TO GAIL
Without whom this book would never have been possible
and
TO LESLIE, JOHN, KATIE, AND COREY
May I always be a positive influence in their lives.

Brief Contents

Contents

Preface

There has never been a greater need for human sexuality education than in the last decades of the twentieth century. Unfortunately, with the AIDS crisis, much vital information has been presented in a negative way. Although one of my goals in writing this book is for students to understand the relevant facts in order to make responsible decisions in their daily lives, an equally important goal has been to present the information in a warm, non-threatening way that leaves students with positive feelings about sex and their own sexuality.

When I began to write this text I recalled my students' complaints about the dryness, sterility, and length of books I had used previously. So with them in mind, I tried to create a book that was factual and thorough, yet readable and interesting. Thus I have included numerous case studies (most contributed by my own students!) to supplement and make more personal the substantial coverage of scientific studies. Although my writing style is purposely conversational, I have worked hard to maintain the scientific foundation of my presentation. This third edition has well over two thousand references for students who wish to use the book as a resource. In addition, the section on HIV/AIDS is as thorough and up-to-date as can be found in any human sexuality textbook.

The cost of textbooks to students has become an increasing concern to many. Besides the cost of the primary textbook, already prohibitive for some, there is the added cost of a student's study guide. However, because it has been the goal of Prentice Hall and myself to publish a high-quality textbook at the lowest possible expense to the students, the Student Study Guide is included within *Human Sexuality Today*, third edition, *at no additional cost to the student*. This Study Guide includes an interactive review and true-false, matching, and fill-in-the-blank questions. The text also includes thought-provoking questions ("Personal Reflections") at the end of each chapter, and a Glossary and Resources at the end of the book. Special boxes are devoted to two important themes—cross-cultural comparisons, and sexuality and health.

The final test for any textbook is whether or not students will read it, learn from it, and enjoy it. Two classroom-tested versions of this book were used with over 6,000 students before the first edition was even published. Thus, as a result of feedback from students and reviewers, the book has continually evolved in an attempt to create an ever-better product.

ACKNOWLEDGMENTS

I offer my deepest gratitude to Dr. Cameron Camp, who helped write the classroom-testing versions and the first edition of the book. His friendship and support will always be appreciated. Thanks also to Anne Downey for her input on the first edition. Others who helped in the early development of the book include Vic Hughes, Cheryl Stout, Desiree Comeaux, Chantelle Boudreaux, Corbie Johnson, and Kathryn King. I thank Teresa Weysham, who typed the present edition.

Thanks to everyone at Prentice Hall who worked on this book. Special thanks to Bill Webber for having faith in me, Joan Foley for her hard work in production, and Susan Goldfarb for her excellent editing.

My sincere appreciation and thanks to the following colleagues for their valuable input and constructive feedback in reviewing this book: for the *first edition*, Susan Graham-Kresge, University of Southern Mississippi; Kendra Jeffcoat, Palomar College; Deborah R. McDonald, New Mexico State University; Ken Murdoff, Lane Community College; Janet A. Simons, University of Iowa; and Janice D. Yoder, University of Wisconsin-Milwaukee; for the *second edition*, Kendra Jeffcoat, Palomar College; Deborah R. McDonald, New Mexico State University; Ken Murdoff, Lane Community College; and Janet A. Simons, University of Iowa; and for the *third edition*, Donna Ashcraft, Clarion University of Pennsylvania; Robert Clark/Labeff, Midwestern State University; Betty Dorr, Fort Lewis College; and Priscilla Hernandez, Washington State University.

Finally, I cannot thank enough the thousands of students who provided me with chapter reviews and/or case histories. This book was written with students in mind. I hope it helps them to lead healthier, happier, and more fulfilling lives.

Bruce M. King

About the Author

Bruce King received his Ph.D. in biopsychology from the University of Chicago in 1978. He has taught human sexuality to over 30,000 students at the University of New Orleans since 1981. He has conducted human sexuality workshops for physicians, public school teachers, staff members working with the mentally retarded, and parental groups. In addition to conducting research in the field of human sexuality, Dr. King has co-authored a textbook on statistics and has published over 50 papers on the biological basis of feeding behavior and obesity. He is a Fellow in the American Psychological Society and an honorary member of the Golden Key National Honor Society for excellence in teaching.

CHAPTER

1

When you have finished studying this chapter, you should be able to:

1. Explain the need for, and benefit of, sexuality education;
2. Explain why, even though people may have extensive sexual experience, many know very little about human sexuality;
3. Describe cultural perspectives other than our own with regard to sexual behaviors and attitudes;
4. Describe what makes something sexually attractive and how that can change over time or be different in other cultures;
5. Explain the historical influence of Judaism and Christianity on contemporary attitudes about sexuality;
6. Summarize the Victorian era's sexual views and its influence on our sexual values and behaviors today;
7. Explain the contributions of Sigmund Freud, Henry Havelock Ellis, Alfred Kinsey, and Masters and Johnson to the field of sexuality; and
8. Understand the use and limitations of surveys, correlational studies, direct observations, case studies, and experimental research.

Why a Course in Human Sexuality?

Amoebas at the start
Were not complex;
They tore themselves apart
And started sex.

(Arthur Guiterman)

Nearly all colleges and universities around the country now offer courses in human sexuality (Rodriquez et al., 1996). At many of them, the enrollment is over 400 students per semester. Why are so many students taking these courses? **Sexuality** is certainly an important part of our lives. We need only to look at the world population of nearly six billion to see that sexual motivation is quite strong. Although sex is necessary for procreation, it is doubtful that many people think of this on more than just

sexuality All of the sexual attitudes, feelings, and behaviors associated with being human. The term does not refer specifically to a person's capacity for erotic response or to sexual acts, but rather to a dimension of one's personality.

an occasional basis when having sexual intercourse. Sex can be a source of great physical and emotional pleasure, enhancing our sense of health and well-being. It can relieve tensions and anxieties. It can boost self-esteem and make us feel more masculine or feminine. It is also the vehicle through which couples can express their affection for one another.

So why the need for a course in human sexuality? Surely someone in your life took the time and responsibility to educate you about this important topic. However, surveys in my course have consistently revealed that fewer than one-third of the 1,000 students per semester have ever had a serious and meaningful discussion with their parents about sex (see King, Parisi, & O'Dwyer, 1993). Nationally, only 36 percent of teenagers say that they learned "a lot" from their parents about sex (Kaiser Family Foundation, 1996). For many teens whose parents talked to them about sex, it was just a single discussion—one "birds and bees" talk that was supposed to prepare them for life. The students generally received little information from their parents apart from a description of intercourse and the fact that it can result in pregnancy. Here are a few comments that I have received from students regarding their prior sex education. These were selected, not because they are unusual, but because they are typical of the many comments I've received on end-of-semester course evaluations.

> "My father thinks this class is a waste because he feels people instinctively know how to deal with their sexuality. Maybe I'm just a freak of nature, but I've never had any instincts explaining any of this to me."

> "My boyfriend doesn't like me taking the course. He says, 'You don't need a classroom to teach you about sex. I'll teach you everything you need to know.'"

> "Until now my parents never spoke to me about sex. I'm from a very strict family. They made me feel as though it was a sinful subject to talk about."

> "I remember my mother finding a book my sister was reading and screaming at her, so everyone in the house could hear, about what an awful, dirty book it was. It wasn't pornography. It was a book on sex education. She just wanted to learn something correctly."

> "When I was young the word sex was never brought up. My mother had one short talk with me, and that was to explain what a period is."

> "The only time my parents discussed sex with me was after they found some condoms in my dresser drawers."

(from the author's files)

Students also provided the following comments regarding the usefulness of a human sexuality course. Apparently, it's never too late to learn.

> "I'm glad I registered in the class. I sure thought I knew it all and found I knew very little. I've been married six years and knew little about my own body, much less about my husband's."

> "I am 32 years old. I am a paramedic with eight years of street experience. I have personally delivered nine babies and assisted with countless other births. I have personally been sexually active for 15 years. I thought I knew it all. This course has taught me a lot."

> "When selecting this course, the thought came to me that it would be a very easy class because I knew everything about sex because Mother told me. After all, there are 18 of us in the family. Boy, was I wrong. I've learned more in one semester than Mother could teach me in 20 years."

> "Just knowing about sex is just not enough. If I was taught about sex at home or in school in 1954, I may not have had 11 children before I was 28."

> "I was very surprised at the amount of material I learned. I thought it was going to be just about sex. There is a lot more to sex than just sex."

(from the author's files)

Where, then, did most of us learn about sex? You may find it hard to believe, but it has been estimated that during prime time on the major television networks, the average viewer is exposed to 15 instances of sexual behavior per hour (Lowry & Shidler, 1993). On soap operas, sexual references per hour have increased 35 percent since the mid-1980s (Greenberg & Busselle, 1996). The typical viewer sees only one instance of preventive behavior for every 25 instances of sexual behavior. This, of course, would include pre-adult television viewers. Television has been called "the most powerful storyteller in American culture, one that continually repeats the myths and ideologies, the facts and patterns of relationships that define our world and le-

---

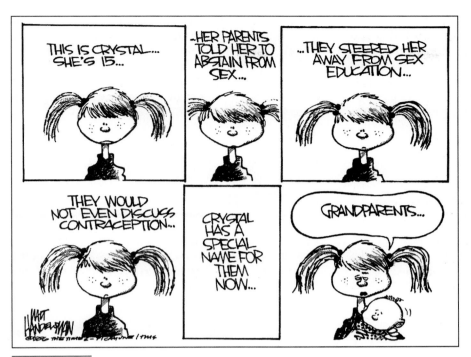

From *The Times-Picayune* (New Orleans), January 31, 1996, page B-6. © Tribune Media Services, Inc. All Rights Reserved. Reprinted with permission.

gitimize the social order" (Brown & Steele, 1996, p. 6). In addition to television, many R-rated movies have sexually explicit scenes with full frontal nudity. Many stores have erotic magazines on open display. Advertisements, such as those for Calvin Klein, use sex to sell products. Sex is everywhere, and children are exposed to it all day long.

In the face of all this, many parents continue to be silent with their children on the subject of sex. However, this too is a source of sex education. Making something mysterious only makes adolescent children want to know more about it. With their parents remaining silent, who can teens turn to? Medical organizations recommend that physicians talk about sexual matters with their teenage patients, but most have not (Schuster et al., 1996). Many young people turn to their friends and the media for their sexual information (Feigenbaum, Weinstein, & Rosen, 1995; Papini et al., 1988; Kaiser Family Foundation, 1996). Much of this information is incorrect, and as a result a majority of Americans are amazingly ignorant about sexual behaviors and sexual health (Reinisch, 1990). Many believe, for example, that people over 60 do not have sex, that masturbation is physically harmful, or that anal intercourse causes AIDS. What this all adds up to is that many people do not fully understand or appreciate the consequences of engaging in sexual relations. Over half of all American teenagers have had sexual intercourse before their seventeenth birthdays (Centers

for Disease Control and Prevention, 1992; Laumann et al., 1994; Seidman & Rieder, 1994), yet only about a fourth of sexually active teenagers in the United States use contraception on a regular basis (Kaiser Family Foundation, 1996). Each year in the United States, over one in ten teenage girls aged 15 to 19 becomes pregnant and over one in four sexually active teens contracts a sexually transmitted disease (Alan Guttmacher Institute, 1996). Most young people do not know the symptoms of sexually transmitted diseases and do not know where to turn if they think they have one.

So, once again, why the need for a course in human sexuality? Probably because you desire to have factual information about a subject that plays, or will play, an important role in your life. Human sexuality courses are designed to do just that. If parents are not going to assume the responsibility, the next best alternative is the schools. Surveys consistently show that the vast majority of Americans support the teaching of sexuality education in schools. One survey found that 90 percent of Americans favor the teaching of sexuality education in high school (Janus & Janus, 1993). A 1985 Gallup poll found that 85 percent of those surveyed believed that sex education should not be limited to reproductive biology and venereal disease, but should include a discussion of birth control as well. Nearly three out of five teenagers say they have not been given enough information about birth control (Kaiser Family Foundation, 1996). Sexuality education curricula are available for elementary and secondary schools (SIECUS, 1996), yet fewer than half the states require that public high schools offer any type of sexuality education (NARAL, 1995).

Canada, England, and many northern European countries have extensive sexuality education programs, and the teenage pregnancy rates are much lower in these countries than in the United States (Dryfoos, 1985; Westoff, 1988). However, the purpose of taking human sexuality courses is much more than just learning about reproduction and sexually transmitted diseases. People want to feel comfortable with their own sexuality and to feel good about themselves. Knowing about their bodies and understanding their feelings and emotions can help people achieve this. No part of our bodies should be shrouded in mystery. Understanding our partners' bodies will help in verbal and nonverbal communication and prevent unnecessary problems. Appreciating that all people are sexual beings can give us a greater understanding of our children, parents, grandparents, and friends. Studies show that sexuality education courses in schools typically result in substantial gains in knowledge and a more tolerant attitude of others (DiClimente et al., 1989; Kirby, 1984). The tolerance that develops with sexuality education is particularly important. Understanding that people are different from ourselves without condemning them is an important part of getting along with others.

A SEXUAL KNOWLEDGE QUIZ

Many of you may already be sexually experienced, and as a result, you may think that you do not need a course in human sexuality. There is more to sexuality, however, than engaging in sexual intercourse. See how well you do on the following 50-question quiz. Do not be afraid to admit that you do not know the correct answer (don't guess)—no one except yourself is going to see the results. The answers are at the end of the quiz.

	True	False	Don't Know
1. Erections in human males result, in part, from a bone that protrudes into the penis.	_____	_____	_____
2. Sperm can be produced only in an environment several degrees lower than normal body temperature.	_____	_____	_____
3. The hymen is a reliable indicator of whether or not a female is a virgin.	_____	_____	_____
4. The inner two-thirds of the vagina is highly sensitive to touch.	_____	_____	_____
5. Many men experience nipple erection when they become sexually aroused.	_____	_____	_____
6. Most men and women are capable of multiple orgasms.	_____	_____	_____
7. Breast size in women is related to the number of mammary glands.	_____	_____	_____
8. Before puberty, boys can reach orgasm, but they do not ejaculate.	_____	_____	_____
9. During sexual intercourse, orgasm in women results from direct stimulation of the clitoris by the penis.	_____	_____	_____
10. Menstrual discharge consists of shredded uterine tissue, blood, and cervical mucus.	_____	_____	_____
11. For hygiene reasons, you should avoid sex during menstruation.	_____	_____	_____
12. Ovulation generally occurs just before menstruation.	_____	_____	_____
13. After a vasectomy, a man can reach orgasm but does not ejaculate.	_____	_____	_____
14. AIDS is the diagnosis for people who have human immunodeficiency virus (HIV).	_____	_____	_____
15. A female can get pregnant as soon as she starts having menstrual periods.	_____	_____	_____

	True	False	Don't Know

16. The combination birth control pill works primarily by preventing implantation of a fertilized egg.

17. Taking the oral contraceptive pill results in fewer serious health problems than do pregnancy and childbirth.

18. Women show their highest levels of sexual desire at the time of ovulation.

19. There are about 12 million new cases of sexually transmitted diseases in the United States each year.

20. The major cause of AIDS is homosexuality.

21. If gonorrhea is not treated, it can sometimes turn into syphilis.

22. Most women do not show symptoms in the early stages of gonorrhea or chlamydia.

23. Gonorrhea, syphilis, and herpes can be successfully treated with antibiotics.

24. In vitro fertilization involves a process where part of fetal development occurs in a test tube.

25. It is usually safe to have sexual intercourse during the seventh and eighth months of pregnancy.

26. "Prepared childbirth" (e.g., Lamaze) refers to delivering a baby without the use of drugs.

27. Most healthy people in their sixties or older continue to engage in sexual behavior.

28. Men's descriptions of orgasm are different from women's descriptions of orgasm.

29. Excessive masturbation can cause serious medical problems.

30. The birth control pill gives women some protection against sexually transmitted diseases.

31. Women who masturbated to orgasm during adolescence generally have less difficulty reaching orgasm during intercourse than women who never masturbated.

32. Couples who have sex very frequently when young generally engage in sex less often than others when old (due to burnout).

33. Adult male homosexuals have lower than normal levels of male hormones.

34. Douching is an effective method of birth control.

35. Recent evidence indicates that environmental factors are most important in determining one's sexual orientation.

36. Prostitutes are generally hypersexual and have pathological sexual needs.

37. Most convicted rapists committed their crimes because of an uncontrollable sex drive.

38. There is a demonstrated link between the availability of pornography and sex crimes.

	True	False	Don't Know
39. It is against the law in many states for a married couple to engage in sexual behaviors other than penile-vaginal intercourse.	_____	_____	_____
40. Most cases of child molestation involve an acquaintance or relative of the child.	_____	_____	_____
41. A pregnant female can transmit syphilis to the unborn baby.	_____	_____	_____
42. Exhibitionists and voyeurs often attempt to rape their victims.	_____	_____	_____
43. Nocturnal emissions ("wet dreams") are often an indication of a sexual problem.	_____	_____	_____
44. Alcohol is a central nervous system excitant that enhances sexual performance.	_____	_____	_____
45. Humans can crossbreed with animals with the use of artificial insemination techniques.	_____	_____	_____
46. Women's sexual desire decreases sharply after menopause.	_____	_____	_____
47. Vaginal infections can be prevented by regular use of feminine hygiene products.	_____	_____	_____
48. A woman's ability to have vaginal orgasms is related to penis size.	_____	_____	_____
49. Oral herpes can be transmitted to another person by oral-genital sexual relations.	_____	_____	_____
50. Unless testosterone is present during embryological development, nature has programmed everyone to be born a girl.	_____	_____	_____

ANSWERS—CHAPTER 1

1. false
2. true
3. false
4. false
5. true
6. false
7. false
8. true
9. false
10. true
11. false
12. false
13. false
14. false (only in the final stages)
15. true
16. false
17. true
18. false (no good evidence for this)
19. true
20. false (homosexuality does not cause AIDS)
21. false
22. true
23. false
24. false
25. true
26. false
27. true
28. false
29. false
30. false
31. true
32. false
33. false
34. false
35. false
36. false
37. false
38. false
39. true
40. true
41. true
42. false
43. false
44. false
45. false
46. false
47. false
48. false
49. true
50. true

These questions were not intended to be tricky or difficult. They are representative of the type of material that is covered in this book. How did you do? Fewer than one-fourth of the students in my classes are able to answer 40 or more correctly at the beginning of the semester. Fewer than half get 35 or more correct. Did you really know all the ones that you got right, or did you just make a correct guess on some of them? If you were not certain of the answers to some of the questions, then that is sufficient reason to read this book.

CROSS-CULTURAL COMPARISONS

It should come as no surprise that people are different. Some people like short hair, while others like it long. Some like blue eyes, while others prefer brown. Some people like to dress up. Some like to dress down. People's sexual attitudes and behaviors differ as well. Some people, for example, have sexual intercourse only in the **missionary position** (i.e., the female lying on her back with the male on top—so called because Christian missionaries instructed people that other positions were unnatural). Others prefer a variety of positions. Some people are most aroused by looking at breasts or a hairy chest. Other people become highly aroused by looking at legs and buttocks.

We learn to accept that other people in our own culture are different from ourselves, and we do not regard them as abnormal when their behavior falls within what we consider the "normal" range of responses. What is normal, however, is defined by the community in which we live. As different as we are, we share much in common. An outsider, such as a person from a different country, is often regarded as very strange by many people. Unfortunately, Americans have a reputation around the world of being egocentric, that is, of viewing our own behaviors and customs as correct or the way things ought to be (a perception leading to the image of the "ugly American"). We must not lose sight of the fact, however, that if we traveled to another country we would probably be regarded as strange. One country's customs and beliefs should not be regarded as correct or normal and another's as incorrect or abnormal (in this book, I will make an exception in cases in which a culture sanctions sexual abuse). Instead, the world can be viewed as one large community, with the behavior and beliefs of all its people falling into the normal range of responses.

The topics covered in some chapters of this book are the same for all peoples of the world (e.g., anatomy, physiology, hormones). When the book covers behaviors and attitudes, it will be primarily from the perspective of people in the United States. Behaviors of people in other cultures will be presented in special boxes, beginning with Box 1–A in this chapter. However, before going on, here is a brief introduction to sexual attitudes and behaviors in a few other cultures around the world. Some of them may seem funny, but remember, *we seem just as strange to them* as they seem strange to us.

missionary position A face-to-face position of sexual intercourse in which the female lies on her back and the male lies on top with his legs between hers. It was called this because Christian missionaries instructed people that other positions were unnatural.

SEXUAL ATTRACTIVENESS

Cultures differ with regard to which parts of the body they find to be erotic. The sight of a navel is considered highly arousing in Samoa. This generally is not true in the United States, but did you know that the TV censors would not allow actress Barbara Eden to have her navel exposed on the television show *I Dream of Jeannie* in the 1960s? A knee is considered to be erotic in New Guinea and the Celebes Islands. When anthropologist Martha Ward visited New Guinea in the early 1970s, she deboarded the plane wearing a miniskirt, blouse, and brassiere, customary clothing for American women at that time. She caused quite a ruckus. She wrote the following letter to one of her colleagues in the United States:

> Dear Len,
> I have had to change my manner of dressing. Thank goodness for the Sears Catalog. The minidresses and short skirts you all are wearing in the States cause quite a stir here. It seems that breasts are regarded as normal female equipment and useful only for feeding babies. Clothing for many women consists of a large towel or three yard length of brightly colored cloth. This is worn around the waist inside the house or in the yard. . . . [In public] if you have on a bra, you don't need a blouse. Bras are considered proper dress for women. Those black Maidenforms with the heavy stitched cups are particularly valued. When Americans are not around, it is sufficient to cover oneself only from the waist down. . . .
> Breasts are not really erogenous, but legs are. Particularly that sexy area on the inside of the knee! No more miniskirts for me. Fitting in and observing local customs means that I have lengthened my skirts to below the knee. . . .
> The American men watch women with nothing on above the waist. The Pohnpeian men comment on American women with short skirts. I am now dressed to please the standards of two cultures. . . .
>
> (From *Nest in the Wind: Adventures in Anthropology on a Tropical Island* by Martha C. Ward, © 1989 by Waveland Press, Inc. Reprinted with the permission of the publishers.)

Many groups of people find body weight to be an important determinant of sexual attractiveness. There is a great deal of pressure in our culture, for example, for men and women to remain thin. It is no coincidence that fashion models are very thin and that movie stars who are considered to be "sexy" have thin waistlines. In many other countries, however, these people would not be considered attractive. For example, women who would be considered obese by most American men are found highly attractive in some other cultures. Adolescent girls are sometimes kept in

Figure 1–1 Rubens's *The Three Graces,* painted in 1630, is a good example of how cultural ideals change over time. Thinness is considered to be physically attractive today, but a thin woman in Rubens's time would have been considered unattractive.

ticularly the mons veneris) as American men are with breasts (Marshall, 1971). In some African cultures, a woman's labia minora are considered to be the most erotic part of her body.

Although walking around naked in public would be considered highly deviant by most people in the United States, there are some cultures in New Guinea and Australia where people go about completely naked. They do, however, have firm rules about staring at other persons' genitals. The Zulus of South Africa also have public rituals that call for people to be naked. They believe that a flabby body results from immoral behavior, and thus if someone refuses to undress for these rituals, it is taken as a sign that the person is trying to hide his or her immorality (Gregersen, 1982). These attitudes about the human body are in marked contrast to those that prevail in Islamic societies, where women must cover their entire body and most of their face when they leave the privacy of their homes.

In some cultures people carve holes in their lips, while in others they stretch their lips or necks, or wear needles through their noses. It is obvious from the lack of universal standards that attitudes about the human body, and what is considered to be sexually attractive, are learned responses.

huts and fed high-calorie diets in order to become more attractive (Gregersen, 1982).

What is considered to be sexually attractive can also change over time. Plump women, for example, were also considered to be very attractive in Western cultures a few centuries ago. If you don't believe this, just look at some famous paintings of nude women that were done two or three hundred years ago (see Figure 1–1).

American men find female breasts very sexually arousing (just recall the large number of magazines for men that show bare-breasted women). There are many areas of the world, however, where naked female breasts have no erotic significance at all (see Figure 1–2). They are important only to hungry babies. On the other hand, Polynesian men are as fascinated with the size, shape, and consistency of female genitals (par-

SEXUAL BEHAVIORS AND ATTITUDES

Kissing is a highly erotic and romanticized part of sexual relations in Western cultures. You will probably be surprised to learn, therefore, that this practice is not shared by all people. "When the Thonga first saw Europeans kissing they laughed, expressing this sentiment: 'Look at them—they eat each other's saliva and dirt'" (Ford & Beach, 1951). Until recently, people in Japan never kissed while interacting sexually. In fact, foreplay before intercourse is entirely unheard of in some cultures.

When you were growing up, many of you probably heard the story of the Hudson Bay Eskimos. It was once the custom for an Eskimo husband to share his wife with any traveler who stayed in the igloo over-

Figure 1–2 This New Guinea girl is dressed up to go courting and find a husband. As in many other areas of the world, bare breasts have no erotic significance. In some areas of New Guinea, males wear only a penis sheath—a piece of bamboo that fits over the penis and attaches around the waist by a string. Some New Guinea tribes wear nothing at all.

night. Western males' eagerness to take advantage of this custom eventually brought it to an end.

Anthropologists believe the most sexually permissive group of people in the world to be the Mangaians, who live on the Cook Islands in the South Pacific (Marshall, 1971). Mangaian boys and girls play together until the age of three or four, but after that they separate into age groups according to sex during the day for the remainder of their lives. When the boys approach adolescence, the arrival of manhood is recognized by superincision of the penis (cutting the skin of the penis lengthwise on top). As the wound heals, the boy is instructed in all aspects of sex, including how to bring a female to orgasm, which is considered important. Girls receive similar instructions from older women. The boy is then given to an experienced woman, who removes the superincision scab during intercourse and teaches the boy an array of sexual techniques. After that, the boy actively seeks out girls at night, having sex an average of 18 to 20 times a week. Mangaian adolescents are encouraged to have sex with many partners and engage in all types of sexual activities, including different positions of intercourse, oral-genital sex, anal sex, and axillary sex (e.g., placing the penis between the armpit or the breasts).

Many other societies in the South Pacific, including Samoa and Pohnpei, similarly encourage their teenage children to enjoy sexual relations (Ward, 1989). In some of these cultures, the boys go into the huts where teenage girls live and have sex with them while the girls' parents are present. The parents ignore

them and act as if the children are invisible. In all of these societies, the physical pleasure of both sexes is emphasized (women have orgasm as often as men) and emotional attachments come later. They regard our custom of emphasizing love before sex as very strange.

The most sexually repressive society in the world is believed to be the Inis Baeg, a fictitious name (coined by anthropologists to preserve anonymity) for a group of people who live on an island off the coast of Ireland (Messenger, 1971). Any mention of sex is taboo, so that children are never told about things like menstruation and pregnancy, which are greatly feared. Nudity is strictly forbidden, to the extent that although they depend on fishing to survive, most of the Inis Baeg never learn how to swim (which requires taking off some clothing) and usually drown if they are in a boating accident. Children wear smocks, and their mothers do not entirely undress them during bathing, but instead wash one arm or leg at a time. Even married adults never see each other completely naked. Sexual relations are not regarded as something positive by either sex. Foreplay is unheard of, and intercourse, which is always done in the missionary position, is completed as quickly as possible because men consider it to be dangerous to their health (women almost never have orgasm).

If the Mangaians and Inis Baeg represent the two extremes, where do we fall on this continuum? In many ways our behavior is permissive—we live during the so-called sexual revolution—but our attitudes about sex are often less than positive (evidenced by the fact that parents and children rarely talk about it together, for example). The constant emphasis on sex on TV, in movies, in magazines, and on the radio give our children one type of message—sex is fun, sex is exciting, sex is great. At the same time, the same children get another type of message from their parents, school, and church—sex is not for you! Is it any wonder that many Americans have ambivalent attitudes about sex? Sex is something good on the one hand, yet bad on the other. We fall somewhere between the Mangaians and Inis Baeg. We are permissive, yet repressed. To see how we arrived at this point in the 1990s, we must examine the history of sexual attitudes in our own culture.

HISTORICAL PERSPECTIVES

JUDAISM

Life for the Biblical Jews was harsh. Living in nomadic tribes that roamed the desert areas of what is now referred to as the Middle East, they considered it a great advantage to have many children. The Jews were directed to do so in the first chapter of the first book of the Old Testament:

> And God blessed them, and God said to them, "Be fruitful and multiply, and fill the earth and subdue it."
>
> (Genesis 1:28, Revised Standard Version)

Having many children assured the survival of the Jewish people. Thus, the primary purpose of sex for the Hebrews was considered to be procreation. This view had a major impact on Western thought about sexuality and is still influential today, even though there is no longer any fear of the world running out of people.

Male children were especially valued because of their dual roles as providers and defenders. In the strongly patriarchal Hebrew society, females were regarded as property (of fathers or husbands), and there were many rules to guarantee that material property was passed on to legitimate offspring. Thus, the Hebrews were very concerned with the social consequences of sex. Sex outside of marriage, for example, was severely condemned and punished. A Jewish woman caught committing adultery was stoned to death, but a man who committed adultery was considered only to have violated the other man's property rights. Rape, too, was considered to be a violation of property rights. The punishment for homosexuality and bestiality was death (Leviticus 18:22–29). Celibacy was looked upon as neglect of one's obligations and was regarded as sinful.

In contrast to those harsh views, the Old Testament presents a positive view of sex within a marriage (Gordis, 1978). A good example of this can be found in the Song of Solomon (Song of Songs):

> How graceful are your feet in sandals,
>   O queenly maiden!
> Your rounded thighs are like jewels,
>   the work of a master hand.
> Your navel is a rounded bowl
>   that never lacks mixed wine.
> Your belly is a heap of wheat,
>   encircled with lilies.
> Your two breasts are like two fawns,
>   twins of a gazelle. . . .
> How fair and pleasant you are,
>   O loved one, delectable maiden!
> You are stately as a palm tree,
>   and your breasts are like its clusters.
> I say I will climb the palm tree
>   and lay hold of its branches.
> Oh, may your breasts be like clusters of the
>   vine,
>   and the scent of your breath like apples,
> and your kisses like the best wine
>   that goes down smoothly,
>   gliding over lips and teeth.
>
> (Song of Solomon 7:1–9, Revised Standard Version)

The genitals were not considered to be obscene, for God had created Adam and Eve in his own image. Sex between husband and wife was cause for rejoicing, a gift from God. A married couple could engage in any sexual activity, with only one restriction—the husband had to ejaculate within his wife's vagina (not doing so was considered "spilling of seed" because it could not lead to having children). Although Jewish women had second-class status, a Jewish wife could take an active role in sexual relations. In fact, her sexual rights were guaranteed in the marriage contract. For example, husbands were not allowed to change jobs without the wife's consent if it meant a change in conjugal rights. The ancient Hebrews did not distinguish between physical and spiritual love.

THE GREEKS AND ROMANS

The ancient Greeks and Romans, like the Jews, placed a strong emphasis on marriage and the family. Although procreation was viewed as the primary purpose of marital sex, a couple had children for the state, not God. Unlike the Biblical Jews, Greek and Roman men were allowed considerable sexual freedom outside marriage. In Greece, sexual relations between adult males and adolescent boys in a teacher-student relationship were not only tolerated, but encouraged as part of the boy's intellectual, emotional, and moral development. Greek women were viewed as second-class citizens, bearers of children with little or no legal status. Roman women were considered to be the property of their husbands, but they enjoyed more social prominence than Greek women.

The Greeks idealized the human body and physical beauty (as is evident in their art), but in the latter part of the Greek era there was a strong emphasis on spiritual development and a denial of physical pleasures. The basis for this change was **dualism,** the belief that body and soul are separate (and antagonistic). Dualism gave rise to an *ascetic philosophy,* which taught that from wisdom came virtue, and that these could only be achieved by avoiding strong passions. Plato, for example, believed that a person could achieve immortality by avoiding sexual desire and striving for intellectual and spiritual love (thus the term *platonic* for sexless love). As you will see next, dualism was a major influence on early Christian leaders.

CHRISTIANITY

Like the theology of the later-period Greeks, Christian theology separated physical love from spiritual love. The period of decline of the Roman empire, which coincided with the rise of Christianity, was marked by sexual excess and debauchery. The views of the early Christians regarding sex were partly the result of an attempt to keep order.

There is little written record about Jesus' views on sex. It is known that except for grounds of adultery, he forbade divorce. It is written in the Gospel of Matthew that Jesus said, "Whosoever looketh on a woman to lust after her hath committed adultery with her already in his heart." Thus, it was not enough for a Christian to conform behaviorally; there was to be purity of inner thoughts as well.

The teachings of the early Christian writers reflect their own personal struggles with sexual temptation. One of the most influential of the early writers was **St. Paul** (about A.D. 5–67), who was a fanatical persecutor of Christians until he had a vision and was converted to Christianity. St. Paul regarded the body as evil and struggled to control it:

> For I know that nothing good dwells within me, that is, in my flesh. I can will what is right, but I cannot do it. For I do not do the good I want, but the evil I do not want is what I do. . . . For I delight in the law of God, in my inmost self, but I see in my members another law at war with the law of my mind and making me captive to the law of sin which dwells in my members.
>
> (Romans 7:18, 7:22–23, Revised Standard Version)

St. Paul blamed Eve for the expulsion from the Garden of Eden and viewed women as temptresses. He strongly believed, as did many others, that the second coming of Christ (signifying the end of the world) would occur in his lifetime and preached that a celibate life-style was the way to heaven. Marriage was only for the weak-willed:

> It is well for a man not to touch a woman. . . . To the unmarried and the widows I say that it is well for them to remain single as I do. But if they cannot exercise self-control, they should marry. For it is better to marry than to be aflame with passion.
>
> (Corinthians 7:1, 7:8–9, Revised Standard Version)

Thus, Paul regarded marriage as a compromise (and a rather poor one at that) to deal with the problems of the flesh. His followers believed that women should be subordinate to men, and therefore should assume the bottom position during sexual intercourse.

St. Jerome (about A.D. 340–420), who spent years fasting in the desert trying to get rid of visions of dancing girls, said a man who loved his wife too passionately was guilty of adultery. Pope John Paul II created some controversy within the Catholic Church when he appeared to echo the beliefs of St. Jerome:

> Adultery in the heart is committed not only because a man looks in a certain way at a woman who is not his wife . . . but precisely because he is looking at a woman that way. Even if he were to look that way at his wife, he would be committing adultery.
>
> (*The New York Times,* October 10, 1980)

Another major influence on Christian beliefs was **St. Augustine** (A.D. 354–430), whose life was full of conflict, beginning with his birth to a pagan father and a Christian mother. He had a mistress and a son at an early age and continued to have numerous sexual affairs into his early thirties, including some with male friends (Boswell, 1980). He is reported to have prayed, "Give me chastity and continence, but do not give it yet" (*The Confessions of St. Augustine,* Book VIII, chap. 7). After reading the works of St. Paul, he converted to Christianity and thereafter led an ascetic life. It was he, more than anyone else, who solidified the Church's antisexual attitude. Augustine believed that all sexual intercourse was sinful, and thus all children were born from the sin of their parents. As a result of the downfall of Adam and Eve, he argued, sex was shameful and equated with guilt. Augustine denounced sex between a husband and wife for the purpose of pleasure, and thus opposed sexual relations during pregnancy or after menopause. He even considered marital sex for the purpose of *procreation* (having children) to be an unpleasant necessity:

dualism The belief that body and soul are separate and antagonistic.

Figure 1–3 St. Augustine (A.D. 354–430). Augustine was a major influence on Christian attitudes about sex. He believed that all sexual acts were driven by lust and were therefore evil, including sex within marriage.

They who marry . . . if the means could be given them of having children without intercourse with their wives, would they not with joy unspeakable embrace so great a blessing? Would they not with great delight accept it?

(Cited in Goergen, 1975)

Augustine not only departed from the Hebrews in denying the pleasures of (marital) sex, but in the process also differed from them by showing complete disgust for the human body: "Between feces and urine we are born."

Augustine's beliefs were systematized by St. Thomas Aquinas (1225–1274) and became doctrine in 1563. Any sexual behavior that was done for pleasure and excluded the possibility of procreation was considered lust and a sin against nature. In descending order of infamy, this included bestiality, homosexuality, intercourse in "unnatural" positions, and masturbation. Although not all Christians today are of the same denomination, they all share the same early history, and thus they all have been influenced by the beliefs of St. Paul and St. Augustine.

Victorian era The period during the reign of Queen Victoria of England (1819–1901). With regard to sexuality, it was a time of great public prudery (the pleasurable aspects of sex were denied) and many incorrect medical beliefs.

In the sixteenth century, Martin Luther (1483–1546) and John Calvin (1509–1564) organized revolts against the Catholic Church. Some of their followers later came to be known as Puritans. The *Reformation* movement, as it was called, included opposition to celibacy. Luther believed, for example, that priests and nuns should be allowed to marry in order to solve the problem of lust. Adultery, homosexuality, and masturbation were still viewed as sinful and severely punished (because of the perceived threat to the family unit), but the linkage of sex with love in marital relationships was emphasized. Thus, the Puritans had a positive view of sex within the marriage, and this attitude was brought over to New England with them. The Puritans proved their worthiness of going to heaven, not by denying the pleasurable aspects of marital sex, but by living frugal and industrious lives.

VICTORIANISM

The nineteenth century is often referred to as the **Victorian era,** after Queen Victoria (1819–1901), who reigned in England for most of the century. It was an era of public prudery and purity. All pleasurable aspects of sex were denied. Women, who just a few centuries before had been considered sexual temptresses, were now viewed as asexual (i.e., having no interest in sex) and as innocent as children. Men were the ones who were viewed as responsible for lust. According to Victorian moralists, a woman's place was in the home, and wives engaged in sex only to perform their "wifely duties." Women's dresses covered the neck, back, and ankles, and the prudery was carried to such an extreme that even piano legs were covered. At the dinner table, it was considered improper to ask for a "breast" of chicken.

The medical views of the nineteenth century generally supported the antisexuality of the era. A prominent Swiss physician named Tissot had published a book in 1741 in which he claimed that masturbation could lead to blindness, consumption, other physical disorders, and insanity. This view had numerous supporters in the medical community, including Benjamin Rush (1745–1813), the father of American psychiatry. The eminent German neurologist Richard von Krafft-Ebing classified sexual disorders in his *Psychopathia Sexualis* (1886), in which the predominant theme was a link between sexuality and disease. A group of British doctors claimed to have evidence that touching a menstruating female could spoil hams (*British Medical Journal,* 1878). To the Victorians, including physicians of that era, loss of semen was thought to be as detrimental to a male's health as loss of blood, a belief that had originated with the Greek physician Hippocrates (Haller & Haller, 1977). As a result of these beliefs, parents often went to ridiculous lengths to prevent masturbation (Hall, 1992) and nocturnal emissions ("wet dreams," or spermatorrhea, as it was called then), including having their male children

"*If sex is dirty and disgusting, why should I save it for someone I love?*"

Reproduced by Special Permission of *Playboy* magazine: copyright © 1993 by Playboy.

while the "surgical appliance" was patented by one J. E. Heyser in 1911. Patents were even issued for devices designed to prevent horses from masturbating.

The amount of misinformation that was distributed by the medical community during this time is really quite appalling. Consider, for example, a book titled *Perfect Womanhood* by "Mary Melendy, M.D., Ph.D.," published in 1903. It was a book of advice for women, and in addition to the usual warnings about masturbation and excessive sex, it contained the following advice about when it was safest to have intercourse (to avoid pregnancy):

It is a law of nature—to which there may be some exceptions—that conception must take place at about the time of the menstrual flow. . . .

It may be said with certainty, however, that from ten days after the cessation of the menstrual flow until three days preceding its return, there is little chance of conception, while the converse is equally true.

castrated or circumcised or making them wear anti-masturbation devices to bed at night. Two such inventions are shown in Figure 1–4. The "spermatorrhea ring" was invented by a Dr. J. L. Milton in the 1800s,

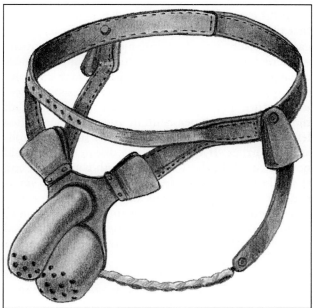

Figure 1–4 Two antisex devices invented and patented during the Victorian era: (A) a spermatorrhea ring to prevent nocturnal emissions; (B) a "surgical applicance" made of leather straps and metal pockets to prevent masturbation.

. . . This law is to the effect that if the conception takes place in the early part of the menstrual period a female child will be the result; if in the latter part, a male child will be born.

These beliefs were based on the fact that most nonhuman mammalian females have a bloody discharge at the time of ovulation (the optimal time for conception to occur). As you will learn in Chapter 3, this is not true for human females (ovulation occurs about two weeks before the menstrual discharge begins); yet some people still believe it.

The private lives of Victorians may not have been as repressed as the image of modesty and prudery they conveyed in public. A recently discovered questionnaire of 45 married women conducted by Dr. Clelia Mosher in 1892 reveals that most of the women desired and enjoyed sex (Jacob, 1981). There was also a great deal of hypocrisy during this era. Prostitution and pornography flourished, and extramarital affairs were common (Trudgill, 1976).

THE TWENTIETH CENTURY

It was not unusual for our great-great-grandfathers, whether they worked on the farm or in the city, to put in 12– to 16–hour days, often six days a week. Child labor was common, so that many children were also working these hours. In addition to puritanical Victorian ideals, the lack of leisure time limited opportunities for sexual relations.

The industrial revolution slowly changed all this. Mechanization and more efficient means of production eventually led to shorter work weeks, freeing people to engage in leisure activities. The invention of the automobile allowed people to get away from the watchful eyes of their parents and neighbors. For the first time, Americans had two things necessary to engage in leisure sex: time and mobility. The growing women's rights movement eventually resulted in more equality for women, and they began to take an active role in sexual matters (they were not the passive, asexual creatures portrayed by Victorian moralists). With the introduction of penicillin in 1940, people worried less about sexually transmitted diseases such as syphilis and gonorrhea. If they got these diseases, they could now be cured. In the 1960s, the birth control pill and the IUD became available, so that having sexual intercourse could be spontaneous and people did not have to worry about unwanted pregnancies. In addition to having more time and mobility, people now believed they were protected from the unwanted side effects of sex.

As a result, we entered the so-called **sexual revolution,** a time when more people than ever in the past were supposedly engaging in more sex, beginning at an earlier age, and with more partners. We have already discussed how movies, TV, and magazines portrayed sexual relations (including premarital and extramarital sex) as exciting and desirable. The newest generation of Americans to come of age was under pressure, not to abstain from premarital sex, but to engage in it.

Although more people are now engaging in premarital sex and oral-genital sex and trying different positions than was the case in past generations, many people have just as many anxieties and feelings of discomfort about their own sexuality as before. The influences of the early Christian church and the Victorian era obviously have not fully disappeared. As a result, many people have ambivalent feelings about sex. We are more permissive, yet still repressive. It is no wonder that so many people still have so many questions, doubts, and fears about sex.

SEX AS A SCIENCE

Until the last few centuries, religious theology was the primary influence in intellectual endeavors. The rise of science was slow and was met with much resistance. Galileo (1564–1642), for example, was forced to publicly recant his support for Copernicus's theory that the earth moved around the sun—and not vice versa, as stated in Church doctrine—or face excommunication. People's desire for facts and knowledge could not be stifled, however, and the following centuries saw great advances in the biological and physical sciences. Science slowly replaced religion as the authority on most subjects.

Unfortunately, sexual behavior as a subject for scientific investigation met with more resistance than most others. As we have already seen, even the medical professionals of the nineteenth and early twentieth centuries were guilty of allowing their personal moral beliefs to interfere with their treatment of sexual subjects. It has not been until the last few decades that researchers interested in studying sexual behavior have been able to apply the two tools necessary for scientific inquiry: observation and measurement. Let us take a brief look at a few individuals who were responsible for making sexual behavior a serious, objective field of study. Further details of their work will be discussed in later chapters.

SIGMUND FREUD (1856–1939)

Sigmund Freud, perhaps more than anyone else, is responsible for demonstrating the influence of sexuality in human life (Figure 1–5). It is all the more remarkable that he did so during the antisexual atmosphere of the Victorian era.

sexual revolution A period in U.S. history, beginning about 1960, of increased sexual permissiveness.

Figure 1–5 Sigmund Freud. The father of psychoanalysis, Freud lived during the Victorian era, a period often described as being indiscriminately antisexual, yet he is given the major credit for showing the importance of sexuality in human behavior and motivations.

In actuality, Freud was not a sex researcher. Rather, he merely discussed sexuality as a primary motivation for behavior. Freud believed that there were three subsystems of the personality. The *id,* Freud said, was present at birth and was directed toward immediate gratification. This was not always possible or socially acceptable, so the *ego* developed as a realistic servant of the id (it postponed immediate gratification). Opposed to both of these was the *superego,* which was the internal representation of all the traditional ideals of society as taught by the parents. These three systems were in constant conflict, a great deal of which was unconscious. Adult behavior, according to Freud, was determined largely by the outcome of these conflicts at different stages of development. Sexual energy—or *libido,* as Freud called it—was said to be channeled into particular areas of the body at different ages. First was the oral stage (in which the center of eroticism was the mouth), then the anal stage, followed by the phallic, latency, and genital stages. You will learn of these stages in greater detail in the chapter on development. Freud claimed that if the conflicts were not properly resolved, neuroses (psychological problems) could result during adulthood. He developed psychoanalysis as a means

Figure 1–6 Henry Havelock Ellis. This English researcher published seven volumes about the psychology of sex between 1896 and 1928. His views were remarkably tolerant compared to those of most Victorian physicians and scientists.

for evaluating and treating unconscious conflicts.

Although many people disagree with Freud's theory today (see, for example, *Time* magazine, November 29, 1993), he remains important for his ideas on sexual motivation and sexuality in infants and children. However, like other Victorian doctors, Freud had many incorrect beliefs about sex, including the belief that loss of semen was as detrimental to a male's health as loss of blood.

HENRY HAVELOCK ELLIS (1859–1939)

When he was a young man in the 1800s, **Henry Havelock Ellis** (Figure 1–6) had frequent nocturnal emissions (wet dreams), which at the time was called spermatorrhea because Victorian physicians believed it was caused by the same thing that caused gonorrhea. The end result, Ellis was told, would be blindness, insanity, and eventual death. He wanted to commit suicide but was too fearful, so instead he kept a diary to document his death by this dreaded "disease." When his eyesight and reasoning did not deteriorate over the passing months, Ellis concluded that loss of semen did not really lead to death, and he became angry about the misconceptions held by the medical profession. In order to spare others the emotional trauma that he had suffered, Ellis devoted the remainder of his life to sexual research (although he reputedly remained a virgin himself).

As a physician practicing in Victorian England, Ellis collected a large amount of information about people's sexual behaviors from case histories and cross-cultural studies. He eventually published six volumes of a series titled Studies in the Psychology of Sex between 1896 and 1910, and a seventh in 1928. He argued that women were not asexual and that men's and women's orgasms were very similar. Ellis was particularly important for his emphasis on the wide range of human sexual behaviors and for his belief that behaviors such as masturbation and homosexuality should be considered normal. His tolerant view of sexuality was in marked

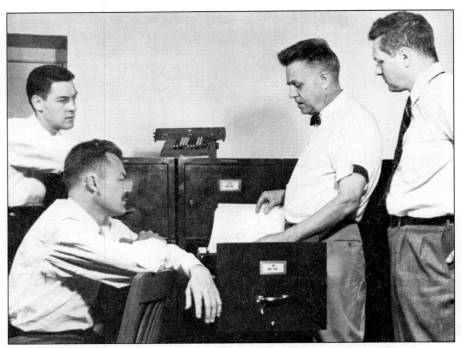

Figure 1–7 Alfred C. Kinsey and colleagues. This college professor's surveys of male sexual behaviors in 1948 and female sexual behaviors in 1953 marked the beginning of the modern era of sexuality research. From left to right: Clyde Martin, Paul Gebhard, Kinsey, Wardell Pomeroy.

1956, but the Institute for Sex Research at Indiana University continues to be a major center for the study of human sexuality.

MASTERS AND JOHNSON

Kinsey paved the way for scientific behavioral studies of sexual behavior, but the study of functional anatomy and physiological studies of sex were still limited to experiments with laboratory animals. Dr. Denslow Lewis, a prominent Chicago gynecologist, presented the first frank discussion of the physiology of sex to the annual meeting of the American Medical Association in 1899. He graphically described erection and vaginal lubrication, and even noted that women could have multiple orgasms (and thus could enjoy sex). His paper was not well accepted. When Lewis had finished with his presentation, Dr. Howard Kelly, the leading gynecologist of the day, responded: "I do hope we shall not have to go into details in discussing this subject. Its discussion is attended with more or less filth and we besmirch ourselves by discussing it in public." Papers read at meetings of the American Medical Association were routinely published in the *Journal of the American Medical Association,* but the journal refused to publish Lewis's.

contrast to that of the Victorian moralists and was a major influence on researchers for several generations.

ALFRED C. KINSEY (1894–1956)

Nearly everyone, of course, has heard of Kinsey's surveys on sexual behavior (1948, 1953). When asked to teach a course on marriage at Indiana University in 1938, **Alfred C. Kinsey** was amazed to find how little objective data there were on sexual behavior. He started collecting his own data by giving questionnaires to his students, but he soon decided that personal interviews resulted in a greater amount of more detailed data. He was joined by Wardell Pomeroy , Clyde Martin, and Paul Gebhard (see Figure 1–7). Their final samples had a total of 5,300 males and 5,940 females.

Kinsey's work opened the doors to a whole new field of research, but this did not come without a price. His findings that most people masturbated, that many people engaged in oral-genital sex, that women could have multiple orgasms, and that many men had had a homosexual experience shocked many people. He was accused of being antifamily and amoral. One Columbia University professor stated that "there should be a law against doing research dealing exclusively with sex" (*New York Times,* April 1, 1948). His work is still the target of impassioned attacks today (e.g., Reisman & Eichel, 1990). Kinsey died an embittered man in

Over half a century later, William Masters, a physician, and Virginia Johnson, a behavioral scientist (Figure 1–8), did not believe that experiments with rats could tell us a great deal about sexual behavior in people. Ford and Beach (1951) had published a book a few years earlier in which they included observations of them having sexual intercourse with their wives, but it lacked objectivity. Therefore, in 1954, **Masters and Johnson** started to directly observe and record the physiological responses in humans engaged in sexual activity under laboratory conditions. Their results, which were based on over 10,000 episodes of sexual activity of 312 men and 382 women, were not published until 1966 (*Human Sexual Response*). Although some people were shocked at Masters and Johnson's observational approach, calling them "scientific peeping Toms," most people in the medical community appreciated the importance of their findings.

As in all other areas of medical science, the understanding of anatomy and physiology led to methods for treating clinical problems and abnormalities. Mas-

Figure 1–8 William Masters and Virginia Johnson. Their laboratory studies of the physiological responses that normally occur during sexual arousal led to the development of modern-day sexual therapy techniques.

ters and Johnson developed the first methods for treating sexual problems and opened a sexual therapy clinic in 1965. Their techniques were described in their 1970 book *Human Sexual Inadequacy*. The behavioral approach to treating sexual disorders, based on Masters and Johnson's work has helped thousands of people to overcome sexual problems, thus improving their sexual relations and their overall satisfaction with life.

THE NATIONAL HEALTH AND SOCIAL LIFE SURVEY

During the 1980s, there were calls from many sources to update what we know about sexual behavior of people in the United States. The Centers for Disease Control and Prevention argued the need for a new national survey (to combat AIDS) in the spring of 1988. In 1989, a panel of the National Academy of Science's National Research Council asked federal agencies to fund new surveys of sexual behavior in order to help with AIDS pre-

vention. In the same year, the U.S. Department of Health and Human Services was getting ready to test a new survey designed to acquire information about the increase in teenage pregnancies, sexually transmitted diseases, and sexual abuse of children. The Department initially approved $15 million for a large survey of 20,000 Americans that would be led by sociologist Edward Laumann. However, U.S. Senator Jesse Helms of South Carolina introduced an amendment in the Senate to eliminate funding for surveys of sexual behavior (saying that they were designed to "legitimize homosexual lifestyles"), and the White House Office of Management and Budget eventually blocked attempts to fund this survey.

Despite the new political obstacles, Laumann and his colleagues Robert Michael, John Gagnon, and Stuart Michaels (see Figure 1–9) eventually received $1.7 million from private foundations, and after scaling down their project, in 1992 they interviewed 3,432 adults (Laumann et al., 1994). Their random sample had a 79 percent participation rate (very high for this type of survey) and was a good representation of all English-speaking adults aged 18 to 59 living in households in the United States. Interviews were conducted face-to-face and lasted 90 minutes. In short, the National Health and Social Life Survey was the most comprehensive nationally representative survey to date, as in-depth as the Kinsey surveys, but with scientifically sound sampling techniques. However, the sample size was not large enough to allow the researchers to draw definite conclusions about small subgroups of the population. Similar national surveys were conducted in the early 1990s in England

Figure 1–9 Robert T. Michael, John H. Gagnon, Stuart Michaels, and Edward O. Laumann. They headed the most scientifically conducted sexual survey to date.

(Wellings et al., 1994) and France (Spira et al., 1994), thus allowing us to make some accurate cross-cultural comparisons.

SCIENTIFIC METHODOLOGY

SURVEYS AND SAMPLES

Throughout this book you will see statistics about sexual behaviors (e.g., percentages of people, frequencies of behavior). In order to know such things, it was necessary for someone like Kinsey, for example, to ask people about their sexual attitudes and behaviors. This involves taking a survey. A **survey** is a study of people's attitudes, opinions, or behaviors. The researcher generally asks a standard set of questions, either in a face-to-face interview or on a paper-and-pencil questionnaire.

The researcher begins by specifying the **population**—the complete set of observations about which he or she wishes to draw conclusions. The population of interest may be quite large (e.g., all adults living in the United States), or it can be small (e.g., all adults living in a town). If the population of interest is large, it may not be possible to obtain responses from everyone. In this case, the researcher must take a **sample** (subset) from the population. There have been many such surveys, some of which you may have heard of: the Kinsey report, the Hunt report (done for *Playboy*), the Hite report, and surveys done by *Redbook, Cosmopolitan,* and other magazines. Sometimes the results of these surveys agree, but in other cases they disagree, sometimes considerably. Which ones should we trust to be true? Are any of them accurate?

It might seem to make sense to some of you that the accuracy of a survey would depend on the size of the sample (i.e., the number of people surveyed). Although sample size is important, there are other factors that, if not considered, can negate any advantage obtained by using a large sample. The classic example that bigger is not necessarily better is a survey conducted in 1936 by a magazine called *Literary Digest*. The magazine sent questionnaires to 10 million people whose names were gathered from lists of telephone and automobile owners. They asked these people whether they were going to vote for Franklin D. Roosevelt or Alfred M. Landon in the upcoming presidential elec-

tion. Over two million people responded, and the results indicated that Landon, a Republican, would beat the Democratic incumbent Roosevelt by a landslide. In fact, Roosevelt beat Landon by a landslide. As a result, the *Literary Digest* lost credibility, which was one of the reasons it went out of business shortly thereafter.

How could a survey that included such a large number of people have been so wrong? The country was in the middle of the Great Depression, and most people were worrying about whether they could continue to buy food to feed their families. Only financially secure people could afford to own cars and have telephones, and they tended to vote Republican. However, most voters were registered Democrats. In other words, despite the large sample size, this survey was grossly inaccurate because of a built-in bias. It surveyed only a small subgroup of the population and was not representative of the entire country. For a sample to be accurate, it must be taken randomly.

What is random sampling? Does it mean blindly picking names from a phone book or stopping people on the street? No, for these types of samples might have the same type of built-in biases as the *Literary Digest* survey. A **random sample** is one in which observations are drawn so that all possible samples of the same size have an equal chance of being selected (see Minium, King, & Bear, 1993). A special type of random sample is a **stratified random sample,** sometimes called a *representative sample* because it pretty accurately represents the target population. This is the type used by the Gallup and Harris polls to predict the outcome of presidential elections today. Before they take their surveys, they break the entire country down by sex, race, education, income, geographical location, and many other factors. They know, for example, exactly what percentage of the population is white, Protestant, college educated, and living in the Northeast. Their sample survey includes this percentage of people with these characteristics. As a result, the Gallup and Harris surveys are rarely off by more than a few percentage points (and because both are done scientifically, they generally agree with each other as well). Yet both polls survey only about 2,500 people out of an estimated 100 million voters.

Kinsey's two studies were quite large, but his samples were not randomly drawn from the U.S. population. They were what is known as *convenience samples*—samples made up of whatever group is available (such as students in a course). Kinsey's samples overrepresented Midwestern, white, college-educated people, and also included a disproportionately large number of inmates from local prisons. Surveys conducted by magazines are also generally quite large, sometimes with over 100,000 respondents, but are they representative of the country as a whole? What kind of people read *Playboy* or *Hustler*? What kind of

survey A study of people's attitudes, opinions, or behaviors. Responses are usually obtained either in a face-to-face interview or on a paper-and-pencil questionnaire.

population The complete set of observations about which a researcher wishes to draw conclusions.

sample A subset of a population of subjects.

random sample A sample in which observations are drawn so that all other possible samples of the same size have an equal chance of being selected.

stratified random sample A sample in which subgroups are randomly selected in the same proportion as they exist in the population. Thus the sample is representative of the target population.

people read *Cosmopolitan* or *Ladies Home Journal*? One-third of the respondents to a 1982 *Playboy* magazine questionnaire said that they had engaged in group sex. Do you really believe this is true for the entire country?

There are more problems in taking a survey of people's sexual behavior than making sure it is representative of the population. Ask yourself the following questions:

1. How often do you masturbate?
2. How many sexual partners have you had in your lifetime?
3. Do you have sexual intercourse in different positions?
4. Do you engage in oral-genital sex?
5. Have you ever had a homosexual experience?
6. Have you ever had an extramarital affair?

Answering some of these questions may have made you feel uncomfortable, for these are very personal questions. How would we know that the people we are surveying are telling the truth? Some people might try to make themselves look good by lying when answering questions about their personal life, so some questionnaires contain "truth items," that is, questions such as "Have you ever told a lie?" It is assumed that everyone has told a lie in his or her life (maybe just a little one), so someone responding negatively to this item would be assumed to be not telling the truth. Incorrect answers, however, are not necessarily the result of intentional deceit, but can also be due to faulty recall. How well do you remember all the events of your childhood, for example? If couples are surveyed, what does the interviewer do if there are discrepancies in the answers given by the two people? How can the interviewer decide who is telling the truth (or is it the case that they both recall events differently)? With regard to sex surveys, two problems are particularly troublesome. Some people may exaggerate their sexual experiences, while others may try to hide certain aspects of their sexuality. Surveys often find, for example, that the number of female sexual partners reported by men is greater than the number of male partners reported by women (see Morris, 1993). Problems like these can sometimes be minimized if the survey is done by questionnaire rather than by interview and the questions can be answered anonymously; that is, the respondent is assured that no one (including the person doing the survey) will ever know who he or she is.

Okay, so now you know enough to design a good sex survey. First you stratify the population of people in which you are interested, and then you randomly sample a specified number of people from each of the subgroups. The questionnaires will be filled out anonymously and contain a few items to check for truthful-

ness. After all this, however, there is another major problem that plagues sex survey research. What do we do when people refuse to participate? In Morton Hunt's survey (1974), which was an attempt to update Kinsey's (1948, 1953) data, only 20 percent of the people contacted agreed to participate. Shere Hite's often-cited surveys (1976, 1981, 1987) had only a 3–6 percent response rate. From a scientific point of view, it is necessary that everyone randomly selected to be in a survey participate in it, but it probably would not surprise you to learn that many people refuse to answer questions like those you just read. We cannot force people to do so. This is the problem of **volunteer bias.** Are there differences between people who agree to participate in a sex survey and those who do not? Should we assume that people who refuse to answer are more conservative in their views and behavior than those who do answer? Or is it possible that the only difference in sexual behaviors between these two groups is that one regards sex as something very private and not to be shared with others? In fact, recent studies have found several differences between people who volunteer to participate in sexual studies and those who refuse. Volunteers had more sexual experiences, were more interested in variety, had a more positive attitude about sex, and had less sexual guilt (Bogaert, 1996; Strassberg & Lowe, 1995). What this means is that the greater the number of people in a survey who refuse to participate, the more cautious we should be in making generalizations about the entire population.

CORRELATION

You are probably aware that there is a relationship between high school grade-point averages and grades earned during the first year of college. The same is true of performance on certain standardized tests (the SAT or ACT, for example) and freshman-year grades. **Correlation** is a mathematical measure of the degree of relationship between two variables. Correlations can be either *positive* (increases in one variable are associated with increases in the other) or *negative* (increases in one variable are associated with decreases in the other), or, for a case in which there is no relationship, zero.

The association between two variables is rarely perfect. We all know students who had mediocre scores on their SAT or ACT test but who did very well in college (perhaps because they are exceptionally motivated and worked hard). On the other hand, a few students who do very well on their entrance tests do rather poorly in college (perhaps because they are not motivated). Generally, however, the association is good enough so that colleges and universities can use the test scores to predict perfor-

volunteer bias A bias in research results that is caused by differences between people who agree to participate and others who refuse.

correlation A mathematical measure of the degree of relationship between two variables.

mance in school. The greater the correlation between two variables, the more accurately we can predict the standing in one from the standing in another.

Correlations are also found in studies of human sexuality. For example, in their recent nationally representative survey, Laumann and his colleagues (1994) found that the more sex a married person has, the more likely he or she is to masturbate (a positive correlation). This may be interesting, but there is a major limitation to the correlational method—correlation never proves causation. In the case of the relationship between sex and masturbation, we cannot tell whether frequent intercourse leads to (causes) frequent masturbation or frequent masturbation leads to frequent intercourse. Or maybe there is a third variable that similarly affects both frequency of intercourse and frequency of masturbation. Laumann's group, for example, believed that frequency of intercourse and masturbation were reflections of a person's overall sex drive. Although we must use caution in drawing conclusions from correlations, they are often very useful in pointing the way to more systematic research.

DIRECT OBSERVATIONS

You have just learned that one of the main problems with conducting a survey is ascertaining the truthfulness of responses. One way around this is to make direct observations of people's behavior. Anthropologists do this when they study the behavior of people of other cultures in their natural settings (this is often called *naturalistic observation* or *ethnographic research*). This is the type of study that Marshall (1971) did of the Mangaians and Messenger (1971) did of the Inis Baeg. Of course, people may not behave normally if they know that they are being observed (this is called the *observer effect*), so it is necessary for the observer to take great care to interfere as little as possible.

For most people, sexual behavior is not done in the presence of others, so some researchers have used what is called the *participant-observer method,* in which the researcher observes while he or she participates. In an early study on "swinging," for example, anthropologist Gilbert Bartell (1970) and his wife attended swingers' parties and observed people engaging in group sex.

Direct observation studies can also be done in the lab. The classic example of this is the work of Masters and Johnson, who, you recall, observed over 10,000 episodes of sexual activity in their laboratory in St. Louis (Masters & Johnson, 1966). For Masters and Johnson, the advantage of observing behavior in the lab was that they could make close observations and also take measures with electrophysiological recording equipment. However, the same limitations apply here as with the previous methods. Were the people Masters and Johnson studied affected by the fact that they knew they were being observed (observer effect)? Are people who would volunteer to have sex under these conditions different from the general population (volunteer bias)? Masters and Johnson were interested in physiological responses, which are probably the same for most people, but we still must be cautious when making generalizations about the entire population.

CASE STUDIES

Clinical psychologists and psychiatrists commonly do in-depth studies of individuals, called **case studies**. They may gather information from questionnaires, interviews with other persons, and even public records, but most of what they learn usually comes from face-to-face interviews conducted over a long period of time. The goal of a case study is to understand a person's behavior and motivations as much as possible. Some of Freud's case studies—such as the case of Little Hans, a 5-year-old boy who had a great fear of horses—are still read today. Freud eventually concluded that Hans's fear of powerful animals grew out of his masturbatory behavior and was actually a disguise for his fear that his father might cut his penis off.

One potential problem with case studies is that the therapist's observations and conclusions might be biased by his or her own emotions and values (called **observer bias**). During the Victorian era, for example, Freud and other physicians believed that masturbation was dangerous and would eventually lead to neurosis (Groenendijk, 1997). In today's world, self-exploration is regarded as normal for young children. Do you suppose that a therapist today would attach great importance to Hans's masturbation?

When working extensively with an individual, a therapist may try different and new techniques, and thus case studies are often the first published reports we have of potentially beneficial treatments. Again, however, great caution must be used in making generalizations about the effect of the treatment for others.

EXPERIMENTAL RESEARCH

The biggest limitation of the research methods you have just read about is that none of them can be used to demonstrate cause-and-effect relationships between two variables. For this, we must use the **experimental method**. With this approach, the researcher systematically manipulates some variable, called an *independent variable,* while keeping all other variables the same. The variable that is measured is called the *dependent variable.* Typically, the researcher compares two groups (sometimes more). One group receives the experimental treatment and the other (called a control group)

case study An in-depth study of an individual.

observer bias The prejudicing of observations and conclusions by the observer's own belief system.

experimental method A study in which an investigator attempts to establish a cause-and-effect relationship by manipulating a variable of interest (the independent variable) while keeping all other factors the same.

does not. If there is any difference between groups in the dependent variable, we can conclude that it was caused by manipulating the independent variable.

I will use one of my own studies as an example (King, Parisi, & O'Dwyer, 1993). As I mentioned earlier, few of my students say that they ever had a meaningful discussion about sex with their parents. The question of interest was whether completion of a college sexuality education course would increase the likelihood that students would talk to their own children about sex. Fifty-two former students who had children that were at least 5 years old were surveyed 2½ to 3 years after they had taken the course. Their responses were compared to those given by a control group of 50 parents on the first day of the course. Here, the completion of a sexuality education course was the independent variable; the incidence of discussions with children about sex was the dependent variable. We found that while only 18 percent of the parents who had not yet begun the course had initiated discussions about sexuality with their children, 86.5 percent of the sexuality-educated parents said that they had done so. We therefore concluded that there was a cause-and-effect relationship: comprehensive college sexuality education increases the likelihood that students will initiate discussions about sex with their children.

While experimental research has advantages over the other methods, it, too, often has limitations. This study, for example, still relied on self-reports. Although I was careful not to survey my former students myself, out of concern that some might give answers they thought would please me, we cannot be sure that everyone answered honestly. In addition, the two groups were convenience samples, so we must be cautious in making conclusions about the effects of college sexuality education courses nationwide. Moreover, researchers simply cannot use experimental designs to address some questions. We may believe, for example, that an infant's hormone levels shortly after birth strongly influence future sexual orientation, but we cannot purposely manipulate hormones in children in order to prove our hypothesis.

In the chapters ahead you will see references to hundreds of studies. Although each has its limitations, it is often the case that a particular subject has been studied in numerous investigations using a variety of methods. The more agreement there is among the results of different studies, the safer we can feel about making general conclusions.

SEXUALITY EDUCATION TODAY

Kinsey's surveys (1948, 1953) were met with public outrage, but surveys of people's sexual behaviors and attitudes have since become commonplace, so that we now have a good idea of what can be considered normal sexual behaviors. Physiological studies with humans rather than animals have become accepted within the scientific community as well. As a result, we now have a large body of data on human sexual responses. The emerging field of sexual therapy has provided us with much information on sexual disorders.

For the first time in our history, we are in a position to give factual information about human sexuality to our children. However, more often than not, sexuality education has become a battleground among groups with different moral and political ideologies. Even if sex educators do not openly talk about values, sex education cannot be value-free (Reiss, 1995), and conflict about values is probably inevitable in democratic societies in which there is a diversity of ethnic, religious, and sexual groups.

On one side of the debate are those who wish to guide their children toward their own view of the world and who regard sexuality education as the exclusive right of the family (thus protecting the children from other points of view). This group is guided by the belief that there has been a decline in morals within our society and that only a narrow set of sexual behaviors is normal and moral. They tend to view sexual diversity as part of the problem of moral decline, and their ideology with regard to sex education is to restrict the curriculum.

On the other side are those who favor what is called comprehensive sexuality education. They believe that children should be exposed to a wide variety of beliefs so that they can make their own choices as to what is best for them. Furthermore, they generally believe that it is the duty of the state to impose this education for the "public good."

Nowhere is the battle over curriculum more heated than with regard to the subject of how to prevent teenage pregnancy and sexually transmitted diseases. At the heart of this debate is the central issue of the purpose or goal of sexuality education. Those favoring comprehensive education argue (often passionately) that students should be taught about birth control methods and the use of condoms (and often advocate making condoms available to teens) in the interest of public health. Most educators who favor a comprehensive curriculum believe that "sex education should reduce teenage pregnancy and help prevent transmission of HIV and other causes of sexually transmitted diseases" (Reiss, 1995). Those favoring a restrictive curriculum, on the other hand, see schools as agents of socialization that shape the moral and sexual norms of future generations, and they want sexuality education to promote family values and abstinence. They frequently cite an article that appeared in the *Atlantic Monthly*, titled "The Failure of Sex Education," which concluded that comprehensive sexuality programs have failed to reduce the number of teenage pregnancies and cases of sexually transmitted diseases (Whitehead, 1994). Advocates of comprehensive education point to other studies. For example, the National Institutes of Health's Consensus panel on AIDS concluded that "abstinence-plus" programs (programs that teach abstinence, contraception, and the prevention of sexually transmitted diseases) are more effective than "abstinence-only" programs (NIH Consensus,

1997). A review of sexuality education programs done for the World Health Organization concluded that the programs did not increase sexual experimentation or activity, and that they often resulted in teens postponing sexual intercourse and/or initiating safer sex practices (Grunseit et al., 1997).

The battle is now being fought in the political arena . In 1995, fewer than half the states required comprehensive sexuality education in schools, and 13 did

Box 1-A   *CROSS-CULTURAL PERSPECTIVES*

Sexuality Education

Children learn about sex from a variety of sources, including parents, brothers and sisters, friends, the media, and school. Parents' attitudes about the sexuality of their children vary considerably across cultures. In many cultures, parents are very permissive, either promoting sexual relations among adolescents and sometimes even among preadolescent children (as in the Mangaian culture or the Muria Gond of India) or at least tolerating it (as with the !Kung of Africa) (Elwin, 1968; Marshall, 1971; Shostak, 1981). In other cultures, parents are very restrictive, never mentioning sex or normal bodily functions such as menstruation to their children and severely restricting premarital sexual relations (as in the case of the Inis Baeg of Ireland and in most Islamic cultures) (Halstead, 1997; Messenger, 1971). In a survey of 141 societies, Broude and Greene (1976) found 25 percent to be highly permissive and another 25 percent highly restrictive. Some anthropologists believe that the more complex a culture is, the more restrictive attitudes about premarital sexual exploration become (Murdock, 1964).

In Europe, the United States, and other English-speaking countries, schools are increasingly becoming the primary source of information about sex (Penland, 1981). The country with probably the longest history of school sexuality education is Sweden, which made it compulsory for all children (starting at age 8) in 1956. Schools in Sweden and Denmark do not have special sex education classes, but instead integrate lessons about sexuality into the rest of the curriculum. Preadolescent Swedish children's knowledge of sexual anatomy and be-

haviors, including the pleasurable aspects, is far superior to that of most North American children (Goldman & Goldman, 1982). People in the United States generally believe that Sweden's sexually permissive culture is a result of Swedish openness, yet in fact, studies show that Swedish people are not promiscuous and that the teenage pregnancy rate in Sweden is substantially lower than it is in the United States (Lindahl & Laack, 1996; Weinberg, Lottes, & Shaver, 1995; Westoff, 1988).

In many parts of the world, government or international agencies are playing a role in sexuality education. In June 1988, then–U.S. Surgeon General Everett Koop's office mailed a brochure entitled "Understanding AIDS" to 107 million American households. In 1993, the World Health Organization called for more sex education in schools and helped to develop innovative programs to promote sexual health education in underdeveloped areas of Africa and Asia (Senanayake, 1992). Government organizations have recently developed sexuality education programs in such countries as Columbia, Nigeria, and India (Mendez, 1996; Nayak & Bose, 1997; Shortridge, 1997). A recent poll in Brazil found that 86 percent of the respondents wanted school curricula to include sexuality education (Egypto, Pinto & Bock, 1996).

The primary source of sexual information may vary from culture to culture, but there is a definite trend in the direction of enlightenment. Throughout the world, the sexual education of children has increasingly become an integrative responsibility shared by parents, schools, and the government.

not even require any education about sexually transmitted diseases (NARAL, 1995). During the 1996–1997 school year, opponents of comprehensive sexuality education introduced 124 measures in 33 states designed to further restrict the curriculum (Mayer, 1997). In 1996, Congress passed legislation, signed into law by President Clinton, that provided $250 million for abstinence-only programs, which would effectively end teaching about other types of birth control.

The result of this battle for control of sexuality education is a great unevenness in programs. In some areas the curriculum reflects a single sexual philosophy, generally ignoring cultural and ideological diversity. Many, perhaps rightfully so, regard this as ideological indoctrination. Other programs, in an attempt to find common ground among differing ideologies, are stripped to the point of offering information only about reproductive biology and some basics of sexually transmitted diseases, and as a result omit important information about sexual health. Almost always left out of the debate are the opinions of young people themselves, most of whom, according to a recent survey, say that they do not receive enough information about sex and birth control and that what they do get comes too late (Kaiser Family Foundation, 1996). Most teens would prefer to receive sexual health education in school (McKay & Holowaty, 1997).

How should we resolve these differences in ideology and construct a sexuality education curriculum? In a thoughtful essay, Alexander McKay (1997) argues for a democratic philosophy of sexuality education that is committed to freedom of belief:

> **In sum, a democratic sexuality education does not insist that students consider the moral dimensions of sexuality from the perspective of a particular sexual ideology. Rather, a democratic sexuality education encourages students to exercise their liberty of thought to deliberate critically between competing ideological perspectives in clarifying their own beliefs and at arriving at new ones. (p. 295)**

As a student at a college or university, you have probably not been required to take this course. When you enrolled, you exercised your freedom of choice. In this text, I will provide accurate information and attempt to make you aware of the diversity in sexual behavior and values. I do not advocate one life-style over another or one set of values over another. My goal for you upon completion of this course is that you feel more comfortable with your own sexuality, and at the same time find a tolerance and respect for the beliefs of others.

Key Terms

ascetic philosophy 11	Masters and Johnson 16	sexual revolution 14
case study 20	missionary position 7	sexuality 1
correlation 19	observer bias 20	stratified random sample 18
dualism 11	population 18	St. Augustine 11
Henry Havelock Ellis 15	procreation 10	St. Paul 11
experimental method 20	random sampling 18	survey 18
Sigmund Freud 14	Reformation 12	Victorian era 12
Alfred C. Kinsey 16	sample 18	volunteer bias 19

Personal Reflections

1. From whom (or from where) did you acquire most of your information about sex (e.g., from parents, from friends, from teachers, from the media)?

2. Did your parents discuss sexuality with you? If not, why do you suppose they did not? If they did, was it in ongoing discussions or just one or two "birds and bees" talks? Do you think that one or two talks can adequately prepare most people for a lifetime of sexual relations and problems?

3. From whom do you hope your children will learn about sexuality? Do you plan to play a role in their sex education?

4. Do you believe that sexuality education should be taught in school? If so, beginning at what level? Why?

5. Christian views about sex were strongly influenced by St. Paul and St. Augustine, who believed that the only legitimate reason to engage in sex was for procreation (to have children). How do you feel about engaging in sex for pleasure?

6. How do you feel about sex as a subject for scientific study? Explain.

Suggested Readings

Abramson, P. (1990). Sexual science: Emerging discipline or oxymoron? *Journal of Sex Research, 27,* 147–165. An excellent review of methodology problems in sex research.

Bullough, V. L. (1994). *Science in the bedroom: A history of sex research.* New York: Basic Books. An in-depth history by one of the premier historians of the field.

Caron, S. L., & Bertran, R. M. (1988, April). What college students want to know about sex. *Medical Aspects of Human Sexuality.*

Davenport, W. (1977). Sex in cross-cultural perspective. In F. Beach (Ed.), *Human sexuality in four perspectives.* Baltimore: Johns Hopkins Press. A comparison of different sexual attitudes and behaviors around the world.

D'Emilio, J., & Freedman, E. (1988). *Intimate matters: A history of sexuality in America.* New York: Harper & Row. A very scholarly history.

Journal of Moral Education, 26(3), 1997. This issue is devoted exclusively to articles about the debate over sexuality education and how to arrive at a curriculum in a culture with a diversity of religious, ethnic, and sexual groups.

Journal of Sex Research, 22, February 1986. This special issue is devoted to problems in methodology.

Marshall, D. S. (1971, February). Too much in Mangaia. *Psychology Today.*

Messenger, J. L. (1971, February). Sex and repression in an Irish folk community: The lack of the Irish. *Psychology Today.*

Michael, R. T., Gagnon, J. H., Laumann, E. O., & Kolata, G. (1994). *Sex in America: A definitive survey.* Boston: Little, Brown. The condensed, popularized version of the landmark survey conducted by the National Opinion Research Center.

Suggs, D. N., & Miracle, A. W. (1993). *Culture and human sexuality.* Pacific Grove: Brooks/Cole. Cross-cultural essays about human sexuality.

CHAPTER 2

Our Sexual and Reproductive Anatomy

Many of the first words that we learn are anatomical terms. Parents often spend hours teaching their young children to point to and name different parts of the body, such as the mouth, eyes, ears, and nose. Unfortunately, when naming body parts, many parents simply skip from arms, chest, and "tummy" to legs, knees, and feet, completely omitting any mention of the genitals and anus. Other parents teach their children the correct anatomical words for all other parts of the body, but substitute such "cute" words as "weeney," "booty," "peanut," and "talleywacker" for the genitals. As for functions of this mysterious body area, many of us are simply taught during toilet training to make "pee-pees" and "poo-poos." Although some parents became sexually experienced without ever knowing the correct anatomical terms, many others who know better nevertheless feel very uncomfortable about teaching their children to say the words "penis" or "vagina."

As we grow older, we learn new words from our peers that describe our sexual anatomy and behavior, but quite often these are slang terms. Men are more likely than women to use slang terms when referring to their own **genitalia,** but many students use no words at all to refer to their partner's or their own genitals (Fischer, 1989). Very few students use the word *intercourse* when referring to the act of copulation. The use of slang terminology and the discomfort that many people feel about using the correct words reflect the ambivalent feelings that many people have about sex. There is perhaps no better example of the distorted negative attitudes that some people have toward sex than the common use of the word "fuck" instead of intercourse. It is frequently used to express displeasure ("What the fuck is going on here?"), trouble ("I got fucked over at work"), or aggression ("Fuck you!"), yet at many other times we think of sexual intercourse as a pleasurable experience. The word "fuck," by the way, probably originated from the Middle English word "fucken." At one time it was not considered improper to use, but it acquired negative connotations in the mid-1600s.

Many of the slang terms used to describe our sexual anatomy also have negative connotations. A commonly used term for the external female genitalia, for example, is "pudendum," which is derived from a Latin word meaning "something to be ashamed of." Similarly, the slang term "prick" is often used in a derogatory manner.

Slang terms can result in misinformation about sexual anatomy and behavior. In response to one of the items on a questionnaire given on the first day of a class in human sexual behavior, many students indicated that they believed that an erection in human males occurs when a bone protrudes from the base of the penis. This belief should not be surprising, for after all, why else would an erection be called a "boner"? Slang terms such as "cherry," "nuts," and "pussy" can also be very misleading.

It is not the intent of the author to teach "dirty" words or slang terminology. In fact, people who take courses in human sexuality often end up using the proper terms more frequently (Fischer, 1989). We must acknowledge the fact, however, that many of us come to class with a sexual vocabulary that consists almost entirely of slang words. The question that many students have, of course, is, "What good will it do to learn the correct anatomical terms, which are almost always Latin or Greek?" The

genitalia The reproductive organs of the male or female.

lack of such knowledge obviously does not hinder most people from becoming sexually experienced. Knowledge of sexual anatomy may help dispel any misleading ideas acquired through use of slang terms, but will it help make us better sexual partners? Learning correct anatomical terms may be very helpful, particularly if it results in our becoming less inhibited about our own bodies and those of our partners. Many people have never even examined themselves (possibly because many parents discouraged or punished such behavior during childhood) and are ashamed of their genitals, rather than viewing them as a source of pleasure and a positive part of their anatomy. This is more often true of women than men (Reinholtz & Muehlenhard, 1995). Here are a few comments that I received from students when I was using a classroom-testing version of this book:

> "As for the illustrations in Chapter 2, I thought they were good and very well illustrated, but they were rather embarrassing to look at."

> "Figure 2–2 made me feel uncomfortable. I felt as if everyone could see me."

> "The diagrams and illustrations are quite to the point. Somehow I still find these photographs repulsive."

> "The pictures in this chapter were quite graphic. I wish they were on another separate page so that we wouldn't have to look at them the entire time we read a particular page. Even just reading this by myself was a little embarrassing."

> "I had difficulties looking at the pictures without embarrassment and shame. I should be able to look without any bad feelings."

> (from the author's files)

No one should feel embarrassed or ashamed about any part of their own body, including their genitals. If you have never done so, I encourage you to examine yourself during this part of the course. By making us more aware of our own bodies and those of our partners (or future partners), a chapter on basic anatomy might result in our becoming more comfortable with exploring other aspects of our sexuality as well.

In the act of sexual intercourse, you share in mutual bodily stimulation with your partner. *Intercourse* means "communication" (*Webster's New World Dictionary*), but communication, whether verbal or tactile, is not possible without some basic knowledge. Should your sexual relations result in having children, it is hoped that someday you will teach them about their bodies without having to resort to slang or cute terminology that avoids the basic facts of life (see Chapter 17).

EXTERNAL FEMALE ANATOMY

You can see the location of the female reproductive system in Figure 2–1. A picture of the external female genitalia, collectively known as the **vulva** (Latin for "covering"), is shown in Figure 2–2. The vulva consists of the mons veneris, the labia majora, the labia minora, the clitoris, and the vaginal and urethral openings. When examining herself (which requires use of a hand mirror), a woman should not consider herself to be abnormal if she does not look identical to the drawing in Figure 2–2. As with all other parts of the body, the female genitalia vary in appearance from person to person, differing in size, shape, and color (see Figure 2–3). Many women have negative feelings

about their genitalia (remember the use of the term *pudendum*). The feelings that a woman has toward her genitalia are directly related to her participation in and enjoyment of sexual activity (Reinholtz & Muehlenhard, 1995).

THE MONS VENERIS

Translated from Latin, **mons veneris** means "mount of Venus." The term refers to the soft layer of fatty tissue overlaying the area where the pubic bones come together. Venus, of course, was the Roman goddess of love, and the mons area is considered to be very erotic after puberty, when it becomes covered with hair. Up until the early 1970s, magazines such as *Playboy* and *Penthouse* considered *pubic hair* too explicitly erotic for publication, and the hair was airbrushed out of photographs. Pubic hair varies considerably in color, texture, and waviness.

Our sensitivity to touch is dependent on the density of nerve endings in a particular area of skin. Areas such as the lips and fingertips have a high density of nerve endings and thus are very sensitive to touch, while areas such as the back have a much lower density of nerve endings and are much less sensitive to touch. Because of numerous nerve endings in the mons area, many women find gentle stimulation of the mons pleasurable.

Although the soft layer of fatty tissue cushions the pubic region during intercourse, one can only speculate as to what purpose nature intended pubic hair to serve. Some researchers believe that this pocket of hair on an otherwise mostly hairless body is meant to be sexually attractive. However, pubic hair also traps vaginal secretions. Males of most mammalian species find vaginal odors erotically stimulating, and similar results have been found for humans under laboratory conditions (Hassett, 1978). Many people in Western cultures, however, are conditioned to have negative feelings about such odors (see the section on the vagina). Shaving the pubic hair is a recent but increasingly popular practice among American women, but it may result in chafing, irritation of the skin, and possible infection from ingrown hairs.

THE LABIA

The labia consist of two outer (**labia majora,** or major lips) and two inner (**labia minora,** or minor lips) elongated folds of skin, which, in the sexually

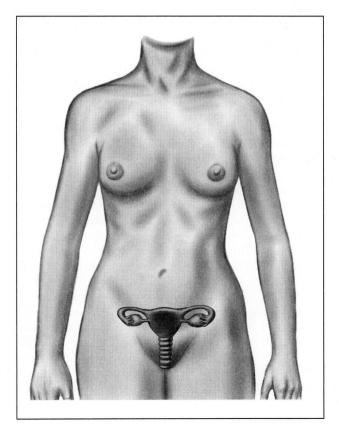

Figure 2–1 Anatomical location of the female reproductive system.

vulva A term for the external female genitalia, including the mons veneris, labia majora, labia minora, clitoris, vaginal opening, and urethral opening.

mons veneris The soft layer of fatty tissue that overlays the pubic bone in females. It becomes covered with pubic hair during puberty.

labia majora Two elongated folds of skin extending from the mons to the perineum in females. Its outer surfaces become covered with pubic hair during puberty.

labia minora Two hairless elongated folds of skin located between the labia majora in females. They meet at the top of the vulva to form the clitoral hood.

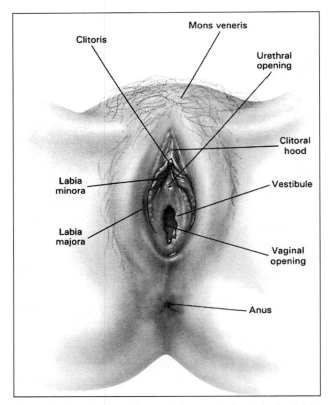

Clitoris

Mons veneris

Urethral opening

Clitoral hood

Labia minora

Vestibule

Labia majora

Vaginal opening

Anus

Figure 2–2 The female genitalia (vulva), with the labia parted to show the vaginal and urethral openings.

unstimulated state, cover (and thus protect) the vaginal and urethral openings. The outer surfaces of the labia majora become covered with hair at the time of puberty. The labia majora extend from the mons to the hairless bit of skin between the vaginal opening and the anus, called the **perineum**. Because the perineum often tears during childbirth, physicians commonly make an incision in it (called an *episiotomy*) prior to delivery (see Chapter 7). When sewing up this area, some old-fashioned doctors add an extra stitch to tighten the vaginal opening (called a husband's knot, as its only purpose is to increase the male's pleasure when sexual intercourse is resumed). After childbirth it is not uncommon for the labia majora to remain apart to some extent in the unstimulated state.

The pinkish and hairless labia minora are located between, and sometimes pro-

perineum Technically, the entire pelvic floor, but more commonly used to refer to the hairless bit of skin between the anus and either the vaginal opening (in females) or the scrotum (in males).

clitoral hood The part of the labia minora that covers the clitoris in females.

Bartholin's glands Glands located at the base of the labia minora in females that contribute a small amount of alkaline fluid to their inner surfaces during sexual arousal.

clitoris A small, elongated erectile structure in females that develops from the same embryonic tissue as the penis. It has no known reproductive function other than to focus sexual sensations.

trude beyond, the labia majora. Among the bush people of Africa, particularly the Hottentots, elongation of the labia minora is considered to be highly erotic, and females are instructed to pull on them from early childhood. Some "Hottentot Aprons," as they are called, can protrude as much as 7 inches.

The labia minora, which meet at the top to form the **clitoral hood** (or *prepuce*), are very sensitive to touch. They have numerous blood vessels that become engorged with blood during sexual stimulation, causing them to swell and turn bright red or wine-colored.

Located at the base of the labia minora are the **Bartholin's glands,** which, during prolonged stimulation, contribute a few drops of an alkaline fluid to the inner surfaces via ducts. The small amount of fluid does not make a significant contribution to vaginal lubrication during sexual intercourse, but it does help to counteract the normal acidity of the outer vagina (sperm cannot live in an acidic environment). Infection or cysts of the Bartholin's glands sometimes occur, and a physician should be consulted if there is any local swelling or irritation.

THE CLITORIS

The **clitoris** (from a Greek word meaning "hill" or "slope") develops from the same embryonic tissue as the penis and has at least as many nerve endings as the much larger penis, making it extremely sensitive to touch. In fact, it is the only structure in either males or females with no known function other than to focus sexual sensations (Masters & Johnson, 1970). The only visible portion is the glans, which in a sexually unaroused woman looks like a small, shiny button located just below the hood of skin formed where the two labia minora meet.

The body, or *shaft,* of the clitoris is located beneath the clitoral hood. It is about one inch long and one-quarter inch in diameter, but divides to form two much larger structures called *crura* (Latin for "legs"), which fan out and attach to the pubic bone. Like the penis, the clitoris contains two *corpora cavernosa,* made up of spongy tissue that becomes engorged with blood during sexual arousal. This causes the clitoris to increase in size; but because of the way it is attached to the pubic bone, the clitoris does not actually become erect like the penis. Any form of sexual arousal may result in engorgement and enlargement of the clitoris. If sexual stimulation continues, the previously visible clitoris pulls back against the pubic bone and the glans disappears from view beneath the clitoral hood.

Contrary to what some people may believe, the sexual pleasure of females is not related to the size of the clitoris (Money, 1970). Another popular myth is that women are more sexually responsive the shorter

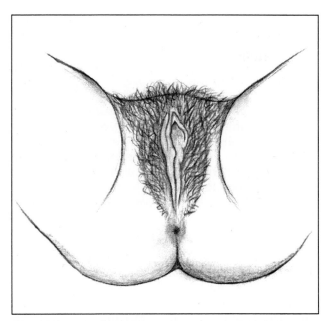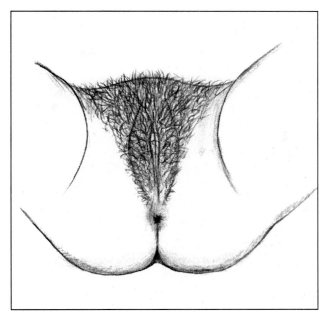

Figure 2–3 Two variations in the appearance of female genitalia.

the distance between the vaginal opening and clitoral glans (so that the penis can more easily stimulate the clitoris); but this, too, is untrue. The penis, in fact, usually does not come into direct contact with the clitoris during sexual intercourse. The thrusting of the penis only indirectly stimulates the clitoris, by causing the clitoral hood to rub back and forth over the glans. Surgical removal of the clitoral hood (*clitoral circumcision*) in order to directly expose the clitoris was once a common practice, done in order to enhance responsivity in wives who were thought to be "frigid." However, problems with sexual desire and trouble attaining orgasm are only rarely the result of an anatomical problem.

Surgical removal of the clitoris was sometimes performed in the United States and Europe during Victorian times in order to prevent female children from masturbating and growing up "oversexed." Clitoridectomy is no longer performed in America, of course, but it is very common in Northern African countries and parts of the Middle East, Malaysia, and Indonesia (see Box 4–A on page 83).

Thickened secretions called *smegma* can accumulate beneath the clitoral hood and result in discomfort during sexual intercourse by causing the glans to stick to the clitoral hood. The secretions can generally be washed off, however, by bathing thoroughly.

THE VAGINAL OPENING

The vaginal and urethral openings are visible only if the labia minora are parted. The area between the two labia minora is sometimes referred to as the **vestibular area** (Latin for "entrance hall") and the vaginal opening as the *introitus* (Latin for "entrance"). The vaginal opening, which is very sensitive to stimulation, is not a permanently open orifice always ready for penetration by the penis. It is surrounded by the **bulbocavernosus muscle,** a ring of sphincter muscles similar to the sphincter muscles surrounding the anus. Sexually experienced women can learn to voluntarily contract or relax these muscles during intercourse, and during childbirth the vaginal opening expands enough to accommodate delivery of a baby. In sexually inexperienced women, on the other hand, these muscles may involuntarily contract as a result of extreme nervousness, making penetration very difficult. The **vestibular bulbs,** which are located underneath the sphincter muscles on both sides of the vaginal opening, also help the vagina grip the penis by swelling with blood during sexual arousal.

In sexually inexperienced females, a thin membrane called the **hymen** (named after the Greek god of marriage) may partially cover the opening to the vagina. This membrane is found only in human females. Until shortly before birth, the hymen separates the vagina from the urinary system. It ruptures after birth,

vestibular area A term sometimes used to refer to the area between the two labia minora.

bulbocavernosus muscle A ring of sphincter muscles that surrounds the vaginal opening in females or the root of the penis in males.

vestibular bulbs Structures surrounding the vaginal opening that fill with blood during sexual arousal, resulting in swelling of the tissues and a narrowing of the vaginal opening.

hymen The thin membrane that partially covers the vaginal opening in many sexually inexperienced women. Its presence or absence, however, is really a very poor indicator of prior sexual experience.

but in humans remains as a fold of membrane around the vaginal opening. It is found in all normal newborn infants (Berenson, 1993; Berenson, Heger, & Andrew, 1991). Although the hymen has no known physiological function, it has recently been suggested that it serves to protect against vaginal infections in human infant females (Hobday, Havry, & Dayton, 1997).

The hymen may have one opening (annular hymen), two openings (septate hymen), or several openings (cribriform hymen) that allow for passage of menstrual flow. In the rare instance that the hymen has no opening (imperforate hymen), a simple surgical incision can be made at the time of first menstruation. The vaginal opening of women who have given birth is sometimes referred to as the *parous introitus* ("passed through").

The presence of the hymen has been used by men throughout history as proof of virginity. In the time of the Biblical Hebrews, a newly wed woman who did not bleed during first intercourse was sometimes stoned to death.

> But if the thing is true, that the tokens of virginity were not found in the young woman, then they shall bring out the young woman to the door of her father's house, and the men of her city shall stone her to death with stones, because she has wrought folly in Israel by playing the harlot in her father's house; so you shall purge the evil from the midst of you.
>
> (Deuteronomy 22:20–21, Revised Standard Version)

Even in many parts of the world today, a newly wed wife can be divorced or exiled if her husband believes her to be unchaste (Ford & Beach, 1951). Many Muslims, Chinese, and Moroccans display a bloodstained sheet on the wedding night as proof of a new bride's chastity. It was once the custom in some cultures for a priest to deviriginate a woman. In other cultures, girls are ritually "deflowered" with stone phalluses or horns. Although these extreme customs are not found in Western cultures, many American men nevertheless expect their female partners to bleed during first intercourse.

> "A girlfriend of mine, a few years ago, told me that she was a virgin, so naturally I figured that after we had sex she would bleed. Then, when she didn't I felt upset and betrayed."

> "When my girlfriend and I first had intercourse it was just understood by listening to my peers that she was supposed to bleed and I should see some kind of skin hanging (hymen). This

didn't happen and for years it bothered me because I thought she lied to me when she told me that I was her first. It wasn't until after taking this course that I realized how ignorant my friends and I were."

> "Being very athletic came natural for me and by the time I was 12, I had won four athletic trophies and many certificates. My mother took me to a doctor right after I made 14 and he told her that I was not a virgin because my hymen was not in place. I cried for 2 weeks. It was not true. My life was very traumatic for the next 6 months. Now at 40 years old, I am still angry."

(from the author's files)

In actuality, the hymen is really a very poor indicator of whether or not a female is sexually experienced, and there is often very little bleeding or discomfort during first sexual intercourse. In fact, pain during first intercourse often reflects a woman's attitudes about sex (Weis, 1985). The hymen sometimes ruptures during vigorous physical activity or insertion of a tampon, and in some women the opening in the hymen only stretches rather than tearing during first intercourse (thus there would be no bleeding). It speaks rather poorly of our attitudes about sex that many newly wed women are expected to bleed and experience great discomfort and pain in order to please their husbands. Untold millions of women throughout history have probably suffered much grief (and even death) for failing to display what many men wrongly believe is the appropriate response to first intercourse.

THE URETHRAL OPENING

Urine passes from the bladder through a small tube called the **urethra** and out the urethral opening, which is located below the clitoris and above the vaginal opening. The male's urethra serves for the passage of sperm as well as urine, but the female's urinary system is not related to her reproductive system.

Cystitis is an inflammation of the bladder with symptoms of frequent and burning urination. Women are more susceptible to this than men. Many factors increase the risk, including recent sexual intercourse and use of a diaphragm with spermicide (Hooton et al., 1996). Rectal bacteria can also cause cystitis. Thus, after a bowel movement, a woman should not wipe herself in the direction from the anus to the vulva (see Chapter 5).

THE BREASTS

Although **breasts** are not part of a woman's reproductive system, they are considered to be highly erotic by most men in Western societies, and therefore we must

---

urethra The passageway from the bladder to the exterior of the body. In males, it also serves as a passageway for semen during ejaculation.

breasts In females, glands that provide milk for infants; located at the front of the chest.

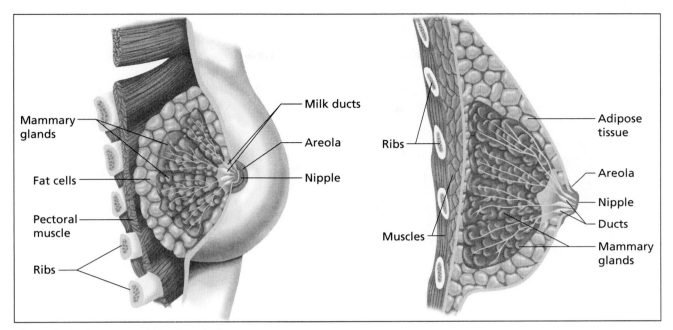

Figure 2–4 Anatomy of the female breast.

consider them part of a woman's sexual anatomy. Recall from Chapter 1, however, that female breasts have no erotic significance in many cultures around the world. (Remember the Tahitian women in the movie *Mutiny on the Bounty* who were naked from the waist up? Their naked breasts were erotic only to the English sailors.)

Breasts develop at puberty as a result of increasing levels of the hormone estrogen, which is produced by the ovaries. Thus, the breasts are really a secondary sex characteristic, just as pubic hair is. Interestingly, it is common for one breast (usually the left) to be slightly larger than the other. Males also have a slight increase in estrogen levels after puberty, and many show a slight increase in breast size (called *gynecomastia*) as a result. This may cause embarrassment for some young teenage boys, but the condition usually disappears within a few years.

Each adult female breast consists of 15 to 20 **mammary** (milk-producing) **glands** (see Figure 2–4). A separate duct connects each gland to the *nipple*, which is made up of smooth muscle fibers and also has lots of nerve fibers (making the nipples sensitive to touch). When a woman becomes sexually aroused, the smooth muscle fibers contract and the nipples become erect. The darkened area around the nipple, called the *areola*, becomes even darker during pregnancy. The small bumps on the areola are glands that secrete oil to keep the nipples lubricated during breast-feeding.

In the late stage of pregnancy, a hormone called *prolactin* from the pituitary gland (at the base of the brain) causes the mammary glands to start producing milk (called *lactation*). When the baby sucks on the nipple, it causes the pituitary to produce the hormone *oxytocin,* which results in the ejection of milk. There can also be secretions from the breasts at other times (a condition known as *galactorrhea*) as a result of certain drugs, birth control pills, stress, or rough fabrics rubbing against the breasts. Some people, including males, have inverted nipples or an extra nipple, but this generally does not pose a health problem.

Breasts vary in size from woman to woman (see Figure 2–5). Breast size is not determined by the number of mammary glands, which is about the same for all women, but by the amount of fatty (adipose) tissue packed between the glands. This is determined primarily by heredity. Although some men believe that large-breasted women have a greater sexual capacity than small-breasted women, there is really no relation between breast size and sensitivity to touch (Masters & Johnson, 1966). Large breasts do not have a greater number of nerve endings than small breasts. Nevertheless, because of the attention that many males give to breast size, some women attempt to enlarge their breasts.

Breast augmentation exercises and the lotions and mechanical devices advertised in women's magazines do not work (remember, breast size is determined by the amount of fatty tissue). In the 1960s, plastic surgeons began implanting soft pouches filled with silicone to increase the size of breasts, and at least two million American women have

mammary glands Milk-producing glands of the breast.

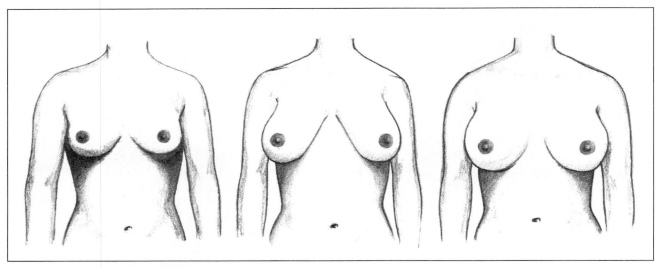

Figure 2–5 Some variations in the appearance of female breasts.

had their breasts enlarged in this manner (see Snyder, 1997). However, complications are common. In about 17.5 percent of women who receive breast implants, a fibrous capsule forms around the implant, making the breast hard, tight, and unnatural in appearance (Gabriel et al., 1997). Other complications include implant ruptures (5.7 percent), hematoma (5.7 percent), and wound infections (2.5 percent).

Ruptured implants initially raised concerns about the possibility of implants causing cancer and autoimmune disease (antibodies attacking the body's own tissues, resulting in arthritis or lupus). However, numerous scientific studies have found no evidence that silicone implants cause cancer or connective tissue disease (see Brinton & Brown, 1997; Marcus, 1996; or Snyder, 1997). Thousands of lawsuits have been filed by women against the manufacturers of implants, but court-appointed experts have also concluded that there is no evidence that implants cause these problems (Nash, 1997).

Although silicone implants apparently do not cause serious health problems, women who are considering implants should not forget about the many cases of painful hardening of the breasts. Talk-show host Jenny Jones has had six operations because of the damage her implants did to her breasts. She was quoted in *People* magazine as saying, "I hate my body a thousand times more now than I ever did before. . . . I would sell everything I own to be able to have the body back that I gave up." Students in my class have relayed similar experiences:

> "My mom had a breast augmentation job done about 2 years ago and has had nothing but problems. She started growing too much scar tissue and lumps in her breasts, and now, after all the pain and money and aggravation

she went through, she's having them removed this summer. I know she did it for my dad, but she can never tell me enough times how sorry she is she ever did it."

> (from the author's files)

In 1992, the Food and Drug Administration (FDA) severely restricted silicone implants for purely cosmetic purposes, but left them available for women with mastectomies (breast removal), for whom they felt the emotional benefits outweighed the other risks. Manufacturers are now making saline-filled implants and experimenting with soybean oil. However, these new implants will not eliminate the problem of encapsulation with hardening of the breasts.

Many naturally large-breasted women would like to be smaller:

> "I currently wear a DD cup. . . . My chest has become a large nuisance. It is hard for me to dance, jog, or do any kind of activity that involves moving around a lot. Finding clothes is very hard. . . . I just recently talked to a friend of mine who had a breast reduction. I really want to have this operation. It would make me feel better about myself."

> (from the author's files)

Breast reduction involves removing some of the fatty tissue, and although it generally has fewer complications, it can cause scars or leave numbness in parts of a breast.

Cancer of the breasts is the second major cause of cancer death in women. Women are advised, therefore, to learn how to examine themselves in order to make an early, possibly life-saving, detection (see Box 2-A).

Box 2-A   *SEXUALITY AND HEALTH*

Breast Cancer and Examination

Apart from skin cancer, cancer of the breasts is the most common type of cancer (and the second leading cause of cancer death) in women, with over 181,000 new cases and 44,000 deaths per year (American Cancer Society, 1997). The odds of getting breast cancer have been increasing over the last few decades, and the latest statistics show that one in eight American women will develop breast cancer in her lifetime (Harris, 1994). A fact that is less well known is that men can get breast cancer too (about 1,400 men will be diagnosed this year).

The cause or causes of most cases of breast cancer are unknown, but there are some factors that put a woman at high risk. These include a family history of breast cancer, being older than 50 years, never giving birth or having the first child after age 30, starting menstruation before age 12, or undergoing menopause after age 55 (see Gail et al., 1989). Breast-feeding lowers the risk of breast cancer (Romieu et al., 1996). About 5 percent of breast cancer cases are caused by an inherited gene mutation (Fitzgerald et al., 1996; Langston et al., 1996).

The rate of breast cancer in North America and Northern Europe is much higher than in Asia or Africa (Henderson & Bernstein, 1991), but children of immigrants from these low-incidence areas have rates of breast cancer similar to those of the high-incidence countries (Kliewer & Smith, 1995). This suggests that lifestyle and environment also play a role. Some of the factors that have been implicated as risks are obesity or lack of exercise (Huang et al.,1997; Stoll, 1996; Thune et al., 1997), alcohol consumption (Longnecker et al., 1992), and smoking (Calle et al., 1994). Estrogen (a hormone released by the ovaries) has long been thought to be a cause of breast cancer (see Toniolo et al., 1995), but results of studies looking at the effects of birth control pills (which contain estrogen) or postmenopausal estrogen replacement therapy have been inconsistent (Cobleigh et al., 1994; Colditz et al., 1995; Sattin et al., 1986; Stanford et al., 1995; Steinberg et al., 1991). The present consensus is that if the birth control pill or hormone replacement therapy does increase the risk, the increase is only very slight, and only with long-term use (Collaborative Group on Hormones, 1996; Stan-

ford et al., 1995; Steinberg et al., 1991; Romieu, Berlin, & Colditz, 1990). How little is known about breast cancer is revealed by the fact that only about half of all cases are associated with any of the known risk factors.

About half of all cases of breast cancer have spread beyond the breast before they are discovered. However, with early detection, there is a 90 percent survival rate (American Cancer Society), and with a little instruction, the female herself is the one most likely to make a life-saving early detection. Some women are reluctant or fearful to examine themselves or to go to a doctor after discovering a lump. When then–First Lady Betty Ford publicly announced in 1974 that she would have surgery for a self-discovered lump, she probably saved thousands of lives by making the subject less taboo. The death of Linda McCartney reemphasized the importance of early detection.

Women should examine their breasts on a monthly basis. The best time is immediately after menstruation ends, when estrogen levels are low and the breasts are not tender or swollen. Women who are pregnant, postmenopausal, or do not menstruate regularly should examine themselves at a regular time each month. First examine yourself in front of a mirror. Do this with your hands at your sides, then with your hands on your hips, and finally with your hands raised above your head. Look for any bulging, flattening, dimpling, or redness. The symptoms are usually painless. In each position, examine one breast with the opposite hand. With fingers flat, press gently in small circular motions, starting at the top, and check the outermost part of the entire breast in clockwise fashion (see Figure 2–6). It is normal to have a ridge of firm tissue in the lower curve of each breast. After you have completed the circle, move your fingers an inch closer to the nipple and make another complete check around the breast. At least three to four complete circles around the breast will be required. Be sure to examine under your arms as well, because this is one of the most common areas for cancer to occur. In addition, examine the nipple by squeezing it and noting any discharge. Repeat the same procedure on the other breast. When you are done, examine

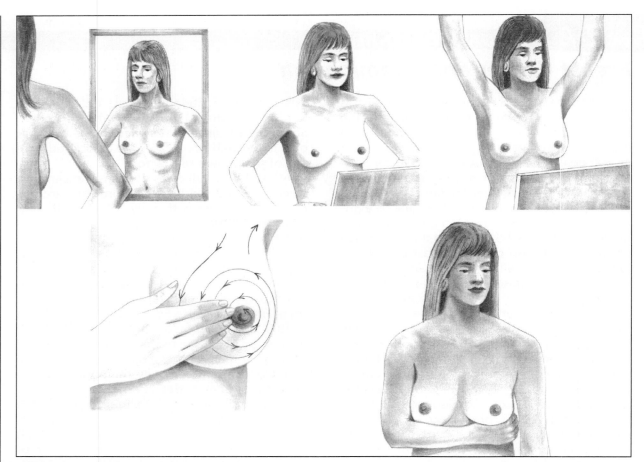

Figure 2–6   Breast self-examination.

yourself again during a bath or shower, as it is often easier to detect a lump, hard knot, or thickening when the skin is wet and slippery. Finally, report any suspected abnormality to your doctor immediately.

> "I am a 33-year-old nurse that has instructed many other women about self-examination, but unfortunately, did not practice what I preached. After one of your lectures I did examine my breasts and was horrified to find a small lump. . . ."
>
> (from the author's files)

Unfortunately, by the time many women discover a lump, the cancer has already spread to other parts of their bodies. As a result, the American Cancer Society has been strongly recommending that women start having a **baseline mammogram** (breast X-ray) to find cancers that are too small to be felt by hand. They suggest one baseline mammogram be taken between the ages of 35 and 40, another every one to two years between the ages of 40 to 49, and an annual breast X-ray after the age of 50. Eleven leading medical organizations endorsed this schedule in 1989. The amount of radiation used in a mammogram is quite small, and the American Cancer Society believes that the long-term benefits far outweigh any potential risks (Mettler et al., 1996). In fact, the number of cancers detected by mammograms has doubled since the mid-1980s. Unfortunately, many women were left confused when in 1993 the National Cancer Institute withdrew its endorsement of mammograms for women under age 50. However, after a review of the evidence, they reversed their decision in 1997.

Eight out of ten lumps that are discovered are benign (noncancerous) cysts (fluid-filled sacs) or fibroadenomas (solid tumors). They are known as *fibrocystic disease* and are believed to be caused by hormones.

baseline mammogram The first of a series of low-radiation X-rays used to detect breast tumors.

"When I was 15 years old, I discovered a breast lump accidentally while bathing. That was one of the scariest days of my life. My doctors told me I had fibrocystic disease. The day before my eighteenth birthday, I had surgery to remove another lump."

(from the author's files)

Therefore, do not panic if you find a lump in your breast, as the chances are good that it is not malignant. However, there is some evidence that having benign tumors increases the risk of future breast cancer among premenopausal women (London et al., 1992).

Biopsies are now being done by needle, which is faster, more exact, and less painful than the conventional surgical biopsy. If you should have a lump and it is diagnosed as cancerous, a number of surgical procedures might be performed: (1) radical *mastectomy*, in which the entire breast, underlying muscle, and lymph nodes are removed; (2) modified radical mastectomy, in which the breast and lymph nodes, but not the underlying muscles, are removed; (3) simple mastectomy, in which only the breast is removed; and (4) *lumpectomy,* in which only the lump and a small bit of surrounding tissue is removed, followed by radiation therapy. Studies show that with early detection, lumpectomy and radiation treatment are as effective as radical mastectomy (Fisher et al., 1995). Preoperative chemotherapy allows even more women to become candidates for breast-saving lumpectomy (Fisher et al., 1997). In cases of cancer that have spread to the lymph nodes, the established procedure is mastectomy with chemotherapy. If one or both breasts have to be removed, breast reconstruction through plastic surgery is often possible (over 50,000 breast cancer patients choose to do this every year), but counseling for both the female and her partner is equally important.

Whatever the therapy, it is common for women with breast cancer to experience serious sexual problems and the need to make adjustments (Kaplan, 1992a; Makar et al., 1997). The reaction of a woman's partner is important. A woman who has had a breast removed needs to be reassured by her partner that she is not going to be desired or loved less than before. Would you expect your partner to love you less if you lost a toe, foot, leg, finger, hand, or arm? Of course not. So why should it be any different if your partner loses a breast? This may seem obvious to you, but your partner will probably want to hear you say it.

"I came from a family consisting of 5 daughters and no one had ever been diagnosed with breast cancer. I had never given it a thought that I would be the one to have a lump in my breast. . . . I had a modified radical mastectomy of my left breast. . . . When my doctor told my husband and me my husband was wonderful about it. All you need to get over the shock is an understanding and supportive husband and family. Emotionally I was able to get over the surgery because he never made me feel like 'half' a woman."

(from the author's files)

For more information and to help find the most appropriate tests and therapies, call the National Cancer Institute hotline:
1–800–4–CANCER.

INTERNAL FEMALE ANATOMY

The vagina, uterus, Fallopian tubes, and ovaries are often referred to as the female's reproductive system (see Figure 2–7). The vagina serves as a depository for the male's sperm. Eggs are produced by the ovaries. During a woman's reproductive years, one or more eggs will mature on a monthly basis and will be released from an ovary and picked up by one of the Fallopian tubes. It is in the Fallopian tube that an egg and a sperm unite to start a new living being. The fertilized egg then travels through the tube and implants itself within the uterus (see Chapter 7).

THE VAGINA

We have already discussed the vaginal opening as part of the genitalia or external anatomy, but the **vagina** (Latin for "sheath") is really an internal structure located behind the bladder and in front of the rectum. It serves not only as a depository for sperm during sexual intercourse, but also as the birth canal, and thus is capable of expanding to about four inches in diameter. However, contrary to what some men may believe, the vagina is not an open orifice always ready to accommodate the insertion of a penis. In the

vagina The sheathlike canal in the female that extends from the vulva to the cervix and that receives the penis during intercourse.

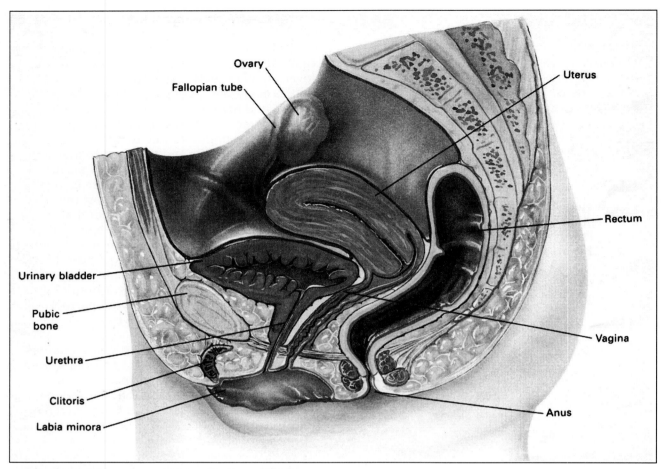

Figure 2–7(A) The internal female reproductive system, viewed from the side.

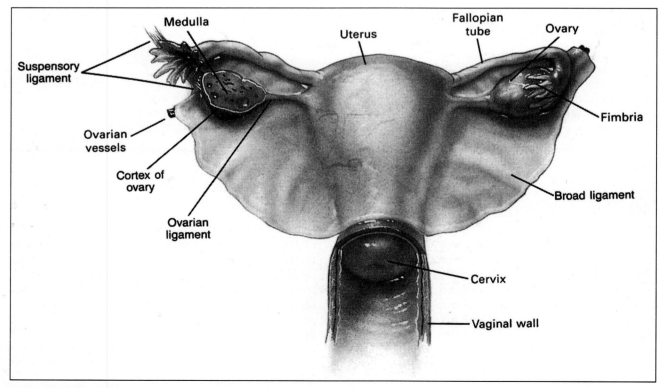

Figure 2–7(B) A frontal view of the internal female reproductive system.

unstimulated state, it is about three to five inches long and its walls are collapsed. It is not until a woman becomes sexually aroused that the walls of the inner two-thirds of the vagina begin to expand to accommodate a penis. The vaginal opening actually narrows during sexual arousal (see Chapter 4).

The vaginal walls have three layers. The inner layer has a soft mucosal surface similar to that of the inside of the mouth. Before puberty, the walls of the vagina are thin and relatively inelastic. The rising levels of female hormones at puberty cause the vaginal walls to thicken and become more elastic and highly vascularized (i.e., having lots of blood vessels). Vaginal lubrication during sexual arousal results from the walls of the vagina becoming engorged with blood, with the resulting pressure causing fluid (super-filtered blood plasma) to be secreted from the mucosal lining. At menopause (the time in life when a female's reproductive capacity ends) the ovaries atrophy, and the consequent loss of hormones causes the walls of the vagina to again become thin and inelastic, similar to their condition before puberty. There is also a decrease in the blood supply to the vaginal walls after menopause, resulting in decreased vaginal lubrication, and thus many older women may require a lubricant during sexual intercourse.

The vagina is a self-cleansing organ. Many potentially harmful bacteria from the outside environment are destroyed by other bacteria that are found naturally within the vagina. In addition, the walls of the vagina continually secrete fluids that help to maintain an acidic environment. A healthy vagina has a musky odor. Vaginal odors act as a sexual stimulant to the males in most mammalian species. Vaginal secretions from female rhesus monkeys, for example, contain fatty acids called copulins, which attract the male monkeys. A similar substance has been found in human vaginal secretions, and one study reported that when married women rubbed perfumes containing human copulin on their breasts before going to bed, many of their husbands showed increased sexual interest (Morris & Udry, 1978).

In laboratory studies, most men do not find musky vaginal odors offensive (Hassett, 1978). Humans, however, are often raised to believe that vaginal odors are offensive. If a person hears enough dirty jokes and stories about vaginal odors before becoming sexually experienced, he or she will come to believe this even though it is not true. How, then, did these stories get started? Probably because most females will occasionally get a vaginal infection that can be accompanied by a foul odor (see Chapter 5 on vaginitis). These infections can be easily and quickly cured. In order to sell their products, however, advertisers of feminine hygiene sprays and douches generally imply that vaginal odors are always offensive and must be masked with a perfumed substance. The fact is that these prod-

ucts do not cure vaginal infections; in fact, they generally make them worse by further changing the chemical environment of the vagina.

> "A year ago I started using a feminine hygiene deodorant spray. I had used this spray for an average of 5 months before I realized that my discharge was abnormal. After talking with my doctor I found that I had a yeast infection. All of this could have been avoided if I had only stayed away from feminine deodorant sprays."
>
> (from the author's files)

Recent studies have found that regular douching increases the risk of pelvic inflammatory disease (a serious infection of a woman's reproduction system; see Chapter 5), ectopic pregnancy, and perhaps even cervical cancer (Zhang, Thomas,& Leybovich, 1997). Women should avoid using douches unless told to do so by a physician.

Although the vaginal opening is very sensitive to touch, the walls of the inner two-thirds of the vagina are relatively insensitive to touch as a result of having few nerve endings. It has been demonstrated that some women cannot even tell when their inner vaginal walls are being touched by a probe (Kinsey et al., 1953). Does this mean that the presence of a penis in the vagina does not contribute to a woman's subjective feelings of pleasure during sexual intercourse? Not necessarily. Although the vaginal walls are sparse in nerve endings, the vagina is surrounded by a large muscle, the **pubococcygeus (PC) muscle,** which is more richly innervated with nerves. It has been found that women who have little vaginal sensation during intercourse (some report that they cannot even feel the penis) have a weak pubococcygeus muscle (Kegel, 1952). As a result, some sex therapists have advocated daily exercises to strengthen this muscle in order to enhance pleasure during sexual intercourse (Barbach, 1976; Graber, 1982; Perry & Whipple, 1981). The exercises are called **Kegel exercises.** They are the same exercises that many physicians instruct women to do after having a baby in order to regain urinary control. The pubococcygeus muscle controls urine flow, but it becomes stretched and weak during pregnancy, causing many women to suffer from urinary incontinence.

For those women interested, here are the steps for doing Kegel exercises: (1) Learn what it feels like to contract the PC muscle by stopping yourself in the middle of

pubococcygeus muscle The major muscle in the pelvic region. Voluntary control over this muscle (to help prevent urinary incontinence or to enhance physical sensations during intercourse) is gained through Kegel exercises.

Kegel exercises Exercises that are designed to strengthen the pubococcygeus muscle that surrounds the bladder and vagina.

urinating with your legs apart; (2) after you know what it feels like to contract the PC muscle, insert a finger into your vagina and bear down until you can feel the muscle squeeze your finger; and (3) once you can do this, start doing daily Kegels without a finger inserted. Physicians and therapists suggest practicing ten contractions of three seconds each three times a day. The best thing about Kegel exercises is that they can be done anytime and anywhere, and no one else knows what you are doing.

There is an exception to the general statement that the vaginal walls are relatively insensitive to touch. Some women have a small sensitive spot on the front wall of the vagina at about the level of the top of the pubic bone (about a third or halfway in from the vaginal opening). This has been named the **Grafenberg (G) spot** after the German doctor who first reported it in 1950 (Perry & Whipple, 1981). Stimulation of the G spot often leads to an orgasm that some women say feels different from an orgasm caused by clitoral stimulation (see Chapter 4). Originally it was claimed that all women had a G spot (Ladas, Whipple, & Perry, 1982), but more recent studies suggest that only 10 percent or fewer women have this area of heightened sensitivity (Alzate & Londono, 1984).

One in every 4,000 to 5,000 girls is born without a vagina (Harnish, 1988). When this occurs, physicians are usually able to surgically construct one.

THE UTERUS

It is within the **uterus,** or womb, that a fertilized egg will attach itself and become an embryo and then a fetus. Resembling a small, inverted pear, the uterus in women who have not had children measures only about three inches long and three inches across at its broadest portion, but it is capable of tremendous expansion. The *cervix,* the narrow end of the uterus, projects into the back of the vagina and can be easily felt (as a slippery bump) by inserting a finger into the back of the vagina. The broad part of the uterus is called the *fundus.* Ligaments hold the uterus in the pelvic cavity at about a 90-degree angle to the vagina, although in some women it may tilt slightly forward or backward.

The uterus has three layers: the innermost **endometrium** where the fertilized egg implants; a strong middle layer of muscles called the *myometrium,* which contracts during labor; and an external cover called the *perimetrium.* The endometrium thickens and becomes rich in blood vessels after ovulation, but it is sloughed off and discharged from the female's body during menstruation if fertilization does not occur. Glands within the cervix secrete mucus throughout the female's menstrual cycle, and a change in the consistency of this mucus is often used to tell when ovulation has occurred. The cervical opening, or *os* (Latin for "mouth"), is normally no wider than the diameter of a matchstick, but it dilates to 10 centimeters (about 4 inches) at childbirth to allow delivery of the baby.

Unfortunately, the cervix is the site of one of the most common types of cancer in women. It is extremely important, therefore, that all females, whether sexually inexperienced or sexually active, go to a physician for an annual examination and Pap smear (see Box 2-B).

THE FALLOPIAN TUBES

Extending 4 inches laterally from both sides of the uterus are the **Fallopian tubes,** or oviducts, as they are sometimes called. There is no direct physical connection between the Fallopian tubes and the ovaries, but the fingerlike projections at the end of the tubes (called *fimbria*) brush against the ovaries. After an egg is expelled from an ovary into the abdominal cavity at ovulation, it is picked up by one of the fimbria. If a sperm fertilizes the egg, it will usually do so within the tube. The fertilized egg then continues its three-to-four-day trip through the tube and normally implants itself in the endometrium of the uterus. Anatomical abnormalities and infections can block the tube and cause sterility (sperm cannot meet egg) or a tubal pregnancy (a fertilized egg implants itself within the tube). The latter poses a serious health threat because the tubes are not capable of expanding like the uterus. Either tying off and cutting or cauterizing the tubes (*tubal ligation* or *laparoscopy*) is a popular means of birth control for couples who already have all the children they want (see Chapter 6).

THE OVARIES

The **ovaries** are the female gonads and develop from the same embryonic tissue as the male gonads, the testicles. They are suspended by ligaments on both sides of the uterus. The ovaries have two functions: to produce eggs, or *ova* (sing. *ovum*) and to produce female hormones (*estrogen* and *progesterone*). A female has all the immature eggs that she will ever have at birth, as many as 300,000 to 400,000, each contained within a thin capsule to form what is called a **primary follicle.** At puberty, one or more of the follicles is stimulated to mature on about a monthly basis (the menstrual cycle). When mature, the follicle is called a *Graafian follicle.* This process ends at menopause, and thus only about 400 of the several hundred thousand ova in a woman's ovaries will ever mature during her lifetime. In contrast, the male's testes produce millions of new sperm every day.

Grafenberg (G) spot A small, sensitive area on the front wall of the vagina found in about 10 percent of all women.

uterus The womb. The hollow, muscular organ in females where the fertilized egg normally implants.

endometrium The inner mucous membrane of the uterus where a fertilized egg implants. Its thickness varies with the phase of the menstrual cycle.

Fallopian tubes The passageways that eggs follow on their way to the uterus.

ovary The female gonad in which ova are produced.

primary follicle An immature ovum enclosed by a single layer of cells.

Box 2-B   SEXUALITY AND HEALTH

Cancer of the Female Reproductive System

Cancer of the Cervix

There are approximately 60,000 new cases of cervical cancer in the United States every year, of which 14,500 are invasive (have spread beyond the surface of the cervix) (American Cancer Society, 1997). The average age of women diagnosed with cervical cancer is 45, but women can have this cancer at almost any age. Women at high risk include those who began having sexual intercourse at an early age or who have had numerous sexual partners (Clarke et al., 1985) and those whose partners have had numerous sexual partners (Slattery et al., 1989). Cancer of the cervix is rare among nuns and celibate women. It is not sex per se that is believed to be the cause of cervical cancer, but viruses that are spread during sexual activity, such as human papillomavirus and herpes virus (Slattery et al., 1989). Regular use of barrier methods of contraception (such as condoms and diaphragms) reduces the risk of getting cervical cancer (Slattery et al., 1989). Women who smoke also have a higher risk of cervical cancer (Daling et al., 1992; Gram, Austin, & Stalsberg, 1992), probably because smoking weakens the body's defenses, making it easier for the cancer-causing viruses to have their effect (Burger et al., 1993).

A pelvic exam and **Pap smear** test (named after Dr. George Papanicolaou, who developed the test) are a necessary part of female health care. The risk of developing cervical cancer is only about one-fourth as great for women who get Pap smears regularly as it is for women who do not (Shy et al., 1989). Most cervical cancer–related deaths occur in women who do not get regular Pap tests. Dr. Patricia Braley, who headed a National Institutes of Health panel on cervical cancer in 1996, said, "In theory, cervical cancer is a cancer we can completely prevent. If we could reach all the women in this country who are not getting regular Pap tests we could eradicate this type of cancer."

Many women are nevertheless reluctant to get a Pap smear because of embarrassment. The first step, therefore, is to find a doctor with whom you are comfortable and who shows concern for his or her patients (as by warming instruments to body temperature before using them). In the examination room, the woman lies flat on her back with her legs apart and her feet in footrests called stirrups. An instrument called a speculum, which is shaped like a duck's bill, is inserted into the vagina and gently opened so that the vaginal walls and cervix may be inspected. A cotton swab or thin wooden stick is used to gently skim the surface of the cervix in order to collect cells, which are then examined under a microscope for abnormalities. During the exam, the doctor will also check the vaginal walls and vulva for possible infections. Some doctors may include a rectal exam at this time also. The entire process generally takes only a few minutes, and although some women may experience discomfort, it is generally not painful.

> "Human sexuality class has made me more educated about my body, and I became more aware. Before I entered this class I was very scared of going to the gynecologist for a visit (age 20). Recently I went for my first visit and found my fears were ridiculous. It was not bad at all and I even felt better to find out I was OK."

> "At the beginning of the semester you mentioned UNO's Health Services and I decided to check them out. When the results of my Pap smear came in, I had tested positive. I had stage II dysplasia. I had cryotherapy. . . . If it had not been for this class, I don't know how long I would have waited to get a Pap smear or what the results would have been."

(from the author's files)

Any detected abnormalities of cells are treated according to severity. An estimated one million women per year are found to have low-grade dysplasia [abnormal cellular changes]. In the precancerous stage, the localized abnormal cells are often destroyed by extreme cold (cryotherapy) or intense heat (electrocoagulation). Because most precancerous abnormalities are slow to develop and have a low risk of becoming cancer, some physicians believe in a watch-and-wait strategy at this stage (Bergstrom et al., 1993; Solomon, 1993). Precancerous abnormalities in

Pap smear A test for cancer of the cervix in females; named for Dr. Papanicolaou, who developed it.

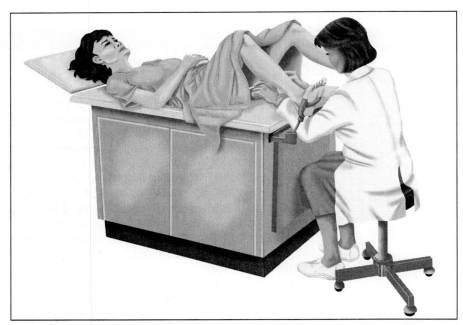

Figure 2–8   The pelvic examination.

Pelvic exams are also done to fit a diaphragm or IUD, to check for vaginal infections, and to examine the uterus during pregnancy. Many women assume that their doctor routinely checks for sexually transmitted disease when taking a Pap smear, but most do not unless specifically asked to do so. If you have had multiple sexual partners (and are thus at high risk to get a sexually transmitted disease), you should consider these additional tests at this time.

cells often take as long as eight years before they start to invade other tissues. If the cancer progresses, warning signs may include bleeding between menstrual periods, excessive bleeding during periods, and bleeding after menopause. However, sometimes there are no symptoms. If cancer of the cervix is found, a woman may have to have a *hysterectomy* (surgical removal of the uterus). This will end her reproductive capabilities, of course, but the surgery does not interfere with sexual desire or make her less feminine in appearance (Schover, 1988).

> "I thought after my hysterectomy my vagina would dry up and I would lose my sexual desire. It has been five years and I still take hormones but I haven't dried up and I have reached my sexual peak. Within these five years I have felt better about my self sexually. . . ."
>
> (from the author's files)

The American Cancer Society advises women that regular checkups should begin when they start having sexual intercourse, or at age 18 even if they are sexually inexperienced. Annual checkups are required at first because the Pap smear test sometimes does not detect precancerous conditions (Solomon, 1993). After three initial negative tests, your doctor may tell you to come back less often.

Cancer of the Endometrium and Ovaries

Cancer of the endometrium is also very common, with an estimated 34,900 new cases and 6,000 deaths per year in the United States (American Cancer Society, 1997). Typically, the first symptom is abnormal vaginal bleeding, so if you should experience this, go to a physician immediately. He or she will make a diagnosis by dilation and curretage (D&C). Should cancer be found, your doctor will probably recommend a hysterectomy, but radiation, drug, or hormonal therapies are also used sometimes.

Ovarian cancer is a rare disease (lifetime risk of less than 1.5 percent), but a deadly one. It causes the death of more American women each year (approximately 14,200) than either cervical or endometrial cancer. Most cases are found in postmenopausal women or women approaching menopause (between 50 and 59 years old). It is more common in women who have never been pregnant, and less so in women who use oral contraceptives. Recently, principal investigators from 12 studies reported that the use of fertility drugs was associated with an increased risk of ovarian cancer (Whittemore et al., 1992a,b). However, the highest risk factor is a family history of ovarian cancer (Lynch & Lynch, 1992). Breast-feeding offers some protection against ovarian cancer (Siskind et al., 1997).

Cancer of the ovaries is usually painless in the early stages. Thus the tumors are often quite large by the time they are discovered, making effective treatment difficult. Today there is a blood test for women at high risk (it tests for CA-125, a protein produced by cancerous ovarian cells). Treatment usually involves removal of the ovaries and radiation.

Many ovarian cysts are benign (noncancerous), but these cysts can become quite large.

Recently, surgeons at Stanford University removed a cyst weighing 303 pounds (the woman herself weighed 210 pounds after the operation). Studies show that in the case of benign cysts, removal of the cyst alone results in fewer problems than does hysterectomy (surgical removal of the uterus).

EXTERNAL MALE ANATOMY

Generally speaking, men have a better idea than women of what their genitals look like. Not only are the male genitalia more visible than the female structures, but boys are taught at a very young age to hold the penis while urinating, and thus may be less inhibited than girls about touching and examining themselves. Although they differ in appearance, many of the male structures develop from the same embryonic tissue as female structures (the penis and clitoris, for example). The male reproductive system and genitalia are shown in Figures 2–9 and 2–10. Penises differ in appearance just like any other part of the body, and men should not consider themselves abnormal if they do not look identical to Figure 2–10.

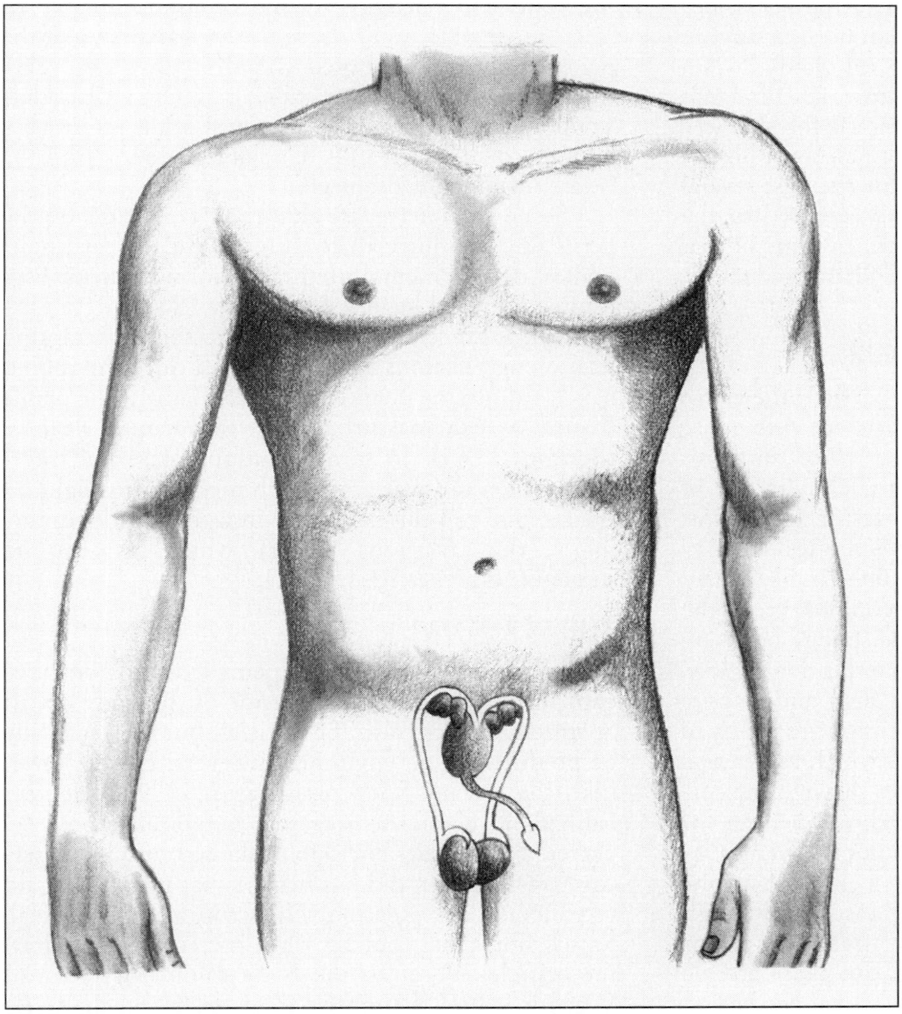

THE PENIS: OUTER APPEARANCE

The **penis** (Latin for "tail") has both a reproductive and a urinary function. It serves to deposit sperm in the female's vagina, and also to eliminate urine from the bladder. When unaroused, the penis is soft and hangs between the legs, but it hardens and becomes erect during sexual stimulation, thus enabling penetration of the vagina. The average size of the penis in the unstimulated (flaccid) condition is 3.75 inches in length and 1.2 inches in diameter, and about 6 inches in length and 1.5 inches in diameter when erect (Jamison & Gebhard, 1988; Sparling, 1997). Some penises are curved to the left or right, and others are curved up or down. In a rare condition called micropenis, the penis never develops to more than 1 inch in length (usually due to low levels of the male hormone testosterone; see Aaronson, 1994). In Chapter 4, we will consider whether penis size relates to sexual satisfaction in women.

Figure 2–9 Anatomical location of the male reproductive system.

> **penis** The male organ for sexual intercourse and the passageway for sperm and urine.

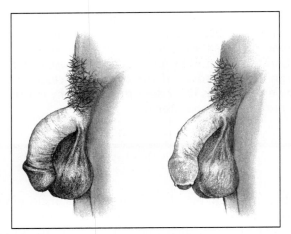

Figure 2–10 A circumcised (left) and uncircumcised (right) penis.

The skin of the penis is very loose to allow for expansion during erection. At birth the skin folds over the glans, the smooth rounded end of the penis, but many males have had their *foreskin* cut off in a surgical procedure known as **circumcision** (see Figure 2–10). If you should have a son, be sure to teach him that good hygiene includes cleaning under the foreskin (see Box 2-C). However, when bathing a newborn, do not pull back forcibly on the foreskin for the first several months, because it is often partially adhered to the penis during this time (Paige, 1978b). In some boys it is still adhered at 3 years of age (Williams, Chell, & Kapila, 1993).

THE PENIS: INTERNAL STRUCTURE

The penis has three parts: the body or shaft; the glans; and the root. Only the first two parts are visible (see Figure 2–11). In cross-section, the *shaft* of the penis can be seen to consist of three parallel cylinders of spongy tissue, two *corpora cavernosa* (or "cavernous bodies") on top and a *corpus spongiosum* (or "spongy body") on the bottom. Each is contained in its own fibrous sheath. The corpora cavernosa do not extend the full length of the penis. Instead, the spongy body expands greatly in front to form the round, smooth *glans*. The raised rim at the border of the shaft and glans is called the *corona* (and is the most sensitive to touch of any part of the penis). The *urethra*, which serves as a passageway for urine and sperm, runs through the corpus spongiosum, and thus the urethral opening (or meatus) is normally located at the tip of the glans. Occasionally it is located to one side of the glans, and although this usually poses no major problem, it can cause a male to have a negative perception of his own genitals (Mureau et al., 1995). For reasons not yet determined, this condition (called *hypospadias*) has nearly doubled in frequency of occurrence since 1970, but it can be corrected by surgery (Paulozzi et al., 1997).

The *root* of the penis consists of the expanded ends of the cavernous bodies, which fan out to form *crura* and attach to the pubic bone (note the similarity in structure to the clitoris), and the expanded end of the spongy body (called the *bulb*). The root is surrounded by two muscles (the bulbocavernosus and the ischiocavernosus) that aid in urination and ejaculation.

In human males, an erection does not result from a bone protruding into the penis, as in some mammalian species (e.g., dogs), nor does it result from voluntary contraction of a muscle. When a male becomes sexually aroused, the arteries going to the penis dilate and the many cavities of the cavernous and spongy bodies fill with blood. Valves in the veins that drain the penis simultaneously close. Expansion of the sinuses (cavities) and contraction of the muscles at the base of the penis may further aid this process by pressing against the veins and partially cutting off the drainage. In brief, an erection results from the spongy tissues of the penis becoming engorged (filled) with blood. Detumescence, or loss of erection, is not as well understood, but probably occurs as a result of the arteries beginning to constrict, thus removing pressure from the veins and allowing drainage.

Some men have an abnormal curvature of the penis due to injury to the uretha and/or cavernous bodies, developmental urethral abnormalities, or Peyronie's disease (Gregory & Purcell, 1989). If too extreme, this can cause pain during erection. Attempting to prolong erections by use of "cock rings" or rubber bands around the penis can severely damage the penile tissues, as can masturbating with vacuum cleaners (Benson, 1985). In some positions of sexual intercourse, men can place too much pressure or weight on their penis, and this can result in a "fracture" (rupturing of the cavernous bodies), which may require surgery (Adduci & Ross, 1991).

THE SCROTUM

The sac located beneath the penis is called the **scrotum**. It holds the testicles outside of the body cavity. Sperm are produced in the testicles, but they can only be produced in an environment several degrees lower than the normal body temperature of 98.6 degrees Fahrenheit. Scrotal temperature is typically about 93 degrees Fahrenheit. The skin of the scrotum is sparsely covered with hair and has many sweat glands that aid in temperature regulation. The scrotum also has small muscle fibers that contract when it is cold to help draw the testicles closer to the body cavity for warmth. When it is hot, the muscle fibers relax and the testes are suspended farther away from the body cavity.

circumcision In males, the removal of all or part of the foreskin of the penis. In females, the removal of the clitoral hood.

scrotum The pouch beneath the penis that contains the testicles.

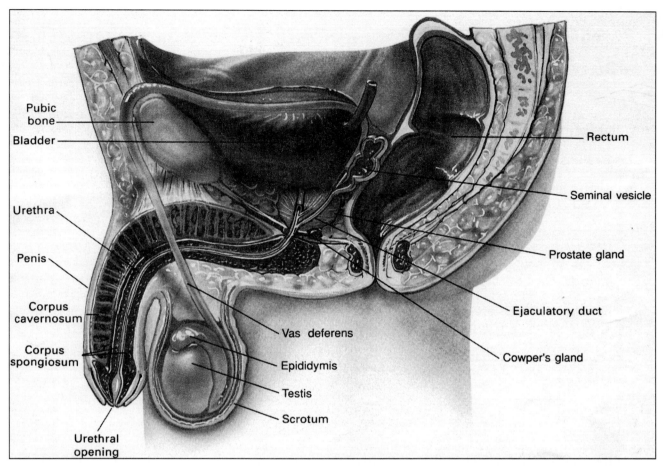

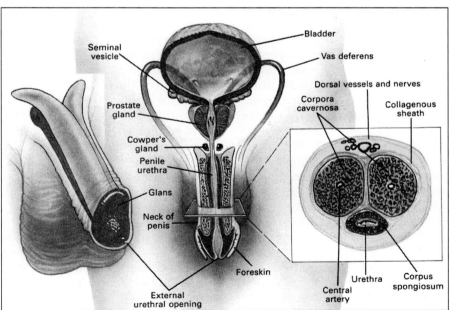

Figure 2–11   The internal male reproductive system: (top) from a side view and cross-section of the penis; (bottom) from a frontal view.

Box 2-C CROSS-CULTURAL PERSPECTIVES/SEXUALITY AND HEALTH

Male Circumcision

Circumcision has been performed on males since at least 4000 B.C. Egyptian mummies dating from this period give evidence of it, and the Greek historians and geographers Herodotus and Strabo later wrote about observing it in their travels (Larue, 1989; Manniche, 1987). The ancient Egyptians and Ethiopians performed circumcisions for cleanliness (Herodotus), and the practice was borrowed by the Hebrews at about the time of Abraham (the custom later fell into disuse, but was revived by Moses). Today, Jewish boys are circumcised on the eighth day of life to symbolize the covenant with God made by Abraham (Genesis 17:9–27). Muslim boys are also circumcised as part of a religious ceremony, often at 13 years of age.

In many cultures today, the operation is done as a puberty rite, and sometimes it involves more than the cutting away of the foreskin. It is done with crude instruments and without anesthesia. The Mangaians and other South Pacific islanders cut lengthwise along the top of the penis (superincision) all the way to the stomach (Marshall, 1971). Primitive tribes in central Australia make a slit to the depth of the urethra along the entire length of the lower side of the penis (subincision) so that males urinate at the base of the penis, requiring that they sit like women (Gregersen, 1982). Still other cultures insert rods with brushes, bells, or pins attached to the end (D. Brown, 1990). Many anthropologists believe that these customs are performed as symbolic pledges of loyalty to the group (Paige & Paige, 1981).

In the United States and England, circumcision became popular during the nineteenth century because Victorian physicians believed that circumcised penises were less sensitive and that the operation would therefore prevent masturbation and hypersexuality. Remember from Chapter 1 that Freud and other physicians of that day believed that loss of semen was just as detrimental to a male's health as loss of blood (Haller & Haller, 1977) and that an array of maladies, including insanity, awaited the masturbator. Circumcision reached its height in popularity between World Wars I and II, but then fell out of favor in most English-speaking countries. Nowadays only one-quarter of all male babies or fewer are routinely circumcised in countries such as England, Canada, New Zealand, and Australia, leaving the United States the only country in which circumcision is still performed for nonreligious reasons on a majority of boys. About 60 percent of all newborn baby boys are circumcised yearly in the United States (National Center for Health Statistics), but the practice is much more common among white males.

Circumcising boys' penises, of course, did not stop them from masturbating. Masters and Johnson do not believe that circumcision affects a male's sexual function in any way whatsoever (Masters, Johnson, & Kolodny, 1992). In fact, a recent study found that circumcised men had greater sexual satisfaction and fewer sexual problems than uncircumcised men, and more of them masturbated (Laumann, Masi, & Zuckerman, 1997).

Today, the reason for circumcision most commonly given by American physicians is to ensure proper hygiene, but there is considerable debate about this, even within the medical community. Glands located beneath the foreskin secrete an oily substance that gets mixed with dead cells to form a cheesy substance called **smegma,** which can cause the foreskin to adhere to the underlying glans of the penis. Without proper hygiene, the smegma serves as a possible breeding ground for bacteria. Several studies have found that many young boys have poor genital hygiene (Schoen, 1990). Cancer of the penis is rare, but some studies reported that it was more common in uncircumcised males (Rotolo & Lynch, 1991). However, penile cancer is no more common among uncircumcised men who regularly clean under their foreskins than among circumcised men.

Occasionally there are serious health risks associated with circumcision, including hemorrhage, infection, and accidental amputation (Poland, 1990; Snyder, 1989; Thompson, 1990). In addition, circumcision is usually done on infant boys without the use of anesthesia, and although some physicians advocate the use of local painkillers (Taddio et al., 1997), for most infants circumcision is a painful operation that

smegma The cheesy secretion of sebaceous glands that can cause the clitoris to stick to the clitoral hood or the foreskin of the penis to stick to the glans.

causes them to cry for hours (Howard, Howard, & Weitzman, 1994; Tyler, 1988). Neonatal circumcision may increase responses to all painful stimuli for several months (Taddio et al., 1995).

> "When I had my son in the 70s I thought after his circumcision he was set in that area. After a couple of weeks I had to take him to the doctor's office because his penis had become red and inflamed. . . ."

> "When I was 10 years old, I saw my baby brother get circumcised. I hated to see the pain that he was going through. I could not see why a baby should go through it at all. So I have decided to not let my baby boy go through that kind of pain. I just won't!"

> (from the author's files)

As a result of these concerns, both the American Academy of Pediatrics and the American College of Obstetricians and Gynecologists took a joint stand against the routine circumcision of newborns (*Guidelines for Perinatal Care,* 1983). However, since that time concerns have been raised about increased risk of urinary tract infection in uncircumcised boys (Fergusson, Lawton, & Shannon, 1988; Herzog, 1989; Wiswell et al., 1987). Some studies have also found that circumcision makes it less likely that a male will be infected with a sexually transmitted disease (including HIV, the virus that causes AIDS) if he has unprotected intercourse with an infected partner (Bongaarts et al., 1989; Caldwell & Caldwell, 1996; Cook, Koutsky, & Holmes, 1994; Royce et al., 1997; Urassa et al., 1997). Women whose partners are uncircumcised also have a greater risk of sexually transmitted diseases, as well as cervical cancer (secondary to the sexually transmitted human papillomavirus) (Agarwal et al., 1994; Hunter et al., 1994). However, a review of 26 studies found no convincing evidence of a relationship between circumcision and susceptibility to sexually transmitted diseases (De Vincenzi & Mertens, 1994).

Largely because of the urinary tract infection studies, the American Academy of Pediatrics in 1989 decided to take a neutral stand on the matter, allowing parents and physicians to decide for themselves (Schoen, 1990). There are still many physicians, however, who feel that the benefits of circumcision do not outweigh the risks (Poland, 1990). In fact, as recently as 1996 the Canadian Paediatric Society, after a review of 671 studies, concluded that "the benefits have not been shown to clearly outweigh the risks and costs" and recommended against routine circumcision of newborns.

There is one circumstance in which there is little argument as to whether circumcision is justified medically, and that is when the foreskin is too tight and cannot be pulled back over the glans. This condition, known as *phimosis,* causes pain during erection.

All of these medical arguments are no doubt very confusing to most parents. Many probably choose circumcision just so that their baby boys will look like Dad.

> "I will have my son circumcised because I don't want him to feel different."

> (from the author's files)

Today, on the other hand, some circumcised men are paying surgeons to reconstruct their foreskins (Greer et al., 1982). My advice to parents is that you should educate yourself about both sides of the argument (I suggest consulting Poland, 1990, and Canadian Paediatric Society,1996, in addition to considering the information in this box) and then make a decision that feels right for you.

In tropical areas of the world, it is not uncommon for men to suffer from elephantiasis of the scrotum (caused by a parasite), a condition in which the scrotum can get as large as a basketball.

INTERNAL MALE ANATOMY

The male internal reproductive system consists of the testicles, a duct system to transport sperm out of the body, the prostate gland and seminal vesicles that produce the fluid in which the sperm are mixed, and the Cowper's gland (see Figure 2–11).

THE TESTICLES

The **testicles** (the male gonads, or *testes*) develop from the same embryonic tissue as the ovaries (the female gonads). Like the ovaries, the testicles have two functions: to produce *sperm* and to produce male hormones. Millions of new sperm start to be produced each day in several

> **testicles** The male gonads that produce sperm and male hormones.

hundred *seminiferous tubules* (the entire process takes about 70 days). Each tubule is 1 to 3 feet in length, and if they were laid together end to end they would measure over one-quarter of a mile in length. In between the seminiferous tubules are cells called *interstitial cells of Leydig,* which produce male hormones. Male (or masculinizing) hormones are called *androgens.* The most important is *testosterone,* and you will find many references to this hormone in later chapters. Another hormone produced in the testicles (by Sertoli cells) is called *inhibin.* I will discuss this hormone in Chapter 3.

The testicles are in the abdominal cavity for most of fetal development, but about two months before birth they descend through the inguinal canal into the scrotum. They fail to descend in about 3 to 4 percent of male births, a condition known as *cryptorchidism* (McClure, 1988). This usually corrects itself, but if it persists it can result in sterility and increased risk of hernia and cancer of the testicles (McClure, 1988; Rao, Wilkinson, & Benton, 1991).

Each testicle is suspended in the scrotum by the *spermatic cord,* a tubelike structure that contains blood vessels, nerves, the vas deferens, and a muscle that

Box 2-D SEXUALITY AND HEALTH

Testicular Cancer and Self-Examination

*C*ancer of the testicles is not very common, accounting for only 1 percent of all cancers in males (about 7,200 new cases per year), but it is the most common type of cancer in men aged 20 to 35 (American Cancer Society, 1997). This type of cancer recently received much attention when United States cyclist Lance Armstrong and Olympic Gold Medal–winning skater Scott Hamilton were both diagnosed with it. Men at highest risk are those whose testicles did not descend into the scrotum before the age of 10, those who had an early puberty, and those who exercise little (United Kingdom Testicular Cancer Study Group, 1994). White males have a higher incidence of this type of cancer than do African-American or Hispanic men. Cancer of the testicles is not related to masturbation, frequent sex, or sexually transmitted diseases.

The most common symptom of testicular cancer is a small lump or a testicle that is slightly enlarged. In the first stages of testicular cancer, the tumors are painless; thus it is important that all men (particularly those in a high-risk group) examine themselves regularly—once a month is advised.

Examine yourself after a warm bath or shower. This causes the testicles to descend and the skin of the scrotum to relax. Examine your testicles one at a time by placing both thumbs on top and the middle and index fingers on bottom (see Figure 2–12). Gently roll your testicle between the thumbs and fingers. When doing so, you will also feel your epididymis,

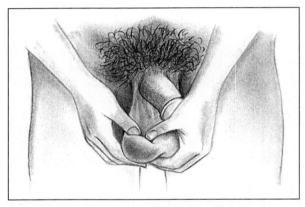

Figure 2–12   Testicular self-examination.

which feels like a spongy rope on the top and back of each testicle.

Not all abnormal lumps are cancerous, but if a lump should be, treatment usually involves surgical removal of the affected testicle and chemotherapy. The cancer rarely spreads to the other testicle, and the unaffected testicle can produce enough male hormones for normal sex drive and enough sperm to father a child. Synthetic implants can be inserted into the scrotum to give the male genitals a normal appearance.

At one time cancer of the testicles was usually fatal, but today it is one of the most curable cancers *if it is detected early and treated right away.* A self-examination takes only about 3 minutes once a month. Isn't it worth your time?

helps to raise and lower the testicles in response to changes in environmental temperature and level of sexual arousal (the testicles are drawn closer to the body during sexual arousal). Each testicle is enclosed in a tight fibrous sheath.

In right-handed men it is normal for the left testicle to hang lower than the right one, and vice versa in left-handed men (don't ask why). Both testicles, however, should be about the same size. The testicles are very sensitive to pressure, and men differ as to whether they like to have them touched during sexual relations.

Men and parents of male children should be aware of several health problems associated with the testicles. The virus that is responsible for mumps, for example, causes the testes to swell painfully, and the pressure against the fibrous sheath can crush the seminiferous tubules, resulting in sterility if mumps occurs after puberty. In an uncommon condition called *testicular torsion,* the spermatic cord twists and cuts off the blood supply to the testicles, necessitating removal if the problem is not corrected in time. This condition is generally accompanied by severe pain, and males should go to a doctor immediately if such pain is experienced. The testes are also the site of a relatively rare form of cancer (see Box 2-D).

The word *testes,* by the way, is derived from a Latin word for *witness* ("to testify"). People shake hands when completing a deal or sealing a promise today, but in Biblical times a man held the testicles of the man to whom he gave an oath. The idea was that if he broke his oath, the children of the other man had the right to take revenge.

THE DUCT SYSTEM

After sperm are produced in the seminiferous tubules, they pass through a four-part duct system (epididymis, vas deferens, ejaculatory duct, and urethra) before being expelled from the penis during an ejaculation. The seminiferous tubules converge to form the **epididymis,** coiled tubes (about 20 feet long if uncoiled) that can be felt on the top and back of each testicle. Sperm mature as they travel through the epididymis, which takes up to 6 weeks (Cooper, 1990). From here the sperm pass into the paired 14– to 16–inch **vas deferens.** The vas begin in the scrotum, then travel through the spermatic cord and enter the abdominal cavity through the inguinal canal. From there, they go up and over the bladder toward the prostate gland.

Many sperm are stored in the expanded end of the vas, called the *ampulla,* prior to ejaculation. Tying off and cutting the vas (*vasectomy*), the sterilization procedure that is performed in men, is similar to tubal ligation in women.

During orgasm, rhythmic muscular contractions force the sperm into the **ejaculatory ducts,** short (about 1 inch long) paired tubes that run through the prostate gland, where they are mixed with fluids from the prostate and seminal vesicles to form *semen.* The two ejaculatory ducts open into the **urethra,** which passes through the spongy body of the penis. Sphincter muscles surround the part of the urethra coming from the bladder. These involuntarily contract during an erection so that urine does not get mixed with semen.

THE PROSTATE GLAND AND SEMINAL VESICLES

Although an average ejaculation contains about 200 to 300 million sperm, they make up very little of the total volume. Most of the volume is seminal fluid from the **seminal vesicles** (accounting for about 70 percent of the fluid) and the prostate gland (about 30 percent). Among other substances, the seminal vesicles (misnamed because anatomists once thought that was where sperm were stored prior to ejaculation) secrete fructose, prostaglandins, and bases. Fructose (a sugar) activates the sperm and makes them mobile, while prostaglandins cause contractions of the uterus, aiding movement through that structure. Bases neutralize the normal acidity of the vagina so that sperm will not be destroyed on their way to the more favorable alkaline environment of the uterus. The **prostate gland** secretes all of these substances and another substance (fibrinogenase) that causes semen to temporarily coagulate after ejaculation, thus helping to prevent spillage from the vagina. More recently, it has been found that the prostate also secretes a powerful antibiotic, possibly to protect the male and female reproductive systems from infection. The prostate is the site of the second most common type of cancer found in males and should be checked annually as men reach the age of 40 (see Box 2-E).

THE COWPER'S GLANDS

The **Cowper's,** or bulbourethral, **glands** are two pea-sized structures located beneath the prostate. They secrete a few drops of an alkaline fluid (a base) that may

---

epididymis The elongated cordlike structure on the back of the testicle. It is the first part of the duct system that transports sperm out of a male's body.

vas deferens The second part of the duct system that transports sperm out of a male's body.

ejaculatory ducts One-inch-long paired tubes that pass through the prostate gland. The third part of the duct system that transports sperm out of a male's body.

urethra The passageway from the bladder to the exterior of the body. In males, it also serves as a passageway for semen during ejaculation.

seminal vesicles Two structures in the male that contribute many substances to the seminal fluid.

prostate gland A gland in males that surrounds the urethra and neck of the bladder and contributes many substances to the seminal fluid.

Cowper's glands Two pea-shaped structures located beneath the prostate gland in males that secrete a few drops of an alkaline fluid prior to orgasm.

Box 2-E   SEXUALITY AND HEALTH

Prostate Problems and Examination

*C*ancer of the prostate is the most common non-skin cancer among men in the United States, with approximately 334,500 new cases and 41,800 deaths each year (American Cancer Society,1997). An average of one in every eight men will be diagnosed with prostate cancer by age 85. The rates are considerably higher among African-American men than among white men. Although the cause of prostate cancer is still unclear, two suspected risk factors are high testosterone levels and dietary fat. The good news is that cancer of the prostate is usually a very slow-growing type of cancer. The 5-year survival rate for all patients is 78 percent. Many men show no symptoms in the early stages. As the cancer advances, symptoms may include difficulty in urination or a frequent need to urinate (especially at night), blood in the urine, and continuing pain in the lower back or pelvic region.

When then-President Ronald Reagan underwent prostate surgery in 1987, he probably did a lot of men a favor by bringing their attention to this chestnut-sized gland that surrounds the urinary tract at the base of the bladder. Since that time, both the diagnosis and treatment of prostate cancer have undergone revolutionary changes. The traditional method to diagnose prostate cancer has been the rectal examination, in which a physician feels the prostate for any abnormalities. Although this method will continue to be widely used, nearly 70 percent of the cancers detected by it are already in an advanced stage. In 1994, the Food and Drug Administration approved the first blood test for detecting prostate cancer. The test measures levels of prostate-specific antigen (PSA), a substance produced by prostate cells. PSA levels are higher in men with prostate cancer (Catalona et al., 1991). A recent review concluded that a combination of diagnostic techniques is better than any single technique (O'Dowd et al., 1997). The American Cancer Society has recommended that all men have a yearly PSA test starting at age 50.

Prostate cancer traditionally has been treated by either radical prostatectomy (surgical removal of the prostate, seminal vesicles, and other tissues), radiation, or castration. However, side effects of these treatments often include impotence and urinary incontinence (inability to control the bladder), and many men feel that the cures are worse than the disease (Litwin, 1993). Because localized prostate cancer is slow to develop and surgery and radiation have limited benefits (Adolfsson, Steineck, & Whitmore, 1993) and unwanted side effects, many experts are now advising that the best treatment is "watchful waiting"—doing nothing except regular monitoring until there is a threat that the cancer will spread to surrounding tissue (Johansson et al.,1997).

Cancer is not the only problem a man can have with his prostate. Inflammatory infections (e.g., prostatitis), sometimes resulting from untreated sexually transmitted diseases, can occur. In addition, as men grow older it is common for the prostate to enlarge as a result of overgrowth of normal prostate tissue. This condition, called *benign prostatic hyperplasia,* affects up to 80 percent of men over the age of 60 (Roehrborn & McConnell, 1991). The overgrowth pinches off the urethra, causing increase in frequency of or difficulty in urination. The traditional treatment has been transurethral prostatectomy, in which the prostate tissue is removed by an instrument inserted through the urethra. However, from about 10 to 15 percent of men who have had this operation suffer complications, including impotence (although some cases of impotence may be psychological in origin). Drugs are available to shrink the prostate, but the most promising new treatments are devices that use microwaves or radio waves to kill the excess tissue. Today there are an increasing number of physicians who believe that unless the urethra is completely blocked or there are other bothersome symptoms (such as urinary difficulties), the best approach, again, is watchful waiting (Denis et al., 1992; Wasson et al., 1995).

If you are male, to be safe you should start going in for a prostate exam at least once a year, starting at age 40.

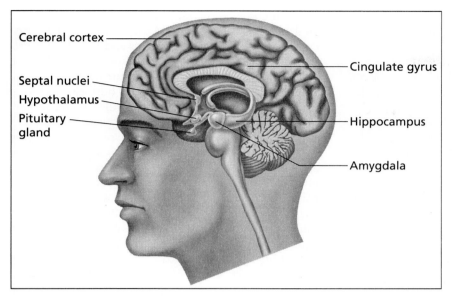

Figure 2–13   The human brain.

appear at the tip of the penis prior to orgasm. The fluid neutralizes the normal acidity of the urethra, and thus sperm are not destroyed when passing through the penis during ejaculation. Many couples use withdrawal as a contraceptive technique (i.e., the male withdraws his penis from his partner's vagina just prior to orgasm), but the fluid from the Cowper's glands often contains sperm, making this an unreliable method of birth control (see Chapter 6).

THE BRAIN

The brain has been called the most sexual organ in the human body. That is probably true. It is the structure that translates nerve impulses from the skin into pleasurable sensations; controls the nerves and muscles used in sexual behavior; regulates the release of hormones necessary for reproduction; and last, but not least, is the origin of our sexual desires. You will read about the brain in many of the chapters to come. Here is a brief introduction to the parts of the brain most directly involved in sexual behavior and reproduction (see Figure 2–13).

The *cerebral cortex* is the outer layer of nerve cell bodies that covers most of the brain. In nonhuman mammals it is relatively smooth in appearance, but in humans it has numerous folds, or convolutions, that allow for more surface area (and therefore more cells). The cortex is the part of the brain that allows us to think and reason. It is also the anatomical origin of our sexual thoughts and fantasies.

Beneath the cortex are several groups of cell bodies and fiber systems that are known collectively as the *limbic system*. This includes structures with some rather peculiar names, such as the amygdala, hippocampus,

cingulate gyrus, and septal area. The limbic system structures are where emotions and feelings originate. These structures apparently are also important for sexual behavior. For example, damage to the amygdala in monkeys, dogs, and cats causes hypersexuality. Male animals with this kind of brain damage will persistently mount not only females, but males and animals of other species. For obvious ethical reasons, researchers cannot destroy parts of the brain in humans, but in a few cases in which the amygdala has had to be removed for medical reasons, the patients afterwards engaged in persistent masturbation (Terzian & Ore, 1955).

Perhaps the most important part of the brain for sexual functioning is the **hypothalamus,** a small area at the base of the brain consisting of several groups of nerve cell bodies. The hypothalamus receives its major input from the limbic system structures. Destruction of certain areas of the hypothalamus completely eliminates sexual behavior in laboratory animals. On the other hand, if electrodes are placed in certain areas of the hypothalamus or limbic system of rats and the animals are allowed to press a lever to deliver electrical stimulation, they will press continuously for hours until they drop from exhaustion (Olds, 1958). Researchers have referred to these brain areas as "pleasure centers."

The hypothalamus is also important for sexual behavior and reproduction because of its relation with the **pituitary gland** directly below it. The pituitary gland secretes chemical substances called hormones into the bloodstream. Some of these hormones are produced in the hypothalamus and others are produced in the pituitary gland, but release of all of the pituitary hormones is controlled by the hypothalamus. You have already learned about two of the pituitary hormones: oxytocin and prolactin (see the section on the breasts). You will learn about two more pituitary hormones—follicle-stimulating hormone and luteinizing hormone—and their role in reproduction in Chapter 3.

OUR SEXUAL BODIES

Does our sexual anatomy include only the genitals and breasts? Of course not. Most of us are attracted to and

hypothalamus A part of the brain that regulates the release of hormones from the pituitary gland.

pituitary gland A gland located at the base of the brain that secretes eight hormones, including follicle-stimulating hormone and luteinizing hormone.

aroused by many other parts of our partner's body. For example, the lips and mouth, which have numerous nerve endings, are very sensitive to touch and are certainly erogenous areas. Kissing is an important part of sexual relations for most people, and it is no accident that women try to make themselves more attractive by wearing lipstick.

Although not sensitive to touch, hair is a very important part of our sexual anatomy, and people spend a great deal of money getting it cut and groomed in the latest attractive styles. Balding men often spend a great deal of money for expensive toupees, and many men love the touch and smell of women's soft and shiny hair. Many women find a man's hairy chest a sexual turn-on. Eyes, too, can be an erogenous part of the body. Do women spend a great deal of time putting on eye makeup because it will improve their eyesight? Eye contact between two people can be an initial and ongoing form of sexual communication (eyes have been called the "windows of our souls").

The buttocks are considered highly erotic by a lot of people. They are really just a large bunch of muscles covered by fat and skin with relatively few nerve endings, yet people wear tight and skimpy bikini bottoms or swim trunks to the beach in order to look more attractive. Incidentally, the males of other primate species mount the female from behind during sex, and the buttocks of the female become swollen and red when she is sexually receptive, which acts as a sexual stimulant to the males. Male-from-behind is the preferred position of sexual intercourse for many humans as well.

In addition to specific parts of the body, one's overall physical appearance is an important part of sexual attractiveness. Unfortunately, in today's society, magazines, television, and movies overemphasize a person's visual appearance to the point that people can't

help but compare themselves to those who are considered physically beautiful. Many Americans have become obsessed about their bodies and have become so image-conscious that they pay thousands of dollars for face-lifts, breast or penis enlargements, hair transplants, and even padding for their buttocks and legs. But is sex only for the very beautiful? Of course not. Physical appearance might make us sexually appealing to others at first, but a good healthy sexual life involves having *sensuality,* which encompasses all of our senses and who we are as a total person. A person's voice, feel, taste, smell, and personality can all be sensual and contribute to his or her sexuality. For example, an overweight person who appreciates the importance of touching and taking his or her time may be a much better sexual partner than one who physically may be a perfect "10."

Think how boring the world would be if we all looked the same, even if all our bodies were perfect. We each have different ideas and standards as to what we find sexually arousing. Some people are attracted to outdoorsy types, while others are more aroused by a person's intellect. Some men are preoccupied with becoming bald, but many women find bald men sexy. We are all unique, and if we can appreciate that uniqueness and love ourselves, then we also have the capacity to truly care for others. Chances are that our partners have (or will have as they grow older) as many unique physical features as we do. In the long run, human tenderness and caring, being interesting and interested, and loving and being loved are more important than physical appearance. In the excitement of sexual relations with someone we truly enjoy and care for, a paunchy midline, a bald spot, or a small penis or small breasts are not important. We all have the capacity to be sensual beings, and we all have bodies that are sexual.

Key Terms

Personal Reflections

1. Do you use sexual slang terms (a) to refer to your genitalia, or (b) when you are upset? Why? If you do, list below the terms you use, then write the correct word next to each. Say each correct term out loud a few times. Are you comfortable doing this, or uncomfortable? Could you use these terms in normal conversation with others without being embarrassed or uncomfortable? (If you cannot, explain why not—after all, you probably use the slang terms.)

2. What are your feelings when you look at drawings such as those in Figures 2–2 and 2–10? If you experience any discomfort or anxiety, explain why.

3. Jews of biblical times believed that humans were created in the image of God, and therefore they were not ashamed of any part of their bodies, including their genitals. How do you feel about this? How do you feel about your own body? (Don't just respond "Good" or "Bad," but explain in some detail.)

4. As part of this chapter, I have suggested that you examine your own genitals. Were you able to do this without feelings of guilt or anxiety? If not, try to ex-plain why you have negative feelings about looking at this part of your body.

5. (a) Women: Have you been to a physician for a pelvic examination? (b) Women aged 35 or older: Have you had a mammogram? (c) Men aged 40 or older: Have you been to a physician for a prostate examination? If you have not, explain why (lack of time is not an acceptable excuse).

6. Do you regularly examine your breasts or testicles for abnormal lumps? If the answer is no, why not? (Neither procedure takes long, so once again lack of time is not a valid excuse.)

7. If in the future you should have a male baby, do you plan to have him circumcised? Why or why not?

8. What body parts of others do you find to be most sexually arousing? Why? Do you think this is a learned response, or is it biologically determined? If the latter, how do you account for the fact that people in other cultures may not be aroused by parts of the body that you find to be arousing (see Chapter 1)?

Suggested Readings

Boston Women's Health Book Collective (1992). *The new our bodies, ourselves* (4th ed.). New York: Simon & Schuster. For women, with an emphasis on health care.

Hoffman, S. A. (1989, April). UTIs: Everything you need to know about urinary tract woes. *American Health.* Urinary tract infections are very common in women. This article discusses causes and preventive steps.

Kiester, E., Jr., & Kiester, S. V. (1996, January). The circumcision decision. *Reader's Digest.* Reviews the evidence for and against circumcision.

Love, S. (1990). *Dr. Susan Love's breast book.* Reading, MA: Addison-Wesley. All about breast care, with many illustrations.

Morgentaler, A. (1993). *The male body: A physician's guide to what every man should know about his sexual health.* New York: Simon & Schuster. The title says it all.

Zilbergeld, B. (1992). *The new male sexuality.* New York: Bantam Books. An excellent source of information about male sexuality and health, this book also provides information about female sexual anatomy.

CHAPTER 3

When you have finished studying this chapter, you should be able to:

1. Identify and describe the hormones of the endocrine system that are involved in reproduction;

2. Describe the four phases of the menstrual cycle;

3. Explain the difference between the menstrual and estrous cycles;

4. Understand the reasoning behind negative attitudes about having sex during menstruation and realize their lack of factual basis;

5. Define and explain menstrual problems such as amenorrhea, premenstrual syndrome (PMS), dysmenorrhea, endometriosis, and toxic shock syndrome (TSS);

6. Explain the regulation of sperm production and testosterone in men;

7. Understand the dangerous effects of taking anabolic steroids; and

8. Understand the relationship between hormones and sexual desire.

Hormones and Sexuality

What are hormones? Do you know what they do? Although most people have heard of hormones, very few are certain of their function. Here are some typical responses by students enrolled in my course. Many had no idea what hormones do:

"Before I was enrolled in this course I did not know what hormones were. All I knew was that they were part of our body and were there for some reason." (female)

"Because everyone talks about hormones 'kicking in' at puberty, I knew it had something to do with growing up, but not exactly the specific roles. I didn't even know that hormones were carried by the blood." (female)

"I thought hormones were the building blocks of our body, like DNA, except hormones were related to attitudes of an individual." (male)

"I thought hormones only made your body develop. They had nothing to do with reproduction, especially the menstrual cycle—that was a real surprise." (female)

"I knew hormones made you grow and sweat. But I didn't know they had anything to do with menstruation. I didn't even know they had any purpose besides just being there." (female)

Many thought that hormones controlled our moods:

"I thought that hormones controlled my moods. Whenever I'm happy, sad, mad, angry, or even 'in the mood' it was all hormones. Any unusual behavior was because of hormones." (female)

"I thought that they gave people mood changes and personality changes. I knew they were liquids of some kind, but I didn't know where they came from." (male)

"I believed that hormones were really only in women and they were only in use when a female was on her period. When women had cramps, I've always heard people say, 'She's on her period. Her hormones are acting up!'" (female)

Others thought that hormones had only to do with sexual behavior:

"I thought hormones made guys horny and go after girls. The girls would always tell us 'The hormones must be flowing' whenever we would do anything to try to impress them. So I figured only guys had them." (male)

"I thought that hormones were something that only males had. Back in middle school my teachers used to tell the males (when we harassed the girls) to control our raging hormones." (male)

"I always thought hormones made a person horny. I always heard people say that when a person was interested in sex 'their hormones were going crazy.'" (female)

"I thought hormones made you horny, like sweat glands made you sweat. Hormones make you look at girls when they pass . . ." (male)

"I always thought that hormones determined how feminine or masculine someone can be. I thought it was what made people gay or straight or in-between." (male)

(from the author's files)

Let us now see what role hormones play in human sexuality.

THE ENDOCRINE SYSTEM

Hormones play a crucial role in both males and females. They are important for growth (growth hormone), metabolism (thyroid hormone), water retention (antidiuretic hormone), reaction to stress (cortisol), and many other functions. They are also important for the development and maintenance of our reproductive system, and they influence our sexual behavior.

Hormones are chemical substances that are released into our bloodstream by ductless *glands* found throughout the body (see Figure 3–1). They are carried in the blood to other parts of the body, where they exert their effects. Hormones released by one gland often cause another gland to release its own hormones. This network of ductless glands is called the **endocrine system.**

Not all glands in the body are endocrine glands. Salivary glands and sweat glands, for example, are not endocrine glands. They secrete their products (saliva and sweat) through ducts to the surface. The testicles and ovaries, on the other hand, are part of the endocrine gland system. The testicles manufacture and release **testosterone,** which is often referred to as the "male hormone," while the ovaries produce the "female hormones" **estrogen** and **progesterone.** However, testosterone is also produced in small amounts by the ovaries, and estrogen in small amounts by the testicles. The *adrenal glands* (located near the kidneys) also produce small amounts of these hormones, so that all three of them are found in both males and females.

---

hormones Chemical substances that are secreted by ductless glands into the bloodstream. They are carried in the blood to other parts of the body, where they exert their effects on other glands or target organs.

endocrine system A network of ductless glands that secrete their chemical substances, called *hormones*, directly into the bloodstream, where they are carried to other parts of the body to exert their effects.

testosterone A hormone that is produced by the testicles (and in very small amounts by the ovaries and adrenal glands).

estrogen A hormone that is produced by the ovaries (and in very small amounts by the testes and adrenal glands).

progesterone A hormone that is produced in large amounts by the ovaries after ovulation. It prepares the endometrium of the uterus to nourish a fertilized egg.

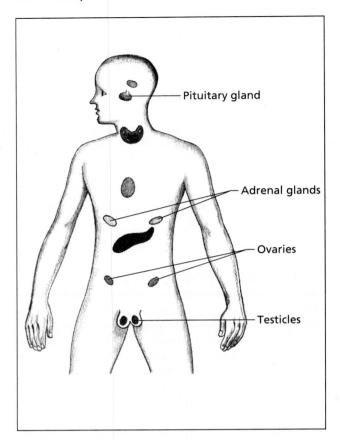

Figure 3–1 Location of the endocrine glands.

the body because many of its hormones (like FSH and LH) influence the activities of other glands (like the ovaries and testicles). We now know, however, that the brain controls the release of pituitary hormones, so maybe it is the brain that should be called the master gland. The release of gonadotropin hormones from the pituitary, for example, is under the control of another hormone, called **gonadotropin-releasing hormone (GnRH),** which is manufactured in the *hypothalamus* (see Figure 3–2). GnRH, the gonadotropins, and the gonadal hormones (testosterone, estrogen, and progesterone) are regulated by negative feedback loops. To better understand a negative feedback loop, think of how a thermostat works. When the room is cold, the thermostat triggers the furnace to go on. An increase in room temperature eventually causes the thermostat to turn the furnace off. Now look at Figure 3–2. In women, increases in blood levels of GnRH from the hypothalamus cause an increase in the release of FSH from the pituitary gland, which in turn results in increased blood levels of estrogen from the ovary. The increased estrogen levels then work to decrease the pro-

What causes the testicles and ovaries to secrete their hormones? The immediate answer is that it is the *pituitary gland,* which, you recall, is located at the base of the brain. The pituitary gland releases eight different hormones into the bloodstream. Two of these have their effect on the ovaries and testicles, and are thus called *gonadotropins* ("gonad-seeking"). In women, **follicle-stimulating hormone (FSH)** stimulates the maturation of a *follicle* (an immature egg) in one of the ovaries. This was its first known function, and thus how it received its name. However, FSH is also released by the male's pituitary to stimulate the production of sperm in the testicles.

A second pituitary gonadotropin hormone is called **luteinizing hormone (LH).** In women, LH triggers *ovulation* (release of an egg). In men, it stimulates the testicles to produce male hormones.

The pituitary was once called the "master gland" of

follicle-stimulating hormone (FSH) A gonadotropin hormone released by the pituitary gland that stimulates the development of a follicle in a woman's ovary and the production of sperm in a man's testicles.

luteinizing hormone (LH) A gonadotropin hormone released by the pituitary gland that triggers ovulation in women and stimulates the production of male hormones in men.

gonadotropin-releasing hormone (GnRH) A hormone released by the hypothalamus in the brain that causes the pituitary gland to release the hormones FSH and LH.

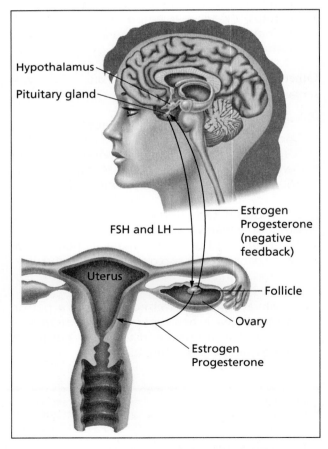

Figure 3–2 The brain-pituitary-gonad feedback loop in females.

duction of GnRH. There is a similar feedback loop among GnRH, LH, and progesterone.

These are just the basics. The complete story, of course, is more complicated. For example, another ovarian hormone, called **inhibin,** helps to regulate FSH production, and input from the brain affects release of GnRH. The feedback-loop model of hormonal regulation is made even more complicated by the fact that female pituitary and gonadal hormones vary cyclically on about a monthly basis. This is known as the **menstrual cycle,** which will now be described in more detail. We will consider the regulation of male hormones later in the chapter.

THE MENSTRUAL CYCLE

A female, you recall, has about 400,000 immature ova at birth. Each egg is surrounded by other cells within a thin capsule of tissue to form what is called a **follicle.** A female does not produce any new eggs during her lifetime, but starting at puberty one (and occasionally more) of the eggs will start maturing on about a monthly basis. Each month, the endometrium of the uterus thickens and becomes highly vascularized (supplied with lots of blood vessels) in preparation for implantation of a fertilized egg. If the egg is not fertilized by a sperm, uterine tissues are discharged from the body, accompanied by bleeding (**menstruation**), and a new ovum starts to mature. This process will end at *menopause,* when the ovaries shrivel up to the point that no eggs can mature (usually around the age of 50), and thus an average woman will have about 400 menstrual cycles in her lifetime.

The average length of an adult human menstrual cycle is about 28 days. I will describe the hormonal events that occur in a 28–day cycle as consisting of four phases. *In numbering the days, it is customary to refer to the start of menstruation as day one.* This is because menstruation is the only event of the cycle that all women can recognize. In actuality, however, menstruation is really the last of the four phases of the cycle, so we will begin with the events that occur after menstruation, on about day five (see Figure 3–3). Try not to be confused by the system used for numbering days. As you will soon learn, consistent 28–day cycles are not very common, so the numbers of the days in parentheses will be different for shorter and longer cycles.

PREOVULATORY PHASE (DAYS 5 TO 13)

This phase is also referred to as the *follicular* or *proliferative phase*. The pituitary secretes relatively high levels of follicle-stimulating hormone (FSH), which, as the name indicates, stimulates the development of a follicle in the ovary. The growing follicle, in turn, be-

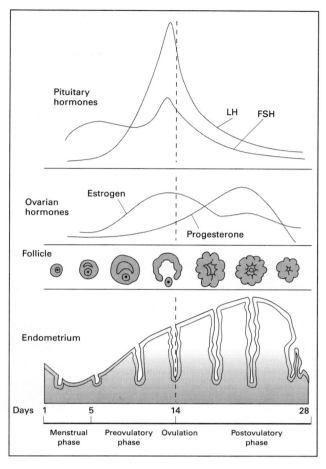

Figure 3–3 Hormonal, ovarian, and uterine changes during a 28-day menstrual cycle.

comes a temporary endocrine gland and secretes increasingly higher levels of estrogen. Estrogen is carried in the bloodstream back to the brain and pituitary, where it inhibits further release of FSH. Estrogen also stimulates release of luteinizing hormone (LH) from the pituitary and promotes proliferation (growth) of the endometrium of the uterus. Near the end of this phase, estrogen levels reach their peak, which triggers a release of LH from the pituitary (called the *LH surge*).

OVULATION (DAY 14)

The LH surge (along with a sharp increase in FSH) signals the onset of ovulation within 12 to 24 hours. By this time, the mature follicle (called a Graafian follicle) has moved to the surface of the ovary (see

inhibin A hormone produced by the testicles and ovaries that inhibits release of follicle stimulating hormone from the pituitary gland.

menstrual cycle The monthly cycle of hormonal events in a female that leads to ovulation and menstruation.

follicle A sac in the ovary containing an ovum and surrounding follicular cells.

menstruation The monthly discharge of endometrial tissue, blood, and other secretions from the uterus that occurs when an egg is not fertilized.

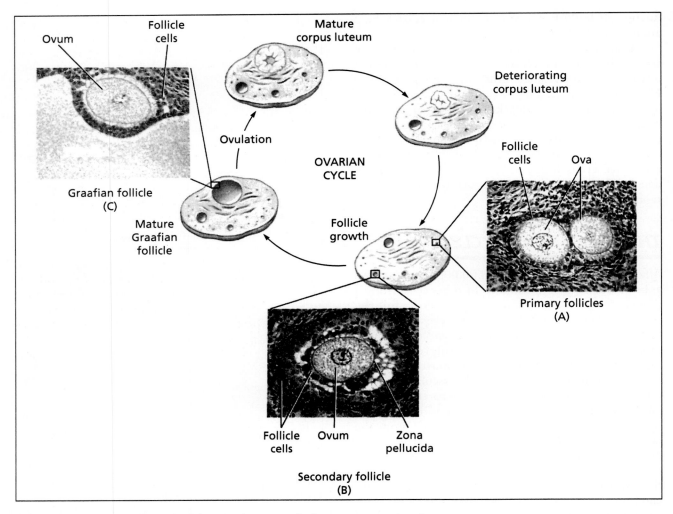

Figure 3–4 Changes that take place in the ovary during a menstrual cycle.

Figure 3–4). At **ovulation,** the follicle ruptures and the ripe ovum is expelled into the abdominal cavity, where it will soon be picked up by a Fallopian tube. Only the ovum is expelled during ovulation. The cells that had surrounded it in the follicle remain in the ovary and now have a new name, the **corpus luteum.** In addition, the hormonal changes induce a change in the quantity and consistency of cervical mucus—from white and sticky to clear and slippery like an egg white—to provide a more hospitable environment for sperm.

Some women experience lower abdominal cramps for about a day during ovulation. This is sometimes referred to as *Mittelschmerz* (German for "middle pain"). It is more common among young females than older women.

ovulation The expulsion of an egg from one of the ovaries.

corpus luteum The follicular cells that remain in the ovary after the follicle expels the ovum during ovulation. They begin to secrete progesterone in large quantities in the postovulatory stage.

POSTOVULATORY PHASE (DAYS 15 TO 28)

The postovulatory phase is also referred to as the *luteal* or *secretory phase*. The cells of the corpus luteum, which are already secreting estrogen, also begin to secrete large levels of progesterone after ovulation. Progesterone inhibits further release of LH from the pituitary and further prepares the thickened endometrium in case the egg is fertilized. The endometrium develops small blood vessels, and its glands secrete nourishing substances. If the egg is fertilized by a sperm, it continues its trip through the Fallopian tube and implants itself in the endometrium. The developing placenta produces the hormone *human chorionic gonadotropin (HCG),* which ensures that the corpus luteum will continue to secrete estrogen and progesterone to maintain the endometrium (thus, HCG serves the same role as the pituitary hormones).

But how often in a female's reproductive lifetime will one of her eggs be fertilized and result in pregnancy? Not too often. Most of the time fertilization does not occur, and in the absence of implantation, the corpus luteum degenerates. There is a sharp decline in estrogen and progesterone near the end of this phase.

MENSTRUATION (DAYS 1 TO 4)

With the decline in levels of estrogen and progesterone, there is loss of the hormones that were responsible for the development and maintenance of the endometrium. The endometrium is sloughed off and shed over a 3-to-6-day period. Menstruation is the discharge of shredded endometrial tissue, cervical mucus, and blood (about 4–6 tablespoons over the entire menstrual period). The loss of estrogen (which was inhibiting the release of FSH, you recall) also results in the pituitary gland once again secreting FSH, and a new cycle begins.

LENGTH OF THE MENSTRUAL CYCLE

In the previous section, you learned that the average length of a woman's menstrual cycle is 28 days. However, most cycles do not last exactly 28 days. There is considerable variability from female to female, and from cycle to cycle in the same female. Most women have cycles that vary in length by 8 days or more (Harlow & Ephross, 1995). The variability in cycle length is greatest in the 2- to 5-year periods after a girl's first menstruation and before a woman's menopause. There is more variability in the preovulatory (follicular) phase than in the postovulatory (luteal) phase. Many things can affect the length of the menstrual cycle, including stress, nutrition, illness, drugs, and—as you will see next—other women's cycles.

MENSTRUAL SYNCHRONY

The sexual behavior of animals is greatly influenced by their sense of smell. If you have ever owned a female cat or dog, you know about this. When your pet was "in heat," male animals from all over the neighborhood started showing up in your yard. They were attracted by natural body scents called **pheromones** produced by substances secreted by your pet. Many people use perfumes and colognes to enhance their sexual attractiveness, but is there any evidence that human sexuality is influenced by natural body odors?

The answer might be yes. In a study of 135 college coeds, Martha McClintock (1971) found that roommates came to have similar menstrual cycles while living together over the course of a year. At the beginning of the year, their cycles were an average of 8.5 days apart, but by the end of the year their cycles were an average of only 4.9 days apart. Mothers and daughters who live together also display menstrual synchrony (Weller & Weller, 1993). McClintock suggested that it was some undetermined natural body odor that synchronized women's cycles. Results of later studies support her hypothesis. Women who were regularly smeared under the nose with sweat collected from the underarms of other women began to have menstrual cycles that coincided with those of the sweat donors (Preti et al., 1986). In a similar experiment, it was found that women with abnormal menstrual cycles who were regularly smeared under the nose with underarm secretions collected from men began to show more normal cycles (Cutler et al., 1986), as is usually the case for women who have sex regularly with men (Veith et al., 1983).

It should be noted that some studies have not found any evidence of menstrual synchrony among women (Wilson, Kiefhaber, & Gravel, 1991) and that other researchers argue that it is common environmental influences (the same life-style) that cause menstrual synchrony, not pheromones (Little et al., 1989). In today's world, the use of deodorants and perfumes would probably mask any human pheromones.

MENSTRUAL VERSUS ESTROUS CYCLE

Does a woman's interest in sexual relations differ, depending on the phase of her menstrual cycle? Well, apparently some people think so:

Dear Abby:

You advised "Mismatched," the woman whose sexual appetite didn't match her husband's, to seek therapy. Bad advice, Abby. It's just another example of the guilt trip that has been laid on women ever since Freud and Kinsey came along with their crackpot theories about sex.

It is a biological truth that female mammals, which includes human beings, have a brief period of sexual desire when the ovum is ready for impregnation by the male.

This period is easily observed in wild mammals and is familiar to owners of female dogs and cats. We humans have chosen to ignore its existence, and try to make our females feel guilty because they desire sex far less frequently than males, whom nature created to be always ready for action.

A woman who

pheromones Chemical substances secreted externally by animals that convey information to, and produce specific responses in, members of the same species.

loves her husband will willingly accommodate his need for sexual pleasure even if she does not desire it at the time. For this she will receive another kind of pleasure—the pleasure of pleasing her husband.

Happily Married

Dear Married: The female of the human species need not be in her fertile phase to desire sex. The biological "truth" you cited is a misconception. (No pun intended.)

(From the "Dear Abby" column by Abigail Van Buren. Copyright 1984, Universal Press Syndicate. Reprinted with permission. All rights reserved.)

That is correct, Abby! There are some obvious differences between women and cats and dogs. Human females, for example, do not walk on all fours or bark or meow. So why is it so difficult for some people to believe that sexual desire in human females may be determined differently than in nonhuman species? Only females of the human species and a few other primate species (mostly apes) have menstrual cycles. The females of other species, including cats and dogs, have **estrous cycles.** Although owners of female cats and dogs may have noticed some periodic estrous discharge, these animals do not have a menstrual period. In some species the endometrium is reabsorbed rather than shed if fertilization does not occur. The slight discharge (called spotting) occurs during ovulation, and it is only at this time that the female is sexually receptive ("in heat," or in estrus) to the male.

Numerous studies have investigated whether human females' sexual interest changes with different stages of the menstrual cycle. Some found a slight peak at about the time of ovulation, but other studies reported peaks at other times of the cycle. Results are influenced by which aspect of sexual activity is measured (e.g., interest, fantasy, dreams, physiological arousal, masturbation, or intercourse). However, a comprehensive review of these studies concluded that the large majority found that peak sexual activity occurs somewhere between a few days after the end of menstruation and the ovulation phase of the cycle (Hedricks, 1994). Some studies have found that the peak period of sexual interest coincides with a midcycle peak in testosterone levels (Alexander et al.,1990; Van Goozen et al., 1997).

Although there appears to be no consistent, distinct peak in sexual interest during any part of the menstrual cycle, the important point is that women, unlike the females of most other mammalian species, can be-

---

estrous cycles The cyles of hormonal events that occurs in most nonhuman mammals. The females are sexually receptive ("in heat," or in estrus) to males only during ovulation.

---

come sexually aroused at any time of the cycle. There is a drop in sexual activity for many women when they menstruate (e.g., Hedricks, 1994), but this is probably more the result of negative social and cultural attitudes about sex during menstruation than the result of hormones.

Why have a few species evolved so that females have menstrual rather than estrous cycles? What purpose does menstruation serve? Recently it has been proposed that the adaptive significance of menstruation is to rid the body of bacteria carried into the uterus by sperm (Profet, 1993) or to eliminate abnormal embryos (Clarke, 1994), but there is little evidence yet to support either theory.

ATTITUDES ABOUT MENSTRUATION: HISTORICAL PERSPECTIVES

The biblical Hebrews regarded a menstruating woman (*Niddah* in Hebrew) as "unclean" or "impure," and believed that she could transfer her condition of uncleanliness to a man during sexual intercourse:

> When a woman has a discharge of blood which is her regular discharge from her body, she shall be in her impurity for seven days, and whoever touches her shall be unclean until the evening.
>
> And everything upon which she lies during her impurity shall be unclean; everything also upon which she sits shall be unclean. And whoever touches her bed shall wash his clothes, and bathe himself in water, and be unclean until the evening. . . . And if any man lies with her, and her impurity is on him, he shall be unclean seven days; and every bed on which he lies shall be unclean.
>
> (Leviticus 15:19–21, 24, Revised Standard Version)

> If a man lies with a woman having her sickness, and uncovers her nakedness, he has made naked her fountain, and she has uncovered the fountain of her blood; both then shall be cut off from among their people.
>
> (Leviticus 20:18, Revised Standard Version)

Sexual intercourse was strictly forbidden for seven days from the time a woman first noticed menstrual blood. If no blood was noticed on the eighth day, she immersed herself in a *mikvah* (ritual bath) and was again considered "pure." The menstrual taboo later was

Box 3-A CROSS-CULTURAL PERSPECTIVES

Attitudes about Menstruation

In our own culture, we have a long history of negative attitudes about menstruation. The Book of Leviticus in the Bible instructed that no one touch a woman while she was in her "impurity," and as recently as the Victorian era physicians advised that menstruating females be kept away from the kitchen or pantry so that they would not spoil hams (*British Medical Journal,* 1878). Menstrual blood was regarded as a pollutant, and menstruating women as impure or unclean. Social isolation was thought to be necessary in order to protect others, as if a menstruating woman had a contagious disease.

Today, **menstrual taboos** (as they are often called) persist in many cultures around the world, including such geographically diverse places as central Africa, Alaska, and the South Pacific. For example, the Havik Brahmins of southern India believe that if a woman touches her husband while she is menstruating, it will shorten his life (Ullrich, 1992). This is similar to the belief held by Sambian (Melanesian) males that menstrual blood endangers their health (Herdt, 1993). The Lele of the Congo believe that if a menstruating woman enters the forest, it will ruin hunting for the men (Douglas, 1966), while the Thonga of South Africa believe that sex with a menstruating woman will cause them to be weak in battle (Ford & Beach, 1951).

In order to ensure that a menstruating woman has no contact with men, many cultures require that she remain in a small menstrual hut, usually on the outskirts of the community, for an entire week. Figure 3–5 shows a group of such huts in New Guinea (Ward, 1989). This practice is carried to an extreme among the Kolish Indians of Alaska, who lock young girls in a small hut for as long as a year after their first menstruation (Delaney et al., 1988). Women of the Dyak tribe (Southeast Asia) must live for a year in a white cabin while wearing white clothes and eating nothing but white foods. Other cultures do not use special huts, but do require that a woman sleep outside of the house while she is menstruating (Ullrich, 1992).

It is a common practice in these cultures to introduce menstrual taboos to young girls dur-

Figure 3–5 Menstrual huts in New Guinea where menstruating females are isolated for seven days.

ing initiation ceremonies held about the time of their first menstruation (Schlegel & Barry, 1979, 1980). These ceremonies are rites of passage into adulthood, signifying that the girls are ready for sexual activity or marriage. Anthropologists believe that the taboos serve to emphasize the low status and inferiority of women in these cultures, where the social organization (including division of labor) is based on gender (Harper, 1964). Formal menstrual taboos are most frequently observed in small food-gathering or horticultural societies (Schlegel & Barry, 1979, 1980). The taboos decrease in importance with increased age of marriage and with increased educational and professional opportunities for women, resulting in a less dominant role for men. These changes can occur rather rapidly (over a period of only 20 years, for example, among the Havik Brahmin of India; Ullrich, 1992).

Larger societies generally do not have formal adolescent initiation ceremonies, and there are even some cultures in which people do not have negative attitudes about menstruation. In Japan, for example, parents celebrate a daughter's

menstrual taboos Incorrect negative attitudes about menstruating females.

first menstruation in a positive manner (Delaney et al., 1988).

> "Growing up, I was not told much about sex by my parents. As a result, I felt uncomfortable about beginning menstruation. My best friend's mom, on the other hand, who was from Japan, told her about menstruation early and openly. When she began menstruation, she felt comfortable telling her mom,

and according to the custom of her native country, a celebration dinner was planned to celebrate her entrance into womanhood."

(from the author's files)

Which girl do you suppose felt better about herself as a blossoming woman during adolescence?

---

extended to all physical contact, and a *Niddah* was required to leave her home and stay in a special house called a house for uncleanliness. Orthodox Jews still adhere to a literal interpretation of the passages in Leviticus. (See also Box 3–A.)

Attitudes about menstruation were even more negative during the time of the Roman Empire. Consider, for example, this passage from the Roman historian Pliny (A.D. 77) describing the possible effects of menstrual blood (Delaney, Lupton, & Toth, 1988):

> Contact with it turns new wine sour, crops touched by it become barren, grafts die, seeds in gardens are dried up, the fruit of trees falls off. . . . the edge of steel and the gleam of ivory are dulled, hives of bees die, even bronze and iron are at once seized by rust, and a horrible smell fills the air, to taste it drives dogs mad and infects their bites with an incurable poison.

As you will see in the next section, many people in the United States still have very negative attitudes about menstruation today.

ATTITUDES ABOUT MENSTRUATION TODAY

Although our country no longer has formal menstrual taboos, there are still many people who believe menstrual myths—incorrect negative beliefs about menstruation. Here are a few of the many provided by students in my class:

> "When I was about ten, I was told that if a woman was on her menstrual cycle, she shouldn't visit friends who had infant boys because the baby would get a strain."

> "My mother told me to tie a piece of black thread around my baby's foot when I went

out in public. She claimed that women menstruating would give the baby colic."

> "When you had your period not to cook or go around the stove while cooking red gravy because it would curdle."

> "All my mother said was, 'Never wash your hair when you are on your period. It will give you cramps.' Not being able to wash my hair for almost a week was pure torture. The grease was unbelievable."

(from the author's files)

Myths that women should avoid bathing, swimming, and exercise during menstruation are common (see Cumming, Cumming, & Kieren, 1991). Bathing, washing their hair, and swimming are not the only things many women do without during menstruation. Results of a survey published in 1978 indicated that the vast majority of Americans aged 55 or older had never had sexual intercourse while the female was menstruating (Paige, 1978a). More recent studies also find that American men and women seldom initiate sex during menstruation (Harvey, 1987; Tanfer & Aral, 1996; see Hedricks, 1994, for a review), although initiating sex is more common among better-educated young white women than among others. This abstinence is a direct reflection of negative attitudes about menstruation. When people can be persuaded to try sexual intercourse during menstruation, they are often surprised by their reaction:

> "My girlfriend sometimes wanted to have sex while she was menstruating. However, I was always reluctant. After taking this course I decided to give it a shot. Much to my surprise, I didn't mind it at all. In fact, she enjoys herself more than normal."

> "The first time my girlfriend and I had sex during menstruation I told her that she was

going to have to remove the condom afterwards without me looking. The thought of it disgusted me. But as we started, we found that we enjoyed it much more because we did not have to worry about getting her pregnant, it relaxed her afterwards, and because of the added fluids, she was much more lubricated. Now we look forward to sex during menstruation."

"I enjoy making love to my boyfriend while I'm having my periods. It might be messy, but it gives me pleasure, and most important it makes my period light the next day. Anyone who hasn't tried making love this way is crazy. If their only excuse is that it's messy, haven't they ever heard of a shower?"

(from the author's files)

A 1981 survey by the manufacturers of Tampax found that one-quarter of Americans believed that menstruation was not an acceptable topic of conversation in the home, and two-thirds believed it was not an acceptable topic in social situations. Menstruation was never even mentioned on television until 1973 (in an episode of *All in the Family*), and the episode drew considerable criticism when it was.

In her book *Outrageous Acts and Everyday Rebellions* (1983), feminist Gloria Steinem asked what our attitudes about this subject would be if men were the ones who menstruated:

Clearly, menstruation would become an enviable, boastworthy, masculine event: Men would brag about how long and how much.

Young boys would talk about it as the envied beginning of manhood. Gifts, religious ceremonies, family dinners, and stag parties would mark the day. . . .

Sanitary supplies would be federally funded and free. Of course, some men would still pay for the prestige of such commercial brands as Paul Newman Tampons, Muhammad Ali's Rope-a-Dope Pads, John Wayne Maxi-Pads, and Joe Namath Jock Shields— "For those Light Bachelor Days!"

(Copyright © 1983 by East Toledo Productions, Inc. Reprinted with the permission of Henry Holt and Company.)

There is nothing dirty or nasty about menstruation, and with a little knowledge, it is not even mysterious, but rather a basic, important biological function. The discharge, you recall, consists of nothing more than some shredded endometrial tissue, mucus, and a small amount of blood. From a contraceptive perspective,

menstruation is the safest phase of the cycle in which to have otherwise unprotected intercourse, although conception is still remotely possible.

What are your attitudes about menstruation? Do you regard it as a normal physiological event, or is it something about which you or your sexual partner feel ashamed or embarrassed? Test your attitude by answering one of the following two questions: (1) Women, would you feel comfortable asking your boyfriend or husband to buy you some tampons or sanitary napkins at the store? (2) Men, if your girlfriend or wife asked you to do so, would you feel comfortable buying her some tampons or sanitary napkins at the store?

Some women not only would not ask a man to buy them tampons or sanitary napkins, but are even embarrassed to buy them for themselves:

"Before taking this class I didn't realize what effect attitudes (about sexuality) in the U.S. had on me. When I went to the store to buy sanitary napkins I would try to hide them until I was getting ready to pay for them. I would not want a man to see me pay for them. I felt that they would think or say she's on her period."

(from the author's files)

If you would be uncomfortable about buying tampons or sanitary napkins, ask yourself why. No one feels uncomfortable about buying toilet paper at the store. The reason for this, of course, is that we all know that everyone has bowel movements. It is a normal

"Oh, don't be such a wimp, Edwards. Surely you can pick me up a package of tampons on your lunch hour."

physiological event. Well, nearly all women will menstruate on a monthly basis for 30 to 40 years of their lives. Menstruation is also a normal physiological occurrence.

> "One class session you asked how many men would buy tampons for their girlfriend. I don't think that is a fair question, because I wouldn't ask my girlfriend to buy condoms for me."
>
> (from the author's files)

When a person buys condoms, it is with the intent of having sex. Tampons, sanitary napkins, and new products such as Instead (which holds menstrual flow rather than absorbing it) are not used for sexual relations. In fact, menstruation has nothing to do with whether or not a female is sexually experienced. Sexually inexperienced women menstruate, just as sexually experienced women do. No one should feel ashamed or embarrassed, or be made to feel so, about a normal bodily function.

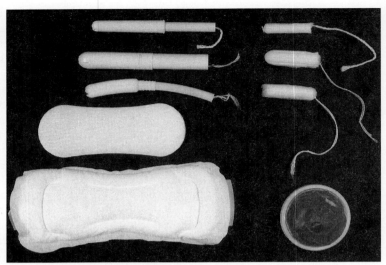

Figure 3–6 Menstrual products. Top: Examples of three types of tampons: with applicators (left) and without applicators (right). Bottom: Examples of two types of menstrual pads: underpant liner for light flow and larger one for heavier flow (left) and a newer product (Instead) that fits under the cervix and holds menstrual flow for up to 12 hours (right).

MENSTRUAL PROBLEMS

Many women experience problems that are associated in some manner with the menstrual cycle.

AMENORRHEA

Amenorrhea refers to the *absence of menstruation* for 6 months or longer. Oligomenorrhea is the absence of menstruation for shorter intervals, or unevenly spaced menstrual periods. Amenorrhea is normal, of course, when a female is pregnant, and it is not unusual during breast-feeding or in the first few months after a woman discontinues use of birth control pills. It is also common for menstrual cycles to be irregular during the first few years after **menarche** (a girl's first menstrual period). With the exception of these reasons, women should be having menstrual periods on about a monthly basis from their late teens until they approach menopause, at which time their cycles will again become irregular. However, about 3 percent of women in their reproductive years have amenorrhea, and oligomenorrhea is even more common (Harlow & Ephross, 1995; Schachter & Shoham, 1994).

A woman's health can affect her hormone levels. There is a relationship between body fat levels and age of menarche. Girls with high body fat levels generally start menstruating at a younger age than girls with low levels of body fat. Amenorrhea is common among female athletes (Shangold, 1985), although intense exercise and stress may also contribute to this (Prior & Vigna, 1991). Women with the eating disorder anorexia nervosa are often considerably underweight, and as a result rarely menstruate (Schweiger, 1991). On the other hand, extreme obesity can also result in amenorrhea (Friedman & Kim, 1985). Adult females who are not menstruating regularly should not assume, however, that this is due to diet or abnormal body fat levels. A failure to menstruate (or menstruate regularly) may also be due to pituitary or ovarian problems.

While poor health can result in amenorrhea, amenorrhea can also affect a woman's health. For example, the low levels of estrogen usually associated with amenorrhea often result in long-term problems such as osteoporosis and cardiovascular disease (Schachter & Shoham, 1994). Thus, a premenopausal woman who is not menstruating regularly should consult a physician.

PREMENSTRUAL SYNDROME (PMS)

> "I suffer from terrible PMS. . . . My boyfriend of four years broke up with me. . . . I go from being angry, to upset, to having my feelings hurt by nothing, to crying. I have a terrible self-image during this time of the month, and I usually have a bubbly, vivacious personality . . ."
>
> (from the author's files)

amenorrhea The absence of menstruation.

menarche The term for a female's first menstrual period.

premenstrual syndrome (PMS) A group of physical and/or emotional changes that many women experience in the last 3 to 4 days before the start of a menstrual period.

In recent years, **premenstrual syndrome (PMS)** has received more attention than any other menstrual problem. PMS refers to a group of physical and/or emotional changes that many women experience in the last 3 to 14 days *before the start of their menstrual period.* Physical symptoms may include bloating, breast tenderness, abdominal swelling, swollen hands and feet, weight gain, constipation, and headaches. Emotional changes may include depression, anxiety, tension, irritability, and an inability to concentrate. Women with PMS usually do not have all of these symptoms. Some women are more bothered with the physical symptoms than the emotional changes, while for others it is just the reverse.

Dr. Katharina Dalton has published numerous articles claiming that marital conflicts (including spouse and child abuse), job absenteeism, accidents, suicide attempts, and criminal acts are more common in the 2 or 3 days before the start of menstruation than at other times of the menstrual cycle (Dalton, 1959, 1960, 1964, 1980). However, many do not believe that Dalton's studies are valid (Fausto-Sterling, 1985; Koeske, 1983).

These inconsistencies have led some researchers to doubt whether or not there really is such a thing as PMS. In addition, recent studies have shown that the term *PMS* is widely overused—and misused (Hardie, 1997a, 1997b). Most women who perceive themselves as having PMS do not really have a recurrent pattern of premenstrual mood changes and mistakenly attribute high stress, poor health, and work problems to PMS. When asked to complete surveys about these problems, men score just as high as most women who perceive themselves as having PMS (Hardie, 1997a). So what distinguishes PMS from other conditions with similar symptoms? In order to be classified as true PMS, the symptoms must regularly occur in a cyclic fashion before menstruation and must end within a few days after the start of menstruation.

For some women, premenstrual mood changes are severe. In fact, the American Psychiatric Association recently decided to classify some of the more severe symptoms of PMS as **premenstrual dysphoric disorder,** or **PMDD** (1994 edition of the *Diagnostic and Statistical Manual of Mental Disorders*). To be diagnosed with PMDD, a woman's symptoms must "markedly interfere" with social relations, work, or education, and include at least one of the following symptoms: markedly depressed mood, marked anxiety or tension, persistent and marked irritability or anger, or "marked affective lability" (extreme changes in mood, such as sudden sadness) (see Steiner, 1997). It is estimated that 3 to 8 percent of all women in their reproductive years have PMDD.

"I am a 29–year-old female who is suffering with PMS. It starts exactly 2 weeks before my menstrual period. I can tell when my PMS starts because I start to get very frustrated for no reason at all. Little things that I can normally handle seem impossible at this time. I start to regret that I have children and I desire a divorce from my husband. It seems like the walls are closing in on me and I suddenly get the urge to run away. I can't make any sense out of myself at this time and a lot of times I think of suicide. I know that I am going thru PMS but I have no control of my emotions and actions, and thinking. I hate myself at this time. As soon as a drop of blood is released from my body (menstruation) I feel like all my stress and anxieties are released with that blood. All of a sudden I feel like myself again"

(from the author's files)

Women with PMS or PMDD generally have normal hormone levels and normal estrogen/progesterone ratios (Steiner, 1997). PMS clinics once prescribed progesterone as a therapy, but studies have shown that this results in no significant improvement and may even make it worse (Freeman et al.,1990). The current consensus of opinion is that PMDD is not the result of a hormone imbalance, but is cyclically triggered by normal ovarian hormone changes (Rubinow & Schmidt, 1995; Steiner, 1997). Considerable evidence has accumulated showing that the ovarian hormones act in combination with a brain chemical called serotonin to produce the symptoms. Several studies have found that the antidepressant drug Prozac (fluoxetine) substantially reduces the tension and irritability in many women with PMDD (e.g., Steiner et al., 1995). Prozac blocks the reuptake of serotonin into brain cells. Recently, other serotonin reuptake–inhibiting drugs such as sertraline have proven to be very effective in reducing the symptoms of PMDD (Yonkers et al., 1997).

Is there anything that women with milder symptoms of PMS can do? Some of the symptoms can often be relieved by medication and changes in diet. Diuretics ("water pills"), for example, can be prescribed to eliminate excess body fluid, which causes bloating and swelling; cutting down on salt intake can also achieve this. Caffeine (found in coffee, tea, many soft drinks, and chocolate) should be avoided because its stimulating effect will worsen problems like nervousness, irritability, and anxiety (Rossignol et al., 1989).

Although a hormonal basis for PMS is suspected, social and cultural factors can also play a role. Several researchers have noted that PMS is greatest in women who have a very negative attitude about menstruation (often due to their husbands' behavior). With so many peo-

premenstrual dysphoric disorder (PMDD) A severe form of PMS that markedly interferes with social relations, work, or education.

ple still treating menstruation as a taboo, it is no wonder that some women are depressed during the few days before it begins. A negative attitude like this will only exaggerate any hormonally caused emotional problems (Brooks, Ruble, & Clarke, 1977).

Feminists, who for years denied that menstruation makes any difference in behavior, worry that the new attention to PMS will revive allegations of biological inferiority and keep women out of positions of responsibility. Former vice president Hubert Humphrey's personal physician publicly stated that women were unfit to hold high public office because of their "raging hormonal influences." At most, this would apply to only a small percentage of women (those with PMDD that cannot be helped with medication), and there are probably just as many men with severe emotional problems that would make them unfit for office. Generalizations about all women are certainly unfounded.

DYSMENORRHEA

Many women experience *painful abdominal cramps during menstruation,* although perhaps not with every menstrual period. Painful menstrual cramps are called **dysmenorrhea,** and symptoms can also include backaches, headaches, a feeling of being bloated, and nausea. Up to 60 percent of women of reproductive age have experienced menstrual pain, and in 7 to 15 percent it is severe. It is most common in young adult women 17 to 24 years old, and more common among women who smoke than those who do not (Harlow & Ephross, 1995).

There are two types of dysmenorrhea—primary and secondary. In *primary dysmenorrhea,* the symptoms are not associated with any pelvic abnormalities; whereas in *secondary dysmenorrhea* there are pelvic abnormalities. Secondary dysmenorrhea is often caused by endometriosis (see next section), pelvic inflammatory disease (see Chapter 5), or ovarian cysts.

Primary dysmenorrhea is not a psychological condition, as was once believed. It is the result of an overproduction of *prostaglandins,* substances that cause contractions of the uterus. The symptoms may be severe enough to cause many women to occasionally miss work or school.

"My periods used to ruin five days out of the month. I could barely get out of bed with extreme cramps to the point of nausea. Nothing helped, not even over-the-counter medicine. It just put me to sleep and didn't help with the pain. I have been on birth control pills for almost a year. It is the best thing that ever happened. I haven't experienced any pain."

dysmenorrhea Painful menstruation.

endometriosis The growth of endometrial tissue outside of the uterus.

"The cramps kept me doubled over for days at a time, feeling like I was going to die. Only recently did I find out about drugs like Advil. . . ."

"Sex actually seems to relieve the severity of the cramps for a while . . ."

(from the author's files)

All three of these personal experiences reflect common (and usually successful) ways of alleviating primary dysmenorrhea. Drugs like ibuprofen (trade names Motrin, Advil, and others) inhibit the production of prostaglandins. Many physicians prescribe birth control pills for women who suffer from dysmenorrhea (Nabrink et al., 1990). Finally, Masters and Johnson (1966) reported that orgasms relieve the symptoms, at least temporarily.

ENDOMETRIOSIS

The endometrium of the uterus grows and thickens during each menstrual cycle and is then shed and discharged during menstruation. **Endometriosis** refers to a condition that occurs in 15 percent of premenopausal women in which the *endometrial tissue also grows outside the uterus.* This may involve the Fallopian tubes, ovaries, external surface of the uterus, vagina, pelvic cavity, and other places. Like endometrial tissue in the uterus, the out-of-place tissue grows and then breaks apart and bleeds during the menstrual cycle, but the blood cannot drain normally like the inner uterine lining. The abnormal bleeding becomes surrounded by inflammation and scar tissue forms, in many cases causing adhesions (abnormal growth that binds organs together). Endometriosis is often found in women who are being examined for infertility (Gruppo Italiano, 1994).

It is not fully understood how endometriosis develops, but those who have it suffer severe cramps and abdominal soreness, especially at menstruation, and often there is excessive menstrual bleeding. Some women experience deep pelvic pain during sexual intercourse as well:

"I have had problems with severe menstrual cramps since I was a junior in high school. The pain got worse and worse over the years. I would have to wear a tampon and 2 sanitary napkins just to catch the blood. It was awful. I remember missing many days of school because the pain was so severe I could not walk. I went to my OB-GYN and he prescribed all types of medication for the pain, but nothing worked. Then it was discovered that I had endometriosis."

(from the author's files)

However, many women with endometriosis show no symptoms, and thus there is some debate as to whether it should be regarded as a disease (see Moen, 1995). For women who have symptoms, relief can often be achieved with hormone therapy that reduces estrogen levels (Barbieri & Gordon, 1991). In some cases, surgical removal of the abnormal abdominal tissue is required, and the use of laparoscopy shows particular promise (Cook & Rock, 1991).

TOXIC SHOCK SYNDROME (TSS)

"I work at a local hospital. . . . One evening a patient came in with flu-like symptoms. The woman got progressively sicker as time went by and her color was actually turning purple in tint. The doctor looked her over from head to toe and then stated that he had found an old tampon in the back of her vagina. He diagnosed her with TSS. The patient's symptoms were getting worse so we went ahead with an emergency transfer by helicopter. After she left, the clerk answered the phone and told us it was the helicopter. Our patient had died. . . ."

(from the author's files)

Toxic shock syndrome (TSS) is caused by toxins produced by a bacterium (*Staphylococcus aureus*). The symptoms, which are often mistaken for the flu, include a sudden high fever, vomiting, diarrhea, fainting or dizziness, low blood pressure, and a red rash that resembles a bad sunburn. It sometimes results in death. Although children and men can get TSS, over 85 percent of all TSS cases are related to menstruation (Bryner, 1989).

There were over 300 cases of TSS in 1980, when the Centers for Disease Control released the first published report of the disease. Most of the cases were in women who had used extra-absorbent tampons, particularly the Rely brand. The number of cases of TSS fell dramatically after highly absorbent tampons were removed from the market, but there are still close to 200 cases reported nationwide each year, and probably many unreported cases.

The acid level in the vagina drops during menstruation, and the air pockets in tampons contain oxygen. If the *Staphylococcus* bacteria are in the vagina and the person does not have antibodies to them (most people do by the time they are adults), the conditions are right for the development of toxic shock. Tampons do not cause toxic shock, but women who use tampons should change them three or four times a day and switch to sanitary napkins at night before going to bed. If any of the previously described symptoms should develop during menstruation, be sure to see a doctor immediately.

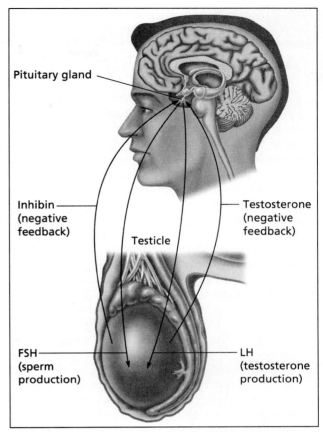

Figure 3–7 The brain-pituitary-gonad feedback loop in males.

REGULATION OF MALE HORMONES

In males, the hypothalamus, pituitary gland, and testicles operate in a feedback loop much like the feedback loop in females (see Figure 3–7). Let us first consider sperm production. Release of gonadotropin-releasing hormone (GnRH) from the hypothalamus causes release of follicle-stimulating hormone (FSH) from the pituitary. Increases in blood levels of FSH stimulate production of sperm in the seminiferous tubules of the testicles. FSH production is inhibited by increases in blood levels of inhibin, produced in males by Sertoli cells in the testicles (Plant et al., 1993).

Testosterone production increases when GnRH from the hypothalamus stimulates release of luteinizing hormone (LH) from the pituitary gland. Higher blood levels of testosterone (produced by the Leydig's cells in the testicles, you recall) then inhibit the production of GnRH.

toxic shock syndrome (TSS) A syndrome with symptoms of high fever, vomiting, diarrhea, and dizziness; caused by toxins produced by the *Staphylococcus aureus* bacterium.

Box 3-B SEXUALITY AND HEALTH

Anabolic Steroids

In their pursuit to be the best, athletes are always looking for an edge over their competitors. Megavitamins, pain killers, stimulants—all have been used in the desire to win. Today there is another substance that is widely used for this purpose by athletes, from high school football players to world-class Olympians: synthetic **anabolic steroids** (the more correct name is *anabolic-androgenic steroids*). These steroid hormones are derivatives of testosterone. They have masculinizing (androgenic) effects and also promote growth by enhancing protein uptake by the muscle cell (the anabolic effect).

When administered in proper dosages by a physician, anabolic steroids often have therapeutic value for people recovering from illness or surgery. They are also used for treatment of osteoporosis, healing of fractures and severe burns, stimulation of appetite, and treatment of muscular dystrophy. However, in an attempt to gain weight and build muscle mass and strength, many athletes started taking many times (often 100 times) the normal dose (Strauss, 1991). More than 300,000 people a year use steroids in this country, including many nonathletes who do it for cosmetic purposes (to look good) (Buckley et al., 1988; Yesalis et al., 1993).

The misuse of anabolic steroids can have serious harmful effects (Strauss, 1991). These include high blood pressure and other serious cardiovascular diseases; tumors of the liver and prostate gland; decrease in size of the testicles and impaired reproductive function in men; and lumps in the breast tissue. Steroids can also stunt growth in boys who have not yet reached their full height. Taking anabolic steroids will not make a male's penis bigger, and steroids are likely to cause acne.

Women who take steroids also risk sterilization, masculinization of the fetus should pregnancy occur, and a much-greater-than-normal chance of breast cancer. Anabolic steroids also have a marked masculinizing effect in women (e.g., causing facial hair growth, male pattern baldness, and deepening of the voice), which is not reversed by discontinuing use of the drugs.

Although anabolic steroids often boost energy and self-esteem, at least temporarily, other behavior changes may include anxiety, irritability, depression, impaired judgment, and paranoid delusions. One study found that one-third of steroid abusers suffered mild to severe mental disorders as a result of using the synthetic hormones (Pope & Katz, 1988). Steroid use also increases aggressiveness and antisocial behavior (Yesalis et al., 1993).

Here are some personal experiences related by students taking my course:

"My friends and I worked out in high school. It was a goal of mine to be bigger and stronger than the others. I decided to take anabolic steroids. Not sure how to do it, I had another inexperienced friend help me inject it into my buttocks. About two weeks later I noticed my testicles were getting smaller. In another week I started to get severe abdominal pains. . . ."

"I am 19 years old and started using steroids last summer. I wanted to bench press 500 pounds. I gained about 40 pounds. These steroids tended to make me irritable and edgy. I occasionally got bad acne on my back and my testicles shrunk a good bit. . . ."

"When I was playing football in college I had a roommate who also played football. This friend of mine felt that he needed that extra edge on all the other players, so he decided to start taking steroids. After taking them for a couple of years he started to develop lumps in his chest. His attitude started to change in a more violent way. By this time the steroids had started to control his actions. He went through many ups and downs with his attitude. Finally, a year before I left, he was unable to function anymore on the field and in the classroom."

(from the author's files)

Many high school athletes and some college athletes probably take anabolic steroids without knowing much about their side effects, but many world-class athletes who know better continue to do so. Fifteen athletes were stripped of 23 medals at an international meet in Caracas, Venezuela, in 1983 because they had used steroids, and another 12 American ath-

anabolic steroids Synthetic steroid hormones that combine the growth (anabolic) effects of adrenal steroids with the masculinizing effects of androgenic steroids.

letes withdrew rather than be tested. In the most publicized case, Canadian sprinter Ben Johnson was stripped of his 1988 Olympic gold medal in the 100-meter dash for taking steroids.

Many champions (or probably most) have earned their medals without ever taking steroids. Why, then, do some persist in light of the dangers? In addition to physical training, most athletes get themselves "psyched up" when competing. Some athletes are easily "psyched out," especially if they believe that the drugs their competitors are taking gives them an edge. In two independent surveys, world-class athletes were asked if they would take a drug that would make them Olympic champions but kill them within 1 to 5 years. A majority responded that they would (Goldman, Bush, & Klatz, 1987; Mirkin & Hoffman, 1978). This may be a sad commentary on our attitude about winning. An anabolic steroid is not a miracle drug that will turn someone into a champion. It may, however, cause serious harm.

Unlike what occurs in females during their reproductive years, the levels of FSH, LH, and testosterone in males are relatively stable. There is a diurnal rhythm of testosterone production (higher levels during the day than at night) as well as some seasonal variation (higher levels in the fall than in the spring), but testosterone levels do not change dramatically in a monthly cycle (Bremner, Vitiello, & Prinz, 1983; Reinberg & Lagoguey, 1978). A few early studies claimed to have found cycles in male testosterone levels (e.g., Doering et al., 1978), but there is presently little evidence to support this.

What accounts for the fact that females of reproductive age have monthly cycles of fluctuating hormone levels while men do not? Are female pituitary glands different from male pituitary glands? Research with animals has shown that when male pituitary glands are transplanted into females, these glands also begin to secrete FSH and LH cyclically. Thus, the difference is in the hypothalamus of the brain, not in the pituitaries. A woman's hypothalamus is, in fact, different in both structure and connections from a man's hypothalamus, and it is for this reason that women show cycles in FSH and LH (and consequently, the gonadal hormones) whereas men do not (Whalen 1977). I will have more to say about this later in the book.

HORMONES AND SEXUAL DESIRE

Is the degree of our sexual desire related to our hormone levels? Do people with high sex drives have high levels of hormones while people with low sex drives have low levels? Do some hormones influence sexual desire more than others?

Research on sex and hormones relies heavily upon two types of studies: those investigating the effects of administering hormones, and those investigating the effects of a loss of hormones (usually after surgical removal of the ovaries or testicles). We will restrict our discussion to human studies, because as you have already learned in reading about menstrual versus estrous cycles, the differences between human sexual behavior and the sexual behavior of most nonhuman mammalian species is so great as to make generalizations from studies with laboratory animals very questionable.

As you will see in Chapter 10, the loss of estrogen and progesterone in women at menopause (or after surgical removal of the ovaries) does not reduce sexual desire or responsivity in most women (Kinsey et al., 1953; Masters & Johnson, 1966), and some women show an increase in sexual behavior because they no longer have to worry about getting pregnant. In those postmenopausal women who do show a decrease in sexual desire, estrogen replacement therapy alone generally has no effect (L. Myers et al., 1990).

There is stronger evidence for testosterone playing a role in sexual desire. In normal young men, 6 to 8 milligrams of testosterone are produced every day (95 percent of it by the testicles and 5 percent by the adrenal glands), but there is a gradual decline in testosterone production as a man grows older. Men who are sexually active often have high testosterone levels, but we do not know which came first (the old chicken-and-egg question; see Udry et al., 1985). Men exposed to sexually explicit pictures show a short-term increase in testosterone levels (Rowland et al., 1987).

Hypogonadal men (men who have abnormally low levels of testosterone) show very little interest in sex. Hormone replacement therapy generally restores sexual desire (Bancroft, 1984). It was once believed that elevating testosterone levels beyond a minimum threshold had no further effect on sexual desire (Bancroft, 1988), but more recent studies have found that administration of testosterone enhances sexual desire and activity, even in men with normal hormone levels (Moss, Panzak, & Tarter, 1993; Alexander et al., 1997; Anderson, Bancroft, & Wu, 1992; Bagatell et al., 1994).

At one time, in some parts of the world, men who were used as harem attendants were forced to have their testicles removed (castration, or *orchiectomy*) so that they would not be interested in the women they were supposed to protect. In many cases this worked, especially if the men were not yet sexually experienced at the time of castration, but there were also stories of some eunuchs (as they are called) having the time of their lives, like roosters guarding a henhouse. Modern studies of castrated men show, however, that sexual drive and behavior are drastically or completely suppressed in a high percentage of cases. In most men, the decline is rapid, but castrated men may continue to show some interest in sex for years (Bancroft, 1984; Davidson et al., 1982). One cannot rule out the possibility that a decrease in sexual interest after castration might be due in part to a negative psychological reaction to having mutilated genitals.

Castration has been performed on sex offenders in some European countries, and most of the subjects have shown a decrease in sexual desire and behavior. In Europe and in the United States, some judges now offer male sex offenders a choice between prison and chemical castration with an antiandrogen drug called Depo-Provera (medroxyprogesterone acetate). This drug and others like it severely reduce testosterone production, and often reduce interest in sex (Cooper & Cernovovsky, 1992; Money, 1987c). Some experts feel that antiandrogen drugs may be useful when combined with psychotherapy, but since castration does not always completely eliminate sexual desire, this is certainly an unproven "cure" (especially since the primary motivation for most rape is nonsexual, as you will learn in Chapter 15).

It appears that testosterone is also the most important hormone for sexual desire in women. Small amounts of testosterone (0.5 milligram) are normally produced daily by a woman's ovaries and adrenal glands. Several studies have found that testosterone levels in women are directly related to measures of sexual behavior and desire (Alexander & Sherwin, 1993; Persky et al., 1982; Sherwin, 1988). A deficiency in testosterone produces a marked decrease in sexual desire and responsiveness in women (Kaplan & Owett, 1993). Estrogen generally has no effect on sexual activity in postmenopausal women (see above), but women administered testosterone after menopause or removal of the ovaries often show enhanced sexual desire (Bachmann, 1992; Sherwin et al., 1985).

In conclusion, a stronger case can be made for the influence of testosterone on sexual desire in both sexes than for estrogen and progesterone. However, human sexual desire appears to be less affected by changes in hormone levels than that of nonhuman species, as many people continue to show some sexual behavior after loss of all gonadal hormones. Once we have become sexually experienced, the brain plays an equal—if not greater—role in our level of desire.

Key Terms

adrenal glands   53
amenorrhea   62
anabolic steroids   66
corpus luteum   56
dysmenorrhea   63
endocrine system   53
endometriosis   64
estrogen   38, 53
estrous cycle   58
follicle   38, 54
follicle-stimulating hormone
   (FSH)   54
follicular (proliferative) phase   55
gonadotropin-releasing hormone
   (GnRH)   54

gonadotropins   54
hormones   53
human chorionic gonadotropin
   (HCG)   56
inhibin   55, 56
luteal (secretory) phase   56
luteinizing hormone (LH)   54
menarche   62
menopause   55
menstrual cycle   55
menstrual taboo   59
menstruation   55
Mittelschmerz   56
orchiectomy   67
ovaries   38, 53

ovulation   55
pheromones   57
pituitary gland   49, 54
premenstrual dysphoric disorder
   (PMDD)   63
premenstrual syndrome
   (PMS)   62
progesterone   38, 53
prostaglandins   64
testicles   45, 53
testosterone   46, 53
toxic shock syndrome
   (TSS)   64

Personal Reflections

1. How was menstruation first described to you: as something positive (e.g., "becoming a woman"), as just a fact of life, or as something negative? What effect do you think these early experiences have had on your present attitudes about menstruation?

2. Are you able to talk about menstruation in your family? If not, why not?

3. (a) Men: Could you go to the store and buy some sanitary napkins or tampons for your girlfriend or wife without embarrassment or anxiety? If not, why not? (b) Women: Could you ask your boyfriend or husband to go to the store to buy you some sanitary napkins or tampons without embarrassment or anxiety? If not, why not?

4. For those of you having male-female or female-female sexual relations, do you have sexual intercourse with your partner during menstruation? If not, why not?

5. Do you regard menstruation as a normal physiological process that is experienced by all women in good health (between puberty and menopause)? If your answer is yes but you answered no to question 2, 3, or 4, you are not being consistent—examine your feelings more carefully.

6. Do you or your partner suffer from premenstrual syndrome? If the answer is yes, is it possible that any negativity that you or your partner have about menstruation may contribute to it? What do you do about PMS?

Suggested Readings

Anthony, J. (1996, May). Endometriosis: The hidden epidemic. *American Health.*

Boston Women's Health Book Collective. (1992). *The new our bodies, ourselves* (4th ed.). New York: Simon & Schuster.

Delaney, J., Lupton, M. J., & Toth, E. (1988). *The curse: A cultural history of menstruation.* Urbana and Chicago: University of Illinois Press.

Gallagher, W. (1988, March). Sex and hormones. *Atlantic Monthly.*

Gastell, B. (1986, January/February). A gift from Mother Nature. *St. Raphael's Better Health.* Discusses myths about menstruation.

Golub, S. (1992). *Periods: From menarche to menopause.* Newbury Park, CA: Sage Publications. Excellent and very thorough.

Harrison, M. (1985). *Self-help for Premenstrual Syndrome.* New York: Random House.

Steinem, G. (1978, October). If men could menstruate. *Ms.* A humorous and political view of menstruation.

U.S. Food and Drug Administration. *Anabolic steroids: Losing at winning.* To receive a copy, write to Dept. 533T, Consumer Information Center, Pueblo, CO 81009.

CHAPTER 4

When you have finished studying this chapter, you should be able to:

1. List and describe the phases of the sexual response cycle in order of their occurrence;

2. Understand the process of vaginal lubrication and penile erection;

3. Describe the orgasmic platform and the "sex-tension flush";

4. Describe the two stages of muscular contractions involved in male orgasm and differentiate between the various kinds of female orgasm;

5. Describe the female multiple orgasm and the male refractory period;

6. Explain and discuss several controversies about orgasm;

7. Understand the role of penis size in sexual pleasuring;

8. Discuss the validity of reputed aphrodisiacs; and

9. Explain and discuss the practice of female genital cutting in other cultures.

Similarities and Differences in Our Sexual Responses

*T*he primary tools of behavioral scientists are observation and measurement. The mechanisms involved in most human behaviors have been under investigation since the last century, but because of the negative attitudes that have prevailed since the Victorian era, human sexual behavior was considered off limits as a subject for scientific inquiry. The first major surveys of human sexual behavior were not published until the middle of this century (Kinsey et al., 1948, 1953). Then, in 1966, William Masters and Virginia Johnson published their book *Human Sexual Response,* based on direct laboratory observation of hundreds of men and women engaged in sexual activity on thousands of occasions. Human sexuality as a topic for research was thereafter no longer taboo. In this chapter, you will learn about the physiological responses that occur during sexual arousal. The emotional and psychological aspects of sexuality will be discussed in later chapters.

Prior to Masters and Johnson, men and women were thought to be quite different in their sexual responses. Men, for example, were thought to heat up like light bulbs, while women, who many thought were incapable of orgasm, were believed to respond only after long periods of foreplay. Masters and Johnson focused on physiological responses during sexual activity and emphasized the similarities, rather than the differences, between the two sexes. Some researchers now believe, however, that Masters and Johnson were too intent on showing that men's and women's sexual responses were similar (Tiefer, 1991). Research conducted in the last 30 years has shown that although men's and women's sexual responses are a lot more similar than once believed, there are some important differences between the sexes, as well as some individual differences within each gender.

MEASUREMENT OF SEXUAL RESPONSES

You were introduced to the work of Masters and Johnson in Chapter 1, where you learned that they used the technique of direct observation to study sexual responses. But how did they actually make their observations and do their recording?

Masters and Johnson recruited their sample of 312 men and 382 women from the local community. The ages of their volunteers ranged from 18 to 89. The laboratory was a plain room that had a bed and recording equipment. The subjects were first allowed to have sex with no one else present so that they would feel comfortable, but after that there was always one or more investigators present. Masters and Johnson recorded over 10,000 sexual episodes leading to orgasm. This included people engaged in masturbation, intercourse, and oral-genital sex. Many subjects were observed dozens of times in order to determine the amount of variability in their responses.

The subjects were hooked up to equipment that measured physiological responses such as heart rate, blood pressure, muscle tension, respiration, and brain waves. Rather than rely on subjects' self-reports of degree of sexual arousal, Masters and Johnson had special equipment made to record the volume of blood in the genitals. Two of these devices are shown in Figure 4–1. The *penile strain gauge* is a thin rubber tube filled with mercury that fits over the base of the penis and transmits a small electric current that can record even a slight change in the circumference of the penis. The *vaginal photoplethysmograph* fits like a tampon into the vagina and has a light and photocell to record blood volume in the vaginal walls (by measuring changes in the reflection of light).

The main disadvantage of studies like Masters and Johnson's is the unnatural setting of the laboratory. Were the subjects affected by the presence of observers or by the recording equipment? Furthermore, how normal were the subjects themselves? Would you volunteer to participate in a study like this? However, it is generally assumed in medical research that studies of normal physiological processes (like digestion, for example) do not require a random sample of subjects because these processes will be more or less the same from person to person. Most importantly, Masters and Johnson's results have stood the test of time. Their findings have been applied with great success in sexual therapy, counseling and other areas.

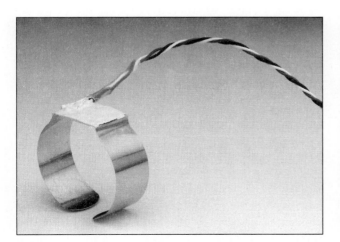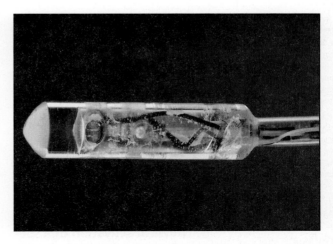

Figure 4–1 A penile strain gauge (left, the Geer gauge) and a vaginal photoplethysmograph (right, the Barlow gauge) used to measure vasocongestion in the genitals.

MODELS OF SEXUAL RESPONSE

Masters and Johnson describe the physiological responses that take place in men and women as occurring in four phases: (1) excitement, (2) plateau, (3) orgasm, and (4) resolution. They refer to this response pattern as the **sexual response cycle.** Although there is little argument concerning specific responses, it must be emphasized that the phases are only a convenient model. Even Masters and Johnson admit that the phases are arbitrarily defined, and rather than indicating a noticeable shift, they often flow together. Hardly anyone, for example, would be aware of the precise moment of leaving the excitement phase and entering the plateau phase.

Other researchers and therapists have proposed models that divide the sexual responses into fewer or more phases than the model used by Masters and Johnson. Many sex therapists have found, for example, that some people never become sexually excited (the first phase in Masters and Johnson's model) because of a lack of sexual desire or an aversion to sex (see Chapter 13). As a result, some therapists have suggested that desire should be considered the first phase of the sexual response cycle (Kaplan, 1979; Zilbergeld & Ellison, 1980). Helen Kaplan, a well-known sex therapist, has proposed a model for sexual responses that has only three phases: desire, excitement, and orgasm. From a subjective point of view, her model is appealing because most people would be able to distinguish the change in these phases. However, because of the historical importance of Masters and Johnson's studies, this book will follow their four-phase model and simply add Kaplan's desire phase to the beginning. None of these models, however, should be viewed as being right or wrong. They are merely different ways of looking at the same experience.

THE SEXUAL RESPONSE CYCLE

DESIRE

Desire, according to Helen Kaplan (1979),

is experienced as specific sensations which move the individual to seek out, or become receptive to, sexual experiences. These sensations are produced by the physical activation of a specific neural system in the brain. When this system is active, a person is "horny"; he may feel genital sensations or he may feel vaguely sexy, interested in sex, open to sex, or even just restless.

The addition of a desire phase has been very popular with many sex therapists, particularly those who feel that the subjective aspects of sexual responsiveness (how one thinks and feels emotionally) are as important as the physiological responses. A person can have the physiological response (e.g., erection) without feeling aroused, or can feel highly aroused but fail to have the physiological responses. Clearly, emotional and cognitive factors play a strong role in sexual motivation. What about the need for intimacy and emotional connectedness as a driving force to have sexual relations?

Researchers have recently distinguished between sexual desire (defined as a global, nonspecific sensation) and eight distinct sexual motivations that result in sexual expression (Hill, 1997; Hill & Preston, 1996). These are "desire for (a) feeling valued by one's partner, (b) showing value for one's partner, (c) obtaining relief from stress, (d) providing nurturance to one's partner, (e) enhancing feelings of personal power, (f) experiencing the power of one's partner, (g) experiencing pleasure, and (h) procreating." The investigators found that individuals differ in their level of interest in each motivation and that sexual arousal in response to different sexual scenarios was most strongly associated with an individual's specific sexual motivations rather than with global sexual desire.

Regardless of what specifically motivates an individual, once a person has sexual desire, stimulation of any of the five senses, or the imagination, can lead to sexual excitement. The physiological responses that occur during excitement and the remaining phases do not differ with different stimuli. The responses are the same regardless of whether the arousing stimuli are tactile, visual, auditory, olfactory, or taste, and whether the source of arousal is another person, of the opposite sex or of the same sex, or oneself (fantasy and/or masturbation).

We can describe most of the responses that occur in terms of two basic physiological processes: vasocongestion and myotonia. **Vasocongestion** refers to tissues becoming engorged (filled) with blood, while **myotonia** refers to a buildup of energy in nerves and muscles, resulting in involuntary contractions.

EXCITEMENT

MALES. The first physical sign of arousal in males, of course, is *erection of the penis* resulting from the spongy tissues of the corpora cavernosa and the corpus spongiosum becoming engorged with blood (see Figure 4–4 on page 77). This vasocongestive response begins within 3 to 8 seconds after stimulation begins but usually does not result in a full erection right away. Many men, however, worry if they do not have a "steel-hard" erection immediately upon stimulation. Remember, the **excitement phase** is just the initial phase of the response cycle, and it is not unusual for men to be easily distracted at first by either mental or external stimuli, resulting in fluctuation in the firmness of the erection.

Vasocongestion of the penis results from nerve impulses causing dilation of the arteries that carry blood to the penis. To better understand this, look at Figure 4–2. There are two centers in the spinal cord responsible for erection. The more important of the two is located in the lowest (sacral) part of the spinal cord. Nerve impulses caused by stimulation of the penis travel to this center, which then sends nerve impulses back to the arteries in a reflex action to initiate dilation of the arteries (and vasocongestion). Many men with spinal cords cut above the sacral erection center can still get erections in this reflexive manner, although

they cannot feel stimulation of the penis (Weiss, 1978). This is true for quadriplegics as well as paraplegics (Alexander, Sipski, & Findley, 1993). In men with intact spinal cords, nerve impulses from the brain (generated by such things as sights, sounds, and fantasies) are carried to the erection center and result in vasocongestion. A second erection center located higher in the spinal cord (thoracolumbar) also receives impulses originating in the brain, and thus it too contributes to psychologically caused erections. Although vasocongestion of the penis sounds like a mechanically easy process, a number of things can interfere with it, including physical or emotional stress and/or fatigue (see Chapter 13).

Vasocongestion of the penis is not the only response that happens in men during the excitement phase. Any time we become excited there is an increase in heart rate and blood pressure. During sexual excitement, the scrotum thickens (the muscle layer contracts) and the spermatic cord shortens, thus elevating the testicles toward the body. Late in the excitement phase the testicles also start to become engorged with blood and slightly enlarge (see Figure 4–4). In addition, nipple erection occurs in some men, but this is not related to a man's masculinity or femininity.

FEMALES. The first sign of sexual arousal in females is also a vasocongestive response, and just as in men, it begins within seconds after the start of stimulation. The vaginal walls become engorged with blood, and the pressure soon causes the walls to secrete drops of fluid (superfiltered blood plasma) on the inner surfaces. This is called *vaginal lubrication*. However, while it is difficult for a man not to notice that his penis is getting firm, many women may not be aware that their vagina is lubricating until several minutes have passed.

Many men mistake the presence of vaginal lubrication as a sign that their female partner is ready to begin intercourse. Although lubrication makes vaginal penetration easier and prevents irritation during thrusting, remember that this is just the first physiological response experienced by a woman and does not mean that she is emotionally, or even physically, ready to begin sexual intercourse. Recent studies have found that there is no relationship between vaginal vasocongestion (physiological arousal) and subjective sexual arousal. For example, nearly all women have genital arousal while watching erotic films, including those women who have a negative reaction or no feeling of being aroused (Laan & Everaerd, 1995).

In addition to vaginal lubrication, the labia majora, which normally cover and protect the vaginal opening, flatten and begin to move apart. The walls of the vagina,

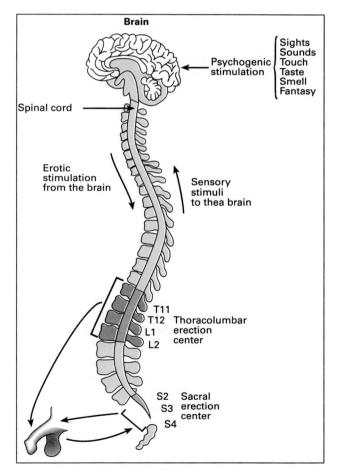

Figure 4–2 Neural mechanisms involved in an erection.

excitement The first stage of the sexual response cycle as proposed by Masters and Johnson. The first signs are erection in the male and vaginal lubrication in the female.

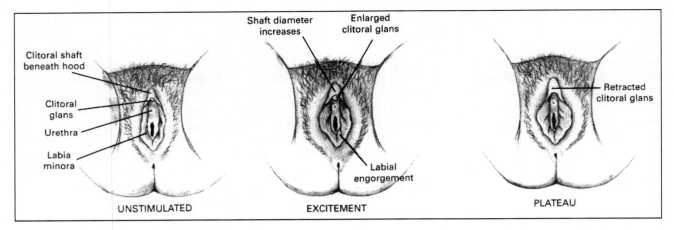

Figure 4–3   Changes in the clitoris and labia during sexual arousal.

which are collapsed in the unstimulated state, begin to balloon out, and the cervix and uterus pull up, thus getting the vagina ready to accommodate a penis (see Figure 4–4). The clitoris, which is made up of two corpora cavernosa, becomes engorged with blood, resulting in an increase in diameter of the shaft (and length in 10 percent of women) and a slight increase in diameter of the glans. As a result, the clitoris is more prominent during the excitement phase than at any other time (see Figure 4–3).

The nipples also become erect during the excitement phase, but this is as a result of contraction of small muscle fibers, not vasocongestion. Vasocongestion of breast tissues is responsible, however, for a slight increase in breast size late in the excitement phase.

PLATEAU

The **plateau** phase is a period of high sexual arousal that potentially sets the stage for orgasm. In some people this phase may be quite short; in others it may last a long time and be as satisfying as orgasm.

MALES. In men, the diameter of the penis further increases during the plateau phase, especially near the corona. In addition, the testes become fully engorged with blood (increasing in size by 50 to 100 percent) and continue to elevate and rotate until their back surfaces touch the perineum, an indication that orgasm is near (see Figure 4–4). The Cowper's glands secrete a few drops of clear fluid that may appear at the tip of the penis. About 25 percent of men experience a sex-tension flush on various areas of their skin (see below).

plateau The second stage of the sexual response cycle proposed by Masters and Johnson. Physiologically, it represents a high state of arousal.

orgasmic platform The engorgement and consequent swelling of the outer third of the vagina during the plateau stage, causing the vaginal opening to narrow by 30 to 50 percent.

FEMALES. In women, many things occur during this period that make the plateau phase distinct from the excitement phase (see Figure 4–4). While the inner two-thirds of the vagina continues to expand like an inflated balloon and the uterus continues to elevate (creating a "tenting effect" in the vaginal walls), something quite different happens in the tissues of the outer third of the vagina. These tissues become greatly engorged with blood and swell, a reaction that Masters and Johnson call the *orgasmic platform*. The swelling of tissues results in a narrowing of the vaginal opening by 30 to 50 percent. Thus, the tissues of the outer third of the vagina grip the penis during this stage (negating the importance of penis size, according to Masters and Johnson).

Changes in blood flow cause a measles-like rash to appear on areas of the skin early in the plateau stage (or late in the excitement stage). This *sex-tension flush,* as it is called, occurs in 50 to 75 percent of all women and generally appears on the breasts, upper abdomen, and buttocks, although it can appear on any part of the body.

There are three additional responses occurring in women during the plateau phase that are sometimes misinterpreted by men to mean that the female is no longer aroused, when in fact she is at a very heightened stage. First, the clitoris pulls back against the pubic bone and disappears beneath the clitoral hood (see Figure 4–3). Second, the breasts, and particularly the areola, become engorged with blood and swell, increasing in size by 20 to 25 percent in women who have not breast-fed a baby and to a lesser extent in those who have. This obscures the nipple erection that was so prominent in the excitement phase. Third, the secretion of fluids from the vaginal walls may slow down if the plateau phase is prolonged (the Bartholin's glands secrete a few drops of fluid during this stage, but not enough to contribute to lubrication). Some men may worry about the disappearance of some responses that were so prominent during the initial excitement phase, but once again, these are the normal physiological responses when women become highly aroused. The glans of the clitoris, for example, is extremely sensitive to touch at this stage, and indirect stimulation is generally sufficient to maintain arousal. In fact, some women find direct stimulation of the clitoris during this phase to be too intense, and less pleasurable than indirect stimulation.

The labia minora also become greatly engorged with blood during the plateau phase, resulting in a doubling or tripling in thickness and a vivid color change (from pink to rosy red or wine color). The thickening of the labia minora helps to further push the labia majora apart and expose the vaginal opening. Masters and Johnson claim that the color change in the labia minora means that a woman is very close to orgasm.

The changes that occur in the plateau phase in males obviously are not as distinct as the changes that occur in females. Although the responses that occur in women prior to orgasm can justifiably be divided into two phases, there are those who disagree with such a division for men (e.g., Robinson, 1976). They claim that the plateau phase in males is really just an advanced stage of excitement, and that Masters and Johnson included a plateau phase for males only because they were determined to show the similarities between men and women.

ORGASM

THE ESSENCE OF ORGASM. With continued and effective stimulation, people experience intense physical sensations lasting a few seconds that are referred to collectively as **orgasm** (climax, or "coming"). Masters and Johnson define orgasm as a sudden discharge of the body's accumulated sexual tension, while *Webster's Dictionary* defines it as "the climax or culmination of a sexual act." But do either of these definitions describe the real essence of orgasm? If a Martian were to land on earth and ask you what an orgasm is, how would you describe it? Hemingway wrote that during orgasm, "the earth moved," but except for a few people in California who might have been having sex when an earthquake struck, I doubt if this is an experience shared by many. Here are some responses students in my class gave me when I asked them to describe their orgasms for someone who had never had one:

"You know those little party poppers that force out streamers and confetti? Well, it's like when you are pulling the string real slowly and feel the tension leading to the explosion and the type of almost shock you feel at the end when it pops—like pleasurable slow motion rhythmic shock." (female)

"It feels like all the tension that has been building and building is released with an explosion. It is the most pleasurable thing in the whole world. It makes my body tingle and I don't want to even move after; I just want to enjoy the rhythmic sensations." (female)

". . . like a bomb but feels great. When you reach your max, you feel like you are going to burst open. Then you feel like you are floating in the sky." (male)

". . . like a roller coaster. At first you feel excited, your whole body feels tingly. My heart feels like it's falling in my chest. Then my stomach and leg muscles tighten up, and all tension is then released in a moment." (male)

"It is at first a force that feels as if you're on a roller coaster ride and are about to be propelled off. Then come the contractions that feel as if the roller coaster ride is hitting every bump on the track . . ." (female)

"A feeling that kind of feels like waves of electricity rushing through your whole entire body. It kind of ripples through and makes your toes and fingers curl." (female)

"Like a bolt of lightning strikes you in the back of your spine. You get stiff, curling your toes and making ludicrous faces." (male)

". . . like you have been drinking 10–15 beers on a road trip and the person driving won't pull over for you to relieve yourself. The feeling is like when they finally do pull over, and the first few seconds of relief." (male)

". . . like when you have to urinate extremely bad and when you finally relieve yourself, you sometimes shake. Compare this at 100 times greater and it feels like an orgasm." (female)

(from the author's files)

Obviously, it is difficult to adequately describe orgasm in words, although we generally might be able to express the feeling that "it sure feels good!" I guess that you have to have had one to really know what it is.

Interestingly, men and women describe orgasm similarly. When physicians and

orgasm The brief but intense sensations (focused largely in the genitals but really a whole body response) experienced during sexual arousal. During orgasm, rhythmic muscular contractions occur in certain tissues in both the male and female. The third stage of the sexual response cycle proposed by Masters and Johnson.

psychologists were presented with descriptions of orgasm such as those above, they were unable to distinguish those that had been written by men from those written by women (Vance & Wagner, 1976; Wiest, 1977).

Using electrophysiological recording equipment, Masters and Johnson (1966) found that orgasm consisted of rhythmic muscular contractions in specific parts of the body that initially occurred every 0.8 seconds, but then diminished in intensity and regularity. Others have reported that the first contractions are only 0.6 seconds apart (Bohlen et al., 1980), but the really important thing is that for the first time definite physiological responses could be identified in specific tissues.

Is an orgasm really nothing more than rhythmic muscular contractions in a few genital tissues? Did Masters and Johnson record what an orgasm actually is, or merely what it looks like? Recent studies have been unable to find a relationship between a person's perception of orgasm and the muscular contractions. Contractions often begin before a person's perception of orgasm begins and end before the person reports orgasm to be over. Moreover, strong contractions are sometimes seen with "mild" orgasms, while weak contractions are sometimes associated with orgasms that are perceived as intensely pleasurable (Bohlen et al., 1980, 1982; Levin & Wagner, 1985). In short, researchers can measure the contractions, but they can't measure the pleasure. The perception of the orgasm can be measured only by the individual person. Subjectively, people report a feeling of pleasure focused initially in the genitals and then spreading throughout the entire body. This has led some experts to conclude that the essence of orgasm lies not in the genitals, but in the brain (see Gallagher, 1986). Evidence for this is provided by the recent finding that some women can experience orgasm in response to imagery alone, in the absence of any physical stimulation (Whipple, Ogden, & Komisaruk, 1992).

Additional proof lies in the fact that some paraplegics and quadriplegics, people whose spinal cord has been transected, report having orgasms (Alexander et al., 1993; Kettl et al., 1991; Sipski & Alexander, 1995; Whipple, Gerdes, & Komisaruk, 1996). They display the typical heart rate changes, muscle tensions, and sex flushes that people with undamaged spines show. A male paraplegic might be capable of having erections during arousal but would not be able to feel them, so the sexual response is clearly in the brain. Stimuli such as emotions, visual and nongenital physical input, and fantasy obviously come to play a major role with disabled people (as well as most others).

In summary, it appears that orgasm is a perceptual experience (generated in the

ejaculation The expulsion of semen from the body.

brain, thus not always requiring genital stimulation) and that its occurrence is subjective (Whipple et al., 1992).

MALES. In males, orgasm occurs in two stages. In the first stage, called *emission,* rhythmic muscular contractions in the vas deferens, prostate gland, and seminal vesicles force the sperm and the prostate and seminal fluids into the ejaculatory ducts, thus forming *semen.* Sphincter muscles (circular bands of muscle fibers) contract and close off the part of the urethra that goes through the prostate, and the semen causes it to swell to two to three times its normal size (called the urethral bulb). These initial contractions give men a feeling of ejaculatory inevitability (the famous "I'm coming")— the feeling that orgasm is about to happen and cannot be put off.

In the second stage, called *expulsion,* these contractions are joined by contractions in the urethra and muscles at the base of the penis to force the semen from the penis—**ejaculation.** During this stage, the sphincter muscles surrounding the part of the urethra coming from the bladder are tightly contracted so that urine is not mixed with semen. A few men have a medical problem known as *retrograde ejaculation* in which the sphincter muscle that allows passage of semen through the penile urethra closes (instead of opening) and the sphincter muscle surrounding the part of the urethra from the bladder opens (instead of closing), thus forcing semen into the bladder instead of out of the body.

Although ejaculation occurs at about the same time as orgasm in adult males, *they are really two different events.* Orgasm refers to the subjective pleasurable sensations, while ejaculation refers to the release of semen from the body. The passage of semen through the penis has nothing to do with the sensation of orgasm or the intensity of orgasm. Think about it: if the passage of fluid through the uretha were the cause of the pleasurable sensations, at which other times during the day would males have a very intense orgasm (but do not)? Orgasm and ejaculation, in fact, do not always occur together. Before puberty, for example, boys can have orgasm, but do not ejaculate (the prostate and seminal vesicles do not enlarge until puberty). Ejaculation in the absence of orgasm has also been reported with some types of illness and medications.

FEMALES. Incredible as it may seem, until just a few decades ago many people (including many physicians and therapists) did not believe that women had orgasms. In fact, as late as 1976, a graduate school professor of mine was still telling students that women were not capable of orgasm. This mistaken belief was due in part to attitudes left over from the Victorian era. Women at that time were not supposed to enjoy

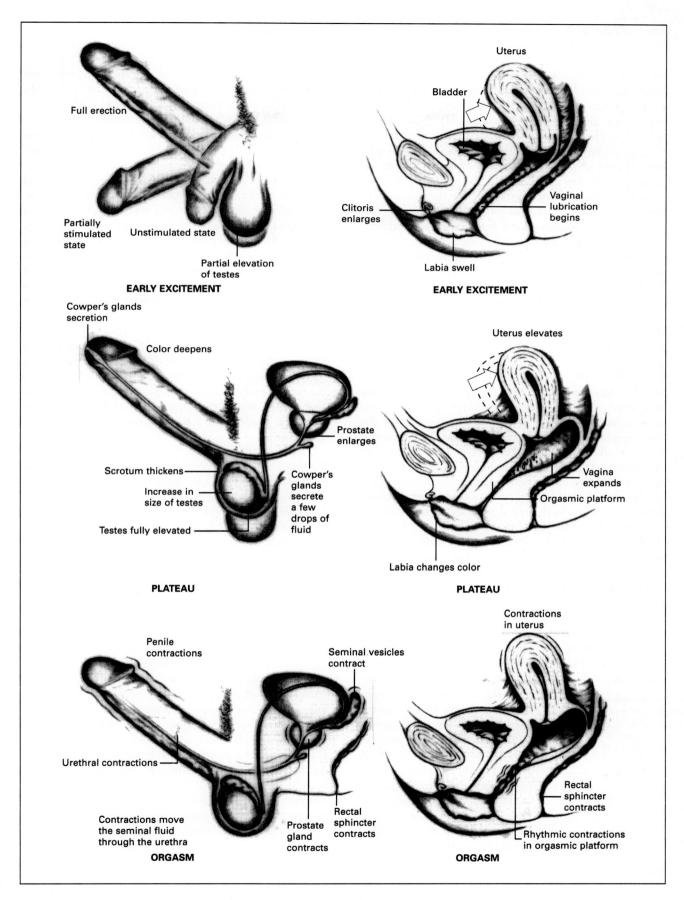

Figure 4–4 Sexual response comparison between male and female.

sex and were thought to participate only to "do their wifely duties." Sex was something that men did to women when they could not control their own desires; thus, the female's role was merely passive. This belief was also partly based on comparative studies. Symons (1979) found that nonhuman female primates living in the wild rarely, if ever, experience orgasm (even if they are physiologically capable of experiencing it, the males mate too quickly for it to occur). It was not until Masters and Johnson recorded electrophysiological responses from females that the debate was finally ended. Like men, women also had rhythmic muscular contractions in specific tissues that were initially 0.8 seconds apart. These tissues included the outer third of the vagina (the orgasmic platform), the uterus, and the anal sphincter muscles. Unlike male orgasms, female orgasms occur in a single stage, without the sense of inevitability (the sense of "coming") that men have.

RESOLUTION

MALES. Resolution is defined as a return to the unaroused state. In men, this involves a loss of erection, a decrease in testicle size and movement of the testicles away from the body cavity, and disappearance of the sex flush in those who have it. Loss of erection (detumescence) is due to the return of normal blood flow to the penis. Some of the excess blood is pumped out by the orgasmic contractions, and normal blood flow then returns as the arteries begin to constrict and the veins open. The testicles had become swollen with blood during the plateau phase, and if orgasm is delayed or not achieved, males may now experience testicular aching (what is sometimes known as "blue balls"). The discomfort is only temporary and is not dangerous—and the testicles and scrotum do not turn blue.

Many men, of course, are capable of reaching orgasm two or more times while having sex. If emotional and/or physical stimulation continues after an orgasm, a man's physiological responses may not fall all the way to preexcitement-phase levels, but they will drop below plateau level for some period of time. Partial or full erection may be possible, but the male will be unable to have another orgasm until his responses build up to plateau level again. (Remember, it is not possible to have an orgasm until one's responses have built up to the high intensity level Masters and Johnson call plateau.) The period of time after an orgasm in which it is physiologically

impossible for a male to achieve another orgasm is called the **refractory period** (see Figure 4–5). The length of the refractory period increases with age and differs from individual to individual and from occasion to occasion. If a man continues to have sexual stimulation after reaching orgasm, the refractory period will generally get longer with each orgasm.

Increased stimulation, of course, can shorten the duration of the refractory period. Successive introductions of new female partners, for example, can keep male laboratory animals performing for prolonged periods of time. This is known as the "Coolidge effect," named after American president Calvin Coolidge. President and Mrs. Coolidge were being conducted on separate tours of a farm. When Mrs. Coolidge reached the henhouse, she asked the farmer whether the continuous and vigorous sexual activity of the hens was really the work of just one rooster. After the farmer acknowledged that it was, Mrs. Coolidge responded, "You might point that out to Mr. Coolidge." Her statement was relayed to the President, who then asked the farmer whether the rooster had sex with a new hen each time. When the farmer responded "Yes," Mr. Coolidge replied, "You might point that out to Mrs. Coolidge."

FEMALES. A female's physiological responses, like those of males, generally drop below the plateau level after orgasm and return to the unaroused state (see the first solid line in Figure 4–5). Some women, however, are capable of having true **multiple orgasms**—two, three, four, or numerous (limited only by the point of physical exhaustion) full orgasms in quick succession without dropping below the plateau level (see the dotted line in Figure 4–5). They must be capable, of course, of having the first orgasm, and additional orgasms are possible only with continued sexual desire and effective stimulation. As many as 10 percent of sexually experienced women have never had an orgasm (see the wavy line in Figure 4–5). Only about 14 to 16 percent of women regularly have multiple orgasms (Athanasiou et al., 1970; Kinsey et al., 1953), but perhaps as many as 40 percent have experienced them occasionally (Darling, Davidson, & Jennings, 1991). Thus, the female sexual response cycle is quite variable. In fact, it is best not to think of *the* human sexual response cycle (a universal model that applies to everybody on all occasions), but rather to *a* human sexual response cycle that differs from person to person and from occasion to occasion (Tiefer, 1991).

Among those women who have experienced multiple orgasms, only a few do so in all of their sexual encounters. Many women report that multiple orgasms are most easily experienced during masturbation, when they are not distracted by the partner and when sexual fantasy is maximal.

resolution The fourth and final stage of the sexual response cycle proposed by Masters and Johnson. It refers to a return to the unaroused state.

refractory period In males, the period of time after an orgasm in which their physiological responses fall below the plateau level, thus making it impossible for them to have another orgasm (until the responses build back up to plateau).

multiple orgasms Having two or more successive orgams without falling below the plateau level of physiological arousal.

"Sometimes [I get] two or three orgasms right after each other. I can have multiple orgasms as much as eight times in a row, but this is only during masturbation."

(from the author's files)

Many physicians and psychoanalysts (mostly men) refused to believe Kinsey when he first reported multiple orgasms in women in 1953. In a book entitled *Kinsey's Myth of Female Sexuality,* for example, it was stated that "one of the most fantastic tales the female volunteers told Kinsey (who believed it), was that of multiple orgasm. . . ." (Bergler & Kroger, 1954). Fourteen percent of the women in Kinsey's famous survey reported having multiple orgasms, yet it was not until Masters and Johnson (1966) physiologically recorded multiple orgasms that many people accepted them as truth. Unfortunately, the pendulum may now have swung too far, with some people misinterpreting Masters and Johnson's findings to mean that all women who have learned to have single orgasms can (or should) have multiple orgasms.

Multiple orgasms, as well as simultaneous orgasms (orgasm at the same time as one's partner), have become the one and only goal in sex for some men and women. This can be self-defeating, as these people feel like failures when the goal is not achieved. Many individuals are completely satisfied by a single orgasm, and they should not put additional pressure on themselves by treating sex like an Olympic event. When people watch the Olympics, it never occurs to them that they should be able to run a 4–minute mile or perform incredible feats on the parallel bars. Many of the same people, however, feel that they have to go for a gold medal every time they have sex. This sexual Olympics mentality can only detract from having a good, healthy, and sexually satisfying experience and can lead to sexual problems as well (see Chapter 13).

At some point after a single orgasm or multiple orgasms, a woman's responses will start to fall below

plateau level and return to normal. The blood drains from the breasts and the tissues of the outer third of the vagina, the sex flush disappears, the uterus comes down, and the vagina shortens in width and length. The clitoris returns to its normal position within seconds, but the glans may be extremely sensitive to touch for several minutes. Perspiration may appear on both the female and male but is often unrelated to the degree of physical exertion.

CONTROVERSIES ABOUT ORGASMS

ARE ALL WOMEN CAPABLE OF ORGASM DURING SEXUAL INTERCOURSE?

"I have had sex but never actually had an orgasm. If I have I did not know it. Sex feels good, but I think I never had one because of the way people say an orgasm feels."

"Unfortunately, I have yet to learn how to have an orgasm with a partner during intercourse, but I have no problem reaching orgasm during oral sex."

(from the author's files)

About 5 to 10 percent of American women have never had an orgasm under any circumstances (Spector & Carey, 1990). Shere Hite (1976) reported that only 30 percent of adult women experienced orgasm regularly during intercourse without simultaneous stimulation of the clitoris. Others put the figure at 40 to 50 percent (Fisher, 1973; Kinsey et al., 1953; Wilcox & Hager, 1980). However, in their more recent, well-conducted study, Laumann and colleagues (1994) found that 75 percent of married women and 62 percent of single women usually or always had an orgasm during sexual intercourse. Even so, this still leaves a sizable minority of women who do not.

Both psychosocial and biological factors appear to play a role in women's ability to reach orgasm (Raboch & Raboch, 1992). From years of experience in her own clinical practice, sex therapist Helen Kaplan (1974) concluded, "There are millions of women who are sexually responsive, and often multiply orgasmic, but who cannot have an

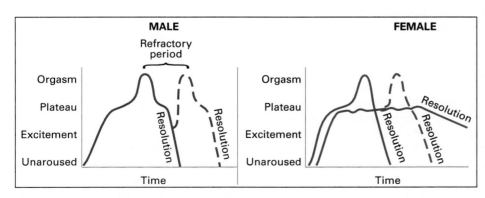

Figure 4–5 Variations in the male and female sexual response cycle.

orgasm during intercourse unless they receive simultaneous clitoral stimulation." Kaplan and others reached this conclusion after interviewing many women who reported that they desired and enjoyed sex and became highly aroused, but had never had an orgasm with a male partner (and some reported having had many partners) without simultaneous clitoral stimulation. On the other hand, Masters and Johnson (1966) and others (Barbach, 1980) claim to have successfully treated similar women and believe that all women in good health are capable of orgasm during intercourse without additional stimulation. We will return to this issue in Chapter 13.

HOW MANY TYPES OF FEMALE ORGASM ARE THERE?

Sigmund Freud, unlike many of his Victorian colleagues, not only believed that women had orgasms, but started another controversy when he said that there were two types of female orgasms, one due to clitoral stimulation and one resulting from vaginal sensations. Moreover, he claimed that clitoral orgasms were infantile and a sign of immaturity (because they were caused by noncoital acts), whereas vaginal orgasms were "authentic" and "mature" and a sign of normal psychosexual development (the result of stimulation by the penis). "The elimination of clitoral sexuality," he wrote, "is a necessary precondition for the development of femininity." This belief was adopted by his followers and caused many women to feel bad or even seek analysis because they could not experience a vaginal orgasm. Freud's views, which were made without any physiological data, have since been refuted by modern sex researchers. It is not true that just because the vagina provides pleasure for the penis, the reverse also must be the case.

Kinsey (1953), who doubted the authenticity of vaginal orgasms, had five gynecologists touch the vaginal walls of 879 women with a probe. Most women could not tell when their vagina was being stimulated, but were very much aware when their clitoris was touched. Physiologically speaking, the inner two-thirds of the vagina has few nerve endings, but it is possible, of course, that the vagina is more sensitive to stimulation during sexual arousal, which Kinsey did not test.

Masters and Johnson (1966) later reported that all female orgasms were physiologically the same. No matter whether the source of stimulation was clitoral or vaginal, identical rhythmic muscular contractions were recorded in the same structures, and the focus of subjective sensation was the clitoris. The clitoris is indirectly stimulated by the penis during vaginal penetration, you recall (by the clitoral hood), but the same contractions were recorded in women who had reached orgasm from breast stimulation alone. According to Masters and Johnson's early physiological research, an orgasm is an orgasm, regardless of the source of stimulation. In other words, Masters and Johnson initially denied the existence of vaginal orgasms.

As you learned earlier, while scientists can measure contractions, they can't measure pleasure. Several studies have found that women subjectively distinguish between orgasms caused by clitoral stimulation during masturbation and orgasms during intercourse (which were described as "fuller, but not stronger," "more internal," and "deeper" [e.g., Bentler & Peeler, 1979; Butler, 1976; Seaman, 1972]):

> "I believe I've had clitoral orgasms, vaginal orgasms, and the blended type. All of these feel different. . . . For me the best type of orgasm is the blended type because I feel it all over and when it is over my body trembles as if I had just stopped running."
>
> (from the author's files)

In fact, some women say that they have never had an orgasm from clitoral stimulation—that their orgasms always result from vaginal stimulation:

> "I am a 30 yr. old woman who has been sexually active since my late teens. I've always been very responsive sexually, resulting in multiple, intense orgasms during intercourse. The significant point is, I have orgasms during intercourse, but never during oral sex or direct clitoral stimulation. My orgasms are strictly vaginal, they are never clitoral. I'd always felt fine about this until I had a relationship with a man who said there was something wrong with me because I did not reach orgasm during oral sex. He made me feel so bad that I literally lost all of my self-esteem and self-confidence. I truly believed I was abnormal. It took me a long time to get over these feelings of inadequacy. I know now that I am okay. I may not be the same as most other women, but there is nothing 'wrong' with me. I consider myself lucky . . . not different."
>
> (from the author's files)

If the walls of the inner two-thirds of the vagina are so sparse in nerve endings, what could be the anatomical basis for vaginal orgasms? Recall from Chapter 2 that the muscle surrounding the vagina, the pubococcygeus (PC) muscle, is more richly innervated with nerve endings than the vaginal walls, and that women who report vaginal sensations during intercourse tend to have strong PC muscles (Kinsey et al.,

1953). In 1982, Ladas, Whipple, and Perry suggested in their book *The G Spot* another possible basis for a vaginal orgasm. The **Grafenberg (G) spot** was first described by the German gynecologist Ernst Grafenberg (1950) in the 1940s. It was described as being a very sensitive area about the size of a dime located on the front wall of the vagina, about halfway between the back of the pubic bone and the front of the cervix (just under the bladder). It swells when stimulated, and a woman's first reaction (lasting a few seconds) may be an urge to urinate.

Ladas, Whipple, and Perry described three types of orgasm. In the *tenting type* described by Masters and Johnson (1966) and Helen Kaplan (1974), orgasm results from clitoral stimulation (which travels via the pudendal nerve) and is characterized by the buildup and discharge of myotonia in the orgasmic platform (which closes) and the pubococcygeus muscle. In the *A-frame type,* orgasm results from stimulation of the Grafenberg spot (which travels via the pelvic nerve) and is characterized by the uterus pushing down instead of elevating, no orgasmic platform, and the buildup and discharge of myotonia in the deeper vaginal muscles. G spot stimulation was often associated with the emission of fluid from the urethra at orgasm. A third type was described as *blended.* Remember, though, that some recent studies suggest that only 10 percent or less of women even have a distinct G spot (Alzate & Londono, 1984; Masters, Johnson, & Kolodny, 1992).

Ladas, Whipple, and Perry's classification is somewhat similar to the three types of female orgasm suggested earlier by Singer and Singer (1972). They distinguished between *vulval orgasms,* which are characterized by rhythmic contractions of the orgasmic platform (induced by direct clitoral stimulation or intercourse), and *uterine orgasms,* which are caused by repeated penis-cervix contact (deep stimulation) that displaces the uterus and stimulates the membrane lining of the abdominal wall. The latter type are characterized by gasping breathing and involuntary breath-holding and, unlike the vulval type, are followed by a refractory period. A *blended type* was also noted. It has been found that orgasmic contraction of the pelvic muscles lasts less than a second, while a uterine contraction can last for as long as 30 seconds.

Is all this just a question of semantics, that is, of people confusing the stimulus with the response? The most effective stimulus for erection in men is touching the glans of the penis (light pressure stimulates the pudendal nerve), but the orgasmic contractions come from the base of the penis and perineum (deep pressure stimulates the pelvic nerve). Yet no one argues whether there are glans or perineal orgasms in males. Instead, men are viewed as having a total-body orgasm, with different ranges of the same experience (some are

mild and others intense). Although most women do not have a distinct G spot, for many the front wall of the vagina is a more sensitive area than other parts of the vagina (Darling, Davidson, & Conway-Welch, 1990) and may be "part and parcel of the female's clitoral/vaginal sensory arm of the orgasmic reflex" (Schultz et al., 1989).

So do women have different types of orgasms, or do they experience different ranges of one orgasm that has both clitoral and vaginal components? Circumstances and emotional climate (e.g., masturbation versus having sex with a partner) can certainly affect the experience. As psychiatrist and sex therapist Helen Kaplan (who believes that there is just one type of orgasm) says: "You use the same muscles to eat a hamburger that you use to eat a dinner at Lutèce, but the experiences are very different" (note: Lutèce is a fancy, expensive New York restaurant; quoted by Gallagher, 1986). As for the opinion of the author of this text, a male, I prefer the view of William Masters, who said: "Physiologically, an orgasm is an orgasm, though there are degrees of intensity. But psychologically—don't ask me about women. I'm a man, and I know that I don't know" (Gallagher, 1986).

Even if it should be proven beyond doubt that there are two or more different types of female orgasm, there is one thing that all sex therapists agree on today—that one type should not be viewed as infantile, immature, or less important and another type as mature, authentic, or more important. In short, a woman should not worry about what kind of orgasm she has. As Barbara Seaman (1972) writes, "The liberated orgasm is any orgasm a woman likes."

DO WOMEN EJACULATE DURING ORGASM?

Adult males ejaculate during orgasm, but do women also ejaculate? Crude novelettes written for the purpose of arousing men often describe female orgasm as including "great gushes of fluid flowing down her thighs." References to female ejaculation have also appeared in some better-known fiction (e.g., D. H. Lawrence's *Lady Chatterley's Lover,* 1930) and in marriage manuals (Van de Velde, 1930). Actually, the emission of a fluid by some women during orgasm was noted as early as 300 years ago by the Dutch embryologist Regnier de Graff, who described the fluid as coming from the small glands surrounding the urethral opening. This was confirmed in the modern scientific literature by Ernst Grafenberg (after whom the G spot was named) in 1950. The emission of fluid in some women was described as a dribble and in others as a gushing stream.

"I have always achieved vaginal orgasms with fluid. The fluid is clear

Grafenberg (G) spot A small, sensitive area on the front wall of the vagina found in about 10 percent of all women.

and white. It has no odor and is not unpleasant. I thought all women have this type of orgasm."

"The first time I emitted a fluid, I thought I had urinated on myself and my boyfriend. I mean it was all over. My boyfriend kept looking at me real strange and I couldn't explain it. After the first time I kept making sure that I went to the bathroom before we had intercourse. When it happened after I knew I couldn't have that much fluid in my bladder. I got a little worried, but my boyfriend enjoyed watching me 'let loose' and got very excited."

(from the author's files)

Prior to the 1980s, women who emitted a fluid during orgasm were almost always told by physicians that they suffered from urinary incontinence, which is the inability to control urination during stress (which also happens to some women during a sneezing bout). In 1981, however, three articles on female ejaculation appeared in *The Journal of Sex Research,* and this was followed in 1982 by a book about the Grafenberg spot and ejaculation (Ladas, Whipple, & Perry). They confirmed that the fluid was emitted not from the vagina, but from **Skene's glands** located in the urethra (Belzer, 1981; Perry and Whipple, 1981). These glands were thought to develop from the same embryological tissue as the male prostate (Heath, 1984; Sevely & Bennett, 1978), and the fluid, which was said to differ in color (clear to milky white) and odor from urine, was found to contain prostatic acid phosphatase, an enzyme found in prostate secretions (Addiego et al., 1981). A few researchers have even begun to refer to the Skene's glands as the female prostate— "a small, functional organ that produces female prostatic secretion and possesses cells with neuroendocrine function, comparable to the male prostate" (Zaviacic & Whipple, 1993). Female ejaculation, it is claimed, is seen most often with stimulation of the Grafenberg spot and in women who have a strong pubococcygeus muscle (Perry & Whipple, 1981; Zaviacic et al., 1988). However, as yet there is no evidence of an anatomical link between the front wall of the vagina and the Skene's glands (Alzate & Hoch, 1986).

How many women emit a fluid (ejaculate) during orgasm? Perry and Whipple claimed in their initial article that "perhaps 10 percent of females" did so, but later claimed in their book that perhaps as many as 40 percent of females had experienced this

Skene's glands Glands located in the urethras of some women that are thought to develop from the same embryological tissue as the male prostate, and that may be the source of a fluid emitted by some women during orgasm.

on occasion. More recent studies agree with this (Darling, Davidson, & Conway-Welch, 1990; Zaviacic et al., 1988). Masters and Johnson, on the other hand, found that only 14 of 300 women they surveyed had experienced emission of fluid at orgasm, but in those who did, the fluid was again found not to be urine (Masters, 1982). However, a later study (which included Beverly Whipple on the research team) of six female "ejaculators" found that the fluid emitted could not be distinguished from urine (Goldberg et al., 1983). It would appear, therefore, that while it can no longer be doubted that some women emit a fluid from the urethra during orgasm, there is some question as to the nature of the fluid. No doubt some women do suffer from stress-induced incontinence, but in others it may actually be a true ejaculation.

CAN MEN HAVE MULTIPLE ORGASMS?

Masters and Johnson originally took the position that only women could have true multiple orgasms—two or more orgasms within a short time without dropping below plateau. Whenever they recorded orgasm with ejaculation from men, it was always followed by a refractory period. Only when a man's level of arousal built back up to the plateau level could he have another orgasm. However, two survey studies question whether there is a true gender difference in the ability to have multiple orgasms. The first study reported 13 men who said that they had multiple dry "mini-orgasms" by withholding ejaculation before having a full wet orgasm (i.e., with ejaculation). The wet orgasm was always followed by a refractory period (Robbins & Jensen, 1978). In the second study, 21 men reported experiencing a variety of multiple orgasms (Dunn & Trost, 1989). Some had one or more "dry" orgasms before ejaculating, while others had an orgasm with ejaculation followed by one or more dry orgasms. As you have learned by now, scientists have always been the last ones to believe people's claims about experiencing different (from most others) types of orgasm. Until someone records rhythmic muscular contractions during multiple dry orgasms, that will probably be the case here as well.

PENIS SIZE: DOES IT MATTER?

In an article written for *Cosmopolitan* (1984), Michael Barson stated that "for a man, the one emotion more disquieting than his own sense of not being big enough is the terror of finding himself in bed with a woman who shares his opinion." It should not be surprising that many men have anxieties about the size of their penis, for penis size has been a focus of attention throughout history. Priapus, the Greco-Roman god of procreation, was always displayed as having an enor-

Box 4-A   CROSS-CULTURAL PERSPECTIVES

Cutting of Females' Genitals

Cutting of females' genitals has been a common practice in many cultures since antiquity. The Greek geographer Agatharchides, for example, made note in the second century B.C. of the clitoral excision of Ethiopian women (Widstrand, 1964). In England and the United States, various forms of clitoral circumcision were used in the late nineteenth and early twentieth centuries as a medical treatment for "frigidity," "nymphomania," and masturbation.

Today, cutting of female genitals is still performed in many parts of the world (see Figure 4–6). Some cultures practice *clitoridectomy,* removal of the clitoris, while others subject females to *infibulation,* genital mutilation that involves cutting away the entire clitoris and labia minora and parts of the labia majora, and then sewing together (sometimes with thorns) the raw sides of the vulva except for a small opening for urine and menstrual flow (Giorgis, 1981). The procedures are usually done by an older village woman with a knife or razor and without anesthesia, requiring that the girl be tied or held down. It is usually done during early childhood, but in some cases, as in Nigeria, it is performed just prior to marriage (Messenger, 1971). Many of the girls suffer hemorrhage or shock, and some die. Long-term effects include chronic uterine, vaginal, and pelvic infections; cysts; and abscesses (Abu-el-Futuh Shandall, 1967; Toubia, 1994).

The World Health Organization estimates that 85 to 114 million women have had their genitals cut away in this manner. This includes most women in Somalia, Ethiopia, Sudan, and Egypt and about half the women in Kenya, with smaller numbers in other African countries

(Lightfoot-Klein & Shaw, 1991). The practice is also common in several Middle Eastern countries and in parts of Indonesia, Malaysia, and the India-Pakistan subcontinent. Most of the cultures in which genital cutting is practiced are Muslim, but it is also known to have been committed by Christians and some Ethiopian Jews.

Why are girls subjected to this mutilation? In the words of Raqiya Dualeh Abdalla, a Somali woman who denounced the practice in her 1982 book *Sisters in Affliction,*

> the various explanations and mystifications offered to justify the practice . . . all emanate from men's motive to control women economically and sexually and as personal objects.

Genital cutting substantially reduces sexual pleasure. Most genitally mutilated women do not enjoy sex, and for many sexual intercourse is painful (Abu-el-Futuh Shandall, 1967; Lightfoot-Klein, 1989). The men in these cultures, for whom sexual pleasure is reserved exclusively, believe that the elimination of sexual

Figure 4–6 Seita Lengila, a 16-year-old Kenyan girl, is held down by village women as her clitoris is cut away with a razor. The wound was then cleaned with cow urine and smeared with goat fat to stop the bleeding.

84 *Chapter 4*

desire in women "frees them to fulfill their real destiny as mothers." Marriage and motherhood are generally the only roles allowed for women in cultures where genital cutting is performed, and for a girl to refuse to submit is to give up her place in society.

A second, but related, reason is economic. The countries where genital mutilation is practiced are strongly patriarchal, and thus there is great concern about guaranteeing the inheritance of property from fathers to legitimate sons. Women are regarded as property and are often sold in marriage. The sewing together of the vulva guarantees that a new wife is a virgin on the wedding night (the vulva is partially cut open by the male), and the elimination of sexual desire reduces infidelity, so that any offspring are almost assuredly the husband's. Infibulated women are fully cut open at childbirth and then partially restitched afterwards.

In 1992, the World Health Organization unanimously approved tougher action to end the practice of female genital cutting, as had the Organization of African Unity in 1990. The Centers for Disease Control and Prevention estimate that at least 168,000 girls of African ancestry living in the United States have been cut or are at risk of being cut (Jones et al., 1997). In 1995, the American College of Obstetricians and Gynecologists officially opposed the performance of these procedures by physicians in the United States. A new federal law went into effect in 1997 making it a crime (punishable by up to 5 years in prison) for parents to arrange to have this done to their children.

Attempts by outsiders to put an end to genital mutilations have resulted in a clash of cultural values. Westerners generally cite humanitarian and feminist values, but many in Western cultures ritually follow the practice of cutting away the foreskins of boys, usually without the use of anesthesia (see Box 2–C). There is legitimate concern that Western people's negative reaction to the practice will generalize to the cultures where it is practiced. In the words of Nahid Toubia, a Sudanese author:

> Over the last decade the . . . West has acted as though they have suddenly discovered a dangerous epidemic which they then sensationalized in international women's forums creating a backlash of over-sensitivity in the concerned communities. They have portrayed it as irrefutable evidence of the barbarism and vulgarity of underdeveloped countries . . . [and the] primitiveness of Arabs, Muslims, and Africans all in one blow. (1988, p. 101)

Recent studies have found that many African and Arabic groups have switched from extreme infibulations and clitoridectomies to more symbolic genital cuts (e.g., Asali et al., 1995; Gruenbaum, 1996). This allows group cohesiveness and identity without subjecting girls to the greater risk and health consequences of the more severe procedures.

---

mous penis. Many writers have also described the penis in exaggerated terms. In the novel *Justine* by the Marquis de Sade, for example, the hero's penis was described as "of such length and exorbitant circumference that not only had I never clapped eyes upon anything comparable, but was absolutely convinced that nature had never fashioned another as prodigious; I could scarcely surround it with both hands, and its length matched that of my forearm." In D. H. Lawrence's *Lady Chatterley's Lover*, the gamekeeper's penis was even given its own name, "John Henry." More recently, penis size has been emphasized in magazine centerfolds and erotic X-rated movies. Bernie Zilbergeld (1978), a noted sex therapist, has said that "it is not much of an exaggeration to say that penises in fantasyland come in only three sizes—large, gigantic, and so big you can barely get through the front door." All of this tends to leave the impression that penis size is related to virility, that is, the ability of a man to sexually please a woman, so that even men with normal-sized penises may wish they were larger.

> Penises come in a variety of shapes and sizes . . . and about the only thing most penises have in common is that they are the wrong size or shape as far as their owners are concerned. In the many hours we have spent talking to men in and out of therapy, we have heard every conceivable complaint about penises. They are too small (the most common complaint), too large, too thin, too thick, stand up at too small (or great) an angle when erect, bend too much to the right, or left, or in the middle, or don't get hard enough when they are erect.
>
> (Zilbergeld, 1978)

"I have an inferiority complex about my penis size. Even though my penis is larger than normal (about 7 inches) . . . no matter how many women tell me that my penis is large, I still have that complex."

(from the author's files)

Breast size in women is relatively easy to assess, but for most heterosexual males the only opportunity to compare themselves to others is in the locker room, where they may get a false impression. The average size of a flaccid penis is about 3.5 to 3.75 inches in length and 1.0 to 1.2 inches in diameter. Researchers at the Kinsey Institute have found, however, that there is considerably more variation in penis size in the flaccid condition than in the erect state. Penises that appear small when flaccid display a proportionately greater increase in size upon erection than penises that appear

Figure 4–7 The fertility gods of the Greeks and Romans were always sculptured to have an enormous penis.

large in the unaroused state (Jamison & Gebhard, 1988). Frequent intercourse does not cause a penis to get larger. Penis size is also not related to height, weight, build, or race (Money et al, 1984; see Eldridge Cleaver's book *Soul on Ice* for a discussion of how the story that black men have larger penises got started).

This is not to say that all penises are the same size when erect. Kinsey's group found that the average size of an erect penis (measured along the top) was 6.2 inches in length and 1.5 inches in diameter (Jamison & Gebhard, 1988). However, that study relied on self-measurement, and some men may have exaggerated. Two more recent studies that did not rely on self-reports found the average length of an erect penis to be smaller, with many men in the 4.5-to-5.75-inch range (Sparling, 1997; Wessells, Lue, & McAninch, 1996).

Today, some men are paying thousands of dollars each to have their penises made larger by injecting fat from other parts of the body. However, the American Society for Aesthetic Plastic Surgery has issued a statement warning that the safety of this procedure has not yet been established. Before any man considers having this operation, let us examine whether the size of a male's penis really is important in pleasing a woman during sexual intercourse.

In order to answer this question, we must first re-examine some basic facts about female anatomy. A woman's vagina, you will recall, is neither a bottomless cavern nor an open hole always ready to accommodate a penis. In the unaroused state, the vagina measures 3 to 5 inches in length and its walls are collapsed. When a woman becomes sexually aroused, the walls of the inner two-thirds of the vagina expand and its depth averages 6 inches, which perhaps not coincidentally is about the same length as the average penis (isn't nature wonderful?). In its capacity to physically stimulate the vagina, therefore, there is no real advantage to having a penis any longer than about 6 inches, for there is no place for it to go. And don't forget that the walls of the inner two-thirds of the vagina are relatively sparse in nerve endings anyway. The tissues of the vaginal opening are highly sensitive to touch, but recall that the vaginal opening narrows when a woman is highly sexually aroused, minimizing the importance of penis size. Most important, remember that the focus of physical sensations in most women is not the vagina, but the clitoris, which usually doesn't come into direct physical contact with the penis during intercourse. Thrusting of the penis stimulates the clitoris only indirectly by causing the clitoral hood to rub back and forth over the clitoral glans.

Do women consider the size of a penis important? In one study, it was found that depictions of large penises produced no more arousal in women than depictions of smaller ones (Fisher, Branscombe, & Lemery, 1983). In another study, several hundred women were asked what was most important to them during sexual intercourse. Not one mentioned penis size (Zilbergeld, 1978). Some expressed a personal preference when questioned further (e.g., circumcised or uncircumcised), but it was never a high priority. In surveys conducted in my class, only once has a woman expressed a preference for large penises.

> "I love men who have big dicks. They don't have to be hudge [sic] or anything, but I much prefer the larger size."
>
> (from the author's files)

For the large majority of women, penis size was not important. Here are some typical responses:

> "Penis size—of course it doesn't matter!"
>
> "And about penis size, we don't care!"
>
> "Also, penis size doesn't mean anything, especially to women."
>
> "It's not what you have that's so important, but how you use it."
>
> "It's not the size, but the performance that counts."
>
> "It is not the size of the ship that matters, but the motion of the ocean."
>
> (from the author's files)

In fact, *Forum* editor Philip Nobile found that most women preferred an average-size penis. The women in these surveys were more concerned with their male partner taking his time and his total response to her during sexual intercourse. The quality of the entire sexual experience—not the length of the penis—was the important factor in sexual enjoyment and satisfaction.

APHRODISIACS: DO THEY HELP?

People have searched for **aphrodisiacs**—substances that enhance sexual desire or prolong sexual performance (named after the Greek goddess of love, Aphrodite)—for centuries. At one time or another, over 500 substances have been believed to be aphrodisiacs, including bull's testicles, powdered rhinoceros horn, elephant sperm, turtle eggs, ginseng roots, and vitamin E,

aphrodisiacs Substances that enhance sexual desire or performance.

anaphrodisiacs Substances that suppress sexual functioning.

as well as common foods like bananas, potatoes, tomatoes, asparagus, garlic, and radishes (Stark, 1982). Many people like to believe that oysters are an aphrodisiac. However, the fact is that none of these substances has any physiological effect on sexual responsivity. Any temporary improvement in sexual functioning is purely a psychological effect (believing that something will work can be arousing), and the effect will wear off shortly.

Spanish fly (*cantharides*), made from powdered beetles that are found in Spain, is perhaps the most famous reputed aphrodisiac. Taken orally, the active substance causes inflammation and irritation of the urinary and genital tracts, which some may interpret as lust. In males, Spanish fly can cause a painful, persistent erection (priapism), even when there is no desire for sex. This is a dangerous drug, a poison, that can result in ulcers of the digestive and urinary tracts, diarrhea, severe pain, and even death.

Alcohol is commonly believed to enhance sexual desire and responsivity. One of its initial effects is dilation of blood vessels in the skin, which gives us a feeling of warmth and well-being. A drink or two depresses certain areas of your brain, making you less inhibited about engaging in sex or other behaviors that you might not ordinarily do (Crowe & George, 1989). In fact, many college students "let themselves drink more than normal in order to make it easier for them to have sex with someone" (Anderson & Mathieu, 1996). The loss of inhibitions can even result in a person's engaging in sexual behaviors that are associated with a high risk of contracting HIV (the virus that causes AIDS) and other sexually transmitted diseases (Avins et al., 1994). Although alcohol may make a person less inhibited about engaging in sex, anything more than a single drink can impair the nervous system responses needed for engaging in sex. As Shakespeare says of drink in *Macbeth,* "it provokes the desire, but it takes away from the performance." Alcohol, in fact, is an **anaphrodisiac,** a suppressant of sexual functioning. In even moderate amounts it can cause erectile problems in men (Farkas & Rosen, 1976; Rosen, 1991) and difficulty in reaching orgasm in both sexes (Malatesta, 1979, 1982; Rosen, 1991). Chronic alcoholics almost always have sexual problems (Rosen, 1991).

Some illegal drugs also have reputations as sexual stimulants. Cocaine is a central nervous system stimulant, and thus may give the user a temporary burst of energy and a feeling of self-confidence, but the drug's effects are often equally due to heightened expectations and the social situation in which it is used (i.e., like alcohol, drugs give people an excuse that they can use afterwards for having had sex). The use of cocaine, however, can often result in erectile failures and difficulties in reaching orgasm (Cocores,

Dackis, & Gold, 1986). Heavy, chronic use of cocaine generally leads to nearly total sexual dysfunction (impotence and inability to reach orgasm), although abusers may still try to engage in sex compulsively (Washton, 1989). Some people have tried rubbing cocaine on the clitoris to increase female responsivity, but cocaine is in fact a topical anesthetic (i.e., it deadens nerve endings). A few men have suffered permanent nerve and tissue damage, requiring amputation, as a result of injecting cocaine directly into their penises.

Amphetamines ("speed" or "uppers") in low doses can also give the user a burst of energy (sometimes used for marathon sexual encounters), but will decrease sexual functioning in higher doses (Buffom et al., 1981) and can result in severe paranoia if used chronically. Amyl nitrate ("poppers" or "snappers"), which is used medically for treatment of coronary disease (it dilates blood vessels), is inhaled just prior to orgasm by some people in order to intensify the sensations (Cohen, 1979). This can be especially dangerous for people with heart or blood pressure problems.

Most marijuana users report that the drug improves sexual pleasure (Goode, 1972; Kolodny, Masters, & Johnson, 1979). It does not increase desire or arousal, nor does it intensify orgasms, but it enhances relaxation, thus allowing increased awareness of touch. Men who use marijuana chronically, however, may experience erectile problems, lowered testosterone levels, and decreased sperm production (Kolodny, 1981; Kolodny et al., 1974).

Among young people, a drug known as Ecstasy (3, 4-methylenedioxymethamphetamine, or MDMA) has become popular as an aphrodisiac. Proponents claim that it enhances a user's sense of well-being and empathy without the visual distortions of other hallucinogens. MDMA was originally a "designer drug," designed to be very similar to banned drugs but just different enough in structure to escape the law. However, it too is now outlawed. Research has shown that it causes permanent brain damage, and some users have experienced psychotic episodes. What's more, designer drugs are manufactured in clandestine labs without stringent quality control, and sometimes just a small unintended alteration in the chemical structure has devastating effects. A botched batch of designer drugs in the early 1980s, for example, caused permanent Parkinson's disease in numerous young people in California.

Most recently, studies have focused on pharmacological agents that affect central nervous system neurotransmitters—the chemical substances that allow brain cells to "talk" to one another (Rosen & Ashton, 1993). Of these, *yohimbine hydrochloride,* a substance derived from the sap of an African tree, has received the most attention because of a report that it caused intense sexual arousal in male rats (Clark, Smith, & Davidson, 1984). Although initial studies suggested that it might also have a positive effect on sexual desire in humans, a review of the work found that over half the people tested showed no response (Rosen, 1991). There is some evidence, however, that yohimbine is of some help in men with psychologically caused impotence (see Segraves, 1991).

Another psychopharmacological agent that might affect sexual responses is oxytocin, a pituitary hormone that promotes ejection of milk. It was recently reported that a woman who was prescribed this hormone as a nasal spray experienced intense sexual desire afterwards (Anderson-Hunt & Dennerstein, 1994), but it is too early to make general conclusions from a single case study.

Despite the lack of scientific evidence that most of the chemicals tried as aphrodisiacs really work, some people will probably continue to search for a problem-free drug that enhances sexual performance. Their efforts would probably be better rewarded if they instead took the time to learn how to communicate and be intimate with their partners. We will discuss how to do this in later chapters.

AND AFTERWARDS?

You have just read Masters and Johnson's description of the sexual response cycle as ending with orgasm and the return of physiological responses to normal, but an individual does not have to stop paying attention to his or her partner as soon as he or she reaches orgasm (e.g., by rolling over and going to sleep, or by lighting up a cigarette). This is one of the things that distinguishes human beings from other species. It is a rare couple who can experience simultaneous orgasms on any more than an occasional basis, which means that most of the time, someone will get there first. If it is the male, this will often mean loss of erection. Stimulation of one's partner, however, is not totally dependent on the functioning of the genitals. Sexual intimacy is a whole-body response, involving lips, tongue, hands, and emotional responsivity. Even if your partner has reached orgasm, the time afterwards can be a period of mutual touching, hugging, and emotional intimacy. When Ann Landers asked her female readers in 1985 if they would forego "the act" if they could be held closely, 100,000 women responded, of whom 72,000 of them said yes. Touching is necessary for a baby's emotional well-being, and we do not lose that need as we grow older. Sexual intimacy is more than genital thrusting and orgasm. It is having someone to touch and to touch you. It is a sense of closeness, a sense of partnership. This next section is a good example of that.

SEXUALITY AND THE HANDICAPPED

Is sexual pleasure reserved only for the young and beautiful? Judging by the number of jokes about the sexual behavior of the elderly and obese, and of other people whom our society often labels as "unattractive," many people apparently think so. These negative attitudes are often extreme in the case of the handicapped. What about people who are blind, deaf, spinal cord–damaged, or suffer from other physical disabilities? What about the mentally retarded? Do these groups of people have sexual desires and feelings? If so, should they be allowed to express those desires?

In this chapter you have learned about the effects of one kind of physical disability on sexual functioning: the effects of spinal cord damage on the ability to achieve erection and/or orgasm. However, as difficult as a physically handicapped person's disabilities may make it to engage in sex, a greater barrier to most handicapped individuals' having a sexual relationship is society's attitude about them. For example, I recently received the following note from a spinal cord–damaged student:

"Recently my boyfriend's mother wanted him to explain to me that she wants grandchildren. I guess she assumed that because I'm in a wheelchair that I'm incapable of having children. Her ignorance amazed me. People who are in wheelchairs, whether they be paraplegics or quadriplegics, are no less sexually able. Sexuality is basically psychological, in terms of its influence upon a person's overall behavior. Thus, the fact that one is paralyzed from the neck down or from the hips down doesn't discount from one's sexual drive or fertility."

(from the author's files)

Many people treat physically and mentally handicapped adults as if they were asexual or childlike and think it ridiculous that these groups should even have any interest in sex (Gardner, 1986). Parents of handicapped children sometimes try to protect them by withholding sexual information, while friends and acquaintances of handicapped adults often try to protect them by never discussing the subject in their presence. Even the medical profession, which has perfected many ways to assure the reproductive capabilities of disabled persons, has only recently begun to realize the importance of sexual education and counseling in helping patients and staff deal with the emotional, self-esteem, and social issues that may impair sexual functioning (Alexander et al., 1993; Rieve, 1989).

The well-intended but incorrect assumption of those who would protect the handicapped by denying their sexuality is that the handicapped have more important things to worry about. For example, in the case of spinal cord–damaged people, who generally have no genital sensations, it is often assumed that they would

Figure 4–8   We are all sexual beings.

(or should) not be interested in sex, but instead should be most concerned about physically rehabilitating the functional parts of their bodies.

Are sexual relations really so unimportant that they should be given such a low priority? To answer this, we must first ask why it is that people engage in sex. Is it possible for a person to enjoy sexual relations even if he or she cannot have sexual intercourse? Is sexual pleasure restricted to genital sensations (e.g., is a man's sexuality focused exclusively in his penis)? Recall that paralyzed individuals often become highly sensitive to touch in their functional body parts. There are many reasons that people engage in sexual relations besides the physical pleasure of genital sensations. Sexual relations are important to many people for the shared intimacy. Equally important to the physical intimacy is the opportunity for emotional intimacy provided by the closeness of sexual relations. Hugging, holding, kissing, and shared affection are as important in sexual relations as intercourse for many people. Physical closeness can be very psychologically exciting (see Figure 4–8). Here is part of a letter from the husband of a woman who had undergone three years of surgeries and rehabilitation for severe spinal cord damage suffered in an automobile accident:

"In July 'Susan' felt well enough to try once more. Sex was very good that day . . . we took it slow and easy. . . . We compensated with lots of holding and touching, something that often had been lacking in our pre-accident relationship. Incredibly, just holding each other in bed became very satisfying, and it actually seemed to bring us closer together emotionally."

(from the author's files)

People with spinal cord damage often show a decrease in sexual desire and satisfaction after the injury, but the sexual desire of the partner can help overcome this. Spinal cord–injured individuals may show a decrease in sexual intercourse, but often compensate for this with other sexual activities (Alexander et al., 1993). There are, in fact, many ways to enjoy sexual relations besides intercourse.

What about the mentally handicapped? Severe mentally retarded individuals display sexual behavior, including touching, kissing, and masturbation, but are unable to differentiate between appropriate times and places. This "inappropriate" (by normal standards) behavior results in many people, even special education teachers and administrators, having negative views of sexual behaviors in developmentally handicapped persons (e.g., Scotti et al., 1996; Wolfe, 1997). However, denial of sexuality in (and for) the mentally handicapped is not a realistic solution. "Regardless of their specific disabilities, persons with mental retardation are individuals with sexual feelings who develop physically at a rate comparable to that of normal young adults and respond to many of the same sexual stimuli and situations as do persons without mental retardation" (Scotti et al., 1996). A more realistic approach is to provide sexuality education to both the mentally handicapped and to their service providers (Ames, 1991).

Different handicaps present different problems in one's ability to have sexual relations, but none of the physical problems are as great an obstacle to healthy sexuality as society's attitude about handicapped people. Remember, regardless of what we look like or our physical or mental capabilities, we are all sexual beings.

Here are some helpful references regarding sexuality and the handicapped. See Chipouras et al., 1979, for a sexual bill of rights for disabled people. Two issues of *SIECUS Report* (March 1986 and April/May 1995) are devoted to meeting the needs of people with disabilities. For resources about sexuality and persons with spinal cord injuries, see Tepper and Lawless (1997).

Key Terms

anaphrodisiac 86
aphrodisiac 86
clitoridectomy 83
desire 72
ejaculation 76
erection 42, 72
excitement 72
Grafenberg (G) spot 38, 81
infibulation 83

Helen Kaplan 72
Masters and Johnson 16, 71
multiple orgasms 78
myotonia 72
orgasm 75
orgasmic platform 74
plateau 74
refractory period 78
resolution 78

retrograde ejaculation 76
semen 47, 76
sex-tension flush 74
sexual response cycle 72
Skene's glands 82
vaginal lubrication 73
vasocongestion 72

Personal Reflections

1. Do you think it is important that we know the physiological responses that normally occur during sexual arousal? Is it important that physicians and therapists have this information? What do you think about studies, such as those by Masters and Johnson, that require observing and recording individuals engaged in sex? If your opinion is negative, how else could we learn about the normal physiological responses during sexual arousal?

2. List the physiological responses that you experience during sexual arousal (e.g., vaginal lubrication, nipple erection, secretion from Cowper's glands, sex-tension flush, etc.). How do you feel about your responses? Are you comfortable with them, or do some cause you anxiety? Why? What about your partner's responses?

3. Do you know what factors influence your continued arousal as you go through the excitement, plateau, and orgasm phases of the sexual response cycle? Do you know which kinds of stimulation will give you the most pleasure? Do you know what pleases your partner most? Is it assumption or fact? Have you talked with your partner about how best to please each other? If not, why not?

4. Do you judge your sexual experiences by your ability to reach, or your frequency of, orgasm? Your partner's orgasm? Why? Many women do not experience orgasm during intercourse. How do you think they feel about it? How would you feel if it were you? Is it necessary to have an orgasm to enjoy sex?

5. Do you wish you had different-sized or different-shaped breasts or a different penis? How would you like to be different? Why? How important is the size of a person's penis or breasts in sexually pleasing a partner compared to a person's mind (his or her attitude about sex)?

6. What was your reaction in reading about genital cutting in African and Arab cultures? What is your view about male genital circumcision in this country?

7. We do not mutilate women's genitals in this country to deny them their sexuality. Do we deny women's sexuality in other ways? How?

8. If you were to have an accident and become physically handicapped, do you think you would still want to hug, hold, kiss, and share affection (and have sex, if possible) with your partner? How do you presently react when you see a handicapped person engaging in these behaviors?

Suggested Readings

Furlow, F. B., & Thornhill, R. (1996, January/February). The orgasm wars. *Psychology Today.* Considers the function of female orgasms from the view of evolutionary psychology.

Gallagher, W. (1986, February). The etiology of orgasm. *Discover.*

Kaufman, M. (1995). *Easy for you to say: Q & A's for teens living with chronic illness or disability.* Toronto: Key Porter Books. A lengthy, excellent chapter on sexuality.

Konner, M. (1988, March/April). Is orgasm essential? *The Sciences.*

Lightfoot-Klein, H. (1989). *Prisoners of ritual: An odyssey into female genital circumcision in Africa.* Binghamton, NY: Haworth Press.

Masters, W., & Johnson, V. (1966) *Human sexual response.* Boston: Little, Brown. The classic work.

Raphael, B. J. (1974, January). The myth of the male orgasm. *Psychology Today.* (Originally printed in the *Village Voice.*) A satire on male attitudes about female orgasm.

Tiefer, L. (1995). *Sex is not a natural act and other essays.* Boulder, CO: Westview. Offers some criticisms of the Masters and Johnson model; is also well written.

Toubia, N. (1993). *Warrior marks: Female sexual mutilation and the sexual blinding of women.* New York: Harcourt Brace.

White, D. (1981, September). Pursuit of the ultimate aphrodisiac. *Psychology Today.* Reviews the scientific evidence.

Yoffe, E. (1994, July/August). The truth about women and sex. *Health.* Discusses why some women have difficulty reaching orgasm.

CHAPTER

5

When you have finished studying this chapter, you should be able to:

1. Explain what causes sexually transmitted diseases, how they are spread, and who can get them;

2. Describe the symptoms for gonorrhea, chlamydia, syphilis, chancroid, herpes, hepatitis, venereal warts, AIDS, trichomoniasis, pubic lice, and scabies;

3. Understand that with many STDs it is possible to be infected and have no symptoms (for particular STDs, name which gender is more likely to be asymptomatic, and at what stage);

4. Describe the method of diagnosis and treatment for each STD;

5. Summarize the complications that can occur if STDs are not treated early;

6. Discuss the possible effects on the fetus and newborn if a pregnant woman has various STDs;

7. Describe the three types of vaginitis and discuss vaginal health care; and

8. Discuss safer sex practices.

Sexually Transmitted and Sexually Related Diseases

*T*he National Institute of Medicine has declared that sexually transmitted diseases (STDs) are presently at epidemic proportions (see Donovan, 1997b). Every year approximately 12 million Americans are diagnosed with an STD. In fact, *one in four of all Americans presently between the ages of 15 and 55 will have at least one sexually transmitted disease sometime in their lives* (Centers for Disease Control and Prevention). Of the nation's 10 most frequently reported infections, 5 are STDs (chlamydia, gonorrhea, hepatitis B, syphilis, and AIDS). The total cost to the country is over $17 billion a year. It is to everyone's advantage, therefore, that we educate ourselves about the causes, symptoms,

TABLE 5–1

STDS IN THE UNITED STATES

Type of Sexually Transmitted Disease	Estimated New Cases per Year in the U.S. (1998)
Chlamydia	4,000,000
Trichomoniasis	3,000,000+
Gonorrhea	800,000
Human Papillomavirus Infection	500,000–1,000,000
Genital Herpes	>500,000
Syphilis	<100,000
Hepatitis B (sexually transmitted)	50,000+
AIDS	60,000
HIV infection	?

Source: Centers for Disease Control and Prevention and studies cited in text.

modes of transmission, and treatments for these diseases. If left untreated, some of these diseases can cause sterility, blindness, and even death, not to mention ruined relationships.

WHAT ARE THEY AND WHO GETS THEM?

Some people believe that sexually transmitted diseases are punishment for having sinful or immoral sex and that only promiscuous persons get STDs. They talk as if they believe that sexual behavior causes the diseases. These beliefs are not only incorrect, they are also cruel, for tens of thousands of loyal and faithful men and women have caught these diseases from unfaithful partners; and a similar number of innocent newborn children have contracted them from infected mothers.

"I got gonorrhea from my boyfriend. I, myself, thought that I was engaged in a monogamous relationship. However, it was not monogamous for him. . . ."

"I had never had sex with anyone else before I got married. Well, I was married 5 years and had never cheated. I came down with genital herpes. . . ."

"Over the past two years, I have had one steady boyfriend. I have never slept around and I thought he was faithful to me. I would have put my life on the line to prove this. During these two years I have contracted chlamydia three

times, PID, and trichomoniasis. . . ."

"My first time having sex was at 16. My only fear was getting pregnant. My boyfriend said he was infertile so I couldn't get pregnant. We had sex. Approximately one month later I found out I had gonorrhea. I had no idea I could be a victim of an STD."

"I am a 48-year-old woman who contracted a sexually transmitted disease from my husband. As a result, I had to have a total hysterectomy. . . ."

(from the author's files)

These people were not promiscuous, yet they still became infected with sexually transmitted diseases. What causes STDs? It is not sexual behavior per se; the behavior is merely the mode of transmission for bacteria, viruses, or parasites that must be present for the diseases to be transmitted. *Sexually transmitted diseases are spread, for the most part, by sexual contact (including vaginal and anal intercourse and oral-genital contact) with someone who has the bacteria, viruses, or parasites that cause the disease.*

Bacteria are very small single-celled organisms. They lack a nuclear membrane but have all the genetic material (RNA and DNA) and metabolic machinery to reproduce themselves. **Viruses** are just a protein shell around a nucleic acid core. They have RNA or DNA, but not both, and thus cannot reproduce themselves. They invade host cells that provide the material to manufacture new virus particles. The bacteria and viruses responsible for STDs have been reported to live outside the body for only a short period of time. The common belief that you can catch these diseases from a toilet seat is greatly exaggerated, although it might be possible to expose yourself to the bacteria or virus by using a damp towel soon after someone with a disease uses it (and in this case, you are more likely to spread it to your eyes than to your genitals).

Some people believe (or at least would like to believe) that sexually transmitted diseases occur only in, or primarily in, the poor, the uneducated, and minority groups. Sexually transmitted diseases, however, are nondiscriminating. The bacteria or viruses that cause these diseases do not care whether you are white, black, yellow, or brown; whether you are on welfare or a millionaire; whether you bathe every day or only once a week; or whether you are a grade school dropout or college-educated. In fact, at least one of these diseases

bacteria Small, single-celled organisms that lack a nuclear membrane, but have all the genetic material (RNA and DNA) to reproduce themselves.

virus A protein shell around a nucleic acid core. Viruses have either RNA or DNA, but not both, and thus cannot reproduce themselves. They invade host cells that provide the material to manufacture new virus particles.

(chlamydia) is at epidemic proportions among college students today. The point, here and in the previous case histories, is that anyone having sex is at some risk for contracting a sexually transmitted disease.

Some of these diseases were once called **venereal diseases** (after the Roman goddess of love, Venus), but this term generally referred to diseases that are spread almost exclusively by sexual contact (gonorrhea, syphilis, and three other lesser-known diseases). Today, the description **sexually transmitted diseases** is generally preferred and refers to these diseases plus other infectious diseases that can be, but are not always, transmitted by sexual contact.

There are still other diseases of the sexual organs that are caused by overgrowths of yeast or fungal organisms found naturally in the body, and these too can sometimes be passed on during sex. I will refer to these as **sexually related diseases.** A few other sexually transmitted diseases are not really diseases at all, but rather infestations of parasites (pubic lice and scabies) that are transmitted from person to person during sexual contact.

WHERE DID THEY COME FROM?

Some people argue that humans first contracted sexually transmitted diseases by having sex with animals, but no animal species is known to have gonorrhea or syphilis. It is more likely that previously harmless bacteria or viruses mutated (they are highly susceptible to change) into strains causing disease. Although where and how STDs originated remains unknown, some of these sexually transmitted diseases have been around for at least a few thousand years, as many famous Greeks, Romans, and Egyptians (including Cleopatra) were described as having the symptoms. In fact, public kissing was banned for a period in Roman times because of an epidemic of a disease that was described as being identical to what we now know as herpes. An epidemic of syphilis also swept through Europe in the early 1500s, and it is still hotly debated whether Columbus and his sailors brought it back from the New World or vice versa (e.g., Luger, 1993; Rosebury, 1971). In fact, Columbus died of advanced syphilis.

The fact is that it does not really matter where or how STDs originated; they are here to stay and we must deal with them. In our own country, there have been major outbreaks of STDs from time to time, most recently after World War I and again after the war in Vietnam. The present epidemic situation can probably be attributed to several factors:

1. the discovery of penicillin and other antibiotics (about the time of World War II), which made people less fearful about contracting an STD;

2. a large drop between the end of World War II and 1955 in the amount of federal funds available for STD prevention (possibly due to overconfidence in antibiotics);
3. the new sexual freedom that many people believed possible as a result of the birth control pill and other reliable means of birth control;
4. the shift from use of the condom to use of the birth control pill; and
5. lack of education, or just plain ignorance.

Surveys of teenagers indicate that many know very little about sexually transmitted diseases, although they believe themselves to be well informed. Many people are unaware, for example, that they can have an STD but show no symptoms and still be contagious. As a result, they do not seek treatment as early as they should. I will discuss some precautionary measures that can help you avoid contracting sexually transmitted diseases and will outline what to do in case you should get one. First, let us examine the symptoms and treatment for some of the most widely known diseases.

GONORRHEA

Gonorrhea (from the Greek *gonos*, "seed," and *rhoia*, "flow"; also known as "the clap" or "the drip") is probably the oldest of the STDs. Its symptoms are described in Leviticus in the Old Testament, which dates back to about 1500 B.C., and were also described in detail by Greek physicians. The number of reported cases of gonorrhea has declined to about 400,000 a year, but many go unreported, and the Centers for Disease Control and Prevention estimates that there will be 800,000 new cases in the United States this year. The rate of infection is considerably higher among African Americans than it is among Hispanics or whites (Coutinho, 1994).

SYMPTOMS AND COMPLICATIONS

Gonorrhea is caused by a bacterium (*Neisseria gonorrhoeae*, often referred to as gonococcus) named after Albert Neisser, who discovered it in 1879. It lives on warm, moist mucous membranes in the urethra, vagina, rectum, mouth and throat, and eyes. A person gets gonorrhea by having his or her mucous membranes come into contact with an-

venereal diseases Term originally used to refer to gonorrhea, syphilis, and three other diseases that are spread almost exclusively by sexual contact.

sexually transmitted diseases Diseases that can be, but are not necessarily always, transmitted by sexual contact. This term is generally preferred to the term *venereal diseases*.

sexually related diseases Diseases of the reproductive system or genitals that are not contracted through sexual activity. Often involve overgrowths of bacteria, yeasts, viruses, or fungal organisms that are found naturally in sexual and reproductive organs.

gonorrhea A sexually transmitted disease caused by the *Neisseria gonorrhoeae* bacterium (often referred to as "the gonococcus"), which lives on mucous membranes.

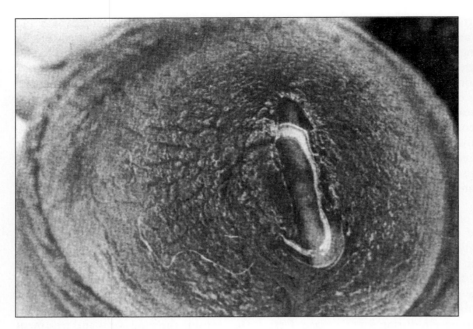

Figure 5–1 The pus-like discharge from the penis caused by gonorrhea.

show no symptoms (are *asymptomatic*), while in others the early symptoms disappear. These asymptomatic males are still infected and can pass the gonococcus to new partners. If left untreated, gonorrhea spreads up the male's reproductive system and causes inflammation of the prostate, seminal vesicles, bladder, and epididymis. By this time, the male usually experiences severe pain and fever. In 1 to 3 percent of cases, the disease gets into the bloodstream and causes inflammation of the joints (*gonococcal arthritis*), heart, or brain covering, or it may be transmitted to the eyes by a contaminated hand (Braverman & Strasburger, 1994). However, it rarely results in death.

In females, the gonococcus initially invades the cervix, but unlike males, *most women show no symptoms during the early stages.* They are often unaware of their infection unless told about it by an infected male partner, and thus can unknowingly pass gonorrhea on to new partners. In the 20 to 40 percent of women who show symptoms, there may be an abnormal vaginal discharge and irritation of the vulva and urethra, causing burning during urination. Because so many women are asymptomatic, they often go untreated, and the gonococcus spreads through the uterus into the Fallopian tubes. In up to 40 percent of untreated women, the tubes become swollen and inflamed, a condition known as **pelvic inflammatory disease** (**PID**), with symptoms of severe abdominal pain and fever (see pages 96–97). Hospitalization may be necessary at this stage. PID can cause scarring of the Fallopian tubes, which blocks passage of the sperm and egg—a common cause of sterility and tubal pregnancies in women (see Chapter 7).

If a pregnant woman contracts gonorrhea, the fetus will not be affected because the gonococcus normally is not carried in the bloodstream. However, a baby's eyes can become infected at delivery as it passes through the infected cervix and vagina (a condition called *gonococcal opthalmia neonatorum;* see Figure 5–2). This can rapidly lead to blindness. In some developing countries, 3 to 15 percent of babies are at risk, but the rate for gonorrhea in pregnant women in the United States is less than 1 percent, and most states require that an antibiotic or silver nitrate drops be put into all babies' eyes at birth to prevent possible infections (Laga, Meheus, & Piot, 1989).

other person's infected membranes. Because of the location of these membranes in the body, this normally occurs only during intimate contact. However, your chances of catching gonorrhea from having vaginal intercourse once with an infected person are not 100 percent. A man's risk of infection during intercourse with an infected woman is about 30 to 50 percent, while a woman has about a 50 to 60 percent chance of catching it the first time she has intercourse with an infected man (Platt et al., 1983; Rein, 1977). The gonococcus can also infect the anus and rectum (during anal intercourse) and the throat (during oral-genital sex, particularly fellatio), but less is known about the relative risk by these means of transmission. The bacteria have been found to survive for a short time outside the body, but your chances of picking them up from nonsexual contact, particularly a dry toilet seat, are extremely remote.

In males who become infected during intercourse, the gonococcus bacteria invade and cause inflammation of the urethra (*gonococcal urethritis*). This usually results in a thick, pus-like white or yellowish discharge from the urethra ("the drip"), starting 2 to 10 days after infection (see Figure 5–1). There is generally an irritation or a burning sensation at the urethral opening, and urination is often frequent and painful. Because the symptoms are painful and obvious, most men seek treatment immediately. However, *about 5 to 20 percent of men*

pelvic inflammatory disease (PID) A bacterially caused inflammation of a female's reproductive tract, particularly the Fallopian tubes, that can result in sterility. The most common (though not the only) cause is untreated gonorrhea and/or chlamydia.

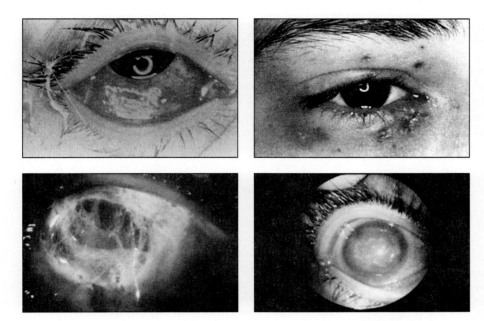

Figure 5–2 Eye infections caused by gonorrhea (left) and herpes (right).

DIAGNOSIS AND TREATMENT

Most individuals with symptoms of gonorrhea will seek treatment immediately. Remember, however, that many people—particularly females—have no early symptoms. Therefore, if you suspect that you have had sex with someone who has gonorrhea, see a doctor as soon as possible. Males who have a urethral discharge can be diagnosed quickly by an examination of the pus under a microscope. This is not 100 percent accurate, however, and the only sure way to test for gonorrhea is to have a culture test. Culture tests are almost always done for women with suspected gonorrhea. A cotton swab is inserted into the area of suspected infection (about one-half inch up the urethra in males, into the cervix in females, or in the throat or rectum of either sex), and the cells are grown in culture for a couple of days and then tested.

There was little that could be done for people with gonorrhea prior to the discovery of antibiotics. The urethral inflammation caused by gonorrhea often caused blockage in men, and because the male's urethra also serves as the passageway for urine, it was often necessary to clean out an infected urethra with a pipe cleanerlike instrument. Fortunately, penicillin was discovered around the time of World War II and proved to be highly effective. Unfortunately, the bacterium developed resistance to penicillin over time. A dosage of 150,000 units was given to people for gonorrhea in the 1940s, but by the 1980s the standard dosage was nearly 5 million units, and there were some strains that could not be destroyed with any dose (*penicillinase Neisseria gonorrhoeae*). There were only about 8,000 re-

ported cases of penicillin-resistant gonorrhea in 1985, but the number of cases had increased to 56,616 in 1989. Many physicians had switched to tetracycline by the early 1980s, but by the end of the decade the bacterium was becoming resistant to it as well (Plasma-mediated, 1990). To complicate matters, many people who are diagnosed with gonorrhea are also found to have another sexually transmitted disease called chlamydia, which is treated with tetracycline or an erythromycin. To combat these multiple problems, most physicians today treat gonorrhea with a drug called ceftriaxone (which is presently effective against all strains), followed by a week of tetracycline or an erythromycin (Handsfield, 1990; Schwarcz et al., 1990). In the future, it may be possible to combat the spread of gonorrhea by vaccinating people (Sparling et al., 1994).

Should you ever be diagnosed with gonorrhea, be sure to tell all your recent partners that they may have gonorrhea so that they, too, can be treated. Gonorrhea is not like chicken pox, where you become immune after having had the disease. You can catch gonorrhea again if you again have sex with someone infected with the gonococcus.

CHLAMYDIA AND NONGONOCOCCAL URETHRITIS

Any inflammation of the urethra not caused by the gonococcus is called **nongonococcal** (or **nonspecific**) **urethritis** (**NGU**). This term is often reserved for men with an inflamed urethra, but the organisms that cause it cause infection in women (*cervicitis*) as well. There are several organisms that can cause infection (including *Ureaplasma urealyticum* and *Trichomonas vaginalis*), but the one that is most frequently responsible today is *Chlamydia trachomatis*. The infection caused by this sexually transmitted bacterium is called **chlamydia.**

Chlamydia is much more common than gonorrhea, with an estimated 4 million new cases in the United States this year (Centers for Disease

nongonococcal (nonspecific) urethritis (NGU) Any inflammation of the urethra that is not caused by the *Neisseria gonorrhoeae* (gonococcus) bacterium.

chlamydia A sexually transmitted disease caused by the *Chlamydia trachomatis* bacterium, which lives on mucous membranes.

Control and Prevention). It is especially prevalent in teenagers and young adults and is more common among whites than African Americans (Zimmerman et al., 1990). A recent study of nearly 150,000 sexually active females aged 15–19 years found that 10 percent of them had chlamydia (Mosure et al., 1997).

Because many students have never heard of chlamydia, they think that it cannot be all that bad. In fact, although its initial symptoms are usually milder than those of gonorrhea, if left untreated it is more likely than gonorrhea to cause damage to the reproductive organs.

SYMPTOMS AND COMPLICATIONS

Like the gonococcus, *Chlamydia trachomatis* lives only on mucous membranes. Chlamydia is spread when the infected membranes of one person come into contact with the membranes of another person.

If a person with chlamydia shows early symptoms, he or she will start doing so within 1 to 3 weeks after infection. The early symptoms are milder than, but similar to and sometimes mistaken for, those of gonorrhea: irritation and burning of the urethra (although usually not painful) and a discharge (usually thin and clear rather than pus-like). However, most women and a large proportion of men have no symptoms at all in the initial stage.

> "I found out that I had chlamydia when my girlfriend handed me a bottle of pills and told me to follow the instructions. I didn't even know what chlamydia was. . . . I had no idea I had an STD. . . ."

> "When I was attending LSU, I went to see the doctor for a regular gynecological exam. The nurse suggested that I consider some other tests for STDs because some of them had few or no symptoms. At first I declined, thinking that I had not slept with many people and, also, that if I had anything I would surely know from symptoms. But I changed my mind and was tested. When I called for results, I was shocked to find out that I was positive for chlamydia."

(from the author's files)

If left untreated, chlamydia spreads through the reproductive system, which in men can cause infection of the prostate and epididymis, and possible sterility (Greendale et al., 1993). In women the spread of chlamidia can cause pelvic inflammatory disease,

lymphogranuloma venereum (LGV) A sexually transmitted disease common in tropical countries that is caused by chlamydia. If left untreated, it causes swelling of the inguinal lymph nodes, penis, labia, or clitoris.

which can leave the Fallopian tubes scarred and result in sterility or increased risk of tubal pregnancy (Brunham et al., 1988; Sherman et al., 1990). At this stage the infection can be quite painful and may require hospitalization. Research has shown that *a single attack of chlamydia is about three times more likely than gonorrhea to cause sterility in women.* Babies born to women who have chlamydia at the time of delivery can get eye infections (*chlamydia conjunctivitis*) and/or nose-throat infections, or even pneumonia (an estimated 182,000 cases per year; Schacter et al., 1986). There is an increased risk of stillbirths, premature deliveries, and death as well. In underdeveloped countries, the chlamydia bacterium often causes an eye infection called *trachoma*, which is spread to other people's eyes by flies. It is the leading cause of blindness in those areas. In tropical countries, chlamydia is also responsible for a serious sexually transmitted disease called **lymphogranuloma venereum (LGV)** (there are fewer than 400 cases annually in the United States). If left untreated, LGV causes the inguinal lymph nodes to swell, followed by swelling of the penis, labia, or clitoris (sometimes enormously so, resulting in elephantiasis of the external genitalia).

DIAGNOSIS AND TREATMENT

Until recently, the only way to diagnose chlamydia was with culture tests that were expensive and technically demanding, sometimes taking up to a week to complete. In the 1990s, several nonculture tests (that detect chlamydial antigens or nucleic acids) became available that could give results in a matter of minutes. However, studies have found that these tests are not as sensitive as the culture tests, and thus often miss people who really have chlamydia (Lin et al., 1992; Schachter et al., 1992).

If chlamydia is confirmed, it is usually treated with tetracycline, doxycycline, or erythromycin, but because in some cases the infection persists, some doctors are using newer antibiotics such as azithromycin (Martin et al., 1992). Remember, many people have no symptoms in the early stages, so if you think someone you have had sex with has chlamydia, seek treatment immediately.

PELVIC INFLAMMATORY DISEASE (PID) IN WOMEN: A LIKELY CONSEQUENCE OF UNTREATED CHLAMYDIA OR GONORRHEA

In the previous sections on gonorrhea and chlamydia, I referred to pelvic inflammatory disease in women as a possible consequence if the diseases went untreated. PID is a general term for an infection that travels from the lower genital tract (vagina and cervix) to the Fallopian tubes, and sometimes to surrounding structures such as the ovaries and pelvic cavity (Berger, Westrom, & Wolner-Hanssen, 1992). It results from infection with a variety of organisms, but 50 to 75 percent of cases are caused by *Chlamydia trachomatis* or *Neisseria gonorrhoeae* (Brunham et al., 1988; Rice & Schachter, 1991).

Over one million cases of PID occur each year in the United States (Donovan, 1997), requiring about 182,000 hospitalizations (Rolfs, Galaid, & Zaidi, 1992). One in seven women of reproductive age have had PID (Aral, Mosher, & Cates, 1991).

Symptoms may include tenderness or pain in the lower abdomen (sometimes to the extent that standing up straight is impossible), high fever, and chills. However, three times as many women experience a persistent low-grade infection, either with no symptoms or with intermittent abdominal cramps (Wolner-Hansen, Kiviat, & Holmes, 1989). In these "silent" infections (i.e., the women do not go to a physician for diagnosis), the long-term inflammation can result in pelvic adhesions and abscesses.

The inside diameter of a Fallopian tube is only about that of a human hair; thus, even a small amount of scar tissue can result in a life-threatening ectopic (tubal) pregnancy or in permanent sterility (Weinstock et al., 1994). The chances of sterility increase from 11 percent of women after one case of PID to 54 percent of women with three episodes (Weström, 1980). Largely because of the increase in sexually transmitted diseases, the rates for ectopic pregnancies and infertility have increased dramatically in young women during the last two decades. Over 40 percent of ectopic pregnancies are directly related to STDs and pelvic inflammatory disease (Coste et al., 1994).

Because PID is so closely associated with the acquisition of chlamydia and gonorrhea, it should not be surprising that the major risk factors for PID are related to sexual behavior: early age for first experience of intercourse and history of multiple partners (Aral et al., 1991; Lidegaard & Helm, 1990). Interestingly, there is also an increased risk from douching (probably by forcing the bacteria into the uterus) and smoking (by impaired tubal ciliary function or impaired immune response; Forrest et al., 1989; Wolner-Hanssen et al., 1990; Scholes, Daling, & Stergachis, 1992).

Is there anything that women can do to minimize the chances of getting PID? One option, of course, is to reduce the number of one's sexual partners. But many of us may be sexually monogamous and still have several (monogamous) partners over time. Spermicides and barrier methods of contraception, such as male or female condoms and the diaphragm, reduce the risk of PID by reducing the risk of sexually transmitted diseases (Lidegaard & Helm, 1990; Washington, Cates, & Wasserheit, 1991; see Box 6–A on pages 154–155). The birth control pill increases the risk of chlamydial infections of the lower genital tract, but decreases the risk of PID, while IUDs increase the risk

of PID (see Chapter 6). It is very important that women having sexual relations have regular pelvic examinations (and perhaps ask their doctor to test for gonorrhea and chlamydia if they have had new partners since their last exam).

SYPHILIS

The consequences of untreated gonorrhea or chlamydia can be serious, but untreated **syphilis** is far worse, being responsible for the deaths of over 100 million people worldwide in this century alone (Chiappa & Forish, 1976). As stated earlier, some people have speculated that Columbus brought the disease back from the New World (Catterall, 1974; but see Luger, 1993, for another explanation), while others have speculated that the disease evolved from the treponema organism, which causes the skin disease yaws in tropical areas. They believe that when humans migrated to cooler climates, the skin provided a less desirable environment and the organism retreated to the genitals and anus, becoming a sexually transmitted disease. Whatever the origin, there is no written description of the disease's symptoms prior to the time of Columbus.

It was once thought that gonorrhea and syphilis were the same disease. In the 1700s, an English doctor named John Hunter tried to prove this by inoculating his own penis with the pus from a patient with gonorrhea. He thought he had proven his hypothesis when he developed the symptoms of both diseases, but in fact the patient had simply had both diseases. *Treponema pallidum,* the spiral-shaped bacterium (spirochete) that causes syphilis, was finally identified in 1905.

The United States has made great efforts to eradicate syphilis by tracking down the sexual contacts of infected persons, and most states require premarital and prenatal blood tests. With the discovery of antibiotics, the number of new cases had dropped to around 7,000 in the late 1950s. The rate started rising again in the 1960s and there was another increase in the middle to late 1980s, but syphilis has declined dramatically since then: there were only 11,624 new cases reported to the Centers for Disease Control and Prevention in 1996. However, the actual rate is certainly much higher. The disease is found mainly in southern states and large urban areas and is much more common among African Americans than whites (Nakashima et al., 1996). These ethnic differences are not direct risk factors for syphilis, but are instead related to more specific risks. Syphilis is particularly bad in the inner cities because of widespread unemployment, poverty, poor education and health care, and a greater frequency of prostitution.

> **syphilis** A sexually transmitted disease caused by the *Treponema pallidum* bacterium (spirochete), which can also pass directly through any cut or scrape into the bloodstream.

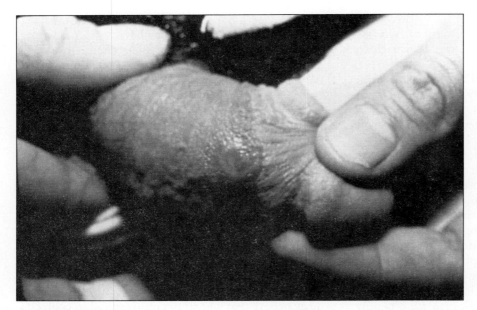

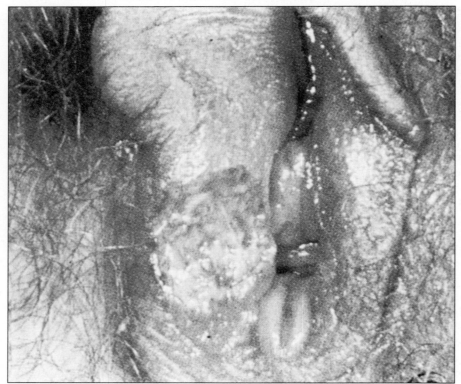

Figure 5–3 A primary-stage syphilis chancre on the penis (top) and labia (bottom).

SYMPTOMS AND COMPLICATIONS

The vast majority of cases of syphilis are transmitted by sexual contact. However, unlike the bacteria that cause gonorrhea and chlamydia, which require contact of mu-

chancre The painless sore that is the main symptom of the primary stage of syphilis.

cous membranes for transmission, the spirochete that causes syphilis can also pass directly through any little cut or scrape of the skin into the bloodstream. This means that one could potentially get syphilis by merely touching the sores of another person, as has happened to some dentists and physicians. If a person has been infected by the spirochete, the first symptoms generally appear 2 to 4 weeks later. If many bacteria are transmitted, the incubation period can be as short as 10 days, but if only a few are transmitted, the sores can take as long as 90 days to appear.

The symptoms of syphilis generally occur in four stages. The **primary stage** begins with the appearance of a very ugly ulcerlike (raised edges with a crater) sore called a *chancre* (see Figure 5–3) at the site where the spirochete entered the body (usually the penis, cervix, lips, tongue, or anus). The chancre is full of bacteria and highly infectious, but even though the chancre looks awful, it is usually painless, and thus some people do not seek treatment. Because the painless chancre generally appears on the cervix in women, they usually are unaware that they are infected during the initial stage unless they get a pelvic exam. Even if a person is not treated, the sore will disappear in a few weeks. However, the spirochete continues to be carried in the person's bloodstream, and thus individuals remain infected *and* infectious even after the chancre disappears.

Symptoms of the **secondary stage** of syphilis appear within a few weeks to a few months after the chancre heals. The main symptom is an *itchless, painless rash* that appears all over the body, including the palms of the hands and the soles of the feet (see Figure 5–4). In the moist areas around the genitals, the rash appears as large sores called *condylomata lata*, which break and ooze a highly infectious fluid full of

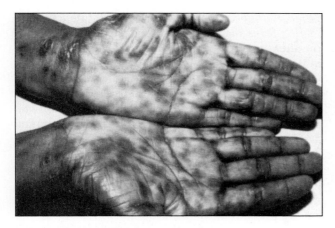

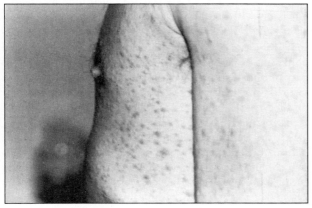

Figure 5–4 Secondary-stage syphilis rash on the palms and abdomen.

bacteria. Other symptoms can include a sore throat, a persistent low-grade fever, nausea, loss of appetite, aches and pains, and sometimes even hair loss. These symptoms are bothersome enough that many infected people will seek treatment, but many people mistake the symptoms for the measles, an allergic reaction, or some other disease (syphilis is sometimes called "the great imitator"). If treatment is not obtained, these symptoms, like the chancre of the primary stage, will also disappear after a few months and not return.

An infected person now enters the symptomless **latent stage,** and after about a year is no longer contagious because the spirochete is no longer found on the mucous membrane surfaces. This stage may last for years with no noticeable signs, but all the while the bacteria are attacking the internal organs of the body, particularly the heart and blood vessels and the brain and spinal cord.

About two-thirds of the people with untreated syphilis will not experience any

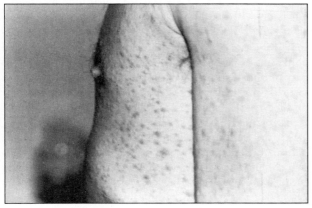

Figure 5–5 A baby born with congenital syphilis.

more problems with the disease, but one-third will develop serious complications as a result of the accumulated organ damage. They are now in the **late (or tertiary) stage**. Large ulcers called *gummas* often appear on the skin and bones. Damage to the heart and blood vessels frequently results in death, while damage to the central nervous system can lead to paralysis, insanity, and/or deafness.

Particularly heartbreaking is the fact that an *unborn baby can catch syphilis from an infected mother.* This happens because the spirochete travels form the pregnant woman's blood into the placental blood system. If it is not diagnosed and treated, the fetus will be aborted, stillborn, or born with a rather advanced stage of the disease called **congenital syphilis.** Complications of congenital syphilis can include deformation of the bones and teeth, blindness, deafness, and other abnormalities that can result in an early death (see Figure 5–5). Blood tests for syphilis are done routinely during pregnancy tests, and if detected the disease can be treated. Congenital syphilis is most common in babies born to mothers who never went for a prenatal checkup (Ricci et al., 1989).

DIAGNOSIS AND TREATMENT

The spirochete taken from a chancre or sore can be identified under a microscope, but blood tests are almost always done as well (some conditions, such as lupus and some types of arthritis, and some flu vaccines can cause a false positive test result). The saddest thing about any suffering or deaths that occur from syphilis is that the spirochetes are so easily eradicated with antibiotics. Penicillin is very effective, with the amount and duration of treatment depending on the stage of infection (Hook, Sondheimer, & Zenilman, 1995). Although antibiotics can eradicate the bacteria at any stage, they cannot reverse any organ damage that may have already occurred if the disease was left untreated until the late stage. It is important, therefore, that people seek treatment as early as possible.

LESS COMMON BACTERIAL STDS

There are other types of bacterially caused STDs that are common in tropical climates but rare in the United States. I have already mentioned lymphogranuloma venereum (see the section on chlamydia); two additional examples are chancroid and granuloma inguinale. **Chancroid** is caused by *Hemo-*

> **gumma** A soft, gummy tumor that appears in late (tertiary) syphilis.
>
> **chancroid** A sexually transmitted disease, caused by the *Hemophilus ducreyi* bacterium, which is characterized by small, painful bumps.

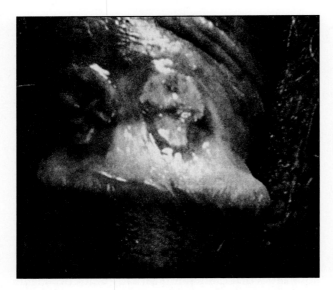

Figure 5–6 Two adjacent chancroid lesions on the penis.

philus ducreyi and is characterized by small bumps on the genitals or other sites that rupture into soft crater-like sores (see Figure 5–6). The incubation period from the time of infection to the appearance of the first sores is only 1 to 8 days. Unlike the syphilis chancre, chancroid is painful. The lymph nodes in the groin area often become inflamed and swollen. There are only about 1,000 to 2,000 new cases of chancroid per year in the United States, where it is spread mainly by sexual contact, but in moist tropical areas it can be spread by nonsexual physical contact as well. Chancroid is treated with erythromycin or ceftriaxone (Schmid, 1990).

Granuloma inguinale is rare in the United States, with fewer than 50 cases per year. The main symptom is a painless pimple that then ulcerates and spreads to surrounding areas, permanently destroying the tissue and causing death if not treated. It is treated with tetracycline.

Shigellosis is another uncommon sexually transmitted disease. It is contracted from exposure to feces infected with the Shigella bacteria (a result of behavior such as oral stimulation of the anus). Symptoms include acute diarrhea, fever, and pain. Thirty percent of the cases in the United States occur in male homosexuals. Shigellosis is treated with tetracycline or ampicillin.

granuloma inguinale A rare (in the United States) sexually transmitted disease that is characterized by ulceration of tissue.

shigellosis A disease that can be contracted during sexual activity by exposure to feces containing the *Shigella* bacterium.

HERPES

Before all the attention given to AIDS, herpes was the media's favorite STD, appearing on the cover of *Time* and *Newsweek* and even the subject of a feature article in *Rolling Stone* magazine. Although it does not receive the attention it used to, herpes is still with us, infecting more than 500,000 Americans every year (Catotti, Clark, & Catoe, 1993). Unlike gonorrhea, chlamydia, and syphilis, there is no cure for herpes. The men and women infected this year will be added to those who already have it, making the total number of genital herpes sufferers at least 45 million—one in five Americans age 12 or older (Fleming et al., 1997). This is a 30 percent increase since the late 1970s. While genital herpes is much more common among blacks than whites or Hispanics, the incidence rate has quintupled among white teens since the late 1970s.

Herpes (from a Greek word meaning "to creep") has been with us for a long time. It is *spread by direct skin-to-skin contact* (from the infected site on one person to the site of contact on another person). A terrible epidemic in Roman times led the emperor Tiberius to ban public kissing. Herpes was finally identified as a virus in the 1940s, with two different types (I and II) of the virus identified in the mid-1960s. The type I virus is much more common than type II (see Burke, 1993, for a review of many studies). Both types cause painful blisters, but in the mid-1960s there was a 95 percent site specificity: 95 percent of people with fever blisters or a cold sore around the mouth had *herpes simplex virus type I* (and thus this type was called "oral herpes"), and 95 percent of people with blisters around the genitals had *herpes simplex virus type II* (which thus came to be called "genital herpes"). Only 5 percent of individuals with blisters on the mouth had type II, and vice versa. Thus, it was commonly believed that HSV type I was acquired by kissing and HSV type II only by sexual intercourse.

By 1980, there was no longer a strong site specificity, particularly in young people. In some clinics, 20 to 35 percent of those persons who had blisters on the genitals had the type I virus rather than type II, and 30 percent who had blisters on the mouth had type II herpes rather than type I (Corey et al., 1983; Peter et al., 1982). Some believe that this is due to the increased popularity of oral-genital sexual behavior in recent decades. Whether or not this is true, it is certain that *both the type I and type II viruses can be transferred from mouth to genitals, and vice versa. The two viruses are not different clinically, as both cause the same painful symptoms.* Nevertheless, many persist in their mistaken belief that oral herpes can only be caught by kissing and genital herpes only through intercourse. Perhaps this letter to Ann Landers will be more persuasive:

> Dear Ann Landers:
>   I got genital herpes from my boyfriend of two years from an innocent fever blister on his lip. I could not believe a woman could get

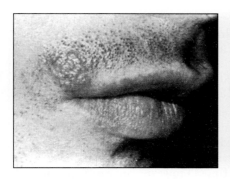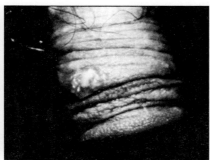

Figure 5–7 Herpes blisters on the mouth (left) and penis (center). Photo (right) shows ulcerated herpes lesions on the penis.

a venereal disease from indulging in oral sex with someone who had a simple cold sore. My doctor cultured it and sure enough it came back Herpes Simplex Type I (the oral kind). But now I have genital herpes.

I developed a high fever after being exposed to the virus on his lip. That virus caused two fever blisters on my genitals. They reappear every two months and are quite painful.

My boyfriend dumped me when he learned I was infected. He remains "pure" and does not come down with a sexually transmitted disease, while I am alone and scared. There is no justice! Please, Ann Landers, inform the public at once. Tell them what can happen if a person has contact with those innocent-looking sores near the mouth.

Sign Me Dumb and Angry

Dear Angry: I have printed this information in my column more than once. Several readers wrote to say I was ill-informed, stupid, and just plain nuts. Thanks for your testimony.

(Reprinted with permission of Ann Landers and Creators Syndicate.)

A person with blisters on the mouth can not only transfer the virus to another person's genitals, and vice versa, but a person with herpes can transfer it by touch to other parts of his or her own body—including the eyes—as well. This is referred to as *autoinoculation.* I will refer to herpes blisters or ulcerations on the mouth as **oral herpes** and blisters or ulcerations on the genitals as **genital herpes,** without worrying whether the blisters were caused by the type I or type II virus (see Figure 5–7). Although oral herpes can be contracted through oral-genital sex, experts believe most cases of oral herpes in adults to be reactivations of latent infections acquired during childhood (see Fleming et al., 1997). The virus can be easily spread by even a quick,

casual kiss, and thus *it should not be assumed that a person suffering an attack of oral herpes was recently involved in sexual activity.* By the way, not all mouth ulcers are caused by the herpes virus; they can also be caused by bacteria, allergic reactions, or autoimmune responses. Canker sores, for example, are not caused by herpes virus. However, "fever blisters" and most cold sores are herpes.

There are actually seven human herpes viruses, but only types I and II are sexually spread. The herpes simplex virus types I and II are very closely related to the viruses that cause chicken pox in children and shingles in adults (varicella zoster virus), infectious mononucleosis (Epstein-Barr virus), and cytomegalic ("large-cell") inclusion disease in infants (cytomegalovirus).

SYMPTOMS—PRIMARY ATTACK

Symptoms of the initial herpes attack normally appear about 2 to 20 days after contact with an infected person, with an average of 6 days. The symptoms occur in three stages. In stage one, the **prodromal stage,** the person feels a tingling, burning, itching, or anesthetic-like sensation on the skin surface where he or she came into contact with the virus. Sometimes the symptoms are more diffuse; in the case of genital herpes, for example, the infected person may feel pain running down the buttocks and thighs. These initial signs indicate the beginning of *viral replication and shedding* (i.e., viruses traveling from the source, free to incorporate themselves into other cells). Therefore, the individual should cease immediately all intimate contact involving the infected area.

Within a few hours the skin surface will break into a rash, which is followed a short time later by the appearance of fluid-filled *blisters* called *vesicles* (see Figure 5–7). These can be excruciatingly painful. After a while, the vesicles break open, resembling pustules and then ulcerated

oral herpes A herpes infection in or around the mouth. It can be caused by herpes simplex virus types I or II.

genital herpes Herpes infection in the genital region. It can be caused by herpes simplex virus types I or II.

running sores. Although herpes lesions generally resemble blisters, many people have sores that are more ulcerative in appearance. These symptoms mark the second, or **vesicle stage,** which can last from 2 to 3 weeks. The first, or primary, herpes attack includes more than just painful blisters, however. The person often has flulike symptoms, suffering muscle aches and pains, headache, fever, and swollen lymph glands. Urination can be quite painful for those with genital herpes (catheterization is sometimes required), and even walking or sitting can be extremely painful sometimes.

The sores eventually begin to develop scales and form scabs. This is the third, or **crusting-over stage.** Concomitant bacterial infection of the sores sometimes occurs during this stage, but the lesions eventually heal without leaving scars. Viruses are still being shed during the crusting-over stage; thus, intimate contact with the infected area should be avoided until all the sores are fully healed. The average duration of a primary attack is about 16 days (Beutner, 1996).

The primary attack is severe because the person has not yet built up antibodies to the virus. A person can have more than one primary attack of herpes, however. An individual could have a primary attack after exposure to the type I virus, for example, and later be exposed to the type II virus and have another primary attack (Al Samarai et al., 1989). Primary attacks to different strains of the same virus are also possible. A person with genital herpes, therefore, should not assume that it is safe to have intercourse with someone who also has herpes.

RECURRENT ATTACKS

During the primary herpes attack, some viruses migrate up sensory nerves from the infected site to the nerve's ganglion (the nerve cell center): the sacral nerve ganglion in the case of genital herpes and the trigeminal nerve ganglion in the case of oral herpes. The viruses become dormant (inactive) when they reach the ganglion. About one-third of those who have suffered primary herpes symptoms never again have another attack, but the remaining cases are evenly divided between those who suffer occasional attacks and those who suffer regular recurrent attacks. A recurrent attack occurs as a result of the virus reactivating and traveling down the sensory nerve to or near the same area as the initial sores, where it again starts multiplying. Genital herpes infections are six times more likely to recur as oral infections (Lafferty et al., 1987), and people with genital herpes suffer an average of four recurrent attacks a year. Recurrent attacks are usually less severe than the primary attack (because the individual now produces antibodies) and generally last only from 5 to 10 days.

What causes recurrent attacks? *Stress* to the immune system, either of a physical or emotional nature, seems to be the major cause. A recurrent attack can be brought on by illness, fatigue, menstruation, too much sunlight, and/or anxiety.

ASYMPTOMATIC AND UNRECOGNIZED INFECTIONS

The herpes viruses are easily spread when sores (blisters or ulcerations) are present. The sores are an obvious sign of viral shedding, and infected persons can take measures to prevent the spread of the virus to other persons at this time. However, in many people the symptoms are mild, while still other individuals are asymptomatic shedders of the virus. In fact, as many as 90 percent of people infected with herpes simplex type II in the genital region are unaware of their infectious potential (Fleming et al., 1997). Half or more of these individuals have clinically recognizable symptoms, but the symptoms are so mild that the individuals do not recognize them as herpes. Experts new believe that most transmission of genital herpers occurs during periods when there are no recognizable symptoms (Conant, 1996; Pereira, 1996).

> "The guy I had been dating for two years told me that he had 'jock itch' and showed me a sore at the base of his penis. I didn't know what jock itch even looked like, much less herpes, so I believed him. His sore went away. About a month later another sore came up, so he went to the doctor . . . [who] told him he had herpes. Two days later I had a sore on my labia. . . . The doctor diagnosed the sore as herpes."
>
> (from the author's files)

SERIOUS COMPLICATIONS

We generally think of herpes blisters as occurring only on the mouth and the genitals or anal area, but *the blisters may appear on any part of the body.* The virus can be spread to the eyes (by touching blisters that appear elsewhere on the body and then rubbing the eyes, or by using saliva to insert contact lenses), resulting in *herpes keratitis,* or ocular herpes. This is the *leading cause of blindness resulting from infection* in the United States today (see Figure 5–2). Symptoms escalate from feeling as if you have something in your eye to marked irritation and pain, often accompanied by conjunctivitis (inflammation of the mucous membrane). Small lesions will often be visible on the eyeball and always result in scarring as they heal (it is the scars that eventually cause blindness). So if you have herpes, be sure to wash your hands if you touch the sores.

Occasionally, dormant oral or facial herpes migrates to the brain rather than to the face, resulting in *herpes encephalitis,* a rare but often fatal condition. Primary attacks of herpes sometimes include inflammation of the membranes covering the brain and spinal cord (*herpes meningitis*), but full recovery usually follows.

Women with genital herpes run four to eight times greater risk of getting cancer of the cervix than women who have never had herpes (Graham et al., 1982). The relationship between herpes and cancer is poorly understood (herpes is not necessarily the cause, but perhaps a cofactor in susceptibility). Women who have genital herpes should get a Pap smear test twice a year instead of once.

Herpes can result in serious complications during pregnancy. A primary attack during pregnancy is generally more severe than usual, and recurrent attacks tend to occur more frequently. The stress of delivery may also bring on a recurrent attack. The greatest risk is to the unborn or newborn baby, particularly for women who have a primary attack near the time of labor (Brown et al., 1997). The virus ordinarily does not get into the bloodstream, but there is a higher-than-normal rate of premature births, spontaneous abortions, and congenital malformations (e.g., brain damage or mental retardation), suggesting the possibility that the virus can cross the placenta (Stone et al., 1989).

Neonatal herpes is devastating. If a woman has an active case of genital herpes during delivery, there is at least a 50 percent chance that the baby will catch it. The first signs appear 4 to 21 days after birth. Many babies who get herpes at birth die or suffer permanent and severe neurological problems (see Brown et at., 1997). Eye infections, widespread skin eruptions, and damage to internal organs can occur as well. The disease is so virulent in newborns because they have no defenses to fight off the infection. Unfortunately, it is not possible to assess the risk of spreading infection to the baby by looking for symptoms in the mother prior to birth. Most babies who become infected are born to mothers who do not display genital blisters at the time of delivery (Stone et al., 1989). It is for these reasons that most doctors prefer to deliver babies by cesarean section to women with genital herpes. Because many people with the herpes virus have no symptoms, or have symptoms they do not recognize to be herpes (see previous section), pregnant women are advised to get a blood test (Kulhanjian et al., 1992). An active case of oral herpes is not a risk factor at the time of delivery, but parents (or anyone else) with an active case of oral herpes should avoid kissing or fondling a baby.

DIAGNOSIS AND TREATMENT

First, the bad news: as yet, there is no cure for herpes. Nothing you can do or take will rid your body of the herpes virus. The virus lies dormant in the ganglion cells between active outbreaks, and the potential for another attack is always there. Herpes sufferers nevertheless remain persistent in their attempts to find a home cure. Treatments such as applying baking soda, bleach, ether, yogurt compresses, seaweed, and even peanut butter or snake venom to the blisters have all been tried. None of these has proven effective in scientific tests, but just attempting to do something probably makes some people feel better. However, some of these home remedies (e.g., ether) have proven to make matters even worse in some cases. Creams and ointments should also be avoided because they help to spread the virus to adjacent areas and prevent the skin from drying. Many people attempt to control herpes through diet, such as eating foods high in L-lysine (e.g., fish, meat, milk, beans, cheese, eggs) and avoiding foods high in arginine (e.g., nuts, chocolate, whole wheat bread, raisins, oatmeal, brown rice). The Centers for Disease Control and Prevention indicates that there is no merit to these specific diet claims.

Now the good news. Although there is no known cure for herpes, there is a drug presently available that relieves symptoms and speeds up the healing process during the primary attack. The drug is called *acyclovir* (marketed under the name Zovirax) and is available both as an ointment and as a tablet. People who have frequent recurrent attacks can take tablets 5 times daily (Beutner, 1996; Conant et al., 1996). A newer version of the drug (*valaciclovir*) that requires only 2 pills daily should be available soon (Bodsworth et al., 1997). Another antiviral cream, penciclovir, has recently been found to help heal cold sores (oral herpes) (Spruance et al., 1997). Acyclovir is not a cure, however, and recurrent attacks will return once the drug is discontinued.

There is real hope for a herpes vaccine in the near future (Pereira, 1996). In the meantime, it is recommended that sufferers avoid stress, wash their hands regularly to prevent autoinoculation, and avoid wearing nylon or tight-fitting underwear. Furthermore, because the herpes virus can live outside the body for a while, such items as towels and toothbrushes should not be shared. Intimate contact with the infected area should certainly be avoided during an active attack; and because some people shed the virus when they have no symptoms, the use of condoms for those with genital herpes (or for individuals whose partner has herpes) is recommended between active attacks.

THE PERSONAL SIDE OF HERPES

Most people with oral herpes continue to lead normal lives. A person with oral herpes can engage in sexual intercourse without worrying about infecting his or her partner, but kissing and oral-genital relations during an active attack should be avoided, of course.

Genital herpes is almost always contracted by intimate sexual contact. Recall, too, that people with genital herpes usually have frequent recurrent attacks. Besides the physical symptoms, individuals suffering with genital herpes almost always have some psychological difficulties adjusting to their infection. They often pass through several stages: shock, emotional numbing, isolation and loneliness, and sometimes severe depression and impotence. Initially there is often a frantic search for a second or even third doctor's opinion (denial); a lot of "Why me?" responses; and guilt and anger. The anger is sometimes directed at the medical profession for its inability to cure herpes, but almost always at the partner responsible for the infection and/or the opposite sex in general. Others invest a great deal of time and effort in experimental drugs, fad diets, and even faith healers.

After the initial shock, many genital herpes sufferers have feelings of ugliness, self-loathing, and guilt. The reactions of others may make them feel like "social lepers." As a result, some may have sexual problems, and others may just swear off sex. A 1993 survey of readers of *the helper* (a magazine for people with herpes) found that in the previous 12 months, 52 percent had suffered depression and fear of rejection, 36 percent had feelings of isolation, and 10 percent had self-destructive feelings (Catotti, Clarke, & Catoe, 1993).

> "In my early 20's I felt a painful sore on my vagina. I had no idea what it was. At the time I hadn't had many partners. Later I had another eruption. My doctor told me I had herpes. Needless to say, I was shocked. I went through a denial phase that followed many nights of wet pillows. It took me years to accept it. . . ."
>
> (from the author's files)

So how should you deal with personal relationships if you have herpes? First of all, accept the fact that although there is no cure, you do not have the plague. *The use of condoms will minimize the chances of infecting a partner* (in case there is viral shedding without symptoms). You should try to minimize the stress in your life, and if you have frequent or long-lasting attacks, talk to your doctor about taking acyclovir. Lastly, tell new partners that you have herpes in advance of having intimate relationships with them. You should not just tell them that you have herpes; you should educate them—teach them about herpes in order to dispel unfounded fears and false beliefs. You might still be rejected occasionally, but chances are that this type of

person would run off during any other type of crisis as well. Many couples have enjoyed long-lasting relationships where one partner had herpes and had not infected the other. Here is the address of an organization that can provide the herpes sufferer with current up-to-date information on herpes research, therapy, and care:

American Social Health Association
P.O. Box 13827
Research Triangle Park, NC 27709

HEPATITIS

Hepatitis is an inflammation of the liver that can have many causes. The incubation period (time period before symptoms appear) is 2 to 6 weeks for hepatitis A and 6 weeks to 6 months for hepatitis B. The symptoms can range from mild (poor appetite, diarrhea) to severe (fever; vomiting; pain; fatigue; jaundiced, or yellow-tinged, skin and eyes; dark urine).

Hepatitis A (infectious hepatitis) is caused by a small virus (HAV) that is spread by direct or indirect oral contact with contaminated feces. There are about 130,000 new cases each year in the United States (Horstman, 1997). The disease can be spread sexually (by contact with the anus of an infected person during sex) and has a higher incidence in male homosexuals than in other groups. However, it is most often contracted through nonsexual means, such as eating food handled by infected individuals, or shellfish taken from contaminated waters. A vaccine is available for those who want protection from this virus.

Hepatitis B (serum hepatitis) is caused by a different, larger virus (HBV) and is transmitted by infected blood or body fluids such as saliva, semen, and vaginal secretions (Zuckerman, 1982). It can cause a number of liver diseases, including cancer and cirrhosis. There are more than 300 million people worldwide who are infected with this virus (Ayoola, 1988). At least 25 percent, and perhaps as many as 50 percent, of the 128,000 new cases each year in the United States are spread during sex (Brandt, 1982; Centers for Disease Control and Prevention, 1994). The disease is most common in homosexual males (particularly those who engage in anal sex) and drug users who share needles. One can also get the virus during a blood transfusion, and over two-thirds of all pregnant females infected with hepatitis B pass the infection to their babies. The antiviral drug interferon is effective in about half of all cases (Horstman, 1997). About 90 percent of those with hepatitis recover, but up to 10 percent remain chronically infected and thus become carriers of the disease. The Centers for Disease Control and Prevention and the American Academy of Pedia-

hepatitis A, B, and C Liver infections caused by viruses. Type A is spread by direct or indirect contact with contaminated feces. Type B is transmitted by infected blood or body fluids, with 25 to 50 percent of the cases contracted during sex. Type C is spread mainly by contaminated blood, but may possibly be spread during sexual intercourse in some cases.

tricians have recommended that *all* children (babies as well as adolescents) be vaccinated against the hepatitis B virus. Vaccination has resulted in dramatic reductions in the number of cases of liver cancer in other parts of the world. (Chang et al., 1997).

Hepatitis C, which is caused by still another virus (HCV), was not identified as a specific virus until 1988. About 150,000 Americans are infected each year, and as many as 4 million are presently infected. About half of them remain chronically infected and about one-fifth will develop cirrhosis of the liver, resulting in 8,000 deaths a year. The hepatitis C virus is most commonly spread through contact with contaminated blood, and thus it is frequently found in people who inject illicit drugs and share needles. Although the virus can be spread during sex, most researchers believe that this is uncommon (MacDonald, Crofts, & Kaldor, 1996). There is no vaccine available for protection against hepatitis C.

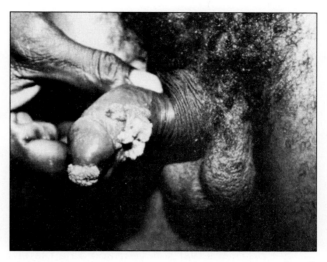

Figure 5–8 Venereal warts on the penis caused by human papillomavirus.

GENITAL HUMAN PAPILLOMAVIRUS INFECTION

Have you ever heard of genital human papillomavirus (HPV) infection? Most college students haven't (Vail-Smith & White, 1992), yet this is a very common and often life-threatening sexually transmitted disease. It is estimated that 24 to 40 million Americans are already infected with HPV and that there are 500,000 to 1 million new cases every year (Donovan, 1993). Perhaps as many as 1 in 3 sexually active adults have HPV (Aral & Holmes, 1991).

Human papillomaviruses infect epithelial cells (the covering of internal and external surfaces of the body). Over 70 different types have been identified (de Villiers, 1989). HPV infections are called *clinical infections* if there are symptoms visible to the naked eye and *subclinical infections* if there are no visible symptoms. Subclinical HPV infections are 10 times more common than clinical infections (Rudlinger & Norval, 1996).

We can divide clinical HPV cases into three broad categories: (1) those that cause a rare skin condition (epidermodysplasia verruciformis) that appears as reddish plaques, (2) those that cause nongenital warts (including common, plantar, and flat warts), and (3) those that cause warts in the genital and anal area (although these can also cause problems in the mouth, pharynx, and larynx). There are about 25 different human papillomaviruses that can cause anogenital warts, but the large majority of genital warts are caused by HPV types 6 and 11. It was once thought that the only type of genital wart was *condyloma acuminata,* the type shown in Figure 5–8, but recent studies have

shown that HPV can result in a wide variety of anogenital lesions and that a single individual can have many different-appearing warts (Rudlinger & Norval, 1996).

For those people who develop *genital (venereal) warts,* the first symptoms generally appear anywhere from 3 weeks to 8 months after contact with an infected person. The warts can cause itching, irritation, or bleeding. In males, the warts generally appear on the penis, scrotum, and/or anus, but they can also grow within the urethra. In females, the warts can appear on the cervix and walls of the vagina as well as the vulva and anus, and thus the virus can be transmitted to a newborn baby during vaginal delivery. Women are much more likely than men to develop visible warts (and serious complications—see below) because the cervix has a zone of cells that are continually undergoing cell division (which makes it easier for the viruses to exploit them). The skin of the penis, on the other hand, resists transformation by the virus.

"I contracted venereal warts from a married woman I frequently slept with. I then gave them to my girlfriend. I did not know I had the affliction until she was diagnosed during a semi-annual exam with her gynecologist. The removal procedure for me as a male was rather simple as opposed to her treatment. Her first treatment involved cryogenically freezing her cervix. After that unsuccessful attempt they resorted to laser surgery. She is

human papillomaviruses (HPV) Viruses that cause abnormal growths in epithelial cells. There are over 70 types. A few (types 6 and 11) cause genital warts, while others (types 16, 18, 31, 33, and 45) can lead to cancer of the cervix.

still not cured. If I could change things I would. We're not together anymore, but I still pray for her. I surely will never forget and I practice safe sex with my new girlfriend."

"When I was 17 years old I went to the doctor for a Pap smear. He found pre-malignant warts on my cervix. When he informed me that they were sexually transmitted, I didn't want to believe him. I went to an all-girls Catholic high school and the guys I slept with were usually from a boy's Catholic high school. I would never think that I would get a STD. I always thought STDs happened to lower class people.

"My doctor performed laser surgery to get rid of the warts. A year and a half later I had another outbreak, had surgery, and within seven weeks after the surgery they came back. These last two times they weren't just on the cervix, they spread all the way to the vulva. . . . I have abstained from sex since the beginning of the second outbreak."

(from the author's files)

Women with HPV infections have a greatly increased risk of developing cervical and/or vulvar cancer (Koutsky et al., 1992b). Approximately 90 percent of all women with cervical cancer are found to have HPV infections (Lowy, Kirnbauer, & Schiller, 1994). The latency period between infection with HPV and detection of cancer is 5 to 25 years (Schiffman & Brinton, 1995). However, only a few kinds of human papillomaviruses are found in cervical cancers: HPV 31, HPV 33, HPV 45, HPV 18 (in 10 to 20 percent of cases) and, in particular, HPV 16 (in at least 40 to 60 percent of cases). Note that these types of HPV are different from the two types that cause most genital warts (types 6 and 11). What this means, of course, is that *most women with HPV infections who are at risk of developing cervical cancer will have no visible symptoms.* Because of this, a regular pelvic exam and Pap smear are an essential part of female health care. Women who test positive for HPV should have Pap smears at least every 6 months. Although HPV has been associated with cancer of the penis, it is much less likely that the viruses will lead to cancer in men (as is the case for visible warts, you recall).

Two of the major risk factors for HPV infection in young women are the number of lifetime male sexual partners and the male partner's number of lifetime sexual partners (Burk et al., 1996). Eighty percent of the male partners of infected women have no visible symptoms (Rudlinger & Norval, 1996).

molluscum contagiosum A sexually transmitted virus whose symptom look like small pimples filled with kernels of corn.

Small external warts can be treated by applying a solution of podophyllin (sold in stores as Condylox), which causes them to dry up. Larger warts can be removed by cauterization or minor surgery. Internal warts are often removed by laser surgery. For HPV infections of the cervix, cryotherapy (freezing) is the preferred therapy. However, none of these methods attack the virus directly, so recurrences of the warts are common (see previous case histories; see also Kraus & Stone, 1990). Injection with the antiviral drug interferon has shown some promise. It is not surprising that people with HPV infections are also affected emotionally by the disease and, in the case of women, have anxieties about cancer (Persson, Dahlof, & Krantz, 1993).

Because millions of Americans infected with HPV have no symptoms, and thus are unaware that they can infect others, the best way to eliminate the disease would be to vaccinate people before they are exposed to the virus (see Frazer, 1996). The first tests of a vaccine in humans began in 1997, but it will probably take at least six years before a vaccine is ready for large-scale use.

MOLLUSCUM CONTAGIOSUM

Molluscum contagiosum is a painless growth that is caused by a virus. The virus is spread by direct skin-to-skin contact, and thus it can be transmitted during sex. There is only one case of this disease for every 100 cases of gonorrhea in the United States (Hatcher et al., 1994). The growths are usually 1 to 5 mm in diameter and look like small pimples filled with kernels of corn. They can be treated with podophyllin or by freezing, but they usually disappear on their own within 6 months.

HIV INFECTION AND AIDS

When a person has sex, they're not having it just with that partner. They're having it with everybody that partner has had it with for the past ten years.

(Otis Bowen, former U.S. Secretary of Health and Human Services)

In the early 1980s, doctors in California were puzzled when five men were diagnosed with *Pneumocystis carinii pneumonia*, a rare respiratory infection usually seen only in people with depressed immune systems. Several young men in Miami were simultaneously diagnosed as having *Kaposi's sarcoma*, a rare cancer of the capillary system that appears as purple blotches on the skin (see Figure 5–9). It, too, is a sign of a depressed immune system. More and more cases of depressed immune systems began to appear, but it was not until September 1982 that the new disease was given a name: **acquired immunodeficiency syndrome (AIDS).**

It was not until the spring of 1984 that the cause of AIDS was identified as a virus. Dr. Robert Gallo of the U.S. National Cancer Institute called the virus *human T-cell lymphotropic virus-III* (HTLV-III) and said that it was one of the family of HTLV viruses. Dr. Luc Montagnier of the Pasteur Institute in Paris simultaneously and independently identified the virus and named it *lymphadenopathy-associated virus* (LAV), after the swollen lymph nodes that are among the early AIDS symptoms. There is a controversy about which group discovered the virus first, but after two decades in which several million people have died of AIDS, that hardly seems important anymore. The virus is now referred to by everyone as **human immunodeficiency virus,** or **HIV.**

In technical terms, HIV is a Lentivirus, a subfamily of retroviruses. Retroviruses reverse the normal pattern of reproduction in cells. There are two major types of HIV: *HIV-1,* which is the most common type worldwide, and *HIV-2,* found mainly in West Africa. HIV-2 wasn't observed in the United States until 1987. HIV displays considerable genetic variability (Janssens, Buve, & Nkengasong, 1997). To date, researchers have identified 10 major HIV-1 subtypes (classified as group M subtypes A–J) and several very aberrant subtypes (group O). The most common type found in the United States is subtype B (group O was not observed here until 1996), but different subtypes can be found in different people living in the same area, and even within one infected individual (Hu et al., 1996). HIV has obviously undergone extensive mutation, but examination of the genetic make-up of various subtypes suggests that all the subtypes evolved from a single virus.

HIV AND THE BODY'S IMMUNE SYSTEM

The **immune system** is that part of your body that defends you against bacteria, viruses, fungi, and cancerous cells. White blood cells, or *lymphocytes,* are the main line of defense in this system. There are several types. One type, called *CD4+ lymphocytes* (also called *helper T cells* and *T4 lymphocytes*) have the job of recognizing the disease-causing agents and then signaling another type of white blood cell (*B cells*) to produce antibodies. The antibodies bind to the recognized agents so that they can be identified by yet another type of white blood cell, the *killer T cells* (*T* is for the thymus gland, where the cells develop). The CD4+ cells then signal the killer T cells to destroy the identified disease-causing agents. In addition to CD4+ cells, there are also *suppressor T cells* that suppress the activity of the B cells and killer T cells. A person normally has twice as many CD4+ (helper) cells as suppressor T cells.

Recall now that viruses must invade and live in normal body cells (called host cells) in order to replicate themselves. The human immunodeficiency virus infects CD4+ cells and reproduces itself, causing death of the cell in about one-and-a-half days and the release of more of the virus. The replicated virus then invades other CD4+ cells. As many as 2 billion CD4+ cells are killed each day (Perelson et al., 1996). Thus, HIV destroys the cells that ordinarily would work to fight it off. The body tries to replace the dead cells, but the number of CD4+ cells begins to decline early in the course of HIV infection (Lang et al., 1989). The victim eventually has fewer helper (CD4+) T cells than suppressor T cells, leaving the body defenseless against viruses, bacteria, and other infection-causing agents.

CD4+ white blood cells are not the only body cells invaded by HIV. The virus has also been found in cells of the brain, gastrointestinal tract, kidney, lungs, cerebrospinal fluid, and other blood and plasma cells. However, it is the destruction of CD4+ cells that leads to the disease we call AIDS.

PROGRESSION OF HIV INFECTION

After invading the body, HIV reproduces unchecked during the first few weeks, reaching enormous levels in the bloodstream. Between 30 and 60 percent of newly infected persons experience flulike symptoms during this time. This is called **primary HIV infection.** The body's immune system then launches a huge counterattack that kills the virus by the billions, reducing it to very low levels (Daar et al., 1991). In time, however, the virus regains the upper hand by slowly killing off the CD4+ cells.

A normal, healthy person has between 1,000 and 1,200 CD4+ cells per cubic mm of blood. Although CD4+ cell counts decline with time, people with HIV show no visible symptoms in the first stage of the chronic infection, called **asymptomatic HIV infection.** This stage often lasts for years.

Infected persons generally start to show symptoms of a weakened immune system when their CD4+ cell count falls below 500 per cubic mm of blood. Symptoms may include fatigue, persistent headaches, loss of appetite, recurrent diarrhea, loss of body weight, low-grade fever (often accompanied by "night sweats"), swollen lymph nodes, and "colds," "flus," and yeast infections that linger on and on. At this point, the infection is called **symptomatic HIV infection** (formerly called *AIDS related complex,* or *ARC*).

acquired immunodeficiency syndrome (AIDS) An often fatal disease caused by a virus (HIV) that destroys the immune system. It is spread by intimate sexual activity (the exchange of bodily fluids) or contaminated blood.

human immunodeficiency virus (HIV) A virus that kills CD4+ cells, eventually resulting in AIDS.

immune system The bodily mechanisms involved in the production of antibodies in response to bacteria, viruses, and cancerous cells.

Figure 5–9

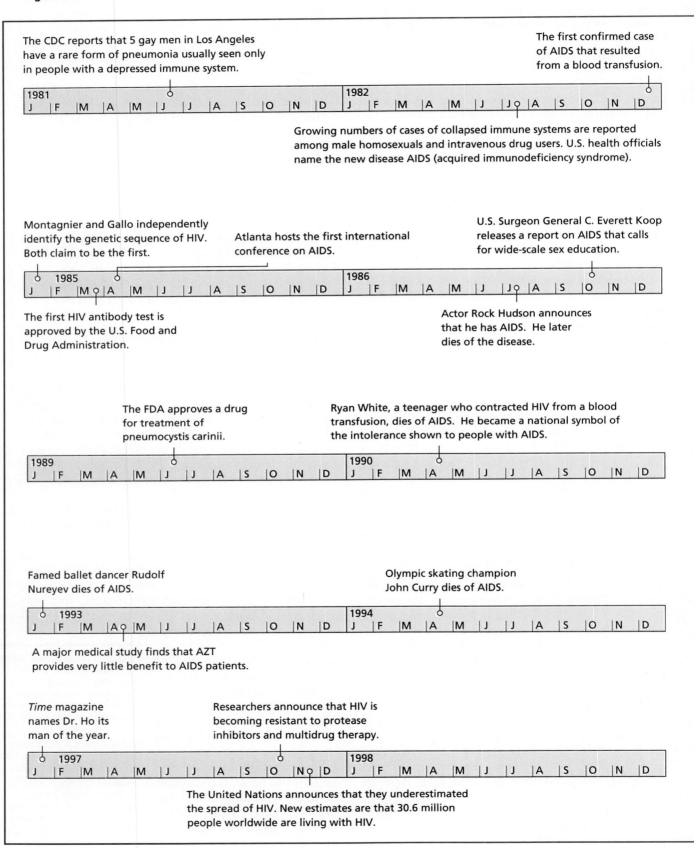

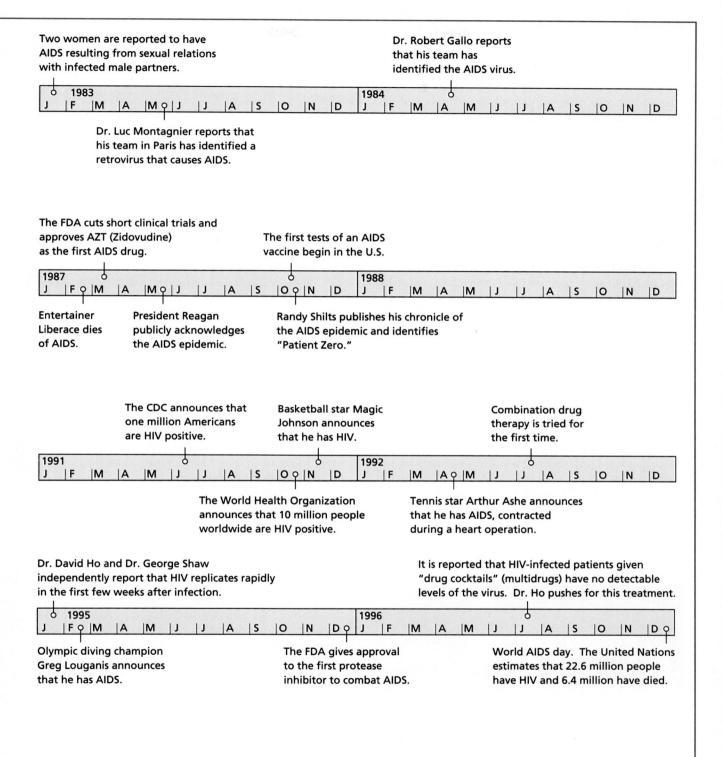

Two women are reported to have AIDS resulting from sexual relations with infected male partners.

Dr. Robert Gallo reports that his team has identified the AIDS virus.

1983
J |F |M |A |M |J |J |A |S |O |N |D

1984
J |F |M |A |M |J |J |A |S |O |N |D

Dr. Luc Montagnier reports that his team in Paris has identified a retrovirus that causes AIDS.

The FDA cuts short clinical trials and approves AZT (Zidovudine) as the first AIDS drug.

The first tests of an AIDS vaccine begin in the U.S.

1987
J |F |M |A |M |J |J |A |S |O |N |D

1988
J |F |M |A |M |J |J |A |S |O |N |D

Entertainer Liberace dies of AIDS.

President Reagan publicly acknowledges the AIDS epidemic.

Randy Shilts publishes his chronicle of the AIDS epidemic and identifies "Patient Zero."

The CDC announces that one million Americans are HIV positive.

Basketball star Magic Johnson announces that he has HIV.

Combination drug therapy is tried for the first time.

1991
J |F |M |A |M |J |J |A |S |O |N |D

1992
J |F |M |A |M |J |J |A |S |O |N |D

The World Health Organization announces that 10 million people worldwide are HIV positive.

Tennis star Arthur Ashe announces that he has AIDS, contracted during a heart operation.

Dr. David Ho and Dr. George Shaw independently report that HIV replicates rapidly in the first few weeks after infection.

It is reported that HIV-infected patients given "drug cocktails" (multidrugs) have no detectable levels of the virus. Dr. Ho pushes for this treatment.

1995
J |F |M |A |M |J |J |A |S |O |N |D

1996
J |F |M |A |M |J |J |A |S |O |N |D

Olympic diving champion Greg Louganis announces that he has AIDS.

The FDA gives approval to the first protease inhibitor to combat AIDS.

World AIDS day. The United Nations estimates that 22.6 million people have HIV and 6.4 million have died.

HIV infection is not called **AIDS** until it has become life-threatening. As the CD4+ cell count approaches 200 per cubic mm of blood, infected persons' immune systems have become so weakened that they fall prey to what are referred to as *opportunistic infections*—diseases such as lymphomas, Kaposi's sarcoma, (see Figure 5–10), *Pneumocystis carinii* pneumonia, recurrent bacterial pneumonia, pulmonary tuberculosis, cryptococcal meningitis, wasting syndrome, and invasive cervical cancer (Katlama & Dickinson, 1993; Mocroft et al., 1997). These are often accompanied by severe mental disorders and/or emaciation (Maj, 1990). Many AIDS patients are infected with a mycoplasma (a primitive microorganism that doesn't even have cell walls) known to kill people who are not even infected with HIV (Lo et al., 1989). Dr. Montagnier has speculated that the mycoplasma makes HIV more powerful (see Wright, 1990). In 1993, the Centers for Disease Control and Prevention expanded the definition of AIDS to include *anyone whose CD4+ cell count was below 200,* whether or not they had yet fallen victim to an opportunistic disease. Some people whose CD4+ cell counts have dropped below 200 have gone three years or longer without developing AIDS-related illnesses (Hoover et al., 1995). In the end, however, AIDS patients die as a result of the opportunistic infections (not as a direct result of HIV).

A few researchers have doubted the belief that AIDS is caused by the depletion of CD4+ cells by HIV. One researcher, for example, proposed that it is not HIV infection per se that causes AIDS, but high-risk behaviors (such as drug use and promiscuous homosexual activity) associated with HIV infection (Duesberg, 1989, Duesberg & Ellison, 1996). However, others have shown that individuals who engage in high-risk behavior but are not infected with HIV do not develop AIDS (Ascher et al., 1993; Schecter et al.,

1993). In the summer of 1992, several individuals were discovered to have AIDS-like symptoms (including depleted CD4+ cells) without HIV (e.g., Laurence et al., 1992), causing a scare that a new, unknown virus could also cause AIDS. It was later found that this condition is very rare, has numerous causes, and is not evidence against HIV as the cause of AIDS (Fauci, 1993).

Once an individual is infected with HIV, how long does it take to develop a full-blown case of AIDS? About 5 percent will develop AIDS within the first 3 years after infection with HIV-1; about 20 percent within 5 years; and 50 percent in 10 years. Only 12 percent of infected persons will not yet have been diagnosed with AIDS after 20 years (Buchbinder et al., 1994; Gauvreau et al., 1994; Rutherford et al., 1990; Munoz & Xu, 1996). About one in 100 have genes that appear to completely protect them from AIDS (Liu et al., 1996).

What accounts for the fact that some HIV-infected people progress to AIDS in just a few years while others take many years? The type of HIV may be one factor. For example, persons infected with HIV-2 take longer to develop AIDS than those infected with HIV-1 (Whittle et al., 1994). People who already have a weakened immune system (as a result of age or illness, for example) have a more rapid progression to AIDS. Psychosocial processes (e.g., amount of stress) may also play a role (Capitano & Lerche, 1991). People who experience acute primary HIV infection (flu-like symptoms or worse upon initial exposure) are more likely to develop AIDS earlier than people who are asymptomatic upon initial infection (Lindback et al., 1994; Sinicco et al., 1993).

Several recent studies have shown that the best single predictor for progression to AIDS is not an individual's CD4+ cell count, but his or her plasma HIV RNA levels, which show the amount of virus in the blood (Mellors et al., 1996, 1997; T.R. O'brien et al., 1996; W.A. O'brien et al., 1996; Shearer et al., 1997). This is called the *viral load.* The higher the viral load, the sooner the person will develop AIDS. HIV-infected persons can have the same CD4+ cell count but very different levels of the virus. The combination of HIV RNA levels and CD4+ cell count is an even better predictor of the progression to AIDS (Mellors et al., 1997). Today, drug therapies have been designed to lower the viral load, and thus in the future it should take longer for an HIV-infected person to develop AIDS.

How long can an HIV-infected person expect to live after being diagnosed with AIDS? The news is not good, but it's improving. The average survival time after a diagnosis of AIDS in the early 1980s was only 11 months. Today it is 12 to 18 months (Mocroft et al., 1996), and new drug therapies offer hope that it

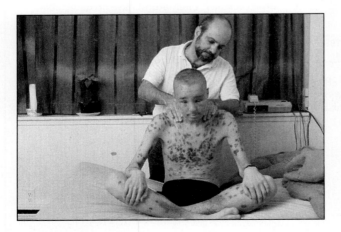

Figure 5–10 Kaposi's sarcoma lesions on a patient with AIDS.

will increase even more. The major determining factors for length of survival are age, CD4+ cell counts at the time of diagnosis, and the type of opportunistic disease (Mocroft et al., 1997). In individuals whose CD4+ cell counts drop below 200 but who do not initially have clinical diseases, the average survival rate is about 3 years (Keet et al., 1994).

WHERE DID HUMAN IMMUNODEFICIENCY VIRUS COME FROM?

There is no question that HIV-1 and HIV-2 are new to the human species, but where did they come from? "Conspiracy" theories are popular. At one time, the Soviets blamed it on U.S. biological experiments gone awry. Filmmaker Spike Lee publicly said that AIDS was "a government-engineered disease targeted at gays, African Americans, and Hispanics" (November 1992). Another person circulated a booklet claiming that WHO, the World Health Organization, "murdered Africa with the AIDS virus" (Douglass, 1988). He, as well as others, said that the virus was injected into people with various vaccines (e.g., against smallpox and hepatitis B). *Rolling Stone* magazine claimed that HIV was introduced to humans during testing of a polio vaccine in Africa in the late 1950s (the polio virus was grown in monkey kidney cells, and the article suggested that the cells were contaminated with a monkey virus).

Scientific studies of the origins of HIV have indeed implicated nonhuman primates but do not support conspiracy theories (see Desrosiers, 1990; Gao et al., 1992; Myers, MacInnes, & Korber, 1992). There is a diverse genetic pool of nonhuman *simian* (primate) *immunodeficiency viruses* (SIV). The viruses are found in rhesus monkeys, sooty mangabeys, African green monkeys, and chimpanzees. Some individuals have been found to be infected with strains of HIV-2 that are more closely related to SIV than to any previously known human virus (Gao et al., 1992; Hahn, 1990). HIV-2 can infect baboons and rhesus macaques (Castro et al., 1991). In other words, there is considerable evidence that humans acquired HIV-2 through cross-species transmission; that is, that it "may simply be simian immunodeficiency viruses residing in and adapting to a human host" (Myers, MacInnes, & Korber, 1992). Researchers have yet to find another SIV with the same degree of similarity to HIV-1, but some strains with notable similarities have been discovered.

Another possibility is that the virus has been present in humans in a nonvirulent form for a long time. HIV mutates very quickly (about one percent of the genetic material changes each year), and it may simply have mutated to its present very deadly form.

Whether HIV originated in monkeys or mutated into its present form, there is no question that it first appeared in Africa. Tests of old blood samples have established that HIV-1 was in central Africa in 1959,

suggesting that the virus first entered humans in the 1940s or early 1950s (Zhu et al., 1998). Kaposi's sarcoma was showing up at a high rate in central Africa in the 1970s. Features of modern society such as worldwide travel, promiscuous sexual activities, and sharing of syringes and hypodermic needles probably greatly helped the spread of the virus.

HOW IS HIV SPREAD?

What is the evidence concerning the transmission of HIV? Some people worry about catching it from just breathing the same air as, or touching objects previously held by, AIDS victims. Some neighbors of AIDS victims worry about mosquitoes. Some people have even quit donating blood. Can you get AIDS by casual contact with an HIV-infected person?

If HIV could be contracted from casual contact, you would expect that family members living with AIDS patients would be at great risk. However, no family member of an AIDS patient has ever become infected by sharing household facilities and items, and no child has ever become infected by just playing with or living with an AIDS-infected brother or sister (Friedland et al., 1990; Gershon, Vlahov, & Nelson, 1990). HIV has sometimes been found in body fluids such as tears and saliva, but not in high enough concentrations to pose a threat. A protein found in saliva helps block HIV from infecting human cells (McNeely & Wahl, 1995). In fact, the concentrations in saliva are so low that your chance of acquiring HIV by being bitten by an HIV-infected individual is extremely slim (Richman & Rickman, 1993).

In the final stages of AIDS, patients cannot hold food down, often have uncontrollable diarrhea, and generally sweat profusely. Yet except for a few people who accidentally stuck themselves with a contaminated needle and a few others whose skin was exposed to a large amount of contaminated blood (which may have entered the body through a small cut), none of the doctors, nurses, or other medical personnel who have cared for the most seriously ill AIDS patients on a daily basis have ever developed an HIV infection.

Malaria and some other diseases are spread by mosquitoes, but it is certain that HIV is not spread in this manner (Castro et al., 1988). If it were, we would be seeing a much larger number of children and elderly people with AIDS.

There is no need for panic—*HIV is not spread by casual contact*. There are three main ways in which HIV is transmitted: (1) sexual contact with an infected person; (2) exposure to infected blood (mainly needle sharing among intravenous drug users); and (3) mother-to-infant transmission. Thus, HIV is transmitted almost exclusively by intimate sexual contact and contaminated blood. Let us examine each of these means of transmission.

There are still some individuals who claim that it is a myth that HIV can be spread by sex between heterosexuals (Brody, 1995). Believe me, it is no myth. HIV has been found in the semen and vaginal fluids of HIV-infected individuals (see Mostad & Kreiss, 1996), although not in all infected persons. Remember, the concentrations of HIV are greatest at the very beginning of the infection, then decrease dramatically, and then increase again as the disease progresses toward the AIDS stage (Daar et al., 1991). This means that the risk of transmission of HIV is greatest in the first 60 days and in the symptomatic HIV and AIDS stages (De Vincenzi et al., 1994; Jacquez et al., 1994). Recent studies also indicate that some of the subtypes of HIV are easier to transmit during sex than others (Royce et al., 1997; Soto-Ramirez et al., 1996). A major factor is whether HIV is transmitted onto an uninfected person's mucous membranes or directly into his or her blood. Any sexual activity in which bleeding occurs would greatly increase the risk (e.g., anal intercourse; having an open sore on the penis, vagina, or cervix; menstruation) (see Royce et al., 1997). Sores on the genitals caused by other sexually transmitted diseases such as syphilis, chancroid, or herpes greatly increase the risk of HIV transmission (see Wasserheit, 1991). Studies show that HIV is at least twice as easy to transmit from an infected male to a female partner than vice versa (European Study Group, 1992; Nicolosi et al., 1994; Padian, Shiboski, & Jewell, 1991).

Although there have been a few isolated reports of heterosexual or male homosexual partners becoming infected during oral sex, most experts believe that the risk of HIV transmission through oral-genital sex is very small (see Ostrow et al., 1995). Female-to-female transmission of HIV during oral-genital sex has not been observed (Petersen et al., 1992). There has been only one case of transmission by kissing, and this was through the man's blood, not saliva (both partners had gum disease) (CDC press release, 1997).

Let me summarize what we know about HIV transmission during sex (see Royce et al., 1997). Stage of infection and bleeding are factors that increase the risk. In terms of behavior, HIV is spread most easily during anal intercourse (because of bleeding, even if it is slight; see Caceres & Griensven, 1994), followed by vaginal intercourse (especially when there are open sores), and occasionally during oral sex.

As previously mentioned, a second major way in which HIV is transmitted is by exposure to contaminated blood. Needle-sharing drug users are at especially high risk (Des Jarlais et al., 1996). But what about our chances of getting HIV should we ever need a blood transfusion? At one time this was a major concern. Several thousand people contracted HIV from blood transfusions before screening tests were begun in 1985. Today, because all blood is tested, only 2 of every 1 million units are likely to be infected with HIV (Schreiber et al., 1996). A few units of infected blood still get through because the test cannot detect HIV if the blood donor has just been infected (see page 116). Unfortunately, many people in the United States have stopped donating blood out of fear of getting AIDS, and as a result, there is a severe shortage of blood in many parts of the country. At blood donation centers in the United States, needles are sterilized and used only once. You will not get AIDS from donating blood!

The third major way that HIV is transmitted is from an infected mother to her infant (see John & Kreiss, 1996). Worldwide, more than 1,000 children are born with HIV every day (Sharland et al., 1997). The Centers for Disease Control and Prevention estimates that in the United States, 7,000 women a year who become pregnant are infected with HIV and 1,000 to 2,000 of their babies become infected (a transmission rate of 15–30 percent). The infections can occur either very late in pregnancy, during delivery, or by breast-feeding (Bertolli et al., 1996; European Collaborative Study, 1992, 1994b). The risk is greatest in women who have high levels of the virus in their blood and when fetal membranes rupture several hours before delivery (Dickover et al., 1996; Landesman et al., 1996; Sperling et al., 1996). The rapidity with which children infected with HIV progress to AIDS is directly related to their viral load (Shearer et al., 1997). Most infected children show symptoms of HIV infection by 6 months of age (European Collaborative Study, 1991). By 1 year of age, about a fourth have AIDS and many have died. With immature and weakened immune systems, nearly all eventually fall prey to various opportunistic diseases.

Because of the attention it has received in the media, one other potential means of transmission needs to be discussed. Many Americans have become afraid of contracting HIV from an infected health care professional (e.g., a physician or dentist). These fears developed when it was widely reported that seven patients of an HIV-infected Florida dentist (who later died of AIDS) had become infected with HIV (Ou et al., 1992). A 1994 segment of the investigative TV news show *60 Minutes* later cast doubt on whether the patients had actually acquired HIV from the dentist, but the case nevertheless raised the important issue of patients' rights.

What are the facts about your risk of contracting HIV from a health care professional? Studies of thousands of patients of HIV-infected surgeons and dentists reveal that the risk is practically nonexistent (Dickinson et al., 1993; Mishu et al., 1990; Rogers et al., 1993). Nevertheless, because chemical disinfectants do not always kill the virus (Sattar & Springthorpe, 1991), surgeons and dentists are now required to steam-clean their instruments and wear protective gloves. Expert review panels will decide on a case-by-case basis whether HIV-infected health professionals may continue performing surgical procedures (American College of Physicians, 1994; Lo & Steinbrook, 1992). Conclusion: there is no good reason to avoid getting regular health and dental checkups.

WHO HAS HIV/AIDS?

HIV infection is not confined to the United States; it is a true pandemic disease. In late 1997, the United Nations announced that they had grossly underestimated how quickly HIV had spread worldwide. The new estimates were that there are 16,000 new infections a day (5.8 million a year) and that there are 30.6 million people currently living with HIV. About 2.3 million people were predicted to die of AIDS in 1997. These estimates include 1,600 new infections and 1,200 deaths each day for children. The U.S. Census Bureau's World Population Profile said that 121 million people could be infected by the year 2020.

At present, 90 percent of persons infected with HIV live in developing countries (Quinn, 1996). Two-thirds of the world's cases are in sub-Saharan Africa, where HIV is believed to have originated. Countries such as Uganda, Zambia, and Rwanda have been devastated. In some places, as many as one out of every four people are infected with HIV, resulting in an average reduction in life expectancy of over 20 years (Caldwell & Caldwell, 1996; Quinn, 1996).

HIV may have originated in Africa, but the World Health Organization says that "the most alarming trends of HIV infection are in south and southeast Asia where the disease is spreading in some areas as fast as it was a decade ago in sub-Saharan Africa." In places like Thailand, HIV was once confined to users of intravenous drugs and prostitutes, but it has now spread into the general population in epidemic form (Mertens & Low-Beer, 1996).

India is another country where HIV has reached epidemic proportions. (Jain, John, & Keusch, 1994). The World Health Organization estimates that already over one-third of all the prostitutes in Bombay are infected with HIV, and similar numbers are being found among patients in some of the city's STD clinics.

In South America, about 2 million people are believed to have HIV (World Health Organization, 1995). Spread of the infection may initially have been helped by unregulated blood banks, but once in the population it spread rapidly as a result of cultural attitudes that encourage male promiscuity. HIV spread rapidly into the heterosexual community. In Brazil, for example, there was only one woman with AIDS for every 124 men with AIDS in 1984, but by 1993 the ratio was down to four men to every AIDS-infected woman.

In Africa, HIV infection is a true heterosexual disease (see Fig. 5–11). Approximately half of the millions of people presently infected with HIV are women (Mertens & Low-Beer, 1996). As a result, nearly 2 million children have been born with HIV. Heterosexual transmission is helped by high numbers of cases of other sexually transmitted diseases that cause genital ulcerations (e.g., syphilis, chancroid, and herpes; WHO Consensus Statement, 1989). HIV has already killed over 1 million children worldwide and ophaned another 5 to 10 million (Quinn, 1996).

What about the United States? The Centers for Disease Control and Prevention reported that there were 58,443 new cases of AIDS in 1997 (HIV/AIDS Surveillance Report, 1998). This is an underestimate, due to delays in reporting and unreported cases. Nevertheless, this is a 23 percent decrease from the number of reported cases in 1993. This may mean that the number of new cases is leveling off, but part of the decrease is certainly due to new drug therapies that have slowed the progression of the disease.

At the start of January 1998, there had been a cumulative total of 619,690 reported cases of AIDS in the United States. Over 60 percent of these had already died. The Centers for Disease Control and Prevention estimates that at least 1 million more Americans are presently infected with the human immunodeficiency virus. By the mid-1990s, AIDS had become *the leading cause of death among Americans aged 25 to 44,* ahead of accidents, cancer, and heart disease (Centers for Disease Control and Prevention, 1995).

Although HIV infection is a heterosexual disease worldwide, to date AIDS has been largely confined to specific subgroups in the United States (see Figure 5–12). Of the cases reported to the CDC in 1997, 35.3 percent of the new cases of AIDS in adolescents and adults occurred among homosexual or bisexual

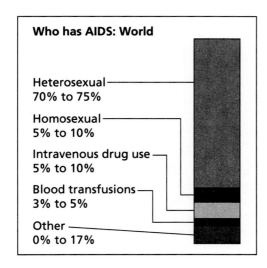

Figure 5–11 AIDS cases worldwide, 1997.

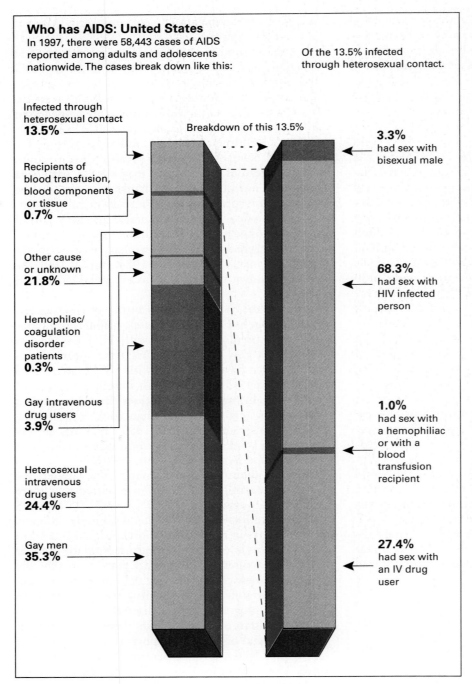

Who has AIDS: United States
In 1997, there were 58,443 cases of AIDS
reported among adults and adolescents
nationwide. The cases break down like this:

Of the 13.5% infected
through heterosexual contact.

Infected through
heterosexual contact
13.5%

Breakdown of this 13.5%

3.3%
had sex with
bisexual male

Recipients of
blood transfusion,
blood components
or tissue
0.7%

Other cause
or unknown
21.8%

68.3%
had sex with
HIV infected
person

Hemophilac/
coagulation
disorder
patients
0.3%

1.0%
had sex with
a hemophiliac
or with a
blood
transfusion
recipient

Gay intravenous
drug users
3.9%

Heterosexual
intravenous
drug users
24.4%

27.4%
had sex with
an IV drug
user

Gay men
35.3%

Figure 5–12   AIDS cases in the United States, 1997.

men, and 24.4 percent occurred among heterosexual
intravenous (injecting) drug users. Another 3.9 percent were gay intravenous drug users. Nearly 22 percent acquired HIV by unknown means. At least 13.5
percent acquired HIV through heterosexual contact
(HIV/AIDS Surveillance Report, 1998).

The large majority of new cases still occurred
among men (a reflection of the high prevalence of the
disease among homosexual and
bisexual men). However, the
number of new AIDS cases who
are women is increasing dramatically, from only 7 percent of all
new cases in 1985 to 21.8 percent in 1997 (HIV/AIDS Surveillance Report, 1998). A dis-
proportionate number of AIDS
cases occurred among African
Americans (44.2 percent) and
Hispanics (about 20.5 percent),
particularly those in low-income
groups. Among women diag-
nosed with AIDS, over three-
fourths were African American
or Hispanic. At least 50 percent
of women with AIDS became infected with HIV during heterosexual contact (Neal et al., 1997;
Wortley & Fleming, 1997).

Why have so many cases in
the United States shown up in
the male homosexual community? Political commentator
Patrick Buchanan was quoted as
saying: "The poor homosexuals.
They have declared war on nature, and now nature is exacting
an awful retribution." However,
you have already learned that a
person's sexual behavior, or sexual orientation, does not cause
HIV infection. To become infected, you must come into contact with the human immunodeficiency virus. The virus cannot
tell whether you are homosexual
or heterosexual. The reason it
first spread so rapidly among the
male homosexual community
here is threefold: many gay men
had numerous, often hundreds,
of sexual partners; the practice of
unprotected anal intercourse was
common; and AIDS was a new
disease with a long incubation
period. Before anyone even
knew that HIV or AIDS existed, the virus had already
been spread to tens of thousands of male homosexuals.

Keep in mind that those who are diagnosed with
AIDS are not the only people who have HIV. Remember, *people diagnosed with AIDS first became infected
with HIV 5 to 15 years before.* Individuals who become
infected with HIV now probably will not reach the
stage of infection we call AIDS until the years 2005 to

2015. So where is HIV now? Is it spreading into the heterosexual intravenous drug–free community?

The answer is yes. Masters, Johnson, and Kolodny (1988) reported that among heterosexuals with six or more sexual partners a year, 7 percent of the women and 5 percent of the men tested positive for HIV. Similar results have been reported by others (Kelen et al., 1988).

The most likely pathways by which HIV first spread into the heterosexual community was through bisexual men and intravenous drug users infecting their heterosexual partners (Chu et al., 1992; Lewis, Watters, & Case, 1990; van den Hoek, van Haastrecht, & Coutinho, 1990). Another likely route of infection is through prostitutes, many of whom are intravenous drug users. A 1990 study of 1,305 female prostitutes across the United States found that 6.7 percent tested positive for HIV (Khabbaz et al., 1990). In some cities the infection rate is as high as 25 percent for female prostitutes and 50 percent or higher for male homosexual prostitutes (Boles & Elifson, 1994; McKeganey, 1994).

Remember, at least 13.5 percent of the new cases of AIDS in 1997 were people who became infected with HIV by heterosexual transmission, up from only 1.9 percent in 1985. The Centers for Disease Control and Prevention says this is a conservative estimate. Already, over half of all new AIDS cases in U.S. women are due to heterosexual transmission. AIDS is already the leading cause of death among African-American women aged 15 to 44 in New York State and New Jersey and is the fourth leading cause of death among all U.S. women aged 25 to 44 (CDC, 1996). More than 50,000 U.S. children have been left motherless because of AIDS.

Despite the tremendous amount of publicity about AIDS, including the fact that worldwide the large proportion of cases are acquired by heterosexual transmission, many heterosexuals in the United States still do not believe that they can get HIV. If you still are not convinced, there is no better evidence than what happened in Haiti, where, as one researcher stated, you can see "how rapidly and dramatically the AIDS virus can spread in a promiscuous heterosexual population." In 1983, 71 percent of the cases of AIDS in Haiti occurred among male homosexuals and intravenous drug users, about the same percentage as presently found here. By the late 1980s, the large majority of new AIDS patients were heterosexuals (Pape & Warren, 1993). Only 15 percent of patients with AIDS in 1979–1982 were females, but by 1991–1992, nearly half the new AIDS patients were females. In urban areas, 7.5 percent of the adult population is infected with HIV, and AIDS is the leading cause of death among sexually active adults (Mertens et al., 1995). One of the reasons HIV spread so rapidly in

Haiti is the common acceptance of multiple sexual partners (in one study it was not uncommon to have had more than 10 partners a year), along with the common practice of prostitution (Pape et al., 1990).

Perhaps there aren't as many people with as many multiple sexual partners in the United States as in Haiti, or as many men who solicit the services of prostitutes, but that doesn't mean that HIV will not spread among heterosexuals here, only that it will spread more slowly. Dr. Ward Cates of the Centers for Disease Control and Prevention said, "Anyone who has the least ability to look into the future can already see the potential for this disease being much worse than anything mankind has seen before."

THE HUMAN SIDE OF AIDS

The deaths of people like actor Rock Hudson and tennis star Arthur Ashe, and the news that basketball great Magic Johnson was infected with HIV, made the disease more personal for many people—it had affected "someone they knew." But with each passing year, there are more and more people who do not have to read a newspaper to get a firsthand experience with AIDS. AIDS is more than a disease; it is a personal tragedy, not only for the patient but for the patient's family and loved ones as well. Some students in my course have asked me to share their experiences. Here are a few:

"My reaction to a low T-cell test was devastating. It was like a bomb exploding inside or a knife through the heart. My time is near and each day is a gift to cherish. . . . Finding out I was HIV positive was not as shocking as finding out that what I have changed to improve the quality of my life makes little difference in my life expectancy. . . ."

"My father has AIDS. You really don't know the severity of a problem until you're faced with it. Watching my father die has been extremely painful and stressful. . . ."

"We are discussing STDs and AIDS in class and although these are issues that are in the media constantly, I still believe that most students don't think that AIDS can happen to them. I have seen AIDS happen firsthand. People need to know that this is a reality in all of our lives. I learned in one of the most painful ways I can imagine. Two years ago my older brother died of AIDS. I was only 19, he was 24. It had been said that everyone's lives would be touched by AIDS, and when I was younger I didn't believe it. Now I do. I was with my brother when he died, as was my whole family. We did not know he had AIDS until two weeks before he died. He had hidden it from us. He came home to die. It changed my life.

"Believe me, AIDS is real. Don't wait for someone close to you to die before you take it seriously. Don't wait to test positive before you believe in AIDS. I think my brother thought he was invincible. I used to think I was, too, until the night he died."

(from the author's files)

The emotional impact of finding out that you have AIDS or are infected with HIV can be devastating. Severe depression is common in the early phases of HIV infection and again at the symptomatic stage, and the suicide rate is 17 to 66 times greater in individuals diagnosed with AIDS than in the general population (see Kalichman & Sikkema, 1994). On the other hand, many people have led productive lives for years after being diagnosed with AIDS. There is increasing evidence that the immune system can be affected by a positive or negative attitude (Siegel, 1989). It is important, therefore, that HIV-infected individuals not give up hope and that they receive as much positive support as possible.

TESTING FOR HIV

The standard tests for HIV do not test for the virus directly, but instead detect antibodies to the virus that are produced by an infected person's immune system. Blood is first checked with the ELISA (enzyme-linked immunosorbent assay) test. However, the test can result in "false positives" (i.e., uninfected blood testing positive). A recent flu vaccination can sometimes produce a false positive, for example (Mac Kenzie, et al., 1992). To ensure against false positives, the ELISA test is repeated if the results are positive. If there is a second positive indication, the blood is tested again with either the Western Blot or IFA (immunofluorescent assay) tests. These are more expensive and demanding tests that give information about particular antibodies. A negative test result means only that a person *probably* has not been infected with HIV. Remember, it often takes several months for an infected person's body to start producing antibodies to the virus; a person who has just recently been infected will not test positive.

The antibody tests were based mainly on HIV subtype B, the most common strain in North America and Europe. The tests have had to be modified as new subtypes appeared and slipped past the tests.

New tests are now available that measure the presence of antibodies to HIV in saliva or urine. The sample is checked with the ELISA test and positive results are then confirmed with the Western Blot blood test.

These tests are less accurate than the blood test, but may be more appealing to some people than blood tests.

The HIV-antibody tests are very effective in detecting HIV infection in individuals who are already showing symptoms (and thus have high levels of HIV and antibodies to HIV). They are not as reliable in screening HIV-infected persons who are still healthy. Today, there is a technique that directly tests for HIV RNA. It is called the polymerase chain reaction (PCR) and it can detect a single HIV-infected white blood cell among 100,000 healthy ones (Piatuk et al., 1993). However, this technique is expensive and time-consuming.

There is considerable controversy over home HIV testing kits (Schopper & Vercauteren, 1996). Purchasers of the kits send dried blood (from a finger prick), saliva, or urine to a lab. They learn by telephone whether or not they are infected with HIV (identification is by a number on the kit, and thus is anonymously done). Infected persons talk to a counselor. Supporters, which include the Centers for Disease Control and Prevention, say that the kits will increase the number of people who are aware of their infection and can then stop spreading the virus. Opponents worry that employers might (illegally) use the tests, or that some people might kill themselves if they learned by phone that they had HIV. Then there are questions about lab accuracy, and the question of how the home tests will affect tracking of the epidemic (physicians currently report new cases to the CDC).

TREATMENT FOR HIV/AIDS

There are many drugs approved for treating patients with HIV (see Cohn, 1997). One group of antiretroviral drugs is technically called *nucleoside analogs*. These include zidovudine (AZT), didanosine (ddI), zalcitabine (ddc), stavudine (d4T), and lamivudine (3TC). Another drug is nevirapine (Viramune; a nonnucleoside inhibitor). These drugs slow the progression of HIV infection by blocking an essential enzyme (called reverse transcriptase) needed for the virus to replicate itself (this occurs at step 2 in Figure 5–12). However, these drugs do not rid an infected individual of HIV.

The first antiviral drug used was zidovudine. The initial studies were very promising, indicating that the drug delayed the onset of AIDS (e.g., Fischl et al., 1989; Volberding et al., 1990), but subsequent studies delivered more sobering news. For many HIV-infected patients, AZT's side effects (e.g., anemia) made them so ill that it offset any advantage to postponing the onset of AIDS (Lenderking et al., 1994; Oddone et al., 1993). Several large-scale studies found that while AZT delayed the onset of AIDS when given early, it did not prolong life (Hamilton et al., 1992;

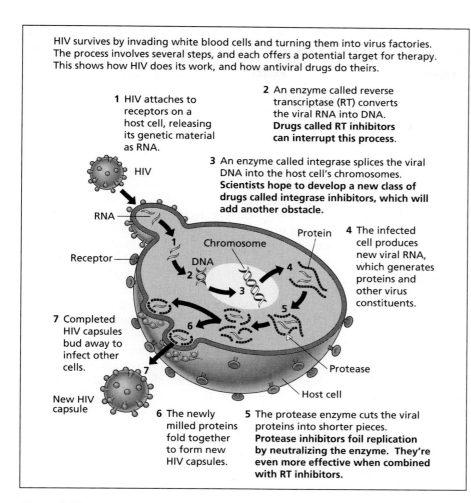

HIV survives by invading white blood cells and turning them into virus factories. The process involves several steps, and each offers a potential target for therapy. This shows how HIV does its work, and how antiviral drugs do theirs.

1 HIV attaches to receptors on a host cell, releasing its genetic material as RNA.

2 An enzyme called reverse transcriptase (RT) converts the viral RNA into DNA. **Drugs called RT inhibitors can interrupt this process.**

3 An enzyme called integrase splices the viral DNA into the host cell's chromosomes. **Scientists hope to develop a new class of drugs called integrase inhibitors, which will add another obstacle.**

HIV

RNA

Receptor

Protein

Chromosome

DNA

4 The infected cell produces new viral RNA, which generates proteins and other virus constituents.

7 Completed HIV capsules bud away to infect other cells.

New HIV capsule

Protease

Host cell

6 The newly milled proteins fold together to form new HIV capsules.

5 The protease enzyme cuts the viral proteins into shorter pieces. **Protease inhibitors foil replication by neutralizing the enzyme. They're even more effective when combined with RT inhibitors.**

Figure 5–13

Nordic Medical Research Council, 1992; Simberkoff et al., 1996). It is now the general consensus that the benefits of AZT, when used alone, fade very rapidly. In fact, the virus quickly becomes resistant to any of the drugs used alone (Cohn, 1997; Craig & Moyle, 1997).

You learned earlier that over 1,000 babies a day catch HIV from their mothers. However, treatment of HIV-infected pregnant women with zidovudine reduces the transmission of the virus to babies by about two-thirds, from about 24 percent to 8 percent (Connor et al., 1994; Sperling et al., 1996). As a result, the Centers for Disease Control and Prevention has recommended that all pregnant women be tested for HIV. Zidovudine, didanosine, or both are also used to slow the progression of HIV infection in children with the disease (England et al., 1997).

Recall that a variety of opportunistic diseases develop as HIV destroys the body's immune system. Over 40 drugs have been approved to treat these AIDS-related infections.

The newest drugs available to fight HIV are called *protease inhibitors*. This includes saquinavir (Invirase), Ritonavir (Norvir), indinavir (Crixivan), and nelfinavir (Agouron). These drugs block an enzyme called protease that is critical to the last stages of HIV replication (step 5 in Figure 5–12) (see Cohn, 1997).

The most exciting advance in the treatment of HIV has been use of a drug "cocktail"— the use of several drugs in combination (see Carpenter et al., 1997; Collier et al., 1996; Deeks et al., 1997). Administration of nucleoside analogs and protease inhibitors together simultaneously attack the replication of HIV at two different places (early in the process at step 2 and late in the process at step 5; see Figure 5–13). In many patients, this approach initially knocks blood HIV levels down to undetectable levels. The U.S. Department of Health and Human Services officially recommended this treatment in 1997. As a result of this new combination drug therapy, deaths form AIDS dropped in 1996 for the first time (by 13 percent).

The initial optimism that the combination drug approach might be a cure for HIV infection has been followed quickly by some sobering reality. Some patients cannot tolerate the side effects, while many others cannot afford the treatment (costs are between $10,000 and $12,000 a year). The worst news is that many physicians are reporting that the virus is already showing resistance to the new therapy. Many patients who initially show a dramatic decrease in their viral loads start to show increases again after several months (Deeks et al., 1997; Finzi et al., 1997; Wong et al., 1997). However, by continually checking HIV RNA levels and CD4+ cell counts, physicians may be able to switch the combinations of drugs as resistance develops (Carpenter et al., 1997). Even if combination therapy should not prove to be a cure, it will certainly allow HIV-infected individuals to lead much longer and more productive lives.

In late 1997, a team of researchers led by Dr. Robert Gallo reported they had identified a molecule called a chemokine that blocks the receptor used by the virus to invade CD4+ cells (Pal et al., 1997). The molecule must be tested in monkeys first, but the discovery could lead to powerful new drugs to prevent HIV infection.

The final stages of AIDS are so debilitating that the person must be hospitalized or cared for entirely by others. In the absence of community-based treatment resources for AIDS patients, hospitals have had to assume the major burden for medical and psychosocial support services. The average hospital bill for an AIDS patient is about $70,000. The total direct lifetime costs of treating a person infected with HIV from time of infection until death is about $119,000 (Hellinger, 1993). Indirect costs, such as lost productivity, are estimated to be at least 10 times higher (Bertozzi, 1996). In addition, AIDS affects everyone in the household, not just the patient. Many middle-class households will be pushed into poverty, while households in poverty will be pushed to the edge of survival.

Because HIV is presently incurable, the prevention of HIV infection depends on preventing transmission of, and avoiding exposure to, the virus. An ideal way to combat HIV would be to develop a vaccine (see Bolognesi, 1996). Scientists are identifying subunits of the virus that, when injected into the body, will stimulate the production of protective antibodies. Large-scale human testing began on two vaccines in 1994. Both vaccines are derived from a protein (gp 120), located on the surface of the virus, that attaches to human cells. When injected, the vaccines cause the body to make antibodies that target the gp 120 protein, thus interfering with the protein's ability to attach to the CD4+ cell. About two dozen other vaccines are presently being tested with humans on a smaller scale. Unfortunately, there are some major problems with developing an HIV vaccine. First, it will be difficult to develop a vaccine that is effective against all the different genetic subtypes of HIV. Second, because the virus continues to mutate, a vaccine developed today may become ineffective over time. Dr. Anthony Fauci of the National Institutes of Health said, however, that the threat of AIDS is so great that a less-than-perfect vaccine would be acceptable. The best hope presently is that one of the vaccines undergoing testing might prove to be 85 percent effective.

PUBLIC REACTIONS TO AIDS

Despite the laws of the land, discrimination continues to play a brutal and important role in the lives of those infected with HIV. The extent and the cruelty of such discrimination have been brought forth in heartrending testimony to the National Commission at each of its many hearings. As a colleague and I wrote recently, "The pain, suffering and despair of the disease alone are dreadful enough. The added stigma makes it virtually unbearable. You lose not only your life, but also your pride, your job, your insurance, your friends and your family. Posterity remembers you for dying of AIDS, not for having lived."

(Dr. David Rogers [1992], vice chair of the National Commission on AIDS)

The 1993 movie *Philadelphia*, starring Tom Hanks, focused attention on the pervasiveness of the stigma related to AIDS and the discrimination often directed at HIV-infected individuals. A sizable minority of the public have feelings of anger, disgust, or fear toward persons with AIDS, and over a third of Americans feel that "people with AIDS should be legally separated from others to protect the public health" (Herek & Capitanio, 1993). In 1993 Congress did, in fact, pass a bill that banned immigration by HIV-infected foreigners. President Clinton signed a defense bill in 1996 that contained a provision requiring that the Pentagon discharge any military personnel with HIV.

The public's fear of AIDS has been fueled by well-publicized cases of individuals who knowingly put others at risk. In 1992, health officials in Michigan confined a woman after she was accused of having sex with six men without telling them she was infected with HIV. In 1997, police in New York arrested a 21-year-old HIV-positive man who may have infected over 100 female sexual partners (Gega, 1997).

But the negativity directed at people with AIDS is not limited to the few who would knowingly spread the disease. Some children who contracted AIDS, such as Ryan White of Kokomo, Indiana (who became infected with HIV from a blood transfusion and later died of AIDS), have been temporarily barred from attending school. In Florida, the home of three hemophiliac children with AIDS was burned to the ground when they attempted to go to public school.

How should we as a society deal with the HIV epidemic? First of all, it is important to educate the public that the virus is not spread by any type of casual contact, and thus there is no medical basis for avoiding nonsexual interactions with infected individuals. Most states now have laws that protect HIV-infected persons against discrimination in the workplace and in housing. Ensuring the privacy and equal treatment of persons with HIV infection is essential to the success of testing, treatment, and outreach programs. Public health officials will not gain the confidence and cooperation of infected individuals if these persons are not legally protected from stigma and irrational prejudice. As former president Ronald Reagan said, "It is the disease that is frightening, not the people who have it."

Some medical journals have argued that there should be more routine testing of people's blood for HIV (for example, testing of all hospital patients, health care workers, pregnant women, and newborns; Angell, 1991). Others have called for more widespread testing (Kutchinsky, 1992). A recent *Time* / CNN poll found that 38 percent of adult Americans say they have been tested for HIV, up from only 6 percent in 1988 (*Time,* January 6, 1997).

Those who want more testing believe that most people would not knowingly spread the virus if they knew they were infected. Routine testing is already done for syphilis (e.g., when people apply for a marriage license and during prenatal exams). However, some people oppose mandatory testing because of possible civil rights violations (positive test results might be passed on to employers and insurance companies).

Some communities have initiated needle exchange programs in order to eliminate HIV among injecting drug users. Although well-intentioned, these programs have had only limited success (Strathdee et al., 1997).

One likely way to reduce the spread of HIV is education, but there is even controversy about that. Former U.S. Surgeon General C. Everett Koop, a noted conservative, startled many people when he advocated the teaching of sex education, including facts about AIDS and condoms, in all public schools. Former U.S. Secretary of Education William Bennett, on the other hand, opposed Koop and wanted to stress moral values and abstinence from sex until marriage.

A big question, of course, is whether or not education about AIDS will convince people to change their life-styles. The Surgeon General's office has been educating and warning people about the long-term dangers of smoking since 1964 (is there really anyone who hasn't heard that smoking can cause lung cancer?), yet many people continue to smoke.

THE EFFECT OF AIDS ON SEXUAL BEHAVIOR

Have AIDS and AIDS prevention programs resulted in long-term changes in sexual behavior in the general population? Two reviews have concluded that the answer is unequivocally yes (Choi & Coates, 1994; Kalichman, Carey, & Johnson, 1996). In 1995, the Centers for Disease Control and Prevention reported that among sexually active teens, 53 percent reported using condoms the last time they had sex. Condom use was highest among men who believed that condoms are effective in preventing HIV infection (e.g., Hingson et al., 1990; Siegel et al., 1991). Among heterosexuals, African-American men were much more likely than white men to have increased their use of condoms and/or reduced their number of sexual partners (Catania et al., 1993; Melnick et al., 1993; Tanfer et al., 1993).

However, not everyone is practicing safer sex behaviors, and several studies have reported some very disturbing trends. For example, in the male homosex-

ual community there has been some relapse from safer sex practices, particularly among young gay men (Handsfield & Schwebke, 1990; Stall et al., 1990). A 1993 study by the Department of Health in San Francisco concluded that "AIDS education campaigns have failed to make an impression on San Francisco's gay youths, who continue to contract HIV by engaging in dangerous sex practices" (Krieger, 1993).

Among heterosexuals, a very large national survey conducted in the 1990s found that while use of condoms may have increased, only a small minority of sexually active adults used them all the time (Catania et al., 1995). Another study concluded that "roughly 85 percent of the adult population aged 18 to 50 were estimated to be at some risk for behavioral exposure to HIV virus with regard to reported sexual behavior" and that "over time, those who could be considered at no risk have declined as a proportion of the population" (Campostrini & McQueen, 1993).

Most disturbing are findings that there are some people who apparently have no regard for their own health or the health of others. For example, many people who have had a sexually transmitted disease in the past still do not take precautions to prevent another infection in the future (O'Campo et al., 1992). In fact, a sizable minority of people who discover that they are infected with HIV continue to engage in unsafe sexual activity (Cleary et al., 1991). Many keep their infections secret from their partners (Marks, Richardson, & Maldonado, 1991). King and Anderson (1994) reported that some people might be willing to accept death by AIDS for an opportunity to engage in a few years of having unlimited sexual partners. Results like these indicate strongly that HIV will continue to spread for some time to come. What precautions are you taking to prevent becoming infected? Be sure to read the section on practicing safer sex.

PARASITIC INFESTATIONS

Viral and bacterial diseases are not the only things that can be transmitted from one person to another during sex. It is also possible to pick up parasites from an infested person. Condoms will do little good in preventing these types of infestations.

Pubic lice (*Pthrius pubis,* or "crabs") are 1-to-2-mm-long, grayish (or dusky red after a meal of blood), six-legged parasites that attach themselves to pubic hair and feed on human blood (see Figure 5–13). Infestation with the lice is technically called *pediculosis pubis.* The lice cause intense itching.

"I was a senior in high school and I was sleeping with a guy who I

pubic lice An infestation of the parasite *Pthrius pubis,* which attach themselves to pubic hair and feed on blood. Also known as "crabs."

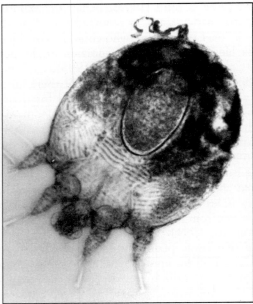

Figure 5–14 Top: *Pthirus pubis* (crab lice). Bottom: Gravid female *Sarcoptes scabiei mite.* (David Toplin & Terry Meinking: Scabies, lice, and fungal infections. *Primary Care, 16*(3), 551–568. Dept. of Dermatology and Cutaneous Surgery, and Epidemiology and Public Health, University of Miami School of Medicine, Miami, Florida.)

had been involved with for a year. For about two weeks I had been itching in the pubic area. The longer it went on, the more irritating it got. Finally, I examined my pubic hair and found little crab-looking things. I was shocked and disgusted. . . ."

"I noticed an itching down in my pubic area. It got extensively annoying so I took a look to see what was happening down there. I almost started to scream when I noticed little flat insects crawling around that area. . . ."

(from the author's files)

The lice do not live for more than 36 hours away from the human body, but any eggs that fall off can survive and hatch up to 10 days later (Meinking & Taplin, 1996). You could, therefore, pick them up from the sheets, towels, or clothing of an infested person. Normal bathing will not wash the lice off, but you can eliminate them by applying pyrethrin products (e.g., 1 percent permethrin cream rinse) or 1 percent lindane (Kwell) lotion, cream, or shampoo to the infested area *and all other hairy body areas.* Repeat the treatment in seven to 10 days (Meinking & Taplin, 1996). Be sure also to clean all clothing, sheets, and towels that might be infested in very hot water (or have them dry-cleaned) to avoid reinfestation.

Scabies is a contagious infestation of 0.3-to-0.4-mm-long, pearly white, parasitic itch mites (*Sarcoptes scabiei*; see Figure 5–14). The mites burrow under the skin to lay their eggs, resulting at first in itchy, red, pimplelike bumps when the eggs hatch. Large patches of scaling skin result if the mites are not immediately destroyed. Secondary bacterial infection is common. The itch mite is acquired by close contact with infested persons (sexual contact is not necessary), and thus is most common in people living and sleeping in crowded conditions. However, the mite "is notorious for its lack of respect for person, age, sex or race, whether it be in the epidermis of an emperor or a slave, a centurion or a nursling, it makes itself perfectly at home with undiscriminating impudence and equal obnoxiousness" (Friedman, 1947). Scabies has traditionally been treated with 1 percent lindane lotion or permethrin 5 percent cream, but recent studies have found that a single oral dose of ivermectin is also highly effective (Meinking et al., 1995). Should you ever get scabies, be sure also to wash all your sheets, towels, and clothing to avoid reinfestation.

Pinworms (*Enterobius vermicularis*) live in the large intestine and are generally gotten through non-sexual contact with the eggs. The female pinworms leave the rectum at night and lay their eggs around the anus, which causes intense itching (see Weber, 1993). Pinworms are common in children, who pass them from one to another by hand-to-mouth contact while playing. Pinworms can also be transmitted sexually in adults by manual or oral contact with the anus of an infected person. This mode of transmission occurs most commonly among homosexual males. Such sexual practices can also result in the transmission of a one-celled animal (*Entamoeba histolytica*) that causes intestinal **amebiasis,** or amebic dysentary.

scabies A contagious infestation of parasitic mites *(Sarcoptes scabiei)* that burrow under the skin to lay their eggs.

amebiasis Dysentery caused by infestation of amoebae, one-celled organisms.

VAGINITIS

Vaginitis is a general term that refers to any inflammation of the vagina. There are basically three types, and it is rare that a female does not experience at least one of them in her lifetime. Vaginitis has many causes, including hormone imbalances, allergic reactions, chemical irritation, and infection. Symptoms generally include discomfort or pain during urination, along with a vaginal discharge that often has a disagreeable odor.

TRICHOMONIASIS (TRICHOMONAL VAGINITIS)

Only one type of vaginitis, trichomoniasis, is usually contracted by sexual contact. The Centers for Disease Control and Prevention estimates that there will be 3 million new cases of "trich" this year, which is more than gonorrhea, syphilis, and genital herpes combined. Other researchers estimate that there may be as many as 8 million new cases each year (Martens & Faro, 1989).

Trichomoniasis is caused by a one-cell protozoan named *Trichomonas vaginalis* that lives in the vagina and urethra. Symptoms in women generally appear from 4 days to 1 month after exposure and include a copious, foamy, yellowish-green discharge with a foul odor accompanied by severe vaginal itching. However, not all women show symptoms.

> "About two weeks after becoming sexually involved with my boyfriend, I began to have severe pain during intercourse. Soon I noticed a discharge, itching, and an extremely foul odor. After many weeks of shame I finally went to the infirmary with my roommate, who was having urinary problems. Imagine how surprised we were to find out that we both had trichomoniasis. How could we catch such a thing? The only person she had sex with was her boyfriend and the only person I had sex with was my boyfriend! In the end we found out that our boyfriends went out together one night and had unprotected sex with a girl that they had known less than 6 hours!"
>
> (from the author's files)

T. vaginalis can survive in urine or tap water for hours or sometimes days, so it is possible to pick it up by using a wet toilet seat or by sharing towels. Nonetheless, the vast majority of cases are transmitted by sexual intercourse. Infected males generally show no symptoms, and in those who do, it is usually only a slight urethral discharge. Thus, *men tend to act as asymptomatic carriers,* spreading it to new female partners—just the opposite of what we often see in the early stages of gonorrhea, chlamydia, and syphilis.

Trichomoniasis can be very irritating (due to itching and burning), but until recently most authorities did not believe that there were any serious long-term consequences if it was left untreated, and thus regarded trich as a "minor" STD. However, a recent study found that women with trichomoniasis were as likely to develop tubal adhesions leading to infertility as women with gonorrhea (Grodstein, Goldman, & Cramer, 1993). The authors concluded that trichomoniasis "should be considered a potentially important sexually transmitted disease in women."

The diagnosis of "trich" is made by examining any discharge under a microscope and then growing the parasite in culture. However, the tests have not been totally reliable. If trichomoniasis is suspected, both the infected woman and her male partner (who is usually asymptomatic) should be treated. Otherwise, he will just reinfect her when intercourse resumes. "Trich" is very easily eradicated by a drug called metronidazole (its trade name is Flagyl), but it should not be taken during pregnancy.

MONILIASIS (OR CANDIDIASIS)

Moniliasis is caused by an overgrowth of a microorganism that is normally found in the vagina (*Candida albicans* in 80 percent of cases and other *Candida* species in 20 percent). It is a fungus or yeast infection that is very common in women. Three-fourths of women will experience it at least once in their lifetimes, and 40 to 45 percent will experience it at least twice. About 10 to 20 percent of women have what is called *recurrent candidiasis*—four or more episodes a year (see Lauper, 1996).

Symptoms include a thick, white, cheesy discharge accompanied by intense itching. Unfortunately, many women do not realize how common normal yeast infections are, and feel alone or dirty when it happens to them.

> "The first time I got a yeast infection I was terrified. I didn't know what it was. All I knew was that the itching was intolerable and I had this discharge that wasn't normal. . . ."
>
> "The first time I had it was when I was 16. I started to have an uneasy itchy feeling. I took a shower and thought it would go away. It not

vaginitis A general term that refers to any inflammation of the vagina. There are three general types: trichomoniasis, moniliasis, and bacterial vaginosis.

trichomoniasis A type of vaginitis caused by a one-celled protozoan that is usually transmitted during sexual intercourse.

moniliasis Sometimes called candidiasis. A type of vaginitis caused by the overgrowth of a microorganism (*Candida albicans*) that is normally found in the vagina. Moniliasis is a fungus or yeast infection and usually is a sexually related, rather than a sexually transmitted, disease.

only continued, but it became worse. The itchiness was now accompanied by a foul smelling discharge. . . . I felt very dirty. . . ."

(from the author's files)

Yeast infections are sometimes transmitted sexually, but the vast majority of cases are not contracted in this manner. *Children, even babies, can get this, too* (Robinson & Ridgway, 1994). Thus, monilial infections are not really considered a sexually transmitted disease, such as trichomoniasis, but are a sexually related disease. The microbe can also invade the mouth (where it is called *thrush*), anus and skin, as well as internal organs.

Anything that changes the chemical environment of the vagina can result in an overgrowth, including hormone changes, diabetes, heavy use of antibiotics (which kill off the "friendly" bacteria in the vagina that keep the yeast in check), and even overly zealous hygiene (e.g., the use of perfumed feminine hygiene products). Many women first experience a monilial infection during pregnancy, for example, or when they first start taking oral contraceptives.

"Once when I had an infected tooth, my dentist prescribed an antibiotic. About the third day I developed a yeast infection. No one had ever told me I could get a yeast infection from an antibiotic. When I told some of my friends, they told me this had happened to at least two of them before. . . ."

"I had my first yeast infection when I was pregnant with my first child. I was so embarrassed about having this terrible itch in the genital area that I couldn't tell anyone. Finally it was time for my regular prenatal checkup, and when the doctor examined me and saw that I had scratched myself raw he wanted to know why I had not called him. I was just too embarrassed to say I had an itch in the genital area."

(from the author's files)

Women with immune deficiencies (e.g., HIV infection and AIDS) are particularly susceptible to yeast infections. The question of what causes recurrent infections is still controversial (see Lauper, 1996). Reduction of intestinal sources of *Candida albicans* and treatment of partners do not prevent recurrences. Some people believe that women can suffer from a chronic candidiasis syndrome (also called candidiasis hyper-sensitivity syndrome or the "yeast connection") characterized by severe PMS, gastrointestinal problems, depression and anxiety, and fatigue (Crook, 1983), but this has been challenged by the American Academy of Allergy and Immunology (1986), and treatment to remove the fungus does not reduce internal or psychological symptoms (Dismukes et al., 1990).

Yeast infections are often treated with antifungal creams or suppositories. There are over-the-counter drugs that you can purchase in the drugstore (e.g., Monistat and Gyne-Lotrimin). Drugs that you can take by mouth (Fluconazole, itraconazole) are also available (Lauper, 1996). However, if the symptoms do not go away within a few days, see your doctor immediately, for you might have one of the other types of vaginitis.

BACTERIAL VAGINOSIS

This type of vaginitis is caused by the interaction of several vaginal bacteria (particularly *Gardnerella vaginalis*) that can also live in the male's urethra. It is the most common of the three types of vaginitis (Martius, 1996a). **Bacterial vaginosis** used to be called by a variety of names (haemophilus, Gardnerella, or nonspecific vaginitis), but the new name more correctly reflects what we now know about the microbiology of the disease (Eschenbach, 1993). Whether or not bacterial vaginosis can be transmitted sexually is still controversial. It is more common in women with many sexual partners, but treating partners does not prevent recurrences of the disease (see Nilsson et al., 1997).

The main complaint of patients is a vaginal odor (sometimes described as fishy), but there is also an abnormal discharge (grayish and nonclumpy). We now know that this type of infection can lead to serious upper reproductive tract infection (Eschenbach, 1993). The antibiotics of choice are metronidazole or clindamycin, but the recurrence rate is very high (Martius, 1996a, 1996b).

CYSTITIS AND PROSTATITIS

Cystitis refers to a bacterial infection of the bladder. Because the female's urethra is considerably shorter than the male's, and thus bacteria have a shorter distance to travel, cystitis is much more common in women than men. It is especially common in sexually active young women and is strongly associated with recent intercourse, especially after using a diaphragm with spermicide (Hooten et al., 1996). However, many cases are unrelated to sexual activity. The bacterium *Escherichia coli,* for example, is often transmitted from the rectum to the urethral opening by wiping forward from the anus after a bowel movement. Women should always wipe themselves from front to back. Cystitis can

---

bacterial vaginosis A type of vaginitis caused by the interaction of several vaginal bacteria (particularly *Gardnerella vaginalis*).

cystitis A bacterial infection of the bladder (often caused by the bacterium *Escherichia coli*).

Box 5-A SEXUALITY AND HEALTH

Vaginal Health Care

For women, vaginal infections are probably second in frequency only to the common cold. Here are some hygiene tips to help minimize your chances of getting vaginitis:

- Bathe regularly, avoiding deodorant soaps and bubble baths, and do not share washcloths and towels.
- Dry the vulva thoroughly and wear cotton panties (synthetic fabrics retain heat and moisture, conditions in which bacteria thrive).

"I used to get yeast infections several times a year. Every time I would visit or call my doctor. . . . I finally figured it out on my own with a little help from one of the hottest days in June. I would always wear silk or polyester blend underpants and nylon hosiery. Does any of these materials allow absorption? Of course not. I ran out and bought cotton underpants and hosiery with garter belts. Five years have passed and not one yeast infection yet."

(from the author's files)

Certainly never wear synthetic underpants to bed. Also, the yeast that results in monilial infections can survive normal laundering in hot water. As a result, a recent study suggested that women put their laundered but damp cotton underwear in a microwave for 5 minutes (do not try this with dry or synthetic fabrics).

- Avoid feminine hygiene products.
- After a bowel movement, wipe the anus from front to back, not back to front, as this can spread rectal bacteria to the vagina.
- Allow time for adequate vaginal lubrication during intercourse, and if you use a lubricant, use a water-soluble one such as K-Y jelly (petroleum-based lubricants can not only harbor bacteria, but disintegrate condoms as well).
- If you and your partner engage in any type of anal stimulation during sex, be sure to wash the hands or penis before touching the vaginal area again.
- If your partner has sexual relationships with people other than yourself, make sure he uses a condom during intercourse.
- If you have frequent yeast infections, ask your doctor about douching with a mild acidic solution.
- Keep stress to a minimum, eat well, and get adequate sleep.
- See a doctor as soon as you notice any of the previously mentioned symptoms. Remember, vaginal infections are very common and nothing to be ashamed of, and they can be treated very easily.

---

also be caused by the *N. gonorrhoeae* and *Chlamydia trachomatis* bacteria, but it is generally not considered a sexually transmitted disease.

Symptoms may include a frequent urge to urinate, painful urination, and lower abdominal pains. Cystitis is treated with either sulfa drugs or antibiotics. Vigorous intercourse, especially in women first becoming sexually active, can also result in urinary tract infection (hence the term *honeymoon cystitis*) by inward friction on the urethra, allowing nonsexually transmitted bacteria to ascend. For those women who experience postcoital urinary tract infection, trimethoprim and sulfamethoxazole taken together within 2 hours of intercourse prevents infections (Stapleton et al., 1990).

The *E. coli* bacteria can also be transmitted to a male's prostate during sexual activity, resulting in **prostatitis.** Symptoms may include lower back and/or groin pain, fever, and burning during ejaculation. This, too, is treated with antibiotics.

PRACTICING SAFER SEX

Add them up: 4 million new cases of chlamydia per year, at least 3 million new cases of trichomoniasis, nearly 1 million new cases of gonorrhea, one-half million to 1 million new cases of human

prostatitis Inflammation of the prostate gland.

papillomavirus infection, over 500,000 new cases of herpes, an unknown number of new HIV infections—plus syphilis, sexually transmitted cases of hepatitis, chancroid, and others. There will be about 12 million new cases of sexually transmitted diseases this year—about 35,000 new cases every day—with most cases occurring within your age group (late teens through 40). Of course, some people who are sexually active will be infected with the same STD two or more times in a single year, or even get two or more different STDs at once. Yes, having one does not prevent you from getting others, so it is possible to hit the jackpot and have more than one at once.

What are you doing to avoid sexually transmitted diseases? It appears that many college students are doing little or nothing. In fact, many continue to engage in high-risk behaviors. There is a strong association between number of sexual partners and having, or having already had, a sexually transmitted disease (e.g., a woman with five or more lifetime partners is eight times more likely to have had an STD than a woman with only one partner; Joffe et al., 1992); yet, according to surveys on college campuses, many students have had multiple lifetime sexual partners. At a midwestern university, for example, females reported an average of 5.6 partners; males reported 11.2 partners (Reinisch et al., 1992). At a heavily Catholic southern university, one-fourth of the students had had at least four partners (King & Anderson, 1994), and at a university in the northeast, over 20 percent of the coeds reported having had three or more sexual partners in the past year (DeBuono et al., 1990). A national survey of heterosexuals living in large cities found that while 39 percent reported having multiple sexual partners or casual sex, only a third used condoms consistently (Catania et al., 1995). Thirty-eight percent of young homosexual men have engaged in unprotected anal intercourse (CDC press release, 1996).

Why do so many people continue to engage in high-risk sexual behaviors (see Box 5–B)? Many young men and women who have had multiple sexual partners believe "it won't happen to me" (Kusseling et al., 1996; Reitman et al., 1996). Recall, however, that the Centers for Disease Control and Prevention says that for one in four Americans, it *will* happen to them. So let's get serious and consider how you can minimize the chances of it ever happening to you.

The only way to completely avoid any chance of ever getting an STD is never to have sex, but few of us wish to live the life of a Tibetan monk. There is no such thing as safe sex with another person; anyone having sex is at some risk. We can, however,

safer sex Sexual behaviors involving a low risk of contracting a sexually transmitted disease. These include consistent use of condoms and/or abstaining from sex until one enters a long-term monogamous relationship.

practice **"safer sex"**—behaviors that minimize the chances of contracting an STD.

One safer-sex solution is *to restrict your sexual activity to a mutually faithful long-term monogamous relationship*. If neither you nor your partner have an STD at the start of the relationship and neither of you have other partners, you do not have to worry about STDs. However, there are problems with this approach. First, while many of us limit our sexual behavior to monogamous relationships, most monogamous relationships do not last forever. Many of us have life-styles of what is commonly called *serial monogamy*. We break up and eventually enter into another monogamous relationship. You may have been having sex with only one person for a certain period of your life, but how many monogamous relationships have you had in your lifetime? If just one of those partners had an STD, he or she could have transmitted it to you. When we have new sexual partners, we can rarely be sure of their previous sexual history. One study, for example, found that many people have lied to new partners about their past sexual experiences (Cochran & Mays, 1990).

> "A friend of mine was going out with a guy from ———. They had a sexual relationship and used condoms most of the time. He only recently mentioned that he had tested positive for the AIDS virus in the past. My friend is under tremendous psychological distress. . . ."
>
> (from the author's files)

A second problem is that few of us can be 100 percent certain that our partner will remain monogamous. Read again the personal case histories at the beginning of the chapter.

A second solution is to *always use condoms*. A study of 256 heterosexual couples in which one partner was infected with HIV found that for those couples who used condoms consistently, none of the noninfected partners became infected during a total of 15,000 episodes of intercourse (De Vincenzi et al., 1994). It is certainly best to use condoms if you are going to engage in "casual" sex.

> "While [I was] working in the emergency room one evening a 17–year-old patient came in complaining of spotting. She informed us that she was HIV positive while [we were] taking her case history. This has become a common thing so I didn't think too much about it. A nurse took it upon herself to talk about the importance of sexual responsibility. [The patient] stated that she knew she had the infection a year ago and had slept with half of her high school football team. When we ques-

tioned her on whether condoms were used she said no. We asked her if she had told the boys beforehand and she said no. . . ."

(from the author's files)

We would all like to believe that most people would not knowingly give others a sexually transmitted disease, but so many people are asymptomatic during the early stages of various STDs that many people are not even aware that they have an infectious disease. For women whose partners refuse to wear condoms, there are some birth control techniques that reduce the risk of HIV (or other STD) infection (see Box 6-A in Chapter 6).

There are problems with condoms (as STD prevention) as well. Even if you and your partner or partners use them consistently and properly, they sometimes tear (see Chapter 6). Even microscopic tears are large enough to allow the passage of viruses. In other words, as is the case with practicing monogamy, condoms cannot guarantee that you will never get a sexually transmitted disease. Which method is better? In a study that used probability modeling to test the amount of risk reduction, the authors concluded that

Figure 5–15 A giant condom displayed at the Fifth International AIDS Conference in Montreal.

"consistent and careful condom use is a far more effective method of reducing HIV infection than is reducing the number of sexual partners" (Reiss & Leik, 1989).

Whichever behavioral strategy you use—condoms or the practice of monogamy (in a *long-term* relationship)—will greatly reduce your chances of ever getting a sexually transmitted disease. Allow me to review some recommendations for safer sex:

1. Limit the number of sex partners. The more sex partners you have, the higher your risk of contracting a sexually transmitted disease, because chances increase that sooner or later you will have sex with an infected individual.
2. Be selective and avoid having sex with anyone who has had lots of other sex partners (e.g., prostitutes or promiscuous individuals). You are at high risk of getting an STD if your partner is promiscuous—even if you are not—so take the time to get to know your partner before you start a sexual relationship.
3. Avoid having sex with anyone who uses intravenous drugs or whose partner uses them.
4. When you are about to have sex with a new partner, leave the lights on and be observant during foreplay, or take a shower together. If you notice anything peculiar, such as sores or a discharge, it will not be too late to change your mind. However, looking for symptoms is not enough by itself. Remember, many people with STDs do not show any symptoms. People with the AIDS virus may not show any symptoms for 5 years or longer.
5. Use a condom, or have your partner use one. This is not a 100 percent guarantee that you will not contract an STD (condoms sometimes tear), but it substantially reduces the risk. The risk can be reduced even further by using condoms in combination with spermicides, which kill many of the bacteria and viruses that cause STDs.
6. If you are sexually active, have checkups regularly. Remember, women are not usually tested for STDs during their routine pelvic exams unless they ask their doctor to do so.

WHAT TO DO AND WHERE TO GO IF YOU HAVE AN STD

You would think it would be obvious to someone what to do if they thought they might have, or thought they were at risk for, a sexually transmitted disease—go to a doctor! However, many people hesitate to do this out of fear, shame, guilt, or just plain denial. "Among groups with high-risk behaviors, only 34 percent of sexually experienced teenagers were screened for STD

Box 5-B SEXUALITY AND HEALTH

Impediments to Practicing Safer Sex

Until an effective vaccine is developed, health officials must rely on convincing people to change their behavior in order to prevent the spread of HIV. For most of the population, this means convincing them to practice safer sex behaviors. During the 1980s, it was hoped that educating people about AIDS and the transmission of HIV would lead them to practice safer sex. Schools began teaching about HIV; the surgeon general mailed an informational brochure to all U.S. households; and hardly a day went by that newspapers did not include an article about HIV or AIDS.

Studies in the 1990s have found that most young people today are, in fact, knowledgable about the transmission of HIV and the consequences of AIDS (Anderson & Christenson, 1991; Ku, Sonenstein, & Pleck, 1992). However, AIDS and sex education programs have had, at best, only a modest success in convincing people to reduce their number of sexual partners and/or to use condoms regularly (Ku et al., 1992). Even college courses have had only minimal success in convincing students to reduce high-risk sexual behaviors (e.g., Baldwin, Whiteley, & Baldwin, 1990; King & Anderson, 1994). Even scare tactics have not produced behavior change (Ross et al., 1990; Sherr, 1990). In fact, researchers find that there is no relationship at all between people's knowledge of AIDS and their sexual behavior (Andersson-Ellstrom et al., 1996; DiClemente et al., 1990; Klepinger et al., 1993; Ku et al., 1992). This has led many researchers to conclude that, by itself, knowledge is "not sufficient for compliance to safer-sex guidelines" (Ahia, 1991; Helwig-Larsen & Collins, 1997).

If it is not lack of knowledge that prevents many people from practicing safer sex, what is it? Actually, there are many reasons. We will consider a few of the most important.

The most common reason that college students give for not using condoms is that one was not available when they were in the mood to have sex (Franzini & Sideman, 1994). About one-third of heterosexual adults are embarrassed about buying condoms (Choi, Rickman, & Catania, 1994). Condoms are also less likely to be used if the male is "high" from alcohol or drugs prior to sex (Ku, Sonenstein, & Pleck,

1993; Mahoney, 1995; Weinstock et al., 1993). People are more likely to have a condom available if they have used them in the past, especially if they are with a partner with whom they have agreed to use condoms.

The major problem here is that to have come to an agreement, a couple has to have talked about using condoms, and most people, including those who are college educated, do not know "how to engage in the interpersonal negotiations that have to occur in order for a condom to be used" (Morrison, quoted in Gladwell, 1992). Practicing safer sex means that you must be able to talk with partners about using condoms, as well as about their previous sexual history, and that you must be able to put a stop to sexual activity if your partner will not use one. This requires some degree of assertiveness. Young people who have these behavioral skills are much more likely to practice safer sex (Fisher & Fisher, 1992; Lear, 1995; Mahoney, Thombs, & Ford, 1995; Yesmont, 1992).

One factor that prevents some women from acting in their own best interest is the perceived power difference between men and women. If a woman's predominant experience of male-female relations is of men exerting power over women and always getting their way, they are not likely to insist that a male use a condom if he doesn't want to (Browne & Minichiello, 1996; Holland et al., 1991; Seal, 1996). This may be common in cultures that emphasize the ideal of the virile, macho male (as in Hispanic culture).

Even when there is little or no power difference, communication between a man and a woman about safer sex is often hindered by cultural double standards and sexual "scripts" (Ehrhardt, 1992; Metts & Fitzpatrick, 1992; Muehlenhard & Quackenbush, 1988). In American culture, men are given more sexual freedom than are women. Thus, a woman may be viewed negatively by a partner and by others if she openly expresses a desire for sex, and she may be considered respectable only if she refuses a partner's sexual advances. Women who believe that their partners endorse this standard are not likely to provide a condom or suggest that their partner use one. As you will learn in Chapter 15, many women engage in "scripted refusal" with new partners—first telling them no (in order to appear respectable) and then consenting (Muehlenhard & McCoy, 1991). Women who engage in this behavior are very unlikely to look after themselves by carrying and providing their partners with condoms.

The most common HIV-prevention behavior used by college students is to have sex only with individuals they believe to be "safe" (Maticka-Tyndale, 1991). They look for partners who appear to have no symptoms (Baldwin & Baldwin, 1988; Ferguson, 1997) and then restrict their sexual relations to a monogamous relationship. Many have life-styles of what I have previously referred to as serial monogamy. However, sexual scripts often work against the use of condoms when one starts dating a "respectable" and "safe" partner where the goal is the formation of a relationship (Browne & Minichiello, 1996; Metts & Fitzpatrick, 1992). There is a negative relational significance associated with condoms: if they are not being used for birth control, their only other use is for the prevention of STDs with an "unsafe" partner (Pilkington, Kern, & Indest, 1994). When the focus of "getting to know you" is on the other person's suitability as a relational partner, the subject of condoms may be viewed negatively, suggesting promiscuity and possible infection (by oneself or the other) and increasing the likelihood of rejection (see Choi et al., 1994). In such situations, a couple are likely to avoid directly asking each other about past sexual behavior or the use of condoms and just talk about AIDS and safer sex in general rather than in terms of the relationship. People (incorrectly) take a partner's willingness to talk about AIDS as a sign that he or she is safe. Even when condoms are used during early sexual episodes, they are likely to be abandoned once the two individuals perceive themselves as starting a relationship (Ku, Sonenstein, & Pleck, 1994). But without previous explicit, honest discussion about prior sexual experiences, you cannot be certain. Or, put another way, "You know if you're using a condom or you're not. You don't know if you're picking the right partner" (Heterosexual AIDS, 1988).

Our culture also has a taboo against seriously discussing any sexual topic (Wight, 1992). For example, dating partners generally do not ask each other directly if the other wishes to have sex, but instead communicate their intentions to one another indirectly via ambiguous body language and other behaviors. Indirect communication also serves as a way of avoiding rejection—cues and signals that go unacknowledged are generally considered to be less rejecting than a direct refusal to verbal requests (O'Sullivan & Byers, 1992). But you can't communicate about the use of condoms with indirect gestures.

In conclusion, the number of individuals practicing safer sex will probably not increase substantially beyond the present levels until health educators begin to teach behavioral skills and address gender issues. In the meantime, you might benefit by engaging in some role playing (Kelly et al., 1989). With a friend of your particular sexual orientation, pretend that the two of you are on a date and that there is a sexual attraction. Take turns bringing up the subject of condoms and responding to your partner's bringing up the subject of condoms. After you discuss your reactions, try the discussion again, and keep doing so until you are comfortable. This may sound silly, but unless you are already comfortable with the idea of talking about condoms, how do you think you will feel with a date in a real-life situation?

---

in the past year, as were only 43 percent of women with a positive STD history and 32 percent of women with 10 or more lifetime partners" (Mosher & Aral, 1991). Now that you know the consequences of untreated STDs, I hope that this will not include you. So, if you think you have an STD (or if you have put yourself at high risk of getting one):

1. See a physician immediately.
2. If he or she diagnoses you as having as STD, abstain from having sex until you are cured (or use condoms if there is no cure).

3. If you have an STD, tell your partner or partners so that they, too, can be treated and not infect others or yourself again. You do not want to play STD Ping-Pong with your partner!

There are two other reasons that someone might not go to a doctor immediately. Some (such as poor college students, for example) may not be able to afford the doctor's charges and lab test costs, and others might not know where to go even if they had the money (they might not have a doctor, or they may wish to avoid the family doctor). Fortunately, the U.S.

"It's not that I don't trust <u>you</u>, Kevin, I just
don't trust the women who've been with the men who've been
with the women who've been with the men who've
been with the <u>women</u> you've <u>been</u> with."

Public Health Service provides clinics for the diagnosis and treatment of sexually transmitted diseases—often free of charge! The government is committed to eradicating STDs. So there is no excuse not to be treated. Contact the Public Health Service clinic in your area.

The Centers for Disease Control and Prevention also have an STD informational National Hotline toll-free number: 1-800-227-8922.

POSITIVE SEXUALITY IN THE ERA OF AIDS

AIDS, human papillomavirus infections, herpes—the attention being given to these and other sexually transmitted diseases is making many people afraid of sex. My intent in this chapter has been to educate you about some of the *possible* consequences of sexual behavior, not to scare you so badly that you will want to avoid ever having sexual relations.

There are many activities human beings engage in that involve risk. Take skiing, for example. No one has to tell us that going down a snow-and-ice-covered mountain on a pair of skis is a dangerous activity. If trees, boulders, and other skiers are not avoided, the result can be terrible injury, yet this doesn't detract from the tremendous enjoyment many people derive from this winter sport. Those who ski learn that there are limitations to how they seek that pleasure. Skiing out of control has its consequences.

The same is true of driving an automobile. Many people own sports cars, and many others would like to. They get a certain thrill by accelerating quickly and driving fast. But, again, there are limitations. Taking a corner at too high a speed can be fatal. At high speeds, we must be able to react to unanticipated events (e.g., another car cutting in front of us) in a fraction of a second. Driving while having impaired judgment and motor coordination due to alcohol consumption is certainly an often fatal high-risk behavior. The fact that each year thousands of people are injured or killed because they foolishly took that risk does not detract from the pleasure that the rest of us often get from driving our cars.

AIDS is not the first sexually transmitted disease, only the most recent. Syphilis and gonorrhea were greatly feared before the discovery of antibiotics. Many millions more have died of syphilis than of AIDS, yet this hasn't stopped people from enjoying sex. Why? *Because sex per se does not cause any of these diseases.* Like driving an automobile, if we avoid foolish, high-risk behaviors, we can still enjoy the pleasures of sexual relations. The few seconds it takes to put on a seat belt does not detract from your enjoying your automobile. Similarly, the few seconds it would take you or your partner to put on a condom will not detract from the enjoyment of sex. Like drinking and driving, making sexual decisions while under the influence of alcohol is also foolish.

AIDS does not strike randomly—it only affects those who are exposed to the virus. If you make just a few adjustments to reduce the risk of exposure, sex can be as pleasurable and exciting as always.

Key Terms

acquired immunodeficiency syndrome (AIDS)   107
amebiasis   120
asymptomatic   94, 107
autoinoculation   101
bacteria   92
bacterial vaginosis   122
cervicitis   95
chancre   98
chancroid   99
chlamydia   95
cystitis   122
genital herpes   101
genital (venereal) warts   105
gonorrhea   93
granuloma inguinale   100
gummas   99

hepatitis A   104
hepatitis B   104
hepatitis C   105
human immunodeficiency virus (HIV)   107
human papillomavirus (HPV) infection   105
immune system   107
lymphogranuloma venereum (LGV)   96
molluscum contagiosum   106
moniliasis   121
nongonococcal (nonspecific) urethritis (NGU)   95
opportunistic infections   110
oral herpes   101

pelvic inflammatory disease (PID)   94
pinworms   120
primary HIV infection   107
prostatitis   123
pubic lice   119
safer sex   124
scabies   120
sexually related disease   93
sexually transmitted disease   93
shigellosis   100
syphilis   97
trichomoniasis   121
vaginitis   121
venereal disease   93
virus   92

Personal Reflections

1. How do you think you would feel if you found out you had a sexually transmitted disease (or, if you have previously been infected, how did you feel)? Would you feel differently than you would if you had contracted the flu? Why? How would you tell your partners of their possible infection?

2. How would you feel if (a) your best friend told you that he or she had an STD? (b) your teenaged son told you that he had an STD? (c) your teenaged daughter told you that she had an STD? (d) your partner told you that he or she had an STD acquired from a previous relationship?

3. Many babies are born with sexually transmitted diseases as a result of their mothers being infected during pregnancy. Is having an STD an indication of immorality, irresponsibility, both, or neither? Explain

(you may wish to reconsider your answers to questions 1 and 2 at this point).

4. What do you presently do to protect yourself from sexually transmitted diseases? Now that you have learned more about STDs, what changes in your sexual life-style do you plan to make to better protect yourself?

5. Have you ever engaged in sex with a new partner while under the influence of alcohol? If your answer is yes, did your use of alcohol result in your engaging in unprotected sex? If it did, what will you do in the future to prevent this?

6. Do you and your new partner(s) discuss safer sex? Why or why not?

Suggested Readings

Ahmed, P. I. (Ed.) (1992). *Living and dying with AIDS*. New York: Plenum Press. Many personal as well as practical essays on HIV and AIDS.

Anderson, P. B., de Mauro, D., and Noonan, R. J. (Eds.) (1996). *Does anyone remember when sex was fun? Positive sexuality in the age of AIDS*. Dubuque, IA. Essays on positive sexuality from many different perspectives.

Boston Women's Health Book Collective. (1992). *The new our bodies, ourselves* (4th ed.) New York: Simon & Schuster. The landmark book on women's sexuality and health.

Brandt, A. M. (1985). *No magic bullet: A social history of venereal disease in the United States Since 1880*. New York: Oxford University Press.

Ebel, C. (1994). *Managing herpes: How to live and love with a chronic STD.* Research Triangle Park, NC. The best book ever written about herpes.

Gordon, A. N. (1990, February). New STD menace: HPV infection. *Medical Aspects of Human Sexuality.* Discusses the human papillomavirus and its link to cervical and penile cancer.

Planned Parenthood. *Vaginitis: Questions and answers.* 16 pages.

Shilts, R. (1987). *And the band played on: Politics, people, and the AIDS epidemic.* New York: St. Martin's Press. An investigative reporter's look at the incompetence and infighting that prevented an early effective campaign against AIDS.

Time Magazine (1997, January 7). Several articles devoted to the AIDS pandemic and new treatments.

Walker, L. A. (1995, September). Dangerous liaisons. *New Woman.* Women talk frankly about condoms.

Westheimer, R. (1992). *Dr. Ruth's guide to safer sex: Exciting, sensible, sexual directions for the 90s.* New York: Warner Books. Easy to read.

Zilbergeld, B., and Barbach, L. (1988). *How to talk with your partner about smart sex!* This excellent 1988 tape can be purchased from the Fay Institute, 7242 Ariel Avenue, Reseda, CA 91335.

CHAPTER 6

Birth Control

*T*he teenage pregnancy rate in the United States is the highest of any developed country in the world. This is not a recent phenomenon. Teenage births in the United States reached a peak in 1957 and have declined a bit since then. What is new is the proportion of births to unwed teens. Only 15 percent of teenage births were out of wedlock in 1960, but this figure rose to 29 percent in 1970, to 49 percent by 1980, and to 76 percent in the 1990s. The greatest increase in pregnancy rates has been among young teens aged 14 to 17 without a high school education (Williams, 1991). About 85 percent of births to unwed teens are unintended or unwanted, and nearly one-third are terminated by abortion (Alan Guttmacher Institute, 1996; Henshaw, 1997).

Recently there has been some improvement. The teenage birthrate declined for the fourth straight year in 1995, and births to

TABLE 6–1

YEARLY TEENAGE PREGNANCY RATE (PER 1000 WOMEN) IN WESTERN NATIONS

United States	105
England and Wales	58
Norway	45
Canada	41
Finland	38
Sweden	33
Netherlands	15

Source: Westoff, 1988. Reprinted with permission of the publishers.

unwed teens that year dropped for the first time in 20 years (Rosenberg et al., 1996). Nevertheless, the teenage pregnancy rate in the United States is still twice as high as in England, over three times as high as in Sweden, and nine times as high as in Japan (Jones et al., 1986; Westoff, 1988; see Table 6–1).

The rate for out-of-wedlock births is much higher for nonwhite women than it is for white women. In 1995, 70 percent of African-American women who gave birth were single, compared to 41 percent for Hispanics and 25 percent for whites (Rosenberg et al., 1996), and these numbers are higher for teenagers. However, white teens are much more likely than black teens to "hide" an unintended pregnancy by getting married (39 percent versus 7 percent, respectively, of nonmaritally conceived births were legitimated in 1981; Jones et al., 1986). Because the divorce rate for couples married as teenagers is much higher than average, many of these prematurely married white teens end up as young single parents anyway.

The high pregnancy rate among U.S. teens reflects widespread nonuse of **contraception.** One-fourth of U.S. women use no contraceptive method at their first premarital sexual intercourse (Alan Guttmacher Institute, 1996). Use of contraception at first intercourse is lowest among young teens. Teens' self-esteem, ego development, and educational aspirations often predict whether or not they use contraception (Goodson, Evans, & Edmundson, 1997). Teens with repeat pregnancies generally have a history of behavior problems (school problems, drug and alcohol use, etc.) (Gillmore et al., 1997).

For young teens, ignorance about reproduction and contraception is common.

"As a Paramedic I've delivered more babies than I can count. Once I responded to a person with 'rectal bleeding.' We expected to find a typical older individual and prepared for such. To our surprise, a young girl greeted me saying that 'she has something coming out her rectum and she is bleeding all over.' The fifteen-year-old female was scared to death, still clothed in underwear, bleeding, with a large mass between her legs, inside the underwear. Immediately I removed obstructions, assisted with the remainder of the delivery, resuscitated the infant and treated the mother for shock. During this ordeal I gained information from the family that no one (including the patient, who in this case I had adequate reason to believe was totally honest), had any inclination that this fifteen-year-old was even pregnant.

"In this case the child and 'child mother' lived. What an education for a fifteen-year-old. Is this really what we want for our children, or do we want to provide the availability of more classes such as Human Sexuality. I vote for the latter."

(from the author's files)

However, most teens know about contraception, and also know that if they have sex without contraception they are taking a chance. But if they have already taken that chance without having it result in pregnancy, they are likely to take the chance again. Eventually, they begin to feel that they can't get pregnant—that it won't happen to them (Luker, 1975).

"I have been having sex since I was 16. I have had different partners and unprotected sex with all of them. I always believed that I couldn't get pregnant. It only happens to bad girls and people who wanted it to happen. There were a couple of late periods and worrying, but it never made me want to try birth control. . . . I found out I was two months pregnant. I was completely shocked."

(from the author's files)

Who are the male partners responsible for these teenage pregnancies? Half of the fathers of babies born to women aged 14 to 17 are at least 20 years old, and 27 percent are at least 5 years older (Lindberg et al., 1997; Taylor et al., 1997). Nevertheless, many teenage males also become fathers, albeit usually absentee fathers (see Thornberry, Smith, & Howard, 1997). Males who do poorly in school, lack long-term goals, come from impoverished backgrounds, and engage in other socially deviant behaviors (e.g., drugs) have the most irresponsible attitudes about reproduction. In fact, many of these men consider getting a woman pregnant a sign of masculinity. This attitude is more

contraception The prevention of conception.

common among African-American men than among whites (Marsiglio, 1993a). Another problem is that many men take no responsibility if they consider their partner an acquaintance (as opposed to a girlfriend): oftentimes "girls think they have boyfriends. Boys think they have acquaintances" (L. Smith, 1993).

"I am an admitted 'serial monogomist.' One of these types of relationships led to pregnancy right after high school. . . . My ex and I no longer speak to each other and he has never been a part of his own child's life (by choice). . . ."

(from the author's files)

The teens most likely to have babies are generally the ones least prepared to take care of them (Williams, 1991). The cost to the nation is staggering. The U.S. government spends over $30 billion a year on welfare support for families started by teenagers (Center for Population Options, 1992) and the cost would be even higher if it were not for publicly funded expenditures for contraceptive services (Forrest & Singh, 1990). But the cost to the individual can be even more tragic. Teenage mothers are much more likely to have high-risk babies with health problems, a problem that is often made worse because many pregnant teens never seek prenatal care (Hein, Burmeister, & Papke, 1990). As a result, the United States has one of the highest infant mortality rates among developed countries (Singh & Yu,1995). Although some of these problems can be

attributed to the socioeconomic status of teen mothers (Makinson, 1985), children born to teenagers are twice as likely to be premature and three times as likely to have brain or nervous disorders and score significantly lower on achievement and IQ tests. In addition to not being economically or physically ready to have children, teenagers are generally not emotionally ready to have children either. Children of teenage parents are much more likely than other children to be victims of child abuse (Zuravin, 1991).

Being an unwed teenage mother does not guarantee that a girl will be locked into poverty with no job skills and little education, but most will experience long periods of hardship because of early childbearing (Furstenberg et al., 1990). Females born to teenage mothers are twice as likely to become teenage mothers themselves (Manlove, 1997), and those who do are even more likely to experience economic dependence and inability to escape poverty (Furstenberg et al., 1990; Horwitz et al., 1991). The children of teen parents who later become young teen parents themselves often report that they experienced emotional deprivation at an early age, have had significant depressive symptoms, and "seek emotional closeness through sexual activity and early parenthood" (Horwitz et al., 1991).

Perhaps the most unfortunate part of all this is that the high rate of births to teens continues to occur in a day and age when highly effective means of birth control are available. This was not always the case. The prevailing attitude toward sex during Victorian times and throughout the history of the Christian church, as you recall, was that it was for procreation only, and thus birth control was opposed. In many states it was against the law to sell or distribute contraceptive devices. As a result, women generally had several or even many children, and many poor women who worried that they could not feed another child died during crude and often self-attempted abortions. It was as a result of watching one of these poor tenement women die from a self-induced abortion attempt in 1912 that Margaret Sanger gave up a nursing career and founded Planned Parenthood. The work of the reformers and the changing attitude about the role of women in general gradually had its effect.

Two-thirds of Americans surveyed in a 1936 Gallup poll said that information about birth con-

Margulies/ *The Journal,* VA/Rothco.

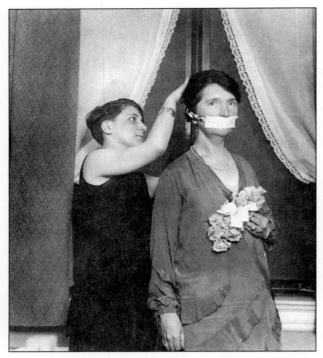

Figure 6–1   Margaret Sanger, a founder of the birth control movement, was forbidden to talk about birth control, so in 1929 she voluntarily had her mouth taped and wrote about the subject on a blackboard.

trol should be available to those who wanted it. The last of the state laws prohibiting the sale of contraceptive devices was finally repealed in 1965 (by a U.S. Supreme Court decision in the case of *Griswold v. Connecticut*), and by then 81 percent of the public approved of the availability of birth control information. The birth control pill was introduced in 1960, but until recently condoms and other birth control devices that did not require a visit to the doctor were kept behind the counter in drug stores, which inhibited many people from purchasing them. Today, male and female condoms and spermicides are sold openly in most drug stores, yet the number of teenage pregnancies continues to be high.

WORLD POPULATION

In addition to personal reasons, there are reasons why a society as a whole should practice birth control. Imagine two shipwrecked couples on a small South Pacific island in the year 1700. The island has a small supply of fresh water, some animal life but no predators, edible vegetation, and a small amount of land suitable for farming. For these four people, the island is Utopia—it has more than they will ever need. Each couple is fertile

and has four children (probably an underestimate without birth control), and the children in turn pair off in their late teens to form four couples, each of which has four more children. Thus, in each generation there would be twice as many people as in the previous generation. If there were no accidental deaths and the people lived until age 65, there would be at least 224 people living on the island by the year 1800. For those people, the island is no longer Utopia. The animals have all been killed, the edible plant life is consumed within days of becoming ripe, and there is not nearly enough farmland to support a population of this size. What's worse, even the water supply must be rationed, and the population continues to increase. Soon there will be large-scale death from starvation and dehydration.

Wildlife experts are used to seeing this happen to populations of deer in areas where there are no predators, but is this scenario too far-fetched for humans? The world is really just an island. We have a limited amount of natural resources, including water. Should they become exhausted, we would find ourselves in the same situation as the inhabitants of our hypothetical island. The world population has been increasing at an alarming rate (see Figure 6–2), and it is now more than 25 times greater than in the year A.D. 1. It was about 800 million in 1760, reached 5 billion in 1987, and reached 5.85 billion at the start of 1998. Each year, 87 million people are born. The United Nations (Population Division) estimates that the world population could rise to 8.5 billion by the year 2025 and reach 10 billion before the middle of the next century. In some countries, such as Rwanda, the fertility rate is as high as eight children per mother.

Many experts predict that unless this world population explosion is substantially slowed, the end result will be widespread starvation. Robert Repetto of the World Resources Institute estimates that if the present population doubles, all of the current cropland in the world would have to produce 2.8 tons of grain per acre per year, equivalent to the most productive American farms. Paul and Anne Ehrlich, authors of *The Population Explosion* (1990), give the following warning:

> **We should not delude ourselves: the population explosion will come to an end before very long. The only remaining question is whether it will be halted through the humane method of birth control, or by nature wiping out the surplus. We realize that religious and cultural opposition to birth control exists throughout the world; but we believe that people simply do not understand the choice that such opposition implies. Today, anyone opposing birth control is unknowingly voting to have the human population size controlled by a massive increase in early deaths.**

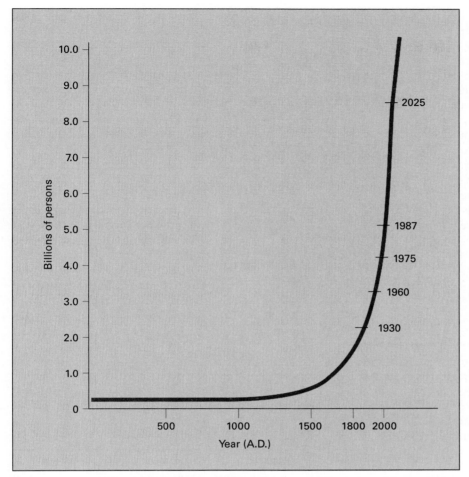

Figure 6–2 World population growth. Notice how rapidly the world population has increased since the beginning of this century.

In some parts of the world, mass starvation is no longer hypothetical, but already a reality. Most recently, our attention has been focused on Ethiopia and Somalia, where thousands of people have died of starvation because the land can no longer support the population. The sight on TV of little children dying a slow, horrible death is not pretty, but it is not just happening in Somalia and Ethiopia. It is happening in many parts of Africa and Asia, and in Brazil alone there have been an estimated 2 million children abandoned by their parents.

China, the world's most populated country with about 1 billion people, has much less farmland than the United States. There is only one-fifth of an acre of land per person, which is just enough to provide food for one person. Any additional increase in the population will mean starvation for millions. As a result, the Chinese government was forced in 1982 to order mandatory birth control. For the next 100 years, couples will not be allowed to have more than one child. Although this will eventually reduce their population

by several hundred million, there are presently so many people under the age of 30 that even if the policy is strictly adhered to, the population will still increase by the amount of the U.S. population. To oversee this policy, each district and town has a Family Planning Council. Birth control courses are given before marriage. Newlywed couples then sign a contract limiting themselves to one child, and those who adhere to the policy earn "model worker" status in their factory, which entitles them to many benefits, including bonus income and free education for their children. Factories with 100 percent participation become "model factories." Older women are assigned to "look after" small groups of females in each neighborhood. If a second pregnancy is detected, tremendous pressure from community leaders is put on the couple to have an abortion and sterilization.

These measures seem very harsh in terms of our own standards of freedom, and with an abundance of farmland in America it is hard for us to imagine such a thing ever happening here. During the "baby boom" years (following the end of World War II through the mid-1950s), there was an average of 3.5 babies born to each American couple. If that rate had continued, the population of the United States today would have been around 500 million, nearly twice what it actually is. Just imagine what the congestion and pollution would have been like. Would our natural resources have been strained? I do not know, but our life-style could not possibly have been as good as it is today.

What prevented the United States from having this population explosion? Quite simply, the use of birth control. Reversible birth control devices to allow couples to postpone having children, and sterilization techniques for couples who already have children, have made it possible for people to limit the number of children they have. The birthrate started to fall in the early 1960s, and by 1970 Americans were averaging about two babies per couple. From 1970 to the mid-1980s, the U.S. fertility rate averaged about 1.8 births per woman, below the rate needed to replace the parents' generation. It rose slightly in the late 1980s, but by 1995 the fertility rate had fallen again to its lowest level since 1986 (Rosenberg et al., 1996; see Figure 6–3). As a result, the population of the United States has stabilized, with only a small increase due to immigration.

Stop and think about it for a minute. Would you really like to see us in a situation where the U.S. population doubles every generation? We might be able to continue to live comfortably during our lifetime, but what would life be like for our grandchildren?

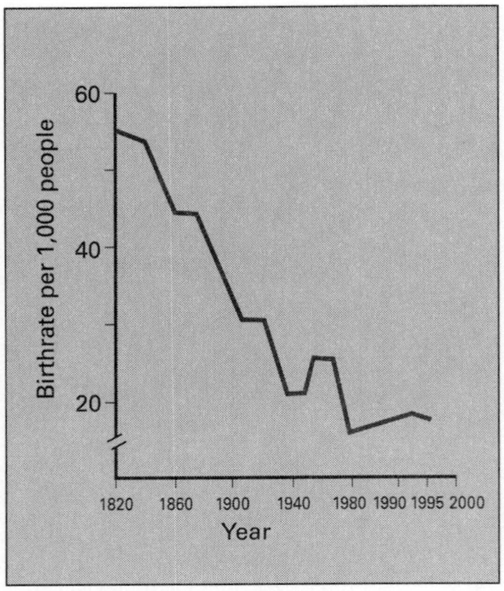

Figure 6–3   The U.S. birthrate.

EVALUATING DIFFERENT BIRTH CONTROL METHODS

If all fertile couples practiced unprotected sexual intercourse (no pills, condoms, diaphragms, etc.) and did not abstain from having sex at any time of the month (no fertility awareness methods), 85 to 90 percent of them would conceive within a year. In judging the effectiveness of the various birth control techniques, we need to know how many couples using a particular contraceptive method would conceive in a year's time. There are actually two figures here that are important: the perfect-use pregnancy rate and the typical-use pregnancy rate. Older studies use the term "failure rates." A failure rate is not the same as pregnancy rate, and it has been recommended that the term not be used (Steiner et al., 1996), but I occasionally will have to use it when referring to results of older studies. For a particular contraceptive technique, the **perfect-use pregnancy rate** is the percentage of couples who would conceive if all couples who were using the technique used it consistently and properly. The **typical-use pregnancy rate** is usually worse than the perfect-use rate because not all people use a method consistently or properly. The birth control pill, for example, was designed to be taken every day, but not all women remember to do that, and this results in some pregnancies. Although all the birth control techniques that you will learn about are better than using nothing at all, some are more effective than others. Ideally, a couple wishing to postpone or prevent pregnancy would like their birth control technique to be 100 percent effective (zero pregnancies in a year's time). Some contraceptive methods approach this, but others do not.

Pregnancy rates are almost always stated in terms of the number of pregnancies that occur *in the first year of use* for every 100 couples using the technique (see Table 6–2). Although I will follow that tradition, most young people want to have several years of sexual relations before their first planned birth. If their chosen contraceptive technique has a pregnancy rate of 5 percent in the first year of use, what are the chances of an unplanned pregnancy if they continue to use the technique over a much longer time period?

Estimates differ, depending on whether one assumes that the pregnancy rate remains constant or declines over time, but even the most conservative estimates show that over a long period of time the risk of pregnancy is high for most contraceptive techniques (Ross, 1989). For a technique with a first-year pregnancy rate of only 5 percent, for example, the 10-year pregnancy rate is between 23 percent (this assumes that the risk will have declined to zero by the end of that time span) and 40 percent (this assumes that the risk remains constant). For a contraceptive technique with a one-year pregnancy rate of 10 percent, the 10-year pregnancy rate is between 40 percent and 66 percent. Some of the contraceptive techniques that we will consider have a first-year pregnancy rate of 20 percent. Over a 10-year time period, the chance of having an unplanned pregnancy while using these techniques is between 67 percent and 90 percent. Thus, while contraceptive methods reduce the number of unplanned pregnancies, most are far from being 100 percent effective over a long period of time.

In evaluating birth control techniques, we must look at more than just their effectiveness. Some techniques are more suitable for certain subgroups of the population than others. The birth control pill is highly effective, but I would not recommend it for women with diabetes or cardiovascular problems, for example. Some people only wish to postpone pregnancy and therefore require a reversible technique (i.e., one that can be discontinued at any time), while others wish to totally prevent any more pregnancies and require a permanent technique. Factors such as spontaneity and aesthetics sometimes play a role, and some people may

perfect-use pregnancy rate   For a particular birth control technique, the percentage of pregnancies during the first year of use by couples who use the technique properly and consistently.

typical-use pregnancy rate   For a particular birth control technique, the percentage of pregnancies during the first year of use by all couples who use the technique, regardless of whether or not they use it properly and consistently.

be looking for simultaneous protection against sexually transmitted diseases. In many cases, religious beliefs must be considered. I will provide as much information about each of the techniques as possible, pro and con, so that whenever you decide to use birth control, you will be able to decide which method is best for you.

CONTRACEPTIVE MYTHS

There is a lot of peer pressure to engage in sex during the adolescent years. In addition to the pressure to conform, teens are often given a great deal of misinformation about contraception by friends and potential partners. Some of the information is so incorrect that it can be considered myth. For example, how many of you were told "You can't get pregnant the first time you have sex" or "You can't get pregnant if you don't kiss"? Well, it just is not true. A girl is capable of reproduction as soon as she starts having menstrual cycles (recall that in the United States, the first menstrual period occurs at an average age of 12½). This is evidence that she is ovulating, i.e., producing eggs. Some of you may have been told that sexual intercourse is safe if the female is on top, or if you have sex while standing up, but these statements are also untrue. It is also not true that a woman must have an orgasm during sex in order to get pregnant. No doubt there are other stories that some of you have heard, but unless they are included in the remainder of the chapter (i.e., they have a basis in fact), chances are they just are not true.

RELATIVELY INEFFECTIVE METHODS

One of the most popular ways in which young people try to prevent pregnancy is **withdrawal (coitus interruptus),** where the male withdraws his penis just before reaching orgasm and ejaculates outside his partner's vagina. In a Gallup poll taken in the early 1990s, 32 percent of college women believed this to be an effective means of birth control. It sounds logical, for if the male does not ejaculate in the female's vagina, there will be no sperm there—right? Wrong! The Cowper's glands, you recall, secrete a few drops of fluid before a man reaches orgasm, and this fluid can contain stray sperm. Although a few studies have reported relatively low failure rates for this method (Vessey, Lawless, & Yeates, 1982), most find that nearly 20 out of every 100 couples using it conceive within the first year of use (see Hatcher et al., 1994).

"I was one that became pregnant at an early age using the withdrawal method. My boyfriend and I normally used condoms during sexual intercourse but sometimes we wouldn't bother with them."

(from the author's files)

Withdrawal is better than nothing, but it is ineffective compared to some other types of contraception. What's more, withdrawal may not be very physically or emotionally satisfying to the male or his partner. Even in cases where a couple does not actually begin intercourse, if the male ejaculates on or near his partner's vaginal opening (during heavy petting, for example), it is possible that some sperm could get into the vaginal secretions and make their way into the vagina.

Some women believe that **douching** is an effective method of birth control. Again, it sounds logical. If you wash out the contents of the vagina, there should be no sperm. The problem is that no matter how quickly a woman douches after intercourse, some sperm will have already made it into the cervix. In fact, the pressure caused by douching can actually force sperm into the cervical opening. Over 40 out of every 100 couples who use this as their only means of birth control will conceive in a year's time. Some women have tried douching with ice water or Coca Cola after intercourse in the belief that these will kill sperm, but this, too, is ineffective because of the previously mentioned reasons (and, because of the sugars in it, Coca Cola is likely to cause a fungus or yeast infection as well). In addition, recall from Chapter 2 that frequent douching can increase the likelihood of pelvic inflammatory disease and also ectopic pregnancy (Zhang, Thomas, & Leybovich, 1997).

It is commonly believed that **breast-feeding** a newborn baby protects a woman from getting pregnant. Breast-feeding has been called "nature's contraceptive." There is some truth to this. The sucking response of the baby on the mother's nipple inhibits the release of gonadotropin hormones (FSH and LH) from the pituitary. Before modern contraceptive techniques were introduced, breast-feeding was the major factor that determined the length of the interval between pregnancies. Women normally do not start cycling regularly immediately after giving birth, and one study found that each month of breast-feeding adds, on average, about 0.4 months to the birth interval (Jain & Bongaarts, 1981). Studies in rural areas of China and India have found that when breast-feeding is frequent and prolonged, the birth interval is 5 to 10 months longer than for

withdrawal (coitus interruptus) Withdrawal of the male's penis from his partner's vagina before ejaculation in order to avoid conception. It is sometimes ineffective because fluids from the Cowper's glands may contain sperm.

douching A feminine hygiene practice of rinsing out the vagina, usually with specifically prepared solutions.

breast-feeding In reference to contraception, the sucking response by a baby on the mother's nipple that inhibits release of follicle-stimulating hormone, thus preventing ovulation.

women who do not breast-feed (Jain, 1969; Potter et al., 1965). A recent large-scale study found that if a mother is fully breast-feeding *and is not menstruating,* the chance of pregnancy during the first 6 months is less than 2 percent (Labbok et al., 1997). *The risk of ovulation rises if a woman continues breast-feeding beyond 6 months* (Gray et al., 1990).

The problem is that for this to work, the baby must be fed "on demand," that is, whenever it wants to feed, at any hour of the day or night. There can be no substituting a bottle (not even with water), and even a pacifier must be avoided as it might satisfy the baby's need to suck. Breast-feeding has been shown to work in less developed countries, where women remain at home all the time or can breast-feed in public (Jain, 1969; Potter et al., 1965), but the typical-use pregnancy rate in the United States (where most women cannot feed on demand) is very high (Hatcher et al., 1994). In other words, American couples who rely on this method quite often have babies 10 to 12 months apart.

ABSTAINING FROM SEX

The most effective way to avoid an unwanted pregnancy is to abstain from sex. Individuals who make this decision can be 100 percent worry free about getting pregnant or getting someone pregnant.

What about sex education programs that teach abstinence-only as birth control? In 1996 the U.S. government provided $250 million to give to the states to start abstinence-only programs. This approach is popular with many Americans, most of whom believe that unplanned pregnancies are a big problem and that a decline in moral standards has contributed to the problem (Mauldon & Delbanco, 1997). But how effective are these programs? California has led the way in implementing abstinence-only programs. The largest statewide program, called Education Now and Babies Later, encouraged preteens to abstain from sex until they were older. Despite the good intentions of its founders and teachers, evaluation of the program has found it to have no influence on teens' sexual decisions (Kirby et al., 1997; see Roan, 1995). "Ab-

stinence-plus" programs that teach abstinence *and* contraception, on the other hand, have often proved to be very effective in reducing teenage pregnancy rates (e.g., Tiezzi et al., 1997). Recall from Chapter 1 that most teens say they do not have enough information about birth control and that what information they do receive comes too late (Kaiser Family Foundation, 1996).

Some comparisons with other countries might be helpful here. Denmark, for example, has comprehensive sexuality education in its schools. Danish teenagers become sexually active at as young an age as American teens (half by age 17), but the vast majority use contraception, and the teenage birth rate in Denmark is among the lowest in the world (Wielandt & Knudsen, 1997). Switzerland recently promoted condoms for teens to combat AIDS. Condom use increased dramatically and did not lead to any major changes in number of sexual partners or frequency of sex (Dubois-Arber et al., 1997).

Although abstinence-only programs have not proven to be very successful, it remains true that individuals who choose to abstain will avoid unwanted pregnancies. Apparently, more teens are, in fact, making that choice. Results of a national survey released by the National Center for Health Statistics in 1997 found that, for the first time since the 1970s, there was a decline in the number of teenagers engaging in sexual intercourse.

TABLE 6–2

NUMBER OF PREGNANCIES PER 100 WOMEN DURING THE FIRST YEAR OF CONTINUOUS USE

Method	Perfect Use	Typical Use
Male sterilization	0.1	0.15
Female sterilization	0.2	0.4
Norplant	0.2	0.2
Depo-Provera	0.3	0.3
Combination birth control pill	0.1	3.0
Progestin only (mini) pill	0.5	4.0
IUD (copper)	0.8	1.0
IUD (Progestasert)	1.5	2.0
Male condom (with spermicide)	<1.0	12.0
Male condom (without spermicide)	3.0	12.0
Female condom	5.0	21.0
Diaphragm (with spermicide)	6.0	15.0
Cervical cap (for women who have not given birth)	8.0–10.0	18.0
Spermicides alone	6.0	21.0
Withdrawal	4.0	18.0
Fertility Awareness:		
Calendar	9.0	20.0+
BBT	2.0	20.0+
Billings	3.0	16.0+
No method	85.0	85.0

Source: Various sources cited in text.

FERTILITY AWARENESS: ABSTAINING FROM SEX DURING OVULATION

The **fertility awareness method** (also called the *rhythm method* or *natural family planning*) may be the only method acceptable to many of you because of your religious beliefs. These are the only birth control methods that are not opposed by the Catholic Church and evangelical Protestant denominations. There are three variations to this method, but all are based on predicting when ovulation occurs. *A female can only get pregnant when an egg is present during the first 24 hours or so after ovulation.* After that, the egg is overly ripe and a sperm cannot fertilize it. Rhythm methods involve the identification of "safe days" in a woman's menstrual cycle and the abstinence of sex during the "unsafe period." It would be almost impossible for a woman who is not having regular menstrual cycles to use these techniques. For those who do cycle regularly, they require a great deal of training and motivation.

CALENDAR METHOD

Sperm can live in a Fallopian tube for several days. Although most do not live for more than 3 days, a recent study found that some can live for as long as 5 days. Wilcox, Weinberg, and Baird (1995) calculated the following probabilities of conception (in relation to the day of ovulation):

Day	Probability
−5	0.10
−4	0.16
−3	0.14
−2	0.27
−1	0.31
0	0.33

In a woman with perfect 28-day cycles, ovulation would be at midcycle, and the safest time to have sex to avoid pregnancy would be during menstruation (yet 20 percent of college women in a recent Gallup poll believed it was dangerous to have sex during menstruation). The problem is that few women have cycles of the same length month after month. It is normal for a woman's cycle to differ in length over time by several days (see Chapter 3). Most of the differences are due to fluctuations in the preovulatory phase of the cycle. Ovulation generally occurs about 14 days before the start of menstruation, but it doesn't do any good to count back 14 days from the start of menstruation (it is too late by then, anyway). We need to know when ovulation is going to occur.

So the first thing a female must do if she is going to use this method is keep track of the length of her menstrual cycles for a minimum of eight cycles (more would be better). Let us take a hypothetical example of a woman who has kept track of ten cycles and found that although her average cycle length was 28 days, she had one that was only 24 days and another that lasted 32 days. The **calendar method** uses a formula to calculate the unsafe period. Subtract 18 from the length of the shortest cycle (24 − 18 = 6) and 11 from the length of the longest cycle (32 − 11 = 21). The unsafe period would thus be from days 6 through 21 (by the numbering system explained in Chapter 3; day 1 is the start of menstruation). This is a period of over two weeks that the woman and her partner must abstain from having sexual intercourse. If the woman had had a cycle shorter than 24 days or longer than 32 days, the period of abstinence would have been even longer.

The calendar method has proved to be very unreliable, with some studies reporting a typical-use failure rate of as high as 40 to 45 percent in one year (Ross & Piotrow, 1974). Most studies report that the typical-use failure rate is about 20 percent per year (Hatcher et al., 1994; Trussell et al., 1990). The calendar method has been called the Russian roulette of birth control: the losers get pregnant right away and the winners get pregnant later and worry in the meantime. In addition to severely testing human motivation, all it takes for the method to fail is just one cycle that is shorter or longer than previously observed. There are two variations of the rhythm method that use biological markers to better identify the time of ovulation.

BASAL BODY TEMPERATURE METHOD

A woman's basal (resting) body temperature rises by a few tenths of one degree Fahrenheit 24 to 72 hours after ovulation. In the **basal body temperature method,** couples abstain from having sexual intercourse from the end of menstruation until about four days after a temperature rise is noted. The basal body temperature should be taken at the same time every day, preferably first thing in the morning (before going to the bathroom, eating, drinking, or smoking). Note that this method still involves a long period of abstinence, so it again severely tests human willpower. Even those women who attempt to use it properly often get pregnant, for its success depends on being able to detect a very small tempera-

fertility awareness methods Methods of birth control that attempt to prevent conception by having a couple abstain from sexual intercourse during the woman's ovulation.

calendar method A fertility awareness method of birth control that attempts to determine a woman's fertile period by use of a mathematical formula.

basal body temperature The temperature of the body while resting. It rises slightly after ovulation.

ture increase (and, more importantly, is based on the assumption that the increase in body temperature is due to ovulation). If no increase is noted, a couple must go an entire cycle without sexual relations. What's more, small temperature increases of a few tenths of a degree are often caused by low-grade infections that we are not even aware of, by stress, or by other factors. Some women have a history of irregular basal body temperature. The typical-use failure rate is about 20 pregnancies per every 100 couples per year (Hatcher et al., 1994; Trussell et al., 1990).

BILLINGS METHOD (CERVICAL MUCUS OR OVULATION METHOD)

The fertility awareness method that is most widely used today is the **Billings method,** a method that attempts to pinpoint the time of ovulation by noting changes in the consistency of a woman's **cervical mucus.** It is named after two Australian doctors who developed it (Billings, Billings, & Catarinch, 1974). Mucus is discharged from the cervix throughout a female's menstrual cycle, and its appearance changes from white (or cloudy) and sticky to clear and slippery (like that of an egg white) a day or two before ovulation. Until recently, the only way to note this change was a finger test, that is, placing a finger in the vagina and then examining the mucus. Women who use this method should not douche, for that makes it more difficult to notice the changes in consistency of the mucus. A couple is instructed to abstain from sexual intercourse from the end of menstruation until four days after the mucus has changed consistency.

If you wish to use this method, do not rely on the preceding simple instructions. There are organizations that will give you more detailed instructions. Be warned, however, that some of these are advocate organizations (i.e., they strongly favor this form of birth control over others) that may exaggerate the effectiveness of the Billings method. Theoretically, it has only a 2 percent pregnancy rate, but the typical-use failure rate is much higher. The Alan Guttmacher Institute reported that the accidental failure rate for married women in 1982 was 16 percent, but most studies have found about 20 pregnancies per year for every 100 couples using this method (Hatcher et al., 1994; Medina et al., 1980; Trussell et al., 1990; Wade et al., 1980). Similar high failure rates obtained in carefully designed studies done in five different countries led the World Health Organization to conclude that the Billings method was "relatively ineffective in general use for preventing pregnancy" (Seventh Annual Report, 1978; see also World Health Organization, 1981).

Advocate groups generally dismiss these high failure rates as due to improper instructions and unmotivated participants (i.e., user failures due to people having sex during unsafe periods). Most of the pregnancies do, in fact, result from conscious deviations from the rules (World Health Organization, 1981), no doubt because the average period of abstinence is 17 days per cycle. Recent data from a World Health Organization clinical trial indicate that the first-year failure rate is, in fact, only 3.1 percent during *perfect use* if women also abstain from sex during stress, but 86.4 percent during imperfect use (Trussell & Grummer-Strawn, 1990). Thus, the Billings method can be very effective if used perfectly, but perfect compliance is difficult for many couples who desire sexual intercourse.

Some couples combine the basal body temperature and Billings methods (along with other signs of ovulation, such as ovulatory pain) to enhance the effectiveness of these fertility awareness methods. This is often referred to as the *sympto-thermal method* and has a lower pregnancy rate than any of the other methods used alone (Medina et al., 1980; Wade et al., 1980). However, one of the problems with any method based on changes in cervical mucus is that perhaps as many as one-third of all women do not have the typical mucus pattern. Even in those who do, the finger test is obviously not very scientific. Not surprisingly, fewer than 2 percent of all Americans use fertility awareness for birth control (Peterson, 1995).

There is real hope for those who want to use the fertility awareness method. Chemical testing kits that will more accurately pinpoint the time of ovulation (by measuring a vaginal mucus enzyme or sodium and potassium levels in saliva) may soon be available in stores for home use. Because sperm can live in the Fallopian tubes for 2 to 5 days, the chemical tests would have to predict ovulation at least 5 days in advance to be useful as a contraceptive technique (there are already some products available that predict ovulation 12 to 36 hours in advance, but these are useful only for couples who want to conceive).

SPERMICIDES: SUBSTANCES THAT KILL SPERM

Spermicides are chemicals that kill sperm (nonoxynol-9 or octoxinol-9). They are sold in a variety of forms. Spermicidal foams, jellies, creams, and film (VCF, or *vaginal contraceptive film,* is a square of film that is placed over the cervix) help hold sperm-killing chemicals in the vagina against the cervix and simultaneously act as a barrier to sperm. Suppositories and tablets are

Billings method A fertility awareness method of birth control in which changes in the consistency and quantity of cervical mucus are used to tell when ovulation has occurred.

cervical mucus The slimy secretion of mucous membranes located inside the cervix.

spermicides Chemicals that kill sperm. In most products, the chemical is nonoxynol-9.

also available. Specific instructions differ for each type of spermicide, but all of them must be placed in the back of the vagina shortly before sexual intercourse begins. They lose their effectiveness over time, so new spermicide must be inserted shortly before each time a woman has intercourse. Be very careful not to confuse spermicidal products with feminine hygiene products. They are often displayed together in drug stores, so if you want the contraceptive material, be sure that the package says "spermicide" or specifies that the product "kills sperm."

Spermicides came on the market before companies were required to prove the effectiveness of new drugs. In 1996 the Food and Drug Administration threatened to make manufacturers prove the effectiveness of spermicides. To date, surveys have found the typical-use rate to be about 21 pregnancies per 100 couples per year (Hatcher et al., 1994; Trussell et al., 1990), but many of these pregnancies are due to improper or inconsistent use (user rather than method failures). One survey found that when spermicides are used consistently and properly, the failure rate can be as low as 6 percent (Harrison & Rosenfield, 1996). However, it is believed (but not proven) that foams and suppositories are more effective than jellies and creams. Women should consider using spermicides in combination with a blockade or barrier method (diaphragm or condom). Some brands of condoms already contain a spermicide.

In addition to their contraceptive benefit, spermicides containing *nonoxynol-9* reduce the risk of contracting gonorrhea and chlamydia by killing the bacteria that cause them (see Box 6–A). The use of spermicides may also reduce the risk of cervical cancer, probably by helping to kill human papillomavirus (see Chapter 5; Celentano et al., 1987). One study suggested that spermicides were as effective as condoms in preventing sexually transmitted diseases (Rosenberg & Gollub, 1992), but other researchers seriously doubt this. They worry that although spermicides kill HIV in test tubes, they might irritate the vagina and make transmission of HIV easier (e.g., Bird, 1991; Cates, Stewart, & Trussell, 1992).

On the negative side, several studies have found that spermicides containing nonoxynol-9 (including condoms with spermicide) increase the risk of urinary tract infection (Fihn et al., 1996; Hooten et al., 1996; McGroarty, Reid, & Bruce, 1994). However, another study found that condoms with spermicide reduced the risk compared to unlubricated condoms (Foxman et al., 1997). Even if it should turn out to be true that spermicides increase the risk of urinary tract infections in women, avoiding spermicides for this reason must be weighed against the risk of unwanted pregnancy and exposure to sexually transmitted diseases. A few early studies suggested a small increase in the risk of birth defects if spermicides are used at the time of conception or during pregnancy. However, the authors of the most frequently cited study (Jick et al., 1981) repudiated their own results (Watkins, 1986), and most experts believe that spermicides do not cause birth defects (Einarson et al., 1990; Lovik et al., 1987).

Most college women rate spermicides as one of their least preferred types of contraception (Sarvela et al., 1992). The most common complaint is that, like the condom and the diaphragm, inserting the spermicidal material just before intercourse interferes with the spontaneity of sexual relations and detracts from the mood. Others complain that spermicides are messy or detract from oral-genital sex. A few people (1 in 20) find that spermicides burn or irritate the vagina or penis. However, as a backup to barrier techniques, and because of their antiviral and antibacterial activity, spermicides have advantages that for many women may outweigh the disadvantages.

BARRIER METHODS: PREVENTING SPERM FROM MEETING EGG

Barrier methods of birth control are designed to prevent pregnancy by placing a blockade between the penis and cervix so that sperm cannot reach the egg if ovulation has occurred. Thus, these methods do not require you to abstain from sexual intercourse, but they may interfere with the spontaneity of sexual relations by requiring you to stop and put something on or put something in.

MALE CONDOMS

Condoms ("rubbers," "safes," or prophylactics) are thin sheaths made of latex rubber, lamb intestine, or polyurethane that fit over the penis and thus trap the sperm (see Figure 6–4). They also serve as a barrier between the male's membranes and female's membranes, and therefore *are also highly effective in preventing the spread of sexually transmitted diseases*. In fact, condoms were originally invented for this purpose in the 1500s. A French writer in 1761 described condoms as "armor against love, gossamer against infection." Mass production of condoms became possible in the mid-1800s with the discovery of how to make rubber more flexible and durable. Today, condoms are the most frequently used method of birth control in Japan, England, and some northern European countries, but for a variety of reasons (to be discussed later), American men have been reluctant to use them.

As a birth control technique, condoms are highly effective. They present no major

barrier (blockade) methods General term for contraceptive methods that block the passage of sperm.

condom *For males*, a thin sheath made of latex rubber, lamb intestine, or polyurethane that fits over the penis.

health hazards and are available without a prescription. What's more, their use requires no mathematical calculations and no subjective evaluation of biological processes. There are only a few simple rules to remember:

1. Put the condom on before you start having intercourse. Otherwise, it is no more effective than withdrawal. If the condom does not have a nipple tip, leave a little extra space at the tip of the penis to catch the ejaculate. If you are uncircumcised, pull back the foreskin while putting on the condom.
2. Hold on to the base of the condom when you are finished and withdraw. Otherwise, you might leave it and its contents in the female's vagina.
3. Use a condom only once. If you have intercourse more than one time while having sex, use a new condom each time. Some people say that you can wash and dry them, powder them with cornstarch, and then roll them up for reuse, but they will eventually tear, so for maximal effectiveness use them only once. After all, even condoms made from animal membranes are not terribly expensive.
4. Do not store rubber condoms for long periods of time in a warm place (such as your wallet) or where they are exposed to light. This will eventually deteriorate them. Store them someplace cool, dark, and dry. Also, *do not use rubber condoms in combination with mineral oil, baby oil, vegetable oil, hand lotions, Vaseline, or other petroleum jellies,* for these, too, can damage them.

"During a recent sexual experience, my partner and I used oil from one of those fancy 'bath and body stores' for lubrication in conjunction with a condom. During intercourse, the condom disintegrated."

(from the author's files)

A man should never begin intercourse until his partner is lubricating, but if additional lubricant is needed, use a water-based one (do not confuse "water-based" with "water-soluble") like K-Y Jelly or Today Personal Lubricant. Better yet, use a spermicide. (The new polyurethane condoms can be used with oil-based lubricants.)

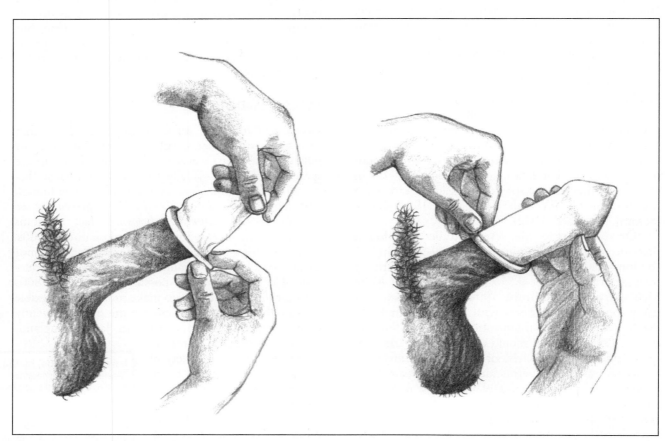

Figure 6–4   Use of the condom.

Condoms come rolled up, dry or lubricated, and until recently they came in only one size. However, there is now a large size for men who find the usual type too tight. (American manufacturers had been reluctant to make different-sized condoms for fear of insulting a male's ego—can you imagine a man asking for a small condom?) A man may wish to practice putting a condom on his penis if he has never used one before. They are unrolled on the penis, not yanked on like a sock.

If these very simple instructions are followed, the theoretical failure rate is only about 3 percent (Hatcher et al., 1994; Trussell et al., 1990). If always used with a spermicide, the theoretical failure rate is less than 1 percent (Kestelman & Trussell, 1991). However, the typical-use rate is about 12 pregnancies per 100 couples per year. Most of these pregnancies are due to inconsistent use, and teens are more likely than older persons to use condoms inconsistently. One study found that the failure rate for women over 30 (almost all of whom used condoms consistently) was only 1–4 percent, but for women under the age of 25 the failure rate was between 10 percent and 33 percent (Schirm et al., 1982).

Not all pregnancies are due to user failures. *Consumer Reports* (May, 1995) found that several brands of condoms failed their laboratory strength tests, including six varieties of the best-selling brand, Trojans. Many brands, however, were sufficiently strong (see *Consumer Reports*). Several studies have found that during human use, the breakage rate is about 1–3 percent during vaginal intercourse (see Cates & Stone, 1992a, 1992b; Grady & Tanfer, 1994; Messiah et al., 1997), but others have reported breakage rates of over 5 percent (Russell-Brown et al., 1992; Steiner et al., 1993; Trussell, Warner, & Hatcher, 1992). The breakage rate may be considerably higher during anal intercourse (Silverman & Gross, 1997). It must be noted that condom failure due to breakage is most common among men who do not use condoms regularly (Messiah et al., 1997; Steiner et al., 1993), indicating that inexperienced men are probably not putting them on properly. The use of two condoms during sex reduces the breakage rate to near zero (Rugpao et al., 1997). Slippage of condoms occurs in about one percent of cases (Messiah et al., 1997), often due to penetration beyond the end ring or improper withdrawal (see rule no. 2). The breakage and slippage rates for the new polyurethane condoms are about the same as for rubber condoms (Rosenberg et al., 1996).

Remember, the failure rate of condoms can be even further reduced by using them in combination with a spermicide. Many brands of condoms come with a spermicide in their lubricant, but using a vaginal spermicide with condoms is much more effective (*Consumer Reports*). When used with spermicide, condoms are the most effective of the nonprescription birth control techniques.

Condom use by teenagers more than doubled during the 1980s (Sonenstein, Pleck, & Ku, 1989; see also Peterson, 1995), but most Americans at risk for getting a sexually transmitted disease still do not use them regularly (Catania et al., 1995).

If condoms are so effective, why aren't all sexually active men using them? Many men do not want to use condoms because they say condoms reduce their sensitivity ("like taking a shower with a raincoat on").

> **"Rubbers seem to make getting laid not feel as good as bare skin sex. Is this the price to pay for not having an STD?"**
>
> (from the author's files)

Here are some lines that women in my class have told me that men commonly use to keep from using a condom:

> **"I promise I will not come in you."**
>
> **"I will pull out before I nut."**
>
> **"Nothing is going to happen."**
>
> **"It doesn't feel the same."**
>
> **"I can't feel anything with it on."**
>
> **"I don't have a disease."**
>
> **"If you're not having sex with anybody else, why should I use it? If you trust me, you wouldn't make me use it."**
>
> **"We know each other deeply. I love you."**
>
> (from the author's files)

Before they have ever tried one, most men have heard many times that condoms will reduce their sensitivity. Their preconceived notions probably exaggerate the problem. Condoms will reduce a male's sensitivity—slightly—but *Consumer Reports* (1989) found that many men eventually find this to be something positive because it makes sex last longer. The new polyurethane condoms are only half as thick as rubber condoms and transmit heat better than rubber. Men generally prefer these to rubber condoms in regard to comfort, sensitivity, lack of smell, and natural feel (Rosenberg et al., 1996).

Others complain that they are allergic to rubber. About 1 to 3 percent of men and women are allergic (Hatcher et al., 1994), but again, some condoms are made of natural animal membranes or polyurethane. Still others complain about the loss of spontaneity; that is, that it is a nuisance to stop and put a condom on in the middle of having sex. How long does it take to put on a condom? The answer is, only a few seconds. These reasons for not wearing condoms are in many cases just excuses. The real reason many men won't use condoms is that they are too embarrassed to go into a store and purchase them.

> "During high school, I was actually embarrassed to buy condoms. I only would get them in those bathroom machines. Perhaps if they were discretely available, then more teens would buy them."
>
> (from the author's files)

A few women in my classes have said that they have talked a partner out of using a condom, but the large majority have to talk partners into using them. By the way, here are some lines that women say they have used successfully to get a male partner to use a condom:

> "No condom, no sex."
>
> "If you choose not to use a condom, you're choosing not to have sex with me."
>
> "If you don't use one, you won't get none."
>
> "Boy, I'm not having your baby."
>
> "Do you want your nickname to be Daddy?"
>
> "It's not that I don't trust you; I don't trust who you've been with."
>
> "There will be no action in the ballpark until the player with the bat puts his glove on."
>
> (from the author's files)

However, if a man refuses to use condoms or get a vasectomy (usually done only on older men who want no more children), this generally means that his female partner must assume the responsibility for birth control. In fact, 40 to 50 percent of condoms are purchased by women today. Putting on a condom takes less time than putting in a diaphragm or cervical cap, and it isn't as messy as most vaginally used spermicides. The latter three must be purchased at a store just like condoms, and use of a diaphragm requires that a woman have a pelvic exam. Although many men may complain about loss of spontaneity and reduced sensitivity, these are just minor annoyances compared to the

side effects often associated with use of the pill or an IUD. Whatever excuses may be used, there is no excuse for not using a condom if the only alternative is unprotected intercourse that may result in unwanted pregnancy, not to mention the possible spread of sexually transmitted diseases.

There are certain subgroups of the population for which I highly recommend use of the condom. If you are sexually active and have multiple partners over time, or if your partner has had multiple partners or is an intravenous drug user, you are in a high-risk category for contracting a sexually transmitted disease. STDs are in epidemic proportions today. The walls of latex condoms prevent the bacteria and viruses that cause sexually transmitted diseases from passing through them. Although skin condoms are stronger (and thus are very effective for contraception), they have small pores that may allow passage of the hepatitis B virus and the virus that causes AIDS. Thus, the latex condom is highly effective for both birth control and STD prevention (see Chapter 5 and Box 6–A for additional information).

MALE RESPONSIBILITY IN AVOIDING UNWANTED PREGNANCIES

In many countries around the world, male contraceptive techniques are the most popular. Thus, men as well as women take responsibility for birth control. In the United States, however, the responsibility of avoiding an unwanted pregnancy has traditionally fallen on women. Recently, the Society for Black Psychologists in New Orleans developed an attitude questionnaire for adolescent males designed to get them to assume more of the responsibility. The students are assembled and given the following statements, to which they respond "True" or "False":

1. Getting a girl pregnant makes you a man.
2. Many adolescent males feel that if a girl becomes pregnant it is her fault and responsibility.
3. Many males believe that if a girl has sex with him she would have sex with other boys.
4. Many males feel that any boy who refuses to have sex with a girl who is willing is a chump.
5. If an adolescent male remains a virgin, his friends think something is wrong with him.
6. All teenage mothers return to school and graduate.
7. A teenage father's education is interrupted as much as the teenage mother's education.
8. Most teenage fathers take care of their babies.
9. Males have the greatest respect for girls who become pregnant.
10. Most teenage males don't care if their teenaged sisters get pregnant.
11. Most teenage males don't care if your teenaged sister gets pregnant.   (Reprinted by permission.)

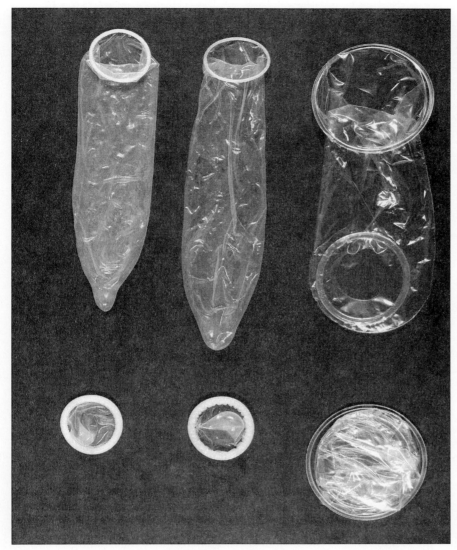

woman inserts it with her fingers by pushing the closed-ended ring against the cervix, as she would a diaphragm. It is held in place by the outer ring, which fits on the outside of the vagina, partially covering the labia (see Figure 6–6). It can be inserted up to 8 hours before having sexual intercourse.

Many people are put off by the appearance of the female condom, but it is thinner than male condoms, feels softer than rubber, and transfers heat, and thus many of those who try it say that it feels more like unprotected intercourse than sex with male condoms.

"It looked so unromantic I was sure I wasn't going to like it. Were we surprised! My boyfriend quit using male condoms now. There's a lot more friction and it feels more natural. . . ."

(from the author's files)

As with the diaphragm, the female condom raises no major health concerns and does not have adverse effects on the vaginal walls (Soper, Brockwell, & Dalton, 1991). However, female condoms are considerably more expensive than male condoms and, like male condoms, are used for only one act of intercourse.

Many women know about HIV and other sexually transmitted diseases but are nevertheless reluctant to talk about safer sex (e.g., using condoms) on dates. For those women, and for those whose partners refuse to wear a male condom, the female condom provides them with the opportunity to protect their own health. The typical-use pregnancy rate for female condoms was initially reported to be about 25 percent per year (FDA hearing, 1992), due largely to improper use. Women apparently are receiving better instructions now, for recent studies have found that the pregnancy rate during typical use was the same as for the diaphragm, sponge, or cervical cap (Farr et al., 1994; Trussell et al., 1994). For those who use the female condom correctly, the current one-year

Figure 6–5 Male condoms (left and middle) are sold rolled and in two sizes. Female condoms (right) are intravaginal pouches that are held in place by two flexible rings.

A response of "True" to any of the statements is then followed by further discussion. Rates for teenage pregnancy will probably remain high as long as males continue to pressure females to engage in sex without any regard for long-term consequences.

THE FEMALE CONDOM

Only recently, in 1993, did we see the introduction of the first woman-controlled barrier method of contraception designed to give simultaneous protection against sexually transmitted diseases (see Gollub & Stein, 1993). The **female condom** (trade name: Reality), as it is called, is actually an intravaginal pouch—a 7-inch-long polyurethane bag that is held in place in the vagina by two flexible rings (see Figure 6–5). A

condom For females, a polyurethane intravaginal pouch held in place by two flexible rings. Condoms are effective as contraception and for prevention of sexually transmitted diseases.

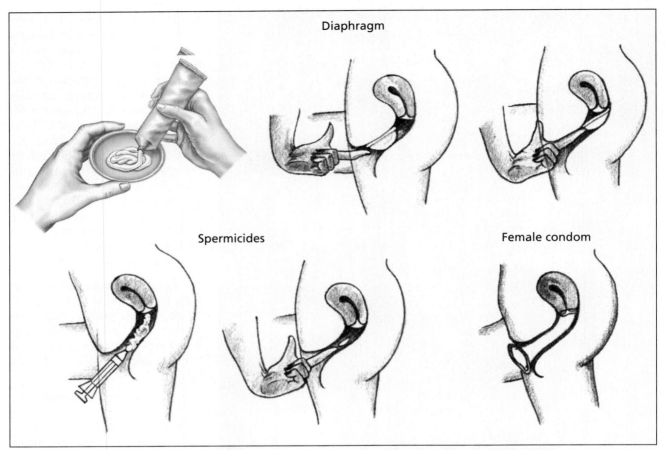

Figure 6–6 Use of the diaphragm (with spermicide), spermicides (foam and suppository), and female condom.

pregnancy rate is about 5 percent (Harrison & Rosenfield, 1996).

The Reality Female Condom is just the first of several new female barrier techniques to reach the market. Many others are in various stages of development (see section on future technology).

THE DIAPHRAGM

A **diaphragm** is a large, dome-shaped rubber cup with a flexible rim that fits over the cervix. It thus works by preventing the passage of sperm into the uterus. In biblical times, women sometimes inserted camel or crocodile dung into the back of the vagina to prevent the sperm's passage. Fortunately for today's women, the rubber cup was invented in 1882 in Germany. However, because of laws that prohibited the mailing or transportation of "obscene" materials (this included birth control devices), the diaphragm was not used in the United States until the 1920s. By the 1930s, the diaphragm was being used by one-third of all couples who practiced

diaphragm A dome-shaped rubber cup with a flexible rim that fits over the cervix and thus acts as a contraceptive device by serving as a barrier to the passage of sperm into the uterus.

birth control. With the introduction of the birth control pill in 1960, the use of the diaphragm became almost nonexistent. There was a small resurgence in its popularity in the 1970s, but its use among single women declined sharply again in the 1980s (Mosher, 1990).

A woman has to be fitted for a diaphragm by a physician or health care worker (the distance between the cervix and pubic bone is measured). She will then be given a prescription for a diaphragm that fits her. It is not something you borrow from a girlfriend for a date. You may have to be refitted if your body undergoes any major changes (e.g., pregnancy or weight changes of 10 pounds or more). A woman getting her first diaphragm is also going to need instructions and coaching to learn how to insert it properly and quickly (see Figure 6–6). This may be one reason there was a decrease in the use of the diaphragm, for it was easier for doctors to prescribe pills than to take time to teach a patient how to use a diaphragm.

After you have learned to insert your diaphragm properly, there are only a few simple rules to remember:

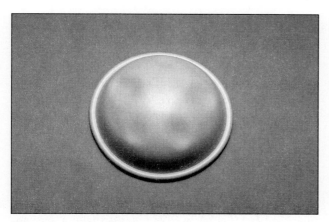

Figure 6–7 The diaphragm.

The cervical cap.

1. As with condoms, diaphragms should be used with a spermicidal jelly. This recommendation is almost universal (Hatcher et al., 1994). Although two studies have found that spermicides do not further reduce the pregnancy failure rate (Craig & Hepburn, 1982; Ferreira et al., 1993), spermicides are nevertheless beneficial in helping to kill bacteria and viruses that cause sexually transmitted diseases (see Box 6–A). Spread spermicidal jelly on the inside part of the diaphragm (the part in contact with the cervix) and rim before insertion.

2. Insert the diaphragm no more than 2 hours in advance of starting intercourse. Otherwise, the spermicide dissipates and adds little additional protection. For maximal effectiveness, the diaphragm should be put in immediately prior to intercourse, although this may interfere with spontaneity.

3. Leave the diaphragm in for 6 to 8 hours after intercourse to make sure that there are no more live sperm. If you have intercourse more than once while having sex, place more spermicide in the vagina before each time. Do not remove the diaphragm to do this (sperm may get into the cervix while the diaphragm is removed), but instead use a plunger or some other means of application. Do not leave a diaphragm in for more than 12 hours.

When used in combination with a spermicide, the perfect-use pregnancy rate for the diaphragm is usually stated as about 6 percent per year (Harrison & Rosenfield, 1996). The typical-use pregnancy rate is about 15 percent per year (Trussell, Strickler, & Vaughn, 1993). Most of these pregnancies, particularly among young women, are due to inconsistent use. Because of inconsistent use, one study found that the pregnancy rate among young women was 21 percent in one year (Kovacs et al., 1986), but in a group of older married

women it was well below 5 percent (Vessey, Lawless, & Yeates, 1982). A recent study found that when the diaphragm was used continuously (without spermicide, removing it only during a daily shower to wash it, with immediate reinsertion) the failure rate was only 2.8 percent (Ferreira et al., 1993). Some of the failures are also due to improper insertion or to the diaphragm's becoming dislodged during vigorous intercourse (this happens most often when the female is on top).

Women with a history of urinary tract infections should not use a diaphragm, especially with a spermicide (Hooten et al., 1996). The diaphragm also should not be used during menstruation or periods of abnormal vaginal discharge. A few women may have allergies to rubber or spermicide. The most common complaint about the diaphragm is aesthetics. Some complain that insertion is a nuisance, interfering with spontaneity, and that spermicides can be messy and detract from such sexual behaviors as oral-genital sexual relations. These complaints are not heard as often, however, in women who are sexually experienced and in a monogamous relationship.

"For years I used a diaphragm. My partners always said that they could not feel it in me and that it did not take away from their pleasure. I also could not feel it if it was inserted properly. The only problem I found was that I felt like I needed a suitcase to carry everything when I went out—the diaphragm with case, gel, and the applicator. . . ."

(from the author's files)

The most important thing to consider is that the diaphragm offers an inexpensive, safe, and relatively effective means of birth control when used consistently and properly.

THE CERVICAL CAP

The **cervical cap** is another barrier device that is designed to prevent passage of sperm from the vagina into the uterus. Made of latex rubber or plastic, it is smaller and more compact than a diaphragm and resembles a large rubber thimble (see Figure 6–7). *It is used with a small amount of spermicide* (for added effectiveness) and fits over the cervix by suction. Insertion and removal of the cap are more difficult than for the diaphragm, but once in place it is more comfortable than the diaphragm and can be left in for 24 hours. Women should check each time after having sexual intercourse to make sure that the cap has not been dislodged.

Although the cervical cap has been popular in England for years, it has not been used very much in this country. In fact, it did not receive approval for marketing in the United States until 1988 because of concerns about prolonged exposure to secretions trapped by the cap (including cervical mucus and menstrual flow). However, after a seven-year trial process, the Food and Drug Administration concluded that the cap does not cause toxic shock syndrome. It is presently recommended that the cap be used only by women who have normal Pap smears because of possible adverse effects on cervical tissue (Mishell, 1989).

The one-year typical-use failure rate for the cervical cap is listed as 18 pregnancies per 100 women, but two studies (Cagen, 1986; Powell et al., 1986) found the typical-use rates to be 8 percent and 14 percent. The cap offers substantially better protection to women who have not yet given birth (Trussell et al., 1993b). For these women, the pregnancy rate is only 8 to 10 percent when the cervical cap is used correctly.

BARRIER METHODS AND SPONTANEITY

All of us, I suppose, would like our sexual relations to be passionate and spontaneous, with no interruptions. Putting on a condom or inserting a diaphragm or spermicide does not take very long for many sexually experienced people—some individuals even integrate the use of barrier methods, e.g., putting on a condom, as part of foreplay—but some couples occasionally take chances and have unprotected intercourse because they do not want to stop and possibly ruin the mood. This is probably the most common cause of failure for barrier methods. The contraceptive techniques that we will cover next get around this problem by not requiring anything be done at the time when you want to have sex (except getting your partner in the mood, of course).

cervical cap A contraceptive device that fits over the cervix by suction, thus blocking the passage of sperm into the uterus.

intrauterine device (IUD) A birth control device, usually made of plastic with either a copper or progesterone coating, that is placed in the uterus to prevent conception and implantation.

THE IUD

The **IUD (intrauterine device),** like the birth control pill, came on the market in the 1960s and helped to open a new era of what people thought was going to be spontaneous and worry-free sex. However, there were numerous health problems caused by a few of the early IUDs, and as a result of the subsequent lawsuits, most were taken off of the market. Before we continue with this part of the story, let us examine how the IUD works.

The IUD is a small plastic and/or metal device (of various shapes) that is placed into the uterus by a physician (see Figure 6–8). This is actually a very ancient idea—camel drivers used to insert a stone into a female camel's uterus to prevent pregnancy (don't ask me how they came up with this idea). It has long been believed that IUDs work as birth control by preventing implantation of a fertilized egg, and thus this method was not acceptable to Roman Catholics and others for moral reasons. However, several studies have found that IUDs work primarily by preventing fertilization, mostly by their effect on sperm transit through the uterus (Bromham, 1993; Senanayake, 1990; Sivin, 1989). It is very possible that IUDs work for both of these reasons (Spinnato, 1997). The IUDs sold in the United States today have either a copper or proges-

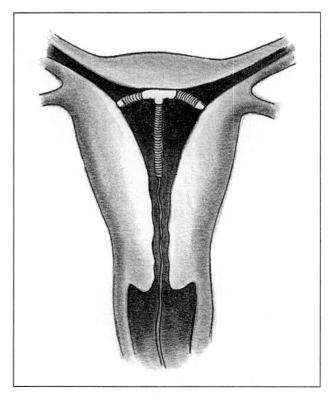

Figure 6–8 An IUD (intrauterine device), inserted.

terone coating, both of which further impair the passage of sperm. These IUDs are highly effective, with typical-use failure rates of less than 2 percent per year (Hatcher et al., 1994). The copper T 380 A IUD can be used effectively for 10 years (Population Council, 1994).

Now back to the story of the early IUDs. Insertion of an IUD requires dilation of the cervical opening, and insertion of some types of IUDs was particularly difficult (the uterus was occasionally punctured during insertion). Once inserted, the woman's uterus often attempted to expel the foreign object. Because some women did not notice that their IUD had been expelled, manufacturers added a thread to the IUD that protruded from the cervix into the vagina so that women could check to see if their IUD was still in place. However, the thread served as an easy path for bacteria and viruses to follow (Vessey et al., 1981). The spread of gonorrhea and chlamydia into the upper reproductive tract, causing pelvic inflammatory disease, was especially rapid in women with IUDs. Pelvic inflammatory disease left many women sterile or with an increased chance of a tubal pregnancy. Another major problem with some older types of IUDs was that implantation of a fertilized egg did occasionally occur. Although not as serious as these problems, the most common complaint of users of some older types of IUDs was heavier and longer menstrual flow, with more cramps and spotting.

Most of these health risks were limited to one type of IUD, the Dalkon Shield (see Sivin, 1993). The company that manufactured it had had no previous experience in contraceptive technology, and despite early warnings from physicians that it was difficult to insert and complaints from users, the company marketed it anyway. Many pregnancies and infections, and even some fatalities, followed. The company filed for bankruptcy after hundreds of millions of dollars in lawsuits were filed. Most of the other IUDs available then were not associated with major health problems, but their manufacturers nevertheless removed them from the market.

The two types of IUDs on the market today are regarded by most physicians and researchers as safe products (e.g., Grimes, 1992; Newton & Tacchi, 1990; Petitti, 1992). The progesterone (Progestasert T) and copper-coated (Copper T) IUDs used today have a single-filament polyethylene string and are *not* associated with an increased risk of pelvic inflammatory disease (PID), ectopic pregnancy, endometrial cancer, infertility, or increased bleeding or pain, and there have been no fatalities (Chi, 1993b; Cramer et al., 1985; Datey et al., 1995; Ebi et al., 1996; Rosenblatt et al., 1996; Sivin, 1993). They also have a lower expulsion rate. The only real risk of PID associated with the IUD is at insertion (due to bacteria), and this risk can be elimi-

nated by better sterilization procedures (Farley et al., 1992; Lee et al., 1983). Of those women who use an IUD, 97 percent have a favorable opinion about it, the highest for any contraceptive technique (Forrest & Fordyce, 1988).

> **"After the birth of my second child, my doctor strongly advised no more children for a while. I was only 24 years old, so my doctor suggested a three-year IUD. It was very easily done in the doctor's office and my doctor advised me it was safe to have sex right away. It was a wonderful method of birth control...."**
>
> (from the author's files)

To be absolutely safe, many physicians advise women who have multiple sexual partners over time to use a contraceptive technique other than the IUD (Mishell, 1989). The IUD cannot protect a woman from her own, or her partner's, sexual behavior. But for women in a stable, monogamous relationship who have completed their families, the IUD is a safe, highly effective method of contraception that also permits spontaneity in sexual relations.

HORMONAL METHODS OF CONTRACEPTION

THE BIRTH CONTROL PILL (ORAL CONTRACEPTION)

Of the approximately 35 million women who practice birth control in the United States, nearly 14 million use the birth control pill (Hatcher et al., 1994), making it by far the most popular of the reversible methods of contraception. It is also the most effective of reversible nonsurgical techniques, with a perfect-use pregnancy rate of close to zero. The typical-use pregnancy rate is three pregnancies per year for every 100 couples using the pill (Hatcher et al., 1994; Trussell et al., 1990). Most of the pregnancies that do occur are due to user, not method, failures. If a woman consistently follows just one simple rule, she really does not need to worry about getting pregnant. What is that? Take the pill every day. However, *the birth control pill is not effective immediately,* and a backup method of birth control should be used during the first month of use. After that, even if a woman forgets occasionally, taking two pills the next day will probably compensate for the missed pill, but any further forgetfulness will certainly result in unprotected intercourse. If you forget to take the pill for a few days, you again must use a backup method of birth control until you have taken daily an entire package of pills. Women, as well as their doctors, should also be aware that *the pill is rendered less effective by simultaneous use of many types of antibiotics, barbiturates, analgesics, and tranquilizers.*

"During our senior year in high school, my cousin came to me because she was afraid that she was pregnant. She was on the pill, and she was taking them correctly. We could not figure out what went wrong. . . . About a month before she became pregnant, she went to the doctor for a bad cold. It turned out that the medicine which was prescribed for the cold made the birth control less effective. "

(from the author's files)

There is really not just one pill, but several different types. In fact, there are over 30 types marketed in the United States. The most widely used and most effective are the **combination pills,** which contain synthetic estrogen and synthetic progesterone (**progestins**). Most contain a fixed amount of the two hormones in each capsule, but there are also combination pills that adjust the levels of progestins in an attempt to mimic the natural hormonal phases of the menstrual cycle (although different manufacturers of this type of pill cannot agree on what is "natural"). Regardless of the type, the pill is taken for 21 days and then discontinued for 7 to permit menstrual bleeding. Most companies package the 21 pills containing hormones with seven additional pills that contain no hormones (placebos), so that women will not get out of the habit of taking a pill every day.

The combination pill works primarily by preventing ovulation. The hormones in the pill prevent the pituitary from releasing FSH and LH (see Figure 6–9). Without an egg present, it is impossible to get pregnant. Even if ovulation should somehow occur, the progestin inhibits the development of the uterine lining (making implantation difficult) and keeps the cervical mucus thick (making passage of the sperm difficult). When a woman decides to stop using the pill, her menstrual cycles may not become regular right away, but there is no long-term effect on fertility (Chasen-Taber et al., 1997; Huggins & Cullins, 1990).

There is also a **minipill** that contains only progestins. It works primarily by inhibiting development of the endometrium and by keeping cervical mucus thick. This pill is for women who are breast-feeding (estrogen inhibits milk production) or who cannot tolerate the side effects of the estrogen-containing pill. It is less effective than the combination pill (1.4–4.3 pregnancies per 100

combination pill An oral contraceptive that contains both synthetic estrogen and synthetic progesterone.

progestin A synthetic form of progesterone.

minipill An oral contraceptive that contains only progestins.

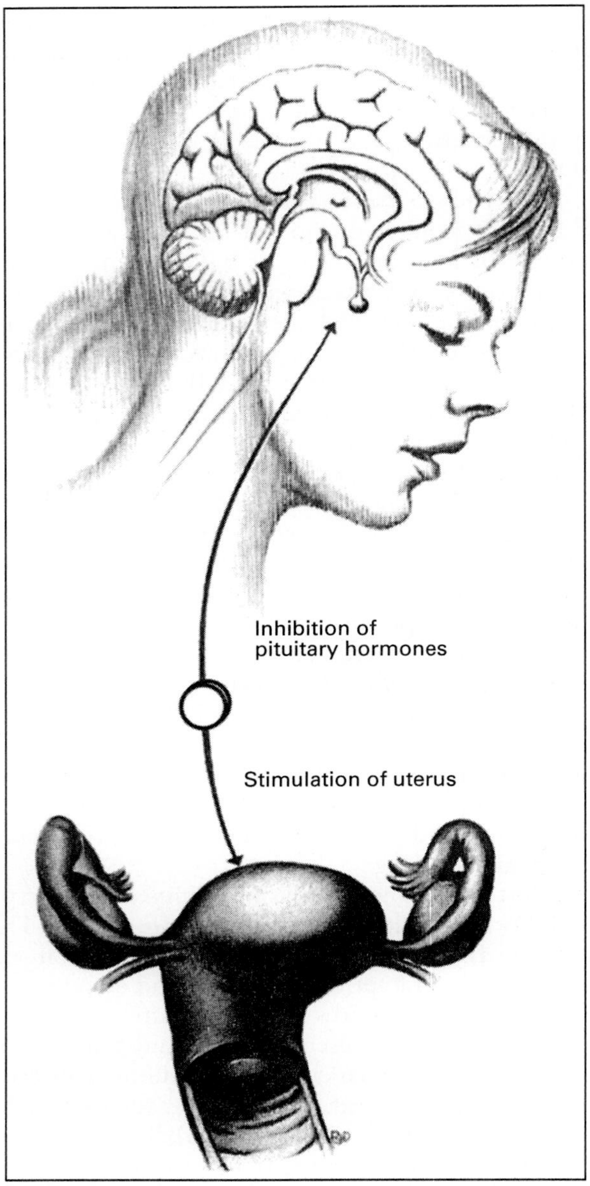

Figure 6–9 How oral contraceptives work.

Inhibition of pituitary hormones

Stimulation of uterus

women in a year), but just as safe, if not safer (Chi, 1993a). However, in some women these pills may cause menstrual problems.

There was a decline in the use of the combination pill during the 1970s and early 1980s when discoveries of serious health risks were reported (Mosher, 1990). However, many of these studies were conducted with the first-generation, high-dosage pill containing more than 150 micrograms of estrogen. Manufacturers of oral contraceptives stopped marketing high-dosage pills in 1988. Today's "third generation" pills have less than 50 micrograms of estrogen, and therefore have less risk associated with them.

The combination pill also has some significant health benefits. Nevertheless, there are specific subgroups of women who should not use the combination pill. Let us examine the risks and benefits in detail.

NEGATIVE SIDE EFFECTS AND HEALTH RISKS. There were some early reports of a few women showing decreased sex drive as a result of taking the combination pill (Masters & Johnson, 1970). These reports have been questioned, and more recent studies have found that pill users have less vaginal lubrication, but engage in sex more frequently, have greater sexual interest, and report more sexual satisfaction than do nonusers (Alexander et al., 1990; McCoy & Matyas, 1996). Studies with chimpanzees suggest that oral contraceptives might adversely affect sexual desire only in women who already had low sexual desire or preexisting problems in sexual adjustment (Nadler et al., 1993).

It is common for women who are first starting to take the pill to experience some symptoms that mimic those of early pregnancy. These may include nausea, constipation, weight gain, breast tenderness (the bloated feeling is due to water retention), and emotional depression. These responses usually occur only during the first few days of each cycle and usually diminish within a few months. If they continue or are bothersome, the woman's doctor may be able to adjust the dosage of the hormones. Occasionally, use of the pill results in brownish spots on the face (chloasma). Because the hormones in the pill alter the chemical environment of the vagina, there is also an increased susceptibility to yeast infection. As a result of the lowered dose of hormones in today's pill, breakthrough bleeding (i.e., bleeding during the 21 days of hormone use) sometimes occurs, but it usually disappears during the first four months of use. Despite these problems, 95 percent of women who use the pill have a very favorable opinion of it as a contraceptive technique (Forrest & Fordyce, 1988).

The previously mentioned problems may be bothersome, but they are not considered serious health risks. However, there are other, more serious problems that have been associated with use of the pill. Studies conducted in the 1970s, when the dosage of estrogen in the birth control pill was much higher than it is today, showed use of the pill to be associated with (though not necessarily the cause of) high blood pressure (4 percent of users), diabetes (1 percent), migraine headaches (0.5 percent), and liver cancer (rare; Palmer et al., 1989; Rinehart & Piotrow, 1979). The pill was once thought to be the cause of gallbladder disease, but recent studies do not support this belief (Vessey & Painter, 1994). Prior use of the pill is not associated with birth defects, but a woman should not continue to take the pill after she is pregnant (Huggins & Cullins, 1990).

Older studies with pills that contained high dosages of estrogen suggested that their use could cause serious cardiovascular problems such as blood clots, heart attacks, and strokes (Mann & Inman, 1975; Rinehart & Piotrow, 1979; Vessey, 1973). Users of the pill who were in their thirties and smoked, for example, were seven times more likely to have a heart attack than users who did not smoke. More recent studies have shown that other risk factors, primarily smoking, account for most of the cardiovascular problems (Archer, 1990; Goldbaum et al., 1987). In fact, the Food and Drug Administration voted in late 1989 to allow women over the age of 40 who do not smoke to continue taking the pill (Kaeser, 1989).

What about today's "third generation" low-dose pills? Pill users who are younger than 35 and do not have high blood pressure have very little or no risk of stroke or heart attack (Carr & Ory, 1997; Petitti et al., 1996; WHO Collaborative Study, 1996a, 1996b). The newest pills slightly increase the risk of a nonfatal blood clot, but reduce the risk of heart attacks as compared to the risk for users of the second-generation pill (Farley et al., 1995; Jick et al., 1995, 1996; Lewis et al., 1996; Spitzer et al., 1996). The authors of the blood clot studies said that their findings were "no cause for alarm," and more recent studies have found no increase in risk at all (Farmer et al., 1997). Oral contraceptive users over the age of 35 who smoke still have an unacceptably high risk of heart attack and stroke and should not use the combination pill (Carr & Ory, 1997).

Users of the birth control pill may have an increased risk of cervical cancer (Ursin et al., 1994), but there is a great deal of uncertainty as to whether this is due to the pill or to other risk factors (see Mishell, 1989). There were also concerns for many years that use of birth control pills containing estrogen would increase the risk of breast cancer. Breast cancer is the second most common type of cancer in women, affecting one in eight women at some point in their life (see Box 2–A). Although some studies claimed to have found a link between the pill and breast cancer, there have been over 20 studies, including one by the Centers for Disease Control and Prevention (Sattin et al., 1986), that have found that pill users are at no greater risk than nonusers (e.g., Romieu et al., 1989). A recent review of 54 studies found that women currently taking the combination pill have only a slighlty increased risk of breast cancer and that there was no increase in risk 10 years after stopping the pill (Collaborative Group on Hormonal Factors, 1996).

When discussing the health risks associated with use of the pill, you should put things into proper perspective and compare the risks to what would happen if nobody used the pill. The health risks (including deaths) *are significantly less than those caused by preg-*

TABLE 6–3

RISK OF DEATH IN A YEAR'S TIME RESULTING FROM SELECTED ACTIVITIES

Activity	Chance of Dying
Motorcycling	1 in 1000
Automobile driving	1 in 6000
Power boating	1 in 6000
Playing football	1 in 25,000
Using tampons (toxic shock), age 15–44	1 in 350,000
Having sexual intercourse (pelvic infection), age 15–44	1 in 50,000
Using Contraception	
Oral contraception—nonsmoker	1 in 63,000
Oral contraception—smoker	1 in 16,000
Using IUDs	1 in 100,000
Undergoing sterilization	
Laparoscopic tubal ligation	1 in 67,000
Vasectomy	1 in 300,000
Deciding about pregnancy	
Continuing pregnancy	1 in 11,000
Terminating pregnancy	
Illegal abortion	1 in 3,000
Legal abortion	
Before 9 weeks	1 in 260,000
9–12 weeks	1 in 100,000
13–15 weeks	1 in 34,000
After 15 weeks	1 in 10,200

Source: Hatcher et al. (1994). Reprinted with permission of the publishers.

nancy and childbirth (DaVanzo, Parnell, & Foege, 1991; see Table 6–3). We can also substantially reduce these risks by identifying certain groups of women who should not use the pill. These include women with a history of circulatory problems, women in their thirties who smoke (or younger users who are heavy smokers), women with sickle-cell anemia (because of the risk of clots), women with diabetes or high blood pressure or a family history of these disorders, very obese women, women with hepatitis or other liver problems, and those with severe focal migraine headaches. Young women who do not fall into any of these groups have little to worry about. In fact, some health researchers believe that today's birth control pills are safe enough to be made available without prescription (Trussell et al., 1993a).

HEALTH BENEFITS. After 35 years of use, it is now known that women who take the birth control pill are at less risk for some health problems than nonusers (see Drife, 1989). Some of these findings were unexpected and could properly be called fringe benefits. The major benefits include a decreased risk of cancer of the endometrium (women on the

Depo-Provera Market name for medroxyprogesterone acetate, a chemical that when injected suppresses ovulation for 3 months.

pill have only one-third the risk that nonusers have) and cancer of the ovaries (60 percent of the risk that nonusers have; Hulka et al., 1982; Rosenberg et al., 1994). (Cautionary note: It has not yet been demonstrated that today's low-dosage oral contraceptives offer the same level of protection against reproductive cancers; see Goldzieher, 1994). Not only is there no reliable evidence that use of the birth control pill is linked to breast cancer, but in fact there is a substantial decrease in the number of benign breast tumors (one-fourth as likely). Use of the pill also results in reduced rates of ovarian cysts (one-fourth as likely), rheumatoid arthritis (one-half as likely; Ory, Rosenfeld, & Landman, 1980), and pelvic inflammatory disease (one-half as likely; Kols et al., 1982; Panser & Phipps, 1991). The birth control pill is also known to reduce premenstrual syndrome, reduce menstrual pain and bleeding (anemia is only two-thirds as likely; Milsom et al., 1990; Mishell, 1982), and improve acne. In fact, some doctors prescribe the pill to women primarily to reduce menstrual problems, and only secondarily as a contraceptive technique.

> "I am now on the birth control pill. . . . My cramps were minimized a great deal and so was the length and heaviness of my period."
>
> (from the author's files)

For older women who do not smoke, are not obese, and do not have hypertension or cardiovascular problems, today's low-dosage pills are also being prescribed to regulate menstrual cycle disturbances that are common before menopause (Volpe et al., 1993).

"THE SHOT"

In early 1993, a new type of contraception became available to women in the United States: **Depo-Provera,** an injectable drug containing a progestin (medroxyprogesterone acetate, or DMPA, a synthetic form of progesterone). The injectable contraceptive was already being used by nearly 4 million women in

90 different countries. The injection is good for 3 months and prevents the problem of pregnancies resulting from missing a daily dose, as with the pill. It works by preventing ovulation (by blocking pituitary hormones) and has a yearly failure rate of less than 1 percent (World Health Organization Task Force, 1983). After a woman stops injections, it takes from several months to a year for fertility to return.

Depo-Provera was denied approval by the Food and Drug Administration twice in the 1970s because of concerns that it might be linked to cancers of the breast, cervix, and liver. However, the World Health Organization gave its approval to Depo-Provera in 1993 after concluding that it did not increase the risk of these cancers, and that it actually reduced the risk of cancer of the uterus (Klitsch, 1993a). But there are some possible side effects that might cause some women to discontinue this method. These include menstrual irregularities (including an increase or decrease in menstrual bleeding), weight gain (an average of 15 pounds), tiredness, weakness, nervousness, dizziness, and/or headaches. In fact, these problems are so bothersome that only about one-third of women who try "the shot" are still using it after a year (Sangi-Haghpeykar et al., 1996; Westfall, Main & Barnard, 1996).

In 1993, the World Health Organization also approved two other injectable contraceptives: Cyclofem and Mesigyna. These drugs contain estrogen and progestin, and are injected once a month. They are nearly 100 percent effective and have less severe side effects than Depo-Provera.

NORPLANT: THE FIVE-YEAR HORMONE IMPLANT

By now, many of you have heard of **Norplant,** the hormonal implant that offers contraceptive protection for up to 5 years. Approved for use in the United States in 1990, it consists of six flexible silicone tubes, each about the size of a match, that are inserted under the skin in a fanlike pattern on the inside of a woman's upper arm (see Figure 6–10). The procedure takes only 15 to 30 minutes and costs about the same as a 5-year supply of birth control pills. The tubes contain a progestin (levonorgestrel), a synthetic form of progesterone that is slowly released over time and prevents pregnancy by inhibiting ovulation and thickening cer-

Figure 6–10   Norplant.

vical mucus. The tubes are nonbiodegradable and can be removed at any time.

Norplant's failure rate is said to be about 2 in 1,000 (Trussell et al., 1990), as compared to 30 in 1,000 for the pill (most of which are user failures). This makes it nearly as reliable as sterilization. Only about 11 to 15 percent of women who choose Norplant decide to have it removed by the end of one year, some because they plan to have another child, but many because of unhappiness with the side effects (Frank et al., 1993; Peers et al., 1996). About three-fourths of women with Norplant implants experience side effects in the first year. The most common side effects (as with all progestin forms of contraception) are spotting or irregular bleeding (this can include prolonged menstrual bleeding or no bleeding at all), weight gain, and headaches, but some other side effects include nervousness, dizziness, nausea, breast tenderness, and acne (Sivin, 1994). However, only one-third of the women who experience side effects are bothered by them. Over two-thirds of Norplant users are very satisfied (Krueger, Dunson, & Amatya, 1994).

"I chose the Norplant birth control method after I had my first child at age 20. My life, of course, changed 100 percent, something I was not ready for, mentally or financially. I am now back in school as a single mother. I decided to try Norplant when I was still married and didn't want any more 'accidents.' . . . The in-office procedure

Norplant   Six flexible silicone tubes containing a progestin that are implanted under the skin to prevent conception; they remain effective for up to 5 years.

Box 6-A **SEXUALITY AND HEALTH**

Contraceptive Methods That Help Prevent Sexually Transmitted Diseases

Sexually active individuals who are not in monogamous relationships must (and should) worry about sexually transmitted diseases in addition to pregnancy. Unfortunately, contraceptive development has concentrated almost entirely on preventing pregnancy and has ignored STDs. What do we know about the effectiveness of currently available contraceptive devices in preventing STDs?

Fertility awareness methods, sterilization, Norplant, Depo-Provera, the IUD, and the birth control pill offer no protection against STDs (Cates & Stone, 1992b). The IUD, you recall, increases a woman's chances of getting pelvic inflammatory disease should she get a bacterial infection of the lower genital tract (e.g., gonorrhea or chlamydia). The birth control pill offers some protection against pelvic inflammatory disease (e.g., Panser & Phipps, 1991) but increases replication of HIV in the vagina (Mostad et al., 1997) and also increases the risk of cervical infections with chlamydia (e.g., Oriel, Johnson, & Barlow, 1978; also see Cates & Stone, 1992b). Even worse, many women stop using condoms when they start using the shot or Norplant (e.g., Sangi-Haghpeykar et al., 1997).

That leaves barrier methods and spermicides. Let us consider spermicides first. The active ingredient in most spermicides, nonoxynol-9, not only kills sperm, but also is effective against most of the bacteria and viruses that cause STDs, including gonorrhea, chlamydia, trichomonas, hepatitis, and HIV (the virus that causes AIDS). There are not enough data to make conclusions about whether spermicides protect against herpes, syphilis, or chancroid. While spermicides reduce the risk, *they do not reduce the risk to zero.* At best, there is a reduction in risk of about 50 percent. (There are far too many studies for me to cite here; for an excellent review, see Rosenberg & Gollub, 1992; or see Howe, Minkoff, & Duerr, 1994, for a review specific to HIV.)

For men, of course, use of rubber condoms will reduce their risk of getting an STD; "skins" (condoms made of sheep intestine) have pores that are large enough for the passage of HIV and other viruses (Liskin, Wharton, & Blackburn, 1990). However, not all the news is reassuring. Rubber condoms may not have any holes before use, but under conditions that resemble intercourse, about a third of these condoms develop very small holes that might be large enough to allow passage of HIV (Carey et al., 1992). What's more, condoms break from about 1 to 3 percent of the time during vaginal intercourse, and more often during anal intercourse (Messiah et al., 1997; Silverman & Gross, 1997; see Cates & Stone, 1992a for a review of numerous studies). Incorrect and inconsistent use also reduces the effectiveness of condoms against STDs. Overall, under actual use conditions, *condoms reduce the rates for STD infection an average of about 50 percent* (Rosenberg, Hill, & Friel, 1991). Of course, we would expect that if condoms were always used correctly and consistently, and always in combination with a spermicide, their effectiveness at STD prevention would be much greater. In one study of 124 serodiscordant couples (only one member was HIV-positive) who used condoms consistently, none of the HIV-negative partners became infected during a total of 15,000 acts of intercourse (De Vincenzi et al., 1994).

What about women? What contraceptive methods can they use that will simultaneously reduce the risk of contracting STDs? At first glance, an obvious choice might be to have their partners always wear condoms, but this "may be unrealistic because traditional sex roles in most cultures do not encourage a woman to talk about sex or to initiate sexual practices or otherwise control an intimate heterosexual encounter" (Rosenberg & Gollub, 1992). Michael Rosenberg and his colleagues (1992) reported that under actual use, women using the diaphragm had lower rates of infection for gonorrhea and trichomonas than women whose partners used condoms. The diaphragm was also more effective than condoms against chlamydia. Other researchers believe that the authors "overstated the scientific case for these methods, especially in comparison with the condom" (Cates, Stewart, & Trussell, 1992). However, there is no question that the barrier contraceptives (diaphragm and the new female condom), if used consistently, properly, and with spermicide, offer women some protection against many STDs (Cates & Stone, 1992a; Rosenberg & Gollub, 1992). There is not yet enough data to tell whether use of the diaphragm with spermicides offers any protection against HIV, but some initial results were not promising (Kreiss et al., 1992). We do know, of course, that the male condom helps protect both the male and his partner against this deadly disease. We must wait and see how effective the female condom is in preventing STDs under actual-use conditions.

lasted only thirty minutes. . . . I felt no pain. . . . I started to gain a little weight and had some slight dizziness. Then my menstrual periods stopped. Now 10 months later they are normal again and I have no other symptoms. My Norplant will last for 5 years. This is just enough time for me to finish school. I would recommend this method of birth control to anyone."

(from the author's files)

Like this student, most women who use Norplant do so as a way to time having children (Frank et al., 1992). The Food and Drug Administration and World Heath Organization believe it is a safe form of contraception because it contains no estrogen. Norplant had already been used by 500,000 women worldwide before it was approved for use in the United States. Nevertheless, Norplant should not be used by women with liver disease or tumors, breast cancer, or a history of blood clots. The television news program *Eye to Eye* reported in 1994 (May 8) that removal of Norplant was difficult and painful in some women, but the manufacturer claims that this was due to improper implantation by doctors who were not qualified for the procedure.

VOLUNTARY STERILIZATION

Over one million **sterilization** procedures are performed in the United States every year. In fact, more Americans rely on sterilization for birth control than on any other birth control technique (L. S. Peterson, 1995). For a couple to rely on this method, it is not necessary that they both be sterilized, so this figure is derived by adding male and female sterilizations (approximately 30 percent of contraceptive users rely on female sterilization and 12 percent on male sterilization). More than half of all married American couples eventually use sterilization after the birth of their last planned child. The use of sterilization has increased most during the past decade among less-educated low-income African-American and Hispanic women, especially those who had never married (Mosher, 1990). In many other countries, sterilization is even more popular than here (Mumford & Kessel, 1992).

In men, the sterilization procedure is called **vasectomy.** In Western cultures, it has traditionally been done by making a small incision or two in the scrotum (under local anesthesia), then tying off and cutting the vas deferens (see Figure 6–11). The entire procedure takes only about 20 minutes. More recently, "no-scalpel" vasectomy has been introduced from China, where it has been performed safely on millions of men (Shun-Quiang, 1988). With this technique, the skin of the scrotum is simply pierced with a sharp instrument. It takes less than 10 minutes, produces less bleeding, and results in fewer infections (Jow & Goldstein, 1994; Nirapathpongporn et al., 1990). Despite the ease of these procedures, there is some evidence that there may have been a decline recently in the number of men seeking a vasectomy (Adler, 1994).

A vasectomy does not interfere with the production of hormones or sperm in the testicles. New sperm simply cannot get past the point where the vas have been tied and cut, and eventually they are destroyed by other cells called phagocytes. Sperm that are already past this point at the time of operation, however, must be eliminated before a man is "safe." Other means of contraception should be used until the man has had 12 to 16 ejaculations. Some men are reluctant to get a vasectomy because they think that they will no longer be able to ejaculate (obviously confusing ejaculation with orgasm), but recall from Chapter 2 that nearly all of the fluid in a man's ejaculation comes from the prostate gland and seminal vesicles, which are unaffected by the procedure. A man with a vasectomy will still ejaculate during orgasm, but his ejaculation will contain no sperm.

In women, the term **tubal ligation** is often used for a variety of techniques, but it really refers just to the tying of the Fallopian tubes. More commonly, the procedure involves cutting as well as tying the tubes (see Figure 6–11).

sterilization A general term for surgical techniques that render an individual infertile.

vasectomy The male sterilization technique in which the vas deferens is tied off and cut, thus preventing passage of sperm through the male's reproductive tract.

tubal ligation A female sterilization technique that originally referred only to the tying of the Fallopian tubes, but which is now often used as a general term for a variety of female sterilization techniques.

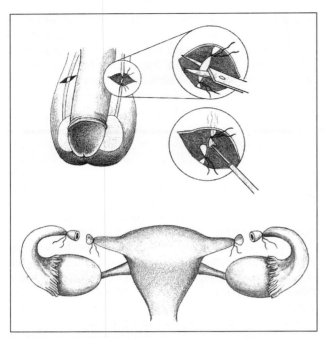

Figure 6–11 A vasectomy and a tubal ligation.

Both methods, of course, prevent passage of the egg and sperm. The egg is absorbed by the woman's body. In the past, these procedures required hospitalization, general anesthesia, and large abdominal incisions.

Today, most tubal ligations involve small incisions. In a **minilaparotomy,** 1–inch incisions are made in the abdomen and the Fallopian tubes are pulled to the opening and cut, tied, or blocked with clips. A **laparoscopy** is a procedure in which a laparoscope, a long, tubelike instrument that transmits TV pictures, is inserted through a small incision in the navel. Once the tubes are located, surgical instruments are inserted through the laparoscope (or other tiny incisions) and the tubes are cut or cauterized. Both procedures are usually done with general anesthesia, but do not require lengthy hospitalization. Most women go home the same day. Unlike the male technique, a woman may safely engage in sexual intercourse as soon as she feels the desire to do so. Instead of using abdominal incisions, some physicians approach the Fallopian tubes through small incisions in the vagina (a procedure called **culpotomy**). However, minor complications (infection or hemorrhage) are twice as high as with the other procedures.

minilaparotomy A female sterilization technique in which small incisions are made in the abdomen and the Fallopian tubes are then pulled to the opening and cut, tied, or blocked with clips.

laparoscopy A technique that involves inserting a slender, tubelike instrument (laparoscope) into a woman's abdomen to examine (via fiberoptics) her reproductive organs or a fetus. The procedure is often used to perform female sterilization.

culpotomy A female sterilization technique in which the Fallopian tubes are approached through a small incision in the back of the vagina and then are cut or cauterized.

A nonsurgical female sterilization has been tested on thousands of women around the world and shows much promise (see Mumford & Kessel, 1992). Blockage of the tubes is achieved by inserting a pellet of quinacrine (a drug) through a modified copper T IUD inserter. This method is 98 to 99 percent effective but has not yet been approved for use in the United States.

COMPLICATIONS. Surgical complications resulting from vasectomy are uncommon (less than 5 percent) and generally minor. However, there have been some concerns about the long-term effects. Rather than being destroyed, some sperm leak into the bloodstream, and as a result, at least half of the men who have had a vasectomy produce antibodies to their own sperm. Some initial research with monkeys suggested that this could lead to accelerated arteriosclerosis (hardening of the arteries; Clarkson & Alexander, 1980), and a small study with humans suggested a link between vasectomy and testicular cancer (Cale et al., 1990). But several large studies conducted with thousands of humans have found no link between the surgery and later development of arteriosclerosis or testicular cancer (e.g., Massey et al., 1984; Moller, Knudsen, & Lynge, 1994; Perinn et al., 1984).

More recently, some studies have raised concerns that vasectomy may increase men's risk of prostate cancer (Giovannucci et al., 1993a, 1993b; see Klitsch, 1993b). However, after a scientific panel at the National Institutes of Health met to discuss these findings, they concluded that the evidence was not strong. Subsequent studies supported this conclusion (Moller, Knudsen, & Lynge, 1994; Zhu et al., 1996). An accompanying editorial stated: "There is no known biologic mechanism to explain a causal relationship." Cancer of the prostate is 50 times more common in the United States than in China and other developing countries, and the factors causing the prevalence of prostate cancer in the United States are so poorly understood that this complicates any research about vasectomy as a cause (Farley et al., 1993).

Some physicians have suggested that there is a "post–tubal ligation syndrome" characterized by heavier-than-normal menstrual bleeding and more severe menstrual cramps. One study reported that these symptoms occurred in one in every seven to ten cases of tubal ligation (Rulin et al., 1989). This has sometimes been used as a justification for performing hysterectomies (Goldhaber et al., 1993), but pathology exams usually reveal nothing abnormal and studies to date have found no support for a biological basis for hysterectomy (Stergachis et al., 1990). Studies conducted by the World Health Organization found no increase in menstrual bleeding associated with sterilization (World Health Organization, 1984, 1985), but a replication study found a small increase (Richards et al., 1991). A more recent study found that half of all sterilized women were experiencing heavy menstrual flow five years after surgery (Wilcox et al., 1992). However, most studies find that the large majority of women are satisfied with their choice of surgical contraception. What's more, sterilization in women has some definite health benefits: it substantially decreases their risk of getting cancer of the ovaries (Miracle-McMahill et al., 1997). Overall, the long-term benefits outweigh the risks (Kawachi, Colditz, & Hankinson, 1994).

CAN STERILIZATIONS BE REVERSED? Sterilization procedures are highly effective, with failure rates of 1 to 4 in 1,000 (Hatcher et al., 1994; Trussell et al., 1990). Today, they should also be considered permanent. Few people who decide to be sterilized ever wish to have the procedure reversed (about 4 percent change their mind), but for those who do there is no guarantee of success. The problem is that the sewing together of the vas deferens or Fallopian tubes requires very skilled microsurgery techniques (the inside diameter of a Fallopian tube, for example, is no greater than a human hair). The success rate for vasectomy reversal has been steadily improving (a new technique using laser surgery appears to be very promising), with some doctors now reporting restored sperm flow in over 50 percent of their patients. However, even with restored sperm flow, the chance of a man impregnating a woman decreases markedly if the reversal is not performed within a year or two, probably due to the increased levels of the man's antibodies to his own sperm. The pregnancy rate after a reversal is about 50 percent (Hatcher et al., 1994). Successful reconstruction of the Fallopian tubes is very difficult because the slightest scarring can impair passage of the egg (and result in tubal pregnancy even if fertilization occurs). Success rates range from 43 percent to 88 percent (Hatcher et al., 1994; Kim et al., 1997). So while there have been successful reversals in both males and females, the success rates are still far less than 100 percent, and therefore only people who are quite sure that they do not want any more children should opt for these procedures.

Most physicians refuse to perform these procedures on young unmarried people, and even married couples with children often are told to consider their decision for a few days to be sure. Some states have a mandatory waiting period (30 days in New York, for example). Sometimes there is more than just the question of children to be considered before making the decision. People who worry about conflicts with their religious beliefs or think that they will not be a "whole person" after sterilization should work these problems out beforehand. Men who have a history of erectile problems are also not good candidates (the procedure might contribute to psychologically caused impotence). Once done, however, many people report an increase in sexual desire because they no longer have to worry about pregnancy or contraceptive side effects, and sexual relations can be completely spontaneous.

WHICH PARTNER SHOULD BE STERILIZED? Remember, if a couple agrees that they do not wish to have any more children and choose sterilization as their contraceptive technique, only one of them need have surgery. Which one should it be—the man or the woman? One study that compared expenses, complication rates, and the number of people who regretted their operations concluded that the male sterilization technique was preferable in every way (Kjersgaard et al., 1989). Whatever the choice may be, it is important that the couple make an educated decision so that both persons will be happy with the decision.

UNWANTED PREGNANCIES: THE FINAL OPTIONS

There is no reversible method of contraception that is 100 percent effective (although Norplant, Depo-Provera, and the pill come close); thus, unwanted pregnancies sometime occur even for couples who are using the most reliable forms of contraception. If an unwanted pregnancy should occur, a woman (or a couple) really has only three options: to continue with the pregnancy and keep the baby; to continue with the pregnancy and put the baby up for adoption; or to terminate the pregnancy (abortion). In many cases, none of these is an easy choice.

The responsibilities involved with **keeping and raising a baby** force many women to drop out of school and remain uneducated and unskilled, locking them into a lifetime of poverty and reliance on state funds and services to raise the child. The rate of child abuse resulting from the emotional frustration in these situations is much higher than normal (Zuravin, 1991).

Some women or couples choose to put an unplanned baby up for **adoption** by others who are better able to care for it. In many cases, this may be best for the child, but nevertheless, it often results in long-lasting adjustment problems for those who have surrendered their child. The following is a typical case:

"I was one of the teenage pregnancies. I was 14 years old and was pregnant. My boyfriend, now my husband, was also 14 years old and thought that he could not produce sperm yet, and I believed him. It has caused me a very hard life because of the guilt I feel for having to give up my child. We now have two more children, but it still doesn't fill the void. I only wish we both could have been aware of this then. All of the agony could have been spared."

(from the author's files)

For many women or couples, **abortion** is no less difficult a choice.

"At the age of 18 I became pregnant. My boyfriend of four years and I had recently broken up, so I was alone. I suppose I could have told my parents, but I was too scared of their reactions. At three months I terminated my pregnancy. It hurt very badly emotionally. Occasionally I think of how old my baby, boy or girl, would be today."

(from the author's files)

However, the large majority of women who have first-trimester abortions do not experience emotional or psychological problems afterwards (Adler et al., 1990; Russo & Dabul, 1997). Many feel a sense of relief because of compelling health or economic considerations (see Lewis, 1997). The few women who do suffer emotionally after an abortion generally had emotional problems before (Russo & Dabul, 1997).

Of these three options, only abortion can be considered a method of birth control. However, many people have strong negative views of the use of abortion as a method of birth control. Let us discuss this topic in more detail.

ABORTION

There are an estimated 36 to 53 million abortions performed annually throughout the world, of which only about 26 to 31 million are legal (Henshaw, 1990). Abortion is the main form of birth control in China and Russia, with the average Russian woman having six or seven abortions in her lifetime. Although the U.S. abortion rate is well below the worldwide rate (28 versus 37 to 55 per 1,000 women aged 15 to 44), nearly a third of all pregnancies in this country end in abortion. Most abortions performed in this country are for middle-class white women (Henshaw & Kost, 1996). Over 1.2 million abortions were performed in the United States in 1995, down from the high of 1.43 million in 1990 (Centers for Disease Control and Prevention, 1997), but increases in some

abortion Termination of pregnancy. Depending on how far the pregnancy has advanced, this can be done by taking a pill (Ovral, birth control pills in high dosage, RU 486); scraping the uterine lining (dilation and curettage); removing the uterine lining by suction (dilation and evacuation); or inducing labor (by injecting hypertonic saline or prostaglandins).

areas of the country suggest that the numbers may rise again.

Over three-fourths of all women who obtain abortions do so within 10 weeks of conception (Koonin et al., 1992). Death resulting from a legal abortion is very uncommon—less than one per 100,000 abortions—and, in fact, is much less common than deaths resulting from childbirth (Grimes & Cates, 1980; Koonin et al., 1992). Modern legal abortions do not increase a woman's subsequent risk of infertility, miscarriage, ectopic pregnancy, or having a low-birthweight baby (Frank et al., 1993; Holt et al., 1989). A recent review of 23 studies concluded that abortion did increase a woman's lifetime risk of breast cancer by 30 percent (Brind et al., 1996). However, a more recent very large study concluded that there was no risk of breast cancer associated with abortion (Melbye et al., 1997), and an editorial in the prestigious *New England Journal of Medicine* agreed (Hartge, 1997).

What are some of the possible consequences of legalizing abortion? Of abolishing abortion? It has been estimated that adolescent childbearing in New York City is 14.1 percent (for white adolescents) to 18.7 percent (for African-American adolescents) lower than it would be if abortion were illegal (Joyce & Mocan, 1990). In Romania, the abolishment of legalized abortion led to the highest maternal mortality rate in Europe and the abandonment of tens of thousands of children in institutions (Stephenson et al., 1992).

There are several methods used to terminate a pregnancy:

1. *Emergency birth control:* There are several methods that can be used to prevent implantation after unprotected intercourse (see Glasier, 1997; Trussell et al., 1997). At the start of 1997, doctors could prescribe the "morning-after pill" (Ovral), which contains ethinyl-estradiol (a synthetic estrogen) and norgestrel (a progestin). When two pills are taken within 72 hours of intercourse followed by two more 12 hours later, they prevent implantation of the fertilized egg in 98 to 99 percent of cases. In early 1997, the Food and Drug Administration gave approval for doctors to prescribe several brands of traditional birth control pills in high dosages as emergency birth control (U.S. Food & Drug, 1997). Seven brands of birth control pills contain the same hormones as Ovral, and almost all women could safely use the pills for this short period (Trussell et al., 1997). However, in high dosages the pills may cause nausea and vomiting. Insertion of a copper IUD within 7 days after intercourse is also a highly effective way of preventing implantation (see Trussell et al., 1997).

It is believed that emergency birth control could prevent as many as 1 to 2 million unintended pregnancies each year (Trussell & Stewart, 1992). Only about one-fourth of all doctors tell their teenage patients about the availablity of emergency birth control, but the large majority will prescribe it if asked to (Gold, Schein, & Coupey, 1997).

2. *Medical (nonsurgical) abortion:* In 1988, the French, Chinese, and Swedish governments approved a pill (*RU 486*, or mifepristone, made by the French company Roussel-Uclaf) that can be used to chemically induce an abortion up to eight weeks after a woman's last menstrual period. Since then, this pill has been used by over 200,000 European women. One-third of all abortions performed in France are presently done by this method. It is about 95 percent effective when taken in combination with oral prostaglandins (misoprostol), which cause contractions of the uterus (Grimes, 1997). RU 486 works by blocking the body's use of progesterone, the hormone necessary for maintaining the endometrium of the uterus, thus causing shedding of the uterine lining (Baulieu, 1989). Another drug combination (methotrexate and misoprostrol) has also proved to be effective to terminate early pregnancies and is currently undergoing clinical trials (Creinin et al., 1996). The main side effects are nausea, cramping, and prolonged uterine bleeding (see Grimes, 1997). Contrary to popular belief, RU 486 is not taken at home and must be administered in a clinic. A woman must visit a doctor three times in order to have a drug-induced abortion.

Scientific advisors to the Food and Drug Administration recommended in 1996 that RU 486 be approved for abortions. However, the European company that was going to market the pill in the United States hesitated because of the abortion controversy here, and the earliest the drug could be available is 1998.

3. *Dilation and curettage (D & C):* Once the standard procedure for unwanted pregnancies of 15 weeks or less, D & C procedures are now used less and less for purposes of abortion. Under general anesthesia, a female's cervix is dilated, and the lining of the uterus is then scraped with a metal instrument called a curette. In addition to the risks associated with general anesthesia, this procedure results in more bleeding and discomfort than D&E (see below) procedures.

4. *Dilation and evacuation (D & E):* In this procedure, which does not require general anesthesia, a tube is inserted through the cervix and the fetal material is removed by suction. This can be done without anesthesia in the first four to six weeks of pregnancy, but requires local anesthesia and dilation of the cervix for pregnancies terminated dur-

ing weeks 7 through 12. Later-stage abortions may require gentle scraping of the uterine walls in addition to suction. The older D&C procedure was associated with a risk of subsequent sterility, but early first-trimester abortions performed by the vacuum technique do not appear to have this risk (Frank et al., 1993).

5. *Induced labor:* This method is used exclusively for termination of pregnancies that have proceeded beyond 16 weeks (which accounts for only 5 percent of all abortions; Koonin et al., 1992). A solution of hypertonic saline or prostaglandins (hormones which cause smooth muscle contractions) is injected into the amniotic sac and induces labor within 12 to 36 hours. The fetus is born dead.

6. *Intact dilation and evacuation (intact D&E):* This procedure, also known as *dilation and extraction (D&X),* is performed in very-late-term pregnancies and supposedly only when the mother's health is in danger. The fetus' head is partially delivered and its brain is then removed by suction (thus the frequently used description "partial-birth abortion"). The Centers for Disease Control and Prevention estimate that about 1.3 percent of abortions are by intact D&E (see *AMA Bulletin,* March 3, 1997).

THE PRESENT STATUS AND FUTURE OF ABORTION IN THE UNITED STATES. Abortion is a subject that has polarized people throughout the nation. Those opposed to abortion ("right-to-lifers") express concerns about protecting human life. "Pro-choicers," on the other hand, wish to retain the right of individuals to make decisions about something that will have a major impact on their own lives. In a 1996 survey conducted by the Gallup Organization, 83 percent of Americans supported a woman's right to abortion in at least some circumstances (Moore, Newport, & Saad, 1996). However, a 1998 New York Times/CBS News Poll found that while a majority of the American public still supports legalized abortion, they also feel that it should be harder to get and less readily chosen (Goldberg & Elder, 1998). Although half of those surveyed regarded abortion as murder, a third of those took the position that it is sometimes "the best response to a bad situation."

The laws regarding abortion have been in a state of transition for over two decades. Prior to 1973, each state could allow or prohibit abortion as it wished, but in that year a 7-to-2 Supreme Court decision (*Roe v. Wade*) prohibited the states from interfering with decisions made between a woman and her doctor during the first three months of pregnancy. The ruling also prevented states from prohibiting abortion in the second trimester of pregnancy, but did allow for more regulations designed to protect a woman's health (hospitalization, for example). Antiabortion forces won a victory in 1977 when the Supreme Court upheld the

Hyde Amendment, which disallowed the use of federal Medicaid money to pay for abortions. Without public funding, obtaining a legal abortion became difficult for women with limited incomes.

Although the decision in *Roe v. Wade* giving a woman the right to choose is still the law, subsequent court decisions have given states greater leeway in regulating abortion. In 1989, a sharply divided Supreme Court upheld by a 5–4 vote (*Webster v. Reproductive Health Services*) a Missouri law that further restricted public funds and facilities for abortions and that also required physicians to test for fetal viability (the potential for the fetus to survive) at 20 weeks. The earliest point of fetal viability is 24 weeks (the law allowed for a 4-week margin of error in determining the time of conception), and the vast majority of abortions have already been performed before the twentieth week. In 1990, the Court voted 6–3 to uphold state laws that ban abortions for girls under the age of 18 unless a parent is notified. In a 1992 case (*Planned Parenthood v. Casey*), the Court, in another 5–4 vote, upheld a Pennsylvania law that imposed a 24-hour waiting period, required doctors to tell women about other options, and also required parental notification in the case of minors. In his first two weeks in office, President Clinton abolished the "gag rule," which had prevented counseling about abortion in federally funded clinics. In 1996, Clinton twice vetoed Congress's attempts to ban intact D&E abortions. Legal decisions regarding abortion are likely to continue to change as the members of the Supreme Court are replaced, either by more conservative or more liberal justices.

Religious beliefs regarding abortion have often changed as well. Within the Catholic Church, for example, the first pope to declare that abortion was murder was Sixtus V in 1588, but that was reversed just three years later by Pope Gregory XIV. For the next three hundred years, abortion was allowed up to the time of what was termed "animation by a rational soul" (up to 40 days after conception in the case of a male embryo and for the first 80 days in the case of a female fetus; Connery, 1977; Grisez, 1970). In 1869, however, Pope Pius IX again declared that abortion was murder, and this has remained the belief of every pope since.

With the diversity of opinion regarding this subject, religious and legal beliefs about abortion will probably continue to change for many decades to come.

FUTURE TECHNOLOGY

In more and more societies, rich with technological diversity and replete with seemingly endless permutations of consumer goods, it is odd indeed that [the present array of contraceptive technologies], central to the lives of so many individuals, families and societies, is so limited.

(Harrison and Rosenfield, 1996. Reprinted with permission of the publisher.)

The female condom became available to women in the United States in 1993. This probably was the first of what will be a whole new generation of female barrier contraceptives. In addition to the condom, there are two other vaginal pouches under development. The *Woman's Choice Female Condomme* is inserted into the vagina with an applicator and has a latex rubber umbrella-like cap that is about a third thicker than the male condom. The *unisex condom garment* is a polyurethane item resembling a bikini bottom with an attached sheath that can be worn by either men or women (as a vaginal liner or a penis covering) for contraception and protection against sexually transmitted diseases.

Researchers are also testing variations of the cervical cap. The *Lea's Shield* is made of silicone rubber and has a loop that makes it easy to insert and remove. It is a one-size-fits-all barrier contraceptive that allows passage of cervical secretions and is intended to be sold over the counter (Mauck et al., 1996). The *Fem Cap* is also made of silicone rubber. It conforms to the vaginal walls by means of a surrounding "brim." The *Oves Cervical Cap*, unlike the others, is disposable, and will probably be available across the counter.

Research into a *male pill* continues, but do not expect it on the market any time soon (see Comhaire, 1994). The problem is trying to find a pill that inhibits sperm production without simultaneously lowering sexual desire or a man's ability to get and maintain an erection. There was some success with gossypol, a Chinese derivative of cottonseed oil that disables sperm-producing testicular cells, but research was abandoned because as many as 20 percent of men who took it became permanently sterile. Present research is focusing on the administration of hormones to block the pituitary hormones responsible for sperm production (see Chapter 3), while simultaneously delivering other hormones to assure that there will be no loss of sex drive (Meriggiola et al., 1997; Harrison & Rosenfield, 1996).

Perhaps the most promising future means of birth control is *vaccination* (Alexander, 1994). It might be possible to prevent pregnancy the same way you can the flu. Researchers are trying to develop vaccines that will cause the body to produce antibodies to proteins in either sperm or eggs. Scientists in India are already clinically testing a vaccine that stimulates the production of antibodies to HCG (human chorionic gonadotropin), a hormone necessary to maintain pregnancy. With this vaccine, a fertilized egg cannot implant and is discharged during menstruation. U.S. scientists are also working on a contraceptive vaccine.

For example, if women could be inoculated with a vaccine that triggered production of antibodies to a protein found only in sperm, this would prevent fertilization. A booster shot would be needed every couple of years to continue the effectiveness of such vaccines.

A report issued in 1990 by the National Research Council's Institute of Medicine warned that the United States is far behind other countries in offering choices of birth control. The biggest deterrent to new contraceptive methods is not technology or lack of research, but an insurance–product liability problem. American consumers are accepting of the fact that contraceptive methods sometimes fail to prevent pregnancy, but they are less forgiving about side effects. However, as Dr. Carl Djerassi (one of the developers of the birth control pill) stated to a congressional committee in 1978, "there is no such thing as a wholly safe contraceptive." Some methods, such as Depo-Provera, went through more than 20 years of FDA hearings before they won final approval, yet were available to people around the world all during that time.

When a legal case involves an individual versus a large corporation, juries often award large settlements to the individual in the belief that the corporation can afford it. A jury awarded $5 million, for example, to a couple who claimed that their child's birth defects were caused by spermicides, even though the authors of the study that claimed to link spermicides with birth defects had repudiated their own results. A few more cases like this and we may see the end of spermicides and the diaphragm. There has never been a contraceptive device that was not associated with some risk, but unless consumers change their attitudes and stop unrealistically expecting perfection, we may retreat to the days when there was nothing available except the condom.

CHOOSING A CONTRACEPTIVE METHOD

It is important that men and women choose a method of birth control with which they and their partner are comfortable. A woman may like the diaphragm, but if she is in a relationship it will not be good if her partner resents her stopping in the middle of having sex to put it in. A man may prefer that his female partner take the pill, but it will not be good for the couple's relationship if she resents having to suffer possible side effects. A couple can work these kinds of problems out only by talking about the matter and communicating with each other on a regular basis.

Choosing a particular birth control method does not mean that you have to stick with it if you are unhappy about it for any reason. As you know by now, there are many alternatives. For example, about one out of seven married women attempting to avoid pregnancy decide to abandon their chosen method within the first year of use. Unfortunately, half of these women use no type of contraception at all for at least a month afterwards (Grady, Hayward, & Florey, 1988). If you do not presently wish to have children and are unhappy with the birth control method you use, decide on an alternative method *before* you stop using your present one.

When deciding which type of birth control to use, ask yourself the following questions; they were developed to help you decide whether the method you are considering is a good choice or a poor choice *for you* (adapted from Robert Hatcher et al. (1994). Reprinted with permission of the publishers.)

Method of birth control you are considering using: _____

Length of time you used this method in the past: _____

Answer "yes" or "no" to the following questions:	**Yes**	**No**
1. Have I had problems using this method before?	_____	_____
2. Have I ever become pregnant while using this method?	_____	_____
3. Am I afraid of using this method?	_____	_____
4. Would I really rather not use this method?	_____	_____
5. Will I have trouble remembering to use this method?	_____	_____
6. Will I have trouble using this method correctly?	_____	_____
7. Do I still have unanswered questions about this method?	_____	_____
8. Does this method make menstrual periods longer or more painful?	_____	_____
9. Does this method cost more than I can afford?	_____	_____
10. Could this method cause me to have serious complications?	_____	_____

	Yes	No
11. Am I opposed to this method because of any religious or moral beliefs?	_____	_____
12. Is my partner opposed to this method?	_____	_____
13. Am I using this method without my partner's knowledge?	_____	_____
14. Will using this method embarrass my partner?	_____	_____
15. Will using this method embarrass me?	_____	_____
16. Will I enjoy intercourse less because of this method?	_____	_____
17. If this method interrupts lovemaking, will I avoid using it?	_____	_____
18. Has a nurse or doctor ever told me *not* to use this method?	_____	_____

Total Number of "Yes" Answers: _____

Most individuals will have a few "yes" answers. "Yes" answers mean that potential problems may lie in store. If you have more than a few "yes" responses, you may want to talk to your physician, counselor, partner, or friend. Talking it over can help you to decide whether to use this method, or how to use it so it will really be effective for you. In general, the more "yes" answers you have, the less likely you are to use this method consistently and correctly.

This chapter has only introduced the various contraceptive methods. If additional information is desired, you are urged to contact your local Planned Parenthood organization or your state family planning clinic. The people who work there will be glad to help you.

Sex is for people who are ready and willing to take a lifetime responsibility for the outcome of sex.

(Sue Finn, as quoted by L. Smith, 1993)

Many parents are opposed to the teaching of contraception in school because they worry that it will lead to sexual activity. Research has shown, however, that young people who have had sex education are no more likely to engage in sexual intercourse than those who have not had it—and more importantly, that they are less likely to get pregnant. Look again at the table at the beginning of the chapter. Danish, Dutch and Swedish children have sex just as much as American teens, but far fewer get pregnant or get someone pregnant (e.g., Wielandt & Knudsen, 1997). In those countries, children are taught about contraception at an early age, in an open and frank manner. Family planning programs that include discussions with older students on resisting peer pressure are particularly effective in reducing teen pregnancies (Howard & Mc-Cabe, 1990).

Key Terms

Personal Reflections

1. Are you prepared at the present time to assume the responsibilities of being a parent? Why or why not?

2. How would your life change if you or your partner were pregnant (consider your relationship, finances, career goals, etc.)?

3. If you are having sexual relations, what do you do about contraception?

4. Whose responsibility is it for birth control—your responsibility or your partner's? Why? If you answered that it is your partner's responsibility, do you think that you share *any* of the responsibility for the consequences of engaging in sexual intercourse?

5. Considering all factors (including your sexual lifestyle), which birth control technique is best for you? If you are not presently using it, how can you obtain it?

Suggested Readings

Babies who have babies: A day in the life of teen pregnancy in America. (1994, October 24). *People* magazine. If you want a good idea of what it's like to be a young unwed teenager, read this.

Consumer Reports. (1995, May). How reliable are condoms? Tests the strength of over 40 types of condoms, along with giving consumer ratings.

Goldberg, M. S. (1993, September). Choosing a contraceptive. *FDA Consumer.* Accurate information about contraception and the prevention of sexually transmitted diseases.

Gordon, L. (1990). *Woman's body, woman's right: A social history of birth control.* New York: Penguin. An excellent history of birth control.

Hatcher, R., et al. (1994). *Contraceptive technology.* New York: Irvington. This is the most comprehensive and technical review of contraceptive methods. It is revised every two to three years.

Jarrow, J. P. (1987, December). Vasectomy update: Effects on health and sexuality. *Medical Aspects of Human Sexuality.* Discusses the effects of vasectomy on the endocrine and immune systems.

Levathes, L. (1995, January/February). Listening to RU 486. *Health.* Discusses the emotional experiences of taking the "abortion pill."

Patient Guide: How to use a condom. (1987, July). *Medical Aspects of Human Sexuality.* Gives instructions on how to use condoms correctly.

Planned Parenthood. (1989). *Facts about birth control.* Easy to understand; written at a sixth-grade level.

Tribe, L. (1992). *Abortion: The clash of absolutes.* New York: W.W. Norton. An even-handed treatment of the controversy over abortion.

Trussell, J. (1995). *Emergency contraception: The nation's best kept secret.* Atlanta: Bridging the Gap Press. How to use the pill as emergency birth control.

Wattleton, F. (1989, July 24/31). Teen-age pregnancy: The case for national action. *The Nation.* Discusses the high teen pregnancy rate in the United States and suggests possible causes and solutions.

CHAPTER 7

When you have finished studying this chapter, you should be able to:

1. Explain the process of conception, including the terms primary follicle, Graafian follicle, capacitation, and zygote;

2. Trace the process of development from the zygote through implantation, and discuss some problems that sometimes occur with implantation;

3. Describe what occurs in each trimester of pregnancy for both mother and embryo/fetus;

4. Summarize the possible effects of smoking, drug use, environmental hazards, diseases, and anesthetics on the embryo and fetus;

5. Compare and contrast several methods for detecting fetal abnormalities;

6. Discuss traditional and alternative birthing practices, including Lamaze, Leboyer, home births, and the use of midwives;

7. Explain the three stages of labor and the events that occur in each;

8. Identify and explain the problems that can occur during breech deliveries, cesareans, and preterm deliveries;

9. Describe the hormones necessary to produce milk for breast-feeding and discuss the benefits of breast-feeding;

10. Discuss recommendations concerning sexual intercourse during and following pregnancy; and

11. Compare and contrast the various methods for dealing with infertility.

Pregnancy and Childbirth

Pregnancy and childbirth are intense experiences, remembered for a lifetime. The most potent human emotions—hope, fear, love, loneliness, depression, and joy—are often felt at their maximum levels. The first time that a woman feels a baby move inside of her, her first labor contractions, the first time that a child is viewed by its parents after birth—these are often the things that make life worth experiencing. Giving birth means assuming a great responsibility, for the newborn is totally dependent upon his or her parents. Relationships change with the addition of a new life and new responsibilities within a family. This requires that many adjustments in living be made. As you will see, it is important to begin making these adjustments before the baby is born—sometimes even before the baby is conceived.

CONCEPTION AND IMPLANTATION

As you learned in Chapter 2, the ovaries in females store eggs (**ova**). Each egg is surrounded by a small sac, forming what is called a **primary follicle.** As a result of the release of FSH from the pituitary gland, one (or more) of these primary follicles matures to become a **Graafian follicle.** About midway through a 28-day menstrual cycle, the Graafian follicle breaks open and the ovum is released into the abdominal cavity, where it is picked up by one of the Fallopian tubes. The ovum moves through the Fallopian tube to the uterus. It will take 3 to 7 days to reach the uterus, but *the ovum can only be fertilized during the first 24 hours after it leaves an ovary* (Moore, 1982). If a sperm does not penetrate the ovum during this time, the ovum becomes overly "ripe" and eventually disintegrates in the uterus, leaving the woman's body during her menstrual flow.

At orgasm during sexual intercourse, a man will ejaculate 200–300 million **sperm** into the vagina. They then attempt to pass through the cervix and uterus to the Fallopian tubes. This is not an easy journey. Some sperm are stopped by the force of gravity, some by the acidity of the woman's reproductive tract, some by clumping, and some by simply taking wrong turns (because usually only one egg is released, only one Fallopian tube will be the right path for the sperm to follow). Only a few thousand will live long enough to reach the Fallopian tubes, and fewer than 50 will reach the egg within the Fallopian tube (Moore, 1982). Most sperm live for only 3 days inside a woman's reproductive tract, but a few may live for as long as 5 days (Wilcox et al., 1995). Thus, intercourse can result in conception during only 6 days out of every 28-day menstrual cycle (5 days before ovulation and the day of ovulation). If you want to know the probability of conception on any one of these 6 days, go back to the section on fertility awareness methods in Chapter 6.

While the sperm are in the female's reproductive tract, they undergo a process called **capacitation** in which their membranes become thin enough so that an enzyme (called hyaluronidase) can be released to soften the egg's outer layers. Only sperm that undergo capacitation are able to unite with an egg. For capacitation to finish, sperm need to spend several hours in the reproductive tract. The egg sends out tiny projections

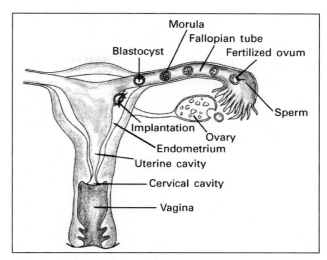

Figure 7–1 Ovulation, fertilization, and implantation.

and pulls one sperm to its surface, the *zona pellucida.* That sperm secretes the enzyme and penetrates the egg's surface, and **conception** takes place. An instantaneous chemical reaction within the ovum then prevents other sperm from penetrating the ovum. Within 24–30 hours, the nuclei of the sperm and the ovum fuse to form a one-celled organism called a **zygote.** This single cell contains the complete genetic code or blueprint for a new human being.

Shortly afterward, the zygote splits into two separate cells, then four, then eight, etc. As cell division occurs, each individual cell gets smaller. This collection of cells is called a **morula.** While cell division is rapidly taking place, the morula is slowly continuing its trip toward the uterus. When the conceptus has about 100 cells, it has developed a fluid-filled center and is then called a **blastocyst.** After the blastocyst reaches the uterus, it may float there for several days. At this point it is still smaller than the head of a pin.

The endometrium is receptive to **implantation** for a 4- or 5-day period (Rogers, 1995). At about 11 to 12 days after conception, the blastocyst attaches itself to the endometrium via hairlike roots called *villi.* By this time the endometrium has a large supply of blood that can serve as a source of nutrients and oxygen. By about 14 days after conception, implantation is usually complete (see Figure 7–1). The blastocyst is

---

ovum An egg; the female reproductive cell.

primary follicle An immature ovum enclosed by a single layer of cells.

Graafian follicle A mature follicle.

sperm The germ cell of a male.

capacitation A process that sperm undergo while traveling through the female's reproductive tract in which their membranes become thin enough so that an enzyme necessary for softening the ovum's membrane can be released.

conception The union of an egg and a sperm.

zygote The one-celled organism created from the fusion of a sperm and egg.

morula The collection of cells formed when the zygote begins rapid cell division.

blastocyst The fluid-filled sphere that reaches the uterus after fertiliza-

tion and which was created by the transformation of the morula through continued, rapid cell division.

implantation The process by which the blastocyst attaches itself to the wall of the uterus.

now called an embryo; it is called a fetus at about 8 weeks (see section below on pregnancy).

The outer cell layers of the embryo are called the **trophoblast** (from the Greek *trophe,* meaning "to nourish"). The trophoblast begins to grow rapidly and forms four protective layers or membranes, each with a special function. One produces blood cells for the embryo until it can produce its own. Another cell layer forms the **umbilical cord**—the major link between the developing baby and its mother. The third membrane is called the **amnion.** It is a thick-skinned sac filled with water that surrounds the embryo. By surrounding the developing baby with liquid, protection is provided against bumps, sudden movements by the mother, and changes in temperature. The fourth membrane, called the **chorion,** will develop into the lining of the placenta. The **placenta** is an organ that serves as a connection or interface between the embryo's systems and those of the mother. The embryo is connected to the placenta by the umbilical cord. Through it, the embryo receives nourishment from its mother. It also receives antibodies for protection against infection. Wastes from the embryo's system travel from the umbilical cord to the placenta, and from there are taken up by the mother's excretory system.

Over 98 percent of term pregnancies result in the birth of a single baby. What causes twins (or other multiple births)? If two different ova are fertilized by two different sperm, the result is *dizygotic* (fraternal or nonidentical) twins. *Monozygotic* (identical) twins result when a fertilized ovum subdivides before it implants in the uterus. Interestingly, the chance of having multiple births increases with the age of the mother (Lazar, 1996).

PROBLEMS WITH IMPLANTATION

It has been estimated that about three-fourths of conceptions either fail to implant or are spontaneously aborted within the first 6 weeks (Boklage, 1990). Over 20 percent fail before pregnancy can be detected by chemical tests (Wilcox et al., 1988). Failure to implant may be a means of "weeding out" or preventing the further development of blastocysts that are not completely healthy or normal.

One type of problem with implantation is called **ectopic pregnancy** (from the Greek *ektopos,* meaning "out of place"). In this case, conception is successful, but implantation takes place outside the uterus. In about 96 percent of these cases, implantation occurs in a Fallopian tube. In rarer cases, implantation can occur on an ovary, on the cervix, or in the abdomen. When implantation occurs in a Fallopian tube (often called a *tubal pregnancy*), there is simply not enough room for the embryo to grow. Unlike the uterus, the Fallopian tubes are not capable of much expansion. The embryo may be dissolved by the body's immunological system, but if it continues to grow after the eighth week, it may break the tube. This is a serious complication of pregnancy, and it can be fatal. Although the mortality rate has been dropping due to earlier detection, ectopic pregnancies still account for about 9 percent of all pregnancy-related deaths (Yao & Tulandi, 1997).

The rate of ectopic pregnancies increased fourfold from 1970 to the mid-1980s, and they now account for about 2 percent of all pregnancies (Yao & Tulandi, 1997). Ectopic pregnancies are caused by conditions that block or slow passage of the conceptus to the uterus. The most common cause is scarring of the Fallopian tubes due to pelvic inflammatory disease resulting from untreated chlamydia or gonorrhea (Chow et al., 1990). Use of condoms reduces the risk of tubal pregnancies in sexually active women. Other factors include anatomical malformations of the tubes, use of (older types of) intrauterine devices, increasing age at time of conception, smoking, and douching (Ankum, 1996). If a woman has an ectopic pregnancy, she has an increased risk of having more ectopic pregnancies in the future, because many of the causes of the original problem, such as scarring, will still be present in later pregnancies (Ankum et al., 1996a).

> "I got pregnant only to learn it was tubular. . . . One year later I got pregnant again and again it was tubular. The specialist told me I needed surgery to remove scar tissue that was choking my Fallopian tube. . . ."
>
> (from the author's files)

Until recently, the only way to diagnose an ectopic pregnancy was to perform surgery and directly examine the internal reproductive organs (*laparoscopy*). Today, the vast majority of ectopic pregnancies can be diagnosed by transvaginal sonography (which uses sound waves to create a picture of the inside of a woman's reproductive tract) in combination with blood HCG levels (Ankum et al., 1996b). In the past, women diagnosed with tubal pregnancy always had part of the affected tube surgically removed, which greatly reduced the chances of pregnancy in the future. Now, if an ectopic pregnancy is confirmed, it can be treated by laparoscopic surgery rather than tubal sectioning (Maruri & Azziz, 1993). The probability of subsequent successful pregnancies is just as high with the conservative

trophoblast The outer four cell layers of the embryo.

umbilical cord The cord that connects the embryo or fetus to the mother's placenta.

amnion A thick-skinned sac filled with water that surrounds the fetus.

chorion The fourth membrane of the trophoblast; it develops into the lining of the placenta.

placenta An organ that serves as a connection or interface between the fetus's systems and those of the mother.

ectopic pregnancy The implantation of a fertilized egg outside of the endometrium of the uterus.

therapy as with the radical therapy (Yao & Tulandi, 1997). Alternatively, some physicians are treating ectopic pregnancies by administering a drug called methotrexate (Korhonen, Stenman, & Ylostalo, 1996). This field is undergoing rapid change, and there is increasing optimism that more and more women will be treated with these conservative procedures.

PREGNANCY

Pregnancy lasts an average of 260 to 270 days. "Due dates," the dates that physicians set as the "expected time of arrival" for babies, are rarely the dates of actual birth. This is because the exact date of conception is hard to measure with complete accuracy and because there is great variability among births, even with the same parents. Because pregnancy lasts about nine months, this time is divided into three-month periods called **trimesters** for descriptive purposes.

THE FIRST TRIMESTER—THE MOTHER

Recall from Chapter 3 that a missed menstrual period is not a sure sign of the start of pregnancy. There are many things that can cause a missed period, including stress and illness. Women athletes, especially track stars and gymnasts, commonly miss one or more periods in the course of several months. In addition, a woman can miss a period without knowing it, because implantation of the blastocyst into the uterine wall can cause bleeding that may be mistaken for menstruation.

When a period is missed, women often want quick and accurate information about whether or not they are actually pregnant. To get this information, they take a "pregnancy test." These tests are conducted by physicians and medical technicians, and some are even sold in drugstores for use at home. They work by determining if a hormone secreted by the placenta (**human chorionic gonadotropin**, or **HCG**) is present in a woman's urine. To test for HCG, a drop of urine is mixed with chemicals in a tube or glass. If the mixture does not coagulate (lump together), HCG is present and the woman is pregnant. This test is most accurate (95–98 percent) two weeks or more after a missed period. Tests that measure HCG in blood can detect if a woman is pregnant even sooner than this. Even when home pregnancy tests are used properly, their accuracy rate is only 70–90 percent, depending on the brand.

The most common symptom of early pregnancy is nausea. It is called "*morning sickness*" (medically referred to as *nausea and vomiting of pregnancy*), though it can occur at any time of the day. Seventy percent of pregnant women experience mild or moderate morning sickness. It usually begins by 4 to 6 weeks after conception, reaches its peak by 8 to 12 weeks, and then spontaneously disappears by the twentieth week. Some women eat crackers, drink soda water, and try a variety of other "home remedies" to cope with morning sickness. Although there are a number of speculations as to its cause (particularly rising estrogen levels), nothing has been proven conclusively (Deuchar, 1995). Because the cause of morning sickness is not known, it should come as no surprise that there is no known safe yet effective way of treating it. Ask your mother or an older female friend if she has ever had morning sickness (and if she did, how she tried to deal with it). Just remember to tell her that you are asking because the author of this book suggested the idea. Also, ask your father if he ever suffered from it—many men experience some of the same symptoms of pregnancy as their partners. In the United States this is sometimes called "sympathy pains," but the medical term for it is **couvade syndrome** (from a French term for the birthing process).

Other symptoms that women may exhibit during the first trimester include enlarged and tender breasts, prominent veins on the breasts, darkened areolas, enlarged nipples, increased frequency of urination, and irregular bowel movements. In addition, many women feel tired and run-down.

THE FIRST TRIMESTER—THE EMBRYO/FETUS

In the first trimester of pregnancy, a substantial amount of development takes place (see Figure 7–2). When the blastocyst implants, it is less than one-third of an inch in diameter. After implantation, it is known as an **embryo.** Cell differentiation begins to take place now, and development occurs in orderly ways. Growth in the embryo occurs from the head downward (cephalo-caudal, or "head to tail," development) and from the center (spine) outward (proximal-distal development). As mentioned earlier, the outer layers of the blastocyst form structures designed to nourish and protect it. In the embryo, three inner cell layers will form specific parts of the body. The *ectoderm* forms the nervous system, skin, and teeth. The *mesoderm* forms the muscles, skeleton, and blood vessels. The *endoderm* forms the internal organs (lungs, liver, digestive system, etc).

In the third week of pregnancy, a central structure called a "neural tube" becomes a dominant feature. This will become the central nervous system. By the end of the fourth week, the umbilical cord, heart, and digestive system begin to form. At 6 weeks, the embryo has a "tail," which will become the tip of the spine. It also has structures that look like gills but are actually parts of the neck and face that have not yet fully devel-

trimesters Three-month periods of pregnancy (months 1–3, 4–6, and 7–9), so labeled for descriptive purposes.

human chorionic gonadotropin (HCG) A hormone secreted by the chorion that stimulates the ovaries during pregnancy.

couvade syndrome The experiencing of pregnancy symptoms by male partners; sometimes called "sympathy pains."

embryo The term given to the blastocyst after it has implanted.

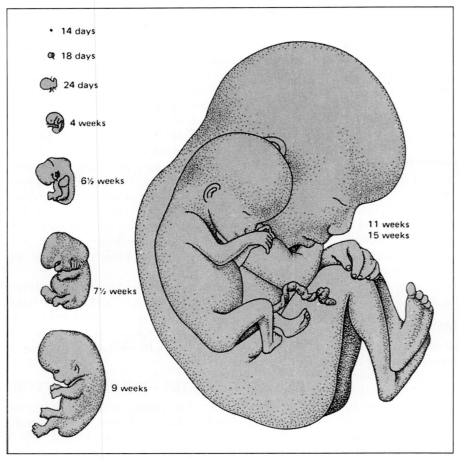

14 days

18 days

24 days

4 weeks

6½ weeks

7½ weeks

9 weeks

11 weeks
15 weeks

Figure 7–2   The embryonic period. This is a life-size illustration of the growth of the human embryo and fetus from 14 days to 15 weeks.

"During my fifth month of pregnancy, I felt the baby move for the first time. It almost felt like the rumbling a stomach makes when you're hungry. It could be described as a 'fluttering,' and it is very exciting. Unfortunately, it doesn't happen that often at first, so you have to 'catch it' if you want your partner to share in the experience."

(from the authors' files)

Women generally begin to get over their morning sickness in the second trimester. They respond to quickening, as just described, but they also notice that their figures are changing, and they may worry about no longer being attractive to their partners. As her abdomen expands, red lines, or "stretch marks," may develop on a mother-to-be. The breasts begin to swell and may start to leak *colostrum,* a thick, sticky liquid that is produced before milk starts to flow. Water retention may cause edema (swelling) in the ankles, feet, and hands. As a result, it becomes harder to remove rings from fingers. Women may begin to develop varicose veins and/or hemorrhoids. Interestingly, the cessation of morning sickness often brings an increase in appetite, and some women may experience heightened sensuality. Freed of the worry about getting pregnant (it's already happened!), some women begin to express their sexuality fully for the first time. Masters and Johnson (1966) report that some women experience their first orgasms in the second trimester. I will discuss sexual behavior during pregnancy at length a little later.

oped. At 8 weeks, the embryo is about 1⅛ inches long. All organs have begun to develop. The heart is pumping, and the stomach has begun to produce some digestive juices. The embryo, although small, is well on its way toward developing into a unique human being. After 8 weeks and until birth, the developing organism is called a **fetus.**

THE SECOND TRIMESTER—THE MOTHER

In the fourth or fifth month of pregnancy, the movements of the fetus can be felt by its mother. This first experience of movement is called **quickening** (Figure 7–3). Although a woman generally knows by this time that she is pregnant, this knowledge is usually an abstract or "intellectual" thing. Once a woman begins to feel a new life moving inside of her, the fetus begins to be viewed as a person, and an emotional attachment to her unborn baby generally begins to form.

fetus The term given to the embryo after the eighth week of pregnancy.

quickening The first time a pregnant woman experiences movement of the fetus, usually in the fifth month.

THE SECOND TRIMESTER—THE FETUS

At this time, the fetus begins to make sucking motions with its mouth. In the fifth month, the fetus has a detectable heartbeat (see Figure 7–5). It will respond to sound, and it begins to show definite periods of sleep and wakefulness. If born at this time, the fetus has only a 1-in-10,000 chance of survival. In the sixth month, the fetus can open its eyes. It will suck its thumb and respond to light. At the end of the second trimester, the fetus is about 1 foot long and weighs about 1 pound.

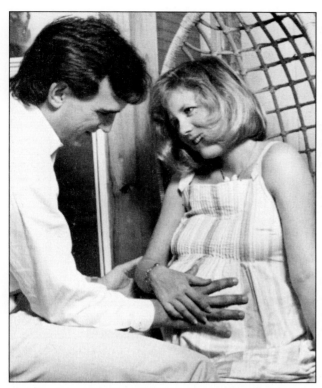

Figure 7–3 A woman can feel her unborn baby moving for the first time (called quickening) in the fourth or fifth month of pregnancy.

THE THIRD TRIMESTER—THE MOTHER

For the expectant mother, walking, sitting, and rising become more difficult. Pregnant women often have to learn new ways to sit down and to get up out of a chair. Expectant mothers often experience back pain from carrying a new weight load in front of them, which shifts their center of gravity. The uterus, enlarging with the rapidly growing fetus, puts pressure on the bladder and stomach. As a result, urination becomes more frequent—a woman may have to urinate four or five times a night. Indigestion, heartburn, gas, and constipation are also common complaints. The active movements of the fetus may prevent restful sleep, especially if the fetus develops hiccups. Finding a comfortable sleeping position is not easy, and a woman may have to change positions several times each night. Leg cramps may occur, a condition sometimes treated by taking extra calcium. The woman's navel may push out. Often she experiences a low energy level (Leifer, 1980).

Until recently, most authorities recommended an overall weight gain of 22 to 27 pounds at term (American College of Obstetricians and Gynecologists, 1985). However, a later study found that 80 percent of women with good pregnancy outcomes had had gains of 22 to 46 pounds (Abrams & Parker, 1990). In 1990, the Institute of Medicine made new recommendations based on initial weight and height: 28–40 pounds for underweight women, 25–35 pounds for normal-weight women, 15–25 pounds for overweight women, and at least 15 pounds for obese women. If weight gain is too low, there is concern about the fetus having proper levels of nourishment. If weight gain is too high, blood pressure may rise and there is an increased risk of a difficult labor and the need for emergency cesarean delivery (Johnson, Longmate, & Frentzen, 1992). Also, excess weight may be hard to lose after the baby is born. Weekly checkups by a physician are usual when the due date gets close.

THE THIRD TRIMESTER—THE FETUS

By the end of the seventh month of pregnancy, the fetus is about 15 inches long and weighs about 1½ pounds (see Figure 7–6). Fatty tissues begin to develop under the skin. A lack of these tissues is what causes premature infants to look so skinny. In the eighth month, the fetus's weight begins to increase dramatically. At the end of the eighth month, it will weigh about 4 pounds and will be 16 to 17 inches long. From this point on, the fetus will gain about ½ pound per week. In the ninth month, the fetus will grow to about 20 inches in length and weigh an average of 7 to 7½ pounds (see Figure 7–7). There is considerable variation in these measurements, however.

The fetus is covered with light hair (**lanugo**) and a waxy bluish substance (**vernix caseosa**). These will make the newborn look very interesting, but they serve as protective devices. Today, the vernix is often allowed to be absorbed into the skin after birth rather than being immediately washed off.

SEXUAL INTERCOURSE DURING PREGNANCY

Feelings about sexual behavior during pregnancy vary from culture to culture. In some cultures, a man's penis is seen as intruding on the territory of the fetus if sexual intercourse occurs during pregnancy. In other cultures, pregnancy is closely associated with sexuality, and women may even experience orgasms while giving birth. There is a wide variety of beliefs (many of which are mistaken) about having sexual intercourse during pregnancy.

In North America, two patterns of sexual activity have been observed during pregnancy. The first is a steady decline in sexual activity throughout pregnancy (Alder, 1989; Call, Sprecher, & Schwartz, 1995). However, a recent large-scale study found that 90 percent of couples were still having intercourse in the fifth month of pregnancy (Hyde et al., 1996).

lanugo Light hair that covers a newborn baby.

vernix caseosa A waxy bluish substance that covers a newborn baby.

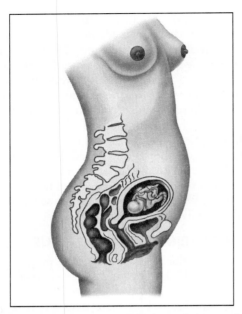

Figure 7–4 THIRD MONTH. The fetus is now about 3 inches long and weighs about 1 ounce. It may continue to develop in the position shown or may turn or rotate frequently. The uterus begins to enlarge with the growing fetus and can now be felt extending about halfway up to the umbilicus.

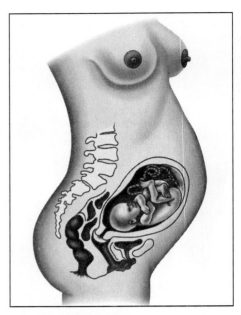

Figure 7–5 FIFTH MONTH. The fetus measures about 10–12 inches long and weighs from ½ to 1 pound. It is still bright red. Its increased size now brings the dome of the uterus to the level of the umbilicus. The internal organs are maturing at astonishing speed, but the lungs are insufficiently developed to cope with conditions outside of the uterus.

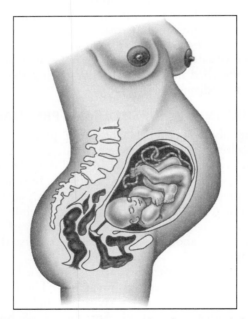

Figure 7–6 SEVENTH MONTH. The fetus's weight has about doubled since last month, and it is about 3 inches longer. However, it still looks quite red and is covered with wrinkles that will eventually be erased by fat. At 7 months the premature baby has a fair chance for survival in nurseries cared for by skilled physicians and nurses.

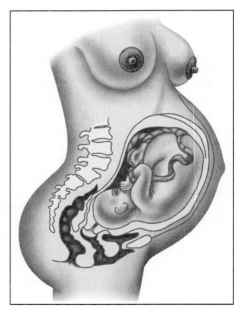

Figure 7–7 NINTH MONTH. At birth or full term the baby weighs on average about 7¼ pounds if a girl and 7½ if a boy. Its length is about 20 inches. Its skin is coated with a creamy coating. The fine downy hair has largely disappeared. Fingernails may protrude beyond the ends of the fingers.

"The fact that it's impossible to get any more pregnant, it makes sex during pregnancy very exciting. My senses seemed much more heightened and both my husband and I found it to be extremely pleasurable."

(from the author's files)

The second pattern, discussed by Masters and Johnson (1966), is an increase in sexual activity from the first to the second trimester, followed by a decrease in the third trimester (Bogren, 1991; Hart et al., 1991). As you might expect, attitudes about sex during pregnancy are related to attitudes about sex before pregnancy (Fisher & Gray, 1988).

Women (and men) list three general reasons for a declining interest in sexual intercourse during pregnancy. One reason is physical discomfort. The traditional man-on-top position can be increasingly uncomfortable for the woman as pregnancy progresses. This discomfort can be alleviated through the use of alternate positions that lessen pressure on the woman's abdomen during intercourse. The female-on-top position is probably most comfortable, and it has even been suggested that having sex in this position is a good exercise for women preparing for "natural" childbirth (Bradley, 1981). After the fetus drops, the penis can make contact with an unyielding cervix during sexual intercourse, so shallow penetration is recommended for intercourse late in pregnancy.

A second reason for decreased interest in sex is that women may feel that they no longer appear attractive. They fear that their partners will see them as "ugly," lose interest, and begin to look for new sex partners. Interestingly, men generally do not list this as a cause of decreased sexual activity.

"We never had sexual intercourse after the first 5 months. I felt disgusted, embarrassed and upset with the way my body had changed. . . ."

(from the author's files)

The third reason is fear about the pregnancy or about the possibility of harming the fetus (Bogren, 1991).

"At about the middle of my pregnancy, my husband had a decrease of interest in sex while my interest had increased. I was already feeling fat and ugly, and this just made things worse. Many nights I stayed up crying because not only did I think I was fat and ugly, but I thought my husband did too. Then I realized his problem was that he was scared he was going to hurt me or the baby. I tried to con-

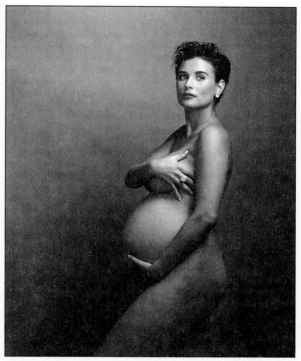

Figure 7–8 In a pose for *Vanity Fair* magazine (August 1991), actress Demi Moore shows that pregnancy is a time when a woman can still be proud of her body.

vince him that he wouldn't, but I guess he just couldn't stop thinking about it while we were having sex."

(from the author's files)

This leads to an important question: How far into pregnancy is it safe to engage in sexual intercourse? There is currently some debate about having sexual intercourse during the last month of pregnancy. Masters and Johnson claim that there is no problem with having sexual intercourse throughout the ninth month, except for certain cases that will be discussed shortly. Grudzinskas et al. (1979) compared women who were sexually active in the ninth month of pregnancy with women who were not. Sexual intercourse during this time was associated with more fecal matter (excrement) in the amniotic fluid of fetuses, along with lower scores for newborns on tests of reflexes and development. Naeye (1979) compared similar groups of women and found that sexual intercourse during the ninth month was related to more infections in amniotic fluid and more breathing problems and jaundice in newborns. White and Reamy (1982) cite several flaws in these studies, but it should be remembered that *even in the flawed studies no problems were associated with having sexual intercourse before the ninth month of pregnancy.* The consensus of opinion is summed up in the highly regarded *Williams Obstetrics:* "it has been generally accepted that in healthy pregnant women, sexual intercourse usually does no harm before the last four weeks or so of pregnancy" (Cunningham et al., eds., 1993, p. 263).

In any case, sexual behavior is not limited to intercourse, and in the latter stages of pregnancy alternative forms of sexual stimulation may be preferred. Hugging, touching, and other displays of affection are important ways of communicating attraction. It is important to speak to a pregnant woman about love and devotion and to express pride in her and hope for the future. Telling her that she is beautiful may be a far more effective way of showing that you care than having sexual intercourse (see Figure 7–8).

There are some circumstances in which sexual intercourse should be practiced carefully or avoided altogether during pregnancy. Sexual intercourse with a partner who has a sexually transmitted disease should, of course, be avoided, especially because some of these diseases can be transmitted to the fetus. Intercourse should not occur if there is vaginal bleeding, or if there are signs that the amniotic sac has ruptured. If the cervix is dilated or if there are other signs of premature labor, sexual intercourse should be avoided.

COMPLICATIONS OF PREGNANCY

Although pregnancy is a complex process, about 97 percent of all births result in a healthy, normal infant (Heinonen et al., 1977). However, complications of pregnancy can occur. Many of these can be prevented if parents-to-be are aware of them.

Substances that can harm an embryo or fetus are called **teratogens** (from the Greek word *tera,* meaning "monster"). The placenta, until recently, was thought to be a perfect filter that kept out all harmful substances. Now we know of hundreds of teratogens that can cross the "placental barrier." Three things determine the harm caused by teratogens—the amount of the teratogen, the duration (amount of time) of exposure to the teratogen, and the age of the embryo or fetus. Each part of the developing body has a time, called a **critical period,** when it is most susceptible to damage. These are the times when it is undergoing most of its formation. During critical periods, body parts are sensitive to "signals" or "messages" about how to

teratogens Substances that can harm an embryo or fetus.
critical period In relation to teratogens, the time during embryological or fetal development during which a particular part of the body is most susceptible to damage.
rubella A disease, also known as German measles, caused by a virus. If contracted during pregnancy, it can cause fetal abnormalities.

develop. Usually, these signals come from the genetic code that guides normal development. However, this sensitivity means that the developing embryo or fetus can be easily influenced by teratogens. Although teratogens should be avoided at all times, most body parts are maximally susceptible to damage during the first 8 weeks of development (see Figure 7–9).

Remember that in the first 8 weeks of pregnancy, many women won't be sure if they are pregnant. There is an important lesson to be learned from this. If a woman wants to change her behavior to improve the chances of having a healthy baby (to stop drinking alcoholic beverages, for example), it's better for her to do this when she is trying to become pregnant rather than after she finds out that she is pregnant. Still, if a woman wants to change, it is better to do so late rather than never. The woman's partner can encourage healthier habits by also changing his habits at the same time.

DISEASES

Even the "weakened" disease organisms of certain vaccines can be harmful if administered just before or during early pregnancy (O'Brien & McManus, 1978). Some strains of the flu, mumps, chicken pox, and other common diseases can also cause harm to the embryo or fetus. One of the first teratogens to be discovered was the **rubella** virus, or "German measles." A baby exposed prenatally to rubella may be born blind, deaf, and/or intellectually impaired. This is especially true if the mother contracts this virus in the first 3 months of pregnancy. A study by Michaels and Mellin (1960) found that 47 percent of the babies born to mothers who had rubella in the first month of pregnancy were abnormal, as compared to 22 percent of the babies whose mothers contracted the disease in the second month and 7 percent of the babies whose mothers caught the disease in the third month. In the second trimester, rubella can cause hearing problems and language retardation (Hardy et al., 1969). Rubella is spread by direct contact or through touching droplets from an infected individual. Spring and early summer are times of the greatest risk of contracting this disease. During an outbreak in 1964–1965, 20,000 women gave birth to abnormal infants, and another 11,000 had abortions after contracting the illness (Franklin, 1984). Administering a rubella vaccine to schoolchildren has greatly decreased the prevalence of the disease, and the number of cases has been reduced by about 99 percent since 1969 (CDC, 1997b). A woman can safely be inoculated against rubella any time up to three months before becoming pregnant. It is estimated that over 5 million American women are not immune to rubella (Behrman & Vaughan, 1983). Other diseases that can cause harm to the fetus but that can be inoculated against include mumps, polio, and measles.

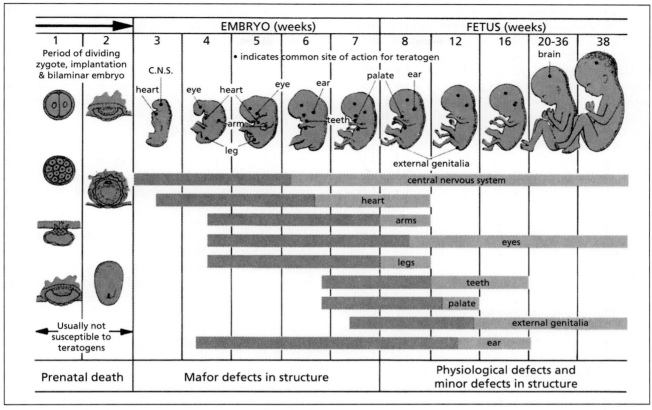

Figure 7–9 Critical periods in prenatal development. Darker shading represents highly sensitive periods; lighter shading represents less-sensitive periods.

"Unbeknown to the doctor or myself, I had chlamydia. Since the baby was in the birth canal he contracted it in the eyes. It was bad! He almost went blind."

(from the author's files)

As you learned in Chapter 5, most types of *sexually transmitted diseases* can affect a fetus or newborn baby. There is a one-in-four chance that a pregnant woman with an HIV infection (the virus that causes AIDS) will pass the virus to the fetus. If a woman has an active case of genital herpes at delivery, you recall, it can result in the death of or neurological damage in the newborn. Blindness in newborns can result if the mother has either gonorrhea or chlamydia at the time of childbirth. Untreated syphilis is particularly bad, because it can be passed to the fetus, resulting in considerable damage by the time the baby is born.

TOXEMIA OF PREGNANCY

A pregnant woman can also have a disease called **toxemia of pregnancy.** In its early stages, this disease is called *preeclampsia,* and up to 10 percent of pregnant women will experience it. Its symptoms include protein in the urine (which allows it to be detected by

physicians), high blood pressure, weight gain, and swollen joints due to excessive water retention. The high blood pressure results from overactivity of what is known as the sympathetic nervous system (Schobel et al., 1996). In about 5 percent of cases, the disease advances to *eclampsia,* characterized by convulsions and coma. In the Western world, eclampsia is one of the major causes of maternal death.

"During my first pregnancy, they discovered that I had toxemia. I was immediately put on a low-sodium diet and had to stay in bed for the rest of my pregnancy. When she was born, my baby weighed only 4 lbs. 10 oz. and her weight then fell to 4 lbs. 1 oz. She had respiratory problems and severe jaundice. They kept her in an incubator for about four weeks. She was a very pretty baby, but took a long time to start crawling, walking, and talking. I didn't get toxemia in my second or third pregnancies."

(from the author's files)

toxemia of pregnancy A disease of pregnancy characterized in its early stage (preeclampsia) by high blood pressure, weight gain, swollen joints, and protein in the urine. Unless it is corrected, it can lead to convulsions and coma.

Toxemia generally occurs in the last trimester of pregnancy. It is most likely to occur in younger women, especially if it is the first birth. Women who develop toxemia in their first pregnancy often do not have it again in later pregnancies. Also, women whose partner(s) had used condoms prior to pregnancy have a much higher risk of preeclampsia (Klonoff-Cohen et al., 1989)—in other words, both these findings indicate that the greater the exposure to semen, the less the likelihood a woman will develop preeclampsia. These results suggest that the disease is due to altered immune responses (see Beer, 1989). Calcium supplementation and bed rest are the usual treatments for preeclampsia (Witlin & Sibai, 1997). However, if these don't work and eclampsia develops, labor will be induced prematurely in order to keep the disease from being life-threatening.

SMOKING

About one-fifth of all pregnant women smoke (Williamson et al., 1989). However, cigarette smoking during pregnancy is associated with a substantial increase in the risk of having a low-birth-weight baby (Nordentoft et al., 1996) and increases the risk of miscarriages, ectopic pregnancies, preterm births, and infant mortality (see DiFranza & Lew, 1995, for a review). Babies that are born to women who smoked while pregnant have decreased respiratory function (Stick et al., 1996). Mothers who smoke during pregnancy double (9 cigarettes or less a day) or triple (10 cigarettes or more a day) the risk that their baby will die of sudden infant death syndrome, also known as "crib death" (Haglund et al., 1990). Only about a third of women who smoke quit smoking when they learn that they are pregnant (Fingerhut et al., 1990).

Interestingly, smoking can also affect male sperm production, with men who smoke having higher numbers of abnormal sperm than nonsmokers (Evans et al., 1981). This might be a factor behind the higher rates of infertility found in male smokers as compared with nonsmokers.

ALCOHOL

A recent survey from the Centers for Disease Control and Prevention found that in recent years there has been a big increase in drinking by pregnant women (CDC, 1997a). Drinking during pregnancy can have dire consequences. Approximately two out of every 1,000 children born in the United States have physical deformities and/or mental retardation because of damage caused by the mother's use of alcohol during pregnancy (*Newsweek*, July 31, 1989). The condition is known as **fetal alcohol syndrome (FAS)** and is six times more common now than it

fetal alcohol syndrome (FAS) A condition common to infants born to alcoholic mothers, involving physical and nervous system abnormalities.

was in 1979 (CDC, 1995). Children born with FAS are underdeveloped and have facial deformities, abnormally spaced eyes, and small heads. They also have damage to the central nervous system, and occasionally they have damage to other organs. FAS is the leading cause of mental retardation in the United States, with an economic impact of over $300 million per year (Abel & Sokol, 1987).

Research with animals indicates that alcohol probably has its maximum effect on the developing fetus in the third week of pregnancy (Rossett & Sanders, 1979; Sulik et al., 1981), but it can have negative effects throughout pregnancy. Animal research also shows that alcohol can cause the umbilical cord to temporarily collapse, cutting off oxygen to the fetus and causing a condition known as *minimal brain damage* (Mukherjee & Hodgen, 1982), which has been associated with hyperactivity and learning disabilities in humans. Even moderate alcohol consumption can result in children with emotional problems and inability to cope in school (known as *fetal alcohol effect*). These data indicate that *it is probably best to avoid alcohol entirely during pregnancy.* If you wouldn't give your newborn baby an alcoholic drink, why would you give it one before it is even born? In fact, because maximal damage might occur before a woman knows that she is pregnant (the third week), it might be best for a woman to stop drinking if she is trying to become pregnant. Even after the baby is born, a mother should not drink if she is breast-feeding, because this can result in the baby having a low rate of motor development and coordination (Little et al., 1989).

COCAINE

The use of cocaine, especially "crack" cocaine, increased dramatically during the 1980s. Sadly, many of the casualties of this epidemic are children born to women who used cocaine during pregnancy. Ten to 15 percent of babies born in large urban areas are affected by their mothers' use of cocaine—over 100,000 babies a year (Chasnoff et al., 1990; U.S. General Accounting Office, 1990; Gomby & Shiono, 1991). The immediate hospital costs alone (at time of delivery) are $5,000 to $12,000 more per child than for unexposed babies (Calhoun & Watson, 1991; Phibbs et al., 1991).

The most widely reported effects of cocaine exposure during pregnancy are diminished growth and preterm birth (both resulting in low birth weight) and decreased head circumference (Bateman et al., 1993; Singer, Arendt, & Minnes, 1993). "Crack babies" have a variety of sensory-motor and behavioral deficits, including irritability and disorientation. Although studies have not yet followed these children in the long term, there is a real potential of later learning problems and behavioral disorders (see Singer, Farkas, & Kliegman, 1992, for a review). Several states have begun to prosecute women who used cocaine during pregnancy. However, the American Medical Association's Board of Trustees suggests that in most cases rehabilitative treatment and education will be more effective (Cole, 1990).

OTHER DRUGS

Many drugs can cross the placental barrier. These include illegal drugs, prescription drugs, and "over-the-counter" drugs. Women who are addicted to heroin (or methadone) while pregnant will give birth to infants who are already addicted as well. These infants must go through withdrawal after being born, showing symptoms such as fever, tremors, convulsions, and difficulty in breathing. This is extremely painful and can be life-threatening (Brazelton, 1970). Heavy use of marijuana during pregnancy can result in problems that resemble fetal alcohol syndrome (Hingson et al.,1982).

Numerous prescription drugs, such as antibiotics and tranquilizers, are clearly harmful to the fetus. In addition, commonly used drugs such as antihistamines (in allergy medication and cough medicine) and megadoses of certain vitamins have proved to have harmful effects. Even aspirin can be harmful. The Food and Drug Administration has issued a warning against taking over-the-counter aspirin products during the last 3 months of pregnancy because they can affect fetal circulation and cause complications during delivery. Caffeine (found in coffee, tea, many soft drinks, and chocolate) has been found to result in reduced birth weights (Vlajinac et al., 1997). Even too much vitamin A during pregnancy can cause birth defects (Rothman et al., 1995). The best advice, which is being given by increasingly large numbers of physicians, is not to take any drugs during pregnancy if at all possible.

ENVIRONMENTAL HAZARDS

Pollutants, such as heavy metals in drinking water, can cause damage to the fetus. One example of this is the high levels of lead and cadmium that can be found in the drinking water of houses in older neighborhoods. These high levels are caused by the passage of water through old water pipes. If the plumbing in your house is old, it is best to let the water run for a few minutes before a pregnant woman drinks it, especially in the early morning, after water has been "sitting still" in the pipes overnight. Mercury-contaminated fish eaten by pregnant women can cause deformities and mental retardation in their children (Milunsky, 1977).

Radiation and X-rays are also powerful teratogens. Radiation from the atomic bombs exploded at Hiroshima and Nagasaki in World War II caused pregnant women exposed to radiation from the bombs to have spontaneous abortions, stillborn babies, and severely handicapped children (Haley & Snider, 1962). X-rays are especially damaging to the fetus in the first trimester. The use of X-rays has been linked to increased risk of leukemia (Stewart & Kneale, 1970). Some researchers claim that X-rays given at low dosages late in pregnancy (for example, an emergency X-ray or an X-ray to determine if the baby's head is too large to pass through the mother's birth canal) cause no harm to the fetus (Moore, 1983). Others feel that ultrasound can give essentially the same information at no risk to the fetus.

ANESTHETICS

Many people reading this book came into the world in a drugged state. This is because anesthetics were used during their mother's labor to relieve pain. Under general anesthesia, a mother is unable to push during contractions. As a result, infants born under these conditions have to be pulled into the world with forceps, which resemble tongs. Anesthetics cross the placental barrier and enter the fetus. Because the fetus's organs are not all fully developed at birth, it cannot get rid of drugs from its system as quickly as its mother. This is one reason why some infants have to be slapped in order to get them to cry and begin breathing. Babies born without the use of anesthetics generally come into the world with eyes open and ready to breathe without having to be slapped into consciousness. Children whose mothers are given anesthetics may be more susceptible to anoxia, an inability to get enough oxygen to the brain. This problem can influence many aspects of development. Horowitz et al. (1977) found no permanent problems in the children of mothers who were given anesthetics at birth. However, these researchers (studying births in Uganda and Israel) noted that the amount of anesthetic given in their study was generally less than the amount of anesthetic given in the United States. Brackbill (1979) reviewed studies involving the use of anesthetics in labor and concluded that using these drugs was related to lower scores on developmental measures in newborns.

Today, most births in which anesthetics are used do not involve general anesthetics. The mother may be anesthetized from the waist down, or any of a number of other possible alternatives. Thus, it is possible to be anesthetized and still be conscious and able to push. These procedures also lessen the effects of anesthetics on the baby. It is important to discuss different anesthetizing procedures with the physician before going into labor, especially to determine the physician's preferred methods.

RH INCOMPATIBILITY

There is a protein in the blood of a majority of people called the **Rh factor.** It got its name from the Rhesus monkey, the species in which this blood

Rh factor A protein in the blood of most individuals. If a mother does not have it and the fetus she carries does, she can develop antibodies against the Rh factor. This is called *Rh incompatibility problem.*

protein was first discovered. If a person has the protein, he or she is "Rh positive." If not, the person is "Rh negative." A blood type of "O positive" means that the person is blood type O and has the Rh factor. Each person has two genes that determine whether or not they have the protein. They receive one of these genes from each parent. The Rh positive gene is dominant, which means that if a person has one Rh positive and one Rh negative gene, the blood will have this protein. A person can have Rh negative blood only when both genes are Rh negative.

In about 8 percent of all pregnancies in the United States, the mother will be Rh negative and the fetus will be Rh positive. When this occurs, there is an Rh compatibility problem. For the first birth, this situation usually isn't dangerous. But during labor and delivery, blood of the fetus and the mother, kept separate in the placenta during pregnancy, may mix together. If this happens, the mother's blood treats the Rh factor in the baby's blood as if it were a dangerous foreign substance, forming antibodies that will attack and destroy the "invader" in the future. If the mother then has another Rh positive fetus, these antibodies can cross the placental barrier and attack the fetus's blood supply, causing a miscarriage, possible brain defects in the fetus, or even death. Fortunately, there is an injection that can be given to the mother to prevent these antibodies from forming. The drug used for this is called *Rhogam*. It consists of Rh negative blood that already has these antibodies in it. Since Rhogam already has antibodies, the mother's system doesn't produce very many of its own. The antibodies in Rhogam eventually die off, leaving the mother capable of carrying another Rh positive fetus. It is important that this drug be given immediately after the birth of the first child or after a first miscarriage of a fetus with Rh positive blood. If there is a delay, the mother's body will build up too many antibodies of its own before the drug can be given, making the drug ineffective.

If a child is born with Rh antibodies from its mother, it may survive if given an immediate and complete blood transfusion. It is possible to give a blood transfusion to a fetus in the uterus if it is too small to survive an induced premature birth, but this is a very new, and still experimental, technique.

chorionic sampling A technique used for detecting problems in a fetus during the eighth to tenth weeks of pregnancy. Hairlike projections (villi) of the chorion are collected and examined.

celocentesis A technique used for detecting fetal abnormalities in which a needle is inserted between the placenta and the amniotic sac to retrieve cells.

amniocentesis A technique for detecting fetal problems, involving collection of amniotic fluid after the fourteenth week of pregnancy.

fetoscopy A procedure in which a tube containing fiberoptic strands is inserted into the amniotic sac to allow inspection of the fetus.

ultrasound A noninvasive technique for examining the internal organs of a fetus; it uses sound waves like a radar or sonar scan to create a picture.

DETECTION OF PROBLEMS IN PREGNANCY

In addition to X-rays, which are generally used only late in pregnancy, if at all, there are several other techniques for detecting problems in the fetus while it is inside the uterus. The first is called **chorionic sampling** or *chorionic villus biopsy*. As you learned earlier, the chorion surrounds the embryo and develops early in pregnancy. In chorionic sampling, a small tube is inserted through the vagina and cervix. Guided by ultrasound (also mentioned earlier), the tube will remove some hairlike cells (villi) from the chorion by gentle vacuum suction. These cells can be examined to detect chromosomal problems (such as Down syndrome) or abnormalities in the fetus, to look for evidence of certain diseases, and even to find out the sex of the fetus. One of the major advantages of chorionic sampling is that it can be used to detect problems early in pregnancy (from the eighth to tenth week of pregnancy), but after 12 weeks of pregnancy all of the villi of the chorion are gone. Initial results indicate that although the safety of chorionic sampling compares favorably with that of other methods of detecting fetal abnormalities (Green et al., 1988), serious complications, including fetal or neonatal death, can occasionally result (Rosenthal, 1991). Researchers are currently testing a newer, "safer" prenatal technique called **celocentesis** in which a needle is inserted between the placenta and amniotic sac to retrieve cells (Jurkovic et al., 1993).

Another technique that provides similar information is **amniocentesis.** This technique has been in use longer than chorionic sampling. In amniocentesis, a hollow needle is inserted through the abdomen and uterus of the mother, through the membrane of the amniotic sac, and into the amniotic fluid surrounding the fetus. Ultrasound is often used to guide the needle. Some of the amniotic fluid is then extracted, and fetal cells that have dropped off and have been floating in the fluid are examined for evidence of abnormalities. This precedure can only be used after the fourteenth week of pregnancy, because before that time there may not be enough fetal cells in the amniotic fluid to make it useful and the fetus may take up too much room in the amniotic sac for the procedure to be used safely. Even so, there is a 1 percent chance that the procedure will cause a spontaneous abortion.

A relatively new technique being used is called **fetoscopy.** A tube that contains fiber optic strands—wires that can send back "pictures"—is inserted into the amniotic sac. The "fetus-scope" can inspect the skin of the fetus and can be used to get a blood sample from the fetus. This technique also carries a risk of damaging the fetus, however.

X-rays, chorionic sampling, amniocentesis, and fetoscopy are called "invasive" techniques, meaning that they invade, or get inside of, the uterus and/or the fetus. **Ultrasound** is considered a "noninvasive" technique because it only bounces sound waves off of the uterus and fetus. Ultrasound can be used to diagnose many problems, though they primarily concern structural difficulties such as a malformation of the skeletal system of the mother or fetus. It cannot detect genetic defects as amniocentesis can. Even ultrasound is recommended *only if the expectant mother has an increased risk factor for a problem* with a pregnancy or delivery. Use of ultrasound as a routine screening procedure has no clinical benefits (Ewigman et al., 1993; LeFevre et al., 1993). Detection procedures should never be used simply to find out in advance if the baby will be a boy or a girl. A balance must be found between detecting potential problems and creating new problems when these procedures are used.

FETAL SURGERY

One of the newest methods in treating a problem pregnancy is **fetal surgery**—performing an operation on the fetus while it is still in the uterus of the mother. Thus far, fetal surgery has been performed primarily for urinary tract obstruction and diaphragmatic hernia. With urinary tract obstruction, urine backs up into the kidneys and damages them. When this problem is detected through ultrasound, the surgeon can insert a needle through the mother's abdomen to help place a little tube (catheter) into the fetus's bladder. Urine then drains through the catheter and bypasses the urinary tract. A newer technique involves making a surgical incision through the fetus's abdomen directly to the bladder. This is done because catheters can get pulled out or plugged up, whereas a surgical incision usually will last until birth, when it can be repaired. The use of surgical incisions rather than catheters is called open fetal surgery. The technological advances in this field have been rapid. In 1989, physicians performed the first successful major surgery on a fetus, repairing a hernia of the diaphragm that if left untreated probably would have resulted in death shortly after birth (Harrison et al., 1990). In 1994, surgeons performed the first endoscopic fetal surgery, using tiny needles guided by miniature cameras without cutting open the mother's uterus. In 1996, a 4-month-old fetus was given a bone marrow transplant while still in the uterus (Flake et al., 1996).

Several factors must be taken into account when deciding whether to do fetal surgery. This type of surgery involves a risk of causing the uterus to go into labor prematurely (Longaker et al., 1991). Excessive bleeding in the mother's uterus due to surgery is also a risk, because the uterus has a huge blood supply in mid-pregnancy. As the technology of fetal surgery evolves, it is expected that risks associated with the procedures will decline and that a greater variety of problems will be remedied in this way.

MISCARRIAGES (SPONTANEOUS ABORTIONS)

Earlier in this chapter you learned that about three-fourths of all conceptions fail to survive beyond 6 weeks. Of the remaining quarter, about 10 percent fail to survive to term (Boklage, 1990). There are many possible causes for **miscarriage,** including genetic, anatomic, and hormonal (too-high levels of luteinizing hormone, for example) causes; infections; and maternal autoimmune responses (Branch, 1990; Regan, Owen, & Jacobs, 1990). But routinely, physicians will not attempt to diagnose causes of miscarriage until the woman has had three of them. *Recurrent miscarriages,* as these are called, occur in 0.5–1.0 percent of women (Alberman, 1988). There are two major reasons for this. The first is that most women who have a miscarriage will carry the next fetus to full term. (Some insurance companies don't want to pay for tests after only one or two miscarriages.) The second is that even after three miscarriages have occurred, 60 percent of these women will later have healthy babies, even without treatment.

For many couples, a miscarriage can be a heavy emotional disappointment:

> "I have been married for 14 years. I conceived 7 times, and every time I had a miscarriage at about 4½ months. My doctor diagnosed that my uterus is too weak to carry the baby after 4 months and no matter what and how I try I cannot have a baby of my own. I get very depressed sometimes and cry. Lately I feel very uncomfortable around other people's babies. . . ."

> "Some people may think only women are upset because of losing a baby due to miscarriage . . . but this hurt me in ways unimaginable. . . . First of all, I felt as if it was my fault for not being there. It hurt the most because a part of us had not yet had a chance to grow and mature into our baby. Our relationship is just now returning to normal. . . ."

(from the author's files)

It is important to allow adults who experience miscarriage to express their emotions and grieve (Cole, 1987). It often takes several months to recover.

After reading this section on the possible dangers the fetus is exposed to, you may begin to believe that it is impossible to protect any unborn baby from these dangers and that most babies will be harmed by something as a re-

fetal surgery A surgical operation on the fetus while it is still in the uterus.

miscarriage A spontaneous abortion.

sult. At this point I would like to emphasize that hundreds of millions of babies are born without serious problems. A fetus is incredibly resistant to harm. Eating well, avoiding toxic substances, getting early and regular care from a physician with whom you can communicate and who shares your point of view, and preparing for the birth of your child through study and training (which I will describe shortly)—these all go a long way toward making pregnancy, labor, and birth an enriching, happy experience.

NUTRITION DURING PREGNANCY

One of the best ways to increase the chances of having a healthy baby is to eat well during pregnancy. A poor diet can increase the risk of complications such as toxemia, anemia, and low birth weight.

Remember, if you are pregnant, you are now eating for two. You will need to eat more *good-quality* foods than you usually do. This especially applies if you are a teenager because of the high incidence of low-birth-weight babies in this age group. Anyone expecting twins must also eat plenty of healthy foods. Here are some suggestions for foods you should eat every day to assure a healthy diet (Brewer & Brewer, 1985):

- 4 glasses of any kind of milk (e.g., whole, low fat, skim). If you do not like milk, substitute yogurt.
- 2 eggs.
- 2 servings of meat, including chicken or fish. If you are a vegetarian, substitute beans and rice.
- 2 servings of leafy green vegetables.
- 1 yellow vegetable (five times a week).
- 5 servings of whole breads or cereals (part of your milk intake can be with cereal).
- 1 potato.
- 2 pieces of fresh fruit.

While this diet will provide you with most of the necessary nutrients, the U.S. Public Health Service recommends that pregnant women *supplement their intake of folic acid* by taking a multivitamin.

Your physician may make additional recommendations if you have certain conditions (such as adjusting salt intake to ease toxemia). Some women have negative feelings about gaining weight during pregnancy, but remember, this is not only normal but necessary. Never put yourself on a low-calorie diet during pregnancy.

PREPARING FOR CHILDBIRTH

prenatal examination A health checkup by a physician during pregnancy.

The World Health Organization estimates that worldwide nearly 600,000 women a year die as a result of childbirth (WHO, 1996). Death rates vary, of course, depending on where one lives and the availability of health care. Maternal death rates are lower in the United States than in developing countries, but even here nearly 20 percent of pregnant women never bother to go in for a **prenatal examination** (Rosenberg et al., 1996). Prenatal exams are an essential part of health care during pregnancy. However, there is more to giving birth than assuring the safety of the mother and baby. What about a woman's psychological well-being and emotional adjustment? Modern maternity care in developed countries is aided by many technological advances, but, as Ulla Waldenström (1996) asks, "Does safety have to take the meaning out of birth?" In this section you will learn about several alternatives to the traditional medical model of birthing—methods that are concerned with safety, but also try to make childbirth a meaningful experience.

The use of general anesthesia, labor rooms where women waited for hours (usually separated from their male partners) before giving birth, delivery rooms, waiting rooms for men, and other aspects of "modern" birth techniques were all instituted to make it easier for health professionals to deliver their services (Mead, 1975). Women and their male partners were usually given very little education about what was to come, with only the expectation that it would be painful. For many women, labor and delivery were not positive experiences.

"In 1970 at age 20 I prepared to give birth to my first child. 'Induced labor' was a convenient form of delivery for the physician, and widely practiced in the 70s. At my last visit to his office, he 'broke the water bag,' handed me five Kotex pads, and told me to walk across the street to ——— hospital and admit myself. He said he would check on me later that evening. After having called my husband for me at the hospital, the delivery room nurse started a 'drip' to begin contractions. I now know why the old ——— hospital had bars on the labor room windows. I cannot describe that intense pain, especially not having been prepared for natural childbirth. To jump would have been a blessing."

"My experience with childbirth was very traumatic. The pain I had to endure. . . . I remember reaching out to punch the nurse because the prepping solution she used on me was so cold. The next thing I knew, the gas mask covered my face and I was out cold."

"I woke up in recovery. It was over. 'Ms. G.,' the nurse said, 'You had a baby boy.'"

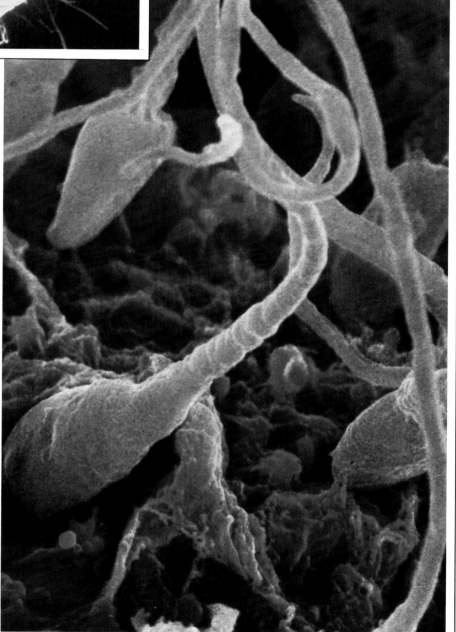

A close-up photo of sperm on
the surface of an ovum

Many sperm attempting to
penetrate the membrane of an
ovum. Only one will be
successful.

A FIBEROPTIC VIEW OF EMBRYONIC AND FETAL DEVELOPMENT

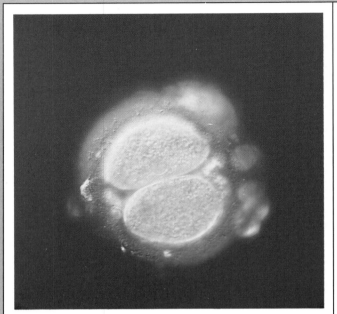

(a) One day (two cells)

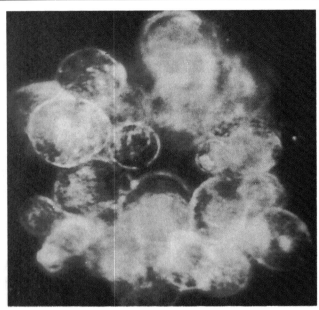

(b) 3 1/2 days

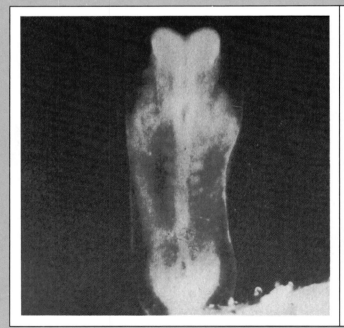

(c) 21 days (note the primitive spinal cord)

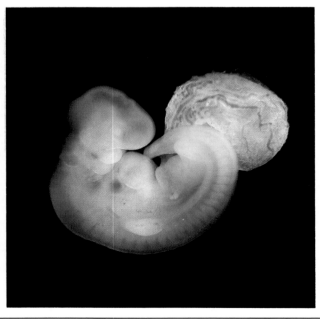

(d) Four weeks

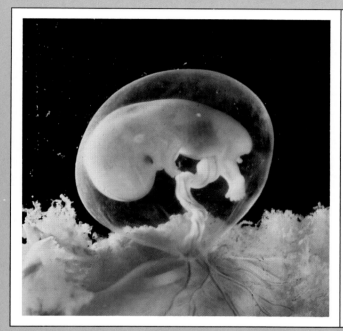

(e) Eight weeks

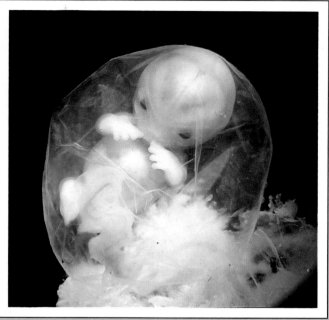

(f) Twelve weeks

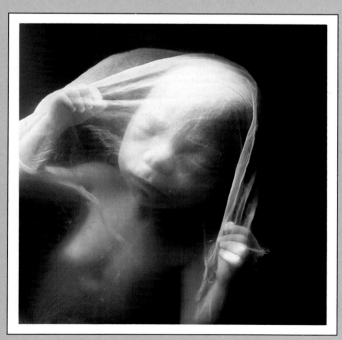

(g) Sixteen weeks

Stages of prenatal development. The heart and nervous system begin to function by the end of four weeks. The fetus is recognizably human by 16 weeks, but will stand very little chance of survival if born before 24 weeks. Note the umbilical cord connection with the placenta in the photo taken at eight weeks.

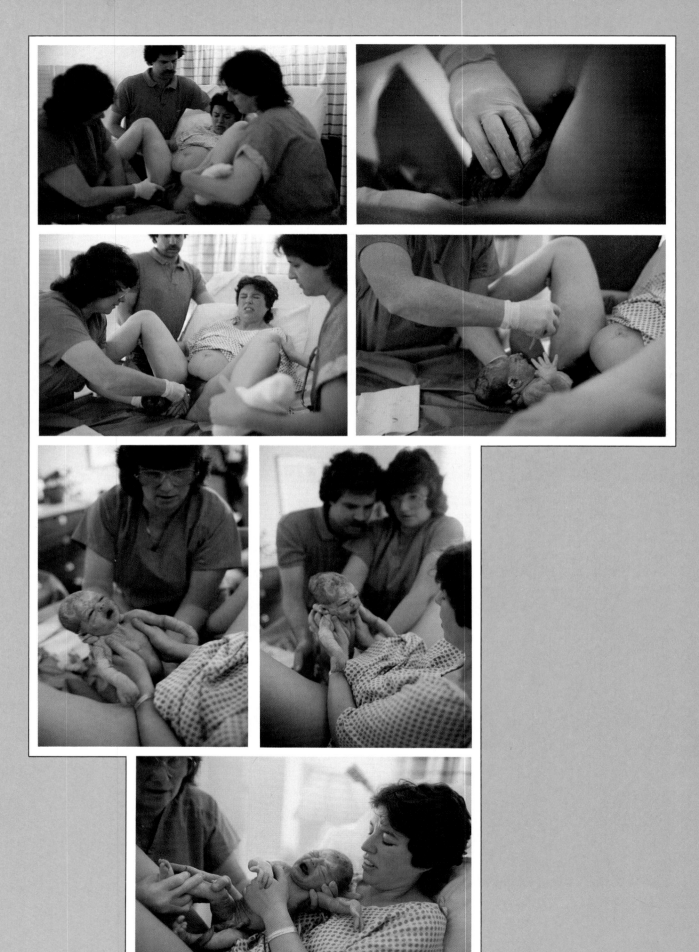

"It is good the way births are being handled now. When I had babies the hospital treated you like the baby belonged to them. My second child wasn't even in the same building with me and I wasn't allowed to see her until the next day. It was torment wondering how she looked."

(from the author's files)

Fathers were generally the forgotten person when a baby was born. They seldom were allowed to be with their wives during any part of labor or delivery.

"I went out of the room to get a candy bar while the nurse gave my wife an exam (neither I nor my wife had ever heard of Lamaze then). When I returned to her room, I saw my wife being taken away. The nurse explained that my wife was giving birth. I'd been awake for several nights with my wife, so I laid down on her bed to rest and fell asleep. Shortly after 6 P.M., the phone next to the bed rang and the nurse told me I was the father of a baby girl."

(from the author's files)

In 1932, a British physician named Grantly Dick-Read wrote a book in which he advocated the practice of *natural childbirth* (Dick-Read, 1972). He became the founder of a movement away from "modern" techniques, a movement that has become increasingly popular in America. Dr. Dick-Read came to the conclusion that modern birthing techniques, in which pregnant women were regarded as "patients" (implying that pregnancy is an illness to be treated), led to a **fear-tension-pain cycle.** He reasoned that women were being put into strange environments, not knowing what was going to happen next and surrounded by strangers, and that this created fear. He believed that fear could cause physiological changes associated with tension, such as tightened muscles. These things were believed to substantially increase the sensation of pain felt by mothers during labor and delivery. Instead, he advocated educating mothers-to-be about pregnancy, labor, birth, and birth procedures to relieve fear of the unknown. He also advocated the training of relaxation techniques to relieve the physiological effects of tension. Drugs were permissible if thought necessary, but were to be avoided if possible. Today, people more properly refer to these techniques as **prepared childbirth** rather than natural childbirth (Bing, 1990).

THE LAMAZE METHOD

A French physician, Fernand Lamaze, traveled to the Soviet Union in 1951 to study birthing techniques in that country. The Soviets had extended the condition-ing techniques of Ivan Pavlov into the area of labor and delivery. They had reasoned that if conditioning could create new associations between food and bells, associations between labor and pain could be replaced with new associations. Women were conditioned to learn pleasant new associations for the sensations of labor and birth. They were also taught relaxation and breathing techniques to aid in making labor easier. Throughout the Soviet Union, Lamaze saw that women were giving birth without either using anesthetics or experiencing great pain. This was a stunning sight to Lamaze. He returned to France, modified the Soviet techniques slightly, and began using this new approach to childbirth in his practice (Lamaze, 1970). An American woman, Marjorie Karmel, used this approach under Lamaze's care and wrote about her experience in a best-selling 1959 book titled *Thank You, Dr. Lamaze.* Since then, the *Lamaze method* has become the most popular form of childbirth in the United States. Compare the following childbirth experiences from two individuals who went through Lamaze training with the four case histories that were presented at the beginning of this section. Which persons do you think remember giving birth as a positive experience?

"I have gone through three different ways of giving birth—the first being totally drugged out. I didn't get to see my baby until the next morning. The next two deliveries were natural childbirth using a birthing room and I didn't have any restraints on me. You are given the freedom and dignity of childbirth. I used a birthing chair with my last baby. It was a more natural way of delivery and I found pushing to be less of an effort. The relaxation and breathing techniques taught with Lamaze training really kept me focused and prepared. You know what you want—this is the way you have chosen to give birth. . . . When the nurse came to check, I was already fully dilated and ready. Unfortunately, the doctor was not, so I was asked to 'choo choo' puff through contractions until the doctor arrived. My husband was there the whole time encouraging me. It was such a joy to see what love had produced. The smile on my husband's face. We could hold the baby after its birth—there bonding begins."

"We had our children using Lamaze. The thing I remember most is how in control my

fear-tension-pain cycle According to prepared childbirth advocates, the cycle of events that women experience during labor when they are not properly educated about labor and childbirth.

prepared childbirth Courses or techniques that prepare women for labor and childbirth, with the goal of making it a positive experience.

wife was, even at the very end. She had drawn her own focal point a few weeks before and concentrated on it every contraction while doing her breathing and effleurage. You could really tell the women in the labor room who hadn't gone to Lamaze. I remember one in particular—she had just gone into labor and was very frightened and yelling for the doctor with every contraction. My wife was actually trying to comfort her between her own contractions.

The absolutely best thing about the births was both of us being there to see our children the moment they were born. Each of them were so wide-awake and looking all around. My wife was yelling, 'It's a girl! It's a girl!' To be able to hold them right away was great. When the last child was born, her big brother and big sister were able to put on gowns and hold her after she was born. It was important to be there when my wife was giving birth, to be part of the experience. I cannot remember any other moment when I felt so connected to her and so alive."

(from the author's files)

The Lamaze approach resembles that of Dr. Dick-Read and other forms of prepared childbirth in that it tries to determine the specific causes of pain and provides methods for dealing with each cause. What are the causes of perceived pain in childbirth? Basically, these break down into (1) anxiety and fear; (2) muscle tension; (3) stretching of muscles; (4) too little oxygen getting to muscles; and (5) pressure on nerves. Here are some ways that Lamaze training deals with each of these.

1. **Anxiety and fear.** There are a variety of factors that increase the sensation of pain. It is important to remember that pain is a subjective experience. A football player may break an arm or leg and yet be unaware of it and continue to play until after the game is over (when the pain may suddenly become unbearable). Some people, on the other hand, wince and scream when they get a blood test before the needle even touches them. These examples illustrate two things that influence the experience of pain. The first is the anxiety and fear associated with the experience, and the second is the attention paid to the source of pain.

   In Lamaze training, fear and anxiety are first dealt with through education. The unknown is always frightening, and if a woman in labor does not know what to expect, her anxiety will increase her experience of pain. This is counteracted by teaching expectant mothers about pregnancy and childbirth. They learn about the different stages of labor and medical procedures associated with childbirth. They also are encouraged to preregister at the hospital (so that this doesn't have to be done later when the mother-to-be has more important things on her mind) and to take a tour of the maternity ward and birthing facilities. In this way, the expectant mother knows what to expect and is more familiar with the environment where labor and birth will take place.

   Expectant mothers also are not left alone during labor and birth. Throughout their training, labor, and delivery they are accompanied by a *"coach."* The coach is often the father of the baby, but can also be a friend (Figure 7–10). Some women get a second, "back-up" coach just in case the primary coach cannot be available when they go into labor. The coach helps the mother utilize her training, providing support, encouragement, and instruction. It is the coach's responsibility to keep the mother-to-be as comfortable as possible, while at the same time keeping her concentration focused and her spirits up. The coach learns all of the exercises that the expectant mother does, practices with her, and "paces" her during labor.

   Another way that the subjective experience of pain is reduced is through learning to focus the attention or concentration. The mother-to-be picks out something outside of her body, such as a favorite doll or a piece of furniture, and learns to focus all of her attention on that thing, or "focal point." By concentrating on the focal point, she is not concentrating on her pain, and so the perception or subjective experience of pain is reduced.

2. **Muscle tension.** When people are anxious or fearful, muscles tighten up. This is a natural reaction and can be adaptive, because it allows quick movement in a "fight or flight" situation. However, keeping muscles tense over long time periods can cause them to ache and be used inefficiently. Labor is just such a situation, lasting several hours, with long periods of inactivity between contractions when tense muscles serve no useful purpose. One of the first exercises in Lamaze training involves learning to know when muscles are tense and how to relax them. Expectant mothers will be told to tighten the muscles in one arm, keeping the muscles in the other arm and the legs relaxed. They are told to notice the difference between the "tense" arm and the other limbs. Then, the coach lightly strokes the tensed arm saying "relax, relax" while the mother-to-be slowly relaxes the arm. The coach also learns to distinguish between the relaxed and tensed muscles of the mother. This does

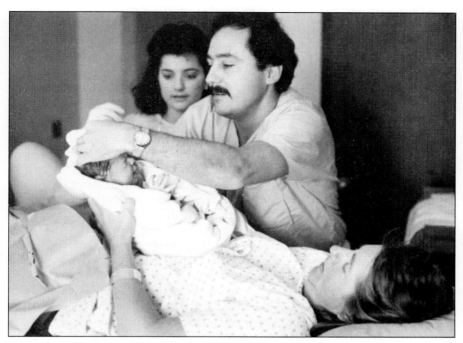

Figure 7–10 In recent years, the birth of a baby has become a family event, with husbands helping their wives during delivery.

begin to ache. The same is true in labor. In order to help deal with this, expectant mothers and coaches learn a variety of breathing techniques. These different types of breathing are generally used during contractions, while the woman's attention is fixed on her focal point, as previously described. During the first part of labor, when contractions are mild and far apart, the expectant mother breathes deeply and slowly (called "low slow" breathing). During transition, she uses a type of breathing called "shallow-chest" or "pant-blow," which keeps the diaphragm (the muscle under the lungs) from pressing down on the top of the uterus during contractions, thus reducing pain. There is even a type of breathing that keeps her from being able to push if the urge to push comes too early or if the physician needs more time.

two things. First, expectant mothers and coaches learn to control the tension of the woman's muscles to prevent her from tensing up, especially between contractions. Second, through lots of practice and repetition, the woman is conditioned to immediately relax her muscles when the coach says the word "relax."

3. **Stretching of muscles.** The muscles used in labor have to go through many hours of exertion. Like any set of muscles, they will get tired and begin to ache if they are overworked, especially if they are out of condition. The idea that a woman is "sick" when she is pregnant and should "take it easy" and rest as often as possible is a thing of the past in a normal pregnancy. In Lamaze training, women are taught a number of exercises so that muscles to be used in labor and delivery can be toned up and strong when the baby decides to arrive. Though weight lifting is not part of Lamaze training itself, some hospitals are experimenting with weight-lifting programs for pregnant women. There are even workout videos for pregnant women.

4. **Too little oxygen getting to muscles.** If you put down this book now, got up to run five miles, and then picked it up again, you would probably have some aching muscles when you resumed your reading. One reason for this is that your muscles use so much energy during vigorous exercise that they often cannot get enough oxygen, and so they

In addition to helping keep oxygen supplied to muscles, these breathing exercises also force the expectant mother to think about something other than the source of pain. This, like using a focal point, reduces the subjective experience of pain. After labor has been going on for some time, the coach may have to breathe along with the expectant mother, "pacing" her and making her use his (or her) face as the focal point. Coaches are more than cheerleaders and must take command if the mother panics during transition.

5. **Pressure on nerves.** Pressure on nerves caused by the fetus moving through the birth canal is dealt with through the use of *effleurage,* or massage. One type of effleurage involves having the expectant mother lightly touch her abdomen with her fingertips, tracing large ovals on her skin during contractions. This light massage generally is a pleasant sensation, and by continually pairing the discomforts of contractions with a pleasant sensation, the subjective experience of pain is reduced. More recently, light massage is being replaced by deep muscle massage. The back, especially the small of the back, is also massaged. If the baby is coming through the birth canal face first rather than crown (back of the head) first, pressure is put on nerves in the lower back (back labor). By applying counter-pressure to the lower back, a coach can reduce some of this discomfort.

These techniques are a sample of some of the components of Lamaze training. The shared training and birth experience can help bring a couple closer together. However, while Lamaze's techniques were a major advance—especially in allowing women and their partners more control over an important event in their lives—he is not without his critics. Some, for example, feel that even his methods are too controlling and that the end result is that women give birth "by direction, as if it were a flight plan" (Armstrong & Feldman, 1990). Today, there are several other birthing techniques that couples can use in place of the traditional hospital–general anesthesia approach. Here are a few of them.

THE BRADLEY METHOD

The second most popular type of prepared childbirth in the United States is based on the work of Dr. Robert Bradley in the 1940s. He, too, was inspired by Dr. Dick-Read and believed that pain during childbirth resulted from culturally learned fear. He argued against the "meddlesome interference with nature's instinctual conduct and plans" and felt that women could learn to deal with labor pain in the same manner as animals do (Bradley, 1981). He emphasized childbirth without medication and had husbands play an even greater role than in Lamaze. Compared to women who have received Lamaze training, women trained in the *Bradley method* are more likely to be critical of the conventional medical model of hospital birth, more likely to question the judgment of their doctors, and more likely to experience less medical intervention during childbirth (Monto, 1996).

THE LEBOYER METHOD

Another French physician, Frederick Leboyer, wrote an influential book in 1975 called *Birth Without Violence*. He felt that traditional techniques of delivery were traumatic for babies, and instead advocated "gentle birth." In the *Leboyer method,* the baby is delivered and immediately placed on the mother's abdomen. The room is dimly lit to simulate the lighting inside of the mother's body. The umbilical cord is not cut for several minutes, until it stops pulsating. The baby is then put in a warm bath (to simulate the amniotic fluid). The idea behind these procedures is to allow the transition of the baby from life inside the mother's body to life in the outside world to be as smooth as possible. Leboyer claims that babies born in this manner immediately smile, are calm, and even develop more rapidly. Critics claim that babies at this age are incapable of experiencing trauma, that dim lighting can interfere with the ability of hospital staff to detect problems in the baby or mother, and that there are no data to confirm Leboyer's claims. At present, this birthing method is still controversial, and it is much more difficult to find a physician willing to use this method as opposed to the Lamaze approach.

HOME BIRTH, BIRTHING ROOMS AND CENTERS, AND MODERN MIDWIFERY

In 1870, fewer than 1 percent of women gave birth in hospitals (Tew, 1990). A century later, only about 1 to 2 percent of women gave birth outside of a hospital. Today, in Europe and the United States, there is a growing movement back to having births in the home. This movement has been driven, in part, by economic considerations. Because of the cost of malpractice insurance, one in eight obstetricians have stopped delivering babies, and one-fourth of the counties in the United States do not have a clinic to provide prenatal care (Nazario, 1990). Of course, home births are also much less expensive. There are also personal considerations. Many couples feel that having a baby in familiar surroundings is more natural than a hospital birth. It also provides an opportunity for the entire family to be present and participate. For pregnant women who are at low risk for complications, home birth is a safe alternative to hospital delivery (Olsen, 1997).

To compete with this movement, many hospitals today have "birthing rooms," which are designed to look like bedrooms rather than hospital rooms. Labor and delivery take place in the same bed, and often a crib is present so that the baby can sleep in the room with the mother after birth. Sometimes even brothers and sisters are allowed to put on hospital gowns and hold the newborn shortly after birth.

In many areas, a closely related option is *birthing centers,* institutions that offer delivery in a homelike atmosphere with the assistance of skilled individuals such as midwives or physicians. Although new to the United States, birthing centers have been popular in Europe for a long time. They have proved to be just as safe for women as traditional hospital care (Waldenström & Nilsson, 1997), but this may be partly because birthing centers are really an option only to women with no foreseeable complications (Rooks et al., 1989). Should there be an emergency, all centers have procedures for a quick transfer to a (usually nearby) hospital.

Whether the choice is to deliver at home, in a birthing center, or in a hospital, there is also a growing trend to use **nurse-midwives** rather than physicians. In many European countries, over two-thirds of all babies are delivered by midwives (Nazario, 1990). Midwives are generally registered nurses trained in obstetrical techniques who are well qualified for normal deliveries (and minor emergencies). They are less expensive than physicians, and they provide personal care during the latter parts of pregnancy, labor, and deliv-

nurse-midwife A registered nurse, trained in obstetrical techniques, who delivers babies, often at the expectant mother's home.

ery. Studies indicate that the outcome for babies (and mothers) is just as good when they are attended by nurse-midwives as when they are attended by physicians (Declercq, 1992; Tew, 1990; see also Nazario, 1990). The American College of Obstetricians and Gynecologists officially approved the use of nurse-midwives in 1971. The use of lay midwives (persons who are not nurses and are generally trained by nurse-midwives) is also increasing in popularity.

Nurse-midwives are certified by the American College of Nurse-Midwives, but in some states there are no certifying agencies. So it is important to learn about the midwife's background, years of experience, and references. I should point out that a midwife may be either male or female, as evidenced by a man who, after practicing this trade for many years, legally had his name changed to "Mr. Midwife."

Women who have some form of prepared childbirth and/or who are assisted during labor and delivery by a midwife generally require less medication for pain, need fewer cesarean sections, and have more positive feelings about the experience (Kennell et al., 1991; McNiven et al., 1992; Turnbull et al., 1996). However, a word of caution: Do not start thinking of childbirth as a contest to see if you can get through with the least amount of medical help. This attitude can make the use of anesthetics or a C-section seem like a "failure." That should not be the case. Any birth that results in a healthy mother and baby is a good one, regardless of the technique used. That is always the goal of childbirth, and should not be forgotten.

CHILDBIRTH

Before birth, the fetus will generally rotate its position so that its head is downward. This can happen weeks or hours before birth and is what is meant when people say "the baby dropped." It is also called **"lightening"** because once the fetus's head (its largest part) has lowered in the uterus, pressure on the mother's abdomen and diaphragm is greatly reduced. Being able to breathe more easily makes a mother feel like the load that she has been carrying (the fetus) weighs less (is "lighter").

Shortly before a woman begins to give birth ("goes into labor"), she may experience a burst of energy. A woman may feel better than she has in months, and may start to clean house or do some other type of work as a means of "burning off" this energy. It is probably a good idea to save some of this energy for later, when it certainly will be needed.

TRUE VERSUS FALSE LABOR
Throughout the latter stages of pregnancy, the uterus will undergo contractions—tightenings and relaxations of the muscles. These are involuntary, and for most of the pregnancy will go unnoticed. These contractions, called **Braxton-Hicks contractions,** are a type of natural "exercise program" for the uterus. Sometimes these contractions become so strong that they are noticed and mistaken for true labor. Parents-to-be sometimes worry and go to a hospital, only to be told that they have had a **false labor** and should return when the woman is in "real" labor. Here are some rules of thumb that help distinguish false labor from true labor.

One important sign of labor is the time between contractions. The uterus is basically a muscle in the shape of a bag. Its job is to push the fetus into the world. Like most muscles engaged in prolonged, heavy labor, it must repeatedly contract, or tighten. If contractions are coming 10 minutes apart on a regular basis, this is a good sign that labor has begun. False labor contractions are usually less regular and will eventually cease. The duration of each contraction also should be measured. If contractions are 30 seconds long on a regular basis, this is a good sign that labor has begun. False labor contractions are usually shorter than contractions marking true labor. With true labor, discomfort caused by contractions is usually felt in the back and abdomen. With false labor, discomfort is more likely to be felt in the groin and lower abdomen.

Two important signs of the beginning of labor involve the cervix. The first sign is **dilation.** In the last month of pregnancy, the cervix will dilate to about 1 centimeter in diameter. When a woman goes into labor, the uterine contractions help cause the cervix to dilate more. It must eventually dilate to 10 centimeters (about 4 inches) in diameter before the baby can be born. In the first stage of labor, the cervix will dilate to about 2 or 3 centimeters. The second sign is **effacement.** The cervix must stretch until it is "thinned out," or "effaced," so that it won't block the baby's passage into the birth canal (vagina). A good sign of true labor is that the cervix is 70 percent or more effaced. At birth, it will be 100 percent effaced. Although both of these cervical changes are good signs that labor has begun, they are only general rules of thumb.

Labor for the first child is usually longer than labor for subsequent children. A mother who has had many children may be asked to go to the hospital when contractions are still 20 minutes apart, because she could go directly from 20 minutes apart to a much shorter interval. It is not uncommon for women to wait too long before going to a hos-

lightening Rotation of the fetus prior to childbirth so that its head is downward, resulting in decreased pressure on the mother's abdomen.

Braxton-Hicks contractions Uterine contractions experienced during the last trimester of pregnancy that are often incorrectly interpreted as the beginning of labor; also called *false labor.*

false labor A set of temporary contractions of the uterus that resemble the start of labor.

dilation Becoming wider or larger.

effacement The thinning of the cervix during labor.

pital and to give birth in taxis, hospital parking lots, or hospital elevators.

False labor can fool even the professionals. I know of one woman who had regular contractions of 30 seconds duration 2 minutes apart for several hours, with 70 percent effacement and 3 centimeters dilation of the cervix. After being examined in the emergency room and admitted, both the nurse and the attending physician assured the parents-to-be that labor was in full swing. Ten hours later the couple checked out. Actual labor and birth came one week (and two more false alarms) later.

STAGES OF LABOR

Labor is divided into three stages. The initial stage of labor is a **start-up stage** and involves the woman's body making preparations to expel the fetus from the uterus into the outside world. This stage usually lasts from 6 to 13 hours. At this time, uterine contractions begin to push the fetus downward toward the cervix. The cervix begins to dilate and efface, as described earlier. At first, contractions are far apart (one every 10 to 20 minutes) and don't last long (15 to 20 seconds), but eventually they begin to come closer together (1 to 2 minutes) and last longer (45 to 60 seconds or longer).

During pregnancy, the cervix becomes plugged up with a thick layer of mucus. This prevents the amniotic sac (bag of water) and the fetus from having direct contact with the outside environment (through the vagina). During labor, this mucous plug will come out. A discharge of mucus and blood (known as a "bloody show") is a sign of labor. However, for some women the mucus comes out a little at a time, so that this sign is not clearly seen. In other women, the plug literally pops out like the cork in a bottle of champagne. As usual, every variation in between is also seen.

For about 10 percent of pregnancies, the amniotic sac will break before labor begins. In some women, the break is a small tear and the fluid seeps out. In other women, the water gushes out. This is what happens when a woman's *"water breaks."* Labor usually begins within a day after the amniotic sac breaks. If not, most physicians will induce labor through use of drugs in order to prevent contact with the outside world from causing infection in the fetus. A woman can also go into labor without having the water break. Physicians sometimes break the amniotic sac on purpose in order to speed up labor. It is even possible to give birth with the fetus en-

closed in an unbroken amniotic sac. Physicians in this country don't allow this to happen and will break the sac first. However, in the Far East, children born in this manner are believed to be blessed with good fortune.

The last part of the first stage of labor is called the **transition phase.** It is called this because it marks the end of the initial stage of labor and the beginning of the next (birth) stage. This phase takes place when the cervix is almost fully dilated (8–10 cm). Contractions are severe, and the woman may feel nauseated, chilled, and very uncomfortable. She will often start to think that there is no end to this process, and many women at this point will announce that they quit and are going home. Fortunately, transition usually lasts 40 minutes or less (compared to the often hours-long duration of the earlier part of the first stage).

During labor, it is important to monitor the fetus's vital signs. A sudden drop in heart rate may indicate a life-threatening situation. Nurses or physicians can monitor heart rate with a stethoscope or by a device called a fetal monitor. With a monitor, a recording device is placed on the expectant mother's abdomen (or, in the second stage of labor, directly on the fetus's scalp) and heart rate is shown on a machine. However, fetal monitors are no better than stethoscopes in detecting fetal distress and reducing the incidence of perinatal deaths, and they frequently give false indications of abnormal heartbeats that can lead doctors to perform cesarean sections unnecessarily (Nelson et al., 1996; Shy et al., 1990).

The **second stage** of labor begins when the cervix is fully dilated and the fetus begins moving through the birth canal (Figure 7–11). It ends with birth. Contractions during this stage of labor are accompanied by an intense desire to push or "bear down." The contractions cause the opening of the vagina to expand, and expectant mothers often get a "second wind" because they now can see an end to their work. Expectant mothers (and fathers) begin to get excited about the fact that they will soon be able to see their baby, and this makes the contractions a little easier to bear. This stage lasts from 30 to 80 minutes on the average, depending on whether it is the first child for a woman or a later child. Just before the actual birth, physicians often use a surgical procedure called an episiotomy (see next section).

The first sign of the fetus is usually the sight of its head at the opening of the vagina. This is called **crowning,** because ideally the crown of the head is leading the way. If the head is turned around so that its nose and forehead are leading, it is called a "back labor" because this will cause the crown of the fetus's head to put pressure on nerves in the lower back, making the labor more painful than normal. The person (usually a physician) delivering the baby will tell the expectant mother when and how hard to push during contractions in order to

start-up stage of labor The stage of labor that begins with uterine contractions pushing the fetus downward toward the cervix and ends when the cervix is fully dilated.

transition phase The last part of the start-up stage of labor, during which the cervix dilates to 10 cm in order for the baby to be able to enter the birth canal.

second stage of labor The stage of labor that begins when the cervix is fully dilated and the fetus begins moving through the birth canal. It ends with birth.

crowning The appearance of the fetus's head at the vaginal opening during childbirth.

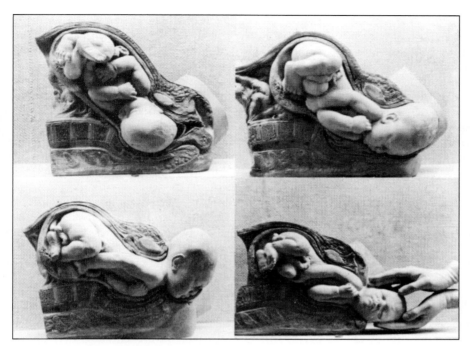

Figure 7–11 The stages of childbirth.

the abdomen is sometimes massaged, and/or drugs may be given. This helps prevent excess bleeding (hemorrhaging) in the uterus.

EPISIOTOMY AND PREPPING

In the United States, over 60 percent of women who deliver vaginally in hospitals are given an **episiotomy** just before they give birth. A episiotomy is an incision made from the bottom of the vagina (through the perineum) toward the anus, thus enlarging the vaginal opening. An episiotomy is a useful procedure if the baby is in distress, and it also reduces tearing of the perineum (which takes longer to heal than an incision). However, many authorities believe that most episiotomies are unnecessary and are done routinely only for the benefit of the physician (Lede, Belizan, & Carroli, 1996). Episiotomies are uncommon in midwife-assisted deliveries (see Henriksen et al., 1994). In fact, a recent study found that women having vaginal deliveries who did not have episiotomies had fewer problems than women who did have them (Argentine Episiotomy Trial, 1993). When sewing up the incision, a few physicians add a "honeymoon stitch" (or "husband's stitch"), a stitch in the lower part of the vaginal opening. Both of these procedures should be abandoned as routine birthing practices.

Another procedure that is done less frequently than in the past is **prepping,** that is, shaving the pubic hair prior to delivery. Many physicians now think that this is unnecessary and can even be a source of infection when the hair begins to grow back.

get the baby into the world. The head is delivered first, with the physician reaching beneath the infant's chin to be sure that the umbilical cord is not wrapped around the baby's neck before the head is brought out. Once the head is delivered, suction is applied to the mouth and nose with a small rubber bulb so that the baby can breathe more easily. The head then turns, and the shoulders and rest of the body come out rather quickly. The mouth and nose are suctioned again.

A newborn will usually cry at birth. If not, the baby's back will be rubbed to start the baby breathing. The umbilical cord will be clamped and cut about 1½ inches from the baby's body. This stub will fall off in a few days, leaving the navel. (The navel is either an "inny" or an "outty," but the type of navel you have seems to depend on heredity, not on how the umbilical cord was cut.)

In the **third stage** of labor, the placenta detaches from the uterus and leaves the mother's body (along with the other materials). This is what is called the **afterbirth.** This stage usually lasts only from 10 to 12 minutes. The physician examines the afterbirth to make sure that it is in one piece; then it is discarded. If even small pieces of the placenta stay in the uterus, infection and bleeding can occur. When this takes place, physicians use a procedure called *D&C* (dilation and curettage), in which the cervix is dilated to allow access to the uterus, which is then scraped clean. If any tears in the vagina have occurred, they are repaired at this time.

After the third stage of labor, the uterus should contract and remain constricted. To help this happen,

PROBLEMS WITH CHILDBIRTH

BREECH BIRTHS

Babies, you recall, are generally born head first. This occurs in about 96 percent of all births, and ideally the back of the head, or crown, leads the way. Sometimes, a fetus will try to come through the birth canal feet or buttocks first. This is called a **breech birth**

---

third stage of labor Detachment and expulsion of the placenta from the uterus after childbirth.

afterbirth The expulsion of the placenta from the uterus after a baby is born.

episiotomy A surgical procedure performed just before birth to reduce the risk of tearing the perineum.

prepping The shaving of a woman's pubic hair prior to delivery of a baby.

breech birth A birth in which the baby is born feet or buttocks first.

and occurs in 2 to 4 percent of all births. A fetus can even try to come out back first, shoulder first, or in any other conceivable position, but these cases are rare. Sometimes a fetus will spontaneously shift from feet to head first just before delivery. Sometimes during labor a woman will be asked to try lying in different positions, such as on her side, in the hope that this will cause the fetus to change to a head-first position. A physician can sometimes turn the fetus by grasping the mother's abdomen and manually trying to change the fetus's position.

Today, some physicians may try to perform a vaginal ("normal") delivery even if a breech birth is unpreventable. This will depend on the circumstances and the physician. If the physician feels that the risks of a vaginal delivery are too great, he or she may perform a C-section.

PLACENTA PREVIA

The blastocyst normally implants in the upper part of the uterus. However, for about one out of every 200 pregnant women, implantation occurs in the lower uterus (Iyasu et al., 1993). As the placenta grows, it often blocks the cervical opening, a condition called **placenta previa.** This does not affect the fetus, but it may prevent a vaginal delivery.

CESAREAN SECTIONS

A **cesarean section** (also spelled *Caesarean*), or **C-section**, is an incision through the abdominal and uterine walls to deliver a baby. Although the name is often attributed to the way in which Julius Caesar was born, in actuality it is derived from *lex cesarea,* a Roman law (715 B.C.) that allowed a fetus to be saved in the case of its mother's death.

Cesarean sections accounted for 5 percent of live births in the United States in 1970, 14 percent in 1979, 19 percent in 1984, and reached a peak of 24.7 percent in 1988. The rate has since declined a bit, to 20.8 percent in 1995 (Curtin & Kozak, 1997). If this number looks large to you, you are not alone in your perceptions. Many individuals, including many health professionals, have begun to call for changes in the way physicians decide when to use this procedure (Young, 1997). Cesareans are supposedly done to reduce the risk to the mother or baby, but most are done on middle- and upper-income women who can afford it (see Porreco & Thorp, 1996). Recall, too, that C-sections are uncommon in deliveries assisted by midwives (Kennell et al., 1991; McNiven et al., 1992). Thus, there has been a suspicion that many C-sections have been performed for the economic benefit of the physician or the hospital, or to avoid the risk of malpractice lawsuits in case of complications during vaginal deliveries.

There are many legitimate reasons for performing a cesarean section. These include *some* cases of breech birth or placenta previa; the baby's head being too big for the mother's pelvis; maternal illness or stress; and fetal stress. It used to be the case that once a woman had a C-section, all later pregnancies would have to end with C-section deliveries (because the cuts were made vertically through the horizontal abdominal muscles and through the contracting part of the uterus). These repeat cesarean sections account for 35 percent of the C-sections performed each year (Curtin & Kozak, 1997). Today, however, incisions are made horizontally through the abdomen and then through the noncontracting part of the uterus. In most cases, it is considered safe for a woman who has previously had a cesarean section to deliver vaginally (Flamm et al., 1990; Rosen et al., 1991) with only a small (but higher than usual) risk of major complications (McMahon et al., 1996). Obstetricians have been slow to respond, but the number of vaginal births after a C-section have quadrupled since the mid 1980s (Porreco & Thorp, 1996). However, C-sections are still the most frequently performed major operation in the United States.

Because C-sections are so common, it is advisable for a pregnant woman to talk with her physician about his or her attitudes regarding the use of cesarean sections before she goes into labor. She should also talk with the physician about episiotomies, use of anesthetics, attitudes about prepared or "natural" childbirth, and all other details of the birth process at this time. This talk should take place when the expectant woman first begins to see the physician. If she doesn't like what she hears, she should find a new doctor.

PRETERM INFANTS

Nine out of every 1,000 births in the United States result in the death of the baby after it is delivered. The leading cause of infant mortality is from preterm birth. An infant is considered preterm if it weighs less than 5½ pounds and the mother was pregnant for less than 37 weeks. Sometimes people use the term "premature infants" to describe such children, but others prefer an alternative term, such as "low-birth-weight infants" (Miller, 1985). I will compromise and use the phrase **preterm infants.** Preterm births account for about 7 percent of all U.S. births. Conditions such as malnutrition and toxemia increase the risk of having a preterm birth, but in 50 percent of preterm births the cause of the problem is unknown. If a woman goes into labor prematurely, drugs can be given to relax the uterus and postpone labor. This must be done early in labor to be effective, however.

placenta previa A condition in which the placenta blocks the cervical opening.

cesarean section (C-section) The surgical removal of a fetus through the mother's abdomen.

preterm infant An infant born weighing less than 5½ pounds and before week 37 of pregnancy.

Advances in neonatology (the care of newborn babies) have reduced the limit of viability to about 23 to 24 weeks after conception (Allen, Donahue, & Dusman, 1993). One of the biggest problems facing a fetus born at this time is that it has great trouble breathing. Before the age of 23 weeks (around 4 months), the fetus cannot produce a liquid called *surfactin.* Surfactin lets the lungs transmit oxygen from the atmosphere to the blood. After 4 months of age, the fetus can produce some surfactin, but it cannot maintain the liquid at the necessary levels until about 8 or 9 months of age. Newborns who cannot maintain proper levels of surfactin develop **respiratory distress syndrome** and die.

> "When I was exactly eight months pregnant my amniotic sac broke and labor was induced. When I saw my baby in the nursery, she was under oxygen, had all kinds of wires connected to her, an IV in her foot, and she had a lot of trouble breathing. I wasn't ready for this. However, I feel that if I was more informed about premature births I would have understood better."
>
> (from the author's files)

Before 1980, there was little hope of survival for babies born weighing less than 2 pounds. Today, about 17,000 infants per year born weighing less than 2 pounds are placed in *intensive care nurseries (ICNs),* where new techniques and equipment allow these infants a good chance of survival. Infants weighing just over 1 pound have about a 30 percent chance of survival, while those weighing about 2½ pounds have about a 90 percent chance of survival (see Phelps et al., 1991). The cost of such technology is very high. For example, the cost of caring for a 4–month premature baby can be over $300,000. In addition, premature infants run a high risk of having handicaps that can require treatment for many years. Babies born weighing less than 750 grams (a little over 1½ pounds) are much more likely than others to have poor psychomotor skills, very low IQ scores, and poor performance in school (Hack et al., 1994). Early childhood intervention improves the IQ and verbal performance of preterm infants born weighing 4½ pounds or more, but is of little or no help to very-low-birth-weight babies. Most of them still remain cognitively and behaviorally impaired after 8 years (McCarton et al., 1997).

AFTERWARDS

BREAST-FEEDING THE BABY

Milk production (lactation) begins about three days after a woman gives birth (Figure 7–12). Recall from Chapter 2 that two hormones produced by the pitu-

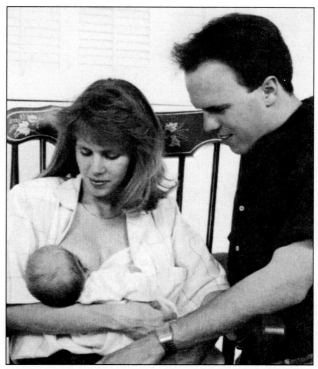

Figure 7–12 Mother's milk contains many infection-fighting proteins and is easier for a baby to digest than cow's milk.

itary gland are responsible for milk production. **Prolactin** stimulates the breasts to produce milk, while **oxytocin** causes the breasts to eject milk. When the baby sucks on the breasts, a milk-flow reflex is created. In fact, over time merely hearing the baby cry can cause the breasts to begin ejecting milk. At first, the breasts give out **colostrum,** mentioned earlier, which is high in protein and also helps immunize the infant against diseases. After about three days, milk begins to flow. The average amount of milk produced is 2½–3 cups per day, but there is a lot of variation among individuals. Some women do not produce enough milk, and some produce too much.

Mother's milk contains many infection-fighting proteins that protect the infant from gastrointestinal illnesses, respiratory infections, meningitis, skin diseases, and other problems (Howie et al., 1990; Institute of Medicine, 1991; Yolken et al., 1992). There is even some evidence that breast-fed babies have higher IQs (see Glick, 1997). Breast-feeding has benefits to mothers as well. When a mother be-

respiratory distress syndrome An illness common in premature infants; caused by insufficient levels of surfactin.

prolactin A hormone released from the pituitary gland that stimulates milk production in the breasts.

oxytocin A hormone released from the pituitary gland that causes the breasts to eject milk.

colostrum A thick, sticky liquid produced by a mother's breasts before milk starts to flow.

gins nursing, a hormone is released that helps the uterus to contract and helps stop internal bleeding. Nursing mothers may also have a reduced risk of breast cancer (Romieu et al., 1996). Breast-feeding does not interfere with sexual desire (Fisher & Gray, 1988; Masters & Johnson, 1966). Despite these benefits, the number of women in the United States who nurse after childbirth is decreasing. You should know that both the American Academy of Pediatricians and the Institute of Medicine endorse breast-feeding. Keep in mind, however, that drugs and pollutants also can be transmitted to the infant through mother's milk. Before taking any medication (including birth control pills), therefore, it is important for nursing mothers to consult their physicians or pharmacists.

Nursing does not come naturally to everyone:

> "Hospitals have a way to reduce an adult to a child with regard to telling you what is and what is not allowed. I assumed that a nurse would instruct me in the method of breast-feeding. Every feeding the nurse would bring the baby and hand me a baby bottle filled with water and a medicine cup with a water-soaked cotton ball in it.
>
> I had no idea what the cotton ball was used for. For every feeding I gave the baby the water bottle and then dabbed her mouth with the cotton ball afterwards. On the second day after delivery I asked the nurse when did she want me to start breast-feeding? Her expression was priceless! 'You mean you haven't?' she said.
>
> I said, 'No, you didn't tell me to; you just keep bringing me this water bottle and cotton ball.' She wondered why the baby had taken so much supplemental milk 'back-stage.'"
>
> (from the author's files)

A mother's breasts and nipples may get sore when she first starts to nurse. Because of reduced estrogen production, her vaginal tissues may also be drier and cause irritation during intercourse (this can be corrected, of course, with use of a lubricant). A nursing mother may not begin her menstrual cycle for several months after birth, which gives her some protection against pregnancy (see Chapter 6). This is especially true when feeding is done "on demand" rather than according to a fixed schedule. It should be remembered, however, that a woman will ovulate and be able to conceive again before she has her next menstrual flow, so this is certainly not a perfect means of contraception.

postpartum blues A mild, transient emotional letdown experienced by a majority of women after giving birth.

Despite the benefits, many women choose not to breast-feed their babies, often because of a career outside the home. If a woman stops breast-feeding (or never begins), milk production will stop within days. This can be speeded up by an injection of estrogen. Be aware, however, that human milk is much easier to digest than cow's milk, and if bottle-fed, a baby should be given formula.

Whether a baby is breast-fed or bottle-fed, fathers should not ignore this new responsibility. A father can sit with his partner while she breast-feeds in the middle of the night, or a mother can use a breast pump to extract her milk for a later feeding (allowing her to sleep while Father feeds the baby).

POSTPARTUM DEPRESSION

The first three months after childbirth are called the *postpartum period.* Although most women look forward to having a baby, many experience a period of negative emotions during this time. In some cases their distress can be extreme:

> "After I had my son I spent weeks upon weeks of sitting alone on the kitchen floor with him on my lap and crying and crying uncontrollably. . . ."
>
> "I was so depressed after giving birth to a little girl. If somebody touched her, I screamed. When my baby was 3 days old I had a fight with my mother-in-law and husband. Actually, they tried to help me out, but I was so depressed that I always fought with them. Things got worse and I thought I was going to get divorced. . . . My doctor explained to me it is because of hormone changes. . . . I am thankful to my husband, who always tried to understand me at that time. . . ."
>
> "I can remember being depressed for a couple of months. . . . I felt so helpless. . . . Sometimes I resented my baby because I felt so bad and was so exhausted. I hated myself for having these feelings. . . ."
>
> (from the author's files)

There is a range of negative emotions that women can experience after childbirth, and it is customary to distinguish three types of depression on the basis of severity (O'Hara, 1997). Many women experience a letdown in the first week or two after delivery, called **postpartum blues.** Symptoms may include a depressed mood, confusion, anxiety, and crying, but the condition is mild and does not last long. However, about 10 percent of women experience **postpartum depression,** characterized by a deeper depression and anxiety, guilt, fatigue, and often obsessive-compulsiveness. The stricken mother may feel that she cannot face daily events, is overwhelmed by child care, and feels ashamed and guilty about her inadequacy as a mother. About one in 1,000 women experience **postpartum psychosis,** characterized by delusions and hallucinations in what often resembles a manic depressive state. I will focus here on postpartum depression.

Biological factors, particularly hormone changes, may contribute to postpartum depression (O'Hara, 1997). The placenta secretes high levels of estrogen and progesterone during pregnancy, and there is a sudden drop in the levels of these hormones after birth. But social and psychological factors certainly play a major role (Gruen, 1990; O'Hara, 1997).

Marital unhappiness and lack of support by the spouse and family, for example, can contribute to the depression. Newborn infants need frequent feeding, which can cause mothers to lose lots of sleep. A woman can feel abandoned if she feels she is not receiving support from her partner. High levels of stress during pregnancy and afterwards (such as unemployment and pressing financial responsibilities) also contribute.

Many studies indicate that if postpartum depression lasts several weeks, it can affect the infant's emotional, behavioral, and cognitive development (Cogill et al., 1986; Murray, 1992). It is probably a mother's style of interacting with her child (brought on by the depression) that results in these negative consequences (Murray & Cooper, 1997). However, it is important to note that postpartum depression is not always a risk factor for later problems in the mother-child relationship (Campbell & Cohn, 1997).

A simple screening questionnaire has been developed to identify mothers with postpartum depression and is a useful tool to suggest who might need support during this difficult time (Cox, Holden, & Sagovsky, 1987). Women with postpartum depression generally are treated in the same manner as other depressed patients (O'Hara, 1997). However, it has been found that postpartum-depressed women also benefit greatly by support from others with the same condition (DeAngelis, 1997).

SEXUAL INTERCOURSE AFTER BIRTH

"We tried to make love again when our first baby was about 3½ weeks old. The pain from the episiotomy made this impossible and we both wound up very frustrated and wondering if our sex life would ever be back to normal. We finally did succeed at about 6 weeks but still it wasn't very comfortable. After a while we got used to the new normal. . . . After my second and third babies were born everything was very different. I didn't have an episiotomy, or tear at all, even though they were both very large. We made love just a few weeks later without any discomfort whatsoever."

(from the author's files)

Labor often leaves a woman tired and sore for some time, and stitches from an episiotomy may leave her uncomfortable. The loss of estrogen results in vaginal dryness and irritation during intercourse for some time. Most physicians advise that sexual intercourse not be resumed until 4 to 6 weeks after the birth.

Women also may worry about their sexuality because of stretch marks, extra weight left over from pregnancy, and "flabby" stomach muscles. The stretch marks soon become paler, and problems with weight and stomach muscles can be helped through exercise. Some women also worry that their vagina has become too large after birth, but the vaginal muscles generally become firm and tight again, especially if the woman continues to practice Kegel exercises (see Chapter 2).

The best time to resume sexual relations depends on many factors. In all cases, however, a man and a woman must learn to relate to each other as both sexual partners *and* as parents. Making this adjustment to new roles and feelings requires time and patience. Differences in sleep patterns with an infant that must be fed at night, demands on time and energy made on parents by children, and difficulty finding periods of privacy all force couples to make adjustments in their sexual behaviors.

Despite these concerns, people do resume sexual intercourse after childbirth. A recent study found that couples start again an average of 7 weeks afterwards (Hyde et al., 1996). Women who are breast-feeding show less sexual activity than others, and women who had a cesarean delivery tend to start earlier than those who delivered vaginally. As you might expect, sexual attitudes in general play a role. Couples who had a positive attitude about sex before pregnancy tend to resume sexual relations earlier than couples with a less than totally positive attitude (Fisher & Gray, 1988).

NEW RESPONSIBILITIES

Becoming a parent is like being asked to start a new job with little or no training. Just because someone has given birth or fathered a child does not mean that she or he has natural parenting ability. Learning how to deal with children effectively and to provide the right balance of discipline and love is not easy. Although 80 percent of women say they would choose to be a parent again if they had

postpartum depression A deep depression experienced by about 10 percent of women in the first few months after giving birth.

postpartum psychosis A psychotic state experienced by about one in 1,000 women after giving birth.

"Pretend you're carrying your golf clubs, and you won't mind it."

Reprinted courtesy of Bunny Hoest and Parade Magazine © 1987.

it to do all over again (Genevie & Margolies, 1987), most find the demands of new (and multiple) roles in their lives to be stressful (Ventura, 1987). Studies find that after the transition to parenthood, happiness within the marriage is directly related to the sharing of child care and housekeeping responsibilities (Hackel & Ruble, 1992). Many new fathers, for example, complain about a diminished sex life and their wives' mood swings, yet do little to relieve their partners in their new responsibilities. Men, do you value your relationship enough to share these responsibilities?

It is useful to acquire some training and knowledge about parenting, just as it is useful to acquire training and knowledge about giving birth. Some excellent references are listed at the end of the chapter. In addition, the Family Resource Coalition has published a listing of programs to improve parenting skills and strengthen families. To learn about such programs in your area, contact the Family Resource Coalition at 200 South Michigan Avenue, 16th floor, Chicago, IL 60604.

INFERTILITY

Infertility is usually defined as the inability of a couple to conceive during 12 months or more of unprotected sexual intercourse. This assumes, of course, that the woman is of reproductive age. Most people think of an infertile couple as a couple who have never been able to

infertility The inability of a couple to conceive.

conceive, called *primary* infertility. However, the inability to conceive after having had one child, called *secondary* infertility, is even more common (Mosher & Pratt, 1990). Many people believe that infertility is increasing, and popular magazines make reference to an infertility "epidemic" (e.g., Ames et al., 1991, in *Newsweek*), but statistics show that in actuality there was no increase during the 1980s (W. D. Mosher & Pratt, 1991). During this time, about one in 12 couples in the United States had difficulty conceiving (Abma et al., 1997). What there has been is a big increase in the number of couples seeking help for infertility, which has led to the perception of an epidemic.

The causes of infertility are many and varied because so many organs of both sexes must function properly for conception to take place. About 40 percent of the problems involved with infertility are due to the male partner, with another 40 percent due to the female. About 20 percent of the cases result from problems in both partners (Hudson, Pepperell, & Wood, 1987; Moore, 1982).

Before we proceed, you should be aware that there is a national organization that offers updated information and support to infertile couples. If interested, contact RESOLVE, 1310 Broadway, Somerville, MA 02144–1731, or call 617–623–0744.

INFERTILITY IN MALES

As you may remember, only a very few of the millions of sperm cells released during ejaculation survive the trip to the egg. If a man's reproductive system cannot produce enough sperm, and/or if the sperm he produces are too slow or too weak, the probability that fertilization can occur is substantially reduced. Low numbers of sperm are caused by a variety of factors, including endocrine problems (low levels of pituitary or gonadal hormones), drugs (including alcohol and many antibiotics), marijuana, radiation, and infections. If a man gets mumps after puberty, for example, it often spreads to the testicles and causes severe swelling that destroys the seminiferous tubules where sperm are produced. Other infections, including some sexually transmitted diseases, can result in blockage of a male's duct system (just as some STDs can lead to blockage of the Fallopian tubes in women). Varicosity of the veins in the spermatic cord (*varicocele*) can also lead to a low sperm count (see Howards, 1995).

Because sperm can only be produced at a temperature several degrees lower than normal body temperature, sperm count is sensitive to temperature variations. For example, athletes who regularly wear athletic supporters or metal cup protectors for long periods of time (resulting in increased testicular temperatures) risk having a temporarily low sperm count. Frequent use of saunas or hot tubs can also cause a temporarily low sperm count. If a couple is trying to conceive, the male may be told to switch to loose-fitting underwear so that his testes will hang farther from his body and thus have less exposure to body heat. Sperm count also decreases during the summer months (Levine et al., 1990), but we are not yet sure that this is due to the heat (Snyder, 1990).

A review of 61 studies conducted worldwide concluded that the average sperm count per ejaculation had dropped by more than half in the last 50 years (Carlsen et al., 1992). What could cause such a dramatic change? Dr. Neils Skakkeback, one of the authors of the paper, said, "It would have to be something in the environment or life style. Changes that occur within a generation could hardly be due to a change in genetic background" (quoted by Epstein in AP release, September 1992). Further evidence is found in the fact that the rates for testicular cancer tripled during the same time period. Could industrial chemical pollutants have such effects? Boys born to (and breast-fed from) mothers exposed to high levels of pollutants can have testicular malformations and dramatically undersized penises, and the same is true for many species in the wild (see Begley & Glick, 1994). Many of the suspected pollutants resemble the female sex hormone estrogen in structure. Since the initial report, some scientists have found no evidence that sperm counts are lower than in the past (Bromwich et al., 1994; Farrow, 1994; Fisch et al., 1995; Paulson, Bermen & Wang, 1995), while others confirmed that there has been a substantial decline (Auger et al., 1995; Irvine et al., 1996). There will certainly be much research in this area in the future.

Infertility is not just a physical problem; it also affects men emotionally:

"Several months ago we began trying to have kids unsuccessfully. When I was younger I had an . . . operation and my doctor said that because of this I would not be able to have children. My wife and I wanted children, and this situation has really put a strain on our marriage."

"Being sterile is not an easy thing to discuss. On top of diminishing dreams and goals, it has made finding the right girl to begin a life-long relationship with very difficult. . . . My future wife and I will never be able to have children of our own. . . . I've always wanted to teach my son to play sports or take my daughter to dancing lessons. . . . Hopefully someday I will find the right woman and we can adopt."

(from the author's files)

What are the medical treatments for male infertility? Infertility due to varicocele (8–23 percent of cases) can often be corrected with surgery (Howards, 1995). If the cause is hormonal, treatment with pituitary hormones (or brain hormones that cause release of pituitary hormones) or gonadal hormones (androgens) is often successful (Menchini-Fabris et al., 1992). A blocked duct system in men is much easier to correct through surgery than are blocked tubes in women.

When a man is found to have a low sperm count, a procedure called **artificial insemination** is often used. In this procedure, sperm are collected via masturbation and inserted into the female partner's vagina or uterus immediately after ovulation. Several ejaculations are usually collected and stored by freezing (frozen semen is just as effective as fresh). For example, if a man's average ejaculation were found to contain 50 million sperm, six to eight ejaculations (totaling 300 to 400 million sperm) might be collected and frozen. When enough sperm have been collected, they are gathered together for a single insertion. Sperm can be safely preserved for more than 10 years, and some men make deposits in sperm banks before having a vasectomy, undergoing chemotherapy, working a hazardous job, or going off to war (Ames et al., 1991).

If a man's sperm count is so low that combining several ejaculations will not help, a couple can choose to use sperm from another, usually anonymous, male. In these cases, artificial insemination is often called *donor insemination*. Donors are usually matched with the physical characteristics of the male. Over 30,000 babies a year are conceived by this method in the United States (see Barratt et al., 1990).

The newest technique involves injecting a single sperm directly into an egg (called **intracytoplasmic sperm injection**). The first U.S. baby conceived by this technique was born in 1993. This procedure is now being used more and more frequently and with good results, and has resulted in decreased demand for donor insemination (Dozortsev et al., 1995).

Sperm with X or Y chromosomes can be partially sepa-

artificial insemination A method of treating infertility caused by a low sperm count in males. Sperm are collected during several ejaculations and then inserted into the partner's vagina at the time of ovulation.

rated (see Chapter 8), and thus artificial insemination also allows couples who wish to do so to attempt to select the sex of their child (Beernink et al., 1993).

INFERTILITY IN FEMALES

The discovery that she is infertile can be emotionally devastating to a woman and her partner (see Rosenthal & Goldfarb, 1997):

"I used various forms of birth control during the first six years of marriage so that my husband could finish college and medical school. We were so excited when we were finally able to start a family. But then I didn't become pregnant! Each month was pure agony as I would start my period. Most of our close friends were getting pregnant and were always asking us why we weren't, which didn't help at all. It was a very bad time for our marriage, also, as we argued about which one of us might be 'at fault.' We began a series of fertility tests after about a year and a half and they were very degrading. . . . After many tests and keeping temperature charts for months, I was given Clomid. I'm happy to report that the second month of Clomid did the trick. . . ."

"My husband and I have been unsuccessful in our attempts to have children. We started trying to have a family 7 years ago. We waited a year and a half before even seeing a doctor. My doctor did the preliminary infertility workup and gave me some low doses of Clomid. After 6 months of no success I went to a specialist. He gave me some industrial strength fertility drugs and after about 4 months I got pregnant and miscarried. We tried the same drugs again with no success and finally threw in the towel. All of this took about 5 years.

The experience is terribly isolating. People don't understand what you're upset about—after all, you didn't lose anything that they can see. All of our friends have had their kids. That's a lot of baby showers and first birthday parties that we didn't show up for. The funny thing about new parents is that it is perfectly normal for them to shamelessly go on and on about their new kids and problems—so even when we got together with our friends for other reasons the main topics were breast-feeding and details about their pregnancy and delivery. It was heartbreaking not to be able to be happy for them because at the time we just happened to be in a lot of pain.

You can also add to the list the ignorant people who think you are selfish and self-centered for not having any kids. You don't tell them you are having problems because it's none of their business and they get to assume the worst. Dr. King—I just think most people don't realize that when you come to realize that you can't have your own kids . . . the couple really does grieve about an imaginary child that everyone has dreamed about. . . . We plan on adopting this summer."

(from the author's files)

For women, problems with fertilization generally come from two sources—structural problems in the Fallopian tubes or uterus, and failure to release eggs (ova). Probably the most common cause of female infertility is blockage of the Fallopian tubes, accounting for about 40 percent of all cases (Paterson & Petrucco, 1987). Blockage can be due to anatomical malformations, growths (e.g., polyps), endometriosis (see below), infections, and scar tissue. Cigarette smoking and douching also increase the risk of infertility by increasing the risk of pelvic inflammatory disease and blocked tubes (Baird et al., 1996; Buck et al., 1997). Scar tissue from sexually transmitted diseases, especially chlamydia and gonorrhea (resulting in pelvic inflammatory disease), accounts for at least half the cases of infertility from blocked tubes. Given the epidemic levels of these diseases, this will continue to be a primary cause for both ectopic pregnancies and infertility in the near future (Weström et al., 1992; see Chapter 5).

The diameter of the passageway that an egg takes through the Fallopian tube is only about half that of a human hair. Sometimes surgeons are able to reopen these tiny structures, but the success rate for microsurgery is less than 50 percent, even for the best of physicians (Patton, Williams, & Coulam, 1987). In recent years, laser beams have been used successfully to unblock tubes. An even newer surgical technique to reopen blocked Fallopian tubes involves inserting a small balloon into the tubes and then inflating it (balloon tuboplasty). A similar technique has previously been used to unclog arteries leading to the heart. This technique is less expensive than microsurgery, and early results suggest that it may be the most effective of all techniques for women with blocked Fallopian tubes (Confino et al., 1990).

Endometriosis can cause infertility by the growth of endometrial tissue in the Fallopian tubes and/or around the ovaries. Many women today are postponing pregnancy, and endometriosis is known as the "career women's disease" because it is most common in women in their mid-twenties and over, especially if they have postponed childbirth. Hormone treatment to control the growth of endometrial tissue or surgery to remove it has commonly been used. Today, laparoscopic surgery (with cauterization) is being used with greater frequency for endometriosis-related infertility (Murphy et al., 1991), and use of a drug (pentoxifylline) is presently undergoing clinical trials.

If a woman is having problems becoming pregnant because she is not having regular menstrual cycles, and thus eggs are not being released, her physician may try **fertility drugs.** There are two major types. One type stimulates the pituitary gland to secrete FSH and LH, the hormones necessary to start the cycle (see Chapter 3). Clomiphene is an example of this type of drug. A second type stimulates the ovaries directly (HMG, or human menopausal gonadotropin, is an example). The success rate for fertility drugs is about 50 to 70 percent in cases where the problem is entirely hormonal.

Although they are often successful, there have been two major concerns with the use of fertility drugs. First, they increase the chance of multiple births from 1 or 2 percent to 10–20 percent. From the early 1970s to the mid-1990s, twin births rose 65 percent and multiple births of three or more increased by nearly 400 percent, largely because of fertility drugs (Luke et al., 1994; Ventura et al., 1997). Multiple-birth babies have a much greater risk of being born prematurely and of having low birth weight. Second, there has been some concern that fertility drugs may increase the risk of ovarian cancer (Whittemore et al., 1992a, 1992b). However, rates for ovarian cancer are already much higher in women who have never given birth, and when this is taken into account there is little evidence that fertility drugs increase the risk (Bristow & Karlan, 1996; Mosgaard et al., 1997).

ASSISTED REPRODUCTIVE TECHNOLOGY

The techniques for overcoming infertility entered a new era with the development of **in vitro** ("in glass") **fertilization** (**IVF**) in England in the 1970s. Here eggs and sperm are combined in a Petri dish, and after fertilization takes place, several fertilized eggs (usually at about the four-cell stage) are placed into the woman's uterus with the hope that at least one will then attach itself to the uterine wall. Fertility drugs are often used to stimulate egg production; the eggs are then surgically removed from the ovaries when mature and sperm is collected via masturbation. The first baby born through use of this technique was Louise Brown of England in 1978 (Steptoe & Edwards, 1978). Chil-

Figure 7–13 A reunion of families who produced babies through in vitro fertilization.

dren conceived with in vitro fertilization are often called "test-tube babies" by the press and others, but as you can see, the procedure does not even use test tubes (Figure 7–13).

The overall success rate for in vitro fertilization is low, 20.7 percent in 1994 (Society for Assisted Reproductive Technology et al., 1997). The average cost of a successful delivery is $67,000 to $114,000 (Neumann, Gharib, & Weinstein, 1994). Many institutes were freezing fertilized eggs in case the first attempt failed, but research has shown that the success rate is higher with fresh fertilized eggs (Levran et al., 1990). The major problem, however, was that eggs fertilized outside a woman's Fallopian tubes simply did not attach to the uterine walls as often as eggs fertilized in a normal manner (i.e., within a Fallopian tube).

In the mid-1980s Dr. Ricardo Asch and his colleagues modified the IVF procedure (Asch et al., 1984). They bypassed the Petri dish by putting eggs and sperm into a pipette and inserting them directly into a Fallopian tube. They called the technique **gamete intrafallopian transfer** (**GIFT**). The fertilized eggs implanted in the uterus at a higher rate than with IVF, and the overall success rate is about 27 percent—40–50 percent in the best clinics (Society for Assisted Reproductive Technology, 1993). But that is not the end of the story. In order to assure fertilization, Asch's

fertility drugs Drugs that improve the chance of conception in women who are infertile owing to hormone deficiencies. The drugs either stimulate the pituitary gland to secrete FSH or LH or stimulate the ovaries directly.

in vitro fertilization A process in which a mature ovum is surgically removed from a woman's ovary, placed in a medium with sperm until fertilization occurs, and then placed in the woman's uterus. This is usually done in women who cannot conceive because of blocked Fallopian tubes.

gamete intrafallopian transfer (GIFT) A procedure for treating female infertility in which sperm and eggs are gathered and placed directly into a Fallopian tube.

group had fertilization take place in a Petri dish (as with IVF) and then placed the fertilized eggs directly into a Fallopian tube, a procedure called **zygote intrafallopian transfer (ZIFT)**. The success rate is presently slightly lower than for GIFT. However, after several years, the pregnancy rates for GIFT and ZIFT have not proved to be much better than for in vitro fertilization. Because GIFT and ZIFT require major surgery, most physicians have returned to using the simpler, less invasive IVF precedure (Menezo & Janny, 1996).

These techniques can also be used to screen potential embryos (preembryos) for genetic defects. In 1993, for example, 8-cell preembryos were tested for the deadly Tay-Sachs disease by removing one cell, and three "clean" fertilized eggs were then placed back in the mother. The couple's first child had died of the disease, but they now know that their new baby, born in 1994, is free of the disease.

These techniques are used for a variety of infertility problems, male or female. If it is necessary, donated eggs or sperm can be used for any of the procedures. However, there are some problems. As with fertility drugs, IVF, GIFT, and ZIFT all involve an increased chance of multiple births (Bollen et al., 1991). And in the end, at least half of all couples who undergo assisted reproductive techniques still will not conceive.

The biggest obstacle to advancement in this field is social, political, and legal. Many people oppose these procedures to treat infertility. The Roman Catholic Church, for example, rejects both artificial insemination and in vitro fertilization, proclaiming that procreation must be "a physically embodied love act." Many physicians believe that GIFT will be ethically acceptable to most religious groups (see Mastroyannis, 1993), but that remains to be seen. The U.S. government has refused to fund research on IVF for the past 20 years. Joseph Schulman, a former section director at the National Institutes of Health, said, "I can think of no other examples where for a decade and a half a whole area of investigation that has in fact proven itself to be of massive value to thousands and thousands of couples has been pigeonholed" (quoted in Baker, 1990). Studies indicate that most people have favorable opinions of infertility treatments that produce a child that is biologically related to both members of the couple (they have less favorable opinions of interventions that use donor eggs, donor sperm, or surrogacy; Halman, Abbey, & Andrews, 1992). How do you feel about this?

zygote intrafallopian transfer (ZIFT) A method of treating infertility caused by blocked Fallopian tubes. An ovum taken from the infertile female is fertilized by sperm from her partner and then transferred to the unblocked portion of a Fallopian tube.

surrogate mother A woman who carries a fetus to full term for another couple, agreeing to give the infant to the other couple after it is born. The infant generally represents a union of the sperm from the man and either his partner's ovum or the ovum of the surrogate mother.

SURROGATE MOTHERS

One of the most controversial fertility solutions is **surrogate motherhood,** sometimes resorted to when a woman cannot conceive or carry a fetus during pregnancy. In such a case, a couple may create a zygote through in vitro fertilization and then have the fertilized egg implanted in the uterus of another woman. In a second method, the surrogate mother is impregnated with the male partner's sperm, usually through artificial insemination. The surrogate mother gives birth, and the infant is raised by the infertile couple. The couple will generally pay for the surrogate mother's medical expenses, along with necessary legal fees. They also will pay the surrogate mother a fee for providing this "service" (usually about $10,000). However, some surrogate mothers have changed their minds after delivery, and a 1987 case involved the rights of surrogate mothers to "back out of" such contracts (the judge awarded custody to the infertile couple). Some states have passed laws that allow surrogacy, while others have attempted to outlaw it. Court tests of these laws continue today. This is certainly one area where laws have not kept pace with the new technology.

It is important that couples attempting to deal with infertility receive counseling about the choices they make, the risk of failures (which is high), the consequences of their decisions, and the specific laws of their state. Some parents who use surrogate mothers may pretend that the child was born to them—a decision that could lead to pain and embarrassment later. Women who want to become surrogate mothers should also consider the emotional cost of having to give up a child they have carried through pregnancy and birth.

DELAYED CHILDBEARING AND ASSISTED REPRODUCTIVE TECHNOLOGY

As women grow older, their fertility declines. For most women, this starts happening when they are as young as 31 years of age (van Noord-Zaadstra et al., 1991). After age 30, the chances of having a healthy baby decrease by more than 3 percent each year. The problem is not with the uterus, but with the decreased quality of eggs as they age. At menopause, you recall, the ovaries stop working entirely and women are no longer capable of reproduction—or at least that was what was thought until recently.

As more and more women have entered careers and chosen to delay pregnancy, they have often been faced with fertility problems when they do decide they are ready for motherhood. Today, however, in vitro fertilization and the related procedures have allowed even some postmenopausal women to become pregnant and have children. Eggs are donated by a young woman (and thus are usually of high quality), and then in vitro fertilization is attempted to achieve fertilization and implantation. The uterus is kept functional with injections of estrogen and progesterone. One of the first women to become pregnant with this technology was a 42-year-old former cancer patient who had lost her ovaries because of chemotherapy (see Sauer, Paulson, & Lobo, 1992).

More recently, postmenopausal women over the age of 50 have elected to have children by egg donation and in vitro fertilization (see Sauer, Paulson, & Lobo, 1993). In 1994, a 62-year-old Italian woman gave birth after receiving donated eggs, followed in 1997 by a 63-year-old California woman (Kalb, 1997). In the words of Dr. Marcia Angell, executive editor of the *New England Journal of Medicine,* "The limits on the child-bearing years are now anyone's guess."

Many people, including many physicians, are debating the ethics and health-related issues of assisted delayed childbearing. The older a woman is during pregnancy, the greater the risk of death to the fetus (or to the baby shortly after birth), and the greater the risk of very low birth weight (Cnattingius et al., 1992). There are also risks to the woman herself—diabetes and high blood pressure, for example. These problems might be minimized by careful screening, but what about the ethical considerations? For those who oppose assisted reproductive technology, proponents offer two arguments. First, there is often a double standard, with many people admiring older men who father children. Why shouldn't women have the same opportunity? Second, qualities like love, experience, and responsibility are more important than age in determining who will be a good mother. As the chairperson of the Council of the British Medical Association said, it is better to have "a fit, healthy 59-year-old than an unfit, unhealthy 19-year-old" mother. What qualities do you think make a good mother?

ADOPTION

When a couple cannot conceive and medical technology cannot correct the problem, there are still some options available. Many couples become parents by adopting a child. One difficulty with the option of **adoption** is that the demand for children to be adopted is very high. Long waits—sometimes months or even years—may be required before a couple is able to find and adopt a healthy child. Some couples choose not to wait, and adopt children of other nationalities or disabled children. For caring people, adoption benefits everyone—the couple who so badly want a child, and a child who needs a home, love, and affection.

For information about adoption (including a bimonthly magazine called *Ours*) and support, you can contact Adoptive Families of America, 3333 Highway 100 North, Minneapolis, MN 55422, or call 612-535-4829.

SUPERFERTILITY

For some women, having children is apparently very easy. According to the *Guinness Book of World Records,* the most prolific mother alive is Leontina Albina of Chile, who gave birth to 59 children—always triplets or twins. She was one of three triplets sent to an orphanage, but while her two brothers were adopted, she was not. She promised herself that if she ever had children, she would never give them away. Leontina married at age 12 and continued to have children into her fifties.

A word of caution: Studies show that women who have had six or more pregnancies are at higher risk for coronary heart disease and cardiovascular disease (Ness et al., 1993). Please take this into consideration if you are planning a large family.

> **adoption** Taking a child into one's family by legal process and raising the child as one's own.

Key Terms

adoption   157, 195
afterbirth   185
amniocentesis   176
amnion   166
artificial insemination   191
blastocyst   165
Bradley method   182
Braxton-Hicks contractions   183
breech birth   185
capacitation   165
celocentesis   176
cesarean section (C-section)   186
chorion   166

chorionic sampling   176
colostrum   168, 187
conception   165
couvade syndrome   167
critical period   172
crowning   184
dilation (cervix)   183
eclampsia   173
ectopic pregnancy   166
effacement (cervix)   183
embryo   167
episiotomy   185
false labor   183

fear-tension-pain cycle   179
fertility drugs   193
fertilization   165
fetal alcohol syndrome (FAS)   174
fetal surgery   177
fetoscopy   176
fetus   168
gamete intrafallopian transfer (GIFT)   193
Graafian follicle   38, 55, 165
human chorionic gonadotropin (HCG)   56, 167
implantation   165

Personal Reflections

1. What qualities do you possess that will make you a good parent? Do you have any qualities that might *not* make you a good parent? What things can you do to prepare yourself to be a good parent?

2. How well suited is your partner to be the mother or father of your child?

3. Do you or your partner smoke, drink alcohol, or take drugs? Do you and your partner eat a well-balanced diet and get ample sleep? What decisions about health can you and your partner make to ensure the health of your baby during pregnancy and afterwards?

4. Make a list of all the words that come to your mind when you think of childbirth, labor, and delivery. Examine your list for negative stereotypical concepts. Are there alternatives?

5. Will pregnancy affect your sexual relationship with your partner? How? Why? Will being parents affect your sexual relationship? How? Why? What steps can you and your partner take to ensure that as parents your sexual relationship will continue to be fulfilling?

6. Imagine that you and your partner have tried to conceive for five years. You both have gone through innumerable tests, but nothing seems to work. How do you feel? What would you do next? What options would you consider? What options would you never consider? Why or why not?

Suggested Readings

Beck, M., et al. (1988, 15 August). Miscarriages. *Newsweek.* Discusses the causes of miscarriage and the medical advances to prevent it.

Begley, S. (1995, 4 September). The baby myth. *Newsweek.* An excellent article on reproductive technology for infertile couples and its low success rate.

Bing, E. (1988). Yes, you can. *Childbirth '88,* vol. 5. Discusses sexual intercourse during pregnancy, assuring women that it is okay. If you want more information, see Bing, E., & Coleman, L. (1983). *Making love during pregnancy,* New York: Bantam.

Burns, L. H. (1987, Fall/Winter). Infertility and the sexual health of the family. *Journal of Sex Education and Therapy.* Discusses the effects of infertility on a couple's sexual health.

Cowan, C. , & Cowan, P. (1992). *When partners become parents.* New York: Basic Books. Addresses the many challenges that couples will face when they become parents. An excellent book that is based on 10 years of research.

Kime, R. (1992). *Wellness: Pregnancy, childbirth and parenting.* Guilford, CT: Dushkin Publishing. Aimed at teaching parents-to-be to think about health.

Lasker, J. N., & Borg, S. (1995). *In search of parenthood: Coping with infertility and high-tech conception.* Philadelphia: Temple University Press. Reviews different treatments for infertility and discusses the psychosocial experiences of people undergoing the procedures.

Slawson, M. (1982, March). Drugs and pregnancy: What we know today. *Your Life and Health.* Discusses the potentially serious effects of nonprescription drugs on pregnancy.

CHAPTER 8

When you have finished studying this chapter, you should be able to:

1. Define gender identity and gender role;

2. Explain various biological influences on sexual anatomy and gender identity;

3. Describe the causes of hermaphroditism and pseudohermaphroditism;

4. Define and differentiate the concepts of gender identity disorder, transsexuality, homosexuality, transvestism, and gender dysphoria;

5. Explain the Freudian, learning, and cognitive-developmental theories of gender identity development;

6. Contrast the differences in emphasis between the individualist and the microstructural theories of gender-role development;

7. Discuss the dualistic and unidimensional models of masculinity and femininity and the concept of androgyny;

8. Summarize the development of gender roles in childhood and adulthood; and

9. Describe, from a microstructural viewpoint, how gender roles have evolved in the United States.

Becoming a Woman/Becoming a Man: Gender Identity and Gender Roles

How does a person become a woman or a man? How do we learn that we are male or female? What does it mean to be "masculine" or "feminine"? Is it appropriate for a woman to ask a man for a date? If a father sees his son playing with a doll, what should the father do? Where do our ideas about "acting like a woman" or "acting like a man" come from? Before these questions can be answered, you must learn the distinction between two terms—*gender identity* and *gender role*.

Gender identity is your sense of self as a male or a female, while **gender role** includes everything you feel, think, say, and do that shows to yourself and others that you are, in fact, a male or a female. In other words, gender role is the way you express your gender identity. Gender identity and gender role are not two different things, but rather two different aspects of the same thing (Money & Tucker, 1975). Gender identity is the inward experience of your gender role, and gender role is the outward expression of your gender identity.

Kelly and Worell (1977) view gender roles as involving two different traits or sets of characteristics—"masculinity" and "femininity." These two characteristics were once viewed as bipolar (opposite ends on a scale), and psychological tests for masculinity and femininity measured people along this one scale. It was believed that if an individual scored high on masculine traits, he or she would not show any feminine traits (see Spence & Helmreich, 1978).

Today, researchers believe that the characteristics of masculinity and femininity are present in both men and women, but to different degrees (Spence, Helmreich, & Stapp, 1974). Sandra Bem (1974) developed the Bem Sex Role Inventory to measure the amount of masculinity and femininity within an individual's gender role. She noted that individuals could be characterized as masculine or feminine if they scored high on one dimension and low on the other. For example, a high score on the femininity dimension and a low score on masculinity would classify a person as "feminine." A person having high scores on both dimensions would be classified as *androgynous*. The term **androgyny** comes from the Greek words *andro,* for "male," and *gyn,* for "female." Some see this combination as socially desirable (Spence, Helmreich, & Stapp, 1975). The concept of androgyny implies that it is possible for a person to be both assertive and compassionate, logical and emotional, depending on what is most appropriate for a particular situation. In addition, Bem claimed that some people scored low on both dimensions. Such individuals would be labeled *undifferentiated.* These combinations of femininity and masculinity are summarized in Table 8–1.

Gender role stereotypes are oversimplified, rigid beliefs that all members of a particular gender have distinct behavioral, psychological, and emotional characteristics. A person is considered "sex-typed" if he or she exhibits a relatively high number of sex-stereotyped characteristics and claims not to have or not to value the other type of characteristics (Worell, 1978). For example, a man would be sex-typed if he chose only to display masculinity in himself and preferred not to show signs of femininity. In contrast to the sex-typed individual is the androgynous person. An androgynous person is capable of integrating both traditionally masculine characteristics and traditionally feminine characteristics into his or her gender role.

In this chapter, you will learn about the forces that shape both gender identity and gender roles. These forces are biological and social, internal and external, involving both genetics and learning. Theorists differ in the importance they give to the influence of these different forces. As always, it seems that most theorists agree that it is the interaction of such forces that ultimately determines gender identity and gender role. I will begin the chapter with a discussion of how gender identity is formed, focusing first on biological aspects of this process. The issue of "nature versus nurture" in the formation of gender identity will be explored. You will learn of instances where an individual's gender identity conflicts with his or her biological sex. Psychological theories of gender identity formation are elaborated upon, including Freudian theory, learning theory, and cognitive-developmental theory. Finally, we will discuss the development of gender roles, and the forces shaping our concepts of masculinity and femininity.

TABLE 8–1

BEM'S SEX ROLE CLASSIFICATIONS

	High Femininity	Low Femininity
High Masculinity	Androgynous	Masculine
Low Masculinity	Feminine	Undifferentiated

BIOLOGICAL INFLUENCES ON GENDER IDENTITY

How do we end up as a boy or girl at birth? Genetics certainly plays an important role, but hormones are equally important.

THE ROLE OF CHROMOSOMES

A human cell normally has 23 pairs of **chromosomes,** rod-shaped structures that determine a person's inherited characteristics. Eggs and sperm have only half the normal genetic material (one chromosome from each of the 23 pairs). Thus, when an egg and sperm unite (called *conception*), the resultant single cell will again

---

gender identity One's subjective sense of being a male or a female. This sense is usually acquired by the age of 3.

gender role A set of culturally specific norms concerning the expected behaviors and attitudes of men and women.

androgyny The ability of an individual to display a variety of personality characteristics, both masculine and feminine, depending on whether a trait or behavior is appropriate in a given situation. It is often viewed as a positive characteristic that gives an individual greater adaptability.

chromosomes Rod-shaped structures containing the genetic material that determines a person's inherited characteristics.

have the normal 23 pairs. One of these pairs determines whether a person is genetically male or female. *Females usually have two X chromosomes (XX), while males usually have one X and one Y chromosome (XY).* Thus, it is the sperm from the male that determines the genetic sex of the child, and this is determined at the moment of conception. However, sometimes individuals with an XX combination are anatomically male and other individuals with an XY combination are anatomically female. How is this possible? Researchers have determined that a smidgen of the Y chromosome—a gene called SRY (sex-determining region of the Y chromosome)—determines maleness (Sinclair et al., 1990; see Sultan et al., 1991). This piece is missing from XY females' Y, and it is present on XX males' X. More recently, it has been found that a gene on the X chromosome—a gene called DSS (dosage-sensitive sex reversal)—may help determine femaleness (Bardoni et al., 1994).

In some cultures it was legitimate for a husband to divorce his wife if she did not have a male baby (as King Henry VIII did). Today, some couples attempt to increase their chances of having a boy by having the woman douche with baking soda, by having intercourse as close as possible to the time of ovulation, or even by wearing boots to bed. Needless to say, none of these techniques have proven reliable (Wilcox, Weinberg, & Baird, 1995). More recently, however, selection of biological sex has been made possible by separating the X and Y sperm (by filtering them through liquid albumin, a protein) and then performing artificial insemination (Glass & Ericsson, 1982). This procedure is very controversial for both ethical and religious reasons (e.g., if most couples preferred male offspring, which is the case in many cultures, there would soon be an oversupply of boys if everyone used this technique), but it appears to be gaining acceptance (Jancin, 1988).

CHROMOSOME DISORDERS

There are over 70 known irregularities involving the sex chromosomes, and approximately one in every 426 people is born with sex chromosome abnormalities (Nielson & Wohlert, 1991). For example, in one out of every 500 live male births, there is one or more extra X chromosomes (XXY or XXXY), a condition known as *Klinefelter's syndrome*. These biological males tend to be tall, with long arms; they have poor muscular development, enlarged breasts, a small penis with shrunken testes, low sexual desire, and mental and academic impairment (Mandoki et al., 1991). Males can also be born with an extra Y chromosome (XYY). These men are also very tall, tend to be mentally dull, and are less fertile than other males. One study reported that there were more XYY males in prison for violent crimes than would be expected (Jacobs et al., 1965), but subsequent research has questioned this finding.

Turner's syndrome is a condition in which there is only one X chromosome (XO). It occurs in about one in every 2,000 to 3,000 live births (Gravholt et al., 1996). Because women with Turner's syndrome are missing a chromosome, the ovaries never develop properly, and these women cannot reproduce. Because of the absence of ovarian hormones, they also do not menstruate or develop breasts at puberty. These females are generally short, with a webbed neck and broad chest, but have normal intelligence. *Triple-X syndrome females* (XXX), on the other hand, may not appear abnormal but are often retarded or mentally disturbed.

One of the most interesting sex chromosome disorders involved Stella Walsh, an internationally renowned athlete who won an Olympic gold medal in the women's 100-meter dash in 1932 and a national pentathlon in 1954 (see Figure 8–1). After her death in 1980, it was found that some of the cells in her body were XX, but others were XY. She had been raised and lived as a female, but had nonfunctional male sex organs.

Fortunately, the vast majority of us are born with the correct number and combination of sex chromosomes. However, as you have learned, having an XY

Figure 8–1 Stella Walsh won an Olympic gold medal in the women's 100-meter dash in 1932. After her death, it was discovered that she had both XX and XY cells.

chromosome combination does not guarantee that a baby will be anatomically a male at birth. Hormones are also critically important in the determination of one's sex.

THE ROLE OF HORMONES

As two researchers put it, "Nature's rule is, it would appear, that to masculinize, something must be added" (Money & Ehrhardt, 1972). This something is *testosterone,* which must be produced at a critical stage of embryological development (i.e., during the first 2 months after conception) in order for an XY combination to result in an anatomically male baby.

Embryos cannot be distinguished anatomically as either male or female for the first few weeks of development. A pair of primitive gonads develops during the fifth and sixth weeks that have the potential for developing into either ovaries or testes. At this stage, the embryo also has two duct systems: (1) the **Wolffian duct system,** which if allowed to develop will become male structures; and (2) the **Mullerian duct system,** which if allowed to develop will become female structures (see Figure 8–2). If the primitive gonads were removed at this stage of development, the baby would *always* be born anatomically a female, even if it were genetically XY. Why?

Normally, the SRY gene on the Y chromosome triggers the transformation of the primitive gonads into testicles during the seventh week of gestation (Haqq et al., 1994). The newly developed testicles begin to secrete testosterone, which promotes the development of the Wolffian duct system into the male internal reproductive system. The masculinizing hormone is also responsible for the development of the external genitalia a few weeks later. What happens to the Mullerian duct system in these male embryos? The SRY gene activates the testicles to secrete an additional substance called **Mullerian duct–inhibiting substance,** which causes these ducts to shrink (Haqq et al., 1994).

What happens to embryos with an XX combination (genetically female)? If there is no Y chromosome (and thus no chemical substance to convert the primitive gonads into testicles), the primitive gonads will eventually develop into ovaries. The absence of large levels of testosterone results in the shrinkage of the

Wolffian duct system and the development of the Mullerian duct system into the female reproductive system.

In summary, it would appear that *unless there are high levels of male hormones at a critical stage of prenatal development, nature has programmed everyone for female development.* In the vast majority of births, the baby's genetic and anatomical sex are matched. Occasionally, however, a different pattern of development may be seen.

ABNORMALITIES IN DEVELOPMENT

Hermaphroditism is a condition in which a person is born with both male and female reproductive systems as a result of the primitive gonads failing to differentiate properly during the embryonic stage (see Blyth & Duckett, 1991). The term *hermaphrodite* is derived from the Greek mythological figure Hermaphroditus, who merged with the nymph Salmacis to form a single body with male and female genitals. Hermaphrodites are usually genetic females, and even though a uterus is almost always present, they often have an ovary and Fallopian tube on one side and a testicle and a vas deferens and/or epididymis on the other (see Figure 8–3). The external genitalia are usually ambiguous in appearance, but because the phallus is often enlarged, nearly two-thirds of these individuals are raised as boys, with complications arising at puberty when they begin to develop breasts and menstruate. Fortunately, this is a very rare condition, with fewer than 100 recorded cases.

More common is a condition known as **pseudohermaphroditism,** in which a person with an XX or XY chromosome pattern is born with the proper set of gonads (ovaries or testicles, respectively), but whose external genitalia are either ambiguous or that of the other sex. In females, the most common cause is known as **adrenogenital syndrome (AGS),** also known as *congenital adrenal hyperplasia,* in which the adrenal glands secrete too much masculinizing hormone during fetal development. The internal organs are normal, but the genitals are masculinized (i.e., the clitoris and labia are enlarged), so that even physicians sometimes mistake these individuals for boys at birth (see Figure 8–4). If the problem is not recognized and corrected early, the excess adrenal hormones will continue to masculinize the individual during subsequent development.

Wolffian duct system A primitive duct system found in embryos that, if allowed to develop, becomes the male reproductive system.

Mullerian duct system A primitive duct system formed in embryos, which, if allowed to develop, becomes the female reproductive system.

Mullerian duct–inhibiting substance A chemical secreted by the testes during embryological development that causes the Mullerian ducts to shrink and disappear.

hermaphrodite A person with both male and female reproductive systems as a result of failure of the

primitive gonads to differentiate properly during embryological development.

pseudohermaphroditism A condition in which a person is born with ambiguous genitalia as a result of a hormonal abnormality.

adrenogenital syndrome A condition in females in which the

adrenal glands excrete too much testosterone during fetal development, causing masculinization. This includes enlargement of the clitoris and labia, so that the genitals are sometimes mistaken for those of a male. Also called *congenital adrenal hyperplasia.*

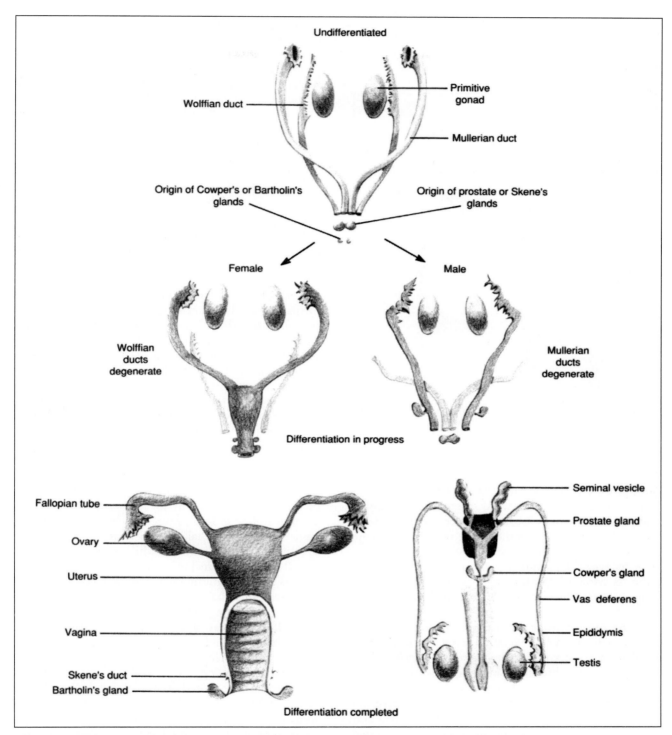

Figure 8–2 Prenatal differentiation of the internal reproductive systems in males and females.

The first treatment for this condition, developed in the 1950s, included administration of the adrenal hormone cortisone and plastic surgery to correct the appearance of the genitals. Early treatment, combined with being raised as a female, generally leads mas- culinized girls to develop female gender identity. However, many of them are reported to show tomboy- ish behavior as children (Berenbaum & Hines, 1992; Ehrhardt & Baker, 1974; Slijper et al., 1992) and to have fewer fantasies about pregnancy and motherhood

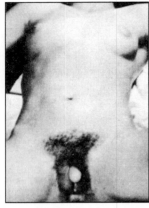

Figure 8–3 Hermaphroditism. This genetic female has one ovary and one testicle and has always lived as a male.

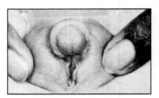

Figure 8–4 Adrenogenital syndrome in an infant female.

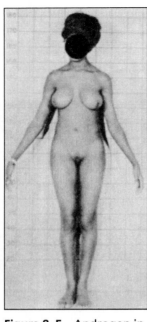

Figure 8–5 Androgen insensitivity syndrome in a genetic male.

during infancy, the condition is often not discovered until the individual fails to start menstruating as a teenager. (Undescended testicles are a risk factor for the development of cancer and are usually removed by surgery.) These individuals can be very feminine in appearance and behavior, and despite the fact that they cannot conceive, they have made excellent adoptive mothers (Money & Ehrhardt, 1972). They are attracted to male sexual partners and clearly have a female gender identity (Hines & Collaer, 1993; Money, 1994).

In 1974, 38 genetic males in the Dominican Republic were discovered who suffered from a type of androgen insensitivity syndrome. Prenatally, their testicles had produced testosterone and Mullerian duct–inhibiting substance, so that their internal structures were male. However, because of an inherited enzyme deficiency, the testosterone was not converted into dihydrotestosterone, which is necessary for proper formation of the external genitals (the condition is therefore sometimes called *DHT deficiency syndrome*). As a result, they had a very small penis that looked like a clitoris, an incomplete scrotum that looked like labia, undescended testicles, and a short closed vaginal cavity.

Eighteen of these children were reared as girls, but at puberty, when their testicles started secreting large amounts of testosterone, their voices deepened, muscles developed, the testicles descended, and their "clitoris" grew to become a penis. Although the urethral opening was located on the perineum and not the tip of the penis, they were capable of ejaculating sperm.

How do you think you would feel if you were being raised as a girl, but suddenly started growing a penis at puberty? The culture in these remote villages has a strong patriarchal "macho" bias. Girls are forced to stay home and do chores, while boys are given considerable freedom, including sexual freedom as adults. Parents of these children generally rejoiced when they discovered that they had an unexpected son. The villagers called them *guevote* ("penis at twelve"). Under these circumstances, it is probably not surprising that 16 of these 18 children decided to become males and subsequently developed a strong sexual interest in girls (Imperato-McGinley et al., 1979).

It should be noted that anthropologist Gilbert Herdt (1990) believes that this interpretation of the adjustment by the DHT-deficient children is an example of observer bias (see Chapter 1). Western culture categorizes people by two genders, but the culture in which the children were raised believes in three genders. In this case, adjustment is easier.

Do you suppose that the transition would have been so easy in a country like the United States? In fact, most American children with this disorder choose to remain female when their bodies become masculinized (Rubin, Reinisch, & Haskett, 1981).

during adolescence (Ehrhardt et al., 1984). Nevertheless, most have a heterosexual orientation as adults, and some who receive treatment later marry and have babies. Many of these women state that the cortisone treatment reduces their sexual desire. Because cortisone reduces the production of androgens (Ehrhardt, Evers, & Money, 1968), it is believed that cortisone is indeed responsible for the decrease in their sexual interest.

The most common cause of pseudohermaphroditism in males is known as **androgen insensitivity syndrome** (sometimes called *testicular feminization syndrome*), in which the testicles secrete normal amounts of testosterone but the body tissues do not respond to it (Kaplan & Owett, 1993). As a result, a clitoris, a short vagina, and labia develop, but the internal female structures fail to develop because the testicles still secrete Mullerian duct–inhibiting substance (see Figure 8–5). These individuals are generally raised as girls. In fact, unless the undescended testicles are felt in the groin

androgen insensitivity syndrome A condition in which the testes secrete normal amounts of testosterone during male embryologic development, but the tissues do not respond to it. As a result, a clitoris, labia, and a short vagina develop, but the internal female structures do not develop because the testes still secrete Mullerian duct–inhibitory substances.

NATURE VERSUS NURTURE IN ACQUIRING GENDER IDENTITY

SEX REASSIGNMENT AT BIRTH

Is gender identity determined mostly by biology or mostly by teaching the values of society (socialization) through child-rearing practices? Much debate has taken place over this question, but recent studies suggest that biology plays a much stronger role than previously believed. Consider, for example, the following case history.

Identical twin brothers were circumcised at 8 months of age, but due to a surgical mistake, one of the twins lost his penis. After considerable deliberation, and upon the recommendation of doctors, the boy's parents decided to raise the child as a girl. They were encouraged to do so because at that time it was believed that at birth people are psychosexually neutral and that normal psychosexual development depends on the appearance of one's genitals.

The child's name was changed, and surgical procedures were done to reconstruct the genitals in order to make them female. The child was dressed in girls' clothing and was encouraged to play with girls' toys and engage in girls' activities.

The published reports during this person's childhood concluded that the sex reassignment had been a success. According to Money (1975), the child readily achieved a female gender identity and gender role. The twin brother developed traditional masculine traits. Largely as a result of the "success" of this often cited case, thousands of children born with ambiguous genitals were given sex reassignment surgery (see Colapinto, 1997).

Later interviews with the "female" twin revealed many problems. The psychiatrist following "her" development found that "she" had great difficulty in her attempt to become a woman (Williams & Smith, 1979). She rejected girls' clothes and toys, preferring boys' clothes and toys. She imitated her father and was regarded as a tomboy. She frequently urinated (or tried to) while standing up. As she entered adolescence, she became depressed and suicidal, which finally resulted in her parents' telling her the truth—"she" was a boy. He chose to have a mastectomy, receive male hormone shots, and undergo surgery to partially restore his penis. Afterwards, he adjusted well, married, and adopted his wife's children. When asked why he had rejected all the attempts to raise him as a female, he said that being a female "did not feel right" (Diamond, 1997; Diamond & Sigmundson, 1997).

This is not an isolated case. Many individuals who underwent sex reassignment as infants because of ambiguous or injured genitals have rejected the socialization process (Diamond, 1997; Cowley, Gideonse, & Underwood, 1997). What can be concluded from these cases? According to Diamond (1997), they show that nature sets a predisposition for gender identity and that "one's sexual identity is not fixed by the gender of rearing."

SEXUAL DIFFERENTIATION OF THE BRAIN

Are male brains different from female brains, and if so, could this account for differences in behavior? In the 1970s, it was discovered that parts of an area of the brain called the hypothalamus (see Chapter 2) were different in male and female rats (Gorski et al., 1978). The difference wasn't noticeable at birth, but became apparent shortly thereafter as a result of differences in testosterone levels (Goy & McEwen, 1980). Studies of humans show that men's and women's hypothalamuses also differ (e.g., Swaab, Gooren, & Hofman, 1992). Apparently, sex hormones very early in life affect not only the anatomy of the genitals, but the anatomy of the brain as well. This would explain why young children given sex reassignment surgery are unable to adjust—sex hormones may have already altered their brains in accord with their genetic sex.

We must be cautious before we accept this idea without hesitation. Because the differences in brain anatomy are not noticeable in newborns, some interesting questions are raised. Do biological (brain) influences determine behavior, or do social stimuli (i.e., the different manner in which people interact with boys and girls) alter both psychology and biology (see Breedlove, 1994, for a review)? This is a fascinating new area of research, and we will return to it in Chapter 9, but it is too early to come to definite conclusions.

GENDER IDENTITY DISORDER

You have just read about a few individuals who had sex reassignment surgery in infancy because of ambiguous-looking or injured genitals. Attempts to raise them as persons of the opposite sex were not successful. Are there ever cases in which individuals with normal genitals are not happy with their gender? The answer is yes.

Some people's gender identity does not match their biological sex. These individuals have normal anatomy, both internally and externally, and yet feel that they are actually a member of the opposite sex. They feel intense distress with their anatomy, a condition known as **gender dysphoria** (American Psychiatric Association, DSM-IV, 1994; Pauly, 1990). People with gender dysphoria feel that they are "trapped in the wrong anatomic body" (Pauly, 1985). This feeling of a mismatch between gender identity and anatomy often begins at a very early age, although in some cases it does not arise until adulthood.

gender dysphoria The feeling of being trapped in a body of the opposite sex.

Adults with this disorder are generally referred to as **transsexuals,** and children with the disorder are diagnosed as having gender identity disorder of childhood. However, in the American Psychiatric Association's latest diagnostic manual, DSM-IV, all conditions have been collapsed into a single diagnosis—**gender identity disorder (GID)** (Bradley & Zucker, 1997). The criteria for gender identity disorder are (a) behaviors that indicate identification with the opposite gender and (b) behaviors that indicate discomfort with one's own anatomy and gender roles. Thus, people with gender identity disorder have an intense desire to belong to the opposite sex, prefer to cross-dress, and participate in the games and pastimes of the opposite sex. Persistent discomfort with their bodies and bodily functions is almost always the case (e.g., boys are disgusted with their penises and testicles and distressed about facial and body hair; girls do not want to grow breasts, menstruate, or urinate in a sitting position).

Among children, boys in therapy for GID outnumber girls 7 to 1, but this may be because our culture has a higher tolerance for girls who display cross-gender behavior (Bradley & Zucker, 1997). The ratio becomes less one-sided in adolescence. Nevertheless, researchers often report that even among adults, male-to-female transsexuals are 3 to 6 times more common than female-to-male transsexuals (Stockard & Johnson, 1980). Among genetic males, transsexualism is found in both heterosexual individuals (most commonly) and homosexual individuals. Among females, it is rare for a heterosexual individual (someone attracted to males) to be transsexual (Dickey & Stevens, 1995). Most female transsexuals attracted to males have a fantasy of being gay.

It is important to understand that transsexualism has to do with gender identity, not sexual orientation—it is not the same as homosexuality. Homosexuals have gender identities that agree with their anatomical sex just as often as heterosexuals. Most male-to-female transsexuals, for example, are attracted to men because they wish to be desired and loved as a woman by a heterosexual male.

Transsexuality is also not the same as transvestism. A transvestite, as you will learn in Chapter 14, dresses in the clothing of the opposite sex in order to achieve sexual arousal. However, a transvestite does not want to change his or her biological sex and does not experience gender dysphoria. In contrast, transsexuals cross-dress as a means of attempting to be more comfortable psychologically with their appearance, not for sexual arousal. Some males are sexually aroused by the thought of themselves with female attrib-utes and may even undergo breast augmentation (Blanchard, 1993), but this is not the same as gender dysphoria either.

The cause(s) of gender identity disorder has not been determined. Children with GID tend to have internalizing psychopathology (e.g., depression, low self-esteem) (Bradley & Zucker, 1997). A recent study found that while adult transsexuals were dysphoric, they did not show signs of other psychopathology and were "normal" in the desired gender (Cole et al, 1997). With regard to biological factors, it is likely (though unproven) that hormones before birth or shortly afterwards have effects on later behavior and gender identity (see previous sections and Collaer & Hines, 1995). Psychosexual factors, particularly the parents' preference for a boy or a girl and parent-child interactions (e.g., whether or not parents discourage cross-sex behavior), may also play a role (Bradley & Zucker, 1997).

What treatment is available? For children with GID, approaches that discourage cross-gender behavior, promote same-sex identification, and encourage parental involvement have had some success (Zucker & Bradley, 1995). Analytical therapy has also been proposed (Coates, 1992). In any case, many children with GID do not become transsexuals as adults (Green, 1987). As for adult transsexuals, many have turned to surgery to deal with the problem (Pauly & Edgerton, 1986).

Prior to undergoing *sex reassignment surgery* (a "sex-change operation"), individuals are carefully screened by psychological testing. In addition, they are usually required to live for one or two years as a member of the "opposite" (anatomical) sex before undergoing surgery (Chong, 1990). For example, a female-to-male transsexual would take on a male identity and wear men's clothing. Sex hormones are also administered to stimulate the growth of secondary sex characteristics of the opposite sex. The male-to-female transsexual is given estrogen, which will result in the development of breasts, the formation of fatty deposits in the hips, and a change in skin tone. The female-to-male transsexual is given testosterone, which will cause a deeper voice, increase facial and body hair, and inhibit menstruation.

During this period, the transsexual can prepare family, friends, and acquaintances for the planned sex change. It is common for there to be a great deal of anxiety, apprehension, and discomfort in the way that family members and friends react to the idea of transsexual surgery. It is best for these feelings to be confronted and dealt with before the surgery. Likewise, obtaining legal recognition of a sex change can be attempted during this try-out period. Many courts in the United States have refused to recognize individuals as really being of the other sex (different from their chromosomal or anatomical sex) after a sex reassignment

transsexual An adult whose gender identity does not match his or her biological sex.

gender identity disorder A disorder whose criteria are (a) behaviors that indicate identification with the opposite gender, and (b) behaviors that indicate discomfort with one's own anatomy and gender roles.

operation (Hurley, 1984). Marriages following the operation may also not be legally recognized if the marriage partner is of the same sex as the transsexual was before the surgery.

The male-to-female transformation can be done in one surgery (Bouman, 1988). The penis and testicles are removed first. Next, a vagina is created using pelvic tissue. Sensory nerves in the skin of the penis are relocated to the inside of the new vagina. Sexual intercourse is possible after the operation, though lubricants may have to be used. There are also reports that a full range of sexual responses, including orgasm, has been achieved after surgery (Money & Walker, 1977), although greater proof of these claims is needed. Male-to-female surgery is a much simpler procedure than female-to-male surgery.

In female-to-male surgery, the uterus, ovaries, and breasts are removed. Some individuals have no further changes made, while others decide to have an artificial penis created. The new penis is made using flaps of skin and muscle (from the groin, forearm, or other places) or by clitoral enlargement (Hage, Bloem, & Suliman, 1993). It is incapable of normal erection, although artificial means of creating erections are possible, such as those used in treating erectile disorders (Hage, Bloem, & Bouman, 1993).

One of the most famous cases of sexual reassignment was that of an ophthalmologist named Richard Raskind (Richards & Ames, 1983). Richard Raskind had attended college and medical school, become a highly ranked tennis player, and married. He hated his maleness and genitalia and reported the agony that he felt in his presurgery dilemma. "As a child . . . I would pray every night that I could be a girl. I knew then that I wanted Renee as my name. It means reborn" (*People* magazine, 1976). He eventually decided to have a sex-change operation and become Renee Richards (Figure 8–6).

After the operation, Renee felt more comfortable with her gender identity, but found that female tennis players did not want to let her compete in the top professional tournaments. They claimed that Renee's physical abilities gave her an unfair advantage. However, a judge ruled that Renee could compete on the professional women's circuit. She did so for several years and later became a tennis coach, most notably for Wimbledon tennis champion Martina Navratilova. In a more recent case (Docter, 1985), a man decided to undergo transsexual surgery at the age of 74. He had been happily married for 37 years, but had had lifelong fantasies of becoming a woman. He finally decided to become a woman after his wife had been dead for several years.

The purpose of sex reassignment surgery is to lessen the feeling of gender dysphoria in the transsexual and to contribute to his or her emotional health and well-being. Most studies have found that the vast

Figure 8–6 Tennis player Renee Richards as she appeared after a sex-reassignment operation (right), and before as Dr. Richard Raskind (left).

majority of transsexuals achieve satisfactory results with the procedure, with only a few expressing feelings of regret (e.g., Blanchard et al., 1989; Pfafflin, 1992; Rakic et al., 1996; see Green & Fleming, 1990 for a review). It is important to remember that *the surgery is used to help confirm an individual's previously established gender identity* (Edgerton, 1984). Such gender confirmation can be heavily influenced by the society within which the transsexual lives. The need for external validation of a person's sense of self varies from individual to individual, with the social acceptance of transsexuality varying across settings and times. It seems reasonable to state that surgery will most help those who can find both internal and external confirmation of their gender identities after the operation.

ARE TWO SEXES ENOUGH?

Consider the following scenario and question:

> **A person who does not start menstruating as expected during adolescence goes to a physician to find out why. The physician discovers that this person—whose outward appearance and sense of self are both female—has no uterus or ovaries, but instead has male (XY) chromosomes. Is this person female or male?**
>
> (McKain, 1996. Reprinted with permission of the publisher.)

Did you hesitate to answer? Most people do. Why? Probably because being male or female is more complicated than having a single characteristic. More importantly, "hesitation in answering the above question suggests how committed people are to the notion that everyone is *either* female *or* male, period. If people were not so wedded to this notion, they would simply respond that the person described above is female in some ways and male in others" (McKain, 1996).

At this point you have been introduced to other examples of individuals with mixed anatomy and/or sexual behavior. We have terms for many of them, such as gender identity disorder, but it has been argued that putting a psychiatric diagnosis on people stigmatizes them (Pauly, 1992). McKain (1992) has argued that we should include people such as hermaphrodites and transsexuals into a single general category called *mixed-sex people.* Recall that some other cultures recognize three sexes (Herdt, 1990). In fact, it has recently been suggested that we should recognize five different sexes (Fausto-Sterling, 1993). The point these authors are trying to make is that our culture's tradition of looking at gender as a dichotomy is not sufficient to include all of the diversity among human beings. What do you think?

PSYCHOLOGICAL THEORIES OF GENDER IDENTITY DEVELOPMENT

In the previous sections, you learned that biology plays a major role in establishing one's gender identity. This is not to say that the environment has no effect. The two interact. It is likely that "nature sets a predisposition for these sexual developments and within such limits the environment works" (Diamond, 1997). Parents are obviously an important part of a child's environment. Three major psychological theories of development claim that the parent is an important role model who influences the development of gender identity.

FREUDIAN THEORY
Sigmund Freud believed that psychological development was influenced primarily by sexual development. He called sexual energy *libido* and thought that the location of the libido (the area of the body responsible for sexual pleasure) changed over time. The part of the body where the libido is focused is called an "erogenous zone." The primary erogenous zone, according to Freud, depends on which *stage of psychosexual development* the individual has reached. In the first year of life, the mouth is the primary erogenous zone, so Freud called this the *oral stage.* Pleasure is derived from sucking and from exploring things with the mouth.

The second stage is called the *anal stage,* which lasts from

---

identification The adoption of the sex roles of the same-sex parent by a child. In Freudian theory, identification resolves the Oedipus complex.

operant conditioning According to learning theory, a type of training thought to be responsible for the acquisition of sex roles by children. It rewards "masculine" behavior in boys and "feminine" behavior in girls and does not reward behaviors associated with the opposite sex.

---

age 1½ to about age 3. According to Freud, the primary source of pleasure in this stage comes from holding in and expelling feces. Conflicts arise when parents attempt to teach their children to control their bowel movements in ways that are socially acceptable. Children at this time may fight parental control and toilet training.

The third stage, called the *phallic stage,* lasts from about age 3 to 5. Freud believed that gender identity was learned at this time. It is during this stage, according to Freud, that boys begin to derive pleasure through masturbation, and as they do so, they begin to sexually desire their mothers and come to view their fathers as powerful rivals who might punish them for these desires by taking away their penises. This fear is called "castration anxiety" and is based on the fact that previous sources of pleasure have already been taken away through weaning and toilet training. In addition, because girls and women do not have a penis, this further suggests to children that the penis may be taken away. Gender identity formation for boys takes place by resolving these conflicts, known as the male *Oedipus complex* (named after the principal character in the Greek play *Oedipus Rex,* in which a man kills the king of Thebes and then marries his queen, who turns out to be his own mother). Following the rule "If you can't beat them, join them," boys try to become like their fathers and adopt their fathers' gender identity. This process occurs unconsciously and is called **identification.** In this way, boys try to assure that for the future, they, too, will be able to fulfill their sexual desires, just like their fathers.

Freud has a more complicated explanation for how gender identity is learned in girls. In the *female Oedipus complex,* according to Freud, when a girl realizes that she has no penis, she feels envious and angry, a state which he referred to as "penis envy." The girl blames her mother for this and wants to possess her father in order to replace the mother. Once she realizes that this cannot happen, she unconsciously identifies with her mother as a means of ensuring that she, too, will be able to fulfill her sexual desires in the future.

Most psychologists do not believe that Freud's theories are accurate descriptions of development. Many feel that his ideas are blatantly sexist and were the result of thinking founded on Victorian ideas of morality and sex roles (e.g., Balmary, 1982; Lerman, 1986). Nevertheless, Freud was one of the first modern thinkers to discuss the possibility that children were sexual and to relate this to gender identity development (see Figure 8–7).

LEARNING THEORY
Learning theory emphasizes the acquisition of new associations. An important type of learning that may be involved in acquiring gender roles is that of **operant conditioning,** which is a type of learning similar to

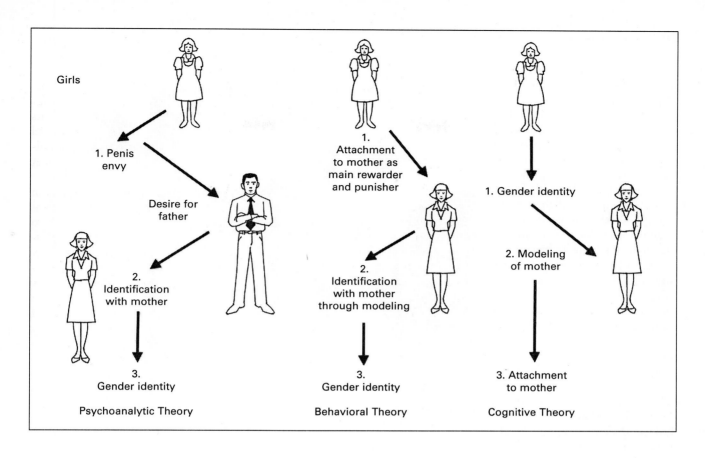

Girls

1. Penis envy

Desire for father

2. Identification with mother

3. Gender identity

Psychoanalytic Theory

1. Attachment to mother as main rewarder and punisher

2. Identification with mother through modeling

3. Gender identity

Behavioral Theory

1. Gender identity

2. Modeling of mother

3. Attachment to mother

Cognitive Theory

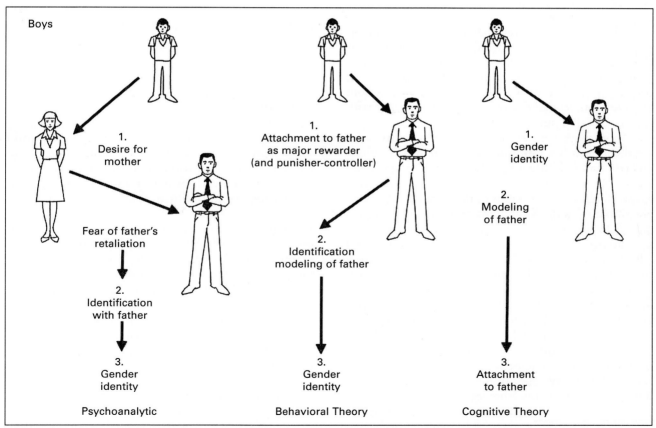

Boys

1. Desire for mother

Fear of father's retaliation

2. Identification with father

3. Gender identity

Psychoanalytic

1. Attachment to father as major rewarder (and punisher-controller)

2. Identification modeling of father

3. Gender identity

Behavioral Theory

1. Gender identity

2. Modeling of father

3. Attachment to father

Cognitive Theory

Figure 8–7 Development of gender identity in girls (top) and boys (bottom), according to Freudian, learning, and cognitive-developmental theories.

training. For example, a boy may be praised by a parent for "acting like a man" and not crying after he falls down; he may be yelled at or punished for putting on his mother's lipstick; or he may be allowed to stop doing his homework in order to help his father fix the car. In each of these three examples, the boy is being "trained" to behave in a "masculine manner." He is learning a gender identity and a gender role, and he will associate good things with "masculine behavior" and negative things (even punishment) with "feminine behavior." For a girl, encouragement and praise may follow the display of "feminine behavior," while "boyish behavior" may result in being ignored or being shown disapproval by parents. Thus, behaviors that are rewarded will continue, and behaviors that are not rewarded will eventually stop.

Figure 8–8   Children ofen learn gender roles and a gender identity by observing and imitating their parents.

Research with children shows that they do not always actually have to perform a behavior to learn it (Bandura & Walters, 1963; Mischel, 1970). Most young boys observe their mothers putting on lipstick, but may never do it themselves because they learn from observing others that the behavior in boys is met with ridicule or other punishment.

Another type of learning described by social learning theory involves **imitation.** Children love to imitate their parents and learn what "daddy" and "mommy" do simply by watching their parents. It is always interesting to ask children, "What kinds of things are daddies supposed to do?" and "What kinds of things do mommies do?" "Do daddies wash dishes?" "Do mommies use a saw and hammer nails?" Answers to these questions will tell a lot about the gender roles that children learn from their parents. (How did you answer these questions?) Children also will try to do the things that their parents do (i.e., imitate them), although the amount of time between when the behavior is observed and when they eventually imitate it can be quite long (Figure 8–8).

Operant conditioning and social learning theory do not have to be competing theories. Imitation may cause a child to try a specific behavior for the first time, but the rewards or punishments that follow that behavior will determine whether the child attempts the behavior again (B. F. Skinner, personal communication).

imitation Following the behaviors of someone taken as one's model.

COGNITIVE-DEVELOPMENTAL THEORY

Cognitive theory views infants as "information seekers." Rather than talking about an "oral stage of psychosexual development" to explain why crawling infants put things into their mouths, cognitive theory claims that this is done because infants can "know" things and acquire information about the world this way. To a crawling infant without language, the world can be divided into things that can and cannot be put in the mouth. It can also be known in terms of things that taste good, bad, or have no taste. Gender is another category into which children can sort themselves.

This theory (Kohlberg, 1966) states that the concept of "male" is first learned by observing others. For example, a boy's father is a male, and because a boy is more like his father than his mother, he begins to think of himself and his father as males. Boys become motivated to seek out information about what a male should do and how he should behave, imitating a ready source of information about male conduct from their fathers. The mother becomes the prime source of information about what is appropriate "female behavior" for girls.

At this early age, children's idea of gender is "concrete," based on physical cues like hairstyle and dress. Cognitive theorists say that while most children have learned that they are a "boy" or a "girl" by the age of three, they do not yet have an understanding of

gender constancy. Gender constancy is the knowledge that one's sex is constant and will not change. For example, a 4-year-old boy may think that since girls have long hair, if he lets his hair grow long he will change into a girl. A 4-year-old girl may believe that when she wears jeans and a football helmet, she is a boy, but when she takes them off, she changes back into a girl again. Because they do not yet have an understanding of gender constancy, young boys may say that they will become mommies when they grow up and girls may say they'll be daddies. This is a natural thing to believe until gender constancy has been acquired, which *usually happens by age 6 or 7.*

Stagnor and Ruble (1987) point out that cognitive theory is similar to psychoanalytic theory in that both describe development as occurring in stages (see Figure 8–7). Both describe a "motivation" to complete development. In psychoanalytic theory, identification is driven by the need to reduce anxiety and maintain a good self-image. In cognitive theory, the motivation is a need to attain mastery over the behaviors typical of one's gender.

Both cognitive theory and social learning theory claim that imitation is extremely important in development. But social learning theory states that behavior is based on the rewards that will follow. Cognitive theory claims that people behave in certain ways because the behavior is rewarding in itself. In other words, social learning theory states that gender understanding occurs because of the learning process. Cognitive theory states that a need to understand one's gender causes a child to want to learn (Stagnor & Ruble, 1987).

GENDER ROLES

THEORIES OF GENDER ROLE

The topics to this point in the chapter have been heavily dominated by discussions of biological causes. Indeed, one point of view, called *biological determinism,* holds that biological influences set predetermined limits to the effects of cultural influences. Some would argue that not only are our anatomy and gender identity heavily influenced by biology, but our behavior is as well. Consider again, for example, girls with adrenogenital syndrome (congenital adrenal hyperplasia) who, as a result of overactive adrenal glands prenatally, are born with masculinized genitals (enlarged clitoris and labia). Most are given corrective surgery in infancy and raised as girls. Although they are, in fact, females and raised as females, many studies have found that as children they are tomboyish in their play activities, prefer male-typical toys, have little interest in babysitting or having children of their own, and score more like males on personality tests (see Collaer & Hines, 1995, for a review). Despite this, their gender identity is consistent with their anatomy—female. The prenatal hormone imbalance in adrenogenital girls, although important in demonstrating the role of biology, no doubt is an extreme example. What about persons with normal hormone levels? A quick reading of Box 8–A makes it obvious that when it comes to "acting like a man" or "acting like a woman," there is a wide range of behaviors that can be affected by culture. In this section you will learn how one's culture works to shape gender.

When people hear the word *sex,* most think of biological features (genitalia, reproductive organs, chromosomes, etc.) or an act (lovemaking, having sex). But what is *gender*? **Gender** (from the Latin *genus* and the Old French *gendre,* meaning "kind" or "sort"), according to Joan Scott (1986),

> is a constitutive element of social relationships based on perceived differences between the sexes, and gender is a primary way of signifying relationships of power. (p. 1067)

Sandra Harding (1986) adds:

> In virtually every culture, gender difference is a pivotal way in which humans identify themselves as persons, organize social relations, and symbolize meaningful natural and social events and processes. (p. 18)

Both Scott and Harding use the term *gender* as an analytical category. The concept of gender was adopted "to distinguish culturally specific characteristics associated with masculinity and femininity from biological features," and thus was used to repudiate biological determinism (Hawkesworth, 1997). In short, gender is the social construction of femininity and masculinity (Komarovsky, 1992).

Gender role is a set of *culturally specific* norms about the expected behaviors and attitudes of men and women. Gender roles vary from culture to culture (see Box 8–A). Another term you will see in this section is **gender-role identity**—"a concept devised to capture the extent to which a person approves of and participates in feelings and behaviors deemed to be appropriate to his or her culturally constituted gender" (Hawkesworth, 1997). For example, girls with adrenogenital syndrome might have a clear sense of themselves as girls (gender identity), but be unhappy with and refuse to behave in accord with our culture's norms of femininity.

---

gender constancy The knowledge that one's sex is constant and will not change. This knowledge is usually acquired by age 6 or 7.

gender The social construction of femininity and masculinity.

gender-role identity The extent to which a person approves of and participates in feelings and behaviors deemed to be appropriate to his or her culturally constituted gender.

Box 8-A CROSS-CULTURAL PERSPECTIVES

Gender Roles

Today in the United States, boys and girls are often treated differently beginning at birth. Consider, for example, the blue or pink name tags that many hospitals use to designate a baby boy or girl. However, this was not always the case. Until the early part of this century, infant boys and girls were treated very much alike in things like dress (both wore gowns) and hair style (long curls). The socialization process of emphasizing masculine or feminine qualities did not begin until the age of 2 or 3 (Garber, 1991).

Some other cultures do not emphasize differences between boys and girls for several years after birth. Some African cultures even have nonmasculine titles (e.g., "woman-child") for boys (Whiting, 1979). In these cultures there is usually a ritualized initiation ceremony for boys (generally between the ages of 7 to 12) to eliminate any identification with female gender roles and emphasize male alliance and gender roles (Munroe, 1980). These rituals often include painful practices such as circumcision of the penis (see Box 2–C). Some of these cultures have rituals for girls as well, and usually focus on their role in fertility (Schlegel & Barry, 1979, 1980). However, there are other cultures that allow even adult males to express their "feminine" side, often in a practice called *couvade* in which males are secluded (like women) and act out giving birth (Paige & Paige, 1981; Whiting, 1979).

Whether emphasis of differences begins in infancy or later in childhood, the large majority of cultures emphasize nurturant behavior in females and achievement and self-reliance in males (Barry, Bacon, & Child, 1957). But even this is not universal. In fact, completely opposite gender roles can be found in cultures living in close proximity to one another. Margaret Mead (1935, 1975), for example, described a tribe in New Guinea, the Tchambuli, in which men were considered too weak to do hard physical labor; were emotional, gossipy, and easily embarrassed; and were primarily artistic in their activities. In that culture, women were independent, aggressive, and made all the important tribal decisions. In another New Guinea culture, the Arapesh, qualities that many Americans would consider "feminine" are reinforced in both males and females. As a result, Arapesh men and women are peaceful, cooperative, and nurturing. Just 80 miles away, in the Mundugumor society, "both men and women are expected to be violent, competitive, aggressively sexed, jealous, and ready to see and avenge insult, delighting in display, in action, in fighting" (Mead, 1935).

Notice that one of the characteristics expected of Mundugumor women is that they be sexually aggressive. This is generally true for women in many cultures in the South Pacific. Recall the Mangaians, for example (Marshall, 1971; see Chapter 1). In Western culture, Victorian morals and ideals expected women to be passive in sexual nature.

What can be concluded from cross-cultural studies such as these? Mead concluded:

> Many, if not all, of the personality traits which we have called masculine or feminine are as lightly linked to sex as are the clothing, the manners, and the form of head-dress that a society at a given period assigned to either sex. . . . The evidence is overwhelmingly in favor of social conditioning.

Risman and Schwartz (1989) describe two major perspectives on gender role development: *individualist theory* and *microstructural theory*. **Individualist theory** is based on the idea that an individual's personality is created relatively early in life. There is a set of personality traits developed early on that do not change much over the life cycle; these traits then determine an individual's expression of masculine or feminine behavior. Thus, an adult female may choose not to enter a "masculine" career, such as welding, because she cannot be motivated to start such a career. Her personality structure and self-concept prevent her from making such a choice.

Microstructural theory, on the other hand, is based on the notion that it is society's expectations of male and female behavior that predict how people will

individualist theory The theory that personality traits develop early in life and do not change much over the life cycle; these traits then determine an individual's expression of masculine or feminine behavior.

microstructural theory A theory that stresses the role of the external environment in shaping gender roles.

behave. Thus, rather than believing that masculinity or femininity is expressed because of internal forces such as rigid personality traits, microstructural theory assumes that men and women would behave identically if society had identical expectations of them. Gender role differences are therefore constructed by the way men and women interact within their environment and with each other. Furthermore, the nature of these interactions can change over time and in different settings.

In the introduction to this chapter, you learned that prior to the 1970s, masculinity and femininity were viewed as opposite ends of a unidimensional continuum (Hathaway & McKinley, 1943; Strong, 1936; Terman & Miles, 1936). In this bipolar model, a person who was high in masculinity would have to be low in femininity, and vice versa.

This conceptualization of masculinity and femininity was revolutionized in 1973 by Anne Constantinople, who proposed that masculinity and femininity were independent constructs. This led to the theory of androgyny, which said that a person could be both masculine and feminine, and that this was the healthiest of all gender roles (Bem, 1974; Block, 1976; Spence, Helmreich, & Stapp, 1975). According to Sandra Bem (1974), "Both historically and cross-culturally, masculinity and femininity seem to have represented two complementary domains of positive traits and behaviors, a cognitive focus on 'getting the job done' (a masculine or *instrumental* orientation) and 'an affective concern for the welfare of others' (a feminine or *expressive* orientation)."

Androgynous individuals are viewed by these researchers as being healthy because of their flexibility, demonstrating instrumental or expressive orientations as the situation demands. They are described as being

> behaviorally flexible with respect to all manner of gender related phenomena. As such, they are willing and able to exhibit masculine behaviors, feminine behaviors, or both as situationally appropriate. Individuals with sex-typed personalities, on the other hand, will tend to avoid or exhibit lower levels of cross-sex-typed behaviors.
>
> (Helmreich, Spence, & Holahan, 1979)

Recently there have been some challenges to the dualistic theory of gender and some questions about androgyny. A recent study, for example, found support for both the dualistic and unidimensional models. Biernat (1991) studied the development of gender stereotypes in social judgments about others and concluded that young children view gender in dualistic terms, but that as they grow older they come to view gender as a unidimensional construct. Others have

even questioned the Bem Sex Role Inventory, the most popular measure of androgyny. One study found that while the items on the inventory did, in fact, measure two different personality characteristics, the two characteristics no longer had anything to do with masculinity and femininity—the authors of the study suggested the terms **instrumental** and **expressive** (Ballard-Reisch & Elton, 1992). One possibility is that people's perceptions of gender roles have changed over the last 20 years. Interestingly, Bem (1979) predicted the destruction of the concept of androgyny:

> If there is a moral to the concept of psychological androgyny, it is that behavior should have no gender. But there is an irony here, for the concept of androgyny contains an inner contradiction and hence the seeds of its own destruction. Thus, as the etymology of the word implies, the concept of androgyny necessarily presupposes that the concepts of femininity and masculinity themselves have distinct and substantive content. But to the extent that the androgynous message is absorbed by the culture, the concepts of femininity and masculinity will cease to have such content and the distinctions to which they refer will blur into invisibility. Thus, when androgyny becomes a reality, the concept of androgyny will have been transcended.

GENDER-ROLE DEVELOPMENT DURING CHILDHOOD

Regardless of the particular theory that researchers support on how gender identity and gender roles are acquired, they do agree that this process occurs and that society's expectations become a part of the individual's own set of attitudes and behaviors. The process of internalizing society's beliefs is called **socialization,** which begins at birth (e.g., different colors are used to represent a baby boy and a baby girl).

What kinds of attitudes do children learn from our society about gender roles? Cognitive theory, described earlier, shows that young children have difficulty understanding complex concepts. Their early learning often consists of dual categories, such as "good and bad" or "big and small." It is not surprising, then, that young children's ideas about gender roles reflect concrete, simple ideas, dividing behaviors or traits into either "masculine" or "feminine." In our society, there are *gender-role stereotypes.* **Stereotypes,** you recall, are oversimplified, pre-

instrumental A personality characteristic; a cognitive focus on "getting the job done."

expressive A personality characteristic; a cognitive focus on "an affective concern for the welfare of others."

socialization The process of internalizing society's beliefs.

stereotypes Oversimplified, preconceived beliefs that supposedly apply to anyone in a given group.

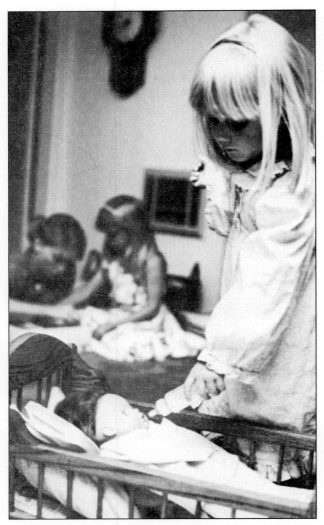

Figure 8–9 Children begin learning roles in infancy. They often show gender-stereotyped behavior even before they have a firm grasp of gender identity or gender constancy.

conceived beliefs that supposedly apply to anyone in a given group.

The stereotype for the female gender role is that of a person who nurtures and cares for others (Figure 8–9); is passive, dependent, and cooperative; seeks social approval; and is expressive in showing her emotions. The male gender-role stereotype emphasizes independence, dominance, competition, aggression, success, achievement, and emotional repression (recall the belief that "big boys don't cry" and the expression "the strong, silent type"). Research done in the early 1970s showed that most people at that time believed men and women differed in these ways (Broverman et al., 1972).

Harvard psychologist Jerome Kagan has said, "If I showed you a hundred kids aged 2, and you couldn't

tell the sex by the haircuts, you couldn't tell if they were boys or girls" (quoted in Shapiro, 1990). If this is true, when do gender roles start developing, and what (or who) is responsible?

Parents who assume "traditional" (stereotypic) gender roles behave differently toward their children (McHale & Huston, 1984). In these families, mothers act as the caregivers, while the fathers act as playmates to their infants (Lamb, 1986; Parke & O'Leary, 1976). Fathers, however, may be more likely to treat their children in gender-stereotyped ways (Leve & Fagot, 1997; Fagot & Leinbach, 1987). Fathers emphasize the strength and coordination of their newborn sons and the beauty and softness of their newborn daughters (Krieger, 1976; Rubin, Provenzano, & Luria, 1974). This is done in spite of there being little difference between the behavior of newborn boys and girls. Fathers are also more likely to offer toys that are stereotypically male to their newborn sons (Jacklin & Maccoby, 1983).

In one study, preschoolers were presented with stereotypic gender-role toys. In one condition, children were given toys that matched stereotypic toys for their own gender. However, in another condition boys were asked to play with dolls or kitchen sets the way girls would, while girls were asked to play with army war toys or cowboy outfits the way boys would. This was done while parents were out of the room. The experimenters were interested in studying how the parents would react when they entered the room and saw the types of toys the children were playing with. Mothers reacted similarly, regardless of whether the toys fit gender-role stereotypes or not, but fathers showed markedly more negative reactions when they found their children playing with toys that did not fit gender roles. This was especially true when boys were playing with "girls'" toys: the fathers often interfered with their sons' play and showed other signs of negative emotions in those situations (Langlois & Downs, 1980) (see Figure 8–10).

"My godson, Mikey, will definitely grow up with the idea that men are the head of the households and that women are 'under' them. My uncle encourages Mikey to play football and run around in the house, yet when he and his sister dance around it is not acceptable. Mikey is only allowed to play 'boy' games. He is not allowed to cook, because that is what a girl does. . . ."

"For as far back as I can remember my father only allowed us to do 'masculine' type things. Sports, cars, outdoor activities, etc. He never allowed us to play with girls. On one occasion I distinctly remember playing with two girl neighbors. My father came looking for me and found his son playing with dolls. For quite a while after that I was known as his 'faggot son.' I couldn't have been more than 10. I never became the 'faggot' son, but I did leave that day with a low self-esteem that I carried with me for quite some time."

(from the author's files)

Other adults treat male and female children in different ways as well. In another study, individuals were shown an infant and asked to make observations about its appearance and behavior. The types of comments made were very different, depending on whether they were told the infant was a boy or a girl (Condry & Condry, 1977). The socialization process in gender-role development is seen everywhere in the child's environment. Children's toy catalogs often use pink pages to show toys for girls' and blue pages for boys' toys (Haffner & Casselman, 1996). The toys available for boys are often action-oriented, while girls are given more passive playthings. Take a walk through a toy store and look at the toys being offered. It is interesting to see how toys are packaged. Many have pictures of children playing with the product on the front of the box. How many toys have pictures of both boys and girls playing with the same toy? (Very few.) Have you ever wondered why boys' bicycles have to be different from girls' bicycles? (They don't.)

Here is a little experiment that you can try on your own. Ask a boy playing with a toy soldier or a plastic model of a superhero if he is playing with dolls. He probably will say "no." (Boys in our culture don't often play with dolls, but they do play with "action figures.") Give the boy a toy soldier and a male "fashion figure," such as the one who usually serves as a "boyfriend" to a famous female doll, and then ask the boy to point to the toys that are dolls. He will probably point to the "fashion figure." Now give the soldier's clothes (and weapons) to the "fashion figure" and make the soldier wear the other toy's clothes (and accessories). Then ask the question again. Which one will the boy now think is the doll? What do you think? Why? Some psychologists think that boys should learn at an early age how to care for infants by playing with baby dolls. Learning at an early age how to put on diapers, etc., can help make a man respond in a more useful way after a real baby is born. Some fathers refuse to change diapers because (1) it is disgusting work, and (2) it is "women's work." It is true that changing diapers can sometimes be disgusting, but it is also necessary. It is not true that this is "women's work." If a man can change the dirty oil in his car, he should be able to manage changing a dirty diaper. Women should also remember that the same holds true for them. Girls can learn to use tools and

Figure 8–10 How would you react if you found your young son playing with a doll? Do you think that playing with dolls can teach boys behaviors that will be useful when they become adults?

change the car oil just as easily as they can learn to change diapers. Dirty oil and dirty diapers are neither masculine nor feminine. They just need to be changed.

Picture books, stories, and television all contribute to the development of gender roles. For example, children's stories often depict boys or men as being adventuresome and courageous, while girls or women are often in need of rescue or are in charge of keeping the house and fulfilling domestic duties (Weitzman et al., 1972). Very few girls or women are the central characters in comics or television cartoons, with female cartoon characters usually taking passive roles, playing the part of comforter and supporter to male characters, and concerned with appearances (Brabant & Mooney, 1997; Thompson & Zerbinos, 1997). A 1997 survey by the Kaiser Family Foundation and Children Now found that this is true for most television shows as well. Female roles on television typically emphasize romance more than work and beauty more than intelligence as the main things by which women and girls are valued. Therefore, it should not be surprising that people who watch a lot of TV generally believe in more stereotyped gender roles (Lips, 1992).

The socialization process begins at an early age. In one study, it was found that boys aged 2½ believed that girls don't fight and that they like to clean house. Girls of the same age believed that boys don't cry and that they like to fight. Boys and girls aged 2 and 3 believed that boys like to help their dads, like to hit, and like to break the rules. They thought that girls liked to help their moms and didn't hit (Kuhn, Nash, & Brucken, 1978). As early as 14 months of age, infants play with toys that are gender stereotyped more often than with other toys (O'Brien & Huston, 1985). Thus, children show gender-stereotyped behavior even before they have a firm grasp of gender identity or gender constancy.

Sandra Bem (1981, 1983) has proposed that children cognitively organize the world according to gender. She calls this a **gender schema**—a set of ideas about gender roles that children create from their interactions with their environment. Objects in the environment are often treated as masculine or feminine. Why, for example, do we often regard dogs as masculine and cats as feminine? Languages such as French or Spanish classify all nouns as either masculine or feminine. Gender schemas can influence decisions about what activities to take part in or learn about, including things like affection, which really has nothing to do with one's sexual anatomy. Unfortunately, many American men regard displays of affection as effeminate, even toward their own children (see Salt, 1991):

> "I know my father cares about his children, but he never displays any kind of affection toward me (male) or my brother, only to my sisters. I think that is one big experience that I missed during my childhood. . . ."
>
> (from the author's files)

Children organize their world according to gender starting at ages 2 or 3 (Hort, Leinbach, & Fagot, 1991). In fact, it may be the first social category used by children (Kohlberg & Ullian, 1974). Bem (1983) concludes that "gender has come to have cognitive primacy over many other social categories because the culture has made it so."

Gender roles are usually learned first in the home, but when children reach school age, peers and teachers become powerful reinforcers of the process of socialization and gender-role development (Fagot, 1995; Moller, Hymel, & Rubin, 1992; Witt, 1997). The activities and chores that teachers assign to children in the classroom are often determined by stereotypic gender roles (Wynn & Fletcher, 1987). Children quickly identify individuals who act "dif-

gender schemas Ideas about gender roles that children create from their interactions with their environment.

ferent." Boys or girls who do not display behaviors similar to the gender roles adopted by most of their peers may be singled out for teasing or ridicule. Interestingly, "tomboy" behavior in girls is usually far more tolerated by peers, teachers, and parents than "effeminate" behavior in boys. It is much easier to carry the label of a "tomboy" than that of a "sissy." Peers will apply great social pressure against boys who show "girlish" behavior (Birns & Sternglanz, 1983). Children generally prefer playmates of the same sex, but if they have to choose, boys prefer girls with masculine play styles to boys with feminine play styles (Alexander & Hines, 1994). As boys and girls approach junior high school age, they tend to participate less in girl-dominated activities and more in male-dominated activities (Sandberg & Meyer-Bahlburg, 1994).

Textbooks and history books in particular have emphasized the central, dominant role of men and the supporting role of women, but there has been some change in recent years. A recent analysis of third-grade basal readers found that male characters are still portrayed as having mostly masculine traits, but that female characters have evolved from having mostly feminine traits to having a balance of masculine and feminine traits (Witt, 1996).

Gender roles in adolescents have also begun to change in recent years. It is now more common for girls to ask boys out. One study, for example, found that while it was once uncommon (and considered improper) for girls to call boys on the telephone, it is now very common (Anderson et al., 1995). The authors were able to identify 1965–1980 as the period in which this gender role changed (Figure 8–11). However, there are still some double standards regarding sexual aggressiveness. Many people, for example, continue to view sexually active females as "sluts" and sexually active males as "studs."

In spite of some changes away from the traditional gender role expectations, stereotypic gender roles are often still in force during puberty and are reinforced by peer pressure (Doyle, 1985). For example, many teenage girls still view being popular and attractive to boys as the primary way to enhance their self-worth. For many teenage boys, achievement in athletics and/or in career-related activities is the primary means of gaining self-esteem (Offer & Offer, 1975).

Recently it was proposed that many of the gender differences in social behavior are due to a difference in self-construals (the manner in which men and women view themselves) (Cross & Madson, 1997). Males, the authors claimed, view themselves as independent, while females define themselves in terms of their relationships with others (interdependency). Others, however, have pointed out that children begin displaying gender-related behaviors before they develop a sense of self (Martin & Ruble, 1997).

Figure 8–11 In the past, traditional gender-role expectations would never have allowed these boys and girls to enjoy these experiences.

I will refer to two recent studies to help summarize what we know about the development of gender roles in childhood and adolescence. There may be little noticeable difference between boys and girls before the age of 2, but there is no question that most children have embraced gender stereotypes by the age of 4. By the age of 3 or 4, children know sex stereotypes about clothing, toys, games, and work, and by the age of 4 or 5 most have stereotyped occupational goals (Huston, 1983). According to Yale psychiatrist Kyle Pruett, "There are rules about being feminine and there are rules about being masculine. You can argue until the

cows come home about whether those are good or bad societal influences, but when you look at the children, they love to know the differences. It solidifies who they are" (quoted in Shapiro, 1990).

However, Pruett's own work shows that children don't necessarily have to develop strongly stereotyped views (see Shapiro, 1990, for a review). He followed the development of children in 16 families where the mothers worked full-time and the fathers were mostly responsible for caring for the children. The children had strong gender identities of being male or female, but more relaxed views of gender roles than their friends. Pruett commented, "I saw the boys really enjoy their nurturing skills. They knew what to do with a baby, they didn't see that as a girl's job, they saw it as a human job. I saw the girls have very active images of the outside world and what their mothers were doing in the workplace. . . ." (quoted in Shapiro, 1990).

Pruett also feels that fathers benefit by assuming more responsibility in raising their children. He says, "The more involved father tends to feel differently about his own life. A lot of men, if they're on the fast track, know a lot about competitive relationships, but they don't know much about intimate relationships. Children are experts at intimacy. After a while the wives in my study would say, 'He's just a nicer guy'" (Shapiro, 1990).

The second study investigated how, and to what extent, young children and adolescents use gender stereotypes to make judgments about the characteristics of individual males and females (Biernat, 1991). It was found that young people rely on gender labels as judgment cues to the same degree from kindergarten to the college years, but that as they grow older, they increasingly rely on more individuating information as well. In other words, as people grow older, they are more able to rely on information about an individual to make judgments, even when the information is not consistent with stereotyped gender roles.

GENDER-ROLE DEVELOPMENT IN CHILDREN RAISED IN SINGLE-PARENT HOUSEHOLDS

The traditional American family has been declining in number since the 1960s. In 1970, 87 percent of children lived with two parents, but in 1996 less than 70 percent did so. This was especially true among African Americans. One-third of all black children lived with one parent in 1970, but by 1996 nearly two-thirds lived with one parent—their mother in all but a few cases (Bianchi & Spain, 1996).

What happens to the development of sex roles in children raised by a single parent? Several early studies concluded that a critical factor in sex-role development was the identification of children with their fathers, by either interacting with their fathers or by watching their parents interact (Biller, 1969; Biller & Bahm, 1971; Roberts et al., 1987). Recall, too, that fathers are more concerned than mothers that their children act in gender-stereotyped ways (Fagot & Leinbach, 1987). Adults in single-parent families have less traditional (less stereotyped) attitudes than adults in two-parent families (Leve & Fagot, 1997). So what happens to children raised in a household in which the father is absent? In brief, children who are raised in single-parent households are more likely than those raised in "intact" households to become androgynous (Russell & Ellis, 1991). This is true not only for boys, but for girls as well.

Androgyny is generally considered to be a positive characteristic (see next section). If so, one study concluded that "parents who model adaptive yet nontraditional roles may play an important role in the healthy development of their children" (Russell & Ellis, 1991).

GENDER-ROLE DEVELOPMENT IN ADULTHOOD

Among adults today, gender roles are in transition (Figure 8–12). Women have not only entered the work force in larger numbers, but have gained entry into ca-

Figure 8–12 What is considered to be masculine and feminine in adult gender roles is far less rigid today than in past generations.

reers and attained access to higher-level positions previously dominated by males. In 1960, fewer than 20 percent of all women with children under the age of 6 were employed outside the home, but in 1996 two-thirds of women with young children were in the labor force and 80 percent of all mothers with school-age children had jobs outside the home (Bianchi & Spain, 1996). In addition, over half of the bachelor's and master's degrees awarded today are to women; 38 percent of all Ph.D.'s are awarded to women, up from only 11 percent in 1960 (Bianchi & Spain, 1996). Thus, the idea that being a mother means giving up having a career is no longer true and is not influencing women to the extent it did in the past.

Nonetheless, gender stereotypes may not have changed much since the 1960s and double standards often still exist, even in two-career families. For example, even if a woman works outside of the home, she generally assumes more responsibility for housecleaning, meals, and child care than does her male partner (Bianchi & Spain, 1996). However, men are spending more time doing housework than in the past, and this is especially true of African-American men (John & Shelton, 1997). Men with traditional stereotyped attitudes are likely to evaluate their marriages negatively if "forced" into egalitarian roles (McHale & Crouter, 1992). Women are also more likely to encounter sexual harassment at work (see Chapter 15). Many women in the work place must be able to balance a feminine appearance with a businesslike appearance, while men are far less concerned with how they dress for work.

Women who work in the home may not have the same chance for self-gratification or self-fulfillment as men, who often serve as both breadwinners and heads of households. Many people still consider work in the home to be "not as important" as work outside the home. In 1990, for example, students at Wellesley College objected to then–First Lady Barbara Bush giving the commencement address because she had been a stay-at-home wife and mother. According to a 1990 survey conducted for *Time* magazine, 86 percent of the young men surveyed wanted a mate who was hard-working and ambitious. While the same survey found that nearly half of the young men wanted to spend more time at home to raise their children, the number of "househusbands," whose primary responsibility is work in the home, remains small, as career achievement is still used to determine the worth of individuals, rather than the ability to maintain a home and care for others.

What is considered to be masculine and feminine in adult gender roles is becoming far less rigid and distinct than was true for past generations (Figure 8–13). Some researchers see these trends as a movement toward androgyny. As you recall, androgyny is the ability of individuals, regardless of gender, to display a variety of personality characteristics, so that the same person can be both assertive and compassionate, both instrumental and expressive, and both masculine and feminine, depending on whether a trait or behavior is appropriate for a given situation. Thus, androgyny is often viewed as a positive characteristic that gives an individual greater adaptability (Spence, Helmreich, & Stapp, 1975). Taylor and Hall (1982) concluded that in American society, androgynous or masculine individuals show greater social adaptation and adjustment, regardless of their gender. They believe, for example, that women who show masculine characteristics may initially not be rewarded for acting that way, but those who continue to do so are eventually given greater rewards and prestige than women with mostly feminine characteristics. In a now classic study, Spence and Helmreich (1972) found that people liked women best who were competent and doing "masculine" things. In a recent review of numerous studies, it was found that the difference between men and women on measurements of masculinity and femininity had, in fact, decreased since the early 1970s (Twenge, 1997). However, the change was not due to men becoming more androgynous (having more feminine traits), but had resulted from women displaying more masculine-stereotyped traits than in the past.

Nevertheless, the push for gender equality has taken its toll on some individuals. Recall, for example, that during the O. J. Simpson trial, prosecutor Marcia Clark's ex-husband went to court to get custody of their children because she was spending too much time at work. Rotheram and Weiner (1983) found that among dual-career couples, dual-career status increased satisfaction with the relationship, but also increased personal stress. By 1990, 80 percent of those in a *Time* magazine survey believed that it was difficult or very difficult to balance marriage, children, and a career (the 1980s "superwoman" ideal) and, in a shift from previous surveys, many women rated a happy marriage as more important to them than a successful career (see Gibbs et al., 1990). It is obvious that different women can have the same roles (mother, wife, worker), yet have different experiences of the quality of each (Baruch & Barnett, 1986).

Some researchers have noted that as individuals develop as adults, they acquire more complex sets of gender roles. For example, the fact of fatherhood may force men to acquire more experience being nurturing, while working outside of the home may force women to become more assertive and independent. Labouvie-Vief (1990) further states that with more complexity in gender-role patterns of behavior comes a greater chance for the integration of gender roles. Labouvie-Vief, Hakim-Larson, and Hobart (1987) found that both men and women who were more developmentally complex also were less likely to approve of imma-

Figure 8–13   Today, the majority of women contribute to their family's income, often in nontraditional roles such as electrical repair and fighter pilot.

ture coping strategies. Although men and women arrive at developmental complexity by different routes, the adaptive end point for both genders is an integration of gender roles; that is, gender roles tend to converge in adulthood (see Blanchard-Fields & Suhrer-Roussel, 1992). Thus, gender-role development extends well into adulthood with cross-gender role characteristics (i.e., behavior of the opposite sex) often emerging among both older men and women. This has been referred to as a *transcendent model* of gender-role development, as opposed to a *traditional model* of gender-role development (Fischer & Narus, 1981). Individuals with a large number of both masculine and feminine characteristics will then be better able to cope with a wide range of life challenges and social situations than will those individuals with a more rigid or narrow range of gender-role patterns of behavior.

GENDER ROLES AND SEXUAL RELATIONS

Victorian-era physicians and sexologists such as Sigmund Freud and Henry Havelock Ellis believed that male and female sexuality were very different. Female sexuality was viewed as more passive, weaker, and less

Box 8-B   CROSS-CULTURAL PERSPECTIVES

The Native American Two-Spirit

Before the influence of Western culture led to their disappearance, there existed among many North American Indian tribes individuals, known until recently as berdaches, who had special gender status. According to Callender and Kochems (1987), a *berdache* was "a person, usually male, who was anatomically normal but assumed the dress, occupations, and behavior of the other sex to effect a change in gender status." Accounts of berdaches date back to the 1500s (Katz, 1976) and they existed in at least 113 groups that extended from California to the Mississippi Valley and upper Great Lakes (list compiled by Callender and Kochems, 1987).

Early observations were heavily influenced by Western values and perceptions and focused primarily on the homosexuality or bisexuality of many of the berdaches. In fact, the word *berdache* is not an Indian term, but rather a medieval French term meaning "male homosexual." According to Callender and Kochems, homosexuality didn't bestow the status of berdache on an individual, but many berdaches became homosexual. Translations of the Indian terms give "halfman-halfwoman" or "man-woman," with no insinuation about sexuality. Native Americans, as well as anthropologists, are presently urging that scholars use the term **two-spirit** or terms within native languages when referring to these individuals (see Jacobs & Thomas, 1994).

The two-spirits were held in awe and highly revered, but were not chiefs or religious leaders (they were not shamans). Individuals became two-spirits by showing a strong interest in the work of the opposite sex during childhood and/or (and more commonly) by having a supernatural vision during adolescence that confirmed the change in gender status. In almost all groups, a male two-spirit dressed and fixed his hair like a woman and assumed a woman's occupations, at which he excelled—part of the reason he was admired. On certain occasions, he dressed like and assumed the role of a man. He was generally responsible for burial rituals, but because he could cross back and forth between genders, he was especially valued as a go-between for men and women. It was these exceptional gender-mixing abilities that led to his being greatly admired. But, as noted, Western values intervened. Lurie (1953), who studied the Winnebago Indians, reported, for example, that "most informants felt that the two-spirit was at one time a highly honored and respected person, but that the Winnebago had become ashamed of the custom because the white people thought it was amusing or evil."

Figure 8–14   We'wha, a nineteenth-century two-spirit from the Zuni tribe.

two-spirit In North American Indian tribes of past centuries, a person, usually a male, who assumed the dress, occupations, and behavior of the other sex in order to effect a change in gender status. The term now preferred by Native Americans and anthropologists over *berdache*.

fulfilling than male sexuality. Many Victorian physicians did not even believe that women had orgasms, as you may recall (see Chapter 4). More recently, Kinsey's surveys (1948, 1953) and the physiological studies by Masters and Johnson (1966) have emphasized that male and female sexuality are very similar. Although some feel that sociopolitical influences may also have contributed to the conclusion that male and female sexuality are similar (Irvine, 1990), a recent study found that there were, in fact, no gender differences in sexual satisfaction (Oliver & Hyde, 1993).

However, there are still some gender differences. As a general rule, men have considerably more permissive attitudes about casual sex than do women (Oliver & Hyde, 1993). The authors of this study suggest that this gender difference helps explain why the same behavior may be interpreted as reasonable by a man, but as harassment by a woman. Also, when people have stereotypic views that men ought to be more sexual than women, it can result in negative labeling of those who don't conform. If a man does not appear to engage in sex as often as others, he may be viewed as inadequate and "unmanly." Conversely, if a woman demonstrates that she really enjoys sex and likes having sex frequently, she may be considered by some to be "easy" or a "slut." Clearly, there is a wide range of individual differences in the way both men and women feel and act with regard to sex. There is no "right" amount of sex drive and/or sexual activity for either men or women.

Many studies have found that males are more interested in the purely physical aspects of sex than are females (Baldwin & Baldwin, 1997; Laumann et al., 1994). Women, on the other hand, are more likely than males to value love and a nurturing relationship. Not surprisingly, then, is the finding that men desire a physically attractive marriage partner more than women do (Regan & Berscheid, 1997), or that women are more likely than men to have sex for emotional intimacy and men are more likely to do so to relieve sexual tensions (Brigman & Knox, 1992).

One commonly held stereotype is that women should be nurturant, loving, and accepting, while men should not express deep feelings and emotions:

"Recently I was involved in a relationship with a very 'sex-typed' guy. . . . Every time I would try to talk to him about how I felt about things or let him know how my emotional needs were going unfulfilled, he would change the subject or dismiss the whole discussion as being 'sissified.'"

(from the author's files)

All people, men and women, need to express their feelings and emotions and be able to communicate them.

The image of man as "strong and silent" robs him of a fundamental part of his humanity and robs his partner of a true companion. Herb Goldberg (1979), author of the book *The New Male,* has written extensively about the harmful effects of men trying to live up to the so-called male role. He states that the entire male socialization process is contrary to the human needs of men and is hazardous to their health and well-being. Good communication leads to happier, healthier, and often more satisfying sexual relations and to greater intimacy (Rubin et al., 1980). Men and women are first and foremost human beings, and we need to strive to live up to human roles rather than rigid male or female roles. It is this shared humanity that should be the focus of our relationships and the strong foundation upon which they are built.

In Chapter 10 you will learn that many parents deny their sexuality by hiding it from their children. As a result, some people view parents, and parenthood, as less sexual than others. For people with extremely stereotyped views of gender roles, this can create a conflict when their own partner becomes a parent:

"My husband and I were in what seemed an ideal marriage. Not only was our coupling wonderful, but we also shared an extremely intense sexual relationship as well. 2½ years into our marriage we were informed that we were going to be parents. My husband was great; he was so attentive and supportive during the time I carried our child. . . .

"He was a proud father. However, things were never the same between us after the birth of our daughter. My husband insisted I stay home and raise her as he did not want her brought up in nurseries or with babysitters. Soon, he was obsessed with work. . . . He was working 18–20 hrs. a day, 6 days a week. . . . There had been little sex between us since the baby, but I attributed it to his long hours. . . .

"I started sensing that my husband was distant. I told him he was growing away from us instead of with us. I asked him what was wrong. . . . He said that ever since I became a mother that he no longer could see me as a sexual partner. I was a mom and that was something sacred that should be put on a pedestal. He said it was impossible to think of me in a sexual way anymore. . . ."

(from the author's files)

Recall from Chapter 4 that in some cultures, females' genitals are mutilated, in part because males view female sexual desire as being inconsistent with their

Figure 8–15 Husbands devote more time to domestic duties today than in the past, but they still do less than their wives (even when the wives are employed outside the home).

roles as wives and mothers. In our own culture, child-bearing often symbolizes the acceptance of traditional gender roles (Nock, 1987). An androgynous person is less likely to have conflicting views about sexuality and parenthood, and more likely to view his or her partner as both nurturant (whether female or male) and sexual.

MICROSTRUCTURAL THEORIES OF GENDER-ROLE DEVELOPMENT

Many adults today were raised with traditional role expectations through socialization in their childhood and adolescence, but are currently being influenced by recent trends in society toward more androgynous gender roles. Such "mixed signals" can result in confusion about what is truly masculine or feminine, with some people claiming that even making such a distinction is unnecessary. In order to better understand how individuals react in times of transition to major social changes that influence traditional gender roles, we shall examine the research based on microstructural theory within the sociological literature.

Sociologists are interested in the processes of group dynamics and change. Microstructural theory, like learning theory in psychology, stresses the role of the external environment in shaping attitudes and behavior. The focus of this type of research is less on the individual and more on large-scale social institutions. In looking at how gender roles have evolved in our country's past, we may better understand what is happening today.

Cancian (1989) and Kimmel (1989) have examined the way in which gender roles have evolved in America. Cancian (1989) argues that love became feminized during the 1800s, when the United States was becoming industrialized. Early American society did not have different gender roles for the expression of love, nurturance, or dependency. According to Cancian (1989), this was because the home was the center of the educational, economic, and social activities of the family, where both men and women shared in all aspects of family life. The church admonished men and women to be nurturant and loving to each other, as this ensured a stable home environment. However, with increased industrialization, men began to earn their livings away from home, and the economic activities of men and women began to split. Men were viewed as breadwinners who left the home to compete in a harsh outside world filled with competition and danger. Women continued to stay at home, and the home began to be viewed as a retreat for the husband after finishing his day's labors. The home became an idealized place of peace and affection, in contrast to the business world. Because the woman maintained the home, the qualities of nurturance, love, and devotion became associated with being feminine. The work that a wife did at home was either ignored or not considered to be as important or meaningful as work done outside the home. Perhaps this was due to the fact that women failed to receive monetary compensation for their labors within the household. Independence, success, achievement, competition, and self-control came to be valued as masculine characteristics as the male pursued work outside the home.

Cancian (1989) believes that our present gender roles evolved as a means of achieving control. Workers in industrial settings were dependent upon companies for their wages, while the individual male was viewed as being independent, tough-minded, and thriving in an impersonal, affection-free business world. This fostered the illusion that the individual male still had control over his life through the display of masculine characteristics, especially in the home. Women were dependent upon wage-earning husbands, but this dependency was hidden by the feminine gender role of women in the form of being nurturant, emotional, passive, and physically weak, though morally pure. Women who worked outside of the home were almost always single, and once married would retire to do the housework of the feminine wife and/or mother. It should be noted that the 1800s saw the opening of the American frontier, and this was seen as a man's job, calling for independence, self-reliance, and the ability to do without female influence.

Kimmel (1989) points out that the end of the 1800s saw a great deal of challenge to traditional masculine and feminine gender roles. Kimmel argues that social changes in the late 1800s gave rise to the "New Woman," who was single, economically independent, and highly educated, and who did not believe that a woman should have to become dependent on a man in order to be feminine. The New Woman was part of a growing women's movement that attempted to challenge traditional gender roles and create a more equal distribution of power within American society. In an attempt to combat traditional gender roles, some women tried to act and dress in a less restrictive manner.

According to Kimmel (1989), men reacted to the rise of the women's movement in one of three ways. The first way was an antifeminist response, with an emphasis on the need to return the woman to the home. A second reaction was a pro-male reaction, which involved segregating boys and girls in separate schools and other social institutions in order to "preserve" boys' masculinity from the corrupting influence of feminism. The third reaction was pro-feminist. Some men joined with feminists to promote equality among the sexes. Kimmel notes that past definitions of masculinity and femininity help reproduce power relationships between the sexes, and when those power relationships are challenged, gender roles are questioned as well and may begin to change.

We have similar forces at work in our society today. Women are seeking more control over job opportunities, the same amount of pay as men receive for comparable work, more freedom in the conduct of their personal affairs, and, in general, more equality between the sexes. Reactions to today's feminist movement are similar to the ones described at the turn of the century, for gender roles are still being seriously questioned and redefined within our society. Qualities that are believed to characterize men and women appear to be undergoing a dramatic change. The initial force behind the changes has come from the women's movement, which asserts that the traditional system of sex-role differentiation is outmoded and destructive (Marecek, 1977). Pleck (1976), a leading authority on the male gender role, notes that women are challenging their traditional gender-role patterns and redefining their roles and their place in society. In turn, men in our society have begun to reexamine their own gender roles. This can cause a conflict within a man between what he was taught was appropriate for his gender role when he was a boy and society's current expectations for the "New Man."

Pleck (1976) refers to this as the *gender* or *sex-role strain perspective.* This perspective contends that the problem with the male gender role is that men in our society are confronted by contradictory demands and expectations from their early socialization process and their adult life experience. For example, men were conditioned to control their emotions and be unexpressive in order to be strong and to be so-called real men, whereas the adult male role (i.e., marital role, parental role) calls for the ability to express feelings and to share affection with their spouses and children.

Perhaps it is more difficult for men to adopt nontraditional role behaviors, as it is harder for men who wish to act in ways that may be labeled feminine. In order for men to move away from traditional gender-role behavior, it has been noted that men not only need to feel such change is desirable and beneficial as far as improving the quality of their lives, but they also need to receive support from their spouses, from other men, and in the workplace (Canavan & Haskell, 1977). Pleck (1976) believes that in redefining their gender roles, men need to make changes in their relationships to women, other men, and their children, and to their work environment.

Many forces have contributed to challenging such traditional attitudes as "a woman's place is in the home" and "a man's home is his castle." These include the rise in the divorce rate, the availability of contraceptives, the large number of women in the work force (especially working mothers), and increased sensitivity to civil rights and personal freedoms. Despite the many gains women have made personally and professionally, feminists are sometimes still considered radical by individuals who are satisfied with the gender roles received from previous generations. Additionally, women often still get paid less to do the same work as men, have fewer opportunities for higher-level leadership roles in corporate or government settings, and must contend many times with conflicting sets of expectations from parents, friends, bosses, and other people about what is appropriate feminine behavior.

Microstructural theories see gender roles as based on gender relationships and power. Reactions to attempts to redistribute power are similar, whether from the turn of the century or today. Risman and Schwartz (1989) argue that the way to change gender roles is not to retrain individuals, but to restructure the social environment. Women will become assertive if their jobs demand it and when they are rewarded for acting assertively. Men will become nurturant when they are held responsible for taking care of others (Chodorow, 1978).

Key Terms

adrenogenital syndrome (AGS) 200
androgen insensitivity
    syndrome 202
androgyny 198
biological determinism 209
chromosomes 198
DHT deficiency syndrome 202
expressive 211
gender 209
gender constancy 209
gender dysphoria 203
gender identity 198
gender identity disorder
    (GID) 204

gender role 198, 209
gender-role identity 209
gender-role stereotype 211
gender schema 214
hermaphroditism 200
identification 206
individualist theory 210
instrumental 211
imitation 208
Klinefelter's syndrome 199
microstructural theory 210
Mullerian duct system 200
Mullerian duct–inhibiting
    substance 200

Oedipus complex 206
operant conditioning 206
phallic stage 206
pseudohermaphroditism 200
socialization 211
stereotype 198, 211
testosterone 46, 53, 200
transsexual 204
Triple-X syndrome 199
Turner's syndrome 199
two-spirit 218
undifferentiated 198
Wolffian duct system 200

Personal Reflections

1. In your opinion, what traits (other than physical characteristics) make a man masculine? What traits make a woman feminine? Why?

2. Compare your answers to question 1. Were there any traits shared by both men and women? Why or why not? Can a man be gentle, caring, affectionate, emotional, and nurturing to his children, and still be masculine? Why or why not? Can a woman be assertive, independent, ambitious in a career, and still be feminine? Why or why not?

3. Do you consider yourself to be sex-typed or androgynous? Why?

4. Have you ever considered marriage? If so, what traits do you want in a husband or wife? Are these the same traits you will want in your partner after you become parents? Is your answer consistent with your answers to the previous questions?

5. Men: Your wife has decided to resume postgraduate studies leading to a career. This will require that she be away from home more than before and that you assume more responsibilities for care of the house and children. Your reaction?

6. Women: Your husband has decided to pass up a career opportunity (leading to higher rank and salary) to spend more time at home with the children. Your reaction?

7. How would you react if you found your young son playing with Barbie dolls? Why? How would you react if you found your young daughter playing with toy soldiers and footballs? Why?

8. How would you react if your teenage son wanted to take home economics instead of machine shop or auto mechanics? Why? How would you react if your teenage daughter wanted to take machine shop or auto mechanics instead of home economics? Why?

9. Look through popular magazines or watch several television shows. Notice the gender roles that are presented for men and for women. What do you feel should be changed, if anything?

10. Visit your local toy store and look at the toys that are displayed. Do manufacturers present different toys in ways that appeal more to girls or more to boys? Do these toys help create stereotypes, or do they simply reflect our society's stereotypical ideas? Would you change anything?

Suggested Readings

Cowley, G., Gideonse, T., & Underwood, A. (1997, May 19). Gender limbo. *Newsweek.* Good article on the lack of success of sex reassignment surgery at birth.

Colapinto, J. (1997, December 11). The true story of John / Joan. *Rolling Stone.* The complete, terrible story of the twin boy whom doctors tried to surgically change to a girl—from the patient's perspective.

Fagot, B. I. (1995). Psychosocial and cognitive determinants of early gender-role development. *Annual review of sex research,* vol. VI. Mount Vernon, IA: Society for the Scientific Study of Sexuality. The title says it all; a very thorough review.

Hyde, J. S. (1996). *Half the human experience: The psychology of women* (5th edition). Boston: Houghton-Mifflin. Comprehensive and well-written review of women and gender roles.

Kipnis, A. R., & Herron, E. (1993, January/February). Ending the battle between the sexes. *Utne Reader.* Excellent article in which the authors try to negotiate an end to the gender battle and an end to gender stereotypes.

Laqueur, T. (1992). *Making sex: Body and gender from the Greeks to Freud.* Cambridge, MA: Harvard University Press (paperback).

Lips, H. (1992). *Sex and gender.* Mountain View, CA: Mayfield. This is a good, easy-to-read introduction to gender roles.

Morris, J. *Conundrum* (1974). New York: Harcourt Brace Jovanovich. The story of a transsexual's experiences.

Segell, M. (1989, January/February). The American man in transition. *American Health.* Discusses the results of a new survey that shows that most people believe that men need to change.

Shapiro, L. (1990, May 28). Guns and dolls. *Newsweek.* An excellent article on the development of gender roles during childhood.

Welbourne-Moglia, A., & Calderwood, D. (1984, November). Gender, gender role, and sexual health. *SIECUS Report.* Compares and contrasts male and female sexual health throughout the life span.

CHAPTER 9

When you have finished studying this chapter, you should be able to:

1. Define the terms sexual orientation, heterosexual, homosexual, and bisexual;

2. Summarize what we know about the prevalence of homosexuality and bisexuality in the United States;

3. Understand the difference between sexual orientation, gender identity, and gender roles;

4. Discuss the origins of sexual orientation, including psychoanalytic, social learning, and biological explanations;

5. Trace the history of attitudes about homosexuality in Western culture;

6. Compare our attitudes about homosexuality with those in other cultures;

7. Understand what it involves for an individual to "come out";

8. Understand homosexual life-styles and the ability of homosexuals to parent; and

9. Discuss the causes of homophobia.

Sexual Orientation

Are you sexually attracted to men? To women? To both men and women? These are questions about **sexual orientation**—one's sexual attraction to members of the opposite sex, the same sex, or both sexes. Attraction to members of the opposite sex is called **heterosexuality,** while attraction to same-sex individuals is called **homosexuality. Bisexuality** refers to sexual attraction to people of either sex.

At one time it was believed that people were either one or the other: heterosexual or homosexual (Gordon & Snyder, 1989). Today, sexual orientation is viewed as a continuum, with people who are exclusively heterosexual at one end and those who are exclusively homosexual at the other end. Kinsey and his colleagues (1948) devised a

heterosexual An individual with a sexual orientation primarily to members of the opposite sex.

homosexual An individual with a sexual orientation primarily to members of the same sex.

bisexual An individual with a sexual orientation to both men and women.

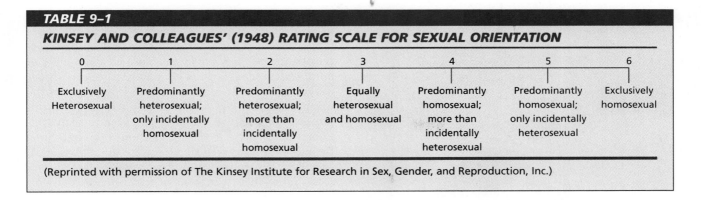

TABLE 9–1

KINSEY AND COLLEAGUES' (1948) RATING SCALE FOR SEXUAL ORIENTATION

0	1	2	3	4	5	6
Exclusively Heterosexual	Predominantly heterosexual; only incidentally homosexual	Predominantly heterosexual; more than incidentally homosexual	Equally heterosexual and homosexual	Predominantly homosexual; more than incidentally heterosexual	Predominantly homosexual; only incidentally heterosexual	Exclusively homosexual

(Reprinted with permission of The Kinsey Institute for Research in Sex, Gender, and Reproduction, Inc.)

7-point rating scale to measure sexual orientation, with 0 representing people who are exclusively heterosexual and 6 representing people who are exclusively homosexual (see Table 9–1). Kinsey's group regarded anyone as bisexual who had a rating of 2, 3, or 4. Kinsey's scale relied heavily on overt behavior, but there are problems with defining a heterosexuality–homosexuality continuum only in terms of behavior. However, before we address these problems, let us examine the results by Kinsey and others regarding behavior.

PREVALENCE OF HOMOSEXUALITY AND BISEXUALITY

The large-scale surveys by Kinsey and his associates discussed in Chapter 1 (Kinsey et al., 1948, 1953) were the first source of information about the prevalence of homosexuality in this country. They found that 4 percent of their sample of white males had been exclusively homosexual for life, 10 percent had been exclusively homosexual for at least a three-year period in their lives, and 37 percent had had at least one homosexual experience to orgasm. This last figure shocked America, and Kinsey's survey was attacked as either biased or totally false (for example, see Reisman & Eichel, 1990). An associate of Kinsey's later claimed that the 37 percent figure was, in fact, too high, but lowered the estimate only to 25–33 percent (Pomeroy, 1972). For women, Kinsey's group found that 13–19 percent had had at least one homosexual experience and that 1–3 percent were predominantly or exclusively homosexual after age 15.

A survey taken over 20 years later by Morton Hunt (1974) for *Playboy* magazine found that 20–25 percent of the males surveyed had had at least one homosexual experience in their lifetimes. However, only 12–13 percent had had a homosexual experience after age 15, and only 2–3 percent were found to be predominantly or exclusively homosexual after that age. A survey taken by the Kinsey Institute at about the same time (1970) found that 3.3 percent of adult males had homosexual experiences "occasionally" or "fairly often"

(Fay et al., 1989). Hunt found that 20 percent of single women reported having had at least one homosexual experience in their lifetimes, but only 10–12 percent had done so after age 15. About 2 percent of single women were predominantly homosexual. For married women, 10–11 percent of Hunt's sample said that they had had at least one homosexual experience, but only 3 percent of them said they had had a homosexual experience after age 15, and fewer than 1 percent were predominantly homosexual. Thus, in the case of both men and women, most people who said that they had had a homosexual experience or experiences had done so as adolescents and were exclusively heterosexual as adults.

> "At the age of about 12, I experienced an act of homosexuality. My cousins and I performed anal intercourse with each other. It never occurred to me that I might be gay when I got older. I am now grown up and am not a homosexual. Maybe because it was just a game to me or exploring different functions."
>
> (from the author's files)

Hunt, like many others, felt that Kinsey overestimated the prevalence of homosexuality in the United States, but at the same time he felt that his survey may have underestimated the number of homosexual people. What do more recent surveys indicate?

In a national cross-sectional survey, Janus and Janus (1993) found that 22 percent of the men and 17 percent of the women surveyed said that they had had homosexual experiences, and that for 9 percent of the men and 5 percent of the women, the experiences were either "frequent" or "ongoing." However, a survey done for *Parade* magazine found that only 3 percent of the men and 1 percent of the women identified themselves as homosexual, and another 3 percent of the men and 0.4 percent of the women said they were bisexual (Clements, 1994). Another survey found that only 2 percent of the men in the sample reported having had a homosexual experience within the last 10 years (Billy et al., 1993).

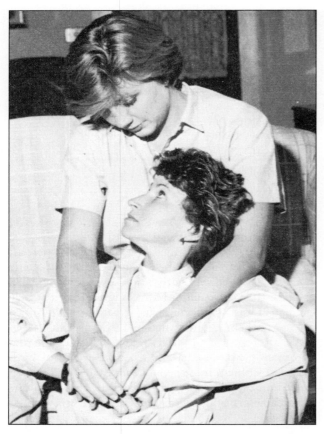

Figure 9–1   Millions of American men and women are either predominantly or exclusively homosexual.

In the most scientifically conducted sexual survey to date (see Chapter 1), Laumann and colleagues (1994) found that 2.8 percent of the men and 1.4 percent of the women surveyed identified themselves as homosexual or bisexual. This was consistent with behavior: 2.7 percent of the men and 1.3 percent of the women in the survey said that they had had sex with someone of their own sex in the past year. In a large national survey done in France at about the same time, the numbers were even smaller (Messiah et al., 1995). However, Laumann's study found that in large urban areas, more than 9 percent of men identified themselves as homosexual.

Two recent reviews are especially relevant here. One reviewed three surveys conducted by the National Opinion Research Center in 1988, 1989, and 1990 (Rogers & Turner, 1991). The other relied on seven surveys, including five conducted after 1987 (Seidman & Rieder, 1994). What did they conclude? As many as 20 percent of men have had a homosexual experience, but only about 5 percent since the age of 18, and only 1–6 percent have done so in the last 12 months (most of the surveys put this last figure at around 2 percent).

After all this research, what is the overall conclusion? *About 3–5 percent of adult American men and about 1 percent of adult American women are homosexual.* Many more people, however, have had sexual experiences with individuals of the same sex at one time in their lives.

DEFINING SEXUAL ORIENTATION: ANOTHER LOOK

As I implied earlier, we must look at more than behavior when characterizing people as heterosexual, homosexual, or bisexual. For example, what about sexual fantasies? Most heterosexuals have had fantasies of homosexual experiences (and vice versa; Bell & Weinberg, 1978). Although this does not mean that they are homosexual or bisexual, what sexual orientation is a person who has been exclusively heterosexual in behavior, but whose sexual fantasies are *predominantly* homosexual in content?

What about a person's sexual self-identity—how someone *feels* about himself or herself? In fact, you do not have to have ever had sex to consider yourself heterosexual or homosexual. Also, some people who are predominantly heterosexual in behavior may feel that they are homosexual or bisexual, but limit themselves because of perceived social constraints and engage in same-sex sexual experiences only on an occasional basis. For example, Laumann et al. (1994) found that more than twice as many people (5.5 percent of women and 6 percent of men) said that they found the idea of having sex with someone of their own sex appealing as had actually engaged in homosexual behavior in the past year or who identified themselves as homosexual or bisexual. Other studies have found that 8–12 percent of heterosexuals report awareness of some attraction to the same sex (McConaghy et al., 1994; Sell, Wells, & Wypij, 1995). However, it is far more common for exclusively homosexual men to be attracted to women than it is for heterosexual men to be attracted to men (Messiah & Mouret-Fourme, 1996).

How, then, shall we define **sexual orientation?** I will use a modification of the definition by Ellis and Ames (1987): *distinct preferences consistently made after adolescence in the presence of clear alternatives.* Preferences can be reflected in desire and/or behavior, and, as noted by Ellis and Ames, "isolated instances of sexual behavior may or may not reflect one's sexual orientation."

Perhaps the biggest confusion, and controversy, is with the term *bisexual.* Many people still insist on viewing sexual orientation as a dichotomous choice (either/or) and have difficulty accepting a third orientation. Many homosexuals (as well as heterosexuals) view people who have engaged in bisexual behavior as really homosexuals who have not yet "come out of the closet" (Gordon & Snyder, 1989; Lever, 1994). One recent study found that people's bisexual activity remained relatively stable over time (McKirnan et al., 1995), while another found that about one third of male bisexuals moved toward homosexuality over time (Stokes, Damon, & McKirnan, 1997).

Is bisexuality a behavior, a feeling, or an identity? In a study of nearly 7,000 men who acknowledged sexual experiences with both men and women, it was found that 2 percent described themselves as homosexual, 69 percent as heterosexual, and only 29 percent as bisexual (Lever et al., 1992). In the survey by Janus and Janus (1993), 22 percent of the men surveyed said that they had had a homosexual experience, but only 4 percent regarded themselves as homosexual and another 5 percent as bisexuals. Thus, most people who have had sexual experiences with both opposite- and same-sex individuals do not regard themselves as bisexual. For example, in an interesting study, Humphreys (1970) examined a behavior called "tearoom trade." This involves men giving oral sex to other men in public restrooms, often without exchanging names. He found that a majority of those who allowed oral sex to be performed on them viewed themselves as heterosexual (54 percent were married and had children), and many did not believe that they had engaged in a homosexual act because they had only been the passive partner. In the study by Lever et al. (1992), those men who identified themselves as bisexual said that they were predominantly or sometimes homosexual, whereas those who identified themselves as heterosex-

ual were predominantly heterosexual in behavior. Similar results were found in a smaller study (Doll et al., 1992). These studies point out the shortcomings of relying exclusively on either behavior or self-identity as a definition.

A commonly used definition of **bisexuality** was suggested by MacDonald (1981, 1982):

> To be bisexual means that a person can enjoy and engage in sexual activity with members of both sexes, or recognizes a desire to do so. Also, although the strength and direction of preference may be constant for some bisexuals, it may vary considerably for others with respect to time of life and specific partners.

Recall that by Kinsey's definition (1948), the sexual behavior (or desire to behave) must be more than incidental. However, Kinsey's scale implies that the farther away an individual gets from one of the extremes (exclusively heterosexual or exclusively homosexual), the weaker that orientation becomes. A bisexual person, therefore, would have only moderate degrees of both orientations. This may not be true. One study found that bisexuals had the same amount of opposite-sex fantasies as heterosexuals and the same amount of same-sex fantasies as homosexuals (Storms, 1980). In conclusion, most researchers today agree that bisexuality is a legitimate classification of sexual orientation, distinct from both heterosexuality and homosexuality.

SEXUAL ORIENTATION, GENDER IDENTITY, AND GENDER ROLES

It has been my experience that many students are uncertain (or mistaken) about the relationship between a person's sexual orientation and his or her gender identity and gender roles. The latter two are independent of sexual orientation (Lips, 1992). The *gender identity* of the vast majority of homosexuals and bisexuals is just as strong and consistent with their anatomical gender as is found among heterosexuals. Homosexuality and bisexuality are not the result of gender dysphoria, and being homosexual is not the same as being transsexual (see Chapter 8). Also, do not confuse homosexuality or bisexuality with transvestism; as you will learn in Chapter 14, most transvestites are heterosexual males.

The real misunderstanding concerns *sexual orientation* and *gender roles.* Early sexologists like Freud and Henry Havelock Ellis thought that homosexuality resulted from reversed gender roles (Ross, 1983). Many people today still hold to this belief

sexual orientation A distinct preference for sexual partners of a particular sex in the presence of clear alternatives.

(De Cecco & Elia, 1993), which is reinforced often by the media's portraying male homosexuals as effeminate and female homosexuals as masculine. While many homosexuals do not conform to gender roles that society considers "normal" for their sex (Snyder, Weinrich, & Pillard, 1994), that is also true for many heterosexuals (Bell, Weinberg, & Hammersmith, 1981). Whether a person conforms or does not conform to gender stereotypes does not always predict sexual orientation.

However, because of these beliefs, many people are quick to label others who do not appear to be "typical." For example, people are much more likely to assume that a man is homosexual if he works as a nurse or secretary, or if his face has some "feminine" features (Deaux & Lewis, 1983; Dunkle & Francis, 1990). Do you remember the 1997 movie *In and Out* starring Kevin Kline? In the movie, many people believe that Kline's character, Howard, is homosexual because he teaches poetry, dresses well, likes to watch Barbara Streisand movies, and is "a totally decent human being." Conversely, many people are quick to attribute feminine characteristics to a man they are told is (but in actuality is not) homosexual (Weissbach & Zagon, 1975).

Probably the major reason for beliefs like these is that for many people, masculinity and femininity are basic assumptions of heterosexuality: men are masculine and attracted to women, women are feminine and attracted to men. If you believe this, then it follows that the less "masculine" a male is, the more likely that he will be perceived as homosexual; and if a man is homosexual, it is more likely that he will be perceived as "feminine" (Ross, 1983). However, as you learned in the previous chapter, gender roles change over time, and this is true within the homosexual community as well as the heterosexual community. Before the 1970s, for example, effeminate qualities were emphasized in the male homosexual community, but since then masculinity has been emphasized (Garber, 1991). Consider, too, the butch and femme roles assumed by many female homosexuals (behavior most prevalent prior to 1960)—their behaviors were quite different, but their sexual orientation was the same. Again, there is no strong evidence that a homosexual or bisexual orientation must be associated with atypical gender roles. A great many homosexuals and bisexuals do not walk, talk, dress, or act any differently than anyone else (Kirk & Madsen, 1989). Consider the following:

> He had attended a Catholic high school, was a star athlete on school teams, and was engaged to be married to his high school sweetheart. They had been going steady for several years. He was handsome, masculine looking, well-liked—an "All American" type of young man. Raised in a supportive family environment along with a younger brother and sister, he had come from as normal a background as could be imagined. He, too, had always thought of himself as normal.
>
> His first awareness of his attraction to men came in adolescence. . . . He felt as if he was the only person in the world with this "problem." He loved his girlfriend, but he knew that something wasn't right and after high school he broke off the engagement. . . . He does not apologize for his homosexuality, for he feels that there is absolutely nothing to apologize for (it would be like apologizing for breathing or the color of his eyes).
>
> He does not explain his homosexuality as the result of following role models, for he had none. He felt that he alone had these feelings during his adolescence. He does not think that he acted out of rebellion against his parents. They were a close, warm family and his parents treated his brother and sister in a similar manner. Neither of them are homosexual. He knows of no close relative who is homosexual. He was just like everyone else in high school and today is indistinguishable from any of the other male students in college. . . .
>
> (from the author's files)

THE ORIGINS OF SEXUAL ORIENTATION

Why are some people heterosexual, others homosexual, and still others bisexual? Researchers and clinicians used to direct their efforts to the question "What causes homosexuality?" However, any study of the causes of homosexuality is inevitably about the causes of heterosexuality, and vice versa—you can't consider one without considering the other (Friedman, 1992).

Is sexual orientation biologically determined? Is it the result of the environment (family background, peers, etc.) shaping individuals? There are probably few topics in the study of human characteristics in which the nature-versus-nurture question has been so strongly debated. I will review this work and then come to two conclusions, which I give in advance: (1) neither heterosexuality nor homosexuality nor bisexuality have a single cause; and (2) biological, psychological, and social/cultural influences interact to produce sexual orientation (see Byne & Parsons, 1993; De Cecco & Elia, 1993; Friedman, 1992).

PSYCHOANALYTIC EXPLANATIONS

Sigmund Freud (1905) viewed all people as capable of becoming either heterosexual or homosexual—neither orientation was assumed to be innate. He believed that a person's sexual orientation depended on how the *Oedipus complex* was resolved (see Chapter 8) and that heterosexuality was the "normal" outcome. Homosexuality was the result of fixation at (or regression to) this stage of development, when a particular set of circumstances occurred.

More specifically, Freud believed that male homosexuality resulted when a boy had a domineering, rejecting mother and turned to his father for love, and later to men in general. According to Freud, female homosexuality develops when a girl loves her mother and identifies with her father and becomes fixated at this stage. As an adult, the woman continues to seek the love of women (mother figures). Freud also felt that some individuals managed to keep their homosexual tendencies hidden from their conscious mind, or "repressed," until these tendencies emerged sometime later. He called these persons latent homosexuals.

In support of Freud, Bieber and colleagues (1962) studied male homosexuals and claimed that they typically had domineering, overly protective mothers and weak, detached fathers who did not provide masculine models. The mothers were seductive (in that they were jealous of, and discouraged, any interest their boys had in girls) and destroyed their boys' masculine behavior. Bieber claimed that the boys became fearful of heterosexual relations and turned to men for the love that they never received from their fathers. In a related study in support of Freud, Wolff (1971) claimed that female homosexuals came from homes with absent fathers and cold and rejecting mothers. The girls then spent the remainder of their lives searching for the love that their mothers never gave them by having other women as sexual partners.

These studies were done with a small number of people who were in therapy. While some other studies have found similar results (Evans, 1969; Whitam, Daskalos, & Mathy, 1995), a much larger study of better-adjusted persons failed to support their findings. Bell, Weinberg, and Hammersmith (1981) could find little evidence that male homosexuals had dominant mothers and weak fathers, or that female homosexuals had chosen their fathers as role models.

What can we conclude? Although there is some evidence that persons with a family background like those described by Bieber or Wolf "have a greater than average likelihood of becoming homosexual" (Marmor, 1980a, 1980b), it is important to keep in mind that homosexuals have all kinds of parents, both good and bad, caring and cold (Masters & Johnson, 1979). The same is true of heterosexuals. The psychoanalytic theories cannot possibly begin to explain the sexual orientation of all individuals.

Interestingly, although Freud's followers viewed homosexuality as a mental illness (see section on homosexuality), Freud himself apparently had a less harsh attitude, as evidenced by this letter to the mother of a homosexual son:

> Dear Mrs. _____ ,
> I gather from your letter that your son is a homosexual. . . . Homosexuality is assuredly no advantage, but it is nothing to be ashamed of, no vice, no degradation, it cannot be classified as an illness; we consider it to be a variation of the sexual function produced by a certain arrest of sexual development.
> Many highly respectable individuals of ancient and modern times have been homosexuals, several of the greatest men among them (Plato, Michelangelo, Leonardo da Vinci, etc.).
> It is a great injustice to persecute homosexuality as a crime, and cruelty, too. . . .

OTHER PSYCHOSOCIAL EXPLANATIONS

Social learning theory explains homosexuality and heterosexuality as learned behaviors (Gagnon & Simon, 1973; Masters & Johnson, 1979). This theory would therefore treat homosexuality and heterosexuality similarly, with each type of sexual orientation being dependent on the learning history of an individual. Homosexuals would be expected to have had good or rewarding experiences with same-sex individuals and bad experiences with members of the opposite sex in their early social experiences. Once initiated, the sexual act itself, being pleasurable, becomes rewarding. Over time, they would therefore learn to associate homosexuality with more pleasure than heterosexuality. Also, they may have had role models to pattern themselves after in their youth.

What is the evidence in support of this explanation? Many people assume that homosexuals are people who are not attractive (or masculine or feminine) enough to "make it" with the opposite sex. However, the data do not support this. Most female homosexuals, for example, have had pleasurable sexual relations with men. They simply prefer sexual relations with women (Martin & Lyon, 1972). In their study, Bell, Weinberg, and Hammersmith (1981) found that "homosexual orientation among females reflects neither a lack of heterosexual experience nor a history of particularly unpleasant heterosexual experiences." However, there is some evidence that an extremely bad negative experience, such as being the victim of rape, can result in an individual's avoiding heterosexual relations (Grundlach, 1977).

There is also some evidence that if a person's first adolescent (or preadolescent) sexual experiences, such as mutual masturbation, are with an individual of the same sex, then there is a greater-than-usual chance that the person's sexual orientation will be homosexual (Van Wyk, 1984, but see Box 9–A on page 242). Storms (1980) has suggested that children who begin puberty early are particularly more likely than others to experiment with adolescent homosexual experiences (for which there are more opportunities than heterosexual experiences), and these early pleasurable experiences make it more than likely that a homosexual orientation will develop. As a general rule, male homosexuals do begin puberty at an earlier age than heterosexual males (Blanchard & Bogaert, 1996a). There is no evidence to support the myth that adolescents became homosexual because they were seduced by an older homosexual (Bell, Weinberg, & Hammersmith, 1981).

Even if the studies about adolescent homosexual experimentation are true, that still leaves the question of which came first: Did homosexual experimentation lead to a homosexual orientation, or did homosexual tendencies lead to homosexual experimentation? Bell, Weinberg, and Hammersmith (1981) found that a person's sexual orientation was usually well established by adolescence and that gender (role) nonconformity during childhood was often, but not always, associated later with a homosexual orientation. Similar results have been reported by others (Phillips & Over, 1995; Saghir & Robins, 1973; Whitam, 1977a).

The relation between childhood gender nonconformity and later homosexuality is greater for boys than it is for girls ("tomboyish" behavior by girls is accepted more than "sissy" behavior by boys). One study found that for boys, the absence of masculine traits was a better predictor of later homosexuality than the presence of feminine traits (Hockenberry & Billingham, 1987). Another study found that for boys who persistently play with girls' dolls and display other stereotypically feminine traits, there is about a 75 percent chance of their being homosexual as adults (Bailey & Zucker, 1995).

Richard Green (1987) studied a small sample of extremely effeminate boys (who he called "sissy boys") over a 15–year period. A few families reinforced the behavior, but most of the families simply tolerated (rather than initiated) it. The boys had poor relations with their fathers, but this could be the result of the boys' "unmasculine" behavior rather than a cause. Most of the boys engaged in homosexual fantasizing and behavior as adolescents, but some developed a clear heterosexual orientation. Thus, even in cases of extreme childhood gender-role nonconformity, one cannot absolutely predict future sexual orientation. Also, keep in mind that homosexual males who were

"sissies" as boys comprise only a portion of homosexuals; many showed typical gender-role behavior as children (Bell et al., 1981; see section on gender roles on page 227).

Green's subjects had been different since early childhood. Bell et al. (1981) suggested that their results, along with those of Green (1987) and others, indicated a possible biological basis for sexual orientation. Let's now consider the evidence for that.

DEMOGRAPHICS

Several studies have found that, on average, homosexual men had a later birth order than heterosexual men (Blanchard & Bogaert, 1996a, 1996b; Blanchard et al.,1996). Moreover, effeminate homosexuality was also associated with a greater number of older brothers, but not older sisters (see Blanchard & Bogaert, 1996b). No birth order effect was observed in homosexual women (Bogaert, 1997). At present, it is unclear just what this means, for neither psychosexual nor biological explanations are satisfactory. Blanchard and Bogaert (1996b) have suggested that the birth order effect in males may be due to a reaction of the mother's immune system triggered by the previous male fetuses.

BIOLOGICAL EXPLANATIONS

There is no question that males and females are structurally different. In Chapter 8, you learned that there are also some differences in male and female brains. Is it possible that brain difference could account for differences in sexual orientation?

GENETIC FACTORS. In 1952, Kallman found nearly 100 percent concordance for homosexual orientation in identical twins (i.e., if one twin was homosexual, the other almost always was too). The concordance for fraternal (nonidentical) twins was only about 10 percent. Similar results were found by a German researcher (Schlegel, 1962), and together these findings challenged the prevailing view that homosexuality was determined by social influences.

The results of these first studies with twins were questioned when several reports appeared in the 1960s and 1970s of single cases or small numbers of identical twins who were discordant for homosexual orientation (see Whitam, Diamond, & Martin, 1993, for a review). However, several recent larger-scale studies have reestablished a possible genetic basis for sexual orientation. Bailey and Pillard (1991) studied 167 pairs of brothers and found that 52 percent of the identical twin brothers of homosexual males in the study were also homosexual, compared with 22 percent of nonidentical twins and 11 percent of adoptive brothers. In another study of 143 pairs of sisters, they found that 48 percent of identical twins of homosexual females were also homosexual, compared to 16 percent of nonidentical twins and 6 percent of adoptive sisters (Bailey et al., 1993). Another study found a 65.8 percent concordance rate for homosexual orientation in identical twin brothers, compared to 30.4 percent in nonidentical twins (Whitam, Diamond, & Martin, 1993). However, not everyone finds these results. A recent study found a high concordance rate in female identical twins (60 percent), but not in male twins (Hershberger, 1997).

Still another group of researchers found that 33 of 40 pairs of homosexual brothers shared the identical DNA pattern on the tip of one arm of their X chromosomes (inherited from the mother; Hamer et al., 1993), but another group of scientists could not duplicate Hamer's findings (Rice et al., 1995) and there have been accusations that the author did not include data that did not support his conclusion (Marshall, 1995). More recently, researchers have presented evidence that male (but not female) homosexuality is determined by a gene in a specific region of the X chromosome (Hu et al., 1995; Turner, 1995). Although it is far removed from human studies, research with flies indicates that sexual orientation can be altered by inserting an extra gene (Zhang & Odenwald, 1995).

What does all this mean? Although the studies suffered from several methodological flaws (Baron, 1993), they leave little doubt of a biological contribution to sexual orientation. But how much? Remember, not all of the identical twins were homosexual, and in the study by Hamer et al. (1993), seven pairs of homosexual brothers did not share the same gene pattern. Michael Bailey, author of several twin studies, concluded that "the degree of the genetic contribution (to homosexuality) could range from 30 percent to 70 percent depending on the assumptions used." Even if his upper estimate is correct, that still leaves a lot of room for the influence of social and environmental factors. Social interactions might well mediate aspects of any genetic role in sexual orientation (Byne & Parsons, 1993). Recall, for example, the extreme gender nonconformity displayed by children in the "sissy boy" syndrome (see previous section). Even if this is due to genetics, treating the children differently from other children could reinforce later nonconformity in gender roles (Hines & Green, 1991).

ANATOMICAL FACTORS. At about the same time that researchers were finding a genetic component to sexual orientation, others were reporting differences between the brains of heterosexuals and homosexuals. Simon LeVay, a neuroscientist and an avowed homosexual with both a homosexual brother and a lover who had died of AIDS, performed autopsies on the brains of 19 homosexual men and 16 heterosexual men. He found that a part of the hypothalamus (at the base of the brain) known to influence sexual behavior (specifically,

the interstitial nucleus of the anterior hypothalamus-3, or INAH-3) was more than twice as large in heterosexuals (LeVay, 1991). This area of the hypothalamus is affected by sex hormones (see next section).

LeVay's study followed a report that another part of the hypothalamus, the suprachiasmatic nucleus, had twice as many cells in homosexual men as in heterosexual men (Swaab & Hofman, 1990). A subsequent study found that a major fiber bundle that connects the two halves of the brain (the anterior commissure) was 34 percent larger in homosexual males than in heterosexual males and no different from the relative size of the same structure in the brains of heterosexual females (when brain weight differences between men and women were adjusted; Allen & Gorski, 1992).

What do the anatomical findings mean, and how are they related to the genetic research? Dr. Dean Heimer of the National Institutes of Health concluded, "The simplest interpretation would be that there is a gene, or genes, that control the growth of [the INAH-3 and other brain structures]. Both point to the same idea, that there is a biological component of sexual orientation" (cited by Maugh in the *Los Angeles Times*, 1991). However, these results do not prove causation; a recent study found that adult sexual behavior in rats can change the anatomy of the brain (Breedlove, 1997).

HORMONAL FACTORS. How might genes affect brain anatomy, and if that can be determined, when? Before birth? Shortly after birth? During childhood? At puberty?

It is likely that the critical link is hormones. Gorski and colleagues (1978) were the first to show differences in brain (hypothalamus) structure between male and female rats. Gorski's group later manipulated the levels of sex hormones in newborn rats and, as a result, were able to get male rats that exhibited the sexual behavior of female rats. Several studies with nonhuman species have shown that treatment with hormones prenatally (before birth) can result in homosexual behavior (Dorner, 1968; Money & Ehrhardt, 1972).

In humans, one thing we do know is that the level of circulating male sex hormones (testosterone) in adult heterosexual and homosexual males does not differ (see Gladue, 1988). Furthermore, giving sex hormones to adult homosexuals (testosterone to homosexual males, for example) does not change their sexual orientation. However, there is evidence for differences in hormonal responses. When injected with the female hormone estrogen, women show an elevation of the pituitary hormone LH (luteinizing hormone; see Chapter 3), while men normally show no elevation. Studies have found that male homosexuals show an intermediate elevation in LH after an injection of estrogen (Dorner et al., 1975; Gladue, Green, & Hellman,

1984). Does this mean that homosexual men have a "female" hypothalamus? These and other studies suggest that if hormones do cause a difference in brain structure, it would have to be caused early in life.

Ethical considerations do not allow manipulation of hormones in human fetuses or infants, but there is some evidence that what has been found in nonhuman species is also true in humans. For example, studies have found that women who were exposed to high levels of masculinizing hormones before birth (see the discussion of congenital adrenal hyperplasia in Chapter 8) are much more likely than others to prefer boys' toys and activities as children (Berenbaum & Snyder, 1995) and develop a homosexual or bisexual orientation as adults (Ehrhardt, Evers, & Money, 1968; Meyer-Bahlburg et al., 1995; Money & Schwartz, 1977).

Ellis and Ames (1987), in an extensive review of both human and animal research, concluded that sexual orientation is primarily determined before birth. According to their theory, all mammals begin life with a single sexual orientation. It is the presence of certain (though yet undetermined) amounts of hormonal factors that direct sexual development and create a variety of sexual orientations, heterosexuality being the most frequent orientation. The presence of testosterone during the second month of embryological development, you recall, is critical for determining one's sexual anatomy (see Chapter 8). Ellis and Ames claim that the critical period for the development of sexual orientation comes between the middle of the second month and the middle of the fifth month of pregnancy; others claim that hormones have their effect in the first year or two after birth (recall the work by Gorski). But notice that all agree that the effect of hormones on sexual orientation occurs very early. Ellis and Ames list several potential factors that could influence an unborn child's development of sexual orientation, including genetic-hormonal factors, drugs, stress in the mother, and immune system dysfunctions.

CONCLUSIONS

Although there is a growing body of evidence for a biological contribution to sexual orientation, this does not mean that one's sexual orientation is destined by biology (Byne & Parsons, 1993; Gladue, 1994). Most researchers agree that biological and social influences *both* contribute to the development of sexual orientation. The question is no longer nature versus nurture, but to what extent each influences orientation. They certainly interact in some complex yet undetermined manner. At the moment, the most we can say for biological factors is that they probably *predispose* (this is not the same as *cause*) an individual to a particular sexual orientation. What this means is that given a certain genetic background and a particular set of social and environmental influences, it is more likely than not that a person will assume a heterosexual or homosexual orientation.

Psychologist Daryl Bem (1996) recently presented an intriguing hypothesis. He proposed that biology (genes, anatomy, hormones) "[does] not code for sexual orientation per se but for childhood temperaments that influence a child's preferences for sex-typical or sex-atypical activities and peers." For example, boys typically prefer rough-and-tumble play, but a boy who is temperamentally not aggressive may shy away from these activities and prefer girl-typical activities. According to Bem, children's preferences make them feel different from opposite-sex or same-sex playmates. In a large-scale study, it was found that 70 percent of adult homosexuals had, in fact, felt different from same-sex children when young (compared to 38–51 percent for heterosexuals; Bell, Weinberg, & Hammersmith, 1981). Bem says that children view the sex they feel different from as "exotic" and later become physiologically aroused (erotic attraction) by that sex. In other words, biology does not code for sexual orientation directly, but for temperaments which the environment then steers in the direction of a particular orientation.

The recent genetic and anatomical findings have created debate in the homosexual community. On the one hand, there are those who feel that proof of a biological determinant of sexual orientation will lead to greater tolerance. As stated by Dr. Richard Pillard, coauthor of the twin studies, "A genetic component in sexual orientation says, 'This is not a fault, and it's not your fault'" (quoted in Gelman et al., 1992). Or, as stated by George Neighbors Jr. of the Federation of Parents and Friends of Lesbians and Gays, "It may help people look at this as a genetic issue. . . . If you believe in God or nature, that's what homosexuality develops from" (quoted by Recur, 1993).

On the other hand, Neighbors and others worry that if a biological basis can be found, some people with an antihomosexual bias will press for medical changes in sexual orientation, as was tried in the recent past (Byne & Stein, 1997). Perhaps the best perspective was presented by David Barr, assistant director of policy for the Gay Men's Health Crisis in New York, who said, "It doesn't really matter why people are gay or not gay. That's not the important question. What's really important is how they're treated. I haven't spent that much time thinking about where my sexuality comes from. I've spent a lot more time thinking about how I fit into this world I have to live in" (quoted by Angier, 1991).

It is certainly the case that homosexual men and women are often treated negatively by others. Let us examine what it means to be homosexual in our culture.

HOMOSEXUALITY

Homosexuality is a word derived from the Greek *homo,* meaning "same" (do not confuse this with the Latin *homo,* meaning "man"). The term was first used by a Hungarian physician in a pamphlet published in Germany in 1869 (Gregersen, 1982) and was popularized in the English language two decades later by Henry Havelock Ellis. The prefix *hetero-* by the way, is also derived from a Greek word and means "different."

Homosexual behavior has been reported in nearly all cultures throughout the world (see Box 9–A). Homosexual behavior (male-male or female-female mounting) has also been observed and studied in a variety of nonhuman species, including many primates and dolphins. Among male animals, it is often done as a display of dominance, and otherwise generally serves as a social function (reinforcing relations) rather than a sexual one (see Denniston, 1980; Ford & Beach, 1951; and Linden, 1992). However, in some cases sexual arousal may also serve as a stimulus (Srivastava, Borries, & Sommer, 1991).

Over the years there have been many derogatory terms used to describe homosexuals (e.g., "faggot," "queer," "fairy," "dyke"). Most homosexuals prefer the term **gay** for a male homosexual (although in some places the term is used for homosexuals of either sex) and **lesbian** for a female homosexual (derived from the name of the Greek island Lesbos, where, in about 600 B.C., the poet Sappho led a circle of young female disciples). Heterosexuals are often referred to as "**straights**." Among some gay and lesbian organizations, the term *homophile* is preferred when referring to a homosexual person.

HISTORY OF ATTITUDES ABOUT HOMOSEXUALITY

Western attitudes about homosexuality have varied considerably throughout history. In ancient Greece, homosexuality was widely accepted as an alternative to heterosexuality (Boswell, 1980). Plato praised homosexual relations in his *Symposium.* Homosexual relations between Greek scholars (e.g., Socrates, Plato) and their students were considered to be a natural part of the young boys' education. Marriage between individuals of the same sex was legal and commonly accepted among the upper class during the time of the Roman Empire.

Homosexual activities were practiced—often as part of religious rituals—by many groups of Hebrews prior to the seventh century B.C. As part of a reformation movement designed to unify the many Hebrew groups, homosexuality was thereafter condemned (along with other acts that had been part of older rituals) and was made punishable by death:

> **If a man lies with a male as with a woman, both of them have committed an abomination; they shall be put to death, their blood is upon them.**
>
> (Leviticus 20:13, Revised Standard Version)

Female homosexuality was dealt with less harshly by the Jews because it didn't involve "spillage of seed" or violation of male property rights.

European Christians, following the Roman tradition, initially were tolerant of homosexuality (Boswell, 1980). It wasn't until the time of St. Thomas Aquinas (A.D. 1225–1274), who emphasized the view of St. Augustine that the only purpose of sex was procreation, that homosexuality came to be viewed as unnatural or "against the laws of nature." However, Boswell (1980) notes that while same-gender sex was considered a sin in the Middle Ages, people who committed the sin were not then viewed as a *type* of person who was different from others. This remained true during the 1600s and 1700s, when homosexual acts became a crime as well as a sin.

It wasn't until the late 1800s and the rise of modern medicine (particularly psychiatry) that people who engaged in homosexual acts came to be viewed as a particular class of people—people with a mental illness. As a result, there has never been a period in Western culture with as

Figure 9–2 Socrates. This Greek philosopher and teacher believed that any evil committed by human beings is done out of ignorance. He devoted his life completely to seeking truth and goodness.

gay A term generally used to refer to male homosexuals, although in some places it is used to refer to homosexuals of either gender.

lesbian A female homosexual.

straight A term used by homosexuals for a heterosexual.

much intolerance of homosexuality as there has been in the twentieth century. Because they were viewed as mentally ill, the medical profession subjected homosexuals to frontal lobotomies, forced castrations or hysterectomies, electric shock treatments, hormone injections, and endless hours of psychotherapy (Katz, 1976; also, see James Harrison's 1992 documentary film *Changing Our Minds*).

Until the 1960s, homosexuals had never been studied outside of a medical or prison setting. Then, in 1957, Evelyn Hooker published a paper showing that nonclinical heterosexual and homosexual males could not be distinguished by the best psychological tests (i.e., the homosexual males were as well adjusted as the heterosexual males). Finally, in 1973, the American Psychiatric Association decided to discontinue classifying homosexuality as a mental illness (see Hooker, 1993, for a review of some of the events leading to this decision). The decision was not unanimously applauded; 37 percent of APA members were opposed to change, and some are still opposed (e.g., Socarides, 1996).

The APA decision has had far-reaching implications. If homosexuality is no longer considered an illness, there is no need to cure it. As you can see, the way individuals deal with homosexuality is based on what they believe. However, attitudes often change very slowly. Simply *being* homosexual is no longer a criminal offense, but many states still have old laws on the books that prohibit the sexual expression of homosexuality. For example, in 20 states there are laws against anal or oral sex that are used usually only to prosecute homosexuals (see Chapter 16). A police raid on the Stonewall Inn in Greenwich Village in 1969 led to three days of public demonstrations, an event that many say was the beginning of the gay rights movement.

Similarly, the Catholic Church no longer considers homosexuality per se a sin, but it continues to condemn homosexual behavior. A letter to bishops worldwide written by Cardinal Joseph Ratzinger and approved by Pope John Paul II stressed that homosexual acts are sins. Homosexuals instead were urged to lead "a chaste life." In another letter to U.S. Catholic bishops in 1992, the Vatican called homosexuality "an objective disorder" and encouraged discrimination against homosexuals in public housing, family health benefits, and the hiring of teachers. The June 1988 convention of the Southern Baptists adopted a resolution that said that homosexuals, "like all sinners, can receive forgiveness through personal faith in Jesus Christ," but added that homosexuality "is not a normal lifestyle and is an abomination in the eyes of God." The Catholic Church and the Southern Baptists are not alone. While there are some

exceptions, most Christian churches continue to have a rejecting or punitive attitude toward homosexuality (Nelson, 1980).

At the present time, many individuals still have a harsh or negative attitude about homosexuality. Recent surveys indicate that about 60 percent of all Americans feel that sex between two adults of the same sex is always wrong and that nearly half are opposed to homosexuals teaching in elementary schools (Yang, 1997). In another survey, 89 percent of 15- to 19-year-old males believed that sex between two men was "disgusting" (Marsiglio, 1993b). Only 12 percent said they could be friends with a gay person. Even teaching about homosexuality is a controversial issue, with many people opposed to it (see Reiss, 1997).

Today, these widespread negative attitudes are spilling over into the voting booth. By 1993 more than 130 states, counties, or cities had passed laws to protect gays and lesbians against discrimination, but there has been a backlash. Oregon voters narrowly rejected an anti-homosexual measure in 1992, but a similar ballot measure in Colorado passed in the same year (the Colorado Supreme Court subsequently overturned it). In 1998, voters in Maine became the first to repeal a homosexual rights law.

The U.S. military has always banned and discharged homosexuals. At the urging of President Clinton, a "compromise" was enacted in 1993: new recruits can no longer be asked if they are homosexual, but they can still be discharged for engaging in homosexual acts.

For people with a homosexual orientation, the battle for acceptance continues.

COMING OUT

With so many people in our society having negative attitudes about homosexuality, it should come as no surprise that acknowledging to oneself that one is homosexual, and then publicly declaring it, is often very difficult. The process of disclosing one's homosexuality is called **coming out**. There are several stages to this process. I will follow the stages outlined by the Boston Women's Health Book Collective (1984) and simply note that other authors' descriptions may differ slightly (e.g., Morris, 1997).

The first stage is *admitting to oneself that one has a homosexual orientation*. Some researchers believe that sexual orientation is established by the age of 4 (Gordon & Snyder, 1989), and many homosexuals knew that they were "different" at a very young age. Attraction to individuals of the same sex generally occurs well before sexual activity begins (Bell, Weinberg, & Hammersmith, 1981).

"I discovered my sexuality at a surprising six years old. Naturally I didn't know at that young age that I was a lesbian, but I can re-

coming out Disclosing to others one's homosexual orientation, from the expression "coming out of the closet."

member having 'crushes' on my teachers and female upperclassmen."

"I think that the most important thing you lectured on was the predisposition of sexual orientation before birth. I, too, felt different as a child as early as by 10."

(from the author's files)

For others, self-recognition is often a confusing and anxiety-producing discovery. Some do not recognize their sexual orientation until well into adolescence, or even later (recall the case history in the section on gender roles).

Dear Ann Landers:

I am a married woman in my late 20s. (We have children.) I never had a homosexual experience until after I was married. I sought counseling, and I am trying to keep my head together.

The feelings I have toward a certain woman are more surprising to me than to anyone. They are more intense than just the physical desire. This is the deepest friendship I have ever known. . . .

I'm sure millions of women who are reading this are thinking "How gross!" "How sick!" A year ago I would have been one of them. I did not go looking for this. It just happened. . . . I am not prepared to throw away my marriage and family, but this new relationship has certainly changed my life. My husband doesn't know about it and hopefully he never will. Most of all I would hate to do anything that would hurt my kids. . . . Please give me some insight.

Staying in the Closet for Now, Illinois

(Reprinted with permission of Ann Landers and Creators Syndicate.)

Because so many people have a negative opinion of it, recognizing that one is homosexual is often a lonely and painful experience.

"As a young gay person, you feel all of the things every other young person feels. You feel all of the normal doubts, uncertainties, questions and frustrations common to that time of life. And you also begin to have feelings and thoughts as precious and deep as any can be: of self, and worth, and love, and life, and purpose. But when you are young and gay, you also face a world of pain and isolation few people have to face.

"We all know how vulnerable and fragile a life can be in those early years. And yet few people seem ready or willing to consider how it must feel to be young and gay. In a time when nurturing and affirmation are so important, the young gay person is told by his church that he is evil. The law says that she is a criminal. Society says that they are misfits. The media displays freaks. And the world seems to say they are less than a whole person. And so, when you are young and gay, you can feel so much hatred, so much rejection, so unwanted, so unwelcome, so worthless, so unaccepted and so misunderstood.

"Other people throughout time have known the kind of pain and persecution which reaches deep into a person's heart. There have always been people who were made to feel hated and demeaned because of their color, their religion, their heritage, or their gender. But as much pain and suffering as these people have known, they also knew they were not alone. They always had someone else they could see who was just like them; someone to identify with. And most important of all, they had someone to talk to; someone to share the pain with.

"When you are young and gay, you feel the same terrible depth of pain. But you look around and there is no one to talk to; no one to share the pain with. There is no one who has feelings like yours or who understands how deeply you hurt. You feel so isolated and so alone. And at times, that vulnerability and pain and isolation can be so terribly overwhelming."

(Steve L., personal communication, New Orleans)

In the next stage, the individual *gets to know other homosexuals,* thus ending the sense of isolation. It is easier to come out if the individual finds acceptance in others, and those most likely to be accepting are other homosexuals. Most homosexuals were raised in the same anti-homosexual environment as others, and by coming into contact with other homosexuals, they can replace these negative stereotypes with a more positive attitude about homosexuality—and about themselves.

"As I was growing up, I heard all of the jokes and stories about the lives gays lived. I saw all of the negative images that the media could dish up for public consumption. . . . I wanted to find someone I could share the rest of my life with.

"And then I looked for that life and tried to find it. . . . The only place society had allowed gays to meet was in one small bar-filled pocket of the city. And so I turned to where I was accepted and where I could breathe. But then I heard society condemn me because of the life-style I 'chose' to lead.

"In time, I grew to learn that there was more of a life to be had than just the narrow world which society tries to impose upon gays. I met many happy and healthy gay people who helped me to understand and learn and grow."

(Steve L., personal communication, New Orleans)

Once contact is made with other homosexuals, a new way of expressing sexuality has to be developed. In many ways, what happens next is quite similar to what happens when heterosexuals begin to date. There are initial relationships and breakups, and often lasting relationships are eventually formed. Contact with other homosexuals may also lead to an increased interest in gay rights and other political activities to support the rights of homosexuals.

In the third, very difficult, stage, the individual tells family and friends of his or her sexual orientation. Many hide their sexual orientation from others for long periods.

"But as bad as you hurt, you keep the pain to yourself. You fear being discovered by the people you love and need the most; for everything your world has taught you makes you afraid you will lose their love and friendship if they find out you are gay. The world gives you the pain, and then makes it impossible for you to talk about it because of fear.

"So not only does the world isolate you from learning about your feelings and others who are like you, but you begin to isolate a part of yourself from family and friends. You begin to develop techniques of telling half-truths and skirting around questions. You learn how to hide inside of yourself and smother feelings. You learn how to be around other people without letting them get too close. You learn how to survive.

"And then, you grow tired of hiding. You grow tired of not being able to share the beauty as well as the pain that fills your life. You grow tired of not being as close as you want to be to the people you care most about.

"But the fear is still very strong. You find it is easier to avoid the questions by just staying away more often. And you don't have to find as much strength to hide and be silent. So you drift further and further from the people you love the most and you feel the pain and isolation building up inside."

(Steve L., personal communication, New Orleans)

Some, because of social constraints and possible punitive actions, feel that they must keep their orientation a secret.

"I've come to terms with my sexuality and live with it. However, I don't choose to show that side of me. I am in the military and must remain discreet. As a matter of fact, I'm in charge of a number of troops. Some members of my unit are in this class, which keeps me from relaxing. Although the new policy in the military is 'Don't ask,' 'Don't tell,' 'Don't pursue,' if found to be gay I would be discharged."

(from the author's files)

Others may choose not to disclose because they are parents (Green & Clunis, 1988). When an individual does finally disclose his or her homosexuality to the family, it often creates a crisis (Cramer & Roach, 1988; Zitter, 1987).

"My parents became aware of my sexual preference when I was eighteen. Since that time, I have endured numerous nasty conflicts with them. My mother told me that she would rather me be a prostitute, and my father told

Figure 9–3   Leonardo da Vinci. This Italian Renaissance painter (*Mona Lisa, The Last Supper*), anatomist, astronomer, and engineer was one of the most versatile geniuses in history.

me I would never be anyone as long I was a 'faggot.' They threw me out immediately after graduation from high school and forbade me to enter their house. They took me back on their terms—that I would go straight and live what is in their opinion a good, decent, moral life-style. I agreed and abided for a few months, but found myself extremely unhappy. I ended up moving out after a fight with my father.

"The continual attempts to persuade me to change drove me to a feeling of no self worth. My parents do not understand that my life-style is not something I can just change—turn on and off like a light switch. They think that if they make me miserable enough, that I'll want to change."

(from the author's files)

In most cases, the family eventually adjusts and accepts their family member's homosexuality (Holtzen & Agresti, 1990). Usually, after the family has been told, the next step is to tell friends, but this too is often met with different reactions.

"I am a 25-year-old female. I've been close friends with 'Darla' since 7th grade. 'Darla' and I used to talk about girl things all the time (boys, dating, sex). I felt there was nothing we couldn't discuss.

"Two years after high school graduation, 'Darla' and I along with another mutual friend got together for pizza and to catch up on old times. 'Darla' then said she had something very important to tell us. She told us she was gay. She said she had those feelings since she was a child and only recently acted on them. I handled her news supportively, but our other friend told 'Darla' that she was 'disgusting' and 'gross.' I'm sure without a doubt that was the hardest day in 'Darla's' life."

(from the author's files)

The emotional issue of revealing one's homosexuality to family received national attention in the newspaper cartoon comic strip series "For Better or For Worse" (by Lynn Johnston, March 1993), in which the 17–year-old character Lawrence, after much agonizing, tells his parents that he is gay. The series is normally carried by 1,400 newspapers, but, not surprisingly, some rejected it during that episode. For those of you who are homosexual and have not yet revealed your orientation to your family, Weinberg (1973) has some excellent suggestions.

The final stage is *complete openness about one's homosexuality*. This includes telling people at work and completes the process, making one's homosexuality a total life-style. The important task for all adults, regardless of their sexual orientation, is to develop an acceptance of their sexuality (i.e., of themselves), to find acceptable ways to express it, and to learn to incorporate their sexuality into their total life-style and personality.

HOMOSEXUAL LIFE-STYLES AND RELATIONS

Finding companions and sexual partners can be difficult for many homosexuals, especially in rural areas and small towns, where there may be little tolerance of homosexuals meeting together in public. In large urban areas there is usually a much larger number of people who are homosexual, as well as a much greater acceptance of their life-style. Prior to the 1970s, gays and lesbians met in "secret" bars, bathhouses, or public restrooms. Over time, gay bars became more than places in which to have sex. Because homosexuality was not openly tolerated, the bars provided an atmosphere where one's sexual orientation was accepted without criticism.

Over the last three decades, the homosexual urban community has grown from a small collection of bars and bathhouses to entire neighborhoods of gay and lesbian stores, restaurants and coffeehouses, churches, and counseling and activity centers. The typical member of the gay community today is well-educated, in a professional career position, and openly homosexual. In San Francisco, homosexuals have achieved positions of political leadership as well (Shilts, 1987). This description, however, fits only those who are openly homosexual. For homosexual individuals who have not yet come out, homosexual relations are still dominated by secret meetings in bars and bathrooms, often with strangers (see Humphreys, 1970, for a description of this "tearoom trade" life-style). Bisexuals, in particular, are unlikely to openly participate in the homosexual community (McKirnan et al., 1995).

In a 1978 study of gays and lesbians in the San Francisco Bay Area, Bell and Weinberg found that 71 percent of homosexual life-styles fell into one of the following five categories:

1. *Close-coupled* (male—10 percent, female—28 percent): In this relationship, individuals have only one partner and generally report that they are happy with their sexual orientation.
2. *Open-coupled* (male—18 percent, female—17 percent): In this relationship there is a primary partner but additional sexual partners are sought outside the relationship. These people are generally happy with their sexual orientation.
3. *Functional* (male—15 percent, female—10 percent): This relationship resembles the "swinging

single." Individuals are comfortable with their homosexuality, enjoy their independence, and have multiple partners.

4. *Dysfunctional* (male—12 percent, female—5 percent): These people are sexually active, but are not happy with their homosexuality and often have sexual problems.

5. *Asexual* (male—16 percent, female—11 percent): These people are single, have few partners, and are not interested in seeking partners.

The types of relationships found by Bell and Weinberg for homosexuals are similar to the types of relationships found among heterosexuals. Many heterosexuals believe that homosexual couples are less committed than heterosexual couples, but studies have found no differences in the amount of satisfaction and personal commitment between cohabiting heterosexuals and cohabiting homosexuals (Kurdek, 1995). In fact, homosexual couples may be better at being "best friends" than heterosexual couples (Peplau, 1988). The smaller number of close-coupled pairs (in the Bell and Weinberg study) as compared to married heterosexuals may be a function of the negative attitudes that our society often has regarding homosexual couples. How would you feel if a homosexual couple who were openly affectionate with each other moved next door to you? Many people would have negative or even hostile reactions. Also, young gays are not likely to see role models of monogamous, close-coupled relationships in gay bars (just as married heterosexual couples are rarely seen in "swinging singles" bars together). Furthermore, TV and the other communications channels of our society do not give monogamous homosexual relationships much coverage or encouragement.

We will now focus specifically on the sexual relations of gays and lesbians. Before we begin, let me put things in proper perspective. The term "homosexual" tends to focus attention only on a person's sex life. However, gays and lesbians spend only a very small part of their lives engaged in sexual activities. The same is true of heterosexuals. Homosexuals and heterosexuals alike spend the large majority of their lives pursuing careers, enjoying recreational activities with family and friends, and exploring social, cultural, and political interests. Most heterosexuals would not want to be judged solely on their sexual behavior! The same is true for homosexuals.

Let us consider males first. Bell and Weinberg (1978) found that over 50 percent of their sample of openly homosexual men living in San Francisco had had at least 500 sexual partners in their lifetimes and 28 percent said that they'd had over 1,000 partners. Their data have been criticized for not being representative of all homosexuals and, in fact, two earlier studies found that only 22 percent (Kinsey et al., 1948) or

8 percent (Hunt, 1974) of gays had had more than 11 sexual partners in their lifetimes. Still, the sexual lifestyle reported by Bell and Weinberg may not have been atypical for openly homosexual males living in a large urban homosexual community in 1978 (*before* we knew about AIDS). Many researchers have found that males, as a general rule, seek physical gratification and do not place as high a value on monogamy as do women. When large numbers of male homosexuals share this set of beliefs and have easy opportunities to meet, the large numbers of partners reported by Bell and Weinberg may not seem surprising. However, again, Bell and Weinberg's study (1978) was before anyone knew about AIDS.

Studies conducted since then indicate that many gay men now place greater emphasis on relationships and emotional commitment than before (Deenen, Gijs, & van Naerssen, 1994). AIDS emphasized the importance of relationships. As one researcher put it, "The relational ethos fostered new erotic attitudes. Most men now perceived coupling, monogamy, and celibacy as healthy and socially acceptable" (Levine, 1992). Unfortunately, recent surveys have found that many young gay men have abandoned safer sex practices (Boxall, 1995). This has been especially true of bisexually active men (McKirnan et al., 1995; Messiah et al., 1995).

The Kinsey (1953) survey found that only 4 percent of the lesbians sampled had had more than 11 partners, and the Hunt survey found only 1 percent. Even in the Bell and Weinberg survey (1978), a majority (70 percent) of lesbians reported having had fewer than ten partners, while 50 percent said that they did not like to have one-night stands. Females, whether straight or lesbian, seem to value commitment and romantic love. This is reflected in the longer-lasting relationships and the smaller number of partners they have compared to men (Leigh, 1989; Tuller, 1988). As usual, males and females are very different in their sexual behaviors and attitudes, regardless of their sexual orientations.

In their sexual relations, homosexual couples do everything that heterosexual couples do (Blumstein & Schwartz 1983; Masters & Johnson, 1979). They kiss, caress, engage in oral-genital sex, and, in the case of men, engage in anal intercourse (very few lesbians practice this). Lesbian couples generally take a great deal more time touching and caressing one another than heterosexual couples, for whom sexual relations are often dominated by the goal-oriented male (Hite, 1976; Masters & Johnson, 1979). Manual stimulation of the genitals and oral-genital sex are the preferred ways of reaching orgasm for lesbians (Bell & Weinberg, 1978; Masters & Johnson, 1979). Many lesbians engage in tribadism, the rubbing of genitals against another's genitals or other body parts, but very few use dildos (objects shaped like a penis). Perhaps because they take more time and understand each other's bodies better, lesbians have orgasms far more regularly during sex than do married heterosexual women (Kinsey et al., 1953). Among gays, oral-genital sex was once the most common sexual practice, followed by manual stimulation; anal intercourse was least common (Bell & Weinberg, 1978).

A common stereotype that many heterosexuals have of homosexuals is that they adopt a single role in a sexual relationship, with one being the "masculine" partner and the other being the "feminine" partner. This, of course, is based on their own heterosexual relations, but in fact is not true. Most homosexuals change positions and roles or use simultaneous stimulation during sex (McWhirter & Mattison, 1984). Hooker (1965) found, for example, that only 20 percent of male homosexuals preferred only one sexual role.

HOMOSEXUALS AND MARRIAGE

Studies estimate that one-fifth of all gay men and one-third of all lesbians have been married (Bell & Weinberg, 1978; Masters & Johnson, 1979). Millions of bisexual men and women are partners in heterosexual marriages (Hill, 1987). While some of these men and women were not aware of their sexual orientations when they married, many others were aware. Why would someone who knows they are homosexual or bisexual choose to marry a heterosexual? There are many possible reasons, including societal or family pressures, affection for the partner, desire to have children, or even negative feelings about leading a homosexual lifestyle (Bozett, 1987; Coleman, 1981/1982).

Of those married people who are consciously aware of their homosexuality or bisexuality, very few admit it to their partners (Stokes et al., 1996). Many experts find that there is little chance of a married homosexual or bisexual person's being monogamous with a heterosexual partner (Clark, 1987; Masters, Johnson, & Kolodny, 1992). The secret life led by the homosexual or bisexual partner can cause them considerable anxiety and stress. When the partner does discover the truth, he or she often experiences considerable pain and anger, which is heightened by feeling stupid and deceived. Most say they would have never married had they known the truth (Hays & Samuels, 1989).

Many gays and lesbians establish long-lasting relationships. Should they be allowed to legally marry? Polls indicate that most Americans oppose the idea of homosexual marriage (Yang, 1997). Why? Many heterosexuals say there is no point to it because homosexual couples can't have children (but see next section). Is

Figure 9–4 Walt Whitman. Many scholars consider his *Leaves of Grass* to be the greatest collection of poems ever written by an American.

the only purpose of marriage to have children?

In 1993, Hawaii's Supreme Court ruled that the state's ban against homosexual marriages might be unconstitutional and sent the case back to a lower court, where, in 1996, a judge ruled that the state could not forbid same-sex marriages. The Constitution requires that states honor each other's laws, but in the same year President Clinton signed the "Defense of Marriage Act" that defines marriage (in federal law) as the union of a man and a woman and allows states to refuse to honor same-sex marriages performed elsewhere. At the start of 1997, 18 states had banned same-sex marriages. The next few years will probably see many legal challenges, both pro and con, related to this issue.

HOMOSEXUALS AS PARENTS

There are, of course, several million married or formerly married homosexual (or bisexual) people who also are parents. Once the "secret" is unveiled, should an openly gay or lesbian person and his or her partner be allowed to raise a child? Should an openly homosexual couple be allowed to adopt a child? Before we continue, read the following real-life story of Kerry and Val, a lesbian couple. It is told in Kerry's own words.

"Val and I have been together for six years. We have two children, a daughter, 5, Nicole, and a son, 2, Nicholas. We both wanted a family but were a little afraid of the effects that our life-style would have on the children. We finally decided that in spite of our sexual orientation, we would still be loving, caring parents with much to offer our children.

"Next arose the problem of how to go about it. (We could not very well do it on our own.) We ended up choosing a mutual friend of ours to father the children. This took much understanding on Val's part to have her lover sleep with someone else. We were stable enough in our relationship that this did not cause us problems. The man we chose is divorced with three children. We wanted someone who already had children so that we would not have any custody problems. He already knew that we were gay and we were honest with him about our plans and wishes. He and I were both checked medically beforehand and the most opportune time for conception was determined.

"I got pregnant easily both times. Val and I went to the doctor together from the beginning. We paid for all the expenses. Val took care of me while I was pregnant. We experienced all of the changes together: hearing the unborn baby's heartbeat, feeling it move, staying up with me when it was too uncomfortable to sleep. Val was with me all through labor and delivery. She got to see the babies being born and was the first to hold them. She was there from beginning to end. It was a wonderful experience for us to share and it brought us much closer.

"We both believe that a lot of love and a lot of time spent with our children is the key. We try to never be too busy to listen to them. We both are involved with Nicole's school (Val especially, she is a room parent for Nicole's class), friends, and activities. This is very difficult because we both work. I am self-employed and attend school full-time, and Val is planning to return to school next semester. We also spend a lot of time involved in family activities such as going to the park or zoo, bike riding, camping, etc. We almost always include the kids in everything we do.

"We both share all the responsibilities for the children, the house (cooking, cleaning, etc.), and finances. Val helps me tremendously with my work so I can devote more effort to school. We are very happy together and can't imagine life without one another. We both wish that legally we could be recognized as a couple. It would be nice to be able to get insurance as a family, file taxes together, etc., but most importantly are the kids. If anything ever happens to me, what will happen to them? Val loves them as if they were her own and they love her and are very attached to her. I would want them to stay with her, but there is nothing I can do legally.

"My parents both know my wishes (they are divorced and both remarried). My dad adores Val and is very proud of the life we have made together and the stable home we have provided for our children. He would do his best to assure that the children remain with Val. On the other hand, my mom hates my life-style. She refuses to have anything to do with Val and because of this we rarely speak. If anything ever happened to me, she would try her best to take the children from Val and would raise the children to hate me (she has told me this). She is not proud of me at all, regardless of how well I have done. She doesn't accept the fact that without Val, I would be nothing. Yet, she is proud of my sisters, who are bums. . . .

"Val and I believe that as far as the chil-dren are concerned, we will never lie or hide any information about our life-style or their pasts. Along with their everyday education, we will also include the information about our life-style as they are able to understand it. We do not sexually do anything in front of the children—no parents should. We do hug, hold hands and kiss goodbye. It is healthy for the children to see people displaying affection for each other.

"We are honest with everyone about our life-style. Our friends, families, and employers and co-workers all know we are gay. To us, honesty is the best policy. We do not flaunt anything in front of them or do anything to embarrass anyone, nor would we, but we don't lie about it either. For the most part we have gotten nothing but respect from our straight friends. Sometimes they are curious, but that is natural. Mostly they respect our privacy and admire our honesty. Once people really get to know us they realize that we are no different from anyone else. At first we were a little worried about the reception that we would get from other parents that our children were associated with. Would they shun our children because of how we live? We wanted to be honest with them because their children were involved and we didn't want them to find out later and think badly of us because we lied. As it turns out, honesty was best.

"Val and I plan to grow old together. We are working together for the future. We have plans and dreams for the future as all couples do. We have and are continuously fighting to better ourselves and be accepted. We want to be treated normal just as everyone else. We are good people. The only difference is what we do behind closed doors. We do the best we can for our children. We want them to grow up normal, happy, and healthy. We are not pushing our life-style on our kids nor do we want them to grow up gay. We want them to have good careers and families of their own. We want grandchildren who will visit us in our old age. But, no matter what they choose, we will always love them."

(from the author's files)

The topic of homosexuals as parents is viewed very negatively by many heterosexuals. In a 1996 poll conducted for *Newsweek*, 31 percent of the Americans surveyed said they did not think homosexuals could be as good at parenting as heterosexuals and 47 percent opposed the right of homosexuals to adopt (Kantrowitz

Figure 9–5 Martina Navratilova. This Czechoslovakian-born tennis player was perhaps the greatest female athlete of the 1980s.

et al., 1996). Other surveys find that nearly two-thirds of Americans oppose adoption by homosexuals (Yang, 1997). This negativity has been reflected in court custody cases as well. In a highly publicized case (*Bottoms v. Bottoms*), a Virginia judge denied a woman custody of her 2–year-old son because her lesbian relationship was "immoral" and illegal. The Virginia Supreme Court upheld the decision in 1995, saying that the mother was an unfit parent (see Swisher & Cook, 1996). Many other gays and lesbians have had similar experiences in court. Why the negativity? According to child psychiatrist Betty Ann June Muller, who debated on the subject at the 1994 annual meeting of the American Bar Association, "judges often have cited long-held stereotypes in justifying awarding custody to the heterosexual parent. Those myths include the view that gays and lesbians are mentally ill, have weak parental instincts or are more likely to raise children who will be homosexual or be sexually abused" (Theim, 1994).

What does the scientific literature show? I have already reviewed the literature on mental illness; most homosexuals are as well adjusted as heterosexuals. As for sexual abuse, several studies have found that very few cases of parental sexual abuse are committed by homosexuals; most are committed by heterosexuals (Barrett & Robinson, 1990; Cramer, 1986). What about openly homosexual parents serving as role models for their children's gender identity and gender-role development? Children raised in a household where the parent is openly homosexual do not show gender identity conflicts, do not have problems with emotional adjustment, and have a heterosexual orientation as often as children raised by heterosexual parents (see Allen & Burrell, 1996; Brewaeys & van Hall, 1997;

and Patterson, 1992, for reviews of many studies). (This last finding is perhaps further evidence that biology is more important than environment in determining sexual orientation.) Openly homosexual parents also generally have positive relationships with their children (Turner et al., 1985).

Perhaps when the public becomes educated about these findings, the negativity will subside. Meanwhile, the controversy continues. When the New York City school board proposed a "Children of the Rainbow" curriculum in 1992 that included the young children's books *Daddy's Roommate* (about a boy who spends weekends with his father and his father's male lover) and *Heather Has Two Mommies,* it ignited a public outcry. Should good parenting be based on sexual orientation, or are there other qualities that make a person a good or bad parent?

MEDIA PORTRAYAL OF HOMOSEXUALS

The top male box office star in Hollywood in the late 1950s was Rock Hudson, who was generally cast in leading romantic roles, but who in his personal life was exclusively homosexual. But to admit that one was homosexual in those days was the kiss of death for one's career, so Hudson's studio arranged a marriage (on paper only) so he could keep his orientation secret. Other leading actors of that day, including Cary Grant and James Dean, were also known to have had sex with men, but they were always portrayed (on screen and off screen) as straight (Leland, 1995).

When Hollywood did star a person as being homosexual, he or she was almost always portrayed as a stereotype, still sometimes seen today in movies such as 1996's *The Birdcage.* However, the media and arts have slowly begun to deal with the subject more honestly. In the 1980s, two Broadway plays, *Torch Song Trilogy* and *La Cage aux Folles,* and movies such as *Kiss of the Spider Woman* and *Welcome Home, Bobby* dealt openly and sympathetically with the lives of male homosexuals. This was followed in 1993 by *Philadelphia,* starring Tom Hanks. Lesbianism as a subject for movies or television has met greater resistance, but when actress Ellen DeGeneres's character declared her homosexuality on the April 30, 1997, episode of *Ellen,* the TV show became the first to be built around (and star) a lesbian.

Hollywood has been slower to actually show romantic or sexual behavior between homosexuals, but even that is changing. Two actresses playing the role of attorneys kissed on NBC's *LA Law* in 1991, followed by Roseanne kissing another woman on her show in 1994. Gay and lesbian couples were portrayed on episodes of *Roseanne* and *Friends* in 1996. Things may have been slow to change, but today, "the mass media that tell most of the stories to most of the people most of the time are slowly becoming more inclusive and accepting of diversity" (Gross, 1996).

Box 9-A   CROSS-CULTURAL PERSPECTIVES

Homosexuality in Other Cultures

Homosexuality is viewed very differently in different cultures. Because there has been very little study of female homosexuality in other cultures, we will restrict ourselves here to attitudes about male homosexuality.

In a review of anthropological studies of 294 societies, Gregersen (1982) found that of 59 societies that had a clear opinion of homosexuality, 69 percent approved of it and only 31 percent condemned it. In an earlier study that included 76 societies in which homosexuality was noted, Ford and Beach (1951) found that 64 percent approved and 36 percent did not. Among major Western industrialized countries, only the United States and Britain bar known homosexuals from the military (Walsh, 1992). Canada, Australia, Germany, France, and the Netherlands are all more tolerant. In Russia and China, on the other hand, homosexuality is still a crime.

Recall that in ancient Greece sexual relations between men was highly idealized (Plato's *Symposium*). It was believed that a boy could acquire knowledge by swallowing semen from an intellectual tutor. Homosexual *activity* with an older male during childhood and adolescence is common in at least 50 Melanesian (part of the South Pacific islands) societies. In perhaps the most studied group, the Sambians (Herdt, 1987, 1991), boys are separated from their mothers about the age of 7 and live with males only. They regularly perform oral sex on older (but not yet adult) males; swallowing semen is a practice that the Sambians believe assures maleness (makes them strong, virile, and good warriors and hunters). The boys continue this for 6 to 8 years and then reverse roles, becoming semen donors to younger males. At 16 years of age, a Sambian boy marries, and for the next year or two is given oral sex by his new bride as well as by younger boys. Once his wife gives birth, however, the homo-

sexual behavior stops, and the new Sambian adult male is exclusively heterosexual (in desire as well as behavior) thereafter. Similar practices have been observed in some groups in South America and Africa (Herdt, 1991).

Several important points can be learned from these cross-cultural studies. First, we must distinguish between homosexual practices and a homosexual orientation. For example, although all Sambian males engage in homosexual practices during a specified time of their lives, Herdt (1981) estimates that fewer than 5 percent have a homosexual orientation as adults. In Mexico, Brazil, Turkey, Greece, Morocco and other northern African countries, it is not uncommon for males to engage in oral-genital sex or anal intercourse, but only the passive partner is regarded as homosexual (Carrier, 1980). Thus, what is considered homosexuality is, in part, culturally defined. The fact that fewer than 5 percent of Sambian males develop a homosexual orientation as adults also says much about the role of the environment and biology. If environment were more important than biological factors, you would certainly expect to see a larger proportion of homosexual adults. On the other hand, the fact that homosexual practice is very age-structured demonstrates the role of the culture and the social surroundings in sexual orientation.

The percentage of homosexuals across cultures that have been extensively studied is about the same: from 1 to 7 percent (Diamond, 1993; Whitam, 1983). Homosexuals in different cultures greatly resemble one another in lifestyle as well, further evidence for a biological basis (Whitam & Mathy, 1986). What does vary from culture to culture are attitudes about homosexuality. That is a learned response, and what is learned depends on the time and the place in which a person is raised.

HOMOPHOBIA

"During my junior year in high school I made a new male friend. The more I got to know him, the more I began to suspect that he was

gay. This scared me. As a result of my growing fears I discontinued our friendship."

"When I was in high school I worked at a flower shop. I didn't realize the owner's son

Figure 9–6 Television star Ellen DeGeneres and actress Anne Heche. In 1997 the television show "Ellen" was the first to portray a homosexual in a leading role.

was gay until my junior year. He acted totally normal. . . . One afternoon I saw him driving somewhere with his boyfriend. Ever since that day I am kind of afraid and don't like to even have conversation with him. I don't know why I fear him so much because he has never done any harm to me."

"I don't even want to be near them or share the same things. . . . I feel like I might catch their 'gay' germs. . . ."

(from the author's files)

Many people are prejudiced against others whom they perceive to be different. People discriminate on the basis of skin color, gender, and religion, but hostility toward homosexuals is probably far more accepted among Americans than any other type of bias (Goleman, 1990). Recall, for example, the study that found that 89 percent of young males consider male homosexuality to be disgusting (Marsiglio, 1993b). AIDS, which many people believe (incorrectly) is a homosexual disease, has increased prejudice against homosexuals (Lewes, 1992). The National Gay and Lesbian Task Force report hundreds of antigay-motivated physical assaults every year. Over 90 percent of gay men have been verbally abused or threatened (Herek, 1989). In such a harsh world, is it any wonder that 30 percent of the suicides committed each year by people aged 15 to 24 are by homosexuals (Health and Human Services Department, 1989)?

Homophobia is an irrational fear of homosexuality (some researchers prefer the term "antigay preju-

dice"; Haaga, 1991). It is greater among males than it is among females (see Kite & Whitley, 1996, for a review). Why do some people have such fear of and hostility toward homosexuals? Studies find that among males, homophobic views are greatest in men with stereotypic male gender-role attitudes, a religious fundamentalist attitude, little education, and/or who regard homosexuality as a choice (Herek & Capitanio, 1995; Johnson, Brems & Alford-Keating, 1997; Marsiglio, 1993b; Van de Ven, 1994). However, this tells us nothing of the psychological basis of homophobia.

Some psychologists believe that homophobia stems from "self-righteousness in which homosexuals are perceived as contemptible threats to the moral universe" (Goleman, 1990). However, there is a major problem with this explanation. Heterosexual males generally have much more hostile attitudes toward male homosexuals than they do toward female homosexuals (Herek, 1984). In fact, many sexually explicit male-oriented magazines regularly devote several pages to female-on-female sexual activity, and female homosexual sex scenes are often shown in hard-core porno films made for male heterosexual audiences. Research shows that heterosexual men find lesbianism highly erotic and that this improves their attitudes toward lesbians (Louderback & Whitley, 1997).

Obviously, male homosexuality must be more threatening to many men than female homosexuality. Some of Freud's followers believe that it is the result of remnants of homosexuality in the "normal" (i.e., heterosexual) resolution of the Oedipus complex—identification with one's father involves love for one's father (de Kuyper, 1993).

Most researchers have a slightly different explanation. For most people, adolescence is a confusing time as they try to establish their own identity and sexual orientation. Many heterosexuals have homosexual fantasies on occasion (see Chapter 11), and lack of understanding of this and other aspects of sexuality enhances the confusion. Gregory Herek, the leading authority on homophobia, believes that antigay prejudice is an attempt to suppress any attraction to the same sex (see Herek, 1988). Many homophobic men, in fact, are physically aroused (show penile enlargement) when shown male

homophobia An irrational fear of homosexual individuals and homosexuality.

© 1977 Dayton Daily News and Tribune Media Services, Inc. Courtesy of Grimmy, Inc.

CAN (SHOULD) SEXUAL ORIENTATION BE CHANGED?

Recall that in the not-too-distant past, homosexuality was viewed by the psychiatric profession as a form of psychopathology that required medical treatment. Many homosexuals were involuntarily subjected to "treatments" such as lobotomies, castration, and electroconvulsive shock therapy. Rejection of the psychopathology model of homosexuality led to the abandonment of the idea that all homosexuals should be "cured," whether voluntarily or involuntarily. There are still therapists who attempt to convert homosexuals to heterosexuality, but the only legitimate reason for attempting that today is personal choice (see Murphy, 1992).

homosexual videos (Adams, Wright, & Lohr, 1996). Herek says, "The teens and early 20's [are] a time of identity consolidation, struggling with issues of manhood and masculinity, how one becomes a man. By attacking a gay man or a lesbian, these guys are trying symbolically to affirm their manhood" (Associated Press, July 1993).

Homophobia affects relations among heterosexuals as well. Many Americans were probably shocked at seeing eastern European and Asian male athletes hug and kiss each other in international competitions such as the Olympics. This is normal in these countries, where they have learned to differentiate affection from sex. Here, homophobia is so widespread that it prevents many same-sex family members (most commonly, fathers and sons) and friends from being affectionate with one another (Britton, 1990; Garnets et al., 1990). It prevents many people, especially men, from enjoying intimate relations with others. Are you able to hug your same-sex parent? Good friends of the same sex? If not, examine which fears prevent you from doing so. Homophobia hurts everyone, including those who display it.

NATIONAL SUPPORT GROUPS

You will find the addresses and phone numbers of several organizations in the "Resources" section following the text. Also, be sure to contact local support groups and organizations. Most college campuses, for example, have gay and lesbian alliances.

Nearly half of all Americans still believe that homosexuals choose their sexual orientation (Yang, 1997). However, when homosexuals were asked if they would choose to take a "magic heterosexual pill" to change their sexual orientation, only 14 percent of gays and 5 percent of lesbians said that they would do so (Bell & Weinberg, 1978). Note that these results indicate that the large majority of homosexuals feel comfortable with their sexual orientation and do not wish to change. It is a myth held by heterosexuals that most homosexuals want to change.

For those homosexual individuals who choose to change their orientation, there are therapists who practice a variety of techniques, including psychoanalysis (Bieber, 1965; Marmor, 1980b) and behavioral therapy (Cautela & Kearney, 1986; Masters & Johnson, 1979, see Murphy, 1992, for a review). However, others state that changing an individual's sexual orientation is not possible (Acosta, 1975; Marciano, 1982). This conclusion has gained strength with the recent findings of a possible biological role in sexual orientation.

Although homosexuals no longer have to fear unwanted medical treatment, it is still questionable whether those gays and lesbians who seek reorientation truly do so "voluntarily" (Halleck, 1976; Murphy, 1992). Most probably do so because of years of dealing with family pressure, job discrimination, legal hassles, and other experiences resulting in self-loathing. According to Murphy (1992), "If this is true, it would appear that 'treating' the 'homosexual' would still be a form of blaming the victim, i.e., treating the person who suffers rather than ameliorating those social forces which devalue homoeroticism."

In 1994, the American Medical Association finally reversed a long-standing policy of recommending that physicians make efforts to turn unhappy homosexuals into heterosexuals. The new policy states that most of the emotional problems of homosexuals are "due more to a sense of alienation in an unaccepting environment" and that "through psychotherapy, gay men and lesbians can become comfortable with their sexual orientation and understand the social responses to it" (AMA, 1994).

Key Terms

bisexual 224, 227	gender roles 209, 227	lesbian 233
coming out 234	heterosexual 224, 227	Oedipus complex 206, 229
gay 233	homophobia 243	sexual orientation 227
gender identity 198, 227	homosexual 224, 227	straight 233

Personal Reflections

1. Are you comfortable with your sexual orientation? If not, why not? What steps can you take to become more comfortable?

2. Homosexuals and bisexuals: Have you told your family and friends of your sexual orientation? Why or why not?

3. How would you react if someone at work told you that he or she was homosexual? What if your best friend told you that he or she was homosexual? Your brother or sister? Your teenage son or daughter? Why?

4. Heterosexuals: Do you ever judge others exclusively, or almost exclusively, by their sexual orientation? How do you suppose you would feel if others ignored your accomplishments in school, in your career, in your sports and hobbies, and in your community and judged you exclusively on some aspect of your sex life?

Suggested Readings

Bailey, J. M., & Pillard, R. C. (1995). Genetics of human sexual orientation. In R. C. Rosen, C. M. Davis, & H. J. Ruppel, Jr. (Eds.), *Annual Review of Sex Research,* vol. VI. Mount Vernon, IA: Society for the Scientific Study of Sexuality.

Bendet, P. (1986, August/September). Hostile eyes. *Campus Voice.* Discusses homophobia on college campuses.

Byne, W. (1994, May). The biological evidence challenged. *Scientific American.* A rebuttal of the popular theory that sexual orientation is influenced by biology.

Clark, D. H. (1987). *The new Loving someone gay.* Berkeley, CA: Celestial Arts. Written for parents of gays.

De Cecco, J. (Ed.). (1988). *Gay relationships.* New York: Haworth Press. Essays on many aspects of gay relations.

Gladue, B. A. (1994, October). The biopsychology of sexual orientation. *Current Directions in Psychological Science,* 3. An excellent review of the biology of sexual orientation.

Helminiak, D. A. (1994). *What the Bible really says about homosexuality.* San Francisco: Alamo Square Press. Man people are convinced that God condemned homosexuality, but this scholarly work shows that He really did not.

Kantrowitz, B. (1996, November 4). Gay families come out. *Newsweek.* Discusses the controversy about homosexuals as parents.

Leland, J. (1995, July 17). Bisexuality. *Newsweek.* Many personal experiences.

Leland, J., Rosenberg, D., & Miller, M. (1994, February 14). Homophobia. *Newsweek.*

Loulan, J. (1984). *Lesbian sex.* San Francisco: Spinsters Ink. Examines the lesbian life-style and sexual relationships.

Plummer, K. (Ed.) (1992). *Modern homosexualities: Fragments of lesbian and gay experience.* London: Routledge. An excellent book that looks at the issue from many perspectives.

Remafedi, G. (1989, July). The healthy sexual development of gay and lesbian adolescents. *SIECUS Report.* Discusses characteristics of healthy homosexual development.

Toufexis, A. (1992, August 17). Bisexuality: What is it? *Time.* A short introduction to a very misunderstood subject.

CHAPTER 10

When you have finished studying this chapter, you should be able to:

1. Explain infants' exploration of their bodies and their need for close physical comfort;

2. Discuss masturbation and sex play in young children;

3. Discuss the development of modesty and inhibitions during the early school-age years;

4. Summarize the changes that occur in boys' and girls' bodies at puberty and identify the hormones that are responsible;

5. Discuss sexual behavior among adolescents and the effects of peer pressure;

6. Explain how the sexual behavior of young adults often differs from that of younger and older age groups;

7. Discuss the sexual life-styles of adults, including marriage, cohabitation, and extramarital sex;

8. Describe the characteristics that men and women look for when choosing a short-term or long-term mate;

9. Describe the sexual behavior of middle-aged individuals and the physical changes that come with aging; and

10. Explain the realities of aging that can directly affect sexuality in the elderly, and discuss the prevalence of sexual relations among healthy elderly couples.

Life-Span Sexual Development

People are sexual beings. To deny our sexuality is to deny our humanity. It should come as no surprise, therefore, that our sexuality begins from the moment of birth (and perhaps before) and lasts until our death, even if we live a very long life.

EARLY INFANCY (AGES 0–1)

Sexuality starts in the womb. Ultrasound recordings have discovered that male fetuses have erections months before they are born (Calderone, 1983b). After birth, male babies often have erections before the umbilical cord is cut. Similarly, female babies can have vaginal lubrication in the first 24 hours after birth (Langfeldt, 1981; Masters, 1980).

An important part of sexual development involves the amount of hugging and cuddling that an infant has with its caregivers. Many physicians and psychologists strongly encourage parents to hold their babies immediately after birth and to continue to give large amounts of hugging and cuddling throughout childhood. Some psychologists and physicians believe that lots of hugging and cuddling between parent and infant allows a strong attachment to be formed between them, which is a process called "bonding." Monkeys raised in isolation and deprived of the opportunity of close physical comfort have difficulty forming relationships later. The same seems to be true for children (Harlow & Harlow, 1962; Money, 1980). Fifty years ago, a physician reported that one in three human infants in foundling homes died during the first year of life, even though they were well fed and provided with adequate medical care. What these babies were not receiving was cuddling, hugging, fondling, and kissing—they apparently died from emotional neglect (Spitz, 1945, 1946). Today, some hospitals are recruiting volunteers to hold and talk to babies (especially premature ones) so that they will not be deprived of early physical contact if their parents cannot provide it. When nursing, infant boys may get an erection and infant girls may display vaginal lubrication. These are normal reflexive reactions, and parents should not be alarmed when they happen.

As soon as infants gain sufficient control over their movements, they begin to touch all parts of their bodies. As part of this exploration, infants sometimes touch their genitals. They can feel pleasure from this and may continue to stimulate themselves. This happens because the nerve endings in the genitals are already developed at birth and transmit sensations of pleasure when stimulated. However, it is important to remember that infants do not comprehend adult sexual behavior. Their behavior is aimed at finding pleasurable physical sensations, not expressing sexual desire. Although the initial sexual self-stimulation of children may be random, once they discover its pleasurable component, such stimulation quickly becomes purposeful (Lidster & Horsburgh, 1994). Taking a child's hand away from his or her genitals may cause the child to make a face or noises (indicating irritation), and the child may quickly try to resume self-stimulation (Bakwin, 1974). Kinsey and his colleagues reported that 32 percent of boys less than 1 year of age were able to have orgasms (though ejaculation during orgasm is not possible until puberty is reached). He also reported a number of cases of girls 3 years old or younger masturbating to orgasm (Kinsey et al., 1948, 1953).

EARLY CHILDHOOD (AGES 2–6)

Before age 2, children are highly *egocentric,* a term coined by psychologist Jean Piaget in his theory of cognitive development (the growth of knowledge in

human beings). This means that they cannot consider other people's points of view. Egocentric children do not play games together, because most play requires cooperation, and cooperation in turn requires respecting or anticipating someone else's point of view. Bodily exploration during this time is confined mainly to self-exploration (Figure 10–1). After the age of 2, however, children increasingly play together, and their natural curiosity now extends not only to their own bodies, but to those of others.

Studies have shown that sexualized behaviors increase dramatically in children after age 2 and reach a peak in the 3-through-5-year age period, and then decrease until puberty (Friedrich et al., 1991; Rutter, 1971). Interest in the genitals is very common for boys and girls in this age group, as are undressing and sexual exploration games that involve showing one's body and genitals to others, or vice versa. Let us examine some of the common, normal behaviors frequently observed in young children.

As just mentioned, by the age of 2, children are very interested in their bodies. Because of this, they will try to watch their siblings and parents bathing and urinating.

"When I was a very young girl, my parents brought me with them to visit some friends of theirs who had two young boys. One boy was my age and one was a couple of years older. I went to the bathroom with them and watched them urinate. My mother got very, very angry. This totally surprised, shocked, and confused me. Maybe if she had had a course like this one, she could have handled the situation a little better."

(from the author's files)

The fact that boys stand and girls sit while urinating is fascinating to children. They will also take their clothes off together and will play games that allow for sexual exploration, such as "playing doctor" and "playing house." Most of the time, these games are limited to viewing and touching genitals. More aggressive sexual behaviors like oral-genital contact and inserting objects into the vagina or anus are uncommon (Friedrich et al., 1991) and may be an indication of abuse. Sexual exploration games will not harm a child's development. What may be harmful is if parents react too strongly when they "catch" their children engaging in sex play.

"I can still remember the day my mother caught me playing 'doctor' with my neighbor. I was five years old. She got so mad at me that she locked me up in the bathroom. I don't remember how long I was in there. I didn't understand what I had done wrong. I ended up crying myself to sleep on the bathroom floor."

"When I was 4, my cousin (age 5) and sister (age 3) had a slumber party. When we were all laying down we started asking questions and being curious. We then started 'playing doctor.' Well, my parents caught us and I was spanked and punished for a long time. I have very vivid memories of this event and now I do not have very positive feelings about sex or my sexuality."

"When I was about 4 or 5 years old I got caught playing doctor with a little boy and girl. My mom was very angry and she put me on my knees with no pants or underwear. She made me stay there with no clothes on until my dad came home, so that he could see me like this. I believe that this really has affected me in a negative way towards sex even now."

(from the author's files)

Sexual exploration games are often played with playmates of the same sex. This is probably more common than sexual contact between sexes, because even at this age children usually play more with playmates of the same sex. Engaging in same-sex contact at this age does not mean that a child will become homosexual as an adult. Adult homosexuals may or may not have been involved in same-sex play in early childhood. The same is true of heterosexual adults.

Figure 10–1 It is normal for children to explore their own bodies.

Recall that Freud said the genitals were the focus of children's pleasure during the *phallic stage* (approximately ages 3–7) of psychosexual development (see Chapter 8). It is true that nearly all young children touch and explore their genitals, although young boys engage in masturbatory behavior more often than young girls (probably because the male genitals are more visible; Friedrich et al., 1991). Here, too, the way parents respond is very important.

"My sister has a five-year-old little boy. . . . She was all worried that something was wrong with him. He would constantly play with his penis. She scolded him one day and told him that if little boys play with themselves like that it would fall off. . . . I am afraid that this could scar him for life. . . ."

"I was raised in a strict Catholic family. I remember distinctly how my initially unhealthy attitude towards my sexuality and that of others began. As a small child, I used to masturbate. My mother would hit the ceiling when she would catch me because my favorite time to do this was at nap time. She would yell and scream and slap my hands and tell me that it was nasty. Her contempt for my actions was so obvious that I could not help but develop a negative attitude towards my sexuality. I worried for a long time that I was the only person who did this and felt very guilty about it. . . ."

(from the author's files)

Most parents respond negatively to finding their young child masturbating (Gagnon, 1985). However, when you become a parent, try not to overreact. Allowing children to explore and satisfy their curiosity enables them to become comfortable with their bodies, both as a child and as an adult (Renshaw, 1988). Punishment or negative messages may lead to a poor body image and later sexual problems (Masters & Johnson, 1970; Money, 1980). It is okay to teach your children that it is not appropriate (as opposed to "bad") to touch their genitals in certain situations; in public, for example. According to one child expert, "The attitude of the parent should be to socialize for privacy rather than to punish or forbid" (Calderone, 1983a). If your child touches herself or himself at home, the best thing to do is ignore it. If you feel that you have to stop it because it bothers you, do not scold; simply distract your child by getting her (or him) involved in another activity.

By now, you should understand that parents' responses to normal behaviors have very important consequences for children's later sexual development and

Figure 10–2 Children will attempt to satisfy their natural curiosity about the human body whenever they get a chance. If this were your child, would you pull her away, or would you let her explore and try to answer any questions she might have about the difference between boys and girls?

their attitudes about their bodies. According to one group of researchers, "From the earliest period in life, the child's family will reciprocally influence the sexual characteristics of the child. The family establishes a 'psychosexual equilibrium' that is a function of the parents' sexual adjustment, the child's developing sexuality, and the impact of the child's sexual development on parental sexual development. . . ." (Friedrich et al., 1991). The way parents react to the sexuality of their children is often an indication of the way they feel about their own sexuality.

It is important that parents be comfortable with their own sexuality, because parents are the major source of sexual information for young children. By age 4, for example, children often have asked parents where babies come from. At this point, parents have a decision to make. Children will either learn about sex from parents or from other people. Generally, those "other people" will be other children who may not

share your values and who may give bad advice or misinformation. Moreover, schools, even those with sex education courses, may not always deal with values. Sex education is largely absent in schools before students reach puberty (except for lessons like "Don't get into a car with strangers" or "Don't let strangers touch you in your private places"). Chapter 17 describes *how to talk to children of different ages about sexuality*—both your sexuality and theirs. The ability to communicate with your children about sex and other topics is a critical part of good parenting.

THE INITIAL SCHOOL-AGE YEARS (AGES 7–11)

Freud believed that this time of life was a period of "latency" when children were not concerned with sexuality. Studies have found that children in fact do engage in less overt sexual behavior at this time (Friedrich et al., 1991; Money & Ehrhardt, 1972). This is probably due to learning cultural standards. By the time they begin kindergarten or first grade, children generally have developed a sense of modesty and inhibition about exposing their bodies in public (Money & Ehrhardt, 1972). They may be very shy about undressing in front of their parents, and for the first time may start demanding privacy in the bathroom.

Some researchers believe that the amount of sex play does not slow down at this stage but is simply hidden more from parents (Goldman & Goldman, 1982; Kinsey, Pomeroy, & Martin, 1948; Kinsey et al., 1953). Children tend to segregate by sex by age 9. Thus, much of their sexual play is with same-sex children (Elias & Gebhard, 1969). This is a normal part of growing up, and parents should not interpret it as homosexual activity as they would activity between two adults.

When caught engaging in sexual exploration games, girls are often treated more harshly than boys. This establishes a double standard—as a general rule, boys are allowed more freedom to explore their sexuality than girls. Consider the following: Suppose the principal of your child's school called and said that your son had been caught telling a female classmate, "You can touch my thing, if I can touch yours." How would you react? Would you react differently if your daughter had said this to a boy? Why?

Although the amount of overtly sexual play may decrease during the initial school-age years, curiosity about sex does not, and children will ask their parents questions such as "Where do babies come from?" Bernstein (1976) claimed that most children in the United States had learned about sexual intercourse and its connection to pregnancy by the age of 8 or 9. However, many continue to have rather strange ideas about

Figure 10–3 When they enter kindergarten or first grade, children are still curious about sexuality, but are also developing a sense of modesty and inhibition about exposing their bodies in public.

Figure 10–4 Boys and girls often play together during early childhood, but tend to segregate by sex by age 9.

sex and birth. Goldman and Goldman (1982) found that at least 50 percent of children 11 years of age or younger in their North American sample believed that children were born through the anus. Many children believed that sex was primarily carried out by insertion of the penis into the anus. Again, how parents respond will affect their children's responses (and willingness to ask questions of their parents about sex) in the future. Unfortunately, many parents are evasive or even negative.

> "My mother damaged my attitude about sexuality by responding to my question about the origins of babies by telling me that when a man and a woman kiss, it's called making love and this is where babies come from. Even today, she still refuses to discuss sex or sexuality. . . ."

(from the author's files)

puberty The time in life when an individual first shows sexual attraction and becomes capable of reproduction.

PUBERTY (AGES 7–15)

Puberty (from the Latin *puber,* meaning "of ripe age, adult") is the time of life when we first show sexual attraction and become capable of reproduction. It does not happen overnight, but instead is *a process lasting several years.* In fact, recent studies have found that puberty begins earlier and lasts longer than previously believed.

Traditionally, the physical and sexual changes that occur during puberty have been attributed to the maturation of the testicles or ovaries. However, puberty is at least a two-part maturational process (see McClintock & Herdt, 1996). In the first part, the adrenal glands start to mature (called *adrenarche*) when children are between the age of 6 and 8. The adrenal glands secrete the androgen hormone DHEA (dehydroepiandrosterone), which is then converted to testosterone and estrogen. Thus, the first increase in sex hormones is not due to the maturing gonads, but to the maturing adrenals. Notice that at this stage girls as well as boys experience an increase in androgens ("male" hormones).

In the second stage, the testicles and ovaries mature (called *gonadarche*), usually several years after adrenarche. In this stage, the pituitary gland begins to secrete FSH in high doses, stimulating the production of sperm in seminiferous tubules in boys and the maturation of ova in girls (see Chapter 3). Prior to this time, boys do not ejaculate (though they can have orgasms) and girls do not have menstrual cycles. In girls, the maturing ovary produces estrogen and progesterone, while increased levels of luteinizing hormone from a boy's pituitary stimulates production of testosterone in the testes. The increased levels of hormones lead to a variety of physiological changes in girls and boys (e.g., breast development, facial hair).

These are often referred to as **secondary sex characteristics.** The changes in physical appearance create psychological changes in the way children think about themselves and others. Let us examine these changes in more detail and discuss how they affect adolescents' behavior.

CHANGES IN GIRLS

One of the first noticeable changes in girls is a growth spurt starting at about the age of 12. Many sixth- and seventh-grade boys suddenly realize one day that a lot of the girls in their classroom are taller than they are (a fact they may not discover until those first school dances). The reason for this is that the growth spurt in boys does not start, on average, until about age 14. Most of the boys eventually catch up with and pass the girls, because girls generally stop growing by age 16 and boys keep growing until about 18; but those first couple of years can be awkward.

The development of breasts in girls during puberty (a result of increased levels of estrogen) is viewed with great attention by both girls and boys. Many girls rush out to purchase training bras, which probably do nothing except make them feel mature. Girls worry that if they are not as "stacked" as their friends, they will be left behind in the rush to be popular with boys. Not all girls develop at the same rate, so insecurities may develop early. Some decide to speed things up on their own, as eleven-year-old Margaret did in Judy Blume's novel *Are You There God? It's Me, Margaret* (Copyright © 1970 by Judy Blume. Reprinted with permission of Simon & Schuster Books for Young Readers.):

I tiptoed back to my room and closed the door. I stepped into my closet and stood in one corner. I shoved three cotton balls into each side of my bra. Well, so what if it was cheating! Probably other girls did it too. I'd look a lot better, wouldn't I? So why not!

I came out of the closet and got back up on my chair. This time when I turned sideways I looked like I'd grown. I liked it!

The increase in estrogen levels also causes an increase in fatty deposits in the hips and buttocks.

Pubic hair generally starts to appear shortly after breast development begins, followed in a couple of years by the appearance of axillary (underarm) hair. The growth of hair on new parts of the body results from increased levels of male hormones (mainly from the adrenal gland). These hormones also cause the sweat glands and sebaceous glands to develop, so that body odor and acne often become a new source of concern.

Puberty may be a process that takes several years, but in many people's minds, a girl's first menstruation marks a turning point in her maturity. The vagina and uterus begin to enlarge about the time that pubic hair appears, and the first menstrual period (called **menarche**) occurs at an average age of about 12.5 for American girls. There is considerable variability, however, and it is normal for the cycles to be very irregular during the first couple of years after menarche. The rise in estrogen levels causes the vaginal walls to become thicker and more elastic and also results in lubrication during sexual arousal.

The average age for menarche has been dropping over the last few centuries. A study of New York City females in 1934 found that the average age was 13.5 years, while the average age in Western Europe in the early 1800s was at least 14.5 years (Bullough, 1981) and perhaps as high as 17 in some places (Tanner, 1962). Many researchers believe that the reason for this decline is better nutrition resulting in an earlier acquisition of some minimally required amount of body fat (the putative signal that tells the brain to start releasing FSH). Women suffering from anorexia nervosa (an eating disorder in which a person has an aversion to food) never menstruate, and very irregular cycles are common in female long-distance joggers who have lower-than-normal levels of body fat.

The reaction of a young woman to her first menstrual cycle will be influenced primarily by what she has been told. Menarche can be viewed as something very positive (a sign of "becoming a woman"), "a simple fact of nature," or "a curse" and something disgusting. Mothers generally discuss menstruation with their daughters and pass on their own attitudes and beliefs about it.

"My mother told me that menstruation was a 'nasty' experience. As a result, I always did my utmost to totally hide

secondary sex characteristics Bodily changes that occur during puberty and differentiate males and females.

menarche The term for a female's first menstrual period.

the fact that I was menstruating. I didn't even want other women to know that I was having my period. I would refuse to purchase sanitary products out of the machines in restrooms if anyone else was in the restroom. . . ."

". . . Then, when my big moment came, instead of treating me as though something natural and exciting were happening, my mother reacted with embarrassment. I was very hurt when she told me to get my sister to show me how to wear a sanitary napkin. We never talked about any of these things. The sanitary napkins were even kept locked. . . . and whenever any of us needed one we would slip it under our shirt and scurry into the bathroom. Menstrual periods were regarded as a shameful secret in our home."

"My mother told me when I was very young that I would have a period around age 13 or so. . . . Finally, one month before my 15th birthday, I started my period. I remember being so excited 'it' finally came. I called my mom in the bathroom and told her. Well, she was so excited she started calling my grandmothers and her friends. When my dad came home he got all excited too. They were so proud I'd finally 'become a woman.'"

(from the author's files)

Which of these girls do you suppose began adolescence with good self-esteem about her body?

Older sisters, friends, sex education courses in schools, and fathers may also give information about menstruation. As you can see from the previous case histories, for some parents the subject is still a taboo, and some girls are never told anything about menstruation until it happens. This is especially true among African-American girls (Scott et al., 1989). In a study done by the Tampax Corporation in 1981, it was found that one-fourth of all people surveyed felt that menstruation was an unacceptable topic for discussion even in the home. Thus, some girls experience menarche with no prior knowledge, which makes it a frightening experience.

For boys, menarche is rarely a topic of discussion with their parents. Most learn about it "on the street," and it seems to be another mysterious difference between men and women. Menarche often is viewed as the topic of crude jokes, though most boys remain ignorant about the basic facts regarding menstruation until they reach adulthood (if then).

nocturnal emission An ejaculation that occurs during sleep in males; a "wet dream."

Figure 10–5   Puberty and the development of secondary sex characteristics mark dramatic increases in interest in sexuality.

CHANGES IN BOYS

As mentioned earlier, pubertal development in boys lags about two years behind development in girls. The first noticeable change in boys is usually growth of the testes and scrotum, the result of increased levels of testosterone. Testosterone then stimulates growth of the penis, prostate gland, and seminal vesicles. The growth of the genitals begins, on average, about the age of 11½ and is completed, on average, by about the age of 15. Boys generally become capable of ejaculation about a year after the penis begins to grow.

The first experience many males have with ejaculation is a **nocturnal emission,** or "wet dream." Some boys have nocturnal emissions frequently, while others have only a few experiences. Fewer fathers prepare their sons for nocturnal emissions than mothers prepare girls for menstruation, and as a result many boys are ashamed, or even frightened, by these experiences.

"When I was 13 years old, I had a 'wet dream' for the first time. I was shocked to later find out that I did not have a disease. I honestly thought I had one."

(from the author's files)

Because they are ashamed, some boys may hide their underwear and sheets or try other ways to "hide the evidence." Because nocturnal emissions are not under voluntary control, there is no reason for shame. It is therefore important to educate children about nocturnal emissions (and for other aspects of puberty) *before* they begin to occur.

Many boys may also be initially frightened by the emission of a strange fluid from the penis that now occurs during masturbation. A case cited by sex therapist Bernie Zilbergeld (1978) is probably not unusual:

© Lynn Johnston Productions Inc. Reproduced with permission. Distributed by United Feature Syndicate, Inc.

".... I kept stroking, my penis got hard, and the sensations felt better and better. Then I was overcome with feelings I had never before felt and, God help me, white stuff came spurting out the end of my cock. I wasn't sure if I had sprung a leak or what. I was afraid but calmed down when I thought that since it wasn't red it couldn't be blood. I kept on stroking and it hurt. I didn't know if the hurt was connected with the white stuff (had I really injured myself?) or if the event was over and my penis needed a rest. But I decided to stop for the moment. Of course, I returned the next day and did it again...."

Many boys also develop temporarily enlarged breasts during puberty, called **gynecomastia.** This results from increased levels of estrogen (Mathur & Braunstein, 1997).

"When I was about 13-years-old something started to happen to my chest. It seemed as though I was beginning to grow breasts. I went to my mother and showed her so she brought me to the doctor. I was a wrestler in middle school and all I could think about was how could I avoid taking off my shirt in public for one to two years...."

(from the author's files)

Gynecomastia usually disappears by the mid-teens, but unless boys are told why this is happening and that it is a normal, temporary condition, they may feel confused, embarrassed, or ashamed. Some fear that they are turning into girls and may stop swimming in public because they are embarrassed to display their male "breasts."

Pubic hair starts to grow about the same time as the genitals start to develop, the result of increasing testosterone levels, but underarm and facial hair generally do not appear for another 2 years. The amount of body hair, however, is also determined by heredity. As in girls, testosterone also causes development of the sebaceous and sweat glands with all their accompanying problems (acne, body odor).

The appearance of facial hair is often as important to boys as the development of breasts is to young girls, for it is one of the few outward signs that they are becoming men. They often start shaving long before it is necessary, only to wish a few years later that they could avoid it altogether. Another change that is obvious to others is a deepening of the voice, a result of testosterone stimulating the growth of the larynx (voice box). This occurs today at an average age of about 13½, but it may have occurred at a later age in past centuries. It was not uncommon in Europe at one time for boys who sang in the great church choirs to be castrated before puberty (with parental approval) in order to preserve their soprano voices. (If castration is done after puberty, by the way, it does not substantially raise a male's voice because the larynx has already undergone its change.)

PRECOCIOUS AND DELAYED PUBERTY

Although the changes that occur during puberty generally begin about the age of 11 or 12, they have been known to occur much earlier. When sexual development begins before the age of nine, it is called **precocious puberty.** The youngest girl known to have given birth was only 5 years old: a Peruvian Indian girl who delivered a baby by cesarean section in 1939. This means, of course, that she was ovulating and having menstrual periods by the age of 4 or 5. The youngest couple on record who became parents were a Chinese boy and girl, aged 9 and 8, respectively.

There have been isolated outbreaks around the world of very young children developing breasts. This has happened to boys and girls as young as 6 months old. A recent study of thousands of American children found that nearly half of the black girls and 15 percent of the white girls studied had begun to develop breasts, pubic hair, or both by age 8 (Herman-Giddens et al., 1997).

gynecomastia Excessive development of the male breasts.

precocious puberty A condition in which puberty begins before the age of 9.

First menstruation still occurred after they had turned 12. It is speculated that the early development of secondary sex characteristics is due to what are called enviromental estrogens, chemical pollutants that resemble the female hormone estrogen.

In some children, the appearance of secondary sex characteristics and physical growth do not begin at the same age as in most children. This is called **delayed puberty.** Most clinical referrals are boys who have failed to grow (Kulin, 1996). The usual treatment is to administer gonadotropin-releasing hormone or androgens (male hormones).

SEXUAL ATTRACTION

Until recently it was believed that children did not experience sexual attraction to others until the testicles or ovaries matured. Several studies have found, however, that children's first sexual attraction occurs at age 10, about the fourth or fifth grade (Hamer et al., 1993; Herdt & Boxer, 1993; Pattatucci & Hamer, 1995). This is well before gonadarche and is true for both heterosexual and homosexual attraction. It coincides instead with rising androgen levels due to the maturation of the adrenal glands (McClintock & Herdt, 1996). This is further evidence that Freud's conception of a latency period between the ages of 6 and 11 is incorrect. In the developmental process, sexual attraction is followed in order by sexual fantasy and sexual behavior. By age 13, the sexual exploration games commonly played by young teens (such as "spin the bottle" and "post office") have a greater erotic content than the games of early childhood.

ADOLESCENCE (AGES 13–18)

Adolescence (from the Latin *adolescens*) refers to the time of life between puberty and adulthood. The term was first used in the English academic literature in 1904 by psychologist G. Stanley Hall. In our culture, this is a transition period before adulthood, and a rather extended one at that, but in many nonindustrial cultures individuals are considered to be adults when they reach puberty.

For most adolescent boys and girls, the most important issue in their lives is self-identity (Erikson, 1968). Because of their rapidly changing bodies, the search for self-identity first focuses on body image and physical characteristics. According to Maddock et al. (1983), "There is an intensification of body awareness . . . based upon the fact that the body is a primary 'symbol of self' in which feelings of personal worth, security, and competence are rooted." An adolescent's self-esteem is generally based on his or her physical attractiveness. Broad-based self-esteem derived from one's accomplishments develops later. Thus, in the adolescent world there is usually a strong relationship between one's physical attractiveness and one's social acceptance (Kleck, Richardson, & Ronald, 1974). For those adolescents who fit the football star or prom queen image, adolescence is often a positive experience. For the rest of us, it is a difficult time, and more and more so if you fall farther and farther away from what is considered attractive. We will return to this issue shortly.

MASTURBATION

For most (but certainly not all) people, the first experience with orgasm occurs during masturbation (for others it occurs during mutual touching, petting, sexual intercourse, or dreams). The first data on masturbation during adolescence came from Kinsey's group (1948, 1953). He found that by age 15, 82 percent of boys and 20 percent of girls had masturbated to orgasm. By the age 20, these figures had risen to 92 percent for males and 33 percent for females. The frequency of masturbation by age 15 was 1.8 times per week on the average for boys and 0.5 times per week on the average for girls. Seventeen percent of the boys interviewed had masturbated from four to seven times per week. Haas (1979) interviewed individuals aged 16 to 19 and found that more than two-thirds of the males and one-half of the females masturbated to orgasm at least one time per week or more.

What do these numbers tell us? For one thing, most adults have masturbated and started doing so during adolescence. This is more true for men than women, but there seems to be a trend indicating that more women also are masturbating to orgasm during adolescence. Kolodny (1980) interviewed women aged 18 to 30 and found that 75 percent reported masturbating during adolescence. Nevertheless, males still have more experience with orgasm by the time they become adults, as a result of their higher frequency of masturbation. There may be cultural differences as well. Some studies indicate, for example, that masturbation is more common among white adolescents than among African-American or Latino teens (Belcastro, 1985; Cortese, 1989).

Santrock (1984) found that only 15 percent of adolescents believe that masturbation is wrong. However, this does not prevent many adolescents from experiencing guilt and other negative feelings about masturbation. Laumann and colleagues (1994) found that about half of men and women who masturbated felt guilty about it.

Many sex therapists believe that masturbation serves important functions in adolescence. Because

---

delayed puberty A condition in which the appearance of secondary sex characteristics and physical growth do not begin until well after they have begun in most children.

adolescence The time of life between puberty and adulthood.

sexual activity among adolescents is not condoned in our society, masturbation can serve as an outlet for sexual tensions. Masturbation also can serve as a way to sexually experiment, gain sexual self-confidence, and control sexual impulses (Barbach, 1980). Kinsey thought that masturbation to orgasm was especially important for girls. He felt that familiarity with orgasm during masturbation made it easier for them to experience orgasm during sexual intercourse as adults. Sexual intercourse does not always automatically result in orgasm for women, and therefore a lack of awareness of their bodily or physical responses to erotic stimulation (or the fear of such responses) may further impede the ability to experience orgasm during intercourse. We will discuss this in Chapter 13.

John Gagnon (1977), a sociologist and psychologist who studies communication between generations, notes that masturbation is the single most powerful predictor of adult sexuality. People who masturbated during adolescence generally engage in sexual activity more frequently and have more positive attitudes about sex than people who did not. However, there are few programs or parents attempting to communicate anything positive about masturbation or other forms of sexual expression. Most communication about sex is negative ("Don't do this"; "Don't let this happen to you"). Gagnon's point is that parents should at least stop attaching guilt to the practice of masturbation in their children (an attitude that reflects the negative Victorian attitudes toward sex described in Chapter 1). If nothing else, it is surely the "safest" sexual outlet available. However, attitudes are hard to change; former U.S. Surgeon General Joycelyn Elders was fired for expressing similar views about masturbation. Masturbation as a normal adult behavior will be discussed in Chapter 11.

SEXUAL FANTASIES

Sexual fantasies in adolescence become very focused and explicit, compared to fantasies in childhood. Male fantasies are often focused on specific sexual activities (e.g., an orgy, having sex with a movie star or cheerleader), while sexual fantasies of females are more likely to involve romance and be socially oriented. Fantasies often accompany masturbation (Haas, 1979). Sexual fantasies can serve many useful purposes, including substituting for unavailable experiences, allowing mental rehearsal of sexual techniques, inducing arousal, and enhancing pleasure. Because many adults use mental imagery to enhance sexual pleasure, fantasizing in adolescence can be a useful prelude to an adult use of sexual imagery (see Chapter 11). Sexual fantasy is only part of the increased use of all types of fantasy in adolescence. By fantasizing, teenagers can experience things that are denied to them in real life. Sex is only one such experience.

PETTING

Kinsey's team of researchers defined *petting* as physical contact between males and females attempting to produce erotic stimulation without sexual intercourse. Such a definition would include a variety of behaviors, such as kissing, deep (French) kissing, stimulation of the female's breasts, touching of the genitals, and oral-genital sex. Many sex researchers today define *petting* as noncoital sexual contact below the waist. Any other physical contact is called *necking*. Kinsey et al. (1953) found that by age 18, over 80 percent of the boys and girls in their sample had engaged in petting. Twenty-one percent of the boys and 15 percent of the girls had petted to orgasm by this age. Typically, adolescents' sexual experiences proceed through the following sequence: holding hands, embracing, necking, feeling breasts (through clothing and then directly), feeling genitals (through clothing first, usually by the male first), and, for many, sexual intercourse (Miller, Christopherson, & King, 1993; Smith & Udry, 1985).

Kolodny (1980) interviewed first-year college students about their sexual experiences in high school and found that 82 percent had engaged in genital touching. Fifty percent of the men and 40 percent of the women had petted to orgasm during adolescence. A recent study found that about half of high school students were virgins, but of these, nearly a third had engaged in masturbation with a partner in the past year and about 10 percent had engaged in oral-genital sex (Schuster, Bell, & Kanouse, 1996). Interestingly, approval of these behaviors by teenagers is greater than the actual number of adolescents engaging in them (Haas, 1979).

SEXUAL INTERCOURSE

The number of unmarried adolescents engaging in sexual intercourse has risen in recent years, and the average age at which they first have intercourse has been steadily declining. Kinsey's group, in 1953, found that only 3 percent of 15–year-old girls had engaged in sexual intercourse. Twenty years later, Jessor and Jessor (1975) found that 55 percent of high school senior girls reported that they had engaged in intercourse. According to the recent survey by Laumann and colleagues (1994), over one-half of American teenagers have engaged in sexual intercourse before their seventeenth birthday (by age 15 for African-American males).

Yearly surveys conducted by the Centers for Disease Control and Prevention and others have confirmed that for both boys and girls, most have had sexual intercourse by age 17, and nearly three-fourths by the twelfth grade (CDC, 1992; see Seidman & Rieder, 1994). However, more recent surveys by the CDC (1995) show that the number of sexually active teens is leveling off, and other studies show that the level of sexual activity is relatively moderate—for males, an average of 1.9 partners in the last year, with an average frequency of intercourse of 2.7 times in the last month (Sonenstein, Pleck, & Ku, 1991).

Teenage boys and girls generally have different reasons for engaging in sex. For teenage girls, love and a committed relationship that is supposed to last are important before engaging in sex, but these things are not important for most teenage males, for whom it is more important that they "never miss an opportunity" (Taris & Semin, 1997).

How do teenagers feel about their first experiences of sexual intercourse? Weis (1983) found that women experience a range of emotions during their first experience of sexual intercourse.

Figure 10–6   The biological changes that occur in adolescence lead to an interest in, and the development of, sexual behavior and sexual identity.

About one-third have strongly negative emotional reactions (e.g., guilt, anxiety, shame, fear, regret), while another third have highly positive emotional experiences. For the final third, their experience is neither positive nor negative, but a "Is that all there is?" response (e.g., disappointment, boredom). Interestingly, more American women have had negative emotional reactions to their first experience of sexual intercourse than women in Sweden, a less sexually repressive culture (Schwartz, 1993). Men generally experience more pleasure and less guilt than women during first intercourse, perhaps because they are more likely to have an orgasm (Sprecher, Barbee, & Schwartz, 1995).

Another study found that about one-third of the women surveyed had experienced severe physical pain at first intercourse, while 28 percent said they had experienced no pain (Weis, 1985). Women with positive experiences and minimal pain generally had an extensive childhood and adolescent history of sex play (kissing, petting, masturbation, and often oral sex), had positive attitudes about their sex play, and had their first intercourse in a safe environment (no fear of discovery) with a caring partner. Those who had experienced negative reactions and pain had little previous experience with sex play, had not planned it, and "gave in" to the male's pressure (see also Thompson, 1990).

Women who have feelings of guilt after first intercourse are more likely than others to be sexually dissatisfied (e.g., have less enjoyment) as adults (Moore & Davidson,

1997). In contrast, individuals who had positive feelings about their early sexual experiences have greater enjoyment of sex as adults (Bauserman & Davis, 1996).

PEER PRESSURE

Studies have shown that about half of all males are motivated by curiosity the first time they have sex, while nearly one-fourth to one-third of females say they "went along" with it or that it was "voluntary but not wanted" (Abma, Driscoll, & Moore, 1997; Laumann et al., 1994). Although many teens engage in sex because they have positive feelings about it, many others who are sexually active are motivated by a desire to boost self-esteem and a desire for acceptance (Hajcak & Garwood, 1988). For most teens there is enormous peer pressure to engage in sex. What is **peer pressure**? It is your peer group's expectations of how you are supposed to behave.

If a teenager believes that most of his or her friends are having sex, there can be intense pressure—internally as well as externally—to conform. In Lakewood, California, in 1993, a group of high school male athletes calling themselves the "Spur Posse" received a lot of national attention when they told investigators, and then the media (including TV talk shows), that they had a contest of who could "bed" the most girls. Several of these teens had already had sex with over 50 girls, some of whom were greatly underage (thus the investigation). To be part of the group, these males were more concerned about scoring than they were about sex. According to sociologist Donna Elder,

peer pressure   Expectations by one's peer group about how one is supposed to behave.

"There's quite a bit of research that adolescent males use sex as another arena to achieve and score. They take the competitive sense and move it into the realm of sexuality" (quoted in Gelman & Rogers, 1993).

Gender roles tend to be very rigid in adolescence, and peer pressure makes failure to conform even more difficult. Most young people believe that the average person engages in sexual intercourse after several dates (Cohen & Shotland, 1996). What about those males who don't wish to become sexually active yet?

> "You were right about the fact that there is great pressure to engage in sex. I didn't have sex until I got married at the age of 22. My friends thought it was dumb for me to want to wait so I lied to them and told them that I had sex and made up a bunch of lies about it. I realize now it was dumb of me. If they were really my friends they would have understood my decision to wait."

> "I am a 19-year-old sophomore and I have not yet had sexual relations. I made a promise to myself long ago that I would not have sex until I am married. I made this promise because of my own moral convictions and I am content with it. I have been the target of a great deal of laughter and ridicule because of this decision. . . . I have been accused of being gay by some and of being sexually repressed by others. I don't see what the fuss is all about. If I am content with my sexuality and I decide to wait until I'm married to have sex, why should I be considered abnormal?"

> (from the author's files)

Many teenage girls also feel pressured to engage in sex:

> "I had sex at 17 in my senior year. . . . So many of my girlfriends would talk about doing it with and for their boyfriends, and would ask me about my experience. I was so ashamed for never having done anything throughout my entire life. . . . I was pressured by myself, him, and all my friends to do it, so I did. For no other reason than curiosity, I went through pain. . . ."

> "My first sexual experience was when I was 15. Everyone else was 'doing it' so I figured I should too. When it was over I cried. I wish I would have waited. Maybe not till I was married, but until I was more mature."

> "I had my first sexual relationships at 14. . . . I realize now that my reason for these relation-ships was to be accepted, wanted, have a boyfriend and pleasing him to keep him."

> (from the author's files)

For many girls, the need for emotional intimacy is an important part of finding their self-identity (see Shaughnessy & Shakesby, 1992). Unfortunately, many mistake sexual intimacy for emotional intimacy. It should not be surprising, therefore, that far fewer teenage girls than boys (fewer than half of sexually active girls, in fact) find sex to be a pleasurable experience (survey by Roper Starch Worldwide, 1994). In many peer groups there is a double standard that also results in a negative sexual experience for many girls. While men all over the world enjoy female promiscuity, in most cultures males prefer women (as mates) who are virgins (Buss & Schmitt, 1993, see Box 10–A).

> ". . . I would let some of the guys from school touch my breasts because of the attention I got from them. I only got a bad reputation instead. . . ."

> "Please talk about labelling women. When I was 15, I had some experiences with some older guys around 17. Well, I did some things that I regret. Word got around at my all-girl Catholic high school that I was a slut. You don't know what it is like to walk down a hall and have girls call you a slut. Guys wanted to go out with me for just one thing. I was depressed and paranoid of what people thought. I even thought of killing myself. I am 21 and have a 'steady' boyfriend who does not know what I did six years ago. I still feel guilty. When I see girls from my high school, some of them still call me a slut. I am very hurt. Please mention this. It is hard to imagine the six years of hurt."

> (from the author's files)

Just as is the case with some teenage boys, many teenage girls decide not to have sexual relations until they are older and are content with their decision:

> "I am writing this letter to let others like myself know that it is okay to be a virgin. I am a 19-year-old college student and a virgin. There is nothing wrong or strange about it. . . . You do not have to give in to sex just to please others or to simply fit in. . . . You should feel good about yourself, whatever and whenever you decide. When the moment is right, so will you be ready. . . ."

> (from the author's files)

If you engage in sex it should be for the right reasons, and you should feel good about yourself afterwards. Dr. Elizabeth Allgeier (1985) modified an eight-question list (Lieberman & Peck, 1982) that is designed to get teens to think about whether they are engaging (or about to engage) in sex for the right reasons. You are ready for sex if . . .

1. You feel guiltless and comfortable about your present level of involvement.
2. You are confident that you will not be humiliated and that your reputation will not be hurt.
3. Neither partner is pressuring the other for sex.
4. You are not trying to:
   a. prove your love for the other person;
   b. increase your self-worth;
   c. prove that you are mature;
   d. show that you can attract a sexual partner;
   e. get attention, affection, or love;
   f. rebel against parents, society, etc.
5. It will be an expression of your current feelings rather than an attempt to improve a poor relationship or one that is growing cold.
6. You can discuss and agree on an effective method of contraception and share the details, responsibilities, and costs of the use of the method.
7. You can discuss the potential of contracting or transmitting sexually transmitted diseases.
8. You have discussed and agreed on what both of you will do if conception occurs, because no contraception method is 100 percent effective.

(Reprinted with permission of SIECUS.)

TEENAGE PREGNANCY

In Chapter 6 you learned about the problem of teenage pregnancy in the United States. As you recall, over 1 million teenage girls become pregnant in the United States every year. Eighty to 90 percent of these pregnancies are unplanned, and about one-third of them end in abortion (Alan Guttmacher Institute, 1996; Henshaw, 1997). Few teen pregnancies lead to marriage (Robinson, 1988), yet fewer than 5 percent of unwed teenage mothers give their children up for adoption (Voydanoff & Donnelly, 1990). With all the responsibilities involved in raising a child, teenage mothers are at high risk for dropping out of school, and therefore becoming low wage earners. Teenage marriages that are the result of unwanted pregnancies often end in divorce or deser-

tion, and the suicide rate among teenage girls in this group is high (Furstenberg et al., 1981).

YOUNG ADULTHOOD (AGES 19–24)

The average age at marriage has increased dramatically in recent decades. The U.S. Census Bureau reported that the median age at which people first got married in 1994 was 24.5 for women and 26.7 for men, the highest levels in this century. This is an increase of more than 2 years of age since 1980 and more than 4 years since 1960. As a result, some psychologists and sociologists think that a new stage of life has developed in today's society—*young adulthood*. This is an extended period of being a single adult that occurs between adolescence and parenthood. Young adults have more freedom from parental restraints and more opportunity for privacy than in the past.

According to Seidman and Rieder (1994), who reviewed the results of numerous recent sexual surveys, the sexual activity of young adults is generally different from that of younger and older age groups. By the age of 20, approximately 90 percent of young adults are sexually experienced and having sex regularly, and most young adults have had multiple serial sexual partners during their young lifetimes. Let us review some of these studies.

Kinsey and his colleagues (1948) found that 68 percent of men surveyed with a college education, 85 percent of men with a high school education, and 98 percent of men with only an elementary school education had had premarital sex. The numbers for females were much lower, but more recent data indicate a striking increase in the number of women who have had premarital sex. Surveys conducted in the late 1980s and early 1990s found that among unmarried persons aged 18–24, 84 percent of the men and 81 percent of the women had engaged in premarital sex with a heterosexual partner (Anderson & Dahlberg, 1992; Gagnon, Lindenbaum, & Martin, 1989; unpublished data from 1988–1990 General Social Surveys). Janus and Janus (1993) found that 91 percent of the men and 83 percent of the women they studied had had premarital sex. Males are much more likely than females to have permissive attitudes about premarital sex, especially "casual" sex (Oliver & Hyde, 1993). Janus and Janus also found that how religious one was did not appear to be a major factor.

As previously mentioned, the lifetime sexual activity of young adults often includes more than one partner. As many as one-third of young women and one-half of young men aged 18 to 19 have had two or more partners in the past year (Forrest & Singh, 1990b) and probably 10 percent or more of young unmarried women have had two or more sexual partners in the past three

Figure 10–7 Young adulthood is an extended period of being a single adult. The large majority of young adults are sexually experienced and have sex regularly.

His two roommates were supposed to be sleeping, so I agreed. He tied me up spreadeagle to the bedposts and blindfolded me. After having sex on and off for about 30 minutes, I was untied and unblindfolded. I saw his two roommates standing in the room wearing underwear. I was then told it was planned to happen this way. They had all 'done me.' I admitted I enjoyed it, and still have group sex to this day."

"This past year I have become very sexually active. I have a steady boyfriend. We love one another to death, but are involved in threesomes and sometimes foursomes. We enjoy sex and have tried everything imaginable. The persons involved in our 'parties' are his roommates. We consider ourselves 'one big happy family'. . . ."

(from the author's files)

months (Seidman, Mosher, & Aral, 1992). As for college students, a study at a private New England university found that three-fourths of the coeds had had more than one sexual partner, and over one-quarter had had three sexual partners in the last year (DeBuono et al., 1990). Over 20 percent of college undergraduates at a Southern university reported having four or more sexual partners (King & Anderson, 1994). Among Canadian undergraduates, over 20 percent of the men and nearly 9 percent of the women had already had at least 10 sexual partners in their lifetimes (MacDonald et al., 1990).

Many of these surveys focused on women in particular. These studies find that the earlier a female becomes sexually experienced, the more likely it is that she will have had multiple sexual partners. Also, the more experienced she becomes, the less time she spends between sexual relationships (Seidman et al., 1992; Tanfer & Schoorl, 1992). Less religious women are more likely to have had multiple partners than religious women, especially Catholic women, and less educated women were found to have had more lifetime partners than college-educated women (except for divorced or separated white women, for whom the reverse was true).

In summary, for adolescents and young adults, the number of sexual partners steadily increases the longer the individual has been sexually experienced (see Tanfer & Schoorl, 1992). For some young adults, this is a period of sexual experimentation in which they engage in "one-night stands" or even group sex:

"After going out and becoming wasted, me and my boyfriend decided to have kinky sex.

However, many adolescents and young adults probably begin a sexual relationship with the hope and expectation that it will last. Reiss (1981) found that most people limited their premarital sexual activity to "serious relationships." More recently, Clements (1994) reported that 86 percent of the women and 71 percent of the men he surveyed said that it was difficult for them to have sex without emotional involvement.

"Dates often leave me feeling bad that I don't engage in the one night stand scenes. . . . I prefer to make love. Sex is not a sport to me. Personally, it's too emotional to take it that lightly. . . ."

(from the author's files)

Thus, for many young adults, their sexual life-style can be called **serial monogamy**—a series of relationships in which sex is reserved for just one other person.

The period of young adulthood is also a time in which most people gradually become less influenced by peer pressure and gain a better understanding of their own sexual motivations.

"I now make my own choices and give my

serial monogamy The practice of having a series of monogamous sexual relationships.

opinions without being intimidated about what someone else might think. But most of all I have stopped confusing sex for the attention I was striving for. For example, I used to think that if I give him sex then that means he must like me or is my friend. . . ."

(from the author's files)

ADULTS AGED 25–39

In their review of sexual surveys, Seidman and Rieder (1994) concluded that relatively long-lasting monogamy was the norm for most adults aged 25 and older. For people in this age group, about 80 percent of sexually active heterosexual men and 90 percent of sexually active heterosexual women report having had only one sex partner within the last year (data largely from the National Surveys of Family Growth and the General Social Surveys). Nearly identical findings were reported by Laumann's group (1994).

For this age group, the best predictors of having more than one partner are marital status, gender, age, and race. Not only are men, particularly younger unmarried men, more likely than women to have had multiple partners in the preceding year, but for unmarried women abstinence becomes more common with increasing age (see Seidman & Rieder, 1994). African Americans are less likely to marry than whites and thus are more likely to have had more than one partner recently (Billy et al., 1993; Laumann et al., 1994; Seidman et al., 1992; Tanfer & Schoorl, 1992).

MARRIAGE

As mentioned in the introduction to this section, many people today are postponing marriage. Still others never marry. This is much truer for African Americans than it is for Americans in general (Saluter, 1994; Tucker, Taylor, & Mitchell-Kernan, 1993). However, because of the history of negative attitudes about sexuality in Western culture (see Chapter 1), for many people sexual relations are legitimized only within marriage. Moreover, numerous studies have found that adults who get and stay married have higher levels of psychological well-being (and in the case of men, less depression) than individuals who stay single (e.g., Horwitz, White, & Howell-White, 1996; Mookherjee, 1997).

The frequency of sexual intercourse in the first year of marriage is usually high, an average of about 15 times a month in the first year (Greenblatt, 1983). The frequency of sexual relations generally declines thereafter. However, for the entire population, the frequency of sexual relations is highest for married couples in their mid-twenties to mid-thirties (Clements,

1994; Janus & Janus, 1993; Laumann et al., 1994). Married women are also more likely than single women to have orgasms during sexual intercourse (Laumann et al., 1994).

The decrease in sexual relations after the first year or two of marriage occurs for a variety of reasons. Sex has to compete with other time demands, such as career advancement. Parenthood means less privacy and more demands, and often results in exhaustion at the end of the day. It is also normal for sexual relations to decrease somewhat as the novelty wears off and individuals within a relationship become accustomed to one another (Blumstein & Schwartz, 1983).

Studies find that married couples today are more likely to use a variety of sexual techniques than in the past. This is true for positions of sexual intercourse and for oral-genital sex (Hunt, 1974). Hunt found that a majority of young people also continue to masturbate after getting married. In fact, people with sexual partners are much more likely to masturbate than people who do not have a partner (Laumann et al., 1994; see Chapter 11). However, a majority of both men and women report marital sex to be very satisfying, more so than do singles (Clements, 1994; Laumann et al., 1994). I will have more to say about marriage and sex in a later section of the chapter.

LIVING TOGETHER (COHABITATION)

A Census Bureau survey showed that 3.5 million unmarried heterosexual couples were living together in 1993. This represents an increase of 120 percent in *cohabitation* since 1980 and well over a 700 percent increase since the 1970 census. Forty percent of couples who live together have children. The increase in the number of couples living together may be one reason for the older age of couples today who marry for the first time.

"I totally believe that it is a good idea to live with someone first if you plan to marry them. . . . I was very uncertain on where our relationship was and where it was headed. I know now after living with someone, just for three months, that he and I had totally different needs in our lives."

(from the author's files)

Living together can be a test or trial period before marriage (there is much more to living together happily and successfully than simply enjoying sex). Fifty-six percent of college students interviewed in a 1989 Gallup survey said that they approved of living together in trial marriage. Living together can also be an alternative to marriage. The amount of commitment that couples make to each other while living together

Figure 10–8 Relatively long-lasting monogamy is the norm for most adults aged 25 and older.

Although most people think that cohabitation is something done mostly by college students, cohabitation is most common among the least educated people in the population, particularly those from less religious families (Bumpass et al., 1991; Thornton, Axinn, & Hill, 1992). On the average, the level of sexual activity is slightly higher among couples who live together than among either married couples or unmarried noncohabitants (Bachrach, 1987; Risman et al., 1981).

Interestingly, studies find that couples who live together before getting married are more likely than others to get divorced (Axinn & Thornton, 1992; Bumpass & Sweet, 1989). The higher divorce rate is not due to the greater time that cohabitors have spent together (DeMaris & Rao, 1992), but instead is the result of a lowered enthusiasm about marriage and a greater tolerance of divorce (Axinn & Barber, 1997). Axinn and Thornton

can vary a great deal (Bumpass, Sweet, & Cherlin, 1991). Many people who live together do not see their relationship as a planned attempt at a trial marriage (Macklin, 1978), but most expect to marry their partner (Bumpass et al., 1991). Studies show that in cohabitation there is generally less certainty about the relationship than in marriage (Bumpass et al., 1991). One advantage people claim for living together over marriage is that since there has been no formal contract, such as a marriage license, it is easy to simply walk away from a relationship that does not seem to be working. One disadvantage in living together is that it is easy to walk out on a relationship that does not seem to be working. (Think about it.)

(1992) believe that cohabitation "may reinforce the idea that intimate relationships are fragile and temporary" and this might "reduce the expectation that marriage is a lifetime relationship." Living together, they said, "probably attracts people who are, on average, more accepting of the termination of intimate relationships." Other studies support their conclusion (Booth & Johnson, 1988; Thomas & Colella, 1992). Although more research is needed to determine why people who live together premaritally are more likely than others to divorce when married, we are able to conclude that cohabitation is not the answer to a stable relationship that many have hoped it would be.

SINGLE PARENTHOOD

In 1994, there were 11.4 million single parents raising children (9.9 million women and 1.5 million men), nearly triple the number in 1970 (U.S. Census Bureau). About 25 percent of white children and 65 percent of black children lived with a single parent. Some adults become single parents through divorce or separation, but nearly 25 per-

Reprinted with special permission of King Features Syndicate.

Box 10-A / CROSS-CULTURAL PERSPECTIVES

How Do People Select a Mate?

The boxes on cross-cultural perspectives presented throughout this book generally show that cultures differ widely in their attitudes and behaviors regarding sexuality. Recently, however, David Buss and David Schmitt (1993) surveyed people throughout the world, from Zambia to China, and found that people's mating preferences and attitudes were very much alike. The only major differences were between men's and women's desires and whether or not people were interested in a short-term relationship (casual sex) or a long-term mate.

Short-term relationships dominate human mating patterns, but men and women engage in short-term liaisons for different reasons, according to the authors of the study. Men tend to engage in casual sex to test their virility, while a woman's short-term relationships are largely experimental in nature, either to test her market value (to see how desirable she is) or to determine what her desires are. On average, men wanted about eighteen women as sexual partners in their lifetime, while the average woman wanted four men.

Other studies confirm that, as a general rule, men are much more likely than women to engage in casual sex with someone they hardly know, or don't know at all (see Oliver & Hyde, 1993). In one study, for example, researchers had an attractive man or an attractive woman approach college students and ask them if they would have sex with him or her (Clark & Hatfield, 1989). Three-fourths of the men said yes, but none of the women agreed to have sex with an attractive stranger. Buss and Schmitt found that most men were willing to have sex with a woman after knowing her for only a week, whereas the typical woman preferred to wait six months.

Buss and Schmitt found that in all cultures the idealistic goal was a lasting relationship. Across all cultures, over 90 percent of people eventually marry (Buss, 1985; Epstein & Guttman, 1984). However, this does not mean that people spend their lives with one partner. About 80 percent of all societies allow men to have multiple wives or mistresses (although only about 20 percent of men actually do so; Ford & Beach, 1951; Murdock, 1967). According to Buss and Schmitt, Western cultures may have outlawed polygyny, but people get

around this by having serial marriages (divorce and remarriage) and by having extramarital affairs. In the United States, the divorce rate is around 50 percent (H. Fisher, 1987), and from 25 to 75 percent of people are estimated to have engaged in adultery (H. E. Fisher, 1987; Hite, 1987; Kinsey et al., 1948, 1953; Symons, 1979).

Buss and Schmitt believe that for males promiscuity is a primal instinct that maximizes reproduction potential. How do men balance their short- and long-term desires? Buss says, "If a man could have his fantasy, he would sequester and monopolize all the attractive women in the country. Indeed, men who are in a position to get what they want—kings, tycoons, celebrities—often do things like that" (quoted by Gura, 1994). Men may enjoy female promiscuity in a short-term relationship, but when looking for a long-term mate they seek a chaste woman, someone who has not yet had sex. Of 37 countries studied by Buss and Schmitt, in more than two-thirds of them the men valued chastity in choosing a mate more than the women did. This was especially true among Palestinian Arabs, Iranians, and male Indonesians.

> "I am of Mexican descent. In our culture, it is 'okay' for boys to engage in sex before marriage. In fact, it is encouraged. Girls, however, are not to have sex until after marriage. We are made to believe that a man will not want to marry a girl who is not a virgin. In most cases, it turns out to be true."
>
> (from the author's files)

Remember from Box 4–A that in large parts of northern Africa and the Middle East, girls' genitals are sewn together to assure chastity at the time of marriage. According to the study, chastity at marriage was not important to most men in Scandinavia and the Netherlands. However, regardless of culture, men placed a high value on physical attractiveness in choosing a long-term mate (Buss, 1989; Buss & Barnes, 1987; Regan & Berscheid, 1997) and preferred mates who were younger than they were (Buss, 1989; Kendrick & Keefe, 1992). Zambian men preferred wives who were at least 7 years younger, while Italian men preferred an age difference of 3 years and U.S. men a difference of about 2 years.

Women's mating strategies are different from men's, according to Buss and Schmitt. Because a woman has a limited number of reproductive years and must invest time and energy in childbearing, women prefer mates who can offer them economic and physical protection. Women, more so than men, prefer men who are good financial prospects (Buss, 1989). Some researchers believe that dating and marriage frequently involve an exchange of sex for financial support (e.g., Muehlenhard & Schrag, 1990). Factors like a promising career, ambition, and education are important factors in women's decisions across all cultures. Women prefer older mates, but are not as concerned about physical attractiveness as are men. Chastity of a potential mate is important to women only in some Oriental cultures.

In line with their mating strategies, men and women use different strategies to keep a mate interested (Buss & Shackelford, 1997). Women are more likely to make themselves more attractive, while men are more likely to spend money on a mate and give gifts.

Buss and Schmitt believe that there is evolutionary value for what men and women seek in a relationship. Whether or not you agree with this, one thing is clear from this large cross-cultural study—love is not really blind.

---

cent of never-married women are also mothers. This trend involves all socioeconomic groups. For example, among never-married managerial or professional women, the proportion who were mothers increased from 3.1 percent in 1981 to 8.2 percent in 1992.

I discussed the effects of single parenthood on children's gender identities and gender roles in Chapter 8. There has been very little research on the sexual behavior of single parents as a group.

EXTRAMARITAL SEX—IN SUPPOSEDLY MONOGAMOUS MARRIAGES

In their original surveys (1948, 1953), Kinsey and his colleagues found that about 50 percent of married men and 24 percent of married women had engaged in extramarital sex on at least one occasion. Surveys by magazines such as *Cosmopolitan, Playboy, Woman's Day,* and *Redbook* have found equally high, and often higher (especially for women), percentages. However, the Kinsey surveys have been criticized for being nonrepresentative, and magazine surveys are almost always nonrepresentative (see Chapter 1).

What do more scientifically conducted current surveys indicate? In their 1990 "New Report on Sex," the Kinsey Institute found that 37 percent of the husbands and 29 percent of the wives surveyed had had at least one extramarital affair. Janus and Janus (1993) found that 35 percent of married men (and 56 percent of divorced men) and 26 percent of married women (and 59 percent of divorced women) in their survey admitted to extramarital affairs. Some therapists believe the figure is even higher (see Mehren, 1991), but some other recent surveys find a lower percentage of people who admit to having been unfaithful to their spouses (Choi, Catania, & Dolcini, 1994; Clements, 1994; Laumann et al., 1994). In their nationally representative sample, Laumann's group found that 25 percent of men and 15 percent of women admitted to having had an affair. This agrees with another recent nationally representative sample (Wiederman, 1997). Keep in mind, however, that if we look only at middle-aged persons (who have had more opportunity), the percentages increase to about 35 percent of married men and 20 percent of married women. By the way, the rate of extrarelational affairs is just as high for cohabitating couples as it is for married couples (Blumstein & Schwartz, 1983).

These statistics on adultery refer to extramarital *sexual* relations. But, according to psychologist Shirley Glass, "That's a male definition [of adultery]. If you look at emotional involvement and sexual involvement short of intercourse, you add another 20 percent" (quoted in Mehren, 1991). For example, it is not uncommon for married persons to have emotional affairs over the telephone or the internet (see Shaw, 1997). Indeed, Thompson (1984) found that extramarital affairs that involved a deep emotional involvement, but were nonsexual, were about half as common as sexual affairs.

Nearly three-fourths of adults in their forties regard adultery as "always wrong." Seventy percent of adults believe that affairs are always harmful to a marriage, although 22 percent believe that affairs can sometimes be good (Adler et al., 1996). Interestingly, Kinsey and his colleagues found that though men were more likely to engage in extramarital sex than women, they were also more likely to rate affairs as harmful to a relationship—if their wives were unfaithful to them. Men are more upset at the idea of their partner's getting physical with someone else than they are about an emotional attachment. Women, on the other hand, are more concerned about emotional infidelity on the part of their mates and are much more likely than men to forgive a partner for sexual infidelity (Buss, 1994).

Why do so many people engage in extramarital sex (or sex outside of any supposedly monogamous relationship)? Some are simply curious to see what sex with another person would be like, while some want more variety in their lives. Some may seek out other partners in order to prove that they are still desirable or young (the famous "midlife crisis" or "middle-aged crazy" cause of extramarital sex), while others may be looking for the companionship no longer found in their marriages. And, of course, there is the excitement factor. As John Gagnon (1977) points out,

> Most people find their extramarital relationships highly exciting, especially in the early stages. This is a result of psychological compression: the couple gets together; they are both very aroused (desire, guilt, expectation); they have only three hours to be together. . . . Another source of attraction is that the other person is always seen when he or she looks good and is on best behavior, never when feeling tired or grubby, or when taking care of children, or when cooking dinner. . . . Each time, all the minutes that the couple has together are special because they have been stolen from all these other relationships. The resulting combination of guilt and excitement has a heightening effect.

Blumstein and Schwartz (1983) found that men generally look for casual sex and have more extramarital partners than women. Women are more likely to seek an emotional attachment with their new partner. Still, in most cases affairs are primarily short-term, or "one-night stands" (Gagnon, 1977).

There is very little relationship between social background—including political leaning (ultraconservative vs. ultraliberal) and religion or lack of it—and whether or not one engages in extramarital affairs (Janus & Janus, 1993; Thompson, 1983). Whitehurst (1972) indicates that two things contribute to having "outside sex." The first is simply opportunity. For example, Levin and Levin (1975) found that while 27 percent of all housewives surveyed reported having had an extramarital affair, 47 percent of married women who worked at least part-time away from home had had sex outside of marriage. Women who work have more opportunity to meet other men and to have affairs than women who stay at home all day (Adler et al., 1996). As psychotherapist Marcella Weiner puts it, "Women are in the work force to stay now, and they have many more opportunities [to meet men]. It has always been okay with men, but it is becoming more and more a phenomenon with women" (quoted in Mehren, 1991). With the rising number of wives entering the work force outside of the home, the number of women having extramarital sex may continue to in-

open marriage A marital relationship in which the couple agrees that it is permissible to have sexual relations outside of the marriage.

Figure 10–9 The addition of children in a relationship may change the pattern of sexual relations, but not necessarily the frequency. Married couples in their mid-twenties to mid-thirties have sex more often than any other group.

crease. The second thing mentioned by Whitehurst is alienation. People can grow apart and develop different interests over time. Thus, if you meet someone else who shares your interests, seems attractive, and seems attracted to you, this can lead to sharing other parts of your life as well (see Chapter 12). An interesting thing about Whitehurst's argument is that situational factors are seen as responsible for the likelihood of extramarital sex, rather than some personality flaw. This would also explain why married individuals generally have affairs in the daytime—it is simply easier to get away and cover one's tracks during the day than at night.

EXTRAMARITAL SEX— CONSENSUAL ARRANGEMENTS

Some couples agree to have sex outside of the marital relationship. One such arrangement is **open marriage,** where both partners agree that it is okay to have sex with others. Even here, however, there are usually some agreed-upon restrictions, such as not having sex with a mutual friend or not having sex with the same person twice (Blumstein & Schwartz, 1983). Couples make these restrictions so that their primary commitment is to the marriage and thus the nonmarital partner will not be viewed as competition. According to Nena and George O'Neill, authors of the 1972 best seller *Open Marriage,* this type of marriage challenges the idea that a single partner can meet all of a person's needs and allows personal growth for both individuals. Blumstein & Swartz (1983) found that perhaps as many as 15 percent of married couples had some sort of "understanding" that allowed for extramarital sexual relations under some conditions. Although this type of extramarital arrangement is supposed to add role equality and flexibility to a marriage (O'Neill & O'Neill, 1972), the divorce rate among open-marriage couples is just as high as for sexually monogamous couples (Rubin & Adams, 1986). However, for those open-marriage couples that did break up, sexual jealousy was rarely the reason.

Another arrangement that allows consensual extramarital sex is **swinging** (often called "wife swapping"), where a married couple has extramarital relations together. Couples get together by answering ads in newspapers or swingers' magazines, or by going to commercial clubs that encourage recreational sex. In some cases, several couples get together for swinging parties. The sexual activity often includes sex between the women while the husbands watch, but male homosexual activity is much less common, and in many cases forbidden. It's been estimated that 2–4 percent of all married couples have engaged in swinging on at least an occasional basis (Hunt, 1974; Weiss, 1983), but these statistics were collected in the 1970s, when swinging was at its peak in popularity. Some couples abandon swinging after giving it a short try, usually because they find that it creates feelings of jealousy, sexual inadequacy, or rejection for them (Masters & Johnson, 1976; Murstein, 1978). Except for the fact that they tend to be less religious than others and to have had more premarital sexual experience, swingers are no different from other couples, and many are middle class and conservative (Jenks, 1985; Murstein, 1978; Weiss, 1983).

At this point it should be emphasized that unless a person *always* uses condoms, anytime a person has sexual partners outside of a monogamous relationship—whether consensual or nonconsensual—his or her chances of contracting a sexually transmitted disease is increased; see Chapter 5.

MIDDLE AGE (AGES 40–59)

Christian and Victorian views of sex emphasized that sex within a marriage was for procreation only, and excluded sexual activities for pleasure. Thus Western cul-

ture came to view older individuals who were through conceiving children as asexual. For example, many students believe that their own parents and grandparents love each other, but how many students think of their parents and grandparents as being sexually active? Studies have found that many college students have difficulty accepting their parents' sexuality and underestimate how often their parents have sex by about half (Allgeier & Murnen, 1985; Pocs & Godow, 1977). Here are a variety of responses from students in my course:

"I was about 14 when my mom told me she and my dad have sex. I was shocked and thought it was disgusting. I didn't want to hear it any more."

"My parents don't have sex. They have other things to do."

"The thought of my parents actually possessing sexuality or having sex mortified me."

"The thought of my parents having sex made me sick, not because of their age, but because of the fact that they were my parents."

"And plus they are in their late 40's and early 50's—I just can't imagine them having sex."

"When I was a teenager, it would make me uncomfortable when they would go in their room and shut and lock the door because I knew what they were going to be doing and I just felt it was weird because to me, they were 'old.' Now that I'm in my twenties, it makes me feel good to know that my parents still love and care about each other, and it helps me to believe that I will have a positive sexual attitude well into my 40's and 50's."

"Sometimes I tease them a little when I catch my dad home late in the morning because I know he stayed home so they could have sex. I think it's great that they still love each other and express themselves sexually after being married for 25 years."

(from the author's files)

Pocs and Godow (1977) concluded that "many parents may appear to be nonsexual because they hesitate to discuss the topic of sexuality in any way with their children or because they are not inclined to exhibit loving, affectionate responses, let alone

swinging A type of open marriage relationship in which a couple has extramarital relations together with other couples.

sexual behavior, in the presence of their children." Allgeier and Murnen (1985) later found that students whose parents had discussed sex with them gave more accurate estimates of their parents' sexual frequency.

Ask yourself these questions: "Do my parents enjoy sexual intercourse? How often? Do they enjoy using different positions? Do they enjoy oral-genital sex? Do my grandparents enjoy sex? How often? Do they enjoy using different positions? Do they enjoy oral-genital sex?"

Think about how you answered these questions, and then ask yourself: "What are my attitudes about sex? Is it something good that two people share?" For those of you who are sexually experienced, also ask yourself, "Do I like sex? How often do I enjoy it?" If you enjoy sex, why wouldn't your parents or grandparents? Sex therapists have found that the way that you feel about sex now is the best estimate of how you will feel about it in the future (Masters & Johnson, 1966). If you enjoy sex now, why should you stop enjoying sex later in life? *Adults of all ages are sexual beings.* Sex can be healthy and enjoyable throughout life. Problems with fully enjoying sex are more often due to one's beliefs and attitudes than to physiological changes (Kellett, 1991). If you want to have a good healthy sex life in the future, you need to be aware now that older adults are still enjoying the pleasurable aspects of sex. From the previous case histories, which students do you think will enter parenthood with more positive attitudes about sexuality?

Figure 10–10 The responsibilities of raising children and having careers may result in fewer opportunities to have sexual relations, but sexual interest does not decline. Do you believe that your parents still enjoy sex? When you become a parent, will you be embarrassed if your children know that you and your partner enjoy sexual relations?

Negative attitudes about sexual activity in older adults can create many unnecessary barriers for people. Today, for example, people are not surprised when two young people decide to live together without getting married. However, the same young people might be shocked if a widowed parent or grandparent decided to move in with an older adult. Barring a tragedy, we will all grow old someday. These negative attitudes must change if today's young people want to be treated decently when they are older.

FREQUENCY OF SEXUAL INTERCOURSE

It is a common belief that sexual activity is highest for young people. As mentioned earlier, two recent national cross-sectional surveys found that sexual activity was highest for people in their mid-twenties to mid-thirties (Clements, 1994; Janus & Janus, 1993). Is there a sharp drop-off after that? Not really, at least not until the sixties. Although most recent surveys do show a decline in sexual activity for people in their forties, and a further decline in the fifties, the drop-off is very gradual (Blumstein & Schwartz, 1983; Clements, 1994; Segraves & Segraves, 1995; Seidman & Rieder, 1994; Smith, 1991). At most, the frequency of sexual intercourse for married couples decreases by only a couple of times a month from the thirties through the fifties (from 7–8 times a month to 5–6 times a month). In fact, Janus and Janus (1993) recently found very little evidence of any decline in sexual activity up through the early sixties. People in all age groups say that they would like to have sex more often (an average of 13 times a month; Clements, 1994), but lack of time and being physically tired prevent them from doing so (Blumstein & Schwartz, 1983).

A lot of people also believe that the sex lives of older married people are dull compared to those of singles. Actually, the surveys indicate that married people have sex more often than singles (never married or divorced) and are happier with their sex lives than are singles (Clements, 1994; Laumann et al., 1994; Smith, 1991). When asked if sex is important to them, more

"Actually, Dad knows quite a bit about sex for a man his age."

Reprinted from *Good Housekeeping* by permission of Rex May.

people in their thirties, forties, and early fifties say yes than do people in their late teens or early twenties (Clements, 1994).

Older couples who have sex frequently are also generally found to have the happiest marriages. Of course, we can't know what came first, the chicken or the egg. According to Andrew Greeley, "Their sex lives may be better because their lives are more satisfying, or the other way around" (cited in Coleman, 1992). We do know that as people grow older, they tend to regard such behaviors as hugging and kissing as much more important and pleasurable than they did when they were younger (Clements, 1994).

LOSS OF A MATE BY DIVORCE OR DEATH

Over half of all marriages in the United States now end in divorce (Martin & Bumpass, 1989). Between 1970 and 1994, the number of people who were divorced increased from 4.3 million to 17.4 million (Saluter, 1996). Other people become single again as a result of their partners' death. When this happens, reentry into the singles world and dating are often difficult, particularly if the relationship had been a long-lasting one. Many divorced and widowed persons have no idea of what the current expectations are in dating, and thus, learning how to interact with potential partners may take some time and at first may be very awkward (Spanier & Thompson, 1987). New rules have to be created or discovered. For example, should a 42-year-old divorced woman have sex with her date (*a*) on the first date, (*b*) after they've gone out a few times, (*c*) only after she thinks they love each other, or (*d*) not at all?

The early sex surveys by Kinsey and colleagues (1953) and Hunt (1974) reported that divorced individuals were very sexually active. However, more recent studies have found that while some people become more sexually active after divorce, about one-fourth to one-third show a decrease in activity (Simenauer & Carroll, 1982). Three-fourths of divorced persons have either a single sexual partner or no partner in the first year (Stack & Gundlach, 1992). Still, almost all divorced men and women will return to an active sex life eventually. Resuming sexual activity is an important part of letting go of the former bonds. For men in particular, the resumption of sexual activity is associated with their well-being and self-confidence (Spanier & Thompson, 1987).

Many divorced or widowed men and women remarry, but there are more opportunities for men. By the time people reach their early forties, there are over 200 single women for every 100 single men (Blumstein & Schwartz, 1983). There is also a double standard in our society. While many people frown upon women having sexual relationships with men much younger than themselves, most feel that it is okay for men to have relationships with younger women. Have you ever noticed that whenever Hollywood puts a middle-aged leading man in a romantic role it is almost always with a woman 10 to 20 years younger (or even more) than he is? Because of the male-female ratio and the double standard, women over age 40 generally have a more difficult time finding new partners than older men. For example, about one-half of widowers, but only one-quarter of widows, remarry (Lown & Dolan, 1988).

About three-fourths of divorced people remarry, usually within three to four years (Coleman & Ganong, 1991; Lown & Dolan, 1988), but do they do a better job of choosing a partner? Apparently not, for half or more of second marriages also end in divorce (Blumstein & Schwartz, 1983; Ganong & Coleman, 1989). Since the early 1970s, the percentage of people who have married three or more times has doubled, from 4 percent to 8 percent (National Center for Health Statistics). For some, marriage is just another form of serial monogamy.

FEMALE SEXUALITY: PHYSICAL CHANGES WITH AGE

As a woman grows older, her ovaries do not respond as well to the pituitary hormones FSH and LH (see Chapter 3), so that menstrual cycle irregularities are common after the age of 35 (cycles become more frequent or less frequent than before). Eventually her ovaries atrophy (wither) and she quits having menstrual cycles entirely, thus ending her ability to have children. The cessation of menstruation is referred to as **menopause** (Note: menopause is defined differently in some other cultures—see Lock, 1994). It usually occurs in a woman's late 40s or early 50s, with the average age being 51.5 (Bromberger et al., 1997). This is an increase of about 4 years over the past century, probably due to improvements in health and socioeconomic status (Flint, 1997). The changes that occur in the few years that precede and follow menopause are called the **climacteric.** How a woman and her partner react to these changes can affect other aspects of a relationship, so it is important for everyone to understand their physical basis (see Nachtigall, 1994 for a review).

As the ovaries atrophy, they produce less and less gonadal hormones, and eventually produce no progesterone and only minute amounts of estrogen. Most women will experience some symptoms as a result of these hormonal changes. Estrogen and progesterone, you recall, inhibited release of FSH and LH from the pituitary. At menopause, the pituitary hormones are no longer inhibited and are released in large amounts. In addition to their reproductive role, these hormones affect the

menopause The term for a female's last menstrual period.

climacteric The changes that occur in women in the few years that precede and follow menopause.

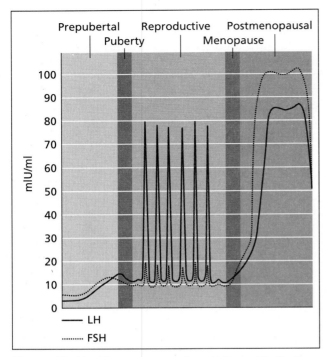

Figure 10–11 The pattern of secretion of pituitary hormones during a female's lifetime. The large increases in FSH and LH after menopause are due to the loss of estrogen and progesterone from the ovaries.

diameter of blood vessels, causing them to dilate. Subjectively, this physiological change is perceived as a *hot flash,* a sudden sensation of warmth or intense heat that spreads over the upper body, face, and head (see Kronenberg, 1994). The hot flash is usually accompanied by flushing and sweating and is often followed by a feeling of being chilled. Pressure in the head or chest, nausea, or feelings of suffocation can also be experienced during a hot flash. The sensations last from a few seconds to 15 minutes or more and are very common at night, thus interfering with sleep. This can lead to fatigue, irritability, and depression. Hot flashes generally disappear within a few years, but 20 percent of affected women may experience hot flashes for 5 years or longer.

The loss of estrogen also causes the vagina to become thinner and less elastic, with a marked *decrease in the amount of vaginal lubrication* during sexual arousal. If the vagina is not properly lubricated, sexual intercourse will be painful, so that some postmenopausal women may start avoiding sexual relations (see Sherwin, 1991). This is unnecessary, however, if a woman and her partner accept vaginal dryness after menopause as a normal physiological response and not as an indication of lack of desire. The dryness can be countered with a lubricant, such as K-Y jelly.

A lot of attention has recently been given to the role of decreased estrogen levels in *osteoporosis,* a condition in which the bones become more brittle with age (see Rozenberg et al., 1994). There are over 250,000 hip fractures a year due to osteoporosis, resulting in up to 50,000 premature deaths in women (National Institutes of Health Consensus Panel).

As women grow older, a few will experience severe weakening of the ligaments that hold the vagina and uterus in place, a condition called prolapse. This condition can usually be corrected, most commonly by reconstructive surgery (Kaminski et al., 1993).

THE MEDICALIZATION OF MENOPAUSE. There are over 50 million women in the United States older than 50, most of whom will live another 20 years or more. Thus, postmenopausal women have become a large segment of society that seeks health care, or, depending on your perspective, that is targeted by health care professionals. There were only 80 publications about menopause in scientific journals in 1965, but 800 in 1995, a clear indication of how much menopause has become a central issue in health care (Palmlund, 1997).

Today, the medical profession treats osteoporosis, hot flashes, vaginal dryness, and other less common problems associated with menopause (e.g., decreased clitoral sensitivity, recurrent urinary tract infections), with *hormone (estrogen) replacement therapy* (in pill, cream, or skin patch form). Numerous studies have shown that estrogen is beneficial for preventing osteoporosis (e.g., Schneider, Barrett-Connor, & Morton, 1997; Writing Group, 1997), heart disease (e.g., Stampfer et al., 1991; Grodstein et al., 1996), and Alzheimer's disease (e.g., Kawas et al., 1997; Tang et al., 1996). Androgen replacement therapy has been used for those postmenopausal women who show a decrease in sexual desire (Sherwin, 1991; but see Casson & Carson, 1996). On the down side, there is concern that estrogen replacement therapy might increase the risks of breast and endometrial cancer. However, results of studies are mixed. Several studies have found that estrogen replacement therapy does not increase the risk of breast cancer (e.g., Dupont & Page, 1991; Stanford et al., 1995), while others have reported an increase with long-term use (e.g., Colditz et al., 1995; Collaborative Group, 1997; Steinberg et al., 1991). In 1997 the Food and Drug Administration approved a "designer" estrogen (raloxifene) that mimics estrogen's beneficial effects for osteoporosis, but without increasing the risk of breast cancer (however, the drug does nothing for hot flashes and increases the risk of blood clots). Progestins are presently given with estrogen to protect against endometrial cancer.

When considering estrogen therapy, a woman must weigh the risks against the benefits. The overall death rate for elderly women taking estrogen is lower than for elderly women not taking estrogen (Grodstein et al., 1997). More women die each year from broken hips, for example, than from cancer of the endometrium. Estrogen should not be given indiscriminately, but many medical experts believe that the benefits outweigh the risks when it is prescribed in small doses for a short period of time (Harlap, 1992; Theriault, 1996).

Medical intervention in treating menopause has not been universally applauded. Many physicians and feminists argue that menopause is a natural part of being a woman that should be allowed to run its course

(see van Hall, 1997). They point out that most women go through menopause with few or no complaints. In fact, a 1997 Gallup poll found that a majority of women aged 45 to 60 regarded menopause as the start of a new and fulfilling stage of life (Utian, 1997). While most women experience a decline in well-being in the first two years after menopause, feelings improve spontaneously after that (Dennerstien, Dudley, & Burger, 1997). Women in some other cultures or ethnic groups report few problems during menopause (see Box 10–B).

Box 10-B CROSS-CULTURAL PERSPECTIVES

Menopause in Different Cultures

Menopause as a biological event is universal—all women worldwide will eventually stop having menstrual cycles and experience a decrease in estrogen levels. But what about the *experience* of menopause? Western health practitioners have medicalized menopause and portray the postmenopausal years negatively, as a time in which the body breaks down (but can be repaired by hormone replacement therapy). Western feminists, on the other hand, portray menopause not as a disease, but as a normal part of aging that can bring liberation and joy. Which of these views is more realistic?

Richters (1997) recently reviewed cross-cultural studies of menopause and found many differences. For example, while a majority of American and European women experience hot flashes after menopause, hot flashes are uncommon among postmenopausal women in Japan or the Mayans of Mexico (Beyne, 1989; Lock, 1994). The Mayans do not even have a word for it. In some areas of Greece, women experience hot flashes, but do not think of them as medical symptoms that require going to a physician (Beyne, 1989). Like the Mayans, traditional Navajo women apparently do not experience the same symptoms as Western women and view menopause as a positive change of life status (e.g., being free of menstrual taboos) (Wright, 1983). Women in northeastern Thailand also view menopause positively as a natural event (Chirawatkul & Manderson, 1994).

Some, but not all, of these differences can be explained by the social position of women

in different cultures (Richters, 1997). In cultures where women gain prestige and power in middle age, menopause may likely be viewed positively, in contrast to the United States, where older persons are often viewed as a burden. Migration from one country to another can rapidly change the perception of menopause. Older women in Italy generally have positive roles, but Italian women who migrated to Australia came to view menopause negatively when those roles were no longer valid (Gifford, 1994).

There are ethnic variations in the experience of menopause within our own country. The average age at menopause among African-American women is two years younger than for white women, with psychosocial stress being the main predictor (Bromberger et al., 1997). African-American women are also more likely to have hot flashes, while Hispanic women are more likely to experience a racing heart. Asian-American women, on the other hand, experience few symptoms (see DeAngelis, 1997). This may reflect ethnic differences in the social status of aging women, but may also be due to diet (Adlercreutz et al., 1992). Smokers in all ethnic groups experienced greater symptoms.

Richters (1997) concluded that many factors might result in the differences in menopausal symptoms. However, one thing can be concluded—the experience of menopause is not the same for all women, either across or within cultures.

So why the big push in our culture to "treat" women during and after menopause? Some claim that postmenopausal women have become a targeted consumer group aimed at making profits for pharmaceutical companies (Palmlund, 1997). Medicalization refers to "the process whereby the normal processes of pregnancy, childbirth, menstruation and menopause have been claimed and redefined by medicine" (Kaufert & Lock, 1997). Opponents of the medicalization of menopause claim that the negative expectations of deteriorating health are the result of culturally determined attitudes, promotions by drug companies, and media pressure, and point out (correctly) that older women in our country are already healthier than older men and outlive men by many years (van Hall, 1997).

These many conflicting results and opinions have left many women confused about what to do (Azar, 1997). In the midst of all this, I will dare to offer an opinion. Hormone replacement therapy should not be mass-marketed as beneficial to all, but instead, the decision to use it should be individualized depending on a woman's own family history and personal risk factors.

FEMALE SEXUALITY: BEHAVIOR

Does a woman's sexual desire end with the climacteric? I have found that many students believe that women no longer have any interest in sex after menopause. This is generally untrue.

> "I was surprised to find out that a woman still had a sexual desire after menopause. I always believed it to be a time of depression and no sex. Well, you learn something new every day."

> "I was totally surprised by the fact that older people still have sex. I often wondered what I was going to do at that age because I enjoy sex tremendously."

> (from the author's files)

For most women, menopause causes no change in feelings of anxiety, anger, depression, stress, excitability, or nervousness (Matthews et al., 1990). The same is true of interest in sex (Hallstrom & Samuelsson, 1990; Segraves & Segraves, 1995). Some women actually show an increased interest in sex after menopause, possibly because they no longer have to worry about getting pregnant. Others show no change, but some do show a decline in both interest and orgasmic capacity, often due to changes like vaginal dryness, as previously noted.

The best predictor of sexual activity for middle-aged women is their sexual activity when younger. Positive personal attitudes about sexuality and past positive experiences regarding the importance of sex are important determinants of sexual functioning in aging

women (Sherwin, 1991; Haavio-Mannila & Kontula, 1997). Some studies have found that women who have regular sexual activity throughout life experience less vaginal atrophy, and thus less of a decline in vaginal lubrication during sexual arousal, after menopause (Bachman, 1989; Leiblum et al., 1983). However, data on sexual activity in middle-aged and older women are limited. For married women, the frequency of and satisfaction with sex are closely associated with how happy they are in the marriage (Hallstrom & Samuelsson, 1990; Hawton, Gath, & Day, 1994), but many middle-aged women do not have a sexual partner. Thus, many researchers feel that frequency of masturbation is the best index to use in measuring female sexuality in the middle-aged and elderly years. Christenson and Gagnon (1965), for example, reported that women aged 50 to 70 who were deprived of a sex partner masturbated twice as often as married women of the same age. Brecher (1984) found that nearly half of the women aged 50 to 59 that he studied masturbated (an average of three times a month). Janus and Janus (1993) recently obtained the following results for the relation between female masturbation and age:

Frequency	Age 18–26	Age 27–38	Age 39–50	Age 51–64
Daily	2%	3%	2%	1%
Several times weekly	6%	11%	12%	7%
Weekly	10%	16%	16%	12%
Monthly	9%	17%	17%	16%
Rarely	29%	33%	37%	34%
Never	44%	20%	16%	30%

(Reprinted with permission of John Wiley & Sons, Inc.)

As you can see, there is very little change through the mid-sixties.

Estimates of the percentage of older women (and men) who masturbate are probably much lower than the true numbers because many older adults do not want to give this information, even if they are willing to divulge the frequency with which they have sexual intercourse. (How would *you* answer a survey question such as "How often do you masturbate in a week?") Sexuality in married women also was related to the age of a husband. Women with younger husbands were found to have more active sex lives than women with older husbands (Brecher, 1984).

For those few middle-aged women who do show a loss of interest in sex, it may not be due to lack of opportunity or any physiological change. Some women suffer from nonsexually related anxiety and depression at about the time of menopause and this, of course, can affect a person's interest in sex. Sociologist Pauline

Bart (1971) has pointed out, for example, that this is about the age that most women have to adjust to their children growing up and leaving home. Bart found that it is the traditional "supermother" type who is most affected by this problem of the "empty nest." It is also a time when a previously family-oriented woman may find that a lack of skills prohibits her from finding meaningful employment. Moreover, some women may feel that they have grown apart from their husbands over the years and may not be interested in having sex with their husbands (this is not to say that they have lost all interest in sex). In short, it is difficult to determine whether a decline of interest in sex is due to a hormonal change or to other changes in a female's life.

MALE SEXUALITY: PHYSICAL CHANGES WITH AGE

At the 1992 meeting of the North American Menopause Society, Dr. Christopher Longcope summarized the conclusions of most of his colleagues: Men normally do not experience menopause. Except in cases where males lose their testicles from surgery (in which case they do show symptoms, including fatigue and hot flashes), there is no sudden loss of hormones as experienced by women. Although the term "male menopause" is commonly used, this generally refers to the midlife psychological and emotional reactions that some men have as a response to changing family relationships (e.g., children growing up) and/or an inability to achieve earlier career goals. Society's attitudes may contribute to this, for many people in our culture treat elderly people like little kids or as nuisances. In cultures where elderly people are highly respected and revered (Oriental cultures, for example), there is no term that is equivalent to male menopause.

Men do show a gradual decline in testosterone levels that begins in their late teens. By age 70, testosterone levels are only 60–70 percent of what they were in early adulthood. This decrease in testosterone will result in some physiological changes as men age. The changes include a longer time to become erect and a less firm erection, shrinkage of the testicles, less forceful ejaculation (often resembling a dribble), and a longer refractory period after orgasm (the length of time before a male is able to have another orgasm), a decline in muscle mass and strength, a reduction in bone calcium, sparse beards, and slightly higher voices (Masters & Johnson, 1966). The production of sperm continues into old age, but it, too, declines after age 40. Nevertheless, many men in their seventies and eighties have fathered children.

Many men show a decline in erectile capacity as they reach late middle and old age (Davidson et al., 1983). This is often true even when there is no decrease in health or other impairment. Although this, too, can be related to decreasing testosterone levels,

some studies have found no relation between testosterone levels and sexual function and behavior (Rowland et al., 1993; Schiavi, 1990). One study suggested that a decline in erectile capacity may also be related to decreased sensory/neural and autonomic functioning (Rowland et al., 1993).

In the mid-1990s there was new interest in the age-related changes in men and calls for medical treatment of the previously described "male climacterium problems." A 1996 cover article in *Newsweek*, for example, was about whether testosterone replacement therapy can keep men young and virile (Cowley et al., 1996). Health food stores began pushing DHEA (dehydroepiandrosterone, a hormone that declines with age) to build muscle mass and increase energy and libido. We may be on the verge of a medicalization of the male "climacterium" similar to what has happened with women (see Oudshoorn, 1997).

If a man is not aware that the previously mentioned changes are normal, he may overreact to them and start to doubt his sexuality, and this performance anxiety can lead to serious problems. Dr. Paul Costa Jr. of the National Institute on Aging recently said, "Some

Figure 10–12 Most surveys have found that there is a gradual decline in sexual activity for people in their 50s. The best predictor is sexual activity when younger.

older men may say their erections aren't as big as they recall them once being. But then their partners say, 'Well, dear, you overestimated them back then, too'" (quoted in Angier, 1992).

MALE SEXUALITY: BEHAVIOR

For most men, there is a gradual decline in sexual activity and interest after age 50 (see Segraves & Segraves, 1995). A man's sexual performance can be affected by many things, including declining health, anxiety over performance, and various medications. However, as with women, the *best predictor of a middle-aged man's level of sexual activity is his level of activity when younger* (e.g., Haavio-Mannila & Kontula, 1997).

As was the case with women, frequency of masturbation can be used as one indicator of sexual activity. Brecher (1984) found that two-thirds of men in their fifties masturbated (an average of five times per month), but that only half of men in their sixties masturbated (an average of four times per month). More recently, Janus and Janus (1993) found very little change in the masturbatory behavior of men up through their mid-sixties:

	Age			
Frequency	**18–26**	**27–38**	**39–50**	**51–64**
Daily	6%	9%	7%	9%
Several times weekly	18%	19%	16%	23%
Weekly	19%	25%	19%	22%
Monthly	8%	9%	12%	4%
Rarely	23%	26%	30%	18%
Never	26%	12%	16%	24%

(Reprinted with permission of John Wiley & Sons, Inc.)

Remember, for many men orgasms experienced during masturbation are in addition to orgasms experienced during sex with a partner (see previous section). Still, some researchers have found that the frequency of orgasm does drop off for some men in their fifties and sixties (Martin, cited in Pesman, 1993). However, this could be due to being accustomed to one's long-term partner (Blumstein & Schwartz, 1983). There is no decline with age in the number of erections men experience at night (see Pesmen, 1993).

For those middle-aged and elderly men who are showing physiological changes, such as a longer time to become fully erect, there is a positive side—more time spent in foreplay, which generally translates into greater sexual happiness for the partner. Even if age interferes with quantity, there is no reason it has to interfere with quality. Sexual relations involve more than just using the penis—they include using the lips, tongue, hands, and other body parts, as well as one's

emotional responsivity. Many men who experience declining erectile capacity with age report that their overall sex life is still just as good (Rowland et al., 1993). Good sexual communication is a total human sexual response.

As a parting thought, here is a quiz (Pesmen, 1993):

> *Question:* What is the most powerful sex organ?
> *Answer:* The brain.
> *Question:* What, then, is the second most powerful sex organ?
> *Answer:* The skin.

As Pesmen points out, "Fortunately for most men, the brain and the skin are stable, healthy partners in mid-life and beyond. Setting penis power aside, no matter what normal changes beset a man's sex organs later in life, a continuing capacity for sexual pleasure remains."

THE ELDERLY YEARS (AGE 60+)

"I didn't think that elderly people thought about sex. I thought when you got older you just lost all interest. . . ."

(from the author's files)

In our culture elderly people are regarded as asexual human beings—neuters. Sex is something enjoyed only by the young and healthy (Gochros, Gochros, & Fisher, 1986). According to Rose and Soares (1993), "Evidence of sexuality among the aged is often joked about, dismissed, or defined as troublesome behavior. A sexually active older man is seen as a 'dirty old man' and a sexually active older woman is considered preposterous or embarrassing. This culturally embedded denial may be rooted in people's difficulty accepting the sexuality of their own parents."

In the 1980s, some behavioral scientists and physicians began to challenge our culture's stereotype of old people as over the hill and sexless. Several studies showed elderly people continuing to enjoy sex well into their sixties, seventies, eighties, and beyond. Among the first was a study done for *Consumer Reports*. Of the people aged 70 and older in the survey, two-thirds of the women (including 81 percent of the married women) and 80 percent of the men said that they were still sexually active (Brecher, 1984). They all had intercourse with their partners, and over half engaged in oral-genital sex. The survey included over 4,000 people, but this was a response rate of only 0.2 percent (so we do not know if the other 99.8 percent

Box 10-C SEXUALITY AND HEALTH

Sex and Heart Attacks

Victorian-era physicians, you recall, believed that sex was harmful to one's health (see Chapter 1). Even Freud believed that loss of semen was as detrimental to a man's health as loss of blood (Haller & Haller, 1977). We now know, of course, that this is not true, but are there other sex-related health problems? An article in a 1986 issue of *National Safety and Health News* startled many people, particularly older ones. The article reported the results of a Japanese study in which investigators looked at the autopsies of 5,000 people who had died suddenly. Thirty-four had died during intercourse. Are people with heart disease, especially former heart attack victims, putting themselves at risk by having sex?

Many people apparently think so. About one-quarter of all men who have had heart attacks never resume sexual activity, and half do so only occasionally (Kolodny et al., 1979). However, physicians and sex therapists say the greatest barrier is not the physical condition of the heart patient, but anxiety and fear. Wives of former heart attack victims also worry that sex might cause another attack (Gould, 1989). This kind of anxiety and stress can cause sexual problems. Female heart patients are less likely than men to cease sexual activity or develop sexual problems (Kolodny, Masters, & Johnson, 1979; Schover & Jensen, 1988).

A recent study from Harvard Medical School of nearly 2,000 patients who had suffered heart attacks found that the chance of having another heart attack resulting from sex was extremely small, only 20 in 1 million (Muller et al., 1996). During sex with one's usual partner, the maximum heart rate increases to about 120 beats per minute (Tobis, 1977). This is no greater, and perhaps less, of an increase than one would experience during a brisk walk or some other form of moderate exercise (Masur, 1979; McLane, Krop, & Mehta, 1980). Extreme physical or emotional stress and things like alcohol or drugs might be risks, and so might sex involving emotional tension or extreme physical exertion. It must be noted, therefore, that of the 34 people that the Japanese study found to have died during sexual intercourse, all were intoxicated or near intoxication and 30 were having extramarital relations with a partner who, on average, was 18 years younger.

Most physicians believe that sexual activities can be resumed within a few weeks after a heart attack (see Tardif, 1989). Generally, masturbation can be resumed first. A general rule of thumb used by some physicians is that sexual activity should not be resumed until the patient can engage in moderate physical exercise (such as climbing two flights of stairs) without exhaustion.

For most people, daily exercise in moderate amounts is recommended for strengthening cardiovascular endurance. A daily brisk walk of at least 20 minutes is highly recommended for older people (Kligman & Pepin, 1992). Regular exercise also improves mental health (Taylor, Sallis, & Needle, 1985). The American College of Sports Medicine recommends any physical activity that raises heart rate at least to 60 to 80 percent of your range. Sexual activity with one's usual partner produces heart rate changes like this. It is very possible, therefore, that regular sexual activity is beneficial in maintaining cardiovascular endurance. Sounds like a lot more fun to me than a stationary bike.

---

were as sexually active). Another study reported that sexually active elderly people found their sexual pleasure to be as good as or better than when they were younger (Starr & Weiner, 1981). Similar results were reported for elderly homosexual individuals (e.g., Berger, 1982; Raphael & Robinson, 1980).

Did the researchers find any age at which sexual activity stopped? Well, in a study of 202 *healthy* men and women aged 80 to 102, Bretschneider and McCoy (1988) found that three-fourths of them still fantasized about sexual relations and enjoyed touching and caressing their partners. Nearly half were having sexual intercourse and a third were engaging in oral-genital sex. Sexual activity was limited by living arrangements and the availability of partners, not by a lack of interest. Here is something to think about: Studies with animals have found that sexually active rats live a great deal longer than sexually inactive rats (Knoll, 1997) and a recent study reported the same for men (Smith, Frankel, & Yarnell, 1997). Could it be that these very elderly people lived as long as they did partly because they enjoyed sex regularly?

The conclusion reached by the behavioral scientists who conducted these studies was a very positive one—sexuality does not have to end at any given age. In the words of Myrna Lewis and Robert Butler (1994), authors of *Love and Sex After 60,* "Relatively healthy older people who enjoy sex are capable of experiencing it often until very late in life."

So which is correct—the cultural stereotype of the aged as sexless, or the optimistic picture painted by the recent surveys of the elderly as a still sexually active group? The truth may be somewhere in between. One recent report, for example, accused the earlier reports and their conclusions of sounding like "pep talks," trying to convince everyone that there would be no loss of sexual pleasure with aging (Rose & Soares, 1993). The authors said the earlier reports "glossed over certain realities of aging which directly affect sexuality": the much greater ratio of women to men, the greater number of people suffering from chronic illness and disability, and the greater number of people dependent on others for daily personal care. Let us consider these problems.

The earlier studies were largely of married men and women—older people who had partners. The ratio of single women to single men that we discussed in the section on middle age becomes even more unfavorable to women over the age of 60. In the United

Figure 10–13 From the moment of birth to the day we die, we are all sexual beings.

States, women outlive men by an average of 8 years. For people over 65, there are 147 women for every 100 men, and only 100 single women for every 27 single men. By age 85, there are 256 women for every man (Robinson, 1983; Weg, 1996). Thus, many women are faced with involuntary sexual retirement. The recent survey by Laumann and colleagues (1994) found that 7 out of 10 women in their seventies were no longer having partnered sex.

In addition to having partners, most of the respondents in the early sex surveys (reporting elderly people having sexual relations) were people in good health. However, many people experience serious medical problems and disabilities as they grow older. Problems like heart disease, stroke, arthritis, diabetes, prostate problems, and problems due to alcohol, smoking, and overeating are common (Segraves & Segraves, 1995; Weg, 1996). By age 65, about a third of all men are experiencing erectile difficulties, mostly due to medical problems, and by age 70 more than half have erectile dysfunctions (Diokno, Brown, & Herzog, 1990; Mulligan et al., 1988; Weizman & Hart, 1987; see Chapter 13).

In the early 1990s, about 13 percent of all persons living in nursing homes and over the age of 65 were married (Kaplan, 1996). However, opportunities for sexual relations are very limited for elderly people with disabilities who are dependent on others for their care. Administrators and staff in most nursing homes are interested in institutional efficiency and tend to view the elderly as sexless (Kaplan, 1996; Weg, 1996). Some nursing homes require married couples to live in separate rooms (when both are institutionalized), and most deny them privacy when the healthy partner of an institutionalized spouse comes to visit (Kaplan, 1996; Wasow & Loeb, 1979). Nursing home residents often say that a lack of privacy prevents them from sexual relations, or even opportunities for masturbation (Spector & Fremeth, 1996). In short, elderly people with sexual needs who live in nursing homes are "truly members of a forgotten, neglected population" (Wasow & Loeb, 1979).

More recent large-scale studies find a steady decline in sexual activity with age (see Segraves & Segraves, 1995, for a review). Marsiglio & Donnelly (1991) found that two-thirds of the people they studied aged 60 to 65 had had sex during the previous month, as compared to 45 percent of people aged 71–75 and 24 percent of people aged 76 and older. Clements (1996) recently reported that 55 percent of people aged 65–69 were sexually active, compared to 28 percent of people aged 75–79 and 13 percent of people over 85. Another recent study of people over 70 found that 30 percent had had sex in the last month (Matthias et al., 1997). An Australian study reported higher percentages of elderly people having sex, but still a decline with age (Minichiello, Plummer, & Seal, 1996). Clements found that of those elderly persons not having sex, two-thirds of the men and nearly a third of the women say they would like to have sex. As previously discussed, the main things keeping older persons from having sex were physical (or mental) health and lack of a partner. However, these are not the only reasons. Other factors include education (more educated people tend to remain more active), religion (more religious people tend to be less sexually active), societal pressures to conform to (asexual) stereotypes, and, of course, one's attitude about sex when younger.

What can we conclude about sexuality in the elderly years? Sexual opportunities are limited for single women, but for healthy men and women with healthy partners, sexual activity will probably continue throughout life if they had a positive attitude about sex when they were younger. Sexual interests may continue even for those with serious medical problems and disabilities. Sex surveys tend to measure sexual behavior in terms of physical acts such as sexual intercourse and masturbation, but there is more to sexuality than this. Elderly people often report great sexual satisfaction with emotionally erotic experiences (Shaw, 1994). Sexual pleasure, for example, includes touching, caressing, physical closeness, and intimacy. These can often be achieved even in the most extreme cases:

> Mr. N was deaf, blind, and unable to talk due to a recent laryngotomy. He was quite dependent on Mrs. N who had severe arthritis and had a mastectomy years ago. They typically sat close together, in part to aid communication due to Mr. N's impairments. But the way they held hands revealed great tenderness and affection. They slept in twin beds and indicated they no longer engaged in intercourse. Yet, they had a nightly ritual in which Mr. N helped his wife bathe and then gave her a massage in bed, which they described with some embarrassment and blushing. Interestingly, through this sensual routine, Mr. N, usually the more dependent of the two, was able to offer something of value to his wife.

> (From Rose and Soares, 1993. Reprinted with permission of Madeline Rose and the publisher.)

Perhaps the biggest problem for elderly people interested in physical relations is our culture's negative stereotypes of the aged as sexless (Barrow & Smith, 1992). Many people poke fun at, or discriminate against, other people who are different from themselves, and this often includes the elderly. But unless there is a tragedy, all of us will be elderly some day. Do you enjoy sexual relations now? Do you ever want to give it up? If not, how do you think you will feel if one day younger people think you are silly, preposterous, or even lecherous or perverted for wanting the pleasure of someone else's touch? The time to address these attitudes is now, not later. We are all sexual human beings, from birth to death.

Key Terms

adolescence 254
climacteric 267
cohabitation 260
delayed puberty 254
egocentricity 247
gynecomastia 31, 253
hot flashes 268

menarche 62, 251
menopause 55, 267
necking 255
nocturnal emission 252
open marriage 264
peer pressure 256
petting 255

phallic stage 206, 248
precocious puberty 253
puberty 250
secondary sex characteristics 251
serial monogamy 124, 259
swinging 265
young adulthood 258

Personal Reflections

1. Does the thought of young children examining and touching their own genitals make you uncomfortable? If so, why? How would you feel if you found your 5-year-old boy with his pants down playing "doctor"? Why? Would your reaction be any different if you found your 5-year-old girl with her pants down playing "doctor"? Why?

2. Try to recall the feelings you experienced during puberty as your body changed. Would better understanding of these events (via parents or sex education classes) have made it a more positive experience?

3. What kinds of pressure to engage in sex have you experienced (from dates, friends, the media, etc.)?

Have you ever engaged in sex because of pressure (to please a boyfriend or girlfriend, to feel older, to be "like everyone else," etc.)? How have you dealt with the pressure? How could you deal better with the pressure?

4. What would you tell a young teenager about sex? Why? Pretend you have an adolescent child who has just told you he or she is considering having sexual intercourse. What is your reaction? Does your answer change depending on the sex of the child? Why?

5. Does the thought of your parents *enjoying* sexual relations make you comfortable or uncomfortable? Why? If you have difficulty viewing your parents as sexual human beings, what effect do you think that will have on your own sexual relationship when you and your partner become parents (and view one another as parents)? Do you think it is healthy or unhealthy for parents to hide their sexuality from their children? Why?

6. Does the thought of elderly people *enjoying* sexual relations make you comfortable or uncomfortable? Why? Do you hope to still enjoy sexual relations when you are in your fifties and older? If not, at what age do you think that you will stop enjoying sex? Why?

Suggested Readings

Adler, J., et. al. (1996, September 30). Adultery: A new furor over an old sin. *Newsweek.*

Alan Guttmacher Institute (1994). *Sex and American teenagers.* New York: Author. A comprehensive survey of the sexual behavior of teens.

Allgeier, E. R. (1985, March). Perceptions of parents as sexual beings: Pocs and Godow revisited. *SIECUS Report.* Discusses the view many children have of their parents as married but asexual.

Barbach, L. (1994). *The pause.* New York: Signet Books. All about menopause.

Beck, M., et al. (1992, May 25). Menopause: The search for straight talk and safe treatment. *Newsweek. Newsweek's* cover story covers both the medical and feminist perspectives on menopause.

Bulcroft, K., & O'Conner-Roden, M. (1986, June). Never too late. *Psychology Today.* Concludes that dating patterns of people over 65 are similar to those of younger adults.

Clements, M. (1996, March 17). Sex after 65. *Parade Magazine.* Discusses sexual activity and satisfaction in the elderly.

Gordon, S. (1986, October). What kids need to know. *Psychology Today.* Discusses some topics about sexuality that children need to know and gives advice to parents and schools about how to teach these subjects.

Janus, S., & Janus, C. (1993). *The Janus report.* New York: John Wiley & Sons. Results of a nationwide survey of sexual behavior.

Michael, R., et al. (1994). *Sex in America.* Boston: Little, Brown. Results of the Laumann et al. survey, the most scientifically conducted survey ever done. Written for the general public.

Segraves, R. T., & Segraves, K. B. (1995). Human sexuality and aging. *Journal of Sex Education and Therapy, 21.* A very thorough review of sexuality and health in aging persons.

Seidman, S. N., & Rieder, R. O. (1994, March). A review of sexual behavior in the United States. *The American Journal of Psychiatry.* An excellent review of the numerous sex surveys.

Sprecher, S., & McKinney, K. (1993). *Sexuality.* Newbury Park, CA: Sage. Sexual activity with dating, marital, and cohabiting relationships.

Willis, J. (1986, July/August). Demystifying menopause. *FDA Consumer.* Discusses the normal experiences of menopause.

CHAPTER 11

When you finish studying this chapter, you should be able to:

1. Explain what is meant by "normal" sexual behaviors;
2. Identify the major historical, as well as current, sources of negative attitudes toward masturbation, and discuss recent findings about this sexual activity;
3. Describe what occurs during the normal sleep cycle that can result in nocturnal orgasms;
4. Summarize the research findings about the frequency, quality, and role of sexual fantasies in enhancing sexual expression;
5. Explain why different cultures, as well as individuals within a culture, use different positions of sexual intercourse;
6. Discuss ways to keep the sexual aspects of a relationship from becoming ritualized;
7. Define fellatio and cunnilingus and describe the current research findings on these activities; and
8. Define a sexually healthy person.

Adult Sexual Behaviors and Attitudes

For the human species to survive, males and females must have sexual intercourse. We need only to look at the large number of people in the world to know that this is not an uncommon sexual practice. But what about other behaviors—that is, behaviors that are not necessary for procreation? How many people engage in oral-genital sex? Is your partner unusual for wanting to try different positions? Is it normal to have sexual fantasies? Is masturbation unnatural and bad for your health? Many people are unsure about the answers to these questions, and thus may have anxieties about whether the sexual behaviors they or their partners wish to engage in are "normal."

WHAT IS NORMAL?

Just what is "normal" sexual behavior? The answer is, it depends. Many factors are involved, including the age of the individual and the society and period in history in which the person lives. What is considered normal changes over the life span of an individual and from generation to generation.

In this chapter you will read about several sexual behaviors that can be considered normal from a statistical point of view. If a behavior is *statistically normal*, it means that a large number of people engage in it. However, you should not consider yourself to be "abnormal" if you have not engaged in all of the behaviors discussed here. It is best to think of normal behavior as a *range of behaviors*. Today, sexual intercourse, masturbation, and oral-genital sex (to name a few examples) are considered to be within the range of behaviors labeled normal for adults by mental health professionals.

For any particular behavior, there is also the question of how frequently people engage in it. How often is normal? Averages can be (and are) computed, but you should not interpret "normal" as meaning only the average value. Again, normal involves a *range of values*. For example, what is the normal number of times per month that married women masturbate? The answer is a range (from 0 to some higher number). Most behaviors are considered normal as long as the individual does not suffer physical and/or psychological damage, does not inflict physical and/or psychological damage on others, and the behavior does not interfere with being able to deal with the activities of daily life.

The incidence and frequency of the various sexual behaviors you will read about in this chapter were obtained by taking sex surveys. Be sure to review the section on surveys and sampling procedures in Chapter 1.

MASTURBATION

"Until I became sexually active with a partner, I masturbated every day. Even after marriage, I continue to masturbate." (male)

"The guilt I feel right after masturbating is so painful that I cry at times." (male)

"I have a steady girlfriend who satisfies my sexual needs. But sometimes I am not able to satisfy hers. This is the primary reason why I enjoy masturbating—because I don't have to worry about trying to satisfy anyone but me." (male)

"I didn't masturbate until after I was sexually active,

masturbation Self-stimulation of one's genitals.

which was in college. In between sex with my boyfriends, I didn't know there were ways I could 'pleasure' myself with sex—and without a partner. Now, after finding it, I truly enjoy it." (female)

"I had masturbated long before I became sexually active and, as long as I kept it to myself, I never seemed to have a problem with it. Now, when my boyfriend asks me if I ever masturbated, I can't help but feel guilty." (female)

"I've masturbated since about the age of 14. I still do, with and without my husband present. I have no guilty feelings about it, not even when I was younger. There are times when I would rather masturbate than make love. Not that I would give up intercourse. It's just nicer sometimes." (female)

"All through my childhood I thought that masturbation was something only guys could do. But Cosmo started printing all kinds of articles telling women to masturbate so they can show their partner what to do, and I got confused. I know I shouldn't be ashamed of my body, but I still can't bring myself to do it." (female)

"I am 36 years old and never in my wildest dreams would I have thought I would be masturbating, but I did and I think it's great because I can stimulate parts that a male partner would not have the slightest idea about." (female)

"When my boyfriend and I engage in sexual activities, I get absolutely aroused when he masturbates in front of me." (female)

(from the author's files)

ATTITUDES ABOUT MASTURBATION

When you were growing up, what kinds of things did you hear might happen to you if you masturbated? Many of us were told (by friends, acquaintances, and sometimes parents) that **masturbation** could cause us to go blind or deaf; have nosebleeds, heart murmurs, acne, or painful menstruation; cause hair to grow on the palms of our hands; and finally, if we didn't stop, result in insanity. Others of us were simply told that we had better not do it to excess. The consequences of doing it to excess were never quite spelled out, but the possibilities that it was bad for our health or could lead to insanity were always implied. And scarier yet, what was considered excessive? Whatever it was, most of us convinced ourselves that it was more than we did it.

It is obvious that masturbation is considered bad by some people, but where did these stories come from? Negative attitudes about masturbation have, in fact, been handed down from generation to generation for centuries. The Greek physician Hippocrates believed that overindulging in sex was harmful to one's health. Ancient Chinese cultures condemned male masturbation as a waste of *yang* (male essence), while the Biblical Hebrews considered it "spillage of seed," punishable by death (recall, from the Bible, the divine slaying of Onan for spilling his seed). The early Roman Catholic Church, you recall, considered the only legitimate purpose of sex to be for procreation (St. Augustine called intercourse an unpleasant necessity), and sexual behaviors that did not have this as a goal (i.e., sex performed for pleasure) were considered immoral. Many scholars believe that the word *masturbation* is derived from the Latin words *manus* ("hand") and *sturpore* ("to defile") or *tubare* ("to disturb").

In 1741, a Swiss physician named S. Tissot published a book entitled *Onania, or a Treatise Upon the Disorders Produced by Masturbation*. In his book, Tissot claimed that excessive sex, and especially masturbation, was dangerous because it deprived vital tissues of blood and led to insanity. As proof of this, he said, one only had to look at the many men in insane asylums who sat around openly masturbating. Tissot's views were accepted by many physicians. In Victorian times (the 1800s and early 1900s), many physicians also believed that loss of semen was as detrimental to a male's health as loss of blood. Freud, for example, believed that the loss of one ounce of semen produced the same degree of fatigue as the loss of 40 ounces of blood. The Victorians, you recall, vehemently denied any pleasurable component of sex and were particularly concerned about masturbation (see Hall, 1992). Freud believed that people who masturbated had "poisoned" themselves and that masturbation caused nervous disorders or neurosis ("neurasthenia") (Groenendijk, 1997). Not only were Victorian-era children forced to wear chastity belts and metal gloves to bed (see Figure 1–4 on page 13), but castration, circumcision, and clitoridectomy became popular during this time as means by which parents tried to prevent their children from masturbating. For an example of extreme Victorian beliefs, read the following passages from a book titled *Perfect Womanhood* written in 1903 by "Mary Melendy, M.D., Ph.D.," which contains the following advice to women about what to tell their male children about masturbation:

> Impress upon him that if these organs are abused, or if they are put to any use besides that for which God made them . . . they will bring disease and ruin upon those who abuse [them]. . . . He will not grow up happy, healthy and strong.
>
> Teach him that when he handles or excites the sexual organs, all parts of the body suffer, because they are connected by nerves that run throughout the system, this is why it is called "self-abuse." . . .
>
> . . . The sin is terrible, and is, in fact, worse than lying or stealing!
>
> If the sexual organs are handled it brings too much blood to these parts, and this produces a diseased condition, it also causes disease in other organs of the body, because they are left with a less amount of blood than they ought to have. . . .
>
> It lays the foundation for consumption, paralysis and heart disease. It weakens the memory, makes a boy careless, negligent and listless.
>
> It even makes many lose their minds; others when grown, commit suicide.

Similarly, Dr. J. H. Kellogg (1888) wrote the following about the effects of masturbation in girls:

> Wide observations have convinced us that a great many of the backaches, sideaches, and other aches and pains of which girls complain, are attributable to this injurious habit. Much of the nervousness, hysteria, neuralgia, and general worthlessness of girls originates in this cause alone.
>
> The period of puberty is one at which thousands of girls break down in health. The constitution, already weakened by a debilitating, debasing vice, is not prepared for the strain, and the poor victim drops into a premature grave.

He wrote the following about masturbation in boys:

> In solitude he pollutes himself, and with his own hand blights all his prospects for both this world and the next. Even after being solemnly warned, he will often continue this worse than beastly practice, deliberately forfeiting his right to health and happiness for a moment's mad sensuality.
>
> (J. H. Kellogg, *Plain Facts for Young and Old*)

This is the same J. H. Kellogg associated with the breakfast cereals. He invented cornflakes to be used as a breakfast food that would curb youthful lust. Gra-

"*Now* they tell us masturbation is harmless!"

Reproduced by special permission of *Playboy* magazine: copyright © 1971 by Playboy.

ham crackers were invented by Dr. Sylvester Graham for much the same purpose (Money, 1985). In his 1834 book, Graham wrote that if a boy masturbated, he would turn into "a confirmed and degraded idiot."

Although not all cultures have had negative attitudes about masturbation (e.g., the ancient Egyptians believed that the world was created when the God Atum ejaculated), negative attitudes are still widely found in Western societies (Gregersen, 1982). For example, the Catholic Church called it an "intrinsically and seriously disordered act" in 1975, and Orthodox Jews regard it as one of the worst sins mentioned in the Torah.

When former Surgeon General Joycelyn Elders was fired in 1994 for talking about masturbation at a conference on AIDS, sociologist John Gagnon was quoted as saying, "It remains a puzzle for me why a relatively innocuous behavior evokes tides of anxiety and fear in otherwise well adjusted people."

INCIDENCE OF MASTURBATION

With this long history of scare tactics, you would expect, then, that very few people masturbate. In fact, the large majority of human beings masturbate. Kinsey and colleagues (1948, 1953) found that 92 percent of the men he surveyed had masturbated by age 20. It was generally the first sexual experience to orgasm for adolescent males. Only a third of the women in Kinsey's study reported that they had masturbated by age 20, but another third had done so by age 40 (a total of 62 percent). Nearly identical results were found 20 years later by Hunt (1974). In other words, men are more likely to start masturbating before they experience sexual intercourse, whereas many women do not start

masturbating until after they start engaging in sexual intercourse (Gagnon, 1977).

Not all people who have masturbated continue to do so regularly. Laumann and colleagues (1994) found that for Americans under the age of 60, only 60 percent of men and 40 percent of women had masturbated in the past year. Both Kinsey's and Laumann's groups found that the better educated a person was, the more likely it was that he or she masturbated. Among today's college students, about twice as many men as women masturbate (Atwood & Gagnon, 1987; Leitenberg, Detzer, & Srebnik, 1993), again reflecting the gender and age differences found by Kinsey (1948, 1953). On the average, men who masturbate do so two or three times more frequently than women who masturbate (Kinsey et al., 1953; Hunt, 1974; Leitenberg et al., 1993). Interestingly, in a recent review of many studies, it was found that although women report a much lower incidence of masturbation than do males, their attitudes toward masturbation do not differ from those of men (Oliver & Hyde, 1993).

Are all of these people risking their health or sanity? No. Masturbation, whether done once a month or three times a day, does not cause any of the problems mentioned previously, and few doctors today, if any, would make such claims. *It is a perfectly normal human behavior.* How can I be so confident as to call it normal when historically it has been so condemned? Just look at the statistics—when the vast majority of people engage in a particular behavior, it has to be considered normal. It is a behavior engaged in by a great many other species as well (Ford & Beach, 1951). Still, there are some who continue to feel guilty or wrong about it—nearly half of all people who masturbate, according to Laumann et al. (1994). Among college students, 40 percent of those who do not masturbate say it is because they believe it is immoral (Atwood & Gagnon, 1987).

METHODS OF MASTURBATION

People differ in the way they masturbate, but men display less variation than do women. Almost all men masturbate by rhythmically stroking up and down the body and glans of the penis with one hand (Hite, 1981). They may differ in how much of the penis they stimulate, and in the rhythm and pressure applied, but it is generally still a stroking motion. Very few thrust. Many men simultaneously stimulate their testicles and other parts of their bodies as well. In order to enhance arousal, many males fantasize and/or look at sexually explicit pictures while masturbating. Hunt (1974) found, for example, that 75 percent of the men he surveyed had fantasies of sexual intercourse with a loved person during masturbation.

Some individuals (mostly adolescent males) try to make their orgasms more intense by performing activities that deprive them of oxygen, such as placing plastic bags over the head or hanging by the neck (Saunders, 1989). This is called *autoerotic asphyxiation.* Although masturbation per se does not cause health problems, these behaviors are dangerous, and 500 to 1,000 people die each year while engaging in them (Byard, Hucker, & Hazelwood, 1990). Aside from simply holding your breath, you should not engage in any behavior that deprives you of oxygen—during sex or at any other time.

Most women who masturbate do so by stimulating the clitoris, labia minora, and/or the entire vulva. Only 20 percent or fewer insert a finger or anything else into their vaginas while masturbating (Hite, 1976; Kinsey et al., 1953). Those who do should not consider themselves abnormal, for remember, "normal" encompasses a wide range of behaviors, not just a single behavior. The manner in which women stimulate themselves is quite diverse. Many prefer to lie on their backs while masturbating, while some prefer a face-down position, with considerable individual preferences in the position of the legs. Some use a single finger, while others use their entire hand to apply pressure to a wider area. Kinsey's group (1953) found that about 11 percent of women stimulate their breasts during masturbation. Although only a few women in his sample reported using a vibrator, the number who do so today has no doubt increased because of the easy availability of these products now. Two percent of the women surveyed in Laumann et al.'s study (1994) said that they had purchased a vibrator or dildo in the past year.

"In the masturbation section I believe you should emphasize the importance of a vibrator for women. Methods of masturbation seem apparent to males, but I think women need more guidance in this area. In my experience a vibrator is the most intensely pleasurable way to masturbate."

(from the author's files)

Like men, most women enhance their arousal by fantasizing during masturbation (Hunt, 1974).

FUNCTIONS OF MASTURBATION

Why do people masturbate? Quite simply, because it feels good. The genitals, you recall, have lots of nerve endings and are thus very sensitive to touch. As you learned in the chapter on development, almost all young children touch their genitals when they have the opportunity (e.g., in the bathtub) because it gives them physical pleasure (Lidster & Horsburgh, 1994). By the early teens, however, genital touching is not just physical. It is often accompanied by erotic thoughts. Al-

though masturbation allows for sexual release when a partner is not available, it does not necessarily cease when one forms a sexual relationship with someone. Hunt (1974) found that 72 percent of the men and 68 percent of the women he studied continued to masturbate after marriage. More recently, Janus and Janus (1993) reported that 66 percent of the men and 67 percent of the women they surveyed agreed or strongly agreed that "masturbation is a natural part of life and continues on in marriage." In fact, Laumann et al. (1994) found high rates of masturbation among young people with sexual partners, and that the more sex a person has, the more likely he or she is to masturbate. In other words, masturbation is not so much an outlet for lack of sex as it is a reflection of a sexually active life-style.

Why do people with sexual partners masturbate? *Variety*—it is a sexual experience that is different (not necessarily better) than sex with their partner.

"I masturbated a lot when I was without a partner, and now I still do. The orgasm is so different than those I get during intercourse. . . . My boyfriend doesn't understand . . . but I love making love with him."

(from the author's files)

Freud thought that adult masturbation was immature. Even today, some people argue that it may be habit-forming and prevent development of "normal" adult sexual relations. However, Kinsey's group (1953) found that women who were able to masturbate to orgasm without guilt or shame during adolescence generally had the least difficulty in making the transition to enjoying sex with a partner. More recent studies report that not only is masturbation during adolescence not harmful to sexual adjustment in young adulthood (Leitenberg, Detzer, & Srebnik, 1993), but that masturbation during marriage is often associated with a greater degree of marital and sexual satisfaction (Hurlbert & Whittaker, 1991). So if you suspect your partner of masturbating occasionally, do not assume that he or she is unhappy with your sex life. There would only be a problem if your partner was masturbating (indicating a need for a sexual outlet) but showing little or no interest or sexual desire for you.

Today, some therapists prescribe masturbation as part of their treatment for orgasm problems in women and for greater ejaculatory control in men (Barbach, 1980; Kaplan, 1974; LoPiccolo & Lobitz, 1972; LoPiccolo & Stock, 1986). Thus, what was once a taboo is now often used as part of sexual therapy. This is not to imply, however, that everyone should masturbate, particularly if they are comfortable with their sexuality. Just as it was once wrong to make people feel bad because they masturbated, it would be equally wrong to make people feel abnormal if they did not. The important point is that those who have a desire to masturbate should be able to do so without feelings of guilt or worries about their health.

NOCTURNAL ORGASMS

"During wet dreams I always seem to wake up when I reach climax and am greeted with a puddle of sperm. I am not embarrassed by such occurrences—it just forces me to wash my sheets." (male)

"I was 17 years old, and one morning I woke up finding my sheets wet. I thought I urinated in bed at first until I realized it wasn't urine. I was ashamed and didn't want anyone to know what happened. I quickly changed my sheets and took a bath." (male)

"I experience nocturnal orgasms 5–6 times a month on the average. They are always corresponding to a sexual dream I am having. The orgasm wakes me up. I consider it a blessing." (female)

"I heard, when I was growing up, about 'wet dreams' but thought that only applied to males. It wasn't until recently I awoke from a deep sleep, very sexually aroused, my heart beating rapidly, my body sweating, and vaginal lubrication had already begun. I don't remember the dream I was having." (female)

(from the author's files)

Many people have **nocturnal orgasms** in their sleep. There are five stages of sleep. Stages I and II are light sleep, and stages III and IV are deep sleep. Adults enter a fifth stage about every 90 minutes. This is called REM, or paradoxical sleep. REM stands for "rapid eye movement" (our eyes move around under the eyelids during this stage), and the term *paradoxical* refers to the fact that the EEG (a measure of the activity of the brain) during this stage is paradoxically identical to that of an alert, awake individual, although we are asleep. Many physiological events normally occur during this stage of sleep, including penile erection in males and vaginal lubrication in females. In fact, when people do not show these responses, it is an indication that something physiologically is wrong. This is also the stage of sleep in which we have our most vivid dreams, but do not assume that the dreams have a

nocturnal orgasm An orgasm that occurs during sleep (sometimes called "wet dreams" in males).

sexual theme just because people are experiencing erection or lubrication. Most nocturnal orgasms are not associated with sexual dreams (Wells, 1986). Remember, unless there is something physiologically wrong, people are going to have these responses regardless of what they are dreaming. An orgasm may result merely from rubbing of the genitals against the sheets and have nothing to do with dream content.

As with attitudes about masturbation, there is a long history of negative opinions about nocturnal orgasms. In medieval times, it was believed that male demons (incubi) had sex with women at night, while female demons (succubi) had sex with men while they slept. In the early 1800s, a French physician named Claude François l'Allernand believed that male nocturnal emissions ("wet dreams") were caused by the same thing that caused gonorrhea. Thus, in the Victorian era, nocturnal emissions were called *spermatorrhea*. Recall from Chapter 1 that as a young man, pioneering sexologist Henry Havelock Ellis thought he was going to die from spermatorrhea. Some Victorian parents made their male children wear spiked "spermatorrhea rings" to bed at night to prevent emissions (see Figure 1–4).

A lot of unfortunate Victorian boys must have experienced punctures. Kinsey et al. (1948) found that 83 percent of the men they surveyed reported having had nocturnal emissions. This occurs most frequently during the teen years and declines substantially after that, but these emissions can occur at any age. Many teenage boys are embarrassed by the wet spot on the sheets, especially if they have not been told in advance by their parents that this may (and is likely to) occur (see Chapter 10 for a more detailed discussion). Parents should not embarrass or ridicule their teenage boys for having wet dreams, particularly since it is a behavior over which they have no control (wet dreams occur during sleep). In fact, many males do not remember having had an orgasm in their sleep and are only aware that they did because of the wet spot.

Forty percent of the women in Kinsey's sample (1953) and 37 percent in a more recent sample (Wells, 1986) also reported having nocturnal orgasms. However, the disparity between the responses of the two sexes may not really be this great. Females do not have a wet spot on the sheets in the morning as proof of a nocturnal orgasm, so only those women who are awakened by the orgasmic contractions can report them.

It is also normal for people to have sexually explicit dreams. Nearly 100 percent of the men and 70 percent of the women in Kinsey's studies reported having such dreams, although in some cases they caused distress. Keep in mind, however, that dreaming about something is not the same as doing it.

SEXUAL FANTASIES

"My sister, who was much older than I, would always have friends over and I would fantasize about making love to them while masturbating." (male)

"They [fantasies] almost always include my wife. It may also include another female . . ." (male)

"I enjoy being male, but have occasionally fantasized about what it is like to be female during lovemaking." (male)

"During sex I have sexual fantasies about making love to a woman. . . . I would view my husband as a woman." (female)

"My sexual fantasy has been the same since age 16. The man uses me, sometimes beats me, and bring his other female partners around me. This fantasy bothers me." (female)

"I have sexual fantasies while making love to my husband, usually about men I'm attracted to from work. I think about how it would be and when I come, it's stronger and harder than it usually is." (female)

(from the author's files)

Nearly everyone has sexual fantasies from time to time. Sometimes the fantasy may be a warm thought of a romantic interaction with a special person, but many times the fantasies are very sexually explicit. It was once thought that only men had sexual fantasies (because women were thought to be less interested in sex), but this is not true. While most men think about sex more often than women (54 percent of men say they do so every day, as compared with 19 percent of women; Laumann et al., 1994), the large majority of women do have sexual fantasies. Kinsey et al. (1953) found that nearly two-thirds of women who masturbated used fantasy to enhance sexual arousal. Masters, Johnson, and Kolodny (1992) found that 86 percent of the women they surveyed had erotic fantasies, and several other studies reported that more than 90 percent of women have had sexual fantasies (Crepault et al., 1977; Leitenberg & Henning, 1995; Pelletier & Harold, 1988). Women's sexual fantasies tend to be more romantic and emotional, while men's fantasies tend to have more explicit and visual imagery (Leitenberg & Henning, 1995). The frequency, length, and explicitness of an individual's sexual fan-

tasies are related to his or her level of sexual experience (Gold & Chick, 1988; Leitenberg & Henning, 1995).

Sexual fantasies can occur at any time. Men have more sexual fantasies than women during nonsexual activity and masturbation, but there are no differences between the sexes during intercourse (Leitenberg & Henning, 1995). One of the most common fantasies during intercourse is the replacement fantasy, where one imagines oneself having sex with someone other than one's sexual partner (Cado & Leitenberg, 1990). If you have ever felt guilty about having these fantasies, as some people do (believing them to be immoral or socially unacceptable), you should be aware that they are so common that chances are that your partner has had them too (Cado & Leitenberg, 1990; Crepault et al., 1977). The fantasized partner can be a friend, a neighbor, a former partner, a celebrity (e.g., a movie star or a musician), or anyone else. Sharing fantasies with one's partner may enhance intimacy and sexual interest in some cases, but may lead to jealousy and self-doubt in other cases. Most people prefer to keep their fantasies private.

The specific content of sexual fantasies is almost limitless, but we can group fantasies into four general categories: (1) exploratory (experimentation with never-before-tried behaviors such as group sex or same-sex activities), (2) intimacy (sexual activities with a known partner), (3) impersonal sex (sex with strangers or watching others have sex; fetishes), and (4) dominance-submission themes (Leitenberg & Henning, 1995; Meuwissen & Over, 1991). For both men and women, fantasies of intimacy with a present, former, or imaginary partner are most common. Dominance-submission fantasies are the least common type, but dominance fantasies are more common for men, while submission fantasies of being overpowered are more common for women.

The content of sexual fantasies does not indicate sexual unhappiness, nor does it signify personality or psychological problems (Davidson & Hoffman, 1986; Leitenberg & Henning, 1995). In fact, sexual fantasies are most common in those individuals who have the fewest number of sexual problems (Leitenberg & Henning, 1995). Most sex therapists view sexual fantasies as something positive to be enjoyed rather than as something bad for one's sexual health. Having a sexual fantasy does not mean that a person actually wants to experience it in real life. Fantasies also provide a safe and private outlet for thoughts that, if actually engaged in, might be considered improper and/or illegal. In fact, in most cases where people have acted out their fantasies, the result has been disappointment. One reason for this is that we are totally in control of the sequence of events in our fantasies, so that they proceed just to our liking; but this is seldom true in real life, where partners can cause distractions and have desires and demands of their own.

Although fantasies are generally viewed as positive, about one-fourth of all people who fantasize feel guilty about it (Leitenberg & Henning, 1995). For the most part, however, fantasies are harmless. Because the vast majority of people have them, they are also in the range of normal behavior.

COITAL POSITIONS AND LOCATIONS

"Before I became sexually active, I believed that everyone who engaged in sexual intercourse did so in the missionary position in the privacy of their own bedroom. Now I look forward to new positions." (female)

"[Our sex] was boring me. We always used the same position—his favorite one." (female)

"My boyfriend has the concept that men should be in control of sex. Well, I say 'What about my pleasure?' My favorite position is 'me on top' because I'm guaranteed to receive an orgasm." (female)

". . . And also I like having sex standing up with my legs wrapped around the male's waist and walking around at the same time." (female)

"I am a mature, responsible, middle-aged female. I am also an experienced diver with a mate who is equally comfortable under water. In my opinion, the most stimulating, exciting, and extraordinary sex on earth happens beneath the surface of the water. Complete sex beyond the grasp of gravity in a quiet, sensuous, foreign world is beyond our limited language to describe." (female)

(from the author's files)

Of those Americans who are having sex, about 95 percent say that the sexual activity always or usually includes vaginal intercourse (Laumann et al., 1994). For most Americans, the first experience with sexual intercourse is in the **missionary position,** that is, a face-to-face position in which the female lies on her back and the male lies on top with his legs between hers. It is called the missionary position because

missionary position A face-to-face position of sexual intercourse in which the female lies on her back and the male lies on top with his legs between hers. It was called this because Christian missionaries instructed people that other positions were unnatural.

coitus Sexual intercourse.

Christian missionaries in foreign countries used to instruct natives that this was the only "proper" way to have sexual intercourse. St. Paul, you may recall, believed that women should be subordinate to men during intercourse (i.e., on the bottom), while St. Augustine said that any other position was unnatural and a sin against nature (see Chapter 1).

For many couples, male-on-top may be the only position in which they ever have sexual intercourse (**coitus**). Kinsey et al. (1948) found, for example, that over two-thirds of the population had never attempted any other position. The preference for this position by many American couples, however, appears to be culturally rather than biologically determined. The female-on-top position is most popular in some other cultures, especially those in which women enjoy high status (see Box 11–A). In most nonhuman primate species, the male inserts his penis into the female's vagina from the rear.

More recent surveys suggest that American couples are practicing a greater variety of sexual positions during intercourse than in Kinsey's time. Hunt (1974), for example, found that of the married people he surveyed, 75 percent had tried female-on-top, 50 percent had tried side-by-side, 40 percent had tried rear vaginal entry, and 25 percent had had intercourse while sitting.

Why do couples experiment with different positions? After all, the missionary position is perfectly adequate for procreational purposes. Well, to answer this, let's first examine the advantages and disadvantages of the missionary position. Perhaps the nicest thing about this position is that it allows for intimacy while having sex. A couple is face-to-face. But for most women, the benefits stop right there. In most couples, the man is considerably heavier than the woman, which means that she has to support a great deal of weight in the male-on-top position (and more and more so as the male gets tired). Some women may even have trouble breathing in this position. Women in my class have additionally pointed out that they receive very little touching or fondling in this position because their male partners are too busy supporting themselves. With a woman's hips "pinned to the mat," the man is in total control (probably the major reason that American men like this position), and it is often difficult to fully enjoy sex when you have little or no control over the movement or tempo.

There are other positions that still involve eye-to-eye contact but allow greater freedom and involvement for the female. As you will see in a later chapter, female-on-top is a position that is recommended by many therapists for optimal sexual arousal in both people. It is generally a lot easier for the male to support the female, and it is also a position in which it is easy for the man or the woman herself to manually stimulate the clitoris (and most other parts of a partner's body) during intercourse. Some couples enjoy rear vaginal entry because in this position the penis stimulates the front wall of the vagina (where some women

Figure 11–1

Figure 11–2

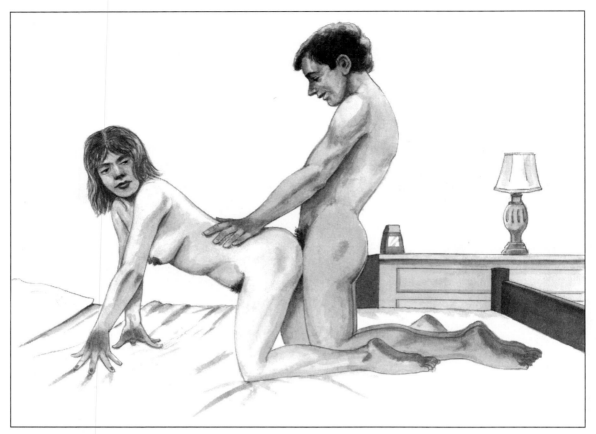

Figure 11–3

Figure 11–4

have a small area of sensitivity called the Grafenberg spot, or G spot, you recall).

Although these considerations are important, probably the main reason that most couples have sexual intercourse in different positions is to bring variety to their sex lives. Many a sexual relationship has become stale and boring by allowing it to become *ritualized,* that is, by only having sex at the same time of day, in the same place, in the same position.

Even the manner in which many couples have sex is highly ritualized. Therapists report that it often proceeds in the following predictable manner: a few fleeting moments of kissing followed very quickly by the man touching and/or kissing the woman's breasts, followed directly by genital touching, followed by intercourse as soon as vaginal lubrication is detected.

Couples try different positions to keep their sex lives exciting, fresh, and fun—the way it was at the beginning of the relationship. Sex manuals are available that show drawings of couples in what appears to be hundreds of different positions, but there are really just a few major variations that a couple can try: they can face each other or the female can turn her back, and they can lie down, sit, or stand up. However, couples should try different positions only because they want to, and not because they feel pressured to do so.

Whether or not a couple ever tries different sexual positions while having intercourse, it is important that they not allow their sex life to become a standardized ritual. The secret to that is to allow your sexual relations to be spontaneous, creative, and fun whenever possible. Some people get locked into a habit of never having sex until right before they go to sleep at night. If you and your partner are in the mood in the morning or afternoon, have sex then. Don't postpone it until later, when you might be tired. It is also normal for couples to have sex in a variety of places, for example, on the couch, on the rug, in different rooms, or in any other place. There is no rule that states you can only have sex in bed.

Dear Ann Landers:

My wife and I have been married 20 years. We have two sweet children and consider ourselves very fortunate. The problem is one we cannot discuss with our minister.

My wife believes that sex anywhere but in the bedroom is sinful according to the Bible. I say a change of setting can add extra pleasure, and so long as there is complete privacy, it is perfectly moral.

The place I have in mind is the car. We have a garage with a sturdy lock on the door. No one could possibly get in.

We are good Christians and want to know what the Bible says about this. Can you contact a religious scholar?

Strictly Confidential in Kentucky

Dear Strictly: Since the Bible predates the automobile by a couple of thousand years, there is no point in bothering a Christian scholar.

If you will settle for my opinion, here it is: It is perfectly all right for a married couple to make love anywhere they choose, providing it is private, safe, and reasonably comfortable.

(Reprinted with permission of Ann Landers and Creators Syndicate.)

If you and your partner are on the couch and get in the mood to have sex, there is nothing immoral, sinful, or indecent in spontaneously doing so right there (assuming, of course, that the shades are pulled, there are no children around, and you are using contraception if pregnancy is unwanted). One of Ann Landers's readers felt compelled to respond to the letter by "Strictly Confidential":

Dear Ann Landers:

I was interested in the letter from the man who enjoyed making love in the car. His wife felt guilty and wanted to know if it was the "Christian" thing to do. You said so long as it was private, not dangerous, and reasonably comfortable, it was nobody's business.

I married one in a million. She was totally uninhibited, willing, and eager to make love any place at any time. I must say, we dreamed up some mighty unusual situations. We traveled quite a bit, and it was not unusual for us to pull off the road in the middle of the day if we ran into a wooded area, a vacant house, a sandy beach, a calm lake, or an inviting motel. On occasion, when the mood came upon us and none of the above was available, we just used the car.

This kept up until we were in our 60s, when my beloved wife passed away. I always felt as if we had the healthiest sex life of anyone I knew because we never stopped turning each other on. Sex was always unpredictable, imaginative, and fun. Our sexual compatibility spilled over into all areas of our life and we were divinely happy.

You can print this letter if you want to but no name or city, please. Just call me . . .

Beautiful Memories

Dear Beautiful: How lucky you were to find each other. It was a perfect match. Lots of readers will be envious.

(Reprinted with permission of Ann Landers and Creators Syndicate.)

Box 11-A CROSS-CULTURAL PERSPECTIVES

Positions of Sexual Intercourse

The males of other primate species always mount the female from behind during sexual intercourse. The buttocks of the females become swollen and red when she is sexually receptive, which acts as a sexual stimulant to the males. With evolution of the two-legged stance in humans, mating occurred primarily face-to-face, with greater opportunity for emotional interaction and communication via facial gestures and vocalization (Lambert, 1995). Some researchers believe that one of the reasons many men find female breasts sexually arousing is that breasts mimic the round contour of the buttocks (Diamond & Karlen, 1980).

Face-to-face is the preferred manner of sexual intercourse in all known cultures, but not necessarily man-on-top. In many African cultures, the most common position is with the partners lying side by side, while in Polynesian and many Asian cultures the most common position is with the woman lying on her back and the man squatting or kneeling between her legs (Gregersen, 1982). (By the way, it was the Polynesians who originally called the male-on-top position taught to them by European missionaries the "missionary position.") Actually, in most cultures there is more than one pre-

ferred position. Mangaians, for example, practice a variety of positions while preferring some variety of male-on-top, but use the male-from-behind position when their partner is pregnant (Marshall, 1971).

Some researchers believe that a culture's preferred position of intercourse is a reflection of women's social status (Goldstein, 1976). In cultures where a woman's sexual satisfaction is considered to be as important as the man's, female-on-top is usually preferred. The ancient Romans preferred this position. In male-dominated cultures, however, the reverse is true (Langmyhr, 1976). There may be no better reflection of the relation between women's social status and coital position than in our own culture. Here, women's gender roles have changed dramatically since World War II. In the 1940s, you recall, Kinsey et al. (1948) found that over two-thirds of the couples studied had never attempted any position other than male-on-top. More and more women began to enter the work force after that, and women's rights evolved rapidly in all areas. Twenty-five years later, Hunt (1974) found that the large majority of couples in the United States had tried the female-on-top position.

However, whenever, and wherever you decide to do it, try to make your sexual relations spontaneous, exciting, enjoyable, and fulfilling. A healthy and satisfying sex life can contribute to our overall physical and emotional well-being.

ORAL-GENITAL SEX

"I was highly against oral sex because I thought it was so nasty and demeaning. But when I experienced it, my attitude changed. I didn't realize what I was against could feel so damn good." (female)

". . . What did surprise me was the fact that I enjoyed giving more than receiving. I think it is a great way for two people to share very intimate moments." (male)

"It is a great way to relax, feel unpressured and not worry about pregnancy or how good you perform." (male)

"I do engage in oral-genital sex, but really prefer it as foreplay. I don't like the idea of a guy coming in my mouth—it gags me!" (female)

"Oral sex is by far the most pleasurable sexual experience I have ever engaged in. It has produced some of the most intense orgasms that I have ever experienced, especially in the 69 position. I also very much enjoy orally stimulating my partner individually as much as being stimulated myself." (female)

(from the author's files)

Sexual relations generally include a great deal of oral stimulation. One of the first sexual behaviors that

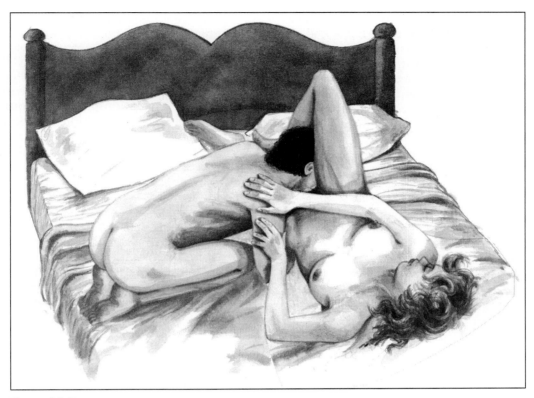

Figure 11–5

most people engage in is kissing, which, of course, involves mouth-to-mouth stimulation. This usually progresses from "dry" kissing (lip to lip) to "wet" kissing (also known as "French" or "soul" kissing, involving the tongue) within a short time. It is also normal while having sex for people to kiss and lick other parts of their partner's body (the neck, for example). For many people, this includes their partner's genitals.

Oral stimulation of the penis is called **fellatio** (from the Latin *fellare,* meaning "to suck"), and oral stimulation of the vulva (clitoris, labia, vaginal opening) is called **cunnilingus** (from the Latin words for "vulva" and "licking") (Figures 11–5 and 11–6). These can be done either as part of foreplay or as the preferred sexual behavior, either by one partner at a time or by both at the same time (the latter is sometimes referred to as "69"). Although the slang term for fellatio is "blow job," the large majority of men prefer a sucking action to blowing. Many women report oral stimulation of the area around the clitoris to be more pleasurable than intercourse (the vaginal walls, you recall, have few nerve endings), and many men find fellatio (particularly orgasm during fellatio) to be more intense than intercourse (Janus & Janus, 1993). The pleasure, however, is not always reserved entirely for the recipient. While it is true that many people prefer receiving oral sex to giving it (Laumann et al., 1994),

many others enjoy orally stimulating their partner (and observing their partner's reaction) nearly or equally as much as being stimulated themselves. For many people, giving oral sex represents a moment of extreme intimacy.

How common are these behaviors? Kinsey's group (1948, 1953) found that about 50 to 60 percent of the married people they studied had engaged in oral-genital sex. However, only 15 percent of singles and only 10 percent of adolescents reported having engaged in it. More recent surveys indicate that about 70 to 90 percent of young adults eventually engage in oral-genital relations (e.g., Billy et al., 1993; Hunt, 1974; Janus & Janus, 1993; Laumann et al., 1994). In another study, nearly 50 percent of the adolescents surveyed reported having engaged in oral-genital sex, with some girls preferring this behavior to intercourse (Newcomer & Udry, 1985).

Although no other behavior has increased as much in popularity since Kinsey's time as oral-genital stimulation (Gagnon & Simon, 1987), there are some notable differences among socioeconomic groups. The appeal of this behavior is positively associated with level of education and negatively associated with how religious one is. Oral-genital sex is mostly

fellatio Oral stimulation of a male's genitals.

cunnilingus Oral stimulation of a female's genitals.

practiced among white, college-educated individuals (e.g., Kinsey et al., 1948; Laumann et al., 1994). In Kinsey's time, oral-genital sex was so uncommon among African Americans that it was almost a taboo, and still today only about 50 percent of African Americans have done it (Laumann et al., 1994). The percentage of Hispanics who have engaged in oral sex is somewhere between the percentages for whites and African Americans.

Some individuals consider oral-genital sex disgusting, and no one should feel pressured to engage in a behavior just because a lot of other people do it. On the other hand, if you expect your partner to do it for you, then you should not have a double standard about doing it for him or her. Double standards in sexual relations can lead to serious problems in other aspects of the relationship.

> "My girlfriend refuses to have oral sex with me, but thinks that it is just fine for me to do it to her! Even though I enjoy doing this to her, I don't think it is fair that she would not even try it." (male)

> "There are people out there who are sexually selfish. I was seeing someone for 7 months. I'll never forget our last encounter. His favorite sexual act was fellatio, so naturally I obliged. After I had done so, I wanted to kiss him. He had a very negative (and immature) reaction. I think if I could have castrated him, I would have!" (female)

> "For almost 18 months I have been involved in what many of my friends consider the 'healthiest' relationship on campus. My boyfriend and I are compatible emotionally, physically, and intellectually. But there is one flaw that tortures me daily: our oral-genital sex relations to this point have been very one-sided, and I am bewildered. My personal hygiene is wonderful but when I finally got the courage to ask him to give me cunnilingus he was mortified. Yet from the beginning I have not hesitated to give him fellatio when he requests. I felt like his pleasure was my reward. When he refused me I felt insulted and I cried for an hour. I just can't lose the feeling of very deep, very personal humiliation." (female)

> (from the author's files)

People often become more accepting of a sexual behavior when they learn that it is not abnormal (or that their partner is not abnormal or kinky for wanting to try it). As an example, try to remember your reaction when you first heard about French kissing. For many of us it seemed disgusting, but we later learned to enjoy it. Nevertheless, some people may have reser-

Figure 11–6

vations about engaging in oral-genital sex because of cleanliness. Certainly, people ought to bathe regularly if they expect their partners to give them oral-genital stimulation (or engage in sexual relations of any kind, for that matter).

Even if a couple has good hygiene habits, however, there may be reservations about the cleanliness and odors of genital secretions. As I mentioned in the chapter on anatomy, the odor of a healthy vagina is not offensive. Some women do not like the taste of semen, but it, too, is relatively clean. In fact, mouth-genital contact is no less hygienic than mouth-to-mouth contact—the mouth harbors as many germs as the genital orifices. This assumes, of course, that neither person has a sexually transmitted disease, for these can be transmitted during oral-genital stimulation as well as during intercourse. On the other hand, there are no health benefits to swallowing genital secretions (such as curing acne or prolonging youthfulness), as some individuals mistakenly believe.

Finally, there are even some people who worry that these are homosexual acts. It is true that the vast majority of homosexuals engage in oral-genital sex, but so do heterosexuals, just as both groups kiss and hold hands. It is not behaviors that determine whether you are heterosexual or homosexual, but the people with whom you regularly practice them.

ANAL STIMULATION

"The first time I had ever heard of anal sex was in 1981. A girl I worked with told me that she and her boyfriend had used this. I knew that male homosexuals had intercourse in that manner, but it never once occurred to me that heterosexuals did also!" (female)

"Last summer my boyfriend decided he wanted to try some of the different types of things. . . . I will never try sex again in the anus because it was so painful. . . ." (female)

"I do very much enjoy anal stimulation, if done properly. For me, anal stimulation during oral sex has produced the 'ultimate' orgasm . . . if it is done gently and not forcefully." (female)

"I like anal sex, probably a lot more than most females. When I first had anal sex, I didn't like it. . . . The difference is in how fast or slow the guy is in penetrating and also how turned on I am before he goes in. . . ." (female)

"My boyfriend and I sometimes use a mini-vibrator while making love. While we're in the missionary position, I insert the vibrator through the anus. At the same time, he continues to penetrate. I receive the ultimate feeling in sexual pleasure." (female)

(from the author's files)

The anus has numerous nerve endings and is very sensitive to touch. Many people enjoy having their anus manually stimulated by their partner's fingers during sexual intercourse. The anal sphincter muscle undergoes rhythmic muscular contractions during orgasm in both males and females, and thus anal stimulation can further enhance the pleasure. Although there is a physical basis for this pleasure, many people feel very negatively about anal sex and consider it abnormal, perverted, or kinky (Janus & Janus, 1993).

Anal intercourse is common among male homosexuals, but how common is it among heterosexual couples? Kinsey et al. (1948) found that 11 percent of the people he surveyed had attempted this behavior (only 8 percent successfully), but surveys conducted more recently indicate that about 20 to 25 percent of young adults have tried anal intercourse (Billy et al., 1993; Hunt, 1974; Laumann et al., 1994; Reinisch et al., 1992). One in 10 heterosexual couples engage in anal intercourse somewhat regularly (Voeller, 1991). Interestingly, Laumann's group reported that only 14 percent of men and 5 percent of women found anal intercourse "very" or "somewhat" appealing. When asked why they participate in this behavior, many women say that it is their least favorite activity, although they occasionally engage in it for variety. But as you can see from the personal testimonials at the beginning of this section, others enjoy it.

For those who engage in this behavior, some words of caution are necessary. First, the anal sphincter muscle contracts in response to attempted penetration, and attempts to force the penis into the anus can result in injury. A water-soluble lubricant (like K-Y Brand Jelly) should be used if you engage in this behavior. The muscle spasms do not relax until 30 to 60 seconds after penetration, so the woman often experiences some initial discomfort, even if her male partner proceeds slowly and gently. Even so, this behavior will almost certainly result in rupturing of small capillaries (unlike the vagina, the rectum does not readily accommodate this kind of stimulation; Agnew, 1986). Anal intercourse itself does not cause AIDS, but the ruptured capillaries maximize the chances of contracting HIV (the virus that causes AIDS) from an *infected* partner. If you are not in a monogamous relationship, use condoms. It is probably the case, however, that condoms tear more often during anal intercourse than during vaginal intercourse (Silverman & Gross, 1997). In addition, bacteria that are normally found in the anus can easily cause infection if introduced into the vagina. Couples who engage in anal stimulation, therefore,

RATINGS OF SEXUAL BEHAVIORS BY AMERICAN MEN AND WOMEN AGED 18 TO 44

	Men				Women			
Sexual Behaviors	Very Appealing	Somewhat Appealing	Not Appealing	Not at All Appealing	Very Appealing	Somewhat Appealing	Not Appealing	Not at All Appealing
Vaginal intercourse	83%	12%	1%	4%	78%	18%	1%	3%
Watching your partner undress	50	43	3	4	30	51	11	9
Receiving oral sex	50	33	5	12	33	35	11	21
Giving oral sex	37	39	9	15	19	38	15	28
Group sex	14	32	20	33	1	8	14	78
Having your anus stimulated by your partner's fingers	6	16	24	54	4	14	18	65
Stimulating your partner's anus with your fingers	7	19	22	52	2	11	16	70
Active anal intercourse	5	9	13	73	—	—	—	—
Using a vibrator or dildo	5	18	27	50	3	13	23	61
Watching others do sexual things	6	34	21	39	2	18	15	66
Same-sex sex partner	4	2	5	89	3	3	9	85
Sex with a stranger	5	29	25	42	1	9	11	80
Passive anal intercourse	3	8	15	75	1	4	9	87

Source: Based on Laumann et al., 1994. Reprinted with permission of the publisher.

should never put anything (e.g., finger, penis, objects) into the vagina that has been in contact with the anus unless it has been washed first.

PREFERRED SEXUAL BEHAVIORS

What are the preferred behaviors for most Americans? In their recent well-conducted study, Laumann and colleagues (1994) found that most Americans were rather traditional (see Table 11–1). All but 5 percent of the people in their survey had engaged in vaginal intercourse the last time they had sex. It was by far the most preferred behavior. Oral-genital sex ranked a distant third, and anal intercourse near the bottom of the list. What was the second most preferred behavior? A rather large proportion of men and women said that it was watching their partner undress.

THE SEXUALLY HEALTHY PERSON

You have now read about several sexual behaviors that a large number of people engage in. They are normal from a statistical perspective. Does this mean that you should try all these behaviors just because a large number of other people practice them? Do we judge sexual normality by the variety of sexual experiences or number of partners a person has had? No, not at all. Sexual health should not be judged like a decathlon event in the Olympics. A **sexually healthy person** is someone who (a) feels comfortable with his or her sexuality (i.e., does not view sex as something naughty, bad, improper, or sinful, and can engage in it without feeling guilty or anxious), and (b) feels free to choose whether or not he or she wishes to try a variety of sexual behaviors. "Feels free" means free of peer pressure, partner pressure, and social pressure. In today's world of epidemic sexually transmitted diseases, sexual health also means freedom from life-threatening diseases. Individuals should not engage in sex simply because they are being pressured by peers or a potential partner. A person can be sexually healthy and still say "no." Always saying "yes" does not prove that one is sexually healthy, any more than it proves that one is a "real man" or a "real woman."

By this definition, someone who has engaged in a wide variety of behaviors with numerous partners, but who gets little fulfillment from his or her sexual relations or who regards sex as dirty, would not be considered sexually healthy. Sometimes we can tell someone's real attitude about sex by how that person regards his or her partner or partners. For example, some people claim to have a positive attitude about sex, but they view their partners as dirty or bad for engaging in sex with them. On the other hand, a person could be sexually inexperienced and still be considered sexually healthy if he or she regarded sex as something good and positive and was choosing not to engage in it for other reasons (e.g., saving sexual relations for someone he or she loved or cared for).

There are many behaviors that a couple can engage in that have not even been mentioned. This chapter has discussed only the most common ones. (I am reminded of the couple who told Dr. Ruth Westheimer [*Playboy*, 1986] that among their favorite practices was having the woman toss onion rings over her partner's erect penis from various distances.) If a couple view their sexual relationship as something good and satisfying and want to experiment with different techniques, should they or anyone else have any concerns? Remember, most therapists take the view that any behavior between consenting adults done in private that does not cause physical, emotional, or psychological harm to anyone involved is okay.

sexually healthy person
Someone who feels comfortable with his or her sexuality and who feels free to choose whether or not to try a variety of sexual behaviors.

Key Terms

anal intercourse 291
autoerotic asphyxiation 281
coitus 284
cunnilingus 289

fellatio 289
masturbation 278
missionary position 7, 284
nocturnal orgasms 282

range of behaviors or values 278
sexually healthy person 293
spermatorrhea 282
statistically normal 278

Personal Reflections

1. What was your emotional reaction, if any, when you first saw the drawings for this chapter? If your reaction was anxiety or disgust, why? Do you regard the behaviors shown as normal?

2. What sexual behaviors do you enjoy the most? The least? Why?

3. Try to recall your reactions when someone first told you about French kissing (also called "wet kissing" and "soul kissing"). Was your reaction somewhat negative? Could you really imagine allowing someone else to put his or her tongue in your mouth? How did your attitude change as you began to realize that most people engage in this behavior? Might this same process eventually change your attitude about any of the behaviors discussed in this chapter to which you presently have negative reactions?

4. Do you have sexual fantasies? (Most people do.) Are you comfortable or uncomfortable with your fantasies—that is, do they cause anxiety or guilt? If your partner told you that he or she had sexual fantasies, what would your feelings be? Analyze the nature of your fantasies (e.g., their content, when they occur). What purpose do you think your fantasies serve?

5. What are your feelings about masturbation? If they are negative, why? Do you have negative feelings when you pleasure yourself by touching (e.g., rubbing, massaging, scratching) other parts of your body? Consider masturbation in context of a larger question: Are genitals reserved only for procreation (to have children), or are they also a source of pleasure?

6. Are you a sexually healthy person? Explain.

7. How do you expect to keep your sexual relationship from becoming ritualized after 6 months, 6 years, 20 years, 60 years? The brain is your largest sex organ. Make sure you continue to use it to your advantage.

Suggested Readings

Atwood, J. D., & Gagnon, J. (1987, Fall/Winter). Masturbatory behavior in college youth. *Journal of Sex Education and Therapy.*

Barbach, L. G. (1992). *For each other: Sharing sexual intimacy.* Garden City, NY: Anchor/Doubleday. Step-by-step exercises to enhance intimacy and improve your sex life.

Blumstein, P., & Schwartz, P. (1983). *American couples.* New York: William Morrow. An excellent survey of married, unmarried, and cohabiting couples.

Comfort, A. (1991). *The new Joy of sex.* New York: Crown. A revised edition of nicely illustrated, well-written sex manual. A best-seller since it was first published in 1972.

Friday, N. (1973). *My secret garden.* New York: Simon and Schuster. All about female sexual fantasies.

Friday, N. (1980). *Men in love.* New York: Delacorte Press. All about male sexual fantasies.

Friday, N. (1991). *Women on top: How real life has changed women's sexual fantasies.* New York: Simon & Schuster. How the changing role of women in society has changed their sexual fantasies.

Janus, S., & Janus, C. (1993). *The Janus report.* New York: John Wiley & Sons. Results of a nationwide survey of sexual behavior.

Kaplan, H. S. (1985, September). Talking frankly about oral sex. *Redbook.* Discusses myths about and problems related to oral-genital sexual relations.

Michael, R., et al. (1994). *Sex in America: A definitive survey.* Boston: Little, Brown. A thinner version of the Laumann et al., (1994) study, adapted for the general public. Probably the most scientifically conducted sex survey ever.

CHAPTER 12

When you have finished studying this chapter, you should be able to:

1. Describe the characteristics that romantic love and friendship have in common, and the characteristics that distinguish romantic love from friendship;
2. Describe companionate love and its relation to romantic love;
3. Discuss the prerequisites for love;
4. Explain what causes jealousy and how best to deal with it;
5. Differentiate between feelings of sexual desire and those of love;
6. Evaluate the place of love without sex in your system of values;
7. Discuss Sternberg's triangular theory of love and how different combinations of three components lead to different kinds of liking and loving;
8. Describe Lee's different styles of love and his advice on how to find a compatible love relationship;
9. Discuss love in other cultures;
10. Explain how a couple can maintain a loving relationship; and
11. Discuss how couples can achieve greater intimacy.

Love and Relationships

Poets write words to rhyme with it. Novelists glorify it in prose. Lyricists praise it in song. Philosophers wonder about its meaning, and politicians all support it.

What is this ubiquitous subject? Why, it's love, of course. Have I exaggerated? Think about it for a moment. How many popular songs that you hear on the radio are about love (first love, new love, broken hearts)? In fact, you may have to listen for quite a while before you find one that is not about love. When was the last time that you read a novel that didn't include at least one romantic subplot?

Love is obviously an emotion that most humans consider to be extremely important to their lives. But what is love? There have probably been as many definitions and thoughts about love as there have been philosophers, behavioral scientists, theologians, and biologists. Freud believed that love was an emotional feeling that resulted from the repression of sexuality during childhood ("aim-inhibited sex") (Freud, 1933/1953). Others have viewed love as a social instinct that satisfies a need for companionship with others. In *The Origins of Love and Hate* (1952), Ian Suttie expressed the belief that "the specific origin of love, in time, was at the moment the infant recognizes the existence of others." In contrast, others, including some noted psychologists, have expressed the belief that love results from conditioning, that is, that it is a learned response. By this account, people come to be desired or loved when others have positive experiences in their presence. Love has also been viewed as a mania (Plato, in *Phaedrus*), a neurosis (Askew, 1965), an addiction (Peele, 1988), a disease (Burton, 1651/1963), and the enshrinement of suffering and death (de Rougement, 1969).

Christian ideals of love are found in St. Augustine's definition: "Love means: I want you to be." Viewing love simply as a desire for another person "to be" honors his or her existence. It transcends sexual desire and acknowledges that this emotion can be felt for a lover, a parent, a child, or a friend of the opposite or the same sex. Noted psychoanalyst Erich Fromm has echoed similar beliefs about love:

> I want the loved person to grow and unfold for his own sake, and in his own ways, and not for the purpose of serving me.

(Erich Fromm, 1956, pp. 23–24)

FRIENDSHIP VERSUS ROMANTIC LOVE

Studies conducted nearly 1600 years after St. Augustine confirm his belief that the structure of love is very similar for various types of close relationships (Sternberg & Grajek, 1984). Let us first ask what distinguishes love from friendship. Nearly 2000 years ago, the Roman statesman and philosopher Seneca wrote, "Friendship always benefits; love sometimes injures." Is this true? Are love and friendship qualitatively different feelings, or is love just a more intense form of the emotion felt in friendship?

Studies by Keith Davis (1985) of the University of South Carolina have revealed several characteristics that are essential for **friendship:** (1) *enjoyment* of each other's company most of the time (although periods of temporary annoyance or anger may occur); (2) *acceptance of one another* as is; (3) a *mutual trust* that each will act in his or her friend's best interest; (4) a *respect for each other* (an assumption that each will use good judgment in making life choices); (5) *mutual assistance* of one another during times of need; (6) *confiding* in one another; (7) an *understanding* of each other's behavior; and (8) *spontaneity* (the freedom to be oneself rather than playing a role).

Davis found that people rated their "spouse/lover" and best friend nearly the same for all of these characteristics except for enjoyment, which more people attributed to their relationship with a lover than to the company of their best friend. What about close friends (rather than best friends)? People tended to rate spouses and lovers higher for enjoyment, respect, mutual assistance, and understanding. Still, close friends also generally fared well on these characteristics.

Are there some characteristics unique to spouses and lovers? Davis found that people generally rated **romantic love** relationships much higher (compared to friends) in *fascination* (a preoccupation with the other person, even when one should be doing other things), *exclusiveness* (not having the same relationship with another person), *sexual desire* (a desire for physical intimacy), and *giving the utmost* when the other is in need. These feelings can be extremely intense. The loved one is perceived as able (and often solely able) to satisfy needs, fulfill expectations, and provide rewards and pleasure.

Notice again that romantic love includes a high degree of sexual desire. That brings us to an interesting question: "What is the difference between the word *love* and the expression *'in love'*?" "The word love is bandied about more promiscuously than almost any other word in the English language," according to Murstein (1988, p. 13). Nevertheless, Meyers and Berscheid (1997) recently found that a large majority perceives a difference between love and "being in love." When they had people place their social relationships into categories, they found that people put far many more people in the "love" category than in the "in love" category, and that people placed in the "in love" category (but not the "love" category) were also included within a "sexual attraction/desire" category. Meyers and Berscheid found that almost all of their subjects knew what it meant if someone told them "I love you, but I'm not in love with you." To them, it meant "I like you, I care about you, I think you're a marvelous person with wonderful qualities and so forth, but I don't find you sexually desirable." Their findings agree with Davis's that the major differences between friendship (love) and romantic love (being "in love") are fascination, exclusiveness, and sexual desire.

friendship A relationship that includes (a) enjoyment of each other's company, (b) acceptance of one another, (c) mutual trust, (d) respect for one another, (e) mutual assistance when needed, (f) confiding in one another, (g) understanding, and (h) spontaneity.

romantic love The combination of passion and liking (intimacy). Romantic love relationships have high levels of fascination, exclusiveness, sexual desire, and "giving the utmost" when the other is in need.

Figure 12–1 Compared to friendship, romantic love rates high in fascination (a preoccupation with the other person), exclusiveness, sexual desire, and giving the utmost when the other is in need.

Davis and others report that most lovers find that their mood depends more on reciprocation of their feelings in romantic relationships than it does in friendships. Lovers tend to view their relationships as more under the control of social rules and expectations. Romantic relationships also rate much higher in ambivalence (mutually existing but conflicting feelings) than friendships. Thus, while romantic relationships are generally more rewarding than friendships, they are also more volatile and frustrating than friendships.

HOW DO I KNOW IF THIS IS REALLY LOVE?

In addition to a preoccupation with the loved person, it has been argued that romantic love involves physiological arousal and the cognitive interpretation of that as being caused by the other person (Berscheid & Walster, 1974). Consider, for example, the following hypothetical scenario:

Susan was in her first year of college. She had gone out a few times, but too many weekend nights had been spent with girlfriends, and she generally felt lonely and bored. She was surprised when David, a handsome junior who had sat next to her in sociology class, called and asked her to go to a movie Friday night. They had a good time and studied together on Sunday. However, David didn't sit next to her in class on Tuesday, and Susan worried all that day and the next if he would call again. She could hardly contain herself when he called Wednesday evening and asked her to a big party at his fraternity.

Susan and David spent a lot of time together during the next two weeks. She was no longer lonely or bored, and, in fact, her girlfriends let her know that they envied her. On the other hand, when she and David were apart, she worried if and when he would call, and her growing sexual attraction to David was keeping her aroused and agitated.

Is Susan in love? She has feelings of extreme happiness (sometimes elation), periods of anxiety and frustration, and sexual desire, and she finds herself thinking about David when she should be concentrating on other things. David fulfills needs and satisfies desires (sexual, physical, and ego). There is no question that she is better off than before. But is this really love, or just infatuation?

Romantic love almost always includes certain *physiological responses,* such as heavy breathing, a pounding heart and increased blood pressure, sweaty palms, and a dry mouth when we are close to or thinking about the loved one. Michael Liebowitz (1983) has found that feelings of romantic love are associated with an increase in three brain chemicals called dopamine, norepinephrine, and phenylethylamine, the last of which is chemically similar to amphetamines. Falling in love, with the release of these chemicals, literally gives the person a natural high. (Some people are known to eat large amounts of chocolate, which is high in phenylethylamine, when suffering the heartache of a broken romance.)

The problem is, of course, that almost any kind of excitement or stress will cause a pounding heart and the other above-mentioned physiological responses. A *cognitive component* is necessary before one can interpret these responses as a particular type of emotion. To prove this, psychologists Stanley Schacter and Jerome Singer (1962) administered adrenalin, which causes increased heart rate and other signs of physiological arousal, to volunteer subjects who were told it was a vitamin shot. The subjects were then instructed to wait in another room with a fellow subject. This other subject, however, was really working for the researchers, and in half the cases acted very happy and in the other half very angry. The real subjects were experiencing strange physiological responses of arousal due to the adrenalin, but it was the environmental cues that determined how they interpreted them. The subjects with the happy person acted happy, while those with the angry person acted angry. When subjects were warned in advance of the physiological changes they would experience, they were not affected by the phony subject's behavior.

If Susan's friends are having similar experiences and say they are "in love," the chances are good that Susan, too, will interpret her newly aroused state as love. Under other circumstances, she might not.

Will Susan's feelings of love for David last? Initially, Susan certainly loves the way David makes her feel. But as many of you have discovered, that initial high usually doesn't last, at least not at the peak it first was at. Just as with amphetamines, the body builds up resistance to phenylethylamine, so that it takes larger amounts to experience the same high. Some people—call them "love junkies," if you like—go from relationship to relationship, ending each one as the initial high (passion) begins to subside. If love is to last

companionate love Love based on togetherness, trust, sharing, and affection rather than passion.

for Susan, she must come to love David for who he is, and not just for what he causes her to experience.

COMPANIONATE LOVE

Companionate love has been defined as "the affection we feel for those with whom our lives are deeply entwined" (Hatfield & Walster, 1978). It is based on togetherness, trust, sharing, affection, and a concern for the welfare of the other (more so than passion). For example, the love between a parent and a child is usually the type we refer to as companionate love. It is also a stable kind of love that is characteristic of lasting adult relationships, for few relationships can sustain the initial level of excitement and sexual passion.

> Young love is a flame; very pretty, often very hot and fierce but still only light and flickering. The love of the older and disciplined heart is as coals, deep burning, unquenchable. . . .
>
> (Henry Ward Beecher)

Scientists have discovered that the continued presence of a partner increases the production of endorphins, natural opiate-like chemicals found in the brain (Fisher, 1992; see Toufexis, 1993). More recent research has found that oxytocin—the hormone released during breast-feeding, labor, and orgasm—might be important for mediating the stage in which couples settle down in a long-term relationship (Fisher, 1997). Perhaps it is these substances that give long-term lovers a sense of calm, peace, and security (in contrast to the high produced by the initial surge of amphetamine-like chemicals).

Many researchers equate the word *love* with companionate love, and the expression "in love" with passionate (romantic) love (Berscheid & Walster, 1978; Fehr, 1994; Hendrick & Hendrick, 1989). The two kinds of love are usually described as a dichotomy. However, Meyers & Berscheid (1997) found that people can experience both simultaneously, and propose instead that the more accurate distinction is passionate/companionate love versus companionate love.

Companionate love, however, does not always come later in a relationship, for some people do not consider romance to be their most important goal when establishing a relationship. Some people desire companionship more than anything else. Companionate love very often includes a good, satisfying sexual relationship as well. This is understandable in that a good overall relationship often leads to a healthy and good sexual relationship. Many people refer to this type of relationship as *realistic love* because it is not based on the fantasies and ideals of romantic love. The predictability (and avoidance of extreme highs and lows) of companionate love offers security, so that people may enjoy their lives outside of the relationship as well as in.

A drab, mundane form of companionship might be called **attachment** (Berscheid, 1982), where one's partner gives few positive rewards for remaining in the relationship aside from predictability (a different use of the term *attachment* is given in the next section). For most people, familiarity is comforting, and this can also be true even when all other aspects of a relationship are poor.

> "I confronted my boyfriend about growing up before we were supposed to get married. I said to him what he had ignored for 29 years of his life—he needs to leave mom and move out on his own. He acted as though I had spoiled all of his high hopes when I told him this. I, like five women before me, chose to confront the issue, yet he is totally bewildered about why his longest relationship with a woman has been merely 3 months long. No matter what confidences I shared with him, he would always have to spill the beans to mom. Any time there was a problem he would slip back into his comfort zone with mom there to cuddle him."

> (from the author's files)

Knowing what to expect may cause less anxiety for some than the thought of leaving and venturing into an unknown future. An extreme example of this is the devotion displayed by some battered wives toward abusing husbands on whom they are economically dependent—the fear of having to support themselves on their own may seem worse than the abuse. I will return to the subject of companionate love in the section on maintaining a relationship.

PREREQUISITES FOR LOVE

SELF-ACCEPTANCE

In order to love another, it is first necessary that one be able to love oneself. As Erich Fromm once stated, "If an individual is able to love productively, he loves himself, too; if he can love only others, he cannot love at all" (1956, p. 60). The first prerequisite for a loving relationship, therefore, is a *positive self-concept* (good **self-esteem**).

Why is a positive self-concept so important? A comedian once said, "I wouldn't want to be a member of any organization that would let me be a member." If you cannot accept and love yourself, it will be impossible for you to accept that someone else might love you. And like the comedian, people who cannot accept themselves generally reject other people.

A positive self-concept does not necessarily mean that someone is self-centered or believes that he or she is always correct, good, and moral and can do no wrong. To accept oneself is to accept one's shortcomings as well as one's strengths, that is, "to accept myself for what I am" (Coutts, 1973). *The manner in which one is raised is important for acquiring a positive self-concept.*

> "I think the reason I don't like myself is because of all the problems I had in the past . . . the absence of Mom and Dad, and when they were around so was abuse. I feel that since my parents couldn't accept me, why should I? I felt like a failure, the ugliest person alive, and basically trash. I had tried to kill myself. . . ."

> (from the author's files)

Figure 12–2 *Cupid and Psyche* by François Gérard conveys many people's idealistic view of romantic love.

attachment A drab, mundane form of companionship where one's partner gives few positive rewards, other than predictability, for remaining in the relationship.

self-esteem The feeling one has about oneself.

It is not unusual for children who have been neglected or abused to have negative self-concepts as adults and sometimes to be unable to be loving to others (in fact, abused children often become abusive parents). Having loving, caring parents during infancy and childhood teaches us not only that other people do nice things that make us feel good (and thus, that they are good), but that we are worthy of having someone else care about us. The trust that is gained by this experience generalizes to other people as we grow older.

Hazan and Shaver (1987) have extended these ideas into what they call an *attachment theory of love*. They claim that the quality of the attachment between parents and child greatly affects the child's ability to form romantic relations as an adult. As adults, people are either *secure lovers* (who don't fear abandonment and find it easy to get close to others), *avoidant lovers* (who have difficulty trusting a partner and letting someone get close to them), or *anxious-ambivalent lovers* (who are insecure in their relations and try desperately to get close to their partners, which often results in their scaring them away). These researchers found that slightly more than half of all adults could be called secure, while about one-fourth were avoidant and one fifth were anxious-ambivalent (Shaver, Hazan, & Bradshaw, 1988).

SELF-DISCLOSURE

Self-acceptance and trust in ourselves give us the potential to trust and love others, but for love to really develop, there must be **self-disclosure** by both parties, resulting in an exchange of vulnerabilities. You cannot really love a person whom you do not really know, and, of course, the same is true for other people. They cannot really love you until they get to know the real you. This is what distinguishes love from infatuation.

Letting others get to know the real you is not always easy. In fact, for some people, emotional intimacy may be more difficult than sexual intimacy. We initially try to look and act our best, but for another person to really get to know you requires that you reveal your needs, feelings, emotions, and values. Unfortunately, some people go through life with a mask on, never allowing themselves to be known by others. A study conducted in the 1950s found that women were much more willing than men to disclose personal information to friends or strangers under laboratory conditions (Jourard & Lasakow, 1958). This is what might be expected in a society where gender roles have traditionally favored emotional repressiveness for men and emotional expressiveness for women, but, as you learned in Chapter 8, gender roles have been changing. Nevertheless, a review of 205 studies of self-disclosure found that there was still a gender difference in disclosure—women disclose slightly more than men in relationships (Dindia & Allen, 1992).

People may be reluctant to reveal what they really think or feel out of fear that other people might judge them or might have less respect for them, or even worse, no longer want them as a lover or a friend. This requires trust, and sometimes that trust is misplaced. Some people may even use another person's trust to fulfill their own selfish desires. If John, for example, feels lust for Joanne, he may show her a lot of attention with the goal of having sex with her. Joanne, on the other hand, may actually be fond of John, interpret his attention to her as love, and subsequently reveal her feelings. John may then say "I love you" to Joanne just to fulfill his own selfish sexual goal. After satisfying his sexual curiosity, John may dump Joanne to pursue other sexual conquests, leaving Joanne emotionally devastated. She may also be more reluctant to trust again.

All of us need the perceptive ability to avoid placing our trust in those who would abuse it, but for love to develop, we must occasionally take that chance or risk. Relationships develop best when two people self-disclose to one another at about the same time. If Joe reveals some of his true thoughts and feelings to Mary, for example, it is expected that she will reciprocate. Well-timed self-disclosure makes a person more likeable to his or her partner (Archer, Berg, & Runge, 1980; Collins & Miller, 1994). Emotional intimacy is achieved only after a couple has shared a reasonable level of self-disclosure (an exchange of vulnerabilities) and each has accepted his or her partner's state of awareness. At this point, they can be said to be interdependent.

JEALOUSY

Shakespeare, in *Othello*, described it as "the green-eyed monster which doth mock the meat it feeds on." He was referring, of course, to the emotion we call jealousy. How would you feel if your partner appeared to be forming an emotional attachment to another person? How would you feel if your partner and another person appeared to be sexually interested in one another? These are situations that typically cause many people to feel jealous. **Jealousy** is an emotional state "that is aroused by a perceived threat to a valued relationship or position and motivates behavior aimed at countering the threat" (Daly, Wilson, & Weghorst, 1982; see also Salovey, 1991; White & Mullen, 1989). The perceived threat can involve more than the actual loss of a partner: it can include loss of face, loss of self-esteem, and loss of feeling special (Buunk & Bringle, 1987). Jealousy is a negative emotion that is hard to

---

self-disclosure Revealing one's thoughts, feelings, and emotions to another.

jealousy An emotional state that is aroused by a perceived threat to a valued relationship or position.

describe, but it usually involves anger, humiliation, anxiety, and depression (Clanton & Smith, 1977).

Some people become jealous if their partner has even casual interactions with other persons. What type of person is most likely to become jealous? Research has shown that people with low self-esteem who are personally unhappy with their lives (Bringle et al., 1977; Pines & Aronson, 1983), and those who place great value on things like popularity, wealth, fame, and physical attractiveness, are more likely than others to be jealous individuals (Salovey & Rodin, 1985). However, cultural factors also play a role. Jealousy is most likely to occur in cultures that consider marriage a means for guilt-free sex, security, and social recognition and that value personal ownership of property (Hupka, 1981). This is true of American culture (as well as many others), and in the United States sexual jealousy by men is the leading cause of beating and murder of wives (Daly & Wilson, 1988). In societies that allow polygyny, jealousy is far less common (Laughlin & Allgeier, 1979).

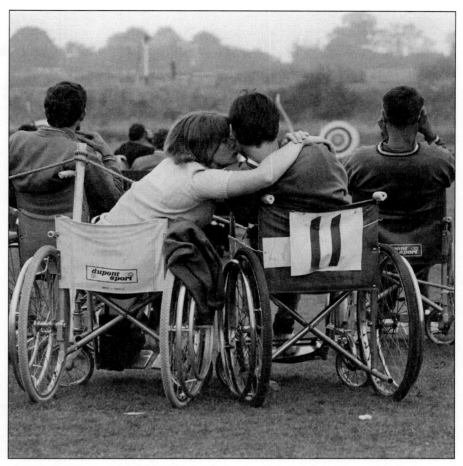

Figure 12–3 For a loving relationship to develop, each person must have good self-esteem and there must be an exchange of vulnerabilities.

Research shows that just as many women as men have feelings of jealousy (Bringle & Buunk, 1985). However, there are important gender differences in jealousy as well. Men are much more likely to become jealous to the perception of a partner's *sexual* infidelity, whereas women are much more likely to experience jealousy as a result of a partner's *emotional* infidelity (Buss et al., 1992; Wiederman & Allgeier, 1993). This difference has been found in different cultures, but to a lesser extent in cultures that are less disapproving of extramarital sex (Buunk et al., 1996). Evolutionary psychologists argue that this difference between men and women is innate, resulting from men's need for certainly about paternity (after the energy spent in courting and mating, he wants to know that he is the father) and women's need for male investment in the children (Buss et al., 1992). Others question the evolutionary explanation and point out that the two types of infidelity, sexual and emotional, are not independent (DeSteno & Salovey, 1996; Harris & Christenfeld, 1996). Men, for example, may be likely to think

that if a partner is emotionally involved, she is also having sex.

Men's and women's reactions to jealousy also tend to differ (Bryson & Shettel-Neuber, 1978; White, 1981a, 1981b). Women tend to experience jealousy when feeling inadequate, and they tend to respond by making themselves more attractive to their partner. Men, on the other hand, tend to initially experience jealous feelings and then have feelings of inadequacy, and they tend to seek solace in outside relationships.

Paradoxically, while jealousy is an emotional reaction to a perceived threat to a relationship and self-esteem, the expression of jealousy is likely to further damage both (Buunk & Bringle, 1987). Withdrawing from or attacking one's partner, for example, may cause him or her to withdraw or attack, thus increasing the fear of loss, which, in turn, may increase or prolong the feeling of jealousy. How, then, should you deal with occasional bouts of jealousy? Acknowledge your feelings to your partner and describe what caused them. Conversely, never purposely try to make your partner jealous, as a few people do in order to get attention (White,

1980a, 1981b). Establish mutual trust, while at the same time respecting your partner's need for some personal freedom. If you can do this, your chances are good of keeping that green-eyed monster at bay.

SEX WITHOUT LOVE

Is it possible to have a life-style of sex without love and still be happy? In the movie *Annie Hall*, Diane Keaton says to Woody Allen, "Sex without love is an empty experience." Woody responds, "Yes, but as empty experiences go, it's one of the best." In his book *Sex Without Love* (1980), Russell Vannoy argues that love is unnecessary and that sex should be enjoyed for its own sake. Masters and Johnson report that they have worked with hundreds of people who weren't in love, but who deeply enjoyed sex. In fact, throughout history there have been many famous people who were noted for their sexual relations with hundreds, often thousands, of partners (e.g., Cleopatra, Empress Theodora, Catherine the Great, Sarah Bernhardt, King Solomon, King Ibn-Saud).

Love and sex share many things in common, including the fact that both may be intensely exciting experiences involving physiological changes that may be expressed in many ways. Feelings of love, however, tend to be more selective than sexual desires, and while the experience of being in love is generally felt all over, sexual sensations are generally confined to a few body parts.

In the United States, most people feel that sexual relations should be reserved for strongly affectionate relationships (Laumann et al., 1994; McCabe, 1987). Women in particular are apt to find sex unenjoyable unless it is within a loving relationship (Christopher & Cate, 1984; McCabe, 1987; Ubell, 1984):

> "I find it much more pleasurable and rewarding to make love with my partner. We have times when we have sex but I enjoy making love much more. My partner caresses me and makes me feel more special and loved."

(from the author's files)

Men, on the other hand, are much more likely than women to enjoy sex without emotional involvement (Carroll, Volk, & Hyde, 1985; Leigh, 1989; Quadagno & Sprague, 1991). In a recent study of people's private wishes (Ehrlichman & Eichenstein, 1992), many more men than women wished "to have sex with anyone I choose." Evolutionary theorists say that because women are the ones who bear children, they have an innate need for (and seek) male partners who will invest in their offspring. Anthropologist John Townsend (1995) found that even highly sexually active women who engage in non–emotionally involved sex are more likely than males to feel emotionally vulnerable and to have anxieties about their partners' willingness to invest.

For most people in our culture, sex without love is a passing stage in relationships. Eventually a large majority of women, and a small majority of men, wish "to deeply love a person who deeply loves me" (Ehrlichman & Eichenstein, 1992). Some people who engage in casual or recreational sex may actually be searching for intimacy. Many people, particularly men, never learn how to express or receive affection (e.g., to hug or cuddle) and mistakenly believe that sex and intimacy are synonymous (see Chapter 10). However, this doesn't mean that a person cannot enjoy a life-style of sex without love. Some people may simply prefer their independence to emotional involvement. And remember, the emphasis in the United States of sex within a loving relationship is a culturally learned value. In some other cultures, people put considerable more emphasis on sexual pleasure (and think that Americans' attitude of "love first" is peculiar) and encourage sexual relations with many partners (see Chapter 1).

LOVE WITHOUT SEX

Keith Davis (1985), you recall, included sexual desire as one of the characteristics that distinguished romantic love from friendship. Is it possible for people to experience romantic love without engaging in sexual relations?

Although sexual deprivation will not kill you like lack of water or food, there is strong evidence that physical contact is important for human beings. Babies who are deprived of physical contact may sicken and die or show arrested social development. Researcher Harry Harlow (1959) raised baby Rhesus monkeys in isolation with make-believe terrycloth mothers. They were well-fed and clung to the terrycloth models as they would their real mothers. When they matured, however, and were placed in cages with other monkeys, the monkeys raised in isolation were antisocial and refused to mate or perform sexually. More than 50 years ago, a physician similarly reported that one in three human infants in foundling homes died during the first year of life, even though they were well fed and provided with adequate medical care. What these babies were not receiving was cuddling, hugging, fondling, and kissing—they apparently died from emotional neglect (Spitz, 1945, 1946). We never outgrow the need to be held and touched, but these needs can be fulfilled without actually engaging in sex.

Love is a feeling, not an act. Therefore, people who have a loving relationship may or may not choose to express their closeness with sex. Although attitudes toward premarital or nonmarital sexual relations are not as stringent as in previous generations, many couples still prefer to reserve their sexuality for marriage (Laumann et al., 1994). This period of voluntary celibacy allows them to devote time and energy to other aspects of their relationship (and allows the intimacy component to grow), perhaps resulting in their finding many nonsexual ways to express their love for one another. Even in sexually consummated relationships, sexual relations may have to be interrupted, sometimes for long periods of time, because of illness or physical separation (due to military duty, etc.), but this doesn't mean that feelings of love are also interrupted.

UNCONDITIONAL LOVE

So far, most of our discussion about love has centered on romantic love, which tends to be **conditional love,** or what Maslow (1968) called *deficiency love* (or *D-love*). We fall in love with someone and remain in love because he or she satisfies certain needs and fulfills desires, and because it is positively reinforcing to be with him or her. We tend to fall out of love when our expectations and needs are no longer met. Our feelings for the other person depend to some extent, perhaps to a large extent, on how he or she makes us feel and contributes to our happiness.

Distinguished from conditional love is **unconditional love** (what Maslow called *being love* or *B-love*), in which one's feelings do not depend on the loved one's meeting certain expectations and desires (Fromm, 1956). People love out of choice, not out of need. Unconditional love is the type of love that many mothers feel for their children (Fromm believed that love is conditional for most fathers), that many grown-up children continue to feel for their parents, and that many people feel for other individuals of the same or opposite sex. How one feels about his or her romantic partner can also eventually transcend the ability of the partner to satisfy needs and fulfill expectations. This is perhaps most closely approximated by what is called companionate love and which is expressed so well in St. Augustine's definition of love, "I want you to be."

THEORIES OF LOVE

Any textbook that includes the topic of love must—as an obligation, I suppose—present some coverage of the theories of love. But which ones? There are as many theories of love as there are definitions, maybe more. Moreover, the theorists cannot even agree on how to organize love. In his preface to a book containing articles by many theorists, Rubin (1988) states:

> Many of the contributors to this volume have developed their own taxonomies of love.

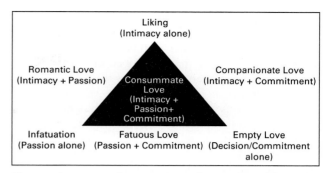

Figure 12–4 Sternberg's triangular model of love. All the different positive emotions that people can have for others are understood by the combinations of intimacy, passion, and decision/commitment.

> Each categorizing scheme differs from the next, and there are no ready translation rules from one chapter's formulation to another's. Just as partners with different views of love may find themselves talking past each other . . . I suspect that some of the contributors to this volume may find it difficult to relate to others' perspectives. (p. ix)

Most theorists touch on only certain aspects of love, and few take the time to support their ideas and conclusions with empirical data. Popular books on love such as Leo Buscaglia's *Love* or M. Scott Peck's *The Road Less Traveled* rarely support their claims with any evidence at all. The works of some of the great thinkers (Plato, Aristotle, Kant), on the other hand, treat love as an academic subject and are often lacking in sensitivity.

Experimental social psychologists are the newest group to offer theories of love. As examples of this approach to love, I have selected the works of two pioneers in the field: Robert Sternberg and John Lee. Other texts may have chosen different theories, but these will give you a good idea of how social psychologists perceive this complex topic we call love.

ROBERT STERNBERG'S TRIANGULAR THEORY OF LOVE

Although most people rate their spouse or lover higher for some characteristics than they do their friends (Davis, 1985), this still does not tell us whether liking and loving are qualitatively different emotions or different regions along a single continuum of emotions. Robert Sternberg (1987) has proposed a *triangular theory of love* in

conditional love Feelings of love that depend on the loved one's satisfying needs and fulfilling desires.

unconditional love Feelings of love that do not depend on the loved one's meeting certain expectations or desires.

which he takes a broader view of liking and loving. He suggests that all the different positive emotions that people can have for other individuals can be understood by the combination of three components. In his model, each component is viewed as the vertex of a triangle, as shown in Figure 12–4. The triangle should not be viewed as a geometric model, but rather as a useful metaphor for visualizing the way in which the three components are related.

The top vertex of the triangle is **intimacy.** Intimacy refers to those feelings in a relationship that promote closeness or bondedness and the experience of warmth. There are many signs of intimacy, including a desire to promote the welfare of the other, experiencing happiness with and having a high regard for the other, receiving and giving emotional support, mutual understanding, and valuing the other person in your life (Sternberg & Grajek, 1984). The left-hand vertex represents the experience of **passion,** or those drives leading to physical attraction, sexual relations, and romance. The right-hand vertex is **decision/commitment,** which includes the decision to love another person and the commitment to maintain the relationship over time. Recent studies have verified that people's concepts of love are made up of these three components (Aron & Westbay, 1996).

Sternberg says that the different combinations of these three components result in different kinds of emotions. When all three components are absent, the result is *nonlove.* This characterizes most of our casual relationships, where there is no love or friendship in any meaningful way. If the intimacy component is expressed alone without passion or decision/commitment, it results in *liking.* The word liking is not used in a trivial manner to refer to casual acquaintances, but instead refers to the feelings of closeness, bondedness, and warmth in true friendships. When passion is felt in the absence of the other two components, the result is *infatuated love,* or what we call "love at first sight." A person feeling passion alone is obsessed with the other person as an ideal, rather than as the individual he or she is in reality. The decision and commitment to love another person without intimacy or passion is experienced as *empty love.* In our society, empty love often occurs at the end of stagnant long-term relationships, but in other cultures where marriages are arranged, it may be the first stage in a long-term relationship.

If you look at the triangle, you can see that there are four possible combinations of the components. If you add passionate arousal to liking (the intimacy component alone,

intimacy Those feelings in a relationship that promote closeness or bondedness and the experience of warmth.

passion The drive leading to physical attraction, sexual relations, and romance.

decision/commitment One of the three basic components in Sternberg's theory of love; the decision that one loves another person and the commitment to maintain the relationship.

you recall), the result is *romantic love.* It results from two people being drawn together both physically and emotionally. There is no commitment, and the lovers may even know that a permanent relationship is not possible, as in "summer love." This is the Romeo-and-Juliet type of love that poets, playwrights, and novelists are so fond of writing about.

The combination of intimacy and decision/commitment without passion leads to *companionate love,* a long-term committed friendship. Sternberg (1988) believes that most romantic love relationships that survive do so by eventually turning into companionate love relationships.

When a commitment is made on the basis of passion without the experience of intimacy, the result is *fatuous love.* This leads to whirlwind romances, the type we often read about involving Hollywood stars. Without intimacy, there is a high risk that the relationship will end once the passion starts to fade (these romances are often over so quickly that there has been no chance for the intimacy component to develop).

Sternberg believes that complete love, what he calls *consummate love,* is found only in relationships that include all three components—passion, intimacy, and commitment. It is his belief that this is the type of love that most of us strive for in our romantic relationships. Thus, other types of relationships are viewed as lacking something (i.e., one or more of the three components).

JOHN LEE'S "MANY COLORS OF LOVE"

The English language gives us only one word—love—to describe a number of interpersonal relationships. Words such as liking, affection, and infatuation are not considered to be synonymous with love. As a result, our different experiences of love tend to be measured as differences in quantity: "Tell me *how much* you love me" or "I love you more than I've ever loved anybody." In our search for a partner, we hope to find someone who loves us *as much* as we love them. When relationships do not work out, we often deny any experience of love (e.g., "After a while, I realized I really didn't love Suzy," or "Suzy really didn't love me"). It's as if our experiences of love are measured in black and white and shades of gray.

Sociologist John Lee (1974, 1976, 1988) believes that there is more than one type of partnering love, with no one of them being singled out as "true love." He suggests that our different experiences are not due to whether or not two persons' feelings are of the same intensity, but rather to how well one person's style of loving matches with another's style. Rather than black and white, Lee uses the analogy of colors to explain love. Different styles of love are portrayed by different colors. Sternberg (1988) believes that his theory of love is most akin in spirit to that of Lee. However, there is a major difference between the two of them. Sternberg's theory implies that there is only one type of true love (consummate love) and that all the others are not really love. Lee's model, on the other hand, does not suggest that one style of love is superior or better than another, but only that there are many different styles. According to Lee, mutual love results from two styles or colors (not intensities) that make a good match.

Just as different colors result from blending red, yellow, and blue, the three primary pigment colors, Lee proposes that different styles of loving arise from blending three primary love-styles—eros, storge, and ludus. Let us first look at each of these and their various combinations, and then we will see what makes a good match.

THE PRIMARY COLORS

Eros (named after the Greek mythological figure) is a highly idealized love based on physical beauty. According to Lee, every erotic lover has a specific ideal physical type that turns him or her on. Erotic lovers look for physical perfection, knowing it is rare, and when they find someone who embodies their ideal, they quickly feel a strong physical attraction and emotions they perceive as love. It is as if they have always known this person that they have just met—their romantic ideal. Thus, the erotic lover is inclined to feel "love at first sight" and wants to have an intimate relationship immediately. Erotic lovers are very affectionate and openly communicate with their idealized partners. However, they are usually quick to find flaws and shortcomings in their new partners, so that many of their relationships are quick to fizzle out because the partner cannot live up to their unrealistic ideal. Therefore, erotic love is generally very transient, and erotic lovers tend to fall in love very often. Erotic lovers desire to have an exclusive relationship with their partners, but they are not jealous and do not try to possess them.

Lee's eros is very similar to the "passionate love" described by Berscheid and Walster (1978). It is quite different from the *eros* of Plato's *Symposium*, which was totally good and wholesome. Lee's choice of terms here is perhaps unfortunate.

Ludus (from the Latin word for "play" and pronounced "loo'-dus") is a self-centered type of love. The ludic lover avoids commitment and treats love like a game, often viewing the chase as more pleasurable than the prize.

"I'm a 23 year old male in your MWF Sex Class. . . . I first had intercourse at age 13. . . . As of this writing, I have had intercourse with 508 women. I have done it 1,823 times in the last 9½ years. My goal is 2,000 times by 1989. I've devoted a notebook strictly for recording names, dates, and places of sexual encounters. . . . I can say that I honestly love all women. . . . "

(from the author's files)

Ludus is similar to Sternberg's (1987) fatuous love. Ludic lovers have no romantic ideal and never see any one person often enough to become dependent on them (or vice versa). Sex is had for fun, not for expressing commitment, for ludic lovers are not very emotional and do not have feelings of falling in love. They avoid jealous partners who might spoil the fun and see nothing wrong with having more than one partner at the same time. They are also more likely than people with other love-styles to use verbal coercion to obtain sex (Sarwer et al., 1993). Ludic lovers can be deceptive with their partners, like Don Juan, but many play the game honestly in order to avoid having feelings of guilt (and thus their partners know exactly what they are getting). This promiscuous type of love-style may seem empty to people with other love-styles, but most ludic lovers look back on their relationships with pleasure.

Storge (from the Greek, and pronounced stor'-gay) is an affectionate type of love that develops from friendship slowly over time. It is essentially the same as Sternberg's (1987) companionate love. The storgic lover does not have a physical ideal and does not go looking for love, but instead develops feelings of affection and commitment with a partner through experiencing activities that they both enjoy. Storgic lovers generally cannot recall a specific point in time when they fell in love. They are more practical than emotional, and their relationships lack the passionate emotional highs (but also avoid the dramatic downturns). In fact, storgic lovers would be embarrassed by having to say "I love you," or by excessive shows of emotion from their partners. They are more interested in talking about their shared interests than their mutual love. A true storgic lover would probably find this whole chapter very silly.

THE SECONDARY COLORS

How many different love-styles are there? You can get numerous colors by blending red, yellow, and blue, and the same is true for mixing the primary love-styles. Look at the color wheel in Figure 12–5. The apexes of the large triangle within the circle represent the three primary colors. Each pair

eros A highly idealized type of romantic love that is based mainly on physical beauty. Erotic lovers look for physical perfection.

ludus A self-centered type of love. Ludic lovers avoid commitment and treat love as a game, often viewing the chase as more pleasurable than the prize.

storge An affectionate type of love that grows from friendship.

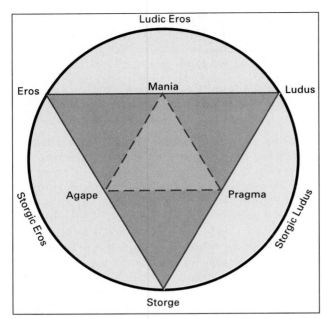

Figure 12–5 John Lee's color wheel of love-styles.

---

can be combined by either going around the circle or along the edge of the triangle. When you combine two things, new properties may emerge and others may be lost. No one, for example, could have predicted the properties of water by looking at the properties of hydrogen and oxygen. Let us now look at a few of the numerous combinations that can result from mixing the primary love-styles.

Pragma (from Greek for pragmatic) is a rational or practical style of loving resulting from combining ludus and storge. Pragmatic lovers have the manipulative confidence of ludic lovers and consciously look for a compatible mate. They are not looking for an exciting romance or affair, but instead (like the storgic lover) wants love to grow out of friendship. However, unlike the storgic lover, the person with a pragmatic love-style has a shopping list of qualities he or she is looking for in a mate and consciously and carefully sorts candidates to see if they have all the desired qualities. Women, for example, give more weight than men to socioeconomic status, ambitiousness, and intelligence in choosing a mate (Feingold, 1992). If a suitable mate cannot be found at work or in the community, the pragmatic lover may join organizations (e.g., church, club) or even try computer dating services to find a compatible partner. Pragmatic lovers want their partners to reciprocate signs of thoughtfulness and com-

pragma A rational or practical type of love. Pragmatic lovers actively look for a "good match."

mania A type of love characterized by an intense emotional dependency on the attention and affection of one's lover.

agape A type of romantic love that puts the interest of the loved person ahead of one's own, even if it means great sacrifice.

mitment, but do not like excessive displays of emotion or jealousy. Sexual compatibility is not unimportant, but it is treated more as a technical skill that can be improved upon if need be rather than as the result of chemistry. A person with this love-style views herself or himself as "in love" and tends to remain loyal and faithful as long as the partner continues to be a good match.

Mania (from Greek meaning love that hits like a lightning bolt) is a love-style characterized by an intense, obsessive emotional dependency on the attention and affection of one's partner. People with manic love-styles are "in love with love" (the excitement of falling in love). Dorothy Tennov (1979) coined the term *limerence* to describe extreme "head over heels" love in which one is obsessed with the loved one (i.e., extreme preoccupation and acute longing for exclusivity). Mania is also similar to Sternberg's (1987) infatuated love. The manic lover is intensely jealous and repeatedly needs to be assured of being loved. The feelings of mania are so intense that this kind of love can be likened to addiction, with the manic lover often not even liking the partner, but at the same time unable to break off the relationship. As a result, manic lovers ride an emotional roller coaster: extreme highs when the partner is showing them attention, and extreme lows when he or she is not. This love-style results from mixing eros and ludus. The manic lover wants an intense, physically stimulating relationship, but usually chooses inappropriate partners and madly projects desired qualities onto them. Everyone but the manic lover can see that the partner does not really have those qualities. The manic lover attempts to manipulate the relationship (by pulling back and not calling the partner, for example), but is too dependent on the partner's love to remain confidently detached like the ludic lover (guess who ends up breaking down and calling first?). Because the manic lover is unable to break off the relationship, it is usually the partner who does so, leaving the manic lover devastated. Although many people have had an experience of manic love, the true manic lover is apt to repeat the same type of relationship again and again.

Agape (from the Greek, and pronounced ah-gah'-pay) is a selfless, altruistic love-style that puts the interest of the loved person ahead of the lover's own interest, even if it means great sacrifice. It is similar to Maslow's (1968) "being love" and Sternberg's (1987) empty love. This is the style of loving proposed by St. Augustine ("Love means: I want you to be") as a goal for all Christians, and the kind of love to aspire to, according to Fromm (1956). It is the result of combining eros and storge. Rather than yearning to find the perfect partner, the agapic lover submits to the perfect will of God and becomes the devoted friend of those who need his or her love. Agapic lovers believe it is their

Figure 12–6 Christian ideals of love as a desire for another person "to be" acknowledge that love can be felt for a lover, a parent, a child, or a friend of the opposite or the same sex.

duty to love, but the feelings are often not directed toward a specific person, but instead to all who need their love.

Lee believes that agape is the least common love-style, and he is probably correct. In fact, some have questioned whether humans even have the capacity to feel true agape, which is selfless (see Nygren's *Agape and Eros,* 1982, and Singer's trilogy *The Nature of Love,* 1984–1987). An example that is often given is found in Charles Dickens' novel *A Tale of Two Cities,* a love story that takes place during the French Revolution. The hero, Sidney Carton, is in love with Lucie Manette, who is in love with Charles Darnay. At the end of the story, Charles Darnay is sentenced to die by guillotine. Sidney's love for Lucie is so strong that he takes Darnay's place and is guillotined so that Lucie can be with the man she loves. But Sidney probably would not have done this if he hadn't felt passionately about her. He certainly would not have done it for anyone else. However, there is no passion or any other type of self-interest in agape. Humans may not be able to achieve agape, but they can strive to achieve a love-style that approaches it.

OTHER COMBINATIONS. Pragma, mania, and agape are not the only possible outcomes of mixing the primary colors. Lee's analogy is mixing flour and water. Try mixing them when the water is hot, and then try it again when the water is cold. You are mixing the same ingredients, but the results are different. Other combinations are possible (e.g., storgic eros, storgic ludus,

ludic eros). For a full description of these, see Lee (1988).

FINDING A GOOD MATCH

Lee's theory has received support from the results of a large-scale study (Hendrick & Hendrick, 1986). Over 1300 students were asked to agree or disagree with various statements about love. Their responses clustered into six different groups that closely fit the six categories just reviewed. There were some interesting gender and ethnic differences as well. Men tended to be more ludic than women, who tended to be more pragmatic, storgic, and manic. Asian-Americans tended to be more pragmatic and storgic than other ethnic groups.

Did you recognize your love-style from the preceding list? There are tests you can take to determine your love-style preference (Hendrick & Hendrick, 1986) but Lee (1988) suggests that you just review your own experiences. When doing this, keep in mind that love-styles are not necessarily fixed. We are not born with a particular love-style that we are stuck with for life, but instead we can learn from our experiences and change our preferences (we cannot learn from our experiences if we look at love as black and white and dismiss failed love as due simply to differences in intensity of feelings). Thus, you may have experienced more than one of the preceding styles. It is not unusual, for example, for a person's first love-style to be mania, or for a person to experience mania for the first time in midlife after a storgic marriage has become dull and uninteresting (conversely, one may prefer a comfortable storgic relationship after experiencing mania). Once you are over mania, it would be a mistake to dismiss it as infatuation (something other than love), for it demonstrates how intense your feelings can be for another person, and this can help you in a later love-style. Rather than accusing your partner of "not really loving me," it is more realistic to tell him or her that "your style of loving is not what I want."

Lee believes that the secret to finding mutual love is to find a good match, not in the amount of love each person gives, but in the style of loving. There are some obvious bad matches—a manic lover with a ludic lover, for example. The ludic lover's vanity will enjoy all the attention given by the manic lover, but the possessiveness and jealousy of the manic lover are not desired. Manic lovers thrive on problems, and the ludic

Box 12-A **CROSS-CULTURAL PERSPECTIVES**

Love and Marriage

Romantic love is not something new; there are references to it throughout history. In Western culture, love became idealized during the Middle Ages with the rise of courtly love (a young man devoting himself to a noble-woman). Even so, romantic love was not considered a good reason to enter into marriage until the 1800s. In colonial times, Americans agreed to marry more or less by arrangement, the main purpose of marriage being to have children (Rothman, 1984). In fact, in a study conducted in the early to mid-1960s, about one-third of the U.S. men and three-fourths of the women surveyed indicated that being "in love" was not a necessity for marriage (Kephart, 1967). By 1976, this had changed; all but about 15 percent of the men and women in one study said that romantic love was a necessary prerequisite for marriage, and the same results were found in 1984 (Simpson, Campbell, & Berscheid, 1986). The change in attitude, mostly by women, was attributed to better economic status—financial independence allowed people to make relationship decisions based on romance.

Is romantic love a creation of highly industrialized cultures where people have much leisure time and many comforts? Anthropologists have found that rules about marriage are, in fact, influenced by economics and politics (Barry & Schlegel, 1984; Hsu, 1981). In industrialized countries, where importance is attached to individuals, monogamy is the standard. In simpler societies, where less importance is given to individuals and more to situations, *polygyny* (allowing a man to have more than one wife) is often the rule. In a study of 862 cultures, 83 percent permitted polygyny and only 16 percent were monogamous (Murdock, 1967). Polygyny is very common in societies with strong fraternal interests (with dowries given at marriage), or where there is warfare for the capture of women, or where there are no constraints on expansion into new lands (White & Burton, 1988).

There are many differences among cultures in the relationship between a husband and a wife. In a great many cultures, marriages are arranged by parents, sometimes at birth. In contrast to the highly romantic beginnings of marriages in the United States, Japanese arranged marriages start off with very little love, but there is no difference in the amount of love within the two types of marriages after 10 years (due largely to the decrease in romantic love over time in American marriages; Blood, 1967). Marriages in India are arranged according to caste, age, and other factors. Many Islamic societies require that a young man marry his father's brother's daughter (first cousin), and husband-wife relations are generally without affection, and are often very hostile (Lindholm & Lindholm, 1980). Pacific islanders and people in Southeast Asia are often required to marry their mother's brother's daughter. Some cultures require the male to capture a wife from an enemy village, as in the case of the Gusii of Kenya (Levine, 1974), who rape their wives whenever they wish to have sex. A male of the Marind Anim of New Guinea must first share his newlywed wife with all the other males of his clan before he is allowed to consummate the marriage (Money & Ehrhardt, 1972).

Does all this mean that romantic love is restricted to a few highly industrialized cultures? No, say William Jankowiak and Edward Fischer (1992), who recently studied love in 166 cultures and found evidence of romantic love in at least 147 of them. Even in cultures where marriages are arranged and rules of marriage are highly structured, romantic love often flowers in clandestine forms (as is often the case with extramarital affairs in our own culture). Romantic weddings may not be universal, but love is apparently "a very primitive, basic human emotion, as basic as fear, anger, or joy" (anthropologist Helen Fisher, quoted by Gray, 1993)—one that escapes all the restrictions and barriers imposed by numerous cultures.

lover is guaranteed to provide them. It is not uncommon, however, to find this pairing, and while not very happy and ultimately doomed, a manic-ludic relationship is always interesting.

With the exception of mania and ludus, *a good match generally results from two styles that are close on the chart* (see Figure 12–5). A pragmatic love-style and storgic love-style make a good match, for example. The farther apart two love-styles are, the less likely it is that they would make a good match. The erotic or manic lover, for example, is likely to find the storgic love-style dull and regard it as friendship without passion and not love at all. The storgic lover will similarly accuse the behavior of the erotic or manic lover as not being love. A relationship between two people with these two different styles is probably doomed within a short time. In the end, each may accuse the other of not having really loved, but in reality both will have loved in his or her own way. It simply was not a good match in love-styles.

Studies have shown that whether or not a relationship is successful can, in fact, be partially predicted by the compatibility of Lee's love-styles (Davis & Latty-Mann, 1987; Hendrick et al., 1988). A recent study found that college students preferred dating partners who had love-styles similar to their own (Hahn & Blass, 1997). One exception noted by this and other studies is the attraction between people with mania and agape love-styles, suggesting that these two types may not be as different as Lee originally believed.

So the next time you find yourself attracted to someone, do not ask yourself how much this person loves you, but instead ask yourself how (in what style of love) they love you and whether or not your colors make a good match. Lee (1988) warns us, however, not to define mutual love in terms of longevity. Two people may have a mutually satisfying relationship, but just because it ends because one of them had a change in love-style does not mean it wasn't wonderful while it lasted.

MAINTAINING A RELATIONSHIP

There is hardly any activity, any enterprise, which is started with such tremendous hopes and expectations and yet which fails so regularly as love.

(Erich Fromm, 1956)

Many factors play a role in the initiation of a relationship. These include (*a*) physical attractiveness—men generally attach more importance to this than women (Buss, 1989; Dion & Dion, 1987; Hatfield & Sprecher, 1986; Walster et al., 1966); (*b*) proximity—people are most likely to fall in love with someone they interact with often (at work, school, etc.) (Berscheid & Walster, 1978); (*c*) similarity—people tend to be attracted to others who have similar love-styles (Lee, 1988), interests, values, intellectual abilities, and degrees of attractiveness (there is no evidence to support the old saying that "opposites attract") (Byrne, Clore, & Smeaton, 1986; Wetzel & Insko, 1982; White, 1980b); and (*d*) reciprocity—we tend to like people who show that they like us (no one likes rejection) (Berscheid & Walster, 1978; Byrne et al., 1986; Curtis & Miller, 1988).

Let us return for a moment to our hypothetical example of Susan and David (see "How Do I Know If This Is Really Love?"). All four factors have played a role in encouraging these two people to begin a relationship. Susan—and David, let's suppose—have experienced physiological arousal in each other's presence and have interpreted their responses as "being in love." The relationship continues to develop if there is mutual self-disclosure, equity in what each wishes to gain from (and give to) the relationship, and finally, commitment. But will it last?

Recent studies show that personality characteristics are important in determining whether or not a relationship lasts. Not only do people prefer mates who are like themselves, but marital and sexual dissatisfaction on the part of one partner can often be predicted if the other partner is lower on agreeableness, emotional stability, and intellect-openness (Botwin, Buss, & Shackelford, 1997). Relationships tend to last when partners idealize one another (Murray, Homes, & Griffin, 1996). However, even when a couple is well matched, they are going to have to learn to deal with change.

The divorce rate is so high not because people make foolish choices, but because they are drawn together for reasons that matter less as time goes on.

(Robert Sternberg, 1985)

If there is anything that is unavoidable in life, it is change. Couples will be faced with a variety of new challenges in life, including parenthood, financial crises, career-related stress, and, inevitably, aging. Two people who believe themselves to be in complete harmony during the dating, passion, and romantic loving period are often surprised to find themselves in disagreement on how to handle these matters. Many couples, for example, report the child-rearing years to be the most stressful and least satisfying in their marriages (Hendrick & Hendrick, 1983). Fixed gender roles—or, conversely, changing gender roles—may contribute to this (see Chapter 8). Today, there are the added chal-

Figure 12–7 Companionate love is a stable kind of love based on togetherness, trust, sharing, affection, and a concern for the welfare of the other.

lenges of two-job families and job-related moves far away from family and friends.

THE DECLINE OF PASSION

The challenges in life are not the only things that change. People also change over time. One of the things that normally occur in a relationship, leading to changes in how we interact with our partners, is **habituation.** Think of your favorite food. What makes it your favorite food? The taste and smell are no doubt important, but I am willing to bet that another factor is that you do not have it very often. If you were served your favorite food every day for a month, do you still think it would be your favorite? Would we look forward to Thanksgiving turkey if we ate turkey every week? It's doubtful. When organisms, including humans, are repeatedly exposed to a stimulus that is initially very positive, it becomes less positive over time. This is why popular weekly TV shows eventually go off the air and

habituation Responding less positively to a stimulus with repeated exposure to it.

why hit songs eventually drop off the Billboard "Top 10" list. We grow tired of seeing and hearing them. The same is true of the ability of other people to stimulate us—socially, emotionally, intellectually, and alas, sexually. Of all the challenges that a couple face in a relationship, habituation is certainly one of the greatest ones (Byrne & Murnen, 1988).

The decline of passion is almost inevitable, for a large component of passion is novelty and fantasy. A new partner cannot continue to arouse strong emotions, and with the passage of time it takes greater and greater stimulation to cause the same response that once was brought about by a mere glance or touch. As the passion subsides and fantasy is replaced with reality, the result is often disappointment. In a study of 62 cultures, Fisher (1992) found that the divorce rate peaked around the fourth year of marriage (seven years if a couple has another child two or three years after the first). The reason most frequently given by couples in the process of divorce is that they had "fallen out of (romantic) love" (see Roberts, 1992) and were bored (Hill, Rubin, & Peplau, 1976). Marital therapy has traditionally focused on improving communication and helping the development of rational skills, and only recently have therapists realized the importance of addressing emotional problems—the loss of sexual attraction and the ebbing of romantic love (Roberts, 1992). Many couples make the mistake of allowing their sex lives to become ritualized, but even when the passion has subsided, a sexual relationship can remain exciting, fun, and fulfilling if the couple work for variety and spontaneity (see Chapter 11).

Although passion will decline with time owing to familiarity and habituation, there is no need for *intimacy* (those feelings and experiences that promote closeness and bondedness) to decline. The key to maintaining a relationship is replacing passion with those things that lead to companionate love.

GROWING TOGETHER/GROWING APART: WILL COMPANIONATE LOVE DEVELOP?

To maintain their relationship, Susan and David must simultaneously deal with the need for similarity in the relationship, the inevitable changes that occur (for example, they will not always be in school together), and the natural process of habituation. Similarity, so important in establishing their relationship, can be maintained and habituation can be kept to a minimum by embarking on new activities together (Byrne & Murnen, 1988). For example, they may find it stimulating to engage in new sports and games together, share new hobbies, *try new sexual experiences together* (see Chapter 11), join dinner groups, plan vacations together, and/or seek new educational opportunities together. One of the major predictors of marital success is the number of shared pleasurable activities

(Ogden & Bradburn, 1968). Couples in happy, long-lasting relationships frequently say that they regard their partner as their best friend (Lauer & Lauer, 1985). Couples who are unhappy in their relationships often report that their children are their only or their greatest source of satisfaction together (Luckey & Bain, 1970). Whether or not Susan and David develop intimacy and companionate love will depend, of course, on their commitment (the third component necessary in Sternberg's model for consummate love). Couples who successfully accomplish this grow together; those who do not grow apart.

> Married lovers grow within love; they develop into better human beings.
>
> (Erich Fromm, 1965)

What is meant by growing together or growing apart? I will try to demonstrate this with the use of intersecting circles (Levinger, 1988). Each person in the relationship is represented by a circle; the degree of intimacy shared is shown by the amount of overlap. The amount of passion felt by one person for the other is represented by plus (+) signs within the circle at the periphery of the intersection.

Look now at Figure 12–8. At the beginning of Susan and David's relationship (two circles at top), both are going to college. Their interests are shown within the circles, and each has a different set of friends: C, D, E, F, and G for Susan, and H, I, J, K, L, and M for David. Susan and David met and began to date, you recall, while taking the same course. Because they have a great deal in common, they have a lot to talk about when they are together: their college courses; their good, bad, and peculiar professors; their campus activities and sporting events; and, of course, their career goals. Let us suppose that they both play tennis and that dancing is their favorite activity on dates. David even goes bowling with Susan, although he doesn't like it much. Over time they establish a mutual set of friends, including other couples (C-D and K-L) and singles (N). They each still have their own sets of friends as well (each has lost contact with some, retained some, and gained some new ones, and each enjoys some activities not shared by the other—bridge and reading for Susan and golf for David). There is a great deal of passion felt by both Susan and David, and before long they tell one another that they are in love (decision). Dating is now more than just a good time, for there is a romantic component to being together. They realize that they share a great deal of each other's lives (i.e., there is a strong intimacy component, as shown by the amount of overlap in the circles in the middle), and neither can imagine life without the other. They make the commitment to have a lifelong relationship.

Let us now jump several years ahead and see what has happened to our imaginary couple. They no longer have the opportunity to spend most of the day together, for college is now just a distant memory. Instead, one of them is pursuing a career and spends 8 to 10 hours a day in an office (in today's world, it could be either Susan or David, but let us assume here that it is David), while the other is a full-time house-and-child caretaker. Yes, our imaginary couple now have children, and they only see each other briefly in the morning, in the evening before they go to sleep, and on the weekends (assuming that our career person does not work weekends as well). How they adjust to this big change will determine whether the relationship remains positive or turns negative.

Unfortunately, many couples **grow apart** over time. Let's assume, for example, that Susan now goes to bowling league, plays bridge and tennis with friends, and enjoys reading and gardening, all without David. David, on the other hand, plays golf and tennis with friends and enjoys an occasional night out with friends, all without Susan. They rarely go dancing anymore, and most of their time together is spent watching TV (with a minimal amount of conversation). Susan and David have children, but they do very little together as a family. David considers it Susan's duty to go to the PTA and drive the children to and from all their activities, even when he is home. Susan and David and their children eat dinner together (quickly) and go to church together, but most of Susan and David's conversations about their children involve school and discipline problems. Raising their children is treated more as a responsibility than as a blessing. Although Susan and David live together and have children, they actually share less together now than when they were dating (there is less intimacy, indicated by less overlap in the circles at the lower left in Figure 12–8). There is no passion left, and very little affection (e.g., hugging, kissing, holding hands) is ever displayed between them. They have grown apart. This is not uncommon.

> "I am married, yet feel as if I do not have a husband. He goes hunting every weekend during hunting season, and goes fishing a lot too. Between his softball league and poker games at the club, most weeknights are taken as well. Most of my time is taken up by the kids' activities and what social life I have is spent with other wives. Most of the time I am too tired to make love when my husband is in the mood, which is not very often. A friend recently described our marriage as 'married singles.'. . ."

"My boyfriend and I rapidly grew apart in

growing apart Having few common interests over time.

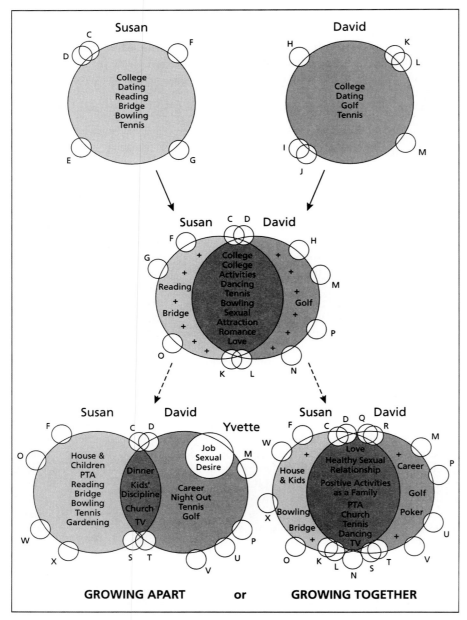

Figure 12–8 An example of growing apart or growing together.

this point, some people may think that it is easier to form a new relationship than to work at their present one. For example, let us suppose that David spends a great deal of time with Yvette while at work (see Figure 12–8). They are together all day at least five days a week. They talk about the boss, fellow workers, job-related problems, and career goals. They even start telling each other about their personal problems at home. Add sexual attraction and feelings of passion, and David may conclude that he has as much in common with Yvette as he once had with Susan when they were in college together. David may decide to have an affair with Yvette, or leave Susan for Yvette. Unless David learns to work at keeping a relationship strong, however, this will just be a short-term solution, for over time David's passionate feelings for Yvette will subside as well and they, too, will grow apart.

Let us now look at another possible outcome for David and Susan. People change over time, but a couple who **grow together** substitute new common interests as old ones drop out (see Figure 12–8). In place of college, which once gave them both a great deal to talk about together, our couple view their children and family activities as something very positive. They go to movies as a family, have picnics together, go to the zoo together, and both parents are involved in Little League, the swimming team, the Boy Scouts and Girl Scouts, the PTA, and church and church social activities. Because they view their children's activities as positive experiences, our two adults have a great deal to talk about. They have substituted new shared activities for some of the ones they used to do together. They still find time (or *make time*) to play tennis on the weekends and go to movies, and they still enjoy dancing together as they did when they were dating. They also have a large number of mutual friends. It is not necessary for Susan and David to do everything together in order to be happy. In addition to being apart during weekdays,

our relationship. We did the same things over and over and did not feel comfortable enough to express our feelings towards one another. We did not accept each other for who we were; we kept trying to change things about each other. . . ."

(from the author's files)

growing together Maintaining common interests (although not necessarily the same ones) over time.

People generally grow apart very slowly, and by the time they realize it, they may have little left in common. At

David still plays golf regularly and enjoys an occasional evening of poker with friends, while Susan has joined a bowling league (David refuses to bowl anymore) and plays bridge with friends. They are different from the way they were when they were dating, but the amount of shared interests is just as great as then. Their feelings of passion for one another have declined, of course (hearts don't beat faster, breathing doesn't get heavier, and palms don't sweat when they see each other), but their sexual interest in one another remains strong. They enjoy each other's bodies and seek variety in the way they stimulate one another. There has been no decline in intimacy. They have grown together.

BECOMING MORE INTIMATE

For most couples, the passion component is generally greatest early in a relationship and then declines over time. Therefore, one of the most important things in maintaining a relationship is developing and maintaining intimacy. Sternberg (1986, 1988) defined *intimacy* as those feelings in a relationship that promote closeness or bondedness and the experience of warmth. Similarly Elaine Hatfield (1984) defines it as "a process in which people attempt to get close to another; to explore similarities (and differences) in the way they think, feel, and behave." It is easy to share our similarities and positive feelings, but what about our differences and negative feelings? In order to become really intimate with someone, you must share your differences as well. Hatfield has made several suggestions designed to teach couples to become more intimate (Hatfield, 1988). Let us briefly examine these.

1. Both individuals in a relationship need to *accept themselves as they are.* I have discussed this earlier in the chapter, but the need for self-acceptance bears repeating. *Your* best may not be *the* best, but it is all that anyone, including yourself, can ask of you. Learn to accept your ideas and feelings as legitimate.

2. Each individual in a relationship needs to *recognize his or her partner for what they are.* As hard as we may be on ourselves, we are often harder on our partners because they are not everything we would want them to be, and we cannot understand why they will not change. Intimacy is not possible with an imaginary "perfect" partner. It can only occur when one recognizes one's mate for what he or she really is—with weaknesses as well as strengths.

3. Each individual must *feel comfortable to express himself or herself.* People are often hesitant to express any doubts, irritation, or anger to their partners in a loving relationship. To do so requires that an individual be capable of independence, for independence and intimacy go hand in hand (Hatfield, 1988). An individual who is totally dependent on his or her mate is unlikely to express any anxieties or fears. It is important that we express positive feelings for our partners, for the ratio of positive to negative interactions is a key factor in maintaining a happy relationship (Byrne & Murnen, 1988). Positive feelings for one's partner can and should be communicated in many ways (for example, by eye contact, holding hands and other physical closeness, thoughtful phone calls, and cards or gifts, as well as sexual interest and verbal responses). However, we must also be able to express our anxieties, fears, and anger to our partners. If we cannot, then we are not really intimate.

4. Learn to *deal with your partner's reactions.* If you express your feelings to your partner, he or she may not like what you say. Your partner may become angry or feel very hurt, but you must learn not to automatically back up or apologize for expressing feelings. Instead, keep calm and keep reminding yourself that your feelings are legitimate and that you are entitled to express them. In other words, keep trying. Only when both of you can express your own feelings and allow the other to do the same is there a chance for real intimacy.

The more intimate two people become, the greater the chance that the relationship will be maintained.

> Love is an active power in man; a power which breaks through the walls which separate man from his fellow men. . . . In love the paradox occurs that two beings become one yet remain two.
>
> (Erich Fromm, 1956)

Key Terms

agape   306
attachment   299
attachment theory of love   300
companionate love   298, 304
conditional love   303
consummate love   304

decision/commitment   304
empty love   304
eros   305
fatuous love   304
friendship   296
growing apart   311

growing together   312
habituation   310
infatuated love   304
"in love"   296
intimacy   304, 313
jealousy   300

Personal Reflections

1. How do you define love? Which of Lee's love-styles best describes you in your present or last relationship? Do you think that you are locked into this love-style for future relationships? Why or why not? Are there changes you need to make so that you can be happy and healthy in your relationships?

2. How important to you is loving and being loved? Do you believe that you could enjoy sex without love? Why or why not? Could you enjoy being in a romantic love relationship for an extended period of time if the relationship did not include sex? Why or why not?

3. Do you have a positive self-concept? Why or why not? If your self-concept is less than totally positive, what effects do you think this has on your ability to participate in a loving relationship?

4. Do you reveal your true needs, feelings, emotions, and values to your partner? Why or why not? Do you think that your partner reveals his or her true needs, feelings, emotions, and values to you? Why or why not? Do you deal with your partner's reactions when he or she does not like you to express your true feelings, or do you back up and/or apologize? Why or why not?

5. Do you ever have feelings of jealousy? Are they so strong that they interfere with your relationship(s)? What are the conditions that usually cause you to become jealous? Does your partner ever try to make you jealous? What things might you do to control these feelings?

6. For those presently in a long-lasting relationship: Have the activities you share with your partner changed since you first began dating? Have new shared activities replaced old ones? Are you growing together or growing apart? If you feel that you are growing apart, what can you do to change this?

Suggested Readings

Ackerman, D. (1994). *A natural history of love.* New York: Random House. Informative and very readable.

Avery, C. S. (1989, May). How do you build intimacy in an age of divorce? *Psychology Today.*

Davis, K. E. (1985, February). Near and dear: Friendship and love compared. *Psychology Today.*

Fromm, E. (1956). *The art of loving.* New York: Harper & Row. An oldie, but a classic.

Gray, P. (1993, February 15). What is love? *Time* magazine.

Hatfield, E., and Rapson, R. L. (1993). *Love, sex, and intimacy: Their psychology, biology, and history.* New York: HarperCollins Publishers. Focuses on romantic relationships.

Hatfield, E., & Rapson, R. L. (1996). *Love & Sex: Cross-cultural perspectives.* Needham Heights, MA: Allyn & Bacon.

Kayser, K. (1993). *When love dies: The process of marital disaffection.* New York: Guilford Press. Explains marital disaffection as going through five phases: disillusionment, hurt, anger, ambivalence, and disaffection.

Rubenstein, C. (1983, July). The modern art of courtly love. *Psychology Today.* Results of a reader survey regarding romance, love, and sexuality reveal some interesting differences between men and women.

Seider, S. (1991). *Romantic longings: Love in America, 1830–1980.* New York: Routledge.

Singer, I. (1984–1987). *The nature of love.* Chicago: University of Chicago Press. A three-volume set for the serious student.

Thayer, S. (1988, March). Close encounters. *Psychology Today.* Discusses the importance of touch in establishing intimacy and examines some gender and cultural differences.

Trotter, R. J. (1986, September). The three faces of love. *Psychology Today.* An easy-to-read review of Sternberg's triangular theory of love.

1. Discuss the common sexual differences couples may encounter in their relationship;

2. Describe some commonly used sexual therapy techniques (the medical history, the sexual history, systematic desensitization, sensate focus, self-awareness, psychosexual therapy, and the PLISSIT model of therapy);

3. Describe sexual problems that can affect both males and females (hypoactive sexual desire, hypersexuality, and dyspareunia);

4. Describe the sexual problems that occur only in males (erectile disorder, premature ejaculation, orgasmic disorder, benign coital cephalalgia, and priapism);

5. Explain the sexual problems that occur only in females (orgasmic disorder and vaginismus); and

6. Explain some similarities and differences in the types of sexual problems experienced by homosexuals.

Sexual Problems and Therapy

In the fantasy world of sex, nothing ever hurts, nothing rubs or chafes, nobody is anxious, no one is tired, there are no menstrual cramps, strange viruses, or pregnancy fears, nothing bad ever happens and everything fits perfectly; in the fantasy world of sex, erections function, orgasms are easy, desire surges, birth-control methods don't interfere with spontaneity, and bodies melt. In the real world, it doesn't always happen this way.

(From *What Really Happens in Bed* by Steven Carter and Julia Sokol. Copyright © 1989 by Steven Carter and Julia Sokol. Reprinted with permission of the publisher, M. Evans and Company, Inc., New York.)

*S*everal studies have found that half or more of all couples have had, or will eventually have, sexual problems at some time in their relationship (e.g., Frank, Anderson & Rubenstein, 1978; Nathan, 1986). In their nationally representative survey, Laumann et al. (1994) found that many Americans had experienced sexual problems lasting several months in just the previous year. Nearly one-third of the women reported a lack of interest in sex, and nearly 30 percent of the men reported continual problems with premature ejaculation, to name just two examples. In this chapter you will learn about the types of sexual problems that men and women may experience. In some cases, individuals have sexual problems regardless of who their partner is. Within a relationship, however, sexual problems frequently arise simply because of individual differences—people are different in what they want and how often they want it. I will begin with a discussion of the types of differences that can lead to sexual problems within a relationship.

INDIVIDUAL DIFFERENCES

DIFFERENT EXPECTATIONS

Men and women often have different ideas about sex and love, and therefore often differ on why they have sex. Many women are looking for affection and are interested in what their partner thinks, while most men engage in sex because of their partner's looks, body, and sexual attractiveness (Baldwin & Baldwin, 1997). Many men equate intimacy with sex. When asked to give six responses to the question "What things do you personally like best in a lover?" for a survey in my course, most of the males listed specific anatomical characteristics (e.g., "nice ass," "big breasts," "sexy legs") or behaviors (e.g., positions, oral sex, no inhibitions). Women, on the other hand, were less concerned about anatomy ("attractive" and "good build" were the only two responses) and specific behaviors and much more concerned about being in a relationship in which they felt cared for, loved, and/or respected. Nearly every woman expressed the desire for her male partner to take his time. In fact, this was often expressed as a complaint—the male partners were not taking their time. These different expectations about a sexual relationship can lead to problems.

> "I got married at a very early age. Sex to me was usually just wifely duties because it was demanded. I used to make up excuses just so I didn't have to do it. Our marriage finally broke up. . . ."
>
> (from the author's files)

Dear Ann Landers:
   Two years ago I found out that my wife of 27 years was having an affair with another man. She was driving 40 miles to a motel twice a week to meet him for "lunch."
   When I learned what was going on, I asked her, "Why? It seemed to me you were getting more than enough sex at home."
   Her reply was, "What I am getting at home is just that. SEX. And sex is no substitute for love."
   Her answer made me stop and think. I had to admit she was right.
   I am writing this letter to all you husbands out there who are making the same mistake. Ask yourselves this question, "Am I making love to my wife, or am I just having sex with her?"

Glad I Got Smart

(Reprinted with permission of Ann Landers and Creators Syndicate.)

Kinsey's group (1948, 1953) found that foreplay for most couples lasted no more than a few minutes. In a later study, Hunt (1974) found that couples averaged 15 minutes of foreplay. However, in the national survey by Laumann et al. (1994), 85 percent of the respondents said that they had spent 15 minutes to more than an hour during their last sexual activity with a partner. This may mean that men are becoming more aware that women are more likely to enjoy sexual intercourse if they take their time.

DIFFERENT ASSUMPTIONS

Men and women often have different initial sexual experiences, which can lead to incorrect assumptions about the opposite sex. The first sexual experience to orgasm for most males is masturbation. Most males masturbate by moving a hand up and down the penis at a continually accelerating rate as they approach orgasm (Masters & Johnson, 1966). Kinsey et al. (1948) found that, on average, men took only 2 to 3 minutes to reach orgasm during masturbation (extremely orgasm-oriented). Women are more variable in their masturbation techniques, but most prefer indirect stimulation of the clitoris (a circular motion or up and down one side). Kinsey found that only 20 percent of the women surveyed masturbated by inserting a finger in the vagina, while Hite (1976) reported that only 1.5 percent of the women she surveyed did so. It takes women longer to masturbate to orgasm on the average than it does for men (Kinsey et al., 1953).

Without any evidence to the contrary, it is logical for people to assume that what feels best for them must feel best for others as well. Men often assume, therefore, that women greatly enjoy having their clitoris and other genitals vigorously stimulated (as by rapid thrusting of a finger in and out of the vagina) during foreplay, when in fact this may not be the case. Similarly, on the basis of her own masturbatory experiences, a woman might assume that her male partner prefers gentle indirect stimulation (Masters & Johnson, 1970), when in fact this may not be the case.

DIFFERENCES IN DESIRE

One of the most common types of problem that a couple might encounter is a difference in the frequency with which sex is desired (Zilbergeld & Kilmann, 1984). In the early 1980s movie *Annie Hall* there is a scene where a man, played by Woody Allen, and a woman, played by Diane Keaton, are shown talking to their respective therapists. Woody says that he and his girlfriend (Diane Keaton) hardly ever have sex, perhaps three times a week. Diane Keaton tells her therapist that they have sex all the time, maybe three times a week. Who has the problem? Is the woman played by Diane Keaton undersexed and to blame, or is the problem that Woody is oversexed? Suppose that Woody, who wants sex once a day, was instead having a relationship with another woman, who wanted sex at least three times a day. Now who is to blame? Would Woody now be considered undersexed, or is his new partner oversexed? It does no good to point the finger and put the blame on one person or the other. People are generally only oversexed or undersexed relative to their partners' desires. When actress Rita Hayworth was married to an Arab prince, she accused him of being oversexed. His response was that he had cut down his level of activity for her. *It is the couple that has the problem.* If the woman played by Diane Keaton had instead had a relationship with someone who also wanted sex once or twice a week, there would be no problem; and the same would be true if Woody had had a relationship with someone who also wanted sex every day.

Sexual desire is not necessarily set at a fixed level. An individual's desire for sex often varies, depending on the circumstances. For example, stress and fatigue can greatly affect our interest in sex. If your partner is showing less interest in sex than he or she once did, this may be due to other things that are happening in his or her life. Our understanding of our partner's feelings would be greatly enhanced, of course, if we could only switch places for a little while.

"When I stayed home and my husband worked, I did not understand his seeming lack of desire. Not that he didn't enjoy it, but it could be placed low on a list of priorities.

"He would say things like, 'Wait, let me finish watching the ball game,' or 'I don't feel well,' or, 'I'm tired.' He would tell me about his problems at work, and I was really interested in making love then. I didn't want to hear how obnoxious, rude, or unintelligent his boss could be.

"I took all of these 'excuses' as a personal put-down. I would say, 'You're home now, forget what happened in the day.' Little did I realize then, he could not separate them. A bad day could not be erased by a few hugs and kisses. At that time, though, I tried to persuade him, which may have turned into a contest of who was going to win.

"After a while, he started telling me that maybe he would be more interested if I didn't push him. After that, I began to let up and things went along fair. I still felt as though I wanted more than he was willing to give. Whenever I would tell him so, he would answer with, 'Quality, not quantity.' Then I would get mad and say, 'You don't understand me and what I need.'

"This period was heightened by his decision to quit his job of 25 years. When he finally decided to do this, I said I would go out to work and became a Kelly Girl. I really only wanted temporary work, but I'd go on short assignments of two or three weeks and stay three or four months with maybe only two or three days off in between. So, it was almost like working at one full time job.

"I started with feeling the pressures of coping with a job. Even though I did not have to stay at a place I did not like, I never turned down an assignment and gave it my all. Now I am receiving tension, problems with bosses, and plain being physically and mentally tired.

"When I came home from work, I'd look for dinner, make remarks about the house, or want to know if he checked on the children's homework. He would be ready to be cuddly and kiss and I had a barrage of questions. I'd also start telling him about problems I had during the day. At first, I realized these were things he said to me, and I didn't say any more. Finally, I told him, 'I can't believe I'm doing and saying the same things you did to me.'

"After a while it became a joke, and I'd say these things and we'd laugh about it. He then knew how it felt and so did I. Well, this same attitude spilled over to the bedroom. I realized I was now telling him, 'I'm tired,' 'I have to get up early,' 'Why didn't you start this earlier?'—all things he had said to me. I knew how he was feeling and he knew how I was feeling. It was not that we didn't want to make love, but we used other issues to cloud our minds. We now are more understanding of each other's feelings, because we know where they are coming from. Even though I'm in school now, my time is even more used up, because of studying, but we are conscious of each other's needs.

"Every so often my mother takes care of the two younger ones and we get away for a few days without any responsibilities or pressures. Just like two lovers on a lost weekend."

(from the author's files)

DIFFERENCES IN PREFERRED BEHAVIORS

Differences in the type of sexual activities each person wants to engage in can also lead to problems. Suppose, for example, that person A (who can be either a man or a woman) wants to have sex only in the missionary position and is in a relationship with person B, who wants to have sexual intercourse in a variety of positions and locations.

"My girlfriend and I are sexually active. I am tired of doing the same old missionary position. Whenever I mention a new position, she doesn't want to do it. I am so tired of the same old position I lose interest in sex. I don't want to even have sex sometimes."

(from the author's files)

Who is to blame for the problem here? Is A inhibited or repressed, or is B perverted? Once again, it is all relative. It does no good to point a finger. Suppose that instead of having a relationship with A, person B was in a relationship with person C, who wanted to try bondage and anal intercourse (which B does not want

to try). Should person B now be viewed as inhibited? Again, *it is the couple that has the problem,* not any one individual.

WHO NEEDS THERAPY?

In summary, there are many differences two people can have that can cause sexual problems. When this happens, it is far too common for one person to get upset and blame the other. It is generally best, however, that a couple view these types of problems as *their* problem, rather than as a particular individual's problem. This means working together to resolve it.

Sexual relations, of course, are only part of a couple's overall relationship. As you learned in the previous chapter, if a couple's overall relationship is good and positive, this generally carries over to sexual relations. Conversely, if a couple is experiencing relationship problems, this generally will affect sexual relations as well.

To work out sexual differences or relationship problems requires that a couple be able to talk about their differences. Most people do not know how to talk about sex comfortably, and Chapter 17 provides you with some guidelines. When couples cannot work out individual differences or relationship problems by themselves, they may need the aid of a counselor. When choosing a counselor, it is important to check his or her credentials. Be sure that you pick one who is properly trained and certified. In addition, just as you should do when choosing a doctor, shop around until you find one that you like. Some couples may prefer male and female co-therapists.

The names and addresses of certified marriage and family therapists in your area can be obtained by writing to the American Association for Marriage and Family Therapy (AAMFT), 1100 17th Street, N.W., 10th Floor, Washington, DC 20036-4601.

Not all sexual problems are the result of individual differences within a relationship. Rather than focusing exclusively on behavioral criteria (e.g., frequency of intercourse), we must also look at a person's basic beliefs, self-concept, and inner feelings and emotions. If a particular sexual behavior is causing an individual a great deal of stress and anxiety and possibly

Figure 13–1 Studies have shown that one-half or more of all couples have had, or eventually will have, sexual problems at some time in their relationship.

interfering with his or her ability to function in a relationship, then that person may be regarded (and probably regards himself or herself) as having a problem. In this case, the individual would probably benefit by seeing a sex therapist. Again, however, it is often beneficial to consider the individual and his or her partner as needing sex therapy together rather than focusing on either one of them.

SEXUAL THERAPY

What is a sex therapist? What do they do? My experience in class has been that many students giggle and laugh when asked this, and some have expressed the mistaken belief that sex therapists have sex with their clients. When you have a toothache you go to a dentist, but does the dentist drill his or her own teeth while you're there? When you have an eye infection you go to an ophthalmologist, and when you have a chest pain you go to a cardiologist. Just as physicians specialize in different kinds of medical problems, therapists often specialize as well. A sex therapist is someone who specializes in helping people with sexual problems. Like any other specialist, a sex therapist has had several years of training before beginning practice.

Sex therapists consider it highly unethical for a therapist to have sex with a client (Lazarus, 1992). The following is an example of an unethical therapist. If anything like this ever happens to you, leave immediately and report the therapist.

> "Six years ago I willingly went to a psychiatrist feeling that I needed someone to talk to that could maybe relinquish some of the thoughts I had about sex that were very disturbing. After two visits the doctor proceeded to show me different sexual positions in his books. The next visit he showed me the positions using me as his partner. I never went back. Two years after this, a friend of mine confided in me and told me she was once a victim of incest (her uncle). She told me the psychiatrist she went to made her opinion of men even worse. That doctor was the same one I went to and he did the same thing to her."
>
> (from the author's files)

Sex therapy is a relatively new field. In fact, the American Psychiatric Association's *Diagnostic and Statistical Manual* (DSM-IV, 1994) did not even list psychosexual disorders until 1980. Until the pioneering work by Masters and Johnson, which was published in 1970, most people with sexual problems went to their family doctor, to a urologist, or, if they could afford

one, to a psychoanalyst. Most physicians were not prepared to deal with such problems. In fact, sex education was not even offered in most medical schools at that time. Psychoanalysis generally involved long-term treatment (sometimes years) and attempted to "cure" the problem by resolving childhood conflicts (believed to be unconscious), viewing the behavioral problem as merely a symptom of some other deeper conflict.

Masters and Johnson (1970) originally believed that most sexual problems were the result of faulty learning and could be undone in a relatively short period of time (often two weeks) by using **cognitive-behavioral therapy.** This kind of therapy focuses on specific behavioral problems, and most therapists work with couples rather than individuals. People with sexual problems are often very distressed and do not want to wait months or years for help. Psychoanalysts often charge, however, that behavioral therapies only treat the symptoms and that if the underlying cause of the problem is not determined, the problem will eventually manifest itself in another manner. No one can deny, however, that behavioral approaches have been enormously successful in treating sexual problems like erectile problems in men and orgasmic problems in women.

Sex therapy is still refining its techniques, and in the years since Masters and Johnson began their clinical practice it has become clear that **psychosexual therapy,** designed to give insight into the historical cause of clients' problems, is often more successful for some types of problems (e.g., low sexual desire or sexual aversion; Kaplan, 1974, 1983). More recently, LoPiccolo (1992) has introduced what he calls *systems therapy* to provide insight to the client about the function of his or her problem in the current relationship. Perhaps the most controversial change in the treatment of sexual problems today has to do with the role of the medical field (see Schover & Leiblum, 1994; Tiefer, 1997). Physicians have increasingly attributed sexual problems to organic causes and are treating them with medical techniques (medications, surgery). I will describe and critique medical approaches in the sections on specific sexual problems later in this chapter.

Sexual therapy programs generally have many things in common. Most follow what is called the **PLISSIT model** (Annon, 1974). PLISSIT is an acronym for *permission, limited information, specific suggestions,* and *intensive therapy*—the four levels of therapy. Each represents a successively deeper level of therapy. In the first level, the therapist "gives permission" for the client to

cognitive-behavioral therapy Therapy that views problems as resulting from faulty learning and that focuses on specific behavioral problems.

psychosexual therapy Therapy that attempts to provide insight into the historical cause of a client's problem.

PLISSIT model An acronym for the four levels of treatment in sexual therapy: permission, limited information, specific suggestions, and intensive therapy.

feel and behave sexually. This is important, because many people and institutions (e.g., parents, some religions) cause people to suppress or repress their sexuality while growing up. In the second level, the therapist gives information to the client (educates him or her about the sexual problem by providing information that relates to the problem, such as teaching the client about clitoral stimulation or discussing the effects of alcohol). This is done in such a manner that the client continues to acquire a positive attitude about sexuality. At the specific suggestion (third) stage, the therapist gives the client exercises to do at home that will help with the specific problem. If the client is still experiencing problems after completing the specific suggestions, then intensive psychosexual therapy will be employed. Only about 10 percent of people who go to sex therapy require this last step.

If you ever think that you may need the help of a sex therapist, be sure to check the credentials of therapists first. In some places, almost anyone can advertise themselves as a "counselor."

The names and addresses of certified sex counselors and therapists in your area can be obtained by writing to the American Association of Sex Educators, Counselors and Therapists (AASECT), 435 North Michigan Avenue, Suite 1717, Chicago, IL 60611-4065.

SEXUAL THERAPY TECHNIQUES

MEDICAL HISTORY

Sexual problems are sometimes caused by physical or medical problems (Rosenstock, 1995). Circulatory problems (e.g., arteriosclerosis), hormone abnormalities (e.g., low testosterone levels), or anything that causes damage to the central nervous system (e.g., diabetes, spinal cord injury) can cause a sexual problem. Alcohol and drugs often cause sexual impairment as well. One common cause of impotence in men, for example, is some prescription medications used to treat hypertension and heart disease (Segraves & Segraves, 1992; Smith & Talbert, 1986). It is important, therefore, that a therapist take a complete *medical history* of the patient (and possibly refer the patient to a physician for a medical exam) before beginning therapy in order to rule out any physiologic basis for the presenting problem. Behavioral therapy will do little good if the problem has an organic cause.

SEXUAL HISTORY

Nearly all sex therapists will take a complete *sexual history* of the client before treatment begins. These histories are very thorough, and the length of time devoted to a history will depend on how candid the client is about his or her past experiences. Some therapists, like Masters and Johnson, prefer to work with couples, because, as they say, "there is no such thing as an uninvolved partner." A person's partner (or partners) is involved in the problem in some manner, and even when they are not the initial cause (in those cases where the problem existed before the relationship started, for example), when one partner has a sexual problem, it is common for the other partner to develop one as well.

SYSTEMATIC DESENSITIZATION

Many patients have severe anxieties about sex in certain situations. Therapists often attempt to reduce this anxiety through **systematic desensitization** involving muscle relaxation exercises or stress reduction techniques. A series of anxiety-producing scenes is presented to the patient, and he or she is told to try to imagine the scene. If this causes anxiety, the relaxation exercises are used until the scene can be imagined without anxiety. The therapist then proceeds to the next scene and repeats the procedure until the patient can complete the entire series without anxiety. Imagining a scene, of course, is not the same as a real-life situation, so a series of homework exercises is usually given as well.

SELF-AWARENESS AND MASTURBATION

As you learned in Chapter 2, some people have never explored their own bodies. As a result, they lack **self-awareness** and are totally out of touch with their own physical responses. If you don't know what your body looks like and how it responds, it will be difficult to communicate your needs and feelings to a partner. Therapists are likely to tell those who are not in touch with their bodies to spend time examining themselves. Some therapists give their clients instructions on how to masturbate (LoPiccolo & Stock, 1986). Many therapists consider it helpful to have a couple masturbate in each other's presence so that each can learn what the other finds most arousing and pleasurable. Isn't it interesting that what once was taboo is now sometimes used as a therapeutic technique?

SENSATE FOCUS

Many people are goal- or performance-oriented during sexual relations (e.g., they focus on orgasm). Others have guilt or anxieties about enjoying sex. As a result, many people never really learn how to give or receive physical pleasure. Most therapists, therefore, instruct couples to use *nondemand mutual pleasuring techniques* when touching each other. They are instructed to go home, get undressed, and take turns touching each other without thinking about the goal of having intercourse or having an orgasm. Touching the breasts and

---

systematic desensitization A therapy technique used to reduce anxiety by slowly introducing elements of the anxiety-producing theme.

self-awareness In sexual therapy, getting in touch with one's own physical responses.

Figure 13–2 Sensate focus is a nondemand mutual pleasuring technique in which a couple learns how to touch and be touched without worrying about performing or reaching a goal.

genitals is forbidden at first, but all other areas of the body are to be explored (see Figure 13–2). The receiver is instructed to focus on the sensations produced by the giver and to provide feedback as to what feels good and what does not. The giver learns what makes his or her partner feel good while simultaneously learning the pleasure of touching. The couple learns to be sensual in a nondemanding situation. Ask yourself the following question: When was the last time you gave your partner a massage without asking or expecting anything in return?

Masters and Johnson, who created these **sensate focus exercises,** often have patients spend several days doing this non-genital-oriented touching. The purpose is to reduce anxiety and teach nonverbal communication skills. Eventually, the couple is instructed to include breast and genital stimulation in their touching, but to avoid orgasm-oriented touching. More and more is slowly added, including mutual touching and *hand-riding techniques* in which the receiver places his

or her hand over the giver's hand and guides it over his or her own body (touching the hand lightly at body spots where gentle stimulation is wanted and pressing harder at spots where firm pressure is preferred). A successful outcome for the treatment of sexual problems is often directly related to the amount of sensate focus that is completed during therapy (Sarwer & Durlak, 1997).

SEXUAL SURROGATES

Sensate focus exercises require a cooperative partner who is willing to devote a great deal of time to it (and also be willing to postpone intercourse and orgasm). What about the client who has no steady partner? What does he or she do? Remember, sexual contact between therapist and client is considered unethical by professional standards, and most prostitutes are not likely to show much sympathetic inter-

sensate focus exercises Exercises designed to reduce anxiety and teach mutual pleasuring through nongenital touching in nondemanding situations.

est. Some sex therapists, therefore, employ surrogate partners who are trained and skilled in sensate focus techniques (as well as being cooperative and sympathetic). *Sexual surrogates* do not advertise their services and do not solicit clients (clients are referred by therapists), and thus their work is not illegal (Nhu, 1997). There are about 200 certified surrogates in the United States, and their services cost about $200 per hour. Although trained surrogates can provide a useful and necessary service, Masters and Johnson and some other therapists have discontinued their use because of emotional attachments that often form between a client and the surrogate.

SPECIFIC EXERCISES

After the sensate focus exercises are successfully completed, therapists generally assign specific exercises to help with the problem for which the person came for treatment. These will be described as we discuss some of the most common types of problems. These exercises usually precede sexual intercourse. When the therapist does allow sexual intercourse, it is almost always in a female-on-top or side-by-side position where neither partner is in total control. Therapists believe that these positions have much more erotic potential than the missionary position that is frequently used by many couples.

SEXUAL PROBLEMS AFFECTING BOTH MALES AND FEMALES

HYYPOACTIVE SEXUAL DESIRE AND SEXUAL AVERSION

"I am 35 years old, my wife is 25, and we have been married for 6 years. From the outside we look like the perfect couple—clean-cut, church-going, the works. The problem is my wife. She devotes all her time to cleaning, cooking, and doing volunteer work for the church. However, we have had sex 3 times in the last 12 months. She says that I should be happy with everything else she does for me. It has even gotten to the point where she wants separate vacations because a hotel room with me is too much pressure for her to stand. I love my wife but can't continue to live this way."

(from the author's files)

The DSM-IV (American Psychiatric Association, 1994) defines a lack of interest in sex as a problem when there is persistent or recurrent absence of sexual fantasies and sexual desire. People with **hypoactive** (or inhibited) **sexual desire** show decreased frequencies of sexual arousal, self-initiated sexual activity, and sexual fantasy (Segraves & Segraves, 1992). Hypoactive sexual desire was first reported in the 1970s (Kaplan, 1977; Lief, 1977), and today many therapists believe that it is the most common problem of couples going into sex therapy (Leiblum & Rosen, 1989). One survey found it to be the most common problem mentioned by women reporting sexual difficulties (Clements, 1994). Recall that Laumann et al.'s (1994) national survey found that nearly a third of the women studied reported a lack of interest in sex in the previous year. Over 15 percent of the men reported the same. A review of the literature indicates that at least 20 percent of all adults, both men and women, have hypoactive sexual desire (Wincze & Carey, 1991).

The problem in diagnosing hypoactive sexual desire, of course, is defining what is meant by "persistent." What is the normal frequency of sexual desire? Note also that *inhibited* is a relative term that must take into account the partner's frequency of desire (see the section on individual differences, p. 317) and the level of desire of other people in the community. Some therapists classify a person as having hypoactive desire if he or she *initiates* an average of two or fewer sexual experiences a month (Schover et al., 1982). This does not have to mean intercourse; it can include masturbation and sex play. Hypoactive sexual desire is more than just a lack of sexual activity, however. These individuals do not desire sex and avoid it even when there is an opportunity to engage in it. Sexually hypoactive persons display less physiological and subjective sexual arousal in response to erotic stimuli than do normal people (Palace & Gorzalka, 1992). Therapist Helen Kaplan's studies (1979) suggest that on some level, inhibited persons suppress their desire because they do not want to feel sexual. Another therapist has called it "sexual anorexia nervosa." In its most extreme form, the avoidance of sex becomes phobic in nature and is called **sexual aversion** (Kaplan, 1987). The anticipation of any kind of sexual interaction (even conversation about sex), or, in some cases, any kind of touch, causes great anxiety, so persons with sexual aversion simply "turn it off" at a very early stage (Ponticas, 1992). Masters, Johnson, and Kolodny (1992) believe that sexual aversion is uncommon, but Helen Kaplan (1987) feels that it is underreported.

It is difficult for people who enjoy sex to imagine that others find it unpleasant. What could cause such a lack of interest in sex? *Primary hypoactive sexual desire* (the case with people who have never had sexual feelings) is much more common in women than men (Wincze & Carey, 1991). *Secondary* or *acquired hypoactive sexual desire* can be attributed to organic factors, such as hormone deficiencies or tumors, in some

hypoactive sexual desire A sexual problem characterized by a persisting and pervasive inhibition of sexual desire.

sexual aversion An extreme form of hypoactive sexual desire in which an individual's avoidance of sex becomes phobic.

cases, but the vast majority of cases are believed to be due to psychological factors (Leiblum & Rosen, 1988). Some cases may be due to a past trauma, such as rape or sexual molestation during childhood (see Chapter 15 on sexual abuse). A person doesn't have to have been a victim of sexual abuse, however, to acquire a negative attitude about sex. Some people simply may never have learned to associate sex with pleasure because of a selfish partner or partners.

> "My husband was drunk on our wedding night. He could hardly walk and practically tore my clothes off. I wanted to kiss and hug and feel like I was loved, but all he wanted was to screw. I wasn't ready and it hurt real bad. He made me do things I didn't want to do. Afterwards he threw up in the bathroom and passed out. I cried all night. Now I still don't want him to touch me."
>
> (from the author's files)

Sometimes, an unwanted pregnancy or contracting (or fear of contracting) a sexually transmitted disease can cause a person to have a negative attitude about sex.

> "When I was 15 I had sex with a prostitute. My friends encouraged me to do so. One week later I noticed two sores on my penis. I couldn't tell my parents, they would kill me. I couldn't sleep. I asked a friend for help. I remember how he looked at me. I felt rejected. . . . I was horrified. . . . His father was a doctor and he treated me. I remember the anxiety when he told me I had chancroid. I didn't know if it was curable. . . . Today, I'm afraid of having sex with anyone. I have not had relations with a woman in nearly five years."
>
> (from the author's files)

Hypoactive sexual desire is also often associated with depression, life stresses, a sexually repressive upbringing, or an aversion to female genitals or to common types of sexual behaviors such as masturbation and oral-genital sex (Donahey & Carroll, 1993; LoPiccolo, 1980). Interestingly, many females with anorexia or other eating disorders have very low sexual functioning and satisfaction (Rothschild et al., 1991).

Today, many therapists believe that problems in a relationship are one of the most important factors contributing to hypoactive sexual desire (McCarthy, 1997; Schwartz & Masters, 1988). Hypoactive women are generally very unhappy with the quality of their relationship with their partner, particularly concerning the expression of affection, while hypoactive men tend to

be overly reliant emotionally on their partner and to have anxieties about finding the right combination of sexual passion and emotional intimacy (Apt, Hurlbert, & Powell, 1993; Donahey & Carroll, 1993; Kaplan, 1988). For probably half or more of the couples who go into therapy for sexual desire problems, one partner or the other has a coexisting sexual problem (see Leiblum & Rosen, 1988). In some cases, hypoactive sexual desire may be a way of avoiding the sexual failure due to the other problem.

Hypoactive sexual desire is probably the most difficult sexual problem to treat (Hawton et al., 1991; Kaplan, 1987; Schover & Leiblum, 1994). Masters and Johnson once reported an 80 percent success rate in treating hypoactive clients with conventional short-term therapy of the type used for other problems, but this claim was severely questioned by other therapists, some of whom reported only a 10–20 percent success rate with conventional techniques. Today, most therapists begin with behavioral and cognitive techniques that are designed to increase pleasure and communication and decrease anxiety, but add couples therapy (to resolve relationship problems and rebuild intimacy) and individual psychosexual therapy (to help the person understand what it is that is causing anxieties about sex) (e.g., LoPiccolo & Friedman, 1988; McCarthy, 1997; Schwartz & Masters, 1988; Zilbergeld & Hammond, 1988).

HYPERSEXUALITY: COMPULSION, ADDICTION, OR MYTH?

At the beginning of this chapter, you learned that two people in a relationship often differ in the frequency with which they desire sex, but that ordinarily neither one should be considered hyposexual or hypersexual. By how much would one's sexual desire have to differ from that of most other people before he or she would be considered extraordinary? This is a matter of subjective judgment, which (as usual) varies across different cultures and times. However, as you have just read, some people can be classified as having hypoactive (inhibited) sexual desire. Can we make the same distinction at the other extreme? Can people be classified as "hypersexual"? Many therapists say no. In fact, the American Psychiatric Association's diagnostic manual, DSM-IV (1994), has no such category. Let us explore this controversial issue further.

History is full of stories of famous people who reportedly had hundreds of sexual partners and often had sex many times a day: Cleopatra, Empress Theodora, Casanova, Catherine the Great, Rasputin, and King Solomon, to name a few. Terms like *nymphomania* and *satyriasis* have long been used to describe people with seemingly insatiable sexual appetites. A recent study suggested that anyone having 7 or more orgasms a week for at least 6 months should be considered hypersexual (Kafka, 1997). However, those therapists who favor a new classification for hypersexuality say that hypersexuality is more than just numbers. Patrick Carnes (1983, 1991) was the first to use the designation sexual addiction. Many therapists objected to this term, saying that there was no evidence of physical addiction as in the case of alcohol or drugs. Goodman (1992) recently redefined *sexual addiction* as follows:

"A disorder in which a behavior that can function both to produce pleasure and to provide escape from internal discomfort is employed in a pattern characterized by (1) recurrent failure to control the behavior, and (2) continuation of the behavior despite significant harmful consequences."

This definition includes aspects of both dependency ("behavior motivated by an attempt to achieve a pleasurable internal state") and compulsivity ("behavior motivated by an attempt to evade or avoid an unpleasurable/aversive internal state") and is similar to the definition in DSM-IV for psychoactive substance addiction.

Opponents of the designation "sexual addiction" argue that it is based on a repressive morality that can be used against anyone engaged in nonmonogamous sex (Barth & Kinder, 1987; Coleman, 1986; Levine & Troiden, 1988). As Coleman (1986) notes, "This concept can potentially be used to oppress sexual minorities . . . because they do not conform to the moral values of the prevailing culture (or therapist). Mental health professionals using such conceptualizations have become simply instruments of such conservative political views and have made people who do not fit into a narrow, traditional sexual life-style feel bad, immoral, and, now mentally ill." These therapists prefer to regard people who show a lack of sexual control as having a *sexual compulsion*.

Note that there is some common ground: both sides admit that it is possible to engage in sex compulsively. I will use the purely descriptive term hypersexual to describe such individuals.

A **hypersexual** individual is distinguished from other people by the compulsiveness with which he or she engages in sex. Hypersexuals engage in sex repeatedly and compulsively to reduce anxiety and distress, usually finding little or no emotional satisfaction. Coleman (1991) and others choose to divide hypersexual (compulsive sexual) behavior into two types: paraphilic and nonparaphilic. I will discuss paraphilic behaviors in the next chapter; this section is devoted to nonparaphilic hypersexuality.

hypersexuality A term used for people who engage in sex compulsively, with little or no emotional satisfaction. Sometimes called *sexual addiction*.

Hypersexual individuals feel driven to have sex, looking for something in sex that they can never find. If they fall in love and establish a relationship, they often cannot stop their promiscuous behavior. Hypersexuality interferes with the ability to carry out normal daily living and results in unhappiness, lack of fulfillment, and the inability to break a pattern of compulsive behavior. The threat of possibly contracting a sexually transmitted disease, including HIV infection and possibly AIDS, does not stop hypersexual people. Although the term sexual addiction may be controversial, it has been found that some hypersexual individuals also display excessive dependence on alcohol, drugs, and/or food to satisfy emotional needs. It is not uncommon, for example, for people who use cocaine to engage in high-risk sexual behaviors (e.g., new partners) when taking the drug (Washton, 1989). The dual addiction becomes a vicious cycle, with one addiction leading to the other.

Coleman believes that hypersexuality ("sexual compulsivity") results from "intimacy dysfunction" during childhood (Coleman & Edwards, 1986). This can occur, for example, if a child is neglected or abused. Such children often have low self-esteem and feelings of inadequacy and unworthiness as adults. They are often very lonely, and their emotional pain causes them to engage in behaviors (e.g., alcohol, drugs, food, sex) that give them pleasure and help them evade their internal discomfort. The relief is only momentary and is followed by a need to engage in the behavior again. However, engaging in the behavior results in greater feelings of unworthiness and shame, which makes it more likely that the person will engage in the behaviors in order to avoid the negative feelings felt at present. Coleman (1991) describes one form of hypersexuality common in women as "compulsive love affairs." These women compulsively lure potential partners to fall in love, but once the partner is "caught" there is no more appeal, and she breaks it off and relentlessly goes looking for the "right" partner.

How would one distinguish hypersexuality from a healthy interest in sex? Remember, it is not so much the frequency of sex as it is one's life-style. Ask yourself some questions as guidelines. Do you *frequently* use sex to alleviate depression or anxiety? Could your sexual behavior abuse or cause injury (physically or emotionally) to yourself or to someone else? Could it possibly damage your relationship, your career, your health, or your self-respect? Have you ever wanted to stop but not been able to do so? A positive response to these questions may make you a candidate for therapy.

Therapy is often twofold: (1) treating the internal discomfort (perhaps initially with antidepressants or stabilizing drugs), and (2) helping the individual find healthy, adaptive ways of dealing with his or her emotions and needs (psychotherapy will probably be used here) (Coleman, 1991). Those therapists favoring the addictive viewpoint probably will also teach cognitive-behavioral strategies to help the individual abstain from the addictive behavior (Goodman, 1992). While therapists continue to argue the validity of the concept of sexual addiction, a number of self-help organizations similar to Alcoholics Anonymous (e.g., Sexaholics Anonymous, Sex Addicts Anonymous) have been created for people who think they need help.

PAINFUL INTERCOURSE

Painful intercourse is called **dyspareunia.** It usually, but not always, is caused by a physical problem. In men, the most common causes are a prostate, bladder, or urethral infection (Davis & Noble, 1991), or the foreskin of the penis being too tight (a condition called *phimosis*). In rare cases, fibrous tissue deposits around the corpora cavernosa of the penis cause curvature of the penis (known as *Peyronie's disease*) and pain during erection (Ansell, 1991; Gregory & Purcell, 1989). In a very rare condition, some men experience persistent and recurring pain in the genitals during or immediately after ejaculation. This is called *postejaculatory pain syndrome* and is caused by involuntary spasms of the genitals that are triggered by psychological conflicts (Kaplan, 1993).

Painful intercourse is much more common in women than men. It has been estimated that 1 to 2 percent of women suffer from chronic dyspareunia, with as many as 15 percent experiencing problems lasting several months or longer (Laumann et al., 1994). One of the most common causes is vaginal dryness. If a woman is not fully lubricated when intercourse begins, the thrusting of the penis will severely irritate the dry vaginal walls. Lack of sufficient lubrication can be due to a partner who doesn't take his time, but it can also be the result of fear or anxiety, which interferes with the vasocongestive process (vaginal lubrication, you recall, is the result of the walls of the vagina becoming engorged with blood). Vaginal dryness can also result from hormonal changes (which occur at menopause), the use of antihistamines and other medications, or even tampons. Use of a water-soluble lubricant can often substantially alleviate this problem.

Dyspareunia in women can also be caused by endometriosis (growth of the endometrium outside the uterus), pelvic inflammatory disease, yeast and other vaginal infections, Bartholin's gland infections, and urinary tract infections. Allergies to semen, feminine hygiene products (deodorants and scented douches), powders, and spermicides can also make sexual intercourse painful (Erard, 1988).

If the physical factors responsible for dyspareunia are not quickly taken care of, dyspareunia can lead to other sexual problems. The anticipation of pain, for example, can become so great that it can lead to erectile problems in men, vaginismus in women, or loss of sexual desire in either sex. The resultant psychological stress can also damage relationships. Remember, pain during intercourse is generally an indication that there is something physically wrong, so see a physician immediately before it leads to greater problems.

MALE SEXUAL PROBLEMS

ERECTILE DISORDER

The term **erectile disorder** refers to a man's inability to attain or maintain an erection. This condition was once called impotence, but this is an unfortunate term because it has negative connotations about a male's manhood. Therapists now prefer the purely descriptive term (National Institutes of Health Consensus Statement, 1992). Erectile disorders can be *primary* (i.e., the male has always had problems) or *secondary* (i.e., the individual has not had erectile problems in the past); and *global* (i.e., the problem occurs in all situations) or *situational* (e.g., a man cannot get an erection with his usual partner, but can with other women or during masturbation).

Historically, and in many cultures, an erect phallis has been a symbol of male power and virility. Having an erectile problem can therefore be psychologically devastating. As one male put it, "I felt I wasn't a full man anymore" (*Newsweek,* June 18, 1984). But the disorder can affect more than the male. It often destroys the self-esteem of the female partner as well, because she often feels as if she is doing something wrong.

Billy and Karen were in their mid-thirties and had been married for three years. They had no children and both worked full-time; he as the manager of a small insurance office and she as a lab technician in a hospital. Their sexual relationship, which began while they were dating, had always been "wonderful."

In their third year of marriage, Karen was promoted to a job selling hospital equipment. About six months later, Billy began to have erectile difficulties. At first, Billy would lose his erection after intercourse had begun, but after about two months he began to have difficulty during foreplay as well. Sometimes he couldn't get an erection at all. He was embarrassed by this (calling

dyspareunia Painful intercourse, usually resulting from organic factors (e.g., vaginal, prostate, or bladder infections).

phimosis A condition in uncircumcised males in which the foreskin of the penis is too tight, causing pain during erection.

erectile disorder A sexual problem in which a male has difficulty getting and maintaining an erection.

himself a "failure") and began reacting by rolling over and going to sleep.

When the problem started, Karen tried to be supportive and didn't say anything, but as Billy's problem worsened she became angered at his silent withdrawal and lack of attempts to satisfy her through other means. As things progressed, Karen, who had been regularly orgasmic in the past, found that she could no longer reach orgasm. Now when they had sex Karen worried so much that Billy would get upset that she, too, lost all focus on sexual pleasure. Billy also began to rush her through foreplay to try and see if he would be able to perform in the "main event." Karen not only missed the intercourse and her own orgasms, but also the tenderness and closeness. Both felt inadequate because they both had sexual difficulty and could not figure out what was causing it.

(Case history courtesy of Dr. David M. Schnarch. Reprinted with permission of W. W. Norton & Company.)

How common are erectile problems? It is a rare male who will not experience an erectile problem at least once in his life. An erection results from the spongy tissues of the corpora cavernosa and corpus spongiosum of the penis becoming engorged with blood (the penile arteries dilate and valves in the veins close). This response is under reflexive control by two centers in the spinal cord (the lower one responds to touch and the upper one responds to erotic thoughts; see Fig. 4–2 on p. 73). The two centers normally work together and depend on the presence of testosterone (the male hormone) and other chemicals. Any number of things can upset the balance and impair functioning, including fatigue, stress, alcohol, and drugs. A drink or two loosens our inhibitions, but most men are unaware that alcohol is also a central nervous system depressant, and it is quite common to experience erectile problems after drinking too much (which sometimes may be only a couple of drinks; R. C. Rosen 1991).

So should we say that all men have erectile problems? No, but it is somewhat arbitrary as to where we draw the line. Masters and Johnson (1970) classify a man as having an erectile disorder if he is having problems in at least 25 percent of his sexual encounters, but others add that the problem must exist for at least 6 weeks to be considered chronic erectile disorder. Laumann et al. (1994) reported that over 10 percent of the men in their national survey said that they had problems lasting at least a few months in keeping an erection during the previous year. Thus, it is estimated that 10 to 20 million American men have an erectile disorder (Leland et al., 1997). Erectile problems are much

more common in elderly men, with only 7 percent of males aged 25 to 55 experiencing such problems, compared to 27 percent of men over 70 (Kravis & Molitch, 1990). However, erectile problems are *not* inevitable with advancing age (i.e., age alone is not the cause), and the problem can be treated with a variety of therapies (NIH Consensus Statement, 1992).

What is responsible for the fact that so many men have an erectile disorder? Masters and Johnson once argued that as many as 80 percent of all cases are due to psychological factors, but physicians now estimate that the large majority of cases have a physical basis (NIH Consensus Statement, 1992; LoPiccolo, 1992). In fact, the biggest change in the treatment of erectile disorders in the past 10 years has been its medicalization—whereby erectile disorders are viewed as a medical problem resulting from organic causes that should be treated with medical solutions (drugs, devices, or surgery). We will focus on the medical approach first, then return to treatment for psychogenic disorders.

Circulatory problems (e.g., arteriosclerosis or sickle-cell anemia), neurological disorders (e.g., resulting from accidents or pelvic surgery), prostate surgery, and hormone imbalances (primary abnormalities or secondary complications due to diabetes) often result in erectile disorders (Kravis & Molitch, 1990; Weinhardt & Carey, 1996). One-fourth of all erectile problems are due to medications prescribed for other conditions, particularly antidepressants and the beta-blocking medications used to treat blood pressure and heart disease (NIH Consensus Statement, 1992). Eighty percent of alcoholics are impaired. Smokers are 50 percent more likely than nonsmokers to suffer from erectile problems (Mannino et al., 1994; see also M. P. Rosen et al., 1991). It is important, therefore, that a client with an erectile problem have a complete medical exam before treatment begins.

Actually, there is an easy test that can often (but not always) determine whether the cause of an erectile disorder is physical or psychological. Males, you remember, get erections at night during REM sleep. If a man gets normal erections during REM sleep, then the problem is most likely psychological. If not, then it is physical.

One medical solution is injection of drugs directly into the penis. Self-injections of phentolamine and papaverine hydrochloride, drugs which cause expansion of vessels that supply blood to the penis and constriction in vessels that carry blood away from the penis, have proven to be successful for many men in reversing erectile problems due to circulatory problems (Althof et al., 1991; Lakin & Montague, 1988). More recently, physicians have been prescribing intracavernosal injection of alprostadil, which has proven to be effective in over 90 percent of all injections (Linet & Ogring, 1996). On the down side, injections can sometimes be painful, cause long-lasting erections, or cause fibrosis (scarring of the soft tissue).

Oral medications (pills) are the next big break-through (see Leland et al., 1997). The first such medication, Viagra, became available in 1998 and is about 80 percent effective. It works by relaxing smooth muscles. The pills work only when a man is sexually stimulated and do not cause erections that last for hours, but the drug can cause headaches and diarrhea. Other types of pills (apormorphine, phentolamine) should be available within a few years.

Another option is a vacuum device that forces blood into the penis, thus causing it to become erect (Witherington, 1990). An elastic ring at the base of the penis keeps it erect. There are few side effects except that this device is clumsy and not very romantic.

For those men for whom medications and therapy do not work, such as those who have suffered permanent nerve or vascular damage, *penile implants* can be used. There are three types. One is made of semirigid plastic rods and is inserted into the erectile chambers of the penis. It is stiff enough to permit intercourse, but flexible enough to be bent downward during daily activity. A second, more sophisticated, type of implant is made of two soft tubes that are implanted in the erectile chambers. They are connected by other tubes to a small pouch containing a saline solution that is implanted under the muscles in the pubic region. The pouch, in turn, is connected to a tiny pump located inside the scrotum. When a man wants to have sex, he gently presses the pump, which forces the saline solution into the tubes in the penis and causes an erection. Squeezing a valve in the pump reverses the process. A third type of implant is a silicone sack that is also inflated by a pump. Tens of thousands of American men are presently using these implants, and although postsurgical levels of sexual satisfaction do not always match presurgical expectations, a large majority of the patients and their partners are satisfied with the results (McCarthy & McMillan, 1990).

The medical model for the treatment of erectile disorders has failed to address some important issues. For example, quite often it is not easy to attribute an erectile problem solely to organic or psychological factors. If a man has a problem due to a physical cause (or a temporary problem due to medication), he can become so apprehensive that he becomes completely impaired (Buvat et al., 1990; NIH Consensus Statement, 1992). Many physicians maintain that the use of injections or pills will reduce psychologically caused problems, but the evidence shows that few men using medical solutions are "cured" (see Schover & Leiblum, 1994). Lastly, it is easier for many men to view themselves as having a disease that can be cured with a magic pill or shot than it is to accept that they have psychological or sexual problems that may require see-

Figure 13–3 Serious stressors, such as financial or job-related problems, can lead to erectile problems or a loss of interest in sex.

ing a sex therapist—even though short-term therapy may result in better long-term adjustment. There are probably psychological problems associated with all erectile problems, even those that are clearly caused by physical problems, and most patients would probably benefit from a combination of medication and therapy. Let us now focus on traditional sexual therapy techniques.

If the cause of an erectile problem is determined to be psychological, the success rate during therapy depends largely on whether it is primary or secondary. Men with primary (psychological) erectile disorders are often found to have had a strict religious upbringing in which sex was equated with sin and guilt (Masters & Johnson, 1970).

"I was raised a _____ throughout my life. . . . I do remember in the 8th or 9th grade my father and I were talking about sex and he told me not to have sex until I was married or I would go to hell—I thought that for the longest time. Then when I discovered masturbation I felt terribly guilty, so I went to talk to a priest. He told me how wrong it was and gave me some prayers to say. From that time on I felt very guilty every time I did it. . . . I haven't had a full erection in 10 years (I'm 26). . . ."

(from the author's files)

An early traumatic experience might also result in long-term erectile problems:

"At 16 years old . . . my best friend thought it was a shame I was still a virgin. . . . In a drunken stupor I was goaded into going downtown to pick up a hooker with my friend. The next day I walked tall and proud (although I felt no different), that is until I noticed a week later a greenish-yellow discharge on my underwear. A doctor diagnosed gonorrhea. I brought it up with my father, who became enraged, and to this day he still teases me about it. My brother laughed at me. . . . I did not date anyone for 3 years. Once I began, it took me 3 or 4 years to have sex. . . . I would get to the point of intercourse, then I would lose my erection."

(from the author's files)

Therapists find that these primary disorders are much more difficult to treat than secondary disorders. Masters and Johnson report a 60 percent success rate, but many other therapists believe that only about 20 percent can be successfully treated (Kaplan, 1979). Men with a primary psychological disorder may be treated by having penile implants.

Situational erectile disorders are quite clearly psychological in origin:

"My husband doesn't stay hard when we make love. Everything is OK at the start, but then he gets soft in a minute or two. He stays hard when I give him oral sex or masturbate him."

(from the author's files)

If this man's erectile problem were caused by something physically wrong with him, he would be unable to maintain an erection with any type of sexual activity.

For men with a secondary erectile disorder due to psychological factors, the most common cause is **performance anxiety**. Performance anxiety is a fear of failure.

performance anxiety A fear of failure during sexual relations that can lead to erectile disorder in males and inhibited orgasm in females.

"My girlfriend and I are sexually active. Since we first started having sex, I would have an orgasm but

she wouldn't. She still hasn't had one, and we have had sex plenty of times. I feel that I am not doing my job as a man. I also feel embarrassed. It makes me feel I do not know much about sex. Lately . . . I have trouble getting an erection. . . ."

(from the author's files)

Men who judge their sexual experiences like Olympic events (e.g., they are unhappy unless their partner has multiple orgasms) put enormous pressure on themselves. The same is true of men who feel they are competing with a real or imagined suitor. In both these cases, the man is likely to become distracted by observing his partner's or his own responses, resulting in an erectile disorder.

As you learned previously, nearly all males will experience an erectile problem sometime in their life, often due to fatigue, drugs, or too much too drink. One of the most common causes is a life change. For example, the stress caused by unemployment frequently leads to erectile problems (Morokoff & Gilliland, 1993). According to Levine and Althof (1991), "Typically, the man does not fully appreciate his response to these life changes, for example, his affects, defenses, coping devices, and adaptational shifts. Neither does he fully realize that the trigger mechanism for his arousal problem is contained in these responses. He avoids recognizing what these affect-laden events mean to him and only focuses on his loss of sexual confidence—that is, performance anxiety. He remains baffled by how he became dysfunctional."

Recall, for example, the case history of Billy and Karen. Let us continue with their story:

When Billy and Karen started therapy, they still had no idea what was causing the problem. They felt they were doing everything the same as before. In actuality, Billy was threatened by Karen's impressive career advancement. Although he tried hard to be a liberated male, he was threatened by her increased income and status and the time she spent entertaining the various physicians and hospital administrators in her territory.

During sexual relations, Billy had begun to worry if Karen found him as appealing as she had when she was more "naive and unworldly." He also had the secret fear that Karen was having an affair, and although she wasn't, he had begun to compete with his own fantasy of Karen's other partner. As with many men, this was sufficient to produce Billy's initial difficulty in maintaining an erection. Once the difficulty began, Billy's antici-

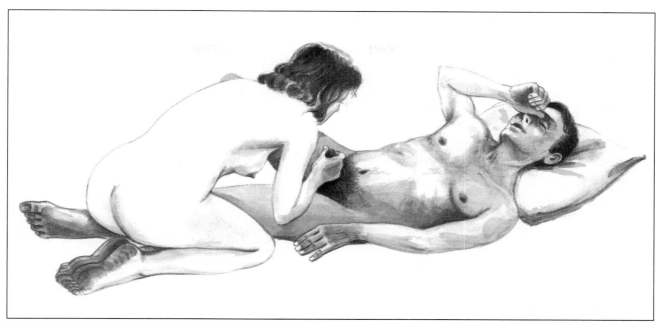

Figure 13–4   A commonly used male training technique.

pation of it happening made it begin to occur earlier in each subsequent sexual encounter.

(Case history courtesy of Dr. David Schnarch. Reprinted with permission of W. W. Norton & Company.)

Although an erectile problem in any of the previously mentioned situations is normal, many men are not aware of that and begin to worry about being "impotent." The next time they have sex, the anxiety caused by fear of failure can result in what is called **spectatoring**—observing and evaluating their own responses rather than experiencing sexual pleasures. Spectatoring is distracting, resulting in a loss of intimacy, and it can lead to loss of an erection, which results in even greater anxiety the next time. The behavior becomes a vicious cycle, and what started as a normal response to fatigue, alcohol, drugs, or other factors can become a chronic psychological disorder.

What is the best therapy for a man with a psychologically caused erectile problem (performance anxiety)? Today, many therapists are emphasizing the importance of the couple's dynamics and interpersonal issues in the cause and treatment of erectile problems: "Clinical experience suggests that couples with strong, committed and supportive relationships generally develop effective coping strategies in the face of even the most severe sexual problems. In contrast, uncommitted couples with unresolved marital conflicts appear to derive little or no benefit from even the most advanced medical, surgical or psychological interventions" (Leiblum & Rosen, 1991). Initial assessment of a couple's motivation and

interactions is important in choosing a therapy approach (Hawton, Catalan, & Fagg, 1992).

After their medical and sexual histories are taken, men with an erectile disorder are usually assigned sensate focus exercises. Many men are amazed to find that they are able to get a full erection during these initial sessions. The reason, of course, is that the touching in sensate focus exercises involves nondemand pleasuring—the man does not have to perform, but instead learns (or relearns) how to experience pleasure. When genital touching is allowed, the female partner is instructed to stop fondling the penis after the male gets an erection (see Figure 13–4). This is intended to teach the man that he does not have to use his erection immediately (many men with erection problems rush their performance for fear that they will lose their erection). With this *"teasing" procedure,* the male's erection comes and goes, and thus he also learns that loss of erection does not mean that he has failed, for it will come back again. When intercourse is finally allowed, it is always in a female-on-top or side-by-side position, and the female is the one who inserts the penis into her vagina. Thus, the male is not put in a position of feeling that he has to perform.

Erectile problems can be so psychologically devastating that many men try to deal with their problem alone. However, there is now an organization called Impotents Anonymous, with 19 chapters nationwide that offer information and support.

spectatoring Observing and evaluating one's own sexual responses. This generally causes distraction, with subsequent erectile and/or orgasm problems.

Wives of men with erectile problems are also welcome to attend meetings.

PREMATURE EJACULATION

> "I have a problem with premature ejaculation. As soon as I start to have sex I ejaculate. I feel like the whole world is laughing at me. . . . All my other friends are having sex and doing it right."

(from the author's files)

When many people think of someone who prematurely ejaculates, they think of a male who reaches orgasm before intercourse even begins. Although this does sometimes happen, it is more often the case that a male ejaculates shortly after the couple begins intercourse. Laumann et al. (1994) reported that 29 percent of the men in their national survey said that they had had problems with premature ejaculation in the previous year, making it the most common male sexual problem. Men who are concerned only with their own orgasms, of course, may not regard this as a problem. In fact, in some cultures around the world where men are very dominant and the sexual pleasure of women is denied, the ability to ejaculate quickly is considered very masculine. It is estimated that only 20 percent of all premature ejaculators in the United States ever seek therapy.

How, then, do we define **premature ejaculation**? The American Psychiatric Association (DSM-IV, 1994) says that its essential feature is "the persistent or recurrent onset of orgasm and ejaculation with minimal sexual stimulation before, on, or shortly after penetration and before the person wishes it," accompanied by "marked distress or interpersonal difficulty." The DSM-IV cautions that one must also consider factors such as age and the novelty of the sexual situation or partner.

How long does a man have to continue in intercourse without being classified as a premature ejaculator? At various times in the past, premature ejaculation has been defined by a specified minimal number of thrusts, a minimal amount of time engaged in intercourse, or in terms of whether or not the female partner reaches orgasm at least 50 percent of the time (see Metz et al., 1997). The problem with the first two types of definition is that they do not consider the partner. Suppose a man always reaches orgasm in only a minute and a half after beginning intercourse, but that his female partner always reaches orgasm within that time period as well. Should he be considered as having a problem? The trouble with the last definition is that it depends exclusively on the partner's responses. What if the partner has difficulties

reaching orgasm regardless of who her partner is? A man could continue intercourse until he's blue in the face and still be classified as a premature ejaculator. Most therapists choose to view premature ejaculation in terms of *absence of reasonable voluntary control* of ejaculation (Kaplan, 1974). By this definition, orgasm should not be a totally involuntary event, and a man should be able to have some control over when it occurs. A man who regularly reaches orgasm before he wants to, therefore, would be a candidate for sexual therapy.

A man who is disturbed by his own inability to exert any control over when he ejaculates may develop low self-esteem and performance anxiety (Kaplan, 1989). This, in turn, can lead to an erectile problem (and commonly does). Others may avoid dating new partners because of fear of embarrassment.

What causes some men to ejaculate too quickly? Traditionally, it has been believed that in most cases premature ejaculation is due to psychological factors (anger or depression) or interpersonal problems (e.g., anger at one's partner). In addition, many therapists believe that the early sexual experiences of many males actually teach them to rush during sex. Male masturbatory behavior is generally orgasm-oriented anyway, but a male may hurry it even more in situations where he fears being caught. Early heterosexual experiences often occur in places like a car, a parent's or friend's house, or some other place where there is a possibility of discovery, and thus sexual intercourse is hurried. However, premature ejaculation does not occur because these men can not monitor their level of sexual arousal (Spiess, Geer, & O'Donohue, 1984).

Recent studies indicate that there may be two different types of premature ejaculation: one caused by psychological factors, and another caused by organic factors (Metz et al., 1997; Tiefer, 1994). Some types of injuries have been known to cause premature ejaculation, as do some medications and withdrawal from some narcotics (see Metz et al., 1997). Moreover, there are several pharmacological agents that delay ejaculation (antidepressants, benzodiazepine anxiolytics, dopamine antagonists).

The following therapeutic exercises are designed to teach a man not to hurry during sexual relations. (Premature ejaculation, however, may sometimes result from an attitude of viewing women as sex objects; such men probably would not seek therapy.) After a couple has completed the sensate focus exercises described earlier in this chapter, the woman is instructed to manually stimulate the man's penis until he gets close to orgasm. She then squeezes his penis firmly between her thumb and index and second fingers for three to five seconds and discontinues stimulation until he calms down. She repeats this three to four times before finally bringing him to orgasm. This *squeeze technique* was de-

premature ejaculation A recurrent and persistent absence of reasonable voluntary control of ejaculation.

veloped by Masters and Johnson (1970) and is still regarded as the most effective treatment (St. Lawrence & Madakasira, 1992). In a variation of this, called *stop-start,* the female partner simply stops manual stimulation and then starts again after the male's feeling of impending ejaculation has diminished (Semans, 1956). Both of these techniques teach the man that he can delay ejaculation, and he can also practice them during masturbation. When intercourse resumes, it is in a side-by-side or female-on-top position (it generally takes longer to reach orgasm if you are not in total control).

Masters and Johnson (1970) have reported success rates of 95 percent. However, long-term follow-up studies show disappointing results. Most men who complete therapy successfully later regress to pretreatment levels (Metz et al., 1997; Schover & Leiblum, 1994). At this point, we do not know whether this means that traditional therapies should be changed (e.g., that follow-up therapy should be instituted) or whether greater effort should be made to identify and treat organic factors.

MALE ORGASMIC DISORDER

This problem can be thought of as just the opposite of premature ejaculation. **Male orgasmic disorder** (formerly called inhibited male orgasm or inhibited ejaculation) refers to a difficulty reaching orgasm and ejaculating in a woman's vagina, and it can be either *primary* or *secondary.* It used to be thought that only 1 or 2 percent of all men suffered from this problem (Apfelbaum, 1980; Masters & Johnson, 1970), but more recent studies indicate that perhaps as many as 4–9 percent of all men have this problem on more than just an occasional basis (Laumann et al., 1994; Spector & Carey, 1990). It is often referred to as **ejaculatory incompetence** if the man is totally unable to ejaculate in a woman's vagina. Although a few cases can be traced to organic causes (e.g., drugs, alcohol, or neurological disorders; Munjack & Kanno, 1979), most of these men are able to reach orgasm either during masturbation or during manual or oral stimulation by the partner (Kaplan, 1979), which indicates that the usual cause is psychological and not physical.

Masters and Johnson (1970) reported that primary orgasmic disorder was often associated with a strict religious upbringing, a fear of getting a woman pregnant, negativity and hostility toward the partner, and/or maternal dominance. Secondary problems are less common and are often associated with some kind of previous trauma. Apfelbaum (1989) more recently found that men with orgasmic disorder focus too much on performance and pleasing their partners, and as a result are subjectively not as aroused (as they are during masturbation, for example).

Inhibited ejaculation is often treated by what's called a *bridge maneuver.* The man is first taught to masturbate to orgasm while alone, and then in the presence of the partner. Once this is successfully completed, the partner manually stimulates the male to orgasm. When this is consistently successful, the female partner stimulates the man until he is very close to orgasm and then quickly inserts his penis into her vagina. The idea is that if this can be done several times, the male will overcome his fears of ejaculating inside her vagina. Apfelbaum additionally encourages men to be more assertive with partners in letting them know what they find to be arousing.

HEADACHES AFTER ORGASM

Several studies have brought attention to a problem suffered by some men during sex: severe headaches that start during or slightly after orgasm (see Diamond & Maliszewski, 1992, for several articles). This condition is called **benign coital cephalalgia** (alternative spelling: *cephalgia*). This is not a new problem, and in fact was first reported by Hippocrates, the father of modern medicine, in the fifth century B.C. It has been estimated that as many as 250,000 Americans may presently suffer from this problem. Most are men who are middle-aged, mildly obese, and somewhat hypertensive. The severe headaches may last for minutes or hours. They are not related to exertion or type of sexual activity, and although physicians sometimes mistake the pains as the sign of an imminent stroke, they are usually not associated with any other serious disorder. The headaches are generally caused by contraction of muscles and/or blood vessels in the head, neck, or upper body during intercourse. For reasons that are not understood, the headaches often disappear for weeks or months at a time and then suddenly reappear. Another study found that the headaches eventually stop recurring in men who do not also suffer tension headaches or migraines (Ostergaard & Kraft, 1992). Treatment may include the teaching of relaxation techniques or medication for high blood pressure.

PRIAPISM

"Sometimes my erection will stay for hours, which is kind of uncomfortable. . . ."

(from the author's files)

To many men, the idea of having a *long-lasting erection* might seem appealing. **Priapism** is a rare condition in which the penis remains erect for prolonged periods of time, sometimes days. It results from damage to the valves regulating penile blood flow and can be caused by tumors, infec-

---

male orgasmic disorder A condition in which a man has difficulty reaching orgasm and ejaculating in a woman's vagina.

ejaculatory incompetence A condition in which a man is totally unable to ejaculate in a woman's vagina during sexual intercourse.

benign coital cephalalgia Severe headaches that start during or slightly after orgasm in males.

priapism A condition in which the penis remains erect for a prolonged period of time, sometimes days.

tion, or chemical irritants (such as Spanish fly; see Chapter 4). It is not usually accompanied by a desire for sex. Priapism is often painful, not to mention inconvenient, and men with this condition would gladly trade it for a normal erection. Many different treatments have been tried, ranging from ice packs to general anesthesia, but results from one study suggest that administration of the beta-agonist drug terbutaline is most effective (Shantha, Shantha, & Bennett, 1990).

FEMALE SEXUAL PROBLEMS

VAGINISMUS

Vaginismus refers to pain experienced during attempted sexual intercourse (intromission). It is caused by *involuntary contractions* of the muscles that surround the outer third of the vagina. Sometimes it is situational and occurs only during attempted intercourse, but in many cases it occurs at attempts to insert anything into the vagina (e.g., a finger or tampon). A pelvic examination may be impossible for women with this condition. To a man attempting intercourse, it feels like his penis is hitting a wall, and forceful attempts at penetration can be quite painful for the woman.

> "I was glad to see this topic addressed in your textbook and to know that it actually had a name. I was a virgin until I was 20 and when I did attempt intercourse, I encountered severe pain. My boyfriend at the time tried to be understanding, but it ended up tearing us apart. No matter how often we tried, we could not get it right. . . . I saw my OB/GYN and she gave me 3 different-sized vaginal dilators. I used these with KY Jelly but this was extremely humiliating. . . . I was helped over my fear by a much older man who took his time and made me feel extremely relaxed."

> (from the author's files)

About 2 percent of all women have vaginismus (Renshaw, 1990). The cause is usually *psychological* (Kaplan, 1974). It is normal for inexperienced women to have anxieties before intercourse that may result in some degree of involuntary muscle contraction, but the vast majority of women soon learn to relax (and even gain voluntary control over these muscles). Persistent and recurrent involuntary muscle spasms are often associated with past trauma (e.g., rape,

vaginismus A sexual problem in females in which pain is experienced during attempted intercourse because of involuntary spasms in the muscles surrounding the outer third of the vagina.

female orgasmic disorder A persistent or recurrent delay in, or absence of, orgasm following a normal sexual excitement phase.

abortion) or negative first sexual experiences, a religious upbringing (where sex is equated with sin), or hostility toward or fear of men (Masters & Johnson, 1970; Van de Wiel et al., 1990). Treatment usually consists of sensate focus relaxation exercises followed by *gradual dilation of the vagina* (with a set of dilators, or increasing from one to three fingers). The use of dilators in an atmosphere that is free of sexual demand, so that there is no performance anxiety, allows the female to slowly overcome her anxieties about having an object in her vagina. Psychotherapy may have to be used as well in order to deal with past sexual trauma, guilt about sex, or relationship problems (Leiblum, Pervin, & Campbell, 1989). Masters and Johnson (1970) report a 100 percent success rate in treating this problem.

FEMALE ORGASMIC DISORDER

The vast majority of women who seek sexual therapy do so because of problems in reaching orgasm. Laumann et al. (1994) found that nearly one-fourth of the women in their national survey said that they had been unable to achieve orgasm for at least several months of the previous year. More than half of all adult women have pretended to have an orgasm during sexual intercourse (Wiederman, 1997). As with erectile problems, **female orgasmic disorder** can be *primary* or *absolute* (never having had an orgasm under any circumstances), *secondary* (having once been regularly orgasmic), or *situational* (e.g., able to reach orgasm during masturbation, but not during intercourse). Some men use the term "frigid" to describe such women, but this is an unfortunate term because it has negative connotations about a woman's emotional responsivity. Most women who have orgasm problems are emotionally responsive to sexual stimulation and otherwise enjoy sex.

One of the very first questions to ask, of course, is whether a woman with problems reaching orgasm during sexual intercourse is receiving sufficient stimulation from her partner. Has there been enough time and foreplay?

> "My problem seems like one that some of my friends have—short foreplay. It takes me a long time to reach orgasm. When we take a long time and work up slowly to lovemaking, I have intense orgasms. It is wonderful for both of us. But this happens very rarely. Because 'John' works a nine hour work day and usually comes home exhausted, our lovemaking is short and to the point. He's satisfied and I lay there feeling nothing because I faked an orgasm for him. Then, when I insist on continuing, I start to feel guilty for wasting his time."

"I have no trouble reaching climax with my husband, but it takes a long time. How can I get him interested in foreplay so we both can experience enjoyment before intercourse? I don't want him to get bored with me."

(from the author's files)

These responses are apparently not unusual. Remember that in response to a survey question asked in my course ("What things do you personally like best in a sexual partner?"), well over half of the women who indicated that they were sexually experienced included as one of their answers that they wanted their partner (or partners) to take his time. Here are some examples:

"Doesn't rush, and takes his time."

"Lover takes his time and is as concerned with fondling and foreplay as he is intercourse."

"Unhurried attitude toward sex and lovemaking in general."

"Be willing to take his time, not hurry."

"Partner who likes to take time with foreplay and not rush into intercourse. A caring and sensitive person who is not out just to get a 'piece of ass.'"

(from the author's files)

A man's first physiological sexual response, you recall, is erection of the penis, which allows him to perform sexually almost immediately, even if he is experiencing very little emotionally. It generally takes women longer to reach orgasm than men (Gebhard, 1966; Kinsey et al., 1953), and many need additional emotional involvement as well, so that lengthy foreplay is very important for most women.

Even if her partner takes his time, many women may still have difficulties reaching orgasm during intercourse. The key for many women is the degree of stimulation to the clitoris, for most women prefer clitoral stimulation to reach orgasm, with or without vaginal stimulation (Darling, Davidson, & Cox, 1991). During sexual intercourse, the penis only indirectly stimulates the clitoris by causing the clitoral hood to rub back and forth over the clitoral glans. Helen Kaplan (1974) and Shere Hite (1976) have estimated, however, that fewer than half of all women are capable of reaching orgasm regularly during intercourse without more direct stimulation of the clitoris. I have seen numerous letters to magazine advice columns written by women who were concerned about

their lack of physical sensations during intercourse, particularly in the missionary position. None could reach orgasm during intercourse, but they had no trouble during masturbation (with more clitoral stimulation), which indicates that there was nothing physically wrong. A recent letter to the Playboy Advisor was written by a woman whose partner refused to give her any type of clitoral stimulation so that she could "learn" how to have a vaginal orgasm.

Unfortunately, many men take the attitude that what feels good to the gander must also feel good to the goose—if it feels good to the man to have his penis in his partner's vagina, then it must also feel equally good to the woman to have his penis in her vagina. The stimulation provided by the vaginal walls to the penis during intercourse is similar to that experienced by men during masturbation, but the reverse is not true for women. The inner two-thirds of the vagina, you recall, has relatively few nerve endings. Most women who masturbate do so by stimulating their clitoris. Men should not feel inadequate if their female partners require manual stimulation of the clitoris during intercourse to reach orgasm.

"Your class has also taught me how to talk to my boyfriend about sex. By stimulation of the clitoris manually it has brought great joy to my sex life. I now even reach orgasm from it during sex."

(from the author's files)

If a man cares for his partner, he will do (or let her do) whatever she says feels best, not what he believes ought to feel best. *Therapists agree that clitoral stimulation during intercourse is not cheating.*

Not all cases of female orgasm problems can be blamed on insufficient stimulation. Sex is more than just the physical rubbing together of tissues. A healthy, positive attitude about sex and pleasure is also very important. Many cases of primary orgasmic disorder, for example, are associated with negative attitudes about sex or feelings of guilt about sex (often due to a very strict religious upbringing, where sexual feelings are associated with sin rather than something positive) (Kelly, Strassberg, & Kircher, 1990; Masters & Johnson, 1970).

"I grew up sexually in marriage. In those days no one discussed sex, and my husband and I both only knew bits and pieces. I understood it to be very natural for women not to have orgasms, therefore, if I did, my private Catholic girl school upbringing made me feel ashamed of my own natural desires and pleasures."

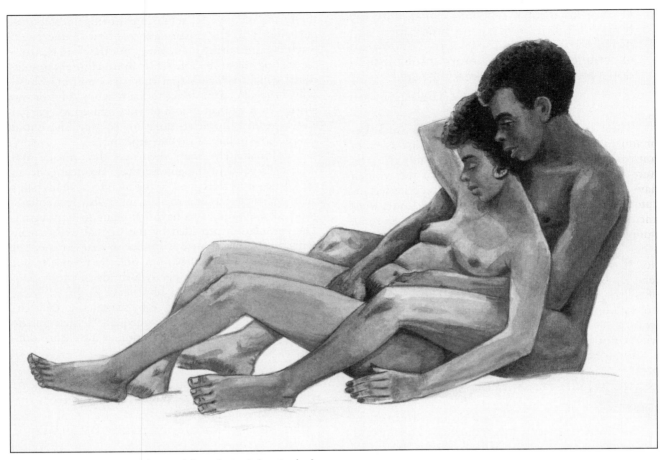

Figure 13–5 A commonly used female training technique.

"I was a Catholic girl all my life. Throughout school there was always a hidden notion that sex was bad and dirty. Those views and ideas were always instilled in me. It made me feel as if I would die if I engaged in any sexual behavior."

(from the author's files)

Even if sex is not associated with sin, it is a fact that most parents allow their daughters less sexual freedom during adolescence than they allow their sons. Female sexuality is more likely to be suppressed during this period than is male sexuality.

"I was told my whole life as a child and as an adolescent that because I was a woman, I was not supposed to feel sexual, or be sexual. My daddy told me that good women who wanted to be loved by a man did not feel sexual. A good man did not look for a woman who was sexual."

(from the author's files)

Sexual suppression generally leads to passivity in sexual relations with a partner. But studies have found that sexually assertive women (those who are active in initiating sex and who communicate with their partners about their likes and dislikes) have higher frequencies of orgasms than less assertive women (Hurlbert, 1991). They also generally report greater satisfaction in their relationships. Multi-orgasmic women are generally more explorative in their sexual activities and more aware of pleasurable sensations of the clitoris via masturbation than are other women (Darling, Davidson, & Jennings, 1991).

In addition, sex involves sweating, underarm and vaginal odors (and often vaginal noises), and facial grimaces that many women were taught are unfeminine. To fully enjoy sex and experience the pleasures of orgasm, one must "let go." Simply telling a young woman that sex is now okay is not going to change these attitudes overnight. It may take years to overcome inhibitions about sex.

"I was raised as good girls don't, and if you do you are used property. After getting married, the morals were supposed to change. I had great difficulty in the transition of why it is OK now, after marriage. I am now divorced. I wish I would have had your class prior to marriage. It would have made the transition a lot easier."

Kinsey's group (1953), you recall, found that half of all women who masturbated did not start doing so until after the age of 20 (and thus probably after they were already sexually experienced). More recent studies have found that married women who have masturbated to orgasm have more orgasms and greater marital and sexual satisfaction than married women who have not masturbated to orgasm (Hurlbert & Whittaker, 1991). In fact, many therapists now include masturbation therapy as a first step in the treatment of inhibited female orgasm in order to make a woman more aware of her own body (Heiman & Grafton-Becker, 1989; LoPiccolo & Stock, 1986). The hope is that once a woman can reach orgasm during masturbation, she will be able to achieve it during sexual intercourse.

Performance anxiety can also contribute to orgasm problems. This can easily happen if a woman's partner expects or demands that she have an orgasm during sex. The subsequent spectatoring then makes the problem even worse.

"About orgasms: this is a subject I have trouble with. Almost every guy I've been with tries

his hardest to make me have one once he finds out I haven't. They want to be the 'first.' Sex can be repulsive when it is done as a forced series of steps."

Just as with erectile disorders, it sometimes takes only a few instances of being unable to reach orgasm for a woman to develop performance anxiety.

"My husband and I have a really good sex life, but I don't reach climax every time. This doesn't bother me, but it really bothers him. I've started worrying about it so much that I'm now having fewer orgasms and starting to fake more."

Women (and their partners) should not expect to reach orgasm with every experience of sexual intercourse. This is an unrealistic expectation that can lead to unnecessary feelings of guilt (Davidson & Moore, 1994).

What procedures are used in therapy to help a woman become more orgasmic (see Heiman & Grafton-Becker, 1989)? As explained before, if a woman has always had difficulty reaching orgasm (and particularly if she has never masturbated), the first thing that a therapist is likely to suggest is self-exploration. You don't necessarily need a therapist, but if you have never explored your body and responses before, some type of instruction will probably be necessary. A book by Heiman and LoPiccolo (1992) is listed at the end of the chapter as a good reference.

The next step is exploration with a partner—the widely used and popular sensate focus exercises. These nondemand exercises allow the woman and her partner to relax and learn what gives each maximum pleasure. The hand-riding technique is particularly useful here (Masters & Johnson, 1970). After several days of this, the therapist will probably instruct the partner to manually explore the woman's genitals, usually while both are sitting in the position shown in Figure 13–5. (Some therapists may first have the woman stimulate her own genitals in front of

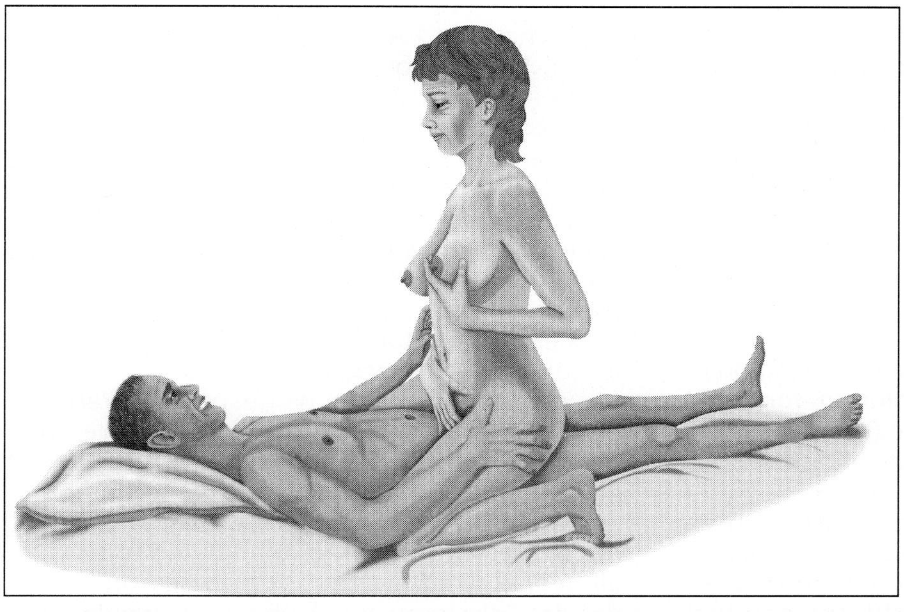

Figure 13–6 For any sexual problem, when the therapist allows the couple to resume sexual intercourse, it is almost always in the female-on-top position. Women with orgasm problems are encouraged to touch their clitoris (or have their partners touch it) during intercourse.

her partner, but this often requires that she get over a great many inhibitions.)

When the couple is finally told that they may resume sexual intercourse, they will almost always be told to use a position that allows the woman a great deal of control, such as woman-on-top. She will decide when intromission begins and initiate pelvic move- ments once they start. By this point in therapy, both she and her partner will be comfortable with manual stimulation of the clitoris during intercourse. Remember, *manual stimulation of the clitoris during intercourse is not cheating.* Helen Kaplan (1974) has suggested that couples at first interrupt intercourse with periods of additional foreplay, but contrary to advice given in a

---

Box 13-A   CROSS-CULTURAL PERSPECTIVES

Sex Therapy

The model of sex therapy presented in this chapter is based primarily on Western middle-class values (Lavee, 1991). The two most important assumptions in the model are that people engage in sex primarily for (physical) pleasure, and that men and women can be viewed as equals. The model also assumes that men and women want to know about sex and that better communication makes for better sex.

Not all cultures share these attitudes. Recall from Box 4–A, for example, that in large parts of northern Africa and the Middle East (with primarily Islamic populations), sexual pleasure is reserved for the males, and women's pleasure is denied by clitoral circumcision and genital mutilation. Many Latin countries are also male-dominant to the point of denying females' sexual feelings (Guerrero Pavich, 1986). In these cultures, a woman's difficulty in reaching orgasm is not considered important, and certainly no grounds for therapy. Even painful intercourse (dyspareunia, vaginismus) is considered unimportant unless it interferes with a woman's ability to engage in sex. For women, the only purposes of sex are procreation and the pleasure of the husband. For men, "premature" ejaculation is not regarded as a problem. Men do regard difficulty in getting and maintaining an erection as a problem, but both men and women reject the very idea of engaging in sensate focus exercises as a solution, for this is a mutual pleasuring technique (Lavee, 1991).

In sharp contrast to these attitudes (and our own) about sex are those held by many Hindu and Buddhist societies (Voigt, 1991). Tantric scriptures urge people to strive for the goal of transcendent unity and harmony, and thus sex is viewed as an opportunity for spiritual growth. While Western therapists emphasize the mutual physical pleasures of sex, these Asian societies emphasize the spiritual union. For example, in this chapter orgasm is presented as resulting from proper physical stimulation and effective technique. Those who follow Tantrism, on the other hand, "understand orgasm as a product of deep relaxation and a profound level of contact between partners" (Voigt, 1991). Western culture emphasizes doing, stimulation, activation, and outcome-focus, while Tantrism emphasizes being, stillness, meditation, and process-immersion. Good sex is not achieved by focusing on one or the other partner, but by focusing on the "between." Couples might be urged to follow five exercises (Voigt, 1991): (1) creation of a ceremonial sex ritual (a simple ceremony that sets the tone for shared sexual expression—for example, candles, flowers, a special room, massage, chanting together or joint meditation); (2) synchronized breathing while embracing or touching; (3) sustained, steady eye contact; (4) becoming motionless at the peak sensual experience during intercourse; and (5) refraining from orgasm at times in order to intensify the sexual-spiritual energy.

Sex therapy as we know it is geared, obviously, to our particular cultural beliefs, and perhaps to only part of our society. Nonwhite people, for example, are much less interested in going to sex therapy for an acknowledged sex problem than are white people (Perez, Mulligan, & Wan, 1993). This may be due, in part, to a distrust of mental health professionals (Wilson, 1986), but it may also reflect different cultural attitudes about sex. Lavee (1991) has suggested that rather than trying to teach all people what healthy sex is (based on middle-class Western values), therapists would be wiser to fit the treatment to the individuals' cultural values.

recent popular book (Eichel & Nobile, 1992), no particular technique of intercourse can guarantee orgasms (Kaplan, 1992b). The couple will have to try different techniques. With some practice, for example, the woman might be able to reach orgasm without manual stimulation by rubbing her clitoris against her partner's pubic bone, or vice-versa (in a "high-ride" position). Recall from Chapter 2 that Kegel exercises may help some women with vaginal sensations and orgasmic responsiveness.

If at this point you are thinking that female orgasm problems have been treated rather mechanically, as if the solution is nothing more than finding the correct manner of physical stimulation, you are probably right. For many women, sexual satisfaction is closely related to how emotionally involved they are with their partners and how happy they are in their overall relationship (Hurlbert & Whittaker, 1991; Newcomb & Bentler, 1983). These things are better predictors of a woman's sexual satisfaction than are frequency of sex and number of orgasms (Hurlbert, Apt, & Rabehl, 1993). Conversely, the better a woman regards her emotional relationship with her partner, the more likely she is to judge her sex life satisfying (Darling et al., 1991; Rosenzweig & Dailey, 1989). In other words, for many women the therapy steps outlined in this section may do little good if the emotional aspects of their relationships are not satisfying.

SEXUAL PROBLEMS AMONG HOMOSEXUALS

Like some heterosexuals, homosexuals sometimes suffer from sexual problems, and they also seek sex therapy to deal with them. However, the types of problems and the focus of therapy is often different for homosexual men and women. Heterosexuals generally go to therapy for problems related to sexual intercourse, but homosexuals seeking sex therapy often have problems that arise from dealing with HIV and AIDS and/or homophobic attitudes within society, which in some cases become internalized (Friedman, 1991; Margolies et al., 1988; Nichols, 1988; Reece, 1988; Shannon & Woods, 1991). These problems can lead to depression (or, in the case of internalized homophobia, repression of sexuality) and consequent loss of sexual desire. Orgasm problems are not common among lesbians, but many lesbians who go to therapy have negative feelings about engaging in cunnilingus (Margolies et al., 1988).

Similarly, many gay men in therapy have negative feelings about anal sex (Reece, 1988).

Masters and Johnson (1979) reported a 93 percent success rate for treating erectile disorders in male homosexuals as well as anorgasmia in lesbians. Sensate focus exercises used with heterosexuals are similarly used with homosexuals. In fact, the success rate for homosexuals is generally higher than that for heterosexuals. This may be due to the fact that members of the same sex may be more understanding of the needs and pleasurable experiences of homosexual partners. In any case, an understanding partner who is sympathetic, willing to communicate, and not overly demanding is crucial to successful sex therapy for both heterosexual and homosexual individuals.

AVOIDING SEXUAL PROBLEMS

What have you learned in this and previous chapters that you could apply to your own life in order to minimize the chance that sexual problems will occur?

First, establish an open, honest line of communication with your partner. Although you should learn to accentuate and reinforce the positive aspects of your relationship, this does not mean that you cannot say no. A sexually healthy person, you remember, is free to make choices, but you must work any differences out with your partner.

In order to communicate, we must be properly educated. Sexual therapy, in fact, is really just sex education combined with a medical exam and counseling. Why did you have to wait until your age to receive a sex education? When you have children, teach them the proper names for their body parts and instruct them about sexuality in such a way that they will grow up to think of the human body as something beautiful and good. A good sex life can contribute to a healthy and satisfying life. You can treat sexuality as something positive while at the same time teaching morals and sexual responsibility. Whatever you do, never equate sex with sin and guilt.

In the relationship with your partner, try to learn and grow together. Remember, a good relationship with your partner (i.e., sharing and caring) often carries over into sexual relations. Keep your sexual relations spontaneous and fun and do not set goals as if sex were an Olympic event. Do not focus on orgasm. Enjoy touching for touching's sake—it doesn't always have to lead immediately to sexual intercourse.

Key Terms

benign coital cephalalgia 331
cognitive-behavioral therapy 319
dyspareunia 325
ejaculatory incompetence 331
erectile disorder 325
female orgasmic disorder 332
global sexual problem 325
hypersexuality 323
hypoactive sexual desire 322
male orgasmic disorder 331

medical history 320
performance anxiety 328
phimosis 325
PLISSIT model 319
premature ejaculation 330
priapism 331
primary sexual problem 325
psychosexual therapy 319
secondary sexual problem 325
self-awareness 320

sensate focus exercises 320
sexual addiction 324
sexual aversion 322
sexual history 320
sexual surrogate 321
situational sexual problem 325
spectatoring 329
systematic desensitization 320
vaginismus 332

Personal Reflections

1. Are you able to comfortably communicate your sexual values and needs to your sexual partner or your potential partner? If not, what barriers are preventing you from doing so? Do you create any barriers that may prevent your partner from comfortably communicating his or her sexual values and needs to you?

2. What do you do when your partner desires sex and you are not in the mood? What would you expect your partner to do if you desired sex and he or she wasn't in the mood? Why? Have you ever pretended to be tired or feeling poorly in order to avoid sex with your partner? Have you ever faked an orgasm? If so, why? Do you think that responses like these are the best way to communicate with your partner about sexual needs?

3. Stop and reflect on the physical interactions you have with your partner(s). In most of your interactions, are you able to engage in extended nongenital touching and caressing of your partner, or are you usually in a hurry to achieve some immediate genital or breast touching? Is your partner able to caress your arms, legs, neck, etc., without your quickly attempting to steer the behavior toward sexual play? Why or why not?

4. Almost all people experience sexual disorders at some point in their lives. If you were in an ongoing relationship and were consistently experiencing a problem, would you go to a sex therapist? Why or why not? If your answer was yes, how long would you wait before you decided to go? Most therapists believe that the earlier therapy is initiated, the better the chance for a cure.

5. Have you ever experienced performance anxiety during sexual relations (e.g., anxiety about pleasing your partner, maintaining an erection, reaching orgasm)? If so, what do you think is the cause of the anxiety?

6. Are you able to "let go" totally during sex (i.e., no anxieties about nudity, sweat, odors, noises, facial expressions, etc.)? If not, why not?

7. If you are in an ongoing sexual relationship, what do you do to prevent the boredom that can arise if sex is allowed to fall into a predictable routine (Saturday night, after the news, same position, etc)?

Suggested Readings

Barbach, L. G. (1982). *For each other: Sharing sexual intimacy.* New York: Anchor/Doubleday.

Belliveau, F., & Richter, L. (1970). *Understanding human sexual inadequacy.* New York: Bantam. An easy-to-read review of Masters and Johnson's work.

The Boston Women's Health Book Collective. (1992). *The new our bodies, ourselves: A book by and for women.* (4th edition). New York: Simon & Schuster.

Cohen, S. S. (1987, April). The power of touch. *New Woman.* Discusses the importance of touching in enhancing intimacy.

Heiman, J., & LoPiccolo, J. (1992). *Becoming orgasmic: A sexual and personal growth program for women.* New York: Fireside Books.

Kaplan, H. (1987). *The illustrated manual of sex therapy.* New York: Brunner / Mazel. Easy to read and tastefully written.

Kaplan, H. S. (1989). *PE: How to overcome premature ejaculation.* New York: Brunner/Mazel.

Kaplan, H. S. (1995). *Sexual desire disorders: Dysfunctional regulation of sexual motivation.* New York: Brunner / Mazel. Extensive coverage of hypoactive sexual desire and hypersexuality from the therapist who pioneered work in desire disorders.

Leland, J., et al. (1997, November 17). A pill for impotence? *Newsweek.*

Schnarch, D. (1991). *Constructing the sexual crucible: An integration of sexual and marital therapy.* New York: W. W. Norton. Written for the professional, this book has had an amazing crossover success with the general public.

Thayer, S. (1988, March). Close encounters. *Psychology Today.* Explores the role of touching in intimate relationships and examines gender and cultural differences.

Zilbergeld, B. (1992). *The new male sexuality.* New York: Bantam Books. A readable book about male sexuality and overcoming problems.

CHAPTER 14

When you have finished studying this chapter, you should be able to:

1. Describe the various ways one can define a behavior as being unconventional (i.e., statistical, sociological, and psychological approaches);

2. Explain the difference between a sexual variant and a paraphilia;

3. Define and discuss the following paraphilias: fetishism, transvestism, exhibitionism and obscene phone calls, voyeurism, zoophilia, pedophilia, sadism, masochism, urophilia, coprophilia, mysophilia, frotteurism, klismaphilia, and necrophilia;

4. Discuss the various theories of how paraphilias develop; and

5. Discuss the treatment (therapy) for individuals with a paraphilia.

Paraphilias and Sexual Variants

There are enormous individual differences among people in every aspect of human behavior. It should be obvious from the many references to other cultures in previous chapters that this includes sexual behavior as well. Variety in behavior is usually considered to be good. We enjoy wearing different clothing, eating different foods, listening to different kinds of music—and, as you learned in Chapter 11, most of us enjoy practicing a variety of sexual behaviors.

What determines whether a behavior is designated "normal" or "conventional" as opposed to "abnormal" or "unconventional"? What is normal or conventional is not always easy to define. Most of us would probably agree that someone who ate beetles was abnormal, but this is common in some African cultures. Obviously, cultural factors must be considered. Even within the same culture, what is believed to be normal today may not have been viewed as normal in the past, and may not be viewed as normal in the future. Here's an example. If a woman wears a string bikini on the beach today, she is considered normal. But if a woman living in 1900 had worn that outfit to a beach, she would have been arrested. The kindest treatment that she might have received would have been treatment for an illness. On the other hand, if a woman wore a full-length swimsuit with bloomers to the beach today (the typical swimwear for American women in 1900), she might be considered sexually inhibited. Similarly, a man or woman can walk naked in New Orleans' French Quarter on Mardi Gras day and get pinched a lot, but not arrested. However, if he or she walked naked down the same streets the day before or the day after Mardi Gras, the person would be arrested.

Even at the same point in time, normal behavior might be a matter of where it is observed. If someone tried to sunbathe in the nude at the local public swimming pool, he or she would probably be arrested. If this same person were sunning himself or herself on

Figure 14–1 Walking on the beach naked might seem unusual and unacceptable to you, but if it is not done in front of unwilling observers, it is a sexual variant, not a paraphilia.

certain beaches on the West or East Coasts, on the other hand, he or she could walk in naked splendor alongside a whole beach full of other naked people. When does such behavior start being normal—before getting to the beach, after getting to the beach, never, or always? Again, the point is that "normal" is not an absolute thing. It can be determined by many factors, including where you are and when you are there.

Researchers have generally attempted to define "normal" or "abnormal" in one of three ways. One way is called the *statistical approach,* in which a behavior is considered normal or conventional if a large number of people engage in it, and abnormal or unconventional if only a few people do it (Pomeroy, 1966). Some professionals who use the statistical approach do not use the terms "abnormal" or "unconventional" to describe behaviors, and instead use the label "sexual minorities" (Gagnon, 1977). There are, of course, some rather large minorities, and there are some behaviors that are so different that they represent a minority of one. However, the fact that only a minority of people have engaged in a behavior does not necessarily make that behavior bad or wrong, or the individuals disturbed or sick. Probably only a small minority of people in our culture, for example, have ever had sex outdoors (in private) at night under the stars, but most would agree that doing so would certainly not be a sign of mental illness. Because of the limitations of the statistical approach—defining normal and abnormal purely in terms of numbers—others have preferred a sociological or psychological approach.

The *sociological approach* calls a behavior unconventional if it is not customary within a society. Customs vary, not only from culture to culture, but also within subcultures. Followers of the sociological approach are concerned with the social conditions that give rise to abnormal behavior as well as the interactions between people labeled abnormal and the rest of society.

The *psychological approach* focuses on the psychological health of the individual and the impact that behaviors have on this health (Buss, 1966). A sexual activity is considered to be a problem behavior if it makes an individual feel distressed or guilty or causes the individual problems in functioning efficiently in ordinary social and occupational roles.

This chapter will focus primarily on the psychological approach. Let us now take a look at how the psychiatric/psychological profession has dealt with the subject of sexual abnormalities.

HISTORICAL PERSPECTIVE

In the Victorian era, sexual behaviors that were considered abnormal were referred to as "perversions," and later as "deviations." The first classification of sexual deviations was made by Richard von Krafft-Ebing in 1886 (*Psychopathia Sexualis*). His work was based on the premise that people are genetically predisposed toward sexual deviation, and although this theory has been discredited, Krafft-Ebing's case histories are still widely cited.

Sigmund Freud believed that any type of sexual behavior that took precedence over heterosexual vaginal intercourse was an indication of impaired psychosexual development. A person's sexual behavior was considered deviant if it was not directed toward an appropriate object (a heterosexual partner) and/or the aim was not sexual intercourse. According to this view, masturbation and oral-genital sex were considered very disturbed. (This is another excellent example of how what is considered abnormal depends on the time and culture, and it also points to a major limitation of the psychological approach.)

Simon (1994) has made an interesting distinction between the terms "deviance" and "perversion." According to him, *sexual deviance* is "the inappropriate or flawed performance of a conventionally understood sexual practice." He uses rape as an example; we rarely refer to it as a perversion. Rape would be considered a perversion only if committed on someone whose inclusion in the act went beyond what we consider to be normal. Simon emphasizes that as sexual practices change, so do our definitions of deviance (oral sex being a good example). *Perversion,* on the other hand, "can be thought of as a disease of sexual desire not only in the sense that it appears to violate the sexual practices of a time and place, but also because it constitutes a violation of common understandings that render current sexual practice plausible. The 'pervert' is disturbing because, at the level of folk psychology, we have difficulty understanding why someone 'might want to do something like that.'" Here, too, Simon emphasizes that our concept of perversion can change with time (masturbation being a good example). Thus, according to Simon, deviance is a problem of control, while perversion is a problem of desire.

Today, the American Psychiatric Association's handbook *Diagnostic and Statistical Manual of Mental Disorders* (DSM-IV, 1994) refers to psychosexual disorders of the type considered in this chapter as **paraphilias,** derived from the Greek *para* (meaning "besides," "beyond," or "amiss") and *philia* ("love"). Not all statistically or sociologically unusual behaviors are paraphilias. For a person's behavior to be called a paraphilia, the distinguishing

feature must be that the person's sexual arousal and gratification *depends almost exclusively* on engaging in or thinking (fantasizing) about the behavior. The essential features of a paraphilia are that the fantasies and behaviors are recurrent (American Psychiatric Association DSM-IV, 1994). As you read through the chapter, there will probably be one or more behaviors covered that you have either fantasized about or engaged in yourself (e.g., voyeurism, exhibitionism, S&M). This does not mean you have a paraphilia. It would be a paraphilia if that was the preferred way in which you become sexually aroused. Otherwise, behaviors and fantasies are generally referred to as **sexual variants** (Stoller, 1977).

Some of the paraphilias covered in this chapter can result in physical harm to oneself (e.g., masochism) or physical or psychological harm to others (e.g., sadism, exhibitionism, voyeurism, pedophilia). Others can lead to arrest and prosecution. However, some would appear to be harmless: transvestism (cross-dressing) in the privacy of one's own home, for example. Some clinicians argue that harmless behaviors (this assumes that the behavior is engaged in without feelings of guilt, shame, or remorse) should not have a negative label, but others say that no paraphilia is harmless because, as the only source of sexual gratification, it deprives the individual of physically affectionate relations with others.

The following sections describe many, but not all, paraphilias. The chapter concludes with a discussion of causes and treatments.

FETISHISM

In anthropology, a fetish is an object that is believed to have magical powers (usually by capturing or embodying a powerful spirit). The term **erotic fetishism** was coined by Alfred Binet (inventor of the first standardized intelligence test) in 1888. It generally refers to achieving sexual arousal and gratification *almost exclusively* by handling or fantasizing about an inanimate object or a particular part of the body. The American Psychiatric Association's DSM-IV (1994) classifies recurrent sexual attraction to specific body parts as a variation of fetishism called *partialism,* but many clinicians do not make the distinction. Some of the more common fetishes involve women's undergarments (panties, bras, or stockings), shoes (usually high-heeled) or boots, hair, objects made of leather or rubber, and feet (Junginger, 1997). Fetishes involving hard objects such as boots, leather, or rubber objects are often associated with sadomasochistic practices and fantasies. Heterosexual male fetishists are aroused by the theme of "femininity" in their preferred object, while homosexual male fetishists are aroused by the theme of "masculinity" in the preferred objects (Weinberg, Williams,

paraphilias General term for a group of sexual disorders in which a person's sexual arousal and gratification depend almost exclusively on unusual behaviors.

sexual variant For an individual, statistically or sociologically unusual behaviors that are engaged in for variety and not as one's preferred manner of becoming sexually aroused.

fetishism A condition in which sexual arousal occurs primarily when using or thinking about an inanimate object or a particular part of the body (such as the feet).

& Calhan, 1994). It is not uncommon for fetishists to be attracted to two or more objects (Junginger, 1997).

The large majority of known fetishists are men, and when their fetish involves clothing or shoes, they greatly prefer articles that have been used and have an odor. A man with a panty fetish, for example, would become highly aroused by a used pair of panties that retained vaginal odor, but would probably not be very aroused by the same panties if they were clean.

> "I work at a department store. One night I was folding towels when a man came over and started asking silly questions like, 'How often do you put rugs on sale?' I noticed that while he was talking, he kept glancing at my feet. He asked me where I got my shoes and if the inside was soft. He also asked if they held an odor. Not thinking, I said 'no' and took off a shoe. By this time, he had knelt down, picked up the shoe, and smelled it several times. He then gave it back and immediately left the store. Since my experience, he has been back in the store twice and has done the same thing to two other girls."

> "I know a married man in his late twenties who has a fetish. He goes around asking women (the ones he has affairs with) for a pair of their dirty underwear."

> (from the author's files)

I chose to cover fetishism first because it provides an excellent example of the difference between a paraphilia and a variant. Most heterosexual males are aroused by female panties, and many men have purchased sheer, lacy lingerie items for their partners from places like Victoria's Secret. Most men have preferences about a partner's anatomy (breasts, buttocks, legs, shoulders), and many have sucked their partner's toes. For the large majority of men, none of this is the end goal; these preferences serve to enhance sexual activity with their partners, not to compete with them. Thus, activities like buying lingerie or sucking a partner's toes are considered normal variants—they fall within the normal range of behavior. The fetishist, on the other hand, is focused exclusively on the object (e.g., panties) or body part (e.g., toes). The partner's personality or body as a whole cannot sustain arousal; her only role is to serve as a vehicle by which to enact the fantasy. In its most extreme form, the fetishist doesn't need or desire the partner at all. The extreme panty fetishist, for example, is completely gratified by stealing (from a laundromat, perhaps) and masturbating with used pairs of women's panties.

Sexual arousal to specific objects or body parts has been explained in terms of a two-step learning process involving both classical and operant conditioning (Junginger, 1997). Other explanations will be considered later. Consider, for example, heterosexual males' arousal by female panties. Most males' first exposure to female panties comes during early necking or petting experiences, at which time they are highly sexually aroused. A few associations of panties with this emotional state, and panties themselves elicit some degree of arousal. This is an example of classical conditioning. A demonstration of this process was provided several years ago (Rachman, 1966). Under laboratory conditions, males were shown pictures of women's boots paired with pictures of nude women. After many pairings, pictures of the boots alone elicited sexual arousal (measured by penile blood volume). Some clinicians have suggested that sexual arousal to an object or body part could develop into a fetish by a person's repeatedly reinforcing the behavior by including the object in fantasies during masturbation. This is an example of operant conditioning. Many men with fetishes were first attracted to their desired objects at an early age (e.g., Gosselin & Wilson, 1980), perhaps as a result of the object's being associated with some of their first experiences of sexual arousal.

How common is fetishism? We don't really know; fetishists rarely are arrested or seek therapy on their own. However, in a recent survey by Janus and Janus (1993) of people's attitudes about fetishes, 22 percent of the men and 18 percent of the women said that they thought fetishes were either "very normal" or "all right" (as opposed to "unusual," "kinky," or "never heard of it"). Fewer than 2 percent said that this was their preferred way to achieve orgasm.

TRANSVESTISM

> "He was an officer in the army (respected by his fellow officers and men), a devoted husband, and a father. He also wore women's clothing at home, often passing as a female relative, and was known as 'aunt' to his child when he assumed his womanly role." (cited in Hirschfeld, 1948, pp. 174–78)

Transvestism (from the Latin *trans*, meaning "across," and *vestia*, meaning "dress") refers to dressing as a member of the opposite sex *in order to achieve sexual arousal and gratification*. Most researchers regard transvestism as very similar to a clothing fetish (and thus it is often called transvestic fetishism), except that the clothing is worn rather than just held and looked at (Zucker & Blanchard, 1997).

transvestism A condition in which sexual arousal is achieved primarily by dressing as a member of the opposite sex.

Cross-dressing is not considered to be an example of transvestism if it is done for fashion (as in the case of most women who dress in "men's" clothing), if the clothes are worn as a party costume or for an act (as with female impersonators, for example), or as part of transsexualism (in which case individuals cross-dress not for sexual arousal, but because they believe themselves to be members of the opposite sex). It also is not considered transvestism if it is done for the purpose of attracting members of the same sex. (Men who have a sexual interest in cross-dressed men constitute a much less common lesser-known paraphilia called *gynandromorphophilia;* Blanchard & Collins, 1993.)

Two surveys of transvestites, one of subscribers to a magazine for transvestites (Prince & Bentler, 1972) and the other of members of a transvestite club (Buhrich, 1976), found that the large majority were heterosexual men and most were married. Similar results have been found in smaller clinical studies (Talamini, 1982; Wise & Meyer, 1980). In fact, the definition of transvestism in the American Psychiatric Association's DSM-IV (1994) focuses exclusively on heterosexual males. However, a more recent study found that 2.4 percent of the male sample identified themselves as homosexual, 10.6 percent as bisexual, and 19.6 percent said that they had no sexual activity other than cross-dressing (Bullough & Bullough, 1997a).

A few men completely cross-dress as women and walk about in public for short periods of time, but most transvestites cross-dress secretly in the privacy of their own homes (Schott, 1995). Many wear only female undergarments under their male clothing during their cross-dressing. The transvestite may masturbate during a cross-dressing episode or during fantasies of cross-dressing, or may use the experience to become aroused enough to have sex with a partner. Although some wives of transvestites support (and may even be aroused by) their husbands' behaviors, most only tolerate it, and many are distressed and resent it, especially if they discovered it only after they were married (Brown, 1994; Brown & Collier, 1989; Weinberg & Bullough, 1988).

How many transvestites are there in the United States? Enough so that there exists an entire transvestite subculture, with clubs, magazines, and newsletters (transvestites generally identify themselves in underground newspapers by the abbreviation TV) (Allen, 1989). In a recent survey, 2 percent of the men felt that cross-dressing was "very normal" and 6 percent thought it was "all right" (Janus & Janus, 1993). Other researchers estimate the prevalence of transvestism to

Figure 14–2 Henry Hyde, Governor of New York (1702–1709), chose to have his portrait painted while dressed in female clothing.

be about 1 percent of the adult male population (Bullough & Bullough, 1993).

Why do some men dress in women's clothing? Transvestites are a very diverse group. Bullough (1991) says transvestism can be looked at "as an attempt for males to escape the narrow confines of the masculine role." Others hold the view that what transvestites have in common is a "comforting fantasy of the self as a girl or woman" (Levine, 1993). Most transvesties started cross-dressing before puberty (Buhrich, 1976; Schott, 1995), but most children have had experiences of cross-dressing (remember, children do not acquire the concept of gender constancy until about age 7—see Chapter 8—so most young boys will dress up as both Mom and Dad). Most research has found that transvestite males were not effeminate during childhood and did not display cross-

Figure 14–3 Transvestite MTV talk-show host RuPaul (right) with comedian Joan Rivers.

gender behaviors or a tendency to play with girls as children (Zucker & Blanchard, 1997).

Is there anything that distinguishes the early childhood cross-dressing experiences of transvestites from those of others? Some transvestites report that they were punished by being made to dress as girls (Stoller, 1977), while some were encouraged to dress as girls (Bullough, Bullough, & Smith, 1983; Schott, 1995). Some clinicians believe that transvestism arises at an early age through association of opposite-sex clothing with sexual arousal (in the same manner that fetishes are acquired). However, explanations for transvestism (or for why there are many more male transvestites than female ones) range from the psychoanalytic (an attempt to conquer castration fears) and the anthropological-economic (Munroe, 1980), to the neurological (right-brain/left-brain differences; Flor-Henry, 1987) and the sociobiological (Wilson, 1987).

EXHIBITIONISM

"One day at sunset, I was walking back to my car in the . . . parking lot. As I was opening my car door I glanced over at the car facing mine. . . . This man was almost in a backbend over his driver seat masturbating at full speed. For a moment I was stunned and shocked. I just got in my car and drove off."

(from the author's files)

In today's society, exposure of the body is highly glamorized and commercialized—in movies and television, books and magazines, and advertisements. Many people, including many celebrities, have posed naked for magazines, and some even star in sexually explicit films or "topless-bottomless" stage shows. Some of you may have gone to a nudist colony at one time or another. However, none of these are examples of exhibitionism because they involve willing observers. **Exhibitionism** (or "indecent exposure") is defined as exposing one's genitals to unsuspecting strangers in order to obtain sexual arousal and gratification. Some of you may have "mooned" someone on occasion or participated in streaking (at one time a popular college prank). In a survey of college men, 21 percent said that they had exposed themselves in public on at least one occasion (Person et al., 1989). Although these acts involve involuntary observers, they are also generally not considered exhibitionism, for another characteristic of exhibitionists is that they engage in the behavior *compulsively.*

Most victims of exhibitionists, or "flashers," are women or children (Murphy, 1997). Some studies have found that as many as 33 to 44 percent of all women have been the victim of exhibitionism on at least one occasion (Cox, 1988; Gittelson, Eacott, & Melita, 1978). One-fifth of the victims found the experience very distressing (one can only assume that a greater percentage of children would be distressed), and another one-fourth said that their attitude about men or sex had been affected.

Although there have been a few reported cases of females exposing their genitals (Grob, 1985; Hollender, Brown, & Roback, 1977), almost all exhibitionists are men. Most started exposing themselves in their midteens or 20s, and half or more either are or have been married (Murphy, 1997). The large majority of these men are described as shy and/or personally immature, but otherwise without serious emotional or mental disorders. They typically feel inadequate and insecure, fear rejection, and have problems forming close, intimate (including sexual) relations (Blair & Lanyon, 1981; Marshall, 1989; Marshall, Eccles, & Barbaree, 1991).

Lance Rentzel was a star wide receiver for the Dallas Cowboys professional football team and had recently married one of the top young actresses in Hollywood. The Cowboys were headed to the Superbowl in 1970, but on November 19, while in the middle of a slump, Lance exposed himself to a 10-year-old girl. The media soon discovered that he had been arrested for the same thing four years earlier while playing for the Minnesota Vikings—and in the midst of another personal slump. Lance Rentzel had always been a hero on the football field and felt that he had to be highly successful to please his family. In his autobiography and account of his exhibitionism, *When All the Laughter Died in Sorrow* (1972), he says that he felt pressure to prove his masculinity over and over again, and as part of this he avoided close, intimate relations with women and instead proved his masculinity to himself by sexual conquests. During times of stress, when he wasn't a "winner," Lance Rentzel resorted to exposing himself as a way to prove his masculinity.

Although most exhibitionists are of normal or above-normal intelligence, some cases of exhibitionism occur among people who are intellectually impaired, either since birth or as a result of a neurological disorder or age. In these cases, the cause is usually an inability to understand what society considers (or the individual previously considered) to be right as opposed to wrong.

Only about half of all exhibitionists have erections while

exhibitionism Exposing one's genitals compulsively in inappropriate settings in order to obtain sexual gratification.

"NO THANKS, I'M TRYING TO QUIT."

Bob Dayton.

Many of you may have made an obscene telephone call as a prank during adolescence. Although inappropriate, this behavior is different from that of the paraphiliac phone caller who makes obscene calls as his primary source of sexual arousal and gratification. **Telephone scatologia** (or *scatophilia,* from the Greek *skopein,* meaning "to view"), as it is called, is often considered a kind of exhibitionism (unique to the twentieth century, of course; Matek, 1988). As with exhibitionists, the scatologist's arousal is proportionate to the victim's reaction of disgust, shock, and fear and the sense of control it gives him. The scatologist often masturbates while making his calls (McConaghy, 1993). One difference between telephone scatologia and exhibitionism is that many scatologists want complete anonymity (Milner & Dopke, 1997).

In April, 1990, the president of American University resigned after police traced obscene phone calls to his office phone. His victims were women who had advertised home day-care services in the local newspaper. He had degrees from Ivy League schools and was a well-known civic activist who had appeared on television hundreds of times as a spokesman for higher education.

(Newsweek, May 7 and June 11, 1990)

What can explain behavior like this on the part of an intelligent (and in this case, prominent) individual? Telephone scatologists feel inadequate and insecure and have difficulty maintaining normal sexual relations, much like exhibitionists (Matek, 1988; Nadler, 1968). Most have anger toward women. They compulsively use the telephone for sexual interactions—it is safe (until recently, anyway), distant, and avoids intimacy. Of those who have been caught, a rather high proportion were sexually abused as children, as was the case with the university president (Gelman, 1990).

There are over one million obscene phone calls reported to phone companies or the police each year in the United States. Some callers boast about themselves and others just breathe heavily, while still others express their sexual intentions, sometimes in the form of threats. Some use the guise of taking a survey to get the victim to talk about herself. Many obscene callers know the names of the people they call, which can make the experience particularly frightening to the vic-

exposing themselves, and they may or may not masturbate during the act (Langevin et al., 1979). But they usually do masturbate shortly afterwards, while fantasizing about the experience, and their erotic turn-on is directly related to the victim's reaction of shock, disgust, and fear (Blair & Lanyon, 1981). Exhibitionists rarely attempt to molest their victims (American Psychiatric Association, DSM-IV, 1994); thus, the best response (if you are ever a victim) is to give no facial or verbal reaction, leave the scene, and report the incident immediately. Interestingly, many exhibitionists compulsively expose themselves in the same location; perhaps the danger of being caught increases the arousal (Stoller, 1977). As a result, they are the most frequently caught paraphiliac offenders, comprising about one-third of all arrests for sexual offense.

OBSCENE PHONE CALLS

"The calls came three to four times a night. The caller would make obscene suggestions about his penis, oral sex, and bondage, and say that he was 'masturbating to the sound of my voice and the thought of my pussy.' Then he would start moaning. I was very shocked and disgusted. . . ."

(from the author's files)

telephone scatologia A condition in which sexual arousal is achieved primarily by making obscene telephone calls.

tim. It is rare, however, for scatologists to molest (or even approach) their victims. They prefer to keep their distance and masturbate either during or after the call.

If you should ever be a victim, the best advice is to say nothing and hang up *immediately* (remember, the scatologist is aroused by your reactions; even slamming the phone down can give him reinforcement). If the calls persist, your telephone company will cooperate in helping the police trace calls. Ask your phone company about Call Trace, a new technology that allows the phone company to instantly record the source of a call made to you. Today, telephone caller ID machines are making obscene phone calls an increasingly risky behavior. When the caller is identified, do not be surprised if he turns out to be someone you know or have met; that is often the case.

VOYEURISM

> "Last month, I was sitting on my bed and I saw a shadow move across my window. I reluctantly peeked through the blinds and there was this man stroking his penis. . . . Last night I got a phone call from a neighbor telling me someone was looking in my window. I reported it. Apparently, they've been trying to catch this guy for a while. I'm very frightened right now."
>
> (from the author's files)

What is a voyeur (from the French *voir,* meaning "to see")? Are men who like to look at naked women in magazines or at topless shows voyeurs? What about people who go to nudist colonies, or men who enjoy looking at women in string bikinis at the beach? What if a man sneaks a peak up a woman's skirt as she is sitting down, or if he is innocently walking down the street, notices a woman undressing in her room, and stops to watch? In their national survey, Laumann et al. (1994) reported that one-third of the men and 16 percent of the women found "watching others do sexual things" to be appealing. Although the word *voyeurism* is often used to describe such behavior, these are really just more examples of normal variants.

To be a **voyeur** (a paraphiliac), one must *repeatedly* seek sexual arousal by observing people undressing or in the nude *without their consent or knowledge.* The voyeur is aroused by the risk of discovery, and thus is not interested in going to nudist colonies or other places where people know that they are being watched (Sagarin, 1977; Tollison & Adams, 1979). A variation of voyeurism is **scoptophilia,** defined as repeatedly seeking sexual arousal by secretly watching sexual acts. In either case, voyeurism is the preferred means of sex-

Figure 14–4 Voyeurs repeatedly seek sexual arousal by secretly watching people undress.

ual arousal. For example, a man was arrested in Louisiana in 1992 for 341 reported cases of peeping in women's windows.

The great majority of voyeurs are males (thus the term *Peeping Tom* applies). Most voyeurs begin their deviant behavior before age 15. They have characteristics that resemble those of exhibitionists: feelings of inadequacy and insecurity, shyness and poor social skills, and a history of great difficulty in heterosexual relations (Gebhard et al., 1965; Kaplan & Krueger, 1997; Tollison & Adams, 1979).

> "I often 'do it' (masturbate) in my car while parked in areas where I can watch women. My social skills are poor, and I do not get on well with people easily."
>
> (from the author's files)

Voyeurs may masturbate while observing their victims, or they may masturbate later while fantasizing about the incident. The large majority of voyeurs are not dangerous, preferring to keep their distance, but there are exceptions (Gebhard et al., 1965; Kaplan & Krueger, 1997; Langevin, 1985). If you are ever a victim, you should consider a voyeur to be potentially dangerous either if he tries to draw attention to the fact that he is watching or if he attempts to approach you or enters your building.

BESTIALITY AND ZOOPHILIA

> "In my early teens my cousins and I penetrated a horse. I did it because of a dare. I've grown up now and have 3 children and a wife. Every now and

voyeurism Repeatedly seeking sexual arousal by observing nude individuals without their knowledge or consent.

scoptophilia The practice of repeatedly seeking sexual arousal by secretly observing genitals and sexual intercourse.

then I think about what I've done and it makes me want to throw up."

"One night several years ago myself and a few acquaintances were sharing some college experiences. One of the group stated that to become a member of _____ he had to have sexual intercourse with a goat."

(from the author's files)

Bestiality is the act of having sexual contact with an animal. Greek mythology provides many stories of gods disguising themselves as animals and having sex with goddesses or humans. But is this type of sexual contact common for humans? Kinsey and his colleagues (1948, 1953) reported that about 8 percent of the men in his sample and over 3 percent of the women had had sexual contact with an animal on at least one occasion. Seventeen percent of boys raised on a farm admitted to such experiences. A more recent study confirms that a fair number of people have experienced sex with an animal on at least one occasion (Alvarez & Freinhar, 1991). For males, bestiality usually involves a farm animal, while household pets are more common for women. Males are most likely to engage in intercourse with the animal; women usually have the animal lick their genitals, or they masturbate a male animal.

For the large majority of people who have engaged in bestiality, it was only one or a few experiences, often during adolescence when no partner was available (Tollison & Adams, 1979). It might have happened as a dare, as an initiation stunt, or out of curiosity. Bestiality becomes a paraphilia, called **zoophilia** (from the Greek *zoon,* meaning animal), only when it is the preferred means of sexual arousal (Perretti & Rowan, 1983). Zoophilia, however, is very rare and is usually a sign of severe psychological problems. Zoophiliacs are often withdrawn with poor interpersonal skills, and frequently have other paraphilias (Milner & Dopke, 1997).

bestiality The act of having sexual contact with an animal.

zoophilia Using sexual contact with animals as the primary means of achieving sexual gratification.

pedophilia A condition in which sexual arousal is achieved primarily and repeatedly through sexual activity with children who have not reached puberty.

sadism A condition in which individuals repeatedly and intentionally inflict pain on others in order to achieve sexual arousal.

masochism A condition in which individuals obtain sexual pleasure primarily from having pain and/or humiliation inflicted on them.

PEDOPHILIA

He was a man loved by children. He played a clown at birthday parties, and little ones loved to see him coming into their homes. Then he was caught molesting a child at a party where he had been hired to entertain. Interviews were conducted with children at parties he had worked before. They revealed that he had used his job as a means to repeatedly engage in sexual behavior with children. He was a pedophiliac.

(summarized from news articles)

Pedophilia (from the Greek, meaning "love of children") is a condition in which an adult's sexual arousal and gratification depend primarily or exclusively on having sex with children. Although legal definitions of a minor differ from state to state (ranging from under 16 to under 21 years of age), sex researchers generally define a child as someone who is younger than 13 years old (Gebhard et al., 1965). Thus, we are discussing sexual relations by an adult with prepubertal children. To be considered an example of pedophilia, the adult must be at least 16 years old and at least 5 years older than the child (American Psychiatric Association, DSM-IV, 1994).

An isolated act of child molestation, although reprehensible, does not necessarily qualify as pedophilia. Remember, to be designated a pedophile, the molester must *repeatedly* seek sexual relations with children.

This is the most serious of the paraphilias, each year resulting in many thousands of victims who suffer long-term emotional and psychological problems. This section is intended only as an introduction to the problem of pedophilia. To learn about pedophiles and the effects their behavior has on the victims, see Chapter 15, "Sexual Abuse."

SADOMASOCHISM

Sadism refers to the infliction of pain on another person for sexual arousal and gratification. The term is named after the Marquis de Sade (1740–1814), a French aristocrat and novelist who wrote stories, supposedly based on his own experiences, of beating and torturing women while being sexually stimulated himself (his works include the novels *Justine* and *120 Days of Sodom*). **Masochism** refers to achieving sexual arousal by experiencing pain. It is named after Leopold von Sacher-Masoch (1835–1895), a practicing masochist who wrote novels (e.g., *Venus in Furs*) about people getting sexual pleasure by having pain inflicted on them. If the partner of a sadist participates willingly, as is usually the case, then he or she is, of course, a masochist, and the linkage of the two behaviors is indicated in the term **sadomasochism** (S&M). The terms sadism and masochism were coined by Krafft-Ebing, who also was one of the first to report clinical cases:

There was a man who habitually attacked young girls on the street with a knife, wounding them in the upper arm. When arrested, he said that he did it because he would have an ejaculation at the moment he cut them.

(cited in Krafft-Ebing, 1951, p. 223)

A 28-year-old man would have a woman tie him up with straps he brought for that purpose. Then he would have the woman take whips (which he had also brought) and beat him on the soles of his feet, the calves, and backside until he ejaculated.

(cited in Krafft-Ebing, 1951, p. 261)

How common are sadism and masochism? In their mildest forms, they are probably very common. Many people, for example, enjoy being bitten, slapped (e.g., on the buttocks), pinched, scratched, or pinned down during sex.

"I consider myself to be a normal sexual person. . . . I love to be spanked." (female)

"I figured I could ask you if it's 'normal' if you like to be bitten during sex? My arousal does not depend exclusively on biting nor does my boyfriend have to bite me hard, I just love it."

(from the author's files)

Kinsey and his associates (1948, 1953) found that over one-fourth of the men and women in their survey had erotic responses to being bitten during sex. Some people are sexually aroused by inflicting or experiencing pain that goes beyond love bites or spanking:

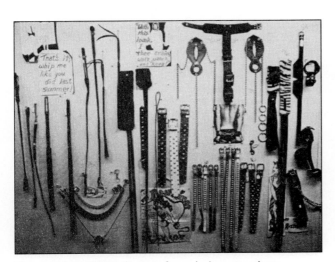

Figure 14–5 S&M devices for sale in a sex shop.

"My best friend always talked about how he wanted to have anal sex with his girlfriend. All he really talked about was how he wanted to hear her let out a cry when he finally got to."

(from the author's files)

Janus and Janus (1993) reported that 16 percent of the men and 12 percent of the women they surveyed agreed or strongly agreed that "pain and pleasure really go together in sex." However, only half as many felt that S&M was very normal or all right, indicating that many people do not regard their own behaviors as S&M. Where then do we draw the line as to what characterizes S&M?

To qualify as a paraphilia, sadomasochism must be the *preferred* means of sexual arousal. Until the 1980s, most of what was known about sadists and masochists was obtained from clinical case studies (such as Krafft-Ebing's) of individuals who had inflicted or experienced tremendous physical pain and harm. In the case of sadists, this often involved unwilling partners as in the case of sadistic murderers (e.g., Ted Bundy, Jeffrey Dahmer, and John Wayne Gacy) or sadistic rapists (Hucker, 1997). Sadomasochism was thus viewed as a psychological abnormality, practiced by people with severe personality disorders. In its extreme form, sadism is less common than masochism (Gebhard et al., 1965), and male sadists far outnumber female sadists (Fiester & Gay, 1991). Sadists who act out their aggressive impulses share many features with other sexual aggressors (Langevin et al., 1988). Some of these features include abuse of alcohol and drugs, poor socialization, frequent history of abuse as children, tendency toward committing nonsexual crimes as well as sexual crimes, and long histories of antisocial behavior. Most experts feel that sadomasochistic activities involving severe pain are relatively uncommon.

Recent studies by sociologists and psychologists in nonclinical settings reveal that the majority of people who participate in sadomasochism differ substantially from individuals who have come to the attention of clinicians. Weinberg and associates (Weinberg, 1987; Weinberg, Williams, & Moser, 1984) found that the distinctive feature in most sadomasochistic relations is not pain, but **domination** (or discipline) and **submission** (D&S). The main features are pain (within well-defined limits), loss of control (e.g., bondage), and humiliation and/or embarrassment (Baumeister & Butler, 1997). Dominators ("master," "mistress," "tops") and submissives ("slaves," "bottoms") act out roles in highly structured scenarios (often involving *bon-*

domination Ruling over and controlling another individual. In sexual relations, it generally involves humiliation of the partner as well.

submission Obeying and yielding to another individual. In sexual relations, it generally involves being humiliated by the partner as well.

dage). This is usually done with a trusted partner (thus allowing them to explore dominance and submission safely), with both agreeing beforehand on the set of activities that follow.

> "My most favorite sexual activity is bondage. This is usually done with a partner with whom I have had sex enough times to develop a feeling of trust. I explain that I will tie her spread-eagle to the bed and do anything to her that I want. After doing so, communication is active and only things that bring pleasure are done. At no time is fear or physical abuse implied. There has never been a bad experience to me or my partner during these adventures."

> "The dominatrix (female dominator) commands me to clean a toilet, then 'rewards' me by allowing me to chew on one of her dirty socks."

> (from the author's files)

The scenarios often include special clothing (e.g., leather garments, high-heeled shoes or boots) and gadgets (e.g., leashes, collars, whips, chains), but pain-inducing behaviors are more symbolic than real. Sadomasochists may go to S&M establishments and pay to experience dominance or submission in structured role-playing.

Dominators and submissives generally are not interested in injury or extreme pain, and pain is erotically arousing *only* when it is part of the agreed-upon ritual (Gosselin & Wilson, 1980). (Male homosexual S&M tends to be more violent than heterosexual S&M; Breslow et al., 1986.) Activities preferred by a majority of both men and women include (in order) spanking, master-slave relationships, oral sex, masturbation, bondage, humiliation, erotic lingerie, restraint, and anal sex (Breslow et al., 1985; Moser & Levitt, 1987). People who participate in dominance and submission have a strong desire to control their environment (Breslow et al., 1986), and the acting out of the highly structured D&S scenarios can become the preferred means of sexual arousal:

> "Growing up I had fantasies about having a partner that would enjoy being on the submissive end of S&M. Well I found one and . . .

Figure 14–6 Participants in S&M and D&S often wear leather garments and boots and use whips, chains, leashes, and collars in highly structured scenarios.

I never knew my tendencies for that could become so everyday. We both fed on that behavior. . . ."

> (from the author's files)

Most nonclinical D&S participants are male (2-to-1 ratio over females), heterosexual, and well educated (Breslow et al., 1985; Moser & Levitt, 1987). Most have a preference for dominant or submissive roles, but many are able to switch between the two. Many female participants do so to please their male partners (Breslow et al., 1985). They, too, tend to be well educated, and while most prefer the submissive role, a sizable minority prefer the dominant role or have no preference (Ernulf & Innala, 1995; Levitt, Moser, & Jamison, 1994). Although most people who engage in D&S are socially well adjusted, close to a third report extreme nervous anxiety (Breslow et al., 1985). Today, some therapists believe that masochism (submission) in particular "does not appear to be itself pathological or a symptom of deeper problems, nor does it generally involve wish for injury, punishment for sexual guilt, or self-destructive impulses (Baumeister & Butler, 1997, p. 237).

Psychoanalytic explanations of sadism often focus on unconscious anger directed at one's mother, while masochism is sometimes explained as sadism directed at oneself. Behavior therapists, on the other hand, contend that S&M results from early associations of pain or aggression with orgasm (e.g., Kernberg, 1991). One study found that 80 percent of male sadomasochists and 40 percent of females attributed their interest in S&M to childhood or adolescent experiences (Breslow et al., 1985), but in another study only 20 percent could recall such experiences (Moser, 1979). Weinberg

(1987) believes that sadomasochism (dominance-submission) can be understood as a social phenomenon. Many men involved in S&M are submissives and many are not only well educated, but successful and prominent as well. The highly structured scripts with a trusted partner allow them to safely play a role that contradicts what our society says is appropriate for males (also see Money, 1984). The dominator role, on the other hand, may be preferred by those who normally have little or no influence or power.

OTHER PARAPHILIAS

The sexual behaviors that have been presented to this point represent only the most common paraphilias. There are less common, more bizarre paraphilias, and I introduce a few of them here. For a description of others, see Money (1984).

Some of these paraphilias might be considered specific types of fetishes (urophilia, coprophilia, mysophilia, klismaphilia), as they involve sexual arousal by specific body discharges (Money, 1986). Behaviors such as urophilia and coprophilia are sometimes included in sadomasochistic acts as well.

"I personally know someone that was a urophiliac. He used to talk to one of my friends. She told me that every time they were together, he used to ask if he could urinate on her belly."

"My boyfriend likes to watch as I urinate. I even let him taste it one time."

(from the author's files)

Urophiliacs (from the Greek *ouron,* meaning urine) are sexually aroused by the act of urination. They enjoy watching others urinate, and many are sexually aroused by urinating on their partner or having their partner urinate on them (often called "golden showers"). Some even drink urine. The noted sexologist Henry Havelock Ellis admitted that he developed a mild form of urophilia during adolescence after witnessing his mother suffer several bouts of urinary incontinence.

Coprophiliacs (from the Greek *kropos,* meaning dung) are sexually aroused by excrement. They may play with, masturbate with, or even eat feces (coprophagia). In some brothels around the world, coprophiliac patrons can pay to watch from under a glass as prostitutes defecate. Although you might think that urophilia and coprophilia are extremely rare, they are common enough that there is a market for videocassettes showing people engaged in these behaviors. Janus and Janus (1993) found that 6 percent of the men in their survey, and 3 percent of the women, be-

lieved that the practice of "golden showers" was very normal or all right. Three percent of the men and 1 percent of the women said that being defecated on was very normal or all right.

Closely related to urophilia and coprophilia is **mysophilia** (from the Greek *mysos,* meaning uncleanliness), sexual arousal caused by filth. For example, there are documented cases of people who are sexually aroused by sweaty or dirty clothing or used menstrual products. Some people are sexually aroused by vomit, a behavior called **vomerophilia** (from the Latin *vomere,* meaning vomit).

"I know a man who has a very high position within a large company. His favorite thing sexually is to be given an enema containing Tabasco sauce. He would hold the fluid in and masturbate. . . ."

(from the author's files)

Another paraphilia is **klismaphilia** (from the Greek *klusma,* meaning enema), obtaining sexual arousal by receiving an enema (Milner & Dopke, 1997). Many klismaphiliacs self-administer the enema, but many others give and/or receive enemas with a partner. Many use warm water, while others use alcoholic solutions or coffee. Unlike most other paraphiliac behaviors, many klismaphiliacs are women.

"A year ago I went with a small group of friends from my high school to London. Well, we were on a very crowded tube [subway] and were packed in so tight that we could barely move. But the guy behind me was able to move. He kept rubbing up and down against me the whole ride. I just thought that someone kept pushing him against me. My friends told me that he looked like he was 'getting off' by doing that. I never wore those pants again because I felt used and dirty."

(from the author's files)

Frotteurism (French, meaning rubbing) involves rubbing one's genitals against other people in public while fully clothed. In a study of college men, one-third admitted to having rubbed up against a woman in a crowded place (Templeman & Stinnett, 1991). Of course, most of these men are not frotteurists because this is not their preferred means of arousal. One study of nonincarcerated para-

urophilia Sexual arousal caused by urine or the act of urination.

coprophilia Sexual arousal caused by feces or the act of defecation.

mysophilia A condition in which sexual arousal is caused primarily by filth or filthy surroundings.

klismaphilia A condition in which sexual arousal is achieved primarily by being given an enema.

frotteurism A condition in which sexual arousal is achieved by rubbing one's genitals against others in public places.

philiacs found that the average number of acts of frottage was over 800 (Abel et al., 1987). Frotteurists seek out crowded buses or other places where "bumping" into strangers is less likely to cause them to be arrested. Again, it isn't unusual to get aroused by close physical contact with someone (as when you are dancing and holding your partner close to you). However, most people hope that they will get together after the dance for more intimate forms of contact. A frotteurist would not be interested in this, and prefers to escape before he is confronted.

Autoerotic asphyxiation (*asphyxiophilia* or *hypoxyphilia*) was introduced in Chapter 11. In this type of paraphilia, which some regard as a type of masochism (DSM-IV, 1994), individuals deprive themselves of oxygen when highly sexually aroused with the hope of intensifying their orgasm. This is often done with a rope, belt, or plastic bag. This behavior results in an estimated 500 to 1,000 deaths every year (Byard, Hucker, & Hazelwood, 1990; Saunders, 1989). Fatal autoerotic asphyxiation almost always occurs with young males masturbating alone (Milner & Dopke, 1997).

Perhaps the most bizarre paraphilia is **necrophilia** (from the Greek *necros,* meaning dead)—obtaining sexual arousal and gratification by having sex with dead bodies. Almost all known cases involve men, although a few cases of female necrophiliacs have been reported (Rosman & Resnick, 1989). Janus and Janus (1993) found that only 1 percent of men, and no women, thought that this behavior was all right. Most experts consider these people to be severely emotionally disturbed or psychotic. Necrophiliacs often seek employment in mortuaries or morgues in order to have access to their "partners." Some necrophiliacs murder in order to obtain corpses (and often mutilate the corpse, as in the case of Jeffrey Dahmer), and thus can be regarded to be both sadists and necrophiliacs (Milner & Dopke, 1997).

MULTIPLE PARAPHILIAS

Many sexual offenders do not wish to describe their "private" lives. However, a study was conducted with outpatient sexual offenders in which the participants were guaranteed immunity from prosecution for any crimes they described in the research interviews (Abel et al., 1987, 1988a). In this study, 62 percent admitted to having committed a previously concealed deviant sexual behavior. Over half reported that they had engaged in more than one type of the previously described behaviors, with the average being three to four types. Most had developed their sexual interests and fantasies by

autoerotic asphyxiation Depriving oneself of oxygen during masturbation-induced orgasm.

necrophilia Sexual arousal by a dead body.

age 12 or 13, but were afraid to discuss their arousal patterns with their parents.

Other studies have also found that individuals in treatment for one type of paraphilia often engage in other paraphilias (Bradford, Boulet, & Pawlak, 1992; Flor-Henry, 1987; Langevin, 1985). In fact, some types of paraphilias tend to cluster together more often than others. Very serious sadistic offenders, for example, tend to engage in transvestism and fetishism as well (Hucker, 1997). Another group of paraphilias often seen together are known as **courtship disorders**—voyeurism, exhibitionism, and frotteurism (Freund, Seto, & Kuban, 1997). These studies indicate that the incidence of multiple paraphilias in individuals may be greater than previously believed, and explanations of the paraphilias must take this into account.

WHAT CAUSES PARAPHILIAS?

What causes people to become paraphiliacs? *Freudian* (psychoanalytic) *theorists* believe that paraphilias are the result of arrested psychosexual development, often due to a failure to resolve emotional conflicts. In this theory, paraphilias serve as defense mechanisms to reduce anxiety. A recent explanation of transvestism, for example, is that the behavior "functions to comfort, soothe, and alleviate unpleasant emotions" (Levine, 1993). Although Freudian theory has the advantage of being very flexible, the effectiveness of psychoanalysis as a therapy for paraphiliacs is very debatable, and is seldom used.

Learning theorists (behaviorists) attribute paraphilias to (usually early) learned associations. Recall, for example, the study by Rachman (1966) regarding acquisition of a "fetish" to boots, or the fact that some sadists (dominators) had early associations of aggression with orgasm (Kernberg, 1991). Some individual case studies also lend credence to learning theory explanations. A man's fetish for plaster casts, for example, was traced to an experience he had at age 11 after he broke his leg (Tollison & Adams, 1979). He got an erection as the nurse held his leg, and later masturbated while fantasizing about it. His repeated fantasy/masturbation episodes led to the fetish.

Although both of these approaches have their advocates, an interesting question still remains: Why are most paraphiliacs males? With the exception of masochism, and perhaps klismaphilia, female paraphiliacs are relatively rare, even among female child molesters, which also is uncommon (Hunter & Mathews, 1997). Part of the answer can be explained by the social constructionist perspective you read about at the beginning of the chapter. What is considered to be abnormal, you recall, depends on the particular time and place. Look, for example, at transvestism and exhibitionism. If women dress in men's clothing, few people today think anything about it—the women are simply being stylish. Similarly, if a woman exposes herself, she is likely to be viewed as "sexy." Thus, in our culture, only men who engage in these behaviors are regarded as behaving inappropriately

Nevertheless, of the nearly 88,000 arrests for sexual offenses other than forcible rape and prostitution in 1993, over 90 percent were men. Most paraphiliacs describe themselves as heterosexual, but they generally have poor social skills, low self-esteem, histories of childhood abuse or neglect (or were raised in families where sex was thought of as evil and normal erotic development was inhibited), and anger at women (see previous sections for specific references). As a result, paraphiliacs have difficulties with intimacy, especially when they attempt to have sexual relationships. These characteristics are very common in men with the courtship disorders (voyeurism, exhibitionism, and frotteurism) and in some kinds of rapists (Freund, Seto, & Kuban, 1997).

Conventional sexual relationships are too complex and threatening for most paraphiliacs, who need to have a great deal of control in order to become sexually aroused. Think for a minute—what do a pair of panties, a dead body, an animal, a person being secretly watched, and a young child—all of them objects of paraphiliacs' fantasies—have in common? These objects are quiet, passive, and *nonrejecting*. Consider, for example the following case history of necrophilia:

> "Two males worked with me at _____. There we worked in the morgue and they used to always want me to watch one morgue while they watched another. Well, one day I went to their morgue after hearing about a young Puerto Rican female body which had been delivered after she died of a drug overdose. There was one guy on top of her and the other was waiting his turn. After talking with them, I found out that they had done this many times before. They tried to get me to do it but I refused, saying they were crazy. They said, 'No, you're crazy because this is the best. We have them any way we want, how we want and do whatever we want to them without any back talk!'"
>
> (from the author's files)

In a *social systems perspective,* the need for some men to dominate women might encourage or sustain attempts to utilize unusual means of becoming sexually aroused in order to ensure against failure. Our society's often subtle but powerful expectation that males should be masculine and dominant may make it difficult for some men to form caring, normal relationships.

Did you notice a common element in all three explanations? In all three, childhood or early adolescent experiences play an important role. The manner in which one is raised is very important for healthy sexual development. John Money (1986) explains the impact of childhood experiences in terms of what he calls "lovemaps." "Lovemaps," according to Money, depict an idealized lover, a love scene, and erotic activities, and develop and are rehearsed during adolescence. For children whose rehearsals have been severely punished or restricted, the "lovemaps" become distorted, with one outcome being the paraphilias.

Today, many researchers are recognizing the similarity between paraphilias and other obsessive-compulsive disorders. Recall from Chapter 13 the cycle of events that drives the individual who is hypersexual (or the sex addict, as some prefer to call these people). The individual is first preoccupied with obsessive thoughts, which lead to ritualistic behaviors causing further sexual excitement, which eventually lead to the sexual act. The act, however, produces despair; and in order to reduce the anxiety, the individual becomes preoccupied again (Carnes, 1983, 1991). Hypersexuality is not recognized as a paraphilia in the American Psychiatric Association's diagnostic manual (DSM-IV, 1994), but the fact that a great many paraphiliacs engage in their behaviors compulsively cannot be denied. Males with paraphilias engage in sexual activities much more often than most people (Kafka, 1997). As sexologist Eli Coleman explains it, the paraphiliac's behavior "may initially be driven by the sexual excitement. . . . But the primary motivation is the reduction of stress and anxiety. There is usually a short-lived feeling of relief, followed by the recurrence of the anxiety" (quoted by Thompson, 1991; see also Coleman, 1991). Although the idea that there are similarities between obsessive-compulsive disorder and paraphilias has merit, preliminary evidence from therapeutic drug studies suggests that the two conditions may not be entirely identical (Kruesi et al., 1992).

Most recently, some researchers have suggested that males with paraphilas have abnormal brain chemistry (e.g., Kafka, 1997). They base this on the observation that antidepressants, psychostimulants, and neuroleptic drugs often have sexual side effects and that many drugs used for psychiatric disorders also alleviate paraphiliac behavior. As you will now see, pharmacological interventions are playing an increasingly greater role in the treatment of paraphilias.

THERAPY

No one type of therapy has proven successful in treating paraphiliacs. However, for those paraphiliacs who are considered to be hypersexual (many) or those who engage in behaviors that cause harm to others (e.g., pedophilia, sadism), *medical approaches* to reduce the sex drive are now commonly being used. These include antiandrogen drugs (e.g., cyproterone acetate or Depo-

Provera) that reduce testosterone levels (Bradford & Greenberg, 1997; Gijs & Gooren, 1996). However, these drugs are not cures; they are used to curb the urge to engage in the behaviors while other therapies can be tried. Drugs used for obsessive-compulsive disorders or generalized anxiety disorders (e.g., Prozac, clomipramine, buspirone) have also proved to be promising in treatment of specific paraphilias such as transvestism, exhibitionism, and fetishism (e.g., Gijs & Gooren, 1996; Kafka, 1991).

Therapy often takes a multifaceted approach. Traditional *psychotherapy* or *group therapy* may be tried to get the individual to become aware of his feelings, while confrontational approaches are sometimes used to force the individual to see the effects his behavior has on others. Behavioral approaches often include *aversion therapy,* in which unpleasant stimuli are presented or associated with the undesired behavior, and *orgasmic reconditioning,* where pleasant stimuli (fantasies) are associated with orgasm.

Therapy is not completed once the paraphiliac's inappropriate urges have been controlled. Remember, paraphiliacs usually have anxieties about intimate relations and poor social skills. Unless these problems can be resolved, the inappropriate behaviors will eventually return. For the anxieties, the most successful approach is *desensitization.* Here, the anxiety-causing stimuli (relations with women, in this case) are slowly introduced to the subject. Fantasies of normal relations with women are attempted first. Once this can be done in a relaxed state, thoughts of normal relations are gradually replaced with actual behaviors (for example, being alone in a room with a woman, sitting next to a woman, and talking to a woman may be introduced in succession). *Social skills training* teaches these men skills, such as how to carry on a conversation and how to cope with rejection, that most of us (but not the paraphiliac) learned during adolescence. Nowadays paraphiliacs are increasingly being treated for hypersexual (compulsive) disorder. Rather than repeat what has already been covered, I refer you back to the section on hypersexuality in Chapter 13.

No type of therapy for *any* problem (not just paraphilias) has a high success rate unless the individual wants to change. This is one reason that therapy ordered by criminal courts are often unsuccessful: the individual does not come to therapy on his own. For many paraphiliacs, the urge to engage in the behavior is so strong that it never seriously occurs to them to give it up (Money, 1986).

The *social systems perspective* advocates a different approach to dealing with paraphilias. Rather than trying to change individuals, those who believe in this perspective advocate changing society's stereotypes about male dominance over females. A prediction of the social systems perspective is that an early socialization process that did not place heavy expectations on males to control or dominate sexual relationships would lead to a sharp decline in the amount of paraphilia in our society. What do you think?

Key Terms

autoerotic asphyxiation 352	klismaphilia 351	scoptophilia 347
bestiality 348	masochism 348	sexual variant 342
coprophilia 351	mysophilia 351	submission 349
courtship disorders 352	necrophilia 352	telephone scatologia 346
domination 349	paraphilia 342	transvestism 343
exhibitionism 345	pedophilia 348	urophilia 351
fetishism 342	sadism 348	voyeurism 347
frotteurism 351	sadomasochism 348	zoophilia 348

Personal Reflections

1. Do you engage in sexual practices that may be physically or psychologically harmful to yourself or others (e.g., exhibitionism, obscene telephone calls, voyeurism, sadomasochism, pedophilia)? If so, have you sought therapy to help with your behavior? If you have not sought therapy, why not?

2. If you are ever victimized by a person with a paraphilia, what do you think you would do? Would you talk about it? To whom? Would you report it? To whom?

3. Do you believe that it is okay for consenting adults to engage in any sexual behavior in private as long as it does not cause physical and/or emotional injury and does not interfere with normal social and occupational activities? Why or why not?

Suggested Readings

Brown, G. R. (1990, June). The transvestite husband. *Medical Aspects of Human Sexuality.*

Bullough, V., & Bullough, B. (1993). *Cross-dressing, sex and gender.* Philadelphia: University of Pennsylvania Press.

Garber, M. (1992). *Vested interests: Cross-dressing and cultural anxiety.* New York: Routledge. Interesting and provocative.

Money, J., & Lamacz, M. (1989). *Vandalized lovemaps.* New York: Prometheus Books. Excellent case histories.

Stoller, R. J. (1977). Sexual deviations. In F. A. Beach (Ed.), *Human sexuality in four perspectives.* Baltimore: John Hopkins University Press. An easy-to-read review of the paraphilias.

Weinberg, T. S. (1995). *S&M: Studies in dominance and submission.* Amherst, NY: Prometheus.

Wilson, G. (Ed.) (1986). *Variant sexuality: Research and theory.* Baltimore: Johns Hopkins University Press. Over 10 years old, but an excellent first source of information.

CHAPTER
15

When you have finished studying this chapter, you should be able to:

1. Discuss, from a historical perspective, attitudes about rape and victims of rape;

2. Understand the prevalence of rape and describe some characteristics of rapists and their victims;

3. Describe acquaintance rape, date rape, sexual coercion, marital rape, gang rape, statutory rape, and the frequency of each;

4. Discuss why rape is common in some cultures and rare in others;

5. Compare and understand the psychodynamic, cultural, social learning, and interpersonal factors that contribute to rape;

6. Identify several myths about rape and appreciate why they are not true;

7. Describe rape trauma syndrome and other reactions to rape;

8. Discuss and evaluate three approaches to rape prevention;

9. Describe sexual harassment in the workplace and the classroom and ways to deal with it;

10. Describe the different types of child molesters and the situations in which sexual abuse of children is most likely to occur;

11. Discuss the various forms of incest and describe the family dynamics that are typically found in cases of father-child incest; and

12. Understand the devastating long-term effects that can result from sexual abuse of children.

Sexual Victimization: Rape, Coercion, Harassment, and Abuse of Children

RAPE

HISTORICAL PERSPECTIVE

In the United States, 16 rapes are attempted every hour; on average, a woman is raped every 6 minutes (U.S. Senate Judiciary Committee hearings, 1990). The United States has a rape rate three times greater than that of Sweden or Denmark, five times greater than that of Canada, seven times greater than that of France, 13 times higher than that of England, and 20 times higher than those of Japan or Israel (U.S. Senate Judiciary Committee, 1990; United Nations/Economic Commis-

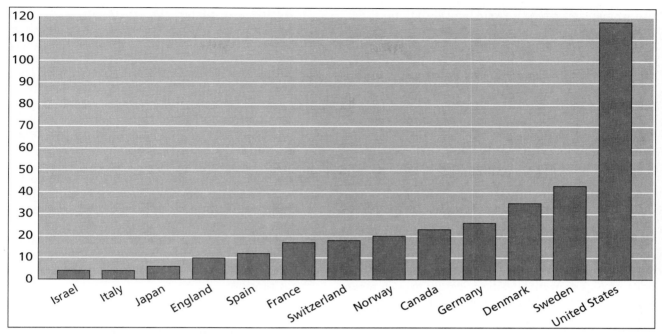

Figure 15–1   Rape rates per 100,000 women aged 15 to 59.

sion, 1995; see Figure 15–1). Compared to men in other cultures, men in the United States obviously have difficulty in their relations with women. Why?

> "The difficult thing to explain is not why gender relations are so violent, but why violence is so gendered, so sexualized."
>
> (Moore, 1994, p. 154)

We will examine rape in some detail and then return to this question.

Rape is "nonconsensual oral, anal, or vaginal penetration, obtained by force, by threat of bodily harm or when the victim is incapable of giving consent" (Koss, 1993b). The word *rape* comes from the Latin word *rapere*, meaning to steal or carry off. A story about the founding of Rome describes how the early Romans abducted women from a neighboring tribe in order to get wives (the "Rape of the Sabine Women"). Many cultures have ritualized the abduction of the bride-to-be by the groom as part of the marriage ceremony. In our own culture, for example, the practice of carrying a new bride across the threshold is a ceremonial re-creation of the Roman legend.

In Western culture long ago (and still true in many parts of the world today), women were treated as if they were property belonging to their father or husband, and fathers and husbands could do as they wished with their property. For example, the Old Testament (Genesis 19:1–8) tells of Lot, who lived in the town of Sodom. Two angels disguised as men came to Sodom, and Lot invited them into his house. Later, the men of Sodom surrounded Lot's house and demanded that he send the angels outside. Lot instead offered to give the crowd his two virgin daughters to do with as they pleased if only the men would leave his guests alone. In Biblical times, rape was viewed as a crime against a man's property rather than against a person (the woman).

The Old Testament (Deuteronomy 22:23–28) also describes different punishments for rape. If a married woman was raped, it was assumed that she had wanted it (and thus violated her husband), and both she and the rapist were put to death. If a betrothed virgin was raped inside the city walls, both the rapist and victim were then stoned to death, because it was assumed that the young woman could have cried out for help and been heard if she had wanted to resist. If a young woman was raped outside the city walls, the punishment was based on whether or not the woman was engaged. If she was, only the rapist was killed because it was assumed that the woman's cries for help could not have been heard. If she was not engaged, the rapist had to pay a fine to the young woman's father (it was his "property" that was damaged) and marry the woman (for who would want to marry "damaged goods"?).

It was not until the thirteenth century that English common law dropped its distinction between raping a mar-

rape Nonconsensual oral, anal, or vaginal penetration, obtained by force, by threat of bodily harm, or when the victim is incapable of giving consent (Koss, 1993b).

ried woman and raping a virgin, and also dropped the practice of forcing the rapist and victim to marry (Brownmiller, 1975). The idea that a wife was property of her husband and that a husband could therefore do as he pleased with her continued for many centuries. You will learn of recent changes in the law regarding this in the section on marital rape.

Throughout history, men of victorious armies have raped the women of defeated nations. When the Japanese army invaded China in 1937, for example, the soldiers raped virtually every woman and girl in the wartime capital of Nanking. They later rounded up tens of thousands of Korean women to serve as sex slaves, with most of the women forced to have sex with dozens of soldiers every day. In the 1990s in Bosnia, Serbian troops raped tens of thousands of Muslim women as part of their "ethnic cleansing" war. Wartime rapes generally are vicious gang rapes, and reports of American soldiers participating in such rapes appear in court-martial records during the Vietnam war.

Why do soldiers (men) engage in such atrocious behavior? It is done to demonstrate power—to destroy the national pride, the manhood, and the honor of the vanquished country. In the words of Susan Brownmiller (1993), the soldier rapes "to give vent to his submerged rage against all women *who belong to other men*. . . . In one act of aggression, the collective spirit of women and of the nation is broken, leaving a reminder long after the troops depart. And if she survives the assault [she becomes a] symbol of her nation's defeat. A pariah. Damaged property."

RAPE STATISTICS

FBI records show that there were 97,769 reported cases of rape in the United States in 1996. Although this is a large number, everyone agrees that the actual number of rapes committed each year is much higher than the number reported. The U.S. Department of Justice (1997) estimates that there are nearly 170,000 rapes and 150,000 attempted rapes committed each year in the United States, while the National Victims Center estimates that there are as many as 683,000 rapes of adult women every year. In the national survey by Laumann et al. (1994), 22 percent of women said that they had been forced to have sex on at least one occasion. Koss (1993b) estimates that a woman living in the United States has a one-in-five chance of being raped in her lifetime.

Women between 16 and 24 years of age are the most frequently reported victims of rape (U.S. Department of Justice, 1997). However, the records show that anyone can be a victim of rape, including young children, elderly women, and men. For example, my grandmother was raped by a teenager when she was 85 years old.

By the age of 80, Hazel's failing health no longer allowed her to make the trip next door to her daughter's house to prepare the family dinner. She was hunched over due to osteoporosis, and her hands were so arthritic that she had to give up playing the organ at church. Her face was heavily wrinkled, and her hair was gray and thin.

When she was 85, a man in his late teens or early twenties broke into her room one night. He forced her to display herself, and he made her lean over his knees while he spanked her. This went on for two or three hours. He beat her up, permanently damaged one of her eyes, and raped her.

There was no known motive. The rapist was a stranger and was never caught. The police said that she had apparently been picked at random.

At least half of all rape victims are girls under the age of 19 (U.S. Department of Justice, 1997). However, these are reported rapes. A study by the National Victims Center, which included both reported and unreported rapes, found that nearly 62 percent of all victims were under the age of 18; 29.3 percent were under the age of 11. Karen Hanna of the National Victims Center says that young people are frequently the target of rape because "the youngest are the least likely to fight back and often don't realize they are victimized" (quoted in Sniffen, 1994).

Even infants can be the victims of rape:

Eight-month-old "Susan" was placed in the care of her mother's boyfriend, "Joe," for a few hours. Upon returning home, her mother found Susan hemorrhaging badly. The baby was rushed to the emergency room of a local hospital, where evidence of vaginal penetration was found. Susan had to have an immediate operation, in which all of her reproductive organs except the vagina were removed in order to stop the bleeding. There were many vaginal lacerations, but the surgeons were able to stitch them up. This was an eight-month-old girl who would not be able to have children in the future and who would have to undergo repetitive reconstructive surgery. She was a victim of rape.

Police found that Joe had abused Susan before, but without vaginal penetration. Susan's mother had refused to contact police because Joe was her boyfriend. After the near death of Susan, her mother reported Joe to the police. He was then arrested for aggravated rape. Because of her failure to report

Joe after his first abuse of Susan, the baby was taken from her mother and placed in a child protection center. Anyone can be a victim of rape.

(summarized from news releases)

The younger the rape victim, the more likely it is that the attacker is an acquaintance or relative (U.S. Department of Justice, 1997). Of rapes of girls under 12 years old, 96 percent were committed by a family member or acquaintance; the figure is 67 percent for women 18 or older. Women are much more likely to report a rape if they are raped by a stranger; thus, most men convicted of rape were strangers to their victims. If, prior to taking this course, you had been asked to describe a "typical" rape, chances are you, too, would have described a scenario in which a victim was raped by a stranger. However, *most rapes are committed by someone the victim knows* (U.S. Department of Justice, 1997; National Victims Center). The rapist can be a friend or acquaintance, but he can also be a husband or boyfriend (or an ex-husband or ex-boyfriend), or a relative (including father or stepfather).

Some rapes are spontaneous, but rapists generally plan their assaults to some degree (see Ward, Hudson, & McCormack, 1997). Even in rape cases committed by a stranger, the rapist often picks a particular place and/or time, with the victim simply being the first person to come along. Thus, being a victim of rape can simply be a matter of being in the wrong place at the wrong time. Rapists, however, often look for vulnerable women, such as those living alone or those who live in a place that can be easily entered (Selkin, 1975). One way that rapists may test for vulnerability is by asking a woman for help (e.g., asking the time, asking for directions). Selkin (1975) found that in 25 percent of the rape cases committed by a stranger, the woman had responded to a request for help. The first interactions between rapists and victims occur indoors as often as outdoors (Amir, 1971), and about a third take place in the victim's home.

SOME CHARACTERISTICS OF RAPISTS

Most rapists are young men who repeat the crime. About 23 percent are reconvicted, but the true repeat offense rate is much higher than this (Polaschek, Ward, & Hudson, 1997; Quinsey et al., 1995). The large majority of rapists have no more history of psychiatric illness than men convicted of other crimes, nor are their IQs any lower (see Polaschek et al., 1997).

There is no compelling evidence to show that rapists, in general, are either oversexed or deprived. Some studies have found higher-than-normal levels of testosterone in sex offenders, but this appears to be true only in the most aggressive of offenders (Rada et al., 1983). But, of course, not all men with high testosterone levels are rapists. Some studies have found that many rapists have sexual partners (wives or girlfriends) that they live with (e.g., Groth, 1979). One study, for example, found that one-third of their sample of rapists were married and having sexual intercourse regularly with their wives (Groth & Burgess, 1977). While convicted rapists who were extremely aggressive have often been found to have had earlier and more frequent childhood sexual experiences (Koss & Dinero, 1989) and to exhibit multiple paraphilias (Marshall, 1996), it is also not unusual for rapists to have had fewer sexual experiences than normal and to have puritanical attitudes about sex (Marshall, 1989). Sexually coercive (nonconvicted sexually aggressive) men have generally had more extensive sexual histories than others (Lalumiere et al., 1996).

Are rapists highly aroused by rape-related stimuli? Early studies suggested this might be so, finding that rapists became sexually aroused while listening to audiotapes of rape while nonrape sex offenders did not (Abel et al., 1977; Barbaree et al., 1979). Another study found that rapists were less aroused by listening to tapes of consenting sex than were nonrapists (Quinsey et al., 1984). Later studies reported mixed results (see Polaschek et al., 1997). In two reviews , it was concluded that these findings held true mainly for the subgroup of rapists that use a great amount of violence in their attacks (Lalumiere & Quinsey, 1994; Marshall, 1992). However, a more recent study found that sexually coercive college men were more aroused than noncoercive men in response to scenes involving verbal pressure and verbal threats (Lohr, Adams, & Davis, 1997).

What about the developmental history of rapists? Studies show that two developmental factors differentiate rapists from other adult males: caregiver inconsistency (e.g., a distant or no relationship with parents) and sexual deviation and abuse in the family (Prentky et al., 1989). An unusually high number of sexual offenders were themselves physically and sexually abused as children (Dhawan & Marshall, 1996). Even if they were not abused themselves, the family environment in which they were raised often included a great deal of verbal or physical abuse by the father, directed at the mother. With this type of family history, a child may well come to view aggression toward women as normal. Psychologist Eugene Porter was quoted as saying, "I haven't seen a rapist who didn't have a childhood horror story" (quoted by Gelman et al., 1990). As a result of this history, it is not surprising that rapists often have low self-esteem and difficulty forming intimate relationships, and even those with partners describe those relationships as lacking intimacy (Marshall, 1989; Marshall, Anderson, & Champagne, 1997; Ward, McCormack, & Hudson, 1997).

Rapists may not have internal controls such as fear, guilt, or sympathy, or they may have learned to suppress them, in contrast to nonrapists. For example, an in-depth study of rapists found that some had fantasized about rape and violence toward women and had committed other sex crimes long before actually committing rape (Abel, 1981). This suggests that the violent behavior escalated over time as their internal controls broke down. The majority of arrested rapists had been drinking prior to the assault (Seto & Barbaree, 1995), thus further releasing any inhibitions or internal controls. Men who force sex on women tend to have adversarial relationships with women in general. According to Dr. Neil Malamuth, "These men feel they have to be in control of their relationships with women, even in conversation" (quoted by Goleman, 1992).

Most rapists have distorted cognitive processes—distorted attitudes and beliefs about sex roles and female behavior, belief in rape myths (see later section), acceptance of violence towards women, and denial of responsibility (for reviews, see Polaschek et al., 1997; Ward, Hudson, & McCormack, 1997). One thing that is clear is that most rapists lack empathy—the ability to see things from the victim's perspective (Marshall, 1996). Most are convinced that their victims wanted to have sex with them, or deserved to be raped. As an example, psychiatrist Gene Abel, one of the foremost experts on rapists, related the story of a rapist (a patient) with a long arrest record for rape who said that he had never raped anyone. When asked how he knew when a woman wanted to have sex with him, Abel's patient responded that he knew a woman wanted sex if she spoke to him (cited in Gelman et al., 1990).

This section is intended only as a general introduction to the subject of rape. Keep in mind that much of what we know comes from studies of convicted rapists, most of whom were strangers to their victims. You will learn more about what makes some men rape in later sections. First let us look at some specific types of rape.

DATE RAPE AND SEXUAL COERCION

One way to classify rapes is by the relationship that exists between the rapist and victim. I will use the term **stranger rape** to refer to a rape committed by someone the victim does not know. But remember that most rapes are committed by someone the victim knows, and thus are often called **acquaintance rape.** Acquaintance rape is called **date rape** if the rape occurs during a social encounter agreed to by the victim. The expression "date rape," though purely descriptive in intent, has turned out

stranger rape Rape committed by someone who is not known by the victim.

acquaintance rape A rape committed by someone who is known to the victim.

date rape A type of acquaintance rape, committed during a social encounter agreed to by the victim.

to be unfortunate because many people believe that it is somehow different from rape by a stranger. Many people attribute more blame, and are less sympathetic, to victims of date rape than they are to victims of rape by a stranger (Barnett et al., 1992; Bridges & McGrail, 1989; Coller & Resick, 1987; L'Armand & Pepitone, 1982). This holds true for police, medical, and social service personnel as well (Holmstrom & Burgess, 1978). *Rape by a date or an acquaintance is no less real than rape by a stranger,* and we should simply refer to it as rape. Here are three examples:

"How many times have we heard our friends say, 'Go out with him, he is such a nice guy'? Well, I'm writing this letter to tell about a terrible occurrence that happened to me by a 'nice guy.'

"I met this guy in one of my college classes. We had four very fun dates. We would go to movies, dancing, parties, and he would always act like a perfect gentleman. He even told me that he did not date many girls at the university because they were only interested in sex. We had a great relationship, and I began to care for him. But my feelings soon changed.

"On our fifth date, he had a little too much to drink. Our friends dropped us at his apartment because he did not want to drive. I knew I was there for the night. I was not worried because he had always treated me with such respect. But that was soon to change. He became very aggressive with me. He told me he wanted to have sex with me because he cared for me so much. I begged him to leave me alone. I tried to fight him off, but I was not strong enough. That turned out to be the worst night of my life, because I was raped.

"I did not think I was leading him on by being in his apartment. I was in shock that such a 'nice guy' could do such a horrible thing to me."

"Robby (a grad student teacher at this university) lived around the corner while I was growing up. I thought of him like a brother, so when I was 17 and he asked me if I was still a virgin, I was not embarrassed to tell him that I was. About a month later a bunch of us went out and after too much to drink, I passed out. I thought I was safe since I was with friends, but he made them leave and I woke up with his fat 270 pound body on top of me. It was horrible and I passed out again. There went 5½ years of friendship and trust. . . ."

Figure 15–2 Rubens's *The Rape of the Daughters of Leucippus* (1617), though a great work of art, has been criticized for glorifying rape.

"When I was 15 I was dating this guy. One night we went to a movie. We got into an argument in the movie theater and he dragged me out of the theater and into the stairwell. We argued some more and he slapped me. When I tried to leave he hit me again. He decided he was going to 'make it up to me' by having sex with me. I repeatedly told him no! I tried to scream and he hit me again. He told me repeatedly that this was what I wanted. I screamed and he just hit me again and again. . . ."

(from the author's files)

How commonly are women forced into unwanted sexual activity on a date (or during a mutually agreed-upon social interaction)? Koss and Oros (1982) surveyed a group of university women and asked if they had ever been raped. Six percent answered "yes" to this question. They were then asked if any of them had had sex with men when they didn't want to, but were pressured into it. Twenty-one percent answered "yes" to this question. In a later study, Koss, Gidycz, and Wisniewski (1987) found that 25 percent of a national sample of college women had been pressured into sexual intercourse when they didn't want to have it. Others have reported similar results (e.g., Laumann et al., 1994; Muehlenhard & Linton, 1987; Russell, 1984).

Only about 5 percent of such rapes are ever reported to the police, and nearly half of the victims have never told anyone (Koss et al., 1988; Muehlenhard & Linton, 1987). Victims of rape often blame themselves (see later section), and this is particularly true when the rape was committed by a date or a close acquaintance (Katz & Burt, 1986). Women who accept extremely stereotypic gender roles often continue in relationships with their violators (Murnen, Perot, & Byrne, 1989).

"When I was 16 I was raped by four guys at a college frat party. In a way I still blame myself for being put in that position. . . . It was my fault for getting drunk. Especially at the age of 16. Guys take advantage of young 16-year-old girls."

"Exactly six months ago today, I was 'date raped.' Though before taking your course, I blamed myself, now I see it was him and not me who was in the wrong. I actually felt responsible and as if I had provoked it."

(from the author's files)

Laumann and colleagues (1994) found that only 3 percent of the men in their national survey felt that they had ever forced a woman into having sex. However, Koss et al. (1987) found that 4.4 percent of the men in their large survey admitted to acts that are legally defined as rape, and another 3.3 percent admitted having attempted rape. One-fourth admitted some form of sexual aggression. Another study found that 15 percent of a sample of college men admitted that they had forced intercourse on a date (Rapaport & Burkhart, 1984). Men often try to rationalize date rape and may claim that "she got me aroused and I couldn't stop," or say "She came to my apartment, so she must have wanted it" (see Cowan & Campbell, 1995). Just because a woman goes out with a man, goes to his home or apartment, or engages in necking or petting does *not* mean that she wants to (or has to) engage in sexual intercourse.

Most people have no difficulty defining sex without consent of the partner as rape if the act involved physical force, violence, or the use of a weapon. But what about physical coercion? What about getting a partner heavily intoxicated (to the point of offering little or no resistance, as in the "Robby" case history) in order to have sex? In the mid-1990s, there was a big increase in the use of Rohypnol ("roofies") and other "date rape drugs" (odorless and tasteless but powerful tranquilizers) that were slipped into unknowing victims' drinks. (Federal law now allows up to 20 years to be added to the sentences of men who use these drugs in order to date-rape.) What about verbal coercion (e.g., verbal pressure, anger, use of psychological power, or lying)?

> "Many times my boyfriend forced me into having sex when I didn't want to. He thought that kissing always led to sex and when I said 'No,' he would become very angry."
>
> (from the author's files)

Koss and colleagues have been criticized for calling experiences like this rape (Gilbert, 1991; Roiphe, 1993), and fewer than half of the victims in Koss' studies viewed their own experiences as rape (Koss et al., 1988). (Women with an exaggerated adherence to stereotypic feminine gender roles are not only the most likely to date men who use force to obtain sex, but also the most likely to rationalize the behavior of these men; McKelvie & Gold, 1994.) A national survey taken for *Time*/CNN found that while over three-fourths of all women *and* men considered sex with an intoxicated partner rape, over half did not classify a male's use of verbal or emotional pressure to have sex as rape (*Time*, June 3, 1991, p. 50). However, as women become more educated about date rape (and especially if they know someone who has been a victim), they are more likely to acknowledge such behavior as rape (Botta & Pingree, 1997). The question of what defines a rape has recently been addressed (Muehlenhard et al., 1992), but whether or not these behaviors are rape is not the real issue. The real issue is whether or not it is right to have sex with people against their will, and the answer to that is "No!"

I will refer to cases in which a person is forced into unwanted sexual activity through physical or verbal coercion as **sexual coercion**. Some instances of coercive sex avoid the label "rape" only because the victims, fearing that they might be raped, "consent" to have sex. Studies have found that about two-thirds of all college men admit to having used sexual coercion of some type in order to have sex with a date (Mosher & Anderson, 1986; Rapaport & Burkhart, 1984). Sexual coercion is also quite common in gay relations between men (Waterman et al., 1989). However, studies have found that women, too, sometimes use sexual coercion with partners. In a study conducted at the University of Northern Iowa, for example, 14.2 percent of the women said that they had forced sexual activity on a partner (Story, 1986). At a West Coast university, 8.1 percent of the women surveyed admitted to such activity (Gwartney-Gibbs et al., 1987), while at Texas A & M University, 13.4 percent said that they had used verbal coercion and 6.5 percent said that they had used physical coercion to pressure a male into sex (Muehlen-

hard & Cook, 1986). Twelve percent of men sampled at the University of South Dakota said that they had been the victims of physical restraint or intimidation to have sex (Struckman-Johnson & Struckman-Johnson, 1994). In a study of several colleges in the New York–New Jersey area, Anderson (1989) found that over 15 percent of women admitted to having abused a partner (e.g., deliberately inducing intoxication) in order to have sex.

In a series of studies, Muehlenhard and her colleagues found that nearly as many men as women had been the victims of sexual coercion (Muehlenhard, 1989; Muehlenhard et al., 1985; Muehlenhard & Long, 1988). However, there was an important difference: the coercion used by males was often physical, whereas the coercion used by women was generally verbal. Verbal coercion often includes tactics such as persuasion, threats to withdraw love, and bribery, but a few women include physical restraint and intimidation as well (Struckman-Johnson & Struckman-Johnson, 1994). Anderson (1991) adds that there is also "the question of whether the male partners in the above studies can properly be viewed as 'victims' in the same manner as female victims of sexual abuse. Even if a female is using verbal or physical force, it requires that he become aroused to engage in sexual intercourse."

Several of the previously cited researchers believe that *some* unwanted sexual experiences with social partners are due to poor communication skills. When it comes to conveying or interpreting sexual intentions, too many people rely on subtle cues (e.g., gestures or body language) that might be misunderstood. Without clear verbal communication in advance, some men misinterpret a woman's desire to engage in kissing and necking as meaning that she wants to "go all the way." When women openly communicate their sexual intentions early during a date, men are less likely to feel "led on" to attempt unwanted behaviors (Muehlenhard, Andrews, & Beal, 1996). On the other hand, Muehlenhard and her colleagues found that over one-third of the college women surveyed admitted to having engaged in **token resistance**—saying "no" when they meant "yes"—during such encounters (Muehlenhard & Hollabaugh, 1988). In the survey for *Time*/CNN, 54 percent of the women indicated that they believed "some women like to be talked into having sex" (*Time*, June 3, 1991, p. 50). Men who have had experiences with women who offer token resistance are likely to push ahead with another partner when she says "no." Some men may even use a negative answer as a rationale to force sex on a partner.

Recent studies have found that just as many men as women have engaged in token resistance to sex, and that men may do so more frequently than women (Muehlenhard, Giusti, & Rodgers, 1993; Muehlenhard & Rodgers, 1993; O'Sullivan & Allgeier, 1995;

sexual coercion The act of forcing another person into unwanted sexual activity by physical or verbal coercion (restraint or constraint).

token resistance Saying "no" to sex when one means "yes."

Sprecher et al., 1995). Most instances of token resistance occur in ongoing relationships, are done as expressive game-playing to increase sexual arousal (not for manipulative reasons), and are perceived as pleasant interactions by both partners. Nevertheless, there is still a danger that men who have had the experience of a partner engaging in token resistance may misinterpret sincere resistance by a future partner.

The use of alcohol no doubt further complicates interpersonal communication. One study found that approximately three-fourths of date rapists *and* their victims had been drinking or taking drugs before the attack (Copenhaver & Grauerho, 1991).

Unfortunately, many people, including many women, believe that it is okay for a male to force a female into unwanted sexual intercourse in some circumstances (Goodchilds & Zellman, 1984; Kanin, 1967; Muehlenhard & Linton, 1987; *Time*/CNN poll, June 3, 1991). These attitudes about relationships and appropriate sexual behaviors are learned early in life. For example, a survey was taken of 1,700 seventh- to ninth-graders who attended an assault awareness program presented by the Rhode Island Rape Crisis Center at schools throughout the state (Kikuchi, 1988). Opinions about male behavior toward a woman on a date included the following:

- Twenty-nine percent of the boys and 20 percent of the girls said that a male has a right to force his date to have sexual intercourse if she has had sexual intercourse with other men.
- Twenty-two percent of the boys and 10 percent of the girls said that a man has the right to have sexual intercourse without the woman's consent if he had spent a lot of money on her.
- Forty-eight percent of the boys believed that it was okay for a man to force a date to have sexual intercourse if she let him touch her above the waist.
- Fifty-seven percent of the boys and 38 percent of the girls believed it was okay for a man to force a date to have sexual intercourse if they had done it before.
- Sixty-seven percent of the boys and 59 percent of the girls said that a husband has the right to force his wife to have sexual intercourse against her will.

Is it any wonder that about one-fifth of all young teenage girls have been forced into unwanted sexual activity (often forced intercourse) by a boyfriend or date (Small & Kerns, 1993)?

It is best to make a firm decision about your sexual behavior for a social occasion *before* it begins (and before you start to drink alcohol), and then to communicate your sexual intentions clearly and verbally to your partner. This will prevent some unwanted sexual experiences. It will not, of course, stop someone who would

use force to have sex, or who would take advantage of a defenseless (e.g., intoxicated) person. When this occurs, it is rape, and no less real than any other type of rape.

RAPE IN MARRIAGE

In the seventeenth century, a chief justice of England named Sir Matthew Hale wrote, "The husband cannot be guilty of rape committed by himself upon his lawful wife. For by their mutual consent and contract, the wife hath given herself up in this kind." There were two ideas in Hale's statement that found their way into the American system of justice (which is based on British common law). The first idea is that when two people marry, the marriage contract implies that the husband has a right to have sex with his wife even if she does not consent to it. The second idea is that this is justified because the wife is the property of the husband.

> "In the fifth year of my marriage I was raped by my husband after a violent physical attack. He blamed me for the attack and also said it was my obligation to provide sex whenever he desired it. . . . My marriage ended quite bitterly. For the last 3½ years I was blaming myself for what happened. . . ."

> "My only form of sexual abuse was a one-time rape by my husband. I remember well; I thought that dying at the moment would be a blessing, anything to get me out of the situation I was in. After ten years, the mental anguish and hurt still remain. Often the memory resurfaces. To think that the man who promised to love, honor, and protect me was capable of such a violent act! How humiliating to have him stand there laughing at me while I lay crying. As far as I am concerned, the man belongs in jail."

> (from the author's files)

Today, citizens and legislatures realize that laws based on such unfair and antiquated ideas must be changed. In 1978, a man named John Rideout was prosecuted in Oregon for the rape of his wife. Although he was found not guilty, the trial inspired several states to change their laws to allow for prosecution of **marital rape.** Today, all 50 states make marital rape a crime, but about half require that there be extraordinary violence before the rape is prosecuted. Only a third provide the same legal protection to wives as in other types of rape (Muehlenhard et al., 1992).

A 1992 report by the National Victims Center estimated that 9 percent of all women had been raped by their husbands or ex-hus-

marital rape Rape of a wife by her husband.

CROSS-CULTURAL PERSPECTIVES

Rape

At the beginning of this chapter, you learned that the rape rate in the United States is much higher than in other Western countries (see Figure 15–1). Rape of women is common in many other societies throughout the world as well. For example, there is a warlike tribe in Africa called the Gusii in which it is the custom for males to obtain brides by stealing young women from neighboring tribes. On the wedding night, the Gusii warrior rapes his new bride as many times as possible to prove his manhood to others (Levine, 1959). Some tribes use rape to punish women (Chappell, 1976). In the Mangaian culture in the South Pacific, for example, women who violate the sanctity of male-only gathering areas are taken into the forest and gang-raped (Marshall, 1971). Throughout history, mass rapes during wartime have routinely been committed by victorious soldiers to punish the defeated nation (see the "Historical Perspective" section).

As you can also see from Figure 15–1, rape is not common in all cultures. Rape is unheard of, for example, among the Arapesh, who raise males to be caring, gentle, and nurturant (Mead, 1935). Anthropologist Peggy Sanday (1981) found over 40 societies that were free of rape. What do these societies have in common that results in this lack of violence by men toward women? Like the Arapesh, boys are raised to be nurturant, not aggressive, and to view women as equals who share power and responsibility within the society.

Rape of women is several hundred times more common in the United States than in societies like the Arapesh, but we are not alone; Sanday (1981) found over 50 "rape-prone" societies. Do these cultures have anything in common? The answer is yes. Sanday found that rape-prone societies like the United States pro-mote and glorify male aggression, treat women as inferior and demean their nurturant roles (from which men remain aloof), view relationships between men and women as adversarial, and, to make matters worse, instill these attitudes in children early in their development.

A recent comparison of university students in the United States and Sweden found that U.S. women were three times as likely as Swedish women to be victims of physical sexual coercion (Lottes & Weinberg, 1996). Very few Swedish men had ever used physical force with a woman. The authors attributed the results to a difference in sex education in the two countries. In Sweden, boys and girls receive sex education in school beginning early in life, and this education emphasizes how to behave ethically. Few American children receive such guidance in school.

American culture is very heterogenous, and the prevalence of rape differs for different ethnic groups. Sexually aggressive behavior is lowest among Asian-Americans (Koss et al., 1987). Note also the low rape rate for Japan in Figure 15–1. Asian-American culture has a collectivist orientation (as compared to the American individualist orientation) that emphasizes group goals over individual goals and social support over individual competition. Hall and Barongan (1997) suggest that a collectivist orientation acts to protect against rape because interpersonal conflict is minimal and rape is viewed as a crime against the group, not just the individual.

The fact that rape is almost nonexistent in some societies proves that it can be eliminated, but for that to happen here we will have to raise boys to have different values than in the past. What type of values will you teach your sons?

---

bands. Other studies have put the estimate even higher (Russell, 1990). Marital rape is often accompanied by severe beatings and other forms of physical abuse (Finkelhor & Yllo, 1985). Frieze (1983) reported that of 137 women beaten by their husbands, 34 percent were also raped as part of their "punishment." Russell (1990) sees marital rape as being motivated by the same things as any other type of rape—power, anger, and sadism. There is no excuse for marital rape or abuse of a spouse—any rape is a real rape, regardless of who the rapist is.

GANG RAPE

In 1989, seven teenagers were arrested in Glen Ridge, New Jersey, for enticing a mentally impaired 17-year-old girl to perform sex acts while other teens watched.

In April 1989, a 12-year-old girl was kidnapped in Los Angeles and assaulted for four days by teenage members of a gang called the Rolling 40s Crips. In 1994, five men were charged with raping a woman in a bar after she had passed out from having too much to drink. In a highly publicized case, six teenagers in Manhattan attacked a lone 26-year-old jogger in Central Park. She was raped, beaten unconscious, and left for dead. This rape was not connected with drugs, robbery, or racial hatred. The only reason given was that it was "something to do" to escape boredom. The youths called their escapade "wilding."

Gang rape refers to cases in which a victim is assaulted by more than one attacker. There are several forms of gang rape. In prisons, weaker men are often gang-raped by other male inmates. With female victims, some cases of gang rape can begin as date rape (O'Sullivan, 1991). On college campuses, as many as 90 percent of all gang rapes are committed by fraternity members (with whom there is often an emphasis on heavy drinking and machismo; Ponce, 1994).

> "When I was 16, I was a very attractive and rebellious teenager. One Friday night my friends and I decided to go to a fraternity party. I did some drugs and I drank a lot of alcohol. I danced with a lot of guys and then a guy that I liked for some time asked me to dance. We danced for some time and he finally asked me to go to a little private party in the back room. When we went into the room, I saw four of his friends standing around a bed. The guy I was with started to kiss me while his friends began removing my clothes. I tried resisting, but with five pairs of hands holding me, it didn't do much good. After they each had a turn, they just left me there and told me if I told anyone, that they would kill me. . . ."
>
> (from the author's files)

In other instances, the victim may be chosen at random. A study by Amir (1971) found that 43 percent of the rape victims surveyed had been attacked by more than one assailant. Of the rapists studied, 71 percent had taken part in rapes with more than one assailant involved. Gang rapes were usually planned, although the specific victim was not always selected in advance. Victims usually lived in the same neighborhood as the rapists.

What motivates individuals to participate in gang rapes? Some of the reasons are the same reasons that single rapes are committed—to assert power and attempt to control others; to express anger; and to conform to the view—often reinforced by movies, televi-

sion, and society in general—that men are the hunters or aggressors and that sexual relations involve conquering female victims. However, gang rapes often have a dynamic of their own. Individuals acting alone may not be capable of sexual assault, but being a member of a group creates special circumstances (O'Sullivan, 1991). For a group member, individuality is lost. This also means that individual responsibility for things a person does can be evaded or diffused while committing a rape as a member of a group. Furthermore, taking part in gang rape may be a way of demonstrating loyalty to the group in the eyes of group members, and hesitant individuals may decide to take part in order not to be seen as cowards (Sanday, 1990). Gang rapes are usually more violent than rapes committed by individuals because each member of the group is challenged to outdo the other (Gidyez & Koss, 1990). They may even bait one another into hideously shocking acts. In the Glen Ridge case, for example, the victim was sexually assaulted with a stick, a broom handle, and a miniature baseball bat.

Individuals who participate in gang rape tend to be younger than rapists who act alone—62 percent are under age 21, as compared to only 20 percent of those responsible for single-offender rapes (United States Bureau of Justice Statistics). The number of 13- and 14-year-old boys accused of rape doubled between 1976 and 1986 (National Center for Juvenile Justice). Peer pressure is intense for teenagers. In a society where movies aimed at teenagers commonly involve ax murder and slashing, and where violence is a commonly portrayed means of interaction between men and women, it is not surprising that gang rapes occur. Only when society expects equality and nonviolence in relations between men and women will gang rape and other forms of rape begin to diminish in frequency.

STATUTORY RAPE

Most states have laws involving the crime **statutory rape,** sometimes referred to as *carnal knowledge of a juvenile.* These laws make it illegal for an adult to have sexual intercourse with anyone under a certain age, even if that person has consented to have sex. The legal "age of consent" varies from state to state, ranging from 14 to 18. Some states also add provisions about difference in age between the adult and the minor, such as stating that a male must be 17 or older and at least 2 years older than the female. The idea behind these laws is that underage individuals are not capable of giving informed consent (i.e., they are not capable of fully understanding the meaning and consequences of sexual relations).

---

gang rape Rape in which a victim is assaulted by more than one attacker.

statutory rape Sexual intercourse by an adult with a (consenting) partner who is not yet of the legal "age of consent." Sometimes called "carnal knowledge of a juvenile."

The mentally handicapped are often protected by these laws as well. A landmark case was the Glen Ridge gang rape (see the section on gang rape). Although the 17-year-old victim allegedly agreed to have sex and had been sexually active since the age of 12, the defendants were convicted after the prosecution showed that she had an IQ of 64 and the social skills of an 8-year-old, and thus could not fully understand her actions. Thus, in some cases, a person can be found guilty of rape even if the victim was consenting.

Recall from Chapter 6 that half of the fathers of babies born to women aged 14 to 17 are at least 20 years old (Lindberg et al.,1997; Taylor et al., 1997). Today, some states are using statutory rape laws to prosecute these men with hopes that it will send a message to others and reduce the rate of teen pregnancies (Donovan, 1997).

MALE RAPE

The FBI's Uniform Crime Report doesn't even acknowledge that **male rape** exists—yet it does. The U.S. Department of Justice (1997) says that over 15,000 men a year report having been victims of rape or attempted rape. There are certainly many more unreported cases.

Can a man be raped by a woman? Many people have difficulty believing that a man could get an erection while feeling great fear or terror, but it does happen. Sarrel and Masters (1982) reported 11 instances in which men had been kidnapped by a woman or a group of women who then forced them (often with use of a weapon) to engage in sexual intercourse. Anderson (1989) reported that some college women admitted to having used physical force or a weapon to have sex with a man. Others have also reported cases where men were victims of female sexual assault (Smith, Pine, & Hawley, 1988). These male victims of rape suffered long-term problems similar to those of female rape victims (see "Reactions to Rape" on pages 371–373) and generally felt guilt, or thought that something was wrong with them, because they were able to respond sexually under the circumstances (Masters, 1986; Sarrel & Masters, 1982).

Most men who are raped are victimized by other men, and many people assume that this occurs only in prison. Indeed, from 0.5 percent to 3 percent of all male inmates are sexually assaulted by other inmates (Moss, Hosford, & Anderson, 1979). The victims are usually gang-raped by men who otherwise consider themselves heterosexual (Cotton & Groth, 1982; Lockwood, 1980). However, males are also raped by other men outside of prison. Laumann et al. (1994) reported that 1.9 percent of the men in their survey had had forced sex with a man. This includes college students (Scarce, 1997).

male rape Rape in which the victim is a male.

"I am a student in your class. I am a 24-year-old male. I am writing this because I feel it can help others in my situation. At 9 years of age, I was raped by three friends. They invited me in[to] their house to play 'strip poker.' I didn't know how to play, so I lost my shirt. Last, but not least, my underwear came off. At this point, the oldest, aged 16, told me to play with his penis. I refused and he punched me in the face. I started to cry and he hit me over and over. I still remember the look on his face. Well, anyway, they all took their penises and proceeded to rape me. They then took turns and at one point they made me perform oral sex on them. I still have nightmares about it. I still remember the blood, the pain, and the shame. What happened to them? Nothing. I never told anyone until this year."

(from the author's files)

Grown men are also victims of rape by other men:

On a stopover in St. Louis, he decided to stroll from the bus station to the Mississippi River. He was standing by the waves looking out at the water when a man approached and asked if he wanted to smoke some marijuana. Senter accepted the offer.

He followed the man, then suddenly found himself being pushed into an empty tractor-trailer rig. . . ."He had his arm on my shoulder and his hand on my belt," Senter said. "I tried to get away once. He slammed me up on the side of the truck and told me not to do that again."

At first, Senter tried to protest. His attacker told him that if he kept quiet, it would be over sooner. So Senter bit down on his thumb to keep from crying out. He bit so hard his thumb was left without feeling for days afterward.

Inwardly, though, his feelings exploded.

(Read, 1993, p. E-3. Reprinted with permission.)

Most rapes go unreported, but this is particularly true when men are raped by other men. However, two groups have published several accounts of male rapes committed outside of prison (Lipscomb et al., 1992; Mezey & King, 1989, 1992). Weapons were used in most of the assaults, and in many cases the assailants were total strangers. Most of the rapists identify themselves as heterosexuals. One study found that there was nothing that distinguished the victims from other men and concluded that "all men are potential victims" (Lipscomb et al., 1992).

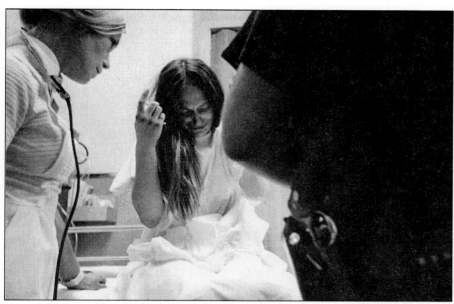

Figure 15–3 If you have ever been a victim of rape or child sexual molestation, just listening to a lecture about sexual abuse can bring back bad memories and upsetting emotions.

In the previous examples the victims, and often the perpetrators, were heterosexual. However, gay men can also be the victims or perpetrators of male rape (Hickson et al., 1994).

EXPLANATIONS OF RAPE

Many years ago it was believed that men raped because of an overpowering sexual urge which women released by their looks, dress, or behavior. Then the pendulum swung so that rape was considered as violence or anger at women expressed by a sexual act. Today, the prevailing view is that there are many types of rapists and that violence and sexual desires are present to differing degrees in different types of rape. Let us review some of the explanations of rape.

PSYCHODYNAMIC THEORIES. Psychodynamic theories view sexually deviant behavior as psychopathology—a character disorder that is very resistant to change. In the 1970s, two such theories about rapists were proposed that have had a substantial impact over the past three decades. Both attempted to classify offenders by their psychology or motivation. The first theory, by Cohen et al. (1971), classified rapists according to whether their aim was aggression, sexual, or sadism. I will focus on the second theory, by Groth, Burgess, and Holmstrom (1977). The two theories, however, have much in common.

Groth, Burgess, and Holmstrom describe three types of rapists. **Power rapists** commit premeditated attacks in order to overcome personal feelings of inse-

curity and inadequacy. The rape "is the means by which he [the rapist] reassures himself of his sexual adequacy and identity, of his strength and potency" (p. 1240). Rapes give power rapists a sense of mastery and control. These rapists seek to humiliate and degrade their victims through language and acts. Power rapists use whatever force is required to overcome the will of their victims, including threats and weapons, and while this sometimes can lead to injury, there is usually little physical harm to the victim beyond the act itself. The rape may last for an extended time. Power rapists usually have great difficulty in relating to women and are often very awkward in their interpersonal relations. Groth (1979) estimates that 55 percent of all rapes are power rapes.

A second type commit what are called anger rapes. An **anger rapist** generally commits unplanned assaults in an attempt to express his hostility or anger about some wrong he feels has been done to him—by life or by his victim. His anger is often directed at women in general. Anger rapists often use more physical abuse than is necessary to commit the rape in an effort to punish their victims. The attack usually is of short duration. Groth estimates that 40 percent of all rapes are anger rapes, and more recent studies suggest that many rapists have problem with anger (see Ward, Hudson, & McCormack, 1997).

Finally, there is the sadistic rape. **Sadistic rapists** are erotically aroused by physical force and derive pleasure from inflicting physical pain on their victims. They may not be sexually satisfied if the victim does not resist and may torture or even murder their victims to satisfy themselves. The attacks usually are of extended duration. Groth estimates that only 5 percent of all rapes are sadistic rapes.

Groth et al. (1977) concluded that although rape involves a sexual act, first and foremost it is an act of violence. Rape is motivated by the need to dominate, humiliate, and exert power over women. According to Groth, "We look at rape as the sexual expression of aggression,

power rapists Rapists who commit their attacks in order to overcome personal feelings of insecurity and inadequacy. Their rapes give them a sense of mastery and control.

anger rapists Rapists who commit their attacks out of hostility or anger directed at their victim or at women in general.

sadistic rapists Rapists who are aroused by physical force and who derive pleasure by harming their victims.

rather than as the aggressive expression of sexuality" (quoted by Gelman et al., 1990). There are many ways in which rape can be used to display dominance or power. In one case, a man known as the "ski mask rapist" committed several rapes in Texas and Louisiana. He always chose a female victim who lived with a male (husband or boyfriend). He would stake out her home, wait until the man left, and then break in when the woman was alone. He would then tie her up, wait until the man returned, and then tie up the man and force him to watch while the woman was raped. If sexual release had been the rapist's primary motivation, he would not have waited until the male partner returned home. In this case, the primary motive was clearly to dominate and humiliate both the woman and her male partner. Gang rapes, either of women or of men, provide another example of the use of rape to demonstrate power or domination.

There are many examples of anger rape as well. The following letter illustrates a clear case of rape used to act out the rapist's hostility toward women:

Dear Ann Landers:

You have printed letters from bartenders, secretaries, lawyers, musicians, housewives, lovesick teenagers . . . , but I have never seen a letter from a rapist. This may be your first, if you have the guts to publish it.

I am 32 years old and started on this rotten road when I was 20. As of last week, I have raped 25 women, mostly in Oklahoma, Arizona, and California. I've never been caught. The system I use is virtually risk-free. I am highly intelligent—a college graduate from an Ivy League school.

I am writing in the hope that you will print the letter and educate the public. They need to know why men rape. Perhaps if the reason were known, today's mothers of young children will do a better job and prevent another generation from growing up to be rapists who terrorize women of all ages. I hate what I am; but I know what caused me to be like this, and hope you will give my views as much coverage as possible.

I have read repeatedly in your column, "Rape is not an act of lust. It is an act of violence." You are so right. Actually, rape has very little to do with sex. It has a lot to do with the way a male relates to females. Almost always, if you put it all together it spells M-O-T-H-E-R.

I came from a well-to-do family and was raised by a domineering, overpowering mother (divorced). My miserable, cruel grandmother lived with us. Almost every day since I can remember, I was slapped, punched, kicked, beaten with a belt or a hairbrush. Once I suffered three broken ribs from being pushed down a flight of stairs. When I was taken to the doctor (Mom thought I might have a broken leg), I was instructed to say I fell off my bike.

I have never had a girlfriend because I despise all women. I know I am sick in the head, but I have this uncontrollable urge to punish all females, and rape is the best way to get even.

Please tell young mothers to love their little boys. If they can't, then do the kid a favor and put him in a foster home or give him in adoption to a family that will treat him like a human being. Children who are whipped and knocked around have the need to get even. . . .

No Name, No City, and No Fingerprints

(Reprinted with permission of Ann Landers and Creators Syndicate.)

Groth et al.'s (1977) classification was based on in-depth research of convicted rapists. But recall that most convicted rapists are stranger rapists and that stranger rapists commit a minority of all rapes. On the basis of what we now know about acquaintance (including date) rapes, some researchers have proposed a fourth type of rapist—the **opportunistic rapist** (Prentky, Knight, & Rosenber, 1988). Here, the primary motivation is sex, but these men have strongly distorted attitudes and beliefs about sex roles and female behavior that negate normal social inhibitions. These men rape impulsively when there is an opportunity, as on a date. They use only enough force to commit the act, and their only anger is at the victim's resistance. Opportunistic rapists probably account for the majority of date rapes.

Recent research has made it clear that rape cannot be explained only by individual psychopathology, for rape is far more common in some cultures than others (see Box 15–A). Cultural, social learning, and interpersonal factors all play a role.

FEMINIST (SEXUAL SCRIPT) THEORIES. Look again at the rape statistics at the very beginning of the chapter. Why is the incidence of rape so much higher in the United States than in many other countries? Studies have found that rape is most common in societies that glorify violence by men, particularly when the society is also sexually repressive (Prescott, 1975; Sanday, 1981). These societies not only encourage boys to be aggressive (see Box 15–A) but have distinct gender

---

opportunistic rapists Rapists motivated by a desire for sex who rape impulsively when there is an opportunity.

roles for men and women, with men regarding women's roles as inferior. In our culture, the stereotypic gender role for male sexuality is "active, aggressive, thrusting and powerful," and for female sexuality it is "passive, powerless, submissive and receptive" (Moore, 1994). Feminist theory holds that the traditional sexual script "supports and condones male sexual coercion against women and that this sexual script remains the normative dating script in our society" (Byers, 1996).

In fact, men who commit rape or engage in sexual coercion are likely to believe in stereotypic gender roles, to view men as dominant, to consider relations with women adversarial (very competitive), and to have male friends who believe that the use of physical force to have sex is okay in some situations (Harney & Muehlenhard, 1991; Kanin, 1985; Malamuth, 1984, 1986; Rapaport & Burkhart, 1984). Henrietta Moore (1994) argues that men's sexual violence towards women does not reflect a breakdown in the social order, but is enacted in order to maintain the existing social order (as a "sign of a struggle for the maintenance of certain fantasies of identity and power").

It has also been found that the amount of sexual violence toward women in a society is greater in cultures that glorify violence by males (e.g., in movies) particularly in times of cultural disorganization (e.g., high unemployment; Baron, Straus, & Jaffee, 1987). You will learn about the effects of pornography in the next chapter. One of the conclusions of research in this area is that men who look at scenes of aggression toward women—and the scenes do not need to be sexually explicit—become less sensitive to the victims of rape and more accepting of sexual violence (Donnerstein & Linz, 1984; Linz, 1989; Malamuth & Check, 1981). Non–sexually explicit rape scenes are exactly the type that are commonly shown on television, despite the warnings of leading researchers in the field (Linz, Wilson, & Donnerstein, 1992).

One study, which found a strong positive correlation between men's attitudes about sexual relations (whether or not they held callous attitudes) and their use of sexual coercion with partners concluded by stating that "the socialization of the macho man, if it does not directly produce a rapist, appears to produce calloused sex attitudes toward women and rape and proclivities toward forceful and exploitive tactics to gain sexual access to reluctant women" (Mosher & Anderson, 1986).

SOCIAL LEARNING THEORIES. These theories have much in common with cultural explanations. You have already read much about social learning research in previous sections, so this will serve mainly as a review. Recall that many rapists were either themselves abused as children or witnessed abuse within the family (Groth,

1979; Seghorn et al., 1987). Whatever type of family life a child experiences while growing up and accepts as "normal," it is likely that he or she will model the same behavior as an adult (Bernard & Bernard, 1983). Interestingly, women with a childhood experience of sexual abuse are more likely than others to be revictimized as adults (White & Humphrey, 1991), perhaps because the experience often lowers self-esteem and makes these women more vulnerable to being easily pressured into sex (Skelton, 1984). Thus, for both men and women, a past experience of sexual abuse (or, to a lesser extent, physical abuse) is predictive of the likelihood of inflicting or receiving such abuse when they are adults (Gwartney-Gibbs et al., 1987; Petrovich & Templer, 1984; Story, 1986).

Learned gender roles are also important. Men who have learned extremely conservative stereotypes and who also believe that relationships between men and women are adversarial tend to believe the myths about rape that legitimize violence against women (see the next section) (Burt, 1980; Koss et al., 1985; Muehlenhard & Falcon, 1990). These men are more likely than others to be sexually violent or aggressive (also see Scully & Marolla, 1984).

INTERPERSONAL THEORIES. Much of this section, too, will be review. Recall that a lack of shared meaning between men and women can lead to unwanted sexual aggression by men (Perper & Weis, 1987). Recall also that many men (and women) believe that rape is justified in some situations, including instances where a man has spent money on the date or feels that she has "led him on" (e.g., by engaging in necking or petting or going to his apartment; Goodchilds & Zellman, 1984; Kikuchi, 1988; Muehlenhard, 1988). Men who accept extreme gender stereotypes of women are most likely to misinterpret behaviors like these as invitations for sex (Muehlenhard & Cook, 1988). The fact that some women engage in token resistance, initially saying "no" when they want the male to continue, compounds the misinterpretation of intentions (Muehlenhard & Hollabaugh, 1989), not only on one date but on future dates with other women as well.

Now recall the research that found that convicted rapists—but not nonrapists—were sexually aroused by listening to tapes of, or viewing, rape scenes (see the section on the characteristics of rapists on page 359). Are there any situations in which normal men might also respond to rape scenes, becoming aroused as rapists do? Barbaree and Marshall (1991) found that many men could become aroused by rape scenes when they had been drinking—and, you remember, most date rapes occur when the male or both people have been drinking (Copenhaver & Grauerho, 1991; Muehlenhard & Linton, 1987). Barbaree's team also found that normal men's reactions to rape scenes be-

Feiffer © 1989 Jules Feiffer. Reproduced with permission of Universal Press Syndicate.

came more like those of rapists if a woman had made them angry just before testing (this was done purposely as part of the research). How might these men behave if they incorrectly believed that a woman had "led them on"? As Dr. Barbaree concluded, "With the right combination of factors, most men can be aroused by violent sex" (cited in Goleman, 1992).

EVOLUTIONARY (SOCIOBIOLOGICAL) THEORIES. Evolutionary theories of rape assume that the inclination of men to rape has a genetic basis (Ellis, 1993; Shields & Shields, 1983; Thornhill & Thornhill, 1992). Basically, these theories state that by having sex with many partners, men maximize their reproductive potential, and that evolution has favored sexually aggressive men by their producing more offspring. Critics have emphasized the dangers of generalizing from studies of animals and insects to humans and contend that other theories do a better job of explaining the specifics of rape behavior (see Polaschek et al., 1997).

MISCONCEPTIONS ABOUT RAPE

By this point in the chapter, you have probably learned some facts about rapes and rapists that have challenged the beliefs you held before en-

rape myths Widespread mistaken beliefs about rape.

rolling in the course. Mistaken beliefs about rape are widespread, so widespread that they are often referred to as **rape myths** (Burt, 1980). These myths serve the function of placing the blame on the victims. In general, men are more accepting of these myths than women—especially young males (Blumberg & Lester, 1991; Johnson, Kuck, & Schander, 1997). This is particularly true of men who accept conservative gender role stereotypes, and the combination of belief in stereotypes and belief in rape myths, you recall, is commonly found in men who engage in sexually coercive behavior (see the section on cultural and social learning theories of rape, pages 368–369). It is important, therefore, that we dispel these myths. Let us examine some of them.

MYTH #1: WOMEN WHO ARE RAPED USUALLY PROVOKED IT BY THEIR DRESS AND BEHAVIOR. Read again the results of the study by Kikuchi (1988) at the end of the section on date rape. Many junior high school boys and girls believed that date rape was okay if the woman had engaged in certain behaviors. Many adults, especially men, continue to hold these beliefs. One-third to one-half of all people believe that a woman is at least partially to blame for rape if she wore a short skirt or went braless, went to the man's home or room, or was drinking alcohol (Burt, 1980; Holcomb et al., 1991; *Time*/CNN poll, 1991). When rape cases go to trial, men are less likely than women to vote for conviction (Schutte & Hosch, 1997). In 1989 a jury in Ft. Lauderdale, Florida, acquitted a drifter on trial for rape because his victim was wearing a lace miniskirt at the time (the jury foreman was quoted as saying, "She asked for it"). The man was a serial rapist who was later convicted of another rape. As a result of this case, the Florida legislature enacted a law that prevents discussion of a rape victim's clothing during court trials. Other states are now passing similar laws. Many women also blame female victims of rape, mistakenly believing that women can prevent themselves from being raped by behaving differently (for example, by dressing conservatively).

When others put the blame for rape on the victim, it makes the emotional trauma all the worse:

> "I lost my virginity through rape at age 15. Even my parents thought it must have been my fault because I was at a party. Needless to say, this made matters worse. I had difficulty having a productive relationship until recently."

(from the author's files)

Why is rape the only crime for which we have a tendency to blame the victims? Consider the following hypothetical scenario:

> Steve wore his finest Italian suit and shoes to the company's reception. He was particularly proud of his new Rolex watch and felt sure it would make an impression on the brunette in Accounting. At the party he had two or three drinks. Steve enjoyed a drink on these occasions because it made him feel more sociable, but he often found that he had had one more than he wanted because of peer pressure—others made him feel like an outcast if he didn't have a drink in his hand. He intermingled freely, and toward the end of the evening finally struck up a conversation with the woman in Accounting. They made plans to get together for lunch the next day, so by the time Steve left, around midnight, he was feeling really good.
>
> On the way home, Steve remembered he was out of bread and eggs, so he stopped at a convenience store, parking his BMW slightly off to the side. When he went to get back in his car, he was stopped by a man who said that he had a gun under his coat and that Steve had better give him his Rolex, wallet, and ring. Steve wanted to avoid injury and quickly handed over his valuables while pleading with the man not to hurt him. The man took the valuables and ran.
>
> Steve called the police immediately, hoping to catch the robber. But when the police arrived, rather than asking for a description of the robber, they started to act suspicious about him. One asked why he was out so late. Didn't he know this was when most muggings take place? The other wanted to know how much he'd had to drink and made him take a breathalyzer test. When Steve finally got to tell them what the robber took, they were not sympathetic, but instead made comments that insinuated that he had invited the robbery by dressing so well and wearing a Rolex in public.
>
> When Steve got home, his two male roommates couldn't believe that he hadn't resisted. They would have put up a fight, they said; it would never have happened to them. Steve went to bed feeling ashamed and guilty. The next day he didn't mention it at the office. . . .

You probably found this scenario ridiculous, and well you should. So, if you have ever thought that it is women's faults that they are raped because of their behavior, stop for a moment and reexamine your attitude.

MYTHS #2 AND 3: WOMEN SUBCONSCIOUSLY WANT TO BE RAPED; NO WOMAN CAN BE RAPED IF SHE TRULY DOES NOT WANT IT. These two closely related ideas are also widely believed (Burt, 1980; Holcomb et al., 1991). Although many women have had fantasies of rape and most are aroused by watching eroticized rape scenes (Stock, 1982, 1983), this does not mean that they actually wish to be raped. Most of us have fantasized about things we would never do in real life; the difference between fantasy and reality is that we are in control of our fantasies. Real rape involves someone else taking away all of your control. Women, unlike many men, are not sexually aroused by watching realistic rape scenes.

A common male phrase is "You can't thread a moving needle." However, most rapists are bigger and stronger than their victims, and many assaults involve more than one male. In our culture, most males are reinforced for their physical aggressiveness while growing up, while many females are reinforced for being passive. The end result is that most women wouldn't even know how to fight back if they were attacked. Nearly a third of all rapists use some kind of weapon (Bureau of Justice Statistics). The use of force and threats can result in anyone's being raped. If the myth were true, then we would certainly never expect men to be raped, but it happens every day, out of prison as well as in.

MYTH #4: WOMEN FREQUENTLY MAKE FALSE ACCUSATIONS OF RAPE. This myth is based on the idea that a woman uses the accusation of rape to get even with a man—for revenge. Many men believe this (Holcomb et al., 1991). However, FBI statistics show that the percentage of reported rapes that are false is actually lower (2 percent) than for most other crimes. (When false accusations do occur, it is usually to provide an alibi, seek revenge, or obtain sympathy; Kanin, 1994.) We do not automatically doubt people when they claim that they were the victim of some other type of crime. So why do you suppose some people are so quick to doubt a woman's accusation of rape?

REACTIONS TO RAPE

In their initial reactions and long-term coping strategies, female victims of rape go through what is called **rape trauma syndrome** (Burgess & Holmstrom, 1974; Burt & Katz, 1988). For most victims, it can be considered a form of posttraumatic stress disorder (Foa & Riggs, 1995). First, there is an *acute phase,* which begins right after the rape and continues for several weeks. During this phase, the victim's body has to recover from the physical damage that may have been caused by the rape. About one-fourth of all rape victims suffer minor injuries, and about 4 percent are seriously injured (National Victim Center, 1992). In addition, the victim must begin to recover from the psychological damage that was done (Arata & Burkhart, 1996; Rynd, 1987). Some victims become very expressive about their feelings and may cry and experience severe depression, become angry or fearful, and/or experience great anxiety. Others may have what is called a controlled reaction, as if they were trying to deny that the rapist had affected them in any way. However, these victims must also work through a great deal of anguish and may become expressive later. Many rape victims experience dissociative symptoms (restricted affect, detachment from other people, concentration problems), and the presence of these symptoms is associated with long-term posttraumatic stress syndrome (Foa & Riggs, 1995). Many rape victims still display symptoms of posttraumatic stress syndrome three months after the attack. Most victims will also experience headaches, sleeplessness, and/or restless behaviors as well in the acute phase. Many victims believed rape myths prior to the rape, and as a result experience guilt and self-blame, which makes the depression even worse (Frazier, 1991). Because the trauma for many victims is

rape trauma syndrome Reactions of a rape victim, including an acute phase (involving initial physical and psychological recovery from the attack) and long-term psychological reorganization (involving attempts to deal with the long-term effects of the attack and to prevent a future rape).

long-term, rape victims may well be referred to as *rape survivors.*

> "The lectures on rape were traumatic for me because I could remember everything so vividly. But I stuck with the lecture so I could find out if it was my fault. Now I don't feel as dirty."
>
> (from the author's files)

Next, the victim enters a *period of long-term reorganization.* In this stage, the rape victim attempts to regain control of her or his life (including taking steps to prevent rape from occurring again) and may move or change jobs. Each of us wants to believe that our world is a relatively safe, secure, and predictable place to live in and to feel that we have control over our lives and what happens to us. This belief allows us to reduce our fears and anxieties about the world we live in so that we can function efficiently and effectively in it. However, when a violent crime such as rape is committed, our belief system is shaken and our security is threatened. Rape victims often feel as if they no longer have any control over what is going to happen to them (Santiago et al., 1985). Many of the reactions to rape are attempts by victims to reassert control over their lives. A basic sense of trust in the goodness of people and society has been damaged, and this damage takes a long time to heal.

Many victims lose their desire for sex and have trouble becoming sexually aroused (Becker et al., 1984; Burgess & Holmstrom, 1979; Gilbert & Cunningham, 1987). Some show a decreased frequency of sexual relations that lasts for years. Reread the story of the teenage girl who was gang-raped. Here is part of her long-term reaction:

> "I talked to other rape victims and I worked through my guilt and blame. I still have a lot of fears about sexual relationships and have not had a sexual relationship since I was raped. . . ."
>
> (from the author's files)

Victims who seek help shortly after being raped generally suffer fewer long-term emotional problems (Stewart et al., 1987). The reactions of the victim's family are very important in this respect. Victims who experience negative social reactions have increased psychological problems (Ullman, 1996). For example, women whose partners respond warmly and sympathetically and do not pressure them to reestablish sexual relations have fewer long-term sexual problems (Howard, 1980). With rape victims who initially suppress their feelings, it is not unusual for symptoms to emerge months or even years after the incident.

Figure 15–4   Most communities now offer rape crisis counseling.

"The first thing I went and did after your class today was run to my car and cry for a half hour. Fourteen years ago I was raped. . . ."

(from the author's files)

Just like female victims, men who are the victims of rape suffer long-term problems (Mezey & King, 1989). They frequently have sexual problems, difficulties in forming close relationships, and a damaged self-image. Unlike women who are victims, they are also likely to have doubts about their sexual self-identity and ask themselves whether the experience means that they are homosexual. Recovery is made more difficult because male victims are even less likely than female victims to tell anyone else, and thus are less likely to seek counseling.

The rates for child molestation, incest, and rape are high in our society, and thus, unfortunately, some victims of sexual abuse are later revictimized.

"When growing up, I was raped by five males in my neighborhood. . . . The speaker and your lectures helped me to realize that it wasn't my fault. . . . Recently, I was raped again . . . and it was very traumatic to me. When I told a close friend of mine, she reacted as if I had done it, asking questions like 'How did you enjoy it?' She just couldn't understand how it could happen to me twice, and to be honest, neither can I."

(from the author's files)

Perhaps as many as 21 to 35 percent of rape victims have been sexually assaulted before (Cohen & Roth, 1987; Ellis et al., 1982; Miller et al., 1978). Studies have found that some (but not all) people attribute greater blame to a repeat victim than to a first-time victim, regarding her as the type of person to get herself into rape situations (Kanekar, Pinto, & Mazumdar, 1985; Schult & Schneider, 1991). Counseling to eliminate self-blame is very important for repeat victims. While rape victims often have sexual problems afterwards, women who have suffered both childhood sexual abuse and rape as adults often have sexual problems that are different from those of other rape victims. Many have histories of brief sexual relations with multiple partners, which often results in unintended pregnancies and abortions. For victims with this kind of history, treatment includes help with decision making (about sexual relations and contraception), as well as help with self-esteem and self-blame (Wyatt et al., 1992).

REACTIONS OF THE PARTNER

A rape victim's partner and family must also deal with their reactions and emotions. When a woman is raped, some male partners blame themselves for not being there to protect her, but it is important that the male put this feeling aside and direct his attention and emotions to the real victim. He should let her recover from the assault at her own pace and allow her to make her own decisions. After all, one of the biggest steps on the road to recovery is regaining control of one's own life.

Perhaps the most devastating thing that can happen to a rape victim *after* the attack is if her partner and/or family blame her because they believe in rape myths. This will make recovery all the more difficult (Davis et al., 1991).

"My girlfriend . . . about 1 year ago (3 years into the relationship) was walking alone on her college campus and was violently raped. I somehow held her responsible for 'letting' this happen to her. Both her parents also blamed her. I know I treated her with a little less respect after this happened. Now I feel like a complete schmuck. I realize that she had no control over what happened. Fortunately, her therapist convinced her that it was not her fault."

(from the author's files)

If your partner is ever the victim of rape, do not ask questions such as "Why did you have to go there?" or "Why were you out so late?" or "What were you wearing?" Questions like these imply that the rape must have somehow been the victim's fault. Both the victim and the partner often find counseling very useful in learning how to cope with rape.

WHAT HAPPENS WHEN A RAPE IS REPORTED?

In most metropolitan areas, the police department has special units for dealing with the sexual abuse of children and rape victims. If a rape is reported, uniformed officers will arrive on the scene and determine whether a case of sexual abuse has occurred. They will search for the attacker, but the investigation of the case will be carried out by a detective from a special rape investigation unit, who will be next on the scene. These detectives are specially trained to deal with victims of rape. Most large cities now have *rape victim advocate pro-*

Box 15-B   SEXUALITY AND HEALTH

Sexual Victimization, Pregnancy, and Sexually Transmitted Diseases

In addition to the immediate physical trauma that may occur during rape, the victim of sexual abuse is also burdened with the fear of possible pregnancy and/or sexually transmitted disease. A national survey found that about 32,000 pregnancies a year result from the rape of women aged 18 and older (Holmes et al., 1996). Many more pregnancies occur from the rape of younger girls. Half of these pregnancies are terminated by abortion.

Although one of the first studies done of rape victims reported that the incidence of STDs was low (Groth & Burgess, 1977), recent studies have found that many rape victims have infections. Between 2 percent and 13 percent of female rape victims have gonorrhea, while 8–26 percent have chlamydia and 6–22 percent have trichomoniasis (Davies & Clay, 1992; Estreich et al., 1990; Glaser et al., 1989, 1991; Jenny et al., 1990; Schwarcz & Whittington, 1990). Fewer than 2 percent have syphilis.

STDs are very common, so not all of the infected victims in the above studies had contracted the diseases during rape. In some, the infections had existed before the assault, while others had been acquired from new partners in the interval between the assault and the medical examination. STDs that are found in the first 24 hours after the assault probably represent preexisting conditions (Blackmore et al., 1982). Nevertheless, recent studies have veri-

fied that many rape victims do, in fact, acquire STDs at the time of the rape (Glaser et al., 1991; Jenny et al., 1990). In child sexual abuse cases, 3 to 16 percent of all victims contract sexually transmitted diseases (Bays & Chadwick, 1993; Paradise, 1990). As a result, authorities are saying that *all* rape victims should be treated with a variety of medications shortly after the rape (e.g., Glaser et al., 1991).

What about the possibility of becoming infected with HIV, the virus that causes AIDS, during a rape? Recall from Chapter 5 that the test for HIV (actually, for antibodies to HIV) cannot confirm the presence of the virus until many weeks after initial infection. So why not test the rapist if he's caught? In most states, victims and prosecutors are not allowed to see the results of rapists' AIDS tests because of a defendant's right to privacy. This seems very unfair to victims (and most others), for worry about the possibility of AIDS adds to the trauma of many rape victims. At least one victim has allowed her attacker to have a reduced sentence in exchange for his AIDS test results (see Salholz et al., 1990). However, remember that it is difficult to spread HIV from person to person—the odds of becoming infected during one act of unprotected vaginal intercourse with an infected partner is only about 1 in 500 (Hearst & Hulley, 1988). Thus, AIDS experts believe that the chance of getting HIV during a rape is extremely low.

grams as well. These programs provide a specially trained counselor to be with the woman from the initial investigation through the prosecution. After the initial investigation, the victim is referred to a hospital for a medical exam.

To maximize the chance of a successful prosecution, it is very important to get a medical exam as soon as possible after a rape. It is crucial not to wash, douche, or change clothing after being raped, as doing so may destroy vital evidence (such as semen in the vagina) needed by the police to catch the rapist and prove that the crime was committed. Some victims are initially reluctant to prosecute, but later change their minds. If a medical exam has taken place, this option is still available. Without such medical attention, a successful prosecution (or even an attempted prosecution) is less likely.

Most hospitals today are much more sensitive to the trauma experienced by rape victims than they were in the past. Their emergency units are educated and trained to deal with rape victims, who are given the second highest priority for treatment (only life-and-death situations are given higher priority). The hospital staff recognizes the need to be very supportive. In addition to possible physical injury, rape also carries the risk of an unwanted pregnancy and/or getting a sexually transmitted disease (see Box 15–B). Rape victims can be given an injection or pills to prevent unwanted pregnancy if they seek medical help shortly after the attack. At most hospitals, social workers are also available to help rape victims in areas that do not have an advocate program. Prosecution of rapists is covered later in this chapter.

PREVENTING RAPE

In general, there are three basic perspectives or approaches regarding the prevention of rape. *The first approach focuses on the rapist.* Malamuth and colleagues (1980, 1981) found that many men indicated that they would commit rape if they were certain they would not get caught. It was noted that watching films that showed violence toward women intensified this attitude. (See Chapter 16, "Sex and the Law.") An approach focusing on the rapist would call for more policemen on patrol, more lights in dark parking lots, tougher laws that are more rigorously enforced, more arrests, and longer prison sentences for rapists. During its 1994 session, for example, Congress passed the Violence Against Women Act, which, among other things, increased penalties for sex offenses. About half of the 50 states have recently passed laws requiring that sex criminals released from prison register their addresses with police.

This first approach attempts to use legal deterrents to prevent rape. Almost everyone is in favor of strict law enforcement and a speedy and just court system.

However, this approach deals with the symptom only (rape behavior), and not the cause (why does a man want to commit rape?). A law-and-order approach, by itself, may only make rapists more cautious in choosing the time, place, and victim.

The second approach focuses on the potential victim. Part of this approach involves decreasing the chances that a woman will be the target of a rape. Women are often urged to avoid settings or people that might increase the likelihood of rape (e.g., avoiding dark alleys and not picking up hitchhikers). They are also taught to keep dead bolts on doors, to avoid advertising the fact that they might live alone, not to trust door chains as a means of identifying strangers, and so on. When going to their cars, women are encouraged to look under and around the vicinity of the car from a distance and to check the back seat before getting in.

This approach also involves reacting to an attempted rape. Women are urged to learn at least a little bit about self-defense. Women who resist decrease the chance of the rape being completed by about 80 percent (Marchbanks, Kung-Jong, & Mercy, 1990). Surveys of victims of attempted rape have found that resistance is related to an increased chance of escape without serious harm (Bart, 1980; McIntyre, 1980; Ullman & Knight, 1991). In our culture, most women are not raised to be physically aggressive, and it has been my experience from talking to women about rape prevention that many at first are uneasy about defending themselves in a manner that might cause physical harm to the rapist. Self-defense courses (such as the ones that are widely offered through the YWCA) prepare women both mentally and physically for how to defend themselves.

Verbal assertiveness and aggression can also help prevent an attack, especially if a woman exhibits them when she is first approached (Warner, 1980). Brodsky (1976) reported that being assertive and fighting back is most likely to be successful when the rapist initially is polite, hesitant, or precedes the attack with conversation. Some women even get themselves to urinate, vomit, defecate, or tell their attacker that they have a sexually transmitted disease in order to make themselves repulsive. Other forms of trickery include trying to talk the rapist into going to a place where someone can recognize what's going on and help.

"I was once approached by a rapist, who dragged me off into an alley. A part of me somehow remained calm. I simply told him that I would rather go to a motel, and that I would pay for it. It blew his mind at first, and he started to choke me. I pushed him away from me and calmly informed him that my husband was a Mafia hit man and that he often checked up on me and would kill us if he found us together. A motel would be the only place where we would be safe alone. And if we went to the phone company where I worked at the time, I would get my check and take the day off. He fell for it.

"I kept calmly talking to him and assuring him how nice it was going to be and how lucky I was to be getting away from my husband. (Any and every lie I could conceive of, I verbalized casually.) At the phone company there was a guard with a gun. When I saw the guard, I began running in circles and screaming for help. The guard quickly pulled out his gun. What amazed me was that the rapist made no attempt to run, but instead told the guard we were going to a motel. His voice had degenerated into blubbering in anticipation. When the police came, he told them he was going to give me a baby in the alley, and I suggested a motel.

"He was a notorious child molester, who had put eight little girls (aged 8 to 11 years old) in the hospital. I was the only one old enough to go to court against him—the kids were too young."

(from the author's files)

There are some problems with this approach to rape prevention as well. The first problem is that by placing the responsibility for preventing rape in the hands of the victim, women who are raped may feel that they have somehow failed and that the rape was their fault. As noted earlier, a woman should never feel guilty for being raped, as she did not instigate or cause the action. Screaming, blowing a whistle, using mace, vomiting, or talking sometimes work very well, but may not work in every case. Techniques that work to drive away one rapist may do the reverse and anger another. There is no perfect defense against rape. Finally, these attempts to prevent rape do not deal with the underlying causes of rape.

A final approach is based on a *social systems perspective*. From this point of view, rape is caused by our society's belief that males should be sexually aggressive and dominant. The explanation of rape as resulting from sexual scripts fits this point of view (look again at the feminist theories of rape). In a society where violence has become both commonplace and acceptable (look at our TV and movie heroes), it should come as no surprise that men use violence as a means of subjugating women. From this perspective, rape can best be prevented by changing social attitudes about the way that men and women relate to each other. Even though this approach attempts to deal di-

sexual harassment Unwelcome sexual advances that persist after the recipient has indicated that they are unwanted. See text for an expanded definition.

rectly with the cause of rape, it may be the most difficult to implement. Nonetheless, although changing the values and attitudes of a society is hard, it is certainly not impossible. Traditional sex role stereotypes have, in fact, been changing over the past two decades in the United States, to the benefit of both men and women. In Mead's (1935) writings about a culture in which males were expected to be nurturant rather than dominant and aggressive toward women, she reported that rape was unknown. Recall, too, that rape is much less common in many other civilized countries.

What do you think? Is a tougher law-and-order approach to rape needed? Should more women become prepared to deal with attempted rape? Is this enough? Will rape ever stop being a serious problem in our society? Why or why not?

SEXUAL HARASSMENT

Anita Hill's accusation of sexual harassment by Supreme Court nominee Clarence Thomas, and similar charges against U.S. Senator Bob Packwood by several of his female staff and interns, brought attention to an issue in the early 1990s that many people (particularly men) had not taken seriously. Was sexual harassment something new? Of course not. Many surveys taken during the 1980s found that a large proportion of women (generally from 40 to 60 percent) in all sorts of occupations felt that they had been the victims of sexual harassment while at work (e.g., Gutek, 1985; Loy & Stewart, 1984; U.S. Merit Systems Protection Board, 1988). Although it is not as common, men too can be the victims of sexual harassment at work (Berdahl, Magley, & Waldo, 1996; Janus & Janus, 1993).

What is sexual harassment? In 1980 the Equal Employment Opportunity Commission (EEOC) established a legal definition of **sexual harassment** in the workplace:

Unwelcome sexual advances, requests for sexual favors, and other verbal or physical conduct of a sexual nature constitute unlawful sexual harassment when (a) submission to such conduct is made either explicitly or implicitly a term or condition of an individual's employment, (b) submission to or rejection of such conduct by an individual is used as the basis for employment decisions affecting such individual, or (c) such conduct has the purpose or effect of unreasonably interfering with an individual's work performance or creating an intimidating, hostile, or offensive working environment.

Behavior is considered harassment when it *persists* after the recipient has indicated that it is unwanted. In 1986, the U.S. Supreme Court expanded the definition by ruling that harassment based on sex, which creates a hostile or offensive work environment, is also illegal (*Meritor Savings Bank v. Vinson,* 1986). In other words, *a victim does not have to prove that sexual favors were demanded in exchange for job benefits.* A hostile or offensive work environment can include such things as nude calendars, lewd jokes, touching, and obsessive staring. This interpretation of sexual harassment was reaffirmed in a 1993 ruling.

In recent years the courts have been struggling to make a decision about whether the definition of sexual harassment encompasses the harassment of males by females (as depicted, for example, in the movie *Disclosure*) (see Franke, 1997). However, in 1998 the Supreme Court ruled that same-sex harassment (e.g., a male employer and male employee) was illegal.

The EEOC and courts have also ruled that a company can be held liable for sexual harassment by its employees if it doesn't take "immediate and appropriate corrective action." Today, many large businesses have explicit policies against sexual harassment, as well as educational programs for their employees. Charges of sexual harassment are usually prosecuted as civil suits (rather than criminal cases) under Title VII of the Civil Rights Act of 1964, which bans discrimination in employment based on sex. The Civil Rights Act of 1991 expanded the limit of punitive and compensatory damages.

Here are two examples of sexual harassment in the workplace provided to me by students:

"My boss started off saying nasty jokes. Then he began to relate the jokes to his own personal experiences. Next he started making comments about how big my breasts were or how my butt looked in jeans. At this time I got real scared. . . ."

"I started working as a legal assistant at the second largest law firm in New Orleans. I was assigned to two young associates. One of the associates was 28, engaged to be married,

Figure 15–5 Many women regard this type of behavior by a male supervisor to be inappropriate and a form of sexual harassment. Do you suppose that this man would behave the same way with a male employee?

and took interest in helping me learn how the trial process is handled. I began to notice little things that 'Rich' did that were making me feel uncomfortable. First, compliments or personal remarks about my looks. Then he would make insinuations in front of others about how late we worked the night before, and how well we worked together. All of these were said in a manner that made me extremely uncomfortable.

"The last straw was the touching, at first on the back as I was typing or reading, but later on the thighs and calves. I went to my superior and we met with the head of personnel and the director of our law department. I quit that day without pressing charges or causing a big fuss. Why? Because the department I worked in was the Labor Department. 'Rich' defended sexual harassment cases and won. He actually defended employees against sexual harassment—he knew better."

(from the author's files)

It is obvious from this last case history that "some men just don't get it." Many more men than women feel that the problem of sexual harassment has been greatly exaggerated (Dietz-Uhler & Murrell, 1992; Safran, 1981), and many men believe that women should not be so quickly offended when men show sexual interest in them (Dietz-Uhler & Murrell, 1992). Although men recognize certain overt behaviors as harassment (e.g., offering favors in return for sex), many men don't view less overt behaviors (e.g., sexual jokes and remarks) as harassment (Burgess & Borgida, 1997; Dietz-Uhler & Murrell, 1992; Powell, 1986). Many men even blame the women for being sexually harassed (Jensen & Gutek, 1982). Thus, men generally take the issue of sexual harassment less seriously than women.

Why is this a serious issue to women? Sexual harassment usually *occurs in a relationship where there is unequal power,* such as employer-employee or teacher-student relationships (MacKinnon, 1979). This makes the victims of harassment very vulnerable, especially if they are economically dependent on their jobs. Victims may feel that there is no recourse. Who do you complain to if it is the boss who is doing the harassing? If a woman refuses sexual advances or objects to an offensive work environment, she might be the target of retaliation (e.g., she may be passed over for raises and promotions, given poor ratings, or even fired). If she quits she may be ineligible for unemployment compensation. Whether she is fired or she quits, she will be unable to ask for references from this employer in the future because he will no doubt invent reasons for her termination having to do with poor performance. The end result is that victims of sexual harassment often are left with feelings similar to those of other victims of sexual abuse—feelings of helplessness (Safran, 1976). In addition, victims often feel humiliation, shame, and anger (Hamilton et al., 1987; Loy & Stewart, 1984).

SOME CAUSES OF SEXUAL HARASSMENT

Carlota Bustelo, of the United Nations Committee on the Elimination of Discrimination Against Women, said that sexual harassment "defines an ancient phenomenon that is frequent in the workplace but that remained undiscussed until the 1970s. Its disclosure, which began in the U.S., was due to changes taking place in the power relationships between men and women in society at large and, more particularly, on the job. . . ." (in *World Press Review,* February 1992). Research has shown that the more a male believes in traditional stereotypical roles for men and women, the more accepting he is of sexual harassment (Dietz-Uhler & Murrell, 1992). Sexual harassment is most common and most severe, for example, in blue-collar occupations in which women are underrepresented (Burgess & Borgida, 1997; Fitzgerald, 1993; Gutek, 1985). Unlike women, most men cannot distinguish sexual attraction from power as motives for harassment (Popovich et al., 1996).

Gender roles have changed considerably as more and more women have entered the work force over the last few decades, and in Chapter 8 you learned that our society has slowly evolved from stereotypical roles to a more androgynous state. Lately, however, there has been a disturbing increase in sexual harassment by young males, which many of them view as an appro-priate gender role. For example, many teenage boys try to "demonstrate their manhood" and gain popularity by yelling explicit propositions to girls or fondling them as they walk by (Henneberger & Marriott, 1993). A survey of teenagers commissioned by the American Association of University Women and conducted by Louis Harris & Associates found that more than two-thirds of the girls (and a fair proportion of the boys) had been fondled, grabbed, or pinched in an explicitly sexual manner (AAUW, 1993). This behavior among young males portends a dim outlook for eliminating sexual harassment in the workplace in the future. "Raising awareness among men," according to one expert, "may be more important than getting women to complain" (Michelle Paludi, quoted in *New York Times,* Oct. 22, 1991, p. C1).

Even in social situations, the signals between men and women are often ambiguous and the possibility of misunderstanding is great. A man may misinterpret how a woman is dressed or behaving as interest in him. Some men, however, construe every smile, gesture, or pleasantry as sexual interest. In our culture, men are also expected to be sexually aggressive, so if a man interprets a behavior as interest (correctly or incorrectly) he is likely to pursue it. Many men interpret initial resistance on the part of a woman as part of the "game," influenced perhaps by the fact that some women first say "no" even when they want to be pursued (Muehlenhard & Hollabaugh, 1988). When, then, does sexual interest become sexual harassment? When a woman has made it clear that sexual advances or comments are unwanted, then any persistence on the part of the male is clearly harassment. This should be true anywhere, but it is particularly true in the workplace.

SEXUAL HARASSMENT OF COLLEGE STUDENTS

Sexual harassment can occur in any setting in which there are power differences between people. College campuses are certainly an example of this. In 1994, the president of a state university in Arkansas was dismissed on grounds of sexual harassment and inappropriate sexual activity. Several studies have found that many college women (17 to 33 percent, depending on the study) have been sexually harassed by an instructor (Adams et al., 1983; Fittzgerald et al., 1988; McCormack, 1985; Popovich & Licata, 1986; Reilly et al., 1986). When one considers all sources, half of all college women have been victims of harassment (Clark, 1993). On college campuses, males also are often the target of sexual harassment (Baier, 1990; Mazer & Percival, 1989; Roscoe et al., 1987).

Harassment can be direct (offering "an A for a lay" or threatening to lower a grade unless the student agrees to sex) or less obvious (e.g., crude comments, hugs, insistence on seeing the student alone). As can happen in the workplace, an instructor can create an offensive environment for a student even if there are no direct propositions or threats:

"It was during my second semester in college when I became the target of sexual harassment from my instructor. The ordeal started when he began to stare at me in class for long periods of time. I grew uncomfortable and more importantly, self-conscious. I couldn't figure out why he was staring at me. He then began to approach me and make small talk. In between 'Hello' and 'How are you?' he would squeeze in a compliment. Eventually his compliments became suggestive, and before long, he was making blatant sexual innuendoes to me in front of the class. He would say 'I can't concentrate on my lectures with you sitting in the front row looking like that,' or, 'You sure do make it hard (emphasis on hard) on a guy.'

"Once, when I wore a skirt to class, he asked, 'Did you wear this for me?' as he fixed his eyes on my legs. On another occasion as I was picking up my school bag he tapped me on the shoulder and said, 'Don't bend over in those jeans like that.' Once he said he'd like to take me to lunch and have me for dessert.

"I finally had enough and I began to sit in the back of the lecture hall. He would come to me after class and tell me he 'missed me in the front row.' That was when I quit going altogether. I had four friends in class who knew what was going on and they insisted after about two weeks of absences that I come back to class. They begged me to press charges against him.

"The day I returned, my professor asked me where I'd been and told me that 'if I knew what he was thinking I could charge him with sexual harassment.' The next week I wore a tie to class and he touched my tie and commented, 'I'd like to tie you up.' A girl I didn't know approached me after class and offered herself as a witness if I wanted to file a formal complaint. I wanted to do just that. However, I was and still am very concerned about my grades; I was sure that if I charged him with this behavior he would have penalized me through my grade.

"This happened two years ago, and I saw this man again last week. (We haven't seen each other since I had his class.) The first thing he said to me—in front of my other professor and many students—was 'Hi honey, my place or yours?' This was the first impression my new professor had of me. This time I'm angry. I don't deserve to be talked to this way, nor does anyone else. My friends were right—he should be stopped!!"

(from the author's files)

It is *never* appropriate for an instructor to make sexual, or even romantic, overtures to a student presently enrolled in his class or under his supervision (as when he is a thesis advisor). Because of the difference in power, all such cases should be regarded as harassment. As in the workplace, students who are the victims of sexual harassment may feel that there is no recourse, and the experience often results in negative emotional effects (Satterfield & Muehlenhard, 1990). Some victims feel so compromised that they change majors or drop out of school (Dziech & Weiner, 1984). Today, most colleges and universities have firm guidelines about harassment and grievance procedures that can be consulted by victims, and the U.S. Supreme Court recently ruled that under Title IX of the Education Amendment of 1972, schools can be held liable for not taking action (*Franklin v. Gwinnett County Public Schools*, 1992). Do you know if your campus has established a policy about sexual harassment?

HOW TO DEAL WITH SEXUAL HARASSMENT

The number of sexual harassment cases reported to the EEOC doubled during the 1980s and more than doubled again between 1990 and 1995 (EEOC). Still, only 5 percent of victims ever decide to take formal action (U.S. Merit Systems Protection Board, 1988).

Sexual harassment can often be prevented if the recipient is assertive at the first sign of inappropriate behavior. Unfortunately, some women are uneasy about being assertive for fear that they will appear "unfriendly" or "impolite." Keep in mind that you are not the one who has behaved improperly. Giving the harasser a firm "no" (or saying that you will not tolerate being talked to like that) is expressing your right to personal dignity. If you are not assertive, some men may misinterpret that as a sign of interest.

If the harasser continues his behavior after you have delivered a clear message that it is not appreciated, then you will have to take other steps to deal with it. Most women choose to avoid it passively (e.g., by ignoring it, dropping the class, or quitting the job) or to defuse it by stalling tactics or making a joke out of it (Cochran, Frazier, & Olson, 1997).

When the harassment is blatant, persistent, and highly offensive, as in the example of the university professor in the case history, stronger measures are necessary. Talk to the harasser, or write a letter to him, relaying the facts and again expressing clearly and firmly that the behavior must stop. Document all the episodes of harassment (dates, places) and list all witnesses. The very next time the behavior occurs, follow the grievance procedure of your company or school and report it. Those companies and schools that have implemented grievance procedures will protect you against retaliation. If you should be treated unjustly, report the problem to the EEOC or file a civil suit (remember, companies and schools can now be held liable for not taking action).

I urge you not to deal with harassment passively, but instead to take assertive actions. Why? Because if you deal with harassment by simply avoiding the harasser, he will start harassing someone else. Case in point: the university professor in the case study continues to teach and continues to treat other coeds in the same manner. Behavior like this will be stopped only when someone decides to press charges.

For more information on how to deal with sexual harassment, contact one of the organizations listed in the "Resources" section following the text.

SEXUAL ABUSE OF CHILDREN

In 1992, a former priest was convicted of sexually molesting 32 children aged 10 to 14, some allegedly in church. In the same year, a former national "teacher of the year" was sentenced to a year in jail for molesting young boys at his cabin home. In 1997, a Nobel Prize–winning scientist was sentenced to jail for sexually molesting boys. In separate cases in the state of Washington, two men with long histories of sexual abuse were convicted of raping many young children and killing several of them, leading that state to pass a "sexual predator" law requiring that career molesters be confined for life.

Our culture has a long history of denial about sexual abuse of children (Olafson, Corwin, & Summit, 1993), but an abundance of research in the past few decades has made it clear that this is a serious, widespread problem. How common is sexual abuse of children? In a comprehensive review of many studies, Peters, Wyatt, and Finkelhor (1986) found that estimates of how many children had been sexually abused ranged from 6 percent to 62 percent for girls and from 3 percent to 31 percent for boys. One of the problems in estimating sexual abuse is how we define it. Within psychology and psychiatry, the term *children* refers to prepubescent children (before age 13), and most professionals also stipulate that the abuse must be by someone at least 5 years older (DSM-IV, 1994). Kinsey et al. (1953) found that 24 percent of the women in their survey qualified as victims of sexual abuse by this definition. In the recent survey by Laumann and colleagues (1994), 17 percent of the women and 12 percent of the men reported having been victims of sexual touching (or worse) by an older person when they were younger than 13. For one-third of the female victims, the abuse occurred (or began) when they were less than 7 years old. The YWCA Rape Crisis Program says that *one out of every four girls and one out of every seven boys will be victims of some form of sexual abuse before they reach the age of 18* (Becker & Coleman, 1988; Wolfe et al., 1988).

As the above data show, nearly twice as many girls as boys are targeted as victims. Kinsey's group found that in 80 percent of the cases they studied, only one incident of molestation had occurred, but a 1985 *Los Angeles Times* poll found that for about 15 percent of all victims, the abuse continues for more than a year. Other studies have found that at least 10 to 15 percent of victims are abused by more than one adult (Conte et al., 1986; Tufts New England Medical Center, 1984).

Laumann et al. (1997) found that sexual contact with children usually involved fondling the genitals but that 10 percent of the girls and 30 percent of the boys who had been molested had been forced to perform oral sex, and 14 percent of the girls had been forced into vaginal intercourse. Eighteen percent of the cases involving boys included anal intercourse.

WHO MOLESTS CHILDREN?

The stereotype of a child molester is often that of a dirty old

---

TABLE 15–1

RELATIONSHIP OF CHILD MOLESTERS TO THEIR VICTIMS, AS RECALLED BY ADULTS WHO HAD BEEN SEXUALLY ABUSED AS CHILDREN (AGED 12 OR YOUNGER)

Abuser	Abused[a]	
	Women (17%)	Men (12%)
Stranger	7%	4%
Acquaintance		
Family friend	29%	40%
Older brother	9%	4%
Stepfather or mother's boyfriend	9%	2%
Father	7%	1%
Other relative	29%	13%
Teacher	3%	4%
Other	20%	21%

[a]Total percentages exceed 100% because some victims were abused by more than one person.

Source: Edward O. Laumann et al. (1994). Reprinted with the permission of The University of Chicago Press; adapted from Table 9.14, p. 343.

man who hangs around playgrounds offering candy to young boys and girls and who snatches the children away when their parents are not present. However, this is usually not the case. Studies have found that in the majority of molestation cases, the molester is an adult known to the victim, and often a relative (Bureau of Justice Statistics for 1992; Laumann et al., 1994; see Table 15–1). In fact, the YWCA Rape Crisis Program estimates that about 80 percent of all molesters are known by their victims (Kikuchi, 1988)

> "When I was 12 years old I was sexually molested by my sister's boyfriend. My mother and sister earlier that day had a doctor's appointment and he had offered to keep an eye on me while they were gone. . . . He dared me to do sexual things. If I refused he would then get forceful toward me and out of being scared I went along with it. He forced me to have oral sex on him. Finally, someone knocked at the door and they saved me from having to go through any more. . . ."

> "When I was little (ages 5–7), I was molested. My parents worked but let me stay with our elderly neighbors. The old man did this to me. He taught me how to kiss and touch his penis. He did this when his wife would either go to the store or leave out of the room. I hated it. He didn't like it when I'd fight him away from me. What I can't understand is why: why me. . . ."

> (from the author's files)

Researchers have attempted to classify molesters in many different ways. For example, Groth, Hobson, and Gary (1982) proposed that molesters could be distinguished as *fixated* (having arrested psychosexual development) or *regressed* (under stress, after normal development). Howells (1981), Lanyon (1986), and others prefer a distinction between **preference molesters** (who have a primary sexual orientation to children and are relatively uninterested in adult partners) and **situational molesters** (whose primary interest is toward adults and who consider their urges toward children, often done impulsively during stress, as abnormal). This last classification has much in common with the one in the American Psychiatric Association's DSM-IV (1994), which classifies molesters as pedophiles or nonpedophiles. You were introduced to the topic of pedophilia (from the Greek, meaning "love of children") in Chapter 14. A **pedophile** is a man for whom children provide the "repeatedly preferred or exclusive method of achieving sexual excitement." Thus, if a man were to commit an act of child molestation but was not repeatedly attracted to children, he would

not be considered a pedophile (here the term *situational molester* is appropriate).

An older classification of incestuous molesters versus nonincestuous molesters, which assumed that the former limited their offenses to family members and were not a threat to the general community, is no longer regarded as a useful distinction. Both groups have similar sexual preference patterns (Abel et al., 1981), and nearly half of all incestuous fathers and stepfathers have sexually abused other children during the same period they were abusing their own children (Abel et al., 1988b).

> "I am a victim of child sexual abuse. I was six years old and my godmother's fiance was the predator. . . . He made me do everything from vaginal penetration to oral-genital sex. . . . Later it was found that he did this to four of his six children, myself, two of my cousins and six other children. . . ."

> (from the author's files)

Pedophiles have been further classified according to their psychological characteristics (Cohen, Seghorn, & Calmas, 1969). The *personally immature pedophile* is the most common. He is attracted to children because he has never developed the social skills necessary to initiate and maintain a sexual relationship with an adult. A relationship with a child allows this individual to be in control, which he cannot achieve with an adult. Notice that the primary motivation here, control, is similar to the motivation in many rape cases. The *regressive pedophile,* on the other hand, is likely to engage in sexual relations in an impulsive manner, often with a child who is a stranger. These people describe their acts as the result of an uncontrollable urge. Unlike the personally immature pedophile, the regressed pedophile has a history of normal adult sexual relations. Regressed pedophiles begin turning to children as a result of developing feelings of sexual inadequacy, often due to stress, and the problem is often complicated by alcohol abuse (Gebhard et al., 1965), which they often use as an excuse for their behavior (MacNamara & Sagarin, 1977). A third type, and fortunately the least common, is the *aggressive pedophile.* These people are not satisfied with just having sex with their young victims, but are aroused by inflicting physical injury. Here, too, notice that the primary motivation—sadistic acts—is similar to the motivation in one of the classifications of rape discussed earlier. In fact, these men can be

preference molesters Child molesters who have a primary sexual orientation to children and who are relatively uninterested in adult partners.

situational molesters Child molesters who have a primary sexual orientation to adults. They have sex with children impulsively and regard their behavior as abnormal.

pedophilia A condition in which sexual arousal is achieved primarily and repeatedly through sexual activity with children who have not reached puberty.

called child rapists (Groth et al., 1982), and they often show a variety of other antisocial behaviors. Seven to 15 percent of all child molestation cases include physical harm to the victims (Conte, 1991; Mrazek et al., 1981).

Pedophiles typically have molested many children before they are caught, if they are caught at all. For example, in one study of male pedophiles who were undergoing treatment, nonincestuous offenders who targeted girls had averaged 20 victims each, while those targeting boys had averaged nearly 200 victims each (Abel et al., 1987). Child molestation committed by men who are not pedophiles is usually within a father-daughter (or stepfather-stepdaughter) relationship and generally involves only one child (Freund, Watson, & Dickey, 1991). Women are sometimes involved in acts of sexual molestation, but pedophilia among women is rare (Finkelhor, 1984; Laumann et al., 1994).

Only a small minority of male sex offenders target both boys and girls (Groth & Birnbaum, 1978). Heterosexual males account for 95 percent of the cases of sexual abuse of girls, while "homosexual" males account for most cases involving boys (Finkelhor & Russell, 1984) (the word homosexual was placed in quotes because most molesters of boys do not have homosexual attractions to adults; Freund & Langevin, 1976). Although the ratio of female to male victims is 2 to 1, because male offenders who target boys generally have many more victims, the ratio of heterosexual pedophiles to homosexual pedophiles has been estimated to be approximately 11 to 1 (Freund & Watson, 1992).

What is the child molester like personally? Most are respectable and otherwise law-abiding citizens (Lanyon, 1986). Many are shy, passive, and unassertive (Langevin, 1983), although incestuous fathers are generally very domineering and controlling within their own families (Meiselman, 1978). However, for the most part there are no reliable personality characteristics correlated with pedophilia (Lanyon, 1986; Okami & Goldberg, 1992). In short, there is no single personality profile; child molesters are a very heterogeneous group (McAnulty, Adams, & Wright, 1994).

We do know that, unlike normal people, child molesters have the capacity to be sexually aroused by young children (Abel et al., 1981; Finkelhor & Araji, 1986; Marshall et al., 1986). This ability is generally acquired when molesters rehearse and fantasize what

Figure 15–6

they will do to children while they masturbate, and this type of rehearsal often goes on for years. A majority of pedophiles were themselves the victims of sexual abuse as children (this is not true of sex offenders against adults; Freund & Kuban, 1994; Groth, 1979), and thus may never have learned that it is wrong to have sex with children. However, some experts believe that the molesters' sexual behaviors are primarily motivated by nonsexual needs (e.g., affection or control; Groth et al., 1982; Lanyon, 1986). The pedophile, unlike most other paraphiliacs, usually does not display other paraphilias (Fedora et al., 1992).

One thing that is well established is that the deviant behavior often begins early in life (Becker & Abel, 1985). One study found that the median age for the first offense was 16 (Groth et al., 1982). In 1992, in Buckley, Washington, a 9-year-old boy was found to have raped or molested eight younger children at knifepoint over a two-year period. Adolescents may be responsible for as many as 30 to 50 percent of child sexual abuse cases (Davis & Leitenberg, 1987; Groth & Loredo, 1981; see Murphy et al., 1992). A recent study found that pedophiles recall having had a much-greater-than-normal curiosity during their own childhoods in seeing naked children, without having had a similar curiosity about adults (Freund & Kuban, 1993). Very young children (6 to 12 years old) who sexually abuse other children have generally been sexually abused themselves and/or have come from highly chaotic families (DeAngelis, 1996).

Tragically, the true pedophile, in order to satisfy his sexual urges, often finds positions where he has access to children (teacher and "other" category in Table 15–1). In New Orleans, a man who worked as a clown at children's birthday parties was convicted of molesting dozens of children who had attended the parties. In a neighboring city, an officer in the child molestation unit was arrested for having sex with the children he was supposed to protect. Pedophiles have been found in positions as teachers, scout leaders, camp counselors, and even ministers or priests.

"I attended a very respectable summer camp during my junior high summers. Every session, about midway through the camp, the camp director took us to _____ Creek to go on the sliding rock. He would tell the kids if they slid down the rock 'skinny sliding' (nude) they would get to be in line first for the dinner buffet. The counselors were sharply divided on this. Some thought it was just funny and others thought it was odd. Two years ago, the director was caught for child molestation. He was convicted and served time in jail."

"I entered a monastery for boys. . . . As a novice, I was told to go to confession daily and the Father Superior would counsel me in both Christian education and my personal life. Confessions were very informal. The Father Superior heard them from 7:30 P.M. until midnight in his room. . . . One night he said, 'And do you know how they took oaths in the Old Testament? One would put his hand on the other man's genitals. But there was no sexual connotation intended. The Jews viewed sex differently than we do.' I was a little puzzled and didn't ask any questions.

"About a week later I went to see one of the brothers for confession, but he told me that the Father Superior wanted to see me. . . . He said, 'Do you want to take an oath like in the Old Testament?' I was stunned and didn't really know what he meant. Finally he took my hand and placed it under his robe on his penis. I was shocked and pulled my hand away immediately, but he said, 'Don't be afraid, there is no sexual intent here. You are keeping a very ancient form of taking an oath.' As he was saying this he took my hand again and placed it on his penis. I wanted to pull back, but I was afraid he would go into another temper tantrum. . . . He didn't do anything else that night, but he talked to me about how much we had to be obedient children. He often stressed obedience—total obedience. (This was how he set us up. We must be totally obedient.)

"Then it happened. One night we were talking and he said, 'Do you want to kiss my penis?'—I was so naive. I didn't know what he was talking about. He said, 'Do you want me to kiss your penis? That's how much I love you.' I said no, but then he said, 'Don't you think I love you? There is no sexual intent. It is a pure act of platonic love.' (He defined platonic love as love without passion—including sexual contact. He emphasized this: without passion. Passion is wrong. All passions. This passionless sex, he explained, could only be between him and his children.) Then he drew me closer to him, he opened my robe, unzipped my pants, put his hands into my underwear, brought my penis into the open and kissed it.

"It later turned into fellatio. . . . He taught me what he wanted done to him. He had orgasms every time that 'I went for confession.' I was instructed never to have orgasm. Never! 'That would not be good for me spiritually.'

"I have some very deep scars and am now receiving therapy."*

(from the author's files)

EFFECTS OF ABUSE ON THE CHILDREN

Because of the strong social stigma against sexual relations between adults and children in our culture, it has traditionally been assumed that the effects on the child were always very negative and long-lasting. Considerable research has now shown, however, that there is no single "post-sexual-abuse syndrome" with a specific course or outcome (e.g., Beitchman et al., 1991, 1992; Lipovsky & Kilpatrick, 1992). In fact, some "victims" recall their experiences as having been positive or pleasurable, leading some investigators to question whether the sexual behavior in these cases can be called "abuse" (e.g., Bauserman & Rind, 1997; Havgaard & Emery, 1989; Okami, 1991). A recent review of several national studies found that child sexual abuse generally was not associated with long-term psychological harm (Rind & Tromovitch, 1997). The children in those studies must have had a different experience than the many others who report anxiety, fear, depression, panic disorder, sexual dysfunctions and other (often long-term) negative reactions (Beitchman et al., 1991, 1992; McCauley et al., 1997; Tebbutt et al., 1997). It would appear that the effects of adult-child sexual relations represent a continuum of experiences.

Several factors contribute to the particular response of a child. The greatest harm in terms of long-lasting negative effects is seen in cases where the abuse is frequent or long-lasting; where the abuse involves penetration, force, threats, or violence; where the abuser is the father or stepfather; and where there is little support from the mother (Bauserman & Rind, 1997; Beitchman et al., 1992; Kendall-Tackett, Williams, & Finkelhor, 1993; Rind & Tromovitch, 1997). Revictimization by others is not uncommon and substantially adds to psychological and emotional distress (Gold, Hughes, & Swingle, 1996; Kellogg & Hoffman, 1997). The most frequent symptoms include fears, poor self-esteem, inappropriate sexual behavior, a greater tendency to have homosexual experiences, and depression. Some victims of severe sexual abuse later show promiscuity and increased sexual risk-taking (Friedrich, 1993; Stock et al., 1997); they confuse sex with love or become sexually preoccupied or aggressive. Some victims of "serious" sexual abuse show

---

*This young man later found out that many other boys had been victimized by the same man.

no apparent symptoms (Kendall-Tackett et al., 1993), but in some cases this might be due to the denial or emotional suppression displayed by many victims (Leitenberg, Greenwald, & Cado, 1992).

Only about a third of all child victims inform their parents or another adult of the abuse (Finkelhor, 1979), and reactions that are perceived by the child as less than supportive can greatly add to the negativity of the experience. Consider, for example, the following three cases:

"When I was younger, I was sexually abused by one of my father's good friends. I told my mother . . . but they acted like they did not believe me. I felt very embarrassed." (female)

"At the age of 9, I was molested by a man right outside of my neighborhood. I told my mother what had happened. I was very hurt and scared, but instead of talking to me, my mother refused to talk about it and acted as if it was my fault. Years later I discovered the man had done the same thing to other children. By then it was too late, I was already emotionally scarred." (female)

"I was about 8 years old when I was sexually abused by a friend of my father. I refused, but he beat me up and made me have sexual intercourse with him. Later the same night I told my father, but my father, instead of going to the police as I thought, beat me very harshly with a wood stick." (male)

(from the author's files)

Like victims of rape, victims of child molestation have a tendency to blame themselves, which can greatly affect self-esteem (Kellogg & Hoffman, 1997).

"I was sexually abused for 9 years as a child by my older sister's husband. For years I believed it had happened because there was something wrong with me."

(from the author's files)

A review of many studies concluded that about two-thirds of all victims of child sexual abuse show

Figure 15–7 This self-portrait by a victim of sexual abuse shows the emotional suffering that is often experienced long after the crime.

substantial recovery during the first 12 to 18 months (Kendall-Tackett et al., 1993). The same has been found for a group of preschoolers who were ritualistically abused (Waterman et al., 1992). Even for these individuals, however, some symptoms, such as sexual preoccupation and aggressiveness, may worsen (Friedrich & Reams, 1987; Gomes-Schwartz et al., 1990). From 10 to 24 percent of all abused female children not only fail to improve over time, but actually get worse (see Kendall-Tackett et al., 1993). As adults, these women display problems (e.g., emotional distress, drug and alcohol abuse, suicide attempts) that are just as great as in women presently being abused (McCauley et al., 1997). Sexual dysfunctions are common among women who experienced childhood abuse involving penetration (Sarwer & Durlak, 1996). Many adults who were sexually abused as children also have serious eating disorders (anorexia nervosa, bulimia) (Laws & Golding, 1996; Wonderlich et al., 1996).

RECOVERED (FALSE?) MEMORY SYNDROME

"For the first time, when I was 17 a boyfriend reached out to touch me in a sexual manner. In that split second, I flashed back to when I was 9 and my uncle (by marriage) was molesting me. Until that moment I had no conscious memory of that event. I began to scream and cry and the poor guy didn't know what to do other than bring me home.

"After 28 years, I still do not remember everything that happened that day. I am convinced that this was not the only occasion that he molested me. For many years I had a deep fear of the Christmas holidays, in particular the family Christmas Eve party.

"I recently found out that this man repeatedly molested all seven of my female cousins. When I finally asked my mother how this could have been going on . . . supposedly she and her three sisters did try to protect us by watching us and trying to keep him from being alone with us. She told me that they didn't tell because they didn't want to hurt their sister (his wife). . . ."

(from the author's files)

One of the most bitter debates in the field of sexual abuse today is whether or not people can forget sexually traumatic past experiences and later remember them. On the one hand, there are therapists who believe that sexual abuse can be so traumatic that individuals, in order to cope, forget or suppress memories of the event, often for many years. Later, either spontaneously or while undergoing therapy, the individual remembers the past events. Many studies have reported cases of sexual abuse victims who said that they had experienced some degree of amnesia of the events for a period of time (e.g., Feldman-Summers & Pope, 1994; Gold et al., 1994; Herman, 1992). These studies find that, on average, about 50 percent of all victims report some amnesia (see Kristiansen, Felton, & Hovdestad, 1996).

On the other hand, others claim that most "recovered" memories are, in fact, false memories created by therapists working with vulnerable people (e.g., Goodyear-Smith, Laidlaw, & Large, 1997; Loftus, 1997a, 1997b; Ofshe & Watters, 1994). Many of the disbelievers have even formed an organization, the False Memory Syndrome Foundation. According to Harvard psychiatrist Harrison Pope, "There are a certain number of therapists who see sexual abuse in every patient they see. And they can lead a patient by a thousand suggestions and implications, and by reinforcing what they do and don't listen to" (quoted by Cole, 1994). Accounts of abuse victims who have retracted their "memories" have been published (e.g., de Rivera, 1997), and some believe that the implanting of false memories has become epidemic.

Proponents of the theory of repressed memories cite evidence that few claims of amnesia are false (Hovdestad & Kristiansen, 1996) and that, in fact, claims of false memories are largely unsubstantiated (Pope, 1996). Some studies have verified that abuse actually occurred (Williams, 1994). It has even been claimed that those who believe that repressed memories are false memories do so because of their own personal needs and sociopolitical interests (Kristiansen, Felton, & Hovdestad, 1996).

If you want to read more from both sides of this debate, see Pope (1996) and the replies in the September 1997 issue of *American Psychologist,* or Volume 5(2) of *Health Care Analysis* (1997).

INCEST

The term incest comes from a Latin word meaning "impure." **Incest** refers to sexual activity between relatives who are too closely related to marry and is illegal in all 50 states. One reason incest is illegal is that this type of mating has the highest probability of producing defective offspring (Baird & MacGillivray, 1982; Thornhill, 1992). Until recently, many people thought that incest was rare, but it happens much more often

than people might (or might want to) believe. Look again at Table 15–1. About 8.5 percent of women and 2.4 percent of men in Laumann et al.'s (1994) national survey said that they had been sexually abused by a family member before the age of 13. Another recent study found that 3.3 percent of the men and 4.5 percent of the women surveyed reported having been molested by a relative as children, and in about 60 percent of these cases the molestation was "often" or "ongoing" (Janus & Janus, 1993). Only about 2 percent of adult-child incestuous relations ever get reported to authorities (Russell, 1984), and some victims do not understand that what happened to them was abuse (Kikuchi, 1988).

Although they have received little attention by researchers, **polyincestuous families** are not uncommon (Faller, 1991). In these families, there are both multiple abusers and multiple victims, both across generations and within the same generation. Victims often become abusers, and in many cases view the activity as acceptable, or even expected. Unlike other cases of child molestation, much of the abuse in polyincestuous families is done by females (Faller, 1987; McCarty, 1986). In over half such families, there are also abusers and victims from outside the family (Faller, 1991).

Much more is known about specific types of incest. After you have read about these, we will return to a discussion of family dynamics in incest.

INCEST BETWEEN SIBLINGS

Brother-sister incest, or in some cases *brother-brother incest,* is believed to be five times more common than parent-child incest (Cole, 1982; Gebhard et al., 1965; Smith & Israel, 1987). Finkelhor (1980) surveyed college undergraduates and found that 15 percent of the women and 10 percent of the men in his sample had had some type of sexual experience with a sibling. This usually involves fondling, but sometimes includes sexual intercourse as well (Meiselman, 1978). In families in which sibling sexual activity occurs, the parents are often distant or inaccessible and/or create a sexual climate in the home (Smith & Israel, 1987).

> "I was a victim from age four until thirteen. My two brothers continually cornered me and forced me. My mother worked and was much too busy or tired to observe. She did not believe me when I tried to tell her. To this day, I still have problems sexually, emotionally, and mentally. My way of solving it was to move away (1200 miles away)."
>
> (from the author's files)

There is considerable debate among professionals about the long-term effects of

incest Sexual contact between two closely related persons.

polyincestuous family Families in which there are both multiple incestuous abusers and multiple victims, across generations and within the same generation.

sibling incest. Some believe that *if* the experience occurs between siblings of nearly the same age (less than a 4-to-5-year age difference), *if* there is no betrayal of trust, and *if* the children are not traumatized by adults in cases where their activity is discovered, then it is just "sex play" and "part of growing up" (Finkelhor, 1980; Forward & Buck, 1978; Pittman, 1987; Steele & Alexander, 1981). However, as you can see from the previous case history, the effects can sometimes be quite negative. This is particularly the case when the incest involves coercion, threats, force, or betrayal of trust (Bank & Kahn, 1982; Canavan et al., 1992; Courtois, 1988). Only about 12 percent of all people with childhood sibling incest experiences have ever told anyone, and "the pain of secrecy" adds to the negative experience (Finkelhor, 1980). In cases where physical pleasure is experienced in exploitive or abusive incest, the result can be confusion and guilt regarding sexuality (Canavan et al., 1992). Other long-term effects can include distrust of men in general, poor self-concept, depression, and confusing intimacy and sex (resulting in thinking that sex is the only way to relate to men; Canavan et al., 1992; Cole, 1982; Russell, 1986).

PARENT-CHILD INCEST

"At the age of five, I was the oldest of two children (my little sister was one year old). My mom would go to church and leave me in the care of my father. He would take out his penis and have me look at it and touch it, but I didn't know what was happening. It stopped for a while, but then he would try to put his tongue into my mouth. At the age of eight, after my mom had another baby, he would touch my vagina and lay on top of me (all 195 lbs.). At the age of 11, he got into my bed trying to take off my panties. I put up a fight but they were removed. He then took out his penis and tried to have sex, but I moved continuously trying to wake up my sister who shared my bed (unsuccessfully). I began to cry; he got up and said he was sorry. I thought 'this bastard!' I was so afraid. I thought if I told he would do it again, and everyone would hate me for breaking up the family. Mothers! Please watch out for your kids. I am 19 years old; I am a sophomore here at UNO, and it still hurts. I try to get over the hating of my father, but I don't really care what he thinks about me." (female)

"My father began to molest me when I was 12—on the night after my mother's funeral. I am an only child and he explained that since my mother was gone I was going to have to

function as the 'woman' of the house. At first I was only required to function as a woman in the evenings (and most mornings too), sharing his bed as my mother had. As I grew older additional chores were added such as cooking and cleaning. It was always explained that if I truly loved him I would never hesitate to do his bidding—oral and anal sex, giving him showers, dressing him, etc. Finally I managed to get the courage to leave my father. . . . Throughout my entire adult life I have never had a successful relationship. I'm not gay, but I don't know how a man should act. I only know how to function as the 'lady of the house.'" (male)

(from the author's files)

Of all the types of incest, parent-child incest probably evokes the greatest emotional response in people. After all, most people believe that it is the role of parents to nurture, provide for, and protect their children, not turn them into sexual objects for their own selfish needs. Yet adults who commit incest with their children come from all walks of life. Former Miss America Marilyn Van Derbur focused much attention on this issue when she revealed that she had been repeatedly assaulted sexually beginning at age 5 by her father, a millionaire businessman, socialite, and philanthropist—a pillar of the community.

Although parent-child incest accounts for only 10 percent of all incest cases (Finkelhor, 1979), nearly 80 percent of arrests involve *father(stepfather)–daughter incest* (Francoeur, 1982). Stepfathers are more likely to victimize their children than are biological fathers (Finkelhor, 1979; Renshaw, 1983), but both types of cases are common. Fathers who commit incest are generally shy and publicly devoted family men who appear to be quite average, but in their own homes they are very domineering and authoritarian—"king of their castle" (Meiselman, 1978; Rosenfeld, 1979). They tend to be impulsive, but generally are not psychotic. In relations with their children, they are often overprotective, selfish, and jealous (Hinds, 1981).

Incestuous fathers, obviously, have the capacity to become sexually aroused by children (remember, nearly half of all incestuous fathers also molest other children; Abel et al., 1988b). Some incestuous fathers try to rationalize their behavior, saying that they are teaching their children in a loving way about life (Hinds, 1981). Virginia Johnson (of Masters and Johnson) has found that incestuous fathers are often very religious and feel that incest is less of a sin than masturbation (*New York Times*, November 1986).

Incestuous fathers who are caught often claim that the influence of alcohol or drugs caused them to commit incest, or that *they* were the ones who were seduced.

Figure 15–8 Former Miss America Marilyn Van Derbur later announced that her father (seen here congratulating her) had sexually abused her when she was a child.

Many of us have had too much to drink at times, but we didn't molest children while under the influence of alcohol. An incestuous father may have a drinking problem, but first and foremost he has a sexual problem.

The role of the mother in father-child incest is critical, and her response, or rather her lack of any, is often as traumatizing to the child as the molestation. Consider, for example, the responses of the following four victims:

". . . My mother didn't believe me. Even at age 19, I still feel resentment for her for not protecting me and for not believing me. At the time she was totally dependent on him (stepfather) and the thought of losing him was unthinkable"

". . . I thought every father did that to their daughter. I thought it was a part of growing up, especially because my mom knew about it"

". . . I eventually told my mother and she told me that it was nothing, that he probably thought it was her and she didn't mention it again"

". . . I grabbed onto her leg saying—screaming, crying—'Don't leave me please. I don't want to be here with him!' She said I had to behave and then left without even looking at me. From this moment on I knew she was not going to listen to me, let alone protect me. This is why I am angry today. . . ."

(from the author's files)

Some mothers do call authorities when they become aware of the incest, but more than two-thirds do not try to protect their daughters (Herman & Hirschman, 1977). Although some mothers of incest victims have normal personalities (Groff, 1987), many are emotionally distant from their children and dependent on their husbands. They fear abandonment more than living with the fact that their children are victims of incest (Finkelhor, 1984; Rosenfeld, 1979). Many were victims of child abuse themselves (Summit & Kryso, 1978), and continue to suffer physical abuse within the marriage (Truesdell et al., 1986). Some even force their daughters to assume the role of sex partner in order to avoid having sexual contact with their husbands (Justice & Justice, 1979).

In some cases, the mother is the abuser. *Mother-son incest* has generally been regarded as rare (see Finkelhor, 1986). However, some believe that it may be underreported because surveys rarely ask specific questions about this type of incest (Lawson, 1993). Mother-son incest typically occurs in disrupted families where the mother seeks emotional support and physical closeness from her son rather than from other adults (Krug, 1989). It generally starts by sharing the boy's bed and may progress to sexual intercourse. Contrary to earlier reports (Meiselman, 1978), most of these mothers are not psychotic or suffering from severe psychological damage (Krug, 1989).

Grandparents as incest perpetrators have also been found to be uncommon (e.g., Cupoli & Sewell, 1988; Kendall-Tackett & Simon, 1987; Russell, 1986), but may be on the increase because of the increased number of working parents leaving their children with the grandparents.

"I only had to see him when we visited in the summer. He would get me downstairs in the cellar alone. . . . He would touch my vagina and I remember it hurting all of the time. He

Box 15-C CROSS-CULTURAL PERSPECTIVES

Incestuous Inbreeding—A Universal Taboo?

Incest has often been called the universal taboo, and there are many theories as to why it is supposedly banned throughout the world. One of the major theories comes from Freud (1913), who believed that most people prefer close kin as mating partners:

> [It is] beyond the possibility of doubt that an incestuous love choice is in fact the first and regular one. . . . (Freud, 1933/1953, pp. 220–221)

But inbreeding can lead to serious genetic defects in offspring (Thornhill, 1992) and would eventually ruin a society if it were practiced widely. Freud believed it was for this reason that societies make rules against incest.

A second, much different, theory was proposed by Edward Westermarck in 1891. He believed that close physical contact between two people during childhood naturally results in loss of sexual attraction. Findings of several anthropological studies have been used as support for this theory (e.g., McCabe, 1983; Pastner, 1986; Shepher, 1971; Wolf, 1970). However, Nancy Thornhill (1991) has proposed that incest rules are not aimed at close family members (because of the Westermarck effect), but instead are directed at more distant kin (cousins) and special kinds of adultery (e.g., a man and his father's wife). Inbreeding between cousins, she says, "can concentrate wealth and power within families to the detriment of the powerful positions of (male) rulers in stratified societies."

Claude Lévi-Strauss (1969) offered yet a fourth explanation. His "alliance theory" says that the incest taboo exists not to prohibit marriage within the family, but is "a rule obliging the mother, sister, or daughter to be given to others." It ensures that women are exchanged between groups. According to Lévi-Strauss, "The incest taboo is where nature transcends itself. It brings about and is the advent of a new order."

There is no question that incestuous inbreeding is viewed very negatively in Western cultures (Ford & Beach, 1951). In the United States, marriage between first cousins is illegal in most states (Bratt, 1984). Fewer than 0.5 percent of all marriages in North America and Western Europe are between first cousins (Lebel, 1983). But what about elsewhere—is inbreeding viewed with equal negativity in non-Western societies? In times past in some cultures (e.g., Egyptian, Inca, Hawaiian), brothers and sisters married within royal families (Gregersen, 1982). For example, the Egyptian pharaoh Rameses II's wives included his younger sister and three of his daughters. Today, in areas that are primarily Muslim (northern Africa; western and southern Asia; Central Asian republics of the former Soviet Union; and northern, eastern, and central India), one-fourth to one-half of all marriages are between persons who are second cousins or closer (see Bittles et al., 1991, for a review). Marriages between parallel first cousins (a male and his father's brother's daughter) are particularly preferred. Among the Hindus of southern India, between 20 and 45 percent of marriages are between close relatives, with the preferred unions being uncle-niece or cross–first cousins (a male and his mother's brother's daughter). One-third to one-half of all marriages in sub-Saharan Africa are believed to be between relatives, and the practice is also thought to be very common in China.

Are there sociological or economic factors that might predict the acceptance of incestuous inbreeding? Yes: in those areas of the world where it is practiced, it is most common among either (a) poor, uneducated rural people, or (b) very rich land-owning families (again, see Bittles et al., 1991, for references).

In conclusion, mother-son incestuous inbreeding is unheard of in any culture, and with rare exceptions, all cultures similarly frown upon father-daughter and brother-sister relations. However, outside Western culture, marriage between first cousins or between uncles and nieces is widely practiced.

would sometimes take his penis out of his pants. . . . I never talked to anyone about it. One time I briefly talked to my sister about it at his funeral. She asked me why I did not seem sad. . . . She said it happened to her also. If it happened to me and my sister, it must have happened to all of the other grandchildren. I wonder if my mother was sexually abused as a child. I wonder if she knew or my grandmother knew. . . ."

(from the author's files)

A recent large study of incestuous grandparents found that nearly all cases were *grandfather-granddaughter incest* (Margolin, 1992). Stepgranddaughters were at a higher risk than biological granddaughters. *Most of the abusive grandfathers had also been sexually abusive fathers.* One has to wonder what kind of mother would leave her children with an adult who had sexually abused her when she was a child. Contrary to earlier studies which typically described the abuse as "gentle," many of the incestuous grandfathers used threats, physical assault, or suddenly grabbed the grandchildren's genitals.

EFFECTS OF INCEST ON CHILDREN

In this section, I will focus on the effects of father-child incest, although much of what you learn will be true of victims of other types of incest as well. Father-child incest generally involves frequent, long-term abuse, and in many cases there is vaginal and/or anal penetration. These are precisely the factors that often cause the greatest long-term harm in child molestation cases (Beitchman et al., 1992; Kendall-Tackett et al., 1993). A lack of support from the mother also substantially affects the victim's outcome.

Another important factor that leads to long-term problems in victims of father-child incest is that many victims blame themselves (often with encouragement from the father; Morrow, 1991). Thus, victims of incest are similar to rape victims in this respect. Consider, for example, the feelings of the following two victims:

"I am 18 now and I believe that it was my fault. No matter how I look at it, I feel as though I could have prevented it. . . ."

"[I] still question myself about if there was something I could have done to prevent it, or worse yet, if there was, somehow, something I did to cause it. . . ."

(from the author's files)

Victims can be very confused if they experienced any degree of physical pleasure during the abuse:

"In my mind, I blamed me, not him. To make matters worse, somewhere along the way my 12-year-old adolescent body had my first orgasm. I hated it, hated myself, hated him. . . . I couldn't understand something I hated feeling good. Heavy, heavy guilt trip. . . ."

(from the author's files)

Sadly, studies have found that many people, particularly men, also *incorrectly* attribute blame to victims, especially if the victims were adolescents at the time of the abuse (Collings & Payne, 1991; Waterman & Foss-Goodman, 1984). If a victim tells someone of his or her experience and is met with this kind of response, this too contributes to a lasting negative outcome.

Children who are victims of incest often have a variety of long-term problems besides self-blame (Lindberg & Distad, 1985; Scott & Stone, 1986). However, as with other types of child molestation, there is no definite or typical "incest syndrome" (Haesevoets, 1997). As adults, victims may have feelings of anxiety, helplessness, and powerlessness. Many victims suffer from low self-esteem and poor self-image (including a sense of separate selves or a loss of memory about the self; Cole & Putnam, 1992).

"I've always felt alone, disassociated. Even when I'm with friends I don't feel like a part of the group. In class, I've never felt like I belonged. . . ."

". . . But deep down I feel really disgusted about my body. . . ."

(from the author's files)

Some victims later develop eating disorders (Hall et al., 1989; Hambridge, 1988), while others suffer from major depressive episodes and alcohol or drug abuse (Burnam et al., 1988).

"By the time I was 13, I had found alcohol and drugs as a safe haven from the immense pain inside me. By the age of 17, I was a full-blown alcoholic experiencing blackouts. . . ."

(from the author's files)

Victims of child incest may also suffer from a variety of sexual problems. Some abused children may even victimize other children in an attempt to work out the distressing events (Johnson, 1989). As adults, they may be unable to form close, trusting relationships because they expect to be betrayed, rejected, and further abused (Cole & Putnam, 1992). This can lead to avoidance of relationships, or, at the other extreme, promiscuity. Victims are also a likely target for revictimization.

> "I find I cannot have open relationships with guys. It is hard for me to trust anyone. For the one time I tried to open up to a guy, I ended up even more hurt. I am 20 years old and I cannot even kiss a guy, much less give a simple hug to someone. . . ."

> "My relationships with men have all been abusive, even to the point that I have been battered and raped on several occasions. . . . I choose men that are very similar to my Dad . . . mainly because that's all I know. . . ."

> (from the author's files)

Boys who are victims of incestuous mothers often experience such anxiety and disgust that as adults they emotionally reject all women (Krug, 1989).

In one study of the backgrounds of women in prison, it was found that over half of the prostitutes, over a third of the felons, and one-quarter of the child molesters had been incest victims (Hinds, 1981). In a more recent study of imprisoned male serial rapists, nearly half had been victims of father-son or stepfather-stepson incest (McCormack et al., 1992).

FAMILY DYNAMICS

Father-child or stepfather-stepchild incest (and in many cases, sibling incest) typically occurs in certain family contexts (Canavan et al., 1992; Larson & Maddock, 1986; Finkelhor, 1978). We have considered some of their characteristics, but let us review them here.

Within the home there is an all-powerful, authoritarian male (who may appear shy but otherwise normal to others) who sets hard, fixed gender roles for everyone and establishes an isolated, rigid system with strong external boundaries that separates the family from the outside world. However, boundaries within the family become blurred, and all emotional needs are met within the family. Independent thoughts and feelings are considered destructive as the family becomes isolated and family members become overdependent on one another. Although marital problems are common, the mother fails to protect her children, adding to their sense of abandonment and fear. Themes associated with insecure parent–child incest (rejection, role reversal, fear) are common within incestuous families (Alexander, 1992). Children who grow up in an environment like this may view it as normal and repeat the pattern when they become parents.

PREVENTING AND DEALING WITH CHILD SEXUAL ABUSE

The first thing to do to minimize the chance of your child's becoming a victim of sexual abuse is to *educate yourself.* You should be able to talk openly and candidly about sexual matters, including sexual abuse, with your child. You should be aware of the fact that young children who tell people about being molested are almost never making the stories up. Remember that most molesters are known to their victims, so do not doubt your child if he or she tells you that the molester is someone you know.

The next thing to do is to *educate your child.* As I have indicated in previous chapters, it is important that children know the correct names for parts of their anatomies (e.g., penis, vagina). If molested, they will then be able to describe to you where they were touched. What words do you use in front of your children to describe body parts? What do your children learn if you refuse to mention certain body parts or functions, or use slang terms?

> Dear Ann Landers:
> I have just spent a week on jury duty, hearing evidence in a sexual battery case. One victim was an 11–year-old girl in the sixth grade. She had difficulty describing what had happened because the poor child did not know the names of the parts of the human body. . . . When asked if she had ever seen a diagram of the human body, she replied, "No, we don't even get that till next year."

> (Reprinted with permission of Ann Landers and Creators Syndicate.)

Children should not be ashamed of their bodies, but they should also know which parts of their bodies are personal or private and should not be touched by other people. You must teach your children to be assertive and to know that it is all right to say "No!" to an adult who attempts to touch them in the genital area or on their breasts or buttocks. Discuss this with all family members present—incest is difficult in families in which there are no secrets. Your child must also know that he or she won't be punished for talking about being molested, no matter what the molester might say. Similarly, children must know that although some molesters will threaten to hurt their parents, these threats are lies. Play "what if" games, letting your child act out what to do if approached by a molester. Treat this as a serious topic and be matter-of-fact, just as you would be when teaching traffic safety. Finally, it is crucial that your child know that he or she can communicate with you about sexual matters in general. If you are always open to discussion about sex, your child will be more likely to confide in you.

Know what to do if your child is molested or approached by a molester. Reassure him or her that talking to you about it was the right thing to do. Let your child know that being molested was not his or her fault and stress that it is the adult who was wrong. Don't confront the molester in front of your child, and if the molester is a family member or friend, let your child know that you want to help the molester get well. Let your child know that he or she will not be harmed by the molester. Finally, contact the police, even if you don't intend to press charges. The police can put you in contact with support groups and give you useful information as well. In reporting the case to the police, you also may help them prevent further child abuse.

There are certain signs to look for that may indicate that a child has been a victim of sexual abuse. A victim may suddenly begin to have episodes of depression, crying, or other out-of-character changes in mood or personality. He or she may begin to want to be alone or become afraid to go to a certain place or see a certain person. A victim may suddenly start to have problems at school or with discipline. Remember, if abuse is occurring, a victim may be unwilling or fearful to ask for help (abusers often threaten their victims). If you are a parent, do not write off these types of behaviors as a phase. Something serious might be hurting your child.

Even if you are not the parent, you can still help. Most states have a special office that people can contact if there is a suspected case of sexual abuse of a child. Staff answering the phone will ask for information, will determine if an emergency exists, and will have a social worker sent to investigate if action seems necessary. There are often several areas of specialization within these offices—for example, abuse and neglect, sexual abuse, and abuse within institutions (e.g., suspected abuse within a child day care center). Investigators from this office generally provide services for the families and children involved, and in extreme cases can have a child removed from a home and put in foster care.

Myths and stereotypes about child abuse are also being refuted by movies, TV, and media stories. As the general public becomes more informed about the topic of child sexual abuse, individuals will feel more inclined to report such cases. In order to focus public attention on this matter, television celebrity Oprah Winfrey, Michael Reagan (Ronald Reagan's son), and U.S. Senator Paula Hawkins have revealed that they were sexually molested as children by an adult. We can all hope that improved communications between children and their parents about human sexuality will result in more widespread knowledge about child molestation (so that it can be prevented), as well as a greater willingness to talk about it and deal with it if it has happened.

PROSECUTION OF SEXUAL OFFENDERS

RAPE

Almost half of all reported rapes are dismissed before trial (U.S. Senate Judiciary Committee report, May 1993). Most of the dismissed cases are acquaintance rapes, for which prosecutors have traditionally been hesitant to bring cases to trial. However, with increased education (of the public, police, and courts) there have been dramatic changes in the manner in which rape victims are treated by the courts. For example, the U.S. Supreme Court has upheld broad **rape-shield laws** that make prosecution easier and prevent the victim from feeling as if she (or he) is on trial. In 1991, the Court ruled that states can enact laws that bar an accused acquaintance rapist from introducing evidence that he and his victim had previously had consensual sex. In addition, a victim's past sexual history is no longer allowed to be introduced during trial, but the accused rapist's past sexual offenses can be. Recall, too, that some states have enacted laws that prevent defense attorneys from talking about the clothing the victim was wearing.

Because many women delay reporting their rapes and many others appear calm and relaxed afterward—two common behaviors associated with rape trauma—many juries in the past have refused to believe victims' stories. Many states now allow expert testimony about rape trauma syndrome to educate jurors about these behaviors. Previously, evidence of emission of semen had to be found before an assault could be called a rape. Today, many states no longer require evidence of ejaculation for rape to be charged. The attacker can be charged and convicted of assault even if he is not charged with rape. In a landmark 1993 case, a rapist was sentenced to 40 years in prison by a jury that disregarded his claim that his victim had consented because she asked him to wear a condom.

However, prosecution of rape is made easier when semen (in the vagina or on pubic hair or clothing) is found during the medical examination, especially now that there is genetic fingerprinting. In the past, semen could only reveal an attacker's blood type, but recent scientific discoveries now make it possible to identify a person from fragments of DNA in their body cells, including semen and blood—

rape-shield laws Laws that make the prosecution of rapists easier and that also prevent the victims from feeling as if they are on trial.

much like a fingerprint. This new identification procedure was first used in a criminal case in England in 1987. A man the police had arrested for two rape-murders was freed after DNA from semen found on the victims did not match his own. The police then took blood samples from 4,000 men who lived in the area, and the real rapist was arrested and convicted when it was found that his DNA matched that found on the victims (see Joseph Wambaugh's *The Blooding,* 1989).

With increasing public awareness of the myths surrounding rape, it is hoped that more attackers will be charged and convicted of rape. If people do not report a rape, it means that someone else will likely be victimized by the same rapist. Even if you choose not to prosecute the crime, the information you provide may be valuable to the police and may prevent future rapes.

CHILD SEXUAL ABUSE

In a case that went before the United States Supreme Court, the Court stated, "Child abuse is one of the most difficult crimes to detect and prosecute, in large part because there often are no witnesses except the victim" (*Pennsylvania v. Ritchie,* 1987, p. 60, cited in Myers, 1993). One of the greatest difficulties was that the Sixth Amendment of the United States Constitution guarantees that in all criminal cases, the accused has the right to confront all witnesses against him. Face-to-face confrontations were often very traumatic for children—who, in addition, were no match for shrewd and manipulative defense lawyers. However, in 1990 the Supreme Court ruled that individuals being prosecuted for child abuse had no guaranteed right to face-to-face confrontations with their young accusers if the children would suffer emotional trauma as a result. The Court upheld state laws that allowed use of video-taped testimony and testimony by one-way closed-circuit television. Most states have eliminated the corroboration rule as well (see Whitcomb, 1992 for a review of these legal changes). Courts are also allowing minors to testify with the aid of anatomically correct dolls so that they can show what happened to them, although there are still some unanswered questions as to the effectiveness of this method of assessment (Lamb et al., 1996).

CONVICTION OF SEX OFFENDERS

Every state has laws that make it illegal to involve children in sexual activity or to fail to protect a child from sexual abuse. The goal of these laws is to protect the child and to see to it that the adult does not repeat the offense (see Conte, 1991). This goal would be best served by a coordinated effort among the police, the courts, and child welfare and community mental health programs, but attitudes in these four institutions often differ with regard to what to do about victims and offenders (Trute, Adkins, & MacDonald, 1992). Without a coordinated effort, many accused adults "fell out of the system" and received little therapy or confinement.

As the extent of sexual abuse of children has become better known, the states and federal government have taken a tougher stand. Washington was the first state to pass a "sexual predator law" that allows the state to confine pedophiles and rapists in mental institutions beyond their prison terms if it can be shown that it is likely that they will commit similar crimes again. Arizona, California, Kansas, Minnesota, and Wisconsin have similar laws. The U.S. Supreme Court upheld these laws in a 1997 decision involving a Kansas man who had sexually abused children his entire adult life. California passed a law in 1996 that allows judges to order chemical castration for repeat sex offenders.

A U.S. Department of Justice study released in 1997 found that there were nearly a quarter of a million convicted sex offenders on probation nationwide. A new federal law passed in 1996 established a nationwide registry to keep track of sex offenders. Part of the law requires states to inform local communities whenever a sex offender moves into the neighborhood. "Megan's Law," as it is called, resulted from the publicity that followed the rape and murder of 7-year-old Megan Kanka in 1994 by a convicted pedophile who was living across the street from the Kanka family. In 1998 the U.S. Supreme Court rejected a challenge to Megan's Law that had been made by a group of sex offenders who claimed that the law amounted to a second punishment. Although well-intentioned, the law has resulted in stigmatizing some people who were convicted of isolated nonviolent sex offenses, often decades ago (Palmer, 1997).

As you will learn in the next chapter, hundreds of thousands of children in poor nations are pushed (and often sold) into prostitution every year and are exploited by pedophile tourists from rich nations. England and other European countries have passed laws that allow the prosecution of their citizens if they are involved in child sex-tourism in other countries. Many people have already been convicted under these laws (Alldridge, 1997).

THERAPY

THERAPY FOR RAPISTS

One of the biggest problems in treating men arrested for rape is their tendency to deny or minimize their offenses. Recall that most rapists have distorted cognitive beliefs and attitudes and that it is not uncommon for rapists to believe that their victims wanted to have sex (see the section on rape). Early programs focused on deviant sexual arousal, but programs today take a multifaceted approach. Successful outcome depends on tailoring the type of therapy to the particular deficits and needs of the individual rapist. It is important, therefore, that there first be a comprehensive assessment of a rapist's developmental history, attitudes and beliefs about women and sex crimes, social competency, empathy, sexual knowledge and preferences, psychiatric history, substance abuse history, and offense variables (Ward, Hudson, & McCormack, 1997).

Progams today generally combine cognitive techniques with the traditional behavioral techniques, and many employ antiandrogen hormone treatment to reduce sex drive (Marshall, Eccles, & Barbaree, 1993; Polaschek, Ward, & Hudson, 1997). Again, however, a big problem is that rapists are reluctant to engage in therapy, and even with those who do, the dropout rate is high (see Polaschek et al., 1997). There is also concern that the recidivisim rate (the percentage of offenders completing therapy who rape again), though improving, is still about 20 percent in the first five years (Prentky et al., 1997). Some have suggested that better success can be achieved by adding empathy training (emotional recognition skills for the self and others) and by treating specific cognitive distortions related to issues of power and the desire to humiliate victims (Marshall, 1993; Polaschek et al., 1997).

THERAPY FOR CHILD MOLESTERS

For the incestuous molester, a family systems approach is often chosen as therapy. Here, therapy for each family member is combined with therapy for each pair of family members and is followed by therapy for all the family members together (Lanyon, 1986). Group therapy with other families is also often included. For positive change to occur, the father must accept responsibility for what he has done and the mother must also accept responsibility (often for what she has *not* done). The child, on the other hand, must learn to believe that he or she is not responsible. The father must also give up his complete authoritarian rule, and each family member must learn to become more independent.

As for the nonpedophile (situational) molester, whose sexual attraction is primarily to adults, many experts believe that society is best served by seeing to it that he receives therapy. Behavioral approaches such as covert sensitization (where aversive stimuli are paired with the deviant sexual fantasies) and orgasmic reconditioning (to reestablish sexual responsiveness to adults) have shown much promise (see Kelly, 1982, and Lanyon, 1986).

The outcome of therapy for pedophiles is less certain. For these men, sexual orientation is primarily and repeatedly directed toward children, and they have little or no sexual interest in adults. As you have already learned, many of these men have victimized dozens, and in some cases hundreds, of children (Abel et al., 1987), yet they typically display considerable denial and defensiveness (Grossman, Haywood, & Wasyliw, 1992). The antiandrogen drugs cyproterone acetate and medroxyprogesterone acetate have proved effective in lowering sex drive, thus allowing clinicians to attempt therapy while the subject is not driven to commit his crimes (Bradford & Greenberg, 1997). As with rapists, a comprehensive cognitive/behavioral/social therapy approach is generally used today (Barbaree & Seto, 1997; Polaschek et al., 1997). However, the judicial system is full of examples of pedophiles who have been caught, confined, and undergone therapy time and time again. A recent study found that 14 percent of child molesters had been rearrested 3 years after release from prison, 30 percent after 10 years, and nearly half after 20 years (Prentky et al., 1997). These types are often referred to as career pedophiles (McCaghy, 1971) or, more recently, as career sexual predators. Prentky argues that simply allowing offenders to complete their prison terms and return to the community is dangerous and that community notification laws such as "Megan's Law" are shortsighted (see DeAngelis, 1997). Instead, he believes that intensive long-term supervision and treatment are best.

As you have learned, several states have passed sexual predator laws, which, you recall, require that molesters identified as career predators be confined for life. Our society will no doubt continue to debate this sensitive issue, weighing the rights of offenders against the rights of victims and future victims, for many years to come. What do you think is best?

THERAPY FOR VICTIMS

You have already learned that victims of rape or child abuse often suffer from a variety of symptoms indicating emotional distress (e.g., depression, anxieties, fears) and that they may exhibit substance abuse, sexual disorders, eating disorders, and relationship problems. Because of our society's attitudes, there is a tendency in part or in full, to blame the victim—a feeling that even the victim may share. Male victims may additionally have questions about their sexual orientation (Gartner, 1997).

For many victims, initial therapy may have to be geared to their posttraumatic stress (Mueser & Taylor, 1997). This often includes a cognitive-behavioral approach to reduce the fear and distress experienced when recalling traumatic events. Eventually, however, the victim must overcome any self-blame, and the involvement of the victim's partner and family can be very beneficial here (Reid, Taylor, & Wampler, 1995). In addition to individual therapy, programs such as those offered by the YWCA emphasize group sessions so that victims have an opportunity to share their experiences with others without the fear of being judged negatively.

Sex is the closest form of communication between two people and can be one of life's most beautiful experiences. It is extremely important, therefore, for a person to get counseling and support from professionals and organizations if he or she is still feeling the effects of incest or child abuse as an adult or is suffering from the effects of being raped. Everyone should have the ability to enjoy sex in ways that are open and honest and that allow them to form healthy, intimate relationships.

RESOURCES FOR THOSE SEEKING HELP

For counseling, contact the *YWCA Rape Crisis Centers* nearest you. Services include a 24-hour rape crisis line; programs for victims of adult and child sexual assault and incest and for families and friends of victims; crisis and long-term counseling; weekly support groups; victim advocacy; information and referral services; training for medical personnel and police; agency consultation; community education, including assault prevention programs for people of all ages, from toddlers to senior citizens; and court monitoring.

For an extensive list of resources, see the "Resources" section following the text.

Key Terms

acquaintance rape   360
aggressive pedophile   381
anger rapists   367
date rape   360
gang rape   365
incest   385
male rape   366
marital rape   363
opportunistic rapists   368

pedophilia   348, 381
personally immature pedophile   381
polyincestuous family   385
power rapists   367
preference molester   381
rape   357
rape myths   370
rape-shield laws   391

rape trauma syndrome   371
regressive pedophile   381
sadistic rapist   367
sexual coercion   362
sexual harassment   376
situational molester   381
statutory rape   365
stranger rape   360
token resistance   362

Personal Reflections

1. Do you ever use sex to express power, dominance, control, or anger? If so, under what circumstances? Why do you use sex as an outlet for these desires and emotions?

2. Have you ever been the victim of rape or sexual abuse? If so, has it affected how you feel about yourself or your relationships with others? If it has affected you in any way (including leaving you with any sense of self-blame), have you ever sought counseling? If you have not sought counseling, why not?

3. Is a man entitled to sexual intercourse when a woman agrees to go out with him, or because he spent money on her, or because she went to his house or apartment, or because she engaged in necking with him? Men: If you answered yes, do you communicate these expectations to your dates before you go out?

4. Have you ever used sexual coercion to have sex with someone when he or she did not want to? This includes (but is not limited to) (*a*) purposely getting someone drunk to have sex; (*b*) physical coercion (e.g., persistent and relentless touching, holding, grabbing);

and (*c*) verbal coercion (e.g., anger and/or threats to break up, other threats, put-downs). If so, why do you suppose you act like this? What effects might your behavior have on those you coerce?

5. Men: Have you ever ignored a date's or acquaintance's saying "no" to your sexual advances? Why? Women: Have you ever said "no" to a man's sexual advances when you did not mean it? Why? If you answered "yes," what are some possible repercussions of your behavior? Is there a better way to communicate with others about your sexual desires?

6. Have you ever been the victim of sexual harassment by a supervisor, employer, or instructor? Did you report it? If not, why not? Are you familiar with your campus's procedures for dealing with sexual harassment? If you are not familiar with the procedures, where can you go to find out?

7. Are you ever sexually aroused by children? If so, have you ever sought counseling? If you have not sought counseling, why not? Would it be better to seek counseling before you acted on an urge, or afterwards?

Suggested Readings

Bass, E., & Davis, L. (1988). *The courage to heal: A guide for women survivors of child sexual abuse.* New York: Harper & Row.

Brownmiller, S. (1975). *Against our will: Men, women, and rape.* New York: Simon & Schuster. A key book in changing our society's attitudes about rape.

Byerly, C. M. (1992). *The mother's book—How to survive the incest of your child.* Dubuque, IA: Kendall/Hunt.

Crewdson, J. (1988). *By silence betrayed: Sexual abuse of children in America.* Boston: Little, Brown.

Davis, L. (1990). *The courage to heal workbook.* New York: Harper & Row. A self-help book.

Finkelhor, D. (1984). *Child sexual abuse.* New York: Free Press.

Gelman, D. et al., (1990, July 23). The mind of the rapist. *Newsweek.* Examines the motivations for rape.

Gilbert, B., and Cunningham, J. (1987, October). Women's post-rape sexual functioning: Review and implications for counseling. *Journal of Counseling and Development.* Discusses the effects of having been raped on long-term sexual functioning.

Groth, A. N. (1979). *Men who rape.* New York: Plenum Press. An excellent psychological and motivational profile of rapists.

Hagens, K. B., and Case, J. (1988). *When your child has been molested.* Old Tappan, NY: Simon and Schuster.

Journal of Social Issues. (1992). Vol. 48, no. 1. A special issue devoted to sexual aggression.

Kohn, A. (1987, February). Shattered innocence. *Psychology Today.* Discusses sexual abuse of girls and boys in the United States.

Koss, M. B., et al. (Eds.). (1994). *No safe haven: Male violence against women at home, at work, and in the community.* Washington, D.C.: American Psychological Association.

Lew, M. (1990). *Victims no longer: Men recovering from incest and other sexual child abuse.* New York: Harper & Row.

Moskal, B. S. (1989, July 3). Sexual harassment, '80's-style. *Industry Week.* Discusses and identifies sexual harassment and its consequences. Includes self-tests.

Newsweek. (1993, October 25). Several articles and essays on date rape and sexual harassment.

Parrot, A., and Bechofer, L. (Eds.). (1991). *Acquaintance rape: The hidden crime.* New York: John Wiley & Sons. Many scholarly articles.

Powell, E. (1991). *Talking back to sexual pressure: What to say, to resist pressure, to avoid disease, to stop harassment, to avoid acquaintance rape.* Minneapolis, MN: Comp Care Publishers. An excellent book; Powell gives many suggestions on how to resist unwanted sexual behavior and devotes much space to behavioral rehearsals.

Sanday, P. R. (1990). *Fraternity gang rape.* New York: New York University Press. Analysis of a seldom-discussed type of sexual violence.

Scarce, M. (1997). *Male on male rape: The hidden toll of stigma and shame.* Insight Books (Plenum).

SIECUS Report, Vol. 23, no. 5 (June/July 1995). Contains a bibliography of readings for children and adolescents.

Volkman, E., and Rosenberg, H. L. (1985, June). The shame of the nation. *Family Weekly.* Discusses child pornography and its link to child abuse.

Yegidis, B. L. (1986, Fall/Winter). Date rape and other forced sexual encounters among college students. *Journal of Sex Education and Therapy,* vol. 12. Discusses date rape from both the victim's and the rapist's point of view.

CHAPTER 16

When you have finished studying this chapter, you should be able to:

1. Define and describe the different types of prostitution;

2. Describe the characteristics of prostitutes and their customers;

3. Evaluate the pros and cons of prostitution from a legal as well as a moral point of view;

4. Trace the history of sexually explicit material in the United States and the legal actions taken to control it;

5. Discuss the uses of sexually explicit material;

6. Compare and evaluate the effects of nonviolent and violent sexually explicit material; and

7. Explain sodomy laws and their implications for heterosexual and homosexual individuals.

Sex and the Law

*S*ex is an important part of most people's lives. Sex provides physical pleasure, it can be given and received as an expression of love and affection, and it can make us feel more masculine or feminine. People engage in a wide variety of sexual behaviors. In Chapter 11 you learned that most therapists take the view that any behavior between consenting adults that is done in private and does not cause physical, emotional, or psychological harm is okay. This text has treated sexual relations between consenting adults as something healthy, good, and positive, but there can be serious consequences if people do not act in a mature, responsible manner (e.g., unwanted pregnancies or sexually transmitted diseases).

There are laws to prevent the sexual abuse of children and nonconsenting adults, as there should be (you learned about these in Chapter 15), but there are also laws that are designed to prevent or regulate sexual activity between consenting adults. As you might suspect, these laws are controversial, with many people objecting to attempts by others to regulate their private lives. Some of these laws were proposed and passed by politicians who were concerned about health-related issues (the spread of STDs, for example), but some appear to be attempts by individuals to force the public to conform to their own moral standards. Many states, for example, once had laws that forbade the sale or distribution of contraceptive devices (these laws were finally ruled unconstitutional by the U.S. Supreme Court in *Griswold v. Connecticut,* 1965).

In this chapter, you will read about several topics for which there has been a long history of individuals attempting to regulate the sexual activity of others. I have purposely attempted not to show my own bias regarding these issues and have tried to take a neutral stand. My intent is to stimulate your own thinking by presenting both sides of an issue. You be the judge as to whether or not there should be laws regulating these activities.

PROSTITUTION

Prostitution is often referred to as "the world's oldest profession." While this may or may not be true, the trade of sex for money or other favors has been around for as long as the recorded history of human beings. In the ancient Hebrew culture, it was not a crime for a father to prostitute his daughter, but if a daughter freely chose to have sex with a man (thus depriving her father of income), she could be put to death. Prostitution was very common in Roman society. In fact, the word **fornication** (sexual intercourse between a man and a woman who are not married to one another) is derived from *cellae fornicae,* the Latin term for the vaulted underground dwellings where prostitutes took their partners after the theater, circus, or gladiator contests.

In the Bible, Jesus told Mary Magdalene to "go, and sin no more," but prostitution flourished in Europe during the Middle Ages, and the Church had a somewhat tolerant attitude about it. St. Augustine believed that prostitution was necessary, as did St. Thomas Aquinas—who, though he considered prostitution a sordid evil, reasoned that just as a palace would become polluted without sewers, the world

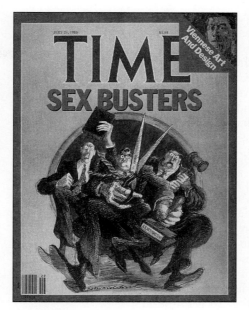

Figure 16–1

would become immersed in lust without prostitution (see Bullough & Bullough, 1997). Parisian prostitutes even adopted Mary Magdalene as their own patron saint.

In Victorian times (mid-nineteenth century to early twentieth century), the upper class maintained a public attitude of purity and prudery, but prostitution was a common sexual outlet for husbands. The poverty and squalor of life in the lower class pushed many young women into prostitution, and it has been estimated that there were as many as 80,000 prostitutes in London at any one time. Prostitution was legalized and regulated by an act of the British Parliament in 1860.

In colonial America and the early years of the United States, Puritanism in the North and the availability of slaves in the South kept prostitution to a minimum. Immigration and industrialization, however, resulted in larger urban areas, and prostitution flourished in the latter half of the nineteenth century. Houses of prostitution (brothels) were commonplace, especially in the West, where men greatly outnumbered women. Ship captains imported thousands of women from the Orient and Central and South America to be forced into prostitution. San Francisco's population reached 25,000 in the early 1850s, and this included 3,000 prostitutes (Tannahill, 1980). Storyville, a 16-block district of legalized prostitution in New Orleans from 1897 to 1907, employed as many as 2,000 prostitutes at any one time (Warner, 1997).

Opposition to prostitution in the United States began to grow during the first part of this century. Illinois became the first state to pass laws against prostitution in 1908. Federal laws passed at the start of World War I shut down many of the houses of prostitution that had existed near service bases (there were no cures for gonorrhea or syphilis at that time, and many men were being rejected for service because of infections).

These laws did not put a stop to prostitution. Kinsey and his colleagues (1948) found that 69 percent of the white males they surveyed had had at least one experience with a prostitute. Twenty-five percent of college-educated men and 54 percent of high school–educated men had had intercourse with a prostitute before marriage. Fewer men have their first sexual experience with

fornication Sexual intercourse between a man and a woman who are not married to one another.

prostitutes today than in the past. In a more recent survey, 20 percent of the male respondents indicated that they had paid for sex on at least one occasion (Janus & Janus, 1993). The percentage was the same for all income groups, and most of the men said that they did this occasionally. In the national survey by Laumann et al. (1994), 17 percent of the men and 2 percent of the women over the age of 18 said that they had paid someone for sex.

DEFINITION AND TYPES OF PROSTITUTION

We generally think of a prostitute as someone who exchanges sex for money. However, by this definition we would have to include people who allow someone to have sex with them for a job or career promotion, to gain a favor, or in return for a trip to Cancun or some other vacation spot. Some prostitutes bitterly complain that what they do is no different from what is done by a woman who remains married to and has sex with a man she cannot stand just for the security. To get around these problems, legal definitions generally distinguish **prostitution** from other activities on the basis of the indiscriminate manner in which it occurs. The Oregon Supreme Court, for example, ruled that "the feature which distinguishes a prostitute from other women who engage in illicit intercourse is the indiscrimination with which she offers herself to men for hire."

How common is prostitution in the United States today? There are generally about 84,000 women working each year as full-time prostitutes (Potterat et al., 1990), not counting the women who have sex in exchange for drugs at crack houses and other such places. However, in a recent cross-sectional national survey, Janus and Janus (1993) found that over 4 percent of the women answered "yes" to the question "Have you ever had sex for money?" In other words, there are many women who have engaged in prostitution part-time or for a short time. Estimates of the number of young teenage prostitutes, for example, range from 100,000–200,000 to over 2 million (U.S. General Accounting Office, 1982).

Prostitution in the United States was transacted primarily in brothels until the turn of the century, but

prostitution The exchange of sex for money. The feature that distinguishes a prostitute from other people who engage in illicit sexual intercourse is the indiscrimination with which she or he offers herself or himself to others for hire.

streetwalker A prostitute who solicits in public.

brothel prostitutes Prostitutes who work in houses of prostitution.

B-girls Prostitutes who work in bars and hotels, generally with the approval of the ownership.

today it occurs in a variety of settings. Prostitutes can be divided roughly into three types according to status. On the lowest tier are **streetwalkers,** who solicit in public. Almost all of these women become prostitutes for financial reasons (e.g., to escape poverty or to support a drug habit) and not because of pathological sexual needs. Very few are forced into prostitution as in

Figure 16–2 Streetwalkers solicit in public and are the lowest paid type of prostitute. They typically have sex with 4 to 5 men a day.

past centuries, but some are recruited. Most streetwalkers are protected by a pimp, who often takes a large share of the earnings. The sexual activity generally takes place in a room in a cheap hotel or apartment where the manager is aware of what is occurring. The typical streetwalker has sex with 4 to 5 men a day, and the sexual activity usually involves giving oral-genital sex (Freund et al., 1989).

The once common house of prostitution is now legal only in some counties in Nevada. The Chicken Ranch, that best little whorehouse in Texas (later moved to Nevada) made famous by the Broadway play and Hollywood movie, once did such a booming business that it had its own FAA-approved airplane strip and operated the Chicken Ranch Airline from Las Vegas. However, business at these legalized houses declined substantially during the 1980s, perhaps because of the fear of AIDS, but probably also owing to the increase in popularity of escort services.

Elsewhere, the **brothel prostitute** has been replaced by **B-girls,** who work in bars and hotels, generally with the approval of the ownership (which may take a cut of the profits or else benefit by the customers the prostitutes attract). Open solicitation, however, is generally frowned upon, so nonverbal messages are often used. The price is usually higher than for a streetwalker's services, and the sexual activity generally takes place in the customer's room.

"Candy" (her own fictitious name) is an attractive 33-year-old African-American B-girl.

She solicits customers in New Orleans' French Quarter, in bars along Bourbon, Chartres, and Decatur Streets. Her minimum charge for sex is $50 for 15 minutes or $200 for a night. The vast majority of her customers are white males in town for conventions. She says that 75 percent of them are married. The sexual activity takes place either in the man's hotel room or in a neutral hotel. Whenever possible, however, she robs her customers rather than having sex with them. Candy says this is usually easy to do before they even get out of the bar. She rubs their genitals, and in their state of excitement they rarely notice that she is pickpocketing them with the other hand. She doesn't worry about returning to the bar in a few days, because by then the "score" has left New Orleans. If she does have sex with a customer, afterwards she tries to relax them so that they fall asleep, and then she robs them of all their money and valuables.

Candy has four children and is on welfare. After high school she worked as a housekeeper and at other minimum-wage jobs. When she was 27, a pimp showed her the expensive nightclubs around town and said that she, too, could enjoy this life-style. After some deliberation, she paid him $300 to be "his lady" and now lives with him (her children live with her mother). She turns all the money that she makes as a prostitute and thief over to him. In return, he gives her protection, buys her nice clothing and jewelry, and lets her drive a Cadillac Seville.

Candy doesn't think about the future. She has been arrested 22 times on various charges in the past seven years, and if convicted of prostitution again, she could be sentenced to two years in prison. She doesn't worry very much about AIDS or other sexually transmitted diseases either. Although she sometimes makes $3,000 to $4,000 in one month, none of it has been saved for when the time comes that she cannot prostitute anymore. She just enjoys the life-style that the money will buy now.

(from the author's files)

More recently, sexual services have increasingly been offered through thinly disguised storefronts such as *massage parlors* (it should be noted, however, that some of these parlors are legitimate). Women who work in massage parlors are often called "hand whores" because toward the end of the massage (often given in the nude) they generally rub the thighs and groin area and let it be known that for a "tip" they will massage the genitals.

The highest-status prostitutes are the **call girls.** New customers are screened (recommendations from trusted sources are often required), and appointments are generally made by telephone. The call girl generally works out of her own apartment, but often offers her company for dinner and social occasions in addition to providing sexual services. The services do not come cheaply, however, with some of the most attractive call girls charging $1,500 per night or more. A more commercialized, and somewhat less expensive, variety of call girl prostitution today is offered through *escort services,* which advertise in newspapers. Judging by the increase in these advertisements, escort service prostitution increased by over 400 percent in the 1980s (Janus & Janus, 1993, p. 132).

Prostitution is not limited to females who cater to heterosexual males. Although not nearly as common as female prostitutes, males sometimes sell their bodies to women as well. These males, called **gigolos,** are the male counterparts of call girls and often work as part of an escort service (generally used by wealthy older single women). For example, the following account is from a 22-year-old student working as a male escort. He said that most of his female clients are 30 to 45 years old and 80 percent are single.

"Being a male escort is not all that people perceive it to be. In this day and age you have women that are successful, and in order for them to stay that way they must constantly work harder than any man they're in competition with. This causes women to be subjected to long hours of work and devotion to their jobs, leaving them with no time at all for a personal life. Every time they attempt to have a relationship, they end up neglecting either work or their partner. That is the point where I come in. I am the individual that can be there for them when they need someone either as a friend or a lover. One of the misconceptions of being a male escort is that they only have sex with their clients. That may be true for females, but in the male's case, out of ten women, only five of them only want sex. Three of them want to be held while watching a movie or listening to music. And as for the other two, they want someone who they can go to events with, someone who can carry on intellectual conversation and stand by their side as their showpiece. "

(from the author's files)

call girls The highest-status prostitutes, who generally work out of their own apartments by appointment only.

gigolos Male prostitutes who cater to females.

Homosexual prostitution is nearly as common as heterosexual prostitution in some large urban areas. Male prostitutes, called **hustlers,** are similar to female streetwalkers in that they generally solicit their customers on the street. Many hustlers do not regard themselves as homosexual (Earls & David, 1989; Lloyd, 1976), because in America the male prostitute generally plays a passive role; that is, the customers pay to give oral sex to the prostitute, and not vice versa. As many as 70 percent of hustlers do this only part-time (Allen, 1980), with the primary motivation being financial and not sexual (Earls & David, 1989).

CHARACTERISTICS OF PROSTITUTES

In many parts of the world, women and girls have no choice; their families sell them into prostitution (Barry, 1995). Even here, the primary reason given by persons engaging in prostitution is economics (Bess & Janus, 1976; Earls & David, 1989; Rio, 1991; Weisberg, 1985). Most prostitutes come from poor families and have little education, or they are young runaways with few or no skills (Bell, 1987; Lloyd, 1976). However, even for call girls or prostitutes who are middle-class housewifes working part-time, the main attraction is still money—their income is higher than that of other women matched for age and education (Exner et al., 1977).

Prostitution and the need for money are often fueled by drug use. The large majority of prostitutes are habitual users of addictive drugs such as heroin and crack cocaine (Feucht, 1993; James et al., 1979; Simon et al., 1992). Pimps often see to it that their prostitutes become addicted just so they will remain economically dependent. For the typical streetwalker or B-girl, prostitution does not turn out to be the gold mine it may have seemed at first, for most end up turning over most of their earnings to the pimp.

Economics may be their main motivation, but many prostitutes are attracted to the thrill and danger of "the life" (Janus & Bess, 1977; Potterat et al., 1985). In fact, Janus and Bess (1977) found that for some prostitutes, the life-style itself was addictive. Most people believe that prostitutes do not really enjoy sex with their customers and perform sex acts mechanically, but one study found that many prostitutes do, in fact, enjoy sex with their customers (including giving fellatio) at least sometimes (Savitz & Rosen, 1988).

Beyond the economic motives and aside from the frequent drug dependency is the fact that most prostitutes have very sad backgrounds. Many prostitutes—including the large majority of teenage prostitutes—were sexually abused as children (Earls & David, 1989; Janus & Bess, 1977; Rio, 1991; Satterfield & Listiak, 1982). Physical abuse by parents is also common, as are parents with alcohol or drug

problems (Hersch, 1988). Many prostitutes had trouble in school, had few friends, and generally felt rejected (Davis, 1978; Lloyd, 1976; Price et al., 1984). As a result, many grew up with severe self-esteem and self-identity problems, and generally are just less psychologically healthy than other people (Gibson-Ainyette et al., 1988).

With this kind of background, it is not surprising that most teenage prostitutes are runaway or throwaway (not wanted even if they return) children who, with no other job options, have turned their first trick by the time they are 14 (Weisberg, 1985).

> "Jean" looks like most other 15-year-old girls. In fact, when her hair is in a ponytail she is often mistaken for 13 or 14 years old. The men, most of whom are married with daughters as old or older than she is, like her very young looks. They want to have sex with a "schoolgirl." Jean fulfills their fantasy by telling them that she is a sophomore at one of the local high schools. She charges them $30 to $50, depending on what they want (oral sex in their car is least expensive). Jean doesn't really go to school at all. She is a runaway. She ran away the last time at age 14 and found herself in New Orleans alone and hungry. That's when she met "Frank." He bought her a hamburger and offered her a place to stay. There were two other girls about her same age staying with Frank. He was particularly nice to Jean and bought her nice dresses. She felt wanted and cared for, and after a few days she and Frank had sex (her first time). He also introduced her to drugs. He told her that she was his special girl. Two weeks later he told her he needed money and asked her to go to bed with another man. The other girls told her it was easy. Jean was surprised to find out how easy it was, and within a short time she was having sex with up to six men a day. All she had to do was stand on a corner and within a short while older men would stop and talk to her.
>
> (from the author's files)

Unfortunately, there are some people who prefer to have sex with young teens, and thus the sexual exploitation of children will continue as long as there is a demand for and supply of destitute young runaway and throwaway children. The 1996 World Congress Against Commercial Sexual Exploitation of Children (Stockholm) estimated that 1 million children a year worldwide are drawn into prostitution (Rios, 1996). In a world dominated by men—and especially in cultures

---

hustler A male prostitute who caters to males.

where there are few opportunities for women to be employed—if a female does not have a father or husband to support her, prostitution may be the only way in which to survive (Bullough & Bullough, 1997b). The following organizations are devoted to helping these children get off the street:

Covenant House
1-800-999-9999
Provides counseling, shelter, and other referral services.

National Network of Runaway and Youth Services
Department P, P.O. Box 8283
Washington, DC 20024
1-800-448-4663
They will put you in touch with the nearest shelter program.

National Runaway Switchboard
1-800-621-4000
They encourage runaway children to reestablish contact with their families.

A recent study of adult male street prostitutes found that nearly all had serious problems with feelings of personal inadequacy, loneliness and social alienation (most said they had no friends), mistrust of others, and depression (Simon et al., 1992). All were drug users and felt that they were locked into their deviate careers. While the authors of the study acknowledge that the psychological problems may reflect the prostitutes' dangerous and chaotic environment, and probably their past, they also suggest that these psychological problems have led them to this particular life-style.

THE CUSTOMER

Prostitution has continued to flourish for centuries despite laws, penalties, and the protests of outraged citizens. Part of the reason it has continued with unabated frequency is the fact that it is perceived as easy income by unskilled, financially destitute women; but without the demand for their services, of course, prostitution could not exist. In order for it to flourish as it has, the demand for sexual services must be great.

Customers are called *johns, tricks,* or *scores* by prostitutes. Most are white, middle-class, middle-aged, and married (James, 1982; Janus & Janus, 1993). There are many reasons why males pay for sex (see Stein, 1977). A few are able to function sexually only with prostitutes (Bess & Janus, 1976), but the reasons for many include lack of a partner (e.g., due to military service, traveling, or being unable to find a partner because of physical or social handicaps) and the desire for a new partner or a novel sexual act. Perhaps the major reason that many men pay for a prostitute's services is to enjoy sex without the time or effort required of an emotional involvement (Stein, 1977). Some men feel that they have to perform (i.e., please their partner) in their usual relationship, but sex with a prostitute allows them to receive pleasure without having to reciprocate. The majority of men who go to streetwalker prostitutes pay to have oral-genital sex performed on them (Leonard, Freund, & Platt, 1989).

Men who pay to have sex with male prostitutes come from all walks of life (Fisher et al., 1982; Kamel, 1983; Lloyd, 1976; Morse et al., 1992). Most describe themselves as heterosexual or bisexual and are married. They are not part of the open gay community and use prostitutes in order to conceal their homosexual activities. Despite their married status, very few use condoms, although the behaviors they engage in put them at risk for sexually transmitted diseases.

PROSTITUTION AND THE LAW

Opponents of prostitution argue that it is often accompanied by drug addiction, a high risk of contracting sexually transmitted diseases (which the john can then spread to his wife or girlfriend[s]), and the possibility of assault and battery by the prostitute and her pimp (what do you suppose the chances are of a victimized john filing charges?). Recent studies in some cities, for example, found that many prostitutes are infected with the virus (HIV) that causes AIDS (see Box 16–A). The primary concern of many people, however, is the moral issue. They believe that prostitution is a sinful and degrading sexual activity, both for the individuals involved and for society as a whole. As a result, there has been a renewed effort by some residents of Nevada to close the legalized houses of prostitution in that state.

Many people, on the other hand, argue that prostitution is a victimless crime (see Rio, 1991). It is certainly true that very few prostitutes are reported by complainants. Instead, most are caught by undercover vice squad policemen posing as johns. Today, many researchers, feminists, and organized groups of prostitutes (e.g., COYOTE) have called for decriminalization or legalization of prostitution (Barry, 1995; Bullough & Bullough, 1997). It is certainly the case that local governments spend a great deal of money on a continually losing battle against prostitution. Proponents advocate the licensing and regulation of prostitutes, including regular required health checks for sexually transmitted diseases. This is already done in Nevada's legalized houses of prostitution, where customers are also now required to use condoms. Proponents say that if prostitution were legalized everywhere, pimps and organized crime would become less influential, and the taxes collected on prostitutes' fees could be used to fight more serious crimes. In addition, some feminists argue that decriminalizing prostitution

Box 16-A   SEXUALITY AND HEALTH

Prostitution and the AIDS Virus

You learned about the human immunodeficiency virus (HIV) and AIDS in Chapter 5. HIV infection, you recall, is the leading cause of death of young adults in at least 64 large U.S. cities (Selik, Chu, & Buchler, 1993). This applies to women as well as men—AIDS is the leading cause of death of young women in at least nine cities. Although once uncommon, heterosexual transmission is now the major cause of most new AIDS cases in women (HIV/AIDS Surveillance Report, 1997a).

Because of their sexual relations with multiple partners, as well as the high percentage who use drugs intravenously, prostitutes are at high risk for contracting HIV and are high-risk sexual partners for those who use them (Des Jarlais et al., 1989; Morse et al., 1991). Fewer than 1 percent of all U.S. citizens are infected with HIV, but what about prostitutes? Here are some examples of results of recent studies: over half the female prostitutes in northern New Jersey are HIV-positive (Turner, Miller, & Moses, 1989); over one-quarter are infected in Miami (Turner et al., 1989); and 6.7 percent tested positively for HIV in a national study (Khabbaz et al., 1990). The risk of HIV infection is greatest for prostitutes working on the street (Pyett & Warr, 1997). Among male prostitutes, 17.5 percent tested positive in New Orleans (Simon et al., 1992), while 27 percent in Atlanta tested positive (Elifson et al., 1989). A recent study reported that in some urban areas, the rate for male prostitutes is 50 percent or higher (Boles & Elifson, 1994). Several of these studies found high rates of infection with hepatitis B and syphilis as well.

You might think that men engaging in high-risk sexual behavior, such as sex with a prostitute, would use condoms. Regular use of condoms does substantially reduce the risk of getting HIV. For example, in legalized brothels in Nevada, where condom use is required, none of the prostitutes had tested positive for HIV by the end of the 1980s, and low rates of infection have also been found among African prostitutes who regularly use condoms (Mann et al., 1987; Ngugi et al., 1988). Condom use by men in the general population is quite low (Catania et al., 1995), but two European studies found that use of condoms was much higher among men who have sex with prostitutes (Barnard et al., 1993; Hooykaas et al., 1989). However, this may not be true in this country (Leonard, Freund, & Pratt, 1989), and even in the European studies there were many clients of prostitutes who did not use condoms. What's more, the clients in one of the studies reported that their condoms tore over 10 percent of the time (Barnard et al., 1993). Alcohol or drug use beforehand may also make it less likely that a client will use a condom with a prostitute (Weinstock et al., 1993).

What can we conclude? By their use of prostitutes, especially when condoms are not used, many men are putting themselves *and their noncommercial partners* at high risk for infection with HIV.

---

would give women more choice and allow many to leave the profession without social stigma.

The United States seems to have very ambivalent attitudes about prostitution. It is illegal in some places and legal in others. Although illegal in most places, many only halfheartedly prosecute it. While all states prosecute the sellers (prostitutes), very few seriously go after the buyers (johns who patronize them). Prostitution is the only sexual offense for which more women than men are prosecuted. If society does not curb the demand, the supply of prostitution will always be there, and thus our society's present treatment of prostitution probably assures that it will continue un-

abated. What do you think we should do to solve this problem: push for greater enforcement (including arrest and severe prosecution of johns), or require the licensing and regulation of prostitutes?

PORNOGRAPHY

Sexually explicit material seems to be everywhere today: on TV, at the movies, in books and magazines, over the telephone, and even available on the internet. This was not always the case, however. There has always been sexually explicit art, pictures, or literature available for those who wanted it. But only in the last few decades has there been such an eruption in the quantity of this material that even those who don't seek it out have difficulty avoiding it. Magazines with sexually explicit covers, for example, are often displayed right behind the cash register at drugstores where children and everyone else cannot help but see them.

How did all this happen? The United States for a long time was a very puritanical country (and believe it or not, still is, compared to some other countries). The first real erotic literature by an American was Walt Whitman's *Leaves of Grass,* written in 1855. However, the **Comstock law,** passed in 1873, made the mailing of this or any other material considered obscene or lewd a felony. Moral crusader Anthony Comstock became a special agent of the U.S. Post Office and personally saw to it that the law named after him was enforced. Even information about birth control was not allowed to be mailed. Importation of James Joyce's novel *Ulysses* was banned when it was first published in 1922. The ban was lifted in 1933, but other novels, such as D. H. Lawrence's *Lady Chatterley's Lover* and Henry Miller's *Tropic of Cancer,* were banned in the United States and most other English-speaking countries until the late 1950s.

Censorship in movies was also very strict during the era. The movie industry was banned from filming erotica in 1935. Clark Gable's final words to Vivien Leigh (who played Scarlet O'Hara) in the movie *Gone with the Wind*—"Quite frankly, my dear, I don't give a damn!"—were shocking to many people. The brief love scene on the beach (with bathing suits on) in the 1950s movie *From Here to Eternity* was considered scandalous. Because it came into people's living rooms, TV was censored even more heavily than movies during the 1950s and 1960s. Many people were outraged when Lucy first used the word "pregnant" on the "I Love Lucy" show. Rock and roll was opposed by many parents in the 1950s, not only for its "suggestive" words (e.g., "Whole Lot of Shakin' Going On," "Let Me Be Your Teddy Bear"), but also because of the performers' gyrations. Elvis Presley was shown only from the waist up when he appeared on the Ed Sullivan Show in 1956.

What about sexually explicit magazines? In the 1950s and 1960s there was really only *Playboy* (on a large scale, anyway). Hugh Hefner spent $600 in 1953 to publish the first issue, which featured nude photos of Marilyn Monroe, and sold it on the street himself. He turned *Playboy* into a multi-million-dollar publishing empire. Hefner really didn't have any competitors until the late 1960s and early 1970s, about the same time that movies and TV became more explicit.

Why the move toward more permissiveness in the middle and late 1960s? The story of Hugh Hefner's success pretty much tells the answer. It became obvious

Figure 16–3 Sexually explicit movies and books are readily available in most parts of the United States.

that despite efforts at censorship, there was a large demand for sexually oriented material. As a result, other magazines (e.g., *Penthouse, Hustler, Playgirl*) started to compete for a share of the market, and in an effort to outdo one another, these magazines continually expanded the borders of sexual license. At the same time, Americans flocked to see erotic European movies starring actresses like Brigitte Bardot, and their box office success led Hollywood to explore sexual boundaries as well.

So where are we in the 1990s? We can easily buy magazines that show hetero- or homosexual couples (or groups of people) having every imaginable kind of sex, using every position, having oral and anal sex, sex with violent themes, or sex with animals, and much more that is probably beyond your imagination. There is a booming $8 billion-a-year market in XXX-rated movies and video rentals (see Figure 16–3). The XXX-rated movie *Deep Throat* cost only $24,000 to make, but it has earned over $25 million in profits. Sexually explicit material has even kept up with technological advances. Today, computer software allows you to watch sex on home computers. This includes CD-ROM "virtual reality" programs that allow some interaction (an estimated $300 million-a-year market). The internet has allowed greater accessibility to sexually explicit material for those embarrassed about going to stores to rent videos or buy maga-

Comstock law A law passed in 1873 that made it a felony to mail any material regarded as obscene or lewd.

zines. Today, such material is readily available in the privacy of your own home at the touch of a mouse.

You no longer have to turn to X-rated material for full frontal nudity and torrid sex scenes. R-rated movies such as 1992's *Body of Evidence* with Madonna and William Dafoe or *Basic Instinct* with Sharon Stone were clearly designed to shock us. But with all the sexually explicit material available, it takes a lot to shock us today, so much so that Madonna's book of photographs, *Sex,* hardly made a ripple when it came out in 1992. Even our ears are being bombarded with explicit material in the 1990s. The 2 Live Crew's album "As Nasty as They Wanna B" in 1990 led to calls for rating CDs, just as is done with movies. New York–based "shock jock" Howard Stern has been fined hundreds of thousands of dollars for his explicitness on the radio. For hard-core fans, there's dial-a-porn (telephone sex). Yes, explicit sex is everywhere today, and you would have to be a hermit living without radio, TV, or magazines to avoid it.

Do we want it? Well, apparently many Americans do. A 1985 Gallop poll found that 40 percent of all VCR owners had watched an X-rated movie in their home within the last year. In fact, when asked if they had seen an X-rated movie in the last year, about a third of all Americans questioned since 1982 said yes (Laumann et al., 1994). Nearly all college men and women have looked at an explicit magazine such as *Playboy* or *Playgirl* by the time they leave high school, and 85 to 90 percent have done so in junior high school (Bryant & Brown, 1989). The average college man spends about 6 hours a month looking at sexually explicit material, and the average college woman about 2 hours (Padgett, Brislutz, & Neal, 1989). In short, a majority of Americans now accept sexually explicit content in the mass media and believe that adults should have the right to obtain it (Winick & Evans, 1994).

USES OF SEXUALLY EXPLICIT MATERIAL

Why do people look at, read, or listen to sexually explicit material? Bryant and Brown (1989) have identified four major reasons. Some people, they say, turn to sexually explicit material to learn about sex. In fact, many college students use sexually explicit material as a source of sexual information (Duncan & Donnelly, 1991; Duncan & Nicholson, 1991):

> "My boyfriend, when I first started dating him, was very misinformed about females, what they like/disliked and about how a woman's anatomy should look like. He now has a more accurate view of things. He got his info from porno movies."
>
> (from the author's files)

A second reason is that it allows people to rehearse sexual behaviors, much as fantasy does. A third, related reason is that today many people consider fantasy and masturbation while looking at or listening to sexually explicit material a form of safe sex (safe from sexually transmitted diseases, for example).

Probably the foremost reason many people use sexually explicit material, however, is that they enjoy the sexual arousal it produces. Many people masturbate while using sexually explicit material. Many couples watch sexually explicit videos together to heighten their arousal, although in a few cases this results in dissatisfaction with a partner's sexual attraction and performance (Zillmann & Bryant, 1988). In short, most people regard sexually explicit material as a form of fantasy and use it to escape from the monotony of their daily routines (Davis, 1983).

EFFECTS OF NONVIOLENT SEXUALLY EXPLICIT MATERIAL

As you will learn shortly, there have been many attempts to ban or censor sexually explicit material. People would not be concerned, of course, if they didn't believe that exposure to sexually explicit material has a negative effect on behavior. In order to determine if this is true, President Lyndon Johnson in 1968 appointed an 18-member commission that spent two years collecting and studying evidence. The findings of the President's Commission on Obscenity and Pornography were released in 1970 and stimulated a decade of research (much of it originally funded by the Commission). Here are some of the major conclusions from that era:

1. Looking at, reading, or listening to sexually explicit material (movies, pictures, books, tapes) produces physiological arousal in a majority of both men and women (Report of the Commission, 1970; Englar & Walker, 1973; Heiman, 1977; Henson, Rubin, & Henson, 1979; McConaghy, 1974; Schmidt & Sigusch, 1970; Wincze, Hoon, & Hoon, 1976). It was once believed that only men were aroused by sexually explicit material, but similar physiological responses were observed in both sexes. Many women may actively avoid sexually explicit material but nevertheless do respond to it when they see or hear it.

2. People are more aroused by looking at sexual behaviors they consider to be normal than they are by unconventional behaviors. Sexually explicit material does not seem to create desires that were not already there (Report of the Commission, 1970).

3. When exposed to sexually explicit material, some people show an increase in sexual behavior, a smaller proportion show a decrease, but the major-

ity of people report no change in their behaviors. Increases (almost always expressed as masturbation or intercourse with the usual partner) are short-lived and disappear within 48 hours (Report of the Commission, 1970).

4. Continued exposure to sexually explicit material leads to a marked decrease in interest (indifference, boredom) in these materials (Report of the Commission, 1970; Lipton, 1983).

5. Contrary to many people's belief that sex offenders are generally obsessed with pornography, convicted sex offenders in prisons (e.g., rapists or child molesters) were often found to have had less exposure to sexually explicit materials (sex education as well as pornography) during adolescence than nonoffenders (Report of the Commission, 1970).

6. Consistent with the previous findings, it was found that when Denmark legalized the sale of hard-core pornographic material to adults in the mid-1960s, there was a lot sold at first as people satisfied their curiosity, but then sales dropped off considerably. There was over a 30 percent drop in sex crimes in the first few years after legalization, suggesting that pornography served as an outlet for some people who might otherwise have committed sex crimes. (Report of the Commission, 1970; Kutchinsky, 1973).

The Commission concluded that there was no significant link between exposure of adults to sexually explicit materials and sex crimes or other deviant or harmful behavior. They recommended the repeal of obscenity and pornography laws and the teaching of sex education in public schools:

> Failure to talk openly and directly about sex has several consequences. It overemphasizes sex, gives it a magical, non-natural quality, making it more attractive and fascinating. It diverts the expression of sexual interest out of more legitimate channels, into less legitimate channels. . . . The Commission believes that interest in sex is normal, healthy, good.

Richard Nixon was president when the Commission's report was released in 1970. He dismissed it as "morally bankrupt" and rejected its findings.

Studies conducted since release of the report have generally supported the Commission's conclusions. For example, studies that looked at the rate of sex crimes in states that had suspended prosecution of antipornography statutes have also concluded there is no link between exposure to sexually explicit material and sex offenses (Winick & Evans, 1996). However, the conclusion that sex offenders have had less exposure to

sexually explicit material has proved to be somewhat controversial. Although most of the studies examined by the Commission found that there was no reliable difference between imprisoned sex offenders and nonoffenders in their rate of exposure to pornography (Cook & Fosen, 1970; Gebhard et al., 1965; Walker, 1970), there was a study that disagreed (Davis & Braught, 1970). A subsequent study supported the Commission's conclusion (Goldstein, 1973), but a still later study found that sex offenders had used sexually explicit material more during adolescence than had nonoffenders (Marshall, 1988). A more recent study found that child molesters had been exposed to sexually explicit material at a later age than were nonoffenders, and that adult sex offenders and nonoffenders did not differ in how often they looked at such material (Nutter & Kearns, 1993). The large majority of sex offenders in these studies did not believe that sexually explicit material caused them to commit their crimes.

The conclusion that women are aroused by sexually explicit material has also been found to be in need of some qualification. Early studies focused on physiological arousal (e.g., vasocongestion of genitals), but subsequent studies have shown that physiological arousal is not necessarily accompanied by subjective feelings of arousal (Laan & Everaerd, 1996). A recent review of many studies found that males report moderately greater arousal to sexual stimuli than do females and that the difference was greatest with very explicit material (Murnen & Stockton, 1997). Other studies have confirmed that most women's arousal to sexually explicit material is greater when the material is female-initiated/female-centered and romantic, as compared to male-initiated/male-centered and not romantic (Laan et al., 1994; Pearson & Pollack, 1997; Quackenbush, Strassberg, & Turner, 1995).

EFFECTS OF VIOLENT AND DEGRADING SEXUALLY EXPLICIT MATERIAL

In the years after the President's Commission of 1970, there were concerns that the amount of sexually explicit material showing women being bound, chained, raped, and tortured had increased (e.g., Check, 1984; Donnerstein & Linz, 1984; Malamuth & Spinner, 1980). Although sexual themes that include violence have actually never accounted for more than a very small proportion (generally less than 10 percent) of sexually explicit videotapes and magazines (Scott & Cuvelier, 1993; Slade, 1984), whether *violent sexually explicit material* is related to sex offenses is nevertheless an important question. The President's Commission did not specifically examine this issue. Can looking at violent pornography cause some men to rape women?

In numerous scientifically conducted studies designed to answer this question, volunteer subjects were shown films or heard tapes of either mutually consenting erotic behavior, scenes of aggression toward women, or scenes with both sexual explicitness and aggression. In many of the studies, violent sexually explicit material was found to cause arousal in male subjects (Malamuth & Donnerstein, 1984). In fact, they were often more aroused by violent sexually explicit scenes (e.g., of naked women bound and appearing distressed) than they were by scenes of women enjoying sex (Heilbrun & Seif, 1988; Malamuth, 1981). After watching violent sexually explicit films, many of the subjects became less sympathetic toward female rape victims, showed greater belief in rape myths (viewing rape victims as more responsible for their assaults), displayed increased hostility to women in the lab, and were more likely to indicate that they would commit rape themselves if they were certain that they would not get caught (Donnerstein, 1982, 1983; Donnerstein & Berkowitz, 1981; Donnerstein & Linz, 1984; Malamuth, 1981; Rosen & Beck, 1988; see Linz, 1989 or Malamuth & Donnerstein, 1984 or Donnerstein, Linz, & Penrod, 1987 for reviews). Donnerstein (1983) was startled to find that by the last day of looking at sexually explicit films showing violence toward women, many men found the material "less debasing and degrading of women, more humorous, more enjoyable, and claimed a greater willingness to see this type of film again." Similar results were found when men looked at *degrading sexually explicit material* (nonviolent material showing women as sex objects where there is a clearly unequal balance of power; Linz, 1985; Zillmann & Bryant, 1982). Both men and women rate themes in which there is active subordination of women to be the most degrading (Cowan & Dunn, 1994). However, not all studies have found that violent pornography causes men to have antifemale thoughts or attitudes, and we are still not certain about the precise conditions that cause some men to react in such a manner (see Davies, 1997; Fisher & Grenier, 1994).

Violent sexually explicit films are often highly eroticized—the female victims are often portrayed as enjoying forced sex after putting up some initial resistance. Interestingly, laboratory studies have shown that many women also are aroused by watching erotic rape scenes (Stock, 1982, 1983). Unlike many men, women are not aroused by watching realistic rape

scenes, but childhood exposure to sexually explicit material results in many women having attitudes more supportive of sexual violence against women (Corne, Briere, & Esses, 1992).

Being aroused by fantasies of rape is not the same as actually committing a rape. None of the previous studies proved that looking at violent pornography really causes sex crimes to be committed (as far as we know, none of the volunteer subjects committed a rape after leaving the lab). To explore whether there really is a causal relationship between violent pornography and sex crimes, Edwin Meese, then the Attorney General of the United States, appointed an 11-member commission on pornography in 1985. They released their conclusions in the spring of 1986, less than a year after they were formed (see Figure 16–4). Their conclusion: a "causal relationship" exists between sexually violent pornography and violence toward women. They stated that these materials help foster the myth that women enjoy being raped and that "sexually explicit materials featuring violence [are] on the whole harmful to society." In contrast to the recommendations of the 1970 commission, they advocated stronger government enforcement of present obscenity laws and additional laws that would result in greater restrictions of sexually explicit material.

The 1986 commission funded no original research, but instead relied heavily on the findings of previously conducted laboratory experiments. However, the scientists who conducted that research accused the Commission of drawing conclusions and making legal recommendations that did not follow from their data (see Linz, Donnerstein, & Penrod, 1987; Mulvey & Havgaard, 1986). Even some of the Commission members were critical of the report. Ellen Levine, editor of *Woman's Day* magazine, and Judith Becker, a clinical psychologist from New York State Psychiatric Institute, issued a formal statement of dissent. They both said that while they found pornography objectionable and offensive to women, the panel's report was flawed and rushed. Others questioned the objectivity of some of the Commission's members, six of whom had previously called for greater government enforcement against pornographic materials. The Commission's chairman, for example, was a U.S. attorney from Virginia

Figure 16–4 With the bare-breasted *Spirit of Justice* in back of him, U.S. Attorney General Edwin Meese accepts the Report of the **1986 Commission on Pornography**. What is erotica to some people is pornography to others.

who had made his reputation by prosecuting owners of adult bookstores. In summary, it would appear that some of the panelists were unable to distinguish between their own personal views and objective research.

Behavioral scientists conducting work in this area were brought together in their own conference in 1986. They agreed with the 1970 commission's conclusion that exposure to nudity alone has no detrimental effects on adult behavior (see Linz et al., 1987). They suggested that the real focus of concern should be depictions of violence toward women, regardless of whether or not it was in a sexually explicit context. Donnerstein, for example, was quoted as saying, "These conclusions [by the 1986 commission] seem bizarre to me. It is the violence more than the sex—and negative messages about human relationships—that are the problem. And these messages are everywhere" (*The New York Times*, May 17, 1986, p. A6). Subsequent studies showed that it is violence-toward-women themes, more than sexual explicitness, that result in potentially harmful attitude and behavioral changes (Linz et al., 1987; Scott & Schwalm, 1988).

This type of potentially harmful material (violence toward women without sexual explicitness) is precisely the type of material that is common in R-rated movies and on TV today. Overall, one out of eight Hollywood films shows a rape scene (National Coalition on Television Violence, 1990). In fact, rape is much more common in R-rated movies, TV shows, and detective magazines than in X-rated movies or magazines like *Playboy* (Dietz et al., 1986; Palys, 1986; Scott, 1986). Studies have consistently found, for example, that viewing nonexplicit movies showing violence toward women results in many men becoming less sensitive toward rape victims (Linz, 1989). So what is the conclusion of the leading behavioral scientists in the field?

> The most clear and present danger . . . is all violent material in our society, whether sexually explicit or not, that promotes violence against women"
>
> (Donnerstein & Linz, 1986).

In order to eliminate possible detrimental effects to society, we would have to suppress all materials showing violence toward women, not just sexually explicit material with violent themes. Scientists suggest that rather than imposing stricter laws, it might be easier to educate the public about the effects of viewing violence so that they can make better choices about which programs, movies, and books they choose to expose themselves and their children to.

SEXUALLY EXPLICIT MATERIAL INVOLVING CHILDREN

"Directly after high school I worked at a photo developing store. One day while sorting some pictures I noticed that the first six shots were of boys that looked to be no older than 10 having fellatio and sodomy with each other. The last eight pictures introduced an older man of about 40 years old. I, of course, showed my manager and he called the police. When the man's wife came to pick up the pictures she was arrested. She was charged with passing pornographic material through our store. Her husband was given a heavier sentence—20 years. He was a prestigious teacher at an uptown school and a career pedophile. In that year alone he had molested 12 boys."

(from the author's files)

The President's Commission of 1970 found no evidence that pornography was harmful to adults and suggested the repeal of laws prohibiting the sale and distribution of sexually explicit material. But what about photographs or films that show children nude or having sex with an adult or another child? Could looking at child pornography cause someone to molest children, or does this type of pornography provide an outlet for those who might otherwise abuse children? We cannot answer this, because for obvious ethical reasons there has been very little research in this area (Jarvie, 1992). Nevertheless, this is one type of pornography that nearly everyone feels should be banned. How can we do this and still protect First Amendment rights?

One way is to ask whether the "stars" of kiddie porn are capable of giving informed consent. Many are runaways who are coerced into participating by financial or other reasons. Some participants are so young that they could not possibly have any idea of what they are doing. They are obviously being exploited by adults (sometimes by their own

"I can watch as much crime and violence as I like as long as they keep their clothes on."

parents or guardians). Children who are used in this manner generally grow up to feel like objects rather than people and often suffer a variety of emotional and sexual problems (Burgess et al., 1984). The U.S. Supreme Court ruled unanimously in 1982 that publishers and distributors of child pornography could be prosecuted for child abuse without requiring the prosecuters to prove that the material was obscene, thus avoiding arguments about First Amendment rights. The passage of the Protection of Children Against Sexual Exploitation Act in 1977, the Child Protection Act in 1984, and the Child Protection and Obscenity Enforcement Act in 1988 made it a federal crime to ship or receive, in interstate or foreign commerce or by mail, pictures of nude children or children (defined as any person under the age of 18) engaged in sexual acts. The maximum punishment for the first offense is 10 years in jail and a fine of $100,000 (except for participating parents and guardians, who can be sentenced to 20 years to life in prison). As a result of the legislation, prosecutions of child pornography increased more than 250 percent in the first few years after the Child Protection Act was passed. Adults who participate in child pornography in any capacity can be prosecuted. In 1990, the Supreme Court upheld an Ohio law that made it a crime even to possess pornography showing children under the age of 18.

This is presently the only type of sexually explicit material that is legally banned. The U.S. Customs Service has established a hotline of the National Center for Missing and Exploited Children to receive tips about pornography: 1-800-843-5678.

Even with this material, however, there is a question of where we should draw the line. What is child pornography? Authorities in programs for sex offenders have noted that there are men who become sexually aroused by watching diaper advertisements on TV. Should we ban all pictures of naked or partially clothed babies and children?

COMPLETE LEGALIZATION OR CENSORSHIP?

What should we do about sexually explicit material? Should we remove all restrictions to the sale and distribution of hard-core material to adults (i.e., make it completely legal), or should we attempt to ban all such material? There are good arguments for both cases.

The President's Commission of 1970, you recall, found evidence that sexually explicit material served as an outlet to people with sexually criminal tendencies and might actually reduce sex crimes. It is probably difficult for you to believe that material you find offensive and repugnant may actually do some good, but that is because you are normal and have no deviant tendencies. Therapist John Money of the Johns Hopkins University has noted that patients in clinics for sex offenders often say that they use pornography to fantasize about abnormal sexual activity so that they will not

have to act out their desires in real life. In fact, most sex therapists still believe that socially unacceptable sexual behaviors often result from attempts to severely repress sexuality. Father Bruce Ritter, founder of Covenant House, is a possible example of this. A member of Meese's commission and an outspoken opponent of both pornography and sex for the purpose of pleasure, even within a marriage ("It wastes the seed," he said), Father Ritter resigned from Covenant House after being accused of sexual molestation by several male residents of the shelter for runaway children.

A recent survey found that over half of all Americans feel that there is too much sexually explicit material in their neighborhood stores (Janus & Janus, 1993), and there have been numerous attempts by citizens, school boards, and local governments to ban sexually oriented materials that they consider offensive. The producers or distributors of the materials, along with others, invariably protest on the grounds of First Amendment rights, and many cases have reached the U.S. Supreme Court. The First Amendment of our Constitution guarantees, among other things, that Congress "shall make no law . . . abridging the freedom of speech, or of the press. . . ." I suspect that the Founding Fathers never intended this to apply to explicit pictures of couples having anal sex, but the problem is deciding where to draw the line. Guess which of the following magazines and books have been targets of censorship by some communities in recent years:

The American Heritage Dictionary
The Bible
Brave New World
The Catcher in the Rye
Cinderella
Club magazine
The Diary of Anne Frank
Gorillas in the Mist
The Grapes of Wrath
Huckleberry Finn
Hustler magazine
Macbeth
Penthouse magazine
Playboy magazine
Romeo and Juliet
Sports Illustrated (swimsuit issues)
Stories about dinosaurs
The Wizard of Oz

If you guessed all of them, you were correct. Yes, there are some people who believe that parts of the Bible are too explicit for others to read. *The American Heritage Dictionary* was banned in some places because it included "bad words." High school teachers in Erie, Pennsylvania, blacked out passages in *Gorillas in the Mist* (by naturalist Dian Fossey) describing the mating habits of apes. Not all of the above materials were

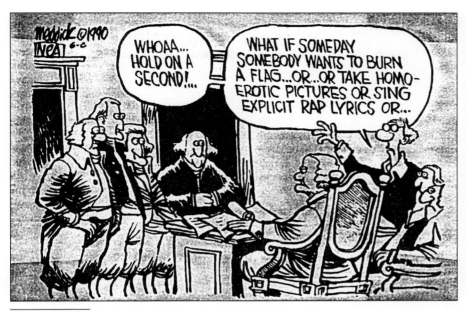

Jim Meddick; reprinted by permission of Newspaper Enterprise Association, Inc.

banned because they were sexually explicit, however. *Cinderella,* for example, was banned from public schools in a Tennessee community in 1986 because it mentioned witchcraft, and *Macbeth* was banned by the same community because it mentioned magic (this was overturned by a decision of the U.S. Supreme Court). Attempts to censor or ban sexually explicit material continue today. The record album *As Nasty as They Wanna B* by the rap group 2 Live Crew became the first ever (in 1990) to be ruled obscene in Federal Court. In July 1990, the curator of Cincinnati's Contemporary Art Center was ordered to go to trial on obscenity charges for putting on an exhibit of homoerotic photos by photographer Robert Mapplethorpe. In 1996, Congress passed the Communications Decency Act, which made it a crime to transmit sexually explicit images through electronic media (the Internet), but a federal district court ruled that it violated the first amendment (Daley, 1996).

With regard to sexually explicit material, the problem the Supreme Court has had to deal with (in cases appealed on First Amendment rights) is trying to distinguish between erotica and pornography. **Erotica** (from the Greek word *erotikos,* meaning "love poem") is any literature or art with a sexual theme, while **pornography** (from the Greek words *porne,* which means "prostitute," and *graphos,* meaning "depicting or writing about") is literature or art with a sexual theme that is designed to cause arousal. Many people use the terms pornography and obscenity interchangeably, while others feel that obscenity is a more limited term (see Penrod & Linz, 1984, for a history of this subject). The key, therefore, is how we define *obscene.* What is erotica to one person is pornography to another. The

Venus de Milo may be regarded as a great work of art by most people (erotica), but to some it is just a statue of a nude woman (obscene and pornographic). Why are old paintings of nude persons or nude couples making love generally considered art, but modern photos of the same thing often considered pornographic? (See Figure 16–5.) Why are drawings of couples having sex considered educational if they appear in sexuality text-books (such as those in Chapter 11 in this book), but pornographic if the same thing is found in magazines (see King & Lococo, 1990)?

After grappling with this issue for many years, the Supreme Court finally came up with a three-part definition of **obscenity** in 1973 (*Miller v. California*). Material was considered obscene if (1) by contemporary community standards it depicts patently offensive sexual conduct; (2) it lacks "serious literary, artistic, political, or scientific value"; and (3) it appeals to prurient interest (lustful craving) in sex.

Even the Supreme Court justices recognized problems with this definition, however. For example, in the following year (*Jenkins v. Georgia,* 1974) the Court ruled that local communities do not have "unbridled discretion" to decide what is obscene (a Georgia community had decided that the Hollywood movie *Carnal Knowledge* was obscene). Who determines what is patently offensive or which art or literature lacks serious value? Former Supreme Court justice Potter Stewart admitted that he could not come up with a good definition of hard-core pornography, but said, "I know it when I see it" (*Jacobelis v. Ohio,* 1965). We all see things differently, however. You know what you find to be obscene, but do you want someone else to make that decision for you?

Even if it should eventually be proven that explicit hard-core material reduces sex crimes, it nevertheless is true that many people find it highly offensive and degrading. Some feminists oppose it on the grounds that it causes men to view women as sex objects and leads to sex discrimination (e.g., Dworkin, 1989). Today's sexually explicit movies do, in fact, emphasize the sexual de-

erotica Any literature or art that has a sexual theme. This is a morally neutral term.

pornography Literature or art with a sexual theme that is designed to cause arousal and excitement and is considered to be obscene.

obscenity In legal terminology, the classification of material that by contemporary community standards depicts patently offensive sexual conduct, lacks serious value, and appeals to prurient interests.

Figure 16–5 Why are many paintings with sexually explicit themes considered great art when modern photos with the same theme are often regarded to be pornographic? (Boucher's *Hercules and Omphale,* 1730s)

sires and prowess of men while portraying women as sexually willing and available (Brosius, Weaver, & Staab, 1993). Catherine MacKinnon (1993) argues that pornography perpetuates sexual inequality and that the antidiscrimination provisions of the Fourteenth Amendment of the Constitution should take precedence over the First Amendment. However, other feminists worry that attempts to ban sexually explicit material on grounds of discrimination reinforce a double standard. "It reasserts that women are sexually different from men and in need of special protection. Yet special protection inadvertently reinforces the ways in which women are legally and socially said to be different from men" (Vance, 1988; also see A. M. Smith, 1993).

We are one of the few countries in the world whose people have freedom of expression. Most people are offended by at least some of the sexually explicit material that is published or filmed today, but many worry that allowing others to have the power to ban certain materials might eventually infringe on their own presently enjoyed freedom of expression (after all, there are those who would ban passages in the Bible, you recall). What do you think should be done?

sodomy laws Laws that prohibit oral and/or anal sex.

LAWS AGAINST FORNICATION, ADULTERY, AND SODOMY

What do you do behind closed and locked doors in your own home or apartment? You're right, it is none of my business. Is it anyone else's business either? Well, you might be surprised to find out, then, that many states have laws prohibiting certain sexual acts, even if done by consenting adults (including married couples) in private.

Seventeen states and the District of Columbia have laws that make *fornication* and/or *cohabitation* a crime (see Table 16–1). The maximum penalty is generally 3 to 6 months in jail and/or a fine. *Adultery,* which is a crime in 26 states, has historically been considered a more serious offense because it violates the sanctity of the family. Also, past attitudes held that women were property belonging to their husbands (and thus adultery was a violation of property rights).

Are people ever prosecuted under these laws? Yes; in states that have laws forbidding adultery, for example, it is not uncommon for one member of a quarreling married couple to press charges against the other to gain an advantage during the settlement, or perhaps just for revenge (Sachs, 1990). In 1997, a North Carolina woman successfully sued her husband's lover for $1 million for breaking up their marriage. You may recall that in the same year the military charged Lt. Kelly Flinn, the first female B-52 pilot, with adultery (Vistica & Thomas, 1997). She subsequently resigned. Some states have begun prosecuting teenagers for fornication in order to prevent teen pregnancy. In Idaho, for example, several teens have been given three-years probation and community service for getting pregnant or getting someone pregnant.

More recently, there has been much attention devoted to **sodomy laws,** which ban specific behaviors between consenting adults. Twenty states have laws that forbid oral or anal sex (see Table 16–1), which are

WAIT... I THINK WE'D BETTER HAVE OUR LAWYER PRESENT.

TABLE 16–1

LAWS REGULATING SEXUAL BEHAVIOR BETWEEN CONSENTING ADULTS

Fornication and/or Cohabitation

State	Punishable Offense	Maximum Penalty
Alaska	Cohabitation	———
Arizona	"Open and notorious cohabitation"	30 days and/or $500
District of Columbia	Fornication	6 months and/or $300
Florida	Cohabitation	60 days and/or $500
Georgia	Fornication	1 year and/or $1000
Idaho	Fornication/cohabitation	6 months and/or $300
Illinois	Fornication/cohabitation	6 months or $500
Massachusetts	Cohabitation	3 months or $300
Michigan	Cohabitation	1 year and/or $500
New Mexico	Cohabitation	———
North Carolina	Fornication/cohabitation	6 months and/or $500
Rhode Island	Fornication	$10
South Carolina	Fornication/cohabitation	1 year and/or $500
Utah	Fornication	6 months or $299
Vermont	Fornication	———
Virginia	Fornication	$100
West Virginia	Fornication/cohabitation	$20 minimum/6 months or $50 minimum
Wisconsin	Fornication	———

Oral or Anal Sex (as of August 1996)

State	Punishable Offense	Maximum Penalty
Alabama	Oral or anal sex with someone other than your spouse	1 year
Arizona	"Infamous crime against nature"	30 days and/or $500
Arkansas	Oral or anal sex with anyone of the same sex	1 year and/or $1000
Florida	"Unnatural and lascivious act"	60 days and/or $500
Georgia	Oral or anal sex	20 years
Idaho	"Infamous crime against nature"	Life (5 years minimum)
Kansas	Oral or anal sex with anyone of the same sex	6 months and/or $1000
Louisiana	"Unnatural carnal copulation"	5 years and/or $2000
Maryland	Oral or anal sex with anyone of the same sex	10 years and/or $1000
Massachusetts	Sodomy and buggery	
Michigan	Abominable and detestable crime against nature	15 years or $2500
Minnesota	Oral or anal sex	1 year and/or $3000
Mississippi	Detestable and abominable crime against nature	10 years
Missouri	Oral or anal sex with anyone of the same sex	1 year and $1000
North Carolina	Crime against nature	10 years
Oklahoma	Detestable and abominable crime against nature with anyone of the same sex	10 years
Rhode Island	Abominable and detestable crime against nature	20 years (7 years min.)
South Carolina	Abominable crime of buggery	5 years and/or $500
Utah	Oral or anal sex with anyone other than your spouse	6 months or $299
Virginia	Oral or anal sex	20 years

Source: National Gay and Lesbian Task Force (August, 1996).

often referred to as "crimes against nature," even though a large number of couples (the vast majority, as far as oral-genital sex is concerned) engage in them (see Chapter 11). These laws reflect the early Judeo-Christian attitude that the only natural sexual act is hetero-sexual intercourse because it is the only sexual behavior that can result in reproduction.

In five states, the sodomy laws are aimed specifically at homosexuals, but in 15 others the laws do not distinguish between heterosexual and homosexual cou-

Box 16-B CROSS-CULTURAL PERSPECTIVES

Illegal Sex Around the World

In the last part of this chapter you learned that many states prohibit and punish sexual behaviors such as fornication, cohabitation, adultery, and oral or anal sex. However, we are not the only country in which this is true. In fact, most countries make it illegal for consenting adults to engage in some forms of sex. The United States is more restrictive than some countries and more permissive than others. Here is a list of some of the countries around the world and the sexual activities that they consider illegal.

	Fornication	Cohabitation	Adultery	All Anal Sex	All Oral Sex	Homosexual Oral or Anal Sex
Mexico	—	—	√	—	—	√
Brazil	—	—	√	—	—	—
Argentina	—	—	√	—	—	—
Chile	—	—	√	—	—	√
England	—	—	—	—	—	—
France	—	—	—	—	—	—
Germany	—	—	—	—	—	—
Sweden	—	—	—	—	—	—
Italy	—	—	—	—	—	—
Russia	—	—	√	—	—	—
China	√	√	√	√	√	—
Japan	—	—	—	—	—	—
India	—	—	√	√	—	√
Australia	—	—	√	—	—	√
Saudi Arabia	√	√	√	√	√	
Iran	√	√	√	√	√	
Egypt	√	√	√	√	√	
Zaire	—	—	—	√	√	
Kenya	—	—	√	√	√	
South Africa	—	—	√	√	√	

ples. The existence of sodomy laws then means, of course, that millions of married American couples are committing crimes for which they could potentially be put in jail (up to 5 years or more in 11 states). However, when the laws are enforced, which isn't too often, they are generally enforced only against homosexuals. Homosexuals regard these laws as "legalized fag bashing" and have legally challenged a number of them.

Does the U.S. Constitution guarantee us the right to privacy? Apparently not. A Virginia law that forbids oral and anal sex between either hetero- or homosexual individuals was upheld by both a state and federal court in 1975. The U.S. Supreme Court refused to grant a hearing on the case (*Doe v. Commonwealth's Attorney*) in 1976 and thus upheld the right of states to regulate the private activity of its citizens. Basically, the ruling was that states have the right to regulate sexual behaviors just as they have the right to require you to

use seat belts or anything else if it is "for your own good."

The case of a Georgia man focused national attention on the legality of sodomy laws. Michael Hardwick was arrested for having oral sex with another man in the privacy of his own bedroom (the arresting officer had come to the house about a traffic violation and was let in by another person). Hardwick, who could have served up to 20 years for this crime, sued the state of Georgia, but the case was thrown out by a U.S. district judge on the basis of the 1976 Virginia ruling. The U.S. 11th Circuit Court of Appeals reinstated the suit (*Bowers v. Hardwick*), and the U.S. Supreme Court this time agreed to hear arguments. The Supreme Court upheld the Georgia law by a 5–4 vote in June 1986. Writing for the majority, Justice Byron White stated: "The proposition that any kind of private sexual conduct between consenting adults is constitutionally insulated from state proscription is unsupportable." In a dissenting opinion, Justice Harry Blackmun said: "What the court really has refused to recognize is the fundamental interest all individuals have in controlling the nature of their intimate associations." Justice Lewis Powell, Jr., later said that he regarded his vote with the majority position to have been a mistake (see Swisher & Cook, 1996).

In their ruling, the Court avoided the larger issue of whether sodomy laws could be used to prosecute heterosexuals—the ruling was limited to "consensual homosexual sodomy." However, there was nothing in the language to cast doubt on the constitutionality of laws that forbid oral or anal sex between married heterosexual individuals. In 1992, Kentucky's Supreme Court struck down that state's sodomy law, which was aimed specifically at homosexuals, stating, "We need not sympathize, agree with, or even understand the sexual preference of homosexuals in order to recognize their right to equal treatment before the bar of criminal justice." Courts in Tennessee and Montana struck down those states' same-sex sodomy laws in 1996.

Interestingly, the American Law Institute's Model Penal Code recommended as far back as 1962 that sexual behaviors between consenting adults (excluding prostitution) be decriminalized. Legislators tend to view *decriminalization* as a safer political route to take than legalization. They worry that any attempt to legalize sexual behaviors will be interpreted by voters as condoning them. Decriminalization simply means either that the acts will no longer be defined as criminal or that they will no longer be subject to penalties, and thus it is morally neutral.

Several states have followed the American Law Institute's recommendations and have decriminalized sexual acts between consenting adults. However, legislators in some states are still proposing restrictive laws. A bill proposed in the state of Washington in 1990, for example, would have made it a prosecutable offense (90 days in jail and a $5,000 fine) for anyone under the age of 18 to engage in necking or petting. Which approach do you believe is more correct? Do you believe that a couple should have the right to decide for themselves which sexual behaviors they engage in (assuming, of course, that it is done with mutual consent and does not cause physical or emotional harm to either person), or that a state should serve as a higher moral authority and regulate the private consensual sexual activities of its citizens?

Key Terms

adultery   263, 410
B-girls   398
brothel prostitute   398
call girls   399
Comstock law   403
degrading sexually explicit material   406

erotica   409
fornication   397
gigolos   399
hustlers   400
johns   401
nonviolent sexually explicit material   404

obscenity   409
pornography   409
prostitution   398
sodomy laws   410
streetwalkers   398
violent sexually explicit material   405

Personal Reflections

1. How do you feel about prostitution: should it remain illegal or should we attempt to legalize and regulate it? Why? If you feel it should be illegal, should the "clients" be prosecuted?

2. Are you ever physically aroused by looking at sexually explicit photos or movies? What is your emotional reaction to this material and any arousal it causes? Why? Have you ever learned anything from photos of naked persons or by watching R-rated or X-rated sex scenes? If you had the power to do it, would you ban this material or allow unrestricted distribution to adults? Why?

3. How would you respond if someone told you that this textbook was obscene and pornographic?

4. Do you think that state legislatures, which represent large groups of people, should have the right to regulate the private (and physically nonharmful) sexual behaviors of consenting adults? Why or why not?

5. If you answered "no" to question 4, do you think that you or any other individual should pass judgment on others for their private consensual sexual behaviors?

Suggested Readings

Baxter, M. (1990, May 5). Flesh and blood. *New Scientist*. A short article about studies of the effects of violent pornography.

Bullough, V., and Bullough, B. (1988). *Women and prostitution: A social history*. Buffalo, NY: Prometheus Books. A sociological, historical, and cross-cultural examination. Very comprehensive.

Donnerstein, E. J., and Linz, D. G. (1986, December). The question of pornography. *Psychology Today*. Authors of pornography research challenge the conclusions of the Attorney General's commission and say that we should be more concerned with violence in the media.

Elmer-Dewitt, P. (1995, July 3). On a screen near you: Cyberporn. *Time* magazine.

Heins, M. (1993). *Sex, sin, and blasphemy : A guide to America's censorship wars*. New York: New Press. Takes the view that sexually explicit material should be protected as a First Amendment right (see MacKinnon for an alternative viewpoint).

Kelly, G. F. (1984). Is it time for sexologists to clarify a position on pornography? *Journal of Sex Education and Therapy*, vol. 10, no. 2.

MacKinnon, C. (1993). *Only words*. Cambridge, MA: Harvard University Press. Takes the view that pornography should be banned as discriminatory against women under the Fourteenth Amendment (see Heins for an alternative viewpoint).

Report of the Commission on Obscenity and Pornography. (1970). New York: Bantam Books. A paperback edition including the results and conclusions of the 1970 commission.

Volkman, E., and Rosenberg, H. L. (1985, June). The shame of the nation. *Family Weekly*. Discusses the problem of child pornography and the factors contributing to it.

Weisberg, K. D. (1984). *Children of the night: A study of adolescent prostitution*. Lexington, MA: Lexington Books.

CHAPTER

17

When you have finished studying this chapter, you should be able to:

1. Take steps that enable you and your partner to become more comfortable talking about your sexual relationship (including agreement on a sexual vocabulary);

2. Take responsibility for your own sexual pleasure;

3. Understand the importance of positive reinforcement and "I" language;

4. Become a better listener;

5. Understand the importance of nonverbal communication;

6. Understand why the sex education of children should be an ongoing process rather than a single "birds and bees" talk;

7. Know at what age children should be told about certain sexuality-related topics by their parents; and

8. Know how to speak to your children in a manner so that they are likely to value your opinion.

Communicating About Sex

You have now nearly completed a comprehensive course in human sexuality. I have chosen to end this text with a chapter on communication skills. Why at the end? **Communication** is the act of making information known, the exchange of thoughts or ideas (to paraphrase the *American Heritage Dictionary*). Until now, you may have lacked factual information about a wide variety of topics in human sexuality. However, when it comes to talking about the topic of sex, it is not enough to have factual information. Many people have trouble talking about sex, even with their partner (Barbach, 1982).

In this chapter I will present some guidelines to help you communicate better about sexual matters. There are a variety of situations in which people might communicate about sexually related

communication The exchange of information, thoughts, ideas, or feelings.

matters, including interactions with a new or short-term partner (dating, initiating sexual activities, contraception, and safer sex); conveying needs and desires to a long-term partner; and conveying information and values to children. Because I cannot possibly cover them all, I have chosen to limit this chapter to sexual communication between long-term partners and communication between parents and their children about sexuality. In Chapter 5 you can find some advice on initiating discussions about safer sex with a dating partner.

The first half of this chapter deals with communication between two people who have made a commitment to a long-term relationship. It assumes that each person genuinely cares for the other; that there is trust and intimacy. As you learned in Chapter 13, most couples are going to experience sexual differences or problems in their relationship (Frank, Anderson, & Rubenstein, 1978; Laumann et al., 1994; Nathan, 1986). How they work those problems out—how they communicate with one another about their differences and problems—will affect their sexual adjustment. In cohabiting couples, sexual satisfaction is associated with understanding (particularly by men) their partner's sexual preferences (Purnine & Carey, 1997). "In many cases, sexual dysfunction problems cannot be addressed until communication improves" (Wincze & Carey, 1991, p. 110). Good communication also leads to a better relationship, and a better relationship generally leads to better sex (Cupach & Metts, 1991; Cupach & Comstock, 1990). The second half of the chapter is designed to help parents (and future parents) take what they have learned in this course and teach it to their children.

In order to better communicate with you about communication, I have elected to abandon the normal academic style of textbook writing. I present to you a few guidelines about improving your communication skills in a question-and-answer format.

TALKING WITH YOUR PARTNER ABOUT SEXUAL DIFFERENCES AND PROBLEMS

WHY IS IT DIFFICULT TO TALK ABOUT SEX?

Intercourse means "communication," but there are few topics that people have more difficulty talking about than sex. One of the major reasons for this is that most people have little experience talking about sex in an intelligent and mature manner. Recall that surveys in my own course consistently indicate that fewer than a third of the 1,000 students I teach every semester have ever had a serious discussion with their parents about sex. Very few ever had a course in sex education during junior high or high school. This means, of course, that most people's only experience with discussing sex is locker room conversation and jokes, usually with individuals of the same sex. Much of this is crude and vulgar and contains misinformation. Adolescents often avoid talking to members of the opposite sex about sex because they fear that this will be interpreted as an invitation or come-on. Thus, males generally have some experience talking about sex with other males, and females with other females, but many people reach adulthood with little or no experience in talking to the opposite sex. Even conversations with people of the same sex may be limited to nonthreatening sexual topics, for to talk about a sexual problem is to admit that you have one. Many people, particularly adolescents, are unwilling to do this because of worries that their friends might think less of them and tell others what they have said. As adults, some people are afraid to talk with their partners about a sexual problem because they are afraid it will threaten the relationship.

Communication about sex is also often hampered when a couple is locked into stereotypic gender roles. For example, one harmful stereotype is that men are experts at sex. This is simply not true. Women and men both *learn* about sex. Individuals differ greatly in the types of sexual experiences that they enjoy. Each person must explore and learn what brings the greatest pleasure to his or her sexual partner. Because sexual behaviors are learned, anyone can become an "expert" if that person is willing to read, to try new behaviors, to listen to feedback and direction from a partner, and to learn as well as to teach.

Another harmful stereotype is that men should not express their feelings and emotions (including sexual feelings), although women do. All people of both sexes need to express their feelings and emotions and be able to communicate them. The image of a man as "strong and silent" robs him of a fundamental part of his humanity and robs his partner of a true companion (see Goldberg's *The New Male,* 1979).

Finally, power differences in a relationship can affect communication. We are most likely to express an opinion, particularly a difference of opinion, to those persons we perceive as having equal status. In some relationships, there is a power imbalance; in heterosexual relationships, it is usually in favor of the male. Although the following suggestions for improving communication should be helpful to everyone, they will be most useful to individuals in an egalitarian relationship. (To resolve a major power imbalance in a relationship may require professional counseling.)

HOW CAN MY PARTNER AND I GET USED TO TALKING ABOUT SEX?

One of the best ways to get used to talking with your partner about sex is to discuss together any sexuality-related topics in your daily newspaper. Hardly a day goes by when there isn't at least one article (and there are often many) in the paper about sexuality—AIDS, birth control, fertility research, and, best of all, results of sex surveys. The "Dear Abby" and "Ann Landers" columns often are about sexually related matters. Show your partner some selected topic in this book and start off with a statement like "I didn't know this" or "This is really interesting."

When you both become more comfortable talking about sex together, then you can more safely talk about your own sexual relationship. Often a good place to start is to talk about why it is difficult for you both to talk about sex. If, for example, your parents had a negative attitude about talking about sex, talk about that and how you would like to be different. Next you might talk about normal bodily functions related to sexuality and how they affect your sexual relationship. Try talking about menstruation, for example. Many people have *learned* (and thus can unlearn) negative attitudes about that (see Chapter 3 for an extensive discussion of learned negative attitudes about menstruation and how they affect sexual relations).

When you and your partner are comfortable enough to talk about sexual differences (in attitudes, preferred behaviors, etc.), you might want to discuss when and how to talk about such matters. For couples interested in exploring their attitudes about different sexual behaviors, looking at a sex manual together might be helpful (e.g., Alex Comfort's *The New Joy of Sex*).

WHAT IF I AM UNCOMFORTABLE WITH THE LANGUAGE OF SEX— WHAT WORDS SHOULD I USE?

Many of the commonly used words for our sexual anatomy and functions reflect the negativity that has prevailed in Western culture since early Christianity. Words like *prick, cock, cunt, pussy, screw,* and *fuck* may be common in conversations among men (Sanders & Robinson, 1979), but many women (and men) might feel uncomfortable using such words with their partners, particularly if they wish to express positive emotions. Medically correct terms such as *penis, vagina,* and *sexual intercourse* may not have any negative connotations but are so unerotic that they, too, may seem inappropriate in conversations between two people in a deeply emotional relationship. Euphemisms such as "go to bed with," "sleeping with," and "making love" may seem more erotic to some and convey that there is a relationship, but they are often inappropriate as well—do we always "make love" when we have sex? Some couples develop a private vocabulary for body parts that is used only in conversations between themselves.

It is not for me to tell you what words to use with your partner when talking about sex. However, it is important that you both be comfortable with a sexual vocabulary. Therefore, as part of the process of getting used to talking about sex, be sure to discuss this with your partner until you both agree on a vocabulary that is explicit and specific enough to convey your feelings and desires and that you also find nonoffensive (yet still erotic).

WHEN (AND WHERE) SHOULD I TRY TO TALK TO MY PARTNER?

When is a good time to tell your partner that you prefer different behaviors, or wish to be touched or made love to in a different manner (or any other preference)? If you do it right after sex, you run the risk of making your comments sound like a report card or performance rating, and perhaps not a very good one at that. If you do it during sex, you might distract or frustrate your partner, and if you express your dissatisfaction before you start, that might cause performance anxiety.

Perhaps a neutral time would be best. Recall from the previous section that one of the general topics you and your partner might first discuss is when to talk about sexual differences. Whenever you do it, make sure you are not frustrated, angry, or highly emotional at the time. It is not so much *when* you discuss your differences but *how* you go about it, and your concerns may not be well received if you begin in a negative emotional state. Try to judge when your partner would be most receptive to talking about your concerns.

Finally, if you have most of your sexual relations in the bedroom, never worry about or discuss your sexual problems while in bed. If you do this too often, the bed can become a stimulus for sexual tension and anxiety; that is, just being in bed can cause you to start worrying. Save the bed for pleasant experiences and pick a neutral site like the living room for worrying about problems.

HOW SHOULD I APPROACH MY PARTNER WITH CONCERNS ABOUT OUR SEXUAL RELATIONSHIP?

Because of their inexperience in talking about sex—particularly sexual problems—many people say nothing and allow a problem to continue, with the end result often being increasingly greater frustration and/or anger. When they finally do get around to saying something, many people begin with statements like "I don't like it when you . . ." or "You don't . . ." These types of statements accentuate the negative and are often met with defensive reactions and counteraccusations (Zimmer, 1983). This can quickly escalate into a full-scale argument.

There are few things that as easily insult a person's ego as criticizing his or her sexual expression or interaction. Nearly everyone would like to believe that they are good sexual partners. To try to solve your own frustra-

Figure 17–1 When talking with your partner, try to avoid negative comments and personal criticism. Accentuate the positive, express your feelings by using "I" instead of "you" or "we," and (unlike this couple) try to make eye contact.

pleasure, and you can do this only if you tell your partner of your needs and desires.

Although I emphasized the use of positive reinforcement in the previous section, that does not mean that you can never complain or express anger. However, when you are angry, learn to control your anger. Express it only after you have calmed down (remember, highly emotional negative accusations will probably be met with similar responses on the part of your partner). Ask yourself what your motivation is for expressing complaints. Is it to hurt your partner, or is it for constructive change? Couples with good relationships accept that there will be differences, and sometimes even conflict, but they resolve those conflicts in constructive ways, not destructive ways. If you criticize your partner, make sure you criticize his or her behavior, not his or her character. There is a big difference between saying "You don't show me often enough how much you care" and saying "You are an uncaring person." If you must criticize your partner's behavior, be sure to praise other aspects of his or her behavior as well.

tions by criticizing your partner, therefore, is asking for trouble. Sometimes people ask for feedback about sex and then get upset with the response. Some people, for example, ask questions like "How was it?" "Did you like it?" or "Wasn't that great?" after having sex. Sometimes they are really looking for a pat on the back or ego reinforcement rather than an honest answer, so this is really not a good time to express any negative reactions. Whenever possible, *accentuate the positive rather than the negative.* This is the most important advice that I can give you. Rather than beginning the conversation with "I don't like it when you . . . ," try "I like it when you . . ." Your partner must do something that you like while having sex, so give him or her some positive reinforcement. Thousands of studies have shown that behaviors increase when they are reinforced, and praise is generally a powerful reinforcer for human beings.

WHAT IF I THINK THAT MY PARTNER IS TO BLAME—CAN I EVER COMPLAIN?

If you haven't expressed your concerns to your partner in a clear, specific manner (because you are uncomfortable talking about sex or for any other reason), do not lay the blame entirely on your partner. The problem has continued because you have allowed it to. Your partner cannot know what you want unless you tell him or her. It is wrong for you to think, or expect, that your partner should know what you want or how to please you. *You must take the responsibility for your own*

HOW SHOULD I EXPRESS MY NEEDS AND DESIRES?

Most communications experts suggest that individuals use *"I" language* when expressing desires to their partners. That means beginning sentences with "I," followed by an expression of *your* feelings, desires, or thoughts. There are some important advantages to this. By directly stating your feelings, you are taking the responsibility for your own well-being. In addition, "I" sentences avoid the blaming or accusatory tone common in statements that begin with "You." Statements such as "You don't like to hug me anymore" or "You don't love me" attack the other person's character and may well be met with a defensive response. The statement "I would like you to hug me more often" more directly expresses your feelings and desires and is less likely to elicit a defensive negative response.

Questions that begin with "Why" ("Why don't you . . . ?") are also often used to criticize a partner. Avoid those as well. Even sentences that begin with "We" can cause problems because they make assumptions about your partner's desires, moods, thoughts, or feelings. Don't talk for your partner.

Dr. Harold Bloomfield and Sirah Vettese, marriage partners and co-authors of *Lifemates: The Love Fitness Program for a Lasting Relationship* (1989; reprinted with permission.), suggest "sexual heart talks" to enhance sexual intimacy. They ask couples to take turns completing the following statements with their spouse:

"The best thing about our sex life is . . ."

"What I find sexually attractive about you is . . ."

"A sexual delight I would like to indulge with you is . . ."

"A sexual fantasy I would like to act out with you is . . ."

"The way I would like to be touched is . . ."

"When I would like to be sexually romantic and you don't respond I . . ."

"I feel like withdrawing from sex when . . ."

"Three sexual fears I have are . . ."

"I feel sexually frustrated when . . ."

HOW CAN I FIND OUT ABOUT MY PARTNER'S DESIRES AND NEEDS?

Yes, it's important that you know and understand your partner's sexual desires and needs, but as a preliminary step, be sure that you are aware of your own. Once you are certain, a good place to start is with self-disclosure. Reveal to your partner (using "I" language) your thoughts, feelings, desires, and needs. This creates an environment of trust and understanding, and most people will self-disclose in return. Compare and discuss how each of you feels about different behaviors. Learn each other's preferences. Here again, a sex manual might be helpful.

Of course, you can also ask your partner about his or her preferences, but make sure that you *ask questions that are open-ended* (e.g., "How do you like me to touch you when . . ."). Avoid questions that can be answered with one word (yes-or-no questions such as "Did you like it?" or "Do you like me to kiss your neck?").

IS LISTENING IMPORTANT? IF SO, HOW CAN I BECOME A BETTER LISTENER?

Effective communication is what holds close relationships together and enables them to grow. However, many people forget that communication requires not just talking, but listening as well. This is as true for sexual relations as it is for any other behavior. How well do you listen?

We may hear with our ears, but *we show how attentive we are with body language*—body position (Do you face your partner while he or she talks to you?), eye contact, facial expressions. Be supportive of your partner's attempts to communicate with you. In addition to positive body language, interact with him or her in a constructive manner. Make comments or paraphrase what your partner is saying so that you both are sure what the other is trying to say—not everyone conveys messages clearly.

To improve your listening skills, try giving your partner a massage. Giving one another a massage strengthens relationships by promoting better communication (Comiskey, 1989). Use some oil and make it a good massage. When starting the massage, it is normal to assume that what feels best to you will feel best to others as well. If you like your shoulders and neck massaged with a lot of pressure, for example, it is only natural that you will start off by attempting to please your partner in this manner. However, not everyone likes the same things you do. Your partner may prefer gentle massaging of the lower back. How will he or she communicate this to you? He or she may tell you, but it is important to realize that a person does not have to use words to communicate. When a massage really feels good, most people emit little noises like "ooh" and "aah." These responses should be just as effective as words in letting you know that what you are doing feels good.

When giving your partner a massage, vary the pressure and explore different areas of your partner's body. Avoid touching genitals and breasts, for the purpose of the massage is not your own sexual pleasure, but to learn how to listen (you may discover, however, that massage is a great form of foreplay and you may want to use it for that purpose in the future). Listen to your partner's responses while you are massaging. If you listen well and care about your partner, you can modify your behavior and give a massage that pleases him or her instead of giving one that might please only you.

If you are a good enough listener to adjust a massage to please your partner, you can easily do the same while having sex. All of us want to believe we are (or will be) a good sexual partner, but to be one we must first be a good listener.

IS IT POSSIBLE TO COMMUNICATE NONVERBALLY?

The answer, of course, is yes. Actually, I have already partially answered this question in my answer to the last question. Things like eye contact, facial expressions, and interpersonal distance (how close you stand or sit next to someone) all express messages. Single persons commonly use these types of messages to convey interest, or lack of interest, in another person. Silence can convey a message (e.g., anger, anxiety). Touch, in particular, is a very powerful means of *nonverbal com-*

munication. As one person who studies communication and intimacy puts it, "If intimacy is proximity, then nothing comes closer than touch, the most intimate knowledge of another" (Thayer, 1986). For example, the manner in which you and your partner go about giving each other a massage may well reflect how much you care about the other.

Perhaps the biggest problem with nonverbal signals is that, because they are not as precise as verbal messages, they can be misinterpreted. This frequently happens in the singles' world.

When two people are communicating verbally, it is important for there to be agreement between the verbal and nonverbal aspects of the communication. How do you suppose your partner would feel if, while telling him or her how much you care, you were inattentive, with little or no eye contact?

IF I WANT MY PARTNER TO CHANGE AND I FOLLOW THESE GUIDELINES, HOW SOON CAN I EXPECT TO SEE A CHANGE?

Even when there is good communication, neither you nor your partner can be expected to change overnight. If there are several things about your partner's attitude or behavior that bother you, do not address them all at once. Attempt to change only one thing at a time. If you present your partner with a long list of criticisms all at once, it will probably be met with defensiveness and retaliatory criticism.

WHAT IF WE CANNOT AGREE?

No one person will always be able to satisfy all the needs and desires of another. That is true even in the best of relationships. People simply have differences of opinion and different preferences. When this happens, good communication can only lead to an understanding of those things that the two of you disagree about. It is often helpful to agree that you disagree. However, if either you or your partner is greatly bothered because your needs and desires are going unfulfilled, then it would be wise to seek professional counseling. Chapter 13 provides the names and addresses of two organizations that will help you locate qualified professionals in your area.

IN THE FUTURE: TALKING WITH YOUR CHILDREN ABOUT SEX

When asked where he received his education, Mark Twain answered, "Throughout my life, except for the years I attended school." This is generally true for most people when it comes to learning about sexually related matters. *Sex education courses* are really just an intervention—a *formal* attempt to provide

sexual learning Knowledge about sex that is received (usually informally) from family, friends, the media, and society in general.

Courtesy of Family Planning Centers of Hawaii.

knowledge. The few people who take comprehensive sex education courses generally do not take them until they are in high school or college. However, **sexual learning** begins in infancy and continues throughout childhood and adolescence. It is informal and is received from family, friends, the media, and the rest of the general environment.

During the preadolescent years, a child's parents are the most powerful influence on his or her beliefs and behavior. Even if parents never discuss sex with their children—and most do not—by their behavior and their attitudes, they still serve as role models from which their children learn.

As children approach their teenage years, their peer groups come to have as much (and often greater) influence on their behaviors and attitudes as their parents. Parents must also contend with the influence of TV and other media. The average child watches several hours of TV every day and is exposed to hundreds of references to sex in a year (Lowry & Shidler. 1993). Between the amount of time that their children spend with peers and watching TV, it is obvious that parents have a difficult time competing as a source of sexual knowledge and values. Unfortunately, the peer group is often misinformed, and TV and movies often display irresponsible sex (e.g., couples having sex without any mention of birth control).

Today, with the threat of AIDS, there has never been a better reason to teach our children responsibility about their health and their bodies. Yet despite all the public debate about AIDS and other sexually related matters, most parents continue to remain silent (*Time*/CNN poll, 1993).

> "Sex was never a topic in my home when I was growing up. It led to embarrassment when I finally had to ask questions, but I could only ask my friends, never my parents. It shouldn't be like that, it shouldn't be embarrassing and I should have been able to talk to my parents."
>
> (from the author's fiiles)

Surveys show that the vast majority of Americans favor sex education in schools (e.g., Janus & Janus, 1993). So why is there reluctance to teach it at home? Probably because most parents never received sex education themselves and thus are uncomfortable with their own sexual knowledge and feelings.

> "Due to lack of knowledge, I could not elaborate enough."
>
> "I always hoped I would be able to handle such topics with my children, but I guess I was unable because it was never really discussed with me."
>
> "If I had been educated earlier, beginning at age 9 or 10, I would have been able to help myself and my children with facts of life much better and much earlier."
>
> (from King & LoRusso, 1997)

Some parents will not discuss sex because they do not want to think about the possibility of their children being sexually active. I even know of some parents who have never talked to their children about sex because of fear that the children would then suspect that they have sexual relations, as if that were wrong.

With the completion of this book, you should be ready to put an end to the cycle of noncommunication that has existed between parents and each new generation of children. I realize, however, that while you have learned the necessary information, you may still feel uncomfortable about discussing sex with someone else. This part of the chapter is intended to help you initiate discussions of sexuality with your children so that they too may acquire factual information. I will try to answer some of the questions that often concern parents.

WHY SHOULD I TALK TO MY CHILD ABOUT SEX?

Unless you are willing to keep your children locked up in a closet until they reach adulthood, *there is no avoiding their learning about sex.* As I have already said, even if they stay home all the time and do nothing but watch TV and listen to the radio, your children will still see and hear thousands of references to sex every year. Do you really want to leave it to the soap operas and "shock jocks" to educate your children about this important subject? What kind of values do you suppose they will learn? Also, don't fool yourself. Children (in all likelihood your own as well) do talk about sex with their friends. Not only may your child's friends be misinformed, but do you want to leave it to your child's peer group to teach sexual values? Even formal sex education courses are generally value-free. By assuming the primary responsibility for your children's sex education, you can not only make certain that they are getting factual information, but at the same time you can stress the social and emotional aspects of sex and teach them values that you consider important. June Reinisch, former director of the Kinsey Institute, said, "If parents are comfortable and have good information, they can be wonderful sex educators. They can contextualize the sexual information into the fabric of family values, ethics, morals and religious beliefs. Sex is embedded in all of these" (quoted by Henderson, 1990).

DOES TELLING CHILDREN ABOUT SEX LEAD THEM TO DO IT?

Some of you may worry that if you educate your children about sex, they will try it and get pregnant or get someone else pregnant. Does driver's education cause automobile accidents? Surveys show that most people engage in premarital sex. The average age of first intercourse today is 16, and in some large metropolitan areas it is much younger (Laumann et al., 1994). Most of these teens have never had formal sexuality education, at school or at home, so sexuality education can hardly be blamed. Studies have shown, you may recall, that children who have had sexuality education are no more likely (and perhaps are less likely) to have sexual relations, to get pregnant, or to contract a sexually transmitted disease than those who have not had sexuality education (Grunseit et al., 1997). Do you need a better reason to talk to your children about sex?

WILL A SINGLE "BIRDS AND BEES" TALK SUFFICE?

Do you really believe that one talk will prepare your child for a lifetime of sexual relations and problems? The answer is no! Experts say that it is a mistake to think that talking about sex means sitting down for one long, serious conversation (see Roan, 1993). Even if you could fit all the necessary information and values into one talk, that one talk can hardly compete with the hundreds of hours of exposure to sexually related matters your child will receive while growing up. As Mark Twain said, education is a lifelong thing. Communication between parents and children about sex and anything else should also be lifelong.

WHEN SHOULD I START TALKING WITH MY CHILD ABOUT SEXUALITY?

When asked why she had not yet begun talking to her 13-year-old daughter about sex, a mother said, "I kept waiting for the right time. But I'm just too afraid to bring it up. I'm not sure what to say" (Roan, 1993).

As I hope you have learned by now, sex education is more than just explaining sexual intercourse. Children should be properly educated about their bodies, for example, and this means teaching them the correct anatomical terms for their genitals at the same time you teach them the names of the other parts of their bodies. "Penis" and "vagina" are not too difficult for young children to pronounce. American parents are notoriously silent about this. One study found that only 23 percent of 9-year-old North American children knew the physical sex differences between newborn baby boys and girls (Goldman & Goldman, 1982). This was a much lower percentage than found for children in some other countries where sex education is provided in school. The following comments were written by a student in my course:

"I disagree strongly with teaching children about sexuality starting from preschool to third grade. Kids at this age don't even know the ABC's. . . . I just don't believe that kids under grade 4 should be told the terms 'penis' and 'vagina.' I do feel that kids under the age of 10 should be taught to know when a stranger is touching them wrong."

(from the author's files)

Think about this for a moment. How will this woman tell her children about inappropriate touching without using the proper words? Will she use cutesy words (e.g., "pee-wee" or "weeney") or make vague references (e.g., "Don't let anyone touch your bottom")? Either of these approaches may convey an attitude to the child that these parts of his or her body are naughty or dirty, and may eventually cause the child to think that the functions of these body parts are also dirty. Teaching boys that they have a penis and girls that they have a vagina is honest and helps them attain a healthy attitude about sexuality.

It is also very important to discuss with your children the physiological changes that start at puberty before they actually begin puberty. Some parents don't talk to their daughters about the menstrual cycle or to their sons about "wet dreams" until after they occur (see Chapter 10). This is very wrong. These changes can be frightening to children, who may think that they are sick or dying.

"There is one thing I would personally put more emphasis on [in the book]. That is the importance of explaining menstruation in advance. From personal experience, I feel this is critical. I had no idea what was happening to me when I started menstruating for the first time. I can remember that I couldn't stop the bleeding and was scared to death and was afraid and embarrassed to tell anyone. I finally told a friend who was in the dark just as I was. She then told her mother. Her mother gave me a pad and said this would happen every month. That was the extent of communication on the subject."

(from the author's files)

It is also important to explain the changes associated with puberty in females to boys and the changes in males to girls. Men should be able to relate to women with empathy and understanding (and vice versa). Ignorance about the physiological (and emotional) changes being experienced by the opposite sex is not a good basis on which to build relationships.

Children themselves are often the best guides as to when to initiate talks about a particular topic. It is common, for example, for young children to ask where babies come from. If they are curious enough to ask questions about some sexually related matter, don't avoid discussing it. Never brush them off. If you do, they may avoid coming to you in the future and may rely instead on getting their information from somewhere else.

Some guidelines for when to start talking about specific topics within human sexuality and family living are presented in Table 17–1. The guidelines are a condensed composite of those suggested by the Sex Information and Education Council of the U.S. (SIECUS National Guidelines Task Force, 1996) and Planned Parenthood

Figure 17–2 The goals of teaching about sexuality are different for children at different ages. At any age, however, be truthful. Don't tell your young child that the baby is in mommy's "tummy." Use correct terms while explaining the body.

TABLE 17–1

GUIDELINES FOR TEACHING CHILDREN ABOUT SEX

Age (School) Level	Concepts to Be Learned
3–5 (preschool)	Correct names for body parts (e.g., penis, vagina); differences between boys and girls; both boys' and girls' bodies are special and each of us can be proud of our special body; adults kiss and hug each other to show how much they care; respect for the privacy of others; their bodies belong to themselves and they have a right to say "no" if they don't want to be touched; how to respond if a stranger offers candy or a ride.
6–9 (early elementary school)	Continue to use a proper vocabulary when talking about body parts; continue to promote self-esteem (their bodies are special); people have body parts that feel good when touched; touching those body parts should be done in private; "sex talk" is done at home; the basics of reproduction and that our genitals and reproductive organs enable this; a fetus is inside the uterus and a baby usually comes out of the vagina; milk comes from breasts; people can have babies only after puberty; meaning of heterosexual and homosexual; awareness of sexuality of people of all ages (parents and grandparents); understanding that there are different types of caring homes and families (e.g., single-parent families); promote non-stereotyped gender roles both inside and outside the home (both mothers and fathers have important jobs and there are no jobs just for boys or just for girls); good hygiene
9–13 (upper elementary school)	Changes in *both* boys and girls during puberty (e.g., girls menstruate; boys ejaculate and have wet dreams) and that the changes occur at different ages in different people; the idea that human sexuality is natural and that sexual feelings are normal; sex is not just for reproduction, but is also pleasurable; masturbation (many people do it, it's not harmful, often a child's first experience with sexual pleasure); intercourse can cause pregnancy and having children is a long-term responsibility that should be planned (include basics on contraception); children are not ready for sexual intercourse; physical appearance does not determine a person's value—liking yourself; communication and assertiveness skills; people can disagree and have different values but still be friends; men and women have equal talents and strengths; recognition of the potential of sexual abuse (including that it is most often committed by someone the child knows) and how to react to it; continued encouragement of good hygiene; encouragement that it is okay to talk to parents about sexuality (*even if our opinions differ*)
13–15 (junior high school)	It is normal to have sexual feelings, desires, and fantasies; sex is pleasurable; the size of one's penis or breasts does not determine whether one will be a good sexual partner; people of all body shapes (including the disabled) have sexual feelings and desire; understanding the differences in sexual behavior (sexual orientation, abstinence, marriage); love is not the same as sexual attraction or desire; understanding the possible consequences of unprotected sexual intercourse (pregnancy); contraception; young teens are not yet ready for sexual intercourse; dating (why, when, how) and relationships (e.g., going steady); peer pressure and your right to say "no" and to disagree with your friends; respecting the limits that are set by your partners; awareness of exploitive relationships (psychological coercion, alcohol, physical force, date rape); physical health (e.g., breast and testicular self-exams); establishing a value system (and consideration of the above topics within that system)
15–18 (high school)	People of all ages are sexual beings; sexuality is only one part of your total personality; integration of sexuality into your value system; feeling good about your body, sexuality, and sexual orientation; how to express your sexual feelings in appropriate ways; communicating your sexual feelings, desires, and limits to partners in an open and honest manner; expressing your sexual feelings without intercourse; further understanding of sexual exploitation among adolescents; for those who engage in sexual intercourse, choosing a method of contraception that will be used consistently; knowledge of the responsibilities of parenthood and child care; understanding that long-term commitments (including marriage and parenthood) require work and that people's sexual and emotional needs change over time

(1991). These are only general guidelines for just a few of the many things your child may want to know about. Remember, if your child asks questions before the age level indicated here, don't wait to respond.

WHAT SHOULD I TELL MY CHILDREN ABOUT AIDS AND OTHER STDS?

The AIDS epidemic, more than anything else, has provided the impetus for bringing frank, explicit discussions about human sexuality to children, both at home and at school. Former U.S. Surgeon General C. Everett Koop, you recall, advocated teaching sex education, including information about AIDS, to children beginning in grade school. Most experts agree that this instruction should begin no later than junior high school.

Figure 17–3 When a pet gives birth, parents have a great opportunity to teach their children about sexuality.

With all the attention devoted to AIDS by the media today, children are naturally going to be curious about it. The first question about sexuality that my son ever asked me was about AIDS. If you hear your child mention AIDS, be sure to take the opportunity to ask what he or she knows about it. Clear up any myths or misunderstandings. If your child never mentions AIDS, do not ignore the subject, for this is a very serious disease that everyone needs to be educated about. Initiate the discussion yourself. Remember, however, that you must teach your child about sexuality on a level that he or she can understand. Whatever their age, teach your children respect for individual differences and acceptance of others.

Most parents fear that in order to explain about AIDS, they must explain behaviors like anal and oral sex. This is really not necessary for preschool and elementary school children. Most preadolescent children are not going to understand anal or oral sex. Young children can be taught to be in charge of their own bodies—sexual responsibility—without being told what anal sex is.

What kind of questions is your child likely to ask? Valentich and Gripton (1989) have listed many questions about AIDS that are frequently asked by children. Early elementary school–age children may want to know answers to questions like: What is AIDS? Can I get AIDS? Can I get it by playing with friends or by touching someone at school? Will it make me sick? Is there medicine I can take to make it go away?

Children aged 9–12 will want to know more about how AIDS is transmitted (e.g., whether they can get AIDS by holding hands, hugging, kissing, having sexual intercourse, or being in the same classroom with someone who has AIDS) and how they can tell if someone has it. Children at this age can also be taught about STDs such as gonorrhea, chlamydia, syphilis, and herpes (and told that they are caused by bacteria and viruses) (SIECUS, 1996). Teenagers will probably want answers to more explicit questions about transmission (e.g., whether they can get AIDS by French kissing, masturbation, genital touching, or sexual intercourse) and prevention (including the use and reliability of condoms). Be sure to include further discussion of other sexually transmitted diseases at this time as well.

You should prepare yourself for questions like these so that you can give your child factual information about means of transmission, prevention, and treatment appropriate for his or her age. When answering questions, you can also discuss values with your child.

No matter what your child's age, *you should not deny the pleasures of sex or resort to scare tactics.* Teaching children that sex is something good and pleasurable is as important as teaching that it must be handled responsibly.

HOW DETAILED SHOULD SEX DISCUSSIONS BE?

Your discussions should be frank and explicit. Children do not want "birds and bees" analogies—they want factual information using real terms. In fact, euphemisms and analogies will only confuse them. At the same time, do not make things too detailed for your child's age level. *Make your answer age-appropriate.* Always check to see if your child understands what you have said. You may have to rephrase your answer to make it clear. Sometimes we give great answers, but not to the question the child asked. Remember, a child may not know what to ask or how to ask it correctly. When a 6-year-old asks where babies come from, for example, many parents worry that they must explain intercourse in detail. Your 6-year-old is not really asking about that. Keep your answer uncomplicated and tell the child only what he or she wants to know. In this case, you could explain that the baby is in a special part of the mother called a uterus (do not avoid explicitness by saying "tummy"—this is incorrect anyway) and that when it is ready to be born it comes out through an opening called the vagina.

Remember that young children are curious about *everything* and that sex is only one part of their world. Also, children don't think about sex the way adults do. Children have not had sexual intercourse (unless they were molested), have not had babies, and may not look at sex as a special topic of conversation. (In fact, chances are you are making a bigger deal about it than your child.) When the 6-year-old son of one of my colleagues asked how babies were born, his father apprehensively explained about the uterus, the entry of the baby into the world through the vagina, etc. After giving detailed, accurate information, the father asked, "Do you have any more questions?" His son yawned and said, "Can we talk about something else now?"

Discussions with your teenage children should definitely be frank and explicit. In a recent study, I asked students and their parents if they had ever had a "meaningful" discussion about sex with one another (King & LoRusso, 1997). Over half of the students responded "no," yet for 60 percent of them one or both parents said that there had been meaningful discussions. Here are some responses by parents who believed that they had had meaningful discussions (but whose children disagreed):

> "I believe children should be taught that total abstinence is the best way. . . . This is so because of God's law and the design for total inner peace and joy."

> "My discussion with _____ dealt with one's soul. A soul is not to be fragmented—not to be given away in bits and pieces. A soul is sacred—sex is sacred."

> "They [the children] knew what was expected, and they behaved accordingly. . . . If you asked about teaching morality to my children, the answer is yes, and if you teach and live a good, healthy, moral life, it usually works."

(from King & LoRusso, 1997)

There is nothing wrong with teaching your children values. However, do not substitute morality for factual information.

WHAT IF I FEEL EMBARRASSED?

Many of you will not be able to avoid embarrassment at times, but the result—your child's education—is worth the discomfort of using real terms and direct language. Be honest. Let your children know that you are a bit embarrassed, but that you are glad they asked to talk with you, that you'll be happy to answer all of their questions, and that it's okay to talk to you about the subject. What you say to your children is important, but it's more important to let them know that you are willing to (and want to) listen to their questions.

HOW SHOULD I TALK WITH MY CHILD?

Most important, *discuss—never lecture.* Also, be sure not to grill your child about his or her behavior or attitudes. Parents who take an assertive approach from a stance of power will probably be met with passivity, while those who take a collaborative approach will probably find that their children actively engage in discussions (Yowell, 1997). In other words, allow your child to have his or her input. Your tone of voice is particularly important, for your child learns attitudes from your tone. Try to sound calm and relaxed at all times. Sol Gordon, author of *Raising Your Child Conservatively in a Sexually Permissive World* (1989), says, "The most important message is that nothing a child does will be made worse by talking to the parent about it. There is no way of dealing with sexuality unless the parent has created the atmosphere of love and caring. Unless that atmosphere is there, nothing works. The child will lie" (quoted by Roan, 1993).

Also, it is not necessary always to have formal talks; this may create tension. You can break the ice by briefly explaining and discussing sexually related matters that your child sees on TV (try watching *Beverly Hills 90210* with your teenage child), hears on the radio, or reads in "Dear Abby."

Remember, *avoid scare tactics.* You must teach sexual responsibility without making your child afraid of sex—children should learn that sex is something good and positive, and not something to feel ashamed or guilty about. Scare tactics probably will not work anyway, and they may cause emotional and psychological problems.

WHAT ABOUT MORALS? AREN'T THEY IMPORTANT TOO?

Most formal sex education is value-free, and that is why it is so important for parents to assume some responsibility for their children's education, even if their children are receiving sex education at school. Sex is more than a biological function. If you want your children to share your beliefs and values about sex, you must talk about them. However, your success in imposing your values upon your children will depend, in large part, on your overall relationship with them. Parents who generally have good interactions with their children are more successful in transmitting values than are parents who generally have poor interactions with their children (Taris & Semin, 1997).

When you discuss moral values with your child, there are some important guidelines you should follow. Most important, *you can emphasize your values, but don't preach.* It is not wrong for you to say how you feel, but if your child has developed some values that are different from yours, you must learn to discuss your differences rather than dictate. Remember, you want to create an atmosphere that allows your child to be comfortable so that he or she will come to you when they have questions. If you always preach and pass judgment, your child may be reluctant to discuss sexuality with you at all, and then you will have no input.

One of the biggest dilemmas facing parents when discussing sexuality with their children is whether to emphasize global moralities or situational ethics. *Global moralities* are values that are supposed to apply to all people under all circumstances. An example is when parents tell their children that sexual relations should be reserved for marriage. As you recall, however, surveys show that well over half of all adolescents engage in premarital intercourse. Can you safely assume that your children will never have premarital sex-

Figure 17–4 When talking to your children about sex, discuss—never lecture. Let them have their input.

ual relations and that you can thus avoid the topic of birth control? Many parents are reluctant to talk about birth control with their children because they think that it promotes promiscuity. The consequences of not talking about it, however, can be devastating (e.g., unwanted pregnancies and/or sexually transmitted diseases).

Situational ethics emphasize the proper thing to do in particular situations (i.e., "We prefer that you not do it, but if you're going to do it anyway, then . . ."). In this case, for example, it could be explained to an adolescent child that condoms substantially reduce the risk of both unwanted pregnancy and sexually transmitted diseases. I cannot tell you which of these two approaches to use. Only you can make that choice.

CAN MY BEHAVIOR AFFECT MY CHILD'S ATTITUDES AND BEHAVIOR?

You should also be aware that *your own behavior can play a role* in your child's sexuality and moral values. It is important for children to learn that their parents share affection for each other. Hugging between parents and between parents and children teaches youngsters that the open display of affection is appropriate and appreciated.

> "My husband and I have always tended to be very private about showing affection. I guess that's because our parents were that way. However, as a result of what I've learned in this class we've been hugging and kissing in front of our kids—who are both teenagers. They've noticed the changes and have even asked us about it, which has given us wonderful opportunities for open and frank discussions with them. Hopefully, it's not too late."
>
> (from the author's files)

Parents who wait until they are alone to kiss or hug are teaching their children that there is something wrong about physical contact between the sexes. *Don't confuse affection with sex.*

At the same time, we live in a day when many parents are divorced and have boyfriends and girlfriends of their own. You must avoid appearing to be hypocritical. If you want your child to abstain from premarital sexual relations, for example, you must avoid premarital sexual relations or be very discreet yourself. What message do you suppose you are sending your child if you tell him or her it is wrong to engage in sex and they know that you are having premarital sexual relations?

Even if you never discuss sexuality with your children, your behavior will still serve as a role model. Unfortunately, many children acquire negative attitudes about sexuality by watching and listening to their parents. Many adults, for example, enjoy "dirty" jokes. What attitude do you suppose is conveyed when children overhear their parents laughing and snickering at jokes involving sex or body parts? Learning that there are "clean" and "dirty" jokes can lead to beliefs that sex is dirty, especially if other sources of information about sex are unavailable.

Children become interested in jokes by the time they reach kindergarten. Although they may not understand the meaning of jokes, they are eager to use them as a source of information. They also want to act older, and seeing older children use obscenities or tell dirty jokes and laugh at them will lead to imitation. Laughing at a child's use of obscenity or joking about it will only reinforce the behavior, making it increase in frequency (remember, you are a role model). If your child uses an obscenity, don't go overboard. Calmly explain what the word means. Often the child will not know. Reacting calmly tells children that they can discuss anything with you.

How do you feel about dirty jokes? What makes something dirty? Is sex dirty? When comedian Woody Allen, who has a bizarre sense of humor, was asked if sex was dirty, he responded, "Only if it's done right." Let's hope your children won't grow up believing this.

HOW DO I KNOW IF I HAVE SUCCEEDED?

You really can judge the success of your attempts to educate your children about sexuality only by their willingness to come to you when they have questions and problems.

> "I am a non-traditional student with three kids. I didn't know that I should start talking with them about sex at such a young age. My boys are 7 and 9 and my daughter is 5 years old. They are very receptive to our talks. And now they always want to talk to me about anything. 'Open lines of communications.' Thanks."
>
> (from the author's files)

All normal adolescents will have an interest in sex and will continue to need information about their sexuality as long as they live. They will want to know about dating, courtship, marriage, living together, starting a family, and many different things at different times in their lives. They will seek information about these and other matters from a variety of sources, and you will want to be one of them. This will only happen, however, if communication between children and parents is open and continuous.

HOPE FOR THE FUTURE

At the beginning of this chapter, I said that fewer than a third of the more than 1,000 students who have taken my course each semester had ever had a serious discussion with their parents about sex. For those who have had one, in many cases it was just a single "birds and bees" talk explaining sexual intercourse. You have learned how this lack of knowledge has been costly to both individuals and society in terms of out-of-wedlock births, sexually transmitted diseases, and sexual abuse. Misinformation about sexual matters is often a contributing factor in sexual disorders and relationship problems as well.

I hope that by expanding people's knowledge, courses like this will result in greater sexual responsibility and fewer of the previously mentioned problems. I hope, too, that this book has helped you feel more comfortable with your sexuality so that you are not ashamed of your body or embarrassed by its functions. Most of all, I hope that when you have children you will be able to discuss sexual matters with them. I've found that students who have completed my course are much more likely to discuss sexuality with their children than are former college students who had not taken a comprehensive human sexuality course (see King, Parisi, & O'Dwyer, 1993).

Let us hope that in the years to come, when I ask my students on the first day of class if they have had discussions with their parents about sex, most of them will say yes.

Key Terms

communication 415
global moralities 426
"I" language 419
nonverbal communication 419
sexual learning 420
situational ethics 426

Personal Reflections

1. Suppose that you are in a long-term monogamous relationship that includes sexual relations. Suppose also that your partner often does not take as much time during foreplay as you would like and does not spend much time holding and touching you afterwards, which often leaves you sexually unfulfilled and frustrated. How would you go about correcting this situation? Be as specific as possible.

2. Suppose that you are unsure of your partner's sexual desires and preferences and you are unsure whether he or she is totally aware of yours. How might you communicate with one another? Suppose your desires and preferences are not exactly the same?

3. Some day you will probably be a parent (if you aren't one already). Do you plan to educate your children about sexuality? If so, how? When will you begin?

4. How do you suppose you would react if:
   (a) your 5-year-old child asked you where babies come from,
   (b) your 9-year-old child asked you what sex is,
   (c) your 10-year-old child asked you what "French kissing" is,
   (d) your 12-year-old son came home from school saying that he had to watch a gross movie about girls having their period,
   (e) your 15-year old son asked you about birth control, and
   (f) your 15-year-old daughter asked you about birth control?

5. Would your reactions to any of the situations in question 4 discourage your child from coming to you in the future and talking to you about sex?

Selected Readings for Couples

A Couple's Guide to Communication by John Gottman. Champaign, IL: Research Press, 1976. The book most recommended by members of the Society for the Scientific Study of Sex (Althrop & Kingsberg, 1992).

For Each Other: Sharing Sexual Intimacy by Lonnie Barbach. New York: Anchor/Doubleday, 1982.

He Says, She Says by L. Glass. New York: G.P. Putnam's Sons, 1992. Designed to break the communications barrier between men and women.

The Love Fitness Program for a Lasting Relationship by Harold Bloomfield and Sirah Vettese. New York: New American Library, 1989. This married couple suggests "sexual heart talks" to enhance sexual intimacy.

The New Peoplemaking by Virginia Satir. Palo Alto, CA: Science and Behavior Books, 1988. Easy to read, this book covers both communication and family relationships.

Self-Disclosure by Valerian Derlega et al. Newbury Park, CA: Sage Publications, 1993.

Why Marriages Succeed or Fail by John Gottman. New York: Simon & Schuster, 1994. Designed to help couples with communication.

You Just Don't Understand: Women and Men in Conversation by Deborah Tannen. New York: William Morrow, 1990 (paperback: Ballantine, 1991). Explains the different styles of conversation used by men and women and gives advice on how to talk to the other sex.

Selected Sex Education References to Help Parents

Ask Me Anything by Marty Klein. New York: Simon & Schuster, 1992. Prepares parents to give straightforward answers to anything their children might ask.

"Everything Your Kids Want to Know About Sex and Aren't Afraid to Ask," article by Alex Packer in *Child* magazine, September 1995.

The Family Book About Sexuality by Mary Calderone and Eric Johnson. New York, Harper & Row, 1989. Very humanistic approach.

How to Talk to Your Children About AIDS by the Sexuality Information and Education Council of the United States (SIECUS), 1994.

How to Talk with Your Child About Sexuality: A Parent's Guide by Planned Parenthood: Faye Wattleton with Elizabeth Keiffer. New York: Doubleday & Co., 1986.

Now What Do I Do? by the Sexuality Information and Education Council of the United States (SIECUS), 1996. Designed to help parents of preteens.

Raising Your Child Conservatively in a Sexually Permissive World by Sol Gordon. New York: Fireside, 1989.

Talking with TV by the Center for Population Options (1025 Vermont Avenue, NW, Suite 210, Washington, DC 20005). An excellent guide for adults and kids on how to make TV a conversation starter for discussions about relationships, values, and sexuality. Updated yearly.

Talking with Your Child About Sex: Questions and Answers for Children from Birth to Puberty by Mary Calderone and James Ramey. New York: Random House, 1983.

"What Kids Need to Know," article by Sol Gordon in *Psychology Today,* October 1986.

What We Told Our Kids About Sex by Betsy and Michael Weisman. New York: Harcourt, Brace Jovanovich, 1987. Gives answers to 102 questions typically asked by preadolescent children.

Selected Sex Education References for Children

(Selections are listed in age-appropriate order.)

Where Did I Come From? by P. Mayle, A. Robins, and P. Walter. Secaucus, NJ: Lyle Stuart, 1973. For young children.

Where Did I Come From? by Peter Mayle. New World. A 27-minute animated videocassette version of the book. Many video rental stores carry it.

Babies Come from People by George and Ann Carpenter. Salt Lake City, Utah: Coldstream, 1975. For preschoolers.

What's Happening To Me? by Peter Mayle. Secaucus, NJ: Lyle Stuart, 1975. For children approaching puberty.

Are You There, God? It's Me, Margaret by Judy Blume. Englewood Cliffs, NJ: Bradbury Press, 1970. An educational story about a girl going through puberty.

Having Your Period by Planned Parenthood (16 pages).

It's Perfectly Normal: Changing Bodies, Growing Up, Sex and Sexual Health by Robbie Harris. Candlewick Press, 1994.

Dr. Ruth Talks to Kids by Dr. Ruth Westheimer. Aimed at children aged 8 through 14. *Time* magazine (May 24, 1993) says that

Dr. Ruth's book provides an excellent middle ground between books that teach abstinence only and those that favor a more comprehensive approach.

What Every Teen-ager Ought to Know by "Dear Abby." (Abigail van Buren). Send a check or money order for $2.50 and a large stamped (50 cents), self-addressed envelope to: Dear Abby, Teen Booklet, P.O. Box 447, Mount Morris, IL 61054.

Teensex? It's OK to Say: No Way! by Planned Parenthood (16 pages, also available on VHS video cassette).

Talk About Sex by SIECUS. For teens, this 46-page booklet "offers clear, honest, straightforward information and instruction about relationships, communication skills, and safer sex behaviors in a very engaging, youth-friendly" manner. 1–4 copies, $2 each. SIECUS, 130 West 42nd Street, Suite 2500, New York, NY 10036.

Easy For You to Say: Q & A's for Teens Living with Chronic Illness or Disability by Miriam Kaufman. Toronto: Key Porter Books, 1995.

Glossary

A

abortion Termination of pregnancy. Depending on how far the pregnancy has advanced, this can be done by taking a pill (Ovral, birth control pills in high dosage, RU 486); scraping the uterine lining (dilation and curettage); removing the uterine lining by suction (dilation and evacuation); or inducing labor (by injecting hypertonic saline or prostaglandins). (Ch. 6)

acquaintance rape A rape committed by someone who is known by the victim. (Ch. 15)

adolescence (ad″ ′l es′ ′ns) The time of life between puberty and adulthood. (Ch. 10)

adoption Taking a child into one's family by legal process and raising the child as one's own. (Ch. 6 and 7)

adrenal (ad- re′ nal) **glands** Endocrine glands located near the kidneys that secrete steroid hormones and small amounts of gonadal hormones. (Ch. 3)

adrenarche (ad″ren-ar′ke) The maturation of the adrenal glands, usually between the ages of 6 and 8. (Ch. 10)

adrenogenital (ah-dre″ no-jen″ y-tal) **syndrome** A condition in females in which the adrenal glands excrete too much testosterone during fetal development, causing masculinization. This includes enlargement of the clitoris and labia, so that the genitals are sometimes mistaken for those of a male. Also called *congenital adrenal hyperplasia*.(Ch. 8)

adultery Sexual intercourse with someone other than one's husband or wife. (Ch. 10 and 16)

afterbirth The expulsion of the placenta from the uterus after a baby is born. (Ch. 7)

agape (ah′ gah pā″) A type of romantic love that puts the interest of the loved person ahead of one's own, even if it means great sacrifice. (Ch. 12)

aggressive pedophile A pedophile who is sexually aroused by inflicting physical injury on victims. (Ch. 15)

AIDS (acquired immunodeficiency syndrome) An often fatal disease caused by a virus (HIV) that destroys the immune system. It is spread by intimate sexual activity (the exchange of bodily fluids) or contaminated blood. (Ch. 5)

AIDS-related complex (ARC) See **symptomatic HIV infection.**

amebiasis (am″ e-bi′ ah-sis) Dysentery caused by infestation of amoebae, one-celled organisms. (Ch. 5)

amenorrhea (ah-men″ o-re′ ah) The absence of menstruation. (Ch. 3)

amniocentesis (am″ ne-o-sen-te′ sis) A technique for detecting fetal problems, involving collection of amniotic fluid after the fourteenth week of pregnancy. (Ch. 7)

amnion (am′ ne-on) A thick-skinned sac filled with water that surrounds the fetus. (Ch. 7)

ampulla (am-pul′ lah) The expanded end of the vas deferens where sperm are stored prior to ejaculation. (Ch. 2)

anabolic (an″ ah-bol′ ik) **steroids** Synthetic steroid hormones that combine the growth (anabolic) effects of adrenal steroids with the masculinizing effects of androgenic steroids. (Ch. 3)

anal intercourse Insertion of the penis into the anus and rectum of a sexual partner. (Ch. 11)

anaphrodisiacs (an″ af-ro-dĭz′ e-ak) Substances that suppress sexual functioning. (Ch. 4)

androgen insensitivity syndrome A condition in which the testes secrete normal amounts of testosterone during male embryologic development, but the tissues do not respond to it. As a result, a clitoris, labia, and a short vagina develop, but the internal female structures do not develop because the testes still secrete Mullerian duct–inhibitory substances. (Ch. 8)

androgens (an′ dro-jens) Hormones that possess masculinizing properties. (Ch. 2)

androgyny (an-droj′ ĭ-ne) The ability of an individual to display a variety of personality characteristics, both masculine and feminine, depending on whether a trait or behavior is appropriate in a given situation. It is often viewed as a positive characteristic that gives an individual greater adaptability. (Ch. 8)

anger rapists Rapists who commit their attacks out of hostility or anger directed at their victim or at women in general. (Ch. 15)

anoxia (ah-nok′ se-ah) An inability to get enough oxygen to the brain. (Ch. 7)

aphrodisiacs (af″ ro-dĭz′ e-ak) Substances that enhance sexual desire or performance. (Ch. 4)

areola (ah-re′ o-lah) The darkened ring surrounding the nipple of a breast. (Ch. 2)

artificial insemination A method of treating infertility caused by a low sperm count in males. Sperm are collected during several ejaculations and then inserted into the partner's vagina at the time of ovulation. (Ch. 7)

ascetic (ah set′ ik) **philosophy** A philosophy that originated among the ancient Greeks. It emphasizes that virtue comes from wisdom and that these qualities can be achieved only by avoiding strong passions. (Ch. 1)

asymptomatic Showing no symptoms. (Ch. 5)

attachment A drab, mundane form of companionship where one's partner gives few positive rewards, other than predictability, for remaining in the relationship. (Ch. 12)

attachment theory of love The theory that the quality of the attachment between parents and child greatly affects the child's ability to form romantic relations as an adult. (Ch. 12)

autoerotic asphyxiation Depriving oneself of oxygen during masturbation-induced orgasm. (Ch. 11 and 14)

autoinoculation The act of infecting oneself. (Ch. 5)

B

bacteria Small, single-celled organisms that lack a nuclear membrane, but have all the genetic material (RNA and DNA) to reproduce themselves. (Ch. 5)

bacterial vaginosis (vaj″ ĭ-no′ sis) A type of vaginitis caused by the interaction of several vaginal bacteria (particularly *Gardnerella vaginalis*). (Ch. 5)

barrier (blockade) method General term for contraceptive methods that block the passage of sperm. (Ch. 6)

Bartholin's (bar′ to-linz) **glands** Glands located at the base of the labia minora in females that contribute a small amount of an alkaline fluid to their inner surfaces during sexual arousal. (Ch. 2)

basal body temperature The temperature of the body while resting. It rises slightly after ovulation. (Ch. 6)

baseline mammogram The first of a series of low-radiation X-rays used to detect breast tumors. Called "baseline" because subsequent mammograms are compared against it.(Ch. 2)

benign coital cephalalgia (sef ″ah-lal′ je-ah) Severe headaches that start during or slightly after orgasm in males. (Ch. 13)

berdache (bar dash′) See **two-spirit**.

bestiality (bes-te-al′ĭ-te) The act of having sexual contact with an animal. (Ch. 14)

beta-endorphins Natural painkillers produced by the body. (Ch. 7)

B-girls Prostitutes who work in bars and hotels, generally with the approval of the ownership. (Ch. 16)

Billings method A fertility awareness method of birth control in which changes in the consistency and quantity of cervical mucus are used to tell when ovulation has occurred. (Ch. 6)

biological determinism The belief that biological influences establish predetermined limits to the effects of cultural influences. (Ch. 8)

bisexual An individual with a sexual orientation to both men and women (Ch. 9)

blastocyst (blas′ to-sist) The fluid-filled sphere that reaches the uterus after fertilization and which was created by the transformation of the morula through continued, rapid cell division. (Ch. 7)

blockade method See **barrier (blockade) method**.

Braxton-Hicks contractions Uterine contractions experienced during the last trimester of pregnancy that are often incorrectly interpreted as the beginning of labor; also called *false labor*. (Ch. 7)

breast-feeding In reference to contraception, the sucking response by a baby on the mother's nipple that inhibits release of follicle stimulating hormone, thus preventing ovulation. (Ch. 6)

breasts In females, glands that provide milk for infants; located at the front of the chest. (Ch. 2)

breech birth A birth in which the baby is born feet or buttocks first. (Ch. 7)

brothel prostitutes Prostitutes who work in houses of prostitution. (Ch. 16)

bulb (of penis) The expanded inner end of the spongy body of the penis. (Ch. 2)

bulbocavernosus (bul″ bo-kav″ er-no′sus) **muscle** A ring of sphincter muscles that surrounds the vaginal opening in females or the root of the penis in males. (Ch. 2)

bulbourethral (bul″ bo-u-re′ thral) **glands** See **Cowper's glands**.

C

calendar method A fertility awareness method of birth control that attempts to determine a woman's fertile period by use of a mathematical formula. (Ch. 6)

call girls The highest-status prostitutes, who generally work out of their own apartments by appointment only. (Ch. 16)

candidiasis (kan″ dĭ-di′ ah-sis) See **moniliasis**.

capacitation A process that sperm undergo while traveling through the female's reproductive tract in which their membranes become thin enough so that an enzyme necessary for softening the ovum's membrane can be released. (Ch. 7)

carnal knowledge of a juvenile See **statutory rape**.

case study An in-depth study of an individual. (Ch.1)

celocentesis (se″ lo-sen-te′ sis) A technique used for detecting fetal abnormalities in which a needle is inserted between the placenta and the amniotic sac to retrieve cells. (Ch. 7)

cephalalgia See **benign sexual cephalalgia**.

cervical cap A contraceptive device that fits over the cervix by suction, thus blocking the passage of sperm into the uterus. (Ch. 6)

cervical mucus The slimy secretion of mucous membranes located inside the cervix. (Ch. 6)

cervicitis (ser″ vĭ-si′ tis) Inflammation of the cervix. (Ch. 5)

cervix (ser′ viks) The narrow lower end of the uterus that projects into the vagina. (Ch. 2)

cesarean (se-sa′ re-an) **section (C-section)** The surgical removal of a fetus through the mother's abdomen. (Ch. 7)

chancre (shang′ ker) The painless sore that is the main symptom of the primary stage of syphilis. (Ch. 5)

chancroid (shang′ kroid) A sexually transmitted disease, caused by the *Hemophilus ducreyi* bacterium, which is characterized by small, painful bumps. (Ch. 5)

child molestation Sexual contact between an adult and a child. (Ch. 15)

chlamydia (klah-mid′ e-ah) A sexually transmitted disease caused by the *Chlamydia trachomatis* bacterium, which lives on mucous membranes. (Ch. 5)

chorion (ko′ re-on) The fourth membrane of the trophoblast; it develops into the lining of the placenta.

chorionic sampling A technique used for detecting problems in a fetus during the eighth to tenth weeks of pregnancy. Hairlike projections (villi) of the chorion are collected and examined. (Ch. 7)

chromosomes (kro′mo-somz)   Rod-shaped structures containing the genetic material that determines a person's inherited characteristics. (Ch. 8)

circumcision (ser″kum-sizh′un)   In males, the removal of all or part of the foreskin of the penis. In females, the removal of the clitoral hood. (Ch. 2)

climacteric (klĭ-măk′ter-ik)   The changes that occur in women in the few years that precede and follow menopause. (Ch. 10)

clitoral hood   The part of the labia minora that covers the clitoris in females. (Ch. 2)

clitoridectomy (klĭ″to-rĭ-dek′to-me)   Removal of the clitoris, often accompanied by removal of the labia minora. In an extreme form, called *infibulation*, the labia majora are also removed, and the sides of the vulva are then sewn together. (Ch. 4)

clitoris (klĭ′to-ris)   A small, elongated erectile structure in females that develops from the same embryonic tissue as the penis. It has no known reproductive function other than to focus sexual sensations. (Ch. 2)

coach   In the Lamaze birthing technique, an individual (usually the woman's partner) who works with the expectant mother during labor to help her use her training and provide support and encouragement. (Ch. 7)

cognitive-behavioral therapy   Therapy that views problems as resulting from faulty learning and that focuses on specific behavioral problems. (Ch. 13)

cohabitation   Living together in an unmarried relationship. (Ch. 10)

coitus (ko′ĭ-tus)   Sexual intercourse. (Ch. 11)

coitus interruptus   See **withdrawal.**

colostrum (ko-los′trum)   A thick, sticky liquid produced by a mother's breasts before milk starts to flow. (Ch. 7)

combination pill   An oral contraceptive that contains both synthetic estrogen and synthetic progesterone. (Ch. 6)

coming out   Disclosing to others one's homosexual orientation, from the expression "coming out of the closet." (Ch. 9)

communication   The exchange of information, thoughts, ideas, or feelings. (Ch. 17)

companionate love   Love based on togetherness, trust, sharing, and affection rather than passion. (Ch. 12)

Comstock law   A law passed in 1873 that made it a felony to mail any material regarded as obscene or lewd. (Ch. 16)

conception   The union of an egg and a sperm. (Ch. 7)

conditional love   Feelings of love that depend on the loved one's satisfying needs and fulfilling desires. (Ch. 12)

condom   *For males,* a thin sheath made of latex rubber, lamb intestine, or polyurethane that fits over the penis. *For females,* a polyurethane intravaginal pouch held in place by two flexible rings. Condoms are effective as contraception and for prevention of sexually transmitted diseases. (Ch. 6)

congenital adrenal hyperplasia   See **adrenogenital syndrome.**

consummate love   According to Sternberg, the complete love found in relationships that include passion, intimacy, and commitment. (Ch. 12)

contraception   The prevention of conception. (Ch. 6)

contraceptive sponge   A contraceptive device made of polyurethane sponge that contains enough spermicide to be effective for 24 hours after being inserted into the vagina. It is no longer sold. (Ch. 6)

coprophilia (kop″ro-fil′e-ah)   Sexual arousal caused by feces or the act of defecation. (Ch. 14)

copulation (kop″u-la′shun)   Sexual intercourse. (Ch. 11)

corona (ko-ro′nah)   The rounded border of the glans of the penis. (Ch. 2)

corpora cavernosa (kor′po-rah kav″er-no′sah)   The two cavernous bodies of the penis or clitoris that fill with blood during sexual arousal. (Ch. 2)

corpus luteum (kor′pus lew′te-um)   The follicular cells that remain in the ovary after the follicle expels the ovum during ovulation. They begin to secrete progesterone in large quantities in the postovulatory stage. (Ch. 3)

corpus spongiosum (kor′pus spun″je-o′sum)   The spongy body of the penis that fills with blood during sexual arousal; the urethra passes through it. (Ch. 2)

correlation   A mathematical measure of the degree of relationship between two variables. (Ch.1)

couvade (koo-vahd′) **syndrome**   The experiencing of pregnancy symptoms by male partners; sometimes called "sympathy pains." (Ch. 7)

Cowper's (kow′perz) **glands**   Two pea-shaped structures located beneath the prostate gland in males that secrete a few drops of an alkaline fluid prior to orgasm. (Ch. 2)

critical period   In relation to teratogens, the time during embryological or fetal development during which a particular part of the body is most susceptible to damage. (Ch. 7)

crowning   The appearance of the fetus's head at the vaginal opening during childbirth. (Ch. 7)

crura (kroo′rah)   The leglike ends of the cavernous bodies of the clitoris of penis that attach to the pubic bone. (Ch. 2)

cryptorchidism (krip-tor′kĭ-dizm)   A defect in which the testes fail to descend into the scrotum. (Ch. 2)

culpotomy (kul-pot′o-me)   A female sterilization technique in which the Fallopian tubes are approached through a small incision in the back of the vagina and then are cut or cauterized. (Ch. 6)

cunnilingus (kun″ĭ-ling′gus)   Oral stimulation of a female's genitals. (Ch. 10)

cystitis (sis-ti′tis)   A bacterial infection of the bladder (often caused by the bacterium *Escherichia coli*). (Ch. 2 and 5)

D

date rape   A type of acquaintance rape, committed during a social encounter agreed to by the victim. (Ch. 15)

decision/commitment   One of the three basic components in Sternberg's theory of love; the decision that one loves another person and the commitment to maintain the relationship. (Ch. 12)

degrading sexually explicit material   Nonviolent sexually explicit material that shows women as sex objects and where there is a clear unequal balance of power. (Ch. 12)

delayed puberty   A condition in which the appearance of secondary sex characteristics and physical growth do not begin until well after they have begun in most children. (Ch. 10)

Depo-Provera   Market name for medroxyprogesterone acetate, a chemical that when injected suppresses ovulation for 3 months. (Ch. 6)

desensitization   See **systematic desensitization.**

desire   A state that "is experienced as specific sensations which move the individual to seek out, or become receptive to, sexual experiences" (Kaplan, 1979). Kaplan and others have suggested that desire precedes excitement as the first stage of the sexual response cycle. (Ch. 4)

detumescence (de″ tu-mes′ ens)   Loss of erection. (Ch. 4)

DHT deficiency syndrome   A type of androgen insensitivity syndrome in which, because testosterone is not converted into dihydrotestosterone, boys' external genitals do not develop properly. Development occurs at puberty with the rise in testosterone.

diaphragm   A dome-shaped rubber cup with a flexible rim that fits over the cervix and thus acts as a contraceptive device by serving as a barrier to the passage of sperm into the uterus. (Ch. 6)

dilation   Becoming wider or larger. (Ch. 7)

dilation and curettage (D&C)   An abortion technique in which the cervix is dilated and the endometrium is removed by scraping it with an instrument called a curette. (Ch.6)

dilation and evacuation (D&E)   An abortion technique in which the cervix is dilated and the endometrium is removed by suction. (Ch.6)

dilation and extraction (D&X)   See **intact dilation and evacuation.**

domination   Ruling over and controlling another individual. In sexual relations, it generally involves humiliation of the partner as well. (Ch. 14)

douching (doosh-ing)   A feminine hygiene practice of rinsing out the vagina, usually with specifically prepared solutions. (Ch. 6)

dualism   The belief that body and soul are separate and antagonistic. (Ch. 1)

dyke   A slang term for a masculine lesbian. This is a stereotype that generally does not reflect how most lesbians look or behave. (Ch. 9)

dysmenorrhea (dĭs″ men-o-re′ ah)   Painful menstruation. (Ch. 3)

dyspareunia (dĭs′ pah-roo′ ne-ah)   Painful intercourse, usually resulting from organic factors (e.g., vaginal, prostate, or bladder infections). (Ch. 13)

E

eclampsia (e-klamp′ se-ah)   An advanced case of toxemia of pregnancy, characterized by convulsions and coma. (Ch. 7)

ectoderm   The layer of the embryo that forms the nervous system, skin, and teeth. (Ch. 7)

ectopic (ek-top′ ik) **pregnancy**   The implantation of a fertilized egg outside of the endometrium of the uterus. (Ch. 7)

edema   Swelling.

effacement   The thinning of the cervix during labor. (Ch. 7)

effleurage (ef-loo-rahzh′)   In the Lamaze birthing technique, a light self-massage of the abdomen that is used to counter the discomfort of labor contractions. (Ch. 7)

egocentric   A term used by Jean Piaget in his theory of cognitive development referring to the inability of young children to consider another person's point of view. (Ch. 10)

ejaculation   The expulsion of semen from the body. (Ch. 4)

ejaculatory ducts   One-inch-long paired tubes that pass through the prostate gland. The third part of the duct system that transports sperm out of a male's body. (Ch. 2)

ejaculatory incompetence   A condition in which a man is totally unable to ejaculate in a woman's vagina during sexual intercourse. (Ch. 13)

embryo (em′ bre-o)   The term given to the blastocyst after it has implanted. (Ch. 7)

emission   The first stage of an orgasm in males. Rhythmic muscular contractions in the vas deferens, prostate gland, and seminal vesicles force the sperm and prostate and seminal fluids into the ejaculatory ducts, thus forming semen. (Ch. 4)

empty love   According to Sternberg, the emotion felt when there is decision and commitment to love another person, but no intimacy or passion. (Ch. 12)

endocrine system   A network of ductless glands that secrete their chemical substances, called *hormones*, directly into the bloodstream, where they are carried to other parts of the body to exert their effects. (Ch. 3)

endoderm   The layer of the embryo that forms the internal organs of the fetus. (Ch. 7)

endometriosis (en″ do-me″ tre-o′ sis)   The growth of endometrial tissue outside of the uterus. (Ch. 3)

endometrium (en-do-me′ tre-um)   The inner mucous membrane of the uterus where a fertilized egg implants. Its thickness varies with the phase of the menstrual cycle. (Ch. 2 and 3)

epididymis (ep″ ĭ-did′ ĭ-mis)   The elongated cordlike structure on the back of a testicle. It is the first part of the duct system that transports sperm out of a male's body. (Ch. 2)

episiotomy (e-piz″ e-ot′ o-me)   A surgical procedure performed just before birth to reduce the risk of tearing the perineum. (Ch. 2 and 7)

erectile disorder   A sexual problem in which a male has difficulty getting and maintaining an erection. (Ch. 13)

erection (of penis)   The state of the penis during sexual arousal (i.e., hard and elongated) due to vasocongestion of tissues. (Ch. 4)

erogenous zone   According to psychoanalytic theory, the part of the body where sexual energy is focused. (Ch. 8)

eros   A highly idealized type of romantic love that is based mainly on physical beauty. Erotic lovers look for physical perfection. (Ch. 12)

erotica   Any literature or art that has a sexual theme. This is a morally neutral term. (Ch. 16)

estrogen (es′ tro-jen)   A hormone that is produced by the ovaries (and in very small amounts by the testes and adrenal glands). (Ch. 3)

estrous cycle The cycle of hormonal events that occurs in most nonhuman mammals. The females are sexually receptive ("in heat," or in estrus) to males only during ovulation. (Ch. 3)

excitement The first phase of the sexual response cycle as proposed by Masters and Johnson. The first signs are erection in the male and vaginal lubrication in the female. (Ch. 4)

exhibitionism Exposing one's genitals compulsively in inappropriate settings in order to obtain sexual gratification. (Ch. 14)

experiment A study in which an investigator attempts to establish a cause-and-effect relationship by manipulating a variable of interest (the independent variable) while keeping all other factors the same. (Ch.1)

expressive A personality characteristic; a cognitive focus on "an affective concern for the welfare of others." (Ch.8)

expulsion The second stage of an orgasm in males. Contractions in the urethra and muscles at the base of the penis force the semen from the penis (*ejaculation*). (Ch. 4)

F

Fallopian (fal-lo' pe-an) **tubes** The passageways that eggs follow on their way to the uterus. (Ch. 2)

false labor A set of temporary contractions of the uterus that resemble the start of labor. (Ch. 7)

false memory syndrome False memories of childhood sexual abuse, often "inadvertently planted by therapists . . ." (Ch. 15)

fatuous love According to Sternberg, the emotion felt when a commitment is made on the basis of passion without the experience of intimacy. (Ch. 12)

fear-tension-pain cycle According to prepared childbirth advocates, the cycle of events that women experience during labor when they are not properly educated about labor and childbirth. (Ch. 7)

fellatio (fe-la' she-o) Oral stimulation of a male's genitals. (Ch. 11)

female orgasmic disorder A persistent or recurrent delay in, or absence of, orgasm following a normal sexual excitement phase. (Ch. 13)

fertility awareness methods Methods of birth control that attempt to prevent conception by having a couple abstain from sexual intercourse during the woman's ovulation. (Ch. 6)

fertility drugs Drugs that improve the chance of conception in women who are infertile owing to hormone deficiencies. The drugs either stimulate the pituitary gland to secrete FSH and LH or stimulate the ovaries directly. (Ch. 7)

fertilization Impregnation; the fusion of a sperm with an ovum. (Ch. 7)

fetal alcohol syndrome (FAS) A condition common to infants born to alcoholic mothers, involving physical and nervous system abnormalities. (Ch. 7)

fetal monitor A device attached to the expectant mother's abdomen during labor to monitor the heart rate of the fetus and the contractions of the uterus. (Ch. 7)

fetal surgery A surgical operation on the fetus while it is still in the uterus. (Ch. 7)

fetishism (fet' ish-ĭ-zum) A condition in which sexual arousal occurs primarily when using or thinking about an inanimate object or a particular part of the body (such as the feet). (Ch. 14)

fetoscopy A procedure in which a tube containing fiberoptic strands is inserted into the amniotic sac to allow inspection of the fetus. (Ch. 7)

fetus The term given to the embryo after the eighth week of pregnancy. (Ch. 7)

fibrocystic (fi" bro-sis' tik) **disease** A benign overgrowth of fibrous tissue in the breast. (Ch. 2)

fimbria (fim' bre-ah) The fingerlike processes at the ovarian end of the Fallopian tubes. (Ch. 2)

focal point In the Lamaze birthing technique, a point outside of the woman's body on which she focuses her concentration during labor contractions. (Ch. 7)

follicle (fol' lĭ-k'l) A sac in the ovary containing an ovum and surrounding follicular cells. (Ch. 2 and 3)

follicle-stimulating hormone (FSH) A gonadotropin hormone released by the pituitary gland that stimulates the development of a follicle in a woman's ovary and the production of sperm in a man's testicles. (Ch. 3)

follicular (proliferative) phase The preovulatory phase of the menstrual cycle, lasting from about day 5 to day 13 in a 28-day cycle. (Ch. 3)

foreskin The loose skin of the penis that folds over the glans and that is sometimes cut away in a procedure known as circumcision. (Ch. 2)

fornication (for' nĭ ka' shun) Sexual intercourse between a man and a woman who are not married to one another. (Ch. 16)

friendship A relationship that includes (a) enjoyment of each other's company, (b) acceptance of one another, (c) mutual trust, (d) respect for one another, (e) mutual assistance when needed, (f) confiding in one another, (g) understanding, and (h) spontaneity. (Ch. 12)

frigidity The common term used by males for females who have orgasm problems, but which is generally not used by therapists because it has negative connotations about emotional responsivity. (Ch. 13)

frotteurism (fro-tur' izm) A condition in which sexual arousal is achieved primarily by rubbing one's genitals against others in public places. (Ch. 14)

fundus (fun' dus) The broad end of the uterus. (Ch. 2)

G

galactorrhea (gah-lak" to-re' ah) The spontaneous flow of milk. (Ch. 2)

gamete intrafallopian transfer (GIFT) A procedure for treating female infertility in which sperm and eggs are gathered and placed directly into a Fallopian tube. (Ch. 7)

gang rape Rape in which a victim is assaulted by more than one attacker. (Ch. 15)

gay A term generally used to refer to male homosexuals,

although in some places it is used to refer to homosexuals of either gender. (Ch. 9)

gender　The social construction of feminity and masculinity. (Ch.8)

gender constancy　The knowledge that one's sex is constant and will not change. This knowledge is usually acquired by age 6 or 7. (Ch. 8)

gender dysphoria (dis-fo′ re-ah)　The feeling of being trapped in a body of the opposite sex. (Ch. 8)

gender identity　One's subjective sense of being a male or a female. This sense is usually acquired by the age of 3. (Ch. 8)

gender identity disorder　A disorder whose criteria are (a) behaviors that indicate identification with the opposite gender, and (b) behaviors that indicate discomfort with one's own anatomy and gender roles. (Ch. 8)

gender role　A set of culturally specific norms concerning the expected behaviors and attitudes of men and women. (Ch.8)

gender-role identity　The extent to which a person approves of and participates in feelings and behaviors deemed to be appropriate to his or her culturally constituted gender. (Ch.8)

gender schemas (ske′ mahs)　Ideas about gender roles that children create from their interactions with their environment. (Ch. 8)

genital herpes　Herpes infection in the genital region. It can be caused by herpes simplex virus types I or II. (Ch. 5)

genital warts　Warts in the genital and anal regions caused by human papillomaviruses (mainly types 6 and 11). The warts are cauliflowerlike growths. (Ch. 5)

genitalia (jen″ ĭ-ta′ leah)　The reproductive organs of the male or female. (Ch. 2)

gigolos (jig′ ah-los)　Male prostitutes who cater to females. (Ch. 16)

gland　An organ or specialized group of cells that separates substances from the blood and secretes them for use by, or elimination from, the body. (Ch. 3)

glans (glanz)　The rounded end of the penis or clitoris. (Ch. 2)

global moralities　Values that are supposed to apply to all people under all circumstances. (Ch. 17)

gonadarche (go″nad-ar′ke)　The maturation of the ovaries and testicles. (Ch.10)

gonadotropin (gon″ ah-do-tro′ pin)　A hormone that has a stimulating effect on the ovaries or testicles. (Ch. 3)

gonadotropin-releasing hormone (GnRH)　A hormone released by the hypothalamus in the brain that causes the pituitary gland to release the hormones FSH and LH. (Ch.3)

gonococcus (gon″ o-kok′ us)　Common name for the bacterium that causes gonorrhea. (Ch. 5)

gonorrhea (gon″ o-re′ ah)　A sexually transmitted disease caused by the *Neisseria gonorrhoeae* bacterium (often referred to as "the gonococcus"), which lives on mucous membranes. (Ch. 5)

Graafian (graf′ e-an) **follicle**　A mature follicle. (Ch. 7)

Grafenberg (graf′ en-burg) **(G) spot**　A small, sensitive area on the front wall of the vagina found in about 10 percent of all women. (Ch. 2 and 4)

granuloma inguinale (gran″ u-lo′ mah ing′ gwĭ-nah le)　A rare (in the United States) sexually transmitted disease that is characterized by ulceration of tissue. (Ch. 5)

growing apart　Having fewer common interests over time. (Ch. 12)

growing together　Maintaining common interests (although not necessarily the same ones) over time. (Ch. 12)

G-spot　See **Grafenberg (G) spot**.

gumma (gum′ ah)　A soft, gummy tumor that appears in late (tertiary) syphilis. (Ch. 5)

gynecomastia (jin″ ĕ-ko-mas′ te-ah)　Excessive development of the male breasts. (Ch. 2 and 10)

H

habituation　Responding less positively to a stimulus with repeated exposure to it. (Ch. 12)

hand-riding　A technique used to teach people how to touch in which the receiver places his or her hand over the giver's hand and guides it over the receiver's body. This is often included as part of sensate focus exercises. (Ch. 13)

HCG　See **human chorionic gonadotropin (HCG)**.

hepatitis (hep″ ah-ti′ tis) **A, B, and C**　Liver infections caused by viruses. Type A is spread by direct or indirect contact with contaminated feces. Type B is transmitted by infected blood or body fluids, with 25 to 50 percent of the cases contracted during sex. Type C is spread mainly by contaminated blood, but may possibly be spread during sexual intercourse in some cases. (Ch. 5)

hermaphrodite (her-maf′ ro-dīt″)　A person with both male and female reproductive systems as a result of failure of the primitive gonads to differentiate properly during embryological development. (Ch. 8)

herpes (her′ pez)　Painful blisters usually found on the mouth (fever blisters; cold sores) or genitals that are caused by a virus (herpes simplex virus type I or II). Genital herpes is almost always contracted by intimate sexual activity, but oral herpes may be spread by casual kissing. (Ch. 5)

heterosexual　An individual with a sexual orientation primarily to members of the opposite sex. (Ch. 9)

HIV　See **human immunodeficiency virus (HIV)**.

homophobia　An irrational fear of homosexual individuals and homosexuality. (Ch. 9)

homosexual　An individual with a sexual orientation primarily to members of the same sex. (Ch. 9)

hormones　Chemical substances that are secreted by ductless glands into the bloodstream. They are carried in the blood to other parts of the body, where they exert their effects on other glands or target organs. (Ch. 3)

hot flashes　A warm feeling over the upper body experienced by menopausal women as a result of increased levels of FSH and LH. (Ch. 10)

HPV　See **human papillomaviruses (HPV)**.

human chorionic (ko″re-on′ ik) **gonadotropin (HCG)**　A hormone secreted by the chorion that stimulates the ovaries during pregnancy. (Ch. 3 and 7)

human immunodeficiency virus (HIV)   A virus that kills CD4+ cells, eventually resulting in AIDS. (Ch. 5)

human papillomaviruses (pap″ ĭ-lo′ mah) **(HPV)** Viruses that cause abnormal growths in epithelial cells. There are over 70 types. A few (types 6 and 11) cause genital warts, while others (types 16, 18, 31, 33, and 45) can lead to cancer of the cervix. (Ch. 5)

hustler   A male prostitute who caters to males. (Ch. 16)

hyaluronidase (hi″ah-lu-ron′ĭ-das)   An enzyme secreted by a sperm that allows the sperm to penetrate the surface of an ovum. (Ch. 7)

hymen (hi′ men)   The thin membrane that partially covers the vaginal opening in many sexually inexperienced women. Its presence or absence, however, is really a very poor indicator of prior sexual experience. (Ch. 2)

hypersexuality   A term reserved for people who engage in sex compulsively, with little or no emotional satisfaction. Sometimes called *sexual addiction*. (Ch. 13)

hypoactive sexual desire   A sexual problem characterized by a persisting and pervasive inhibition of sexual desire. (Ch. 13)

hypothalamus (hi″po-thal′ah-mus)   A part of the brain that regulates the release of hormones from the pituitary gland. (Ch. 2 and 3)

hysterectomy (his″ tĕ-rek′ to-me)   Surgical removal of the uterus. (Ch. 2)

I

identification   The adoption of the sex roles of the same-sex parent by a child. In Freudian theory, identification resolves the Oedipus complex. (Ch. 8)

"I" language   Communicating your desires to another person by beginning sentences with "I," followed by your feelings, desires, or thoughts. (Ch. 17)

imitation   Following the behaviors of someone taken as one's model. (Ch. 8)

immune system   The bodily mechanisms involved in the production of antibodies in response to bacteria, viruses, and cancerous cells. (Ch. 5)

implantation   The process by which the blastocyst attaches itself to the wall of the uterus. (Ch. 7)

impotence   A common term for erectile disorders in males, but generally not used by therapists because it has negative connotations about a male's manhood. (Ch. 13)

incest   Sexual contact between two closely related persons. (Ch. 15)

incontinence   The inability to control one's bladder. (Ch. 4)

individualist theory   The theory that personality traits develop early in life and do not change much over the life cycle; these traits then determine an individual's expression of masculine or feminine behavior. (Ch. 8)

infatuated love   According to Sternberg, the emotion felt when there is passion in the absence of intimacy and commitment. (Ch. 12)

infertility   The inability of a couple to conceive. (Ch. 7)

infibulation   See **clitoridectomy.**

inhibin (in-hib′in)   A hormone produced by the testicles and ovaries that inhibits release of follicle-stimulating hormone from the pituitary gland. (Ch. 2 and 3)

inhibited ejaculation   See **male orgasmic disorder.** (Ch. 13)

inhibited female orgasm   See **female orgasmic disorder**.

instrumental   A personality characteristic; a cognitive focus on "getting the job done." (Ch.8)

intact dilation and evacuation (intact D&E)   An abortion procedure performed in very-late-term pregnancies in which the fetus's head is partially delivered and its brain is then removed by suction. Also called *dilation and extraction (D&X)*. (Ch.6)

intimacy   Those feelings in a relationship that promote closeness or bondedness and the experience of warmth. (Ch. 12)

intrauterine device (IUD)   A birth control device, usually made of plastic with either a copper or progesterone coating, that is placed in the uterus to prevent conception and implantation. (Ch. 6)

introitus (in-tro′ ĭ-tus)   The entrance to the vagina. (Ch. 2)

in vitro fertilization   A process in which a mature ovum is surgically removed from a woman's ovary, placed in a medium with sperm until fertilization occurs, and then placed in the woman's uterus. This is usually done in women who cannot conceive because of blocked Fallopian tubes. (Ch. 7)

IUD   See **intrauterine device (IUD).**

J

jealousy   An emotional state that is aroused by a perceived threat to a valued relationship or position. (Ch. 12)

john   A street term for a male who pays for sex with a prostitute; johns are also called *tricks* or *scores*. (Ch. 16)

K

Kaposi's sarcoma (kap′-o-sez sar-ko′mah)   A cancer of the capillary system that often occurs in people with AIDS. (Ch. 5)

Kegel exercises   Exercises that are designed to strengthen the pubococcygeus muscle that surrounds the bladder and vagina. (Ch. 2)

Klinefelter's syndrome   A condition in males in which there is one or more extra X chromosomes. (Ch. 8)

klismaphilia (kliz″ mah-fil′ e-ah)   A condition in which sexual arousal is achieved primarily by being given an enema. (Ch. 14)

L

labia majora (la′ be-ah majo′ rah)   Two elongated folds of skin extending from the mons to the perineum in females. Its outer surfaces become covered with pubic hair during puberty. (Ch. 2)

labia minora (la′be-ah mino′rah)   Two hairless elongated folds of skin located between the labia majora in females. They meet at the top of the vulva to form the clitoral hood. (Ch. 2)

Lamaze A method of prepared childbirth; named after its founder, Fernand Lamaze. (Ch. 7)

lanugo (lah-nu′ go) Light hair that covers a newborn baby. (Ch. 7)

laparoscopy (lap″ ah-ros′ko-pe) A technique that involves inserting a slender, tubelike instrument (laparoscope) into a woman's abdomen to examine (via fiberoptics) her reproductive organs or a fetus. The procedure is often used to perform female sterilizations. (Ch. 6 and 7)

lesbian (lez′ be-an) A female homosexual. (Ch. 9)

Leydig (li′ dig), **interstitial cells of** The cells in the testes that produce male hormones. (Ch. 2)

libido (lĭ-be′ do) In psychoanalytic theory, the sexual energy of an individual. (Ch. 8)

lightening Rotation of the fetus prior to childbirth so that its head is downward, resulting in decreased pressure on the mother's abdomen. (Ch. 7)

liking According to Sternberg, the emotion felt when there is intimacy without passion or commitment. (Ch. 12)

limerence "Head over heels" love in which one is obsessed with the loved one. (Ch. 12)

love Many possible definitions, and, according to researchers, many different types. For example, see **agape, companionate love, eros, ludus, mania, pragma,** and **storge.** (Ch. 12)

ludic eros According to Lee, the love-style that results from combining the ludus and eros love-styles. (Ch. 12)

ludus A self-centered type of love. Ludic lovers avoid commitment and treat love as a game, often viewing the chase as more pleasurable than the prize. (Ch. 12)

lumpectomy A surgical procedure in which only a breast tumor and a small bit of surrounding tissue are removed. (Ch. 2)

luteal (lu′ te-al) **(secretory) phase** The postovulatory phase of the menstrual cycle, lasting from about day 15 to day 28 in a 28-day cycle. (Ch. 3)

luteinizing (lu′ te-in″ i-zing) **hormone (LH)** A gonadotropin hormone released by the pituitary gland that triggers ovulation in women and stimulates the production of male hormones in men. (Ch. 3)

lymphogranuloma venereum (lim″ fo-gran″ u-lo′ mah vene′ reum) **(LGV)** A sexually transmitted disease common in tropical countries that is caused by chlamydia. If left untreated, it causes swelling of the inguinal lymph nodes, penis, labia, or clitoris. (Ch. 5)

M

male orgasmic disorder A condition in which a male has difficulty reaching orgasm and ejaculating in a woman's vagina. (Ch. 13)

male rape Rape in which the victim is a male. (Ch. 15)

mammary glands Milk-producing glands of the breast. (Ch. 2)

mammogram Low-radiation x-rays used to detect breast tumors. (Ch.2)

mania A type of love characterized by an intense emotional dependency on the attention and affection of one's lover. (Ch. 12)

marital rape Rape of a wife by her husband. (Ch. 15)

masochism (mas′ o-kizm) A condition in which individuals obtain sexual pleasure primarily from having pain and/or humiliation inflicted on them. (Ch. 14)

mastectomy (mas-tek′ to-me) Surgical removal of a breast. (Ch. 2)

masturbation Self-stimulation of one's genitals. (Ch. 10 and 11)

meatus (me-a′ tus) General term for an opening or passageway, as in *urinary meatus.* (Ch. 2)

medical abortion Abortion induced by taking a pill (see **RU 486**). (Ch.6)

medical history One of the first steps used by most sex therapists. A complete medical history or exam is important because many sexual problems have an organic basis. (Ch. 13)

menarche (mě-nar′ ke) The term for a female's first menstrual period. (Ch. 3 and 10)

menopause (men′ o-pawz) The term for a female's last menstrual period. (Ch. 3 and 10)

menstrual cycle The monthly cycle of hormonal events in a female that leads to ovulation and menstruation. (Ch. 3)

menstrual taboos Incorrect negative attitudes about menstruating females. (Ch. 3)

menstruation (men″ stroo-a′ shun) The monthly discharge of endometrial tissue, blood, and other secretions from the uterus that occurs when an egg is not fertilized. (Ch. 3)

mesoderm The layer of the embryo that forms the muscles, skeleton, and blood vessels. (Ch. 7)

microstructural theory A theory that stresses the role of the external environment in shaping gender roles. (Ch. 8)

midwife See **nurse-midwife.**

minilaparotomy A female sterilization technique in which small incisions are made in the abdomen and the Fallopian tubes are then pulled to the opening and cut, tied, or blocked with clips. (Ch.6)

minipill An oral contraceptive that contains only progestins. (Ch. 6)

miscarriage A spontaneous abortion. (Ch. 7)

missionary position A face-to-face position of sexual intercourse in which the female lies on her back and the male lies on top with his legs between hers. It was called this because Christian missionaries instructed people that other positions were unnatural. (Ch. 1 and 11)

Mittelschmerz (mit′ el-shmārts) Abdominal pain at the time of ovulation. (Ch. 3)

molluscum contagiosum (mo-lus′ kum contagio′ sum) A sexually transmitted virus whose symptom looks like small pimples filled with kernels of corn. (Ch. 5)

moniliasis (mon-ĭ-lī′ ah-sis) Sometimes called candidiasis. A type of vaginitis caused by the overgrowth of a microorganism (*Candida albicans*) that is normally found in the vagina. Moniliasis is a fungus or yeast infection and usually is a sexually related, rather than a sexually transmitted, disease. (Ch. 5)

mons veneris (monz ven′ eris) The soft layer of fatty tissue that overlays the pubic bone in females. It becomes covered with pubic hair during puberty. (Ch. 2)

morning sickness   A symptom of early pregnancy involving nausea. (Ch. 7)

morula (mor' u-lah)   The collection of cells formed when the zygote begins rapid cell division. (Ch. 7)

Mullerian (mil-e' re-an) **duct system**   A primitive duct system formed in embryos, which, if allowed to develop, becomes the female reproductive system. (Ch. 8)

Mullerian-duct–inhibiting substance   A chemical secreted by the testes during embryological development that causes the Mullerian ducts to shrink and disappear. (Ch. 8)

multiple orgasms   Having two or more successive orgasms without falling below the plateau level of physiological arousal. (Ch. 4)

myotonia (mi″ o-to' ne-ah)   A buildup of energy in nerves and muscles resulting in involuntary contractions. (Ch. 4)

mysophilia (mi″ so-fil' e-ah)   A condition in which sexual arousal is caused primarily by filth or filthy surroundings. (Ch. 14)

N

necking   Noncoital sexual contact above the waist. (Ch. 10)

necrophilia (nek″ ro-fil' e-ah)   Sexual arousal by a dead body. (Ch. 14)

nipple   The protuberance on the breast that in females contains the outlets for the milk ducts. It consists of smooth muscle fibers. (Ch. 2)

nocturnal emission   An ejaculation that occurs during sleep in males; a "wet dream." (Ch. 10)

nocturnal orgasm   An orgasm that occurs during sleep (sometimes called "wet dreams" in males). (Ch. 11)

nongonococcal (nonspecific) urethritis (NGU)   Any inflammation of the urethra that is not caused by the *Neisseria gonorrhoeae* (gonococcus) bacterium. (Ch. 5)

nonlove   According to Sternberg, the absence of all three components of love: intimacy, passion, and commitment. (Ch. 12)

nonoxynol-9   The active chemical in most spermicides. (Ch. 6)

nonverbal communication   Communicating with another person without using words. (Ch. 17)

nonviolent sexually explicit material   Sexually explicit material that does not show violent interactions between individuals. (Ch. 16)

Norplant   Six flexible silicone tubes containing a progestin that are implanted under the skin to prevent conception; they remain effective for up to 5 years. (Ch. 6)

nurse-midwife   A registered nurse, trained in obstetrical techniques, who delivers babies, often at the expectant mother's home. (Ch. 7)

O

obscenity   In legal terminology, the classification of material that by contemporary community standards depicts patently offensive sexual conduct, lacks serious value, and appeals to prurient interests. (Ch. 16)

observer bias   The prejudicing of observations and conclusions by the observer's own belief system. (Ch. 1)

Oedipus (ed' i-pus) **complex**   According to psychoanalytic theory, the sexual desire that a boy feels toward his mother (male Oedipus complex) or the sexual attraction that a girl feels toward her father (female Oedipus complex) during the phallic stage of psychosexual development. (Ch. 8)

open marriage   A marital relationship in which the couple agrees that it is permissible to have sexual relations outside of the marriage. (Ch. 10)

operant conditioning   According to learning theory, a type of training thought to be responsible for the acquisition of sex roles by children. It rewards "masculine" behavior in boys and "feminine" behavior in girls and does not reward behaviors associated with the opposite sex. (Ch. 8)

opportunistic diseases   A variety of diseases that occur in individuals with a very weakened immune system, as in AIDS. (Ch. 5)

opportunistic rapists   Rapists motivated by a desire for sex who rape impulsively when there is an opportunity. (Ch. 15)

oral contraceptive   The birth control pill. (Ch. 6)

oral herpes   A herpes infection in or around the mouth. It can be caused by herpes simplex virus types I or II. (Ch. 5)

orchiectomy (or″ ke-ek' to-me)   Removal of one or both testicles. (Ch. 3)

orgasm   The brief but intense sensations (focused largely in the genitals but really a whole body response) experienced during sexual arousal. During orgasm, rhythmic muscular contractions occur in certain tissues in both the male and female. The third phase of the sexual response cycle proposed by Masters and Johnson. (Ch. 4)

orgasmic disorder   See **male orgasmic disorder, female orgasmic disorder.**

orgasmic platform   The engorgement and consequent swelling of the outer third of the vagina during the plateau stage, causing the vaginal opening to narrow by 30 to 50 percent. (Ch. 4)

os   The opening of the cervix. (Ch. 7)

ovary (o' vah-re)   The female gonad in which ova are produced. (Ch. 2)

ovulation (o″ vu-la' shun)   The expulsion of an egg from one of the ovaries. (Ch. 3)

ovum   An egg; the female reproductive cell. (Ch. 2 and 7)

oxytocin (ok″se-to'sin)   A hormone released from the pituitary gland that causes the breasts to eject milk. (Ch. 2 and 7)

P

Pap smear   A test for cancer of the cervix in females; named for Dr. Papanicolau, who developed it. (Ch. 2)

paraphilias (par″ ah-fil' e-ahs)   General term for a group of sexual disorders in which a person's sexual arousal and gratification depend almost exclusively on unusual behaviors. (Ch. 14)

passion   The drive leading to physical attraction, sexual relations, and romance. (Ch. 12)

pedophilia (pe″ do-fil′ e-ah)   A condition in which sexual arousal is achieved primarily and repeatedly through sexual activity with children who have not reached puberty. (Ch. 14 and 15)

peer pressure   Expectations by one's peer group about how one is supposed to behave. (Ch. 10)

pelvic exam   A necessary part of female health care to check for cervical and vaginal infections; it includes a Pap smear. (Ch. 2)

pelvic inflammatory disease (PID)   A bacterially caused inflammation of a female's reproductive tract, particularly the Fallopian tubes, that can result in sterility. The most common (though not the only) cause is untreated gonorrhea and/or chlamydia. (Ch. 5)

penis (pe′ nis)   The male organ for sexual intercourse and the passageway for sperm and urine. (Ch. 2)

penis envy   According to psychoanalytic theory, the envy that a girl feels during the phallic stage of psychosexual development, when she realizes that she has no penis (and will never have one). This supposedly causes her to become angry at her mother, whom the girl blames for this condition. (Ch. 8)

perfect-use pregnancy rate   For a particular birth control technique, the percentage of pregnancies during the first year of use by couples who use the technique properly and consistently. See also **typical-use pregnancy rate.** (Ch. 6)

performance anxiety   A fear of failure during sexual relations that can lead to erectile disorder in males and inhibited orgasm in females. (Ch. 13)

perineum (per″ i-ne′ um)   Technically, the entire pelvic floor, but more commonly used to refer to the hairless bit of skin between the anus and either the vaginal opening (in females) or the scrotum (in males). (Ch. 2)

personally immature pedophile   A pedophile who is attracted to children because he never developed the social skills necessary to initiate and maintain a sexual relationship with an adult. (Ch. 15)

petting   Noncoital sexual contact below the waist. (Ch. 10)

phallic (fal′ ik) **stage**   According to psychoanalytic theory, the third stage of psychosexual development, in which a child's sexual desire for the parent of the opposite sex emerges and is resolved through the process of identification. (Ch. 8)

pheromones (fer′ah monz)   Chemical substances secreted externally by animals that convey information to, and produce specific responses in, members of the same species. (Ch. 3)

phimosis (fi-mo′ sis)   A condition in uncircumcised males in which the foreskin of the penis is too tight, causing pain during erection. (Ch. 2 and 13)

PID   See **pelvic inflammatory disease (PID).**

pill   See **oral contraceptive.** (Ch. 6)

pinworms   Small worms (*Enterobius vermicularis*) that live in the large intestine and are generally transmitted through nonsexual contact with the worms' eggs, but which can be transmitted sexually by manual or oral contact with the anus of an infected person. (Ch. 5)

pituitary (pĭ-tu′ ĭ-tār″ e) **gland**   A gland located at the base of the brain that secretes eight hormones, including follicle-stimulating hormone and luteinizing hormone. (Ch. 3)

placenta (plah-sen′tah)   An organ that serves as a connection or interface between the fetus's systems and those of the mother. (Ch. 7)

placenta previa   A condition in which the placenta blocks the cervical opening. (Ch. 7)

plateau   The second phase of the sexual response cycle proposed by Masters and Johnson. Physiologically, it represents a high state of arousal. (Ch. 4)

PLISSIT model   An acronym for the four levels of treatment in sexual therapy: permission, limited information, specific suggestions, and intensive therapy. (Ch. 13)

PMS   See **premenstrual syndrome (PMS).**

pneumocystis carinii (nu″ mo-sis′ tis cari′ nii) **pneumonia**   A rare respiratory infection that often occurs in people with AIDS. (Ch. 5)

polygyny   The practice of having two or more wives at the same time. (Ch. 12)

polyincestuous family   Families in which there are both multiple incestuous abusers and multiple victims, across generations and within the same generation. (Ch. 15)

population   The complete set of observations about which a researcher wishes to draw conclusions. (Ch.1)

pornography   Literature or art with a sexual theme that is designed to cause arousal and excitement and is considered to be obscene. (Ch. 16)

postpartum blues   A mild, transient emotional letdown experienced by a majority of women after giving birth. (Ch. 7)

postpartum depression   A deep depression experienced by about 10 percent of women in the first few months after giving birth. (Ch. 7)

postpartum psychosis   A psychotic state experienced by about one in 1,000 women after giving birth. (Ch.7)

power rapists   Rapists who commit their attacks in order to overcome personal feelings of insecurity and inadequacy. Their rapes give them a sense of mastery and control. (Ch. 15)

pragma   A rational or practical type of love. Pragmatic lovers actively look for a "good match." (Ch. 12)

precocious puberty   A condition in which puberty begins before the age of 9. (Ch. 10)

preeclampsia (pre″ e-klamp′se-ah)   The early stage of toxemia of pregnancy, characterized by high blood pressure, weight gain, and swollen joints. (Ch. 7)

preference molesters   Child molesters who have a primary sexual orientation to children and who are relatively uninterested in adult partners. (Ch. 15)

premature ejaculation   A recurrent and persistent absence of reasonable voluntary control of ejaculation. (Ch. 13)

premenstrual dysphoric disorder (PMDD)   A severe form of PMS that markedly interferes with social relations, work, or education. (Ch. 3).

premenstrual syndrome (PMS)   A group of physical and/or emotional changes that many women experience in the last 3 to 14 days before the start of a menstrual period. (Ch. 3)

prenatal examination A health checkup by a physician during pregnancy. (Ch.7)

prepared childbirth Courses or techniques that prepare women for labor and childbirth, with the goal of making it a positive experience. (Ch. 7)

prepping The shaving of a woman's pubic hair prior to delivery of a baby. (Ch. 7)

preterm infant An infant born weighing less than 5½ pounds and before week 37 of pregnancy. (Ch. 7)

priapism (pri′ ah-pizm) A condition in which the penis remains erect for a prolonged period of time, sometimes days. (Ch. 13)

primary follicle An immature ovum enclosed by a single layer of cells. (Ch. 2 and 7)

primary HIV infection The first few weeks of HIV infection, during which HIV reaches enormous levels in the blood and 30 to 60 percent of infected individuals experience flulike symptoms. (Ch. 5)

primary sexual problem A sexual problem that has existed throughout an individual's life. (Ch. 13)

procreation The act of producing offspring. (Ch. 1)

progesterone (pro-jes′ tĕ-ron) A hormone that is produced in large amounts by the ovaries after ovulation. It prepares the endometrium of the uterus to nourish a fertilized egg. (Ch. 3)

progestin (proj-jes′ tin) A synthetic form of progesterone. (Ch. 6)

prolactin (pro-lak′ tin) A hormone released from the pituitary gland that stimulates milk production in the breasts. (Ch. 2 and 7)

proliferative phase See **follicular (proliferative) phase.**

prophylactic See **condom.**

prostaglandins (pros″ tah-glan′ dinz) Chemical substances in the body that cause uterine contractions. (Ch. 3 and 6)

prostate (pros′ tāt) **gland** A gland in males that surrounds the urethra and neck of the bladder and contributes many substances to the seminal fluid. (Ch. 2 and 5)

prostatitis (pros″tah-ti′ tis) Inflammation of the prostate gland. (Ch. 5)

prostitution The exchange of sex for money. The feature that distinguishes a prostitute from other people who engage in illicit sexual intercourse is the indiscrimination with which she or he offers herself or himself to others for hire. (Ch. 16)

pseudohermaphroditism (su″ do-her-maf′ ro-di-tizm″) A condition in which a person is born with ambiguous genitalia as a result of a hormonal abnormality. See also **adrenogenital syndrome** and **androgen insensitivity syndrome.** (Ch. 8)

psychosexual therapy Therapy that attempts to provide insight into the historical cause of a client's problem. (Ch. 13)

puberty (pu′ ber-te) The time in life when an individual first shows sexual attraction and becomes capable of reproduction. (Ch. 10)

pubic (pu′ bik) **hair** Hair that grows in the genital region starting at puberty; a secondary sex characteristic. (Ch. 2)

pubic (pu′ bik) **lice** An infestation of the parasite *Pthrius pubis,* which attach themselves to pubic hair and feed on blood. Also known as "crabs." (Ch. 5)

pubococcygeus (pu″ bo-kok-sij′ e-us) **muscle** The major muscle in the pelvic region. In women, voluntary control over this muscle (to help prevent urinary incontinence or to enhance physical sensations during intercourse) is gained through Kegel exercises. (Ch. 2)

quickening The first time a pregnant woman experiences movement of the fetus, usually in the fifth month. (Ch. 7)

R

random sample A sample in which observations are drawn so that all other possible samples of the same size have an equal chance of being selected. (Ch. 1)

rape Nonconsensual oral, anal, or vaginal penetration, obtained by force, by threat of bodily harm, or when the victim is incapable of giving consent (Koss, 1993b). (Ch. 15)

rape myths Widespread mistaken beliefs about rape. (Ch. 15)

rape-shield laws Laws that make the prosecution of rapists easier and that also prevent the victims from feeling as if they are on trial. (Ch. 15)

rape trauma syndrome Reactions of a rape victim, including an acute phase (involving initial physical and psychological recovery from the attack) and long-term psychological reorganization (involving attempts to deal with the long-term effects of the attack and to prevent a future rape). (Ch. 15)

recovered memory syndrome The recovery of memories of childhood sexual abuse that were repressed (forgotten) and then later remembered. (Ch. 15)

Reformation Sixteenth-century religious movement that was directed at reforming the Roman Catholic Church. (Ch. 1)

refractory period In males, the period of time after an orgasm in which their physiological responses fall below the plateau level, thus making it impossible for them to have another orgasm (until the responses build back up to plateau). (Ch. 4)

regressive pedophile A pedophile with a past history of normal adult sexual relations. (Ch. 5)

resolution The fourth and final phase of the sexual response cycle proposed by Masters and Johnson. It refers to a return to the unaroused state. (Ch. 4)

respiratory distress syndrome An illness common in premature infants; caused by insufficient levels of surfactin. (Ch. 7)

retarded ejaculation See **male ogasmic disorder.**

retrograde ejaculation An abnormal condition experienced by some males during orgasm in which the semen is forced into the bladder instead of out of the body. (Ch. 4)

Rh factor A protein in the blood of most individuals. If a mother does not have it and the fetus she carries does, she can develop antibodies against the Rh factor. This is called an *Rh incompatibility problem.* (Ch. 7)

Rhogam A drug used to treat Rh incompatibility problems, consisting of Rh-negative blood that already has Rh antibodies. (Ch. 7)

rhythm method See **fertility awareness methods.**

romantic love The combination of passion and liking (intimacy). Romantic love relationships have high levels of fascination, exclusiveness, sexual desire, and "giving the utmost" when the other is in need. (Ch. 12)

root (of penis) The part of the penis consisting of the crura and the bulb. (Ch. 2)

RU 486 A pill that chemically induces an abortion by blocking progesterone. (Ch. 6)

rubella (roo-bel′ ah) A disease, also known as German measles, caused by a virus. If contracted during pregnancy, it can cause fetal abnormalities. (Ch. 7)

S

sadistic rapists Rapists who are aroused by physical force and who derive pleasure by harming their victims. (Ch. 15)

sadism (sad′ izm) A condition in which individuals repeatedly and intentionally inflict pain on others in order to achieve sexual arousal. (Ch. 14)

Sadomasochism (S&M) A term used to indicate the linkage of sadism with masochism. (Ch. 14)

safer sex Sexual behaviors involving a low risk of contracting a sexually transmitted disease. These include consistent use of condoms and/or abstaining from sex until one enters a long-term monogamous relationship. (Ch. 5)

sample A subset of a population of subjects. (Ch. 1)

scabies (ska′ bez) A contagious infestation of parasitic mites (*Sarcoptes scabiei*) that burrow under the skin to lay their eggs. (Ch. 5)

schemas See **gender schemas.**

scoptophilia (skop″ to-fil′ e-ah) The practice of repeatedly seeking sexual arousal by secretly observing genitals and sexual intercourse. (Ch. 14)

scrotum (skro′ tum) The pouch beneath the penis that contains the testicles. (Ch. 2)

secondary sex characteristics Bodily changes that occur during puberty and differentiate males and females. (Ch. 10)

secondary sexual problem A sexual problem occurring in an individual who has not had the problem in past sexual relations. (Ch. 13)

second stage of labor The stage of labor that begins when the cervix is fully dilated and the fetus begins moving through the birth canal. It ends with birth. (Ch. 7)

secretory phase See **luteal (secretory) phase.**

self-awareness In sexual therapy, getting in touch with one's own physical responses. (Ch. 13)

self-disclosure Revealing one's thoughts, feelings, and emotions to another. (Ch. 12)

self-esteem The feeling one has about oneself. (Ch. 12)

semen (se′ men) The fluid expelled from a male's penis during orgasm, consisting of sperm and fluids from the prostate gland and seminal vesicles. (Ch. 4)

seminal vesicles (sem″ ĭ-nal ves′ ĭ-k′lz) Two structures in the male that contribute many substances to the seminal fluid. (Ch. 2)

seminiferous (se″ mĭ-nif′ er-us) **tubules** The tubules in the testicles that produce sperm. (Ch. 2)

sensate focus exercises Exercises designed to reduce anxiety and teach mutual pleasuring through nongenital touching in nondemanding situations. (Ch. 13)

sensuality The state of being sensual. It encompasses all of our senses and who we are as a total person. (Ch. 2)

serial monogamy The practice of having a series of monogamous sexual relationships. (Ch. 10)

Sertoli (ser-to′lēz) **cells** Cells in the testicles that produce the hormone inhibin. (Ch. 2 and 3)

sex roles Behaviors that are associated with being stereotypically masculine or feminine. (Ch. 8)

sex-tension flush The rash that appears on the skin (due to vasocongestion) in 50–75 percent of all women and 25 percent of all men during the plateau phase of the sexual response cycle. (Ch. 4)

sexual addiction See **hypersexuality.**

sexual aversion An extreme form of hypoactive sexual desire in which an individual's avoidance of sex becomes phobic. (Ch. 13)

sexual coercion The act of forcing another person into unwanted sexual activity by physical or verbal coercion (restraint or constraint). (Ch. 15)

sexual deviance "The inappropriate or flawed performance of a conventionally understood sexual practice" (Simon, 1994). (Ch. 14)

sexual harassment Unwelcome sexual advances that persist after the recipient has indicated that they are unwanted. See text for an expanded definition. (Ch. 15)

sexual history One of the first steps in most sexual therapy programs. (Ch. 13)

sexually healthy person Someone who feels comfortable with his or her sexuality and who feels free to choose whether or not to try a variety of sexual behaviors. (Ch. 11)

sexually related diseases Diseases of the reproductive system or genitals that are not contracted through sexual activity. Often involve overgrowths of bacteria, yeasts, viruses, or fungal organisms that are found naturally in sexual and reproductive organs. (Ch. 5)

sexually transmitted diseases Diseases that can be, but are not necessarily always, transmitted by sexual contact. This term is generally preferred to the term *venereal diseases*. (Ch. 5)

sexuality All of the sexual attitudes, feelings, and behaviors associated with being human. The term does not refer specifically to a person's capacity for erotic response or to sexual acts, but rather to a dimension of one's personality. (Ch. 1)

sexual learning Knowledge about sex that is received (usually informally) from family, friends, the media, and society in general. (Ch. 17)

sexual orientation A distinct preference for sexual partners of a particular sex in the presence of clear alternatives. (Ch. 9 and 10)

sexual perversion A problem of sexual desire, "not only in the sense that it appears to violate the sexual practices of a time and place, but also because it constitutes a violation of common understandings that render current sexual practice plausible" (Simon, 1994). (Ch. 14)

sexual response cycle   The physiological responses that occur during sexual arousal, which many therapists and researchers have arbitrarily divided into different phases. (Ch. 4)

sexual revolution   A period in U.S. history, beginning about 1960, of increased sexual permissiveness. (Ch. 1)

sexual surrogate   See **surrogate.**

sexual variant   For an individual, statistically or sociologically unusual behaviors that are engaged in for variety and not as one's preferred manner of becoming sexually aroused. (Ch. 14)

shaft   The long body of the penis or clitoris. (Ch. 2)

shigellosis (she″ gel-lo′ sis)   A disease that can be contracted during sexual activity by exposure to feces containing the *Shigella* bacterium. (Ch. 5)

situational ethics   Values or ethics that emphasize the proper thing to do in particular situations. (Ch. 17)

situational molesters   Child molesters who have a primary sexual orientation to adults. They have sex with children impulsively and regard their behavior as abnormal. (Ch. 15)

situational sexual problem   A sexual problem that occurs only in specific situations. (Ch. 13)

Skene's (skenz) **glands**   Glands located in the urethras of some women that are thought to develop from the same embryological tissue as the male prostate, and that may be the source of a fluid emitted by some women during orgasm. (Ch. 4)

smegma (smeg′mah)   The cheesy secretion of sebaceous glands that can cause the clitoris to stick to the clitoral hood or the foreskin of the penis to stick to the glans. (Ch. 2)

socialization   The process of internalizing society's beliefs. (Ch. 8)

sodomy (sod′ o-me) **laws**   Laws that prohibit oral and/or anal sex. (Ch. 16)

spectatoring   Observing and evaluating one's own sexual responses. This generally causes distraction, with subsequent erectile and/or orgasm problems. (Ch. 13)

sperm   The germ cell of a male. (Ch. 2 and 7)

spermatic (sper-mat′ ik) **cord**   The cord that suspends a testicle in the scrotum. (Ch. 2)

spermatorrhea (sper″ mah-to-re′ ah)   The Victorian name for male nocturnal emissions (wet dreams), so named because Victorian physicians believed it was caused by the same thing that caused gonorrhea. (Ch. 1 and 11)

spermicides   Chemicals that kill sperm. In most products, the chemical is nonoxynol-9. (Ch. 6)

start-up stage of labor   The stage of labor that begins with uterine contractions pushing the fetus downward toward the cervix and ends when the cervix is fully dilated. (Ch. 7)

statistically normal   A range of behaviors or values that encompasses a large number of individuals. (Ch. 11)

statutory rape   Sexual intercourse by an adult with a (consenting) partner who is not yet of the legal "age of con-

sent." Sometimes called "carnal knowledge of a juvenile." (Ch. 15)

stereotypes   Oversimplified, preconceived beliefs that supposedly apply to anyone in a given group. (Ch. 8)

sterilization   A general term for surgical techniques that render an individual infertile. (Ch. 6)

stigmatophilia (stig″ mah to-fil′ e-ah)   Sexual arousal by marking one's own body or inserting foreign objects into the urethra, genitals, or nipples. (Ch. 14)

storge (stor′ ga)   An affectionate type of love that grows from friendship. (Ch. 12)

storgic eros   According to Lee, the love-style that results from combining storge and eros. (Ch. 12)

storgic ludus   According to Lee, the love-style that results from combining storge and ludus. (Ch. 12)

straight   A term used by homosexuals for a heterosexual. (Ch. 9)

stranger rape   Rape committed by someone who is not known by the victim. (Ch. 15)

stratified random sample   A sample in which subgroups are randomly selected in the same proportion as they exist in the population. Thus the sample is representative of the target population. (Ch.1)

streetwalker   A prostitute who solicits in public. (Ch. 16)

submission   Obeying and yielding to another individual. In sexual relations, it generally involves being humiliated by the partner as well. (Ch. 14)

surfactin   A liquid that lets the lungs transmit oxygen from the atmosphere to the blood. (Ch. 7)

surrogate   A person, usually a female, trained and skilled in sensate focus techniques who works under the direction of a therapist with a single partnerless client undergoing sexual therapy. (Ch. 13)

surrogate mother   A woman who carries a fetus to full term for another couple, agreeing to give the infant to the other couple after it is born. The infant generally represents a union of the sperm from the man and either his partner's ovum or the ovum of the surrogate mother. (Ch. 7)

survey   A study of people's attitudes, opinions, or behaviors. Responses are usually obtained either in a face-to-face interview or on a paper-and-pencil questionnaire. (Ch. 1)

swinging   A type of open marriage relationship in which a couple has extramarital relations together with other couples. (Ch. 10)

swish   A slang term for an effeminate gay. This is a stereotype that generally does not reflect how most gays look or behave. (Ch. 9)

symptomatic HIV infection   Formerly called AIDS-related complex (ARC); the early symptoms of HIV infection, which eventually lead to AIDS. (Ch. 5)

sympto-thermal method   A combination of the basal body temperature and Billings fertility awareness methods. (Ch. 6)

syphilis (sif′ i-lis)   A sexually transmitted disease caused by the *Treponema pallidum* bacterium (spirochete), which can also pass directly through any cut or scrape into the bloodstream. (Ch. 5)

systematic desensitization   A therapy technique used to reduce anxiety by slowly introducing elements of the anxiety-producing theme. (Ch. 13 and 14)

T

telephone scatologia (skat″ o-lo′ge-ah)   A condition in which sexual arousal is achieved primarily by making obscene telephone calls. (Ch. 14)

teratogens (teh rat′ ojinz)   Substances that can harm an embryo or fetus. (Ch. 7)

testicles (tes′ ty-k'ls)   The male gonads that produce sperm and male hormones. (Ch. 2)

testicular torsion   An uncommon condition in which the spermatic cord twists and cuts off the blood supply to the testicles. (Ch. 2)

testosterone (tes-tos′ te-ron)   A hormone that is produced by the testicles (and in very small amounts by the ovaries and adrenal glands). (Ch. 2 and 3)

third stage of labor   Detachment and expulsion of the placenta from the uterus after childbirth. (Ch. 7)

thrush   A yeast infection (candidiasis) of the mouth. (Ch. 5)

token resistance [to sex]   Saying "no" to sex when one means "yes." (Ch. 15)

toxemia of pregnancy   A disease of pregnancy characterized in its early stage (preeclampsia) by high blood pressure, weight gain, swollen joints, and protein in the urine. Unless it is corrected, it can lead to convulsions and coma. (Ch. 7)

toxic shock syndrome   A syndrome with symptoms of high fever, vomiting, diarrhea, and dizziness; caused by toxins produced by the *Staphylococcus aureus* bacterium. (Ch. 3)

transition phase   The last part of the start-up stage of labor, during which the cervix dilates to 10 cm in order for the baby to be able to enter the birth canal. (Ch. 7)

transsexual   An adult whose gender identity does not match his or her biological sex. (Ch. 8)

transvestism   A condition in which sexual arousal is achieved primarily by dressing as a member of the opposite sex. (Ch. 14)

triangular theory of love   Sternberg's theory of love, which says that all of our different positive emotions are the result of different combinations of intimacy, passion, and decision/commitment. (Ch. 12)

trichomoniasis (trik″ o-mo-ni′ ah-sis)   A type of vaginitis caused by a one-celled protozoan that is usually transmitted during sexual intercourse. (Ch. 5)

trimesters   Three-month periods of pregnancy (months 1–3, 4–6, and 7–9), so labeled for descriptive purposes. (Ch. 7)

Triple-X syndrome   A condition in which females have an extra X chromosome. (Ch. 8)

trophoblast (trof′ o-blast)   The outer four cell layers of the embryo. (Ch. 7)

tubal ligation   A female sterilization technique that originally referred only to the tying of the Fallopian tubes, but which is now often used as a general term for a variety of female sterilization techniques. (Ch. 6)

tubal pregnancy   Implantation of a blastocyst in a Fallopian tube. (Ch. 7)

Turner's syndrome   A condition in which females have only one X chromosome; as a result, their ovaries never develop properly. (Ch. 8)

two-spirit   In North American Indian tribes of past centuries, a person, usually a male, who assumed the dress, occupations, and behavior of the other sex in order to effect a change in gender status. The term now preferred by Native Americans and anthropologists over *berdache*. (Ch. 8)

typical-use pregnancy rate   For a particular birth control technique, the percentage of pregnancies during the first year of use by all couples who use the technique, regardless of whether or not they use it properly or consistently. See also **perfect-use pregnancy rate.** (Ch. 6)

U

ultrasound   A noninvasive technique for examining the internal organs of a fetus; it uses sound waves like a radar or sonar scan to create a picture. (Ch. 7)

umbilical (um-bil′ ĭ-kal) **cord**   The cord that connects an embryo or fetus to the mother's placenta. (Ch. 7)

unconditional love   Feelings of love that do not depend on the loved one's meeting certain expectations or desires. (Ch. 12)

undifferentiated   Individuals who are low on both masculine and feminine traits. (Ch. 8)

urethra (u-re′ thrah)   The passageway from the bladder to the exterior of the body. In males, it also serves as a passageway for semen during ejaculation. (Ch. 2)

urophilia (u″ ro-fil′ e-ah)   Sexual arousal caused by urine or the act of urination. (Ch. 14)

uterus (u′ ter-us)   The womb. The hollow, muscular organ in females where the fertilized egg normally implants. (Ch. 2)

V

vagina (vah-ji′ nah)   The sheathlike canal in the female that extends from the vulva to the cervix and that receives the penis during intercourse. (Ch. 2)

vaginal lubrication   The first sign of sexual arousal in females; the result of the vaginal walls becoming engorged with blood. (Ch. 4)

vaginismus (vaj″ ĭ-niz′ mus)   A sexual problem in females in which pain is experienced during attempted intercourse because of involuntary spasms in the muscles surrounding the outer third of the vagina. (Ch. 13)

vaginitis (vaj″ ĭ-ni′ tis)   A general term that refers to any inflammation of the vagina. There are three general types: trichomoniasis, moniliasis, and bacterial vaginosis. (Ch. 5)

vas deferens (vas def′ er-enz)   The second part of the duct system that transports sperm out of a male's body. (Ch. 2)

vasectomy (vah-sek′ to-me)   The male sterilization technique in which the vas deferens is tied off and cut, thus preventing passage of sperm through the male's reproductive tract. (Ch. 2 and 6)

vasocongestion (vas′ o kunjes′chun) The engorgement (filling) of tissues with blood. (Ch. 4)

venereal diseases Term originally used to refer to gonorrhea, syphilis, and three other diseases that are spread almost exclusively by sexual contact. (Ch. 5)

venereal warts See **genital warts.**

vernix caseosa A waxy bluish substance that covers a newborn baby. (Ch. 7)

vestibular (ves-tīb′ u-lar) **area** A term sometimes used to refer to the area between the two labia minora. (Ch. 2)

vestibular (ves-tīb′ u-lar) **bulbs** Structures surrounding the vaginal opening that fill with blood during sexual arousal, resulting in swelling of the tissues and a narrowing of the vaginal opening. (Ch. 2)

Victorian era The period during the reign of Queen Victoria of England (1819–1901). With regard to sexuality, it was a time of great public prudery (the pleasurable aspects of sex were denied) and many incorrect medical beliefs. (Ch. 1)

violent sexually explicit material Sexually explicit material that depicts violence, almost always showing males treating females violently. (Ch. 16)

viral load The amount of virus in the blood of an individual. (Ch. 5)

virus A protein shell around a nucleic acid core. Viruses have either RNA or DNA, but not both, and thus cannot reproduce themselves. They invade host cells that provide the material to manufacture new virus particles. (Ch. 5)

volunteer bias A bias in research results that is caused by differences between people who agree to participate and others who refuse. (Ch.1)

voyeurism (voi′ yer-izm) Repeatedly seeking sexual arousal by observing nude individuals without their knowledge or consent. (Ch. 14)

vulva (vul′ vah) A term for the external female genitalia, including the mons veneris, labia majora, labia minora, clitoris, vaginal opening, and urethral opening. (Ch. 2)

W

wet dream An involuntary nocturnal emission of semen. This type of ejaculation often is a boy's first experience with ejaculation, as wet dreams generally begin at the onset of puberty. (Ch. 10 and 11)

withdrawal (coitus interruptus) Withdrawal of the male's penis from his partner's vagina before ejaculation in order to avoid conception. It is sometimes ineffective because fluids from the Cowper's glands may contain sperm. (Ch. 6)

Wolffian duct system A primitive duct system found in embryos that, if allowed to develop, becomes the male reproductive system. (Ch. 8)

Z

zona pellucida (zo′nah pellu′ cĭda) The surface of an ovum. (Ch. 7)

zoophilia (zo″ ofil′ e-ah) Using sexual contact with animals as the primary means of achieving sexual gratification. (Ch. 14)

zygote (zi′ got) The one-celled organism created from the fusion of a sperm and egg. (Ch. 7)

zygote intrafallopian transfer (ZIFT) A method of treating infertility caused by blocked Fallopian tubes. An ovum taken from the infertile female is fertilized by sperm from her partner and then transferred to the unblocked portion of a Fallopian tube. (Ch. 7)

Resources

GENERAL

American College Health Association
P.O. Box 28937
Baltimore, MD 21240-8937
(410) 859-1500

National Self-Help Clearinghouse
25 West 43rd Street
New York, NY 10036
(212) 354-8525
They will help you locate a self-help organization in your area.

SIECUS (Sex Information and Education Council of the U.S.)
130 West 42nd Street, Suite 350
New York, NY 10036
(212) 819-9770

Society for the Scientific Study of Sex
P.O. Box 208
Mount Vernon, IA 52314
(319) 895-8407

ABORTION

National Abortion and Reproductive Rights Action League
1156 15th Street, N.W., Suite 700
Washington, DC 20005
(202) 973-3000
The leading pro-choice abortion group.

National Right to Life Committee
419 7th Street, N.W., Suite 500
Washington, DC 20004
(202) 626-8800
The leading anti-abortion group.

BIRTH CONTROL

Alan Guttmacher Institute
120 Wall Street, 21st Floor
New York, NY 10005
(212) 248-1111

Association for Voluntary Surgical Contraception (AVSC)
79 Madison Avenue
New York, NY 10016
(212) 561-8000

Planned Parenthood Federation of America
810 Seventh Avenue
New York, NY 10019
(212) 541-7800
You can contact the national headquarters to get the address and phone number of the clinic closest to you.

CANCER

American Cancer Society
19 West 56th Street
New York, NY 10019
(212) 586-8700
Local units can be found in your telephone directory.

National Cancer Institute Cancer Information Service
1(800) 4-CANCER

CHILD SEXUAL ABUSE/INCEST

CHILDHELP National Child Abuse Hotline
1(800) 422-4453
Provides information, crisis intervention, and referrals to local services if you wish to report abuse.

Children's Defense Fund
25 E Street, N.W.
Washington, DC 20001
(202) 628-8787
Provides information about child abuse prevention.

Incest Survivors Anonymous
P.O. Box 17245
Long Beach, CA 90807-7245
(562) 428-5599
For incest victims.

Incest Recovery Services of the Family Place
P. O. Box 7999
Dallas, TX 75209
(214) 559-2170
For incest victims.

National Center for Missing and Exploited Children
2101 Wilson Blvd., Suite 550
Arlington, VA 22201
1(800) 843-5678
(703) 235-3900 (office)
For parents who wish to report missing children, and others who may have information about missing children.

RAINN (Rape, Abuse, and Incest National Network)
1(800) 656-HOPE
Contact them for the number and address of the rape crisis program nearest to you.

YWCA Rape Crisis Program
Check your telephone directory for a program nearest to you, or call RAINN (see above).

HIV/AIDS

AIDS Action Council
1875 Connecticut Avenue, N.W., Suite 700
Washington, DC 20009
(202) 986-1300

AIDS Information Exchange
U.S. Conference of Mayors
1620 I Street, N.W.
Washington, DC 20006
(202) 293-7330

American Red Cross AIDS Public Education Program
2025 E Street, N.W.
Washington, DC 20003
(202) 728-6400

ACLU Lesbian and Gay Rights/AIDS Project
American Civil Liberties Union
125 Broad Street, 18th Floor
New York, NY 10004
(212) 549-2500

American Foundation for AIDS
  Research
733 Third Avenue, 12th Floor
New York, NY 10017
(212) 682-7440

National HIV/AIDS Hotline
  (Centers for Disease Control and
  Prevention)
1(800) 342-AIDS

National AIDS Information
  Clearinghouse
1(800) 458-5231
  Service provided by the Centers
  for Disease Control and
  Prevention.

National Association of People with
  AIDS
1413 K Street, N.W.
Seventh Floor
Washington, DC 20005
(202) 898-0414

  See also listings under
  "homosexuality."

HOMOSEXUALITY

ACLU Lesbian and Gay Rights/AIDS
  Project
American Civil Liberties Union
125 Broad Street, 18th Floor
New York, NY 10004
(212) 549-2500

Federation of Parents and Friends of
  Lesbians and Gays (PFLAG)
1101 14th Street, N.W., Suite 1030
Washington, DC 20005
(202) 638-4200

Gay Men's Health Crisis
129 W. 20th Street
New York, NY 10011-3629
AIDS Hotline: (212) 807-6655

Hetrick-Martin Institute
2 Astor Place
New York, NY 10003
(212) 674-2400
  Gay, lesbian, and bisexual youth
  service agency.

Lambda Legal Defense and
  Education Fund, Inc.
6030 Willshire Blvd., Suite 200
Los Angeles, CA 90036-3617
(213) 937-2728

National Gay and Lesbian Task
  Force
2320 17th Street, N.W.
Washington, DC 20009-2702
(202) 332-6483

National Lesbian and Gay Health
  Association
P.O. Box 65472
Washington, DC 20035
 (202) 939-7880

INFERTILITY/ADOPTION

Adoptive Families of America
2309 Como Avenue
St. Paul, MN 55108
(612) 645-9955

American Society for Reproductive
  Medicine
1209 Montgomery Highway
Birmingham, AL 35216
(205) 978-5000
  Formerly called the American
  Fertility Society

National Council for Adoption
1930 17th Street, N.W.
Washington, DC 20009
(202) 328-1200
  Contact them for information
  about adoption services in your
  area

Resolve, Inc.
1310 Broadway
Somerville, MA 02144-1779
(617) 623-0744
  Provides advice and help about
  infertility

PREGNANCY/CHILDBIRTH

American College of Nurse-
  Midwives
818 Connecticut Avenue, N.W.,
  Suite 900
Washington, DC 20006
(202) 728-9860

American College of Obstetricians
  and Gynecologists
409 12th Street, S.W.
Washington, DC 20024
(202) 638-5577

Lamaze International
1200 19th Street, N.W., Suite 300

Washington, DC 20036-2401
1(800) 368-4404
  Contact them for the name and
  address of a certified Lamaze
  instructor in your area.

La Leche League International
1400 North Meacham Road
Schaumburg, IL 60173-4840
(800) LA-LECHE
  Advice about breast-feeding.

March of Dimes Birth Defects
  Foundation
1275 Mamaroneck Avenue
White Plains, NY 10605
(914) 428-7100
  Provides information on birth
  defects and gives referrals to
  genetic counseling and services in
  your area.

RAPE

Center for the Prevention of Sexual
  and Domestic Violence
936 North 34th Street, Suite 200
Seattle, Washington 98103
(206) 634-1903

National Clearinghouse on Marital
  and Date Rape
(510) 524-1582

National Resource Center on
  Domestic Violence
1(800) 537-2238

National Victims Center
2111 Wilson Blvd., Suite 300
Arlington, VA 22201
(703) 276-2880

RAINN (Rape, Abuse, and Incest
  National Network)
1(800) 656-HOPE
  Contact them for the number and
  address of the rape crisis program
  nearest to you.

U.S. Department of Justice
Office of Violence Against Women
10th and Constitution Avenue,
  N.W., Room 5302
Washington, DC 20530
(202) 514-2000

YWCA Rape Crisis Program
  Check your telephone directory

for a program nearest to you, or call RAINN (see above).

RUNAWAY CHILDREN

Covenant House (Crisis Intervention Hotline)
1(800) 999-9999
> Provides counseling, shelter, and other referral services.

National Network of Runaway and Youth Services (Youth Crisis Hotline)
Department P
P.O. Box 8283
Washington, DC 20024
1(800) 448-4663
> They will put you in touch with the nearest shelter program.

National Runaway Switchboard
1(800) 621-4000
> They encourage runaway children to reestablish contact with their families.

SEX/MARITAL THERAPY

AAMFT (American Association for Marriage and Family Therapy)
1133 15th Street, N.W., Suite 300
Washington, DC 20005
(202) 452-0109
> They will provide names and addresses of certified marriage and family therapists in your area.

AASECT (American Association of Sex Educators, Counselors and Therapists)
P.O. Box 238
Mount Vernon, IA 52314
(319) 895-8407
> They will provide names and addresses of certified sex counselors and therapists in your area.

American Board of Sexology, and The American Academy of Clinical Sexologists
1929 18th Street, N.W., Suite 1166
Washington, DC 20009
(202) 462-2122
> They will provide names and addresses of certified sex therapists and sex educators in your area.

SEXUAL HARASSMENT

Center for Women in Government
State University of New York at Albany
Draper Hall, Room 302
135 Western Avenue
Albany, NY 12222
(518) 442-3900

National Organization for Women
1000 16th Street, N.W., Suite 700
Washington, D.C. 20036
(202) 331-0066

SEXUALLY TRANSMITTED DISEASES (OTHER THAN HIV/AIDS)

American Social Health Association
P.O. Box 13827
Research Triangle Park, NC 27709

Herpes Resource Center (hotline)
(919) 361-8488
> Service provided by the American Social Health Association.

STD National Hotline
1 (800) 227-8922
> Service provided by the American Social Health Association, and funded by the Centers for Disease Control and Prevention.

Credits

ABOUT THE AUTHOR: **xxii**, © Peggy Stewart, Saluda, NC.

CHAPTER 1: **8**, Museo del Prado; **9**, courtesy Dr. Martha Ward, Dept. of Anthropology, University of New Orleans; **12**, courtesy St. Augustine and St. Johns County Chamber of Commerce, St. Augustine, Florida; **13 (bottom left)**, Chantelle Boudreaux; **13 (bottom right)**, Don Martinetti; **15 16, 17 (top)**, Corbis-Bettman; **17 (bottom)**, Bruce Powell Photography.

CHAPTER 2: **27**, Precision Graphics/Don Martinetti; **31, 40, 49**, Precision Graphics; **29, 32, 34, 41, 46**, Don Martinetti; **28**, (with Precision Graphics), **36 (top & bottom)**, **43 (top & bottom)** from Frederic Martini, *Fundamentals of Anatomy and Physiology*, © 1989. Reprinted by permission of Prentice Hall, Upper Saddle River, New Jersey; **42**, Don Martinetti/Chantelle Boudreaux.

CHAPTER 3: **54 (left)**, **55**, Don Martinetti/Chantelle Boudreaux; **54 (right)**, **65**, Precison Graphics; **56**, from Frederic Martini, *Fundamentals of Anatomy and Physiology*, © 1989. Reprinted by permission of Prentice Hall, Upper Saddle River, NJ; **59**, courtesy Dr. Martha Ward, Dept. of Anthropology, University of New Orleans; **62**, Bruce King.

CHAPTER 4: **71**, Behavioral Technologies, Inc.; **73, 74, 79**, Don Martinetti/Chantelle Boudreaux; **77**, J. Geer, J. Heiman, H. Leitenberg, *Human Sexuality*, © 1984. Reprinted by permission of Prentice Hall, Upper Saddle River, New Jersey; **83**, Stephanie Walsh/Gamma-Liaison, Inc.; **85**, Joel Gordon/Joel Gordon Photography; **88 (left)**, Ken Magee; **88 (right)**, Joel Gordon/Joel Gordon Photography.

CHAPTER 5: **94, 95 (top), 98 (top), 100, 101 (left)**, Burroughs Wellcome Inc.; **95 (bottom)**, courtesy Bloomfield "Will" Baber; **98 (bottom), 99 (top), 105**, courtesy Dr. Denise Buntin, Memphis Veterans Administration Medical Center; **99 (bottom)**, courtesy of Christopher J. McEwen, M.D., Calais Dermatology Associates; **101 (center & right)**, courtesy of Dr. Denise Buntin & Dr. Larry Millikan, Tulane University Medical Center; **108-109, 113**, Precision Graphics; **110**, A. Ramey/Stock Boston; **114**, Precision Graphics/Don Martinetti; **117**, from *Newsweek* (December 2, 1996). Copyright © 1996 by Newsweek, Inc. Reprinted with the permission of Newsweek; **125**, Jonathan Wenk/Black Star.

CHAPTER 6: **134**, UPI/Corbis-Bettman; **135, 136, 142, 148**, Don Martinetti; **145**, Bruce King; **146**, Precision Graphics/Don Martinetti/Chantelle Boudreaux; **147 (left)**, Frank LaBua/Ortho-McNeil Pharmaceutical, Inc.; **147 (right)**, Frank LaBua/Prentif Lamberts (Dalston) LTD; **150**, from Robert J. Demarest & John Sciarra, *Conception, Birth, and Contraception: A Visual Presentation*, 2/E. Copyright © 1976 by McGraw-Hill, Inc. Reprinted with permission of the publishers; **153**, Wyeth Ayerst Laboratories; **156**, Don Martinetti/Chantelle Boudreaux.

CHAPTER 7: **165**, Chantelle Boudreaux; **168**, Chantelle Boudreaux/Simon & Schuster/PH College; **169, 181, 187**, Ken Magee; **170**, Illustrations from *Pregnancy in Anatomical Illustrations* (1984). Used with permission of Carnation Company, Nutritional Products Division. All rights reserved; **171**, Annie Liebovitz/Contact Press Images; **173**, from Robert M. Liebert, et al., *Developmental Psychology*, 4/E, © 1986. Reprinted by permission of Prentice Hall, Upper Saddle River, NJ; **185**, The American Museum of Natural History; **193**, IVF Australia.

CHAPTER 8: **199, 205**, AP/Wide World Photos; **202 (top & right)**, John Money & Anke A. Erhardt, *Man and Woman, Boy and Girl*. The Johns Hopkins University Press, Baltimore/London, 1973; **202 (bottom)**, The Johns Hopkins University Press; **208**, Ken Karp/Simon & Schuster/PH College; **212**, Spencer Grant/Picture Cube, Inc.; **213**, Suzanne E. Wu/Jeroboam, Inc.; **215 (top)**, Elizabeth Hamlin/Stock Boston; **215 (bottom)**, Cary Wolinski/Stock Boston; **216**, David Woo/Stock Boston; **217 (top)**, Andy Sacks/Tony Stone Images; **217 (bottom)**, Luke Airforce Base; **218**, Smithsonian Institution; **220 (top)**, Robert Brenner/PhotoEdit; **220 (bottom)**, Tony Freeman/PhotoEdit.

CHAPTER 9: **226 (top)**, Ken Magee; **226 (bottom)**, Rose Skytta/Jeroboam, Inc.; **239**, Rare Books Division, The New York Public Library. Astor, Lenox, and Tilden Foundations; **241**, AP/Wide World Photos; **243**, Mark J. Terrill/AP/Wide World Photos.

CHAPTER 10: **248**, Judy S. Geller/Stock Boston; **249**, Frederick D. Bodin/Stock Boston; **250 (left)**, Barbara Rios/Photo Researchers, Inc.; **250 (right)**, **266 (left)**, Ken Magee; **252**, Joel Gordon/Joel Gordon Photography; **256**, John Henley/The Stock Market; **259**, Tony Savino/The Image Works; **261 (top)**, Alfred Jones; **264**, Laura Dwight/Photo Edit; **268**, Precision Graphics; **271**, Will & Deni McIntyre/Photo Researchers, Inc.; **274**, Ariel Skelley/The Stock Market.

CHAPTER 11: **285, 286, 289, 290**, Don Martinetti.

CHAPTER 12: **297, 307**, Ken Magee; **299**, Art Resource, NY; **301**, Lily Solmssen/Photo Researchers, Inc.; **303, 306**, Robert Sternberg, IBM Professor of Psychology and Education, Yale University; **310**, Myrleen Ferguson/PhotoEdit; **312**, Don Martinetti/Chantelle Boudreaux.

CHAPTER 13: **318**, T. Petillot/Explorer/Photo Researchers, Inc.; **321, 329, 334, 335**, Don Martinetti; **327**, Simon & Schuster/PH College.

CHAPTER 14: **341**, Charles Gatewood/The Image Works; **344 (top)**, Corbis-Bettman; **344 (bottom)**, Jim Cooper/AP/Wide World Photos; **347**, David Young-Wolff/PhotoEdit; **349**, Joel Gordon Photography; **350**, Douglas Mason/Woodfin Camp & Associates.

CHAPTER 15: **357**, Precision Graphics; **361**, Art Resource, NY; **367**, Robert Goldstein/Photo Researchers, Inc.; **373**, Bettye Lane/Photo Researchers, Inc.; **377**, Frank Siteman/Stock Boston; **382**, From *Newsweek*, May 14, 1984; copyright Newsweek Inc. All rights reserved. Reprinted with permission; **387**, AP/Wide World Photos.

CHAPTER 16: **397**, copyright 1986 Time, Inc., reprinted by permission; **398**, Adam Scull/Rangefinders/Globe Photos, Inc.; **403**, Eugene Gordon/Simon & Schuster/PH College; **406**, Cynthia Johnson/Time Life Syndication; **410 (top)** SuperStock, Inc.

CHAPTER 17: **418**, Lynne Weinstein/Woodfin Camp & Associates; **422**, Randy Matuson/Monkmeyer Press; **424**, Suzanne Szasz/Photo Researchers, Inc.; **426**, David M. Grossman/Photo Researchers, Inc.

Insert: 1 (top), Gamma-Liaison; **1 (bottom)**, © Lennart Nilsson, *The Incredible Machine*; **2 (top to bottom, left to right)**, © Lennart Nilsson, *Being Born*; Dr. Landrum B. Shettles; Dr. Landrum B. Shettles; © Lennart Nilsson, *A Child Is Born*; **3 (top)**, © Lennart Nilsson, *A Child Is Born*; **3 (bottom)**, © Lennart Nilsson, *Behold Man*; **4**, Heinz Kluetmeier/DOT.

References

AARONSON, I. A. (1994). Micropenis: Medical and surgical implications. *Journal of Urology, 152,* 4–14.

AAUW (1993). *Hostile hallways.* Washington, D.C.: American Association of University Women.

ABDULLA, R. H. D. (1982). *Sisters in affliction: Circumcision and infibulation of women in Africa.* London: Zed Press (Westport, Connecticut: L. Hill)

ABEL, E. L., & SOKOL, R. J. (1987). Incidence of fetal alcohol syndrome and economic impact of FAS-related anomalies. *Drug and Alcohol Dependency, 19,* 51–70.

ABEL, G. G. (1981, October 15). *The evaluation and treatment of sexual offenders and their victims.* Paper presented at St. Vincent Hospital and Medical Center, Portland, OR.

ABEL, G. G., et al. (1987). Self-reported sex crimes of non-incarcerated paraphiliacs. *Journal of Interpersonal Violence, (1),* 3–25.

ABEL, G. G., et al. (1988a). Multiple paraphiliac diagnoses among sex offenders. *Bulletin of the American Academy of Psychiatry and the Law, 16,* 153–168.

ABEL, G. G., et al. (1988b). Predicting child molesters' response to treatment. *Annals of the New York Academy of Sciences, 528,* 223–234.

ABEL, G. G., BARLOW, D. H., BLANCHARD, E. B., & GUILD, D. (1977). The components of rapists' sexual arousal. *Archives of General Psychiatry, 34,* 895–903.

ABEL, G. G., BECKER, J. V., MURPHY, W. D., & FLANAGAN, B. (1981). Identifying dangerous child molesters. In R. B. Stuard (Ed.), *Violent behavior: Social learning approaches to prediction, management, and treatment* (pp. 116–137). New York: Brunner/Mazel.

ABMA, J., et al. (1997). Fertility, family planning, and women's health: New data from the 1995 National Survey of Family Growth. National Center For Health Statistics. *Vital Health Stat, 23(19).*

ABMA, J., DRISCOLL, A., & MOORE, K. (1998). Differing degrees of control over first intercourse and young women's first partners: Data from cycle 5 of the National Survey of Family Growth. *Family Planning Perspectives, 30,* 12–18.

ABRAMS, B., & PARKER, J. D. (1990). Maternal weight gain in women with good pregnancy outcome. *Obstetrics and Gynecology, 76,* 1–7.

ABU-EL-FUTUH SHANDALL, A. (1967). Circumcision and infibulation of females: A general consideration of the problem and a clinical study of the complications in Sudanese women. *Sudan Medical Journal, 5(4),* 178–212.

ACOSTA, F. (1975). Etiology and treatment of homosexuality: A review. *Archives of Sexual Behavior, 4,* 9.

ADAMS, H. E., WRIGHT, L. W., & LOHR, B. A. (1996). Is homophobia associated with homosexual arousal? *Journal of Abnormal Psychology, 105,* 440–445.

ADAMS, J. W., KOTTKE, J. L., & PADGITT, J. S. (1983). Sexual harassment of university students. *Journal of College Student Personnel, 24,* 484–490.

ADDIEGO, F., et al. (1981). Female ejaculation: A case study. *Journal of Sex Research, 17,* 13–21.

ADDUCCI, C., & ROSS, L. (1991, October). Common urethral injuries in men. *Medical Aspects of Human Sexuality,* pp. 32–34.

ADLER, E. (1994, March 4). Vasectomies cheaper, easier—and dropping. (New Orleans) *Times-Picayune,* p. E-10.

ADLER, J., et al. (1996, September 30). Adultery: A new furor over an old sin. *Newsweek,* pp. 54–60.

ADLER, N. E., et al. (1990). Psychological responses after abortion. *Science, 248,* 41–44.

ADLERCREVTZ, H., et al. (1992). Dietary phyto-oestrogens and the menopause in Japan. *Lancet, 339,* 1233.

ADOLFSSON, J., STEINECK, G., & WHITMORE, W. F., JR. (1993). Recent results of management of palpable clinically localized prostate cancer. *Cancer, 72,* 310–322.

AGARWAL, S. S., et al. (1993). Role of male behavior in cervical carcinogenesis among women with one lifetime partner. *Cancer, 72,* 1666–1669.

AGNEW, J. (1986). Hazards associated with anal erotic activity. *Archives of Sexual Behavior, 15,* 307–314.

AHIA, R. N. (1991). Compliance with safer-sex guidelines among adolescent males: Application of the health belief model and protection motivation theory. *Journal of Health Education, 22,* 49–52.

ALAN GUTTMACHER INSTITUTE (1994). *Sex and American teenagers.* New York: Author.

ALAN GUTTMACHER INSTITUTE (1996). Facts in brief: Teen sex and pregnancy. [Brochure]. New York: Author.

ALBERMAN, E. (1988). The epidemiology of repeated abortion. In R. W. Beard & F. Sharp (Eds.), *Early pregnancy loss: Mechanisms and treatment* (pp. 9–17). London: Royal College of Obstetricians and Gynaecologists.

ALDER, E. M. (1989). Sexual behavior in pregnancy, after childbirth and during breast-feeding. *Balliere's Clinical Obstetrics and Gynaecology, 3,* 805–821.

ALEXANDER, C. J., SIPSKI, M. L., & FINDLEY, T. W. (1993). Sexual activities, desire, and satisfaction in males pre-and post-spinal cord injury. *Archives of Sexual Behavior, 22,* 217–228.

ALEXANDER, G. M., & HINES, M. (1994). Gender labels and play styles: Their relative contribution to children's selection of playmates. *Child Development,* 65, 869–879.

ALEXANDER, G. M., & SHERWIN, B. B. (1993). Sex steroids, sexual behavior, and selection attention for erotic stimuli in women using oral contraceptives. *Psychoneuroendocrinology, 18,* 91–102.

ALEXANDER, G. M., SHERWIN, B. B., BANCROFT, J., & DAVIDSON, D.W. (1990). Testosterone and sexual behavior in oral-contraceptive users and nonusers: A prospective study. *Hormones and Behavior, 24,* 388–402.

ALEXANDER, G. M., et al. (1997). Androgen-behavior correlations in hypogonadal men and eugonadal men. *Hormones and Behavior, 31,* 110–119.

ALEXANDER, N. J. (1994). Vaccines for contraception: A summary. *Reproduction Fertility Development, 6,* 417–419.

ALEXANDER, P. C. (1992). Application of attachment theory to the study of sexual abuse. *Journal of Consulting and Clinical Psychology, 60,* 185–195.

ALLDRIDGE, P. (1997, January). The Sexual Offenses (Conspiracy and Incitement) Act 1996. *Criminal Law Review,* pp. 30–40.

ALLEN, D. M. (1980). Young male prostitutes: A psychosocial study. *Archives of Sexual Behavior, 9,* 399–426.

ALLEN, L. S., & GORSKI, R. A. (1992). Sexual orientation and the size of the anterior commissure in the human brain. *Proceedings of the National Academy of Sciences of the U.S.A., 89,* 7199–7202.

ALLEN, M., & BURRELL, N. (1996). Comparing the impact of homosexual and heterosexual parents on children: Meta-analysis of existing research. *Journal of Homosexuality, 32(2),* 19–35.

ALLEN, M. C., DONAHUE, P. K., & DUSMAN, A. E. (1993). The limit of viability—Neonatal outcome of infants born at 22 to 25 weeks gestation. *New England Journal of Medicine, 329,* 1597–1601.

ALLEN, M. P. (1989). Transformations: Crossdressers and those who love them. New York: Dutton.

ALLGEIER, E. R. (1985, July). Are you ready for sex?: Informed consent for sexual intimacy. *SIECUS Report, 13,* 8–9.

ALLGEIER, E. R., & MURNEN, S. K. (1985, March). Perception of parents as sexual beings: Pocs and Godow revisited. *SIECUS Report, 4,* 11–12.

AL SAMARAI, A. M., et al. (1989). Sequential genital infections with herpes simplex virus types 1 and 2. *Genitourinary Medicine, 65,* 39–41.

ALTHOF, S., et al. (1991). Sexual, psychological, and marital impact of self-injection of papaverine and phentolamine: A long-term prospective study. *Journal of Sex and Marital Therapy, 17,* 101–112.

ALVAREZ, W. A., & FREINHAR, J. P. (1991). A prevalence study of bestiality zoophilia in psychiatric in-patients, medical in-patients, and psychiatric staff. *International Journal of Psychosomatics, 38,* 45–47.

ALZATE, H., & HOCH, Z. (1986). The "G-spot" and "female ejaculation": A current appraisal. *Journal of Sex and Marital Therapy, 12,* 211–220.

ALZATE, H., & LONDONO, M. L. (1984). Vaginal erotic sensitivity. *Journal of Sex and Marital Therapy, 10,* 49–56.

AMERICAN ACADEMY OF ALLERGY AND IMMUNOLOGY. (1986). Candidiasis hypersensitivity syndrome. *Journal of Allergy and Clinical Immunology, 78,* 271–273.

AMERICAN ACADEMY OF PEDIATRICS AND AMERICAN COLLEGE OF OBSTETRICIANS AND GYNECOLOGISTS (1983). *Guidelines For Perinatal Care.* Evanston, IL: Author.

AMERICAN CANCER SOCIETY (1997). *Cancer facts and figures—1997.* Atlanta: Author.

AMERICAN COLLEGE OF OBSTETRICIANS AND GYNECOLOGISTS. (1985). *Standards for obstetric-gynecologic services.* Washington, DC: Author.

AMERICAN COLLEGE OF PHYSICIANS AND INFECTIOUS DISEASES SOCIETY OF AMERICA. (1994). Human immunodeficiency virus (HIV) infection. *Annals of Internal Medicine, 120*, 310–319.

AMERICAN MEDICAL ASSOCIATION (AMA) (1994). Health care needs of gay men and lesbians in the U.S. Policy paper adopted December 6. Chicago: Author.

AMERICAN PSYCHIATRIC ASSOCIATION. (1994). *Diagnostic and statistical manual of mental disorder (DSM-IV)* (4th ed.). Washington, DC: Author.

AMES, K., GORDON, J., & MASON, M. (1991, February 18). Savings plan for a generation. *Newsweek*, p. 71.

AMES, K., et al. (1991, September 30). And donor makes three. *Newsweek*, pp. 60–61.

AMES, T. R. H. (1991). Guidelines for providing sexuality-related services to severely and profoundly retarded individuals: The challenge for the nineteen-nineties. *Sexuality and Disability, 9*, 113–122.

AMIR, M. (1971). *Patterns in forcible rape.* Chicago: University of Chicago Press.

ANDERSON, D., & CHRISTENSON, G. (1991). Ethnic breakdown of AIDS related knowledge and attitudes from the National Student Health Survey. *Journal of Health Education, 22*, 30–34.

ANDERSON, J. E., & DAHLBERG, L. L. (1992). High-risk sexual behavior in the general population. *Sexually Transmitted Diseases, 19*, 320–325.

ANDERSON, P. B. (1989). Adversarial sexual beliefs and past experience of sexual abuse of college females as predictors of their sexual aggression toward adolescent and adult males. (Unpublished doctoral dissertation, New York University).

ANDERSON, P. B. (1991). Can men be the victim of unwanted sexual intercourse? In B. M. King, C. J. Camp, & A. Downey (Eds.), *Human sexuality today* (p. 370). Englewood Cliffs, NJ: Prentice Hall.

ANDERSON, P. B., & MATHIEU, D. A. (1996). College students' high-risk sexual behavior following alcohol consumption. *Journal of Sex and Marital Therapy, 22*, 259–264.

ANDERSON, P. B., et al. (1995). Changes in the telephone calling patterns of adolescent girls. *Adolescence, 30*, 779–784.

ANDERSON, R. A., BANCROFT, J., & WU, F. C. (1992). The effects of exogenous testosterone on sexuality and mood of normal men. *Journal of Clinical and Endocrinological Metabolism, 75*, 1503–1507.

ANDERSON-HUNT, M., & DENNERSTEIN, L. (1994). Increased female sexual response after oxytocin. *British Medical Journal, 309*, 929.

ANDERSSON-ELLSTROM, A., et al. (1996). The relationship between knowledge about sexually transmitted diseases and actual sexual behaviour in a group of teenage girls. *Genitourinary Medicine, 72*, 32–36.

ANGELL, M. (1991). A dual approach to the AIDS epidemic. *New England Journal of Medicine, 324*, 1498–1500.

ANGIER, N. (1991, September 1). The biology of what it means to be gay. *New York Times*, sec. 4, p. 1.

ANGIER, N. (1992). A male menopause? Jury is still out. *New York Times*.

ANKUM, W. M. (1996). Is the rising incidence of ectopic pregnancy unexplained? *Human Reproduction, 11*, 238–239.

ANKUM, W. M., MOL, B. W. J., VAN DER VEEN, F., & BOSSUYT, P. M. M. (1996). Risk factors for ectopic pregnancy: A meta-analysis. *Fertility and Sterility, 65*, 1093–1099.

ANKUM, W. M., et al. (1996). Management of suspected ecptopic pregnancy: Impact of new diagnostic tools in 686 consecutive cases. *Journal of Reproductive Medicine, 41*, 724–728.

ANNON, J. (1974). *The behavioral treatment of sexual problems.* Honolulu, HI: Enabling Systems.

ANSELL, J. (1991, September). The crooked penis: Causes and treatment. *Medical Aspects of Human Sexuality*, pp. 32–38.

APFELBAUM, B. (1989). Retarded ejaculation: A much misunderstood syndrome. In S. R. Leiblum & R. C. Rosen (Eds.), *Principles and practice of sex therapy: Update for the 1990's* (pp. 168–206). New York: Guilford Press.

APFELBAUM, B. (Ed.). (1980). *Expanding the boundaries of sex therapy.* Berkeley, CA: Berkeley Sex Therapy Group.

APT, C., HURLBERT, D. F., & POWELL, D. (1993). Men with hypoactive sexual desire disorder: The role of interpersonal dependency and assertiveness. *Journal of Sex Education and Therapy, 19*, 108–116.

ARAL, S. O., MOSHER, W. D., & CATES, W. (1991). Self-reported pelvic inflammatory disease in the United States, 1988. *Journal of the American Medical Association, 266*, 2570–2573.

ARAL, S. O., MOSHER, W. D., & CATES, W. (1992). Vaginal douching among women of reproductive age in the United States: 1988. *American Journal of Public Health, 82*, 210–214.

ARATA, C. M., & BURKHART, B. R. (1996). Post-traumatic stress disorder among college student victims of acquaintance assault. *Journal of Psychology and Human Sexuality, 8*, 79–92.

ARCHER, D. F. (1990, February). Cardiovascular disease and oral contraceptives. *Medical Aspects of Human Sexuality*, pp. 26–33.

ARCHER, R. L., BERG, J. H., & RUNGE, T. E. (1980). Active and passive observer's attraction to a self-disclosing other. *Journal of Experimental Social Psychology, 16*, 130–145.

ARGENTINE EPISIOTOMY TRIAL COLLABORATIVE GROUP. (1993). Routine vs. selective episiotomy: A randomized controlled trial. *Lancet, 342*, 1517–1518.

ARMSTRONG, P., & FELDMAN, S. (1990). *A wise birth.* New York: William Morrow.

ARON, A., & WESTBAY, L.. (1996). Dimensions of the prototype of love. *Journal of Personality and Social Psychology, 70*, 535–551.

ARRILLAGA, A., ERSEK, R. A., BARICOS, W. M., & RYAN, R. F. (1977, November). Method for the prevention of firm breasts from capsular contraction. *Plastic and Reconstructive Surgery*, pp. 752–754.

ASALI, A., et al. (1995). Ritual female genital surgery among Bedouin in Israel. *Archives of Sexual Behavior, 24*, 571–575.

ASCH, R. H., ELLSWORTH, L. R., BALMACEDA, J. P., & WONG, P. C. (1984). Pregnancy after translaparoscopic gamete intra-fallopian transfer [Letter]. *Lancet, 2*, 1034–1035.

ASCHER, M. S., et al. (1993). Does drug use cause AIDS? *Nature* (London), *362*, 103–104.

ASKEW, M. W. (1965). Courtly love: Neurosis as institution. *Psychoanalytic Review, 52*, 19–29.

ASSALIAN, P. (1994). Premature ejaculation: Is it psychogenic? *Journal of Sex Education and Therapy, 20*, 1–4.

ATHANASIOU, R., SHAVER, P., & TAVRIS, C. (1970, July). Sex. *Psychology Today*, pp. 37–52.

ATWOOD, J. D., & GAGNON, J. (1987). Masturbatory behavior in college youth. *Journal of Sex Education and Therapy, 13*, 35–42.

AUGER, J., KUNSTMANN, J. M., CZYGLIK, F., & JOUANNET, P. (1995). Decline in semen quality among fertile men in Paris during the past 20 years. *New England Journal of Medicine, 332*, 281–285.

AVINS, A., et al. (1994). HIV infection and risk behaviors among heterosexuals in alcohol treatment programs. *Journal of the American Medical Association, 271*, 515–518.

AXINN, W. G., & BARBER, J. S. (1997). Living arrangements and family formation attitudes in early adulthood. *Journal of Marriage and the Family, 59*, 595–611.

AXINN, W. G., & THORNTON, A. (1992). The relationship between cohabitation and divorce: Selectivity or causal influence? *Demography, 29*, 357–374.

AYOOLA, E. A. (1988). Viral hepatitis in Africa. In A. J. Zuckerman (Ed.), *Viral hepatitis and liver disease.* New York: Alan R. Liss.

AZAR, B. (1997, October). More study needed on hormone replacement. *APA Monitor*, 33.

BACHMAN, G. A. (1989). Brief sexual inquiry in gynecologic practice. Part 1. *Obstetrics and Gynecology, 73*, 425–427.

BACHMANN, G. (1992). Using androgens to increase libido. *Medical Aspects of Human Sexuality, 26*, 6.

BACHRACH, C. A. (1987). Cohabitation and reproductive behavior in the U.S. *Demography, 24*, 623–637.

BAGATELL, C. J., et al. (1994). Metabolic and behavioral effects of high-dose exogenous testosterone in healthy men. *Journal of Clinical and Endocrinological Metabolism, 79*, 561–567.

BAIER, J. L. (1990). Sexual harassment of university students by faculty members at a Southern research university. *The College Student Affairs Journal, 10* (2), 4–11.

BAILEY, J. M., & PILLARD, R. C. (1991). A genetic study of male sexual orientation. *Archives of General Psychiatry, 48*, 1089–1096.

BAILEY, J. M., PILLARD, R. C., NEALE, M. C., & AGYEI, Y. (1993). Heritable factors influence sexual orientation in women. *Archives of General Psychiatry, 50*, 217–223.

BAILEY, J. M., & ZUCKER, K. J. (1995). Childhood sex-typed behavior and sexual orientation: A conceptual analysis and quantitative review. *Developmental Psychology, 31*, 43–55.

BAIRD, D. D., WEINBERG, C. R., VOIGHT, L. F., & DALING, J. R. (1996). Vaginal douching and reduced fertility. *American Journal of Public Health, 86*, 844–850.

BAIRD, P. A., & MacGILLIVRAY, B. (1982). Children of incest. *Journal of Pediatrics, 101*, 854–858.

BAKER, B. (1990). Birth control. *Common Cause Magazine, 16*(3), 11–14.

BAKWIN, H. (1974). Erotic feelings in infants and young children. *Medical Aspects of Human Sexuality, 8*, 200–215.

BALDWIN, J. D., & BALDWIN, J. I. (1988). Factors affecting AIDS-related sexual risk-taking behavior among college students. *Journal of Sex Research, 25*, 181–196.

BALDWIN, J. D., & BALDWIN, J. I. (1997). Gender differences in sexual interest. *Archives of Sexual Behavior, 26,* 181–210.

BALDWIN, J. I., WHITELEY, S., & BALDWIN, J. D. (1990). Changing AIDS- and fertility-related behavior: The effectiveness of sexual education. *Journal of Sex Research, 27,* 245–262.

BALLARD-REISCH, D., & ELTON, M. (1992). Gender orientation and the Bem sex role inventory: A psychological construct revisited. *Sex Roles, 27,* 291–306.

BALMARY, M. (1982). *Psychoanalyzing psychoanalysis.* Baltimore: Johns Hopkins University Press.

BANCROFT, J. (1984). Hormones and human sexual behavior. *Journal of Sex and Marital Therapy, 10,* 3–21.

BANCROFT, J. (1988). Sexual desire and the brain. *Journal of Sex and Marital Therapy, 3,* 11–27.

BANDURA, A., & WALTERS, R. H. (1963). *Social learning and personality development.* New York: Holt, Rinehart & Winston.

BANK, S. P., & KAHN, M. (1982). *The sibling bond.* New York: Basic Books.

BARBACH, L. G., (1980). *Women discover orgasm.* New York: Free Press.

BARBACH, L. G. (1982). *For each other: Sharing sexual intimacy.* Garden City, NY: Anchor Press/Doubleday.

BARBAREE, H. E., & MARSHALL, W. L. (1991). The role of male sexual arousal in rape: 6 models. *Journal of Consulting and Clinical Psychology, 59,* 621–630.

BARBAREE, H. E., MARSHALL, W. L., & LANTHEIR, R. (1979). Deviant sexual arousal in rapists. *Behavior Research and Therapy, 17,* 215–222.

BARBAREE, H. E., & SETO, M. C. (1997). Pedophilia: Assessment and treatment. In D. R. Laws & W. O'Donohue (Eds.), *Sexual deviance: Theory, assessment, and treatment* (pp. 175–193). New York: Guilford Press.

BARBIERI, R. L., & GORDON, A. M. C. (1991). Hormonal therapy of endometriosis: The estradiol target. *Fertility and Sterility, 56,* 820–822.

BARDONI, B., et al. (1994). A dosage sensitive locus at chromosome XP21 is involved in male to female sex reversal. *Nature Genetics, 7,* 497–501.

BARNARD, M. A., MCKEGANEY, N. P., & LEYLAND, A. H. (1993). Risk behaviors among male clients of female prostitutes. *British Medical Journal, 307,* 361–362.

BARNETT, M. A., et al. (1992). Factors affecting reactions to a rape victim. *Journal of Psychology, 126,* 609–620.

BARON, L., STRAUS, M., & JAFFEE, D. (1987). Legitimate violence, violent attitudes, and rape: A test of the cultural spillover theory. *Annals of the New York Academy of Sciences, SR121,* 1–23.

BARON, M. (1993). Genetics and human sexual orientation. *Biological Psychiatry, 33,* 759–761.

BARRATT, C. L. R., CHAUHAN, M., & COOKE, J. D. (1990). Donor insemination—A look to the future. *Fertility and Sterility, 54,* 375–387.

BARRETT, R. L., & ROBINSON, B. E. (1990). *Gay fathers.* Lexington, MA: Lexington Books.

BARROW, G., & SMITH, T. (1992). *Aging, ageism, and society.* St. Paul, MI: West.

BARRY, H., III, BACON, M. K., & CHILD, I. L. (1957). A cross-cultural survey of some sex differences in socialization. *Journal of Abnormal and Social Psychology, 55,* 327–332.

BARRY, H., III, & SCHLEGEL, A. (1984). Measurements of adolescent sexual behavior in the standard sample of societies. *Ethnology, 23,* 315–329.

BARRY, K. (1995). *The prostitution of sexuality: The global exploitation of women.* New York: New York University Press.

BARSON, M. (1984, March). Penis size: A sexual or political issue? *Cosmopolitan,* pp. 224–226.

BART, P. (1980). *Avoiding rape: A study of victims and avoiders. Final Report.* Rockville, MD: National Institute of Mental Health. (MH29311)

BART, P. B. (1971). Depression in middle-aged women. In V. G. Gornick & B. K. Moran (Eds.), *Women in sexist society.* New York: Basic Books.

BARTELL, G. D. (1970). Group sex among the mid-Americans. *Journal of Sex Research, 6,* 113–130.

BARTH, R. J., & KINDER, B. N. (1987). The mislabeling of sexual impulsivity. *Journal of Sex and Marital Therapy, 13,* 15–23.

BARUCH, G. K., & BARNETT, R. (1986). Role quality, multiple role involvement, and psychological well-being in midlife women. *Journal of Personality and Social Psychology, 51,* 578–585.

BATEMAN, D. A., NG, S. K. C., HANSEN, C. A., & HEAGARTY, M. C. (1993). The effects of intrauterine cocaine exposure in newborns. *American Journal of Public Health, 83,* 190–193.

BAUER, H. M., et al. (1991). Human papillomavirus infection in female university students as determined by a PCR-based method. *Journal of the American Medical Association, 265,* 472–477.

BAULIEU, E.-E. (1989). Ru-486 as an antiprogesterone steroid. *Journal of the American Medical Association, 262,* 1808–1814.

BAUMEISTER, R. F., & BUTLER, J. L. (1997). Sexual masochism: Deviance without pathology. In D. R. Laws & W. O. O'Donohue (Eds.), *Sexual deviance: Theory, assessment, and treatment* (pp. 225–239). New York: Guilford Press.

BAUSERMAN, R., & DAVIS, C. (1996). Perceptions of early sexual experiences and adult sexual adjustment. *Journal of Psychology and Human Sexuality, 8*(3), 37–59.

BAUSERMAN, R., & RIND, B. (1997). Psychological correlates of male child and adolescent sexual experiences with adults: A review of the nonclinical literature. *Archives of Sexual Behavior, 26,* 105–141.

BAYS, J., & CHADWICK, D. (1993). Medical diagnosis of the sexually abused child. *Child Abuse and Neglect, 17,* 91–110.

BECKER, J. V., et al. (1984). Sexual problems of sexual assault survivors. *Women and Health, 9,* 5–20.

BECKER, J. V., & ABEL, G. G. (1985). Methodological and ethical issues in evaluating and treating adolescent sex offenders. In E. M. Otey & G. D. Ryan (Eds.), *Adolescent sex offenders: Issues in research and treatment* (pp. 109–129). Rockville, MD: USDHHS.

BECKER, J. V., & COLEMAN, E. M. (1988). Incest. In V. B. Van Hasselt, R. L. Morrison, A. S. Bellack, & M. Hersen (Eds.), Handbook *of family violence.* New York: Plenum.

BEER, A. E. (1989). Immunology, contraception, and preeclampsia. *Journal of the American Medical Association, 262,* 3184.

BEERNINK, F. J., DMOWSKI, W. P., & ERICSSON, R. J. (1993). Sex preselection through albumin separation of sperm. *Fertility and Sterility, 59,* 382–386.

BEGLEY, S., & GLICK, D. (1994, March 21). The estrogen complex. *Newsweek,* pp. 76–77.

BEHRMAN, R. E., & VAUGHAN, V. C., III. (1983). *Pediatrics.* Philadelphia: Saunders.

BEITCHMAN, J. H., et al. (1991). A review of the short-term effects of child sexual abuse. *Child Abuse and Neglect, 15,* 537–556.

BEITCHMAN, J. H., et al. (1992). A review of the long-term effects of child sexual abuse. *Child Abuse and Neglect, 16,* 101–118.

BELCASTRO, P. (1985). Sexual behavior differences between black and white students. *Journal of Sex Research, 21,* 56–67.

BELL, A. P., & WEINBERG, M. S. (1978). *Homosexualities.* New York: Simon & Schuster.

BELL, A. P., WEINBERG, M. S., & HAMMERSMITH, S. K. (1981). *Sexual preference: Its development in men and women.* Bloomington: Indiana University Press.

BELL, L. (Ed.). (1987). *Good girls/bad girls: Sex trade workers confront feminists.* Seattle, WA: Seal Press.

BELL, T. A., et al. (1992). Chronic chlamydia trachomatis infections in infants. *Journal of the American Medical Association, 267,* 400–402.

BELZER, G. (1981). Orgasmic expulsions of women: A review and heuristic inquiry. *Journal of Sex Research, 17,* 1–12.

BEM, D. J. (1996). Exotic becomes erotic: A developmental theory of sexual orientation. *Psychological Review, 103,* 320–335.

BEM, S. L. (1974). The measurement of psychological androgyny. *Journal of Consulting and Clinical Psychology, 42,* 155–162.

BEM, S. L. (1979). Theory and measurement of androgyny: A reply to Pedhazur-Tetenbaum and Locksley-Colten critiques. *Journal of Personality and Social Psychology, 37,* 1047–1054.

BEM, S. L. (1981). Gender schema theory: A cognitive account of sex typing. *Psychological Review, 88,* 354–364.

BEM, S. L. (1983). Gender schema theory and its implications for child development: Raising gender-aschematic children in a gender-schematic society. *Signs, 8,* 598–616.

BENSON, R. (1985). Vacuum cleaner injury to penis: A common urologic problem? *Urology, 25,* 41–44.

BENTLER, P. M., & PEELER, W. H. (1979). Models of female orgasm. *Archives of Sexual Behavior, 8,* 405–424.

BERDAHL, J. L., MAGLEY, V. J., & WALDO, C. R. (1996). The sexual harassment of men? *Psychology of Women Quarterly, 20,* 527–547.

BERENBAUM, S. A., & HINES, M. (1992). Early androgens are related to childhood sex-typed toy preferences. *Psychological Science, 3,* 203–206.

BERENBAUM, S. A., & SNYDER, E. (1995). Early hormonal influences on childhood sex-typed activity and playmate preferences: Implications for the development of sexual orientation. *Developmental Psychology, 31,* 31–42.

BERENSON, A. B. (1993). Appearance of the hymen at birth and one year of age: A longitudinal study. *Pediatrics, 91,* 820–825.

BERENSON, A. B., HEGER, A., & ANDREW, S. (1991). Appearance of the hymen in newborns. *Pediatrics, 87,* 458–465.

BERGER, G. S., WESTROM, L. V., & WOLNER-HANSSEN, P. (1992). Definition of pelvic inflammatory disease. In G. S. Berger & L. V. Weström (Eds.), *Pelvic inflammatory disease.* New York: Raven Press.

BERGER, R. M. (1982). The unseen minority: Older gays and lesbians. *Social Work, 27,* 236–242.

BERGLER, G., & KROGER, W. S. (1954). *Kinsey's myth of female sexuality.* New York: Grune & Stratton.

BERGSTROM, R., et al. (1993). Detection of preinvasive cancer of the cervix and the subsequent reduction in invasive cancer. *Journal of the National Cancer Institute, 85,* 1050–1057.

BERNARD, M. L., & BERNARD, J. L. (1983). Violent intimacy: The family as a model for love relationships. *Family Relations, 32,* 282–286.

BERNSTEIN, A. C. (1976, January). How children learn about sex and birth. *Psychology Today, 9,* 31–35, 66.

BERSCHEID, E. (1982). Interpersonal attraction. In E. Aronson & G. Lindzey (Eds.), *Handbook of social psychology* (3rd ed.). Reading, MA: Addison-Wesley.

BERSCHEID, E., & WALSTER, E. (1974). A little bit about love. In T. L. Huston (Ed.), *Foundations of interpersonal attraction* (pp. 355–381). New York: Academic Press.

BERSCHEID, E., & WALSTER, E. (1978). *Interpersonal attraction* (2nd ed.). Reading, MA: Addison-Wesley.

BERTOLLI, J., et al. (1996). Estimating the timing of mother-to-child transmission of human immunodeficiency virus in a breast-feeding population in Kinshasa, Zaire. *Journal of Infectious Diseases, 174,* 722–726.

BERTOZZI, S. M. (1996). The impact of human immunodeficiency virus/AIDS. *Journal of Infectious Diseases, 174* (Suppl. 2), S253–S257.

BESS, B. E., & JANUS, S. S. (1976). Prostitution. In B. Sadock et al. (Eds.), *The sexual experience.* Baltimore: Williams & Wilkins.

BEUTNER, K. R. (1996). Genital herpes. In P. Elsner & A. Eichmann (Eds.), *Sexually transmitted diseases. Advances in diagnosis and treatment* (pp. 132–139). Basel: Karger.

BEYNE, Y. (1989). *Menarche to menopause: Reproductive lives of peasant women in two cultures.* Albany, NY: State University of New York Press.

BIANCHI, S. M., & SPAIN, D. (1996). Women, work, and family in America. *Population Bulletin, 51*(3), 1–48.

BIEBER, I. (1965). Clinical aspects of male inversion. In J. Marmor (Ed.), *Sexual inversion* (pp. 248–267). New York: Basic Books.

BIEBER, I., et al. (1962). *Homosexuality: A psychoanalytic study.* New York: Basic Books.

BIERNAT, M. (1991). Gender stereotypes and the relationship between masculinity and femininity: A developmental analysis. *Journal of Personality and Social Psychology, 61,* 351–365.

BILLER, H. (1969). Father absence, maternal encouragement, and sex role development in kindergarten-age boys. *Child Development, 40,* 539–546.

BILLER, H., & BAHM, R. (1971). Father absence, perceived maternal behavior, and masculinity of self-concept among junior high school boys. *Developmental Psychology, 4,* 178–181.

BILLINGS, E. L., BILLINGS, J. J., & CATARINCH, M. (1974). *Atlas of the ovulation method.* Collegeville, MN: Liturgical Press.

BILLY, J. O. G., TANFER, K., GRADY, W. R., & KLEPINGER, D. H. (1993). The sexual behavior of men in the United States. *Family Planning Perspectives, 25,* 52–60.

BING, E. (1990). Lamaze childbirth: Then and now. In S. J. Alt (Ed.), *Lamaze parents* (pp. 14–16). VA: ASPO/Lamaze.

BIRD, K. D. (1991). The use of spermicide containing nonoxynol-9 in the prevention of HIV infection. *AIDS, 5,* 791–796.

BIRNS, B., & STERNGLANZ, S. H. (1983). Sex-role socialization: Looking back and looking ahead. In M. B. Liss (Ed.), *Social and cognitive skills: Sex roles and children's play.* New York: Academic Press.

BITTLES, A. H., MASON, W. M., GREENE, J., & RAO, N. A. (1991). Reproductive behavior and health in consanguineous marriages. *Science, 252,* 789–794.

BLACKMORE, C. A., KEEGAN, R. A., & CATES, W., JR., (1982). Diagnosis and treatment of sexually transmitted diseases in rape victims. *Review of Infectious Diseases, 4*(Suppl), S877–S882.

BLAIR, C. D., & LANYON, R. I. (1981). Exhibitionism: Etiology and treatment. *Psychological Bulletin, 89,* 439–463.

BLANCHARD, R. (1993). The she-male phenomenon and the concept of partial autogynephilia. *Journal of Sex and Marital Therapy, 19,* 69–76.

BLANCHARD, R., & BOGAERT, A. F. (1996a). Biodemographic comparisons of homosexual and heterosexual men in the Kinsey interview data. *Archives of Sexual Behavior, 25,* 551–579.

BLANCHARD, R., & BOGAERT, A. F. (1996b). Homosexuality in men and number of older brothers. *American Journal of Psychiatry, 153,* 27–31.

BLANCHARD, R., & COLLINS, R. I. (1993). Men with sexual interest in transvestites, transsexuals, and she-males. *Journal of Nervous and Mental Disease, 181,* 570–575.

BLANCHARD, R., et al. (1989). Prediction of regrets in postoperative transsexuals. *Canadian Journal of Psychiatry, 34,* 43–45.

BLANCHARD, R., et al. (1996). Birth order and sibling sex ratio in two samples of Dutch gender-dysphoric homosexual males. *Archives of Sexual Behavior, 25,* 495–514.

BLANCHARD-FIELDS, F., & SUHRER-ROUSSEL, L. (1992). Adaptive coping and social cognitive development of women. In E. E. Guice (Ed.), *Women and aging: Now and the future?* Westport, CT: Greenwood Press.

BLOCK, J. (1976). Issues, problems, and pitfalls in assessing sex differences. *Merrill-Palmer Quarterly, 22,* 283–308.

BLOOD, R. O., JR. (1967). *Love match and arranged marriage.* New York: Free Press.

BLOOMFIELD, H., & VETTESE S. (1989). *Lifemates: The love fitness program for a lasting relationship.* New York: New American Library.

BLUMBERG, M. L., & LESTER, D. (1991). High school and college students' attitudes toward rape. *Adolescence, 26,* 727–729.

BLUME, J. (1970). *Are you there God? It's me, Margaret.* Englewood Cliffs, NJ: Bradbury Press.

BLUMSTEIN, P. W., & SCHWARTZ, P. (1983). *American couples.* New York: William Morrow.

BLYTH, B., & DUCKETT, J. W. (1991). Gonadal differentiation: A review of physiological process and influencing factors based on recent experimental evidence. *Journal of Urology, 145,* 689–694.

BODSWORTH, N. J., et al. (1997). Valaciclovir versus aciclovir in patient initiated treatment of recurrent genital herpes: A randomized, double blind clinical trial. *Genitourinary Medicine, 73,* 110–116.

BOGAERT, A. F. (1996). Volunteer bias in human sexuality research: Evidence for both sexuality and personality differences in males. *Archives of Sexual Behavior, 25,* 125–140.

BOGAERT, A. F. (1997). Birth order and sexual orientation in women. *Behavioral Neuroscience, 111,* 1395–1397.

BOGREN, L. (1991). Changes in sexuality in women and men during pregnancy. *Archives of Sexual Behavior, 20,* 35–46.

BOHLEN, J. G., HELD, J. P., & SANDERSON, M. D. (1980). The male orgasm: Pelvic contractions measured by anal probe. *Archives of Sexual Behavior, 9,* 503–521.

BOHLEN, J. G., HELD, J. P., SANDERSON, M. O., & AHLGREN, A. (1982). The female orgasm: Pelvic contractions. *Archives of Sexual Behavior, 11,* 367–386.

BOKLAGE, C. E. (1990). Survival probability of human conceptions from fertilization to term. *International Journal of Fertility, 35,* 75–94.

BOLES, J., & ELIFSON, K. W. (1994). Sexual identity and HIV: The male prostitute. *Journal of Sex Research, 31,* 39–46.

BOLLEN, N., et al. (1991). The incidence of multiple pregnancy after in vitro fertilization and embryo transfer, gamete, or zygote intrafallopian transfer. *Fertility and Sterility, 55,* 314–318.

BOLOGNESI, D. P. (1996). Overview of HIV vaccine development. In G. Giraldo et al. (Eds.), *Development and applications of vaccines and gene therapy in AIDS* (pp. 63–67). Basel: Karger.

BONGAARTS, J., REINING, P., WAY, P., & CONANT, F. (1989). The relationship between male circumcision and HIV infection in African populations. *AIDS, 3,* 373–377.

BOOTH, A., & JOHNSON, D. (1988). Premarital cohabitation and marital success. *Journal of Family Issues, 9,* 255–272.

BOSTON WOMEN'S HEALTH BOOK COLLECTIVE. (1984). *The new Our bodies, ourselves.* New York: Simon & Schuster.

BOSWELL, J. (1980). *Christianity, social tolerance, and homosexuality.* Chicago: University of Chicago Press.

BOTTA, R. A., & PINGREE, S. (1997). Interpersonal communication and rape: Women acknowledge their assaults. *Journal of Health Communication, 2,* 197–212.

BOTWIN, M. D., BUSS, D. M., & SHACKELFORD, T. K. (1997). Personality and mate preferences: Five factors in mate selection and marital satisfaction. *Journal of Personality, 65,* 107–136.

BOUMAN, F. G. (1988). Sex reassignment surgery in male to female transsexuals. *Annals of Plastic Surgery, 21,* 526–531.

BOXALL, B. (1995, September 3). Young gays stray from safe sex, new data shows. *Los Angeles Times,* pp. A1, A24.

BOZETT, F. W. (Ed.). (1987). *Gay and lesbian parents.* New York: Praeger.

BRABANT, S., & MOONEY, L. A. (1997). Sex role stereotyping in the Sunday comics: A twenty year update. *Sex Roles, 37,* 269–281.

BRACKBILL, Y. (1979). Obstetrical medication and infant behavior. In J. D. Osofsky (Ed.), *Handbook of infant development.* New York: Wiley (Interscience).

BRADFORD, J. M. W., BOULET, J., & PAWLAK, A. (1992). The paraphelias: A multiplicity of deviant behaviors. *Canadian Journal of Psychiatry, 37,* 104–107.

BRADFORD, J. M. W., & GREENBERG, D. M. (1997). Pharmacological treatment of deviant sexual behavior. In R. C. Rosen, C. M. Davis, and H. J. Ruppel, Jr. (Eds.), *Annual review of sex research*, vol. vii (pp. 283–306). Mount Vernon, Iowa: Society for the Scientific study of Sexuality.

BRADLEY, R. A. (1981). *Husband coached childbirth*. New York: Harper & Row.

BRADLEY, S. J., & ZUCKER, K. J. (1997). Gender identity disorder: A review of the past 10 years. *Journal of the American Academy of Child and Adolescent Psychology, 36*, 872–880.

BRADLEY, S. J., et al. (1991). Interim report of the DSM-IV Subcommittee on Gender Identity Disorders. *Archives of Sexual Behavior, 20*, 333–343.

BRANCH, D. W. (1990). Autoimmunity and pregnancy loss. *Journal of the American Medical Association, 264*, 1453–1454.

BRANDT, E. N., JR. (1982). Physicians and sexually transmitted disease: A call to action. *Journal of the American Medical Association, 248*, 2032.

BRATT, C. S. (1984). Incest statutes and the fundamental right of marriage: Is Oedipus free to marry? *Family Law Quarterly, 18*, 257–309.

BRAVERMAN, P., & STRASBURGER, V. (1994, January). Sexually transmitted diseases. *Clinical Pediatrics*, 26–37.

BRAZELTON, T. B. (1969). *Infants and mothers*. New York: Delacorte Press/Seymour Lawrence.

BRAZELTON, T. B. (1970). Effect of prenatal drugs on the behavior of the neonate. *American Journal of Psychiatry, 126*, 1261–1266.

BRECHER, E. M. (1984). *Love, sex, and aging*. Mount Vernon, NY: Consumers Union.

BREEDLOVE, S. M. (1994). Sexual differentiation of the human nervous system. *Annual Review of Psychology, 45*, 389–418.

BREEDLOVE, S. M. (1997). Sex on the brain. *Nature, 389*, 801.

BREMNER, W. J., VITIELLO, M. V., & PRINZ, P. N. (1983). Loss of circadian rhythmicity in blood testosterone levels with aging in normal men. *Journal of Clinical Endocrinology and Metabolism, 56*, 1278–1281.

BRESLOW, N., EVANS, L., & LANGLEY, J. (1985). On the prevalence and roles of females in the sadomasochistic subculture. *Archives of Sexual Behavior, 14*, 303–317.

BRESLOW, N., EVANS, L., & LANGLEY, J. (1986). Comparisons among heterosexual, bisexual, and homosexual sadomasochists. *Journal of Homosexuality, 13*, 83–107.

BRETSCHNEIDER, J. G., & McCOY, N. L. (1988). Sexual interest and behavior in healthy 80- to 102-year-olds. *Archives of Sexual Behavior, 17*, 109–129.

BREWAEYS, A., & VAN HALL, E. V. (1997). Lesbian motherhood: The impact on child development and family functioning. *Journal of Psychosomatic Obstetrics and Gynecology, 18*, 1–16.

BREWER, S. S., & BREWER, T. (1985). *What every pregnant woman should know: The truth about diet and drugs in pregnancy*. New York: Penguin Press.

BRIDGES, J. S., & McGRAIL, C. A. (1989). Attributions of responsibility for date and stranger rape. *Sex Roles, 21*, 273–286.

BRIGMAN, B., & KNOX, D. (1992). University students' motivation to have intercourse. *College Student Journal, 26*, 406–408.

BRIND, J., et al. (1996). Induced abortion as an independent risk factor for breast cancer: A comprehensive review and meta-analysis. *Journal of Epidemiology and Community Health, 50*, 481–486.

BRINGLE, R. G., & BUUNK, B. P. (1985). Jealousy and social behavior: A review of personal, relationship and situational determinants. In P. Shaver (Ed.), *Review of personality and social psychology*, Vol. 2 (pp. 241–264). Beverly Hills, CA: Sage.

BRINGLE, R. G., ROACH, S., ADLER, C., & EVENBECK, S. (1977). *Correlates of jealousy*. Paper presented at the annual meeting of the Midwestern Psychological Association, Chicago.

BRINTON, L. A., & BROWN, S. L. (1997). Breast implants and cancer. *Journal of the National Cancer Institute, 89*, 1341–1349.

BRISTOW, R. E., & KARLAN, B. Y. (1996). Ovulation induction, infertility, and ovarian cancer risk. *Fertility and Sterility, 66*, 499–507.

BRITTON, D. M. (1990). Homophobia and homo sociality: An analysis of boundary maintenance. *Sociological Quarterly, 31*, 423–439.

BRODSKY, S. L. (1976). Prevention of rape: Deterrence by the potential victim. In M. J. Walker & S. L. Brodsky (Eds.), *Sexual assault*. Lexington MA: D. C. Heath.

BRODY, S. (1995). Lack of evidence for transmission of human immunodeficiency virus through vaginal intercourse. *Archives of Sexual Behavior, 24*, 383–393.

BROMBERGER, J. T., et al. (1997). Prospective study of the determinants of age at menopause. *American Journal of Epidemiology, 145*, 124–133.

BROMHAM, D. R. (1993). Intrauterine contraceptive devices—A reappraisal. *British Medical Bulletin, 49*, 100–123.

BROMWICH, P., COHEN, J., STEWART, I., & WALKER, A. (1994). Decline in sperm counts: An artifact of changed reference range of "normal"? *British Medical Journal, 309*, 19–22.

BROOKS, J., RUBLE, D. N., & CLARKE, A. E. (1977). College women's attitudes and expectations concerning menstrual-related changes. *Psychosomatic Medicine, 39*, 288–298.

BROSIUS, H.-B., WEAVER, J. B., III, & STAAB, J. F. (1993). Exploring the social and sexual "reality" of contemporary pornography. *Journal of Sex Research, 30*, 161–170.

BROUDE, G. J., & GREENE, S. J. (1976). Cross-cultural codes on twenty sexual attitudes and practices. *Ethnology, 15*, 409–429.

BROVERMAN, I. K., et al. (1972). Sex-role stereotypes: A current appraisal. *Journal of Social Issues, 28*, 59–78.

BROWN, D. (1990). The penis pin. In V. Sutlive (Ed.), *Female and male in Borneo: Contributions and challenges to gender studies* (pp. 435–454). Borneo: Borneo Research Council.

BROWN, G. R., & COLLIER, L. (1989). Transvestites' women revisited: A nonpatient sample. *Archives of Sexual Behavior, 18*, 73–83.

BROWN, G. R. (1994). Women in relationships with cross-dressing men: A descriptive study from a nonclincial setting. *Archives of Sexual Behavior, 23*, 515–530.

BROWN, J. D., CHILDERS, K. W., & WASZAK, C. (1990). Television and adolescent sexuality. *Journal of Adolescent Health Care, 11*, 62–70.

BROWN, J. D., & NEWCOMER, S. F. (1991). Television viewing and adolescents' sexual behavior. *Journal of Homosexuality, 21*, 77–91.

BROWN, J. D., & STEELE, J. R. (1996). Sexuality and the mass media: An overview. *SIECUS Report, 24*(4), 3–9.

BROWN, Z. A., et al. (1997). The acquisition of herpes simplex virus during pregnancy. *New England Journal of Medicine, 337*, 509–515.

BROWNE, J., & MINICHIELLO, V. (1996). Condoms: Dilemmas of caring and autonomy in heterosexual safe sex practices. *Venereology, 9*, 24–33.

BROWNMILLER, S. (1975). *Against our will: Men, women, and rape*. New York: Simon & Schuster.

BROWNMILLER, S. (1993, January 4). Making female bodies the battlefield. *Newsweek*, p. 37.

BRUNHAM, R. C. et al. (1988). Etiology and outcome of acute pelvic inflammatory disease. *Journal of Infectious Diseases, 158*, 510–517.

BRYANT, J., & BROWN, D. (1989). Uses of pornography. In Z. Dolf & B. Jennings (Eds.), *Pornography: Research advances and policy considerations*. Hillsdale, NJ: Erlbaum.

BRYNER, C. (1989). Recurrent toxic shock syndrome. *American Family Physician, 39*, 157–164.

BRYSON, J., & SHETTEL-NEUBER, J. (1978, February). Unbalanced relationships: Who becomes jealous of whom. Cited by James Hasset in *Psychology Today, 11*, 26–29.

BUCHBINDER, S. P., et al. (1994). Long-term HIV-1 infection without immunologic progression. *AIDS, 8*, 1123–1128.

BUCK, G. M., SEVER, L. E., BATT, R. E., & MENDOLA, P. (1997). Life-style factors and female infertility. *Epidemiology, 8*, 435–441.

BUCKLEY, W. E., et al. (1988). Estimated prevalence of anabolic steroid use among male high school seniors. *Journal of the American Medical Association, 260*, 3441–3445.

BUFFUM, J., et al. (1981). Drugs and sexual function. In H. Lief (Ed.), *Sexual problems in medical practice*. Monroe, WI: American Medical Association.

BUHRICH, N. (1976). A heterosexual transvestite club: Psychiatric aspects. *Australian and New Zealand Journal of Psychiatry, 10*, 331–335.

BULLOUGH, B., & BULLOUGH, V. L. (1997a). Are transvestites necessarily heterosexual? *Archives of Sexual Behavior, 26*, 1–12..

BULLOUGH, B., & BULLOUGH, V. L. (1997b). Female prostitution: Current research and changing interpretations. In R. C. Rosen, C. M. Davis, and H. J. Ruppel, Jr. (Eds.), *Annual review of sex research*, Vol. VII (pp.158–180). Mount Vernon, Iowa: Society for the Scientific study of Sexuality.

BULLOUGH, V. L. (1981). Age at menarche: A misunderstanding. *Science, 213*, 365–366.

BULLOUGH, V. L. (1991). Transvestism: A reexamination. *Journal of Psychology and Human Sexuality, 4*, 53–67.

BULLOUGH, V. L., & BULLOUGH, B. (1993). *Cross dressing, sex, and gender*. Philadelphia: University of Pennsylvania Press.

BULLOUGH, V., BULLOUGH, B., & SMITH, R. (1983). Comparative study of male transvestites, male to female transsexuals, and male homosexuals. *Journal of Sex Research, 19*, 238–257.

BUMPASS, L. L., & SWEET, J. A. (1989). Childrens' experience in single-parent families: Implications of cohabitation and marital transitions. *Family Planning Perspectives, 21*, 256–260.

BUMPASS, L. L., SWEET, J. A., & CHERLIN, A. (1991). The role of cohabitation in declining rates of marriage. *Journal of Marriage and the Family, 53*, 913–927.

BURGER, M. P. M., et al. (1993). Cigarette smoking and human papillomavirus in patients with reported cervical cytological abnormality. *British Medical Journal, 306*, 749–755.

BURGESS, A. W., HARMAN, C., McCAUSLAND, M., & POWERS, P. (1984). Response patterns in children and adolescents exploited through sex rings and pornography. *American Journal of Psychiatry, 141*, 656–662.

BURGESS, A. W., & HOLMSTROM, L. L. (1974). Rape trauma syndrome. *American Journal of Psychiatry, 131*, 981–986.

BURGESS, A. W., & HOLMSTROM, L. L. (1979). Rape: Sexual disruption and recovery. *American Journal of Orthopsychiatry, 49*, 648–657.

BURGESS, D., & BORGIDA, E. (1997). Sexual harassment: An experimental test of sex-role spillover theory. *Personality and Social Psychology Bulletin, 23*, 63–75.

BURK, R. D., et al. (1996). Sexual behavior and partner characteristics are the prominent risk factors for genital human papillomavirus infection in young women. *Journal of Infectious Diseases, 174*, 679–689.

BURNAM, M. A., et al. (1988). Sexual assault and mental disorders in a community population. *Journal of Consulting and Clinical Psychology, 56*, 843–850.

BURT, M. R. (1980). Cultural myths and supports for rape. *Journal of Personality and Social Psychology, 38*, 217–230.

BURT, M. R., & KATZ, B. L. (1988). Coping strategies and recovery from rape. *Annals of the New York Academy of Sciences, 528*, 345–358.

BURTON, R. (1963). The anatomy of melancholy. In A. M. Witherspoon & F. Warnke (Eds.), *Seventeenth-century prose and poetry* (pp. 132–133). New York: Harcourt Brace Jovanovich. (Original work published 1651.)

BUSS, A. (1966). *Psychopathology*. New York: Wiley.

BUSS, D. M. (1985). Human mate selection. *American Scientist, 73*, 47–51.

BUSS, D. M. (1989). Sex differences in human mate selection: Evolutionary hypotheses tested in 37 cultures. *Behavioral and Brain Sciences, 12*, 1–49.

BUSS, D. M. (1994, August). *Evolution of desire: Person, environment, interaction and human mating.* Paper presented at the annual convention of the American Psychological Association, Los Angeles.

BUSS, D. M., & BARNES, M. F. (1987). Preferences in human mate selection. *Journal of Personality and Social Psychology, 50*, 559–570.

BUSS, D. M., LARSEN, R. J., WESTEN, D., & SEMMELROTH, J. (1992). Sex differences in jealousy: Evolution, physiology, and psychology. *Psychological Science, 3*, 251–255.

BUSS, D. M., & SCHMITT, D. P. (1993). Sexual strategies theory: An evolutionary perspective on human mating. *Psychological Review, 100*, 204–232.

BUTLER, C. A. (1976). New data about female sexual response. *Journal of Sex and Marital Therapy, 2*, 40–46.

BUUNK, B., & BRINGLE, R. (1987). Jealousy in love relationships. In D. Perlman & S. Duck (Eds.), *Intimate Relationships*. Newbury Park, CA: Sage.

BUUNK, B. P., ANGLEITNER, A., OUBAID, U., & BUSS, D. M. (1996). Sex differences in jealousy in evolutionary and cultural perspective: Tests from the Netherlands, Germany, and the United States. *Psychological Science, 7*, 359–363.

BUVAT, J., et al. (1990). Recent developments in the clinical assessment and diagnosis of erectile dysfunction. In J. Bancroft (Ed.), *Annual review of sex research* (Vol. 2). (pp. 265–308). Mount Vernon, IA: Society for the Scientific Study of Sex.

BYARD, R. W., HUCKER, S. J., & HAZELWOOD, R. R. (1990). A comparison of typical death scene features in cases of fatal male and female autoerotic ashyxia with a review of the literature. *Forensic Science International, 48*, 113–121.

BYERS, E. S. (1996). How well does the traditional sexual script explain sexual coercion? Review of a program of research. *Journal of Psychology & Human Sexuality, 8*, 7–25.

BYNE, W., & PARSONS, B. (1993). Human sexual orientation: The biologic theories reappraised. *Archives of General Psychiatry, 50*, 228–239.

BYNE, W., & STEIN, E. (1997). Ethical implications of scientific research on the causes of sexual orientation. *Health Care Analysis, 5*, 136–148.

BYRNE, D., CLORE, G., & SMEATON, G. (1986). The attraction hypothesis: Do similar attitudes affect anything? *Journal of Personality and Social Psychology, 51*, 1167–1170.

BYRNE, D., & MURNEN, S. K. (1988). Maintaining loving relationships. In R. J. Sternberg & M. L. Barnes (Eds.), *The psychology of love*. New Haven, CT: Yale University Press.

CACERES, C. F., & VAN GRIENSVEN, G. J. P. (1994). Male homosexual transmission of HIV-1. *AIDS, 8*, 1051–1061.

CADO, S., & LEITENBERG, H. (1990). Guilt reactions to sexual fantasies during intercourse. *Archives of Sexual Behavior, 19*, 49–64.

CAGEN, R. (1986). The cervical cap as a barrier contraceptive. *Contraception, 33*, 487–496.

CALDERONE, M. S. (1983a). Childhood sexuality: Approaching the prevention of sexual disease. In G. Albee et al. (Eds.), *Promoting sexual responsibility and preventing sexual problems*. Hanover, NH: University Press of New England.

CALDERONE, M. S. (1983b). Fetal erection and its message to us. *SIECUS Report, 11* (5/6), 9–10.

CALDWELL, J. C., & CALDWELL, P. (1996). The African AIDS epidemic. *Scientific American, 274*, 40–46.

CALE, A. R. J., et al. (1990). Does vasectomy accelerate testicular-tumor? Importance of testicular examinations before and after vasectomy. *British Medical Journal, 300*, 370.

CALHOUN, B. C., & WATSON, P. T. (1991). The cost of maternal cocaine abuse: I. Perinatal cost. *Obstetrics and Gynecology, 78*, 731–734.

CALL, V., SPRECHER, S., & SCHWARTZ, P. (1995). The incidence and frequency of marital sex in a national sample. *Journal of Marriage and the Family, 57*, 639–652.

CALLE, E., MIRACLE-McMAHILL, H. L., THUN, M. J., & HEATH, C. W. (1994). Cigarette smoking and risk of fatal breast cancer. *American Journal of Epidemiology, 139*, 1001–1007.

CALLENDER, C., & KOCHEMS, L. (1987). The North American Berdache. *Current Anthropology, 24*, 443–456.

CAMPBELL, S. B., & COHN, J. F. (1997). The timing and chronicity of postpartum depression: Implications for infant development. In L. Murray and P. Cooper (Eds.), *Postpartum depression and child development*. New York: Guilford Press.

CAMPOSTRINI, S., & McQUEEN, D. V. (1993). Sexual behavior and exposure to HIV infection: Estimates from a general-population risk index. *American Journal of Public Health, 83*, 1139–1143.

CANADIAN PAEDIATRIC SOCIETY, FETUS AND NEWBORN COMMITTEE (1996). Neonatal circumcision revisited. *Canadian Medical Association Journal, 154*, 769–780.

CANAVAN, M. M., MEYER, W. J., III, & HIGGS, D. C. (1992). The female experience of sibling incest. *Journal of Marital and Family Therapy, 18*, 129–142.

CANAVAN, P., & HASKELL, J. (1977). The great American male stereotype. In C. Garney & S. McMahon (Eds.), *Exploring contemporary male/female roles: A facilitator's guide*. La Jolla, CA: University Associates.

CANCIAN, F. M. (1989). Love and the rise of capitalism. In B. J. Risman & P. Schwartz (Eds.), *Gender in intimate relationships: A microstructural approach*. Belmont, CA: Wadsworth.

CAPITANO, J. P., & LERCHE, N. W. (1991). Psychosocial factors and disease progression in simian AIDS: A preliminary report. *AIDS, 5*, 1103–1106.

CAREY, R. F., et al. (1992). Effectiveness of latex condoms as a barrier to human immunodeficiency-sized particles under conditions of simulated use. *Sexually Transmitted Diseases, 19*, 230–234.

CARLSEN, E., GIWERCMAN, A., KEIDING, N., & SKAKKEBACK, N. E. (1992). Evidence for decreasing quality of semen during past 50 years. *British Medical Journal, 305*, 609–613.

CARNES, P. J. (1983). *Out of the shadows: Understanding sexual addiction*. Minneapolis, MN: Compeare Publishers.

CARNES, P. J. (1991). *Don't call it love*. New York: Bantam Books.

CARPENTER, C. C. J., et al. (1997). Antiretroviral therapy for HIV infection in 1997. *Journal of the American Medical Association, 277*, 1962–1969.

CARR, B. R., & ORY, H. (1997). Estrogen and progesterone components of oral contraceptives: Relationship to vascular disease. *Contraception, 55*, 267–272.

CARRIER, J. M. (1980). Homosexual behavior in cross-cultural perspective. In J. Marmor (Ed.), *Homosexual behavior* (pp.100–122).New York: Basic Books.

CARROLL, J., VOLK, K., & HYDE, J. (1985). Differences between males and females in motives for engaging in sexual intercourse. *Archives of Sexual Behavior, 14*, 131–139.

CARTER, S., & SOKOL, J. (1989). *What really happens in bed: A demystification of sex*. New York: M. Evans.

CASSON, P. R., & CARSON, S. A. (1996). Androgen replacement therapy in women: Myths and realities. *International Journal of Fertility, 41*, 412–422.

CASTRO, B. A., et al. (1991). Persistent infection of baboons and rhesus monkeys with different strains of HIV-2. *Virology, 184*, 219–226.

CASTRO, K. G., et al. (1988). Transmission of HIV in Belle Glade, Florida: Lessons for other communities in the United States. *Science, 239*, 193–197.

CATALONA, W., et al. (1991). Measurement of prostate-specific antigen in serum as a screening test for prostate cancer. *New England Journal of Medicine, 324*, 1156–1161.

CATANIA, J. A., et al. (1993). Changes in condom use among Black, Hispanic, and white heterosexuals in San Francisco: The AMEN cohort study. *Journal of Sex Research, 30*, 121–128.

CATANIA, J. A., et al. (1995). Risk factors for HIV and other sexually transmitted diseases and prevention practices among US heterosexual adults: Changes from 1990 to 1992. *American Journal of Public Health, 85*, 1492–1499.

CATES, W., JR., STEWART, F. H., & TRUSSELL, J. (1992). Commentary: The quest for women's prophylactic methods—Hopes vs. science. *American Journal of Public Health, 82*, 1479–1482.

CATES, W., JR., & STONE, K. M. (1992a). Family planning, sexually transmitted diseases and contraceptive choice: A literature update. Part I. *Family Planning Perspectives, 24*(2), 75–84.

CATES, W., JR., & STONE, K. M. (1992b). Family planning, sexually transmitted diseases and contraceptive choice: A literature update. Part II. *Family Planning Perspectives, 24*(3), 122–128.

CATOTTI, D. N., CLARKE, P., & CATOE, K. E. (1993). Herpes revisited: Still a cause of concern. *Sexually Transmitted Diseases, 20*, 77–80.

CATTERALL, R. D. (1974). *A short textbook of venereology.* Philadelphia: Lippincott.

CAUTELA, J., & KEARNEY, A. J. (1986). *The covert conditioning handbook.* New York: Springer.

CELENTANO, D. D., KLASSEN, A. C., WEISMAN, C. S., & ROSENSHEIN, N. B. (1987). The role of contraceptive use in cervical-cancer: The Maryland cervical-cancer case control study. *American Journal of Epidemiology, 126*, 592–604.

CENTER FOR POPULATION OPTIONS. (1992). *Teenage pregnancy and too-early childbearing: Public costs, personal consequences.*

CENTERS FOR DISEASE CONTROL. (1988). Ectopic pregnancies—United States, 1984 and 1985. *Morbidity and Mortality Weekly Reports, 37*, 637–639.

CENTERS FOR DISEASE CONTROL AND PREVENTION. (1992). Revised classification system for HIV infection and expanded surveillance case definitions for AIDS among adolescents and adults. *Morbidity and Mortality Weekly, 41*, 1–19.

CENTERS FOR DISEASE CONTROL AND PREVENTION. DIVISION OF STD/HIV PREVENTION. (1994). Sexually Transmitted Disease Surveillance, 1993. CDC.

CENTERS FOR DISEASE CONTROL AND PREVENTION. (1996). Mortality attributable to HIV infection among persons aged 25–44 years—United States, 1994. *Morbidity and Mortality Weekly Report, 45,* 121–125.

CENTERS FOR DISEASE CONTROL AND PREVENTION (1997a). Alcohol consumption among pregnant and childbearing-aged women—United States, 1991 and 1995. *Morbidity and Mortality Weekly Report, 46* (16), 346–350.

CENTERS FOR DISEASE CONTROL AND PREVENTION (1997b). Rubella and congenital rubella syndrome—United States, 1994–1997. *Morbidity and Mortality Weekly Report, 46* (16), 350–354.

Cervical cap: A birth control method with several advantages. (1989). *Contraceptive Technology Update, 10,* 2.

CHANG, M.-H., et al. (1997). Universal hepatitis B vaccination in Taiwan and the incidence of hepatocellular carcinoma in children. *New England Journal of Medicine, 336*, 1855–1859.

CHAPPELL, D. (1976). Cross-cultural research on forcible rape. *International Journal of Criminology and Penology, 4*, 295–304.

CHASEN-TABER, L., et al. (1997). Oral contraceptives and ovulatory causes of delayed fertility. *American Journal of Epidemiology, 146*, 258–265.

CHASNOFF, I. J., LANDRESS, H., & BARRETT, M. (1990). The prevalence of illicit drug or alcohol use during pregnancy and discrepancies in mandatory reporting in Pinellas County, Florida. *New England Journal of Medicine, 322*, 1202–1206.

CHECK, J. (1984, August). *Mass media sexual violence: Content analysis and counteractive measures.* Paper presented at the annual meeting of the American Psychological Association, Toronto.

CHI, I.-C. (1993a). Safety and efficacy issues of progestin-only oral contraceptives—An epidemiologic approach. *Contraception, 47*, 1–21.

CHI, I.-C. (1993b). What we have learned from recent IUD studies: A researcher's perspective. *Contraception, 48*, 81–108.

CHIAPPA, J. A., & FORISH, J. J. (1976). *The VD book.* New York: Holt, Rinehart & Winston.

CHIPOURAS, S., et al. (1979). *Who cares? A handbook on sex education and counseling services for disabled people.* Washington, DC: George Washington University.

CHIRAWATKUL, S., & MANDERSON, L. (1994). Perceptions of menopause in Northeast Thailand: Contested meaning and practice. *Social Science Medicine, 39*, 1545–1554.

CHODOROW, N. (1978). *The reproduction of mothering: Psychoanalysis and the sociology of gender.* Berkeley: University of California Press.

CHOI, K.-H., CATANIA, J. A., & DOLCINI, M. M. (1994). Extramarital sex and HIV risk behavior among US adults: Results from the National AIDS Behavioral Survey. *American Journal of Public Health, 84*, 2003–2007.

CHOI, K.-H., & COATES, T. J. (1994). Prevention of HIV infection. *AIDS, 8*, 1371–1389.

CHOI, K.-H., RICKMAN, R., & CATANIA, J. A. (1994). What heterosexual adults believe about condoms. *New England Journal of Medicine, 331*, 406–407.

CHONG, J. M. (1990). Social assessment of transsexuals who apply for sex reassignment therapy. *Social Work in Health Care, 14*, 87–105.

CHOW, J. M., et al. (1990). The association between chlamydia trachomatis and ectopic pregnancy. *Journal of the American Medical Association, 63*, 3164–3167.

CHRISTENSON, C. V., & GAGNON, J. H. (1965). Sexual behavior in a group of older women. *Journal of Geriatrics, 20*, 351–356.

CHRISTOPHER, F. S., & CATE, R. (1984). Factors involved in premarital decision making. *Journal of Sex Research, 20*, 363–376.

CHRISTOPHER, F. S., & ROOSA, M. W. (1990). An evolution of an adolescent pregnancy prevention program: Is 'just say no' enough? *Family Relations, 39*, 68.

CHU, S. Y., et al. (1992). AIDS in bisexual men in the United States: Epidemiology and transmission to women. *American Journal of Public Health, 82*, 220–224.

CLANTON, G., & SMITH, L. G. (1977, March). The self-inflicted pain of jealousy. *Psychology Today, 10*, 44–47.

CLARK, D. (1987). *The new Loving someone gay.* Berkeley, CA: Celestrial Arts.

CLARK, J. K. (1993). Complications in academia: Sexual harassment and the law. *SIECUS Report, 21* (6), 6–10.

CLARK, J. T., SMITH, E. R., & DAVIDSON, J. M. (1984). Enhancement of sexual motivation in male rats. *Science, 224*, 847–849.

CLARK, R. D., & HATFIELD, E. (1989). Gender differences in receptivity to sexual offers. *Journal of Psychology and Human Sexuality, 2*, 39–55.

CLARKE, E., HATCHER, J., McKEOWN-EYSSEN, G., & LICKRISH, G. (1985). Cervical dysplasia: Association with sexual behavior, smoking, and oral contraceptive use? *American Journal of Obstetrics and Gynecology, 151*, 612–616.

CLARKE, J. (1994). The meaning of menstruation in the elimination of abnormal embryos. *Human Reproduction, 9*, 1204–1207.

CLARKSON, T. B., & ALEXANDER, N. J. (1980). Long-term vasectomy: Effects on the occurrence and extent of atherosclerosis in rhesus monkeys. *Journal of Clinical Investigation, 65*, 15–25.

CLEARY, P. D., et al. (1991). Behavior changes after notification of HIV infection. *American Journal of Public Health, 81*, 1586–1590.

CLEAVER, E. (1968). *Soul on ice.* New York: Dell.

CLEMENTS, M. (1994, August 7). Sex in America today. *Parade Magazine*, 4–6.

CLEMENTS, M. (1996, March 17). Sex after 65. *Parade Magazine*, 4–6.

CNATTINGIUS, S., FORMAN, M. R., BERENDES, H. W., & ISOTALO, L. (1992). Delayed childbearing and risk of adverse perinatal outcome. *Journal of the American Medical Association, 268*, 886–890.

COATES, S. (1992). The etiology of boyhood gender identity disorder: An integrative model. In J. W. Barron, M. N. Eagle, and D. L. Wolitzky (Eds.), *Interface of Psychoanalysis and Psychology* (pp. 245–265). Washington, D.C.: American Psychological Association.

COBLEIGH, M. A., et al. (1994). Estrogen replacement therapy in breast cancer survivors: A time for change. *Journal of the American Medical Association, 272*, 540–545.

COCHRAN, C. C., FRAZIER, P. A., & OLSON, A. M. (1997). Predictors of response to unwanted sexual attention. *Psychology of Women Quarterly, 21*, 207–226.

COCHRAN, S. D., & MAYS, V. M. (1990). Sex, lies, and HIV. *New England Journal of Medicine, 322*, 774–775.

COCORES, J., DACKIS, C., & GOLD, M. (1986). Sexual dysfunction secondary to cocaine abuse. *Journal of Clinical Psychiatry, 47*, 384–385.

COGILL, S., et al. (1986). Impact of maternal postnatal depression on cognitive development of young children. *British Medical Journal, 292*, 1165–1167.

COHEN, L. J., & ROTH, S. (1987). The psychological aftermath of rape: Long-term effects and individual differences in recovery. *Journal of Social and Clinical Psychology, 5*, 525–534.

COHEN, L. L., & SHOTLAND, R. L. (1996). Timing of first sexual intercourse in a relationship: Expectations, experiences, and perceptions of others. *Journal of Sex Research, 33,* 291–299.

COHEN, M. L., GAROFALO, R., BOUCHER, R., & SEGHORN, T. (1971). The psychology of rapists. *Seminars in Psychiatry, 3,* 307–327.

COHEN, M. L., SEGHORN, T., & CALMAS, W. (1969). Sociometric study of the sex offender. *Journal of Abnormal Psychology, 74,* 249–255.

COHEN, S. (1979). The volatile nitrates. *Journal of the American Medical Association, 241,* 2077–2078.

COHN, J. A. (1997). HIV infections—1. *British Medical Journal, 314,* 487–491.

COLAPINTO, J. (1997, December 11). The true story of John/Joan. *Rolling Stone,* 54–73, 92–96.

COLDITZ, G. A., et al. (1995). The use of estrogens and progestins and the risk of breast cancer in postmenopausal women. *New England Journal of Medicine, 332,* 1589–1593.

COLE, C. M., et al. (1997). Comorbidity of gender dysphoria and other psychiatric diagnosis. *Archives of Sexual Behavior, 26,* 13–26.

COLE, D. (1987, July). It might have been: Mourning the unborn. *Psychology Today,* pp. 64–65.

COLE, E. (1982). Sibling incest: The myth of benign sibling incest. *Women and Therapy, 5,* 79–89.

COLE, H. M. (1990). Legal interventions during pregnancy. *Journal of the American Medical Association, 264,* 2663–2670.

COLE, P. M., & PUTNAM, F. W. (1992). Effect of incest on self and social functioning: A developmental psychopathology perspective. *Journal of Consulting and Clinical Psychology, 60,* 174–184.

COLE, R. (1994, March 26). Lawsuit challenges daughter's memory of sexual abuse. (New Orleans) *Times-Picayune,* p. A-2.

COLEMAN, B. C. (1992, November 1). Sex after 60 makes for happier marriages. (New Orleans) *Times-Picayune,* p. E-1.

COLEMAN, E. (1981–1982). Developmental stages of the coming out process. *Journal of Homosexuality, 7,* 31–43.

COLEMAN, E. (1991). Compulsive sexual behavior: New concepts and treatment. *Journal of Psychology and Human Sexuality, 4(2),* 37–52.

COLEMAN, E., & EDWARDS, B, (1986, July). Sexual compulsion vs. sexual addiction: The debate continues. *SIECUS Report,* pp. 7–10.

COLEMAN, M., & GANONG, L. (1991). Remarriage and stepfamily research in the 1980s: Increased interest in an old form. In A. Booth (Ed.), *Contemporary families: Looking forward, looking back.* Minneapolis, MN: National Council on Family Relations.

COLLABORATIVE GROUP ON HORMONAL FACTORS IN BREAST CANCER (1996). Breast cancer and hormonal contraceptives: Collaborative reanalysis of individual data on 53,297 women with and 100,239 women without breast cancer from 54 epidemiological studies. *Lancet, 347,* 1713–1727.

COLLABORATIVE GROUP ON HORMONAL FACTORS IN BREAST CANCER. (1997). Breast cancer and hormone replacement therapy: Collaborative reanalysis of data from 51 epidemological studies of 52,705 women with breast cancer and 108,411 women without breast cancer. *Lancet, 350,* 1047–1059.

COLLAER, M. L., & HINES, M. (1995). Human behavioral sex differences: A role for gonadal hormones during early development? *Psychological Bulletin, 118,* 55–107.

COLLER, S. A., & RESICK, P. A. (1987). Women's attributions of responsibility for date rape: The influence of empathy and sex-role stereotyping. *Violence and Victims, 2,* 115–125.

COLLIER, A. C., et al. (1996). Treatment of human immunodeficiency virus infection with saquinavir, zidovudine, and zaleitabine. *New England Journal of Medicine, 334,* 1011–1017.

COLLINGS, S. J., & PAYNE, M. F. (1991). Attribution of causal and moral responsibility to victims of father-daughter incest: An exploratory examination of five factors. *Child Abuse and Neglect, 15,* 513–521.

COLLINS, N. L., & MILLER, L. C. (1994). Self-disclosure and liking: A meta-analytic review. *Psychological Bulletin, 116,* 457–475.

COMHAIRE, F. H. (1994). Male contraception: Hormonal, mechanical and other. *Human Reproduction, 9,* 586–590.

COMISKEY, K. M. (1989). Relationship workshops: How "good marriages" can be strengthened. *Contemporary Sexuality, 1,* 10.

COMMISSION ON OBSCENITY AND PORNOGRAPHY. (1970). *The report of the Commission on Obscenity and Pornography.* Washington, DC: U.S. Government Printing Office.

CONANT, M. A., et al. (1996). Genital herpes: An integrated approach to management. *Journal of the American Academy of Dermatology, 35,* 601–605.

CONDRY, J., & CONDRY, S. (1977). Sex differences: A study of the eye of the beholder. *Child Development, 47,* 812–819.

CONFINO, E., et al. (1990). Transcervical balloon tuboplasty: A multicenter study. *Journal of the American Medical Association, 264,* 2079–2082.

CONNERY, J. (1977). *Abortion: The development of the Roman Catholic perspective.* Chicago: Loyola University Press.

CONNOR, E. M., et al. (1994). Reduction of maternal-infant transmission of human immunodeficiency virus type 1 with zidovudine treatment. *New England Journal of Medicine, 331,* 1173–1180.

CONSTANTINOPLE, A. (1973). Masculinity-femininity: An exception to a famous dictum. *Psychological Bulletin, 80,* 389–407.

CONSUMER REPORTS. (1989, March). Can you rely on condoms? *Consumer Reports,* pp. 135–141.

CONSUMER REPORTS (1995, May). How reliable are condoms? *Consumer Reports,* pp. 320–325.

CONTE, J. R. (1991). The nature of sexual offenses against children. In C.R. Hollin & K. Howells (Eds.), *Clinical approaches to sex offenders and their victims* (pp.11–34). New York: Wiley.

CONTE, J. R., et al. (1986). Child sexual abuse and the family: A critical analysis. *Journal of Psychotherapy and the Family, 2,* 113–126.

COOK, A. S., & ROCK, J. A. (1991). The role of laparoscopy in the treatment of endometriosis. *Fertility and Sterility, 55,* 663–680.

COOK, L. S., KOUTSKY, L. A., & HOLMES, K. K. (1994). Circumcision and sexually transmitted diseases. *American Journal of Public Health, 84,* 197–201.

COOK, R., & FOSEN, R. (1970). Pornography and the sex offender: Patterns of exposure and immediate arousal effects of pornographic stimuli. In *Technical report of the Commission on Obscenity and Pornography* (vol. 7). Washington, DC: U.S. Government Printing Office.

COOPER, A. J., & CERNOVSKY, Z. (1992). The effects of cyproterone acetate on sleeping and waking penile erections in pedophiles: Possible implications for treatment. *Canadian Journal of Psychiatry, 37,* 33–37.

COOPER, T. G. (1990). In defense of a function for the human epididymis. *Fertility and Sterility, 54,* 965–975.

COPENHAVER, S., & GRAUERHO, E. (1991). Sexual victimization among sorority women: Exploring the link between sexual violence and institutional practices. *Sex Roles, 24,* 31–41.

CORNE, S., BRIERE, J., & ESSES, L. M. (1992). Women's attitudes and fantasies about rape as a function of early exposure to pornography. *Journal of Interpersonal Violence, 7,* 454–461.

CORTESE, A. (1989). Subcultural differences in human sexuality: Race, ethnicity, and social class. In K. McKinney & S. Sprecher (Eds.), *Human sexuality: The societal and interpersonal context.* Norwood, NJ: Ablex.

COSTE, J., et al. (1994). Sexually transmitted diseases as major causes of ectopic pregnancy: Results from a large case-control study in France. *Fertility and Sterility, 62,* 289–295.

COTTON, D., & GROTH, A. (1982). Innate rape: Prevention and intervention. *Journal of Prison and Jail Health, 2,* 45–57.

COURTOIS, C. A. (1988). *Healing the incest wound.* New York: Norton.

COUTINHO, R. A. (1994, March/April). Epidemiology of sexually transmitted disease. *Sexually Transmitted Diseases,* pp. S51–S52.

COUTTS, R. L. (1973). *Love and intimacy: A psychological approach.* San Ramon, CA: Consensus.

COWAN, C. C., FRAZIER, P. A., & OLSON A. M. (1997). Predictors of responses to unwanted sexual attention. *Psychology of Women Quarterly, 21,* 207–226.

COWAN, G., & CAMPBELL, R. R. (1995). Rape casual attitudes among adolescents. *Journal of Sex Research, 32,* 145–153.

COWAN, G., & DUNN, K. F. (1994). What themes in pornography lead to perceptions of the degradation of women? *Journal of Sex Research, 31,* 11–21.

COWLEY, G., et al. (1996, September 16). Attention: Aging men. *Newsweek,* pp. 68–75.

COWLEY, G., GIDEONSE, T., & UNDERWOOD, A. (1997, May 19). Gender limbo. *Newsweek,* pp. 64–66.

COX, A. D., HOLDEN, J. M., & SAGOVSKY, R. (1987). Detection of postnatal depression: Development of the 10 item Edinburgh postnatal depression scale. *Journal of Personality and Social Psychology, 150,* 782–786.

COX, D. J. (1988). Incidence and nature of male genital exposure behavior as reported by college women. *Journal of Sex Research, 24,* 227–234.

CRAIG, C., & MOYLE, G. (1997). The development of resistance of HIV-1 to zaleitabine. *AIDS, 11,* 271–279.

CRAIG, S., & HEPBURN, S. (1982). The effectiveness of barrier methods of contraception with and without spermicide. *Contraception, 26,* 347–359.

CRAMER, D. (1986). Gay parents and their children: A review of research and practical implications. *Journal of Counseling and Development, 64,* 504–507.

CRAMER, D., & ROACH, A. (1988). Coming out to mom and dad: A study of gay males and their relationships with their parents. *Journal of Homosexuality, 14*, 77–88.

CRAMER, D. W., et al. (1985). Tubal infertility and the intrauterine device. *New England Journal of Medicine, 31*, 941–947.

CREININ, M. D., et al. (1996). Methotrexate and misoprostrol for early abortion: A multicenter trial. I. Safety and efficacy. *Contraception, 53*, 321–327.

CREPAULT, C., et al. (1977). Erotic imagery in women. In R. Gemme & C. C. Wheeler (Eds.), *Progress in sexology* (pp. 267–283). New York: Plenum.

CROOK, W. G. (1983). *The yeast connection: A medical breakthrough.* Jackson, TN: Professional Books.

CROSS, S. E., & MADSON, L. (1997). Models of the self: Self-construals and gender. *Psychological Bulletin, 122*, 5–37.

CROWE, L. C., & GEORGE, W. H. (1989). Alcohol and human sexuality: Review and integration. *Psychological Bulletin, 105*, 374–386.

CUMMING, D. C., CUMMING, C. E., & KIEREN, D. K. (1991). Menstrual mythology and sources of information about menstruation. *American Journal of Obstetrics and Gynecology, 164*, 472–476.

CUNNINGHAM, F. G., et al. (1993). *Williams obstetrics* (19th ed.), Norwalk, CT: Appleton & Lange.

CUPACH, W. R., & COMSTOCK, J. (1990). Satisfaction with sexual communication in marriage. *Journal of Social and Personal Relationships, 7*, 179–186.

CUPACH, W. R., & METTS, S. (1991). Sexuality and communication in close relationships. In K. McKinney & S. Sprecher (Eds.), *Sexuality in close relationships.* Hillsdale, NJ: Erlbaum.

CUPOLI, J. M., & SEWELL, P. M. (1988). One thousand fifty-nine children with a chief complaint of sexual abuse. *Child Abuse and Neglect, 12*, 151–162.

CURTIN, S. C., & KOZAK, L. J. (1997). Cesarean delivery rates in 1995 continue to decline in the United States. *Birth, 24*, 194–196.

CURTIS, R., & MILLER, K. (1988). Believing another likes or dislikes you: Behavior making the beliefs come true. *Journal of Personality and Social Psychology, 51*, 284–290.

CUTLER, W. B., et al. (1986). Human axillary secretions influence women's menstrual cycles: The role of donor extract from men. *Hormones and Behavior, 20*, 463.

DAAR, E. S., MOUDGIL, T., MEYER, R. D., & HO, D. D. (1991). Transient high levels of viremia in patients with primary human immunodeficiency virus type 1 infection. *New England Journal of Medicine, 324*, 961–964.

DALEY, D. (1996). "Interrupting the conversation": Stripping sexuality from modern communication. *SIECUS Report, 25*(1), 13–15.

DALING, J., et al. (1992). Cigarette smoking and the risk of anogenital cancer. *American Journal of Epidemiology, 135*, 180–189.

DALTON, K. (1959, January 17). Menstruation and acute psychiatric illness. *British Medical Journal*, pp. 148–149.

DALTON, K. (1960). Menstruation and accidents. *British Medical Journal, 2*, 1425–1426.

DALTON, K. (1964). *The premenstrual syndrome.* Springfield, IL: Charles C Thomas.

DALTON, K. (1980). Cyclical criminal acts in premenstrual syndrome. *Lancet, 2*, 1070–1071.

DALY, M., & WILSON, M. (1988). Evolutionary social psychology and family violence. *Science, 242*, 519–524.

DALY, M., WILSON, M., & WEGHORST, S. J. (1982). Male sexual jealousy. *Ethology and Sociobiology, 3*, 11–27.

DARLING, C. A., DAVIDSON, J. K., & CONWAY-WELCH, C. (1990). Female ejaculation: Perceived origins, the Grafenberg spot/area, and sexual responsiveness. *Archives of Sexual Behavior, 19*, 29–47.

DARLING, C. A., DAVIDSON, J. K., & COX, R. P. (1991). Female sexual response and the timing of partner orgasm. *Journal of Sex and Marital Therapy, 17*, 3–21.

DARLING, C. A., DAVIDSON, J. K., & JENNINGS, D. A. (1991). The female sexual response revisited: Understanding the multiorgasmic experience in women. *Archives of Sexual Behavior, 20*, 527–540.

DATEY, S., GAUR, L., & SAXENA, B. (1995). Vaginal bleeding patterns of women using different contraceptive methods (implants, injectables, IUDs, oral pills)—An Indian experience. *Contraception, 51*, 155–165.

DA VANZO, J., PARNELL, A. M., & FOEGE, W. H. (1991). Health consequences of contraceptive use and reproductive patterns. *Journal of the American Medical Association, 265*, 2692–2696.

DAVIDSON, J. K., & HOFFMAN, L. E. (1986). Sexual fantasies and sexual satisfaction: An empirical analysis of erotic thought. *Journal of Sex Research, 22*, 184–205.

DAVIDSON, J. K., & MOORE, N. B. (1994). Guilt and lack of orgasm during sexual intercourse: Myth versus reality among college women. *Journal of Sex Education and Therapy, 20*, 153–174.

DAVIDSON, J. M., et al. (1983). Hormonal changes and sexual function in aging men. *Journal of Clinical Endocrinology and Metabolism, 57*, 71–77.

DAVIDSON, J. M., KWAN, M., & GREENLEAF, W. J. (1982). Hormonal replacement and sexuality in men. *Clinics in Endocrinology and Metabolism, 11*, 599–623.

DAVIES, A. G., & CLAY, J. C. (1992). Prevalence of sexually transmitted disease infection in women alleging rape. *Sexually Transmitted Diseases, 19*, 298–300.

DAVIES, K. A. (1997). Voluntary exposure to pornography and men's attitudes toward feminism and rape. *Journal of Sex Research, 24*, 131–137.

DAVIS, B., & NOBLE, M. (1991, April). Putting an end to chronic testicular pain. *Medical Aspects of Human Sexuality*, pp. 26–34.

DAVIS, G. E., & LEITENBERG, H. (1987). Adolescent sex offenders. *Psychological Bulletin, 101*, 417–427.

DAVIS, K. E. (1985, February). Near and dear: Friendship and love compared. *Psychology Today*, pp. 22–28.

DAVIS, K. E., & BRAUGHT, G. N. (1970). Exposure to pornography, character, and sexual deviance: A retrospective study. In *Technical report of the Commission on Obscenity and Pornography* (Vol. 7). Washington, DC: U.S. Government Printing Office.

DAVIS, K. E., & LATTY-MANN, H. (1987). Love styles and relationship quality: A contribution to validation. *Journal of Social and Personal Relationships, 4*, 409–428.

DAVIS, M. S. (1983). *Smut: Erotic reality/obscene ideology.* Chicago: University of Chicago Press.

DAVIS, N. J. (1978). Prostitution: Identity, career, and legal-economic enterprise. In J. M. Henslin & E. Sagarin (Eds.), *The sociology of sex: An introductory reader.* New York: Schocken Books.

DAVIS, R. C., BRICKMAN, E., & BAKER, T. (1991). Supportive and unsupportive responses of others to rape victims: Effects on concurrent victim adjustment. *American Journal of Community Psychology, 19*, 443–451.

DE ANGELIS, T. (1996, October). Project explores sexual misconduct among children. *APA Monitor*, 43.

DE ANGELIS, T. (1997, September). There's new hope for women with postpartum blues. *APA Monitor*, 22–23.

DE ANGELIS, T. (1997, November). Menopause symptoms may vary among ethnic groups. *APA Monitor*, 16–17.

DEAUX, K., & LEWIS, L. L. (1983). Components of gender role stereotypes. *Psychological Documents, 13*, 25.

DEBUONO, B. A., ZINNER, S. H., DAAMEN, M., & McCORMACK, W. M. (1990). Sexual behavior of college women in 1975, 1986, and 1989. *New England Journal of Medicine, 322*, 821–825.

DE CECCO, J. P., & ELIA, J. P. (1993). A critique and synthesis of biological essentialism and social constructionist views of sexuality and gender. Introduction. *Journal of Homosexuality, 24*, 1–26.

DECLERCQ, E. R. (1992). The transformation of American midwifery: 1975 to 1988. *American Journal of Public Health, 82*, 680–684.

DEEKS, S. G., SMITH, M., HOLODNIY, M., & KAHN, J. O. (1997). HIV-1 protease inhibitors: A review for clinicians. *Journal of the American Medical Association, 277*, 145–153.

DEENEN, A. A., GIJS, L., & VAN NAERSSEN, A. X. (1994). Intimacy and sexuality in gay male couples. *Archives of Sexual Behavior, 23*, 421–431.

DE KUYPER, E. (1993). The Freudian construction of sexuality: The gay foundation of heterosexuality and straight homophobia. In *If you seduce a straight person can you make them gay?* (pp. 137–144). New York: Haworth Press.

DELANEY, J., LUPTON, M. J., & TOTH, E. (1988). *The curse: A cultural history of menstruation.* Urbana and Chicago: University of Illinois Press.

DEMARIS, A., & MACDONALD, W. (1993). Premarital cohabitation and marital instability: A test of the unconventionality hypothesis. *Journal of Marriage and the Family, 55*, 399–407.

DEMARIS, A., & RAO, K. V. (1992). Premarital cohabitation and subsequent marital stability in the United States: A reassessment. *Journal of Marriage and the Family, 54*, 178–190.

DENIS, L., et al. (1992). Alternatives to surgery for benign prostatic hyperplasia. *Cancer* (Philadelphia), *70*, 376–378.

DENNERSTEIN, L., DUDLEY, E., & BURGER, H. (1997). Well-being and the menopausal transition. *Journal of Psychosomatic Obstetrics and Gynecology, 18*, 95–101.

DENNISTON, R. H. (1980). Ambisexuality in animals. In J. Marmor (Ed.), *Homosexual behavior* (pp. 25–40). New York: Basic Books.

DE RIVERA, J. (1997). The construction of false memory syndrome: The experience of retractors. *Psychological Inquiry, 8*, 271–292.

DE ROUGEMENT, D. (1969). *Love in the western world*. New York: Fawcett.

DE SADE, D., MARQUIS. (1965). *Justine*. (R. Seaver & A. Wainhouse, Trans.). New York: Grove Press. (Original work published 1791)

DES JARLAIS, J. C., et al. (1989). HIV-1 infection among intravenous drug users in Manhattan, New York City, from 1977–1987. *Journal of the American Medical Association, 261*, 1008.

DES JARLAIS, D. C., et al. (1996). HIV incidence among injecting drug users in New York City syringe-exchange programmes, *Lancet, 348*, 987–991.

DESROSIERS, R. C. (1990). A finger on the missing link. *Nature* (London), *345*, 288–289.

DESTENO, D. A., & SALOVEY, P. (1996). Evolutionary origins of sex differences in jealousy? Questioning the "fitness" of the model. *Psychological Science, 7*, 367–372.

DEUCHAR, N. (1995). Nausea and vomiting in pregnancy: A review of the problem with particular regard to psychological and social aspects. *British Journal of Obstetrics and Gynaecology, 102*, 6–8.

DE VILLIERS, E.-M. (1989). Heterogeneity of the human papillomavirus group. *Journal of Virology, 63*, 4898–4903.

DE VINCENZI, I., & EUROPEAN STUDY GROUP (1994). A longitudinal study of human immunodeficiency virus transmission by heterosexual partners. *New England Journal of Medicine, 331*, 341–346.

DE VINCENZI, I., & MERTENS, T. (1994). Male circumcision: A role in HIV prevention? *AIDS, 8*, 153–160.

DHAWAN, S., & MARSHALL, W. L. (1996). Sexual abuse histories of sexual offenders. *Sexual abuse: A Journal of research and treatment, 8*, 7–15.

DIAMOND, M. (1993). Homosexuality and bisexuality in different populations. *Archives of Sexual Behavior, 22*, 291–310.

DIAMOND, M. (1997). Sexual identity and sexual orientation in children with traumatized or ambiguous genitalia. *Journal of Sex Research, 34*, 199–211.

DIAMOND, M., & KARLEN, A. (1980). *Sexual decisions*. Boston: Little, Brown.

DIAMOND, M., & SIGMUNDSON, H. K. (1997). Sex reassignment at birth: Long-term review and clinical implications. *Archives of Pediatric and Adolescent Medicine, 151*, 298–304.

DIAMOND, S., & MALISZEWSKI, M. (Eds.). (1992). *Sexual aspects of headaches*. Madison, CT: International University Press.

DICKEY, R., & STEPHENS, J. (1995). Female-to-male transsexualism, heterosexual type: Two cases. *Archives of Sexual Behavior, 24*, 439–445.

DICKINSON, G. M., et al. (1993). Absence of HIV transmission from an infected dentist to his patients. *Journal of the American Medical Association, 269*, 1802–1806.

DICKOVER, R. E., et al. (1996). Identification of levels of maternal HIV-1 RNA associated with risk of perinatal transmission. *Journal of the American Medical Association, 275*, 599–605.

DICK-READ, G. (1972). *Childbirth without fear* (4th ed.). New York: Harper & Row. (First published 1932)

DICLIMENTE, R. J., et al. (1989). Evaluation of school-based AIDS education curricula in San Francisco. *Journal of Sex Research, 26*, 188–198.

DICLIMENTE, R. J., et al. (1990). College students' knowledge and attitudes about AIDS and changes in HIV-preventive behaviors. *AIDS Education and Prevention, 2*, 201–212.

DIETZ, P. E., HARRY, B., & HAZELWOOD, R. R. (1986). Detective magazines: Pornography for the sexual sadist? *Journal of Forensic Sciences, 31*, 197–211.

DIETZ-UHLER, B., & MURRELL, A. (1992). College students perceptions of sexual harassment: Are gender differences decreasing? *Journal of College Student Development, 33*, 540–546.

DIFRANZA, J. R., & LEW, R. A. (1995). Effect of maternal cigarette smoking on pregnancy complications and sudden infant death syndrome. *Journal of Family Practice, 40*, 385–394.

DINDIA, K., & ALLEN, M. (1992). Sex differences in self-disclosure: A meta-analysis. *Psychological Bulletin, 112*, 106–124.

DIOKNO, A. C., BROWN, M. B., & HERZOG, A. R. (1990). Sexual function in the elderly. *Archives of Internal Medicine, 150*, 197–200.

DION, K. L., & DION, K. K. (1987). Belief in a just world and physical attractiveness stereotyping. *Journal of Personality and Social Psychology, 52*, 775–780.

DISMUKES, W. E., et al. (1990). A randomized, double-blind trial of nystatin therapy for the candidiasis hypersensitivity syndrome. *New England Journal of Medicine, 323*, 1717–1723.

DOCTER, R. F. (1985). Transsexual surgery at 74: A case report. *Archives of Sexual Behavior, 14*, 271–277.

DOERING, C., et al. (1978). Plasma testosterone levels and psychologic measures in men over a 2-month period. In R. R. Friedman et al. (Eds.), *Sex differences in behavior*. Huntington, NY: Krieger.

DOLL, L. S., et al. (1992). Homosexually and nonhomosexually identified men who have sex with men: A behavioral comparison. *Journal of Sex Research, 29*, 1–14.

DONAHEY, K. M., & CARROLL, R. A. (1993). Gender differences in factors associated with hypoactive sexual desire. *Journal of Sex and Marital Therapy, 19*, 25–40.

DONNERSTEIN, E. (1982). Pornography: Its effect on violence against women. In N. M. Malamuth & E. Donnerstein (Eds.), *Pornography and sexual aggression*. (pp. 53–81). Orlando, FL: Academic Press.

DONNERSTEIN, E. (1983). *Massive exposure to sexual violence and desensitization to violence and rape*. Paper presented at the 26th annual meeting of the society for the scientific study of sex, Chicago, November 20.

DONNERSTEIN, E., & BERKOWITZ, L. (1981). Victim reactions in aggressive erotic films as a factor in violence against women. *Journal of Personality and Social Psychology, 41*, 710–724.

DONNERSTEIN, E., & LINZ, D. (1984). Sexual violence in the media: A warning. *Psychology Today, 18*, pp. 14–15.

DONNERSTEIN, E., & LINZ, D. (1986, December). The question of pornography. *Psychology Today*, pp. 56–59.

DONNERSTEIN, E., LINZ, D., & PENROD, S. (1987). *The question of pornography: Research findings and policy implications*. New York: Free Press.

DONOVAN, P. (1993). *Testing positive: Sexually transmitted disease and the public health response*. New York: Alan Guttmacher Institute.

DONOVAN, P. (1997a). Can statutory rape laws be effective in preventing adolescent pregnancy? *Family Planning Perspectives, 29*, 30–34.

DONOVAN, P. (1997b). Confronting a hidden epidemic: The Institute of Medicine's Report on sexually transmitted diseases. *Family Planning Perspectives, 29*, 87–89.

DORNER, G. (1968). Hormonal induction and prevention of female homosexuality. *Journal of Endocrinology, 42*, 163–164.

DORNER, G., et al. (1975). A neuroendocrine predisposition for homosexuality in men. *Archives of Sexual Behavior, 4*, 1–8.

DOUGLAS, M. (1966). *Purity and danger: An analysis of concepts of pollution and taboo*. New York: Praeger.

DOUGLASS, W. C. (1988). *WHO murdered Africa?* [Circulated booklet].

DOYLE, J. (1985). *Sex and gender*. Dubuque, IA: W. C. Brown.

DOZORTSEV, D., DE SUTTER, P., RYBOUCHKIN, A., & KHONT, M. (1995). Oocyte activation and intracytoplasmic sperm injection (ICSI). *Assisted Reproduction Review, 5*, 32–39.

DRIFE, J. (1989). The benefits of combined oral contraceptives. *British Journal of Obstetrics and Gynaecology, 96*, 1255–1258.

DRYFOOS, J. (1985, November). What the United States can learn about prevention of teenage pregnancy from other developed countries. *SIECUS Report, 14*, 1–7.

DUBOIS-ARBER, F., JEANNIN, A., KONINGS, E., & PACCAUD, F. (1997). Increased condom use without other major changes in sexual behavior among the general population in Switzerland. *American Journal of Public Health, 87*, 558–566.

DUESBERG, P. H. (1989). Human immunodeficiency virus and acquired immunodeficiency syndrome: Correlation but not causation. *Proceedings of the National Academy of Sciences of the U.S.A., 86*, 755–764.

DUESBERG, P. H., & ELLISON, B. (1996). *Inventing the AIDS virus*. Regnery.

DUNCAN, D. F., & DONNELLY, J. W. (1991). Pornography as a source of sex information for students at a private northeastern university. *Psychological Reports, 68*, 782.

DUNCAN, D. F., & NICHOLSON, T. (1991). Pornography as a source of sex information for students at a southeastern state university. *Psychological Reports, 68*, 802.

DUNKLE, J. H., & FRANCIS, P. L. (1990). The role of facial masculinity/femininity in the attribution of homosexuality. *Sex Roles, 23*, 157–167.

DUNN, M. E., & TROST, J. E. (1989). Male multiple orgasms: A descriptive study. *Archives of Sexual Behavior, 18*, 377–387.

DUPONT, W. D., & PAGE, D. L. (1991). Menopausal estrogen replacement therapy and breast cancer. *Archives of Internal Medicine, 151*, 67–72.

DWORKIN, A. (1989). *Pornography: Men possessing women*. New York: Dutton.

DZIECH, B. W., & WEINER, L. (1984). *The lecherous professor: Sexual harassment on campus*. Boston: Beacon Press.

EARLS, C. M., & DAVID, H. (1989). A psychosocial study of male prostitution. *Archives of Sexual Behavior, 18*, 401–419.

EBI, K. L., PIZIALI, R. L., ROSENBERG, M., & WACHOB, H. F. (1996). Evidence against tailstrings increasing the rate of pelvic inflammatory disease among IUD users. *Contraception, 53*, 25–32.

EDGERTON, M. (1984). The role of surgery in the treatment of transsexualism. *Annals of Plastic Surgery, 13*, 473–476.

EGYPTO, A. C., PINTO, M. C. D., & BOCK, S. D. (1996). Brazilian organization develops "sexual guidance" programs defined by long-term communication. *SIECUS Report, 24*(3), 16–17.

EHRHARDT, A. A. (1992). Trends in sexual behavior and the HIV pandemic. *American Journal of Public Health, 82*, 1459–1461.

EHRHARDT, A. A., & BAKER, S. W. (1974). Fetal androgens, human nervous system differentiation, and behavioral sex differences. In R. C. Friedman, R. M. Richart, & R. Van de Wiele (Eds.), *Sex differences in behavior*. New York: Wiley.

EHRHARDT, A. A., EVERS, K., & MONEY, J. (1968). Influence of androgen on some aspects of sexually dimorphic behavior in women with the late-treated androgenital syndrome. *Johns Hopkins Medical Journal, 123*, 115–122.

EHRHARDT, A. A., MEYER-BAHLBURG, H., FELDMAN, J., & INCE, S. (1984). Sex-dimorphic behavior in childhood subsequent to prenatal exposure to exogenous progestogens and estrogens. *Archives of Sexual Behavior, 13*, 457–477.

EHRLICH, P. R., & EHRLICH, A. H. (1990). *The population explosion*. New York: Simon & Schuster.

EHRLICHMAN, H., & EICHENSTEIN, R. (1992). Private wishes: Gender similarities and differences. *Sex Roles, 26*, 399–422.

EICHEL, E. W., & NOBILE, P. (1992). *The perfect fit: How to achieve mutual fulfillment and monogamous passion through the new intercourse*. New York: Donald Fine.

EINARSON, T. P., et al. (1990). Maternal spermicide use and adverse reproductive outcome: A meta-analysis. *American Journal of Obstetrics and Gynecology, 162*, 655–660.

ELIAS, J., & GEBHARD, P. (1969). Sexuality and sexual learning in childhood. *Phi Delta Kappan, 50*, 401–405.

ELIFSON, K. W., et al. (1989). Seroprevalence of human immunodeficiency virus among male prostitutes. *New England Journal of Medicine, 321*, 822–833.

ELLIOT, A. N., O'DONOHUE, W. T., & NICKERSON, M. A. (1993). The use of sexually anatomically detailed dolls in the assessment of sexual abuse. *Clinical Psychology Review, 13*, 207–221.

ELLIS, E. M., ATKESON, B. M., & CALHOUN, K. S. (1982). An examination of differences between multiple and single incident victims of sexual assault. *Journal of Abnormal Psychology, 91*, 221–224.

ELLIS, L. (1993). Rape as a biosocial phenomenon. In G. C. N. Hall et al. (Eds.), *Sexual aggression: Issues in etiology, assessment, and treatment* (pp. 17–41). Washington, D.C. : Taylor & Francis.

ELLIS, L., & AMES, M. A. (1987). Neurohormonal functioning and sexual orientation: A theory of homosexuality-heterosexuality. *Psychological Bulletin, 101*, 223–258.

ELWIN, U. (1968). *The kingdom of the young*. Oxford: Oxford University Press.

ENGLAR, R. C., & WALKER, C. E. (1973). Male and female reactions to erotic literature. *Psychological Reports, 32*, 481–482.

ENGLUND, J. A., et al. (1997). Zidovudine, didanosine, or both as the initial treatment for symptomatic HIV-infected children. *New England Journal of Medicine, 336*, 1704–1712.

EPSTEIN, E., & GUTTMAN, R. (1984). Mate selection in man: Evidence, theory, and outcome. *Social Biology, 31*, 243–278.

ERARD, J. R. (1988). Vaginal discharges: Diagnosis and therapy. *Diagnosis, 10*(6), 47.

ERIKSON, E. (1968). *Identity: Youth and crisis*. New York: Norton.

ERNULF, K. E., & INNALA, S. M. (1995). Sexual bondage: A review and unobtrusive investigation. *Archives of Sexual Behavior, 24*, 631–654.

ESCHENBACH, D. A. (1993). History and review of bacterial vaginosis. *American Journal of Obstetrics and Gynecology, 169*, 441–445.

ESTREICH, S., FORSTER, G. E., & ROBINSON, A. (1990). Sexually transmitted diseases in rape victims. *Genitourinary Medicine, 66*, 433–438.

EUROPEAN COLLABORATIVE STUDY. (1991). Children born to women with HIV-1 infection: Natural history and risk of transmission. *Lancet, 337*, 253–260.

EUROPEAN COLLABORATIVE STUDY. (1992). Risk factors for mother-to-child transmission of HIV-1. *Lancet, 339*, 1007–1012.

EUROPEAN COLLABORATIVE STUDY. (1994). Perinatal findings in children born to HIV-infected mothers. *British Journal of Obstetrics and Gynaecology, 101*, 136–141.

EUROPEAN STUDY GROUP ON HETEROSEXUAL TRANSMISSION OF HIV. (1992). Comparison of female to male and male to female transmission of HIV in 563 stable couples. *British Medical Journal, 304*, 809–813.

EVANS, H. J., FLETCHER, J., TORRANCE, M., & HARGREAVE, T. B. (1981). Sperm abnormalities and cigarette smoking. *Lancet, I*, 627–629.

EVANS, R. B. (1969). Childhood parental relationships of homosexual men. *Journal of Counseling and Clinical Psychology, 33*, 129–135.

EWIGMAN, B. G., et al. (1993). Effect of prenatal ultrasound screening on perinatal outcome. *New England Journal of Medicine, 329*, 821–827.

EXNER, J. F., WYLIE, J., LEURA, A., & PARRILL, T. (1977). Some psychological characteristics of prostitutes. *Journal of Personality Assessment, 41*, 474–485.

FAGOT, B., & LEINBACH, M. (1987). Socialization of sex roles within the family. In D. B. Carter (Ed.), *Current conceptions of sex roles and sex typing*. New York: Praeger.

FAGOT, B. I. (1995). Psychosocial and cognitive determinants of early gender-role development. In R.C. Rosen et al. (Eds.), *Annual Review of Sex Research*, Vol. VI (pp. 1–31). Mount Vernon, IA: Society for the Scientific Study of Sexuality.

FALLER, K. C. (1987). Women who sexually abuse children. *Victims and Violence, 2*(4), 23–27.

FALLER, K. C. (1991). Polyincestuous families: An exploratory study. *Journal of Interpersonal Violence, 6*, 310–321

FARKAS, G. M., & ROSEN, R. C. (1976). Effect of alcohol on elicited male sexual response. *Journal of Studies on Alcohol, 37*, 265–272.

FARLEY, T. M. M., MEIRIK, O., MEHTA, S., & WAITES, G. M. H. (1993). The safety of vasectomy: Recent concerns. *Bulletin of the World Health Organization, 71*, 413–419.

FARLEY, T. M. M., et al. (1992). Intrauterine devices and pelvic inflammatory disease: An international perspective. *Lancet, 339*, 785–788.

FARLEY, T. M. M., et al. (1995). Effect of different progestagens in low estrogen oral contraceptives on venous thromboebolic diseases. *Lancet, 346*, 1582–1588.

FARMER, R. D. T., et al. (1997). Population-based study of risk of venous thromboembolism associated with various oral contraceptives. *Lancet, 349*, 83–88.

FARR, G., GABELNICK, H., STURGEN, K., & DORFLINJER, L. (1994). Contraceptive efficacy and acceptability of the female condom. *American Journal of Public Health, 84*, 1960–1964.

FARROW, S. (1994). Falling sperm quality: Fact or fiction? *British Medical Journal, 309*, 1–2.

FAUCI, A. S. (1993). CD4 1 T-Lymphocytopenia without HIV infection—No lights, no cameras, just facts. *New England Journal of Medicine, 328*, 429–430.

FAUSTO-STERLING, A. (1985). *Myths of gender*. New York: Basic Books.

FAUSTO-STERLING, A. (1993). The five sexes: Why male and female are not enough. *The Sciences, 33*(3), 20–24.

FAY, R. E., et al. (1989). Prevalence and patterns of same-gender sexual contact among men. *Science, 243*, 338–348.

FEDORA, O., et al. (1992). Sadism and other paraphilias in normal controls and aggressive and nonaggressive sex offenders. *Archives of Sexual Behavior, 21*, 1–15.

FEHR, B. (1994). Prototype-based assessment of laypeople's views of love. *Personal Relationships, 1*, 309–331.

FEIGENBAUM, R., WEINSTEIN, E., & ROSEN, E. (1995). College student's sexual attitudes and behaviors: Implications for sexuality education. *Journal of American College Health, 44*(3), 112–118.

FEINGOLD, A. (1992). Gender differences in mate selection preferences: A test of the parental investment model. *Psychological Bulletin, 112*, 125–139.

FELDMAN-SUMMERS, S., & POPE, K. S. (1994). The experience of "forgetting" childhood abuse: A national survey of psychologists. *Journal of Consulting and Clinical Psychology, 62*, 636–639.

FERGUSON, E. (1997). HIV/AIDS knowledge and HIV/AIDS risk perception: An indirect relationship. *Work & Stress, 11*, 103–117.

FERGUSSON, D. M., LAWTON, J. M., & SHANNON, F. T. (1988). Neonatal circumcision and penile problems: An 8-year longitudinal study. *Pediatrics, 81*, 537–540.

FERREIRA, A. E. et al. (1993). Effectiveness of the diaphragm, used continuously, without spermicide. *Contraception, 48*, 29–35.

FEUCHT, T. E. (1993). Prostitutes on crack cocaine: Addiction, utility, and marketplace economics. *Deviant Behavior: An Interdisciplinary Journal, 14*, 91–108.

FIESTER, S. J., & GAY, M. (1991). Sadistic personality disorder: A review of data and recommendations for DSM-IV. *Journal of Personality Disorders, 5*, 376–385.

FIHN, S. D., et al. (1996). Association between use of spermicide-coated condoms and *Escherichia coli* urinary tract infection in young women. *American Journal of Epidemiology, 144*, 512–520.

FINGERHUT, L. A., KLEINMAN, J. C., & KENDRICK, J. S. (1990). Smoking before, during, and after pregnancy. *American Journal of Public Health, 80*, 541–544.

FINKELHOR, D. (1978). Psychological, cultural and family factors in incest and family sexual abuse. *Journal of Marriage and Family Counseling, 4,* 41–49.

FINKELHOR, D. (1979). *Sexually victimized children.* New York: Free Press.

FINKELHOR, D. (1980). Sex among siblings: A survey on prevalence, variety, and effects. *Archives of Sexual Behavior, 9,* 171–194.

FINKELHOR, D. (1984). *Child sexual abuse: New theory and research.* New York: Free Press.

FINKELHOR, D., & ARAJI, A. (1986). Explanations of pedophilia: A four-factor model. *Journal of Sex Research, 22,* 145–161.

FINKELHOR, D., & RUSSELL, D. (1984). The gender gap among perpetrators of child sexual abuse. In D. Russell (Ed.), *Sexual exploitation: Rape, child sexual abuse, and workplace harassment* (pp. 215–231). Beverly Hills, CA: Sage.

FINKELHOR, D., & YLLO, K. (1985). *License to rape: Sexual abuse of wives.* New York: Holt, Rinehart & Winston.

FINZI, D., et al. (1997). Identification of a reservoir for HIV-1 in patients on highly active antiretroviral therapy. *Science, 278,* 1295–1300.

FISCH, H., et al. (1995). Semen analyses in 1,283 men from the United States over a 25-year period: No decline in quality. *Fertility and Sterility, 65,* 1009–1014.

FISCHER, G. J. (1989). Sex words used by partners in a relationship. *Journal of Sex Education and Therapy, 15*(1), 50.

FISCHER, J. L., & NARUS, L. R., JR. (1981). Sex-role development in late adolescence and adulthood. *Sex Roles, 7,* 97–106.

FISCHL, M. A., et al. (1989). Prolonged zidovudine therapy in patients with AIDS and advanced AIDS-related complex. *Journal of the American Medical Association, 262,* 2405–2410.

FISHER, B., et al. (1995). Reanalysis and results after 12 years of follow-up in a randomized clinical trial comparing total mastectomy and lumpectomy with or without irradiation in the treatment of breast cancer. *New England Journal of Medicine, 333,* 1456–1461.

FISHER, B., et al. (1997). Effect of preoperative chemotherapy on local-regional disease in women with operable breast cancer: Findings from national surgical adjuvant breast and bowel project B-18. *Journal of Clinical Oncology, 15,* 2483–2493.

FISHER, B., WEISBERG, D., & MAROTTA, T. (1982). *Report on adolescent male prostitution.* San Francisco: Urban and Rural Systems Associates.

FISHER, H. E. (1987). The four year itch. *Natural History, 10,* 22–29.

FISHER, H. E. (1992). *Anatomy of love: The natural history of monogamy, adultery and divorce.* New York: Norton.

FISHER, H. E. (1997). Paper presented at the meeting of the American Association for the Advancement of Science.

FISHER, J. D., & FISHER, W. A. (1992). Changing AIDS-risk behavior. *Psychological Bulletin, 11,* 455–474.

FISHER, S. (1973). *The female orgasm.* New York: Basic Books.

FISHER, W., & GRAY, J. (1988). Erotophobia-erotophilia and sexual behavior during pregnancy and postpartum. *Journal of Sex Research, 25,* 379–396.

FISHER, W. A., BRANSCOMBE, N. R., & LEMERY, C. R. (1983). The bigger the better? Arousal and attributional responses to erotic stimuli that depict different size penises. *Journal of Sex Research, 19,* 337–396.

FISHER, W. A., & GRENIER, G. (1994). Violent pornography, antiwoman thoughts, and antiwoman acts: In search of reliable effects. *Journal of Sex Research, 31,* 23–38.

FITZGERALD, L. F. (1993). Sexual harassment: Violence against women in the workplace. *American Psychologist, 48,* 1070–1076.

FITZGERALD, L. F., SHULLMAN, S. L., BAILEY, N., & RICHARDS, M. (1988). The incidence and dimensions of sexual harassment in academia and the workplace. *Journal of Vocational Behavior, 32,* 152–175.

FITZGERALD, M. G., et al. (1996). Germ-line BRCA1 mutations in Jewish and non-Jewish women with early-onset breast cancer. *New England Journal of Medicine, 334,* 143–149.

FLAKE, A. W., et al. (1996). Treatment of X-linked severe combined immunodeficiency by in vitro transplantation of paternal bone marrow. *New England Journal of Medicine, 335,* 1806–1810.

FLAMM, B. L., et al. (1990). Vaginal birth after cesarean delivery: Results of a 5-year multicenter collaborative study. *Obstetrics and Gynecology, 76,* 750–754.

FLEMING, C., et al. (1993). A decision analysis of alternative treatment strategies for clinically localized prostate cancer. *Journal of the American Medical Association, 269,* 2650–2658.

FLEMING, D. T., et al. (1997). Herpes simplex virus type 2 in the United States, 1976 to 1994. *New England Journal of Medicine, 337,* 1105–1111.

FLINT, M. P. (1997). Secular trends in menopause age. *Journal of Psychosomatic Obstetrics and Gynecology, 18,* 65–72.

FLOR-HENRY, P. (1987). Cerebral aspects of sexual deviation. In G. D. Wilson (Ed.), *Variant sexuality: Research and theory.* Baltimore: Johns Hopkins University Press.

FOA, E. B., & RIGGS, D. S. (1995). Posttraumatic stress disorder following assault: Theoretical considerations and empirical findings. *Current Directions in Psychological Science, 4,* 61–65.

FONDA, G. (1994). Local oestrogen replacement for local symptoms in older community dwelling women. *Gerontology, 40*(Suppl.), 9–13.

FORD, C. S., & BEACH, F. A. (1951). *Patterns of sexual behavior.* New York: Harper & Brothers.

FORREST, J. D., & SINGH, S. (1990). Public-sector savings resulting from expenditures for contraceptive services. *Family Planning Perspectives, 22,* 6–15.

FORWARD, S., & BUCK, C. (1978). *Betrayal of innocence: Incest and its devastation.* Los Angeles: J. P. Tarcher.

FOXMAN, B., et al. (1997). Condom use and first-time urinary tract infection. *Epidemiology, 8,* 637–641.

FRANCOEUR, R. T. (1982). *Becoming a sexual person.* New York: John Wiley.

FRANK, E., ANDERSON, C., & RUBENSTEIN, D. (1978). Frequency of sexual dysfunction in "normal" couples. *New England Journal of Medicine, 299,* 111–115.

FRANK, M. L., POINDEXTER, A. N., JOHNSON, M. L., & BATEMAN, L. (1992). Characteristics and attitudes of early contraceptive implant acceptors in Texas. *Family Planning Perspectives, 24,* 208–213.

FRANK, M. L., et al. (1993). One-year experience with subdermal contraceptive implants in the United States. *Contraception, 48,* 229–243.

FRANK, P., et al. (1993). The effect of induced abortion on subsequent fertility. *British Journal of Obstetrics and Gynaecology, 100,* 575–580.

FRANKE, K. M. (1997). What's wrong with sexual harassment? *Stanford Law Review, 49,* 691–772.

FRANKLIN, D. (1984). Rubella threatens unborn in vaccine gap. *Science News, 125,* 186.

FRANZINI, L. R., & SIDEMAN, L. M. (1994). Personality characteristics of condom users. *Journal of Sex Education and Therapy, 20,* 110–118.

FRAZER, I. H. (1996). The role of vaccines in the control of STDs: HPV vaccines. *Genitourinary Medicine, 72,* 398–403.

FRAZIER, P. A. (1991). Self-blame as a mediator of postrape depressive symptoms. *Journal of Social and Clinical Psychology, 10,* 47–57.

FREEMAN, E., RICKELS, K., SONDHEIMER, S. J., & POLANSKY, M. (1990). Ineffectiveness of progesterone suppository treatment for premenstrual syndrome. *Journal of the American Medical Association, 264,* 349–353.

FREUD, S. (1905). Three essays on the theory of sexuality. In *Standard edition,* Vol. VII, pp. 125–245. London: Hogarth Press, 1953..

FREUD, S. (1913). *Totem and taboo.* New York: Vintage Books.

FREUD, S. (1953). Contributions to the psychology of love: A special type of choice of objects made by men. In E. Jones (Ed.), *Collected papers* (Vol. 4) (pp. 192–202). London: Hogarth Press. (Originally published 1933).

FREUD, S. (1943). *A general introduction to psychoanalysis.* Garden City, NY: Garden City Publishing. (Originally published 1917).

FREUND, K., & KUBAN, M. (1993). Toward a testable developmental model of pedophilia: The development of erotic age preference. *Child Abuse & Neglect, 17,* 315–324.

FREUND, K., & KUBAN, M. (1994). The basis of the abused abuser theory of pedophilia: A further elaboration on an earlier study. *Archives of Sexual Behavior, 23,* 553–563.

FREUND, K., & LANGEVIN, R. (1976). Bisexuality in homosexual pedophilia. *Archives of Sexual Behavior, 5,* 415–423.

FREUND, K., SETO, M.C., & KUBAN, M. (1997). Frotteurism and the theory of courtship disorder. In D. R. Laws & W. O'Donohue (Eds.), *Sexual deviance: Theory, assessment, and treatment* (pp. 111–130). New York: Guilford Press.

FREUND, K., & WATSON, R. J. (1992). The proportions of heterosexual and homosexual pedophiles among sex offenders against children: An exploratory study. *Journal of Sex and Marital Therapy, 18,* 34–43.

FREUND, K., WATSON, R., & DICKEY, R. (1991). Sex offenses against female children perpetrated by men who are not pedophiles. *Journal of Sex Research, 28,* 409–423.

FREUND, M., LEONARD, T. L., & LEE, N. (1989). Sexual behavior of resident street prostitutes with their clients in Camden, New Jersey. *Journal of Sex Research, 26,* 460–478.

FRIEDLAND, G. H., et al. (1990). Additional evidence for lack of transmission of HIV infection by close interpersonal (casual) contact. *AIDS, 4,* 639–644.

FRIEDMAN, C. I., & KIM, M. H. (1985). Obesity and its effect on reproductive function. *Clinical Obstetrics and Gynecology, 28,* 645–663.

FRIEDMAN, R. (1947). *The story of scabies* (Vol. I). New York: Froben Press.

FRIEDMAN, R. C. (1991). Couple therapy with gay couples. *Psychiatric Annals, 21,* 485–490.

FRIEDMAN, R. C. (1992). Neuropsychiatry and homosexuality: On the need for biopsychosocial interactionism. *Journal of Neuropsychiatry and Clinical Neurosciences, 4,* 1–2.

FRIEDRICH, W. N. (1993). Sexual victimization and sexual behavior in children: A review of recent literature. *Child Abuse and Neglect, 17,* 59–66.

FRIEDRICH, W. N., et al. (1991). Normative sexual behavior in children. *Pediatrics, 88,* 456–464.

FRIEDRICH, W. N., & REAMS, R. A. (1987). Course of psychological symptoms in sexually abused young children. *Psychotherapy, 24,* 160–170.

FRIEZE, I. H. (1983). Investigating the causes and consequences of marital rape. *Signs, 8,* 532–553.

FROMM, E. (1956). *The art of loving.* New York: Harper & Row.

FURSTENBERG, F. F., LEVINE, J. A., & BROOKS-GUNN, J. (1990). The children of teenage mothers: Patterns of early childbearing in two generations. *Family Planning Perspectives, 22*(2), 54–61.

FURSTENBERG, F. F., MENCKEN, J., & LINCOLN, R. (1981). *Teenage sexuality, pregnancy, and childbearing.* Philadelphia: University of Pennsylvania Press.

GABRIEL, S. E., et al. (1997). Complications leading to surgery after breast implantation. *New England Journal of Medicine, 336,* 677–682.

GAGNON, J. H. (1977). *Human sexualities.* Glenview, IL: Scott, Foresman.

GAGNON, J. H. (1985). Attitudes and responses of parents to preadolescent masturbation. *Archives of Sexual Behavior, 14,* 451–466.

GAGNON, J. H., LINDENBAUM, S., & MARTIN, J. L. (1989). Sexual behavior and AIDS. In C. F. Turner, H. G. Miller, & L. G. Moses (Eds.), *AIDS, sexual behavior, and intravenous drug use.* Washington, DC: National Academy Press.

GAGNON, J. H., & SIMON, W. (1973). Sexual conduct: *The social origins of human sexuality.* Chicago: Aldine.

GAGNON, J. H., & SIMON, W. (1987). The sexual scripting of oral genital contacts. *Archives of Sexual Behavior, 16,* 1–25.

GAIL, M. H., et al. (1989). Projecting individualized probabilities of developing breast cancer for white females who are being examined annually. *Journal of the National Cancer Institute, 81,* 1879–1886.

GALLAGHER, W. (1986, February). The etiology of orgasm. *Discover,* pp. 51–59.

GANONG, L., & COLEMAN, M. (1989). Preparing for remarriage: Anticipating the issues, seeking solutions. *Family Relations, 38,* 28–33.

GAO, F., et al. (1992). Human infection by genetically diverse SIVSM-related HIV-2 in West Africa. *Nature* (London), *358,* 495–499.

GARBER, L. (1991). *Vested interests.* Boston: Little, Brown.

GARDNER, N. E. S. (1986). Sexuality. In *The right to grow up. An introduction to adults with developmental disabilities* (pp. 45–66). Baltimore: Paul H. Brookes.

GARNETS, L., et al. (1990). Violence and victimization of lesbians and gay men: Mental health consequences. *Journal of Interpersonal Violence, 5,* 366–383.

GARTNER, R. B. (1997). Considerations in the psychoanalytic treatment of men who were sexually abused as children. *Psychoanalytic Psychology, 14,* 13–41.

GAUVREAU, K., DEGRUTTOLA, U., PAGANO, M., & BELLOCCO, R. (1994). The effect of covariates on the induction time of AIDS using improved imputation of exact seroconversion times. *Statistics in Medicine, 13,* 2021–2030.

GEBHARD, P. H. (1966). Factors in marital orgasm. *Journal of Social Issues, 22,* 88–95.

GEBHARD, P. H., GAGNON, J. H., POMEROY, W. B., & CHRISTENSON, C. V. (1965). *Sex offenders: An analysis of types.* New York: Harper & Row.

GEGAX, T. T. (1997, November 10) The AIDS predator. *Newsweek,* pp. 52–59.

GELMAN, D. (1990, June 11). Was it illness or immorality? *Newsweek,* p. 55.

GELMAN, D., et al. (1990, July 23). The mind of a rapist. *Newsweek,* pp. 46–52.

GELMAN, D., FOOTE, D., BARRETT, T., & TALBOT, M. (1992, February 24). Born or bred? *Newsweek,* pp. 46–53.

GELMAN, D., & ROGERS, P. (1993, April 12). Mixed messages. *Newsweek,* pp. 28–29.

GENEVIE, L., & MARGOLIES, E. (1987). *The Motherhood report: How women feel about being mothers.* New York: Macmillan.

GERSHON, R. R. M., VLAHOV, D., & NELSON, K. E. (1990). The risk of transmission of HIV-1 through non-percutaneous, non-sexual modes—A review. *AIDS, 4,* 645–650.

GIBBS, N., et al. (1990, Fall). The dreams of youth. *Time,* Special Issue, pp. 10–14.

GIBSON-AINYETTE, I., TEMPLER, D. I., BROWN, R., & VEACO, L. (1988). Adolescent female prostitutes. *Archives of Sexual Behavior, 17,* 431–438.

GIDYEZ, C. A., & KOSS, M. P. (1990). A comparison of group and individual sexual assault victims. *Psychology of Women Quarterly, 14,* 325–342.

GIFFORD, S. M. (1994). The change of life, the sorrow of life: Menopause, bad blood and cancer among Italian-Australian working class women. *Culture, Medicine, Psychiatry, 18,* 299–321.

GIJS, L., & GOOREN, L. (1996). Hormonal and psychopharmacological interventions in the treatment of paraphilias: An update. *Journal of Sex Research, 33,* 273–290.

GILBERT, B., & CUNNINGHAM, J. (1987). Women's post rape sexual functioning: Review and implications for counseling. *Journal of Counseling and Development, 65,* 71–73.

GILBERT, N. (1991). The phantom epidemic of sexual assault. *The Public Interest, 103,* 54–65.

GILLMORE, M. R., et al. (1997). Repeat pregnancies among adolescent mothers. *Journal of Marriage and the Family, 59,* 536–550.

GIORGIS, B. (1981). *Female circumcision in Africa.* Addis Ababa, Ethiopia: United Nations Economic Commission for Africa.

GIOVANNUCCI, E., et al. (1993a). A prospective cohort study of vasectomy and prostate cancer in United States men. *Journal of the American Medical Association, 269,* 873–877.

GIOVANNUCCI, E., et al. (1993b). A retrospective cohort study of vasectomy and prostate cancer in United States men. *Journal of the American Medical Association, 269,* 878–882.

GITTELSON, N. L., EACOTT, S., & MELITA, B. (1978). Victims of indecent exposure. *British Journal of Psychiatry, 132,* 61–66.

GLADUE, B. A. (1988). Hormones in relationship to homosexual/bisexual/heterosexual gender orientation. In J. M. A. Sitesen (Ed.), *Handbook of sexology: Vol. 6. The pharmacology and endocrinology of sexual function.* Amsterdam: Elsevier.

GLADUE, B. A. (1994). The biopsychology of sexual orientation. *Current Directions in Psychological Science, 3,* 150–154.

GLADUE, B. A., GREEN, R., & HELLMAN, R. E. (1984). Neuroendocrine response to estrogen and sexual orientation. *Science, 225,* 1496–1498.

GLADWELL, M. (1992, May 16). "Safe–sex" campaign said to be missing the mark. *Washington Post,* pp. A1 & A11.

GLASER, J. B., et al. (1991). Sexually transmitted diseases in postpubertal female rape victims. *Journal of Infectious Diseases, 164,* 726–730.

GLASER, J. B., HAMMERSCHLAG, M. R., & McCORMACK, W. M. (1989). Epidemiology of sexually transmitted diseases in rape victims. *Review of Infectious Diseases, 11,* 246–254.

GLASIER, A. (1997). Emergency postcoital contraception. *New England Journal of Medicine, 337,* 1058–1064.

GLASS, R., & ERICSSON, R. (1982). *Getting pregnant in the 1980s.* Berkeley: University of California Press.

GLICK, D. (1997, Spring/Summer). Rooting for intelligence. *Newsweek* [Special issue], 32.

GOCHROS, H., GOCHROS, J. S., & FISHER, J. (Eds.). (1986). *Helping the sexually oppressed.* Englewood Cliffs, NJ: Prentice Hall.

GOERGEN, D. (1975). *The sexual celibate.* New York: Seabury.

GOLD, M. A., SCHEIN, A., & COUPEY, S. M. (1997). Emergency contraception: A national survey of adolescent health experts. *Family Planning Perspectives, 29,* 15–19, 24.

GOLD, S. N., HUGHES, D., & HOHNECKER, L. (1994). Degrees of repression of sexual abuse memories. *American Psychologist, 49,* 441–442.

GOLD, S. N., HUGHES, D. M., & SWINGLE, J. M. (1996). Characteristics of childhood sexual abuse among female survivors in therapy. *Child Abuse & Neglect, 20,* 323–335.

GOLD, S. R., & CHICK, D. A. (1988). Sexual fantasy patterns as related to sexual attitude, experience, guilt and sex. *Journal of Sex Education and Therapy, 14,* 18–23.

GOLDBAUM, G., et al. (1987). The relative impact of smoking and oral contraceptive use on women in the United States. *Journal of the American Medical Association, 258,* 1339–1342.

GOLDBERG, C., & ELDER, J. (1998, January 16). Public still backs abortion, but wants limits, poll says. A notable shift from general acceptance. *New York Times,* pp. A1 & A16.

GOLDBERG, D. C., et al. (1983). The Grafenberg spot and female ejaculation: A review of initial hypotheses. *Journal of Sex and Marital Therapy, 9,* 27–37.

GOLDBERG, H. (1979). *The new male.* New York: William Morrow.

GOLDHABER, M. K., et al. (1993). Long-term risk of hysterectomy among 80,007 sterilized and comparison women at Kaiser Permanente, 1971–1987. *American Journal of Epidemiology, 138,* 508–521.

GOLDMAN, B., BUSH, P., & KLATZ, R. (1987). *Death in the locker room: Steroids, cocaine, and sports.* Tucson, AZ: Body Press.

GOLDMAN, R., & GOLDMAN, J. (1982). *Children's sexual thinking: A comparative study of children aged 5 to 15 years in Australia, North America, Britain, and Sweden.* London: Routledge & Kegan Paul.

GOLDSTEIN, B. (1976). *Human sexuality.* New York: McGraw-Hill.

GOLDSTEIN, M. J. (1973). Exposure to erotic stimuli and sexual deviance. *Journal of Social Issues, 29,* 197–220.

GOLDZIEHER, J. W. (1994). Are low-dose oral contraceptives safer and better? *American Journal of Obstetrics and Gynecology, 171,* 587–590.

GOLEMAN, D. (1990, July 10). Homophobia: Scientists find clues to its roots. *New York Times,* pp. C1, C11.

GOLEMAN, D. (1992). New studies map the mind of the rapist. *New York Times.*

GOLLUB, E. L., & STEIN, Z. A. (1993). Commentary: The new female condom—Item 1 on a women's AIDS prevention agenda. *American Journal of Public Health, 83,* 498–500.

GOMBY, D., & SHIONO, P. H. (1991). Estimating the number of substance-exposed infants. *Future Child, 1,* 17–25.

GOMES-SCHWARTZ, B., HOROWITZ, J. M., CARDARELLI, A. A., & SAUZIER, M. (1990). The aftermath of child sexual abuse: 18 months later. In B. Gomes-Schwartz, J. M. Horowitz, & A. P. Cardarelli (Eds.), *Child sexual abuse: The initial effects* (pp. 132–152). Newbury Park, CA: Sage.

GOODCHILDS, J., & ZELLMAN, G. (1984). Sexual signaling and sexual aggression in adolescent relationships. In N. Malamuth & E. Donnerstein (Eds.), *Pornography and sexual aggression.* Orlando, FL: Academic Press.

GOODE, E. (1972). Sexual marijuana. *Sexual Behavior, 2,* 45–51.

GOODMAN, A. (1992). Sexual addiction: Designation and treatment. *Journal of Sex and Marital Therapy, 18,* 304–314.

GOODSON, P., EVANS, A., & EDMUNDSON, E. (1997). Female adolescents and onset of sexual intercourse: A therapy-based review of research from 1984 to 1994. *Journal of Adolescent Health, 21,* 147–155.

GOODYEAR-SMITH, F. A., LAIDLAW, T. M., & LARGE, R. G. (1997). Memory recovery and repression: What is the evidence? *Health Care Analysis, 5*(2), 99–101.

GORDIS, R. (1978). *Love and sex: A modern Jewish perspective.* New York: Farrar, Straus & Giroux.

GORDON, S. (1989). *Raising your child conservatively in a sexually permissive world.* New York: Fireside.

GORDON, S., & SNYDER, C. W. (1989). *Personal issues in human sexuality: A guidebook for better sexual health* (2nd ed.). Boston: Allyn and Bacon.

GORDON, T. (1976). P.E.T. (*Parent Effectiveness Training) in action.* New York: Bantam Books.

GORSKI, R., GORDON, J., SHRYNE, J., & SOUTHAM, A. (1978). Evidence for a morphological sex difference within the medial preoptic area of the rat brain. *Brain Research, 148,* 333–346.

GOSSELIN, C., & WILSON, G. (1980). *Sexual variations: Fetishism, sadomasochism, transvestism.* New York: Simon & Schuster.

GOULD, J. B., DAVEY, B., & STAFFORD, R. S. (1989). Socioeconomic differences in rates of cesarean section. *New England Journal of Medicine, 321,* 233–239.

GOULD, L. (1989, April). Impact of cardiovascular disease on male sexual function. *Medical Aspects of Human Sexuality,* pp. 24–27.

GOY, R. W., & MCEWEN, B. S. (1980). *Sexual differentiation in the brain.* Cambridge, MA: MIT Press.

GRABER, B. (Ed.). (1982). *Circumvaginal musculature and sexual function.* New York: Karger.

GRADY, W. R., HAYWARD, M. D., & FLOREY, F. A. (1988). Contraceptive discontinuation among married women in the United States. *Studies in Family Planning, 19,* 227–235.

GRADY, W. R., & TANFER, K. (1994). Condom breakage and slippage among men in the United States. *Family Planning Perspectives, 26,* 107–112.

GRAFENBERG, E. (1950). The role of the urethra in female orgasm. *International Journal of Sexology, 3,* 145–148.

GRAHAM, S., et al. (1982). Sex patterns and herpes simplex virus type 2 in the epidemiology of cancer of the cervix. *American Journal of Epidemiology, 115,* 729–735.

GRAM, I., AUSTIN, H., & STALSBERG, H. (1992). Cigarette smoking and the incidence of cervical intraeptihelial neoplasia, grade III, and the cancer of the cervix uteri. *American Journal of Epidemiology, 135,* 341–346.

GRAVHOLT, C. H., JUUL, S., NAERAA, R. W., & HANSEN, J. (1996). Prenatal and postnatal prevalence of Turner's syndrome: A registry study. *British Medical Journal, 312,* 16–21.

GRAY, P. (1993, February 15). What is love? *Time,* pp. 46–49.

GRAY, R. H., et al. (1990). Risk of ovulation during lactation. *Lancet, 335,* 25–29.

GREEN, G., & CLUNIS, D. (1988). Married lesbians. *Women and Therapy, 8,* 41–49.

GREEN, J. E., et al. (1988). Chorionic villus sampling: Experience with an initial 940 cases. *Obstetrics and Gynecology, 71,* 208–212.

GREEN, R. (1987). *The "sissy boy syndrome" and the development of homosexuality.* New Haven, CT: Yale University Press.

GREEN, R., & FLEMING, D. T. (1990). Transsexual surgery follow-up: Status in the 1990s. In J. Bancroft (Ed.), *Annual review of Sex Research,* (pp. 163–174). Mount Vernon, IA: Society for the Scientific Study of Sex.

GREENBERG, B. S., & BUSSELLE, R. (1996). What's old, what's new: Sexuality on the soaps. *SIECUS Report, 24*(5), 14–16.

GREENBLATT, C. S. (1983). The salience of sexuality in the early years of marriage. *Journal of Marriage and the Family, 45,* 289–299.

GREENDALE, G. A., et al. (1993). The relationship of *Chlamydia trachomatis* infection and male infertility. *American Journal of Public Health, 83,* 996–1001.

GREER, D. M., et al. (1982). A technique for foreskin reconstruction and some preliminary results. *Journal of Sex Research, 18,* 324–330.

GREGERSEN, E. (1982). *Sexual practices.* New York: Franklin Watts.

GREGORY, J. G., & PURCELL, M. M. (1989). Penile curvature: Assessment and treatment. *Medical Aspects of Human Sexuality, 23,* 64.

GRIMES, D. A. (1992). The intrauterine device, pelvic inflammatory disease, and infertility: The confusion between hypothesis and knowledge. *Fertility and Sterility, 58,* 670–673.

GRIMES, D. A. (1997). Medical abortion in early pregnancy: A review of the evidence. *Obstetrics and Gynecology, 89,* 790–796.

GRIMES, D. A., & CATES, W., JR. (1980). Abortions: Methods and complications. In E. S. Hofiz (Ed.), *Human reproduction: Conception and contraception* (pp. 796–813). New York: Harper & Row.

GRISEZ, G. G. (1970). *Abortion: The myths, the realities, and the arguments.* New York: Corpus Books.

GROB, C. S. (1985). Female exhibitionism. *Journal of Nervous and Mental Disease, 173,* 253.

GRODSTEIN, F., et al. (1996). Postmenopausal estrogen and progestin use and the risk of cardiovascular disease. *New England Journal of Medicine, 335,* 453–461.

GRODSTEIN, F., et al. (1997). Postmenopausal hormone therapy and mortality. *New England Journal of Medicine, 336,* 1769–1775.

GRODSTEIN, F., GOLDMAN, M. B., & CRAMER, D. W. (1993). Relation of tubal infertility to history of sexually transmitted diseases. *American Journal of Epidemiology, 137,* 577–584.

GROENENDIJK, L. F. (1997). Masturbation and neurasthenia: Freud and Stekel in debate on the harmful effects of autoerotism. *Journal of Psychology and Human Sexuality, 91,* 71–94.

GROFF, M. G. (1987). Characteristics of incest offenders' wives. *Journal of Sex Research, 23,* 91–96.

GROSS, L. (1996). Lesbians and gays in the broadcast media. *SIECUS Report, 24*(4), 10–14.

GROSSMAN, L. S., HAYWOOD, T. W., & WASYLIW, O. E. (1992). The evaluation of truthfulness in alleged sex offenders' self-reports: 16 PF and MMPI validity scales. *Journal of Personality Assessment, 59,* 264–275.

GROTH, A. N. (1979). *Men who rape.* New York: Plenum.

GROTH, A. N., & BIRNBAUM, H. J. (1978). Adult sexual orientation and attraction to underage persons. *Archives of Sexual Behavior, 7,* 175–181.

GROTH, A. N., & BURGESS, A. W. (1977). Sexual dysfunction during rape. *New England Journal of Medicine, 297,* 764–766.

GROTH, A. N., BURGESS, A. W., & HOLMSTROM, L. (1977). Rape: Power, anger, and sexuality. *American Journal of Psychiatry, 134,* 1239–1243.

GROTH, A. N., HOBSON, W. F., & GARY, T. S. (1982). The child molester: Clinical observations. In J. Conte & D. A. Shore (Eds.), *Social work and child sexual abuse* (pp. 129–144). New York: Haworth Press.

GROTH, A. N., & LOREDO, C. M. (1981). Juvenile sexual offenders: Guidelines for assessment. *International Journal of Offender Therapy and Comparative Criminology, 25,* 31–39.

GRUDZINSKAS, J. G., WATSON, C., & CHARD, T. (1979). Does sexual intercourse cause fetal distress? *Lancet, 29,* 692–693.

GRUEN, D. (1990). Postpartum depression: A debilitating yet often unassessed problem. *Health and Social Work, 15,* 261–269.

GRUENBAUM, E. (1996). The cultural debate over female circumcision: The Sudanese are arguing this one out for themselves. *Medical Anthropology Quarterly, 10,* 455–475.

GRUNDLACH, R. (1977). Sexual molestation and rape reported by homosexual and heterosexual women. *Journal of Homosexuality, 2,* 367–384.

GRUNSEIT, A., et al. (1997). Sexuality education and young people's sexual behavior: A review of studies. *Journal of Adolescent Research, 12,* 421–453.

GRUPPO ITALIANO PER LO STUDIO DELL' ENDOMETRIOS. (1994). Prevalence and anatomical distribution of endometriosis in women with selected gynaecological conditions: Results from a multicentric Italian study. *Human Reproduction, 9,* 1158–1162.

GUERRERO PAVICH, E. (1986). A Chicana perspective on Mexican culture and sexuality. In L. Lister (Ed.), *Human sexuality, ethnoculture, and social work.* New York: Haworth Press.

GUITERMAN, A. (1992). Poem. In *Bartlett's familiar quotations* (15th ed.).

GURA, T. (1994, July 31). Mating study: Men seek variety, women security. (New Orleans) *Times-Picayune,* p. A-7.

GUTEK, B. A. (1985). *Sex and the workplace.* San Francisco: Jossey-Bass.

GWARTNEY-GIBBS, P., STOCKARD, J., & BOHMER, S. (1987). Learning courtship aggression: The influence of parents, peers, and personal experience. *Family Relations, 36,* 276–282.

HAAGA, D. A. (1991). Homophobia? *Journal of Behavior and Personality, 6,* 171–174.

HAAS, A. (1979). *Teenage sexuality: A survey of teenage sexual behavior.* New York: Macmillan.

HAAVIO-MANNILA, E., & KONTULA, O. (1997). Correlates of increased sexual satisfaction. *Archives of Sexual Behavior, 26,* 399–419.

HACK, M., et al. (1994). School-age outcomes in children with birth weights under 750g. *New England Journal of Medicine, 331,* 753–759.

HACKEL, L. S., & RUBLE, D. N. (1992). Changes in marital relationship after the first baby is born: Predicting the impact of expectancy disconfirmation. *Journal of Personality and Social Psychology, 62,* 944–957.

HAESEVOETS, Y.-H. (1997). L'enfant victime d'inceste: Symptomatologie spécifique ou aspécifique? (Essai de conceptualisation clinique). *Psychiatrie de l'enfant, XL* (1), 87–119.

HAFFNER, D. W., & CASSELMAN, M. (1996). Toy story: A look into the gender-stereotyped world of children's catalogs, *SIECUS Report, 24*(4), 20–21.

HAGE, J. J., BLOEM, J. J. A. M., & BOUMAN, F. G. (1993). Obtaining rigidity in the neophallus of female-to-male transsexuals: A review of the literature. *Annals of Plastic Surgery, 30,* 327–333.

HAGE, J. J., BLOEM, J. J. A. M., & SULIMAN, H. M. (1993). Review of the literature on techniques for phalloplasty with emphasis on the applicability in female-to-male transsexuals. *Journal of Urology, 150,* 1093–1098.

HAGLUND, B., et al. (1990). Cigarette smoking as a risk factor for sudden infant death syndrome: A population-based study. *American Journal of Public Health, 80,* 29–32.

HAHN, B. (1990). Biologically unique, SIV-like HIV-2 variants in healthy West African individuals. In M. Girard & L. Valette (Eds.), *Proceedings of the Fifth Colloque des Cent Gandes,* Lyon, France.

HAHN, J., & BLASS, T. (1997). Dating partner preferences: A function of similarity of love styles. *Journal of Social Behavior and Personality, 12,* 595–610.

HAJCAK, F., & GARWOOD, P. (1988). Quick-fix sex: Pseudosexuality in adolescents. *Adolescence, 23,* 755–760.

HALEY, T. J., & SNIDER, R. S. (Eds.). (1962). *Response of the nervous system to ionizing radiation.* New York: Academic Press.

HALL, G. C. N., & BARONGAN, C. (1997). Prevention of sexual aggression: Socioculture risk and protective factors. *American Psychologist, 52,* 5–14.

HALL, L. A. (1992). Forbidden by God, despised by men: Masturbation, medical warnings, moral panic, and manhood in Great Britain, 1850–1950. *Journal of the History of Sexuality, 2,* 365–387.

HALL, R. C. et al. (1989). Sexual abuse in patients with anorexia nervosa and bulimia. *Psychosomatics, 30,* 73–79.

HALLECK, S. L. (1976). Another response to "Homosexuality: The ethical challenge." *Journal of Consulting and Clinical Psychology, 44,* 167–170.

HALLER, J. S., & HALLER, R. M. (1977). *The physician and sexuality in Victorian America.* New York: Norton.

HALLSTROM, T., & SAMUELSSON, S. (1990). Changes in women's sexual desire in middle life: The longitudinal study of women in Gothenburg (Sweden). *Archives of Sexual Behavior, 19,* 259–267.

HALMAN, L. J., ABBEY, A., & ANDREWS, F. M. (1992). Attitudes about infertility interventions among fertile and infertile couples. *American Journal of Public Health, 82,* 191–194.

HALSTEAD, J. M. (1997). Muslims and sex education. *Journal of Sex Education, 26,* 317–330.

HAMBRIDGE, D. M. (1988). Incest and anorexia nervosa: What is the link? *British Journal of Psychiatry, 152,* 145–146.

HAMER, D. H., et al. (1993). A linkage between DNA markers on the X chromosome and male sexual orientation. *Science, 261,* 321–327.

HAMILTON, J. D., et al. (1992). A controlled trial of early versus late treatment with zidovudine in symptomatic human immunodeficiency virus infection. *New England Journal of Medicine, 326,* 437–443.

HAMILTON, J., ALAGNA, S., KING, L., & LLOYD, C. (1987). The emotional consequences of gender-based abuse in the workplace: New counseling programs for sex discrimination. *Women and Therapy, 6,* 155–182.

HANDSFIELD, H. H. (1990). Old enemies: Combating syphilis and gonorrhea in the 1990s. *Journal of the American Medical Association, 264,* 1451–1452.

HANDSFIELD, H. H., & SCHWEBKE, J. (1990). Trends in sexually transmitted diseases in homosexually active men in King County, Washington, 1980–1990. *Sexually Transmitted Diseases, 17,* 211–215.

HAQQ, C. M., et al. (1994). Molecular basis of mammalian sexual determination: Activation of Mullerian inhibiting substance gene expression by SRY. *Science, 266,* 1494–1500.

HARDIE, E. A. (1997a) PMS in the workplace: Dispelling the myth of cyclic dysfunction. *Journal of Occupational and Organizational Psychology, 70,* 97–102.

HARDIE, E. A. (1997b). Prevalence and predictors of cyclic and noncyclic affective change. *Psychology of Women Quarterly, 21,* 299–314.

HARDING, S. (1986). *The science question in feminism.* Ithaca, NY: Cornell University Press.

HARDY, J. B., McCRACKEN, G. H., JR., GILKESON, M. R., & SEVER, J. L. (1969). Adverse fetal outcome following maternal rubella after the first trimester of pregnancy. *Journal of the American Medical Association, 207,* 2414–2420.

HARLAP, S. (1992). The benefits and risks of hormone replacement therapy: An epidemiologic overview. *American Journal of Obstetrics and Gynecology, 166,* 1986–1992.

HARLOW, H. F. (1959). Love in infant monkeys. *Scientific American, 200* (6), 68–74.

HARLOW, H. F., & HARLOW, M. (1962). The effect of rearing conditions on behavior. *Bulletin of the Menninger Clinic, 26,* 213–224.

HARLOW, S. D., & EPHROSS, S. A. (1995). Epidemiology of menstruation and its relevance to women's health. *Epidemiologic Reviews, 17,* 265–286.

HARNEY, P., & MUEHLENHARD, C. (1991). Rape. In E. Graverholz & M. Korlewski (Eds.), *Sexual coercion: A sourcebook on its nature, causes, and prevention.* Lexington, MA: Lexington Books.

HARNISH, S. (1988, July). Congenital absence of the vagina: Clinical issues. *Medical Aspects of Human Sexuality,* pp. 54–60.

HARPER, E. B. (1964). Ritual pollution as an integrator of caste and religion. *Journal of Asian Studies, 23,* 151–197.

HARRIS, C. R., & CHRISTENFELD, N. (1996). Gender, jealousy, and reason. *Psychological Science, 7,* 364–366.

HARRIS, R. (1994). Breast cancer among women in their forties: Toward a reasonable research agenda. *Journal of the National Cancer Institute, 86,* 410–412.

HARRISON, M. R., et al. (1990). Successful repair in utero of a fetal diaphragmatic hernia after removal of herniated viscera from the left thorax. *New England Journal of Medicine, 322,* 1582–1584.

HARRISON, P. F., & ROSENFIELD, A. (Eds.). (1996). *Contraceptive research and development: Looking to the future.* Washington, DC: National Academy Press.

HART, J., COHEN, E., GINGOLD, A., & HOMBURG, R. (1991). Sexual behavior in pregnancy: A study of 219 women. *Journal of Sex Education and Therapy, 17,* 86–90.

HARTGE, P. (1997). Abortion, breast cancer, and epidemiology. *New England Journal of Medicine, 336,* 127–128.

HARVEY, S. (1987). Female sexual behavior: Fluctuations during the menstrual cycle. *Journal of Psychosomatic Research, 31,* 101–110.

HASSETT, J. (1978). Sex and smell. *Psychology Today, 11,* 40–42, 45.

HATCHER, R. A., et al. (1994). *Contraceptive technology* (16th ed.). New York: Irvington Publishers.

HATFIELD, E. (1984). The dangers of intimacy. In V. Derlaga (Ed.), *Communication, intimacy, and close relationships* (pp. 207–220). New York: Academic Press.

HATFIELD, E. (1988). Passionate and companionate love. In R. J. Sternberg & M. L. Barnes (Eds.), *The psychology of love* (pp. 191–217). New Haven, CT: Yale University Press.

HATFIELD, E., & SPRECHER, S. (1986). *Mirror, mirror . . . The importance of looks in everyday life.* Albany: State University of New York Press.

HATFIELD, E., & WALSTER, G. W. (1978). *A new look at love.* Lantham, MA: University Press of America.

HATHAWAY, S. R., & McKINLEY, J. C. (1943). *The Minnesota multiphasic personality inventory.* New York: Psychological Corporation.

HAVGAARD, J. J., & EMERY, R. E. (1989). Methodological issues in child sexual abuse research. *Child Abuse & Neglect, 13,* 89–100.

HAWKESWORTH, M. (1997). Confounding gender. *Signs: Journal of Women in Culture and Society, 22,* 649–684.

HAWTON, K., CATALAN, J., & FAGG, J. (1991). Low sexual desire: Sex therapy results and prognostic factors. *Behavior Research and Therapy, 29,* 217–224.

HAWTON, K., CATALAN, J., & FAGG, J. (1992). Sex therapy for erectile dysfunction: Characteristics of couples, treatment outcome, and prognostic factors. *Archives of Sexual Behavior, 21,* 161–175.

HAWTON, K., GATH, D., & DAY, A. (1994). Sexual function in a community sample of middle-aged women with partners: Effects of age, marital, socioeconomic, psychiatric, gynecological, and menopausal factors. *Archives of Sexual Behavior, 23,* 375–395.

HAYS, D., & SAMUELS, A. (1989). Heterosexual women's perceptions of their marriages to bisexual or homosexual men. *Journal of Homosexuality, 18,* 81–100.

HAZAN, C., & SHAVER, P. (1987). Love conceptualized as an attachment process. *Journal of Personality and Social Psychology, 52,* 511–524.

HEARST, N., & HULLEY, S. B. (1988). Preventing the heterosexual spread of AIDS: Are we giving our patients the best advice? *Journal of the American Medical Association, 259,* 2428–2432.

HEATH, D. (1984). An investigation into the origins of a copious vaginal discharge during intercourse—"enough to wet the bed"—that "is not urine." *Journal of Sex Research, 20,* 194–215.

HEDRICKS, C. A. (1994). Female sexual activity across the human menstrual cycle. In J. Bancroft, C. M. Davis, & H. J. Ruppel, Jr. (Eds.), *Annual review of sex research* (Vol.V) (pp. 122–172). Mount Vernon, IA: Society for the Scientific Study of Sexuality.

HEILBRUN, A., & SEIF, D. (1988). Erotic value of female distress in sexually explicit photographs. *Journal of Sex Research, 24,* 47–57.

HEIMAN, J. (1977). A psychophysiological exploration of sexual arousal patterns in females and males. *Psychophysiology, 14,* 266–274.

HEIMAN, J., & LOPICCOLO, J. (1988). *Becoming orgasmic: A sexual growth program for women.* Englewood Cliffs, N.J.: Prentice Hall

HEIMAN, J. R., & GRAFTON-BECKER, V. (1989). Orgasmic disorders in women, In S. R. Leiblum and R. C. Rosen (Eds.), *Principles and practice of sex therapy: Update for the 1990's* (pp. 51–88). New York: Guilford Press.

HEIN, H. A., BURMEISTER, L. F., & PAPKE, K. R. (1990). The relationship of unwed status to infant mortality. *Obstetrics and Gynecology, 76,* 763–768.

HEINONEN, O. P., SLONE, D., & SHAPIRO, S. (1977). *Birth defects and drugs in pregnancy.* Littleton, MA: Publishing Sciences Group.

HELLINGER, F. J. (1993). The lifetime cost of treating a person with HIV. *Journal of the American Medical Association, 270,* 474–478.

HELMREICH, R., SPENCE, J., & HOLAHAN, C. (1979). Psychological androgyny and sex role flexibility: A test of two hypotheses. *Journal of Personality and Social Psychology, 37,* 1631–1644.

HELWIG-LARSEN, M., & COLLINS, B. E. (1997). A social psychological perspective on the role of knowledge about AIDS in AIDS prevention. *Current Directions in Psychological Science, 6,* 23–26.

HENDERSON, B. E., & BERNSTEIN, L. (1991). The international variation in breast cancer rates: An epidemiological assessment. *Breast Cancer Research and Treatment, 18,* 511–517.

HENDERSON, R. (1990, November 22). Parents need right words, right time to tell kids about sex. *Los Angeles Times,* p. X1.

HENDRICK, C., & HENDRICK, S. (1983). *Liking, loving and relating.* Monterey, CA: Brooks/Cole.

HENDRICK, C., & HENDRICK, S. (1986). A theory and method of love. *Journal of Personality and Social Psychology, 50,* 392–402.

HENDRICK, C., & HENDRICK, S.S. (1989). Research or love: Does it measure up? *Journal of Personality and Social Psychology, 56,* 784–794.

HENDRICK, C., HENDRICK, S., & ADLER, N. (1988). Romantic relationships: Love, satisfaction, and staying together. *Journal of Personality and Social Psychology, 54,* 980–988.

HENNEBERGER, M., & MARRIOTT, M. (1993). What teens call flirting, adults call sex harassment. *New York Times.*

HENRIKSEN, T. B., BEK, K. M., HEDEGAARD, M., & SECHER, N. J. (1994). Methods and consequences of changes in use of episiotomy. *British Medical Journal, 309,* 1255–1258.

HENSHAW, S. K. (1990). Induced abortion: A world review, 1990. *Family Planning Perspectives, 22*(2), 76–89.

HENSHAW, S. K. (1997). Teenage abortion and pregnancy statistics by state, 1992. *Family Planning Perspectives, 29,* 115–122.

HENSHAW, S. K., & KOST, K. (1996). Abortion patients in 1994–1995: Characteristics and contraceptive use. *Family Planning Perspectives, 28,* 140.

HENSON, C., RUBIN, H. B., & HENSON, D. E. (1979). Women's sexual arousal concurrently assessed by three genital measures. *Archives of Sexual Behavior, 8,* 459–479.

HERDT, G. (1990). Mistaken gender: 5-Alpha Reductase hermaphroditism and biological reductionism in sexual identity reconsidered. *American Anthropologist, 92,* 433–446.

HERDT, G. (1991). Commentary on status of sex research: Cross-cultural implications of sexual development. *Journal of Psychology & Human Sexuality, 4*(1), 5–12.

HERDT, G., & BOXER, A. (1993). *Children of horizons.* New York: Beacon Press.

HERDT, G. H. (1987). *The Sambia.* New York: Holt, Rinehart, & Winston.

HERDT, G. H. (1993). Semen transactions in Sambia culture. In D. N. Suggs & A. W. Miracle (Eds.), *Culture and human sexuality* (pp. 298–327.) Pacific Grove, CA: Brooks/Cole.

HEREK, G. M. (1984). Beyond homophobia: A social psychological perspective on attitudes toward lesbians and gay men. *Journal of Homosexuality, 10,* 1–21.

HEREK, G. M. (1988). Heterosexuals' attitudes toward lesbian and gay men: Correlates and gender differences. *Journal of Sex Research, 25,* 451–477.

HEREK, G. M. (1989). Hate crimes against lesbian and gay men: Issues for research and policy. *American Psychologist, 44,* 948–955.

HEREK, G. M., & CAPITANIO, J. P. (1993). Public reactions to AIDS in the United States: A second decade of stigma. *American Journal of Public Health, 83,* 574–577.

HEREK, G. M., & CAPITANIO, J. P. (1995). Black heterosexuals' attitudes toward lesbians and gay men in the United States. *Journal of Sex Research, 32,* 95–105.

HERMAN, J., & HIRSCHMAN, L. (1977). Father-daughter incest. *Journal of Women in Culture and Society, 2,* 735–756.

HERMAN, J. L. (1992). *Trauma and recovery.* New York: Basic Books.

HERMAN-GIDDENS, M. E., et al. (1997). Secondary sexual characteristics and menses in young girls seen in office practice: A study from the pediatric research in office settings network. *Pediatrics, 99,* 505–512.

HERSCH, P. (1988, January). Coming of age on city streets. *Psychology Today, 22,* 28–37.

HERSHBERGER, S. L. (1997). A twin registry study of male and female sexual orientation. *Journal of Sex Research, 34,* 212–222.

HERZOG, L. (1989). Urinary tract infections and circumcision. *American Journal of Diseases of Children, 143,* 348–350.

HETEROSEXUAL AIDS: SETTING THE ODDS. (1988). *Science, 240,* 597.

HHS NEWS (1995, July 7). CDC urges HIV counseling and voluntary testing for all pregnant women.

HICKSON, F. C. I., et al. (1994). Gay men as victims of nonconsensual sex. *Archives of Sexual Behavior, 23,* 281–294.

HILL, C. A. (1997). The distinctiveness of sexual motives in relation to sexual desire and desirable partner attributes. *Journal of Sex Research, 34,* 139–153.

HILL, C. A., & PRESTON, L. K. (1996). Individual differences in the experience of sexual motivation: Theory and measurement of dispositional sexual motives. *Journal of Sex Research, 33,* 27–45.

HILL, C. T., RUBIN, Z., & PEPLAU, L. A. (1976). Breakups before marriage: The end of 103 affairs. *Journal of Social Issues, 32,* 147–168.

HILL, I. (1987). *The bisexual spouse.* McLean, VA: Barlina Books.

HINDS, M. C. (1981, June 15). The child victim of incest. *New York Times.*

HINES, M., & COLLAER, M. L. (1993). Gonadal hormones and sexual differentiation of human behavior: Developments from research on endocrine syndromes and studies of brain structure. *Annual Review of Sex Research, 4,* 1–48.

HINES, M., & GREEN, R. (1991). Human hormonal and neural correlates of sex-typed behaviors. *Review of Psychiatry, 10,* 536–555.

HINGSON, R., et al. (1982). Effects of maternal drinking and marijuana use on fetal growth and development. *Pediatrics, 70,* 539–546.

HINGSON, R. W., STRUNIN, L., BERLIN, B. M., & HERREN, T. (1990). Beliefs about AIDS, use of alcohol and drugs, and unprotected sex among Massachusetts adolescents. *American Journal of Public Health, 80,* 295–299.

HIRSCHFELD, M. (1948). *Sexual anomalies.* New York: Emerson.

HITE, S. (1976). *The Hite report.* New York: Macmillan.

HITE, S. (1981). *The Hite report on male sexuality.* New York: Knopf.

HITE, S. (1987). *Women and love: A cultural revolution in progress.* New York: Knopf.

HIV/AIDS SURVEILLANCE REPORT (1997a). U.S. Department of Health and Human Services, Centers for Disease Control and Prevention, Vol. 8(2).

HIV/AIDS SURVEILLANCE REPORT (1997b). U.S. Department of Health and Human Services, Centers for Disease Control and Prevention, Vol. 9(1).

HOBDAY, A. J., HAURY, L., & DAYTON, P. K. (1997) Function of the human hymen. *Medical Hypotheses, 49,* 171–173.

HOCH, Z. (1986). Vaginal erotic sensitivity by sexological examination. *Acta Obstericia et Gynecolologica Scandinavica, 65,* 767–773.

HOCKENBERRY, S. L., & BILLINGHAM, R. (1987). Sexual orientation and boyhood gender conformity. *Archives of Sexual Behavior, 16,* 475–493.

HOLCOMB, D. R., et al. (1991). Gender differences among college students. *College Student Journal, 25,* 434–439.

HOLLAND, J., et al. (1991). Between embarrassment and trust: Young women and the diversity of condom use. In P. Aggleton, P. Davies, & G. Hart (Eds.), *AIDS: Responses, intervention and care.* Basingstoke, England: Falmer Press.

HOLLENDER, M. H., BROWN, C. W., & ROBACK, H. B. (1977). Genital exhibitionism in women. *American Journal of Psychiatry, 134,* 436–438.

HOLLIN, C. R., & HOWELLS, K. (Eds.), *Clinical approaches to sex offenders and their victims* (pp. 11–34). New York: Wiley.

HOLMES, M. M., et al. (1996). Rape-related pregnancy: Estimates and descriptive characteristics from a national sample of women. *American Journal of Obstetrics and Gynecology, 175,* 320–325.

HOLMSTROM, L. L., & BURGESS, A. W. (1978). *The victim of rape: Institutional reactions.* New York: Wiley.

HOLT, V. L., et al. (1989). Induced abortion and the risk of subsequent ectopic pregnancy. *American Journal of Public Health, 79,* 1234–1238.

HOLTZEN, D. W., & AGRESTI, A. A. (1990). Parental responses to gay and lesbian children. *Journal of Social and Clinical Psychology, 9,* 390–399.

HOOK, E., SONDHEIMER, S., & ZENILMAN, J. (1995). Today's treatment for STDs. *Patient Care, 29,* 40–56.

HOOKER, E. (1957). The adjustment of the male overt homosexual. *Journal of Projective Techniques, 21,* 18–31.

HOOKER, E. (1965). An empirical study of some relations between sexual patterns and gender identity in male homosexuals. In J. Money (Ed.), *Sex research: New developments.* New York: Henry Holt.

HOOKER, E. (1993). Reflections of a 40-year exploration: A scientific view on homosexuality. *American Psychologist, 48,* 1–4.

HOOTEN, T. M., et al. (1996). A prospective study of risk factors for symptomatic urinary tract infection in young women. *New England Journal of Medicine, 335,* 468–474.

HOOVER, D. R., et al. (1995). Long-term survival without clinical AIDS after CD41 cell counts fall below 200 x 106/l. *AIDS, 9,* 145–152.

HOOYKAAS, C. et al. (1989). Heterosexuals at risk for HIV: Differences between private and commercial partners in sexual behavior and condom use. *AIDS, 3,* 525–532.

HOROWITZ, F. D., et al. (1977). The effects of obstetrical medication on the behavior of Israeli newborn infants and some comparisons with Uruguayan and American infants. *Child Development, 48,* 1607–1623.

HORSTMAN, J. (1997, August). Hepatitis C: The hidden epidemic. *Hippocrates,* 31–41.

HORT, B. E., LEINBACH, M. D., & FAGOT, B. I. (1991). Is there coherence among the cognitive components of gender acquisition? *Sex Roles, 24,* 195–207.

HORWITZ, A. V., WHITE, H. R., & HOWELL-WHITE, S. (1996). Becoming married and mental health: A longitudinal study of a cohort of young adults. *Journal of Marriage and the Family, 58,* 895–907.

HORWITZ, S. M., KLERMAN, L. V., KUO, H. S., & JEKEL, J. F. (1991). Intergenerational transmission of school-age parenthood. *Family Planning Perspectives, 23,* 168–177.

HOVDESTAD, W. E., & KRISTIANSEN, C. M. (1996, Summer). A field study of "false memory syndrome": Construct validity and incidence. *Journal of Psychiatry & Law,* 299–338.

HOWARD, C. R., HOWARD, F. M., & WEITZMAN, M. L. (1994). Acetaminophen analgesia in neonatal circumcision: The effect on pain. *Pediatrics, 93,* 641–646.

HOWARD, E. (1980, November). Overcoming rape trauma. *Ms.,* p. 35.

HOWARD, M., & McCABE, J. B. (1990). Helping teenagers postpone sexual involvement. *Family Planning Perspectives, 22,* 21–26.

HOWARDS, S. S. (1995). Treatment of male infertility. *New England Journal of Medicine, 332,* 312–317.

HOWE, J. E., MINKOFF, H. L., & DUERR, A. C. (1994). Contraceptives and HIV. *AIDS, 8,* 861–871.

HOWELLS, K. (1981). Adult sexual interest in children: Considerations relevant to theories of etiology. In M. Cook & K. Howells (Eds.), *Adult sexual interest in children* (pp. 55–94). New York: Academic Press.

HOWIE, P. W., et al. (1990). Protective effect of breast feeding against infection. *British Medical Journal, 300,* 11–16.

HSU, F. L. K. (1981). *Americans and Chinese: Passage to difference* (3rd ed.). Honolulu: University Press of Hawaii.

HU, D. J., et al. (1996). The emerging genetic diversity of HIV. *Journal of the American Medical Association, 275,* 210–216.

HU, S., et al. (1995). Linkage between sexual orientation and chromosome Xq 28 in males but not in females. *Nature Genetics, 11,* 248–256.

HUANG, Z., et al. (1997). Dual effects of weight and weight gain on breast cancer risk. *Journal of the American Medical Association, 278,* 1407–1411.

HUCKER, S. J. (1997). Sexual sadism: Psychopathology and theory, In D. R. Laws & W. O'Donohue (Eds.), *Sexual deviance: Theory, assessment, and treatment* (pp. 194–209). New York: Guilford Press.

HUDSON, B., PEPPERELL, R., & WOOD, C. (1987). The problem of infertility. In R. Pepperell, B. Hudson, & C. Wood (Eds.), *The infertile couple.* Edinburgh: Churchill-Livingstone.

HUGGINS, G. R., & CULLINS, V. E. (1990). Fertility after contraception or abortion. *Fertility and Sterility, 54,* 559–573.

HULKA, B., et al. (1982). Protection against endometrial carcinoma by combination-product oral contraceptives. *Journal of the American Medical Association, 247,* 475–477.

HUMPHREYS, L. (1970). *Tearoom trade: Impersonal sex in public places.* Chicago: Aldine.

HUNT, M. (1974). *Sexual behavior in the 1970s.* Chicago: Playboy Press.

HUNTER, D. J., et al. (1994), Sexual behavior, sexually transmitted diseases, male circumcision, and risk of HIV infection among women in Nairobi, Kenya. *AIDS, 8,* 93–99.

HUNTER, J. A., & MATHEWS, R. (1997). Sexual deviance in females. In D. R. Laws & W. O'Donohue (Eds.), *Sexual deviance: Theory, assessment, and treatment* (pp. 465–480). New York: Guilford Press.

HUPKA, R. B. (1981, August). Cultural determinants of jealousy. *Alternative Life Styles,* p. 4.

HURLBERT, D. F. (1991). The role of assertiveness in female sexuality: A comparative study between sexually assertive and sexually nonassertive women. *Journal of Sex and Marital Therapy, 17,* 183–190.

HURLBERT, D. F., APT, C., & RABEHL, S. M. (1993). Key variables to understanding female sexual satisfaction: An examination of women in nondistressed marriages. *Journal of Sex and Marital Therapy, 19,* 154–165.

HURLBERT, D. F., & WHITTAKER, K. E. (1991). The role of masturbation in marital and sexual satisfaction: A comparative study of female masturbators and nonmasturbators. *Journal of Sex Education and Therapy, 17,* 272–282.

HURLEY, T. (1984). Constitution implications of sex-change operations. *Journal of Legal Medicine, 5,* 633–664.

HUSTON, A. C. (1983). Sex-typing. In E. M. Hetherington (Ed.), *Handbook of child psychology: Vol. 4. Socialization, personality, and social development* (pp. 387–468). New York: Wiley.

HYDE, J. S., DELAMATER, J. D., PLANT, E. A., & BYRD, J. M. (1996). Sexuality during pregnancy and the year postpartum. *Journal of Sex Research, 33,* 143–151.

IMPERATO-McGINLEY, J., PETERSON, R. E., GAUTIER, T., & STURLA, E. (1979). Androgens and the evolution of male-gender identity among male pseudohermaphrodites with 5-reductase deficiency. *New England Journal of Medicine, 300,* 1233–1237.

INSTITUTE OF MEDICINE. (1990). *Nutrition during pregnancy.* Washington, DC: National Academy of Sciences, Subcommittee on Nutritional Status and Weight Gain During Pregnancy.

INSTITUTE OF MEDICINE. (1991). *Nutrition during lactation.* Washington, DC: National Academy Press.

IRVINE, J. M. (1990). From difference to sameness: Gender ideology in sexual science. *Journal of Sex Research, 27,* 7–24.

IRVINE, S., et al. (1996). Evidence of deteriorating semen quality in the United Kingdom: Birth cohort study in 577 men in Scotland over 11 years. *British Medical Journal, 312,* 467–471.

IYASU, S., et al. (1993). The epidemiology of placenta previa in the United States, 1979 through 1987. *American Journal of Obstetrics and Gynecology, 168,* 1424–1429.

JACKLIN, C. N., & MACCOBY, E. E. (1983). Issues of gender differentiation in normal development. In M. D. Levine, W. B. Carey, A. C. Crocker, & R. T. Gross (Eds.), *Developmental-behavioral pediatrics.* Philadelphia: Saunders.

JACOB, K. A. (1981). The Mosher report. *American Heritage, 32*(4), 56–64.

JACOBS, P. A., et al. (1965). Aggressive behavior, mental subnormality, and the XYY male. *Nature* (London), *208,* 1351–1352.

JACOBS, S.-H., & THOMAS, W. (1994). Native American two-spirits. *Anthropology Newsletter, 35*(8), 7.

JACQUEZ, J. A., et al. (1994). Role of the primary infection in epidemics of HIV infection in gay cohorts. *Journal of Acquired Immune Deficiency Syndromes, 7,* 1169–1184.

JAIN, A. K. (1969). Pregnancy outcome and the time required for next conception. *Population Studies, 23,* 421–433.

JAIN, A. K., & BONGAARTS, J. (1981). Breastfeeding: Patterns, correlates and fertility effects. *Studies in Family Planning, 12*(3), 79–99.

JAIN, M., JOHN, T., & KEUSCH, G. (1994). A review of HIV infection in India. *Journal of AIDS, 7,* 1185–1194.

JAMES, J. (1982). The prostitute as a victim. In M. Kirkpatrick (Ed.), *Women's sexual experience.* New York: Plenum.

JAMES, J., GOSHO, C., & WOHL, R. W. (1979). The relationship between female criminality and drug use. *International Journal of the Addictions, 14,* 215–229.

JAMISON, P. L., & GEBHARD, P. H. (1988). Penis size increase between flaccid and erect states: An analysis of the Kinsey data. *Journal of Sex Research, 24,* 177–183.

JANCIN, B. (1988). Prenatal gender selection appears to be gaining acceptance. *Obstetrics and Gynecology News, 23,* 30.

JANKOWIAK, W. R., & FISCHER, E. F. (1992). A cross-cultural perspective on romantic love. *Ethnology, 31,* 149–155.

JANSSENS, W., BUVE, A., & NKENGASONG, J. N. (1997). The puzzle of HIV-1 subtypes in Africa. *AIDS, 11,* 705-712.

JANUS, S. S., & BESS, B. (1977, May 5). *Prostitution: Option or addiction?* Paper presented at the National Convention of the American Psychiatric Association, Toronto, Canada.

JANUS, S. S., & JANUS, C. L. (1993). *The Janus report on sexual behavior.* New York: Wiley.

JARVIE, I. C. (1992). Child pornography and prostitution. In W. O'Donohue & J. H. Geer (Eds.), *The sexual abuse of children: Theory and research,* Vol. 1 (pp. 307–328). Hillsdale, NJ: Erlbaum.

JENKS, R. (1985). Swinging: A replication and test of a theory. *Journal of Sex Research, 21,* 199–210.

JENNY, C. et al. (1990). Sexually transmitted diseases in victims of rape. *New England Journal of Medicine, 322,* 713–716.

JENSEN, J. W., & GUTEK, B. A. (1982). Attributions and assignment of responsibility in sexual harassment. *Journal of Social Issues, 38,* 121–136.

JESSOR, S. L., & JESSOR, R. (1975). Transition from virginity to nonvirginity among youth: A social-psychological study over time. *Developmental Psychology, 11,* 473–484.

JICK, H., et al. (1981). Vaginal spermicides and congenital disorders. *Journal of the American Medical Association, 245,* 1329–1332.

JICK, H., JICK, S., MYERS, M. W., & VASILAKIS, C. (1996). Risk of acute myocardial infarction and low-dose combined oral contraceptives. *Lancet, 347,* 627–628.

JICK, H., et al. (1995). Risk of idiopathic cardiovascular death on non-fatal venous thromboembolism in women using oral contraceptives with differing progestagen components. *Lancet, 346,* 1589–1593.

JOFFE, G. P., et al. (1992). Multiple partners and partner choice as risk factors for sexually transmitted disease among female college students. *Sexually Transmitted Diseases, 19,* 272–278.

JOHANSSON, J. E., et al. (1997). Fifteen-year survival in prostate cancer: A prospective, population-based study in Sweden. *Journal of the American Medical Association, 277,* 467–471.

JOHN, D., & SHELTON, B. A. (1997). The production of gender among black and white women and men: The case of household labor. *Sex Roles, 36,* 171–193.

JOHN, G. C., & KREISS, J. (1996). Mother-to-child transmission of human immunodeficiency virus type 1. *Epidemiologic Reviews, 18,* 149–157.

JOHNSON, B. E., KUCK, D. L., & SCHANDER, P. R. (1997). Rape myth acceptance and sociodemographic characteristics: A multidimensional analysis. *Sex Roles, 36,* 693–710.

JOHNSON, J. W. C., LONGMATE, J. A., & FRENTZEN, B. (1992). Excessive maternal weight and pregnancy outcome. *American Journal of Obstetrics and Gynecology, 167,* 353–372.

JOHNSON, M. E., BREMS, C., & ALFORD-KEATING, P. (1997). Personality correlates of homophobia. *Journal of Homosexuality, 34,* 57–69.

JOHNSON, T. C. (1989). Female child perpetrators: Children who molest other children. *Child Abuse and Neglect, 13,* 571–585.

JONES, E. F., et al. (1986). *Teenage pregnancy in industrialized countries.* New Haven, CT: Yale University Press.

JONES, W. K., SMITH, J., & KIEKE JR., B., & WILCOX, L. (1997). Female genital mutilation/female circumcision: Who is at risk in the United States? *Public Health Reports, 112,* 368–377.

JOURARD, S. M., & LASAKOW, P. (1958). Some factors in self-disclosure. *Journal of Abnormal and Social Psychology, 56,* 91–98.

JOW, W.,& GOLDSTEIN, M. (1994, May) No-scalpel vasectomy offers minimal invasiveness. *Contemporary Ob/Gyn,* pp. 83–96.

JOYCE, T. J., & MOCAN, N. H. (1990). The impact of legalized abortion on adolescent childbearing in New York City. *American Journal of Public Health, 80,* 273–278.

JUNGINGER, J. (1997). Fetishism: Assessment and treatment. In D. R. Laws & W. O'Donohue (Eds.), *Sexual deviance: Theory, assessment, and treatment* (pp. 92–110). New York: Guilford Press.

JURKOVIC, D., et al. (1993). Coelocentesis: A new technique for early prenatal diagnosis. *Lancet, 341,* 1623–1624.

JUSTICE, B., & JUSTICE, R. (1979). *The broken taboo: Sex in the family.* New York: Human Sciences Press.

KAESER, L. (1989). Reconsidering the age limits on pill use. *Family Planning Perspectives, 21,* 273–274.

KAFKA, M. P. (1991). Successful antidepressant treatment of nonparaphiliac sexual addictions and paraphilias in men. *Journal of Clinical Psychiatry, 52,* 60–65.

KAFKA, M. P. (1997). A monoamine hypothesis for the pathophysiology of paraphilic disorders. *Archives of Sexual Behavior, 26,* 343–358.

KAFKA, M. P. (1997). Hypersexual desire in males: An operational definition and clinical implications for males with paraphilias and paraphilia-related disorders. *Archives of Sexual Behavior, 26,* 505–526.

KAHN, J. R., & LONDON, K. A. (1991). Premarital sex and the risk of divorce. *Journal of Marriage and the Family, 53,* 845–855.

KAISER FAMILY FOUNDATION (1996, June 24). Survey on teens and sex: What they say teens today need to know, and who they listen to. [News release].

KALB, C. (1997, May 5). How old is too old? *Newsweek,* p. 64.

KALICHMAN, S. C., CAREY, M. P., & JOHNSON, B. T. (1996). Prevention of sexually transmitted HIV infection: A meta-analytic review of the behavioral outcome literature. *Annals of Behavioral Medicine, 18,* 6–15.

KALICHMAN, S. C., & SIKKEMA, K. J. (1994). Psychological sequelae of HIV infection and AIDS: Review of empirical findings. *Clinical Psychology Reviews, 7,* 611–632.

KALLMAN, F. J. (1952). Comparative twin study on the genetic aspects of male homosexuality. *Journal of Nervous and Mental Disease, 115,* 283–298.

KAMEL, H. (1983). *Downtown street hustlers: The role of dramaturgical imaging practices in the social construction of male prostitution.* San Diego: University of California.

KAMINSKI, P. F., SOROSKY, J. J., PEES, R. C., & PODCZASKI, E. S. (1993). Correction of massive vaginal prolapse in an older population: A four-year experience at a rural tertiary care center. *Journal of the American Geriatric Society, 41,* 42–44.

KANEKAR, S., PINTO, N. J. P., & MAZUMDAR, D. (1985). Causal and moral responsibility of victims of rape and robbery. *Journal of Applied Social Psychology, 15,* 622–637.

KANIN, E. (1967). Reference groups and sex conduct norms. *Sociological Quarterly, 8,* 495–504.

KANIN, E. J. (1985). Date rapists: Differential sexual socialization and relative deprivation. *Archives of Sexual Behavior, 14,* 219–231.

KANIN, E. J. (1994). False rape allegations. *Archives of Sexual Behavior, 23,* 81–92.

KANTROWITZ, B., et al. (1996, November 4). Gay families come out. *Newsweek,* 50–57.

KAPLAN, H. S. (1974). *The new sex therapy.* New York: Bruner/Mazel.

KAPLAN, H. S. (1977). Hypoactive sexual desire. *Journal of Sex and Marital Therapy, 3,* 3–9.

KAPLAN, H. S. (1979). *Disorders of desire.* New York: Bruner/Mazel.

KAPLAN, H. S. (1983). *The evaluation of sexual disorders: Psychological and medical aspects.* New York: Brunner/Mazel.

KAPLAN, H. S. (1987). *Sexual aversion, sexual phobias, and panic disorders.* New York: Brunner/Mazel.

KAPLAN, H. S. (1988). Intimacy disorders and sexual panic states. *Journal of Sex and Marital Therapy, 14,* 3–12.

KAPLAN, H. S. (1989). *How to overcome premature ejaculation.* New York: Brunner/Mazel.

KAPLAN, H. S. (1992a). A neglected issue: The sexual side effects of current treatments for breast cancer. *Journal of Sex and Marital Therapy, 18,* 3–19.

KAPLAN, H. S. (1992b). Does the CAT technique enhance female orgasm? *Journal of Sex and Marital Therapy, 18,* 285–291.

KAPLAN, H. S. (1993). Post-ejaculatory pain syndrome. *Journal of Sex and Marital Therapy, 19,* 91–103.

KAPLAN, H. S., & OWETT, T. (1993). The female androgen deficiency syndrome. *Journal of Sex and Marital Therapy, 19,* 3–24.

KAPLAN, L. (1996). Sexual and institutional issues when one spouse resides in the community and the other lives in a nursing home. *Sexuality and Disability, 14,* 281–293.

KAPLAN, M. S., & KRUEGER, R. B. (1997). Voyeurism: Psychopathology and theory. In D. R. Laws & W. O'Donohue (Eds.), *Sexual deviance: Theory, assessment, and treatment* (pp. 297–310). New York: Guilford Press.

KARMEL, M. (1959). *Thank you Dr. Lamaze.* Philadelphia: Lippincott.

KATLAMA, C., & DICKINSON, G. M. (1993). Update on opportunistic infections. *AIDS, 7,* S185–S194.

KATZ, B. L., & BURT, M. R. (1986, August). *Effects of familiarity with the rapist on postrape recovery.* Paper presented at the meeting of the American Psychological Association, Washington, DC.

KATZ, J. (1976). *Gay American history: Lesbian and gay men in the U.S.A.* New York: Crowell.

KAUFERT, P. A., & LOCK, M. (1997). Medicalization of women's third age. *Journal of Psychosomatic Obstetrics and Gynecology, 18,* 81–86.

KAWACHI, I., COLDITZ, G. A., & HANKINSON, S. (1994). Long-term benefits and risks of alternative methods of fertility control in the United States. *Contraception, 50,* 1–16.

KAWAS, C., et al. (1970). A prospective study of estrogen replacement therapy and the risk of developing Alzheimer's disease: The Baltimore Longitudinal Study of Aging. *Neurology, 48,* 1517–1521.

KEET, I. P. M., et al. (1994). Predictors of disease progression in HIV-infected homosexual men with CD4 + cells—less than 200 X 10(6)/L but free of AIDS-defining clinical disease. *AIDS, 8,* 1577–1583.

KEGEL, A. (1952). Sexual functions of the pubococcygeus muscle. *Western Journal of Surgery, Obstetrics, and Gynecology, 60,* 521–524.

KELEN, G. D., et al. (1988). Unrecognized human immunodeficiency virus infection in emergency department patients. *New England Journal of Medicine, 318,* 1645–1650.

KELLETT, J. M. (1991). Sexuality and the elderly. *Sexual and Marital Therapy, 6,* 147–155.

KELLOGG, J. H. (1888). *Plain facts for young and old, embracing the natural history and hygiene of organic life.*

KELLOGG, N. D., & HOFFMAN, T. J. (1997). Child sexual revictimization by multiple perpetrators. *Child Abuse & Neglect, 21,* 953–964.

KELLY, J. A., ST. LAWRENCE, J. S., HOOD, H. V., & BRASFIELD, T. L. (1989). Behavioral intervention to reduce AIDS risk activities. *Journal of Consulting and Clinical Psychology, 57,* 60–67.

KELLY, J. A., & WORELL, J. (1977). New formulations of sex roles and androgyny: A critical review. *Journal of Consulting and Clinical Psychology, 45,* 1101–1115.

KELLY, M. P., STRASSBERG, D. S., & KIRCHER, J. R. (1990). Attitudinal and experiential correlates of anorgasmia. *Archives of Sexual Behavior, 19,* 165–177.

KELLY, R. J. (1982). Behavioral reorientation of pedophiles: Can it be done? *Clinical Psychology Review, 2,* 387–408.

KENDALL-TACKETT, K. A., & SIMON, A. F. (1987). Perpetrators and their acts: Data from 365 adults molested as children. *Child Abuse and Neglect, 11,* 237–245.

KENDALL-TACKETT, K. A., WILLIAMS, L. M., & FINKELHOR, D. (1993). Impact of sexual abuse on children: A review and synthesis of recent empirical studies. *Psychological Bulletin, 113,* 164–180.

KENDRICK, D. T., & KEEFE, R. C. (1992). Age preferences in mates reflect sex differences in reproductive strategies. *Behavioral and Brain Sciences, 15,* 75–133.

KENNELL, J., et al.(1991). Continuous emotional support during labor in a U.S. hospital. *Journal of the American Medical Association, 265,* 2197–2201.

KEPHART, W. M. (1967). Some correlates of romantic love. *Journal of Marriage and the Family, 29,* 470–474.

KERLIKOWSKE, K., et al. (1995). Efficacy of screening mammography: A meta-analysis. *Journal of the American Medical Association, 273,* 149–154.

KERNBERG, O. F. (1991). Sadomasochism, sexual excitement, and perversion. *Journal of the American Psychoanalytic Association, 39,* 333–362.

KESTELMAN, P., & TRUSSELL, J. (1991). Efficacy of the simultaneous use of condoms and spermicides. *Family Planning Perspectives, 23,* 226–232.

KETTL, P., et al. (1991). Female sexuality after spinal cord injury. *Sexuality and Disability, 9,* 287–295.

KHABBAZ, R. F., et al. (1990). Seroprevalence and risk factors for HTLV-I/II infection among female prostitutes in the United States. *Journal of the American Medical Association, 263,* 60–64.

KIKUCHI, J. J. (1988, Fall). Rhode Island develops successful intervention program for adolescents. *NCASA News,* p. 26.

KIM, S. H., et al. (1997). Microsurgical reversal of tubal sterilization: A report on 1,118 cases. *Fertility and Sterility, 68,* 865–870.

KIMMEL, M. S. (1989). From separate spheres to sexual equality: Men's responses to feminism at the turn of the century. In B. J. Risman & P. Schwartz (Eds.), *Gender in intimate relationships: A microstructural approach.* Belmont, CA: Wadsworth.

KING, B. M., & ANDERSON, P. B. (1994). A failure of HIV education: Sex can be more important than a long life. *Journal of Health Education, 25,* 13–18.

KING, B. M., & LOCOCO, G. C. (1990). Effects of sexually explicit textbook drawings on enrollment and family communication. *Journal of Sex Education and Therapy, 16,* 38–53.

KING, B. M., & LORUSSO, J. (1997). Discussions in the home about sex: Different recollections by parents and children. *Journal of Sex and Marital Therapy, 23,* 52–60.

KING, B. M., PARISI, L. S., & O'DWYER, K. R. (1993). College sexuality education promotes future discussions about sexuality between former students and their children. *Journal of Sex Education and Therapy, 19,* 285–293.

KINSEY, A. C., POMEROY, W., & MARTIN, C. (1948). *Sexual behavior in the human male.* Philadelphia: Saunders.

KINSEY, A. C., POMEROY, W., MARTIN, C., & GEBHARD, P. (1953). *Sexual behavior in the human female.* Philadelphia: Saunders.

KIRBY, D. (1984). *Sexuality education: An evaluation of programs and their effects.* Santa Cruz, CA: Network Publications.

KIRBY, D., KORPI, M., BARTH, R. P., & CAGAMPANG, H. H. (1997). The impact of the Postponing Sexual Involvement curriculum among youths in California. *Family Planning Perspectives, 29,* 100–108.

KIRK, M., & MADSEN, H. (1989). *How America will conquer its fear and hatred of homosexuals in the '90s.* New York: Doubleday.

KITE, M. E., & WHITLEY, B. E., JR. (1996). Sex differences in attitudes toward homosexual persons, behaviors and civil rights: A meta-analysis. *Personality and Social Psychology Bulletin, 22,* 336–352.

KJERSGAARD, A. G., et al. (1989). Male or female sterilization: A comparative study. *Fertility and Sterility, 51,* 439–443.

KLECK, R. E., RICHARDSON, S. A., & RONALD, L. (1974). Physical appearance cues and interpersonal attraction in children. *Child Development, 45,* 305–310.

KLEPINGER, D. H., BILLY, J. O. G., TANFER, K., & GRADY, W. R. (1993). Perception of AIDS risk and severity and their association with risk-related behavior among U.S. men. *Family Planning Perspectives, 25,* 74–82.

KLIEWER, E. U., & SMITH, K. R. (1995). Breast cancer mortality among immigrants in Australia and Canada. *Journal of the National Cancer Institute, 87,* 1154–1161.

KLIGMAN, E. W., & PEPIN, E. (1992). Prescribing physical activity for older patients. *Geriatrics, 47*(8), 33–47.

KLITSCH, M. (1993a). Injectable hormones and regulatory controversy: An end to the long-running story. *Family Planning Perspectives, 25,* 37–40.

KLITSCH, M. (1993b). Vasectomy and prostate cancer: More questions than answers. *Family Planning Perspectives, 25,* 133–135.

KLONOFF-COHEN, H. S., SAVITZ, D. A., CEFALO, R. C., & McCANN, M. F. (1989). An epidemiologic study of contraception and preeclampsia. *Journal of the American Medical Association, 262,* 3143–3147.

KNIGHT, R. A., & PRENTKY, R. A. (1996). A 25-year follow-up of rapists and child molesters. *International Journal of Psychology, 31,* 3073.

KNOLL, J. (1997). Sexual performance and longevity. *Experimental Gerontology, 32,* 539–552.

KOESKE, R. (1983). Lifting the curse of menstruation: Toward a feminist perspective on the menstrual cycle. *Women and Health, 8,* 1–16.

KOHLBERG, L. (1966). A cognitive-developmental analysis of children's sex-role concepts and attitudes. In E. E. Maccoby (Ed.), *The development of sex differences* (pp. 82–173). Stanford, CA: Stanford University Press.

KOHLBERG, L., & ULLIAN, D. Z. (1974). Stages in the development of psychosexual concepts and attitudes. In R. C. Friedman, R. M. Richart, & R. L. Van de Wiele (Eds.), *Sex differences in behavior* (pp. 209–222). New York: Wiley.

KOLODNY, R. C. (1980, November). *Adolescent sexuality.* Paper presented at the Michigan Personnel and Guidance Association annual convention, Detroit.

KOLODNY, R. C. (1981). *Effects of marijuana on sexual behavior and function.* Paper presented at the Midwestern Conference on Drug Use, St. Louis, MO.

KOLODNY, R. C., MASTERS, W. H., KOLODNY, W. H., & TORO, G. (1974). Depression of plasma testosterone levels after chronic intensive marijuana use. *New England Journal of Medicine, 290,* 872–874.

KOLS, A., et al. (1982, May–June). Oral contraceptives in the 1980's. *Population Reports,* Series A, No. 6.

KOMAROVSKY, M. (1992). The concept of gender role revisited. *Gender and Society, 6,* 301–313.

KOONIN, L. M., et al. (1992). Abortion surveillance—United States, 1989. *Morbidity and Mortality Weekly, 41* (Spec. Suppl. 5).

KORHONEN, J., STENMAN, V.-H. H., & YLOSTALO, P. (1996). Low-dose oral methotrexate with expectant management of ectopic pregnancy. *Obstetrics and Gynecology, 88,* 775–778.

KOSS, M. P. (1993a). Detecting the scope of rape: A review of prevalence research methods. *Journal of Interpersonal Violence, 8,* 198–222.

Koss, M. P. (1993b). Rape: Scope, impact, interventions, and public policy responses. *American Psychologist, 48,* 1062–1069.

Koss, M. P., & Dinero, T. E. (1989). Predictors of sexual aggression among a national sample of male college students. In R. A. Prentky & V.L. Quinsley (Eds.), *Human sexual aggression: Current perspectives* (pp. 133–146). New York: New York Academy of Sciences.

Koss, M. P., Dinero, T. E., Seibel, C. A., & Cox, S. L. (1988). Stranger and acquaintance rape: Are there differences in the victim's experience? *Psychology of Women Quarterly, 12,* 1–23.

Koss, M. P., Gidycz, C. A., & Wisniewski, N. (1987). The scope of rape: Incidence and prevalence of sexual aggression and victimization in a national sample of higher education students. *Journal of Consulting and Clinical Psychology, 55,* 162–170.

Koss, M. P., Leonard, K., Beezley, D., & Oros, C. (1985). Nonstranger sexual aggression: A discriminant analysis of the psychological characteristics of undetected offenders. *Sex Roles, 12,* 981–982.

Koss, M. P., & Oros, C. J. (1982). Sexual experiences survey: A research instrument investigating sexual aggression and victimization. *Journal of Consulting and Clinical Psychology, 50,* 455–457.

Koutsky, L. A., et al. (1992). A cohort study of the risk of cervical intraepithelial neoplasia grade 2 or 3 in relation to papillomavirus infection. *New England Journal of Medicine, 327,* 1272–1278.

Kovacs, G. T., et al. (1986). The contraceptive diaphragm: Is it an acceptable method in the 1980's? *Australian & New Zealand Journal of Obstetrics and Gynaecology, 26,* 76–79.

Krafft-Ebing, R. V. (1951). *Aberrations of sexual life.* New York: Staples Press.

Krafft-Ebing, R. V. (1978; originally 1886). *Psychopathia sexualis.* New York: Stein & Day.

Kraus, S. J., & Stone, K. M. (1990). Management of genital infection caused by human papillomavirus. *Reviews of Infectious Diseases, 12,* S620–S632.

Kravis, D., & Molitch, M. E. (1990, February). Endocrine causes of impotence. *Medical Aspects of Human Sexuality,* pp. 62–67.

Kreiss, J., et al. (1992). Efficacy of nonoxynol 9 contraceptive sponge use in preventing heterosexual acquisition of HIV in Nairobi prostitutes. *Journal of the American Medical Association, 268,* 477–482.

Krieger, L. M. (1993, May). Gay youths ignoring AIDS threat, health study says. *San Francisco Examiner.*

Krieger, W. G. (1976). Infant influences and the parent sex by child sex interaction in the socialization process. *JSAS Catalogue of Selected Documents in Psychology, 6,* 36.

Kristiansen, C. M., Felton, K. A., & Hovdestad, W. E. (1996). Recovered memories of child abuse: Fact, fantasy or fancy? *Women & Therapy, 19,* 47–59.

Krueger, S. L., Dunson, T. R., & Amatya, R. N. (1994). Norplant contraceptive acceptability among women in five Asian countries. *Contraception, 50,* 349–361.

Kruesi, M. J. P., et al. (1992). Paraphilias: A double-blind crossover comparison of clomipramine versus desipramine. *Archives of Sexual Behavior, 21,* 587–593.

Krug, R. S. (1989). Adult male report of childhood sexual abuse by mothers: Case descriptions and long-term consequences. *Child Abuse and Neglect, 13,* 111–119.

Ku, L. C., Sonenstein, F. L., & Pleck, J. H. (1992). The association of AIDS education and sex education with sexual behavior and condom use among teenage men. *Family Planning Perspectives, 24,* 100–106.

Ku, L. C., Sonenstein, F. L., & Pleck, J. H. (1993). Young men's risk behaviors for HIV infection and sexually transmitted diseases, 1988 through 1991. *American Journal of Public Health, 83,* 1609–1615.

Ku, L. C., Sonenstein, F. L., & Pleck, J. H. (1994). The dynamics of young men's condom use during and across relationships. *Family Planning Perspectives, 26,* 246–251.

Kuhn, D., Nash, S., & Brucken, L. (1978). Sex role concepts of two- and three-year olds. *Child Development, 49,* 445–451.

Kulhanjian, J. A., et al. (1992). Identification of women at unsuspected risk of primary infection with herpes simplex virus type 2 during pregnancy. *New England Journal of Medicine, 326,* 16–20.

Kulin, H. E. (1996). Extensive personal experience: Delayed puberty. *Journal of Clinical Endocrinology and Metabolism, 81,* 3460–3464.

Kurdek, L. A. (1995). Assessing multiple determinants of relationship commitment in cohabiting gay, cohabiting lesbian, dating heterosexual and married heterosexual couples. *Family Relations, 44,* 261–266.

Kusseling, F. S., Shapiro, M. F., Greenberg, J. M., & Wenger, N. S. (1996). Understanding why heterosexual adults do not practice safer sex: A comparison of two samples. *AIDS Education and Prevention, 8,* 247–257.

Kutchinsky, B. (1973). The effect of easy availability of pornography on the incidence of sex crimes. *Journal of Social Issues, 29,* 163–182.

Kutchinsky, B. (1996). A case for testing: The roles of safer sex and HIV testing in AIDS prevention. In P. B. Anderson, D. de Mauro, & R. J. Noonan (Eds.), *Does anyone still remember when sex was fun?* (3rd ed.). (pp. 44–70) Dubuque, IA: Kendall/Hunt.

Laan, E., & Everaerd, W. (1996). Determinants of female sexual arousal: Psychophysiological theory and data. In R. C. Rosen, C. M. Davis, & H. J. Ruppel, Jr. (Eds.), *Annual review of sex research,* Vol. VI (pp. 32–76). Mount Vernon, IA: Society for the Scientific Study of Sexuality.

Laan, E., Everaerd, W., Van Bellen, G., & Hanewald, G. (1994). Women's sexual and emotional responses to male- and female-produced erotica. *Archives of Sexual Behavior, 23,* 153–169.

Labbok, M. H., et al. (1997). Multicenter study of the lactational amenorrhea method (LAM): I. Efficacy, duration, and implications for clinical application. *Contraception, 55,* 327–336.

Labouvie-Vief, G. (1990). Modes of knowledge and the organization of development. In M. L. Commons, C. Armon, L. Kohlberg, S. A. Richards, T. A. Gropzer, & J. Sannott (Eds.), *Adult development models and methods in the study of adolescents and adult thought.* New York: Praeger.

Labouvie-Vief, G., Hakim-Larson, J., & Hobart, C. J. (1987). Age, ego level, and the life-span development of coping and defense processes. *Psychology and Aging, 2,* 286–293.

Ladas, A. K., Whipple, B., & Perry, J. D. (1982). *The G spot and other recent discoveries about human sexuality.* New York: Holt, Rinehart & Winston.

Lafferty, W. E., et al. (1987). Recurrences after oral and genital herpes simplex virus infection: Influence of site of infection and viral type. *New England Journal of Medicine, 316,* 1444–1449.

Laga, M., Meheus, A., & Piot, P. (1989). Epidemiology and control of gonococcal ophthalmia neonatorum. *Bulletin of the World Health Organization, 67,* 471–478.

Lakin, M. M., & Montague, D. K. (1988). Intracavernous injection therapy in post priapism cavernosal fibrosis., *Journal of Urology, 140,* 828–829.

Lalumiere, M. L., Chalmers, L. J., Quinsey, V. L., & Seto, M. C. (1996). A test of the male deprivation hypothesis of sexual coercion. *Ethology and Sociobiology, 17,* 299–318.

Lalumiere, M. L., & Quinsey, V. L. (1994). The discriminability of rapists from non-sex offenders using phallometric measures: A meta-analysis. *Criminal Justice and Behavior, 21,* 150–175.

Lamaze, F. (1970). *Painless childbirth.* Chicago: H. Regnery. (Originally published 1956).

Lamb, M. E. (1986). *The father's role: Cross-cultural perspectives.* Hillsdale, NJ: Erlbaum.

Lamb, M.E., et al. (1996). Investigative interviews of alleged sexual abuse victims with and without anatomical dolls. *Child Abuse and Neglect, 20,* 1251–1259.

Lambert, J. D. (1995). A legacy of pleasure. In P. B. Anderson, D. de Mauro, & R. J. Noonan (Eds.), *Does anyone still remember when sex was fun?* (3rd ed.) (pp. 13–30). Dubuque, IA: Kendall/Hunt.

Landesman, S. H., et al. (1996). Obstetrical factors and the transmission of human immunodeficiency virus type I from mother to child. *New England Journal of Medicine, 334,* 1617–1623.

Lang, W., et al. (1989). Patterns of T lymphocyte changes with human immunodeficiency virus infection: From seroconversion to the development of AIDS. *Journal of Acquired Immune Deficiency Syndromes, 2,* 63–69.

Langevin, R. (1983). *Sexual strands.* Hillsdale, NJ: Erlbaum.

Langevin, R. (1985). An overview of the paraphelias. In M. H. Ben-Aron, S. J., Hucker, & C. D. Webster (Eds.), *Clinical criminology: The assessment and treatment of criminal behavior* (pp. 177–190). Toronto: M & M Graphics and Clarke Institute of Psychiatry.

Langevin, R., et al. (1979). Experimental studies of the etiology of genital exhibitionism. *Archives of Sexual Behavior, 8,* 307–332.

Langevin, R., et al. (1988). Sexual sadism: Brain, blood, and behavior. *Annals of the New York Academy of Sciences, 528,* 163–171.

Langfeldt, T. (1981). Sexual development in children. In M. Cook & K. Howells (Eds.), *Adult sexual interest in children.* London: Academic Press.

Langlois, J. H., & Downs, A. C. (1980). Mothers, fathers, and peers as socialization agents of sex-typed play behaviors in young children. *Child Development, 51,* 1237–1247.

Langmyhr, G. J. (1976, June). Varieties of coital positions: Advantages and disadvantages. *Medical Aspects of Human Sexuality,* pp. 128–139.

LANGSTON, A. A., et al. (1996). BRCA1 mutations in a population-based sample of young women with breast cancer. *New England Journal of Medicine, 334,* 137–142.

LANYON, R. I. (1986). Theory and treatment in child molestation. *Journal of Consulting and Clinical Psychology, 54,* 176–182.

L'ARMAND, K. L., & PEPITONE, A. (1982). Judgements of rape: A study of victim-rapist relationship and victim sexual history. *Personality and Social Psychology Bulletin, 8,* 134–139.

LARSON, N. R., & MADDOCK, J. W. (1986). Structural and functional variables in incest family systems: Implications for assessment and treatment. *Journal of Psychotherapy and the Family, 2,* 27–44.

LARUE, G. A. (1989, July/August). Religious traditions and circumcision. *The Truth Seeker,* pp. 4–8.

LAUER, J., & LAUER, R. (1985, June). Marriages made to last. *Psychology Today,* pp. 22–26.

LAUGHLIN, C. D. J. R., & ALLGEIER, E. R. (1979). *Ethnography of the So of Northeastern Uganda.* New Haven, CT: Human Relations Area Files, Inc.

LAUMANN, E., MICHAEL, R., GAGNON, J., & MICHAELS, S. (1994). *The social organization of sexuality: Sexual practices in the United States.* Chicago: University of Chicago Press.

LAUMANN, E. O., MASI, C. M., & ZUCKERMAN, E. W. (1997). Circumcision in the United States: Prevalence, prophylactic effects, and sexual practice. *Journal of the American Medical Association, 277,* 1052–1057.

LAUPER, U. (1996). Genital candidosis. In P. Elsner and A. Eichmann (Eds.), *Sexually transmitted diseases. Advances in diagnosis and treatment* (pp. 123–131). Basel: Karger.

LAURENCE, J., et al. (1992). Acquired immunodeficiency without evidence of infection with human immunodeficiency virus types 1 and 2. *Lancet, 340,* 273–274.

LAVEE, Y. (1991). Western and non-Western human sexuality: Implications for clinical practice. *Journal of Sex and Marital Therapy, 17,* 203–213.

LAWRENCE, D. H. (1930). *Lady Chatterley's lover.* New York: W. Faro.

LAWS, A., & GOLDING, J. M. (1996). Sexual assault history and eating disorder symptoms among white, Hispanic and African-American women and men. *American Journal of Public Health, 86,* 579–582.

LAWSON, C. (1993). Mother-son sexual abuse: Rare or underreported? A critique of the research. *Child Abuse & Neglect, 17,* 261–269.

LAZAR, P. (1996). Maturation folliculaire, conceptions gémellaires dizygotes et âge maternal. *C. R. Académie des Sciences, Paris, 319,* 1139–1144.

LAZARUS, J. A. (1992). Sex with former patients almost always unethical. *American Journal of Psychiatry, 149,* 855–857.

LEAR, D. (1995). Sexual communication in the age of AIDS: The construction of risk and trust among young adults. *Social Science Medical, 41,* 1311–1323.

LEBEL, R. R. (1983). Consanguinity studies in Wisconsin. 1. Secular trends in consanguineous marriage, 1843–1981. *American Journal of Medical Genetics, 15,* 543–560.

LEBOYER, F. (1975). *Birth without violence.* New York: Knopf.

LEDE, R. L., BELIZAN, J. M., & CARROLI, G. (1996). Is routine use of episiotomy justified? *American Journal of Obstetrics and Gynecology, 174,* 1399–1402.

LEE, J. A. (1974, October, 8). The styles of loving. *Psychology Today,* pp. 43–50.

LEE, J. A. (1976). *The colors of love.* Englewood Cliffs, NJ: Prentice Hall.

LEE, J. A. (1988). Love-styles. In R. J. Sternberg & M. L. Barnes (Eds.), *The psychology of love* (pp. 38–67). New Haven, CT: Yale University Press.

LEE, N. C., RUBIN, G. L., ORY, H. W., & BURKMAN, R. T. (1983). Type of intrauterine device and the risk of pelvic inflammatory disease. *Obstetrics and Gynecology, 62,* 1–6.

LEFEVRE, M. L., et al. (1993). A randomized trial of prenatal ultrasonagraphic screening: Impact on maternal management and outcome. *American Journal of Obstetrics and Gynecology, 169,* 483–489.

LEIBLUM, S. R., PERVIN, L., & CAMPBELL, E. H. (1989). The treatment of vaginismus: Success and failure. In S. R. Leiblum & R. C. Rosen (Eds). *Principles and practice of sex therapy: Update for the 1990s* (pp. 113–138). New York: Guilford Press.

LEIBLUM, S. R., & ROSEN, R. C. (1989). *Principles and practice of sex therapy: An update for the 1990s.* New York: Guilford.

LEIBLUM, S. R., & ROSEN, R. C. (1991). Couples therapy for erectile disorders: Conceptual and clinical considerations. *Journal of Sex and Marital Therapy, 17,* 147–159.

LEIBLUM, S. R., & ROSEN, R. C. (Eds.). (1988). *Sexual desire disorders.* New York: Guilford.

LEIBLUM, S R., et al. (1983). Vaginal atrophy in the post menopausal woman: The importance of sexual activity and hormones. *Journal of the American Medical Association, 249,* 2195–2198.

LEIFER, M. (1980). *Psychological effects of motherhood: A study of first pregnancy.* New York: Praeger.

LEIGH, B. (1989). Reasons for having and avoiding sex: Gender, sexual orientation, and relationship to sexual behavior. *Journal of Sex Research, 26,* 199–208.

LEITENBERG, H., & HENNING, K. (1995). Sexual fantasy. *Psychological Bulletin, 117,* 469–496.

LEITENBERG, H., DETZER, M. J., & SREBNIK, D. (1993). Gender differences in masturbation and the relation of masturbation experience in preadolescence and/or early adolescence to sexual behavior and sexual adjustment in young adulthood. *Archives of Sexual Behavior, 22,* 87–98.

LEITENBERG, H., GREENWALD, E., & CADO, S. (1992). A retrospective study of long-term methods of coping with having been sexually abused during childhood. *Child Abuse and Neglect, 16,* 399–407.

LELAND, J. (1995, July 17). Bisexuality. *Newsweek,* pp. 44–50.

LELAND, J., et al. (1997, November 17). A pill for impotence? *Newsweek,* pp. 62–68.

LENDERKING, W. R. et al. (1994). Evaluation of the quality of life associated with zidovudine treatment in asymptomatic human immunodeficiency virus infection. *New England Journal of Medicine, 330,* 738–743.

LEONARD, T. L., FREUND, M., & PLATT, J. J. (1989). Behavior of clients of prostitutes. *American Journal of Public Health, 79,* 903.

LERMAN, H. (1986). *A mote in Freud's eye: From psychoanalysis to the psychology of women.* New York: Springer.

LEVAY, S. (1991). A difference in hypothalamic structure between heterosexual and homosexual men. *Science, 253,* 1034–1037.

LEVE, L. D., & FAGOT, B. I. (1997). Gender-role socialization and discipline processes in one- and two-parent families. *Sex Roles, 36,* 1–21.

LEVER J. (1994, August 23). Sexual revelations: The 1994 Advocate survey of sexuality and relationships: The men. *The Advocate,* 661/662, 16–24.

LEVER, J., et al. (1992). Behavior patterns and sexual identity of bisexual males. *Journal of Sex Research, 29,* 141–167.

LEVIN, R. J., & LEVIN, A. (1975, September). Sexual pleasure: The surprising performances of 100,000 women. *Redbook,* p. 51.

LEVIN, R. J., & WAGNER, G. (1985). Orgasm in women in the laboratory—Quantitative studies on duration, intensity, latency, and vaginal blood flow. *Archives of Sexual Behavior, 11,* 367–386.

LEVINE, M. P. (1992). The life of gay clones. In G. Herdt (Ed.), *Gay culture in America: Essays from the field.* Boston: Beacon Press.

LEVINE, M. P., & TROIDEN, R. R. (1988). The myth of sexual compulsivity. *Journal of Sex Research, 25,* 347–363.

LEVINE, R. A. (1959). Gussi sex offenses: A study in social control. *American Anthropologist, 61,* 965–990.

LEVINE, R. A. (1974). Gussi sex offenses: A study in social control. In N. N. Wagner (Ed.), *Perspectives on human sexuality* (pp. 308–352). New York: Behavioral Publishers.

LEVINE, R. J., et al. (1990). Differences in the quality of semen in outdoor workers during summer and winter. *New England Journal of Medicine, 323,* 12–16.

LEVINE, S. B. (1993). Gender-disturbed males. *Journal of Sex and Marital Therapy, 19,* 131–141.

LEVINE, S. B., & ALTHOF, S. E. (1991). The pathogenesis of psychogenic erectile dysfunction. *Journal of Sex Education and Therapy, 17,* 251–266.

LEVINGER, G. (1988). Can we picture "love"? In R. J. Sternberg & M. L. Barnes (Eds.), *The psychology of love* (pp. 139–158). New Haven, CT: Yale University Press.

LÉVI-STRAUSS, C. (1969). *The elementary structures of kinship.* Boston: Beacon Press.

LEVITT, E. E., MOSER, C., & JAMISON, K. V. (1994). The prevalence and some attributes of females in the sadomasochistic subculture: A second report. *Archives of Sexual Behavior, 23,* 465–473.

LEVRAN, D., et al. (1990). Pregnancy potential of human oocytes—The effect of cryopreservation. *New England Journal of Medicine, 323,* 1153–1156.

LEWES, K. (1992). Homophobia and the heterosexual fear of AIDS. *American Imago, 49,* 343–356.

LEWIS, D. K., WATTERS, J. K., & CASE, P. (1990). The prevalence of high-risk sexual behavior in male intravenous drug users with steady female partners. *American Journal of Public Health, 80,* 465–466.

LEWIS, M., & BUTLER, R. (1994). *Love and sex after 60.* New York: Ballantine.

LEWIS, M. A., et al. (1996). Third generation oral contraceptives and risk of myocardial infarction: An international case-control study. *British Medical Journal, 312,* 88–90.

LEWIS, W. J. (1997). Factors associated with post-abortion adjustment problems: Implications for triage. *Canadian Journal of Human Sexuality, 6,* 9–16.

LIDEGAARD, O., & HELM, P. (1990). Pelvic inflammatory disease: The influence of contraceptive, sexual, and social life events. *Contraception, 41,* 475–483.

LIDSTER, C., & HORSBURGH, M. (1994). Masturbation—Beyond myth and taboo. *Nursing Forum, 29*(3), 18–26.

LIEBERMAN, E. J., & PECK, E. (1982). *Sex and birth control.* New York: Schocken.

LIEBOWITZ, M. (1983). *The chemistry of love.* Boston: Little, Brown.

LIEF, H. J. (1977). Inhibited sexual desire. *Medical Aspects of Human Sexuality, 7,* 94–95.

LIGHTFOOT-KLEIN, H. (1989). *Prisoners of ritual: An odyssey into female genital circumcision in Africa.* Binghamton, NY: Haworth Press.

LIGHTFOOT-KLEIN, H., & SHAW, E. (1991). Special needs of ritually circumcised women patients. *Journal of Obstetrics, Gynecology and Neonatal Nursing, 20,* 102–107.

LIN, J.-S. L., et al. (1992). Underdiagnosis of *Chlamydia trachomatis* infection. *Sexually Transmitted Diseases, 19,* 259–265.

LINDAHL, K., & LAACK, S. (1996). Sweden looks anew at ways to reach and teach its young people about sexuality. *SIECUS Report, 24*(3), 7–9.

LINDBACK, S., BROSTROM, C., KARLSSON, A., & GAINES, H. (1994). Does symptomatic primary HIV-1 infection accelerate progression to CDC stage IV disease, CD4 count below 200 x 106/l, and death from AIDS? *British Medical Journal, 309,* 1535–1537.

LINDBERG, F. H., & DISTAD, L. J. (1985). Post-traumatic stress disorders in women who experienced childhood incest. *Child Abuse and Neglect, 9,* 329–334.

LINDBERG, L. D., SONENSTEIN, F. L., KU, L., & MARTINEZ, G. (1997). Age differences between minors who give birth and their adult partners. *Family Planning Perspectives, 29,* 61–66.

LINDEN, E. (1992, August 17). Apes that swing many ways. *Time,* pp. 50–51.

LINDHOLM, C., & LINDHOLM, C. (1980, Summer). Life behind the veil. *Science Digest Special.*

LINET, O. J., & OGRING, F. G. (1996). Efficacy and safety of intracavernosal alprostadil in men with erectile dysfunction. *New England Journal of Medicine, 334,* 873–877.

LINZ, D. (1985). Sexual violence in the media: Effects on male viewers and implications for society. Doctoral thesis, University of Wisconsin, Department of Psychology, Madison.

LINZ, D. (1989). Exposure to sexually explicit materials and attitudes toward rape: A comparison of study results. *Journal of Sex Research, 26,* 50–84.

LINZ, D., DONNERSTEIN, E., & PENROD, S. (1987). The findings and recommendations of the Attorney General's Commission on Pornography: Do the psychological "facts" fit the political fury? *American Psychologist, 42,* 946–953.

LINZ, D., WILSON, B., & DONNERSTEIN, E. (1992). Sexual violence in the mass media: Legal solutions, warnings, and mitigation through education. *Journal of Social Issues, 48,* 145–171.

LIPOVSKY, J. A., & KILPATRICK, D. G. (1992). The child sexual abuse victim as an adult. In W. O'Donohue & J. H. Geer (Eds.), *The sexual abuse of children: Clinical issues* (Vol. 2). Hillsdale, NJ: Erlbaum.

LIPSCOMB, G. H., MURAM, D., SPECK, P. M., & MERCER, B. M. (1992). Male victims of sexual assault. *Journal of the American Medical Association, 267,* 3064–3066.

LIPTON, M. A. (1983). The problem of pornography. In W. E. Fann et al. (Eds.), *Phenomenology and treatment of psychosexual disorders* (pp. 113–134). New York: Spectrum.

LISKIN, L., WHARTON, C., & BLACKBURN, R. (1990). Condoms—Now more than ever. *Population Reports, Series H, 8,* 1–36.

LITTLE, B. B., et al. (1989). Environmental influences cause menstrual synchrony, not pheromones. *American Journal of Human Biology, 1,* 53–57.

LITTLE, R. E., et al. (1989). Maternal alcohol use during breastfeeding and infant mental and motor development at one year. *New England Journal of Medicine, 321,* 425–430.

LITWIN, M. (1993). *Side effects of prostate treatment (impotency and incontinence).* Paper presented at the meeting of the American Urological Association.

LIU, R., et al. (1996). Homozygous defect in HIV-1 coreceptor accounts for resistance of some multiply-exposed individuals to HIV-1 infection. *Cell, 86,* 367–377.

LLOYD, R. (1976). *For money or love: Boy prostitution in America.* New York: Vanguard Press.

LO, B., & STEINBROOK, R. (1992). Health care workers infected with the human immunodeficiency virus. *Journal of the American Medical Association, 267,* 1100–1105.

LO, S. C., et al. (1989). A novel virus-like infectious agent in patients with AIDS. *American Journal of Tropical Medicine and Hygiene, 40,* 213–226.

LOCK, M. (1994). Menopause in cultural context. *Experimental Gerontology, 29,* 307–317.

LOCKWOOD, D. (1980). *Prison sexual violence.* New York: Elsevier.

LOFTUS, E. F. (1997). Creating false memories. *Scientific American, 277,* 70–75.

LOFTUS, E. F. (1997). Memory for a past that never was. *Current Directions in Psychological Science, 6,* 60–65.

LOFTUS, E. F., & KETCHAM, K. (1994). *The myth of repressed memory.* New York: St. Martin's Press.

LOHR, B. A., ADAMS, H. E., & DAVIS, J. M. (1997). Sexual arousal to erotic and aggressive stimuli in sexually coercive and noncoercive men. *Journal of Abnormal Psychology, 106,* 230–242.

LONDON, S. J., CONNOLLY, J. L., SCHMITT, S. J., & COLDITZ, G. A. (1992). A prospective study of benign breast disease and the risk of breast cancer. *Journal of the American Medical Association, 267,* 941–944.

LONGAKER, M. T., et al. (1991). Maternal outcome after open fetal surgery. *Journal of the American Medical Association, 265,* 737–741.

LONGNECKER, M. P., et al. (1992). Risk of breast cancer in relation to past and recent alcohol consumption. *American Journal of Epidemiology, 136,* 1001.

LOPICCOLO, J. (1992). Postmodern sex therapy for erectile failure. In R. Rosen & S. Leiblum (Eds.), *Erectile disorders.* New York: Guilford.

LOPICCOLO, J., & FRIEDMAN, J. (1988). Broad-spectrum treatment of low sexual desire: Integration of cognitive, behavioral, and systemic therapy. In S. R. Leiblum & R. C. Rosen (Eds.), *Sexual desire disorders.* New York: Guilford.

LOPICCOLO, J., & LOBITZ, W. C. (1972). The role of masturbation in the treatment of orgasmic dysfunction. *Archives of Sexual Behavior, 2,* 163–171.

LOPICCOLO, J., & STOCK, W. (1986). Treatment of sexual dysfunction. *Journal of Consulting and Clinical Psychology, 54,* 158–167.

LOPICCOLO, J. (1980). Low sexual desire. In S. R. Leiblum & L. A. Pervin (Eds.), *Principles and practice of sex therapy* (pp. 29–64). New York: Guilford.

LOTTES, I. L., & WEINBERG, M. S. (1996). Sexual coercion among university students: A comparison of the United States and Sweden. *Journal of Sex Research, 34,* 67–76.

LOUDERBACK, L. A., & WHITLEY, B. E., JR. (1997). Perceived erotic value of homosexuality and sex-role attitudes as mediators of sex differences in heterosexual college student's attitudes toward lesbians and gay men. *Journal of Sex Research, 34,* 175–182.

LOVIK, C., et al. (1987). Maternal exposure to spermicides in relation to certain birth defects. *New England Journal of Medicine, 317,* 474–476.

LOWN, J., & DOLAN, E. (1988). Financial challenges in remarriage. *Lifestyles: Family and Economic Issues, 9,* 73–88.

LOWRY, D. T., & SHIDLER, J. A. (1993). Primetime TV portrayals of sex, "safe sex" and AIDS: A longitudinal analysis. *Journalism Quarterly, 70,* 628–637.

LOWY, D. R., KIRNBAUER, R., & SCHILLER, J. T. (1994). Genital human papillomavirus infection. *Proceedings of the National Academy of Sciences, of the U.S.A., 91,* 2436–2440.

LOY, P. H., & STEWART, L. P. (1984). The extent and effects of the sexual harassment of working women. *Sociological Focus, 17,* 31–43.

LUCKEY, E. B., & BAIN, J. K. (1970). Children: A factor in marital satisfaction. *Journal of Marriage and the Family, 32,* 43–44.

LUGER, A. (1993). The origin of syphilis. *Sexually Transmitted Diseases, 20,* 110–117.

LUKE, B. (1994). The changing pattern of multiple births in the United States: Maternal and infant characteristics, 1973 and 1990. *Obstetrics and Gynecology, 84,* 101–106.

LUKER, K. (1975). *Taking chances.* Berkeley: University of California Press.

LURIE, N. O. (1953). Winnebago berdache. *American Anthropologist, 55,* 708–712.

LU-YAO, G. L., MCLERRAN, D., WASSON, J., & WENNBERG, J. E. (1993). An assessment of radical prostatectomy. *Journal of the American Medical Association, 269,* 2633–2636.

LYNCH, H. T., & LYNCH, J. F. (1992). Hereditary ovarian carcinoma. *Hematology/Oncology Clinics of North America, 6,* 783–811.

MACDONALD, A. P. (1981). Bisexuality: Some comments on research and theory. *Journal of Homosexuality, 6,* 21–35.

MACDONALD, A. P. (1982). Research on sexual orientation: A bridge which touches both shores but doesn't meet in the middle. *Journal of Sex Education and Therapy, 8,* 9–13.

MacDonald, M., Crofts, N., & Kaldor, J. (1996). Transmission of hepatitis C virus: Rates, routes, and cofactors. *Epidemiologic Reviews, 18,* 137–148.

Mac Kenzie, W. R., et al. (1992). Multiple false-positive serologic tests for HIV, HTLV-1, and hepatitis C following influenza vaccination, 1991. *Journal of the American Medical Association, 268,* 1015–1017.

MacKinnon, C. (1993). *Only words.* Boston: Harvard University Press.

MacKinnon, C. A. (1979). *Sexual harassment of working women.* New Haven, CT: Yale University Press.

Macklin, E. D. (1978). Review of research on non-marital cohabitation in the U.S. In B. I. Murstein (Ed.), *Exploring intimate lifestyles* (pp. 197–243). New York: Springer.

MacNamara, D., & Sagarin, E. (1977). *Sex, crime, and the law.* New York: Free Press.

Maddock, J. W., et al. (1983). *Human sexuality and the family.* New York: Haworth Press.

Mahoney, C. A. (1995). The role of cues, self-efficacy, level of worry, and high-risk behaviors in college student condom use. *Journal of Sex Education and Therapy, 21,* 103–116.

Mahoney, C. A., Thombs, D. L., & Ford, O. J. (1995). Health belief and self-efficacy models: Their utility in explaining college student condom use. *AIDS Education and Prevention, 7,* 32–49.

Maj, M. (1990). Organic mental disorders in HIV-1 infection. *AIDS, 4,* 831–840.

Makar, K., et al. (1997). Sexuality, body image and quality of life after high dose or conventional chemotherapy for metastatic breast cancer. *Canadian Journal of Human Sexuality, 6*(1), 1–8.

Makinson, C. (1985). The health consequences of teenage fertility. *Family Planning Perspectives, 17,* 132–139.

Malamuth, N. (1981). Rape proclivity among males. *Journal of Social Issues, 37,* 138–157.

Malamuth, N. (1984). Aggression against women: Cultural and individual causes. In N. Malamuth & E. Donnerstein (Eds.), *Pornography and sexual aggression.* New York: Academic Press.

Malamuth, N. (1986). Predictors of naturalistic sexual aggression. *Journal of Personality and Social Psychology, 50,* 953–962.

Malamuth, N., & Check, J. (1981). The effects of mass media exposure on acceptance of violence against women: A field experiment. *Journal of Research in Personality, 15,* 436–446.

Malamuth, N., & Donnerstein, E. (Eds.) (1984). *Pornography and sexual aggression.* New York: Academic Press.

Malamuth, N., & Spinner, B. (1980). A longitudinal content analysis of sexual violence in the best-selling erotic magazines. *Journal of Sex Research, 16,* 226–237.

Malatesta, V. J. (1979). Alcohol effects on the orgasmic-ejaculatory response in human males. *Journal of Sex Research, 15,* 101–107.

Malatesta, V. J., et al. (1982). Acute alcohol intoxication and female orgasmic response. *Journal of Sex Research, 18,* 1–17.

Mandoki, M. W., Sumner, G. S., Hoffman, R. P., & Riconda, D. L. (1991). A review of Klinefelter's Syndrome in children and adolescents. *Journal of the American Academy of Child and Adolescence Psychiatry, 30,* 167–172.

Manlove, J. (1997). Early motherhood in an intergenerational perspective: The experience of a British cohort. *Journal of Marriage and the Family, 59,* 263–279.

Mann, J., et al. (1987). Condom use and HIV infection among prostitutes in Zaire. *New England Journal of Medicine, 316,* 345.

Mann, J. J., & Inman, W. H. (1975). Oral contraceptives and death from myocardial infarction. *British Medical Journal, 2,* 245–248.

Manniche, L. (1987). *Sexual life in ancient Egypt.* London: KPI.

Mannino, D. M., Klevens, R. M., & Flanders, W. D. (1994). Cigarette smoking: An independent risk factor for impotence? *American Journal of Epidemiology, 140,* 1003–1008.

Marchbanks, P. A., Kung-Jong, L., & Mercy, J. A. (1990). Risk of injury from resisting rape. *American Journal of Epidemiology, 132,* 540–549.

Marciano, T. D. (1982). Four marriage and family texts: A brief (but telling) array. *Contemporary Sociology, 11,* 150–153.

Marcus, D. M. (1996). An analytical review of silicone immunology. *Arthritis & Rheumatism, 39,* 1619–1626.

Margolies, L., Becher, M., & Jackson-Brewer, K. (1988). Internalized homophobia: Identifying and treating the oppressor within. In Boston Lesbian Psychologies Collective (Eds.), *Lesbian psychologies.* Urbana: University of Illinois Press.

Margolin, L. (1992). Sexual abuse by grandparents. *Child Abuse and Neglect, 16,* 735–742.

Marks, G., Richardson, J. L., & Maldonado, N. (1991). Self-disclosure of HIV infection to sexual partners. *American Journal of Public Health, 81,* 1321–1323.

Marmor, J. (Ed.). (1980a). *Homosexual behavior.* New York: Basic Books.

Marmor, J. (1980b). Clinical aspects of male homosexuality. In J. Marmor (Ed.), *Homosexual behavior: A modern reappraisal* (pp. 267–279). New York: Basic Books.

Marshall, D. S. (1971). Sexual behavior on Mangaia. In D. S. Marshall & R. C. Suggs (Eds.), *Human sexual behavior* (pp. 103–162). New York: Basic Books.

Marshall, E. (1995). NIH's "gay gene" study questioned. *Science, 268,* 1841.

Marshall, W. L. (1988). The use of sexually explicit stimuli by rapists, child molesters, and nonoffenders. *Journal of Sex Research, 25,* 267–268.

Marshall, W. L. (1989). Intimacy, loneliness, and sexual offenders. *Behavior Research and Therapy, 27,* 491–503.

Marshall, W. L. (1992). Pornography and sex offenders. In D. Zillman & J. Bryant (Eds.), *Pornography: Recent research, interpretation, and policy considerations.* (pp. 185–214). Hillsdale, NJ: Erlbaum.

Marshall, W. L. (1993). A revised approach to the treatment of men who sexually asault adult females. In G. C. N. Hall et al. (Eds.), *Sexual aggression: Issues in etiology, assessment, and treatment* (pp. 143–165). Washington, D.C.: Taylor & Francis.

Marshall, W. L. (1996). Assessment, treatment, and theorizing about sex offenders. *Criminal Justice and Behavior, 23,* 162–199.

Marshall, W. L., Anderson, D., & Champagne, F. (1997). Self-esteem and its relationship to sexual offending. *Psychology, Crime and Law, 3,* 161–186.

Marshall, W. L., Barbaree, H. E., & Christophe, D. (1986). Sexual offenders against female children: Sexual preferences for age of victims and type of behavior. *Canadian Journal of Behavioral Science, 18,* 424–439.

Marshall, W. L., Eccles, A., & Barbaree, H. (1991). The treatment of exhibitionists: A focus on sexual deviance versus cognitive and relationship features. *Behavior Research and Therapy, 29,* 129–135.

Marshall, W. L., Eccles, A., & Barbaree, H. E. (1993). A three-tiered approach to the rehabilitation of incarcerated sex offenders. *Behavioral Sciences and the Law, 11,* 441–455.

Marsiglio, W. (1993a). Adolescent males' orientation toward paternity and contraception. *Family Planning Perspectives, 25,* 22–31.

Marsiglio, W. (1993b). Attitudes toward homosexual activity and gays as friends: A national survey of heterosexual 15- to 19-year-old males. *Journal of Sex Research, 30,* 12–17.

Marsiglio, W., & Donnelly, D. (1991). Sexual relations in later life: A national study of married persons. *Journal of Gerontology, 46,* S338–S344.

Martens, M. G., & Faro, S. (1989). Update on trichomoniasis: Detection and management. *Medical Aspects of Human Sexuality, 23,* 73–79.

Martin, C. L., & Ruble, D. N. (1997). A developmental perspective of self-construals and sex differences: Comment on Cross and Madson (1997). *Psychological Bulletin, 122,* 45–50.

Martin, D., & Lyon, P. (1972). *Lesbian-women.* New York: Bantam Books.

Martin, D. H., et al. (1992). A controlled trial of a single dose of azithromycin for the treatment of chlamydial urethritis and cervicitis. *New England Journal of Medicine, 327,* 921–925.

Martin, T. C., & Bumpass, L. L. (1989). Recent trends in marital disruption. *Demography, 26,* 37–51.

Martius, J. (1996a). Bacterial vaginosis. In P. Elsner and A. Eichmann (Eds.), *Sexually transmitted diseases. Advances in diagnosis and treatment* (pp. 34–39). Basel: Karger.

Martius, J. (1996b). Bacterial vaginosis. In P. Elsner and A. Eichmann (Eds.), *Sexually transmitted diseases. Advances in diagnosis and treatment* (pp. 105–109). Basel: Karger.

Maruri, F., & Azziz, R. (1993). Laparoscopic surgery for ectopic pregnancies: Technology assessment and public health implications. *Fertility and Sterility, 59,* 487–498.

Maslow, A. H. (1968). *Toward a psychology of being* (2nd ed.). Princeton, NJ: Van Nostrand.

Massey, F. J., Bernstein, G. S., & O'Fallon, W. M. (1984). Vasectomy health: Results from a large cohort study. *Journal of the American Medical Association, 252,* 1023–1029.

Masters, W. H. (1980, October 20). *Update on sexual physiology.* Paper presented at the Masters & Johnson Institute's Postgraduate Workshop on Human Sexual Function and Dysfunction, St. Louis, MO.

Masters, W. H. (1982, June 12). *Update on sexual physiology.* Paper presented at the Masters & Johnson Institute's Advanced Workshop on Human sexuality, St. Louis, MO.

MASTERS, W. H. (1986). Sexual dysfunction as an aftermath of sexual assault of men by women. *Journal of Sex and Marital Therapy, 12,* 35–45.

MASTERS, W. H., & JOHNSON, V. E. (1966). *Human sexual response.* Boston: Little, Brown.

MASTERS, W. H., & JOHNSON, V. E. (1970). *Human sexual inadequacy.* Boston: Little, Brown.

MASTERS, W. H., & JOHNSON, V. E. (1976). *The pleasure bond.* New York: Bantam Books.

MASTERS, W. H., & JOHNSON, V. E. (1979). *Homosexuality in perspective.* Boston: Little, Brown.

MASTERS, W. H., JOHNSON, V. E., & KOLODNY, R. C. (1988). *Crisis: Heterosexual behavior in the age of AIDS.* New York: Grove Press.

MASTERS, W. H., JOHNSON, V. E., & KOLODNY, R. C. (1992). *Human sexuality* (4th ed.). New York: HarperCollins.

MASTROYANNIS, C. (1993). Gamete intrafallopian transfer: Ethical considerations, historical development of the procedure, and comparison with other advanced reproductive technologies. *Fertility and Sterility, 60,* 389–402.

MASUR, F. T. (1979). Resumption of sexual activity following myocardial infarction. *Sexuality and Disability, 2,* 98–114.

MATEK, O. (1988). Obscene phone callers. *Journal of Social Work and Human Sexuality, 7,* 113–130.

MATHUR, R., & BRAUNSTEIN, G. D. (1997). Gynecomastia: Pathomechanisms and treatment strategies. *Hormone Research, 48,* 95–102.

MATICKA-TYNDALE, E. (1991). Sexual scripts and AIDS prevention: Variations in adherence to safer-sex guidelines by heterosexual adolescents. *Journal of Sex Research, 28,* 45–66.

MATTHEWS, K. A., et al. (1990). Influences of natural menopause on psychological characteristics and symptoms of middle-aged healthy women. *Journal of Consulting and Clinical Psychology, 58,* 345–351.

MATTHIAS, R. E., LUBBEN, J. E., ATCHISON, K. A., & SCHWEITZER, S. O. (1997). Sexual activity and satisfaction among very old adults: Results from a community-dwelling medicare population survey. *The Gerontologist, 37,* 6–14.

MAUCK, C., et al. (1990). Lea's Sheild: A study of the safety and efficacy of a new vaginal barrier contraceptive used with and without spermicide. *Contraception, 53,* 329–335.

MAULDON, J., & DELBANCO, S. (1997). Public perceptions about unplanned pregnancy. *Family Planning Perspective, 29,* 25–29, 40.

MAYER, R. (1997). 1996–1997 trends in opposition to comprehensive sexuality education in public schools in the United States. *SIECUS Report, 25*(6), 20–26.

MAZER, D. B., & PERCIVAL, E. F. (1989). Students' experiences of sexual harassment at a small university. *Sex Roles, 20,* 1–22.

MCANULTY, R. D., ADAMS, H. E., & WRIGHT, L. W., JR. (1994). Relationship between MMPI and penile plethysmograph in accused child molesters. *Journal of Sex Research, 31,* 179–184.

MCCABE, M. (1987). Desired and experienced levels of premarital affection and sexual intercourse during dating. *Journal of Sex Research, 23,* 23–33.

MCCABE, S. (1983). FBD marriage: Further support for the Westermarck hypothesis of the incest taboo? *American Anthropologist, 85,* 50–69.

MCCAGHY, C. H. (1971). Child molesting. *Sexual Behavior, I,* 16–24.

MCCARTHY, B. W. (1997). Strategies and techniques for revitalizing a nonsexual marriage. *Journal of Sex & Marital Therapy, 23,* 231–240.

MCCARTHY, J., & MCMILLAN, S. (1990). Patient/partner satisfaction with penile implant surgery. *Journal of Sex Education and Therapy, 16*(1), 25–37.

MCCARTON, C. M., et al. (1997). Results at age 8 years of early intervention for low-birth-weight premature infants. *Journal of the American Medical Association, 277,* 126–132.

MCCARTY, L. (1986). Mother-child incest: Characteristics of the offender. *Child Welfare, 6,* 447–458.

MCCAULEY, J., et al. (1997). Clinical characteristics of women with a history of childhood abuse: Unhealed wounds. *Journal of the American Medical Association, 277,* 1362–1368.

MCCLINTOCK, M. (1971). Menstrual synchrony and suppression. *Nature* (London), *229,* 244–245.

MCCLINTOCK, M. K., & HERDT, G. (1996). Rethinking puberty: The development of sexual attraction. *Current Directions in Psychological Science, 5,* 178–183.

MCCLURE, D. (1988, May). Men with one testicle. *Medical Aspects of Human Sexuality,* 22–32.

MCCONAGHY, N. (1974). Penile volume responses to moving and still pictures of male and female nudes. *Archives of Sexual Behavior, 3,* 566–570.

MCCONAGHY, N. (1993). *Sexual behavior: Problems and management.* New York: Plenum Press.

MCCONAGHY, N., BUHRICH, N., & SILOVE, D. (1994). Opposite sex-linked behaviors and homosexual feelings in the predominantly heterosexual male majority. *Archives of Sexual Behavior, 23,* 565–577.

MCCORMACK, A. (1985). The sexual harassment of students by teachers: The case of students in science. *Sex Roles, 13,* 21–32.

MCCORMACK, A., ROKOUS, F. E., HAZELWOOD, R. R., & BURGESS, A. W. (1992). An exploration of incest in the childhood development of serial rapists. *Journal of Family Violence, 7,* 219–228.

MCCOY, N. L., & MATYAS, J. R. (1996). Oral contraceptives and sexuality in university women. *Archives of sexual behavior, 25,* 73–90.

MCGROARTY, J. A., REID, G., & BRUCE, A. W. (1994). The influence of nonoxynol-9-containing spermicides on urogenital infection. *Journal of Urology, 152,* 831–833.

MCHALE, S. M., & CROUTER, A. C. (1992). You can't always get what you want: Incongruence between sex-role attitudes and family work roles and its implications for marriage. *Journal of Marriage and the Family, 54,* 537–547.

MCHALE, S. M., & HUSTON, T. L. (1984). Men and women as parents: Sex role orientations, employment and parental roles with infants. *Child Development, 55,* 1349–1361.

MCINTYRE, J. (1980). *Victim responses to rape: Alternative outcomes. Final report, ROIMH29043.* Rockville, MD: National Institute of Mental Health.

MCKAIN, T. L. (1996). Acknowledging mixed-sex people. *Journal of Sex and Marital Therapy, 22,* 265–274.

MCKAY, A. (1997). Accommodating ideological pluralism is sexuality education. *Journal of Moral Education, 26,* 285–300.

MCKAY, A., & HOLOWATY, P. (1997). Sexual health education: A study of adolescents' opinions, self-perceived needs, and current and preferred sources of information. *The Canadian Journal of Human Sexuality, 6*(1), 29–38.

MCKEGANEY, N. P. (1994). Prostitution and HIV: What do we know and where might research be targeted in the future? *AIDS, 8,* 1215–1226.

MCKELVIE, M., & GOLD, S. R. (1994). Hyperfemininity: Further definition of the construct. *Journal of Sex Research, 31,* 219–228.

MCKIRNAN, D. J., STOKES, J. P., DOLL, L., & BURZETTE, R. G. (1995). Bisexually active men: Social characteristics and sexual behavior. *Journal of Sex Research, 32,* 65–76.

MCLANE, M., KROP, H., & MEHTA, J. (1980). Psychosexual adjustment and counseling after myocardial infarction. *Annals of Internal Medicine, 92,* 514–519.

MCMAHON, M. J., LUTHER, M. P. H., BOWES, W. A., & OLSHAN, A. F. (1996). Comparison of a trial of labor with an elective second cesarean section. *New England Journal of Medicine, 355,* 689–695.

MCNEELY, T., & WAHL, S. (1995). Paper presented at a meeting of the American Society for Microbiology.

MCNIVEN, P., HODNETT, E., & O'BRIEN-PALLAS, L. (1992). Supporting women in labor: A work sampling study of the activities of labor and delivery nurses. *Birth, 19,* 3–8.

MCWHIRTER, D. P., & MATTISON, A. M. (1984). *The male couple: How relationships develop.* Englewood Cliffs, NJ: Prentice Hall.

MEAD, M. (1935). *Sex and temperament in three primitive societies.* New York: William Morrow.

MEAD, M. (1975). *Male and female.* New York: William Morrow.

MEDINA, J. E., et al. (1980). Comparative evaluation of two methods of natural family planning in Columbia. *American Journal of Obstetrics and Gynecology, 138,* 1142–1147.

MEHREN, E. (1991, June 2). What we really think about adultery. *Los Angeles Times,* pp. E1, E8.

MEINKING, T. L., & TAPLIN, D. (1996). Infestations: Pediculosis. In P. Elsner and A. Eichmann (Eds.), *Sexually transmitted diseases: Advances in diagnosis and treatment* (pp. 157–163). Basel: Karger.

MEINKING, T. L., et al. (1995). The treatment of scabies with ivermectin. *New England Journal of Medicine, 333,* 26–30.

MEISELMAN, K. (1978). *Incest: A psychological study of causes and effects with treatment recommendations.* San Francisco: Jossey-Bass.

MELBYE, M., et al. (1997). Induced abortion and the risk of breast cancer. *New England Journal of Medicine, 336,* 81–85.

MELENDY, M. R. (1903). *Perfect womanhood: For maidens-wives-mothers.* Philadelphia: K. T. Boland.

MELLORS, J. W., et al. (1996). Prognosis in HIV-1 infection predicted by the quantity of virus in plasma. *Science, 272,* 1167–1170.

MELLORS, J. W., et al. (1997). Plasma viral load and CD4+ lymphocytes as prognostic markers of HIV-1 infection. *Annals of Internal Medicine, 126,* 946–954.

MELNICK, S. L., et al. (1993). Changes in sexual behavior by young urban heterosexual adults in response to the AIDS epidemic. *Public Health Reports, 108,* 582–588.

MENCHINI-FABRIS, G. F., TURCHI, P., & CANALE, D. (1992). Medical treatment of male infertility. *International Journal of Fertility, 37,* 330–334.

MENDEZ, Z. M. (1996). Columbia's "National Project for Sex Education." *SIECUS Report, 24*(3), 13.

MENEZO, Y. J. R., & JANNY, L. (1996). Is there a rationale for tubal transfer in human ART? *Human Reproduction, 11,* 1818–1820.

MEREDITH, N. (1984, January). The gay dilemma. *Psychology Today, 18,* 56–62.

MERIGGIOLA, M. C., et al. (1997). An oral regimen of cyproterone acetate and testosterone undecanoate for spermatogenic suppression in men. *Fertility and Sterility, 68,* 844–850.

MERTENS, T. E., & LOW-BEER, D. (1996). HIV and AIDS: where is the epidemic going? *Bulletin of the World Health Organization, 74,* 121–129.

MERTENS, T. E., et al. (1995). Global estimates and epidemiology of HIV–1 infections and AIDS: Further heterogeneity in spread and impact. *AIDS, 9* (Suppl.1), S259–S272.

MESSENGER, J. (1971). Sex and repression in an Irish folk community. In D. Marshall & R. Suggs (Eds.), *Human sexual behavior* (pp. 3–37). New York: Basic Books.

MESSIAH, A., & MOURET-FOURME, E. (1996). Homosexuality, bisexuality: Elements of socio-biography. In M. Bozon and H. Leridon (Eds.), *Sexuality and the social sciences: A French survey on sexual behavior* (pp. 177–202) Brookfield MA: Dartmouth.

MESSIAH, A., et al. (1995). Sociodemographic characteristics and sexual behavior of bisexual men in France: Implications for HIV prevention. *American Journal of Public Heath, 85,* 1543–1546.

MESSIAH, A., et al. (1997). Condom breakage and slippage during heterosexual intercourse: A French national survey. *American Journal of Public Health, 87,* 421–424.

METTLER, F. A. (1996). Benefits versus risks from mammography. *Cancer, 77,* 903–909.

METTS, S., & FITZPATRICK, M. A. (1992). Thinking about safer sex: The risky business of "Know your partner" advice. In T. Edgar, M. A. Fitzpatrick, & V. S. Freimuth (Eds.), *AIDS: A communication perspective.* Hillsdale, NJ: Erlbaum.

METZ, M. E., et al. (1997). Premature ejaculation: A psychophysiological review. *Journal of Sex and Marital Therapy, 23,* 3–23.

MEUWISSEN, I., & OVER, R. (1991). Multidimensionality of the content of female sexual fantasy. *Behavior Research and Therapy, 29,* 179–189.

MEYER-BAHLBURG, H. F. L., et al. (1995). Prenatal estrogens and the development of homosexual orientation. *Developmental Psychology, 31,* 12–21.

MEYERS, S. A., & BERSCHEID, E. (1997). The language of love: the difference a preposition makes. *Personality and Social Psychology Bulletin, 23,* 347–362.

MEZEY, G., & KING, M. (1989). The effects of sexual assault on men: A survey of 22 victims. *Psychological Medicine, 19,* 205–209.

MEZEY, G., & KING, M. (1992). Male victims of sexual assault. *AIDS Care, 5,* 247.

MICHAELS, R. H., & MELLIN, G. W. (1960). Prospective experience with maternal rubella and the associated congenital malformations. *Pediatrics, 26,* 200–209.

MILLER, B. C., CHRISTOPHERSON, C. R., & KING, P. K. (1993). Sexual behavior in adolescence. In T. P. Gullot et al. (Eds.), *Adolescent sexuality.* Newbury Park, CA: Sage.

MILLER, C. A. (1985). Infant mortality in the U.S. *Scientific American, 235,* 31–37.

MILLER, J., et al. (1978). Recidivism among sex assault victims. *American Journal of Psychiatry, 135,* 1103–1104.

MILNER, J. S., & DOPKE, C. A. (1997). Paraphilia not otherwise specified: Psychopathology and theory. In D.R. Laws & W. O'Donohue (Eds.), *Sexual deviance: Theory, assessment, and treatment* (pp. 394–423). New York: Guilford Press.

MILSOM, I., SUNDELL, G., & ANDERSCH, B. (1990). The influence of different combined oral contraceptives on the prevalence and severity of dysmenorrhea. *Contraception, 42,* 497–506.

MILUNSKY, A. (1977). *Know your genes.* Boston: Houghton Mifflin.

MINICHIELLO, V., PLUMMER, D., & SEAL, A. (1996). The "asexual" older person? Australian evidence. *Venereology, 9,* 180–188.

MINIUM, E. W., KING, B. M., & BEAR, G. (1993). *Statistical reasoning in psychology and education* (3rd ed.) New York: Wiley.

MIRACLE-MCMAHILL, H. L., et al. (1997). Tubal ligation and fatal ovarian cancer in a large prospective cohort study. *American Journal of Epidemiology, 145,* 349–357.

MIRKIN, G., & HOFFMAN, M. (1978). *The sportsmedicine book.* Boston: Little, Brown.

MISCHEL, W. (1970). Sex-typing and socialization. In P. H. Musses (Ed.), *Carmichael's manual of child psychology,* Vol. 2, (pp. 3–72). New York: Wiley.

MISHELL, D. R., JR. (1982). Non-contraceptive health benefits of oral steroidal contraceptives. *American Journal of Obstetrics and Gynecology, 142,* 809–816.

MISHELL, D. R., JR. (1989). Medical progress: Contraception. *New England Journal of Medicine, 320,* 777–787.

MISHU, B., et al. (1990). A surgeon with AIDS: Lack of evidence of transmission to patients. *Journal of the American Medical Association, 264,* 467–470.

MOCROFT, A., JOHNSON, M. A., & PHILLIPS, A. N. (1996). Factors affecting survival in patients with the acquired immunodeficiency syndrome. *AIDS, 10,*1057–1065.

MOCROFT, A. J., et al. (1997). Survival of AIDS patients according to type of AIDS-defining event. *International Journal of Epidemiology, 26,* 400–407.

MOEN, M. H. (1995). Is mild endometriosis a disease? *Human Reproduction, 10,* 8–12.

MOLLER, H., KNUDSEN, L., & LYNGE, G. (1994). Risk of testicular cancer after vasectomy: Cohort study of over 73,000 men. *British Medical Journal, 309,* 295–299.

MOLLER, L. C., HYMEL, S., & RUBIN, K. H. (1992). Sex typing in play and popularity in middle childhood. *Sex Roles, 26,* 331–335.

MONEY, J. (1970). Clitoral size and erotic sensation. *Medical Aspects of Human Sexuality, 4,* 95.

MONEY, J. (1975). Ablatio penis: Normal male infant sex-reassigned as a girl. *Archives of Sexual Behavior, 4,* 65–72.

MONEY, J. (1980). *Love and love sickness.* Baltimore: Johns Hopkins University Press.

MONEY, J. (1984). Paraphilias: Phenomenology and classification. *American Journal of Psychotherapy, 38,* 164–179.

MONEY, J. (1985). *The destroying angel.* Buffalo, NY: Prometheus Books.

MONEY, J. (1986). *Lovemaps: Clinical concepts of sexual/erotic health and pathology, paraphilia, and gender transposition in childhood, adolescence, and maturity.* New York: Irvington.

MONEY, J. (1987a). Sin, sickness, or status? Homosexual gender identity and psychoneuroendocrinology. *American Psychologist, 42,* 384–399.

MONEY, J. (1987b). Treatment guidelines: Antiandrogen and counseling of paraphiliac sex offenders. *Journal of Sex and Marital Therapy, 13,* 219–223.

MONEY, J. (1994). *Sex errors of the body and related syndromes.* Baltimore: Paul H. Brookes Publishing.

MONEY, J., & EHRHARDT, A. E. (1972). *Man & woman, boy & girl.* Baltimore: Johns Hopkins University Press.

MONEY, J., & SCHWARTZ, M. (1977). Dating, romantic and nonromantic friendships, and sexuality in 17 early-treated adrenogenital females, aged 16–25. In P. A. Lee et al. (Eds.), *Congenital adrenal hyperplasia.* Baltimore: University Park Press.

MONEY, J., & TUCKER, P. (1975). *Sexual signatures.* Boston & Toronto: Little, Brown.

MONEY, J., & WALKER, P. (1977). Counseling the transsexual. In J. Money & H. Musaph (Eds.), *Handbook of sexology* (pp. 1287–1301). Amsterdam: Elsevier/North-Holland.

MONEY, J., et al. (1994). Micropenis: Adult follow-up and comparison of size against new norms. *Journal of Sex and Marital Therapy, 10,* 105–116.

MONTO, M. A. (1996). Lamaze and Bradley childbirth classes: Contrasting perspectives toward the medical model of birth. *Birth, 23,* 193–201.

MOORE, D. W., NEWPORT, F., SAAD, L. (1996, August). Public generally supports a woman's right to abortion. *Gallup Poll Monthly,* pp. 29–35.

MOORE, H. (1994). In P. Harvey & P. Gow. (Eds.), *Sex and violence: Issues in representation and experience.* New York: Routledge.

MOORE, K. (1982). *The developing human: Clinically oriented embryology* (3rd ed.). Philadelphia: Saunders.

MOORE, M. L. (1983). *Realities in childbearing* (2nd ed.). Philadelphia: Saunders.

MOORE, N. B., & DAVIDSON, J. K. (1997). Guilt about first intercourse: An antecedent of sexual dissatisfaction among college women. *Journal of Sex and Marital Therapy, 23,* 29–46.

MOROKOFF, P. J., & GILLILLAND, R. (1993). Stress, sexual functioning, and marital satisfaction. *Journal of Sex Research, 30,* 43–53.

MORRIS, J. F. (1997). Lesbian coming out as a multidimensional process. *Journal of Homosexuality, 33*(2), 1–22.

MORRIS, M. (1993). Telling tails explain the discrepancy in sexual partner reports. *Nature* (London), *365,* 437–440.

MORRIS, N. M., & UDRY, J. R. (1978). Pheromonal influences on human sexual behavior-experimental search. *Journal of Biosocial Science, 10,* 147–159.

MORROW, K. B. (1991). Attributions of female adolescent incest victims regarding their molestation. *Child Abuse and Neglect, 15,* 477–483.

MORSE, E. V., SIMON, P. M., BALSON, P. M., & OSOFSKY, H. J. (1992). Sexual behavior patterns of customers of male street prostitutes. *Archives of Sexual Behavior, 21,* 347–357.

MORSE, E. V., et al. (1991). The male street prostitute: A vector for transmission of HIV infection into the heterosexual world. *Social Science and Medicine, 32,* 535–539.

MOSER, C. (1979). An exploratory-descriptive study of a self-defined S/M (sadomasochistic) sample. Unpublished manuscript, Institute for the Advanced Study of Human Sexuality, San Francisco.

MOSER, C., & LEVITT, E. (1987). An exploratory-descriptive study of a sadomasochistically oriented sample. *Journal of Sex Research, 23,* 322–337.

MOSGAARD, B. J., et al. (1997). Infertility, fertility drugs, and invasive ovarian cancer: A case-control study. *Fertility and Sterility, 67,* 1005–1012.

MOSHER, D., & ANDERSON, R. (1986). Macho personality, sexual aggression, and reactions to guided imagery of realistic rape. *Journal of Research in Personality, 20,* 77–94.

MOSHER, W. D. (1990). Contraceptive practice in the United States, 1982–1988. *Family Planning Perspectives, 22,* 198–205.

MOSHER, W. D., & ARAL, S. O. (1991). Testing for sexually transmitted diseases among women of reproductive age: United States, 1988. *Family Planning Perspectives, 23,* 216–221.

MOSHER, W. D., & PRATT W. F. (1991). Fecundity and infertility in the United States: Incidence and trends. *Fertility and Sterility, 56,* 192–193.

MOSS, C., HOSFORD, R., & ANDERSON, W. (1979). Sexual assault in prison. *Psychological Reports, 44,* 823–828.

MOSS, H. B., PANZAK, G. L., & TARTER, R. E. (1993). Sexual functioning of male anabolic steroid abusers. *Archives of Sexual Behavior, 22,* 1–12.

MOSTAD, S. B., & KREISS, J. K. (1996). Shedding of HIV-1 in the genital tract. *AIDS, 10,* 1305–1315.

MOSTAD, S. B., et al. (1997). Hormonal contraception, vitamin A deficiency, and other risk factors for shedding of HIV-1 infected cells from the cervix and vagina. *Lancet, 350,* 922–927.

MOSURE, D.J., et al. (1997). Genital chlamydia infections in sexually active female adolescents: Do we really need to screen everyone? *Journal of Adolescent Health, 20,* 6–13.

MOOKHERJEE, H. N. (1997). Marital status, gender, and perception of well-being. *Journal of Social Psychology, 137,* 95–105.

MRAZEK, P. B., LYNCH, M., & BENTOUIM, A. (1981). Recognition of child sexual abuse in the United Kingdom. In P. B. Mrazek & C. H. Kempe (Eds.), *Sexually abused children and their families* (pp. 35–49). Oxford: Pergamon.

MUEHLENHARD, C. L. (1989, April). Young men pressured into having sex with women. *Medical Aspects of Human Sexuality,* pp. 50–62.

MUEHLENHARD, C. L., ANDREWS, S. L., & BEAL, G. K. (1996). Beyond "just saying no": Dealing with men's unwanted sexual advances in heterosexual dating contexts. *Journal of Psychology and Human Sexuality, 8,* 141–168.

MUEHLENHARD, C. L., & COOK, S. W. (1988). Men's self-reports of unwanted sexual activity. *Journal of Sex Research, 24,* 58–72.

MUEHLENHARD, C. L., & FALCON, P. (1990). Men's heterosexual skill and attitudes toward women as predictors of verbal and sexual coercion and forceful rape. *Sex Roles, 23,* 241–259.

MUEHLENHARD, C. L., FELTS, A., & ANDREWS, S. (1985, June). *Men's attitudes toward the justifiability of date rape: Intervening variables and possible solutions.* Paper presented at the annual Midcontinent Meeting of the Society for the Scientific Study of Sex, Chicago.

MUEHLENHARD, C. L., GIUSTI, L. M., & RODGERS, C. S. (1993, November 7). *The social construction of "token resistance to sex": The nature and function of the myth.* Paper presented at the annual meeting of the Society for the Scientific Study of Sex, Chicago.

MUEHLENHARD, C. L., & HOLLABAUGH, L. (1989). Do women sometimes say no when they mean yes? The prevalence and correlates of women's token resistance to sex. *Journal of Personality and Social Psychology, 54,* 872–879.

MUEHLENHARD, C. L., & LINTON, M. (1987). Date rape and sexual aggression in dating situations; Incidence and risk factors. *Journal of Consulting Psychology, 34,* 186–196.

MUEHLENHARD, C. L., & LONG, P. (1988, May). *Men's versus women's reports of pressure to engage in unwanted sexual intercourse.* Paper presented at the Western Region meeting of the Society for the Scientific Study of Sex, Dallas, TX.

MUEHLENHARD, C. L., & McCOY, M. L. (1991). Double standard/double blind: The sexual double standard and women's communication about sex. *Psychology of Women Quarterly, 15,* 447–461.

MUEHLENHARD, C. L., POWCH, I. G., PHELPS, J. L., & GIUSTI, L. M. (1992). Definitions of rape: Scientific and political implications. *Journal of Social Issues, 48,* 23–44.

MUEHLENHARD, C. L., & QUACKENBUSH, D. M. (1988, November). *Can the sexual double standard put women at risk for sexually transmitted diseases? The role of the double standard in condom use among women.* Paper presented at the annual meeting of the Society for the Scientific Study of Sex, San Francisco.

MUEHLENHARD, C. L., & RODGERS, C. S. (1993, August 23). *Narrative descriptions of "token resistance" to sex.* Paper presented at the annual meeting of the American Psychological Association, Toronto.

MUEHLENHARD, C. L., & SCHRAG, J. (1990). Nonviolent sexual coercion. In A. Parrot & L. Bechhofer (Eds.), *Hidden rape: Sexual assault among acquaintances, friends, and intimates.* New York: Wiley.

MUESER, K. T., & TAYLOR, K. L. (1997). A cognitive-behavioral approach. In M. Harris & C. L. Landis (Eds.), *Sexual abuse in the lives of women diagnosed with serious mental illness* (pp. 67–90). Amsterdam: Harwood Academic Publishers.

MUKHERJEE, A. B., & HODGEN, G. D. (1982). Maternal ethanol exposure induces transient impairment of umbilical circulation and fetal hypoxia in monkeys. *Science, 218,* 700–702.

MULLER, J. E., et al. (1996). Triggering myocardial infarction by sexual activity: Low absolute risk and prevention by regular physical exertion. *Journal of the American Medical Association, 275,* 1405–1409.

MULLIGAN, T., RETCHIN, S. M., CHINCHILLI, V. M., & BETTINGER, C. B. (1988). The role of aging and chronic disease in sexual dysfunction. *Journal of the American Geriatrics Society, 36,* 520–524.

MULVEY, E. P., & HAVGAARD, J. L. (1986). *Report of the Surgeon General's workshop on pornography and public health.* Washington, DC: U.S. Department of Health and Human Services, Office of the Surgeon General.

MUMFORD, S. D., & KESSEL, E. (1992). Sterilization needs in the 1990s: The case for quinacrine nonsurgical female sterilization. *American Journal of Obstetrics and Gynecology, 167,* 1203–1207.

MUNJACK, D. J., & KANNO, P. H. (1979). Retarded ejaculation: A review. *Archives of Sexual Behavior, 8,* 139–150.

MUNOZ, A., & XU, J. (1996). Models for the incubation of AIDS and variations according to age and period. *Statistics in Medicine, 15,* 2459–2473.

MUNROE, R. L. (1980). Male transvestism and the couvade: A psychocultural analysis. *Ethos, 8,* 49–59.

MURDOCK, G. P. (1964). Cultural correlates of the regulation of premarital sex behavior. In R. A. Manners (Ed.), *Process and pattern in culture* (pp. 339–410). Chicago: Aldine.

MURDOCK, G. P. (1967). *Ethnographic atlas.* Pittsburgh: University of Pittsburgh Press.

MUREAU, M. A. M., SLIJPER, F. M. E., SLOB, A. K. & VERHULST, F. C. (1995). Genital perception of children, adolescents, and adults operated on for hypospadias: A comparative study. *Journal of Sex Research, 32,* 289–298.

MURNEN, S. K., PEROT, A., & BYRNE, D. (1989). Coping with unwanted sexual activity: Normative response, situational determinants, and individual differences. *Journal of Sex Research, 26*(1), 85–106.

MURNEN, S.K., & STOCKTON, M. (1997). Gender and self-reported sexual arousal in response to sexual stimuli: A meta-analytic review. *Sex Roles, 37,* 135–153.

MURPHY, A. A., et al. (1991). Laparoscopic cautery in the treatment of endometriosis-related infertility. *Fertility and Sterility, 55,* 246–251.

MURPHY, T. F. (1992). Redirecting sexual orientation: Techniques and justifications. *Journal of Sex Research, 29,* 501–523.

MURPHY, W. D. (1997). Exhibitionism: Psychopathology and theory. In D. R. Laws & W. O'Donohue (Eds.), *Sexual deviance: Theory, assessment, and treatment* (pp. 22–39). New York: Guilford Press.

MURPHY, W. D., HAYNES, M. R., & PAGE, I. J. (1992). Adolescent sex offenders. In W. O'Donohue & J. H. Geer (Eds.), *The sexual abuse of children: Clinical issues,* Vol. 2, (pp. 394–429). Hillsdale, NJ: Erlbaum.

MURRAY, L. (1992). The impact of postnatal depression on infant development. *Journal of Child Psychology and Psychiatry, 33,* 543–561.

MURRAY, L., & COOPER, P. J. (1997). Postpartum depression and child development. *Psychological Medicine, 27,* 253–260.

MURRAY, S. L., HOLMES, J. G., & GRIFFIN, D. W. (1996). The self-fulfilling nature of positive illusions in romantic relationships: Love is not blind, but prescient. *Journal of Personality and Social Psychology, 71,* 1155–1180.

MURSTEIN, B. I. (1978). Swinging, or comarital sex. In B. I. Murstein (Ed.), *Exploring intimate life styles* (pp. 109–130). New York: Springer.

MURSTEIN, B. I. (1980). Mate selection in the 1970s. *Journal of Marriage and the Family, 42,* 777–792.

MURSTEIN, B. I. (1988). A taxonomy of love. In R. J. Sternberg & M. L. Barnes (Eds.), *The psychology of love* (pp. 13–37) New Haven, CT: Yale University Press.

MYERS, G., MACINNES, K., & KORBER, B. (1992). The emergence of simian/human immunodeficiency viruses. *AIDS Research and Human Retroviruses, 8,* 373–386.

MYERS, J. E. B. (1993). A call for forensically relevant research. *Child Abuse and Neglect, 17,* 573–579.

MYERS, L., et al. (1990). Effects of estrogen, androgen, and progestin on sexual psychophysiology and behavior in postmenopausal women. *Journal of Clinical Endocrinology and Metabolism, 70,* 1124–1131.

NABRINK, M., BIRGERSSON, L., COLLING-SALTIN, A.-S., & SOLUM, T. (1990). Modern oral contraceptives and dysmenorrhea. *Contraception, 42,* 275–283.

NACHTIGALL, L. E. (1994). Sexual function in menopause and post-menopause. In B. A. Lobo (Ed.), *Treatment of the postmenopausal woman: Basic and clinical aspects* (pp. 301–306). New York: Raven Press.

NADLER, R. (1968). Approach to psychodynamics of obscene telephone calls. *New York Journal of Medicine, 68,* 521–526.

NADLER, R. D., DAHL, J. F., GOULD, K. G., & COLLINS, D. C. (1993). Effects of an oral contraceptive on sexual behavior of chimpanzees (Pan troglodytes). *Archives of Sexual Behavior, 22,* 477–500.

NAEYE, R. L. (1979). Coitus and associated amniotic-fluid infections. *New England Journal of Medicine, 301,* 1198–2000.

NAKASHIMA, A. K., et al. (1996). Epidemiology of syphilis in the United States, 1941–1993. *Sexually Transmitted Diseases, 23,* 16–23.

NARAL (1995). *Sexuality education in America: A state-by-state review.* National Abortion and Reproductive Rights Action League.

NASH, J. M. (1997, January 6). Ruling out "junk science." *Time,* p. 102.

NATHAN, S. (1986). The epidemiology of the DSM-III psychosexual dysfunctions. *Journal of Sex and Marital Therapy, 12,* 267–281.

NATIONAL INSTITUTES OF HEALTH (NIH). (1992). Impotence. *NIH Consensus Statement, 10*(4), 1–33.

NAYAK, J., & BOSE, R. (1997). Making sense, talking sexuality: India reaches out to its youth. *SIECUS Report, 25*(2), 19–21.

NAZARIO, S. L. (1990, September 9). Midwifery is staging revival as demand for prenatal care, low-tech births rises. *Wall Street Journal,* pp. B1, B4.

NEAL, J. J., FLEMING, P. L., GREEN, T. A., & WARD, J. W. (1997). Trends in heterosexually acquired AIDS in the United States, 1988 through 1995. *Journal of Acquired Immune Deficiency Syndromes and Human Retrovirology, 14,* 465–474.

NELSON, J. (1980). Gayness and homosexuality: Issues for the church. In E. Batchelor, Jr. (Ed.), *Homosexuality and ethics.* New York: Pilgrim Press.

NELSON, K. B., DAMBROSIA, J. M., TING, T. Y. & GRETHER, J. K. (1996). Uncertain value of electronic fetal monitoring in predicting cerebral palsy. *New England Journal of Medicine, 334,* 613–618.

NESS, R. B., et al. (1993). Number of pregnancies and the subsequent risk of cardiovascular disease. *New England Journal of Medicine, 328,* 1528–1533.

NEUMANN, P. J., GHARIB, S. D., & WEINSTEIN, M. C. (1994). The cost of a successful delivery with in vitro fertilization. *New England Journal of Medicine, 331,* 239–243.

NEWCOMB, M. D., & BENTLER, P. M. (1983). Dimensions of subjective female orgasmic responsiveness. *Journal of Personality and Social Psychology, 44,* 862–873.

NEWCOMER, S. F., & UDRY, R. (1985). Oral sex in an adolescent population. *Archives of Sexual Behavior, 14,* 41–46.

NEWTON, J., & TACCHI, D. (1990). Long-term use of copper intrauterine devices. *Lancet, 335,* 1322–1323.

NGUGI, E. N., et al. (1988). Prevention of transmission of human immunodeficiency virus in Africa: Effectiveness of condom promotion and health education among prostitutes. *Lancet, 2,* 887–890.

NHU, T. T. (1997). Sex surrogate offers healing. (New Orleans) *The Times-Picayune,* October 5, A-20.

NICHOLS, M. (1988). Bisexuality in women: Myths, realities, and implications for therapy. *Women and Therapy, 7,* 235–252.

NICOLOSI, A., et al. (1994). The efficiency of male-to-female and female-to-male sexual transmission of the human immunodeficiency virus: A study of 730 stable couples. *Epidemiology, 5,* 570–575.

NIELSON, J., & WOHLERT, M. (1991). Chromosome abnormalities found among 34,910 newborn children: Results from a 13-year incidence study in Arhus, Denmark. *Human Genetics, 87,* 81–83.

NIH CONCENSUS DEVELOPMENT CONFERENCE STATEMENT, February. 11–13, 1997.

NILSSON, U., et al. (1997). Sexual behavior risk factors associated with bacteria vaginosis and chlamydia trachomatis infection. *Sexually Transmitted Diseases, 24,* 241–246.

NIRAPATHPONGPORN, A., HUBER, D. H., & KRIEGER, J. H. (1990). No-scalpel vasectomy at the King's birthday vasectomy festival. *Lancet, 335,* 894–895.

NOCK, S. (1987). The symbolic meaning of childbearing. *Journal of Family Issues, 8,* 373–393.

NORDENTOFT, M., et al. (1996). Intrauterine growth retardation and premature delivery: The influence of maternal smoking and psychosocial factors. *American Journal of Public Health, 86,* 347–354.

NORDIC MEDICAL RESEARCH COUNCIL'S HIV THERAPY GROUP. (1992). Double blind dose-response study of zidovudine in AIDS and advanced HIV infection. *British Medical Journal, 304,* 13–17.

NUTTER, D. E., & KEARNS, M. E. (1993). Patterns of exposure to sexually explicit material among sex offenders, child molesters, and controls, *Journal of Sex and Marital Therapy, 19,* 77–85.

NYGREN, A. (1982). *Agape and eros.* Chicago: University of Chicago Press.

O'BRIEN, M., & HUSTON, A. C. (1985). Development of sex-typed play behavior in toddlers. *Developmental Psychology, 21,* 866–871.

O'BRIEN, T. E., & MCMANUS, C. E. (1978). Drugs and the fetus. A consumer's guide by generic and brand name. *Birth and the Family Journal, 5,* 58–86.

O'BRIEN, T. R., et al. (1996). Serum HIV-1 RNA levels and time to development of AIDS in the multicenter hemophilia cohort study. *Journal of the American Medical Association, 276,* 105–110.

O'BRIEN, W. A., et al. (1996). Changes in plasma HIV-1 RNA and CD4+ lymphocyte counts and the risk of progression to AIDS. *New England Journal of Medicine, 334,* 426–431.

O'CAMPO, P., et al. (1992). Prior episode of sexually transmitted disease and subsequent sexual risk-reduction practices: A need for improved risk-reduction interventions. *Sexually Transmitted Diseases, 19,* 326.

ODDONE, E. Z., et al. (1993). Cost effectiveness analysis of early zidovudine treatment of HIV infected patients. *British Medical Journal, 307,* 1322–1325.

O'DOWD, G. J., et al. (1997). Update on the appropriate staging evaluation for newly diagnosed prostate cancer. *Journal of Urology, 158,* 687–698.

OFFER, D., & OFFER, J. B. (1975). *From teenage to young manhood.* New York: Basic Books.

OFSHE, R., & WATTERS, E. (1994). *Making monsters.* New York: Charles Schribner's Sons.

OGDEN, S. R., & BRADBURN, N. M. (1968). Dimensions of marriage happiness. *American Journal of Sociology, 73,* 715–731.

O'HARA, M. W. (1997). The nature of pospartum depression. In L. Murray and P. Cooper (Eds.), *Postpartum depression and child development* (pp. 3–31). New York: Guilford Press.

OKAMI, P. (1991). Self-reports of "positive" childhood and adolescent sexual contacts with older persons: An exploratory study. *Archives of Sexual Behavior, 20,* 437–457.

OKAMI, P., & GOLDBERG, A. (1992). Personality correlates of pedophilia: Are they reliable indicators? *Journal of Sex Research, 29,* 297–328.

OLAFSON, E., CORWIN, D. L., & SUMMIT, R. C. (1993). Modern history of child sexual abuse awareness: Cycles of discovery and suppression. *Child Abuse and Neglect, 17,* 7–24.

OLDS, J. (1958). Self-stimulation of the brain. *Science, 127,* 315–324.

OLIVER, M. B., & HYDE, J. S. (1993). Gender differences in sexuality: A meta-analysis. *Psychological Bulletin, 114,* 29–51.

OLSEN, O. (1997). Meta-analysis of the safety of home birth. *Birth, 24,* 4–13.

O'NEILL, N., & O'NEILL, G. (1972). *Open marriage: A new life style for couples.* New York: M. Evans and Co.

ORIEL, J. D., JOHNSON, A. L., & BARLOW, D. (1978). Infection of the uterine cervix with *Chlamydia trachomatis. Journal of Infectious Diseases, 147,* 443–451.

ORR, D. P., BEITER, M., & INGERSOL, G. (1991). Premature sexual activity as an indicator of psychosocial risk. *Pediatrics, 87,* 141–147.

ORY, H. W., ROSENFELD, A., & LANDMAN, L. C. (1980). The pill at 20: An assessment. *Family Planning Perspectives, 12,* 278–283.

OSTERGAARD, J., & KRAFT, M. (1992). Natural course of benign coital headache. *British Medical Journal, 305,* 1129.

OSTROW, D., et al. (1995). A case-control study of human immunodeficiency virus type-1 seroconversion and risk-related behaviors in the Chicago MACS/CCS cohort, 1984–1992. *American Journal of Epidemiology, 142,* 1–10.

O'SULLIVAN, C. S. (1991). Acquaintance gang rape on campus. In A. Parrot & L. Bechhofer (Eds.), *Acquaintance rape: The hidden crime.* New York: Wiley.

O'SULLIVAN, L. F., & ALLGEIER, E. R. (1994). Disassembling a stereotype: Gender differences in the use of token resistance. *Journal of Applied Social Psychology, 24,* 1035–1055.

O'SULLIVAN, L. F., & BYERS, E. S. (1992). College students' incorporation of imitation and restrictor roles in sexual dating interactions. *Journal of Sex Research, 29,* 435–446.

OU, C.-Y., et al. (1992). Molecular epidemiology of HIV transmission in a dental practice. *Science, 256,* 1165–1171.

OUDSHOORN, N. E. J. (1997). Menopause, only for women? The social construction of menopause as an exclusively female condition. *Journal of Psychosomatic Obstetrics and Gynecology, 18,* 137–144.

PADGETT, V. R., BRISLUTZ, J. A., & NEAL, J. A. (1989). Pornography, erotica, and attitudes toward women: The effects of repeated exposure. *Journal of Sex Research, 26,* 479–491.

PADIAN, N. S., SHIBOSKI, S. C., & JEWELL, N. P. (1991). Female-to-male transmission of human immunodeficiency virus. *Journal of the American Medical Association, 266,* 1664–1667.

PAIGE, K. E. (1978a). The declining taboo against menstrual sex. *Psychology Today, 12*(7), 50–51.

PAIGE, K. E. (1978b). The ritual of circumcision. *Human Nature, 1,* 40.

PAIGE, K. E., & PAIGE, J. M. (1981). *The politics of reproductive ritual.* Berkeley: University of California Press.

PAL, R., et al. (1997). Inhibition of HIV-1 infection by the β-chemokine MDC. *Science, 278,* 695–698.

PALACE, E. M., & GORZALKA, B. B. (1992). Differential patterns of arousal in sexually functional and dysfunctional women: Physiological and subjective components of sexual response. *Archives of Sexual Behavior, 21,* 135–159.

PALMER, J. R., et al. (1989). Oral contraceptive use and liver cancer. *American Journal of Epidemiology, 130,* 878–884.

PALMER, L. (1997, October 12). Megan's law often brands wrong people. (New Orleans) *Times-Picayune,* p. A-10.

PALMLUND, I. (1997). The marketing of estrogens for menopausal and postmenopausal women. *Journal of Psychosomatic Obstetrics and Gynecology, 18,* 158–164.

PALMLUND, I. (1997). The social construction of menopause as risk. *Journal of Psychosomatic Obstetrics and Gynecology, 18,* 87–94.

PALYS, T. S. (1986). Testing the common wisdom: The social content of video pornography. *Canadian Psychology, 27,* 22–35.

PANSER, L. A., & PHIPPS, W. R. (1991). Type of oral contraceptive in relation to acute, initial episodes of pelvic inflammatory disease. *Contraception, 43,* 91–99.

PAPE, J. W., & WARREN, W. D., JR. (1993). AIDS in Haiti, 1982–1992. *Clinical Infectious Diseases, 17,* S341–S345.

PAPE, J. W., et al. (1990). Prevalence of HIV infection and high-risk activities in Haiti. *Journal of Acquired Immune Deficiency Syndromes, 3,* 995–1001.

PAPINI, D. R., FARMER, F. L., CLARK, S. M., & SNEL, W. E. (1988). An evaluation of adolescent patterns of sexual self-disclosure to parents and friends. *Journal of Adolescent Research, 3,* 387–401.

PARADISE, J. E. (1990). The medical evaluation of the sexually abused child. In R. M. Reece (Ed.), *The pediatric clinics of North America: Child abuse* (pp. 839–862). Philadelphia: Saunders.

PARKE, R. D., & O'LEARY, S. E. (1976). Father-mother, -infant interaction in the newborn period: Some findings, some observations and some unresolved issues. In K. F. Riegel & J. A. Meacham (Eds.), *The developing individual in a changing world: Vol 2. Social and environment issues* (pp. 653–663). Chicago: Aldine.

PASTNER, C. M. (1986). The Westermarck hypothesis and first cousin marriage: The cultural modification of negative imprinting. *Journal of Anthropological Research, 24,* 573–586.

PATERSON, P., & PETRUCCO, A. (1987). Tubal factors and infertility. In R. Pepperell, B. Hudson, & C. Woods (Eds.), *The infertile couple.* Edinburgh: Churchill-Livingstone.

PATTATUCCI, A. M. L., & HAMER, D. H. (1995). Development and familiarity of sexual orientation in females. *Behavior Genetics, 25,* 407–420.

PATTERSON, C. J. (1992). Children of lesbian and gay parents. *Child Development, 63,* 1025–1042.

PATTON, P. E., WILLIAMS, T. J., & COULAM, C. B. (1987). Microsurgical reconstruction of the proximal oviduct. *Fertility and Sterility, 47,* 35–39.

PAULOZZI, L. J., ERICKSON, J. D., & JACKSON, R. J. (1997). Hypospadias trends in two US surveillance systems. *Pediatrics, 100,* 831–834.

PAULSEN, C. A., BERMAN, N. G., & WANG, C. (1995). Data from men in greater Seattle area reveals no downward trend in semen quality: Fur-

ther evidence that deterioration of semen quality is not geographically uniform. *Fertility and Sterility, 65,* 1015–1020.

PAULY, I. B. (1985). Gender identity disorder. In M. Farber (Ed.), *Human sexuality: Psychosexual effects of disease* (pp. 295–316). New York: Macmillan.

PAULY, I. B. (1990). Gender identity disorders: Evaluation and treatment. *Journal of Sex Education and Therapy, 16,* 2–24.

PAULY, I. B. (1992). Terminology and classification of gender identity disorders. *Journal of Psychology and Human Sexuality, 5*(4), 1–14.

PEARSON, S. E., & POLLACK, R. H. (1997). Female response to sexually explicit films. *Journal of Psychology & Human Sexuality, 9*(2), 73–88.

PEELE, S. (1988). Fools for love: The romantic ideal, psychological theory, and addictive love. In R. J. Sternberg & M. L. Barnes (Eds.), *The psychology of love* (pp. 159–188). New Haven, CT: Yale University Press.

PEERS, T., STEVENS, J. E., GRAHAM, J., & DAVEY, A. (1996). Norplant implants in the UK: First year continuation and removals. *Contraception, 53,* 345–351.

PELLETIER, L., & HAROLD, E. (1988). The relationship of age, sex guilt, and sexual experience with female sexual fantasies. *Journal of Sex Research, 24,* 250–256.

PENLAND, L. R. (1981, April). Sex education in 1900, 1940, and 1980: A historical sketch. *Journal of School Health,* pp. 274–281.

PENROD, S., & LINZ, D. (1984). Using psychological research on violent pornography to inform legal change. In N. M. Malamuth & E. Donnerstein (Eds.), *Pornography and sexual aggression* (pp. 247–275). Orlando, FL: Academic Press.

PEPLAU, L. A. (1988). Research on homosexual couples. In J. De Cecco (Ed.), *Gay relationships.* New York: Haworth Press.

PEREIRA, F. A. (1996). Herpes simplex: Evolving concepts. *Journal of the American Academy of Dermatology, 35,* 503–522.

PERELSON, A. S., et al. (1996). HIV-1 dynamics in vivo: Virion clearance rate, infected cell life-span, and viral generation time. *Science, 271,* 1582–1586.

PEREZ, E. D., MULLIGAN, T., & WAN, T. (1993). Why men are interested in an evaluation for a sexual problem. *Journal of the American Geriatric Society, 41,* 233–237.

PERINN, E. B., et al. (1984). Long-term effect of vasectomy on coronary heart disease. *American Journal of Public Health, 74,* 128–132.

PERLMAN, D., & DUCK, S. (Eds.), *Intimate relationships.* Newbury Park, CA: Sage.

PERPER, T., & WEIS, D. (1987). Proceptive and rejective strategies of U.S. and Canadian college women. *Journal of Sex Research, 23,* 455–480.

PERRETTI, P. O., & ROWAN, M. (1983). Zoophilia: Factors related to its sustained practice. *Panminerva Medicine, 25,* 127.

PERRY, J. D., & WHIPPLE, B. (1981). Pelvic muscle strength of female ejaculators: Evidence in support of a new theory of orgasm. *Journal of Sex Research, 17,* 22–39.

PERSKY, H., et al. (1982). The relation of plasma androgen levels to sexual behavior and attitudes of women. *Psychosomatic Medicine, 44,* 305–319.

PERSON, E. S., et al. (1989). Gender differences in sexual behaviors and fantasies in a college population. *Journal of Sex and Marital Therapy, 15,* 187–198.

PERSSON, G., DAHLOF, L. G., & KRANTZ, J. (1993). Physical and psychological effects of anogenital warts on female patients. *Sexually Transmitted Diseases, 20,* 10–13.

PESMEN, C. (1993, March). The last orgasm. *GQ,* pp. 266–270.

PETER, J. B., BRYSON, Y., & LOVETT, M. A. (1982, March–April). Genital herpes: Urgent questions, elusive answers. *Diagnostic Medicine,* pp. 71–74, 76–78.

PETERS, S. D., WYATT, G. E., & FINKELHOR, D. (1986). Prevalence. In D. Finkelhor (Ed.), *A sourcebook on child sexual abuse* (pp. 15–59). Beverly Hills, CA: Sage.

PETERSEN, L. R., et al. (1992). No evidence for female-to-female HIV transmission among 960,000 female blood donors. *Journal of Acquired Immune Deficiency Syndromes, 5,* 853–855.

PETERSON, J. L., MOORE, K. A., & FURSTENBERG, F. F., JR. (1991). Television viewing and early initiation of sexual intercourse: Is there a link? In *Gay people, sex, and the media.* Binghamton, NY: Haworth.

PETERSON, L. S. (1995, February 14). *Contraceptive use in the United States: 1982–1990. Advance Data, from Vital and Health Statistics* (No. 260). Hyattsville, MD: National Center for Health Statistics.

PETITTI, D. B. (1992). Reconsidering the IUD. *Family Planning Perspectives, 24,* 33–35.

PETITTI, D. B., et al. (1996). Stroke in users of low-dose oral contraceptives. *New England Journal of Medicine, 335,* 8–15.

PETROVICH, M., & TEMPLER, D. (1984). Heterosexual molestation of children who later become rapists. *Psychological Reports, 54,* 810.

PFAFFLIN, F. (1992). Regrets after sex reassignment surgery. *Journal of Psychology and Human Sexuality, 5*(4), 69–85.

PHELPS, D. L., et al. (1991). 28-day survival rates of 6676 neonates with birth weights of 1250 grams or less. *Pediatrics, 87,* 7–17.

PHIBBS, C. S., BATEMAN, D. A., & SCHWARTZ, R. M. (1991). The neonatal costs of maternal cocaine use. *Journal of the American Medical Association, 266,* 1521–1526.

PHILLIPS, G., & OVER, R. (1995). Differences between heterosexual, bisexual, and lesbian women in recalled childhood experiences. *Archives of Sexual Behavior, 24,* 1–10.

PIATUK, M. P., LUK, K.-C., WILLIAMS, B., & LIFSON, J. D. (1993). Quantitative competitive polymerase chain reaction of accurate quantitation of HIV DNA and RNA species. *BioTechniques, 14,* 70–77.

PILKINGTON, C. J., KERN, W., & INDEST, D. (1994). Is safer sex necessary with a "safe" partner? Condom use and romantic feelings. *Journal of Sex Research, 31,* 203–210.

PINES, A., & ARONSON, E. (1983). Antecedents, correlates, and consequences of sexual jealousy. *Journal of Personality, 51,* 108–136.

PITTMAN, F. (1987). *Turning points: Treating families in transition and crisis.* New York: Norton.

PLANNED PARENTHOOD. (1991). *Human sexuality: What children should know and when they should know it.* New York: Author.

PLANT, T. M., WINTERS, S. J., ATTARDI, B. J., & MAJUMDAR, S .S. (1993). The follicle stimulating hormone–inhibin feedback loop in male primates. *Human Reproduction, 8* (Suppl.2), 41–44.

PLASMA-MEDIATED ANTIMICROBIAL RESISTANCE IN NEISSERIA GONORRHOEA. (1990). *Morbidity and Mortality Weekly Report, 39,* 284.

PLATT, R., RICE, P. A., & MCCORMICK, W. M. (1983). Risk of acquiring gonorrhea and prevalence of abnormal adnexal findings among women recently exposed to gonorrhea. *Journal of the American Medical Association, 250,* 3205–3209.

PLECK, J. H. (1976). The male sex role: Definitions, problems and sources of change. *Journal of Social Issues, 32*(3), 155–164.

POCS, O., & GODOW, A. G. (1977). Can students view parents as sexual beings? *The Family Coordinator, 26,* 31–36.

POLAND, R. L. (1990). The question of routine neonatal circumcision. *New England Journal of Medicine, 322,* 1312–1315.

POLASCHEK, D. L. L., WARD, T., & HUDSON, S. M. (1997). Rape and rapists: Theory and treatment. *Clinical Psychology Review, 17,* 117–144.

POMEROY, W. B. (1966). Normal vs. Abnormal sex. *Sexology, 32,* 436–439.

POMEROY, W. B. (1975). The diagnosis and treatment of transvestites and transsexuals. *Journal of Sex and Marital Therapy, 1,* 215–224.

PONCE, P. (1994, April 10). Fraternities working to be campus rape solution. (New Orleans) *Times-Picayune,* p. A-4.

PONTICAS, Y. (1992). Sexual aversion versus hypoactive sexual desire: A diagnostic challenge. *Psychiatric Medicine, 10,* 273–281.

POPE, H. G., JR., & KATZ, D. L. (1988). Affective and psychotic symptoms associated with anabolic steroid use. *American Journal of Psychiatry, 145,* 487–490.

POPE, K. S. (1996). Memory, abuse, and science: Questioning claims about false memory syndrome epidemic. *American Psychologist, 51,* 957–974.

POPOVICH, P. M., et al. (1996). Physical attractiveness and sexual harassment: Does every picture tell a story or every story draw a picture? *Journal of Applied Social Psychology, 26,* 520–542.

POPOVICH, P. M., & LICATA, B. J. (1986). Assessing the incidence and perceptions of sexual harassment behaviors among American undergraduates. *Journal of Psychology, 120,* 387–396.

POPULATION COUNCIL (1994, September 27). Copper T380A Intrauterine Device is effective for 10 years [News release].

PORRECO, R. P., & THORP, J. A. (1996). The cesarean birth epidemic: Trends, causes, and solutions. *American Journal of Obstetrics and Gynecology, 175,* 369–374.

POTTER, R. G., et al. (1965). Applications of field studies to research on the physiology of human reproduction: Lactation and its effects upon birth intervals in eleven Punjab villages, India. In M. Sheps & J. Ridley (Eds.), *Public health and population change.* Pittsburgh: University of Pittsburgh Press.

POTTERAT, J. J., PHILLIPS, L., ROTHENBERG, R. B., & DARROW, W. V. (1985). On becoming a prostitute: An explanatory case-comparison study. *Journal of Sex Research, 21,* 329–335.

POTTERAT, J. J., WOODHOUSE, D. E., MUTH, J. B., & MUTH, S. Q. (1990). Estimating the prevalence and career longevity of prostitute women. *Journal of Sex Research, 27,* 233–243.

POWELL, G. N. (1986). Effects of sex role identity and sex on definitions of sexual harassment. *Sex Roles, 14,* 9–19.

POWELL, M. G., MEARS, B. J., DEBER, R. B., & FERGUSON, D. (1986).

Contraception with the cervical cap: Effectiveness, safety, continuity of use, and user satisfaction. *Contraception, 33,* 215–232.

PREGNANCY + ALCOHOL = PROBLEMS. (July 31, 1989). *Newsweek,* p. 57.

PRENTKY, R. A., KNIGHT, R. A., & ROSENBERG, R. (1988). Validation analysis on a taxonomic system for rapists: Disconfirmation and reconceptualization. *Annals of the New York Academy of Sciences, 528,* 21–40.

PRENTKY, R. A., LEE, A. F. S., KNIGHT, R. A., & CERCE, D. (1997). Recidivism rates among child molesters and rapists: A methodological analysis. *Law and Human Behavior, 21,* 635–659.

PRENTKY, R. A., et al. (1989). Developmental antecedents of sexual aggression. *Development and Psychopathology, 1,* 153–169.

PRESCOTT, J. W. (1975, April). Body pleasure and the orgins of violence. *The Futurist,* pp. 64–74.

PRETI, G., et al. (1986). Human axillary secretions influence women's menstrual cycles: The role of donor extract of females. *Hormones and Behavior, 20,* 474–482.

PRICE, V., SCANLON, B., & JANUS, M.-D. (1984). Social characteristics of adolescent male prostitution. *Victimology, 9,* 211–221.

PRINCE, V., & BENTLER, P. M. (1972). Survey of 504 cases of transvestism. *Psychological Reports, 31,* 903–917.

PRIOR, J., & VIGNA, Y. (1991). Ovulation disturbances and exercise training. *Clinical Obstetrics and Gynecology, 34,* 180–190.

PROFET, M. (1993). Menstruation as a defense against pathogens transported by sperm. *Quarterly Review of Biology, 68,* 335–381.

PURNINE, D. M., & CAREY, M. P. (1997). Interpersonal communication and sexual adjustment: The roles of understanding and agreement. *Journal of Consulting and Clinical Psychology, 65,* 1017–1025.

PYETT, P. M., & WARR, D. J. (1997). Vulnerability on the streets: Female sex workers and HIV risk. *AIDS Care, 9,* 539–547.

QUADAGNO, D., & SPRAGUE, J. (1991, June). Reasons for having sex. *Medical Aspects of Human Sexuality,* p. 52.

QUAKENBUSH, D. M., STRASSBERG, D. S., & TURNER, C. W. (1995). Gender effects of romantic themes in erotica. *Archives of Sexual Behavior, 24,* 21–35.

QUINN, T. C. (1996). Global burden of the HIV pandemic. *Lancet, 348,* 99–106.

QUINSEY, V., CHAPLIN, T., & UPFOLD, D. (1984). Sexual arousal to nonsexual violence and sadomasochistic themes among rapists and non-sex-offenders. *Journal of Consulting and Clinical Psychology, 52,* 651–657.

QUINSEY, V. L., et al. (1995). Predicting sexual offenses. In J. C. Campbell (Ed.), *Assessing dangerousness: Violence by sexual offenders, batterers, and child abusers* (pp. 114–137). Thousand Oaks, CA: Sage.

RABOCH, J., & RABOCH, J. (1992). Infrequent orgasms in women. *Journal of Sex and Marital Therapy, 18,* 114–120.

RACHMAN, S. (1966). Sexual fetishism: An experimental analogue. *Psychological Record, 16,* 293–296.

RADA, R. T., et al. (1983). Plasma androgens in violent and nonviolent sex offenders. *Bulletin of the American Academy of Psychiatry and the Law, 11,* 149–158.

RAKIC, Z., et al. (1996). The outcome of sex reassignment surgery in Belgrade: 32 patients of both sexes. *Archives of Sexual Behavior, 25,* 515–525.

RAO, M., WILKINSON, J., & BENTON, D. (1991). Screening for undescended testes. *Archives of Disease in Children, 66,* 934–937.

RAPAPORT, K., & BURKHART, B. R. (1984). Personality and attitudinal characteristics of sexually coercive college males. *Journal of Abnormal Psychology, 93,* 216–221.

RAPHAEL, S., & ROBINSON, M. (1980). The older lesbian: Love relationships and friendship patterns. *Alternate Lifestyles, 3,* 207–229.

READ, K. (1993, April 27). Male rape: One of the last taboo topics. (New Orleans) *Times-Picayune,* pp. E-1 & E-3.

RECUR, P. (1993, July 17). Study: Male homosexual gene pattern is found. (New Orleans) *Times-Picayune,* p. G-20.

REECE, R. (1988). Special issues in the etiologies and treatments of sexual problems among gay men. *Journal of Homosexuality, 15,* 43–57.

REGAN, L., OWEN, E. J., & JACOBS, H. S. (1990). Hypersecretion of luteinizing hormone, infertility, and miscarriage. *Lancet, 336,* 1141–1144.

REGAN, P. C., & BERSCHEID, E. (1997). Gender differences in characteristics desired in a potential sexual and marriage partner. *Journal of Psychology & Human Sexuality, 9*(1), 25–37.

REID, K. S., TAYLOR, D. K., & WAMPLER, R. S. (1995). Perceptions of partner involvement in the therapeutic process by patients who experienced sexual abuse as children. *Journal of Sex Education and Therapy, 21,* 36–45.

REILLY, M. E., LOTT, B., & GALLOGLY, S. M. (1986). Sexual harassment of university students. *Sex Roles, 15,* 333–358.

REIN, M. F. (1977). Epidemiology of gonococcal infections. In R. B. Roberts (Ed.), *The gonococcus* (pp. 1–31). New York: Wiley.

REINBERG, A., & LAGOGUEY, M. (1978). Circadian and circannual rhythms in sexual activity and plasma hormones (FSH, LH, testosterone) of five human males. *Archives of Sexual Behavior, 7,* 13–30.

REINHOLTZ, R. K., & MUEHLENHARD, C. L. (1995). Genital perceptions and sexual activity in a college population. *Journal of Sex Research, 32,* 155–165.

REINISCH, J. M. (1990). *The Kinsey Institute new report on sex.* New York: St. Martin's Press.

REINISCH, J. M., SANDERS, S. A., HILL, C. A., & ZIEMBA-DAVIS, M. (1992). High-risk sexual behavior among heterosexual undergraduates at a midwestern university. *Family Planning Perspectives, 24,* 116–121, 145.

REISMAN, J. A., & EICHEL, E. W. (1990). *Kinsey, sex and fraud: The indoctrination of a people.* Lafayette, LA: Huntington House.

REISS, I. L. (1981). Some observations on ideology and sexuality in America. *Journal of Marriage and the Family, 43,* 271–283.

REISS, I. L., ANDERSON, R. E., & SPONAUGLE, G. C. (1980). A multivariate model of the determinants of extramarital sexual permissiveness. *Journal of Marriage and the Family, 42,* 395–411.

REISS, I. L., & LEIK, R. K. (1989). Evaluating strategies to avoid AIDS: Number of partners versus use of condoms. *Journal of Sex Research, 4,* 411–433.

REISS, M. J. (1995). Conflicting philosophies of school sex education. *Journal of Moral Education, 24,* 371–382.

REISS, M. J. (1997). Teaching about homosexuality and heterosexuality. *Journal of Moral Education, 26,* 343–352.

REITMAN, D., et al. (1996). Predictors of African American adolescents' condom use and HIV risk behavior. *AIDS Education and Prevention, 8,* 31–48.

RENSHAW, D. (1983). *Incest-understanding and treatment.* Boston: Little, Brown.

RENSHAW, D. (1990). Short-term therapy for sexual dysfunction: Brief counseling to manage vaginismus. *Clinical Practice in Sexuality, 6*(5), 23–29.

RENSHAW, D. C. (1988). Young children's sex play: Counseling the parents. *Medical Aspects of Human Sexuality, 22,* 68–72.

RENTZEL, L. (1972). *When all the laughter died in sorrow.* New York: Saturday Review Press.

REPORT OF THE COMMISSION ON OBSCENITY AND PORNOGRAPHY (1970). New York: Bantam Books.

RICCI, J. M., FOJACO, R. M., & O'SULLIVAN, M. J. (1989). Congenital syphilis: The University of Miami/Jackson Memorial Medical Center experience, 1986–1988. *Obstetrics and Gynecology, 74,* 687–693.

RICE, G., ANDERSON, C., RISCH, N., & EBERS, G. (1995). *Male homosexuality: Absence of linkage to microsatellite markers on the X chromosome in a Canadian study.* Paper presented at the 21st annual meeting of the International Academy of Sex Research, Provincetown, MA.

RICE, P. A., & SCHACHTER, J. (1991). Pathogenesis of pelvic inflammatory disease. *Journal of the American Medical Association, 266,* 2587–2593.

RICHARDS, J. M., KOHLER, C. L., RYAN, W. G., JACKSON, J. R., & GOLDENBERG, R. L. (1991). Contraceptive female sterilization in Alabama: A replication of the WHO study. *Contraception, 43,* 325–333.

RICHARDS, R., & AMES, J. (1983). *Second serve.* New York: Stein & Day.

RICHMAN, K. M., & RICHMAN, L. S. (1993). The potential for transmission of human immunodeficiency virus through human bites. *Journal of Acquired Immune Deficiency Syndromes, 6,* 402–406.

RICHTERS, J. M. A. (1997). Menopause in different cultures. *Journal of Psychosomatic Obstetrics and Gynecology, 18,* 73–80.

RIEVE, J. E. (1989). Sexuality and the adult with acquired physical disability. *Nursing Clinics of North America, 24,* 265–276.

RIND, B., & TROMOVITCH, P. (1997). A meta-analytic review of findings from national samples on psychological correlates of child sexual abuse. *Journal of Sex Research, 34,* 237–255.

RINEHART, W., & PIOTROW, P. T. (1979, January). OCs: Update on usage, safety, and side effects. *Population Reports,* Series A.

RIO, L. (1991). Psychological and sociological research and the decriminalization or legalization of prostitution. *Archives of Sexual Behavior, 20,* 205–217.

RIOS, D. M. (1996, September 15). Suffer the children. (New Orleans) *Times-Picayune,* pp. A1, A8, A10.

RISMAN, B. J., HILL, C., RUBIN, Z., & PEPLAU, L. A. (1981). Living together in college: Implications for courtship. *Journal of Marriage and the Family, 43,* 77–83.

RISMAN, B. J., & SCHWARTZ, P. (1989). Being gendered: A microstructural view of intimate relationships. In B. J. Risman & P. Schwartz (Eds.), *Gender in intimate relationships: A microstructural approach.* Belmont, CA: Wadsworth.

ROAN, S. (1993, February 16). Beyond the birds and the bees. *Los Angeles Times,* pp. E1 & E8.

ROAN, S. (1995, July 12). Are we teaching too little, too late? *Los Angeles Times,* pp. E1, E4.

ROBBINS, M., & JENSEN, G. (1978). Multiple orgasm in males. *Journal of Sex Research, 14,* 21–26.

ROBERTS, C., GREEN, R., WILLIAMS, K., & GOODMAN, M. (1987). Boyhood gender identity development: A statistical contrast of two family groups. *Developmental Psychology, 23,* 544–557.

ROBERTS, T. W. (1992). Sexual attraction and romantic love: Forgotten variables in marital therapy. *Journal of Marital and Family Therapy, 18,* 357–364.

ROBINSON, A. J., & RIDGWAY, G. L. (1994). Sexually transmitted diseases in children: Non viral including bacterial vaginosis, *Gardnerella vaginalis,* mycoplasmas, *Trichomonas vaginalis, Candida albicans,* scabies and pubic lice. *Genitourinary Medicine, 70,* 208–214.

ROBINSON, B. (1988). Teenage pregnancy from the father's perspective. *American Journal of Orthopsychiatry, 58,* 46–51.

ROBINSON, P. (1976). *The modernization of sex.* New York: Harper & Row.

ROBINSON, P. (1983). The sociological perspective. In R. Weg (Ed.), *Sexuality in the later years: Roles and behavior.* New York: Academic Press.

RODRIGUEZ, M., et al. (1996). Teaching our teachers to teach: A SIECUS study on training and preparation for HIV/AIDS prevention and sexuality education. *SIECUS Report, 28*(2), 3–11.

ROEHRBORN, C. G., & MCCONNELL, J. D. (1991). Benign prostatic hyperplasia guideline panel, 1991. Agency for Health Care Policy and Research.

ROGERS, A. S., et al. (1993). Investigation of potential HIV transmission to the patients of an HIV-infected surgeon. *Journal of the American Medical Association, 269,* 1795–1801.

ROGERS, D. E. (1992). Report card on our national response to the AIDS epidemic—some A's, too many D's. *American Journal of Public Health, 82,* 522–524.

ROGERS, P. A. W. (1995). Current studies on human implantation: A brief overview. *Reproduction Fertility Development, 7,* 1395–1399.

ROGERS, S. M., & TURNER, C. F. (1991). Male-male sexual contact in the U.S.A.: Findings from five sample surveys, 1970–1990. *Journal of Sex Research, 28,* 491–519.

ROIPHE, K. (1993). *The morning after: Sex, fear, and feminism on campus.* Boston: Little, Brown.

ROLFS, R. T., GALAID, E. I., & ZAIDI, A. A. (1992). Pelvic inflammatory disease: Trends in hospitalizations and office visits, 1988. *American Journal of Obstetrics and Gynecology, 166,* 983–990.

ROMIEU, I., et al. (1989). Prospective study of oral contraceptive use and risk of breast cancer in women. *Journal of the National Cancer Institute, 81,* 1313–1321.

ROMIEU, I., et al. (1996). Breast cancer and lactation history in Mexican women. *American Journal of Epidemiology, 143,* 543–552.

ROMIEU, I., BERLIN, J. A., & COLDITZ, G. (1990). Oral contraceptives and breast cancer: Review and meta-analysis. *Cancer* (Philadelphia), *66,* 2253–2263.

ROOKS, J., et al. (1989). Outcomes of care in birth centers. *New England Journal of Medicine, 321,* 1804–1811.

ROOKUS, M. A., et al. (1994). Oral contraceptives and risk of breast cancer in women aged 20–54 years. *Lancet, 344,* 844–851.

ROSCOE, B., GOODWIN, M., REPP, S., & ROSE, M. (1987). Sexual harassment of university student and student-employees: Findings and implications. *College Student Journal, 12,* 254–273.

ROSE, M. K., & SOARES, H. H. (1993). Sexual adaptations of the frail elderly: A realistic approach. *Journal of Gerontological Social Work, 19,* 167–178.

ROSEBURY, T. (1971). *Microbes and morals: The strange story of venereal disease.* New York: Viking Press.

ROSEN, M. G., DICKINSON, J. C., & WESTHOFF, C. L. (1991). Vaginal birth after cesarean: A meta-analysis of morbidity and mortality. *Obstetrics and Gynecology, 77,* 465–470.

ROSEN, M. P., et al. (1991). Cigarette smoking: An independent risk factor for atherosclerosis in the hypogastric-cavernous arterial bed. *Urology, 145,* 759–763.

ROSEN, R. C. (1991). Alcohol and drug effects on sexual response: Human experimental and clinical studies. In J. Bancroft (Ed.), *Annual review of sex research,* Vol. 2 (pp. 119–179). Mount Vernon, IA: The Society for the Scientific Study of Sex.

ROSEN, R. C., & ASHTON, A. K. (1993). Prosexual drugs: Empirical status of the "new aphrodisiacs." *Archives of Sexual Behavior, 22,* 521–540.

ROSEN, R. C., & BECK, J. (1988). *Patterns of sexual arousal.* New York: Guilford.

ROSENBERG, H. M., et al. (1996). Birth and deaths: United States, 1995. *Monthly Vital Statistics Report, 45*(3), Suppl. 2.

ROSENBERG, L., et al. (1994). A case-control study of oral contraceptive use and invasive epithelial ovarian cancer. *American Journal of Epidemiology, 139,* 654–661.

ROSENBERG, M. J., & GOLLUB, E. L. (1992). Methods women can use that may prevent sexually transmitted disease, including HIV. *American Journal of Public Health, 82,* 1473–1478.

ROSENBERG, M. J., HILL, H. A., & FRIEL, P. J. (1991, October 7). *Spermicides and condoms in prevention of sexually transmitted diseases: A meta-analysis.* Presented at the meeting of the International Society for Sexually Transmitted Disease Research, Banff, Canada.

ROSENBERG, M. J., WAUGH, M. S., SOLOMON, H. M., & LYSZKOWSKI, A. D. L. (1996). The male polyurethane condom: A review of current knowledge. *Contraception, 53,* 141–146.

ROSENBERG, M. J., et al. (1992). Barrier contraceptives and sexually transmitted diseases in women: A comparison of female-dependent methods and condoms. *American Journal of Public Health, 82,* 669–674.

ROSENBLATT, K. A., et al. (1996). Intrauterine devices and endometrial cancer. *Contraception, 54,* 329–332.

ROSENFELD, A. (1979). Endogamic incest and the victim perpetrator model. *American Journal of Diseases of Children, 133,* 406.

ROSENSTOCK, H. A. (1995). Medical considerations for successful sex therapy. *SIECUS Report, 23*(5), 11–12.

ROSENTHAL, E. (1991). Technique for early prenatal test comes under question in studies. *New York Times.*

ROSENTHAL, M. B., & GOLDFARB, J. (1997). Infertility and assisted reproductive technology: An update for mental health professionals. *Harvard Review of Psychiatry, 5,* 169–172.

ROSENZWEIG, J. M., & DAILEY, D. M. (1989). Dyadic adjustment/sexual satisfaction in women and men as a function of psychological sex role self-perception. *Journal of Sex and Marital Therapy, 15,* 42–56.

ROSMAN, J. P., & RESNICK, P. J. (1989). Sexual attraction to corpses: A psychiatric review of necrophilia. *Bulletin of the American Academy of Psychiatry and the Law, 17,* 153–163.

ROSS, C., & PIOTROW, P. T. (1974). Periodic abstinence: Birth control with contraceptives. *Population Reports,* Series I, No. L.

ROSS, J. A. (1989). Contraception: Short-term vs. long-term failure rates. *Family Planning Perspectives, 21,* 275–277.

ROSS, M. W. (1983). Homosexuality and sex roles: A re-evaluation. *Journal of Homosexuality, 9,* 1–6.

ROSS, M. W., et al. (1990). The effect of a national campaign on attitudes toward AIDS. *AIDS Care, 2,* 339–346.

ROSSETT, H. L., & SANDERS, L. W. (1979). Effects of maternal drinking on neonatal morphology and state regulation. In J. D. Osofsky (Ed.), *Handbook of infant development.* New York: Wiley (Interscience).

ROSSIGNOL, A. M., ZHANG, J., CHEN, Y., & XIANG, Z. (1989). Tea and premenstrual syndrome in the People's Republic of China. *American Journal of Public Health, 79,* 67–69.

ROTHERAM, M. J., & WEINER, N. (1983). Androgyny, stress, and satisfaction: Dual-career and traditional relationships. *Sex Roles, 9,* 151–158.

ROTHMAN, E. K. (1984). *Hands and hearts: A history of courtship in America.* New York: Basic Books.

ROTHMAN, K. J., et al. (1995). Teratogenicity of high vitamin A intake. *New England Journal of Medicine, 333,* 1369–1373.

ROTHSCHILD, B. S., FAGAN, P. J., WOODALL, C., & ANDERSON, A. E. (1991). Sexual functioning of female-disordered patients. *International Journal of Eating Disorders, 10,* 389–394.

ROTOLO, J., & LYNCH, J. (1991, June 15). Penile cancer: Curable with early detection. *Hospital Practice,* pp. 131–138.

ROWLAND, D. L., et al. (1987). Endocrine, psychological and genital response to sexual arousal in men. *Psychoneuroendocrinology, 12,* 149–158.

ROWLAND, D. L., GREENLEAF, W. J., DORFMAN, L. J., & DAVIDSON, J. M. (1993). Aging and sexual function in men. *Archives of Sexual Behavior, 22,* 545–557.

ROYCE, R. A, SENA, A., CATES, W., & COHEN, M. S. (1997). Sexual transmission of HIV: Current concepts. *New England Journal of Medicine, 336,* 1072–1078.

ROZENBERG, S., et al. (1994). Osteoporosis prevention with sex hormone replacement therapy. *International Journal of Fertility, 39,* 262–271.

RUBIN, A. M., & ADAMS, J. R. (1986). Outcomes of sexually open marriages. *Journal of Sex Research, 22,* 311–319.

RUBIN, J., PROVENZANO, F., & LURIA, Z. (1974). The eye of the beholder: Parents' views on sex of newborns. *American Journal of Orthopsychiatry, 44,* 512–519.

RUBIN, R. T., REINISCH, J. M., & HASKETT, R. F. (1981). Postnatal gonadal steroid effects on human behavior. *Science, 211,* 1318–1324.

RUBIN, Z. (1988). Preface. In R. J. Sternberg & M. L. Barnes (Eds.), *The psychology of love* (pp. vii–xii). New Haven, CT: Yale University Press.

RUBIN, Z., et al. (1980). Self-disclosure in dating couples: Sex roles and the ethic of openness. *Journal of Marriage and the Family, 42,* 305.

RUBINOW, D. R., & SCHMIDT, P. J. (1995). The treatment of premenstrual syndrome—Forward into the past. *New England Journal of Medicine, 332,* 1574–1575.

RUDLINGER, R., & NORVAL, M. (1996). Human papillomavirus infections. In P. Elsner & A. Eichmann (Eds.), *Sexually transmitted diseases. Advances in diagnosis and treatment* (pp. 67–76) Basel: Karger.

RUGPAO, S., et al. (1997). Multiple condom use and decreased condom breakage and slippage in Thailand. *Journal of Acquired Immune Deficiency Syndromes and Human Retrovirology, 14,* 169–173.

RULIN, M. C., et al. (1989). Changes in menstrual symptoms among sterilized and comparison women: A prospective study. *Obstetrics and Gynecology, 74,* 149–154.

RUSS, D. M., & SHACKELFORD, T. K. (1997). From vigilance to violence: Mate retention tactics in married couples. *Journal of Personality and Social Psychology, 72,* 346–361.

RUSSELL, C. D., & ELLIS, J. B. (1991). Sex-role development in single parent households. *Social Behavior and Personality, 19,* 5–9.

RUSSELL, D. E. H. (Ed.). (1984). *Sexual exploitation: Rape, child sexual abuse and workplace harassment.* Beverly Hills, CA: Sage.

RUSSELL, D. E. H. (1986). *The secret trauma: Incest in the lives of girls and women.* New York: Basic Books.

RUSSELL, D. E. H. (1990). *Rape in marriage* (rev. ed.). Bloomington: Indiana University Press.

RUSSELL-BROWN, P., et al. (1992). Comparison of condom breakage during human use with performance in laboratory testing. *Contraception, 45,* 429–437.

RUSSO, N. F., & DABUL, A. J. (1997). The relationship of abortion to well-being: Do race and religion make a difference. *Professional Psychology: Research and Practive, 28,* 23–31.

RUTHERFORD, G. W., et al. (1990). Course of HIV-1 infection in a cohort of homosexual and bisexual men: An 11 year follow up study. *British Medical Journal, 301,* 1183–1188.

RYKEN, T. C., HENDERSON, S. T., & MERRELL, A. N. (1990, January). Stigmatophilia. *Medical Aspects of Human Sexuality,* pp. 50A–50B.

RYND, N. (1987). Incidence of psychosomatic symptoms in rape victims. *Journal of Sex Research, 24,* 155–161.

SACHS, A. (1990, October 1). Handing out scarlet letters. *Time,* p. 98.

SAFRAN, C. (1976, November). What men do to women on the job: A shocking look at sexual harassment. *Redbook,* 148–150, 156.

SAFRAN, C. (1981, March). Sexual harassment: The view from the top. *Redbook,* pp. 47–51.

SAGARIN, E. (1977). Power to the peephole. *Sexual Behavior, 3,* 2–7.

SAGHIR, M. T., & ROBINS, E. (1973). *Male and female homosexuality: A comprehensive investigation.* Baltimore, MD: Williams & Wilkins.

SALHOLZ, E., SPRINGEN, K., DE LA PEÑA, N., & WITHERSPOON, D. (1990, July 23). A frightening aftermath: Concern about AIDS adds to the trauma of rape. *Newsweek,* p. 53.

SALOVEY, P. (Ed.). (1991). *The psychology of jealousy and envy.* New York: Guilford.

SALOVEY, P., & RODIN, J. (1985, September). The heart of jealousy. *Psychology Today,* pp. 22–29.

SALT, R. E. (1991). Affectionate touch between fathers and preadolescent sons. *Journal of Marriage and the Family, 53,* 545–554.

SALUTER, A. F. (1994). Marital status and living arrangements: March 1993. *Current Population Reports, 20,* 478.

SALUTER, A. F. (1996). Marital status and living arrangements: March 1994. *Current Population Reports.*

SANDAY, P. (1981). The socio-cultural context of rape: A cross-cultural study. *Journal of Social Issues, 37,* 5–27.

SANDAY, P. (1990). *Fraternity gang rape: Sex, brotherhood and privilege on campus.* New York: New York University Press.

SANDBERG, D. E., & MEYER-BAHLBURG, H. F. L. (1994). Variability in middle childhood play behavior: Effects of gender, age, and family background. *Archives of Sexual Behavior, 23,* 645–663.

SANDERS, J., & ROBINSON, W. (1979). Talking and not talking about sex: Male and female vocabulary. *Journal of Communication, 29,* 22–30.

SANGI-HAGHPEYKAR, H., et al. (1996). Experiences of injectable contraceptive users in urban setting. *Obstetrics and Gynecology, 88,* 227–233.

SANGI-HAGHPEYKAR, H., POINDEXTER, A. N., III, & BATEMANM, L. (1997). Consistency of condom use among users of injectable contraceptives. *Family Planning Perspectives, 29,* 67–69, 75.

SANTIAGO, J. M., McCALL-PEREZ, F., GORCEY, M., & BEIGEL, A. (1985). Long-term psychological effects of rape in 35 rape victims. *American Journal of Psychiatry, 142,* 1338–1340.

SANTROCK, J. W. (1984). *Adolescence: An introduction* (2nd ed.). Dubuque, IA: W. C. Brown.

SARREL, P., & MASTERS, W. (1982). Sexual molestation of men by women. *Archives of Sexual Behavior, 11*(2), 117–131.

SARVELA, P. D., et al. (1992). Connotative meanings assigned to contraceptive options. *Journal of American College Health, 41,* 91–97.

SARWER, D. B., et al. (1993). Sexual aggression and love styles: An exploratory study. *Archives of Sexual Behavior, 22,* 265–275.

SARWER, D. B., & DURLAK, J. A. (1996). Childhood sexual abuse as a predictor of adult female sexual dysfunction: A study of couples seeking sex therapy. *Child Abuse and Neglect, 20,* 963–972.

SARWER, D. B., & DURLAK, J. A. (1997). A field trial of the effectiveness of behavioral treatment for sexual dysfunctions. *Journal of Sex and Marital Therapy, 23,* 87–97.

SATTAR, S. A., & SPRINGTHORPE, V. S. (1991). Survival and disinfectant inactivation of the human immunodeficiency virus: A critical review. *Reviews of Infectious Diseases, 13,* 430–447.

SATTERFIELD, A., & MUEHLENHARD, C. (1990, November). *Flirtation in the classroom: Negative consequences on women's perceptions of their ability.* Paper presented at the annual meeting of the Society for the Scientific Study of Sex, Minneapolis, MN.

SATTERFIELD, S., & LISTIAK, A. (1982, May 15–21). *Juvenile prostitution: A sequel to incest.* Paper presented at the 135th meeting of the American Psychiatric Association, Toronto.

SATTIN, R. W., et al. (1986). Oral-contraceptive use and the risk of breast cancer: The Cancer and Steroid Hormone Study of the Centers for Disease Control and the National Institute of Child Health and Human Development. *New England Journal of Medicine, 315,* 405–411.

SAUER, M. V., PAULSON, R. J., & LOBO, R. A. (1992). Reversing the natural decline in human fertility. *Journal of the American Medical Association, 268,* 1275–1279.

SAUER, M. V., PAULSON, R. J., & LOBO, R. A. (1993). Pregnancy after age 50: Application of oocyte donation to women after natural menopause. *Lancet, 341,* 321–323.

SAUNDERS, E. J. (1989). Life-threatening autoerotic behavior: A challenge for sex educators and therapists. *Journal of Sex Education and Therapy, 15*(2), 77–81.

SAVITZ, L., & ROSEN, L. (1988). The sexuality of prostitutes: Sexual enjoyment reported by streetwalkers. *Journal of Sex Research, 24,* 200–208.

SCARCE, M. (1997). Same-sex rape of male college students. *Journal of American College of Health, 45,* 171–173.

SCHACHTER, J., et al. (1986). Prospective study of perinatal transmission of *Chlamydia trachomatis. Journal of the American Medical Association, 255,* 3374–3377.

SCHACHTER, J., et al. (1992). Nonculture tests for genital tract chlamydial infection. *Sexually Transmitted Diseases, 19,* 243–244.

SCHACHTER, M., & SHOHAM, Z. (1994). Amenorrhea during the reproductive years—Is it safe? *Fertility and Sterility, 62,* 1–16.

SCHACTER, S., & SINGER, J. (1962). Cognitive, social and physiological determinants of emotional state. *Psychological Review, 69,* 379–399.

SCHECHTER, M. T., et al. (1993). HIV-1 and the aetiology of AIDS. *Lancet, 341,* 658–659.

SCHIAVI, R. C. (1990). Sexuality and aging in men. In J. Bancroft (Ed.), *Annual review of sex research,* Vol. 1 (pp. 227–249). Mount Vernon, IA: Society for the Scientific Study of Sex.

SCHIFFMAN, M. H., & BRINTON, L. A. (1995). The epidemiology of cervical carcinogenesis. *Cancer, 76,* 1888–1901.

SCHIRM, A. L., TRUSSELL, J., MENKEN, J., & GRADY, W. R. (1982). Contraceptive failure in the United States: The impact of social, economic and demographic factors. *Family Planning Perspectives, 14*(2), 68–75.

SCHLEGEL, A., & BARRY, H., III. (1979). Adolescent initiation ceremonies: Cross-cultural codes. *Ethnology, 18,* 199–210.

SCHLEGEL, A., & BARRY, H., III. (1980). The evolutionary significance of adolescent initiation ceremonies. *American Ethnologist, 7,* 696–715.

SCHLEGEL, W. S. (1962). Die konstitutionbiologisches grundlagen der homosexualitat. *Zeitschrift fuer Menschliche Vererbungs und Konstitutionslehre, 36,* 341–364.

SCHMID, G. P. (1990). Treatment of chancroid, 1989. *Reviews of Infectious Diseases, 12,* S580–S619.

SCHMIDT, G., & SIGUSCH, V. (1970). Sex differences in responses to psychosexual stimulation by films and slides. *Journal of Sex Research, 6,* 268–283.

SCHNEIDER, D., BARRETT-CONNOR, E., & MORTON, D. (1997). Timing of postmenopausal estrogen for optimal bone mineral density. *Journal of the American Medical Association, 277,* 543–547.

SCHOBEL, H. P., et al. (1996). Preeclampsia—a state of sympathetic overactivity. *New England Journal of Medicine, 335,* 1480–1485.

SCHOEN, E. J. (1990). The status of circumcision of newborns. *New England Journal of Medicine, 322,* 1308–1312.

SCHOOLNIK, G. (1988). *Sex and cancer: Human papilloma virus—the sexually transmitted disease of the 90's.* Paper presented at the seventh annual American Medical Association Science Reporters Conference, San Francisco.

SCHOPPER, D., & VERCAUTEREN, G. (1996). Testing for HIV at home: What are the issues? *AIDS, 10,* 1455–1465.

SCHOTT, R. (1995). The childhood and family dynamics of transvestites. *Archives of sexual behavior, 24,* 309–327.

SCHOVER, L. R. (1988). *Sexuality and cancer: For the woman who has cancer, and her partner.* New York: American Cancer Society.

SCHOVER, L. R., & JENSEN, S. B. (1988). *Sexuality and chronic illness: A comprehensive approach.* New York: Guilford.

SCHOVER, L. R., & LEIBLUM, S. R. (1994). Commentary: The stagnation of sex therapy. *Journal of Psychology and Human Sexuality, 6*(3), 5–30.

SCHOVER, L. R., et al. (1982). Multiaxial problem-oriented system for sexual dysfunctions. *Archives of General Psychiatry, 39,* 614–619.

SCHREIBER, G. B., et al. (1996). The risk of transfusion—Transmitted viral infections. *New England Journal of Medicine, 334,* 1685–1690.

SCHULT, D. G., & SCHNEIDER, L. J. (1991). The role of provocativeness, rape history, and observer sex in attributions of blame in sexual assault. *Journal of Interpersonal Violence, 6,* 94–101.

SCHULTZ, W. C. M. et al. (1989). Vaginal sensitivity to electric stimuli: Theoretical and practical implications. *Archives of Sexual Behavior, 18,* 87–95.

SCHUSTER, M. A., BELL, R. M., & KANOUSE, D. E. (1996). The sexual practices of adolescent virgins: Genital sexual activities of high school students who have never had vaginal intercourse. *American Journal of Public Health, 86,* 1570–1576.

SCHUSTER, M. A., BELL, R. M., PETERSEN, L. P. & KANOUSE, D. E. (1996). Communication between adolescents and physicians about sexual behavior and risk prevention. *Archives of Pediatrics and Adolescent Medicine, 150,* 906–913.

SCHUTTE, J. W., & HOSCH, H. M. (1997). Gender differences in sexual assault verdicts: A meta-analysis. *Journal of Social Behavior and Personality, 12,* 759–772.

SCHWARCZ, S. K., & WHITTINGTON, W. L. (1990). Sexual assault and sexually transmitted diseases: Detection and management in adults and children. *Reviews of Infectious Diseases, 12*(Suppl. 6), S682–S690.

SCHWARCZ, S. K., et al. (1990). National surveillance of antimicrobial resistance in *Neisseria gonorrhoeae. Journal of the American Medical Association, 264,* 1413–1417.

SCHWARTZ, I. M. (1993). Affective reactions of American and Swedish women to their first premarital coitus: A cross-cultural comparison. *Journal of Sex Research, 30,* 18–26.

SCHWARTZ, M., & MASTERS, W. (1988). Inhibited sexual desire: The Masters and Johnson Institute treatment model. In S. R. Leiblum & R. C. Rosen (Eds.), *Sexual desire disorders.* New York: Guilford.

SCHWEIGER, U. (1991). Menstrual function and luteal-phase deficiency in relation to weight changes and dieting. *Clinical Obstetrics and Gynecology, 34,* 191–197.

SCOTT, C. S., ARTHUR, D., PANIZO, M. I., & OWEN, R. (1989). Menarche: The Black American experience. *Journal of Adolescent Health, 10,* 363–368.

SCOTT, J. E. (1986). An updated longitudinal content analysis of sex references in mass circulation magazines. *Journal of Sex Research, 22,* 385–392.

SCOTT, J. E., & CUVELIER, S. J. (1993). Violence and sexual violence in pornography: Is it really increasing? *Archives of Sexual Behavior, 22,* 357–371.

SCOTT, J. E., & SCHWALM, L. (1988). Rape rates and the circulation rates of adult magazines. *Journal of Sex Research, 24,* 241–250.

SCOTT, R. L., & STONE, D. A. (1986). MMPI measures of psychological disturbance in adolescent and adult victims of father-daughter incest. *Journal of Clinical Psychology, 42,* 251–259.

SCOTTI, J. R., SLACK, B. S., BOWMAN, R. A., & MORRIS, T. L. (1996). College student attitudes concerning the sexuality of persons with mental retardation: Development of the perceptions of sexuality scale. *Sexuality and Disability, 14,* 249–263.

SCULLY, D., & MAROLLA, J. (1984). Convicted rapists' vocabulary of motives: Excuses and justifications. *Social Problems, 31,* 530–544.

SEAL, A. (1996). Women, STDs and safe sex: A review of the evidence. *Venereology, 9,* 48–53.

SEAMAN, B. (1972). *Free and female.* New York: Coward, McGann & Geoghegan.

SEGHORN, T. K., PRENTKY, R. A., & BOUCHER, R. J. (1987). Childhood sexual abuse in the lives of sexually aggressive offenders. *Journal of the American Academy of Child and Adolescent Psychiatry, 26,* 262–267.

SEGRAVES, R. T. (1991). Pharmacological enhancement of human sexual behavior. *Journal of Sex Education and Therapy, 17,* 283–289.

SEGRAVES, R. T., & SEGRAVES, K. (1992). Aging and drug effects on male sexuality. In R. Rosen & S. Leiblum (Eds.), *Erectile disorders.* New York: Guilford.

SEGRAVES, R. T., & SEGRAVES, K. B. (1995). Human sexuality and aging. *Journal of Sex Education and Therapy, 21,* 88–102.

SEIDMAN, S. N., & RIEDER, R. O. (1994). A review of sexual behavior in the United States. *American Journal of Psychiatry, 151,* 330–341.

SELIK, R. M., CHU, S. Y., & BUCHLER, J. W. (1993). HIV infection as leading cause of death among young adults in US cities and states. *Journal of the American Medical Association, 269,* 2991–2994.

SELKIN, J. (1975). Rape. *Psychology Today, 8,* 70–76.

SELL, R. L., WELLS, J. A., & WYPIJ, D. (1995). The prevalence of homosexual behavior and attraction in the United States, the United Kingdom and France: Results of national population-based samples. *Archives of Sexual Behavior, 24,* 235–248.

SEMANS, J. H. (1956). Premature ejaculation: A new approach. *Journal of Southern Medicine, 79,* 353–361.

SENANAYAKE, P. (1990). IUDs: A global review. XVIIth Current Fertility Control Symposium. *British Journal of Family Planning, 15*(Suppl.), 8–12.

SENANAYAKE, P. (1992). Positive approaches to education for sexual health with examples from Asia and Africa. *Journal of Adolescent Health, 13,* 351–354.

SETO, M. C., & BARBAREE, H. E. (1995). The role of alcohol in sexual aggression. *Clinical Psychology Review, 15,* 545–566.

SEVELY, J. L., & BENNETT, J. W. (1978). Concerning female ejaculation and the female prostate. *Journal of Sex Research, 14,* 1–20.

SHANGOLD, M. (1985). Causes, evaluation, and management of athletic oligamenorrhea. *Medical Clinics of North America, 69,* 83–95.

SHANNON, J. W., & WOODS, W. J. (1991). Affirmative psychotherapy for gay men. *Counseling Psychologist, 19,* 197–215.

SHANTHA, T. R., SHANTHA, D. T., & BENNETT, J. K. (1990, April). Priapism. *Medical Aspects of Human Sexuality,* pp. 63–67.

SHAPIRO, L. (1990, May 28). Guns and dolls. *Newsweek,* pp. 54–65.

SHARLAND, M., et al. (1997). Paediatric HIV infections. *Archives of Disease in Childhood, 76,* 293–297.

SHAUGHNESSY, M. F., & SHAKESBY, P. (1992). Adolescent sexual and emotional intimacy. *Adolescence, 27,* 475–480.

SHAVER, P., HAZAN, C., & BRADSHAW, D. (1988). Love as attachment. In R. J. Sternberg & M. L. Barnes (Eds.), *The psychology of love* (pp. 68–99). New Haven, CT: Yale University Press.

SHAW, J. (1994). Aging and sexual potential. *Journal of Sex Education and Therapy, 20,* 134–139.

SHAW, J. (1997). Treatment rationale for internet infidelity. *Journal of Sex Education and Therapy, 22,* 29–34.

SHEARER, W. T., et al. (1997). Viral load and disease progression in infants infected with human immunodeficiency virus type 1. *New England Journal of Medicine, 336,* 1337–1342.

SHEPHER, J. (1971). Mate selection among second generation kibbutz adolescents and adults: Incest avoidance and negative imprinting. *Archives of Sexual Behavior, 1,* 293–307.

SHERMAN, K. J., et al. (1990). Sexually transmitted diseases and tubal pregnancy. *Sexually Transmitted Diseases, 17,* 115–121.

SHERR, L. (1990). Fear arousal and AIDS: Do shock tactics work? *AIDS, 4,* 361–364.

SHERWIN, B. B. (1988). A comparative analysis of the role of androgens in human male and female sexual behavior: Behavior specificity, critical thresholds and sensitivity. *Psychobiology, 16,* 416–425.

SHERWIN, B. B. (1991). The psychoendocrinology of aging and female sexuality. In J. Bancroft (Ed.), *Annual review of sex research,* Vol. 2 (pp. 181–198). Mount Vernon, IA: Society for the Scientific Study of Sex.

SHERWIN, B. B., GELFAND, M., & BRENDER, W. (1985). Androgen enhances sexual motivation in females: A prospective crossover study of sex steroid administration in the surgical menopause. *Psychosomatic Medicine, 47,* 339–351.

SHIELDS, W. M., & SHIELDS, L. M. (1983). Forcible rape: An evolutionary perspective. *Ethology and Sociobiology, 4,* 115–136.

SHILTS, R. (1987). *And the band played on.* New York: St. Martin's Press.

SHORTRIDGE, J. L. (1997). Nigerian guidelines for sexuality education introduced at ceremony in Lagos. *SIECUS Report, 25*(2), 4–7.

SHOSTAK, M. (1981). *Kalahari hunter-gatherers.* Cambridge, MA: Harvard University Press.

SHUN-QUIANG, L. (1988). Vasal sterilization techniques: Teaching material for the national standard workshop. Chong-quing, China: Scientific and Technical Literature Press.

SHY, K., et al. (1989). Papanicolaou smear screening interval and risk of cervical cancer. *Obstetrics and Gynecology, 74,* 838–843.

SHY, K., et al. (1990). Effects of electronic fetal-heart-rate monitoring as compared with periodic auscultation, on the neurologic development of premature infants. *New England Journal of Medicine, 322,* 588–593.

SIECUS NATIONAL GUIDELINES TASK FORCE. (1996). *Guidelines for comprehensive sexuality education.* (2nd ed.) New York: Author.

SIEGEL, B. S. (1989). *Love, medicine & miracles.* New York: Perennial/Harper & Row.

SIEGEL, D., et al. (1991). AIDS knowledge, attitudes, and behavior among inner city, junior high school students. *Journal of School Health, 61,* 160–165.

SILVERMAN, B. G., & GROSS, T. P. (1997). Use and effectiveness of condoms during anal intercourse. *Sexually Transmitted Diseases, 24,* 11–17.

SIMBERKOFF, M. S., et al. (1996). Long-term follow-up of symptomatic HIV-infected patients originally randomized to early versus later zidovudine treatment: Report of a Veterans Affairs Cooperative study. *Journal of Acquired Immune Deficiency Syndromes and Human Retrovirology, 11,* 142–150.

SIMENAUER, J., & CARROLL, D. (1982). *Singles: The new Americans.* New York: Simon & Schuster.

SIMON, P. M., et al. (1992). Psychological characteristics of a sample of male street prostitutes. *Archives of Sexual Behavior, 21,* 33–44.

SIMON, W. (1994). Deviance as history: The future of perversion. *Archives of Sexual Behavior, 23,* 1–20.

SIMPSON, J. A., CAMPBELL, B., & BERSCHEID, E. (1986). The association between romantic love and marriage: Kephart (1967) twice revisited. *Personality and Social Psychology Bulletin, 12,* 363–372.

SINCLAIR, A. H., et al. (1990). A gene from the human sex-determining region encodes a protein with homology to a conserved DNA binding motif. *Nature* (London), *346,* 240–244.

SINGER, I. (1984–1987). *The nature of love* (Vols. 1–3). Chicago: University of Chicago Press.

SINGER, J., & SINGER, I. (1972). Types of female orgasm. *Journal of Sex Research, 8,* 255–267.

SINGER, L., ARENDT, R., & MINNES, S. (1993). Neurodevelopmental effects of cocaine. *Clinics in Perinatology, 20*(1), 245–262.

SINGER, L., FARKAS, K., & KLIEGMAN, R. (1992). Childhood medical and behavioral consequences of maternal cocaine use. *Journal of Pediatric Psychology, 17,* 389–406.

SINGH, G. K., & YU, S. M. (1995). Infant mortality in the United States: Trends, differentials, and projections, 1950 through 2010. *American Journal of Public Health, 85,* 957–964.

SINICCO, A., et al. (1993). Risk of developing AIDS after primary acute HIV-1 infection. *Journal of Acquired Immune Deficiency Syndromes, 6,* 575–581.

SIPSKI, M. L., & ALEXANDER, C. J. (1995). Spinal cord injury and female sexuality. In R. C. Rosen, C. M. Davis, & H. J. Ruppel, Jr. (Eds.), *Annual Review of Sex Research,* Vol. VI (pp. 224–244). Mount Vernon, IA: Society for the Scientific Study of Sexuality.

SISKIND, V., GREEN, A., BAIN, C., & PURDIE, D. (1997). Breastfeeding, menopause, and epithelial ovarian cancer. *Epidemiology, 8,* 188–191.

SIVIN, I. (1989). IUDs are contraceptives, not abortifacients: A comment on research and belief. *Studies in Family Planning, 20,* 355–359.

SIVIN, I. (1993). Another look at the Dalkon Shield: Meta-analysis underscores its problems. *Contraception, 48,* 1–12.

SIVIN, I. (1994). Contraception with NORPLANT implants. *Human Reproduction, 9,* 1818–1826.

SKELTON, C. (1984). Correlates of sexual victimization among college women. (No. DER 84-26609). Unpublished doctoral dissertation, Auburn University, Auburn, AL.

SLADE, J. (1984). Violence in the hard-core pornographic film: A historical survey. *Journal of Communication, 34,* 148–163.

SLATTERY, M. L., et al. (1989). Sexual activity, contraception, genital infections, and cervical cancer: Support for a sexually transmitted disease hypothesis. *American Journal of Epidemiology, 130,* 248–258.

SLIJPER, F. M. E., et al. (1992). Evaluation of psychosexual development of young women with congenital adrenal hyperplasia: A pilot study. *Journal of Sex Education and Therapy, 18,* 200–206.

SMALL, S. A., & KERNS, D. (1993). Unwanted sexual activity among peers during early and middle adolescence: Incidence and risk factors. *Journal of Marriage and the Family, 55,* 941–952.

SMITH, A. M. (1993). 'What is pornography?': An analysis of the policy statement of the campaign against pornography and censorship. *Feminist Review, 43,* 71–87.

SMITH, E. A., & UDRY, J. R. (1985). Coital and non-coital sexual behaviors of white and black adolescents. *American Journal of Public Health, 75,* 1200–1203.

SMITH, G. D., FRANKEL, S., & YARNELL, J. (1997). Sex and death: Are they related? Findings from the Caerphilly cohort study. *British Medical Journal, 315,* 1641–1645.

SMITH, H., & ISRAEL, E. (1987). Sibling incest: A study of the dynamics of 25 cases. *Child Abuse and Neglect, 11,* 101–108.

SMITH, L. (1993, January 6). Saying no to birth control. *Los Angeles Times,* pp. A1 & A12.

SMITH, P. I., & TALBERT, R. L. (1986). Sexual dysfunction with antihypertensive and antipsychotic agents. *Clinical Pharmacy, 5,* 373–384.

SMITH, R., PINE, C., & HAWLEY, M. (1988). Social cognitions about adult male victims of female sexual assault. *Journal of Sex Research, 24,* 101–112.

SMITH, T. W. (1991). Adult sexual behavior in 1989: Number of partners, frequency of intercourse, and risk of AIDS. *Family Planning Perspectives, 23,* 102–107.

SNIFFEN, M. J. (1994, June 23). Study: Most rape victims young. (New Orleans) *Times-Picayune,* p. A-9.

SNYDER, J. L. (1989, July/August). The problem of circumcision in America. *The Truth Seeker,* pp. 39–42.

SNYDER, J. W. (1997). Silicone breast implants. *The Journal of Legal Medicine, 18,* 133–220.

SNYDER, P. J. (1990). Fewer sperm in the summer: It's not the heat, it's . . . *New England Journal of Medicine, 323,* 54–56.

SNYDER, P. J., WEINRICH, J. D., & PILLARD, R. C. (1994). Personality and lipid level differences associated with homosexual and bisexual identity in men. *Archives of Sexual Behavior, 23,* 433–451.

SOCARIDES, C. W. (1996). *Homosexuality: A freedom too far.* Phoenix, AZ: Adam Margrave Books.

SOCIETY FOR ASSISTED REPRODUCTIVE TECHNOLOGY, AMERICAN FERTILITY SOCIETY. (1993). Assisted reproductive technology in the United States and Canada: 1991 results from the Society for Assisted Reproductive Technology generated from the American Fertility Society Registry. *Fertility and Sterility, 59,* 956–962.

SOCIETY FOR ASSISTED REPRODUCTIVE TECHNOLOGY, et al. (1997). *The 1995 assisted reproductive technology success rate: National survey and fertility clinics report.* Atlanta: U.S. Department of Health and Human Services.

SOLOMON, D. (1993). Screening for cervical cancer: Prospects for the future. *Journal of the National Cancer Institute, 85,* 1018–1019.

SONENSTEIN, F. L., PLECK, J. H., & KU, L. C. (1991). Levels of sexual activity among adolescent males in the United States. *Family Planning Perspectives, 23,* 162–167.

SOPER, D. E., BROCKWELL, N. J., & DALTON, H. P. (1991). Evaluation of the effects of a female condom on the female lower genital tract. *Contraception, 44,* 21–29.

SORENSON, R. C. (1973). *Adolescent sexuality in contemporary America.* New York: World Publishing Co.

SOTO-RAMIREZ, L. E., et al; (1996). HIV-1 Langehans' cell tropism associated with heterosexual transmission of HIV. *Science, 271,* 1291–1293.

SPANIER, G. B., & THOMPSON, L. (1987). *Parting: The aftermath of separation and divorce.* Newbury Park, CA: Sage.

SPARLING, J. (1997). Penile erections: Shape, angle, and length. *Journal of Sex and Marital Therapy, 23,* 195–207.

SPARLING, P. F., ELKINS, C., WYRICK, P. B., & COHEN, M. S. (1994). Vaccines for bacterial sexually transmitted infections: A realistic goal. *Proceedings of the National Academy of Sciences of the U.S.A., 91,* 2456–2463.

SPECTOR, I., & CAREY, M. (1990). Incidence and prevalence of the sexual dysfunctions: A critical review of the empirical literature. *Archives of Sexual Behavior, 19,* 389–408.

SPECTOR, I. P., & FREMETH, S. M. (1996). Sexual behaviors and attitudes of geriatric residents in long-term care facilities. *Journal of Sex & Marital Therapy, 22,* 235–246.

SPENCE, J. T., & HELMREICH, R. L. (1972). Who likes competent women? Competence, sex-role congruence of interests, and attitudes towards women as determinants of interpersonal attraction. *Journal of Applied and Social Psychology, 2,* 197–213.

SPENCE, J. T., & HELMREICH, R. L. (1978). *Masculinity & femininity: Their psychological dimensions, correlates, and antecedents.* Austin: University of Texas Press.

SPENCE, J. T., HELMREICH, R. L., & STAPP, J. (1974). The personal attrib-

utes questionnaire: A measure of sex role stereotypes and masculinity-femininity. *JSAS Catalog of Selected Documents in Psychology, 4,* 43.

SPENCE, J. T., HELMREICH, R. L., & STAPP, J. (1975). Ratings of self and peers on sex role attributes and their relation to self-esteem and conceptions of masculinity and femininity. *Journal of Personality and Social Psychology, 32,* 29–39.

SPERLING, R. S., et al. (1996) Maternal viral load, zidovudine treatment, and the risk of transmission of human immunodeficiency virus type 1 from mother to infant. *New England Journal of Medicine, 335,* 1621–1629.

SPIESS, W. F. J., GEER, J. H., & O'DONOHUE, W. T. (1984). Premature ejaculation: Investigation of factors in ejaculatory latency. *Journal of Abnormal Psychology, 93,* 242–245.

SPINNATO J. A., II. (1997). Mechanism of action of intrauterine contraceptive devices and its relation to informed consent. *American Journal of Obstetrics and Gynecology, 176,* 503–506.

SPIRA, A., BAJOS, N., and the ACSF GROUP. (1994). *Sexual behavior and AIDS.* Ashgate, England: Avebury.

SPITZ, R. A. (1945). Hospitalism: An inquiry into the genesis of psychiatric conditioning in early childhood. In D. Fenschel & A. Freud (Eds.), *Psychoanalytic studies of the child* (Vol. 1). New York: International Universities Press.

SPITZ, R. A. (1946). Hospitalism: A follow-up report. In D. Fenschel, P. Greenacre, & A. Freud (Eds.), *Psychoanalytic studies of the child* (Vol. 2.) (pp. 113–117). New York: International Universities Press.

SPITZER, W. O., et al. (1996). Third generation oral contraceptives and risk of venous thromboembolic disorders: An international case-control study. *British Medical Journal, 312,* 83–88.

SPRECHER, S., BARBEE, A., & SCHWARTZ, P. (1995). "Was it good for you, too?": Gender differences in first sexual intercourse experiences. *Journal of Sex Research, 32,* 3–15.

SPRUANCE, S. L., et al. (1997). Penciclovir cream for the treatment of herpes simplex labialis. *Journal of the American Medical Association, 277,* 1374–1379.

SRIVASTAVA, A., BORRIES, C., & SOMMER, V. (1991). Homosexual mounting in free-ranging female Hanuman Langurs (*Presbytis entellus*). *Archives of Sexual Behavior, 20,* 487–512 .

ST. LAWRENCE, J. S., & MADAKASIRA, S. (1992). Evaluation and treatment of premature ejaculation: A critical review. *International Journal of Psychiatry in Medicine, 22,* 77–97.

STACK, S., & GUNDLACH, J. H. (1992). Divorce and sex. *Archives of Sexual Behavior, 21,* 359–368.

STAFFORD, R. S., SULLIVAN, S. D., & GARDNER, L. B. (1993). Trends in cesarean section use in California, 1983 to 1990. *American Journal of Obstetrics and Gynecology, 168,* 1297–1302.

STAGNOR, C., & RUBLE, D. N. (1987). Development of gender knowledge and gender constancy. In L. S. Lieben & M. L. Signorella (Eds.), *Children's gender schemata.* San Francisco: Jossey-Bass.

STALL, R., et al. (1990). Relapse from safer sex: The next challenge for AIDS prevention efforts. *Journal of Acquired Immune Deficiency Syndromes, 3,* 1181–1187.

STAMPFER, M. J., et al. (1991). Postmenopausal estrogen therapy and cardiovascular disease. *New England Journal of Medicine, 325,* 756–762.

STANFORD, J. L., et al. (1995). Combined estrogen and progestin hormone replacement therapy in relation to risk of breast cancer in middle-aged women. *Journal of the American Medical Association, 274,* 137–142.

STAPLETON, A., LATHAM, R. H., JOHNSON, C., & STAMM, W. E. (1990). Postcoital antimicrobial prophylaxis for recurrent urinary tract infection. *Journal of the American Medical Association, 264,* 703–706.

STARK, R. (1982). *The book of aphrodisiacs.* New York: Stein & Day.

STARR, B. D., & WEINER, M. B. (1981). *The Starr-Weiner report on sex and sexuality in the mature years.* New York: Stein & Day.

STD-related infertility finds 125,000 victims a year in U.S. (1988). *Reproductive Health Digest* 2(4).

STEELE, B. F., & ALEXANDER, H. (1981). Long-term effects of sexual abuse in childhood. In P. B. Mzazek & C. H. Kempe (Eds.), *Sexually abused children and their families* (pp. 223–233). New York: Pergamon.

STEIN, M. L. (1977). Prostitution. In J. Money & H. Musaph (Eds.), *Handbook of sexology.* Amsterdam: Elsevier/North-Holland.

STEINBERG, K. K., et al. (1991). A meta-analysis of the effect of estrogen replacement therapy on the risk of breast cancer. *Journal of the American Medical Association, 265,* 1985–1990.

STEINEM, G. (1983). *Outrageous acts and everyday rebellions.* New York: Holt, Rinehart & Winston.

STEINER, M. (1997). Premenstrual syndromes. *Annual Review of Medicine, 48,* 447–455.

STEINER, M., et al. (1995). Fluoxetine in the treatment of premenstrual dysphoria. *New England Journal of Medicine, 332,* 1529–1534.

STEINER, M., DOMINIK, R., TRUSSELL, J., & HERTZ-PICCIOTTO, I. (1996). Measuring contraceptive effectiveness: A conceptual framework. *Obstetrics and Gynecology, 88* (3 Suppl.), 24S–30S.

STEINER, M., PIEDRAHITA, C., GLOVER, L., & JOANIS, C. (1993). Can condom users likely to experience condom failure be identified? *Family Planning Perspectives, 25,* 220–223, 226.

STEPHENSON, P., WAGNER, M., BADEA, M., & SERBANESCU, F. (1992). Commentary: The public health consequences of restricted induced abortion—Lessons from Romania. *American Journal of Public Health, 82,* 1328–1331.

STEPTOE, P. C, & EDWARDS R. G. (1978). Birth after the implantation of a human embryo [letter]. *Lancet, 2,* 366.

STERGACHIS, A., et al. (1990). Tubal sterilization and the long-term risk of hysterectomy. *Journal of the American Medical Association, 264,* 2893–2898.

STERNBERG, R. J. (1986). A triangular theory of love. *Psychological Review, 93,* 119–135.

STERNBERG, R. J. (1987). Liking vs. loving: A comparative evaluation of theories. *Psychological Bulletin, 102,* 331–345.

STERNBERG, R. J. (1988). Triangulating love. In R. J. Sternberg & M. L. Barnes (Eds.), *The psychology of love* (pp. 119–138). New Haven, CT: Yale University Press.

STERNBERG, R. J., & GRAJEK, S. (1984). The nature of love. *Journal of Personality and Social Psychology, 47,* 312–329.

STEWART, A., & KNEALE, G. W. (1970). Radiation dose effects in relation to obstetric X-rays and childhood cancers. *Lancet, 1,* 1185–1188.

STEWART, B., et al. (1987). The aftermath of rape. Profiles of immediate and delayed treatment seekers. *Journal of Nervous and Mental Disorders, 175,* 90–94.

STICK, S. M., et al. (1996). Effects of maternal smoking during pregnancy and a family history of asthma on respiratory function in newborn-infants. *Lancet, 348,* 1060–1064.

STOCK, J. L., BELL, M. A., BOYER, D. K., & CONNELL, F. A. (1997). Adolescent pregnancy and sexual risk-taking among sexually abused girls. *Family Planning Perspectives, 29,* 200–203, 207.

STOCK, W. E. (1982, November). *The effect of violent pornography on women.* Paper presented at the national meeting of the Society of the Scientific Study of Sex, San Francisco.

STOCK, W. E. (1983). *Effects of exposure to violent pornography.* Presented at the 26th annual meeting of the Society for the Scientific Study of Sex, Chicago.

STOCKARD, J., & JOHNSON, M. M. (1980). *Sex roles, sex inequality and sex role development.* Englewood Cliffs, NJ: Prentice Hall.

STOKES, J. P., DAMON, W., & McKIRNAN, D. J. (1997). Predictors of movement toward homosexuality: A longitudinal study of bisexual men. *Journal of Sex Research, 34,* 304–312.

STOKES, J., McKIRNAN, D., DOLL, L., & BURZETTE, R. (1996). Female partners of bisexual men: What they don't know might hurt them. *Psychology of Women Quarterly, 20,* 267–284.

STOLL, B. A. (1996). Obesity and breast cancer. *International Journal of Obesity, 20,* 389–392.

STOLLER, R. J. (1977). Sexual deviations. In F. Beach (Ed.), *Human sexuality in four perspectives* (pp. 190–214). Baltimore: Johns Hopkins University Press.

STONE, K. M., et al. (1989). National surveillance for neonatal herpes simplex virus infections. *Sexually Transmitted Diseases, 16,* 152–160.

STORMS, M. (1980). Theories of sexual orientation. *Journal of Personality and Social Psychology, 38,* 783–792.

STORY, M. (1986). Factors affecting the incidence of partner abuse among university students. Unpublished manuscript, University of Northern Iowa, Home Economics Department, Waterloo.

STRASSBERG, D. S., KELLY, M. P., CARROLL, C., & KIRCHER, J. C. (1987). The psychophysiological nature of premature ejaculation. *Archives of Sexual Behavior, 16,* 327–336.

STRASSBERG, D. S., & LOWE, K. (1995). Volunteer bias in sexuality research. *Archives of Sexual Behavior, 24,* 369–382.

STRATHDEE, S. A., et al. (1997). Needle exchange is not enough: Lessons from the Vancouver injecting drug use study. *AIDS, 11,* F59–F65.

STRAUSS, R. H. (1991). Anabolic steroids in the athlete. *Annual Review of Medicine, 42,* 449–457.

STRONG, E. K. (1936). Interests of men and women. *Journal of Social Psychology, 7,* 49–67.

STRUCKMAN-JOHNSON, C., & STRUCKMAN-JOHNSON, D. (1994). Men pressured and forced into sexual experience. *Archives of Sexual Behavior, 23,* 93–114.

SULIK, K. K., JOHNSON, M. C., & WEBB, M. A. (1981). Fetal alcohol syndrome: Embryogenesis in a mouse model. *Science, 214,* 936–938.

SULTAN, C., et al. (1991). SRY and male sex determination. *Hormone Research, 36,* 1–3.

SULTAN, F. E., & CHAMBLES, D. L. (1982). Pubococcygeal function and orgasm in a normal population. In B. Graber (Ed.), *Circumvaginal musculature and sexual function* (pp. 74–87). New York: Karger.

SUMMIT, R., & KRYSO, J. (1978). Sexual abuse of children: A clinical spectrum. *American Journal of Orthopsychiatry, 48,* 237–251.

SUTTIE, I. D. (1952). *The origins of love and hate.* New York: Julian Press.

SWAAB, D. F., GOOREN, L. J. G., & HOFMAN, M. A. (1992). The human hypothalamus in relation to gender and sexual orientation. *Progress in Brain Research, 93,* 205–219.

SWAAB, D. F., & HOFMAN, M. A. (1990). An enlarged suprachismatic nucleus in homosexual men. *Brain Research, 537,* 141–148.

SWISHER, P. N., & COOK, N. D. (1996). Bottoms v. Bottoms: In whose best interest? Analysis of a lesbian mother child custody dispute. *University of Louisville Journal of Family Law, 34,* 843–895.

SYMONS, D. (1979). *The evolution of human sexuality.* New York: Oxford University Press.

TADDIO, A., et al. (1995). Effect of neonatal circumcision on pain responses during vaccination in boys. *Lancet, 345,* 291–292.

TADDIO, A., et al. (1997). Efficacy and safety of lidocaine-prilocaine cream for pain during circumcision. *New England Journal of Medicine, 336,* 1197–1201.

TALAMINI, J. T. (1982). *Boys will be girls: The hidden world of the heterosexual male transvestite.* Washington, DC: University Press of America.

TANFER, K., & ARAL, S. O. (1996). Sexual intercourse during menstruation and self-reported sexually transmitted disease history among women. *Sexually Transmitted Diseases, 23,* 395–401.

TANFER, K., GRADY, W. R., KLEPINGER, D. H., & BILLY, J. O. G. (1993). Condom use among U.S. men, 1991. *Family Planning Perspectives, 25,* 61–66.

TANFER, K., & SCHOORL, J. J. (1992). Premarital sexual careers and partner change. *Archives of Sexual Behavior, 21,* 45–68.

TANG, M. - X., et al. (1996). Effect of estrogen during menopause on risk and age at onset of Alzheimer's disease. *Lancet, 348,* 429–432.

TANNAHILL, R. (1980). *Sex in history.* New York: Stein & Day.

TANNER, J. M. (1962). *Growth at adolescence.* Oxford: Blackwell.

TARDIF, G. S. (1989). Sexual activity after a myocardial infarction. *Archives of Physical Medicine and Rehabilitation, 70,* 763–766.

TARIS, T. W., & SEMIN, G. R. (1997). Gender as a narrative of the effects of the love motive and relational context on sexual experience. *Archives of Sexual Behavior, 26,* 159–180.

TARIS, T. W., & SEMIN, G. R. (1997). Passing on the faith: How mother-child communication influences transmission of moral values. *Journal of Moral Education, 26,* 211–221.

TAYLOR, C. V., SALLIS, J. F., & NEEDLE, R. (1985). The relation of physical activity and exercise to mental health. *Public Health Reports, 100,* 195–201.

TAYLOR, D., et al. (1997). Risk factors for adult paternity in births to adolescents. *Obstetrics and Gynecology, 89,* 199–205.

TAYLOR, M. C., & HALL, J. A. (1982). Psychological androgyny: Theories, methods, and conclusions. *Psychological Bulletin, 92,* 347–366.

TEBBUTT, J., SWANSTON, H., OATES, R. K., & O'TOOLE, B. I. (1997). Five years after child sexual abuse: Persisting dysfunction and problems of prediction. *Journal of the American Academy of Child and Adolescent Psychiatry, 36,* 330–339.

TEMPLEMAN, T. L., & STINNETT, R. D. (1991). Patterns of sexual arousal and history in a "normal" sample of young men. *Archives of Sexual Behavior, 20,* 137–150.

TENNOV, D. (1979). *Love and limerence.* New York: Stein and Day.

TEPPER, M. S., & LAWLESS, B. (1997). Relevant resources for sexuality and spinal cord injury. *Sexuality and Disability, 15,* 203.

TERMAN, L., & MILES, C. C. (1936). *Sex and personality.* New York: McGraw-Hill.

TERZIAN, H., & ORE, G. (1955). Syndrome of Kluver and Bucy reproduced in man by bilateral removal of the temporal lobes. *Neurology, 5,* 373–380.

TEW, M. (1990). *Safer childbirth? A critical history of maternity care.* London: Chapman & Hall.

THAYER, L. (1986). *On communication.* Norwood, NJ: Ablex.

THEIM, R. (1994, August 6). Gay parents' custody rights debated by lawyers. (New Orleans) *Times-Picayune,* p. A-7.

THERIAULT, R. L. (1996). Hormone replacement therapy and breast cancer: An overview. *British Journal of Obstetrics and Gynaecology, 103* (Suppl.13), 87–91.

THOMAS, E., & COLELLA, V. (1992). Cohabitation and marital stability: Quality or commitment? *Journal of Marriage and the Family, 54,* 259–267.

THOMPSON, A. (1983). Extramarital sex: A review of the research literature. *Journal of Sex Research, 19,* 1–22.

THOMPSON, A. (1984). Emotional and sexual components of extramarital relations. *Journal of Marriage and the Family, 46,* 35–42.

THOMPSON, L. (1991). Paraphilia spans extremes of sexual disorders. *The Washington Post.*

THOMPSON, R. (1990). Is routine circumcision indicated in the newborn? An opposing view. *Journal of Family Practice, 31,* 189–196.

THOMPSON, T. L., & ZERBINOS, E. (1997). Television cartoons: Do children notice it's a boy's world? *Sex Roles, 37,* 415–432.

THOMPSON, S. (1990). Putting a big thing into a little hole: Teenage girls' accounts of sexual initiation. *Journal of Sex Research, 27,* 341–361.

THORNBERRY, T. P., SMITH, C. A., & HOWARD, G. J. (1997). Risk factors for teenage fatherhood. *Journal of Marriage and the Family, 59,* 505–522.

THORNHILL, N. W. (1991). An evolutionary analysis of rules regulating human inbreeding and marriage. *Behavioral and Brain Sciences, 14,* 247–293.

THORNHILL, N. W. (1992). Human inbreeding. In N. W. Thornhill (Ed.), *The natural history of inbreeding and outbreeding: Theoretical and empirical perspectives.* Chicago: University of Chicago Press.

THORNHILL, R., & THORNHILL, N. W. (1992). The evolutionary psychology of men's coercive sexuality. *Behavioral and Brain Sciences, 15,* 363–421.

THORNTON, A., AXINN, W. G., & HILL, D. H. (1992). Reciprocal effects of religiosity, cohabitation, and marriage. *American Journal of Sociology, 98,* 628–651.

THUNE, I., et al. (1997). Physical activity and the risk of breast cancer. *New England Journal of Medicine, 336,* 1269–1275.

TIEFER, L. (1991). Historical, scientific, clinical and feminist criticisms of "The human sexual response" model. In J. Bancroft, C. M. Davis, & H. J. Ruppel (Eds.), *Annual review of sex research,* Vol. 2 (pp. 1–23). Mount Vernon, IA: Society for the Scientific Study of Sex.

TIEFER, L. (1994). Might premature ejaculation be organic? The perfect penis takes a giant step forward. *Journal of Sex Education and Therapy, 20,* 7–8.

TIEFER, L. (1997). The medicalization of sexuality: Conceptual, normative, and professional issues. In R.C. Rosen, C. M. Davis, and H. J. Ruppel, Jr. (Eds.), *Annual review of sex research,* Vol.VII (pp. 252–282). Mount Vernon, IA: Society for the Scientific Study of Sexuality.

TIEZZI, L., et al. (1997). Pregnancy prevention among urban adolescents younger than 15: Results of the 'In Your Face' program. *Family Planning Perspectives, 29,* 176–186, 197.

TIME. (1993, May 24). Time/CNN poll: pp. 62–63.

TIME (1997, January 6). Time/CNN poll: "Have you ever been tested for AIDS?" p. 83

TOBIS, J. (1977). Cardiovascular patients and sexual dysfunctions. In R. Marinelli & A. Dell Orto (Eds.), *The psychological and social impact of physical disability.* New York: Springer.

TOLLISON, C. D., & ADAMS, H. E. (1979). *Sexual disorders: Treatment, theory, and research.* New York: Gardner Press.

TONIOLO, P. G., et al. (1995). Prospective study of endogenous estrogens and breast cancer in postmenopausal women. *Journal of the National Cancer Institute, 87,* 190–197.

TOUBIA, N. (1988). Women and health in Sudan. In N. Toubia (Ed.), *Women of the Arab world* (pp. 98–109). London: Zed Books.

TOUBIA, N. (1994). Female circumcision as a public health issue. *New England Journal of Medicine, 331,* 712–716.

TOUFEXIS, A. (1993, February 15). The right chemistry. *Time,* pp. 49–51.

TOWNSEND, J. M. (1995). Sex without emotional involvement: An evolutionary interpretation of sex differences. *Archives of Sexual Behavior, 24,* 173–206.

TRUDEL, G., & SAINT LAURENT, S. (1983). A comparison between the effects of Kegel's exercises and a combination of sexual awareness relaxation and breathing on situational orgasmic dysfunction in women. *Journal of Sex and Marital Therapy, 9,* 204–209.

TRUDGILL, E. (1976). *Madonnas and Magdalens: The origins and development of Victorian sexual attitudes.* New York: Holmes & Meier.

TRUESDELL, D., MCNEIL, J., & DESCHNER, J. (1986). Incidence of wife abuse in incestuous families. *Social Work, 31,* 138–140.

TRUSSELL, J., & GRUMMER-STRAWN, L. (1990). Contraceptive failure of the ovulation method of periodic abstinence. *Family Planning Perspectives, 2*(2), 65–75.

TRUSSELL, J., KOENIG, J., ELLERTSON, L., & STEWART, F. (1997) Preventing unintended pregnancy: the cost-effectiveness of three methods of emergency contraception. *American Journal of Public Health, 87,* 932–937.

TRUSSELL, J., & STEWART, F. (1992). The effectiveness of postcoital contraception. *Family Planning Perspectives, 24,* 262–264.

TRUSSELL, J., STRICKLER, J., & VAUGHAN, B. (1993b). Contraceptive efficacy of the diaphragm, the sponge and the cervical cap. *Family Planning Perspectives, 25,* 100–105, 135.

TRUSSELL, J., STURGEN, K., STRICKLER, J., & DOMINIK, R. (1994). Comparative contraceptive efficacy of the female condom and other barrier methods. *Family Planning Perspectives, 26,* 66–72.

TRUSSELL, J., WARNER, D. L., & HATCHER, R. A. (1992). Condom slippage and breakage rates. *Family Planning Perspectives, 24,* 20–23.

TRUSSELL, J., et al. (1990). Contraceptive failure in the U.S.: An update. *Studies in Family Planning, 21*(1), Table 1.

TRUSSELL, J., et al. (1993a). Should oral contraceptives be available without prescription? *American Journal of Public Health, 83,* 1094–1099.

TRUTE, B., ADKINS, E., & MACDONALD, G. (1992). Professional attitudes regarding the sexual abuse of children: Comparing police, child welfare and community mental health. *Child Abuse and Neglect, 16,* 359–368.

TUCKER, M. B., TAYLOR, R. J., & MITCHELL-KERNAN, C. (1993). Marriage and romantic involvement among aged African Americans. *Journal of Gerontology, 48,* S123–S132.

TUFTS NEW ENGLAND MEDICAL CENTER. (1984). *Sexually exploited children: Services and research project.* Boston: Author.

TULLER, N. R. (1988). Couples: The hidden segment of the gay world. In J. De Cecco (Ed.), *Gay relationships.* New York: Haworth.

TURNBULL, D., et al. (1996). Randomized, controlled trial of efficacy of midwife-managed care. *Lancet, 348,* 213–218.

TURNER, C. F., MILLER, H. G., & MOSES, L. E. (Eds.). (1989). *AIDS: Sexual behavior and intravenous drug use.* Washington, DC: National Academy Press.

TURNER, P. H., et al. (1985, March). *Parenting in gay and lesbian families.* Paper presented at the first meeting of the Future of Parenting Symposium, Chicago.

TURNER, W. J. (1995). Homosexuality, type 1: An Xq28 phenomenon. *Archives of Sexual Behavior, 24,* 109–134.

TWENGE, J. M. (1997). Changes in masculine and feminie traits over time: A meta-analysis. *Sex Roles, 36,* 305–325.

TYLER, D. C. (1988). Pain in the neonate. *Pre- and Peri-Natal Psychology Journal, 3,* 53–59.

UBELL, E. (1984, October 28). Sex in America today. *Parade,* pp. 11–13.

UDRY, J., BILLY, J., MORRIS, N., GROFF, T., & RAJ, M. (1985). Serum androgenic hormones motivate sexual behavior in adolescent boys. *Fertility and Sterility, 43,* 90–94.

ULLMAN, S. E., & KNIGHT, R. A. (1991). A multivariate model for predicting rape and physical injury outcomes during sexual assaults. *Journal of Consulting and Clinical Psychology, 59,* 724–731.

ULLMAN, S. E. (1996). Social reactions, coping strategies, and self-blame attributions in adjustment to sexual assault. *Psychology of Women Quarterly, 20,* 505–526.

ULLRICH, H. E. (1992). Menstrual taboos among Havik Brahmin women: A study of ritual change. *Sex Roles, 26,* 19–40.

UNITED KINGDOM TESTICULAR CANCER SOCIETY STUDY GROUP (1994). Aetiology of testicular cancer: Association with congenital abnormalities, age at puberty, infertility, and exercise. *British Medical Journal, 308,* 1393–1399.

UNITED NATIONS/ECONOMIC COMMISSION FOR EUROPE. (1995). *Trends in Europe and North America: The statistical yearbook of the Economic Commission for Europe.* Author.

U.S. MERIT SYSTEMS PROTECTION BOARD. (1988). *Sexual harassment in the Federal workplace: An update.* Washington, D.C.: U.S. Government Printing Office.

URASSA, M., et al. (1997). Male circumcision and susceptibility to HIV infection among men in Tanzania. *AIDS, 11,* 73–80.

URSIN, G., et al. (1994). Oral contraceptive use and adenocarcinoma of cervix. *Lancet, 344,* 1390–1394.

U.S. ATTORNEY GENERAL'S COMMISSION ON PORNOGRAPHY. (1986). *Final Report of the Attorney General's Commission on Pornography.* Washington, DC: U.S. Justice Department.

U.S. DEPARTMENT OF JUSTICE. *Uniform crime reports for the United States 1993.* Washington, DC: U.S. Government Printing Office.

U.S. DEPARTMENT OF JUSTICE (1997). *Bureau of Justice statistics sourcebook of criminal justice statistics—1996.* Washington, DC: U.S. Government Printing Office.

U.S. FOOD AND DRUG ADMINISTRATION (1997). Prescription drug products; certain combined oral contraceptions for use as postcoital emergency contraception; notice. *Federal Register, 62,* 8610–8612.

U.S. GENERAL ACCOUNTING OFFICE. (1982). *Sexual exploitation of children—A problem of unknown magnitude.* Gaithersburg, MD: Author.

U.S. GENERAL ACCOUNTING OFFICE. (1990). *Drug exposed infants: A generation at risk.* Washington, DC: Author. (Publication GAO/HRD-90-138)

UTIAN, W. (1997, Sept. 4). Women's experience and attitudes toward menopause. Paper presented at the annual meeting of the North American Menopause Society, Boston, MA.

VAIL-SMITH, K., & WHITE, D. M. (1992). Risk level, knowledge, and preventive behavior for human papillomaviruses among sexually active college women. *Journal of American College Health, 40,* 227–230.

VALENTICH, M., & GRIPTON, J. (1989). Teaching children about AIDS. *Journal of Sex Education and Therapy, 15*(2), 92.

VANCE, C. S. (1988). Ordinances restricting pornography could damage women. In R. T. Francoeur (Ed.), *Taking sides: Clashing views on controversial issues in human sexuality.* Guilford, CT: Dushkin Publishing Group.

VANCE, E. B., & WAGNER, N. N. (1976). Written descriptions of orgasm: A study of sex differences. *Archives of Sexual Behavior, 5,* 87–98.

VAN DEN HOEK, A., VAN HAASTRECHT, H. J. A., & COUTINHO, R. A. (1990). Heterosexual behavior of intravenous drug users in Amsterdam: Implications for the AIDS epidemic. *AIDS, 4,* 449–453.

VAN DE VELDE, T. H. (1930). *Ideal marriage: Its physiology and technique.* New York: Couici-Friede Publishers.

VAN DE VEN, P. (1994). Comparisons among homophobic reactions of undergraduates, high school students, and young offenders. *Journal of Sex Research, 31,* 117–124.

VAN DE WIEL, H., JASPERS, J., WEIJMAR-SCHULTZ, W., & GAL, J. (1990). Treatment of vaginismus: A review of concepts and treatment modalities. *Journal of Psychosomatic Obstetrics and Gynecology, 11,* 1–18.

VAN GOOZEN, S. H. M., et al. (1997). Psychoendocrinological assessment of the menstrual cycle: The relationship between hormones, sexuality, and mood. *Archives of Sexual Behavior, 26,* 359–382.

VAN HALL, E. V. (1997). The menopausal misnomer. *Journal of Psychosomatic Obstetrics and Gynecology, 18,* 59–62.

VAN LOOK, P. F. A., & VON HERTZEN, H. (1993). Emergency contraception. *British Medical Bulletin, 49,* 158–170.

VAN NOORD-ZAADSTRA, B., et al. (1991). Delaying childbearing: Effect of age on fecundity and outcome of pregnancy. *British Medical Journal, 302,* 1361–1365.

VANNOY, R. (1980). *Sex without love: A philosophical exploration.* Buffalo, NY: Prometheus Books.

VAN WYK, P. (1984). Psychosocial development of heterosexual, bisexual, and homosexual behavior. *Archives of Sexual Behavior, 13,* 505–544.

VEITH, J., et al. (1983, August 27). *Exposure to men influences the occurrence of ovulation in women.* Paper presented at the 91st Convention of the American Psychological Association, Anaheim, CA.

VENTURA, J. N. (1987). The stresses of parenthood reexamined. *Family Relations, 36,* 26–29.

VENTURA, S. J., et al. (1997). Report of final natality statistics, 1995. National Center for Health Statistics. *Monthly Vital Statistics Report, 45* (11).

VESSEY, M. P. (1973). Oral contraceptives and stroke. *New England Journal of Medicine, 288,* 906–907.

VESSEY, M. P., et al. (1981). Pelvic inflammatory disease and the intrauterine device. *British Medical Journal, 282,* 855–857.

VESSEY, M. P., LAWLESS, M., & YEATES, D. (1982). Efficacy of different contraceptive methods. *Lancet, 1,* 841–842.

VESSEY, M. P., & PAINTER, R. (1994). Oral contraceptive use and benign gallbladder disease; revisited. *Contraception, 50,* 167–173.

VISTICA G. L., & THOMAS, E. (1997, June 2). Sex and lies. *Newsweek,* pp. 26–31.

VLAJINAC, H. D., et al. (1997). Effect of caffeine intake during pregnancy on birth weight. *American Journal of Epidemiology, 145,* 335–338.

VOELLER, B. (1991). AIDS and heterosexual anal intercourse. *Archives of Sexual Behavior, 201,* 233–269.

VOIGT, H. (1991). Enriching the sexual experience of couples: The Asian traditions and sexual counseling. *Journal of Sex and Marital Therapy, 17,* 214–219.

VOLBERDING, P. A., et al. (1990). Zidovudine in asymptomatic human immunodeficiency virus infections. *New England Journal of Medicine, 322,* 941–949.

VOLPE, A., SILFERI, M., GENAZZANI, A. D., & GENAZZANI, A. R. (1993). Contraception in older women. *Contraception, 47,* 229–239.

VOYDANOFF, P., & DONNELLY, B. (1990). *Adolescent sexuality and pregnancy.* Newbury Park, CA: Sage.

WADE, M. E., et al. (1980, September). *A randomized prospective study of the use-effectiveness of two methods of natural family planning.* Paper presented at the International Federation for Family Life Promotion Second International Congress, Navan, Ireland.

WALDENSTRÖM, U. (1996). Modern maternity care: Does safety have to take the meaning out of birth? *Midwifery, 12,* 165–173.

WALDENSTRÖM, U., & NILSSON, C.-A. (1997). A randomized controlled study of birth center care versus standard maternity care: Effects on women's health. *Birth, 24,* 17–26.

WALKER, C. E. (1970). Erotic stimuli and the aggressive sexual offender. In *Technical report of the Commission on Obscenity and Pornography,* Vol. 7 (pp. 91–147). Washington, DC: U.S. Government Printing Office.

WALSH, M. W. (1992, October 29). Canada far ahead of U.S. in recognizing gay rights. *Los Angeles Times,* pp. A1, A8.

WALSTER, E., ARONSON, V., ABRAHAMS, D., & ROTTMAN, L. (1966). Importance of physical attractiveness in dating behavior. *Journal of Personality and Social Psychology, 4,* 508–516.

WAMBAUGH, J. (1989). *The blooding.* New York: Bantam Books.

WARD, M. C. (1989). *Nest in the wind: Adventures in anthropology on a tropical island.* Prospect Heights, IL: Waveland Press.

WARD, T., HUDSON, S. M., & MCCORMACK, J. (1997). The assessment of rapists. *Behaviour change, 14,* 39–54.

WARD, T., MCCORMACK, J., & HUDSON, S. M. (1997). Sexual offenders' perceptions of their intimate relationships. *Sexual abuse: A journal of research and treatment, 9,* 57–74.

WARNER, C. (1997, September 29). Storied history. (New Orleans) *Times-Picayune,* pp. A1, A6–A7.

WARNER, C. G. (Ed.). (1980). *Rape and sexual assault.* Germantown, MD: Aspen Systems Corp.

WASHINGTON, A. E., CATES, W., & WASSERHEIT, J. N. (1991). Preventing pelvic inflammatory disease. *Journal of the American Medical Association, 266,* 2574–2580.

WASHTON, A. M. (1989, December). Cocaine abuse and compulsive sexuality. *Medical Aspects of Human Sexuality,* pp. 23, 32–39.

WASOW, M., & LOEB, M. (1979). Sexuality in nursing homes. *Journal of the American Geriatrics Society, 27,* 73–79.

WASSON, J. H., et al. (1995). A comparison of transurethral with watchful waiting for moderate symptoms of benign prostatic hyperplasia. *New England Journal of Medicine, 332,* 1995.

WATERMAN, C. K., DAWSON, L. J., & BOLOGNA, M. J. (1989). Sexual coercion in gay male and lesbian relationships: Predictors and implications for support services. *Journal of Sex Research, 26,* 118–125.

WATERMAN, C. K., & FOSS-GOODMAN, D. (1984). Child molesting: Variables relating to attribution of fault to victims, offenders and nonparticipating parents. *Journal of Sex Research, 20,* 329–349.

WATERMAN, J., KELLY, R. J., MCCORD, J., & OLIVERI, M. K. (1992). *Behind the playground walls: Sexual abuse in preschools.* New York: Guilford.

WATKINS, R. N. (1986). Vaginal spermicides and congenital disorders: The validity of a study. *Journal of the American Medical Association, 256,* 3095.

WEBER, M. (1993). Images in clinical medicine: Pinworms. *New England Journal of Medicine, 328,* 927.

WEG, R. B. (1996). Sexuality, sensuality, and intimacy. In G. L. Maddox (Ed.), *The encyclopedia of aging* (2nd ed.) (pp. 479–488). New York: Springer Publishing.

WEINBERG, G. (1973). *Society and the healthy homosexual.* New York: Anchor.

WEINBERG, M. S., LOTTES, I. L., & SHAVER, F. M. (1995) Swedish or American heterosexual college youth: Who is more permissive? *Archives of Sexual Behavior, 24,* 409–438.

WEINBERG, M. S., WILLIAMS, C. J., & CALHAN, C. (1994). Homosexual foot fetishism. *Archives of Sexual Behavior, 23,* 611–626.

WEINBERG, M. S., WILLIAMS, C., & MOSER, C. (1984). The social constituents of sadomasochism. *Social Problems, 31,* 379–389.

WEINBERG, T. (1987). Sadomasochism in the United States: A review of recent sociological literature. *Journal of Sex Research, 23,* 50–69.

WEINBERG, T., & BULLOUGH, V. (1988). Alienation, self-image, and the importance of support groups for the wives of transvestites. *Journal of Sex Research, 24,* 262–268.

WEINHARDT, L. S., & CAREY, M. P. (1996). Prevalence of erectile disorder among men with diabetes mellitus: Comprehensive review, methodological critique, and suggestions for future research. *Journal of Sex Research, 33,* 205–214.

WEINSTOCK, H., et al. (1994). *Chlamydia trachomatis* infections. *Infectious Disease Clinics of North America, 8,* 797–819.

WEINSTOCK, H. S. et al. (1993). Factors associated with condom use in a high-risk heterosexual population. *Sexually Transmitted Diseases, 20,* 14–20.

WEIS, D. L. (1983). Affective reactions of women to their initial experience of coitus. *Journal of Sex Research, 19,* 209–237.

WEIS, D. L. (1985). The experience of pain during women's first sexual intercourse: Cultural mythology about female sexual initiation. *Archives of Sexual Behavior, 14,* 421–428.

WEISBERG, D. K. (1985). *Children of the night: A study of adolescent prostitution.* Lexington, MA: Lexington Books.

WEISS, D. (1983). Open marriage and multilateral relationships: The emergence of nonexclusive models of the marital relationship. In E. Macklin & R. Rubin (Eds.), *Contemporary families and alternative lifestyles.* Newbury Park, CA: Sage.

WEISS, H. (1978). The physiology of human penile erection. In A. Comfort (Ed.), *Sexual consequences of disability.* Philadelphia: Stickley.

WEISSBACH, T. A., & ZAGON, G. (1975). The effect of deviant group membership upon impression of personality. *Journal of Social Psychology, 95,* 263–266.

WEITZMAN, L. J., et al. (1972). Sex role socialization in picture books for pre-school children. *American Journal of Sociology, 77,* 1125–1150.

WEIZMAN, R., & HART, J. (1987). Sexual behavior in healthy married elderly men. *Archives of Sexual Behavior, 16,* 39–44.

WELLER, A., & WELLER, L. (1993). Menstrual synchrony between mothers and daughters and between roommates. *Physiology and Behavior, 53,* 943–949.

WELLINGS, K., FIELD, J., JOHNSON, A. M., & WADSWORTH, J. (1994). *Sexual behavior in Britain: The National Survey of Sexual Attitudes and Lifestyles.* London: Penguin Books.

WELLS, B. L. (1986). Predictors of female nocturnal orgasms. *Journal of Sex Research, 22,* 421–437.

WESSELLS, H., LUE, T. F., & McANINCH, J. W. (1996). Penile length in the flaccid and erect states: Guidelines for penile augmentation. *Journal of Urology, 156,* 995–997.

WESTERMARCK, E. A. (1891). *The history of human marriage.* New York: Macmillan.

WESTFALL, J. M., MAIN, D. J., & BARNARD, L. (1996). Continuation rates among injectable contraceptive users. *Family Planning Perspectives, 28,* 275–277.

WESTOFF, C. F. (1988). Unintended pregnancy in America and abroad. *Family Planning Perspectives, 20,* 254.

WESTROM, L. (1980). Incidence, prevalence, and trends of acute pelvic inflammatory disease and its consequences in industrialized countries. *American Journal of Obstetrics and Gynecology, 138,* 880.

WESTROM, L., et al. (1992). Pelvic inflammatory disease and fertility. *Sexually Transmitted Diseases, 19,* 185–192

WETZEL, C., & INSKO, C. (1982). The similarity-attraction relationship: Is there an ideal one? *Journal of Experimental Social Psychology, 18,* 253–276.

WHALEN, R. (1977). Brain mechanisms controlling sexual behavior. In F. A. Beach (Ed.), *Human sexuality in four perspectives.* Baltimore: Johns Hopkins University Press.

WHIPPLE, B., GERDES, C. A., & KOMISARUK, B. R. (1996). Sexual response to self-stimulation in women with complete spinal cord injury. *Journal of Sex Research, 33,* 231–240.

WHIPPLE, B., OGDEN, G., & KOMISARUK, B. R. (1992). Physiological correlates of imagery-induced orgasm in women. *Archives of Sexual Behavior, 21,* 121–133.

WHITAM, F. L. (1977a). Childhood indicators of male homosexuality. *Archives of Sexual Behavior, 6,* 89–96.

WHITAM, F. L. (1977b). The homosexual role: A reconsideration. *Journal of Sex Research, 13,* 1–11.

WHITAM, F. L. (1983). Culturally invariable properties of male homosexuality: Tentative conclusions from cross-cultural research. *Archives of Sexual Behavior, 12,* 207–226.

WHITAM, F. L., DASKALOS, C. T., & MATHY, R. M. (1995). A cross-cultural assessment of familial factors in the development of female homosexuality. *Journal of Psychology and Human Sexuality, 7*(4), 59–76.

WHITAM, F. L., DIAMOND, M., & MARTIN, J. (1993). Homosexual orientation in twins: A report on 61 pairs and three triplet sets. *Archives of Sexual Behavior, 22,* 187–205.

WHITAM, F. L., & MATHY, R. M. (1986). *Male homosexuality in four societies: Brazil, Guatemala, the Philippines, and the United States.* New York: Praeger.

WHITCOMB, D. (1992). Legal reforms on behalf of child witnesses: Recent developments in the American courts. In H. Dent & R. Flin (Eds.), *Children as witnesses* (pp. 151–165). Chichester, England: Wiley.

WHITE, D. R., & BURTON, M. L. (1988). Causes of polygyny: Ecology, economy, kinship, and warfare. *American Anthropologist, 90,* 871–887.

WHITE, E., MALONE, K. E., WEISS, N. S., & DALING, J. R. (1994). Breast cancer among young U.S. women in relation to oral contraceptive use. *Journal of the National Cancer Institute, 86,* 505–513.

WHITE, G. L. (1980a). Inducing jealousy: A power perspective. *Personality and Sexual Psychology Bulletin, 6,* 222–227.

WHITE, G. L. (1980b). Physical attractiveness and courtship progress. *Journal of Personality and Social Psychology, 39,* 660–668.

WHITE, G. L. (1981a). Some correlates of romantic jealousy. *Journal of Personality, 49,* 129–147.

WHITE, G. L. (1981b). Relative involvement, inadequacy, and jealousy: A test of a causal model. *Alternative Lifestyles, 4,* 291–309.

WHITE, G. L., & MULLEN, P. E. (1989). *Jealousy: Theory, research, and clinical strategies.* New York: Guilford.

WHITE, J., & HUMPHREY, J. (1991, August). *Victims of rape, resistance, revictimization, and acknowledgment.* Paper presented at the annual meeting of the American Psychological Association, Washington, DC.

WHITE, S. E., & REAMY, K. (1982). Sexuality and pregnancy: A review. *Archives of Sexual Behavior, 11,* 429–444.

WHITEHEAD, B. D. (1994). The failure of sex education. *Atlantic Monthly, 274,* 55–80.

WHITEHURST, R. N. (1972). Extramarital sex: Alienation or extension of normal behavior. In J. N. Edwards (Ed.), *Sex and society.* Chicago: Rand McNally.

WHITING, B. (1979). Contributions of anthropology to the study of gender identity, gender role, and sexual behavior. In H. Katchadourian (Ed.), *Human sexuality: A comparative and developmental perspective.* Berkeley: University of California Press.

WHITTEMORE, A. S., HARRIS, R., ITNYRE, J. & COLLABORATIVE OVARIAN CANCER GROUP. (1992a). Characteristics relating to ovarian cancer risk: Collaborative analysis of 12 U.S. case-control studies. II: Invasive epithelial ovarian cancers in white women. *American Journal of Epidemiology, 136,* 1184–1203.

WHITTEMORE, A. S., HARRIS, R., ITNYRE, J., & COLLABORATIVE OVARIAN CANCER GROUP. (1992b). Characteristics relating to ovarian cancer risk: The pathogenesis of epithelial ovarian cancer. *American Journal of Epidemiology, 136,* 1212–1220.

WHITTLE, H., et al. (1994). HIV-2 infected patients survive longer than HIV-1 infected patients. *AIDS, 8,* 1617–1620.

WHO COLLABORATIVE STUDY OF CARDIOVASCULAR DISEASE AND STEROID HORMONE CONTRACEPTION (1996a). Haemorrhagic stroke, overall stroke risk, and combined oral contraceptives: Results of an international, multicentre, case-control study. *Lancet, 348,* 505–510.

WHO COLLABORATIVE STUDY OF CARDIOVASCULAR DISEASE AND STEROID HORMONE CONTRACEPTION (1996b). Ischaemic stroke and combined oral contraceptives: Results of an international, multicentre, case control study. *Lancet, 348,* 498–505.

WIDSTRAND, C. G. (1964). Female infibulation. *Studia Ethnographica Upsaliensia, 20,* 95–122.

WIEDERMAN, M. W. (1997). Pretending orgasm during sexual intercourse: Correlates in a sample of young adult women. *Journal of Sex & Marital Therapy, 23,* 131–139.

WIEDERMAN, M. W. (1997). Extramarital sex: Prevalence and correlates in a national survey. *Journal of Sex Research, 34,* 167–174.

WIEDERMAN, M. W., & ALLGEIER, E. R. (1993). Gender differences in sexual jealousy: Adaptationist or social learning explanation. *Ethology and Sociobiology, 14,* 115–140.

WIELANDT, H., & KNUDSEN, L. B. (1997). Birth control: Some experiences from Denmark. *Contraception, 55,* 301–306.

WIEST, W. (1977). Semantic differential profiles of orgasm and other experiences among men and women. *Sex Roles, 3,* 399–403.

WIGHT, D. (1992). Impediments to safer heterosexual sex: A review of research with young people. *AIDS Care, 4,* 11–23.

WILCOX, A. J., WEINBERG, C. R., & BAIRD, D. D. (1995). Timing of sexual intercourse in relation to ovulation: Effects on the probability of conception, survival of the pregnancy, and sex of the baby. *New England Journal of Medicine, 333,* 1517–1521.

WILCOX, A. J., et al. (1988). Incidence of early loss of pregnancy. *New England Journal of Medicine, 319,* 189–194.

WILCOX, D., & HAGER, R. (1980). Toward realistic expectation for orgasmic response in women. *Journal of Sex Research, 16,* 162–179.

WILCOX, L. W., et al. (1992). Menstrual function after tubal sterilization. *American Journal of Epidemiology, 135,* 1368–1381.

WILLIAMS, L. B. (1991). Determinants of unintended childbearing among ever-married women in the United States: 1973–1988. *Family Planning Perspectives, 23,* 212–221.

WILLIAMS, L. M. (1994). Recall of childhood trauma: A prospective study of women's memories of child sexual abuse. *Journal of Consulting and Clinical Psychology, 62,* 1167–1176.

WILLIAMS, N., CHELL, J., & KAPILA, L. (1993). Why are children referred for circumcision? *British Medical Journal, 306,* 28.

WILLIAMS, P., & SMITH, M. (1979). Interview. In The first question [Department film]. London: British Broadcasting System Science and Features.

WILLIAMSON, D. F., et al. (1989). Comparing the prevalence of smoking in pregnant and nonpregnant women, 1985 to 1986. *Journal of the American Medical Association, 261,* 70.

WILSON, G. D. (Ed.), (1987). *Variant sexuality: Research and theory.* Baltimore: Johns Hopkins University Press.

WILSON, G. D., (1987). An ethological approach to sexual deviation. In G. D. Wilson (Ed.), *Variant sexuality: Research and theory.* Baltimore: Johns Hopkins University Press.

WILSON, H. C., KIEFHABER, S. H., & GRAVEL, V. (1991). Two studies of menstrual synchrony: Negative results. *Psychoneuroendocrinology, 16,* 353–359.

WILSON, P. (1986). Black culture and sexuality. *Journal of Social Work and Human Sexuality, 4,* 29–46.

WINCZE, J., & CAREY, M. (1991). *Sexual dysfunction: A guide for assessment and treatment.* New York: Guilford.

WINCZE, J. P., HOON, E. F., & HOON, P. W. (1976). Physiological responsivity of normal and sexually dysfunctional women during erotic stimulus exposure. *Journal of Psychosomatic Research, 20,* 445–451.

WINICK, C., & EVANS., J. T. (1994). Is there a national standard with respect to attitudes toward sexually explicit media material. *Archives of Sexual Behavior, 23,* 405–419.

WINICK, C., & EVANS, J. T. (1996). The relationship between nonenforcement of state pornography laws and rates of sex crimes arrests. *Archives of Sexual Behavior, 25,* 439–453.

WISE, T., & MEYER, J. (1980). Transvestism: Previous findings and new areas for inquiry. *Journal of Sex and Marital Therapy, 6,* 116–128.

WISWELL, T. E., ENZENAUER, R. W., CORNISH, J. D., & HANKINS, C. T. (1987). Declining frequency of circumcision: Implications for changes in the absolute incidence and male to female sex ratio of urinary tract infections in early infancy. *Pediatrics, 79,* 338–342.

WITHERINGTON, R. (1990). External penile appliances for management of impotence. *Seminars in Urology, 8,* 124–128.

WITLIN, A. G., & SIBAI, B. M. (1997). Hypertension in pregnancy: Current concepts of preeclampsia. *Annual Review of Medicine, 48,* 115–127.

WITT, S. D. (1996). Traditional or androgynous: An analysis to determine gender role orientation of basal readers. *Child Study Journal, 26,* 303–318.

WITT, S. D. (1997). Parental influence on children's socialization to gender roles. *Adolescence, 32,* 253–259.

WOLF, A. P. (1970). Childhood association and sexual attraction: A further test of the Westermarck hypothesis. *American Anthropologist, 72,* 503–515.

WOLFE, A. A., WOLFE, V. V., & BEST, C. L. (1988). Child victims of sexual abuse. In V. B. Van Hasselt, H. R. L. Morrison, A. S. Bellack, & M. Hersen (Eds.), *Handbook of family violence.* New York: Plenum.

WOLFE, P. S. (1997). The influence of personal values on issues of sexuality and disability. *Sexuality and Disability, 15,* 69–90.

WOLFF, C. (1971). *Love between women.* New York: Harper & Row.

WOLNER-HANSSEN, P., KIVIAT, N. K., & HOLMES, K. K. (1989). Atypical pelvic inflammatory disease: Subacute, chronic, or subclinical upper genital tract infection in women. In K. K. Holmes et al. (Eds.), *Sexually transmitted diseases* (2nd ed.). New York: McGraw-Hill.

WONDERLICH, S. A., WILSNACK, R. W., WILSNACK, S. C. & HARRIS, T. R. (1996). Childhood sexual abuse and bulimia behavior in a nationally representative sample. *American Journal of Public Health, 86,* 1082–1086.

WONG, J. K., et al. (1997). Recovery of replication-competent HIV despite prolonged suppression of plasma viremia. *Science, 278,* 1291–1295.

WORELL, J. (1978). Sex roles and psychological well being: Perspectives on methodology. *Journal of Consulting and Clinical Psychology, 46,* 777–791.

WORLD HEALTH ORGANIZATION. (1978). *Special Programme of Research, Development and Research Training: Seventh annual report.* Geneva: Author.

WORLD HEALTH ORGANIZATION. (1981). A prospective multicentre trial of the ovulation method of natural family planning. II. The effectiveness phase. *Fertility and Sterility, 36,* 591–598.

WORLD HEALTH ORGANIZATION. (1984). Mental health and female sterilization: Report of a WHO collaborative prospective study. *Journal of Biosocial Sciences, 16,* 1–21.

WORLD HEALTH ORGANIZATION. (1985). Mental health and female sterilization: A follow-up. *Journal of Biosocial Sciences, 17,* 1–18.

WORLD HEALTH ORGANIZATION (1995). AIDS-global data. *Weekly Epidemiology Records, 70,* 353–355.

WORLD HEALTH ORGANIZATION (1996) *Revised 1990 estimates of maternal mortality—A new approach by WHO and UNICEF.* WHO/FRH/MSM 96.11: UNICEF? PLN? 96.1. World Health Organization, Geneva.

WORLD HEALTH ORGANIZATION CONSENSUS STATEMENT. (1989). Sexually transmitted diseases as a risk factor for HIV transmission. *Journal of Sex Research, 26,* 272–275.

WORLD HEALTH ORGANIZATION TASK FORCE ON LONG-ACTING AGENTS FOR THE REGULATION OF FERTILITY (1983). Multinational comparative clinical trial of long-acting injectable contraceptives: Norethisterone enanthate given in two dosage regimens and depotmedroxyprogesterone acetate: final report. *Contraception, 28,* 1–20.

WORTLEY, P. M., & FLEMING, P. L. (1997). AIDS in women in the United States. *Journal of the American Medical Association, 278,* 911–916.

WRIGHT, A. L. (1983). A cross-cultural comparison of menopause symptoms. *Medical Anthropology, 7,* 20–36.

WRIGHT, K. (1990). Mycoplasmas in the AIDS spotlight. *Science, 248,* 682–683.

WRITING GROUP FOR THE PEPI TRIAL (1996). Effects of hormone therapy on bone mineral density. *Journal of the American Medical Association, 276,* 1389–1396.

WYATT, G. E., GUTHRIE, D., & NOTGRASS, C. M. (1992). Differential effects of women's child sexual abuse and subsequent sexual revictimization. *Journal of Consulting and Clinical Psychology, 60,* 167–173.

WYNN, R., & FLETCHER, C. (1987). Sex role development and early educational experiences. In D. B. Carter (Ed.), *Current conceptions of sex roles and sex typing.* New York: Praeger.

YALOM, I. D. (1960). Aggression and forbiddenness in voyeurism. *Archives of General Psychiatry, 3,* 305–319.

YANG, A. S. (1997). Attitudes toward homosexuality. *Public Opinion Quarterly, 6,* 477–507.

YAO, M., & TULANDI, T. (1997). Current status of surgical and nonsurgical management of ectopic pregnancy. *Fertility and Sterility, 67,* 421–423.

YESALIS, C. E., KENNEDY, N. J., KOPSTEIN, A. N., & BAHRKE, M. S. (1993). Anabolic-androgenic steroid use in the United States. *Journal of the American Medical Association, 270,* 1217–1221.

YESMONT, G. A. (1992). The relationship of assertiveness to college students' safer sex behaviors. *Adolescence, 27,* 253–272.

YOLKEN, R. H., et al. (1992). Human milk mucin inhibits rotavirus replication and prevents experimental gastroenteritis. *Journal of Clinical Investigation, 90,* 1984–1991.

YONKERS, K. A., et al. (1997). Symptomatic improvement of premenstrual dysphoric disorder with sertraline treatment. *Journal of the American Medical Association, 278,* 983–988.

YOUNG, D. (1997). A new push to reduce cesareans in the United States, *Birth, 24,* 1–3.

YOWELL, C. M. (1997). Risks of communication: Early adolescent girls' conversation with mothers and friends about sexuality. *Journal of Early Adolescence, 17,* 172–196.

ZAVIACIC, M., & WHIPPLE, B. (1993). Update on the female prostate and the phenomenon of female ejaculation. *Journal of Sex Research, 30,* 148–151.

ZAVIACIC, M., ZAVIACICOVA, A., HOLOMAN, I. K., & MOLCAN, J. (1988). Female urethral expulsions evoked by local digital stimulation of the G-spot: Differences in the response patterns. *Journal of Sex Research, 24,* 311–318.

ZELNIK, M., & KANTNER, J. F. (1979). Reasons for nonuse of contraception by sexually active women aged 15–19. *Family Planning Perspectives, 11,* 289–296.

ZHANG, J., THOMAS, A. G., & LEYBOVICH, E. (1997). Vaginal douching and adverse health effects: A meta-analysis. *American Journal of Public Health, 87,* 1207–1211.

ZHANG, S. D., & ODENWALD, W. F. (1995). Misexpression of the white (omega) gene triggers male-made courtship in drosophilia. *Proceedings of the National Academy of Sciences of the United States of America, 92,* 5525–5529.

ZHU, K., et al. (1996). Vasectomy and prostate cancer: A case-control study in a health maintenance organization. *American Journal of Epidemiology, 144,* 717–722.

ZHU, T. (1998). An African HIV-1 sequence from 1959 and implications for the origin of the epidemic. *Nature, 391,* 594–597.

ZILBERGELD, B. (1978). *Male sexuality: A guide to sexual fulfillment.* Boston: Little, Brown.

ZILBERGELD, B., & ELLISON, C. R. (1980). Desire discrepancies and arousal problems in sex therapy. In S. R. Leiblum & L. A. Pervin (Eds.), *Principles and practice of therapy.* New York: Guilford.

ZILBERGELD, B., & HAMMOND, D. (1988). The use of hypnosis in treating desire disorders. In S. R. Leiblum & R. C. Rosen (Eds.), *Sexual desire disorders*. New York: Guilford.

ZILBERGELD, B., & KILMANN, P. (1984). The scope and effectiveness of sex therapy. *Psychotherapy, 21,* 319–326.

ZILLMANN, D., & BRYANT, J. (1982, Autumn). Pornography, sexual callousness, and the trivialization of rape. *Journal of Communication,* pp. 10–21.

ZILLMANN, D., & BRYANT, J. (1988). Pornography's impact on sexual satisfaction. *Journal of Applied Social Psychology, 18,* 438–453.

ZIMMER, D. (1983). Interaction patterns and communication skills in sexually distressed, maritally distressed, and normal couples: Two experimental studies. *Journal of Sex and Marital Therapy, 9,* 251–265.

ZIMMERMAN, H. L., et al. (1990). Epidemiologic differences between chlamydia and gonorrhea. *American Journal of Public Health, 80,* 1338–1342.

ZITTER, S. (1987). Coming out to mom: Theoretical aspects of the mother-daughter process. In Boston Lesbian Psychologies Collective (Eds.), *Lesbian psychologies: Explorations and challenges* (pp. 177–194). Urbana: University of Illinois Press.

ZUCKER, K. J., & BLANCHARD, R. (1997). Transvestic fetishism: Psychopathology and theory. In D. R. Laws & W. O'Donohue (Eds.), *Sexual deviance: Theory, assessment, and treatment* (pp. 253–279). New York: Guilford Press.

ZUCKER, K. J., & BRADLEY, S. J. (1995). *Gender identity disorder and psychosexual problems in children and adolescents.* New York: Guilford.

ZUCKERMAN, A. J. (1982, November/December). Viral hepatitis. *Practical Gastroenterology, 16,* 21–27.

ZURAVIN, S. J. (1991). Unplanned childbearing and family size: Their relationship to child neglect and abuse. *Family Planning Perspectives, 23,* 155–161.

STUDENT STUDY GUIDE

ASSESSING YOUR RISK FOR GETTING HIV/AIDS AND OTHER SEXUALLY TRANSMITTED DISEASES

This year there will be close to 12 million newly diagnosed cases of sexually transmitted diseases in the United States. One in four Americans between the ages of 15 and 55 will have at least one STD sometime in their lifetimes (Centers for Disease Control). Ask yourself, "Am I at risk of getting a sexually transmitted disease?" What precautions do you take to prevent it from ever happening? Surveys conducted on college campuses indicate that most students do nothing more than look for symptoms in potential new partners. However, many infected individuals, including those with the virus that causes AIDS, do not show any symptoms. We must, therefore, do more than just look for symptoms. As a first step, you should have some idea of your level of risk. Everyone who has sexual relations is at some risk, some more than others. The following questions are designed to help you assess your own level of risk. Be honest when you answer the questions; the questionnaire cannot be of any use to you if you have not answered each item truthfully.

1. How many sexual partners (intercourse or oral-genital sex) have you had (a) in the last seven years if you have not had an HIV antibody test, or (b) since your last HIV antibody test that proved negative?

    4 points ___ 42 or more (an average of 6 or more per year)
    3 points ___ 8–41
    2 points ___ 2–7
    1 point ___ 1
    0 points ___ none

2. Do you use drugs intravenously?

    4 points ___ yes
    0 points ___ no

Note: If you scored zero on the first two questions, you are at extremely low risk of being infected with the human immunodeficiency virus or any other virus, bacterium, or parasite discussed in Chapter 5. You may stop here. If you scored greater than zero, complete the questionnaire.

3. How many sexual partners have you had in the last year?

    4 points ___ 6 or more
    3 points ___ 3–5
    2 points ___ 2
    1 point ___ 1
    0 points ___ none

4. What kinds of sexual contacts have you had? (*Note*: Working from top to bottom, give yourself the points *only* for the *first* category you check.)

    4 points ___ "one-night stands" with relative strangers, sex with a prostitute, or group sex
    3 points ___ (a) "casual sex" (i.e., sex with acquaintances or friends with whom you do not have a "steady" relationship) or (b) steady relationship(s), but you have "cheated" on at least one occasion

    1 point ___ steady relationship(s) without cheating during relationship
    0 points ___ you have not yet had sexual intercourse or oral-genital sex

5. What is your sexual orientation?

    4 points ___ male homosexual, male bisexual, or male heterosexual with occasional male homosexual experiences (*Note*: Homosexuality does not cause STDs, but STDs are very prevalent in these groups.)
    1 point ___ (a) exclusively male or female heterosexual, or (b) female homosexual
    0 points ___ you have not yet had sexual intercourse or oral-genital sex

6. Have you been a victim of forced sexual activity in the last seven years?

    4 points ___ yes, and you have not been tested for STDs (including an HIV antibody test)
    1 point ___ yes, but subsequent STD tests, including an HIV antibody test, were negative
    0 points ___ no

7. What kinds of sexual activities do you usually engage in? (*Note*: Working from top to bottom, give yourself the points *only* for the *first* category you check.)

    4 points ___ anal intercourse, fisting, oral-anal, or any other activities involving contact with feces or urine
    3 points ___ unprotected (no condom) vaginal intercourse
    2 points ___ unprotected oral-genital sex
    1 point ___ protected (condom) vaginal and/or oral-genital sex
    0 points ___ kissing, caressing, massage, masturbation

8. How often have you had sex after using alcohol or drugs?

    4 points ___ very often
    3 points ___ usually at parties only
    2 points ___ sometimes
    1 point ___ hardly ever
    0 points ___ never

9. Where do most of your sexual partners live?

    1 point ___ in a large city or immediate suburb
    0 points ___ in a small city or town
    0 points ___ you have not yet had sexual intercourse or oral-genital sex

10. In the last year, how often have you or your sexual partners used condoms?

    4 points ___ never or rarely, and you have had more than one partner
    3 points ___ usually, but sometimes you have taken chances
    2 points ___ never or rarely, but you have had only one steady partner
    1 point ___ every time (for each act of penetration), including the first time with a new partner
    0 points ___ (a) you have not yet had sexual intercourse or oral-genital sex, or (b) you are an exclusively homosexual female

11. How many sexual partners has your current (or last) partner had prior to you?

    4 points ___ many
    3 points ___ several
    2 points ___ one or a few
    1 point ___ none (you think)
    0 points ___ you have not yet had sexual intercourse or oral-genital sex

12. How certain are you that your current (or last) sexual partner has not had sex with anyone else since you and he or she started having sex?

    4 points ___ you are certain your partner has had sex with others during your relationship
    3 points ___ you suspect your partner of having sex with others during your relationship
    2 points ___ you are uncertain
    1 point ___ you are relatively certain that he or she has not (*Note*: No one can be 100 percent certain.)
    0 points ___ you have not yet had sexual intercourse or oral-genital sex

(*Note*: Answer question 13 only if you are not a homosexual or bisexual male.)

13. Have any of your sexual partners had male homosexual experiences or had sex with someone who has? (*Note*: Homosexuality per se does not cause STDs.)

    4 points ___ definitely yes
    3 points ___ possibly
    1 point ___ do not know
    0 points ___ you are relatively certain that he or she has not

14. Have any of your sexual partners used drugs intravenously?

    4 points ___ yes
    2 points ___ you do not know
    0 points ___ no

15. Do you and your sexual partner(s) discuss sexually transmitted diseases and "safer sex"?

    2 points ___ rarely
    1 point ___ yes, but you have not been tested
    0 points ___ (a) yes, and you have both been tested, or (b) you have not yet had sexual intercourse or oral-genital sex

HOW TO SCORE

If you use condoms every time during vaginal or anal intercourse (be honest—it must be *every* time), for questions 3, 4, 5, 8, 11, 12, and 13 subtract 1 point for each question for which you scored a "2," and subtract 2 points for each question for which you scored a "3" or a "4." If you and your partner are in a monogamous relationship lasting several years and neither of you has ever cheated, you may subtract 2 points if you scored "2" or more for question 7.

Add up your adjusted total points for the 15 questions (a maximum of 4 points on questions 1–8 and 10–14; 2 points on question 15; and 1 point on question 9).

15 points or less   Unless you take drugs intravenously (an automatic high risk), you are at low risk of contracting a sexually transmitted disease.

16–29 points   You are at moderate risk and should adjust your life-style.

30 points or more   You are at high risk. You are strongly advised to change your life-style. If you take drugs intravenously, you are at high risk regardless of your point total.

If you are at moderate or high risk, how can you adjust your life-style? Look back at the questions and see where simple changes would result in subtracting points. The easiest way is by limiting the number of sexual partners you have and/or by always using condoms. Think about it—it could save your life.

CHAPTER 1 *INTERACTIVE REVIEW*

Some people feel that sex education should be the responsibility of parents, yet only about a third of all college students have ever had a meaningful discussion about sex with their parents. Many young people turn to (1) _____ and (2) _____ for information about sex, but much of what they learn is incorrect. The best alternative would be for children to receive factual information in school. Surveys indicate that approximately (3) _____ percent of Americans favor sex education in school, but fewer than half of all junior high schools and high schools in the United States presently offer such courses.

Sex is only a part of (4) _____, which encompasses all of the sexual attitudes, feelings, and behaviors associated with being human. Sexual behaviors and attitudes (such as what is considered to be sexually attractive) vary from culture to culture, and can even change within a culture over time. American sexual attitudes are both permissive and repressed, and as a result many people in the United States have ambivalent feelings about sex.

In order to understand how we arrived at our current values, we must examine the history of sexual attitudes in our culture. The idea that the primary purpose of sex is for procreation (to have children) originally came from (5) _____, but they also had a very positive attitude about their bodies and sexual pleasure between husbands and wives. The early (6) _____ affirmed the procreational purpose of sex, but completely denied its pleasurable aspects. Sexual desire, even within marriage, was now associated with guilt. The two biggest proponents of this view within Christianity were (7) _____ and (8) _____. In Western culture, negative attitudes about sex reached their zenith during the reign of (9) _____ of England. During this time, the medical profession contributed many incorrect negative beliefs about engaging in sex, including the belief that excess sex, particularly masturbation, could lead to serious medical problems and eventually insanity.

The industrial revolution slowly changed Americans' lives, including their sex lives. With shorter workdays and work weeks and greater mobility (e.g., automobiles) than in past generations, people now had more free time to spend together. With the availability of (10) _____ during World War II and the marketing of (11) _____ in 1960, the United States entered the sexual revolution.

Because of the antisexual attitudes of the Victorian era, scientific study of human sexuality was slow to develop. (12) _____, who emphasized the sexuality of all people, including children, and (13) _____, who published seven volumes about the psychology of sex, were two individuals of the Victorian era who attempted to counter antisexual attitudes. However, it has only been within the last 30 to 40 years that the scientific and medical communities have accepted sex as a subject for serious discussion and research. The first large-scale surveys were done by (14) _____ in the 1940s and early 1950s. (15) _____ published their physiological investigations of human sexual behavior in 1966. Even today, however, there is some resistance on the part of governmental agencies to fund research on human sexual behavior.

Researchers use several methods to study sexuality. Surveys are conducted by taking (16) _____ from the population. The way in which a sample is drawn and the manner in which the survey is constructed determine to what extent results can be correctly generalized to the population. Other methods include (17) _____, (18) _____, (19) _____, and (20) _____.

CHAPTER 1 *SELF-TEST*

A. TRUE OR FALSE

[T] [F] 21. The Old Testament presents a positive view of sex within marriage.

[T] [F] 22. American sexual behaviors are considered to be the norm by the rest of the world.

[T] [F] 23. Because of AIDS, all states now require that public schools offer education about sexually transmitted diseases.

[T] [F] 24. The larger the number of people in a survey, the more accurate it always is.

[T] [F] 25. All cultures consider female breasts to be highly erotic.

[T] [F] 26. The Puritans' sexual beliefs were a major influence on current American values.

[T] [F] 27. Kissing is one sexual behavior that is done worldwide.

[T] [F] 28. One method of obtaining a random sample is to randomly pick names from a phone book.

[T] [F] 29. Each year in the United States, one in ten teenage girls aged 15 to 19 gets pregnant and one in four sexually active teens gets a sexually transmitted disease.

T ☐ F ☐ 30. Over half of all American teenagers have had sexual intercourse before their seventeenth birthday.

T ☐ F ☐ 31. The era of permissiveness known as the "sexual revolution" started during the Victorian era.

T ☐ F ☐ 32. About one-fourth of all societies in the world encourage or openly tolerate children engaging in sex.

T ☐ F ☐ 33. A small survey conducted by Dr. Clelia Mosher in 1892 showed that most Victorian women did not desire or enjoy sex.

T ☐ F ☐ 34. Kinsey's surveys are a good example of the use of random sampling techniques.

T ☐ F ☐ 35. A strong positive correlation between two variables is evidence of a cause-and-effect relationship.

MATCHING, PART 1

_____ 36. A Victorian-era physician who emphasized the sexuality of children and adults

_____ 37. He viewed sex for procreation as an unpleasant necessity and equated guilt with sexual desire

_____ 38. Conducted the first large-scale physiological study of human sexual behavior

_____ 39. He systematized Christian thought during the Middle Ages

_____ 40. They believed that the purpose of sex was for procreation, but had a very positive attitude about sexual relations between a husband and wife

_____ 41. Reformation leader who said that priests and nuns should be allowed to marry

_____ 42. They proved their worthiness, not by denying the pleasurable aspects of marital sex, but by living frugal and industrious lives

_____ 43. They believed in an ascetic philosophy: wisdom and virtue come from denying physical pleasures

_____ 44. He conducted the first large-scale survey of American sexual attitudes and behaviors

_____ 45. The first major influence on Christian sexual values, he regarded bodily pleasures as evil and thought it "well for a man not to touch a woman"

_____ 46. A Victorian-era sex researcher with a tolerant attitude about sexuality

_____ 47. They held antisexual attitudes that were reinforced by the mistaken medical beliefs of that time

_____ 48. He headed a recent survey of a nationally representative sample of adults living in households.

a. Henry Havelock Ellis
b. Edwardians
c. St. Paul
d. ancient Greeks
e. Jesus
f. Sigmund Freud
g. Masters and Johnson
h. Biblical Hebrews
i. Puritans
j. Victorians
k. Morton Hunt
l. St. John
m. Martin Luther
n. St. Thomas Aquinas
o. Alfred Kinsey
p. St. Augustine
q. Edward Laumann

MATCHING, PART 2

49. The major method of investigation used by Masters and Johnson: _____.

50. Professor Jones studies in the lab the effects of different amounts of alcohol consumption on men's arousal from viewing sexually explicit films showing violence toward women. Her method of investigation is called _____.

51. In question 50, the different amounts of alcohol consumption are the _____.

52. In question 50, the level of arousal is the _____.

53. Professor Jones's subjects are a _____.

54. If Professor Jones's subjects were volunteers from her class, they would be a _____.

55. Professor Jones finds that a group given alcohol shows more arousal than a control group that does not consume alcohol. If she kept all other factors the same, she can draw a conclusion based on _____.

a. population
b. sample
c. convenience sample
d. stratified random sample
e. survey
f. correlation
g. direct observation
h. case study
i. experimental research
j. volunteer bias
k. observer bias
l. independent variable

56. Professor Jones finds a strong linear relationship between the amount of alcohol consumed and the degree of arousal. This is an example of the use of _____.

57. Professor Jones makes a conclusion about the effects of alcohol in all adult males. She has made a _____.

58. In Professor Jones's study, "all adult males" is the _____.

59. If the men who agreed to participate in Professor Jones's study did so because they enjoyed watching sexually violent movies more than other people do, this would be an example of _____.

60. The major method of investigation used by anthropologists: _____.

61. If an anthropologist's description of another culture's behaviors was influenced by his own (Western) values, this would be an example of _____.

62. Dr. Smith does an in-depth investigation and analysis of a serial rapist. His method of investigation is called a _____.

63. Professor Thomas gives a paper-and-pencil questionnaire to his class asking about their sexual attitudes. This is an example of a _____.

64. If a study samples randomly from targeted subgroups in the same proportion as they are found in the general population, the result is a _____.

m. dependent variable
m. generalization
o. cause and effect

C. FILL IN THE BLANKS

65. Anthropologists believe the most sexually repressive society in the world to be the _____.

66. Four important influences that led to the "sexual revolution" were (a) _____, (b) _____, (c) _____, and (d) _____.

67. A random sample is properly defined as a sample drawn from a population in a manner so that _____ has an equal chance of being selected.

68. _____ believed that intellectual love could lead to immortality.

69. The _____ had a positive view of sex within marriage and brought this view over to the colonies in the New World.

70. A democratic philosophy of sexuality education is committed to _____.

71. A major influence on early Christian thought was the Greek philosophy of dualism, which separated _____ and _____.

72. Freud referred to sexual energy as _____.

73. Victorian physicians called nocturnal emissions _____ because they believed they were caused by the same thing that causes gonorrhea.

CHAPTER 2   INTERACTIVE REVIEW

When referring to their genitals or breasts, many people feel more comfortable using slang terms than the correct anatomical names. The use of sexual slang often reflects (1) _____ feelings about sex and may lead to misinformation.

The external female genitalia are collectively known as the (2) _____. This includes the (3) _____, (4) _____, (5) _____, (6) _____, (7) _____, and (8) _____. The (9) _____ and (10) _____ become covered with hair at puberty. The (11) _____ has no known function other than to focus pleasurable sensations. It is most similar in structure to the male's (12) _____. The (13) _____ meet at top to form the clitoral hood. The area between the labia minora is called the (14) _____. Many sexually inexperienced females have a thin membrane, called the (15) _____, that partially covers the vaginal opening. It has no known physiological function, but its presence or absence is used by men in many cultures to tell whether or not their partner is a virgin. In actuality, it is a (16) _____ indicator of prior sexual experience.

A woman's breasts are not part of her reproductive anatomy, but in cultures where men consider them to be erotic, they are part of her sexual anatomy. One in (17) _____ women will get breast cancer sometime in their lifetimes.

A female's internal reproductive system includes the (18) _____, (19) _____, (20) _____, and (21) _____. The (22) _____ is the depository for sperm, the birth canal, and the exit route for menstrual fluids. It is a self-cleansing organ, and its odor is generally not offensive. The walls of the inner two-thirds of the (23) _____ are relatively insensitive to touch, but about 10 percent of all women have an area of heightened sensitivity on the front wall called the (24) _____. When a mature ovum (egg) is released from an (25) _____, it is picked up by a (26) _____, which then transports it to the (27) _____. If fertilized by a sperm, the egg usually implants in the (28) _____ of the uterus. The female's urinary system is not related to her reproductive system except for the fact that the urethral opening is located between the (29) _____ and the (30) _____. Women should have an annual pelvic exam and Pap smear to test for (31) _____.

The male's external anatomy consists of the penis and the (32) _____, which contains the (33) _____. Erection of the penis occurs when the two (34) _____ and the (35) _____ become engorged with blood. The rounded end of the penis, called the (36) _____, is covered by the (37) _____ unless this excess skin has been surgically removed in an operation called (38) _____. There is great debate today about the merits of doing this procedure routinely on males.

The male's internal reproductive system includes the (39) _____, which produce sperm and the male hormone (40) _____, and a four-part duct system (starting from the testicles) consisting of the (41) _____, (42) _____, (43) _____, and (44) _____, which transports sperm out of the body. During an ejaculation, sperm are mixed with fluids from the (45) _____ and (46) _____ to form (47) _____. A small amount of fluid is released by the (48) _____ before a male reaches orgasm. Cancer of the (49) _____ is the most common type of cancer in men aged 20 to 35. The most common cancer (other than skin cancer) among all men in the United States is cancer of the (50) _____.

CHAPTER 2   SELF-TEST

A. TRUE OR FALSE

T ☐ F ☐ 51. The testicles in adult males normally produce millions of new sperm every day.

T ☐ F ☐ 52. It is normal for one testicle to hang lower in the scrotum than the other.

T ☐ F ☐ 53. Most of the fluid in a male's ejaculation comes from the prostate gland.

T ☐ F ☐ 54. Sperm can only be produced in an environment that is several degrees lower than normal body temperature.

T ☐ F ☐ 55. Most doctors routinely check for sexually transmitted diseases when they perform pelvic exams.

T ☐ F ☐ 56. Most women have a G spot.

T ☐ F ☐ 57. The American Cancer Society advises that women should have baseline mammograms, starting with one between the ages of 35 and 40.

T ☐ F ☐ 58. The ovaries produce hundreds of new eggs every month during a woman's reproductive years.

T ☐ F ☐ 59. The American Academy of Pediatrics presently favors the routine circumcision of males for health reasons.

T ☐ F ☐ 60. The use of feminine hygiene sprays and douches is a recommended part of normal feminine hygiene.

T ☐ F ☐ 61. Females can reduce their chances of getting a urinary tract infection (cystitis) by wiping themselves from front to back after a bowel movement.

T ☐ F ☐ 62. There is no direct physical pathway between the ovaries and the Fallopian tubes.

T ☐ F ☐ 63. The majority of older men experience prostate problems as a result of their prostates becoming smaller with age.

T ☐ F ☐ 64. The labia majora are hairless and meet at the top to form the clitoral hood.

T	F	65. The vagina is an open orifice that is always ready to accommodate insertion of a penis.
T	F	66. Deep, pleasurable vaginal sensations during intercourse are due to sensitive vaginal walls.
T	F	67. It is not uncommon for adolescent males to have enlarged breasts due to increased estrogen levels.
T	F	68. Erections in men are due in part to a bone that protrudes into the penis.
T	F	69. The hymen is a good indicator of whether or not a woman has had sexual intercourse.
T	F	70. A woman's sexual responsiveness is related to breast size.

B. MATCHING

_____ 71. hypothalamus

_____ 72. areola

_____ 73. os

_____ 74. mammogram

_____ 75. mons veneris

_____ 76. cystitis

_____ 77. perineum

_____ 78. introitus

_____ 79. corpora cavernosa

_____ 80. bulbocavernosus muscle

_____ 81. corona

_____ 82. fimbria

_____ 83. glans

_____ 84. Bartholin's glands

_____ 85. smegma

_____ 86. pubococcygeus muscle

_____ 87. circumcision

_____ 88. vestibular bulbs

_____ 89. spermatic cord

_____ 90. pituitary

a. inflammation of the bladder

b. term used to refer to the vaginal opening

c. darkened area around the nipple

d. opening to the cervix

e. X-ray procedure used for early detection of breast cancer

f. hairless area of skin between the vaginal opening and the anus

g. cheesy substance secreted by glands in the foreskin and clitoral hood

h. soft layer of fatty tissue overlaying the pubic bone

i. two spongy cylinders in the penis

j. large muscle that surrounds the vagina and the bladder

k. raised ridge where the glans and shaft of the penis meet

l. tiny fingerlike endings of the Fallopian tubes

m. smooth rounded end of the penis

n. two glands located at the base of the labia minora

o. surgical removal of the foreskin

p. ring of sphincter muscles that surround the vaginal opening and root of the penis

q. structures located on both sides of the vaginal opening that become engorged with blood during sexual arousal

r. tubelike structure that suspends the testicles in the scrotum

s. area of the brain important for sexual functioning

t. gland located at the base of the brain

C. FILL IN THE BLANK

91. The American Cancer Society advises all women aged 18 or over to have regular pelvic exams and _____ tests to check for _____ cancer.

92. All men over the age of 40 should have an annual examination to check for cancer of the _____.

93. In females, the two outer elongated folds of skin that extend from the mons to the perineum are called the _____.

94. The innermost layer of the uterus, which is sloughed off and discharged from the female's body during menstruation, is called the _____.

95. In human males, an erection results from the spongy tissues of the penis becoming engorged with _____.

96. When a male becomes sexually aroused, a few drops of a clear fluid produced by the _____ may appear at the tip of the penis.

97. The soft layer of fatty tissue overlaying the pubic bone in females is called the _____.

98. Breast size in women is determined by the _____.

99. The best time for most women to examine their breasts for abnormal lumps is _____.

100. The sac beneath the penis is called the _____.

101. After sperm travel through the vas deferens, they enter the paired _____.

102. Most of the fluid in an ejaculation comes from the _____.

103. After an egg is released from an ovary, it travels through one of the _____ on the way to the uterus.

104. The best time for a man to examine his testicles for abnormal lumps is _____.

105. Two glands that secrete small amounts of alkaline fluid into ducts at the base of the labia minora are called _____.

106. Sperm are produced in the _____ of the testicles.

107. The male hormone testosterone is produced by the _____ in the testicles.

108. Both the penis and the _____ have corpora cavernosa.

109. The muscle surrounding the vagina and bladder is called the _____ muscle.

110. In women who have a G spot, it is located on the _____ wall of the vagina.

111. _____ is a condition in which the spermatic cord becomes twisted and cuts off the blood supply to the testicles.

112. The corpora cavernosa expand and fan out to form _____, which attach to the pubic bone.

113. The pituitary hormone that causes production of milk is _____.

114. Ejection of milk is caused by the homone _____.

CHAPTER 3   INTERACTIVE REVIEW

Hormones are chemical substances that are released into the bloodstream by ductless (1) _____ glands. The ovaries and testicles are part of this system. The testicles produce the "male hormone" (2) _____ and the ovaries produce the "female hormones" (3) _____ and (4) _____. Hormones from the (5) _____ cause the ovaries and testicles to produce their hormones.

In adult females, an egg matures on an average of every (6) _____ days. The pituitary hormone that starts the menstrual cycle is called (7) _____. This hormone stimulates the development of a (8) _____ in the ovary. During the preovulatory phase of the menstrual cycle (also called the (9) _____ phase), estrogen from the follicle promotes growth of the (10) _____, inhibits release of (11) _____, and stimulates release of (12) _____. The (13) _____ surge signals the onset of (14) _____, at which time the ovum is expelled into the (15) _____ and is picked up by a (16) _____. During the postovulatory phase of the cycle (also called the (17) _____ phase), the (18) _____ secretes progesterone in large amounts. If the egg is fertilized by a sperm, it normally implants in the (19) _____. If the egg is not fertilized, the corpus luteum degenerates and the (20) _____ is shed and discharged in a normal physiological process called (21) _____. Although the average length of the menstrual cycle is 28 days, the large majority of women have cycles that vary in length by (22) _____. In most nonhuman mammalian species, this cycle of hormonal events is called the (23) _____ cycle. Unlike human women, females of species with this type of cycle are sexually receptive to males only during (24) _____. Some cultures have menstrual (25) _____ that prohibit contact with a menstruating woman, and even in our own culture many women and their partners avoid sexual intercourse during menstruation, but this generally reflects inaccurate information and/or negative culturally learned responses. Menstrual discharge consists simply of (26) _____, (27) _____, and (28) _____.

In males, FSH stimulates (29) _____, while (30) _____ stimulates the production of (31) _____ in the Leydig's cells of the testicles. Misuse of anabolic steroids, derivatives of testosterone, can cause serious harmful effects such as (32) _____ (name three effects).

Some women experience menstrual-related problems. The absence of menstruation is called (33) _____.

Emotional and/or physical changes taking place (34) _____ days before the start of menstruation are referred to as (35) _____. The thing that distinguishes this condition from other emotional states is that it ends (36) _____. The major cause of (37) _____, painful menstruation, is an overproduction of (38) _____. Endometriosis refers to a condition in which (39) _____. Women who use tampons but do not change them frequently risk getting a serious bacterial infection called (40) _____.

In men, sexual desire appears to be related to circulating levels of (41) _____. Studies of women after menopause, or after surgical removal of the ovaries, suggest that women's sexual desire is not strongly affected by the hormones (42) _____ or (43) _____. Other studies indicate that the hormone (44) _____ does influence sexual desire in women.

CHAPTER 3   SELF-TEST

A. TRUE OR FALSE

T   F   45. Hormones are important for women's sexual desire, but not for men's.
T   F   46. Testosterone is found only in males, and estrogen is found only in females.
T   F   47. All female mammals have menstrual cycles.
T   F   48. Most women's menstrual cycles are 28 days in length.
T   F   49. It is medically safe for a man to have sexual intercourse with a menstruating woman.
T   F   50. Premenstrual syndrome involves emotional, but not physical, changes before the start of menstruation.
T   F   51. Women who use tampons should change them three or four times a day, even if the tampons are advertised as long-lasting.
T   F   52. Large doses of anabolic steroids can cause mental disorders.
T   F   53. Men can no longer have an interest in sex after castration.
T   F   54. Women show a large increase in sexual desire around the time of ovulation.
T   F   55. Females produce new eggs throughout their lifetimes.
T   F   56. Cervical mucus changes during ovulation.
T   F   57. Near the end of the postovulatory phase of the menstrual cycle, there is an LH surge.
T   F   58. Ovulation almost always occurs 14 days before the start of menstruation.
T   F   59. There is some evidence suggesting that the female menstrual cycle can be altered by certain smells.
T   F   60. Women with PMS generally have normal hormone levels.
T   F   61. Only women can get toxic shock syndrome.
T   F   62. Estrogen, according to most research, is the primary hormone responsible for sexual desire in women.

B. MATCHING

_____ 63. endocrine system
_____ 64. testosterone
_____ 65. luteinizing hormone
_____ 66. estrogen
_____ 67. FSH
_____ 68. follicular phase
_____ 69. corpus luteum
_____ 70. amenorrhea
_____ 71. luteal phase
_____ 72. dysmenorrhea
_____ 73. *Mittelschmerz*

a. substances that cause muscle contractions of the uterus
b. cells surrounding the ovum that are left behind at ovulation
c. emotional and/or physical changes that precede menstruation
d. male hormone produced in testicles, ovaries, and adrenal glands
e. cramps experienced for about a day during ovulation
f. postovulatory phase when the corpus luteum begins to secrete large amounts of progesterone
g. endometrial tissue growing outside the uterus
h. network of ductless glands
i. rupture of the mature ovum into the abdominal cavity
j. female hormone produced in ovaries, testicles, and adrenal glands
k. preovulatory phase when the pituitary secretes FSH and the follicle secretes estrogen
l. pituitary hormone that stimulates production of sperm and maturation of ova

_____ 74. inhibin

_____ 75. ovulation

_____ 76. prostaglandins

_____ 77. premenstrual syndrome

_____ 78. endometriosis

m. absence of menstruation

n. pituitary hormone that triggers ovulation in women and the production of testosterone in men

o. painful cramps during menstruation

p. substance produced in the ovaries and testicles that inhibits production of FSH

C. FILL IN THE BLANK

79. The gland located at the base of the brain that secretes follicle-stimulating hormone and luteinizing hormone is the _____ gland.

80. Natural body scents that can affect the behavior of other members of the same species are called _____.

81. The release of FSH from the pituitary stimulates _____ in the ovary.

82. Female gonadotropin and gonadal hormone levels show cyclic fluctuations referred to as _____.

83. Each immature ovum is surrounded by other cells within a thin capsule of tissue to form what is called a _____.

84. During ovulation, the cells that surround the ovum in the follicle remain in the ovary, and are then referred to as the _____.

85. After ovulation, there is a large increase in circulating levels of the hormone _____.

86. A woman's first menstrual cycle is called _____.

87. The removal of the testicles is known as castration or _____.

88. Once people have become sexually experienced, the _____ plays a role equal to, if not greater than, hormones in their level of sexual desire.

89. _____ is the hormone most influential in influencing sexual desire in women.

90. Toxic shock syndrome is caused by _____.

91. If an egg is fertilized by a sperm and implantation occurs, the corpus luteum is maintained by a hormone from the developing placenta called _____.

92. The hypothalamic hormone that causes release of FSH and LH from the pituitary is called _____.

93. In the female hormone feedback loop, LH production is suppressed primarily by _____.

94. In the male hormone feedback loop, LH production is suppressed primarily by _____.

95. In the male hormone feedback loop, FSH production is suppressed by _____.

CHAPTER 4   INTERACTIVE REVIEW

(1) _____ observed and recorded physiological responses from hundreds of people engaged in sexual activity. This work led them to conclude that males and females are (2) _____ in their responses than previously believed. They divided the physiological responses during sex into four phases: (3) _____, (4) _____, (5) _____, and (6) _____. This pattern of responses is often referred to as the (7) _____. Other researchers have organized the responses into fewer or more phases, and this chapter follows the model of sex therapist (8) _____ and includes (9) _____ in the first phase.

The first response in the excitement phase for both males and females is a (10) _____ response, which results in (11) _____ in men and (12) _____ in women. During the (13) _____ phase in women, the tissues of the outer third of the vagina become swollen with blood, causing the vaginal opening to narrow. This response is referred to as the (14) _____. Also during this phase, the clitoris (15) _____. Early in the plateau phase, (16) _____ percent of all women and about 25 percent of all men experience a rash on the skin called the (17) _____.

At the time of orgasm, both men and women have (18) _____ in specific tissues that initially occur every 0.8 seconds. However, recent studies indicate that the real essence of orgasm is not in the genitals, but in (19) _____. Men's and women's descriptions of orgasm are (20) _____. Male orgasms occur in two stages, (21) _____ and (22) _____. The expulsion of semen from the penis is called (23) _____. After males have an orgasm, their physiological responses generally dip below plateau, during which time they cannot have another orgasm. This period of time is called the (24) _____. Unlike most men, some women can have (25) _____—two or more orgasms in quick succession—without dropping below the plateau level. The return to the unaroused state is called (26) _____.

By most estimates, only (27) _____ percent of adult women experience orgasm regularly during intercourse. Masters and Johnson claim that all women in good health are capable of reaching orgasm during intercourse, but Helen Kaplan and others believe that many women are incapable of reaching orgasm during intercourse without simultaneous (28) _____ stimulation.

Freud believed that there were two types of female orgasms, one caused by (29) _____ stimulation and another by (30) _____ sensations. Masters and Johnson originally claimed that all female orgasms were identical and were focused in the (31) _____. However, some women report having experienced different types of orgasms, and work conducted in the 1980s revealed that many of these women had a sensitive area on the (32) _____ wall of the vagina called the (33) _____. Stimulation of this area sometimes resulted in (34) _____ during orgasm. In some women the fluid was identified as urine, but in others it contained an enzyme found in secretions from the male's (35) _____.

Most women say that the size of a partner's penis is (36) _____ for their pleasure during intercourse. Many people have tried to enhance their sexual desire or performance by taking substances called (37) _____, but there is little evidence that they have any real effect. In many places in the world, sexual pleasure is denied to females by mutilating their genitals in procedures called (38) _____ and (39) _____. The World Health Organization has had little luck in stopping these ancient cultural practices.

CHAPTER 4   SELF-TEST

A. TRUE OR FALSE

T F 40. Once a person has sexual desire, stimulation of any of the five senses can lead to excitement.

T F 41. The sexual responses caused by touching are the same as those caused by fantasy.

T F 42. The first physiological signs of arousal in men occur within seconds, but take several minutes to begin in women.

T F 43. Vaginal lubrication is actually superfiltered blood plasma.

T F 44. The presence of vaginal lubrication means that a woman is ready to begin sexual intercourse.

T F 45. Both men and women can experience nipple erection during the excitement phase.

T F 46. When the clitoris pulls back and disappears beneath the clitoral hood, it means that a woman is less sexually aroused than before.

T F 47. The secretion of fluids from the vaginal walls may slow down if the plateau phase is prolonged.

T F 48. The Bartholin's glands secrete a small amount of fluid during the initial excitement phase.

T F 49. Changes that occur during the plateau phase in males are not as distinct as the changes that occur in females.

T F 50. Generally speaking, descriptions of orgasm written by women can be easily distinguished from those written by men.

T F 51. Orgasm and ejaculation occur at the same time in men and are actually the same event.

T F 52. The rhythmic muscular contractions during orgasm recorded by Masters and Johnson are not strongly related to a person's subjective sensations of pleasure.

T F 53. A full orgasm for the male, with ejaculation, is almost always followed by a refractory period.

T F 54. Some people who are paralyzed, with no sensations from the genitals, can experience orgasm.

T	F	55. The walls of the inner two-thirds of the vagina are very sensitive to touch, thus making penis length an important factor during intercourse.
T	F	56. There is more variation in penis length in the unaroused condition than in the erect state.
T	F	57. Alcohol excites the central nervous system and enhances sexual performance.
T	F	58. Regular long-term use of marijuana can lower testosterone levels and decrease sperm production.
T	F	59. A man's penis size is related to his height, weight, and race.
T	F	60. Regular use of cocaine often leads to erectile problems and difficulties reaching orgasm.
T	F	61. Sexual arousal is more strongly associated with specific sexual motivations than it is to global sexual desire.
T	F	62. There is a strong relationship between vaginal vasocongestion and a woman's subjective sense of sexual arousal.

B. MATCHING (SOME QUESTIONS HAVE MULTIPLE ANSWERS)

_____ 63. desire

_____ 64. excitement phase

_____ 65. plateau phase

_____ 66. orgasm phase

_____ 67. resolution phase

_____ 68. vasocongestive response

a. Cowper's glands secrete a few drops of clear fluid
b. penis starts to become erect
c. labia minora become engorged with blood
d. specific sensations cause the individual to seek out, or become receptive to, sexual experiences
e. rhythmic muscular contractions occur in outer vagina, uterus, and anal sphincter muscles
f. clitoris pulls back against the pubic bone
g. vaginal lubrication begins
h. blood drains from breasts, outer third of the vagina, labia minora, and clitoris
i. sex-tension flush appears
j. nipples become erect
k. clitoris is most prominent at this time
l. scrotum thickens and helps pull the testicles toward the body
m. blood drains from penis and testicles
n. walls of the inner two-thirds of the vagina begin to balloon out
o. return to the unaroused state
p. rhythmic muscular contractions in vas deferens, prostate gland, urethra, and anal sphincter muscles
q. outer third of the vagina becomes engorged with blood
r. nipples appear to be less erect
s. labia majora flatten and spread apart
t. testicles increase in size by 50–100 percent
u. in males, a fluid is emitted from the urethra

C. FILL IN THE BLANKS

69. The two basic physiological responses that occur during sexual arousal are called _____ and _____.

70. Sperm, prostate fluids, and seminal fluids combine to form _____.

71. _____ started the controversy about clitoral versus vaginal orgasms.

72. Some women emit a fluid during orgasm that comes from the _____.

73. Most women and many men get a skin rash called the _____ when they become highly sexually aroused.

74. Substances that suppress sexual functioning are called _____.

75. _____ refers to the buildup of energy in nerves and muscles resulting in involuntary contractions.

76. The first stage of orgasm in males is called _____.

77. A medical problem in which the sphincter muscles dysfunctionally force semen into the bladder is called _____.

78. The engorgement and swelling of the outer third of the vagina was named the _____ by Masters and Johnson.

79. A sensitive area that is found on the front wall of the vagina in about 10 percent of all women is called the _____.

80. After ejaculating, most men have a period of time called the _____ during which it is impossible for them to have another orgasm.

81. In women, the _____ become greatly engorged with blood during the plateau phase, resulting in a doubling or tripling in their thickness and a vivid color change.

82. The type of female orgasm resulting from clitoral stimulation and characterized by the buildup and discharge of myotonia in the orgasmic platform and the pubococcygeus muscles is called _____.

CHAPTER 5   *INTERACTIVE REVIEW*

There are about 12 million new cases of sexually transmitted diseases in the United States every year. For Americans under the age of 55, about one in (1) _____ will have at least one STD sometime in their life.

Sexual behavior does not cause STDs. The behavior is merely the mode of transmission for (2) _____, (3) _____, and/or (4) _____ that must be present for the diseases to be transmitted. These causative agents are nondiscriminating—anyone having sex with an infected partner can get a sexually transmitted disease.

The three most common STDs (in the U.S.) caused by bacteria are (5) _____, (6) _____, and (7) _____. In the early stages of (8) _____ and (9) _____, most women and many men show no symptoms. If they are not treated, the bacteria that cause these two STDs can invade the upper reproductive tract and cause prostatitis and epididymitis in men and (10) _____ in women. (11) _____ passes through several stages, and if left untreated can cause organ damage and death. These types of STDs can be cured with (12) _____.

Sexually transmitted diseases caused by viruses are generally incurable because the viruses live inside normal body cells. Herpes has been around since biblical times and is spread by direct (13) _____-to-_____ contact. Genital herpes is almost always transmitted by genital-to-genital or mouth-to-genital contact; oral herpes is often spread by more casual contact. After the primary attack, many herpes sufferers experience (14) _____ attacks when the virus replicates and sheds. The virus is most easily transmitted during active attacks, but can be transmitted at other times.

Hepatitis is an inflammation of the (15) _____ that can be caused by contact with contaminated (16) _____ (hepatitis A) or infected (17) _____ (hepatitis B). One-fourth to one-half of all cases of hepatitis B are sexually transmitted.

Warts in the genital and anal areas are caused by a few types of the (18) _____ virus. Other types of the virus are associated with cancer of the cervix and penis. Physicians can attempt to remove the warts or abnormal cervical cells, but they cannot attack the viruses directly; thus, in many cases, the condition reappears.

AIDS is caused by the (19) _____. HIV invades and destroys (20) _____ (also called helper T) cells, white blood cells that are a critical part of the body's (21) _____ system. Most people infected with HIV remain asymptomatic for several years. HIV infection is not called AIDS until the immune system is so weakened that the individual's life is threatened. People with AIDS eventually die from a variety of (22) _____ infections. About half of all individuals who become infected with HIV will develop AIDS within (23) _____ years. HIV probably originated in Africa from simian (nonhuman primate) immunodeficiency viruses. It is transmitted by (24) _____ and exposure to infected (25) _____. It is not spread by (26) _____. HIV infection is a true pandemic, with at least 120 million cases expected worldwide by the year 2020. Worldwide, most cases are spread by sexual activity between (27) _____. Presently, there is no cure, but there are several antiviral drugs that slow the progression of the infection.

Pubic lice and scabies are not really diseases, but (28) _____. Pubic lice attach themselves to pubic hair and feed on blood. Scabies is caused by a mite that burrows under the skin to lay its eggs.

Vaginitis is a general term that refers to any inflammation of the vagina. There are three types. (29) _____ is caused by a one-celled protozoan that is transmitted during sexual intercourse. The other two types of vaginitis are not considered STDs because most cases are not acquired during sexual relations. (30) _____ vaginitis is a fungus or yeast infection that is very common in females. Anything that changes the chemical environment of the vagina can cause a yeast infection. Probably the most common type of vaginitis is (31) _____, caused by the interaction of several vaginal bacteria. Vaginal infections are very common and nothing to be ashamed of.

STDs do not strike randomly—they only affect those who are exposed to the bacteria or viruses. You can greatly reduce your chances of ever contracting a sexually transmitted disease by practicing safer sex. This means making one of two choices: (1) using condoms (preferably with spermicide) properly and consistently, or (2) abstaining from sexual relations until you are reasonably confident that you are in a long-term monogamous relationship. See the self-assessment questionnaire at the beginning of this study guide.

CHAPTER 5   SELF-TEST

A. TRUE OR FALSE

T F 32. People who have had a sexually transmitted disease in the past are immune to getting that type of STD again.

T F 33. Untreated chlamydia results in serious complications more often than untreated gonorrhea.

T F 34. Cold sores and fever blisters are symptoms of herpes.

T F 35. Homosexuality is one of the causes of AIDS.

T F 36. In the past century, more people have died from syphilis than from AIDS.

T F 37. If left untreated, gonorrhea can turn into syphilis.

T F 38. Yeast infections (monilial vaginitis) can usually be prevented by using feminine hygiene products.

T F 39. A person with a cold sore on the mouth can give his or her partner genital herpes during oral-genital sex.

T F 40. If neither person has an STD to begin with, a monogamous heterosexual or homosexual couple can engage in oral-genital or anal sex without fear of getting an STD, including AIDS.

T F 41. Gonorrhea, chlamydia, and syphilis are often contracted by people who use toilet seats previously used by infected individuals.

T F 42. Having a sexually transmitted disease is a sign of promiscuity.

T F 43. Chlamydia is more common than gonorrhea.

T F 44. Syphilis is caused by a bacterium.

T F 45. It is possible to contract HIV by hugging, touching, or being close to an infected person.

T F 46. The large majority of women do not show any symptoms in the early stages of gonorrhea or chlamydia.

T F 47. Herpes can sometimes be cured with antibiotics.

T F 48. A person can have only one type of sexually transmitted disease at a time.

T F 49. HIV, the virus that causes AIDS, is sometimes spread by mosquitoes.

T F 50. Viruses lack the genetic material necessary to reproduce themselves.

T F 51. Pelvic inflammatory disease can be caused by infections that were not contracted during sex.

T F 52. Condoms are generally ineffective against the human immunodeficiency virus.

T F 53. The eggs of pubic lice can survive for days on towels or sheets.

T F 54. A person who has no symptoms does not have to worry about having gonorrhea, chlamydia, or syphilis.

T F 55. Only females can contract trichomoniasis and monilial infections.

T F 56. Herpes can be cured with a drug called acyclovir.

T ☐ F ☐ 57. The human papillomaviruses that cause cervical cancer are the same as those that cause genital warts.

T ☐ F ☐ 58. Regular douching decreases the risk of pelvic inflammatory disease.

T ☐ F ☐ 59. The major symptom of the primary stage of syphilis is a rash.

T ☐ F ☐ 60. Recurrent herpes attacks are generally more painful, and last longer, than the first (primary) attack.

T ☐ F ☐ 61. Herpes is the leading infectious cause of blindness in the United States.

T ☐ F ☐ 62. Hepatitis A is most often spread by nonsexual means.

T ☐ F ☐ 63. Worldwide, most cases of HIV infection (infection with the virus that causes AIDS) are contracted by sex between heterosexuals.

T ☐ F ☐ 64. Generally speaking, people who get AIDS this year are those who became infected with HIV 5 to 15 years ago.

T ☐ F ☐ 65. Death by AIDS can generally be prevented by early treatment with the drug AZT.

T ☐ F ☐ 66. To minimize their chance of a vaginal infection, women should wear panties made of synthetic fabrics.

T ☐ F ☐ 67. Cystitis can sometimes be caused by vigorous intercourse.

T ☐ F ☐ 68. People who get HIV are extremely contagious in the first 60 days after exposure.

T ☐ F ☐ 69. Among young Americans aged 25 to 44, AIDS causes more deaths than accidents, cancer, or heart disease.

B. MATCHING

SYMPTOMS

_____ 70. painful, craterlike sores

_____ 71. itchless, painless rash all over the body

_____ 72. pus-like discharge and/or burning during urination

_____ 73. intense itching caused by grayish, six-legged parasites

_____ 74. thick, white, cheesy vaginal discharge and intense itching

_____ 75. loss of appetite, fatigue, slow recovery from colds and flus, continual yeast infections, purple blotches on skin, pneumonia

_____ 76. cauliflower-like growths

_____ 77. severe abdominal pain and fever

_____ 78. scaling skin caused by pearly mites

_____ 79. large, ulcerlike, painless sore

_____ 80. fluid-filled blisters

_____ 81. copious, foamy, yellowish-green vaginal discharge and odor

_____ 82. wart-like growths that look like small pimples filled with a kernel of corn

_____ 83. thin, clear discharge and irritation of the urethra

_____ 84. jaundiced or yellow tinge of skin and eyes

_____ 85. large ulcers (gummas) on the skin and bones; damage to the heart and nervous system

_____ 86. inflammation of the bladder

INFECTION (OR INFESTATION)

a. chancroid

b. chlamydia

c. gonorrhea

d. hepatitis

e. herpes

f. HIV infection (AIDS)

g. molluscum contagiosum

h. moniliasis (yeast infection)

i. PID

j. primary stage syphilis

k. cystitis

l. secondary stage syphilis

m. scabies

n. trichomoniasis

o. venereal warts (human papillomavirus infection)

p. late-stage syphilis

q. pubic lice

C. FILL IN THE BLANKS

87. Trichomoniasis, pubic lice, and scabies are all caused by_____.

88. Inflammation of the vagina is referred to as _____.

89. In women, untreated gonorrhea or chlamydia can lead to _____.

90. A woman who has had _____ is at high risk of getting cancer of the cervix.

91. The sexually transmitted disease that impairs the immune system is called _____.

92. Trichomoniasis, moniliasis, and bacterial vaginosis are all _____.

93. A viral liver infection that can be sexually transmitted is _____.

94. In terms of number of new infections per year, the most common viral sexually transmitted disease in the United States is _____.

95. The drug that is prescribed to relieve the symptoms of herpes is _____.

96. Two serious possible consequences of PID are _____ and _____.

97. Three sexually transmitted diseases that can be passed from an infected pregnant female to the fetus are _____, _____, and _____.

98. _____ is the most common thing that brings on recurrent herpes attacks.

99. The bacteria that cause gonorrhea and chlamydia live on _____.

100. Any inflammation of the urethra not caused by the gonococcus is called _____.

101. In tropical countries, chlamydia is responsible for an STD called _____.

102. _____ is a sexually transmitted disease for which many more women show symptoms than men.

103. Antibiotics commonly lead to _____ infections.

104. _____ live in the large intestine and can be transmitted during sex, but are generally acquired by children through nonsexual means.

105. Many people diagnosed with gonorrhea are also found to have _____.

106. HIV attacks immune system cells called _____.

107. People with HIV are most contagious (when) _____.

108. The most common AIDS test tests for _____.

109. Gonorrhea, chlamydia, syphilis, herpes, HPV infection, HIV infection, trichomoniasis: for which of these diseases do women usually not have visible symptoms in the initial stage? _____

110. For which of the diseases listed in the previous question do men usually not show symptoms in the initial stage? _____

111. The best ways to avoid sexually transmitted diseases are _____ and _____.

112. The newest group of drugs used to prevent replication of HIV are called _____.

CHAPTER 6   INTERACTIVE REVIEW

There are well over one million births to unwed females in the United States every year. Nearly half of these are to teenagers. The teen pregnancy rate in the United States is (1) _____ as high as in most other developed countries. Unfortunately, the teens most likely to have babies are those least prepared to take care of them.

Worldwide, there is need for effective contraceptive technology. World population is expected to double by the middle of the next century, yet already there are many parts of the world experiencing food shortages and mass starvation.

While all the methods of birth control discussed in this chapter are better than using nothing, some are more effective than others. Relatively ineffective methods include withdrawal, douching, and breast-feeding. Fertility awareness methods (the (2) _____, (3) _____, and (4) _____ methods) rely on calculating when (5) _____ will occur and abstaining from intercourse during that period. The pregnancy rates for these methods are higher than for most others. (6) _____ work by killing sperm and are generally used in combination with a barrier method of contraception ((7) _____, (8) _____, or (9) _____). Barrier methods block sperm from getting into the (10) _____. Spermicides and barrier methods often interfere with spontaneity, but for people who are not in a monogamous relationship, a major advantage of these methods (not offered by other methods) is that they (11) _____.

Intrauterine devices (IUDs) have had a reputation for causing serious problems, but today's IUDs are safe as well as highly effective. However, they are recommended only for women in long-term monogamous relationships who have completed their families. The combination birth control pill contains synthetic (12) _____ and (13) _____ and works by preventing ovulation. The pills used today contain low dosages of estrogen and are safer than the high-dosage pills used in the past. Nevertheless, there are certain groups of women who, for medical reasons, should not take the pill. (14) _____ (the shot good for 3 months) and (15) _____ (the implant good for 5 years) contain only progestin and are believed to be safer than the pill. All three hormonal methods of contraception have theoretical failure rates of less than 1 percent.

If you consider male and female sterilization techniques together, more Americans rely on sterilization for birth control than any other method. In a vasectomy, the (16) _____ is tied off and cut. Afterwards, the male still ejaculates during orgasm, but his ejaculation contains no sperm. For a (17) _____ or (18) _____, a woman's Fallopian tubes are cut, cauterized, or blocked by clips. The failure rate is close to zero.

For women who have an unplanned or unwanted pregnancy, one option is abortion. Various techniques can be used to terminate pregnancy, including hormones or chemicals, scraping or suctioning away of the endometrium, and induced labor. The Supreme Court has recently made several decisions affecting the availability of abortion in the United States.

It is important for you to choose an effective contraceptive technique *before* you have sexual intercourse. A questionnaire is provided at the end of the chapter that is designed to help you decide which method is best for you.

CHAPTER 6   SELF-TEST

A. TRUE OR FALSE

T F	19.	The rates for births to unmarried teenage girls have decreased since 1970 because of increased use of contraceptives.
T F	20.	The world population is expected to stabilize by the year 2000 because of increased use of birth control.
T F	21.	The U.S. fertility rate is presently about 2.0 births per woman.
T F	22.	If the perfect-use pregnancy rate for a contraceptive method is 20 percent, then for 100 couples who start using the method, about 20 of them will have conceived after 3 years.
T F	23.	In order for a woman to get pregnant, she must have an orgasm.
T F	24.	More couples rely on sterilization than on the birth control pill.
T F	25.	It is possible for a girl to get pregnant as soon as she starts having menstrual periods.
T F	26.	The combination birth control pill reduces the risk of cancer of the endometrium and ovaries.
T F	27.	A relatively effective method of birth control for women is to wash out the contents of the vagina immediately after sexual intercourse.
T F	28.	The birth control pill reduces the risk of getting a sexually transmitted disease.
T F	29.	Abortion was allowed by the Catholic Church until the late 1800s.
T F	30.	A vasectomy works as birth control by preventing ejaculation.
T F	31.	If carefully washed and dried, condoms can be used safely on more than one occasion.

T	F	32. Taking antibiotics makes the birth control pill less effective.
T	F	33. Legal abortions do not increase a woman's later risk of infertility, miscarriage, or a low-birth-weight baby.
T	F	34. A diaphragm works by being inserted into the uterus, thus blocking sperm.
T	F	35. Because IUDs are so effective, they are highly recommended for young, sexually active women.
T	F	36. The major difference between the cervical cap and the diaphragm is that the cervical cap does not need to be used with spermicide.
T	F	37. Most daughters of teenage mothers become teenage mothers themselves.
T	F	38. Breast-feeding a baby inhibits the release of follicle-stimulating hormone from a woman's pituitary gland.
T	F	39. With perfect use, fertility awareness methods are as effective as the birth control pill.
T	F	40. Spermicides help to kill HIV, the virus that causes AIDS.
T	F	41. Pregnancy and childbirth result in more maternal deaths than the birth control pill.
T	F	42. In the future, it will probably be possible to be vaccinated to prevent pregnancy (as is done for the flu).
T	F	43. The French abortion pill, RU 486, allows women to have abortions in the privacy of their own homes.
T	F	44. Rubber condoms are more effective than "skin" condoms for prevention of STDs.
T	F	45. Nearly a third of all pregnancies to unwed teens in this country end in abortion.
T	F	46. The female condom is actually an intravaginal pouch.
T	F	47. In high dosages, combination birth control pills can be used as emergency birth control.
T	F	48. Most women over the age of 40 can continue to take the combination birth control pill.
T	F	49. Once inserted under the skin, Norplant must remain in place for 5 years.
T	F	50. The diaphragm offers women some protection against gonorrhea and chlamydia.

B. MATCHING (SOME ANSWERS CAN BE USED MORE THAN ONCE)

CONTRACEPTIVE METHOD

_____ 51. breast-feeding
_____ 52. calendar method
_____ 53. basal body temperature method
_____ 54. Billings method
_____ 55. spermicides
_____ 56. male condom
_____ 57. female condom
_____ 58. diaphragm
_____ 59. cervical cap
_____ 60. emergency birth control
_____ 61. IUD
_____ 62. combination pill
_____ 63. Depo-Provera
_____ 64. Norplant
_____ 65. vasectomy
_____ 66. tubal ligation
_____ 67. laparoscopy

a. once believed to work by preventing implantation, but now known also to work by preventing fertilization
b. active ingredient kills sperm
c. prevents release of FSH
d. involves noting changes in cervical mucus in order to determine when to abstain from intercourse
e. prevents sperm deposited in the vagina from getting into the uterus
f. involves tying, cutting, or cauterizing Fallopian tubes, or blocking them by clips
g. contains enough spermicide to be effective for 24 hours
h. blocks sperm from getting into the vagina
i. uses simple mathematical formula to determine when to abstain from intercourse
j. progestins prevent ovulation
k. uses changes in resting temperature to determine when to abstain from intercourse
l. involves tying or cutting off vas deferens
m. synthetic estrogen helps to prevent ovulation
n. prevents implantation

C. FILL IN THE BLANKS

68. The perfect-use pregnancy rates of contraception are higher than the typical-use pregnancy rates mainly because some people do not use the methods _____ or _____.

69. Withdrawal is not a highly effective method of birth control because of secretions from _____.

70. After a vasectomy, a male is not sterile until _____.

71. Fertility awareness methods require that a couple abstain from sexual intercourse from near the end of _____ until several days after _____.

72. For breast-feeding to be effective as a birth control method, a mother must allow her baby to nurse _____.

73. A _____ resembles a large thimble and fits over the cervix by suction.

74. If a woman's shortest recorded menstrual cycle is 25 days and her longest is 31 days, by the calendar method she should abstain from intercourse from day _____ to day _____.

75. Women's basal body temperature _____ 24 to 72 hours after ovulation.

76. Women's cervical mucus becomes _____ at the time of ovulation.

77. The active ingredient in most spermicides is _____.

78. Rubber condoms should not be used in combination with _____ lubricants.

79. The IUD is regarded as a safe contraceptive method for women _____.

80. Two nonprescription barrier types of contraception are the _____ and the _____.

81. The combination birth control pill is made less effective if a woman takes _____.

82. The most common side effect of Depo-Provera and Norplant is _____.

83. Four birth control techniques that offer women some protection against STDs are _____, _____, _____ and _____.

84. One study that compared male and female sterilization techniques concluded that the _____ technique was preferable in every way.

85. In cases of rape, emergency birth control can be provided within the first few days by _____.

86. If used properly and consistently, the most effective nonprescription nonsurgical method of birth control is _____.

87. The most effective reversible method(s) of birth control is(are) _____.

88. Sperm can live in a Fallopian tube for as long as _____.

89. The best way to avoid an unwanted pregnancy is _____.

CHAPTER 7   INTERACTIVE REVIEW

Sexual intercourse can result in conception on only about 6 days of the menstrual cycle (from about (1) _____ days before (2) _____ to about (3) _____ day afterwards). Fertilization usually occurs within a (4) _____, and implantation normally occurs in the (5) _____ of the uterus. Implantation of a fertilized egg at any other location is called an (6) _____ pregnancy, and the most common cause is scarring of the Fallopian tubes due to (7) _____. Pregnancy tests are positive if a woman's urine or blood contains a hormone secreted by the developing placenta called (8) _____.

Pregnancy is divided into three 3-month intervals called (9) _____. After implantation, the conceptus is called an (10) _____ for the first (11) _____ months and then a (12) _____ for the last

(13) _____ months of pregnancy. An expectant mother can first feel her unborn baby moving (an experience called (14) _____) during the (15) _____ month of pregnancy. Sexual intercourse during pregnancy is generally thought not to be harmful to the fetus, although some health professionals recommend that a pregnant woman not engage in intercourse during the ninth month of pregnancy.

There are many possible causes of complications of pregnancy. Substances that can cross the "placental barrier" and harm the unborn baby are called (16) _____. Examples include viruses, smoking, alcohol, cocaine and other drugs, and environmental hazards. Fetal abnormalities can often be detected by ultrasound or by a variety of invasive techniques such as chorionic sampling, (17) _____, celocentesis, and fetoscopy. Today, it is sometimes possible to perform surgery on a fetus while it is still in the uterus.

Prepared childbirth courses attempt to break the (18) _____ cycle through education and preparation. Lamaze and other techniques generally include a "coach" (usually the woman's partner) to help the expectant mother through labor and delivery. These methods hold to a philosophy that the mother-to-be and the coach should be in control of the birthing process and that anesthetics should not be used unless absolutely necessary. Today, there is a growing movement advocating going back to giving birth in the home (or in a homelike environment), usually with the aid of a nurse- (19) _____ rather than a physician.

The actual birthing experience is known as labor. Before giving birth, many women experience "false labor." True labor is distinguished from false labor by the timing and length of (20) _____ and by the (21) _____ of the cervix. Labor proceeds in three stages. The initial or (22) _____ stage usually lasts from 6 to 13 hours, although each stage of labor varies a great deal from woman to woman and even from birth to birth in the same woman. The (23) _____ breaks (or is broken by the physician) just prior to or during this stage. The last part of the first stage of labor is called the (24) _____. It marks the beginning of the birth process. In the second stage of labor, the cervix is (25) _____ and the fetus begins moving through the birth canal. The baby usually enters the world (26) _____ first, and after delivery the (27) _____ is cut. In the third stage of labor the (28) _____ detaches from the uterus and leaves the mother's body. If complications develop during childbirth, the baby can be delivered surgically through the mother's abdomen by a (29) _____. Today, in most cases, it is considered safe for women who have previously had a C-section to deliver vaginally during subsequent deliveries.

An increasing number of U.S. women are breast-feeding their babies. Mother's milk contains many infection-fighting (30) _____ and is easier to digest than cow's milk. After giving birth, many women experience negative emotions, called (31) _____. Hormone changes, lack of sleep, and new demands can all contribute to this. Most physicians advise that women not resume sexual intercourse for (32) _____ weeks after delivery. Studies find that after the transition to parenthood, happiness within a relationship is related to (33) _____.

Infertility affects about one in (34) _____ couples in the United States. About (35) _____ percent of these cases are due to problems with the male, (36) _____ percent to problems with the female, and about (37) _____ percent to problems wirh both partners. In men, infertility is usually caused either by blockage of the duct system or (38) _____. For the latter, a procedure called (39) _____ is often used. In women, infertility is usually the result of blocked (40) _____ or a failure to (41) _____.

Today, infertile couples can be helped by a variety of assisted reproductive procedures in which an egg is fertilized outside of a woman's body and then placed into her uterus (called (42) _____) or Fallopian tube (called (43) _____), or in which both eggs and sperm are placed into a Fallopian tube (called (44) _____). These techniques have allowed some postmenopausal women to have children as well. For many infertile couples, adoption has proved to be a rewarding alternative.

CHAPTER 7 SELF-TEST

A. TRUE OR FALSE

[T] [F] 45. At conception, the one-celled organism is called an embryo.

[T] [F] 46. A primary follicle first begins to mature under stimulation of luteinizing hormone.

[T] [F] 47. About 75 percent of all conceptions either fail to implant or are spontaneously aborted within the first 6 weeks.

[T] [F] 48. "Morning sickness" is most common during the last trimester of pregnancy.

[T] [F] 49. Men sometimes experience symptoms of pregnancy while their partners are pregnant.

[T] [F] 50. An expectant mother cannot feel her unborn baby moving inside her ("quickening") until the second trimester.

[T] [F] 51. Sexual intercourse during the seventh and eighth months of pregnancy is often harmful to the mother and/or fetus.

[T] [F] 52. Drugs can adversely affect the fetus, but smoking, while hazardous to the mother, has little effect.

[T] [F] 53. If a woman is Rh positive and her fetus is Rh negative, the condition is called Rh incompatibility.

[T] [F] 54. The only noninvasive technique for examining a fetus inside the uterus is ultrasound.

[T] [F] 55. One beer or one glass of wine a day is thought to be acceptable when a woman is pregnant.

[T] [F] 56. One of the most common causes of scarring of the Fallopian tubes is untreated gonorrhea and chlamydia.

[T] [F] 57. "Prepared childbirth" means having a baby without the use of anesthetics.

[T] [F] 58. A baby is born during the third stage of labor.

[T] [F] 59. If a woman has an ectopic pregnancy, there is an increased risk that her future pregnancies will also be ectopic.

[T] [F] 60. Women who have delivered babies by cesarean section usually can deliver vaginally in subsequent pregnancies.

[T] [F] 61. Natural filters protect the baby from alcohol, drugs, and pollutants during breast-feeding.

[T] [F] 62. The expression "test-tube babies" refers to the process whereby conception and embryological development (i.e., the first 2 months) occur outside the mother's body.

[T] [F] 63. Secondary infertility (inability to conceive after having had one child) is more common than primary infertility.

[T] [F] 64. Some studies have found that the average sperm count in men today is less than half of what it was in men 50 years ago.

[T] [F] 65. Artificial insemination refers to the fertilization of an egg by synthetically manufactured sperm.

[T] [F] 66. Preembryos can be screened for genetic defects.

[T] [F] 67. With assisted reproductive techniques, women over the age of 60 can have babies if they wish to.

[T] [F] 68. "Morning sickness" can be effectively treated with hormones.

[T] [F] 69. True labor is marked by the onset of Braxton-Hicks contractions.

[T] [F] 70. In Lamaze and some other prepared childbirth programs, pain is controlled by using focal points, effleurage, and breathing techniques.

B. MATCHING (MATCH THE PROBLEM WITH THE TREATMENT OR SOLUTION)

TREATMENT OR SOLUTION

_____ 71. artificial insemination

_____ 72. balloon tuboplasty

_____ 73. cesarean section

_____ 74. chorionic sampling (or amniocentesis, or fetoscopy)

_____ 75. episiotomy

_____ 76. fertility drugs

_____ 77. fetal monitor

_____ 78. fetal surgery

_____ 79. gamete intrafallopian transfer

_____ 80. in vitro fertilization

_____ 81. laparoscopic surgery

_____ 82. surrogate mother

_____ 83. zygote intrafallopian transfer

_____ 84. intracytoplasmic sperm injection

PROBLEM

a. Susan has a tubal (ectopic) pregnancy.

b. Sherry and John cannot conceive because she is not having regular menstrual cycles.

c. Harriet's physician is worried that her perineum might tear during delivery.

d. Carol and David cannot conceive because he has a low sperm count.

e. Maria and John cannot conceive. Eggs taken from Maria are fertilized by sperm from John and the fertilized eggs are placed into one of Maria's Fallopian tubes.

f. Phyllis is not yet fully dilated and her baby is in distress.

g. Diane cannot carry a fetus during pregnancy, but she and Frank want to have a baby that is biologically theirs.

h. Mary is in labor and her physician wishes to keep track of the unborn baby's vital signs.

i. Shelly and Mark cannot conceive. Eggs taken from Shelly are fertilized by sperm from Mark and then placed into Shelly's uterus.

j. Cindy and Bert cannot conceive because she has blocked Fallopian tubes, but they have ruled out assisted reproductive techniques.

k. Linda and Walter's unborn baby has a life-threatening urinary tract infection.

l. Joan and Tom are worried that their unborn baby may have an abnormality.

m. Betty and Sam cannot conceive. Eggs taken from Betty and sperm taken from Sam are placed directly into one of her Fallopian tubes.

C. FILL IN THE BLANKS

85. Zygote, _____, _____, embryo, fetus.

86. The major link between the developing baby and the mother is the _____.

87. A pregnant woman who has gained excessive weight and has high blood pressure and swollen joints may have _____.

88. Before birth the fetus' head drops lower in the uterus. This is called _____.

89. Three rules of thumb to determine if labor has begun are _____, _____, and _____.

90. The _____ trimester is the time when a developing fetus is generally most susceptible to harmful substances.

91. The vast majority of ectopic pregnancies are _____ pregnancies.

92. In 1990, the Institute of Medicine recommended that normal-weight women gain _____ pounds during pregnancy.

93. By the time a fertilized egg has reached the uterus, it is a fluid-filled sphere called a _____.

94. A noninvasive technique for examining a fetus is _____.

95. The part of a baby that normally comes into the birth canal first is _____.

96. The cervix becomes fully dilated during the _____ stage of labor.

97. An egg remains ripe only for about _____ after ovulation.

98. The fetus is protected in a thick-skinned sac called the _____.

99. _____, _____, and _____ are three procedures used by physicians during labor and childbirth that many experts feel are done far too routinely (and unnecessarily).

100. For a baby to have any chance of surviving, it must develop in the uterus for at least _____ weeks.

101. If a woman _____ during pregnancy, that increases the risk of low birth weight, spontaneous abortions, and sudden infant death syndrome.

102. A prepared childbirth method that emphasizes childbirth without medication is called the _____ method.

CHAPTER 8   INTERACTIVE REVIEW

(1) _____ is your subjective sense of being male or female. It is influenced by both biological and social factors. A person's genetic sex is determined at conception by the combination of an egg, which has an X chromosome for sex, and a sperm, which can have either an X or a Y chromosome. (2) _____ combinations usually result in girls, while (3) _____ combinations usually produce boys. There are over 70 known irregularities in chromosome combinations. These include males with one or more extra X chromosomes ((4) _____ Syndrome), females with only one X chromosome ((5) _____ Syndrome), and females with Triple-X Syndrome.

Hormones determine whether an embryo will develop anatomically as a male or female. Unless there are high levels of (6) _____ at this critical stage of prenatal development, nature has programmed the body to develop into a (7) _____. Hormone abnormalities before birth can result in a mismatch between genetic and anatomical sex, or a baby whose external genitalia are ambiguous in appearance. Some individuals feel that they are trapped inside a body of the wrong gender, a feeling called (8) _____. Persons whose gender identity does not match their biological sex are said to have (9) _____. They often elect to undergo sex reassignment surgery.

Three psychological theories, though very different in their explanations, suggest that gender identity is learned. (10) _____ theory emphasizes unconscious identification with the parent of the same sex, while (11) _____ theory emphasizes the role of reinforcement and imitation. Cognitive-developmental theory states that children do not acquire the concept of (12) _____ until the age of (13) _____.

(14) _____ are norms (what is considered appropriate) about the behaviors and attitudes of men and women. They vary from culture to culture and can change over time within the same culture. (15) _____ theory emphasizes how gender roles are integrated within the individual early in life as part of the individual's personality. The development of gender roles begins before a child develops gender identity or gender constancy. (16) _____ theory emphasizes the role of the environment or social context in influencing the way gender roles develop over the life span.

Masculinity and femininity were once viewed as opposite ends of a unidimensional continuum, but today they are generally viewed as (17) _____constructs. This has led to the theory of (18) _____, which says that a person can be both masculine and feminine and that this is the healthiest of all gender roles because of the flexibility it gives individuals in different situations.

Parents, teachers, the media, and peers are all powerful influences on the process of socialization and gender role development. Gender is possibly the first social category learned by children. When children cognitively organize the world according to gender, they create gender (19) _____. Children raised in single-parent households tend to be more (20) _____ than children raised in two-parent households. As individuals become adults, they generally acquire more complex sets of gender roles, and for both sexes there is an integration of gender roles. Continued belief in stereotypic gender roles can adversely affect one's personal and sexual relations with a partner.

Traditional gender roles in our culture evolved over time in response to social forces. Those forces have been changing (an example has been the increased presence of women in the work force), and as a result, gender roles are in a state of transition. This may encourage individuals to base their behaviors, beliefs, and attitudes on human roles rather than being limited by rigid gender-role expectations.

CHAPTER 8   SELF-TEST

A. TRUE OR FALSE

[T]   [F]   21. By the age of 3, children know whether they are a boy or a girl, but do not understand that this cannot change.

[T]   [F]   22. In all cultures studied thus far, males are independent, aggressive, and make the important decisions, while females are emotional, nurturant, and considered too weak to do hard labor.

[T]   [F]   23. In the absence of testosterone during embryological development, we would all be born anatomically female.

[T]   [F]   24. Fathers tend to treat their children in more gender-stereotypic ways than do mothers.

[T]   [F]   25. According to microstructural theory, the way to change gender roles is not to retrain individuals but to restructure the social environment.

[T]   [F]   26. Transsexual is a term for a type of homosexual.

[T]   [F]   27. "Tomboy" behavior in girls is usually better tolerated by peers, parents, and teachers than is "effeminate" behavior in boys.

[T]   [F]   28. Double standards often still exist in two-career families.

[T]   [F]   29. According to social learning theory, imitation is a process by which children may learn gender identity and gender roles.

[T]   [F]   30. Turner's syndrome results from an extra Y chromosome.

[T]   [F]   31. Parents who assume "traditional" gender roles behave differently toward their male and female children.

[T]   [F]   32. The rise of industrialization helped create the male gender role of being independent and unemotional.

[T]   [F]   33. Gender role is the way you express your gender identity.

[T]   [F]   34. The most common cause of pseudohermaphroditism in males is androgen insensitivity syndrome.

[T]   [F]   35. Males' and females' hypothalamus are different.

[T]   [F]   36. A transsexual is someone who cross-dresses for sexual arousal.

[T]   [F]   37. Children do not show gender-stereotyped behavior until they have developed gender constancy.

[T]   [F]   38. A recent study found that young children view gender in dualistic terms, but as they grow older they come to view gender as a unidimensional construct.

[T]   [F]   39. Some cultures have a practice called couvade in which males act out giving birth.

[T]   [F]   40. There is little noticeable difference between the behavior of boys and the behavior of girls before the age of 2.

[T]   [F]   41. As people grow older, they are more able to rely on information about an individual to make judgments, even when the information is not consistent with stereotyped gender roles.

[T]   [F]   42. Fewer than half of all women with young children are in the labor force.

[T]   [F]   43. As a result of converging gender roles, men and women no longer differ with regard to permissive attitudes about casual sex.

[T]   [F]   44. Pleck's perspective on gender-role strain contends that the problem with the male gender role is that men are confronted by contradictory expectations in their early socialization and their adult life experience.

B. MATCHING (SOME QUESTIONS HAVE MORE THAN ONE ANSWER)

_____ 45. Bob is low in instrumental orientation ("getting the job done") and has great affective concern for the welfare of others.

_____ 46. Frank prefers men as sexual partners.

_____ 47. Susan cannot reproduce; is short, with a webbed neck; and has one X chromosome.

_____ 48. Joe is neither instrumental nor expressive, neither assertive nor emotional.

_____ 49. David's body did not respond to testosterone during prenatal development; he has female genitalia and undescended testes.

_____ 50. Sam was born with a very small penis that looked like a clitoris, an incomplete scrotum, and a short, closed vaginal cavity. At puberty his voice deepened, his testes descended, and his "clitoris" grew to become a penis.

_____ 51. Wayne is both instrumental and expressive, assertive and emotional.

_____ 52. Phillip believes he is a woman trapped in a male body.

_____ 53. Sharon is slightly retarded and has an extra X chromosome.

_____ 54. Carol has an enlarged clitoris and labia because of too much masculinizing hormone during her fetal development.

_____ 55. Mike believes that men should be assertive, aggressive, success-oriented, unemotional, and play little role in housekeeping and child-care responsibilities.

_____ 56. Harold is tall, with long arms; he has a small penis, shrunken testes, low sexual desire, and an extra X chromosome.

_____ 57. Barbara is genetically a female, but has both male and female reproductive systems as a result of the failure of her primitive gonads to differentiate during the embryonic stage.

a. adrenogenital syndrome
b. androgen insensitivity syndrome
c. androgynous individual
d. DHT-deficient individual
e. "feminine" on Bem Sex Role Inventory
f. gender dysphoric
g. hermaphrodite
h. homosexual
i. Klinefelter's Syndrome
j. pseudohermaphrodite
k. stereotyped individual
l. transsexual
m. Triple-X Syndrome
n. Turner's Syndrome
o. undifferentiated individual
p. gender identity disorder

C. FILL IN THE BLANKS

58. A child's knowledge that his or her sex does not change is called _____ in cognitive-developmental theory.

59. According to Cancian, gender roles for males and females did not start to differ until the _____.

60. According to _____ theory, an individual's gender role results from society's expectations of male and female behavior.

61. According to Freud, children acquire the gender identity of the same-sex parent through the process of _____.

62. The process of internalizing society's beliefs is called _____.

63. Oversimplified, rigid beliefs that all members of a particular gender have distinct behavioral and emotional characteristics are called _____.

64. On Bem's Sex Role Inventory, a person who scores low on both dimensions is called _____.

65. The duct system in embryos that can develop into the female reproductive system is called the _____ duct system.

66. According to Freud, children acquire their gender identity in the _____ stage of psychosexual development.

67. If an individual scores high on both the femininity dimension and the masculine dimension of the Bem Sex Role Inventory, he or she would be called _____.

68. Money and Ehrhardt (1972) stated, "Nature's rule is, it would appear, that to masculinize, something must be added." That something is _____.

69. According to Freud, gender identity is acquired when children resolve the _____ complex.

70. _____ cross-dress for sexual arousal and gratification, whereas _____ cross-dress because they truly believe they are members of the opposite sex.

71. In social learning explanations of gender identity development, when children watch their mothers and fathers and copy them, it is called _____.

72. The _____ view of gender role development holds that biological influences establish predetermined limits to the effects of cultural influences.

73. According to Bem, masculine is to instrumental as feminine is to _____.

74. According to Kagan, boys' and girls' behaviors do not differ at the age of _____.

75. People who watch a lot of television generally believe in more _____ gender roles.

76. Some languages classify all nouns as either masculine or feminine. This is an example of _____.

77. A recent study found that young people rely on _____ as judgment cues, but as they grow older they increasingly rely on more _____ information.

78. If research on single-parent households is correct, then a greater proportion of African-American children will grow up to be _____ than is true for Caucasian children.

79. The finding that cross-gender-role characteristics often emerge among older men and women has been referred to as the _____ model of gender role development.

80. As a general rule, men have considerably more permissive attitudes about _____ than women do.

81. Among early North American Indian tribes, a _____ was a highly respected male who cross-dressed and assumed the behaviors of a woman.

82. _____ is the extent to which a person approves of and participates in feelings and behaviors deemed to be appropriate to his or her culturally constituted gender.

CHAPTER 9 INTERACTIVE REVIEW

In this chapter, sexual orientation is defined as (1) _____ consistently made after adolescence in the presence of clear alternatives. Isolated instances of sexual behavior may or may not reflect one's sexual orientation. Recent surveys indicate that (2) _____ percent of the adult male population and about (3) _____ of the adult female population have a homosexual orientation. About 20 percent of all males (and a lower percentage of females) have had sex with a same-sex individual on at least one occasion, but fewer than 5 percent regard themselves as bisexual. A true bisexual would have a rating of (4) _____ on Kinsey's 7-point rating scale.

The (5) _____ of the vast majority of homosexuals and bisexuals are just as strong and consistent with their anatomical sex as among heterosexuals. Conformity or nonconformity with gender (6) _____ does not predict one's sexual orientation. A great many homosexuals and bisexuals do not walk, talk, dress, or act any differently than anyone else.

There are three major explanations for the origins of sexual orientation. Psychoanalytic explanations state that heterosexuality is the "normal" outcome and that homosexuality results from problems in resolving the (7) _____ complex. Some of Freud's followers believed that homosexual males had (8) _____ mothers and (9) _____ fathers, while homosexual females had (10) _____ fathers and (11) _____ mothers. However, later research does not generally support these claims. Homosexuals, as well as heterosexuals, have all kinds of parents, both good and bad, caring and cold. (12) _____ regards homosexuality and heterosexuality as learned behaviors; good or rewarding experiences with individuals of one sex and/or bad experiences with individuals of the other sex (particularly during adolescence) would lead to a particular orientation. However, social and cultural studies show that environment alone cannot explain sexual orientation. In the (13) _____ society in Melanesia, all boys engage in homosexual acts for many years, yet the large majority grow up to have a heterosexual orientation.

Recent studies suggest that biology plays a major role in the origin of sexual orientation. Studies with twins indicate a (14) _____ factor, while anatomical studies have found differences in the (15) _____ of heterosexuals and homosexuals. Many researchers believe that differences in (16) _____ before or shortly after we are born predispose us to a particular sexual orientation. Nearly all agree that biological and social/cultural influences (17) _____ to produce sexual orientation.

Anthropologists find that most societies are (18) _____ tolerant of homosexuality than prevailing attitudes in the United States. Attitudes about homosexuality have varied considerably in Western culture. The (19) _____ and (20) _____ accepted homosexuality, as did the early Christians. It was not until St. (21) _____ that homosexuality came to be viewed as unnatural or "against the laws of nature," and not until the 1600s and 1700s that it was regarded as criminal. In the 1800s, the medical (psychiatric) model regarded homosexuality as an illness, and homosexual individuals were regarded as having pathological conditions in need of being "cured." It was not until 1973 that the American Psychiatric Association removed homosexuality as an official mental illness. Nevertheless, most people in the United States continue to have a harsh and negative attitude.

The process of (22) _____, or identifying oneself as homosexual, involves several often painful stages. The first stage is (23) _____. In the final stages, the individual attempts to gain the acceptance and understanding of friends, family, and co-workers. This is difficult in a society filled with (24) _____, an irrational fear of homosexuality. Experts believe that antigay prejudice often reflects a person's attempts to affirm his or her own manhood or womanhood.

Prior to the 1970s, opportunities for homosexuals to meet were limited to a few bars, bathhouses, and public restrooms. Today, most metropolitan areas have well-established homosexual communities where individuals can openly associate. Women, whether straight or lesbian, tend to value (25) _____ in their relations. Prior to the 1980s, many gays engaged in impersonal sex with numerous partners, but today there is a greater emphasis in the gay community on establishing relationships.

At the time this book was written, homosexual marriages were illegal in all 50 states. Most Americans oppose the idea of openly homosexual individuals raising children (even their own), yet research shows that most parental sexual abuse is committed by heterosexuals and that children raised by homosexual parents grow up to have normal gender identities and gender roles, and almost always have a heterosexual orientation. This last finding again suggests a greater role for biology than environment in the origin of sexual orientation, and it is the hope of the researchers conducting biological studies that their findings will promote greater understanding and tolerance.

CHAPTER 9   SELF-TEST

A. TRUE OR FALSE

- [T] [F]   26. In some cultures, homosexual behavior is considered normal for males during adolescence.
- [T] [F]   27. Homosexuals usually act and/or dress differently from heterosexuals.
- [T] [F]   28. Freud said that homosexuality was not an illness.
- [T] [F]   29. It is common for heterosexuals to have homosexual fantasies.
- [T] [F]   30. Bisexuals are people who are afraid to admit to themselves their real homosexuality.
- [T] [F]   31. Most homosexuals are unhappy with their sexual orientation and would like to become heterosexual.
- [T] [F]   32. A homosexual orientation indicates a gender identity problem.
- [T] [F]   33. A male homosexual can be made heterosexual by administering large doses of testosterone.
- [T] [F]   34. People who have had homosexual experiences are by definition homosexual in orientation.
- [T] [F]   35. "Coming out" is regarded as a form of exhibitionism and is generally unhealthy.
- [T] [F]   36. Attitudes about homosexuality are learned, and what is learned depends on the time and place in which a person is raised.
- [T] [F]   37. Although being homosexual is no longer criminal, in many states it is against the law to engage in homosexual behaviors.
- [T] [F]   38. Freud believed that all people were capable of becoming either heterosexual or homosexual, depending on their early childhood experiences.

T ☐ F ☐   39. Researchers have found a higher rate of concordance for homosexuality between identical twins than between nonidentical twins.

T ☐ F ☐   40. Homosexual parents are more likely than heterosexual parents to molest their children.

T ☐ F ☐   41. A bisexual is anyone who has had sex with both men and women.

T ☐ F ☐   42. More recent studies support Kinsey's findings that 10 percent of all American males are predominantly homosexual.

T ☐ F ☐   43. Homosexual experiences before age 15 are a good predictor of adult sexual orientation.

T ☐ F ☐   44. Bisexuals have fewer opposite-sex fantasies than heterosexuals and fewer same-sex fantasies than homosexuals.

T ☐ F ☐   45. Many homosexuals and bisexuals have gender dysphoria.

T ☐ F ☐   46. Most female homosexuals have not had pleasurable sexual relations with men.

T ☐ F ☐   47. Sexual orientation is usually well established by adolescence.

T ☐ F ☐   48. Extremely effeminate ("sissy") boys always develop a homosexual orientation.

T ☐ F ☐   49. If one identical twin is homosexual, the other twin always is too.

T ☐ F ☐   50. Studies have found that areas of the hypothalamus (in the brain) are different in heterosexuals and homosexuals.

T ☐ F ☐   51. Male rats in which sex hormones have been manipulated shortly after birth often show female sexual behavior.

T ☐ F ☐   52. Among major Western industrialized countries, only the United States and Britain bar known homosexuals from the military.

T ☐ F ☐   53. Most homosexuals adopt a single role ("masculine" or "feminine") in sexual relationships.

T ☐ F ☐   54. Children raised by openly homosexual parents are more likely to develop a homosexual orientation than children raised by heterosexual parents.

T ☐ F ☐   55. On average, homosexual males had a later birth order than heterosexual males.

T ☐ F ☐   56. Homophobic males generally have equally negative attitudes about male and female homosexuals.

B. MATCHING

_____ 57. George is a zero on Kinsey's rating scale.

_____ 58. Henry has had a few same-sex sexual experiences, but has had sex only with women since he turned 15 ten years ago.

_____ 59. Joyce has been married and enjoyed sexual relations with her husband, but now prefers sexual relations with women.

_____ 60. Frank has had sex with both men and women.

_____ 61. Michael believes he is a woman and has sex only with men.

_____ 62. Carl sometimes enjoys sex with women, but prefers sexual relations with men.

_____ 63. Tom is 12 years old and his few sexual experiences have been exclusively with boys his age.

_____ 64. Alice enjoys sex with men most of the time, but she often enjoys sex with women as well.

_____ 65. Diane has had sex only with men, but occasionally has homosexual fantasies.

_____ 66. Matthew occasionally goes to bathhouses and lets other men give him oral-genital sex.

_____ 67. Mary's gender identity is not consistent with her anatomical gender.

_____ 68. Steve is 19 years old and his only sexual experiences have been with women, but all his sexual fantasies are about men.

a. probably heterosexual

b. probably homosexual

c. probably bisexual

d. probably transsexual

e. not enough information to make a guess

C. FILL IN THE BLANKS

69. In lesbian relationships, rubbing the genitals together is called _____.

70. In this book, sexual orientation is defined as distinct preferences consistently made after _____ in the presence of clear alternatives.

71. Humphreys (1970) called the practice of men giving oral sex to other men in public restrooms _____.

72. _____ and _____ believed (incorrectly) that homosexuality resulted from reversed gender roles.

73. Freud called individuals who kept their homosexual tendencies hidden from their conscious mind _____.

74. Simon LeVay found that a part of the brain called the _____ was different in heterosexual and homosexual men.

75. Researchers believe that hormones may affect sexual orientation during which time of life? _____

76. According to this chapter, biological factors probably _____ individuals to a particular sexual orientation.

77. Many homosexuals refer to heterosexuals as _____.

78. The four stages of the coming-out process are (a) _____, (b) _____ (c) _____, and (d) _____.

79. Bell and Weinberg (1978) described homosexuals who were happy in their sexual orientation and in a monogamous relationship as _____.

80. Many male homosexuals have changed their sexual life-styles because of _____.

81. Name four nonliving homosexuals who made significant contributions to humankind: (a) _____, (b) _____, (c) _____, and (d) _____.

82. Daryl Bem proposed that biology codes for _____ that influence a child's preferences for sex-typical or sex-atypical activities and peers.

CHAPTER 10 *INTERACTIVE REVIEW*

People are sexual beings from the moment of birth. In infancy, self-stimulation of the genitals is motivated by a general attempt to explore the body and seek pleasure. As children grow, their curiosity about their bodies continues and, during early childhood, reaches a peak in the age period from (1) _____ years old. Early childhood games such as "house" or "doctor" allow exploration of same-sex and opposite-sex individuals. In the initial school-age years, children develop a sense of (2) _____. Although overt sexual play may decrease, curiosity about the human body and sexuality does not.

(3) _____ is the time in life in which an individual first shows sexual attraction and becomes capable of (4) _____. Many bodily changes occur, the result of changing (5) _____ levels, and these physical changes require many adjustments for young adolescents. Unfortunately, many young teens acquire a negative attitude about their bodies as a result of their parents' reactions to normal events such as menstruation and nocturnal emissions.

Interest in sex increases dramatically at puberty, and earlier childhood games evolve into more erotic and consciously sexual games such as "spin the bottle." The most important issue in the lives of most adolescents is (6) _____, and during these years there is a focus on body image and physical characteristics. Almost all teenaged boys and a majority of girls masturbate, which is usually accompanied by sexual fantasies. Sex therapists feel that masturbation is a safe and healthy sexual outlet for teenagers. Adolescents' interpersonal sexual experiences generally begin with holding hands, followed in order by embracing, (7) _____, (8) _____, and, for many, sexual intercourse. Surveys find that most boys and girls have engaged in sexual intercourse by age (9) _____, and nearly three-fourths have done so by the time they complete high school. For some teens, however, having sexual relations is not a positive experience, yet many engage in it because of (10) _____. The teenage pregnancy rate in the United States is one of the highest among industrialized countries.

The average age at marriage has steadily increased and is now in the mid-twenties. Most young adults believe that premarital sex, especially in a "serious" relationship, is acceptable. Most have also had two or more sexual partners, often in a life-style pattern called (11) _____.

By the time people are in their mid-twenties, most people's sexual life-style is characterized by (12) _____, either in marriage or by cohabitation. In all cultures, the idealistic goal is a (13) _____. Throughout the world, men value female promiscuity when pursuing short-term relationships, but when choosing a long-term mate, men in most cultures value (14) _____. When choosing a long-term mate, women generally prefer men who can offer them (15) _____. Despite the ideal, about a third of all married men and about a fifth of all married women eventually engage in extramarital relations. Some do so with the consent of their mate.

Most young adults underestimate the frequency of sex practiced by older adults. Sexual activity is highest for people in their (16) _____, and only gradually declines up through the early sixties. Surveys indicate that married people have sex more often than singles and are happier with their sex lives than are singles. Still, over half of all marriages in the United States end in divorce. About three-fourths of all divorced people remarry, and for some marriage is just another form of serial monogamy.

As women grow older, their menstrual cycles become irregular and eventually stop entirely, which is called (17) _____. This usually occurs in a woman's late forties or early fifties, and the change in hormones can result in hot flashes, a decrease in (18) _____, and osteoporosis. These conditions can be reversed with (19) _____ replacement therapy. Experts feel that men normally do not experience menopause, but they do show a gradual decline in (20) _____ beginning in their late teens. This also results in some physical changes, such as a decline in muscle mass and strength, less firm erections, less forceful ejaculations, and longer (21) _____ periods. However, most elderly people continue to enjoy sexual relations as long as they remain otherwise healthy and have a partner. For healthy couples, the best predictor of whether or not they will enjoy sex in old age is (22) _____.

CHAPTER 10 SELF-TEST

A. TRUE OR FALSE

23. Ultrasound recordings have found that male fetuses have erections while still inside the uterus.
24. Children cannot masturbate to orgasm until they reach puberty.
25. The growth of pubic hair on girls is the result of increased levels of testosterone.
26. If a child is found engaging in sexual exploration games with another child of the same sex before age 7, it means the child is probably gay.
27. Gagnon believes that teenage masturbation is the single most powerful predictor of adult sexuality.
28. Couples who lived together before marriage are less likely than other couples to get divorced.
29. In a study of 202 men and women aged 80 to 102, it was found that nearly half still engaged in sexual intercourse.
30. The lack of vaginal lubrication in women who have undergone menopause is an indicator that their sex life is coming to an end.
31. The average married couple in their twenties or thirties have sex more often than young singles.
32. Adults almost always quit masturbating after they are married.
33. The average age at which girls have their first menstrual cycle has been decreasing over the last two centuries.
34. "Hot flashes" are due, in part, to increasing levels of pituitary hormones.
35. Sexual fantasies are the same for males and females during adolescence.
36. The best predictor of sexual activity for older women is the amount of vaginal lubrication they had when they were younger.

T F	37.	A heart attack almost always causes older males to become impotent.
T F	38.	Boys often develop enlarged breasts during puberty.
T F	39.	Early studies found that babies could die from emotional neglect.
T F	40.	In early childhood it is normal for children to engage in sexual behaviors that imitate adults.
T F	41.	The way parents react to the sexuality of their children is generally an indication of the way they feel about their own sexuality.
T F	42.	Beginning in kindergarten, most U.S. children develop a sense of modesty about exposing their bodies in public.
T F	43.	There is no physical reason why elderly couples cannot continue to have sexual intercourse.
T F	44.	In terms of hormone changes with aging, men and women are very much alike.
T F	45.	In men, the production of sperm generally stops by the early fifties.
T F	46.	Most women show a decreased interest in sex after menopause.
T F	47.	In some cultures, hot flashes are uncommon among postmenopausal women.
T F	48.	Pubertal development in boys lags about 2 years behind development in girls.
T F	49.	If an adult male is castrated, it substantially raises his voice.
T F	50.	Girls as young as 6 or 7 have given birth.
T F	51.	Most teenage girls experience severe pain during their first sexual intercourse.
T F	52.	The sexual activity of young adults is generally different from the activity of younger and older age groups.
T F	53.	Most young adults limit their premarital sexual activity to "serious" relationships.
T F	54.	Men and women are more likely to be upset by their partner's having a sexual relationship with someone else than they are about their having an emotional attachment.
T F	55.	There is a strong relationship between social background (religious and political views) and whether or not one engages in extramarital affairs.
T F	56.	The divorce rate for open-marriage couples is lower than that for sexually monogamous couples.
T F	57.	Children's first sexual attractions occur around age 10.
T F	58.	Hot flashes are a common symptom of menopause in all cultures.

B. MATCHING: WHICH HORMONE CHANGES ARE MOST LIKELY TO BE ASSOCIATED WITH THE FOLLOWING PHYSICAL OR BEHAVIORAL CHANGES?

_____ 59. breast development in girls

_____ 60. vaginal dryness at menopause

_____ 61. hot flashes in women

_____ 62. development of sweat glands (body odor, acne)

_____ 63. gynecomastia

_____ 64. pubic hair and body hair

_____ 65. menarche

_____ 66. less firm erection and longer refractory period

_____ 67. reversal of effects of menopause

_____ 68. decreased sexual desire in women (review from Chapter 3)

_____ 69. decreased sexual desire in men (review from Chapter 3)

_____ 70. children's first sexual attractions

a. increase in estrogen

b. decrease in estrogen

c. increase in testosterone

d. decrease in testosterone

e. increase in FSH and LH

f. increase in DHEA from adrenal glands

C. FILL IN THE BLANKS

71. The decline in sexual activity among older single women is most often due to _____.

72. When sexual development begins before age 9, it is called _____.

73. According to Jean Piaget, before the age of 2 children are _____.

74. The changes that occur in women in the few years that precede and follow menopause are called _____.

75. Before age 2, bodily exploration is usually confined to _____.

76. Freud referred to what this chapter calls the initial school-age years as the _____ stage of psycho-sexual development.

77. Development of breasts and growth of facial hair are examples of _____.

78. The first experience many males have with ejaculation is a "wet dream," or _____.

79. For many teenage girls, the need for _____ is an important part of finding their self-identity.

80. Throughout the world, men tend to engage in casual sex to _____.

81. About 80 percent of all societies allow men to _____.

82. Women are most concerned about _____ infidelity by their mates.

83. Whitehurst believes that the two things most likely to lead to extramarital sex are _____ and _____.

84. From a person's late thirties to the early sixties, the frequency of sexual intercourse _____.

85. Puberty is at least a two-part maturational process. The two parts are called _____ and _____.

86. If a child is late in showing growth and development of secondary sex characteristics, it is called _____.

CHAPTER 11   INTERACTIVE REVIEW

This chapter presents several normal adult sexual behaviors. "Normal" is defined from a statistical point of view, and is best thought of as a (1) _____ of behaviors or values. Thus, you should not regard yourself as abnormal if you have not engaged in all the behaviors discussed here.

Historically, masturbation has been presented as unnatural, immoral, and bad for one's physical and mental health. In his landmark surveys of 1948 and 1953, (2) _____ found that (3) _____ percent of the men and (4) _____ percent of the women surveyed had masturbated, thus demonstrating that masturbation is, in fact, a very normal human sexual behavior. Modern medicine has also shown that masturbation has no negative medical consequences. Most men and women masturbate even when they are in sexual relationships. Most men, and probably most women, have also experienced orgasms during sleep, called (5) _____. They occur during a stage of sleep called (6) _____, during which all men in good health (7) _____ and women experience (8) _____.

Compared to those of men, sexual fantasies of women tend to be more (9) _____. Although at times a fantasy can be very sexually explicit, that does not mean that a person would actually want to do what he or she has fantasized. Fantasies are usually not an indication of sexual unhappiness or personality or psychological disorders.

The most common position for sexual intercourse in the United States is (10) _____-on-top, also called the (11) _____ position. However, there is no "correct" way of having sexual intercourse except what is right for you and your partner. Making your sexual encounters spontaneous, exciting, fulfilling, and not ritualized may mean exploring a variety of times, places, and positions.

Oral-genital sexual relations (properly referred to as (12) _____ and (13) _____) as a form of sexual expression has increased tremendously in popularity since Kinsey's time. In addition, a considerable number of heterosexual couples have tried anal intercourse on at least one occasion. However, just because many people may be exploring a particular form of expression does not mean it fits into everyone's value system.

A healthy and satisfying sexual relationship can contribute to one's overall physical and emotional well-being. (14) _____ individuals consider sex to be a positive and good thing, and feel free to choose when, where, and with whom to engage in a particular sexual activity.

CHAPTER 11   SELF-TEST

A. TRUE OR FALSE

T F 15. People who masturbate to excess run the risk of serious medical problems.

T F 16. Kellogg's cornflakes and Graham crackers were developed as foods to prevent youthful masturbation.

T F 17. Women do not have nocturnal orgasms.

T F 18. Substantially more men have sexual fantasies than women.

T F 19. A person can be sexually healthy and choose not to have oral-genital sex, or even sexual intercourse.

T F 20. Male-on-top is the preferred position of intercourse in all cultures.

T F 21. Masturbation is normal for men, but not for women.

T F 22. Studies have found that males masturbate in more varieties of ways than do females.

T F 23. Assuming that neither person has a sexually transmitted disease, oral-genital sex is no less hygienic than kissing.

T F 24. Most married people who masturbate do so because they are unhappy in their sexual relationships.

T F 25. The content of sexual fantasies usually does not indicate sexual unhappiness or personality or psychological problems.

T F 26. Therapists sometimes instruct their clients on how to masturbate in order to help them overcome sexual problems.

T F 27. "Wet dreams" are dreams with a sexual theme.

T F 28. Fantasizing about another person while having sex with your partner is usually an indication of a serious relationship problem.

T F 29. Hunt's survey found that most couples have had sexual intercourse in more than one position.

T F 30. No other behavior has increased as much in popularity since Kinsey's time as oral-genital sex.

T F 31. Anal intercourse between men and women is often a sign of latent homosexuality.

T F 32. Women's sexual fantasies are just as explicit as men's.

T F 33. Surveys have consistently found that over 90 percent of all men have masturbated by age 20.

T F 34. Generally speaking, men have sexual fantasies more often than women.

T F 35. More people with sexual partners masturbate than people who do not have a sexual partner.

B. MATCHING (ALL QUESTIONS HAVE MULTIPLE ANSWERS)

_____ 36. probably a sexually healthy person

_____ 37. probably a sexually unhealthy person

_____ 38. not enough information to tell

a. a 24-year-old man who has never had sexual relations

b. a 20-year-old woman who has had 15 sexual partners

c. a 20-year-old woman who is abstaining from sex, but is looking forward to sexual relations in a long-term relationship

d. a 20-year-old man who has had 15 sexual partners

e. someone who has had oral-genital sex and sexual intercourse in many positions

f. a male who engages in sex with many partners to prove his sexuality to his peers

g. someone who refuses to try oral-genital sex

h. a man who has had many sexual partners and regards them as sluts

i. someone who doesn't like oral-genital sex but always agrees to it

j. someone who engages in sex because of low self-esteem

k. someone who has had many partners and regards his or her experiences positively

l. someone who masturbates every day

m. someone who has always been faithful to his or her partner but feels guilty about desiring to have sex with him or her

n. someone who enjoys trying new sexual experiences with his or her partner, but chooses not to engage in anal sex

C. FILL IN THE BLANKS

39. Oral stimulation of a woman's genitals is called _____.

40. Approximately every 90 minutes during sleep, females experience _____ and males experience _____.

41. Couples should not engage in anal and vaginal intercourse without washing in between because of _____.

42. Perhaps the most common type of fantasy during sexual intercourse is _____.

43. The main reason that many people continue to masturbate after forming a sexual relationship is _____.

44. Couples who always have sex in the same place and in the same manner risk letting their sex lives become _____.

45. Laumann and his colleagues (1994) found that the two sexual behaviors most preferred by Americans are _____ and _____.

46. Laumann and his colleagues also found that masturbation is mostly a reflection of _____

47. The preferred manner of sexual intercourse in all known cultures is _____.

48. _____ is the position of intercourse usually preferred in cultures where a woman's sexual satisfaction is considered to be as important as the man's.

49. Victorian-era physicians incorrectly believed that nocturnal emissions were caused by the same thing that causes _____.

50. A sexually healthy person is someone who (a) _____ and (b) _____.

CHAPTER 12   INTERACTIVE REVIEW

There have probably been as many definitions and descriptions of love as there have been poets, philosophers, and scientists who study love. Recent work has shown that although friendship and romantic love share many characteristics, romantic lovers differ from friends in terms of their (1) _____, (2) _____, (3) _____, and willingness to give the utmost when the other is in need. The feeling of romantic love initially includes physiological arousal, but a (4) _____ is required to interpret the responses as love. Feelings of passion almost always decrease with time, and long-term happy relationships are based more on togetherness, trust, sharing, and affection, or what is called (5) _____ love.

The ability to love another person requires (6) _____ and (7) _____. (8) _____ is an emotional state that is aroused by a perceived threat to a valued relationship. It is more common in people who have low (9) _____ and/or who are personally unhappy in their lives, or who put great value on things like (10) _____. Love and sex share many things in common, but it is possible to enjoy sex without emotional involvement, and it is also possible to love without having sexual relations. Love can be (11) _____ and depend on another person's ability to satisfy our needs and desires, or it can be (12) _____ and not dependent on the loved one's meeting certain expectations and desires.

Robert Sternberg has suggested that liking, loving, and all the other positive emotions we feel for other individuals can be understood by the combination of three components: intimacy, passion, and decision/commitment. According to this triangular theory, (13) _____ is intimacy alone, without passion or commitment. (14) _____ is liking plus feelings of passion without commitment, while (15) _____ is intimacy and commitment without passion. (16) _____, which Sternberg believes to be true love, requires the presence of all three components together.

In contrast to Sternberg, John Lee does not believe that there is only one type of love that should be viewed as true love. He proposes that there are many styles of loving. (17) _____ is based on an ideal of physical perfection; (18) _____ grows from friendship; (19) _____ is rational and practical; (20) _____ involves intense emotional dependency; (21) _____ is self-centered; and (22) _____ puts the interest of the loved person first. According to Lee, the degree of happiness an individual feels in a loving relationship depends greatly on how well his or her love-style (23) _____ that of the loved one.

Relationships that last are generally more realistic than idealistic and based more on companionship and affection than on passion. In order to maintain a relationship, a couple must substitute new shared activities for old ones as their lives change, and they also must develop skills to achieve greater (24) _____, those feelings and experiences that promote closeness and bondedness.

CHAPTER 12   SELF-TEST

A. TRUE OR FALSE

- T F 25. People rate best friends and lovers similarly on characteristics such as acceptance, trust, respect, mutual assistance, and understanding.
- T F 26. Romantic relationships are generally more rewarding than friendships, but they are also more volatile and frustrating.
- T F 27. According to Lee, a good match generally results from two love-styles that are far apart on his "color wheel" chart.
- T F 28. According to Lee, love-styles are fixed and repetitive.
- T F 29. As a general rule, people are most attracted to opposites.
- T F 30. The decision and commitment to love another person without intimacy or passion is experienced as empty love, according to Sternberg.
- T F 31. In the initiation of a relationship, men attach greater importance to physical attractiveness than do women.
- T F 32. Long periods of sexual deprivation (including masturbation) can result in death for an adult.
- T F 33. Contrary to popular belief, men are no more likely than women to enjoy sex without emotional involvement.
- T F 34. One of the major predictors of marital success is the number of pleasurable activities the couple shares.
- T F 35. Within a monogamous relationship, sexual intercourse and affection are synonymous.
- T F 36. Jealousy is most common in cultures that allow polygyny.
- T F 37. Men and women respond to jealousy in much the same way.
- T F 38. Feelings of romantic love are associated with the release of amphetamine-like chemicals in the brain.
- T F 39. The manner in which one is raised is important for acquiring a positive self-concept.
- T F 40. Self-disclosure is always important in a relationship, and the timing or mutuality of it does not really matter.
- T F 41. Men and women are most likely to become jealous to the perception of a partner's sexual infidelity.
- T F 42. Romantic love tends to be unconditional.

T F   43. Anthropologists have found that romantic love is an idealized notion that is found almost exclusively in industrialized cultures where emphasis is on the individual.

T F   44. Babies who are deprived of physical contact may sicken or die.

T F   45. In order for a couple to become more intimate, it is important for them to withhold their negative feelings about each other.

T F   46. Theoretically, humans are probably not capable of true agape.

B. MATCHING

_____ 47. agapic love

_____ 48. attachment

_____ 49. companionate love

_____ 50. conditional love

_____ 51. consummate love

_____ 52. empty love

_____ 53. erotic love

_____ 54. fatuous love

_____ 55. infatuated love

_____ 56. ludic love

_____ 57. manic love

_____ 58. pragmatic love

_____ 59. romantic love

_____ 60. storgic love

_____ 61. unconditional love

a. highly idealized love based on physical beauty; these type of lovers are inclined to feel "love at first sight," but it usually doesn't last long

b. a love based on the partner's satisfying certain needs and desires; Maslow called it "deficiency love"

c. an affectionate type of love that develops from friendship slowly over time

d. the love that results when commitment is made on the basis of passion without the experience of intimacy; leads to whirlwind romances that usually end when the passion starts to fade

e. a selfless love-style that puts the interest of the loved person ahead of the lover's own, even if it means great sacrifice

f. a love that does not depend on the loved one meeting certain expectations and desires; Maslow called it "being love"

g. commitment without intimacy or passion; often the first stage in arranged marriages

h. a love in which one's partner gives few positive rewards, other than predictability, for remaining in the relationship

i. a self-centered love in which the lover avoids commitment and treats love like a game

j. a love characterized by passionate arousal and liking without commitment; results from two people being drawn together both physically and emotionally

k. a love that results when passion is felt in the absence of intimacy and commitment, where the lover is obsessed with the other person as an ideal; similar to what Lee calls erotic love

l. love characterized by an intense emotional dependency on the attention and affection of the partner; "head over heels" love in which one is obsessed with the loved one

m. a love found only in relationships that include intimacy, passion, and commitment

n. the combination of intimacy and commitment without passion; based on togetherness, trust, sharing, and affection

o. a rational or practical style of love; this type of lover consciously looks for a compatible mate (a "good match")

C. FILL IN THE BLANKS

62. Fatuous love in Sternberg's model most closely resembles _____ in Lee's theoretical model.

63. Companionate love in Sternberg's model most closely resembles _____ in Lee's theoretical model.

64. Infatuated love in Sternberg's model most closely resembles _____ in Lee's theoretical model.

65. St. Augustine's idea of love ("I want you to be") is probably closest to _____ in Lee's theoretical model.

66. In order for lovers to experience complete love in Sternberg's model, a relationship must have _____, _____, and _____.

67. According to Freud, love is the result of _____.

68. Davis and others report that most romantic lovers find that their mood depends on _____ of their feelings more than in friendships.

69. Feelings of romantic love are associated with an increase in three brain chemicals called _____, _____, and _____.

70. Companionate love is often referred to as _____ love because it is not based on the fantasies and ideals of romantic love.

71. In Hazen and Shaver's attachment theory of love, adults who do not fear abandonment and find it easy to get close to others are called _____ lovers.

72. People who put great value on things like popularity, wealth, fame, and physical attractiveness are more likely than others to _____.

73. As a general rule, women are much more likely than men to experience jealousy in response to a partner's _____ infidelity.

74. Dorothy Tennov coined the term _____ to describe extreme "head over heels" love in which one is obsessed with the loved one. This would be experienced in the type of love Lee called _____.

75. According to Lee's model, in the type of love called _____, lovers feel that the act of loving is more important than the partner for whom it is expressed, and thus feel fulfilled even if their love is not reciprocated.

76. Elaine Hatfield has suggested that couples can achieve greater intimacy by (a) _____, (b) _____, (c) _____, and (d) _____.

77. A large component of passion is novelty and fantasy. The decline of passion in a relationship is almost inevitable because of _____.

78. Meyers and Berscheid found that when people placed their social relationships into categories, people placed in the "in love" category were also included within the _____ category.

CHAPTER 13   INTERACTIVE REVIEW

Studies have found that (1) _____ of all couples in the United States will eventually experience sexual problems. While some couples may grow apart and experience a variety of relationship problems (sex being just one indicator that there are other problems), even couples with a good relationship can experience sexual problems. It is common for two people in a relationship to differ in their preferences for frequency and type of sex. This often results from different expectations and assumptions about sex. When this happens, it is best to consider the problem as the (2) _____ problem. If the problem persists and is causing stress and anxiety, it is advisable to seek professional help.

Sex therapists are trained to (3) _____. They consider it unethical to (4) _____. Most sexual problems are treated with (5) _____, developed by Masters and Johnson; psychosexual therapy; or a combination of both. Sexual therapy programs generally have many things in common and follow the (6) _____ model, an acronym for permission, limited information, specific suggestions, and intensive therapy. Each represents a progressively deeper level of therapy. Most therapists will probably take a (7) _____ and (8) _____ history first. Some clients will need self-awareness exercises, but for most couples the first set of general exercises will be nondemand mutual pleasuring techniques called (9) _____. After completing these, the couple will then be assigned specific exercises.

Many therapists believe that the most common problem of couples seeking therapy is (10) _____. This refers to a persistent and pervasive inhibition of sexual desire. In its most extreme form, the avoidance of sex becomes phobic in nature and is called (11) _____. This generally requires psychosexual therapy. At the other extreme, hypersexual individuals are distinguished by the (12) _____ with which they engage in sex. A controversial term for these persons is (13) _____. Painful intercourse, or (14) _____, can occur in either men or women and can have a variety of causes, both organic and psychological.

The most common sexual problem specific to males is probably (15) _____, which is usually defined as a recurrent and persistent absence of (16) _____. Probably the most psychologically devastating male problem is (17) _____, which can have organic and/or psychological causes. The most common psychological cause is (18) _____. Difficulty reaching orgasm and ejaculating in a woman's vagina is called (19) _____ and usually has a psychological cause. Some men suffer from benign coital cephalalgia, or (20) _____, while others suffer from long-lasting (often painful) erections, a condition called (21) _____.

Some women experience involuntary contractions of the muscles surrounding the vaginal opening when they attempt intercourse, which results in pain. (22) _____, as it is called, usually has a psychological cause. The vast majority of women who go to sexual therapy do so because of problems (23) _____. This can be due to a variety of causes, including poor techniques by the partner, sexual repression during the woman's upbringing, and general relationship problems. When one member of a couple has a sexual problem, it is not unusual for the partner to develop a corresponding sexual difficulty.

Homosexuals can have many of the same sexual problems experienced by heterosexuals, but often experience additional problems that arise from dealing with (24) _____ and/or (25) _____.

CHAPTER 13   SELF-TEST

A. TRUE OR FALSE

T	F	
☐	☐	26. Bob wants to have sex twice a week but is in a relationship with Sue, who wants to have sex every day. In this example, it is clear that Sue is hypersexual.
☐	☐	27. Psychoanalysis has proven to be the most effective technique for treating most sexual problems.
☐	☐	28. The most frequent sexual problem for women seeking sex therapy is difficulty reaching orgasm.
☐	☐	29. Stimulation of the clitoris during intercourse should not be necessary, and a woman should be able to be orgasmic through intercourse alone.
☐	☐	30. Sex therapists generally treat sexual problems by having sexual relations with their clients.
☐	☐	31. Vaginismus is usually caused by psychological factors.
☐	☐	32. Performance anxiety can cause sexual problems in men, but not women.
☐	☐	33. The American Psychiatric Association defines a premature ejaculator as any male who usually reaches orgasm within 2 minutes of beginning sexual intercourse.
☐	☐	34. Most women who have difficulty reaching orgasm do not enjoy sex.
☐	☐	35. In order to help their clients, most sex therapists require them to stop masturbating.
☐	☐	36. Dyspareunia is usually caused by a physical problem.
☐	☐	37. Many men with psychologically caused erectile problems get full erections during sensate focus exercises.
☐	☐	38. Sexual problems in one partner in a relationship often lead to sexual problems in the other partner.
☐	☐	39. Masters and Johnson found that the success rate for treating sexual problems in homosexuals was generally higher than that for heterosexuals.
☐	☐	40. Female orgasm problems are always the result of insufficient stimulation.
☐	☐	41. Men and women often have different ideas about sex and affection.
☐	☐	42. If it feels good to you, it must feel good to your partner.
☐	☐	43. Most sex therapists today omit taking a medical history because most sex problems are psychologically caused.
☐	☐	44. Sexual surrogates are mostly former prostitutes.
☐	☐	45. Therapists are in agreement that hypersexuality is a form of addiction similar to addiction to alcohol or drugs.
☐	☐	46. Erectile problems are inevitable as men grow older.

T	F	47. A majority of erectile problems have a physical basis.
T	F	48. Headaches during orgasm occur mainly in men.
T	F	49. The sex therapy model presented in this chapter is applicable to most peoples of the world.

B. MATCHING: FOR EACH OF THE SEXUAL PROBLEMS BELOW, MATCH WHICH TECHNIQUES THERAPISTS WOULD BE LIKELY TO USE (IN ORDER) ACCORDING TO THE TEXT. ANSWERS MAY BE USED MORE THAN ONCE.

_____ 50. hypoactive sexual desire

_____ 51. hypersexuality

_____ 52. dyspareunia

_____ 53. psychological erectile problem

_____ 54. premature ejaculation

_____ 55. male orgasmic disorder

_____ 56. benign coital cephalalgia

_____ 57. vaginismus

_____ 58. female orgasmic disorder

a. sensate focus
b. psychosexual (or psycho-)therapy
c. Kegel exercises
d. bridge maneuver
e. use of dilators
f. resume female-on-top position of intercourse
g. self-exploration and masturbation
h. tease technique
i. treatment of the internal discomfort (antidepressants or stabilizing drugs)
j. treatment of specific organic cause (usually has organic cause)
k. squeeze technique or stop-start technique
l. relaxation techniques or medication for high blood pressure

C. FILL IN THE BLANK

59. When a man with no previous history of sexual problems starts to experience difficulty getting and/or maintaining an erection, it is called _____.

60. Most women need stimulation of the _____ in order to reach orgasm, even during intercourse.

61. Most therapists believe that the _____ position has the most erotic potential for both people during intercourse.

62. Sex therapists generally take medical histories of their clients because some sexual problems have a _____ cause.

63. When a person begins to observe and evaluate his or her own sexual responses during sex, that is called _____, and it can be a cause of sexual problems.

64. PLISSIT is an acronym for _____, _____, _____ and _____.

65. The therapy technique used to reduce anxiety by slowly introducing elements of the anxiety-producing stimulus is called _____.

66. Sensate focus exercises are _____ techniques.

67. Many therapists classify a person as having hypoactive sexual desire if they initiate an average of _____ sexual experiences a month.

68. With probably half or more of couples who seek therapy for sexual desire problems, one or the other partner has _____.

69. People who engage in sex compulsively are called _____.

70. In men, painful intercourse can be caused by the foreskin of the penis being too tight, a condition called _____.

71. In men, a life stress such as unemployment could lead to _____.

72. If a man is totally unable to ejaculate in a woman's vagina, this is called _____.

73. Vaginismus is usually caused by _____.

74. Therapists do not use the term "frigid" to describe women with orgasm problems because _____.

75. Tantrism emphasizes _____ during sex.

76. Probably the most difficult sexual problem to treat successfully is _____.

CHAPTER 14   *INTERACTIVE REVIEW*

This chapter considers three different ways to define a behavior as unconventional or abnormal. The (1) _____ approach defines a behavior as normal if most people do it. Therefore, unconventional behaviors are those that are engaged in by relatively few people. The (2) _____ approach looks at various behaviors to determine if they are customary within a given society. The (3) _____ approach evaluates the behaviors with respect to whether or not they cause an individual to feel distressed or guilty, and/or cause the person problems in functioning efficiently in ordinary social and occupational roles. What is considered unconventional depends on the time and (4) _____ in which the behavior is displayed.

Paraphiliacs are individuals whose sexual arousal and gratification (5) _____ on behaviors (or fantasizing about behaviors) that are outside the usual and accepted sexual patterns for adults. Unusual behaviors that are used to enhance sexual activity with a partner, not to compete with him or her, are called (6) _____. (7) _____ refers to achieving sexual arousal when using or fantasizing about an inanimate object or a particular part of the body. (8) _____ refers to achieving sexual arousal and gratification from dressing as a member of the opposite sex. (9) _____ expose their genitals compulsively in inappropriate settings, while (10) _____ repeatedly make obscene phone calls for sexual arousal. (11) _____ prefer to seek sexual gratification by watching people undress(ed) without their knowledge or consent. Zoophiliacs prefer (12) _____ as sexual partners; pedophiliacs prefer (13) _____. (14) _____ intentionally inflict pain and/or humiliation on others for sexual arousal and gratification, while (15) _____ obtain sexual pleasure from having pain and/or humiliation inflicted on them.

Other paraphilias include (16) _____ (sexual arousal by the act of urination), (17) _____ (arousal by feces), (18) _____ (arousal by filth), (19) _____ (arousal by being given an enema), (20) _____ (rubbing one's genitals against others in public places), and necrophilia (sex with (21) _____). Recent studies have found that people who practice paraphilias often engage in more than one type.

The large majority of paraphiliacs are (22) _____. Most describe themselves as heterosexual, but have difficulty maintaining intimate (particularly erotic) relations. The paraphiliacs' behavior(s) allows them to have control over the objects of their sexual arousal. Therapy is generally unsuccessful in eliminating the paraphilia, often because the individual does not want to give it up.

CHAPTER 14   *SELF-TEST*

A. TRUE OR FALSE

☐T ☐F 23. Freud considered oral-genital sex to be a sign of impaired psychosexual development.

☐T ☐F 24. All abnormal behaviors are called paraphilias.

☐T ☐F 25. Some of the more common fetishes involve leather, rubber, shoes, feet, and women's undergarments.

☐T ☐F 26. Most men who buy scanty, frilly panties for their female partners have a panty fetish.

☐T ☐F 27. Young children who occasionally dress up in clothing of the opposite sex usually grow up to be transvestites.

☐T ☐F 28. Most transvestites are heterosexual men and married.

☐T ☐F 29. Exhibitionists often behave as if they wish to be caught.

☐T ☐F 30. Strippers and nudists are the most common examples of exhibitionists.

☐T ☐F 31. The best way to react to an exhibitionist is to show disgust.

☐T ☐F 32. Most exhibitionists and voyeurs are harmless.

T	F	
T	F	33. Voyeurs often like to go to nudist colonies to observe naked bodies.
T	F	34. A person may commit bestiality and not have a paraphilia.
T	F	35. Many child molesters are not pedophiles.
T	F	36. Sadomasochists act out highly structured scenarios in which pain-inducing behaviors are more symbolic than real.
T	F	37. About half of all sadomasochists are women.
T	F	38. Many men involved in S&M prefer the submissive role, and are well-educated and successful.
T	F	39. Most paraphiliacs have difficulties with intimacy.
T	F	40. Most paraphiliacs wish to be cured.

B. MATCHING (MATCH THE BEHAVIOR WITH THE NAME OF THE PARAPHILIA THAT BEST FITS IT):

_____ 41. bestiality

_____ 42. coprophilia

_____ 43. exhibitionism

_____ 44. fetishism

_____ 45. frotteurism

_____ 46. klismaphilia

_____ 47. masochism

_____ 48. mysophilia

_____ 49. necrophilia

_____ 50. pedophilia

_____ 51. sadism

_____ 52. sexual variant

_____ 53. telephone scatologia

_____ 54. autoerotic asphyxiation

_____ 55. transvestism

_____ 56. urophilia

_____ 57. voyeurism

a. Billy must seek crowded places in order to get away with this paraphilia.

b. Henry is trying to find employment in a morgue in order to act out his paraphiliac fantasies.

c. Joe has volunteered to serve as counselor at the elementary school's summer camp to have access to the objects of his sexual arousal.

d. John is highly aroused by watching his partner have a bowel movement.

e. Jim has sneaked out to the barn to find a four-footed sexual partner.

f. Allen peeks in windows in order to become aroused.

g. Bert is hanging around the laundromat hoping to steal the object of his sexual desire from an unwashed basket of clothing.

h. Carl is crawling on his hands and knees and being pulled by a leash in order to get sexually aroused.

i. Stan asks his partner to urinate on him whenever they have sex.

j. Greg is wearing pink, frilly panties and a bra underneath his suit and tie.

k. Walter is waiting in his car in the grocery parking lot to show his penis to the first woman who walks by.

l. Susan gets highly aroused whenever she ties her partner up and orders him to lick her feet.

m. Sam gets highly aroused when his wife wears a negligee or teddy.

n. David is describing his erect penis to a woman in his class he has called but does not know.

o. Lenny is becoming highly aroused by his partner's nasal mucus.

p. Paul has paid a prostitute to lay him across her lap and give him an enema.

q. Larry puts a plastic bag over his head when he is highly sexually aroused.

C. FILL IN THE BLANK

58. The first classification of sexual deviations was made by _____.

59. Transvestism is most closely related to _____.

60. Mainstream S&M is more accurately described as _____ and _____.

61. The most frequently caught paraphiliac offenders are _____.

62. Telephone scatologia is most closely related to _____.

63. You should consider a voyeur to be potentially dangerous if he _____ or _____.

64. A variation of voyeurism in which a person repeatedly seeks sexual arousal by secretly watching sexual acts is called _____.

65. The act of having sex with an animal is called _____.

66. For pedophilia, sex researchers generally define a child as someone who is younger than _____.

67. Weinberg and others have found that the distinctive features in nonclinical sadomasochism are _____ and _____.

68. Coprophiliacs are aroused by _____ or the act of _____.

69. Urophilia, coprophilia, mysophilia, and klismaphilia involve sexual arousal by _____ and thus might be considered specific types of _____.

70. The objects of paraphiliacs' sexual arousal (e.g., a pair of used panties, a dead body, an animal, a young child) have in common that they are _____.

71. If a paraphiliac's inappropriate sexual urges can be controlled, then the most successful therapy is _____ combined with _____.

72. The _____ explanation for paraphilia is that it arises from society's expectation that men should dominate women.

73. According to Simon, deviance is a problem of _____, while perversion is a problem of _____.

CHAPTER 15 INTERACTIVE REVIEW

Historically, rape has been viewed as a violation of (1) _____ and victims were treated like "damaged goods." Even today, in the United States, many states will not prosecute a husband for raping his wife unless there is extreme physical harm to the victim.

There are about 100,000 reported cases of rape in the United States every year, but the actual number of rapes committed is many times greater. Most rapes are committed by (2) _____. Most victims are women between (3) _____ years of age, but anyone can be a victim of rape, including men. Rapists generally plan their crimes and seek victims who appear to be (4) _____.

Studies of convicted rapists reveal that most have (5) _____ IQs, have no more history of psychiatric illness than other criminals, are not oversexed, and have normal social skills. However, many were physically or sexually abused as children, or come from families in which they witnessed abuse of the mother by the father. Most rapists are (6) _____ men who repeat the crime and who lack internal controls and (7) _____.

Rape by a date or an acquaintance is no less real than rape by a stranger. Rape is rape. About 25 percent of all college women have been forced into sex when they did not want to have it. About 7 to 15 percent of all men admit to using physical force in order to have sex on a date, while two-thirds admit to using (8) _____ to have sex. Women, too, often use coercion to have sex, but the coercion used by females is generally (9) _____, while that used by males is generally (10) _____. Many adolescents and young adults believe that a man has the right to force a date to have sexual intercourse under some circumstances. Poor communication skills regarding sexual intentions also contribute to unwanted sexual activities.

Many rapes involve more than one assailant, and these (11) _____ rapes are usually more violent than rapes committed by individuals. Marital rape, too, is often accompanied by severe beatings and physical abuse. Most states have laws that make it illegal for an adult to have sex with anyone under the age of consent—this crime is generally called (12) _____. The mentally handicapped are often protected by these laws as well.

There are several explanations for rape. Psychodynamic theories explain rape as resulting from psychopathology. Studies of convicted rapists have found that rapists can be divided into three types, according to their motivation: (13) _____, (14) _____, and (15) _____. For these men, rape is not a crime of passion; it is a crime of violence and aggression. On the basis of what we now know of date rapists, some researchers have proposed a fourth type of rapist, the (16) _____ rapist, who has distorted cognitive processes and for whom the primary motivation is sex. Recent studies have shown that cultural factors, social learning, and interpersonal factors (a lack of shared meaning between men and women) also help explain why rape occurs. Rape is almost nonexistent in some societies, but is common in countries like the United States that promote and glorify (17) _____, and where large numbers of people conform to conservative stereotypic gender roles and men view relationships between the two sexes as (18) _____.

Many mistaken beliefs about rape are so widespread that they are called myths. This chapter considered four rape myths: (19) _____, (20) _____, (21) _____, and women frequently make false accusations of rape. None of these myths are true, but people who believe them tend to blame the victim for being raped.

Rape victims experience short- and long-term reactions in what is called (22) _____. In the acute phase, a victim's reactions may be either expressive or (23) _____. In the period of (24) _____, the victim attempts to regain control of her or his life. Long-term sexual problems are not uncommon. The reactions of a victim's partner and family are important to the recovery process, and it is crucial that they do not attribute blame to the victim.

If you are ever raped, it is important to report the crime immediately and get a medical exam as soon as possible. Do not wash, douche, or change clothes until the exam is completed. Today, most police forces have specially trained units, including counselors, to work with rape victims, and the U.S. Supreme Court has approved broad (25) _____ laws to protect victims from feeling as if they are on trial during court proceedings.

There are three basic approaches to rape prevention. One focuses on the rapists and advocates the creation of safer physical environments and tougher laws and punishments. Another focuses on potential victims and suggests educating women to be less vulnerable to assault. The third, a (26) _____ perspective, advocates changing society's values and its attitudes toward how men and women interact with one another.

Unwelcome sexual advances, requests for sexual favors, and other unwanted verbal or physical conduct of a sexual nature in the workplace or classroom is called (27) _____. The Supreme Court has ruled that this also includes the creation of a hostile or offensive work environment, and that a victim does not have to prove that sexual favors were demanded in exchange for job benefits. Companies and schools can be held liable for not taking immediate and appropriate corrective action.

We now know that sexual abuse of children is more widespread than previously believed. Most victims are molested by (28) _____. We can divide child molesters into two classifications: the (29) _____ molester, or pedophile, whose primary sexual orientation is to children, and the (30) _____ molester, whose primary sexual orientation is to adults. Pedophiles have been further classified according to their psychological characteristics as either (31) _____, (32) _____, or aggressive. Of pedophiles who have been caught, those who target girls average about 20 victims each, while those who target boys average nearly (33) _____ victims each. Heterosexual pedophiles outnumber homosexual pedophiles by 11 to 1. The effects of sexual abuse on children are often long-lasting, but there is no single "post-sexual-abuse-syndrome."

Sexual relations between relatives who are too closely related to marry is called incest. (34) _____ incest is five times more common than parent-child incest, although father-daughter and stepfather-stepdaughter incest account for 80 percent of all arrests. Incestuous fathers often appear to be shy and family-oriented publicly, but are dominating and authoritarian in their own homes. In these cases, the mothers are usually aware that the incest is going on. Because the abuse in father-child incest is usually frequent and long-term, the victims generally have a variety of long-lasting problems. It is important for children who have been sexually abused to receive counseling and support from professionals and organizations familiar with sexual abuse.

CHAPTER 15   SELF-TEST

A. TRUE OR FALSE

- [T] [F] 35. An adult woman cannot be raped if she truly does not want to be raped.
- [T] [F] 36. Rapists are generally young men who have a consenting sexual partner.
- [T] [F] 37. Groth et al. (1977) found that rape was the sexual expression of aggression, not the aggressive expression of sexuality.
- [T] [F] 38. Rape was originally considered to be a crime against a man's property.
- [T] [F] 39. In about half of the 50 states, a husband cannot be prosecuted for rape of his wife unless there is extraordinary violence.
- [T] [F] 40. The percentage of reported rapes that are false is lower than for most other crimes.
- [T] [F] 41. Anyone can be a victim of rape, including any man.
- [T] [F] 42. The majority of acquaintance rapes are reported to police, but never prosecuted.
- [T] [F] 43. If a woman willingly engages in necking or petting with a man, she cannot charge him with rape if he then forces her to have sexual intercourse.

T	F	44.	Rapists are more likely to initiate their crimes outdoors than indoors.
T	F	45.	Most rapists prefer victims who are feisty and who will try to physically resist.
T	F	46.	Most rapists believe that their victims want to have sex with them, or deserve to be raped.
T	F	47.	About 55 percent of all rapists are found to be psychotic.
T	F	48.	Date rape is not as real as rape by a stranger.
T	F	49.	Muehlenhard and colleagues found that over one-third of all college women had said "no" to a date when they really meant "yes" (to having sex).
T	F	50.	Well over half of all young adolescent boys and girls believe that a husband has the right to force his wife to have sexual intercourse against her will.
T	F	51.	Males who participate in gang rape tend to be younger than rapists who act alone.
T	F	52.	A man cannot be convicted of rape if the female agreed to have sexual intercourse.
T	F	53.	A man cannot be raped by a woman.
T	F	54.	Women who are raped usually have done something to provoke it.
T	F	55.	Many victims of rape do not show any emotional reactions afterwards.
T	F	56.	A man who tells sexual jokes, makes sexual references, and/or stares obsessively at a female co-worker cannot be prosecuted for sexual harassment unless he demands sexual favors.
T	F	57.	About 80 percent of all child molesters are known by their victims.
T	F	58.	Most pedophiles were themselves sexually abused as children.
T	F	59.	Pedophile is the term used by the American Psychiatric Association for all child molesters.
T	F	60.	Men who molest children generally begin their deviant behavior as teenagers.
T	F	61.	The Supreme Court has ruled that people being prosecuted for child sexual abuse have a constitutional right to face their accuser in court.
T	F	62.	Fathers who commit incest appear to most people to be quite average.
T	F	63.	In families with father-child incest, the large majority of mothers do not try to protect their children.
T	F	64.	Incestuous inbreeding (first cousins or closer in relation) is a universal taboo.
T	F	65.	Victims of sexual abuse tend to blame themselves.

B. MATCHING

_____ 66. aggressive pedophile

_____ 67. anger rapist

_____ 68. opportunistic rapist

_____ 69. personally immature pedophile

_____ 70. power rapist

_____ 71. preference molester

_____ 72. regressive pedophile

_____ 73. sadistic rapist

_____ 74. sexual harasser

_____ 75. situational molester

a. Herbert feels insecure and inadequate. He commits rape to reassure himself of his sexual adequacy, strength, and potency.

b. Frank's sexual urges are primarily toward adults, but he has fondled a child and had the child touch his penis. He considers his behavior abnormal.

c. Jim tells crude sexual jokes and makes sexual references to his female employees.

d. Steve has sex with prepubescent boys and beats them before and afterwards.

e. Joe is sexually aroused and impulsively forces his date to have sex with him when they are alone.

f. Albert loves it when his rape victims try to resist and enjoys hurting them for a prolonged time.

g. David has never had a successful relationship with a woman and seeks out children, whom he can control, for sexual gratification.

h. Mike has no interest in adults as sexual partners. His sexual orientation is to children only.

       i. Robert rapes because he despises all women, whom he blames for his lack of success.

       j. Tom has always had sexual relations with adults, but work and marital problems have led him to drink, and when he feels sexually inadequate, he molests a child.

C. FILL IN THE BLANKS

76. Mentally handicapped people are often protected against sexual abuse by the same laws that protect _____.

77. Russell and Howell (1983) estimated that a woman living in the United States had a one-in-_____ chance of being raped in her lifetime.

78. The majority of rapists had been _____ prior to the attack.

79. Give four examples of sexually coercive behavior: _____, _____, _____, and _____.

80. Give two characteristics of cultures that are "rape-free": _____ and _____.

81. In _____ rape, the individual responsibility for things a person does can be diffused or forgotten.

82. Statutory rape laws make it illegal to have sex with anyone _____.

83. A recent study found that _____ distinguishes the victims of male rape from other men.

84. Psychodynamic theories of rape explain it as resulting from _____.

85. The majority of date rapes are probably committed by _____ rapists.

86. Men who look at scenes of aggression toward women become _____.

87. Men who normally are not aroused by rape scenes are likely to become aroused by such scenes if _____ or _____.

88. A person who believes in rape myths is likely to _____ the victims.

89. In the _____ of the rape trauma syndrome, victims attempt to regain control of their lives.

90. Most large cities now have rape victim advocate programs that provide rape victims with _____.

91. In cases of rape where there is semen, rapists can be positively identified by _____.

92. A problem with approaches to preventing rape that focus on laws and punishment and teaching women how best to avoid rape is that these approaches _____.

93. Charges of sexual harassment are usually prosecuted as civil suits under Title VII of the _____ of 1964.

94. Sexual harassment usually occurs in a relationship where there is _____.

95. Research has shown that the more a male believes in _____, the more accepting he is of sexual harassment.

96. A man who repeatedly prefers children for achieving sexual excitement is called a _____.

97. In the previous question, "children" refers to persons under the age of _____.

98. The least common type of pedophile is the _____ pedophile.

99. The text suggests that the best way to minimize the chances of your child being a victim of sexual abuse is to educate _____ and educate _____.

100. In _____ families there are both multiple abusers and multiple victims, both across generations and within the same generation.

101. _____ incest is probably the least common type of incest.

102. In cases of father-child incest, a _____ approach is often chosen as therapy.

CHAPTER 16 *INTERACTIVE REVIEW*

In addition to laws that protect children and nonconsenting adults from sexual abuse, many states also have laws that attempt to regulate the sexual behaviors of consenting adults. Prostitution is illegal in all parts of the United States except for a few counties in (1) _____. In a recent survey, (2) _____ percent of the male respondents indicated that they had paid for sex on at least one occasion. Female prostitutes can be divided into three types according to status: (3) _____ who solicit in public, brothel prostitutes and (4) _____, and (5) _____, who generally work out of their own apartments. Male prostitutes who cater to homosexuals are called (6) _____. The primary reason most individuals engage in prostitution is (7) _____. Most prostitutes were sexually or physically abused as children and were loners. Most customers are white, (8) _____, and (9) _____. Opponents of prostitution argue that it spreads sexually transmitted diseases and is often associated with other crimes. Proponents of legalization argue that licensing and required health checks would reduce both STDs (including HIV infection) and the influence of organized crime.

The sale of sexually explicit material is a multi-billion-dollar-a-year business. This could not have happened, of course, if there were not a large demand for such material. According to surveys, about (10) _____ of all Americans have seen an X-rated movie or video in the past year. The President's Commission of 1970 and subsequent researchers have generally concluded that exposure to nonviolent sexually explicit material has (11) _____ effect on individuals. Subsequent research has shown that many men are aroused by sexually explicit material depicting violence toward women. Exposure to sexually explicit material with violent themes often results in men becoming (12) _____ toward rape victims, having greater belief in (13) _____, displaying (14) _____ to women, and being more likely to say that they would (15) _____. However, behavioral scientists have concluded that it is the (16) _____ in this material, not the (17) _____, that is potentially harmful.

Most people find some types of sexually explicit material offensive, but attempts to censor or ban this material raise questions about First Amendment rights. What is (18) _____ to one person is pornography to another, and who is to decide where to draw the line? The one exception is sexually explicit material involving (19) _____, for which persons can be prosecuted on the basis of abuse.

Several states have sodomy laws that prohibit (20) _____ and/or (21) _____ sex between consenting adults. Many people believe that they should have the right to decide for themselves what to do in privacy, but the Supreme Court has upheld the constitutionality of these laws.

Additional topics on the legality of sexual behavior are included in Chapters 7 ("Pregnancy and Childbirth"; see the section on surrogate mothers), 9 ("Sexual Orientation"), and 15 ("Sexual Victimization").

CHAPTER 16 *SELF-TEST*

A. TRUE OR FALSE

T F 22. Most people who become prostitutes do so because of abnormal sexual needs.

T F 23. Kinsey found that in the 1940s, a majority of white males had had sex with a prostitute.

T F 24. When exposed to sexually explicit material, most people show an increase in sexual behavior.

T F 25. The 1970 President's Commission concluded that there was no significant link between exposure of adults to sexually explicit materials and sex crimes.

T F 26. In many states, a married couple caught having oral sex in the privacy of their own home can be sent to prison.

T F 27. Prostitution is the only sexual offense for which more women than men are prosecuted.

T F 28. Erotica and pornography are synonyms, and are really the same thing.

T F 29. When Denmark legalized the sale of hard-core pornographic material to adults, there was a marked decrease in sex crimes afterward.

T	F	30.	People are most aroused by looking at sexually explicit material showing behaviors they regard as normal.
T	F	31.	Adultery is a crime in nearly half of the 50 states.
T	F	32.	Prostitution was uncommon during the Victorian era.
T	F	33.	Men who use male prostitutes usually pay to give oral sex to the prostitute.
T	F	34.	Many prostitutes enjoy sex with their customers.
T	F	35.	HIV infection is uncommon among prostitutes, probably because most have their customers use condoms.
T	F	36.	Nearly all junior high school boys and girls have looked at a sexually explicit magazine.
T	F	37.	Many people use sexually explicit material as a source of sexual information.
T	F	38.	A majority of both men and women are aroused by sexually explicit material.
T	F	39.	There is substantial evidence that sex offenders have had more exposure to sexually explicit material than have nonoffenders.
T	F	40.	Exposure to violent sexually explicit material often results in women adopting attitudes that are more supportive of sexual violence against women.
T	F	41.	There is more violence against women shown in X-rated movies than in R-rated movies.
T	F	42.	It is a crime to possess sexually explicit material (e.g., magazines, videos, photos) showing children.
T	F	43.	St. Augustine and St. Thomas Aquinas believed prostitution was necessary.
T	F	44.	Worldwide, one million children are drawn into prostitution every year.

B. MATCHING

_____ 45. B-girl

_____ 46. brothel prostitute

_____ 47. call girl

_____ 48. gigolo

_____ 49. hustler

_____ 50. john

_____ 51. not a prostitute

_____ 52. pimp

_____ 53. streetwalker

a. Susan has sex for money at a legalized house in Nevada.

b. Joan has sex with her boss in the hope that it will get her a promotion.

c. Hank works out of an escort service and is paid by older women to have sex.

d. Stephanie works out of her own apartment, and her clients must make an appointment to have sex with her.

e. David takes 75 percent of the money earned by six women who do tricks on the west side of town.

f. Joe pays women to give him oral-genital sex.

g. Sam accepts money from men on the street who want to have sex with him.

h. Sally solicits for sex in a downtown bar.

i. Bonnie solicits men in public.

C. FILL IN THE BLANKS

54. The feature that distinguishes a prostitute from other women who engage in illicit sexual intercourse is the _____ with which she offers herself to men for hire.

55. Sexual intercourse between a man and a woman who are not married is called _____.

56. A hustler is on the same status level as a female _____.

57. Pornography is distinguished from erotica in that it is literature or art with a sexual theme that is considered to be _____.

58. Laws that prohibit oral or anal sex are called _____ laws.

59. Recent evidence suggests that it is exposure to _____, more than just sexual explicitness, that results in potentially harmful attitude and behavioral changes.

60. In *Miller v. California,* 1973, the Supreme Court ruled that material was obscene if (a) _____, (b) _____, and (c) _____.

61. Most teenaged prostitutes are _____ children who have turned their first trick by the time they are 14.

62. The _____, passed in 1873, made the mailing of material considered to be obscene a felony.

63. Probably the foremost reason many people use sexually explicit material is that _____.

64. With regard to pornography, men and women rate themes in which there is _____ as the most degrading.

CHAPTER 17   INTERACTIVE REVIEW

Communication is the exchange of information. Because of our culture's negative attitudes about sex and the negativity communicated by most parents, many people feel uncomfortable talking about sex, even with their sexual partners. (1) _____ and power differences in a relationship also contribute to this difficulty.

You and your partner can get used to talking about sex together by discussing sexuality-related articles in newspapers, books, and magazines. One of the first things you must do is agree on a comfortable (2) _____ to use when talking about sex. When talking about sexual differences or problems, be sure to emphasize the (3) _____ rather than the (4) _____ things that your partner does. That doesn't mean that you can never complain, but when you do, focus on your partner's behavior rather than on his or her character.

Take responsibility for your own pleasure by expressing your feelings and desires to your partner in a clear, specific manner. Whenever possible, begin sentences with "I" rather than (5) _____, (6) _____, or (7) _____. It is often easier to find out about your partner's sexual desires and needs if you first (8) _____. However, good communication is a two-way street; it requires that you also become a good (9) _____. In addition to communicating verbally, we also communicate nonverbally with [give four examples] (10) _____, (11) _____, (12) _____, and (13) _____. Even with good communication skills, people will not always agree, but it is possible to agree that you disagree.

It is equally important for parents to learn to communicate with their children about sexuality in a positive manner. Your children will learn about sex whether or not you ever talk to them about it. (14) _____ begins in infancy, and children receive it from their peers, the media, and your own behaviors. Children who have had sex education are (15) _____ likely to have sexual relations than those who have not. Avoid that one long "birds and bees talk" approach; communication between parents and their children about sexuality should begin (16) _____ and be ongoing. This chapter presents guidelines for age-appropriate discussions about sexuality-related topics.

Discussions with your children about sexuality should be frank and explicit. You should strive to create an atmosphere of love and caring, so avoid (17) _____ and do not lecture to them. You can emphasize your own sexual values, but do not preach. Your own behavior will affect your child's attitudes and moral values, so don't let them overhear you use "dirty" language or tell "dirty" jokes—and teach them, by example, that affection is something positive. The only way you will be able to know whether or not your attempts to educate your children about sexuality have been successful is by their (18)_____.

CHAPTER 17   SELF-TEST

A. TRUE OR FALSE

| T | F | 19. | You can prevent your children from learning about sex by never talking to them about it. |

19. You can prevent your children from learning about sex by never talking to them about it.

20. Most children can be adequately taught about sexuality with one long, serious talk.

21. Parents should teach the words penis and vagina to their children at the same time they teach them the names for their other body parts.

22. Young adolescent males should be taught about menstruation.

23. When discussing sex with their children, parents should insist that the children reveal their sexual attitudes and behavior.

T	F	24. In order to dissuade your child from engaging in sex, you should emphasize the possible negative consequences (e.g., AIDS).
T	F	25. It is okay to teach your children about values while giving them factual information.
T	F	26. You should avoid open displays of affection with your partner (e.g., hugging or kissing) when your children are present.
T	F	27. Your own sexual behavior and attitudes will likely serve as a role model for your children, whether or not you discuss sexuality with them.
T	F	28. Generally speaking, men are naturally more "expert" at sex than are women.
T	F	29. When talking to your partner about sex, use of the term "making love" is always better than saying "having sex" or "sexual intercourse."
T	F	30. The best time to talk about a sexual problem or difference with your partner is immediately after sex.
T	F	31. The best place to discuss a sexual problem or difference with your partner is in bed.
T	F	32. When you have a sexual difference or problem with your partner, you should never complain or express your anger.
T	F	33. When talking to your partner about your sexual attitudes and preferences, it is good to begin sentences with "we" because it emphasizes the relationship.
T	F	34. Giving your partner a massage can strengthen communication by making you a better listener.
T	F	35. When approaching your partner about differences in sexual attitudes or preferences, it is best to present all your criticisms at once.
T	F	36. In most cases, when parents believe that they have had meaningful discussions with their children about sex, the children also recognize the talks as meaningful.

B. MATCHING: AT WHAT AGE (GRADE LEVEL) SHOULD YOU BEGIN TO TALK TO YOUR CHILD ABOUT THE FOLLOWING TOPICS?

_____ 37. the changes that occur in boys and girls at puberty

_____ 38. correct names for genitals

_____ 39. the idea that human sexuality is natural and that sexual feelings are normal

_____ 40. peer pressure and the right to say "no"

_____ 41. an explanation of AIDS and how it is transmitted

_____ 42. promotion of nonstereotyped gender roles

_____ 43. basics of reproduction and the role of our genitals and reproductive organs in enabling it

_____ 44. integration of sexuality into an individual's value system

_____ 45. awareness of exploitive relations

_____ 46. how to express sexual feelings in appropriate ways

_____ 47. the fact that certain body parts feel good when touched

a. preschool

b. early elementary

c. upper elementary

d. junior high

e. high school

C. FILL IN THE BLANKS

48. When expressing your desires to your partner, you should try to begin sentences with _____.

49. The word intercourse means _____.

50. When communicating with another, it is important that there be agreement between the _____ and _____ aspects of communication.

51. When answering your child's questions about sex, be sure to make your answers _____-appropriate.

52. If you must criticize your partner, make sure you criticize his or her _____, not his or her _____.

53. With regards to sexual relations, you must take responsibility for _____.

54. If you want your child to actively engage in your attempts to discuss sex, you should take a _____ approach.

55. Your success in imposing your values upon your children will depend, in large part, on _____.

ANSWERS—CHAPTER 1

1. their friends
2. the media
3. 85–90
4. sexuality
5. the biblical Hebrews
6. Christians
7. St. Paul
8. St. Augustine
9. Queen Victoria
10. penicillin
11. the birth control pill and IUD
12. Sigmund Freud
13. Henry Havelock Ellis
14. Alfred Kinsey
15. Masters and Johnson
16. samples

17. correlational studies
18. direct observation
19. case studies
20. experimental research
21. true
22. false
23. false
24. false
25. false
26. false
27. false
28. false
29. true
30. true
31. false
32. true
33. false
34. false
35. false

36. f
37. p
38. g
39. n
40. h
41. m
42. i
43. d
44. o
45. c
46. a
47. j
48. q
49. g
50. i
51. l
52. m
53. b
54. c
55. o
56. f

57. n
58. a
59. j
60. g
61. k
62. h
63. e
64. d
65. Inis Baeg
66. leisure time, mobility, birth control, antibiotics
67. each possible sample of that size
68. Plato
69. Puritans
70. freedom of belief
71. body, soul
72. libido
73. spermatorrhea

ANSWERS—CHAPTER 2

1. ambivalent
2. vulva
3. mons veneris
4. labia majora
5. labia minora
6. clitoris
7. vaginal opening
8. urethral opening
9. mons veneris
10. outer surfaces of the labia majora
11. clitoris
12. penis
13. labia minora
14. vestibular area
15. hymen
16. poor
17. eight
18. vagina
19. uterus
20. Fallopian tubes
21. ovaries

22. vagina
23. vagina
24. Grafenberg (G) spot
25. ovary
26. Fallopian tube
27. uterus
28. endometrium
29. clitoris
30. vaginal opening
31. cancer of the cervix
32. scrotum
33. testicles
34. corpora cavernosa
35. corpus spongiosum
36. glans
37. foreskin
38. circumcision
39. testicles
40. testosterone

41. epididymis
42. vas deferens
43. ejaculatory ducts
44. urethra
45. prostate gland
46. seminal vesicles
47. semen
48. Cowper's glands
49. testicles
50. prostate gland
51. true
52. true
53. false
54. true
55. false
56. false
57. true
58. false
59. false
60. false
61. true
62. true

63. false
64. false
65. false
66. false
67. true
68. false
69. false
70. false
71. s
72. c
73. d
74. e
75. h
76. a
77. f
78. b
79. i
80. p
81. k
82. l
83. m
84. n

85. g
86. j
87. o
88. q
89. r
90. t
91. Pap smear; cervical
92. prostate gland

93. labia majora
94. endometrium
95. blood
96. Cowper's glands
97. mons veneris
98. amount of fatty tissue
99. just after menstruation

100. scrotum
101. ejaculatory ducts
102. seminal vesicles
103. Fallopian tubes
104. after a warm bath or shower
105. Bartholin's glands
106. seminiferous tubules

107. cells of Leydig
108. clitoris
109. pubococcygeus (PC) muscle
110. front
111. testicular torsion
112. crura
113. prolactin
114. oxytocin

ANSWERS—CHAPTER 3

1. endocrine
2. testosterone
3. estrogen
4. progesterone
5. pituitary gland
6. 28
7. follicle-stimulating hormone (FSH)
8. follicle
9. follicular (or proliferative)
10. endometrium
11. FSH
12. luteinizing hormone (LH)
13. LH
14. ovulation
15. abdominal cavity
16. Fallopian tube
17. luteal (or secretory)
18. corpus luteum
19. endometrium
20. endometrium
21. menstruation
22. at least 8 days
23. estrous
24. estrus (ovulation)
25. taboos
26. shredded endometrial tissue
27. cervical mucus

28. blood
29. sperm production
30. luteinizing hormone
31. male hormones (e.g., testosterone)
32. high blood pressure; liver, prostate, and breast tumors; impaired reproductive function; masculinization in women; emotional or psychological problems
33. amenorrhea
34. 3 to 14
35. premenstrual syndrome (PMS)
36. with the start of menstruation
37. dysmenorrhea
38. prostaglandins
39. endometrial tissue grows outside the uterus
40. toxic shock syndrome
41. testosterone
42. estrogen
43. progesterone

44. testosterone
45. false
46. false
47. false
48. false (28 days is an average)
49. true
50. false
51. true
52. true
53. false
54. false (there is no strong eivdence for this)
55. false
56. true
57. false
58. true
59. true
60. true
61. false
62. false
63. h
64. d
65. n
66. j
67. l
68. k
69. b
70. m
71. f
72. o

73. e
74. p
75. i
76. a
77. c
78. g
79. pituitary
80. pheromones
81. maturation of a follicle
82. the menstrual cycle
83. follicle
84. corpus luteum
85. progesterone
86. menarche
87. orchiectomy
88. brain
89. testosterone
90. toxins produced by a bacterium (*Staphylococcus aureus*)
91. human chorionic gonadotropin (HCG)
92. gonadotropin-releasing hormone (GnRH)
93. progesterone
94. testosterone
95. inhibin

ANSWERS—CHAPTER 4

1. Masters and Johnson
2. more similar
3. excitement
4. plateau
5. orgasm
6. resolution

7. sexual response cycle
8. Helen Kaplan
9. desire
10. vasocongestive
11. penile erection
12. vaginal lubrication

13. plateau
14. orgasmic platform
15. pulls back against the pubic bone and disappears beneath the clitoral hood

16. 50 to 75
17. sex-tension flush
18. rhythmic muscular contractions
19. the brain
20. similar
21. emission

22. expulsion
23. ejaculation
24. refractory period
25. multiple orgasms
26. resolution
27. 30 to 50
28. clitoral
29. clitoral
30. vaginal
31. clitoris
32. front
33. Grafenberg (G) spot
34. emission of fluid
35. prostate gland
36. not important
37. aphrodisiacs

38. clitoridectomy
39. infibulation
40. true
41. true
42. false
43. true
44. false (not necessarily)
45. true
46. false
47. true
48. false
49. true
50. false
51. false
52. true
53. true

54. true
55. false
56. true
57. false
58. true
59. false
60. true
61. true
62. false
63. d
64. b, g, j, k, l, n, s
65. a, c, f, i, q, r, t
66. e, p, u
67. h, m, o
68. b, c, g, i, k, q, t
69. vasocongestion, myotonia

70. semen
71. Freud
72. urethra
73. sex-tension flush
74. anaphrodisiacs
75. myotonia
76. emission
77. retrograde ejaculation
78. orgasmic platform
79. Grafenberg (G) spot
80. refractory period
81. labia minora
82. the tenting type

ANSWERS—CHAPTER 5

1. four
2. bacteria
3. viruses
4. parasites
5. gonorrhea
6. chlamydia
7. syphilis
8. gonorrhea
9. chlamydia
10. pelvic inflammatory disease (PID)
11. syphilis
12. antibiotics
13. skin, skin
14. recurrent
15. liver
16. feces
17. blood or body fluids
18. human papilloma
19. human immunodeficiency virus (HIV)
20. CD4+
21. immune
22. opportunistic
23. 10
24. intimate sexual activity
25. blood
26. casual contact
27. men and women

28. infestations of parasites
29. trichomoniasis
30. monilial
31. bacterial vaginosis
32. false
33. true
34. true
35. false
36. true
37. false
38. false
39. true
40. true
41. false
42. false
43. true
44. true
45. false
46. true
47. false
48. false
49. false
50. true
51. true
52. false
53. true
54. false
55. false
56. false
57. false
58. false
59. false

60. false
61. true
62. true
63. true
64. true
65. false
66. false
67. true
68. true
69. true
70. a
71. l
72. c
73. q
74. h
75. f
76. o
77. i
78. m
79. j
80. e
81. n
82. g
83. b
84. d
85. p
86. k
87. parasites
88. vaginitis
89. pelvic inflammatory disease
90. human papillomavirus infection

91. human immunodeficiency virus (HIV) infection
92. types of vaginitis
93. hepatitis
94. human papillomavirus infection
95. acyclovir
96. ectopic pregnancy, infertility
97. syphilis, human immunodeficiency virus, hepatitis
98. stress
99. mucous membranes
100. nongonococcal (or nonspecific) urethritis
101. lymphogranuloma venereum
102. trichomoniasis
103. yeast (moniliasis)
104. pinworms
105. chlamydia
106. CD4+ cells
107. during the first 60 days and late in the infection (AIDS)
108. HIV antibodies
109. gonorrhea, chla-

mydia, syphilis (the chancre is often internal), HPV infection,

HIV infection
110. HPV infection, HIV infection, trichomoniasis

111. always use condoms, abstain from sex until you are in a long-term

monogamous relationship
112. protease inhibitors

ANSWERS—CHAPTER 6

1. over twice
2. calendar
3. basal body temperature
4. Billings
5. ovulation
6. spermicides
7. the diaphragm
8. the cervical cap
9. condoms
10. uterus
11. offer some protection against sexually transmitted diseases
12. estrogen
13. progesterone
14. Depo-Provera
15. Norplant
16. vas deferens
17. tubal ligation
18. laparoscopy
19. false
20. false
21. true
22. false
23. false
24. true
25. true

26. true
27. false
28. false
29. true
30. false
31. false
32. true
33. true
34. false
35. false
36. false
37. false
38. true
39. false
40. true
41. true
42. true
43. false
44. true
45. true
46. true
47. true
48. true
49. false
50. true
51. c
52. i
53. k
54. d

55. b, e
56. h
57. h
58. e
59. e
60. n
61. a, n
62. c, j, m
63. j
64. j
65. l
66. f
67. f
68. consistently, properly
69. the Cowper's glands
70. he has had 12 to 16 ejaculations
71. menstruation, ovulation
72. on demand
73. cervical cap
74. 7, 20
75. rises slightly
76. clear and slippery
77. nonoxynol-9
78. oil-based

79. in a stable monogamous relationship who have completed their families
80. male condom, female condom
81. many types of antibiotics, barbiturates, analgesics, and tranquilizers
82. menstrual irregularities
83. male condom, female condom, diaphragm with spermicide, spermicides
84. male
85. Ovral ("morning after" pill)
86. male condom with spermicide
87. the hormonal methods
88. 5 days
89. to abstain from sex

ANSWERS—CHAPTER 7

1. 5
2. ovulation
3. one
4. Fallopian tube
5. endometrium
6. ectopic
7. chlamydia and/or gonorrhea
8. human chorionic gonadotropin (HCG)
9. trimesters
10. embryo
11. 2
12. fetus

13. 7
14. quickening
15. fourth or fifth
16. teratogens
17. amniocentesis
18. fear-tension-pain
19. midwife
20. contractions
21. dilation
22. start-up
23. amniotic sac
24. transition phase
25. fully dilated
26. head
27. umbilical cord

28. placenta
29. cesarean (C-section)
30. proteins
31. postpartum blues or depression
32. 6
33. the sharing of child-care and housekeeping responsibilities
34. 12
35. 40
36. 40
37. 20

38. a low sperm count
39. artificial insemination
40. Fallopian tubes
41. ovulate
42. in vitro fertilization
43. zygote intrafallopian transfer
44. gamete intrafallopian transfer
45. false
46. false
47. true
48. false

49. true
50. true
51. false
52. false
53. false
54. true
55. false
56. true
57. false
58. false
59. true
60. true
61. false
62. false
63. true
64. true
65. false
66. true

67. true
68. false
69. false
70. true
71. d
72. j
73. f
74. l
75. c
76. b
77. h
78. k
79. m
80. i
81. a
82. g
83. e
84. d

85. morula, blastocyst
86. umbilical cord
87. preeclampsia
88. lightening
89. contractions are coming 10 minutes apart (or sooner) on a regular basis; contractions are at least 30 seconds long on a regular basis; cervix is dilated to at least two to three centimeters; cervix is at least 70 percent effaced
90. first

91. tubal
92. 25 to 35
93. blastocyst
94. ultrasound
95. the head
96. end of the first (beginning of the second)
97. 24 hours
98. amnion
99. episiotomies, prepping, cesarean sections
100. 23 to 24 weeks
101. smokes
102. Bradley

ANSWERS—CHAPTER 8

1. gender identity
2. XX
3. XY
4. Klinefelter's
5. Turner's
6. testosterone
7. female
8. gender dysphoria
9. gender identity disorder (transsexuals)
10. Freudian
11. learning
12. gender constancy
13. 6 or 7
14. gender roles
15. individualist
16. microstructural
17. independent
18. androgyny
19. schemas
20. androgynous
21. true
22. false

23. true
24. true
25. true
26. false
27. true
28. true
29. true
30. false
31. true
32. true
33. true
34. true
35. true
36. false
37. false
38. true
39. true
40. true
41. true
42. false
43. false
44. true
45. e

46. h
47. n
48. o
49. b, j
50. b, d, j
51. c
52. f, l, p
53. m
54. a, j
55. k
56. i
57. g
58. gender constancy
59. 1800s (when the United States and other countries became industrialized)
60. microstructural
61. identification
62. socialization
63. stereotypes
64. undifferentiated

65. Mullerian
66. phallic
67. androgynous
68. testosterone
69. Oedipus
70. transvestites; transsexuals
71. imitation
72. biological determinism
73. expressive
74. 2
75. stereotyped
76. gender schema
77. gender labels (stereotypes); individuating
78. androgynous
79. transcendent
80. casual sex
81. two-spirit
82. gender-role identity

ANSWERS—CHAPTER 9

1. distinct preferences
2. 3 to 5
3. one
4. 2, 3, or 4
5. gender identities
6. roles
7. Oedipus

8. domineering, overprotective
9. weak, detached
10. absent
11. cold, rejecting
12. social learning theory

13. Sambian
14. genetic
15. brains
16. hormones
17. interact
18. more
19. ancient Greeks

20. Romans
21. Thomas Aquinas
22. coming out
23. admitting to oneself that one has a homosexual orientation

24. homophobia
25. commitment and romantic love
26. true
27. false
28. true
29. true
30. false
31. false
32. false
33. false
34. false
35. false
36. true
37. true
38. true
39. true
40. false
41. false
42. false
43. false
44. false
45. false
46. false
47. true
48. false
49. false
50. true
51. true
52. true
53. false
54. false
55. true
56. false
57. a
58. a
59. b
60. e
61. d
62. b
63. e
64. c
65. a
66. e
67. d
68. b
69. tribadism
70. adolescence
71. "tearoom trade"
72. Freud, Henry Havelock Ellis
73. latent homosexuals
74. hypothalamus
75. before or shortly after birth
76. predispose
77. straights
78. admitting a homosexual orientation to oneself; getting to know other homosexuals; telling family and friends of one's sexual orientation; complete openness about one's homosexuality
79. close-coupled
80. AIDS
81. Socrates, Leonardo da Vinci, Michaelangelo, Gertrude Stein, Walt Whitman
82. temperaments

ANSWERS—CHAPTER 10

1. 3 through 5
2. modesty (and inhibitions about exposing their bodies in public)
3. puberty
4. reproduction
5. hormone
6. self-identity
7. necking
8. petting
9. 17
10. peer pressure
11. serial monogamy
12. monogamy
13. lasting relationship
14. chastity
15. economic and physical protection
16. mid-twenties to mid-thirties
17. menopause
18. vaginal lubrication
19. estrogen
20. testosterone
21. refractory
22. their sexual activity when they were younger
23. true
24. false
25. true
26. false
27. true
28. false
29. true
30. false
31. true
32. false
33. true
34. true
35. false
36. false
37. false
38. true
39. true
40. false
41. true
42. true
43. false
44. false
45. false
46. false
47. true
48. true
49. false
50. true
51. false
52. true
53. true
54. false (true for men only)
55. false
56. false
57. true
58. false
59. a
60. b
61. e (caused by b)
62. c
63. a
64. c
65. e
66. d
67. a
68. d
69. d
70. f, c (from DHEA)
71. lack of a partner
72. precocious puberty
73. egocentric
74. the climacteric
75. self-exploration
76. phallic
77. secondary sex characteristics
78. nocturnal emission
79. emotional intimacy
80. test their virility
81. have multiple wives
82. emotional
83. opportunity, alienation
84. declines only slightly
85. adrenarche, gonadarche
86. delayed puberty

ANSWERS—CHAPTER 11

1. range
2. Kinsey
3. 92
4. 62
5. nocturnal orgasms
6. REM
7. have erections
8. vaginal lubrication

9. romantic and emotional
10. male
11. missionary
12. fellatio
13. cunnilingus
14. sexually healthy
15. false
16. true
17. false
18. false
19. true
20. false
21. false
22. false
23. true
24. false
25. true
26. true
27. false (not necessarily)
28. false
29. true
30. true
31. false
32. false
33. true
34. true
35. true
36. c, k, n
37. f, h, i, j, m
38. a, b, d, e, g, l
39. cunnilingus
40. vaginal lubrication, penile erection
41. rectal bacteria
42. the replacement fantasy
43. variety
44. ritualized
45. vaginal intercourse, watching one's partner undress
46. a sexually active lifestyle
47. face-to-face
48. female-on-top
49. gonorrhea
50. feels comfortable with his or her sexuality; feels free to choose whether or not he or she wishes to try a variety of sexual behaviors

ANSWERS—CHAPTER 12

1. fascination (preoccupation with the other person)
2. exclusiveness
3. sexual desire
4. cognitive component
5. companionate
6. a positive self-concept
7. self-disclosure
8. jealousy
9. self-esteem
10. popularity, wealth, fame, and physical attractiveness
11. conditional
12. unconditional
13. liking
14. romantic love
15. companionate love
16. consummate love
17. eros
18. storge
19. pragma
20. mania
21. ludus
22. agape
23. matches
24. intimacy
25. true
26. true
27. false
28. false
29. false
30. true
31. true
32. false
33. false
34. true
35. false
36. false
37. false
38. true
39. true
40. false
41. false (true for men only)
42. false
43. false
44. true
45. false
46. true
47. e
48. h
49. n
50. b
51. m
52. g
53. a
54. d
55. k
56. i
57. l
58. o
59. j
60. c
61. f
62. ludus
63. storge
64. mania
65. agape
66. passion, intimacy, and commitment
67. repression of sexuality during childhood
68. reciprocation
69. dopamine, norepinephrine, and phenylethylamine
70. realistic
71. secure
72. experience jealousy
73. emotional
74. limerence; mania
75. storgic eros
76. (a) both individuals accepting themselves as they are, (b) each individual recognizing his or her partner for what they are, (c) each individual feeling comfortable to express himself or herself, (d) learning to deal with the partner's reactions.
77. habituation
78. sexual attraction/desire

ANSWERS—CHAPTER 13

1. at least half
2. couple's
3. help people with sexual problems
4. have sex with a client
5. cognitive-behavioral therapy
6. PLISSIT
7. medical
8. sexual
9. sensate focus exercises
10. hypoactive sexual desire
11. sexual aversion
12. compulsiveness
13. sexual addicts
14. dyspareunia
15. premature ejaculation
16. reasonable voluntary control

17. erectile disorder
18. performance anxiety
19. orgasmic disorder
20. headaches after orgasm
21. priapism
22. vaginismus
23. reaching orgasm
24. HIV/AIDS
25. homophobic attitudes within society
26. false
27. false
28. true
29. false (not necessarily)
30. false
31. true
32. false
33. false
34. false
35. false
36. true
37. true
38. true
39. true
40. false
41. true
42. false
43. false
44. false
45. false
46. false
47. true
48. true
49. false
50. a, b
51. i, b
52. j
53. a, h, f
54. a, k, f
55. g, d
56. l
57. a, e
58. g, a, f (possibly c as well)
59. secondary erectile disorder
60. clitoris
61. female-on-top
62. physical
63. spectatoring
64. permission, limited information, specific suggestions, intensive therapy
65. systematic desensitization
66. nondemand mutual pleasuring
67. two or fewer
68. a coexisting sexual problem
69. hypersexual individuals (some refer to them as sexual addicts)
70. phimosis
71. erectile disorder
72. ejaculatory incompetence
73. psychological problems
74. it has negative connotations about a woman's emotional responsivity
75. spiritual union
76. hypoactive sexual desire

ANSWERS—CHAPTER 14

1. statistical
2. sociological
3. psychological
4. place
5. depend almost exclusively
6. sexual variants
7. fetishism
8. transvestism
9. exhibitionists
10. telephone scatologists
11. voyeurs
12. animals
13. prepubertal children
14. sadists
15. masochists
16. urophilia
17. coprophilia
18. mysophilia
19. klismaphilia
20. frotteurism
21. dead bodies
22. males
23. true
24. false
25. true
26. false
27. false
28. true
29. true
30. false
31. false
32. true
33. false
34. true
35. true
36. true
37. false
38. true
39. true
40. false
41. e
42. d
43. k
44. g
45. a
46. p
47. h
48. o
49. b
50. c
51. l
52. m
53. n
54. q
55. j
56. i
57. f
58. Richard von Krafft-Ebing
59. fetishism
60. domination, submission
61. exhibitionists
62. exhibitionism
63. tries to draw attention to the fact
that he is watching; attempts to approach you or enters your building
64. scoptophilia
65. bestiality
66. 13 years old
67. domination, submission
68. feces, defecation
69. bodily discharges, fetishes
70. quiet, passive, and nonrejecting
71. desensitization, social skills training
72. social systems perspective
73. control, desire

ANSWERS—CHAPTER 15

1. property
2. someone the victim knows
3. 16 and 24
4. vulnerable
5. average
6. young
7. empathy
8. coercion
9. verbal
10. physical
11. gang
12. statutory rape
13. power
14. anger
15. sadism

16. opportunistic
17. violence by men
18. adversarial
19. women who are raped usually provoke it by their dress and behavior
20. women subconsciously want to be raped
21. no woman can be raped if she truly does not want it
22. rape trauma syndrome
23. controlled
24. long-term reorganization
25. rape shield
26. social systems
27. sexual harassment
28. someone they know
29. preference
30. situational
31. personally immature
32. regressive
33. 200
34. brother-sister
35. false
36. true
37. true
38. true
39. true
40. true
41. true
42. false
43. false
44. false
45. false
46. true
47. false
48. false
49. true
50. true
51. true
52. false
53. false
54. false
55. true
56. false
57. true
58. true
59. false
60. true
61. false
62. true
63. true
64. false
65. true
66. d
67. i
68. e
69. g
70. a
71. h
72. j
73. f
74. c
75. b
76. juveniles (persons under the legal "age of consent")
77. four
78. drinking
79. getting a partner intoxicated in order to have sex, physical (relentless) pressure, verbal pressure (e.g., anger), emotional/psychological pressure (e.g., threatening to end the relationship)
80. boys are raised to be nurturant, not aggressive; boys are raised to view women as equals who share power and responsibility
81. gang
82. under the "legal age of consent"
83. nothing
84. psychopathology (a character disorder)
85. opportunistic
86. less sensitive to the victims and more accepting of sexual violence
87. they had been drinking, if a woman had made them angry
88. blame
89. period of long-term reorganization
90. specially trained counselors
91. DNA
92. place the responsibility for preventing rape in the hands of the victim; thus, women who are raped may feel that it was somehow their fault
93. Civil Rights Act
94. unequal power
95. traditional stereotypical gender roles
96. pedophile (preference molester)
97. 13
98. aggressive
99. yourself, your child
100. polyincestuous
101. mother-child
102. family systems

ANSWERS—CHAPTER 16

1. Nevada
2. 17 to 20
3. streetwalkers
4. B-girls
5. call girls
6. hustlers
7. economics (the need for money)
8. middle-class
9. married
10. one-third
11. no harmful
12. less sympathetic
13. rape myths
14. increased hostility
15. commit rape themselves if they were certain they would not get caught
16. depictions of violence toward women
17. sexual explicitness
18. erotica
19. children
20. oral
21. anal
22. false
23. true
24. false
25. true
26. true
27. true
28. false
29. true
30. true
31. true
32. false
33. true
34. true
35. false
36. true
37. true
38. true
39. false
40. true
41. false
42. true
43. true
44. true
45. h
46. a
47. d
48. c
49. g
50. f
51. b
52. e
53. i

54. lack of discrimination
55. fornication
56. streetwalker
57. obscene
58. sodomy
59. material portraying violence toward women

60. (a) by contemporary community standards it depicts patently offensive sexual conduct; (b) it lacks serious literary, artistic, political, or scientific value; and (c) it appeals to prurient interests
61. runaway or throwaway
62. Comstock law
63. they enjoy the sexual arousal it produces

64. active subordination of women

ANSWERS—CHAPTER 17

1. stereotypic gender roles
2. vocabulary
3. positive
4. negative
5. "you"
6. "why"
7. "we"
8. self-disclose
9. listener
10. eye contact
11. facial expressions
12. interpersonal distance
13. touch
14. sexual learning
15. no more

16. in early childhood (preschool years)
17. scare tactics
18. willingness to come to you when they have questions and problems
19. false
20. false
21. true
22. true
23. false
24. false
25. true
26. false
27. true

28. false
29. false (not true for all couples in all instances)
30. false
31. false
32. false
33. false
34. true
35. false
36. false
37. c
38. a
39. c
40. d
41. b
42. b

43. b
44. e
45. d
46. e
47. b
48. "I"
49. "communication"
50. verbal, nonverbal
51. age
52. behavior, character
53. your own pleasure
54. collaborative
55. your overall relationship with your children

Index